ROBOTICS AND MANUFACTURING

RECENT TRENDS IN RESEARCH, EDUCATION, AND APPLICATIONS

VOLUME 4

ASME PRESS SERIES: PROCEEDINGS OF THE INTERNATIONAL SYMPOSIA ON ROBOTICS AND MANUFACTURING

Volume 1, Recent Trends in Robotics: Modeling, Control, and Education, edited by M. Jamshidi, J.Y.S. Luh, and M. Shahinpoor, North-Holland, New York, 1986

Volume 2, Robotics and Manufacturing: Recent Trends in Research, Education, and Applications, edited by M. Jamshidi, J.Y.S. Luh, H. Seraji, G.P. Starr, ASME Press, New York 1988

Volume 3, Robotics and Manufacturing: Recent Trends in Research, Education, and Applications, edited by M. Jamshidi and M. Saif, ASME Press, New York, 1990

Volume 4, Robotics and Manufacturing: Recent Trends in Research, Education, and Applications, edited by M. Jamshidi, R. Lumia, J. Mullins, and M. Shahinpoor, ASME Press, New York, 1992

ASME PRESS SERIES

ROBOTICS AND MANUFACTURING

RECENT TRENDS IN RESEARCH, EDUCATION, AND APPLICATIONS

VOLUME 4

Proceedings of the Fourth International Symposium on Robotics and Manufacturing (ISRAM '92), held November 11-13, 1992 in Sante Fe, New Mexico, U.S.A.

EDITORS

Mohammad Jamshidi
University of New Mexico

Ronald Lumia
National Institute of Standards and Technology

Joe Mullins
University of New Mexico

Mohsen Shahinpoor
University of New Mexico

CAD LAB
CAD Laboratory for Systems/Robotics
University of New Mexico

University of New Mexico

ASME PRESS NEW YORK 1992

The cover art was supplied by Z. Dong, B. Tabarrok, and Z. Qin of the University of Victoria, Department of Mechanical Engineering, and appears in their article, "An Integrated Approach to Shape Design, Stress Analysis, and Cutting Pattern Generation of Tension Structures," on page 939.

345 East 47th Street, New York, NY 10017

ISBN 0-7918-0030-X
ISSN 1052-4150

THIS VOLUME IS DEDICATED TO:

PROFESSOR LOTFI A. ZADEH

Father of Fuzzy Logic

For fifty years of contributions to circuit and network theories, control systems, system theory, artificial intelligence, system identification and inspiring the world with his creative works on fuzzy sets, fuzzy logic, approximate reasoning and fuzzy control.

Lotfi A. Zadeh (1921-   ) is a prominent scientist, engineer, and system theorist from Baku, Azarbayejan. Called the "father of fuzzy logic," he received his M.S. from M.I.T. (1946), his Ph.D. in Electrical Engineering from Columbia University (1949) and several honorary degrees, including an Honorary Doctorate from Paul-Sabatier University in Toulouse, France, and an Honoris Causa from the State University of New York. Currently Professor Emeritus at the University of California, Berkeley, and Director of Berkeley's Initiative on Soft Computing, Dr. Zadeh is a Guggenheim Fellow (1967-68), a Fellow of the AAAS, a Fellow of the World Council of Cybernetics, a Fellow of the IRE (IEEE) (1958), a Member of the National Academy of Engineering (1973), and a Foreign Member of the Russian Academy of Natural Sciences (1992). In addition, he has received Japan's Honda Award, the Berkeley Citation, the IEEE Education Medal, IEEE Centennial Medal, and the IEEE Hamming Medal. Among his major technical contributions are his dissertation of "Frequency Analysis of Time-Varying Networks," his "Generalization of Wiener's Theory of Prediction," and many other meaningful works which have furthered his highly significant contribution of the concept of fuzzy set.

TABLE OF CONTENTS

PREFACE

Robotics and manufacturing problems are among the most important issues facing the nations of the world, particularly the industrialized world. Recently, in many international forums, environmental issues have been raised making the role of manufacturing, automation, and robotics even more important. For example, consider the use of mobile robots for restoration and clean-up of contaminated sites; or hardened robots working in nuclear power plants in place of human beings. Likewise, industrial automation is best achieved by the use of fast, accurate robots for many tasks and sub-tasks in manufacturing plants. These applications contribute to the economic growth and prosperity of our society as a whole.

Organized by the University of New Mexico's CAD Laboratory for Intelligent and Robotic Systems, the Fourth International Symposium on Robotics and Manufacturing (ISRAM '92) is the fourth in a series of symposia that are rapidly becoming common forums for international scientists and engineers to come together and discuss issues of robotics and manufacturing. Gathering leaders of industry, government and academia from more than 15 countries in Sante Fe, New Mexico, the three days of ISRAM '92 provide the opportunity for more than 300 experts on robotics and manufacturing to disseminate the most current knowledge in the field, information necessary to forge ahead in this post Cold War era.

The object of this fourth volume of the ASME Press Series on Robotics and Manufacturing is to report on the technical presentations offered at ISRAM '92. Easily accessible through the organized table of contents and the subject index at the end of the book, these presentations are categorized into nine parts in this volume, as follows:

1. Robot Kinematics, Dynamics, Analysis, and Design
2. Sensing and Sensors
3. Robot Control
4. Parallel and Redundant Robots
5. Telerobotics and Space Applications of Robots
6. Flexible and Mobile Robots
7. Fuzzy Logic Applications in Robotics and Manufacturing
8. Intelligent Systems and Intelligent Manufacturing
9. Design and Economics of Manufacturing Systems

The editors would like to take this opportunity to thank the authors for their contributions to robotics and manufacturing in general, and to this volume in particular. We wish to express our gratitude to all members of the program committee, with special thanks to University of New Mexico's Dr. James Thompson

(Dean, College of Engineering), Professor Nasir Ahmad (Chair, EECE Department), and Professor Fred Ju (former chair, ME Department) for supporting our efforts. Also, we thank Jerry Parkinson (Los Alamos National Laboratory), Doug Bosomworth (Motorola, Inc.), Chuck Weisbin (Jet Propulsion Laboratory), and Bruno Bosacchi (AT&T Bell Laboratories) for their excellent organization of various keynote speeches and their general support of this Symposium. Listed below is the entire team of dedicated organizers and members of our two program committees:

Robotics Program:
Chair: Ron Lumia, National Institute of Standards & Technology, USA
Co-Chair: Charlie Nguyen, Catholic University of America, USA
Members: C. Abdallah, S. Ahmad, M. Ahamdi, J. Albus, W. Amara, J. Anderson, J. Angeles, P. Backes, B. Bavarian, M. Bayoumi, H. Berenji, T. Beugelsdijk, G. Bojadziev, W. Book, J. Burdick, R. Carelli, R. Carroll, D. Cherchas, R. Colbaugh, P. Dauchez, P. Eicker, M. Eslami, D. Fontaine, K. Furuta, W. Gruver, Y. Gu, W. Hamel, R. Harrigan, S. Hayati, H. Hemami, J. Herendon, T. Hsia, R. Inigo, A. Ismail, A. Katbab, K. Khorasani, P. Khosla, S. Kitamura, H. Koivo, T. Kokkini, L. Latre, M. Leahy, M. Mahmoud, R. Mann, S. Ollendorf, R. Patel, F. Pfeiffer, F.Pin, F. Pourboghrat, J. Rico-Martinez, M. Shahinpoor, M. Spong, M. Saif, H. Seraji, G. Starr, A. Steiger, N. Vadiee, G. Varsi, I. Walker, C. Weisbin, G. Widmann, J. Yuh

Manufacturing Program:
Chair: Joe Mullins, University of New Mexico, USA
Co-Chair: Doug Bosomworth, Motorola, Inc., USA
Members: N. Boustany, F. Choobineh, T. Davis, Z. Dong, K. Futukami, Z. Geng, F. Gomide, R. Goodwin, R. Gore, F. Hadaegh, X. Jiang, S. Kamat, S. Kassicieh, R. Klabunde, A. Kusiak, L. Lemke, S. Lu, A. Manesh, A. Miller, A.Nee, H. Parsaei, M. Rabins, V. Salminen, F. Sassani, M. Semens, M. Shahinpoor, B. Tabarrok, J. Tarro

Organizing Co-Chairs:
Mo Shahinpoor, University of New Mexico, USA
Jerry Parkinson, Los Alamos National Laboratory, USA

Finally, we wish to express our most sincere gratitude to Mr. Kishan Kumbla, Ms. Catherine Osborn, Ms. Elizabeth Baldwin, Ms. Nancy Gillan, and Ms. Janet Watts of the University of New Mexico's CAD Laboratory for Intelligent and Robotic Systems for their tireless assistance to the editors as well as their continuous cooperation throughout ISRAM '92.

Mo Jamshidi	Ron Lumia	Joe Mullins	Mo Shahinpoor
Albuquerque, NM	Gaithersburg, MD	Albuquerque, NM	Albuquerque, NM

July 29, 1992

PART 1 : ROBOT KINEMATICS, DYNAMICS, ANALYSIS, AND DESIGN

AUTOMATED HANDLING OF DEFORMABLE OBJECTS BY COORDINATED MANIPULATORS

Ming Z. Chen and Yuan F. Zheng
Department of Electrical Engineering
The Ohio State University, Columbus, OH 43210

ABSTRACT

Coordinating two manipulators to automatically handle deformable objects is studied in this paper. The kinematic constraints are first derived. Then three methods are identified for the manipulators. Analysis shows that the third method results in the minimum forces and moments between the object and the hands. Motion trajectories of the end-effectors are then derived for the third method. Finally, experiments are presented to verify the theoretical studies.

Keywords: Coordinated robots, Automated handling of materials, deformable object, Force and moment minimization, Optimal trajectory

1. INTRODUCTION

In recent years, coordinating two or more manipulators for manufacturing automation has received much attention. Studies of constraints imposed on two robots were conducted by Mason[1]. Zheng and Luh later extended Mason's results[2]. Dynamics of coordinated robots has also been extensively studied[3-7].

While the studies on the coordination of manipulators continue[8-10], one aspect has never received serious attention, i.e., handling deformable objects by multiple manipulators. In reality, however, there is a frequent occurrence of non-rigid objects. For instance, in the automobile industries, etc., flexible plates have long been used in the assembly of vehicles. Traditionally, to handle non-rigid materials, special toolings needed to be designed such as a vacuum pad. The pads are designed to cover the area of object surface. However, such a mechanism needs a large pad when the object has a large area, which often becomes too heavy to be carried by a manipulator. Alternatively, it is possible to handle deformable objects by coordinating two manipulators. This, however, raises a new problem on coordination. It will be studied in this paper.

In reality, there are three types of deformable objects, which are one, two, or three dimensional. If the sheet is picked up in such a way that the end-effectors are positioned at the center of the shorter dimension, the deformation is more likely to occur in the direction of the longer dimension. As a result, two dimensional objects can be treated as one dimensional. For the reasons just cited, we concentrate our study on one dimensional objects, which are simply called beams in the paper.

2. KINEMATIC CONSTRAINTS ON TWO MANIPULATORS HANDLING A DEFORMABLE BEAM

Consider two manipulators each with n joints. Let (x_n^l, y_n^l, z_n^l) and (x_n^f, y_n^f, z_n^f) be the coordinate systems of two end-effectors, respectively(Fig. 1). Since the beam is flexible, one may obtain the following inequality equation:

$$r_{min} < |p(q^f)-p(q^l)| = \left| R_0^n (q^l) r^l \right| = |r^l| < r_{max} \tag{1}$$

where p(q) is a position vector, q is a vector of n joint displacements, R_0^n (q) is the orientation matrix, r^l is a vector and r_{min} and r_{max} are the minimum and maximum length of $|r^l|$. It is clear r_{max} must be less than or equal to the length of the beam.

Assume that twisting of the beam is not allowed. However, the orientation between the vectors $a(q^l)$ and $a(q^f)$ can be modified, where a(q) is unit vector of co-ordinates z_n. The constraints imposed on $a(q^l)$ and $a(q^f)$ can thus be depicted by

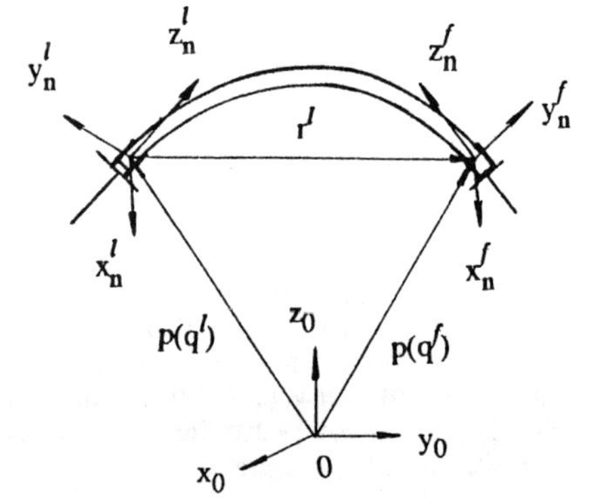

Fig. 1 The coordinate system and notations for two manipulators

$$a_{min} < a^T(q^l)\, a(q^f) < a_{max}. \tag{2}$$

It is clear that a_{min} = -1 when the beam has no deformation, and a_{max} is less than 1.

3. HANDLING A DEFORMABLE BEAM BY TWO MANIPULATORS WITH ONLY POSITION ALTERNATABLE

It can be found that three methods are theoretically possible to handle a flexible beam:

Method A: the relative orientation is allowed to change, but the position is not.
Method B: the relative position is allowed to change, but the orientation is not.
Method C: both the relative position and orientation are allowed to change.

Since the beam is firmly grasped by the end-effectors, one cannot change the relative orientations without changing the relative positions. We therefore can eliminate method A. To evaluate methods B and C, we will select the forces and moments between the beam and the end-effectors as the criterion.

3.1 Deflection Behavior of the Beam under Small Deflections

Consider a beam held by the end-effectors. According to [11], the beam can be deformed in two different cases. Fig. 2 shows the first case in which the deformation has one buckle. In the second case the deformation has two buckles.

In case 1, two inflection points appear on the beam. No moments are exerted on the inflection points, which thus are called the points of zero moment (ZM). If the deflection is small, the force exerted on the point of ZM can be expressed as [11]

$$F_{1S} = \frac{\pi^2 EI_z}{(L/2)^2} = \frac{4\pi^2 EI_z}{L^2} \tag{3}$$

where E is the stiffness of material, I_z is the moment of inertia of the cross-section, and L is the length of the beam. Obviously, the end-effector has to provide the same force to the beam end.

In case 2, it is found [11] that the force exerted on the end-effectors is twice as much as F_{1S}. Therefore, case 2 should be avoided in practice.

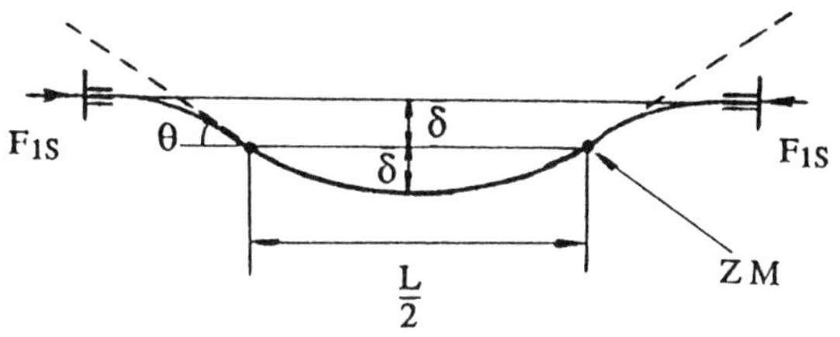

Fig. 2 The beam has one buckle in its deformation

3.2 Deflection Behavior of the Beam under Large Deflections

Introduce a fixed-free ended beam under large deflection, whose behavior is the same as the half of a hinged-hinged beam[11]. Thus, from Fig. 2, it can be seen that the entire beam can be divided into four pieces of fixed-free ended beam, separated by the points of ZM. For every piece, we can determine its behavior of the large deflection. Hence, the behavior of the entire beam can be expressed as follows (here F_{1L} is used instead of F_{1S})

$$F_{1L}=\frac{(4R(\alpha))^2EI_z}{L^2},\ \ \|r^1\| = L(\frac{2P(\alpha)}{R(\alpha)} - 1),\ \ 2\delta = \frac{L}{R(\alpha)}\sin(\frac{\theta}{2}). \tag{4}$$

where α is the deflection angle (here α is the same as θ shown in Fig. 2), $R(\alpha)$ and $P(\alpha)$ can be calculated from the elliptic integrals:

$$R(\alpha,\phi_1)=\int_0^{\phi_1}\frac{d\phi}{\sqrt{1-k\sin^2\phi}},\ \ P(\alpha,\phi_1)=\int_0^{\phi_1}\sqrt{1-k\sin^2\phi}\,d\phi \tag{5}$$

where $k=\sin^2(\alpha/2)$. From (5), one may have $R(\alpha) = R(\alpha,\phi_1)\big|_{\phi_1=\frac{\pi}{2}}$ and $P(\alpha) = P(\alpha,\phi_1)\big|_{\phi_1=\frac{\pi}{2}}$. The moment on the end-effector can be calculated as $M_{1L} = F_{1L}\delta$.

3.3 The Minimum Distance between the Two Grippers

Consider a single piece ended by a ZM point and the end-effector(Fig. 2). It behaves as a fixed-free ended beam, for which the following must always hold [11]

$$\frac{F_{1L}}{A} + \frac{y_m F_{1L}\delta}{I_z} < \sigma_{max} \tag{6}$$

where σ_{max} is the maximum allowable stress for the beam, A is the area of cross section, and y_m is the distance from the center to the edge of the area. The hinged forces and moments thus found still have to satisfy the maximum load that the manipulators can carry. One can determine the minimum distance r_{min} using (6).

4. HANDLING A DEFORMABLE BEAM BY TWO MANIPULATORS WITH BOTH ORIENTATION AND POSITION ALTERNATABLE

In this section, we investigate method C. It is clear that there are many possible combinations of the relative position and orientation between the two end-effectors. Therefore, we should select the best combination from all the possibilities.

4.1 The Best Combination of the Relative Position and Orientation

We propose and prove the following claim.

Claim: When the two end-effectors grasp the beam in such a way that the two end points coincide with the ZM points of the beam, and the orientations of the end-effectors are the same as the deflection angle of the beam (Fig. 3), the end-effectors suffer the minimum forces and moments.

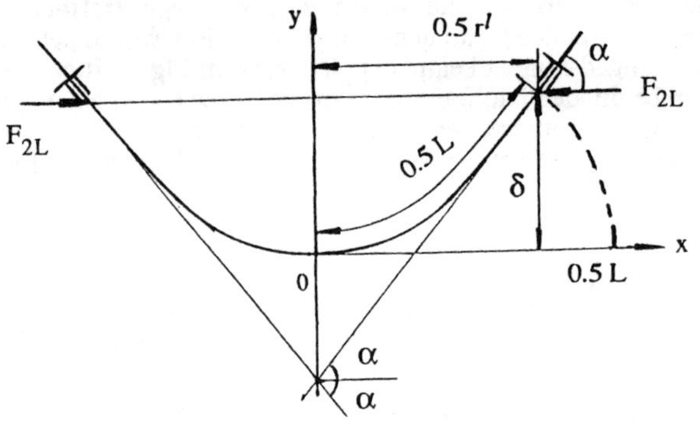

Fig. 3 The end-effectors hold the ZM points when the beam is bent

Proof: The end points of the beam are now at the ZM points. It can be seen that the force exerted on the ZM point is related to the length of the portion between the two ZM points. Longer length results in small forces. Since the length of the beam is the maximum distance that the portion between the two ZM points could be, the force exerted on the ZM point or the center of the end-effectors becomes the minimum.

Next we consider the moments exerted on the end-effectors. Since the orientations of the end-effectors coincide with the deflection angle of the beam, the curvature of the beam at the tip of the end-effectors are null. It follows that the moments exerted on the end-effectors are zero. ♦

4.2 The Minimum Distance and the Maximum Orientation Difference between the Two End-Effectors

The beam now behaves as two pieces of fixed-free ended beam (Fig. 3). The exerted force, the distance between the two end-effectors and the deflection can be written as:

$$F_{2L}=\frac{(2R(\alpha))^2EI_z}{L^2}, \quad |r^l| = L(\frac{2P(\alpha)}{R(\alpha)} - 1), \quad \delta = \frac{L}{2R(\alpha)} \sin(\frac{\alpha}{2}). \tag{7}$$

Obviously, F_{2L} is less than F_{1L}. The minimum distance between the two end-effectors can be determined using the procedure in subsection 3.3 using (7) instead of (4). Essentially, one needs to find a maximum deflection angle, α_{max}. Once α_{max} is found, not only the minimum distance can be found, but also the maximum orientation difference can be obtained as:

$$a^T(q^l)a(q^f) \leq a_{max} = \cos(2\alpha_{max}). \tag{8}$$

4.3 The Relationship between the Positions and Orientations of the End-Effectors while Bending the Beam

The trajectory of the end-effector can be expressed as (Fig. 3):

$$x(\alpha)=L\left(\frac{P(\alpha)}{R(\alpha)} - \frac{1}{2}\right), \quad y(\alpha) = \frac{L}{R(\alpha)}\sin(\alpha/2) \tag{9}$$

There are two problems involved in this method. First, the computation involved in (9) is complicated. Secondly, even if the computation can be handled by the controller, executing such a trajectory is time consuming. In reality, we coarsely divide the trajectories into a few pieces. The positions of the two end points of each piece can be specified by (9) or using a look-up table. The motions between the two end points are simply linear. Now the problem is how the force and moment will be affected by using this approximation method. This will be studied as follows.

4.4 Behavior of the Approximation Method

If we use the line segment instead of the curve, the shape of the beam will be changed and the inflection points will occur on the beam instead at the end points. We may extend the length of the original beam in such a way that the inflection points of the original beam are also at the quarter points of the extended beam[11]. After detailed analysis we obtain the following relationship:

$$\frac{F_x}{F_{2L}} = \left(\frac{2R(\alpha_x) - R(\alpha_x,\phi_1)}{R(\alpha)}\right)^2. \tag{10}$$

where F_x is the force exerted on the ZM point and α_x is the deflection angle at the ZM point. Since $R(\alpha_x)$ is greater than $R(\alpha_x,\phi_1)$, and $R(\alpha_x)$ is greater than $R(\alpha)$, F_x is greater than F_{2L}. However, as long as the difference between the line and the curve is small, the F_x is still manageable. And the force arm can be written as:

$$h = \frac{L\sqrt{2(\cos\varphi - \cos\alpha_x)}}{2(2R(\alpha_x) - R(\alpha_x,\phi_1))} \tag{11}$$

where φ is the angle between the end-effector and ox. Consequently, the moment exerted on the end-effector is $F_x h$.

5. EXPERIMENTAL STUDY

To verify the results, we have conducted experiments in our laboratory. Basically, in the experiment, a number of elastic sheets needed to be assembled to form a book.

For the alignment purpose, each sheet has four holes. Those holes are used to align the sheet with pins which are fixed on a board. Since the tolerance between the pin and the hole is very tight, it is difficult to align all the holes with the pins at the same time. Instead, the manipulators bend the sheet such that the two middle holes can make contact with the pins first. Once the middle holes are aligned with the pins, the manipulators extend the sheet such that the remaining holes are aligned with the pins. In this process, the two manipulators follow the trajectories defined in the previous sections.

We have tested both method B and method C in the experiment. The required bending angle was 40°. When method B was used, we found that some permanent damage was made.

We then turned to method C. In programming the motions of the end-effectors, we divided the 40° deflection into four small pieces with 10° in each piece. Use the end points of the pieces as the destinations for the end-effectors to move. By using method C the assembly was successfully completed without any damage to the sheet.

6. CONCLUSIONS

In this paper we have discussed the mechanism for two coordinated manipulators to handle deformable objects. The main contributions can be summarized as follows:

(1) We have identified two methods for the two manipulators to handle deformable objects. From the analysis, we identified that the second method is better than the first method.

(2) For the second method, we specified the end-effector trajectories and further developed an approximation method. Finally, experiments were given to prove the second method was effective for handling deformable objects.

ACKNOWLEDGEMENT: This work is supported by ONR under grant N00014-90-J-1516, and by a Presidential Young Investigator Award of NSF to Yuan F. Zheng (DDM-8996238).

REFERENCES

1. M.T. Mason, "Compliance and force control for computer controlled manipulators," IEEE Trans. on System, Man and Cyber. Vol. SMC-11, N. 6, June 1981, pp. 418-432.
2. J.Y.S. Luh and Y.F. Zheng, "Constrained relations between two coordinated industrial robots for motion control," Int. J. of Robotics Research, Vol. 6, No. 3, Fall, 1987, pp. 60-70.
3. Y.F. Zheng and J.Y.S. Luh, "Joint torques for control of two coordinated moving robots," Proc. 1986 IEEE Int. Conf. on Robotics and Automation, San Francisco, CA, April 7-10, 1986, pp. 1375-1380.
4. T.J. Tarn, A.K. Bejczy and X. Yun, "Dynamic coordination of two robot arm," Proc. IEEE 25th CDC, Athens, Greece, Dec. 10-12, 1986, pp. 1267-1270.
5. U. Ozguner, S. Yurkovich and F. Al-Abbass, "Decentralized variable structure control of a two-arm robotic system," Proc. 1987 IEEE Int. Conf. on Robotics and Automation, Raleigh, NC, March 31-April 3, 1987, pp. 1248-1254.
6. T. Yoshikawa and X. Zheng, "Coordinated dynamic hybrid position/force control for multiple robot manipulators handling one constrained object," Proc. 1990 IEEE Int. Conf. on Robotics and Automation, Cincinnati, Ohio, May 13-18, 1990, pp. 1178-1183.
7. A.J. Koivo, "Adaptive position-velocity-force control of two manipulators," Proc. IEEE 24th CDC, Fort Lauderdale, FL, Dec. 11-13, 1985, pp. 1529-1532.
8. R. Guptill and P. Stahura,"Multiple robotic devices: position specification and coordination," Proc. 1987 IEEE Int. Conf. on Robotics and Automation, Raleigh, NC, March 31-April 3, 1987, pp. 1655-1659.
9. Y.F. Zheng, J.Y.S. Luh and P.F. Jia, "Integrating two industrial robots into a coordinated system," Computer in Industry, Vol. 12, 1989, pp. 285-298.
10. J.M. Tao, J.Y.S. Luh and Y.F. Zheng, "Compliant coordination control of two moving industrial robots," IEEE Trans. Robotics and Autom., Vol. RA-6, No. 3, 1990, pp. 322-330.
11. S.F. Borg, Fundamentals of Engineering Elasticity, D. Van Nostrand Company, Inc. 1962.

Magnetically Levitated Micro-Robots: Design, Control, Experimentation and Applications

Jerry Robichaux and Shaheen Ahmad

Real-Time Robot Control Lab, School of Electrical Engineering, Purdue University West Lafayette, IN 47905-1285, USA

ABSTRACT

Since the middle 1980's, there has been tremendous development of micro electro-mechanical systems (MEMS) components, such as gears, sensors and motors [1,2,3,4]. MEMS technology has applications in the automotive, semiconductor, medical, security and space industries. Accelerometers, pressure sensors, high resolution positioning systems [5][6], high-density magnetic recording systems [7], micro-surgical systems, and scanning surface microscopes [8] are among the wide variety of applications which may benefit from MEMS technology. To realize these applications, many of the MEMS components will have to be integrated with semiconductor circuits to produce sensors and actuators in discrete small packages. In this paper we describe a magnetically levitated micro manipulation system which was developed for the assembly, machining and general tooling of MEMS devices.

1.0 INTRODUCTION

The micro manipulator in this paper is a rigid body device with permanent magnets attached at the vertices of its triangular frame. The manipulator has three degrees of freedom for micro manipulation in the x-y plane and in-plane rotation. The propulsion of the manipulator is accomplished by the direct application of magnetic forces rather than from mechanically connected motors. This gives the manipulator its small size, small mass, high mechanical bandwidth and high positioning resolution. The magnetic forces result from the interaction of the permanent magnets with actively controlled magnetic fields generated from a five layer printed circuit board.

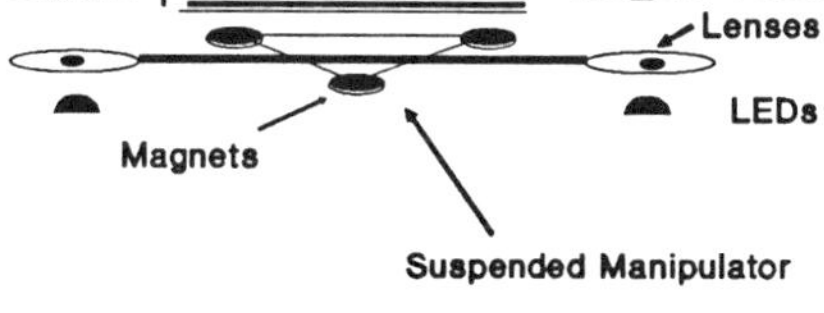

Fig. 1. Sectional view of the position sensing components and manipulator as it is levitated below the multi-layer magnetic array in its precision work area of 1mm x 1mm.

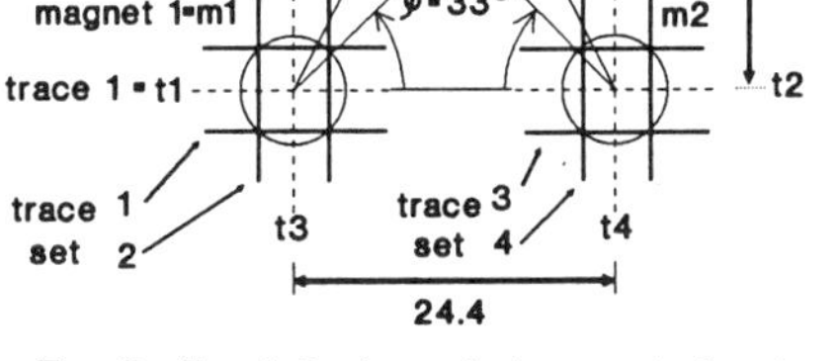

Fig. 2. Detailed view of the manipulator's relationship to the printed circuit board elements giving trace and manipulator dimensions as well as the single trace, trace set, and magnet identifications.

Nonlinear controllers are utilized to achieve a positioning accuracy of 1 micron over a 1000 x 1000 micron work area, and to exert precise forces in two dimensions, in the range of tens of micro-Newtons to 2 milli-Newtons. Fig. 1 illustrates a sectional view of the manipulator levitated below the multiple layer magnetic array along with the sensor components.

The position of the manipulator is sensed with infrared LED's and lateral effect photo diodes (LEPD's) also called position sensing devices (PSD's). The LED's emit light which gets collected through lenses mounted on each side of the manipulator, the light is focussed down to a 200 micron spot onto stationary PSD's located above each lens. The PSD's then detect the location of the focussed emission to determine the manipulator's position. The position sensing resolution of the manipulator is currently .5 microns, however this can be substantially improved.

Fig. 2 provides a more detailed view of the manipulator's relationship to the printed circuit board elements. The bottom magnets (m_1 and m_2) together with the traces in layers 3 and 4 provide fine positioning capability. Layers 1 and 2 have parallel trace sets, providing tangential spring like forces to all three magnets. The tangential spring forces will act on the manipulator by restoring it to the center of the fine motion work area. Three multiple turn air-core solenoids in layer 5 provide the majority of lifting force, while contributing minimal tangential forces for all three magnets within the precision work area.

Each permanent magnet on the manipulator is suspended just below the plane of the magnetic array. The motions of the manipulator are restricted to the plane such that roll and pitch angles are close to zero. To reduced coulomb friction the manipulator is dithered by switching the magnetic field from the air core solenoids. The dither frequency, f_d, ranges from 500 to 2000 Hz. Dithering the manipulator implies that it will periodically make contact with the surface of the magnetic array. While this reduces coulomb friction to nearly zero, because of the increased eddy current losses and the periodic surface contact it also increases the effective viscous friction of the system. For a large enough f_d the manipulator behaves as if it were on a bearing surface with an average z axis height and exhibits smooth planar motion. Therefore, the manipulator behaves as if it were levitated on a viscous bearing at a height of $z_o<0$. Parameter identification experiments are used to determine the effective viscosity of the system.

2.0 MAGNETIC FIELD MODEL TO DETERMINE THE FORCES ON A PERMANENT MAGNET IN A STATIC MAGNETIC FIELD

The magnetic field from a permanent magnet can be modeled as a magnetic dipole. A magnetic dipole is formed from electrons moving in a closed circular path, and can be modeled as a pair of fictitious magnetic charges having opposite polarities of ($\pm q_m$), with units [Am], separated by a dipole length l. Therefore we can model a cylindrical axially magnetized permanent magnet as a cylinder with two oppositely charged surfaces separated by a dipole length l. To compute the force exerted on a magnetic dipole due to an external magnetic field, **B** [Wb/m^2], an analogy of Coulomb's law applied to magnetism is used in conjunction with the surface charge model of a permanent magnet. We have, $\mathbf{f}=q_m\mathbf{B}$. Where the force, **f**, is exerted on a magnetic charge q_m. To calculate the net force on a permanent magnet requires us to integrate over the top and bottom surfaces of the magnet.

2.1 Magnetic Field for a Straight Conductor

As the fields acting on the manipulator are generated by printed circuit board traces, the field due to a trace can be determined from the Biot-Savart law. For a trace of width, w, located in the xy plane parallel to the y axis, we have,

$$d\bar{B} = \frac{\mu_o}{2\pi} i(x_c) \left[\frac{z}{(x-x_c)^2+z^2} \hat{a}_x - \frac{x-x_c}{(x-x_c)^2+z^2} \hat{a}_z \right] dx_c \ , \qquad (1)$$

where x_c is the conductor coordinate along the x axis, and the coordinates x and z are the points of interest for which **dB** is to be evaluated. Integrating (1) over the width of the conductor from x_c=-w/2 to x_c=+w/2 we can compute the total magnetic flux density at point (x,y,z). The surface current $i(x_c)$ [A/m] is the current distribution in the conductor, which can be shown to be uniform in the presence of the external field due to a permanent magnet. An expression for the force acting on a permanent magnet, with two oppositely charged surfaces parallel to the x-y plane, due to a current carrying conductor in the x-y plane, can be written as,

$$\bar{f} = \sum_{j=1}^{2} \int_{x_c} \int_{x_m} \int_{y_m} \rho_{ms} dy_m dx_m d\bar{B}, \qquad (2)$$

where j is used to index the top (j=1) and the bottom (j=2) magnet surfaces. Further, $dx_m dy_m$ is an incremental area on the magnet surface and ρ_{ms} is the magnetic charge density per unit area with units [A/m]. Equation (2) was numerically integrated for $i(x_c)$ = 1A, and $\rho_{ms}=3.24 \times 10^6$ and the resulting normal and tangential profiles are given in Figs. 3 and 4 below. For a small region of $x \in [-.5,.5]$ mm, it is easy to visualize that a parabolic equation could be used to represent the tangential force component, and a linear equation could approximate the normal force component. It is important to note that the magnetic forces attenuate inversely proportional to the distance cubed, therefore only the traces within 5mm of the magnets are considered to generate forces on that magnet. For instance from Fig.2, the fields from trace 1 and 3, trace set 1 and 2 and coil 1 only effect magnet 1.

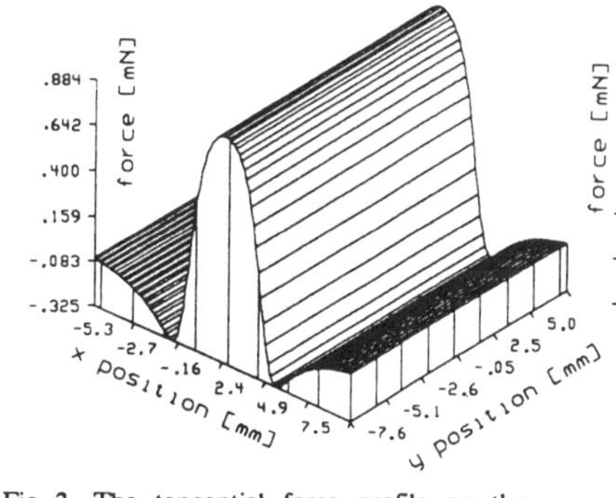

Fig 3. The tangential force profile on the permanent magnet resulting from the magnetic field generated by a single printed circuit board conductor.

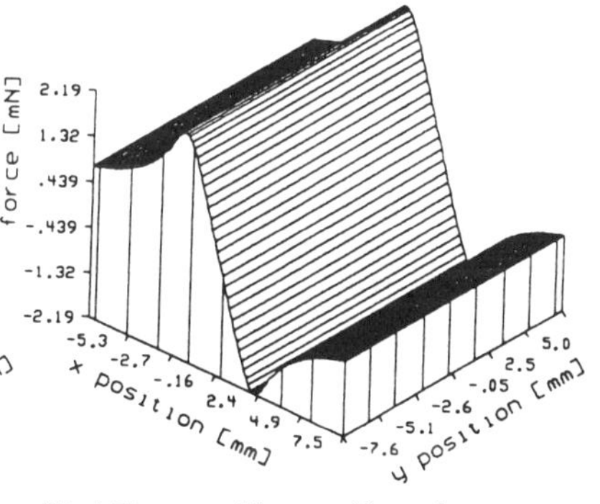

Fig 4. The normal force profile on the permanent magnet resulting from the magnetic field generated by a single printed circuit board conductor.

The parabolic tangential force equation for a trace parallel to the y axis is of the form,

$$\bar{f}_t = I\left[a(x_m-x_c)^2+c\right]\hat{a}_x + 0\,\hat{a}_y \quad [\text{mN}] \ , \qquad (3)$$

where x_m and x_c are the coordinates along the x axis for the magnet and conductor respectively. Further, the parameters a and c are numerically determined for the traces in layers 3 and 4, and I is the current through the trace. The normal forces, f_n, for a trace parallel to the y axis may be approximated with a linear equation of the form,

$$\bar{f}_n = I\left[k(x_m-x_c)+b\right]\hat{a}_z \quad [\text{mN}] \ , \qquad (4)$$

where parameters k and b are also numerically determined for layers 3 and 4. Similar approximations are made for traces parallel to the x axis.

The parallel pairs of traces in layers 1 and 2 of Fig. 2 form trace sets. The tangential forces are linearly proportional to displacements x_c - x_m, and are similar to the forces generated by a spring displaced from its equilibrium position. When $x_m=x_c$ the tangential force is zero, and the normal force is a maximum. The tangential and normal profiles may also be approximated by first and second order polynomials provided $|x_c - x_m| \leq .5$mm and $|y_c - y_m| \leq .5$mm. The tangential forces for a trace set parallel to the y axis, may be approximated by a linear equation, and the normal forces for a trace set parallel to the y axis may be approximated by the parabolic equation.

2.2 Magnetic Fields and Forces from Multiple Turn Coils

Multiple turn coils are used in the magnetic array to provide significant normal forces to the manipulator. A wire wound multiple turn coil may be approximated as multiple layers of concentric coils. Therefore, we vectorially add the **B** field from each single turn coil in the multiple turn set. For a coil at the origin of a spherical coordinate system, the magnetic vector potential, **A**, for point $P(R,\theta,\phi)$ is obtained from,

$$\bar{A} = \frac{\mu_o I}{4\pi} \oint_{C'} \frac{dl'}{R_1}, \tag{5}$$

where the closed line integral is evaluated along the coil length dl', R_1 is the distance from a point on the coil length to P, and I is the current in the coil. We first numerically integrate (5) for 100 coils and then the **B** field is calculated by numerically taking the curl of **A**. Once **B** is obtained equation (2) is numerically integrated to form tangential and normal force profiles between a permanent magnet and the multiple turn coil.

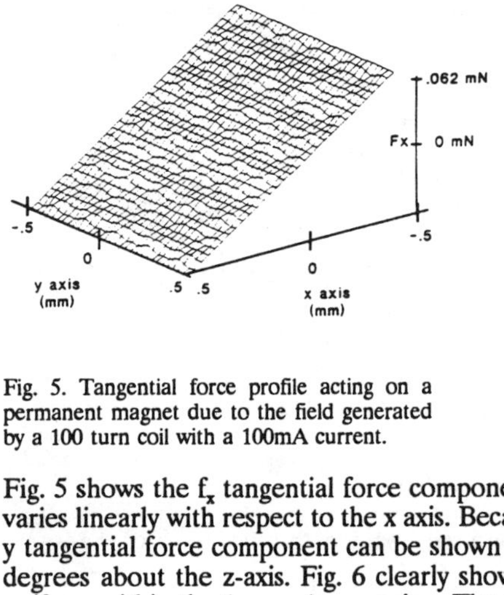

Fig. 5. Tangential force profile acting on a permanent magnet due to the field generated by a 100 turn coil with a 100mA current.

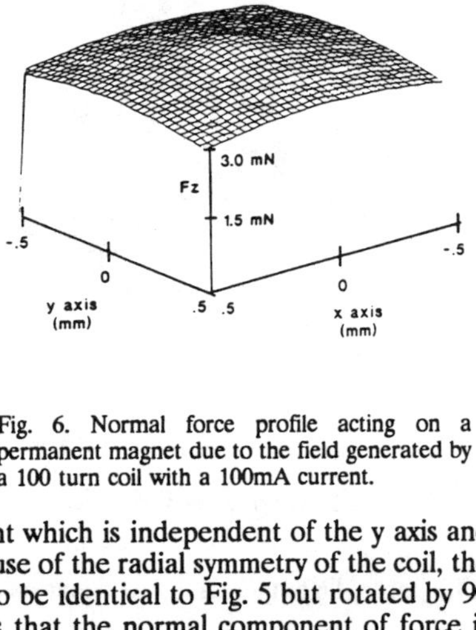

Fig. 6. Normal force profile acting on a permanent magnet due to the field generated by a 100 turn coil with a 100mA current.

Fig. 5 shows the f_x tangential force component which is independent of the y axis and varies linearly with respect to the x axis. Because of the radial symmetry of the coil, the y tangential force component can be shown to be identical to Fig. 5 but rotated by 90 degrees about the z-axis. Fig. 6 clearly shows that the normal component of force is uniform within the 1mm x 1mm region. The normal and tangential forces acting on the permanent magnets due to the external fields generated by the multiple turn coils may be approximated with linear and constant parameter algebraic equations of the form,

$$\bar{f} = I\left[k_c(x_m - x_c)\hat{a}_x + k_c(y_m - y_c)\hat{a}_y + d_c\,\hat{a}_z\right] \ [\text{mN}] \ . \tag{6}$$

Static parameter identification experiments indicate that the parameter values are d_c=30 [mN/A] and k_c=-10 [mN/Amm] for a current input with f_d=2000 Hz and with an 80% duty cycle.

3.0 DYNAMIC EQUATIONS OF MOTION OF THE MICRO MANIPULATOR

In this section we will derive the system of equations describing the sum of the tangential forces acting on the manipulator which result in motions along the xy plane and orientation in the xy plane θ, given that the manipulator coordinates are (x,y,θ). Again the motion of the manipulator is restricted to $x\in[-.5,.5]$ mm, $y\in[-.5,.5]$ mm, $z=z_o$, and the manipulator is not pitched ($\phi=0$), or rolled ($\psi=0$), with respect to the printed circuit board plane. Since tangential forces act on each magnet located at a radial distance from the manipulator center of mass, a net torque will result, rotating the manipulator about the z axis and enabling in-plane, θ, rotation. The net torque from each of the three magnets (i= 1 to 3) is given from the following cross product,

$$\bar{T}=\sum_{i=1}^{3}(\bar{r}_i \times \bar{f}_i), \tag{7}$$

here f_i is the sum of the tangential forces from the single traces, trace sets and multiple turn coils on the permanent magnet (i), and r_i is the vector from the manipulator center of mass to the center of the magnets. The total tangential force/torque expression for the manipulator may be written by combining the single trace inputs $u_1=[i_1\ i_2\ i_3]^T$ the parallel trace set and coil inputs $u_2=[i_5\ i_6\ i_7\ i_8]^T$ and their respective polynomial force/torque expressions.

$$\begin{bmatrix} f_x \\ f_y \\ T \end{bmatrix} = \begin{bmatrix} 0 & 0 & f_{x1}+f_{x2} \\ f_{y1} & f_{y2} & 0 \\ \tau_1 & \tau_2 & \tau_3+\tau_4 \end{bmatrix} \begin{bmatrix} i_1 \\ i_2 \\ i_3 \end{bmatrix} + \begin{bmatrix} k_{x5} & k_{x6} & k_{x7} & k_{x8} \\ k_{y5} & k_{y6} & k_{y7} & k_{y8} \\ \tau_5 & \tau_6 & \tau_7 & \tau_8 \end{bmatrix} \begin{bmatrix} i_5 \\ i_6 \\ i_7 \\ i_8 \end{bmatrix} \tag{8}$$

We will refer to (8) with the matrix notation given in (9), where F(x) is the force/torque matrix for single trace inputs u_1, and K(x) is the spring constant matrix for the trace set and multiple turn coil inputs u_2, or,

$$f = F(x)u_1 + K(x)u_2 \ . \tag{9}$$

Thus from rigid body dynamics, the manipulator dynamic model is given as,

$$D\ddot{x} + B\dot{x} = F(x)u_1 + K(x)u_2 \ , \tag{10}$$

where $D\in\mathbf{R}^{3x3}=\text{diag}[m,\ m,\ I_\theta]$, is the nonsingular positive definite symmetric inertia matrix, and $B\in\mathbf{R}^{3x3}=\text{diag}[b_x,\ b_y,\ b_\theta]$ is the effective viscosity contribution from eddy current and dithering losses. The vector u_2 is set to be constant, and control of the manipulator motion is achieved by the input u_1. Parameter identification experiments indicate that the viscosity values for b_x and b_y are on the order of .20 Ns/m and b_θ = 1.50×10^{-4} Nsm. Step response experiments demonstrate that the manipulator is slightly underdamped, with a high mechanical bandwidth and is capable of moving at velocities of 4000 microns/second.

4.0 FEEDBACK LINEARIZATION CONTROLLER DESIGN

A feedback linearization controller which inverts the plant model is given as,

$$u_1 = F^{-1}(x)\{D[\ddot{x}_d + K_d\dot{e} + K_p e] + B\dot{x} - K(x)u_2\} \tag{11}$$

where we have defined the tracking error to be $e=x_d-x$. The desired trajectories $x_d(t)\in\mathbf{R}^3$ are in the (x,y,θ) space and are twice differentiable $x_d(t)\in\mathbf{C}^2$. Substituting the control law into the manipulator dynamic equation and assuming that we have accurate knowledge of the plant model results in the following error dynamic equation as,

$$D[\ddot{e}+K_d\dot{e}+K_p e]=0 \quad . \tag{12}$$

Where K_p and $K_d \in \mathbf{R}^{3x3}$ are diagonal positive definite feedback gain matrices. The error, $e(t)\rightarrow 0$ exponentially as $t\rightarrow\infty$ if $Is^2+K_d s+K_p$ is Hurwitz polynomial, i.e. the eigenvalues are on the left half plane, and $I\in\mathbf{R}^{3x3}$ is the identity matrix.

Fig. 7 shows experimental results for the diagonal elements of K_p=20 and K_d=.1, and for a desired trajectory along the y axis. In this experiment the control law is updated at 30 Hz. Notice that the initial error of -20 microns settles within .2 seconds and the error remains bounded by ±5 microns over the majority of the trajectory. As our model for the manipulator looses accuracy near the extremities of the fine motion work area, the error near the end of the trajectory also increases. Other trajectory tracking experiments demonstrate the tracking errors for the x and y axis remain bounded by ±5 microns. When the integral of the position error is introduced into the feedback loop, the steady state error is within ±1 micron.

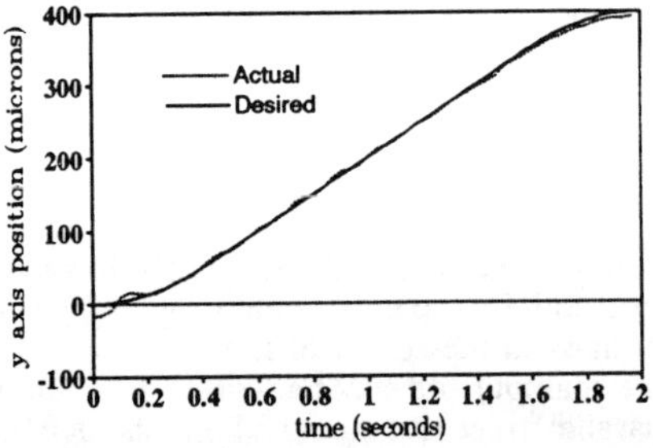

Fig. 7. Experiment of the micro manipulator following a twice differentiable trajectory along the y axis.

5.0 CONCLUSION AND FUTURE WORK

A micro manipulator test bed has been developed in our laboratory, it utilizes feedback linearization control to track desired position and velocity trajectories. Preliminary experiments indicate that force control with micro-Newton accuracy is also achievable with the present test bed. Soon with the ability to exert precisely controlled forces the manipulator will be utilized to perform tasks such as parts placement, assembly, and deburring of MEMS components.

6.0 REFERENCES

1. W. Hoffer,"The Coming Era of Tiny Technology," *Nation's Business*, (January 1989).
2. "A tiny World of Action," *Compressed Air Magazine*, (September 1988).
3. S.C. Jacobsen, "The Wobble Motor: An Electrostatic Planetary-Armature Miroactuator," *Micro Electro Mechanical Systems*, Proc. IEEE, Salt Lake City, Utah, February 1989.
4. K.D. Wise, "Integrated Microelectromechanical Systems: A Perspective on MEMS in the 90's,", *Micro Electro Mechanical Systems*, Proc. IEEE, Napa Valley, CA, February 1990.
5. I.J. Busch-Vishniac, "Applications of Magnetic Levitation Based Micro-Automation in Semiconductor Manufacturing," *IEEE Transactions on Semiconductor Manufacturing, Vol. 3, No. 2*, May 1991.
6. J. Robichaux, "Model Based Maglev Microbotic Motion Control," *IEEE International Conference on Robotics and Automation*, Sacramento, CA, April 1991.
7. H. Zappe, "Technological Challenges in Magnetic Recording," *Micro Electro Mechanical Systems*, Proc. IEEE, Napa Valley, CA, February 1990.
8. C. Quate, "Imaging with the Tunneling and Force Microscopes," *Micro Electro Mechanical Systems*, Proc. IEEE, Napa Valley, CA, February 1990.

ON THE KINEMATICS OF MANIPULATORS WITH GENERAL GEOMETRY : APPLICATION TO THE FEASIBILITY ANALYSIS OF CONTINUOUS TRAJECTORIES

PHILIPPE WENGER
Laboratoire d'Automatique de Nantes - URA CNRS n° 823
Ecole Centrale de Nantes
1, rue la Noë 44072 Nantes Cedex- France

ABSTRACT

This paper presents a novel general framework for analysing the feasibility of continuous trajectories for robot manipulators with arbitrary nonredundant geometry. Robots of general geometry may change posture while not meeting a singularity. To take account of this surprising result, the notion of aspects is generalized and the maximal domains of feasible continuous trajectories, called t-connected subspaces, are defined in the robot's workspace.

1- INTRODUCTION

Many robotic tasks require following a prescribed path in the workspace. For a nonredundant robot manipulator to be able to track the path, it is necessary that no change of posture occurs during motion [1], [2]. Otherwise, the robot's end-effector may have to leave the trajectory, as shown on figure 1. The notion of aspects [1] was introduced in this context. The aspects are the maximal singularity-free domains of the robot's jointspace. For robots with simple geometries (all industrial manipulators belong to this class) it was shown that, for a given point in the robot's workspace, there is at most one inverse kinematic solution in each aspect. Each of these solutions is related to a type of posture (like "elbow up" and "elbow down"). Since the aspects are the maximal uniqueness jointspace domains, their image in the robot's workspace yield the t-connected subspaces (the maximal regions of feasible continuous trajectories). These regions are bounded by surfaces induced by singularities and joint limits [1], [5], [6]. In this paper, it is shown that these results do not hold anymore when the robot's geometry does not have simplifying conditions, because there may be more then one solution to the inverse kinematic problem in each aspect. In other words, some robots may change posture while not meeting a singularity. This surprising result was recently pointed out in [3]. In this paper, new notions are established which lead to the jointspace uniqueness domains for nonredundant robots with arbitrary geometry. Mapping the uniqueness domains into the workspace yields the t-connected subspaces.This paper synthetizes a more extensive work [4].

2- A NEW FRAMEWORK

2.1- Type 1 and type 2 geometries

A robot is said to have a *type 1 geometry* if it must pass through a singularity when changing posture. All usual industrial geometries are type 1. A robot with *type 2 geometry* can change posture while not meeting a singularity. There is more than one solution per aspect to the inverse kinematic problem for robots with type 2 geometry. Characterization of all type 1 geometries is not obvious at all. In [3], some rules are suggested for identifying type 1 regional RRR geometries. It has been shown in [4], using geometric arguments and previous well-known kinematic results [7], that all RPP, PRP, and PPR geometries are type 1. In addition, a non exhaustive list of RRR, PRR, RPR, and RRP type 1 geometries was given, using numerical verification. These geometries remain type 1 when adding a wrist.

2.2- Showing the problem

A typical case of a RRR regional structure with type 2 geometry is presented on figure 2 at configuration q=0. Joint axis 1 (resp. 2) and joint axis 2 (resp. 3) are orthogonal, but not intersecting. The robot is assumed to have unlimited joints. It has 2 aspects A_1 and A_2 and 2 inverse solutions per aspect [4]. Let $WA_1=f(A_1)$ and $WA_2=f(A_2)$ (f is the geometrical operator). It can be shown that $WA_1=WA_2=W$ (W means the entire workspace). Figure 3 shows that any continous trajectory lying in WA_1 or in WA_2 is not necessarily feasible. Figure 4a depicts a section of WA_1. Let T=AB be a continuous trajectory in WA_1. A numerical calculus of $f^{-1}(AB)$ is shown in figure 4b. The preimage of T is composed of 2 separate paths (and not 4, since the opposite sides of the square are actually coincident, due to unlimited joints). When tracking path 1 from $f^{-1}(A)$ to $f^{-1}(J)$, the end-effector follows T from A to J. Then, at point J, the end-effector cannot move forward anymore. If path 1 is tracked further, the end-effector goes back to I, and finally returns to A. Similarly, tracking path 2 from $f^{-1}(B)$ makes the end-effector moving from B to I and back to J and B. Thus, T is infeasible. This example reveals two interesting results : first, the number of inverse kinematic solutions is not necessarily the same for all points in the image of an aspect (all points lying between I and J have two inverse solutions in the aspect, while all point lying on AI and JB, have only 1 solution) ; second, points lying within the image WA_i of an aspect A_i may have singular inverse kinematic solutions in the boundary A_i of the same aspect A_i (this is the case for points I and J). In other words, some configurations of the boundary A_i map onto points that are not necessarily on the boundary of WA_i.
Next sections present a set of proved theorems. For room considerations, their proofs will not be reported here, they can be found in [4].

2.3- Notion of Characteristic Surfaces

The characteristic surfaces (CS) are theoretically defined bellow. It will be stated further that they separate the different inverse kinematic solutions in A_1.

Definition : Let A_i be an aspect in Q. The *characteristic surfaces* of A_i, denoted $S_c(A_i)$, are defined as the preimage in A_i of the boundary $\bar{A}_i$:

$$S_c(A_i) \triangleq f^{-1}(\, f(\bar{A}_i)) \cap A_i \qquad (1)$$

The CS generally depend on q_1. However, if joint 1 is unlimited revolute, since in this case the workspace structure is independant of q_1, the CS are also independant of q_1. Thus, if the robot to be studied has 3 joints, the CS can be represented in plane (q_2, q_3). Figure 4 depicts the CS for the unlimited 3R-joint robot of figure 2. All CS have been obtained using a simple numerical procedure described in [4].

2.4- Basic components and basic regions

The CS induce a partition on each aspect into open connected sets called basic components :

Definition 3 : Let A_i be an aspect. The *basic components* of A_i, denoted $\{Ab_{ik}\ ;\ k\in K\}$, are defined as the connected components of the set $A_i \dot{-} S_c(A_i)$ ($\dot{-}$ means the difference between sets).The following relation holds :

$$A_i = (\cup_{k\in K} Ab_{ik}) \cup S_c(A_i) \quad (2)$$

Definition 4 : Let $WA_i=f(A_i)$ and $WAb_{ik}=f(Ab_{ik})$. WAb_{ik} is a domain of W which will be called *basic region* of A_i. Thus :

$$WA_i = (\cup_{k\in K} Wb_{ik}) \cup f(S_c(A_i)) \quad (3)$$

In addition, it can be easily shown [4] that the following result holds:

Proposition 1 : The basic regions of a given aspect are either coïncident, or disjoint sets of W. They can never overlap each other (but two basic regions of two different aspects may overlap).

Physically, the coïncident basic regions of a given aspect A_i represent the different accessibility "strata" in the image of A_i. Their inverse mapping in the jointspace yields the basic components and separates these strata. This result can be stated as follows :

Theorem 1 : For all points in the workspace, there is one unique inverse kinematic solution in each basic component.

Theorem 1 yields the straightforward following consequence :

The change of posture is done by passing through either a singularity or a characteristic surface.

Figures 5 presents the basic components and corresponding basic regions for the regional geometry of figure 2. Both aspects A_1 and A_2 are composed of 3 basic components $A_1=Ab_{11}\cup Ab_{12}\cup Ab_{13}\cup S_c(A_1)$ and $A_2=Ab_{21}\cup Ab_{22}\cup Ab_{23}\cup S_c(A_2)$ (the reader must not forget to identify the opposite sides of the square, since the joints are unlimited). For all points lying in the inner part of W ($WAb_{11}=WAb_{13}=WAb_{22}=WAb_{23}$), there are four inverse kinematic solutions, respectively in Ab_{11}, Ab_{13}, Ab_{22}, and Ab_{23}.

2.5- Uniqueness domains

Theorem 1 gives sufficient conditions for defining domains of Q (the basic components) where there is one unique solution to the inverse kinematic problem. However, there may exist larger domains in Q where there is still one unique solution. Following theorem 2 states the natural intuitive result that there should be still one unique inverse solution at each point lying in the union of disjoint basic regions. Let define the uniqueness domains as the larger open connected subspaces of Q where f is one-to-one. They are characterized as follows :

Theorem 2 : The uniqueness domains Qu_j are the union of two sets : the (nonconnected) set of adjacent basic components $\cup_{k\in K'} Ab_{ik}$ of a same aspect A_i whose respective images are disjoint basic regions, and the set $s_c(K')$ of the characteristic surfaces which separate the components of $\cup_{k\in K'} Ab_{ik}$:

$$Qu_j=(\cup_{k\in K'} Ab_{ik})\cup s_c(K') \quad (4)$$

with $K'\subset K$ such that $\forall k_1;k_2\in I'$, $f(Ab_{ik_1})\cap f(Ab_{ik_2})=\emptyset$

Proof of this theorem is a straightforward consequence of the two following results (that can be easely verified using continuity arguments [4]) :

- f is one-to-one from the border between two adjacent components of $\cup_{k\in K'}Ab_{ik}$ onto its image;
- Given two pairs of adjacent components of $\cup_{i\in I'}Ab_i$, the two borders separating the two respective pairs map onto disjoint sets in W. In addition, they do not overlap the set $f(\cup_{k\in K'}Ab_{ik})$.

Remarks : Unlike the aspects and the basic components, the uniqueness domains do not necessarily induce a partion on Q. Figure 6 depicts the 4 uniqueness domains of our RRR regional structure.

2.6- Characterizing the t-connected subspaces

The t-connected subspaces WA_t are the regions in W where any continuous trajectory can be followed by the robot's end-effector [2]. In other words, any path in WA_t is the image under f of a path in Q. Thus, it is necessary and sufficient that f be a one-to-one continuous map from some subspace of Q onto WA_t, and that f^{-1} be also continuous on WA_t. This is true as soon as WA_t is the image of any uniqueness domains Qu_j. The fact that f^{-1} is also continuous is due to the compactness of $f(Qu_j)$:

Theorem 3 :
The t-connected subspaces are the mapping of the uniqueness domains :

$$WA_t = f(Qu_j) \triangleq f\Big((\cup_{k\in K'}Ab_{ik})\cup s_c(K')\Big)=(\cup_{k\in K'}WAb_{ik})\cup f(s_c(K')) \qquad (5)$$

Remark :
Manipulators with type 1 geometries have only one basic component per aspect (the entire aspect). Equation (3-5) tells us that the t-connected subspaces of such manipulators are the images of the aspects, which is a well-known result [1], [8].

Figure 7 shows (in a cross section) the four t-connected subspaces of our 3R regional struture. For this example, each t-connected subspace $WA_{tj}=f(Qu_j)$ is the image of an aspect A_i minus a 2-dimensional subspace, that is actually the set $f(s_c(A \div Qu_j))$. This set forms an internal boundary through which the end-effector cannot pass. On the figure, this 2-dimensional set is represented as internal dark lines. This figure clearly expains why the continuous trajectory AB depicted in figure 4a is infeasible (AB crosses the forbidden internal boundary). The t-connected subspaces have been computed using simple procedures described in [4].

3- CONCLUSIONS AND PERSPECTIVES

In this paper, a new general formalism for the kinematic analysis of all nonredundant manipulators has been proposed. Both unlimited and limited joint have been considered in this study (examples with limited joints can be found in [4]). The jointspace uniqueness domains have been characterized for all geometries, their images under f define the maximal regions of feasible continuous trajectories of W. This work brings significant contribution to the comprehension of general manipulators kinematics. Further developments will take into account obstacles in the kinematic analysis. Kinematic performances of type 2 geometries in cluttered environments will be analysed and compared to those of type-1 geometries.

REFERENCES

1. BORREL P. and LIEGEOIS A. , "A study of manipulator inverse kinematic solutions with application to trajectory planning and workspace determination, Proc. IEEE Int. Conf. Rob. and Aut., pp1180-1185, 1986.

2. WENGER Ph. and CHEDMAIL P. : "On the connectivity of manipulator free-workspace", Journal of Robotic Systems, Vol 8(6), Dec 1991.
3. BURDICK J. W., "A classification of 3R regional manipulator singularities and geometries", Proc. IEEE Int. Conf. on Robotics and Automation, pp 2670-2675, Sacramento, California, 1991.
4. WENGER Ph : "Nouveaux résultats sur la cinématique des robots : généralisation des aspects et classification des morphologies", Internal Report No 91-21, Laboratoire d'Automatique de Nantes, E.C.N., Sept 1991.
5. KHOLI D. et SPANOS J., "Workspace analysis of mechanical manipulators using polynomial discriminants", Journal of Mechanisms, Transmissions, and Automation in Design, Vol. 107, June 1985, pp209-215.
6. KHOLI D. et HSU M.S., "The jacobian analysis of workspaces of mechanical manipulators", Journal of Mechanisms and Machine Theory, Vol 22, No 3, pp 265-275, 1987.
7. ROTH B., "Performance evaluation of manipulators from a kinematic viewpoint", Cours de robotique, IRIA, Toulouse, 1976, p : 233-263.
8. WENGER Ph. and CHEDMAIL P. : "Manipulator free-workspace, a new approach for robotic CAD systems", Proc. Int. Symp. on Rob. and Man. Syst. ISRAM 90, Vancouver, Canada, July 90.

FIGURES

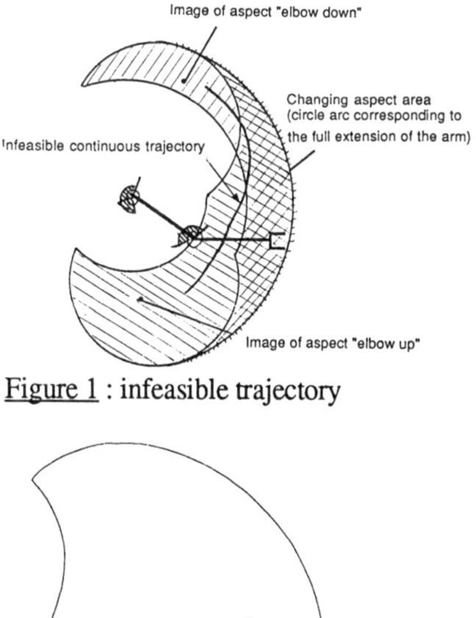

Figure 1 : infeasible trajectory

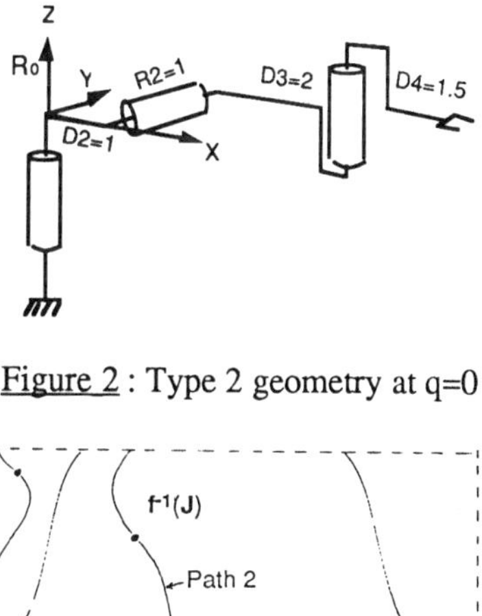

Figure 2 : Type 2 geometry at q=0

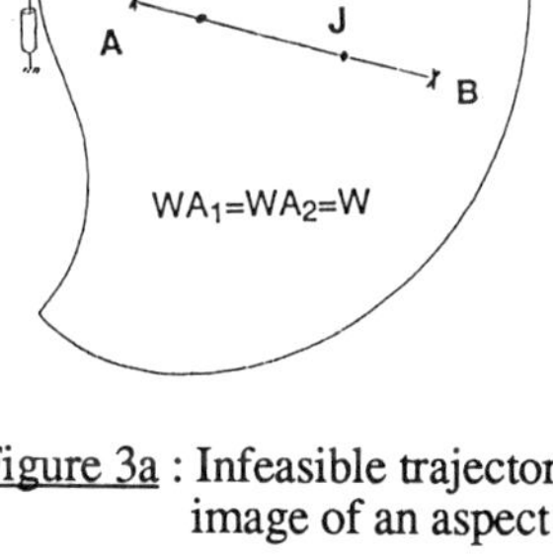

Figure 3a : Infeasible trajectory in the image of an aspect

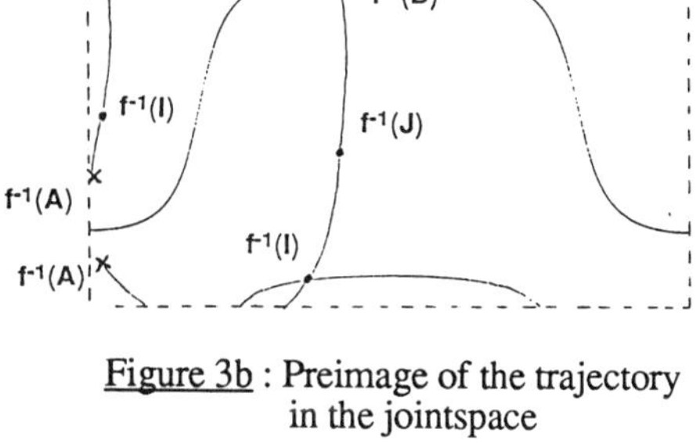

Figure 3b : Preimage of the trajectory in the jointspace

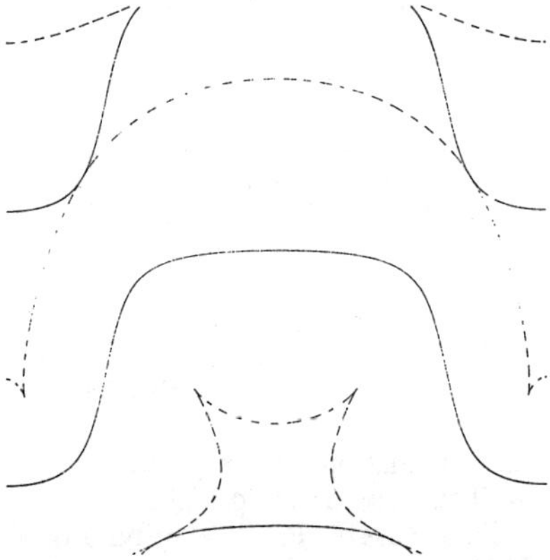

Figure 4 : characteristic surfaces (in dotted lines) and singularities (in full lines)

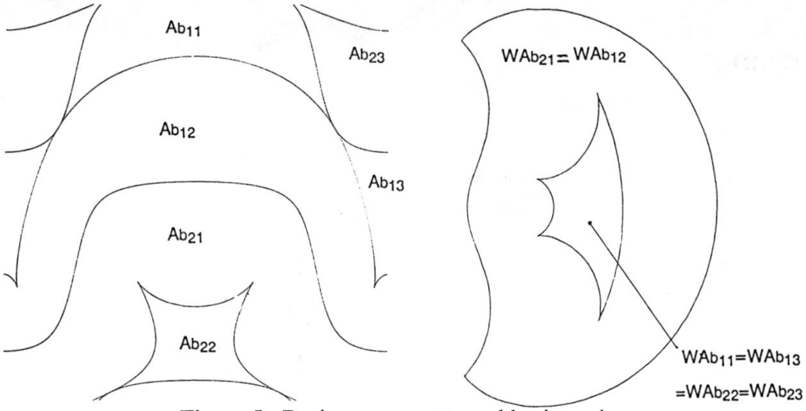

Figure 5 : Basic components and basic regions

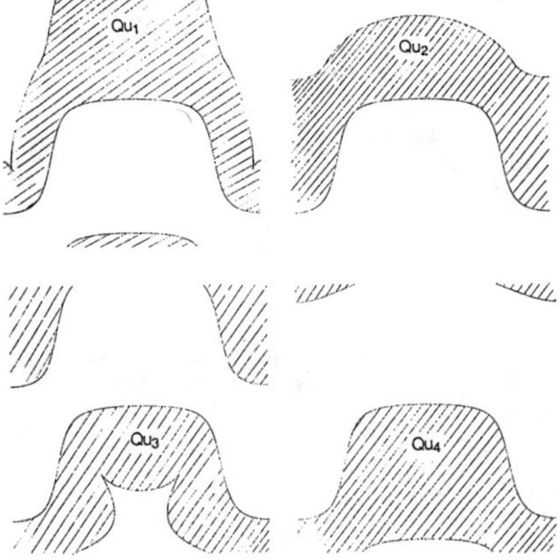

Figure 6 : Uniqueness domains

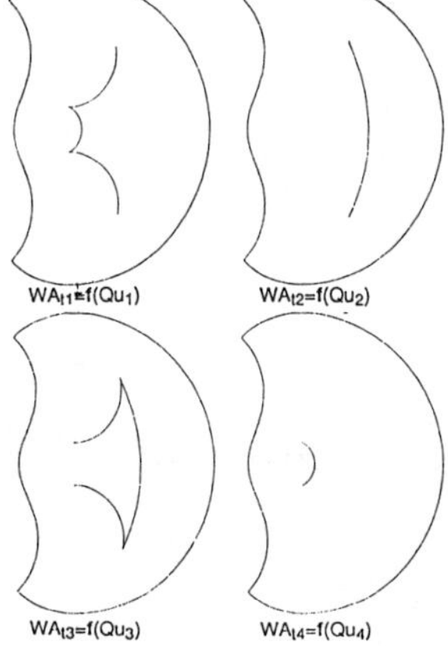

Figure 6 : T-connected subspaces

Torsional Stiffness of Mechanical Joints Using the Roto-Lok® Rotary Drive Technology

Donald G. Carson
Sagebrush Technology Inc.
Albuquerque, NM USA

Diamantis D. Preonas
Sagebrush Technology Inc.
Albuquerque, NM USA

ABSTRACT

Precision positioning of robotic or machine tool elements requires extremely high torsional stiffness. Quite often the limiting factor in the stiffness of a mechanism is the transmission linkage that couples the motor to the output. One solution has been to eliminate this transmission and drive the output axis directly. Another solution has been to add larger, heavier and more complex transmissions to achieve additional stiffness. In the case of robotics this adds mass which is self-defeating as the robot becomes excessively heavy with a poor strength to weight ratio. In the pursuit of precision other alternative approaches have been to let the joints be relatively compliant and use machine vision, tactile sensors or sophisticated controllers to circumvent the difficulties of poor mechanical stiffness.

A patented technology, the Roto-Lok rotary drive, can provide torsional stiffnesses generally higher than the structures to which they are attached. The drive utilizes many precision cables, wound in a unique way between a capstan (input member) and a driven drum (output member) to provide torque with high torsional stiffness and total freedom from backlash. Over the past decade applications for the technology have ranged from a few oz-in of output torque to 60,000 lb-ft. Several of these applications are described and guidelines provided to incorporating this technology into new designs.

INTRODUCTION

The Roto-Lok drive couples two rotating shafts together by means of flexible cables. Invented by one of the authors (Carson) the Roto-Lok drive was initially developed as a smoother, more precise means of pointing large observatory telescopes. The invention was intended to replace costly and delicate precision gears by using the averaging effect of multiple inexpensive cables to provide high torsional stiffness, zero backlash and superb drive smoothness. The technology has turned out to have numerous additional attributes that have made it attractive in a number of diverse applications in science and industry; applications as varied as positioning diamond cutting tools or steering 510 square meter solar collectors.

The Roto-Lok drive is a traction or friction drive. *Figure 1* illustrates the drive concept showing just one cable. Each cable is fixed on one end to a drum (the larger shaft), passes in a figure-eight wrap around the capstan (the smaller shaft), and terminates back on the drum with a tension spring holding it tight. The tension is such that when torque is applied to one shaft the other shaft is turned by the friction coupling of the cable against the shaft surface. The drive ratio is the ratio of pitch diameters of the two shafts. In most applications the cable is wrapped in a helix around both shafts. Usually one of the shafts has a helical groove that positions the cable axially to keep adjacent cables from rubbing where they cross between the drum and the capstan.

Typical Cable Wrap for Multiple Turn Roto-Lok Drive

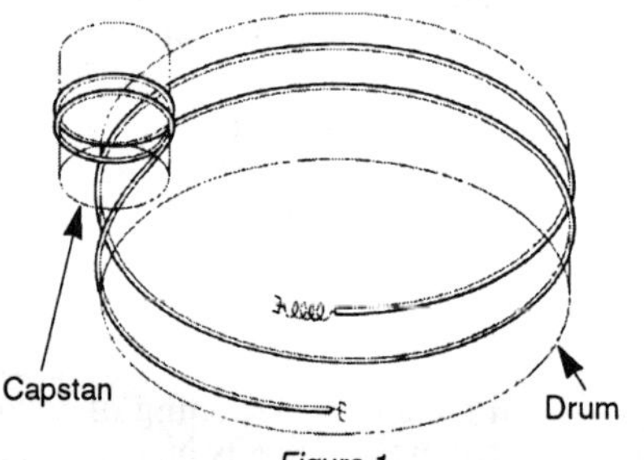

Figure 1

The rotation of the Roto-Lok drive is limited because each end of the cable must be terminated on the drum. The rotation must stop when the capstan gets to the end of the cables. Although most applications require less than 360° of angular travel some systems have travel of up to several revolutions (*Figure 1* illustrates two revolutions). The practical limit is set by the length of the capstan which must be long enough to accommodate the longer helical wrap required by more angular travel. This limited angular range is no problem in most robotics applications where signal or power cable routing dictate rotation angles of under 360°.

Roto-Lok Drive Basic Geometry

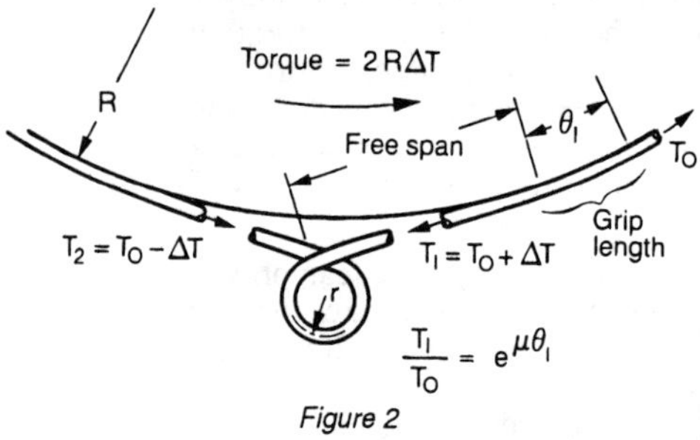

Figure 2

Figure 2 illustrates the basic geometry of the drive. With no torque load the drum and capstan are linked together by several very short pieces of cable (the free span illustrated) that are unsupported by either the drum or capstan. In effect these cable links are very stiff springs. When a torque load is applied, friction gradually picks up the load according to the familiar belt formula:

$$\frac{T_1}{T_0} = e^{\mu\Theta}$$

where μ is the coefficient of friction and Θ is the angle over which the load T_1-T_0 is gradually picked up. A conservative coefficient of friction between the cable and capstan is 0.1; with two turns about the capstan Θ is about 6 thus the maximum tension ratio from one side of the capstan to the other is about 1.8. In practice this means that the tension difference that can be supported before the cable slips is over 50% of the initial cable tension supplied by the spring and a significant torque can be supported.

The spring itself provides several functions. First it preloads its cable. Second it accommodates changes in temperature, wear, and load changes by keeping the cable taut. And third, it keeps the cable in place during assembly and use. Note that a spring is used only at one end of the cable, while the other end is firmly affixed to the drum. In most designs a symmetrical arrangement of cables is used with half the cables hard mounted on the left side of the drum and half on the right side. Thus in load conditions which are felt all the way to the cable ends, one is pulling the drum directly against the hard mounts.

The rotational stiffness of the drive is set by the stiffness of the short cable links between the capstan and drum and the number of links. In virtually every application investigated to date, the stiffness of this linkage can readily be made stiffer than the structure and bearings holding the parts together. One should note that as the torque load increases, more and more cable is affected, thus the stiffness goes down. When the direction of the load change reverses, the stiffness goes up to its original value (established by the cable free span) and again drops as more and more cable is affected. In a typical load carrying application, the deflection at a 10% load is 2.5 times what would be projected from the low load stiffness. This does lead to some hysteresis which is a characteristic of all traction drives. *Figure 3* presents a typical hysteresis curve for a 100 lb-ft drive. The small loop labeled A is the path taken as a load goes from -100 lb-ft to +54 lb-ft, to +21 lb-ft, to +100 lb-ft. The points labeled "knee" occur when the load is sufficient to reach all the way to the cable ends; at this point one is pulling against half hard terminated cable ends and half spring terminated ends.

Figure 3 illustrates one of the principal features of the Roto-Lok drive; it is stiffest at low loads or under a reversing load and becomes more compliant under a monotonically increasing load. The cable links are always in tension so there is no lost motion upon a direction reversal or when the incremental change in load changes direction. This is an ideal situation for a servo system; its gain can be high for fine positioning and homing in, yet compliant enough under high loading that the servo controller does not become unstable. In effect, the drive provides built-in mechanical damping and is truly free of any backlash.

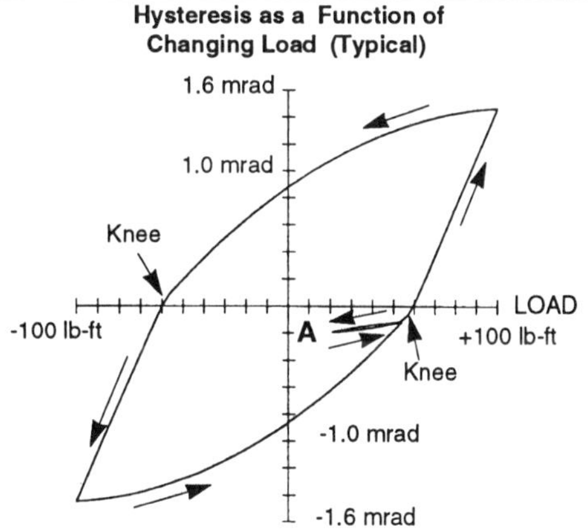

Figure 3

One result of this hysteresis behavior is that the cable remembers past history. When moving, the capstan gently picks up cable at one tension and lays it down on the drum usually at another tension. Thus in some systems that require fine absolute position motion, a feedback sensor on the output drum is often required to maintain absolute position control while a velocity sensor on the faster capstan shaft is usually adequate for velocity feedback. It should be noted that in at least half the systems developed to date, position control of the input capstan shaft alone is more than adequate to do the job.

Hysteresis also implies losses; however, the drive is remarkably efficient. To date we have been unable to separate the cable friction losses from bearing friction losses experimentally. If one integrates that area within the hysteresis curve of *Figure 3* and makes reasonable assumptions on the work being done by the 100 ft-lb torque, the cable losses can be calculated to be less than 1%. This is borne out in practice when one sees that a 20:1 or 32:1 ratio drive is easily back driven. Because there is no rubbing or sliding action or other mechanical action to produce sound, the drive is essentially noiseless. One is simply winding cable on one side of the capstan and off the other; the friction losses come from internal cable friction, bending of any cable jacket material, and the microcreep of hysteresis.

The engineering of cables themselves has progressed remarkably in the past 25 years. Our conjecture is that small cable technology was driven by the "golf-ball" typewriter and the myriad of copy machines, computer printer, and other electro-mechanical gadgets that have become prevalent in the past 25 years. Although cables can be engineered for many applications, so far we have used standard "aircraft design" cables in most of our applications simply because there is more engineering design data available for this style of cable than any other.

Cables do impose some limits on the size of Roto-Lok drives (on the small end). For long fatigue lifetime ($>10^7$ cycles) the capstan diameter should be 25 times the cable diameter. For example, if a cable with a 0.020" diameter is selected a 1/2" pitch diameter capstan is the smallest practical diameter; thus for a 20:1 drive ratio the drum should have diameter of 10". *Table 1* below presents the results of a design study in which 10 cables were used to obtain the maximum torque consistent with long lifetime for a number of drive ratios and drum diameters. In each case a standard commercial cable was selected and spring tensions optimized for maximum torque with long lifetime.

One of the first things that becomes apparent in *Table 1* is that the torque capacity increases dramatically with increasing drum diameter. This extraordinary gain is due to several factors. First, doubling the radius gives twice the torque for the same tangential load. Doubling the capstan size doubles the allowable cable diameter which increases its capacity by a factor of four and allows larger spring tensions. The empirical data of *Table 1* can be fit to the simple equation:

Output Torque as a Function of Diameter & Ratio

Ratio	Diameter (inches) 8"	16"	24"
10	320	3,240	11,000
15	138	1,125	4,850
20	79	640	2,170

Torque
lb-in

Table 1

$$T=81\,\frac{D^{3.16}}{R^{2.23}}$$

where T is the torque in lb-inches, D is the drum diameter in inches, and R is the drive ratio. Over the range of the parameters of *Table 1*, this equation provides a useful estimate of the capacity of a 10-cable Roto-Lok drive.

Torsional stiffness increases even more dramatically with drum diameter than load capability. *Table 2* presents the low load stiffnesses of the same 10-cable drives of *Table 1* where cable stretch and capstan shaft bending are both accounted for but bearing and structural deflections are assumed negligible. In this case doubling the radius halves the tangential load for one-fourth the angular deflection; and quadruples the cable cross section for an even stiffer cable and reduces the capstan shaft bending. This overall gain is partially offset by practical geometric considerations; for example, an increase in the capstan drum spacing to ease assembly which increases the unsupported cable free span.

As for torque, the data of *Table 2* can be fit to an empirical equation:

$$k=1.39 \times 10^5 \frac{D^{3.32}}{R^{1.82}}$$

where k is the rotary stiffness in lb-in/radian, D is the drum diameter in inches, and R is the drive ratio.

Output Stiffness as a Function of Diameter & Ratio

Ratio	Drum Dia. 8"	Drum Dia. 16"	Drum Dia. 24"
10	2.06×10^6	21.9×10^6	75.9×10^6
15	1.01×10^6	9.54×10^6	42.0×10^6
20	0.58×10^6	6.33×10^6	20.8×10^6

Stiffness (at low load) lb-in/radian

Table 2

APPLICATIONS

One of the first applications of the Roto-Lok drive technology was, not surprisingly, an astronomical telescope. In this case two two-axis drives were built for the "Infrared-Optical Telescope Array (IOTA) Consortium", a joint effort of Harvard, University of Massachusetts, MIT, the Smithsonian Astrophysical Observatory, and the University of Wyoming. In the IOTA project the light from two one-meter flat mirrors spaced up to 38 meters apart are combined interferometrically to provide the resolution of a 38 meter optic. Two Roto-Lok drives position each of these mirrors to sub-arc-second tolerances. In this case the design challenge was not load, but extreme stiffness under wind loading with a resonant frequency of 70 Hz or more. A 16-cable drive using 225 lb break strength cables, a 20 inch drum, and a 14.29 drive ratio yielded the requisite stiffness of 5×10^7 in-lb/radian.

The first commercial application was a wafer saw turntable used to dice silicon wafers. The application seems mundane, but after an optical calibration, the wafers must be positioned open-loop to 6 arc-second accuracy to keep the diamond saw in the middle of a 6 mil wide street (between chips) up to 12 inches long. Over 1000 of these wafer saws have been produced.

The largest application to date has been a 60,000 lb-ft drive used to position 5.1 meter by 100 meter solar collectors for a commercial solar power plant where life cycle cost was the deciding feature. In this case the critical element was the peak wind load during sudden thunderstorms. Three half-inch cables, a 7:1 ratio, and a 2 meter drum were used to satisfy this requirement.

The most sophisticated application has been multiple two-axis optical gimbals used as part of a proof of concept demonstration of satellite-to-satellite laser communications. In this case Sagebrush built the specialty servo drive motors right on the capstan shafts and the whole package was integrated to achieve 1.24 microradian tracking.

A number of applications take advantage of the extreme smoothness and stiffness of the Roto-Lok technology to do things that were previously not achievable with conventional technologies. One of these is the control of diamond turning cutters in production equipment used to make opthomological plastic contact lenses. Previously, throughput was limited by vibrations traced to the old drive that caused striations in the lens figure and required significant secondary polishing operations. The Roto-Lok drive yields a four times improvement in smoothness of the optical surface in half the time, thus requiring much less post polishing.

The Roto-Lok technology satisfied another smoothness requirement for the motion picture industry. In this case it replaces another technology in the azimuth drive of a 12 foot automated camera boom used for special effects cinematography.

In some cases the robustness, simplicity, and low cost of the Roto-Lok drive was the deciding factor in its selection. Two examples of this are a diesel engine remote throttle control and a mechanism for positioning the targeting mirror in a surface to air missile.

Framework for Task Based Design of Manipulators

Jin-Oh Kim and Pradeep K. Khosla
The Robotics Institute, Carnegie-Mellon University, Pittsburgh, PA, 15213

ABSTRACT

In this paper, we introduce a framework called Task Based Design (TBD) where we design robot manipulators based on a given task. TBD selects an optimal manipulator which can perform the given task best. With the proposed design framework, given degrees of freedom (DOF) and type, we design dimension, pose and base position simultaneously. New DOF and type can be tested iteratively until we obtain a manipulator that satisfies the given task requirement. Our framework consists of three modules of DOF section, type generation, and optimization algorithm and three inputs of task specification, manipulator specification and dexterity measure. We present an example of simple TBD with a 3 DOF planar manipulator. The framework is so general that it can be applied to path placement, workspace design and manipulator selection problem, just by fixing some variables.

INTRODUCTION

In theory, a robot's task can be changed by simply loading a new program into its controller; however, in practice, this is rarely the case. Each robot has a specific configuration and limited sensing capabilities that support only the applications for which the system was designed. For example, SCARA type manipulators are suitable for table-top assembly operations requiring selective compliance and accuracy but are not good for tasks requiring large workspace. The CMU Reconfigurable Modular Manipulator System (RMMS) was conceived to address the problems associated with conventional fixed-configuration manipulators[11].

The RMMS utilizes a stock of assemblable joint and link modules of different size and performance specifications. The modularity in mechanical, electrical and electronic design allows the user to design the optimal manipulator for the task at hand. As opposed to existing commercial manipulators which are made to perform as many tasks as possible, the RMMS has a unique feature that can provide a special manipulator for the given specific task. For a fully exploiting the above features of the RMMS, we need a framework for design of manipulators based on a given task specification. The main purpose of this research is to develop such a framework for design, called Task Based Design (TBD). The design framework developed in the paper can be applied to the general manipulator design.

The goal of TBD is to obtain *kinematic robot parameters* from a specified task. The robot parameters are DOF (degrees of freedom), the Denavit-Hartenberg parameters (type, dimension and pose) and the base position of a manipulator. Results of TBD include a set of optimal poses at a selected finite number of task points, which can be directly used as input for the planning problem over the whole trajectory. The framework can be extended to its subproblems where we want to use the best dexterity of a given manipulator. For example, path placement problem[9,10] and workspace design problem[1] are subproblems of our TBD.

In the following section, we define our TBD with three assumptions. Next, we introduce our design framework in detail. After showing an example of simple TBD of a 3 DOF manipulator for a crank turning task, we summarize this paper in the last section.

PROBLEM DESCRIPTION AND ASSUMPTIONS

Design variables for robot manipulators include

1. Degrees of freedom (N),
2. Denavit-Hartenberg (D-H)[3] parameters (α_i, a_i, d_i, θ_i) and
3. Base position (B),

where α_i is the i-th link twist angle, a_i the i-th link length, d_i the i-th joint offset distance and θ_i the i-th joint variable (angle). The D-H parameters is composed of type (T), dimension (D) and pose (P) of the manipulator. The type of n DOF manipulator is represented by n sets of triple (α_i, a_i, d_i). The magnitudes of a_i and d_i are determined by dimensional synthesis and the magnitude of θ_i is determined by pose synthesis. But both are made at the same time. The design problem addressed in this paper is to map the specified task onto the above robot parameters. The problem in which the above five variables are unknown can be stated as follows.

Design a manipulator (M) of DOF N, type T, dimension D, pose P and base position B subject to the given task specification **E**, manipulator specification **R** and some optimality criterion **C.**

For this problem, design variables are 5 tuples of (N,T,D,P,B). The input to this problem is a triple (**E**, **R**, **C**). The task specification is related to only a given specific task and the manipulator specification constrains the search space of design variables. In this paper, we use a dexterity measure as an optimality criterion. Even though our ultimate goal is to solve this problem, our optimization algorithm does not solve this problem directly because it is highly nonlinear and tremendously complex. Instead, we first solve its subproblem where DOF and type are given, and repeat this with other DOF and/or type until a manipulator that satisfies given task/manipulator specifications is obtained.

We made the following assumptions in this paper to reduce complexity.

1. The three design variables of (D, P, B) are continuous and the DOF (N) and the type (T) are discrete.
2. Only serial manipulators with revolute joints are considered.
3. The task space can be approximated by a finite number of task points.

The first assumption helps a lot to reduce complexity because treatment of type as continuous variable make the number of variables almost twice as many as discrete case. In the following section, we propose our framework for TBD.

FRAMEWORK

The design framework proposed in this paper is shown in FIG. 1. As mentioned in the previous section, the input to design framework is a triple of task specification, manipulator specification and dexterity measure. In addition to this triple, framework has three modules of DOF selection, type generation, and optimization algorithm. These three modules are related to different design variables of 5 tuples of (N,T,D,P,B), These design variables are decomposed into two groups (continuous and discrete). In our framework, optimization is directly performed in the space of continuous variable and optimal values of discrete variables are obtained by performing the optimization, repeatedly. DOF, one of discrete variable, is selected based on how many DOF is necessary for a given task and all possible candidate types (the other discrete variable) for a selected DOF are generated according to rules developed in this paper. A set of DOF and type is forwarded to the optimization stage where we optimize dimension, pose and base position, simultaneously. Task specification represents all constraints related to task and manipulator specification covers constraints on dimension, pose and base position. Dexterity measure can be used to select one optimal manipulator inside of solution space in which dimension, pose and base position are variables or an approximate manipulator may be chosen in terms of least square concept. We developed an optimization algorithm called Multi-Population Genetic Algorithm (MPGA) that searches a global optimum in parallel over design space which includes lots of local minima and inexplicit and nondifferentiable constraints. In the following subsections, we describe all units in detail.

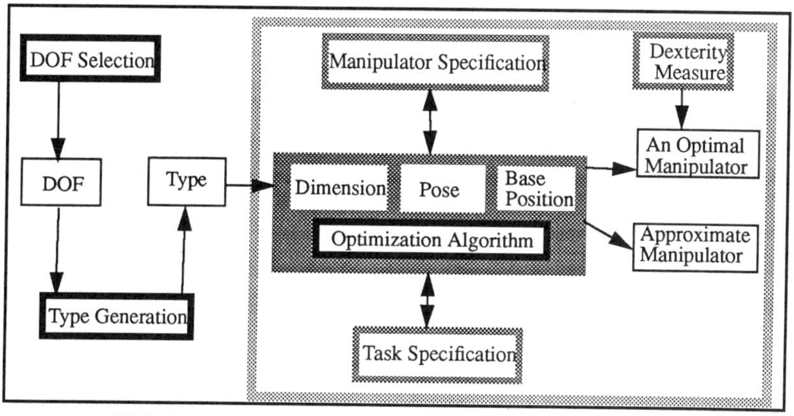

FIG. 1. Schematic Diagram for design framework

Task specification

In general, we can define *task space* as a space of a given task that is required to be reached and *workspace* as a volume reachable by a manipulator. The concepts of task space and workspace are extended and generalized in our design problem. The task specification is a generalization of task space while manipulator specification in the following subsection is that of workspace. Task specification is related to task only. Below is summarized task specifications that are used in our design.

(1) Reachability constraint must be satisfied at the finite number of task points, which were assumed to be able to approximate a given trajectory. The reachability constraint enables us to dispense with inverse kinematics that exists for only special class of manipulators. Typically, the closest and furthest points from a given base space should be included in a set of task points. Selection of task points affects the success of design. However, as the number of task points increases, the risk would be lessened. The number of task points increases when some section of trajectory is found to be not reachable by a designed manipulator in a follow-up checking stage.

(2) Task constraints vary depending on the given task. Tasks such as bracing, obstacle avoidance, singularity avoidance and dexterity may require redundancy of a manipulator. Among these, I consider dexterity and singularity avoidance, which is represented by features of the velocity and force ellipsoids1.

(3) Joint angle change constraint is based on the fact that the joint angle change between two adjacent task points must be kept small. This constraint depends on the distance between two adjacent task points.

Manipulator specification

Manipulator specification, an extended concept of task space of a manipulator, constrains design variables of D, P and B. Like task specification, manipulator specification is expressed in the form of constraint.

(1) Dimension constraint is a constraint on link (link length and joint offset distance). The constraint is that the inner link is always longer than the outer link. This constraint is based on the following observations; the outer link is more appropriate for fine motion and the inner link for gross motion and the dextrous space requires the outer link to be shorter than the inner link.

(2) Pose constraint is about joint limit. The joint limit constraint assumes that the total ranges of all actuators are limited and is formulated by an interval between two limits of range. Only after a structure is determined, joint limit is represented by two absolute boundary joint angles.

(3) Base position constraint is a constraint on the possible location of base position. This is expressed by a bounded space in a 3 dimensional space.

Dexterity measure - optimality criterion

The need of dexterity measure is obvious. It is useful when there exist multitude of solutions that satisfy all task/manipulator constraints. Dexterity measures quantize performance and thus can be called "performance index". There are two requirements that must be satisfied by a well defined dexterity measure; physical meaning and scale independence. Without physical meaning, optimality of a designed manipulator is hard to interpret. If it is scale dependent, we may come up with an awkward design like infinite link length. Dexterity measures that satisfy this two requirements are the relative manipulability, the condition number and the measure of isotropy[5].

DOF selection

In this module, we select a minimum DOF which can match a given task. In general, 6 DOF is necessary for a spacial positioning and orientating. However, not all tasks require 6 DOF. For example, when a tool at the end-effector has an orientational symmetry as in arc welding and grinding tasks, the minimum DOF becomes five. On the other hand, when more dexterity is required or there are obstacles in task space, the minimum DOF may become more than six. In this paper, we define the minimum DOF as the number of joints that matches position and orientation of a given task. When this minimum DOF fails to provide enough dexterity required, we increase DOF by one and test all candidate types of manipulators. We repeat this until we get an optimal manipulator for a given task.

Type generation

All possible candidate types of manipulators for a selected DOF are generated in this module, based on some heuristic rules that have been applied to almost all existing manipulators. The rules we use in this unit are as follows;

(1) Kinematic simplicity: The kinematic simplicity constrains connection between two adjacent joints. The link twist angle (α) is either 0 or $\pi/2$. The link length (a) and joint offset distance (d) are zero or nonzero and at least one of two is zero. Then, there exist four possible connections between two joints; **1**(α=0, a=nonzero, d=0), **2**($\pi/2$, 0, 0), **3**($\pi/2$, nonzero, 0) and **4**($\pi/2$, 0, nonzero).

(2) Planar manipulator: The planar manipulator is defined as a manipulator whose every joint axis is parallel to each other. This planar manipulator cannot be used for 3 dimensional positioning even though it has the DOF that matches a given task.

(3) Mechanical simplicity[4]: The connection that requires two actuators located at the same position should be eliminated because it is hard to be manufactured. In terms of the 4 possible connections derived from kinematic simplicity, this can be restated as "connection **2** must be followed by connection **4** only". Otherwise, a manipulator is said to be mechanically complex.

(4) Spherical wrist: The connection between a spherical wrist and positioning structure must be **2**. Otherwise, it violates the mechanical simplicity.

The above rules are independent of tasks. The application of the above rules to 6 DOF manipulators with a spherical wrist results in the 8 possible candidate types of manipulators as in FIG. 2.

Optimization algorithm: Multi-Population Genetic Algorithm (MPGA)

Our MPGA[6,7] is a parallel implementation of Simple Genetic Algorithm (SGA)[2] that is an adaptive search technique based on mechanics of natural genetics. Different from SGA and other GAs, MPGA uses multiple populations. Each population searches an optimal manipulator for one task point and are connected to each other by connecting constraints. Two connecting constraints are link constraint and base position constraint. Link constraint enforces link lengths for all optimal manipulators corresponding to all task points to be the same and the base position constraint makes all optimal manipulators have the same base

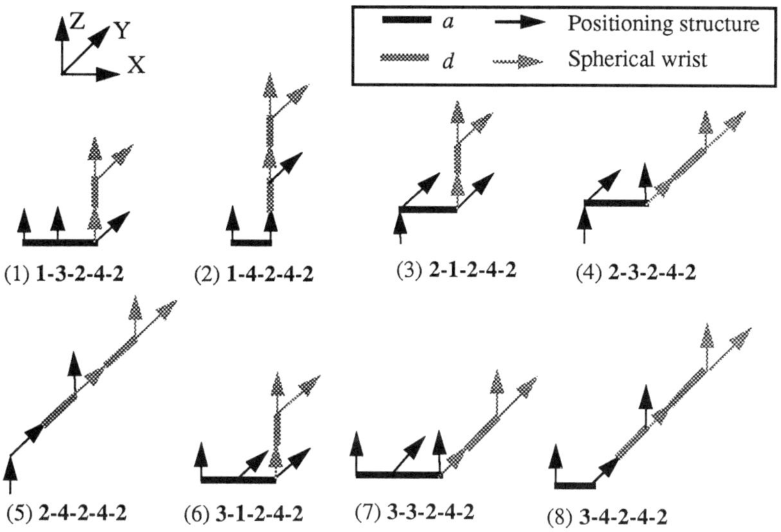

FIG. 2. Eight candidate types of 6 DOF spatial manipulators

position. Without these two constraints, our MPGA is simply a collection of SGAs each of which will design an optimal manipulator for one corresponding task point.

The advantage of MPGA is that the complexity does not increase much as the number of task points increases. On the contrary, if we use other search techniques like SGA and simulated annealing, the search space of task based design increases exponentially as the number of variables increases. This implies that the quality of the solution decreases rapidly as the search space increases. For our MPGA, as task points increases, only the number of populations increases.

Another advantage is that MPGA can provide a general search technique for a manipulator either with or without a prismatic joint by just turning on or off link constraint and for a manipulator with mobile base by adjusting base position constraint.

EXAMPLE OF TASK BASED DESIGN

In this section, we present an example of TBD for crank turning task in FIG. 3(a). For simplicity, the base position is given at the origin. Whole trajectory is in a plane. A 2 DOF planar manipulator is enough for this task. However, we assume that 2 DOF is not enough to perform this task with high dexterity (relative manipulability) and we design an optimal 3 DOF planar redundant manipulator. Thus, 3 DOF and type (planar) are input to the optimization stage in FIG. 1. We design based on the 4 task points from the circular trajectory. The result is shown in FIG. 3(b). We have used the relative manipulability[5] as a dexterity measure in this example. Our design framework has been applied to design of an optimal 7 DOF spatial manipulator for space shuttle tile servicing task. Details can be found in (Kim, 1992).

SUMMARY

In this paper, we have introduced a framework for Task Based Design (TBD) of optimal manipulator which is an ongoing project related to the CMU RMMS. The TBD designs an optimal manipulator that is most appropriate for a given task. The framework for TBD consists of 3 inputs (task/manipulator specification and dexterity measure) and 3 modules (DOF selection, type generation and optimization algorithm). We have shown a simple TBD with a 3 DOF planar manipulator for crank turning task. Our TBD is so general that

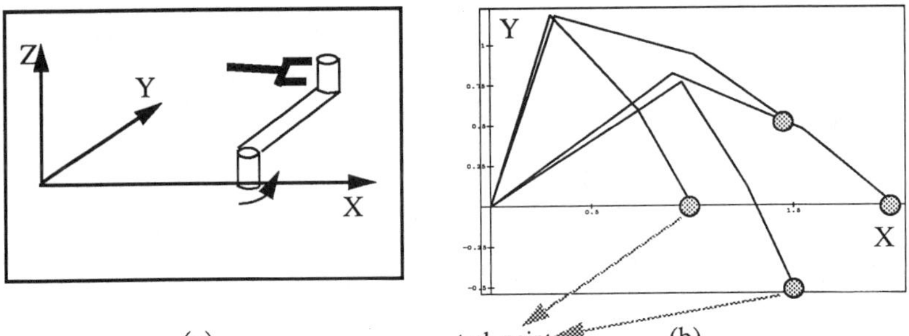

(a) (b)

FIG. 3. Task Based Design of an optimal 3 DOF planar manipulator

all of its subproblems can be solved using only partial of our framework. For example, path placement can be solved without DOF selection, type generation and fixing variables corresponding dimension since a manipulator is given.

ACKNOWLEDGMENT

This research was funded in part by NASA under grant NAG-1-1075, DOE under grant DE-F902-89ER14042, the Department of Electrical and Computer Engineering, and The Robotics Institute, Carnegie-Mellon University.

REFERENCES

1. Chiu, S, "Task Compatibility of Manipulators Postures", *The International Journal of Robotics Research*, vol.7, No. 5, pp. 13-21, (October, 1988).
2. Goldberg, D.E., *Genetic Algorithms in Search, Optimization, and Machine Learning*, (Addison-Wesley publishing company, 1989).
3. Hartenberg, R.S. and Denavit, J., *Kinematic Synthesis of Linkages*, (McGraw-Hill Book Company, 1964).
4. Hollerbach, J.M. "Evaluation of Redundant Manipulators Derived form the PUMA Geometry", Robotics and Manufacturing Automation: The Winter Annual Meeting of the ASME, pp. 187-192, Florida, (November, 1985).
5. Kim, J.-O. and Khosla, P.K., "Dexterity Measures for Design and Control of Manipulators", *IEEE/RSJ int. workshop on Intelligent Robots and Systems (IROS'91)*, Osaka, Japan, (November, 1991).
6. Kim, J.-O. and Khosla, P.K., "A Multi-Population Genetic Algorithm and Its Application to Design of Manipulators", *IEEE/RSJ int. workshop on Intelligent Robots and Systems (IROS'92)*, (July, 1992).
7. Kim, J.-O. and Khosla, P.K., "Task Based Synthesis of Optimal Manipulator Configurations", 1992 Japan-USA Symposium on Flexible Automation - A Pacific Rim Conference- ISCIE, San Francisco, CA, (July, 1992).
8. Kim J.-O. *Task Based Design of Robot Manipulators: Application to the CMU RMMS*, Ph.D. Thesis, The Robotics Ph.D. Program, Carnegie-Mellon University, Pittsburgh, (August, 1992).
9. Nelson, B., Pedersen, K. and Donath, M., "Locating Assembly Tasks in a Manipulator's Workspace", *IEEE int. conf. on Robotics and Automation*, pp. 1367-1372, (1987).
10. Alfonso Pamanes G. J., and Zeghloul, S., "Optimal placement of Robotic Manipulators Using Multiple Kinematic Criterion", *IEEE int. conf. on Robotics and Automation*, pp. 933-938, Sacramento, CA, (1991).
11. Schmitz, D.E., Khosla, P.K., and Kanade, T., "The CMU Reconfigurable Modular Manipulator System", *Proceedings of the 18-th ISIR*, Australia, (1988).

ERROR-MODEL-BASED ROBOT CALIBRATION USING A MODIFIED CPC MODEL

Hanqi Zhuang, Luke K. Wang and Zvi S. Roth
Robotics Center and Dept. of Electrical Engineering
Florida Atlantic University
Boca Raton, FL 33431

I. INTRODUCTION

Calibration is a process by which the accuracy of the manipulator is improved by modifying its control software. In an error-model-based kinematic calibration process, selection of a proper kinematic model is one of the keys to the success of the calibration task[1]. This paper presents a new kinematic model particularly designed for a convenient error model based robot calibration.

A kinematic model suitable for robot calibration is one which is complete and parametrically continuous[1]. A complete kinematic model is one that can completely describe the geometry and motion of any type of serial manipulators. It also allows the arbitrary placement of the world coordinates and assignment of the robot home position. In order to be complete the model has to possess the (minimum) required number of independent kinematic parameters. This number, for a serial manipulator having a low pair joints is 4R+2P+6[1], where R and P are the number of revolute and prismatic joints respectively. Parametric continuity implies no parameter jumps in the model. It means that any continuous change of position and orientation of any robot joint axis will result in continuous changes of the model link parameters.

For DH model, parameter jumps always arise when two neighboring joint axes change from nominally parallel to actually almost-parallel. The assignments of the world and tool frames are also restricted in the DH model. Many kinematic modeling conventions have been proposed to coup with the problems existed in the DH model[2]. Roberts[3] presents a new line representation consisting of four parameters which are the theoretical minimum. The Shape Matrix approach[4] separates the joint variables from the other link parameters. The complete and parametrically continuous (CPC) model[5] has a similar structure to the shape matrix convention. For the internal link transformations it adopts Roberts' line representation. Two extra parameters are introduced in the last link transformation for the arbitrary assignment of tool frame. It was shown[5] that the CPC model is complete and parametrically continuous. The error model in the CPC convention is also singularity free.

This paper presents a modification to the CPC model. Rather than using the direction vectors of joint axes, it employs two angles to accommodate the required rotation for each internal link transformation. Furthermore, three angular and three linear parameters are employed for the last link transformation. By this modification, the resulting error model can be derived in the same manner as that for the DH model, which is more user friendly compared with the CPC error model. The penalty for this modification is that the error model may be singular if the tool axis is perpendicular to the last joint axis. This singularity can be avoided by either changing the z axis of the tool frame or by introducing an intermediate link frame that is fixed in the tool frame and has different orientation with respect to the tool frame.

II. THE MODIFIED CPC (MCPC) KINEMATIC MODEL

The MCPC model uses four parameters to represent the internal link transformation of a robot. It separates joint variables from link parameters. Besides, the MCPC model employs two angular parameters to model the rotation part of each link transformation. The procedure for the assignment of the MCPC internal link parameters is as follows (Figure 1):

1. Construct a plane (**B**$_i$) perpendicular to the z_i axis passing through the origin of $\{i-1\}$.
2. Rotate an angle α_i about x_{i-1} axis until the y axis of the moving frame lies on plane **B**$_i$. This y axis is now parallel to the y_i axis.
3. Rotate an angle β_i about the resulting y axis until the x axis of the moving frame lies on plane **B**$_i$. This x axis is now parallel to the x_i axis. The resulting z axis is also parallel to the z_i axis.
4. Translate along the x_i and y_i axes by x_i and y_i, respectively. The moving frame is now coincident with $\{i\}$.

In summary, the ith link transformation in the MCPC convention is

$$B_i = Q_i\ Rot(x\ ,\ \alpha_i)\ Rot(y,\ \beta_i)\ Trans(x_i\ ,\ y_i\ ,0) \qquad i=0,1,2,...,n-1\ . \tag{2.1}$$

Similar to the CPC model, two additional link parameters are used in the nth link transformation to allow an arbitrary assignment of the tool frame, that is,

$$B_n = Q_n\ Rot(x\ ,\ \alpha_n)\ Rot(y\ ,\ \beta_n)\ Rot(z\ ,\ \gamma_n)\ Trans(x_n\ ,\ y_n\ ,\ z_n)\ . \tag{2.2}$$

For convenience, we sometimes denote $(x_i, y_i, 0)$ by $\boldsymbol{l}_i$ and $(x_n\ , y_n\ , z_n)$ by $\boldsymbol{l}_n$.

III. THE LINEARIZED MCPC ERROR MODEL

The derivation procedure is composed of two steps. First, the differential relationships between cartesian errors and link parameter errors in each link are derived. The MCPC error model is then obtained by substituting the above results into a generic form of a robot kinematic error model[6]. The MCPC error model is given in the following theorem.

Theorem 1: Let $\boldsymbol{\delta}$ and $\boldsymbol{d}$ be respectively the position and orientation error vectors of the robot end-effector. Then the linearized MCPC error model can be written as follows:

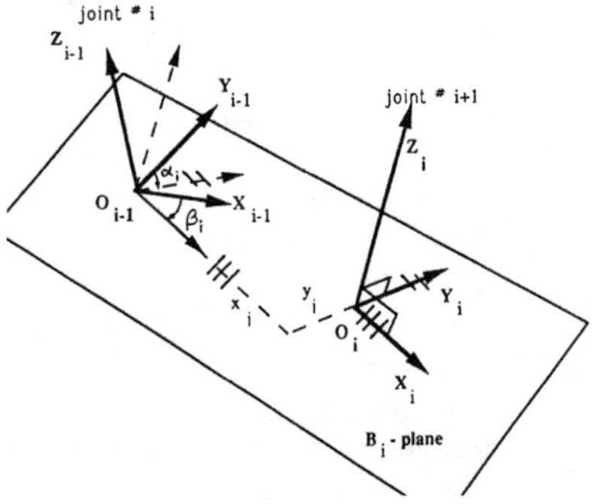

FIG. 1. The MCPC modeling convention.

$$\delta = R_\alpha\ \Delta\alpha + R_\beta\ \Delta\beta + k_{\gamma,n}\ \Delta\gamma_n \tag{3.1}$$

$$d = T_\alpha\ \Delta\alpha + T_\beta\ \Delta\beta + T_x\ \Delta x + T_y\ \Delta y + k_{\gamma,n}\times l_n\ \Delta\gamma_n + k_{z,n}\ \Delta z_n \tag{3.2}$$

where $\Delta\alpha = [\Delta\alpha_0\ \Delta\alpha_1\ ...\ \Delta\alpha_n]^T$, and similar for $\Delta\beta$,Δx, and Δy. $\boldsymbol{R}_\alpha$, $\boldsymbol{R}_\beta$, $\boldsymbol{T}_\alpha$, $\boldsymbol{T}_\beta$, $\boldsymbol{T}_x$ and $\boldsymbol{T}_y$ are all 3×(n+1) matrices whose components are functions of joint variables and link parameters. More specifically, the ith column of each such matrix is as follows:

$$[R_\alpha]_i = \begin{bmatrix} n_{i+1} \cdot k_{\alpha,i} \\ o_{i+1} \cdot k_{\alpha,i} \\ a_{i+1} \cdot k_{\alpha,i} \end{bmatrix} \tag{3.3a}$$

$$[R_\beta]_i = \begin{bmatrix} n_{i+1} \cdot k_{\beta,i} \\ o_{i+1} \cdot k_{\beta,i} \\ a_{i+1} \cdot k_{\beta,i} \end{bmatrix} \tag{3.3b}$$

$$[T_\alpha]_i = \begin{bmatrix} ((p_{i+1}+l_i) \times n_{i+1}) \cdot k_{\alpha,i} \\ ((p_{i+1}+l_i) \times o_{i+1}) \cdot k_{\alpha,i} \\ ((p_{i+1}+l_i) \times a_{i+1}) \cdot k_{\alpha,i} \end{bmatrix} \tag{3.3c}$$

$$[T_\beta]_i = \begin{bmatrix} ((p_{i+1}+l_i) \times n_{i+1}) \cdot k_{\beta,i} \\ ((p_{i+1}+l_i) \times o_{i+1}) \cdot k_{\beta,i} \\ ((p_{i+1}+l_i) \times a_{i+1}) \cdot k_{\beta,i} \end{bmatrix} \tag{3.3d}$$

$$[T_j]_i = \begin{bmatrix} n_{i+1,\,j} & o_{i+1,\,j} & a_{i+1,\,j} \end{bmatrix}^T \tag{3.3e}$$

$$k_{z,n} = \begin{bmatrix} n_{n,z} & o_{n,z} & a_{n,z} \end{bmatrix}^T \tag{3.3f}$$

where $\boldsymbol{k}_{\alpha,i} = Rot^T(y,\beta_i)[1\ 0\ 0]^T$ and $\boldsymbol{k}_{\beta,i} = [0\ 1\ 0]^T$. $\boldsymbol{k}_{z,n}$ is the 3rd row vector of the rotation submatrix in B_n. The $\boldsymbol{n}_i$, $\boldsymbol{o}_i$, $\boldsymbol{a}_i$, and $\boldsymbol{p}_i$ are the 1st, 2nd, 3rd, and 4th column vectors (each is 3×1) of the homogeneous transformation $T_i=B_0B_1\ldots B_i$.

IV. EXPERIMENTAL STUDY

A PUMA 560 robot was used for experimental study. Its MCPC parameters are listed in Table I. A stereo vision system was used as measurement device to collect robot poses at different robot measurement configurations. One set of identified error parameters are listed in Table II. The pose errors using the identified models obtained with different number of measurement configuration are shown in Figures 2-3.

TABLE I. Nominal MCPC link parameters for the PUMA 560.

Link #	x_i(mm)	α_i(deg)	y_i(mm)	β_i(deg)	z_6(mm)	γ_6(deg)
0	-150	90	-960	180		
1	0	-90	-700	0		
2	431.8	0	0	0		
3	-20.32	90	149.09	0		
4	0	-90	-433.07	0		
5	0	90	0	0		
6	0	0	0	0	300	100

The following remarks are given as conclusions:

1. The nominal parameters in Table I were established using "rough" position measurements. Therefore the corrections to the nominal link parameters are very substantial, comparing to those in the Tables II.
2. The error parameters in internal links are small, reflecting the proportionality (or parametric continuity) of the MCPC model.
3. The experimental study confirmed that due to measurement uncertainties of the camera system and the CMM, at least 10 measurements have to be taken for the calibration of the whole manipulator. On the other hand, using more than 10 measurements did not improve much the calibration accuracy.

TABLE II. Identified MCPC link parameter errors (# of measurements = 15).

Link #	x_i(mm)	α_i(rad)	y_i(mm)	β_i(rad)	z_6(mm)	γ_6(rad)
0	40.260292	0.048515	-41.885082	0.004098		
1	-0.154299	-0.000715	1.745687	-0.065828		
2	0.488876	-0.000027	0.093721	-0.000460		
3	-1.116960	0.004340	-0.297157	0.010392		
4	-0.076503	0.001741	2.043655	-0.023318		
5	0.929132	-0.001362	0.603039	-0.078147		
6	-12.263172	0.054457	-6.760906	0.027106	31.2467	0.00233

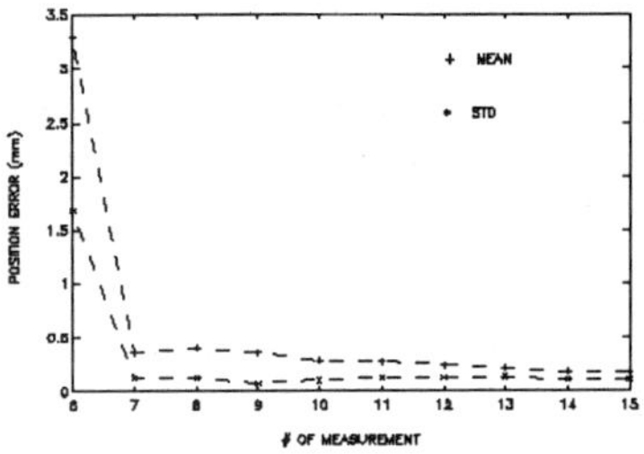

FIG. 2. Position errors after the entire robot is calibrated.

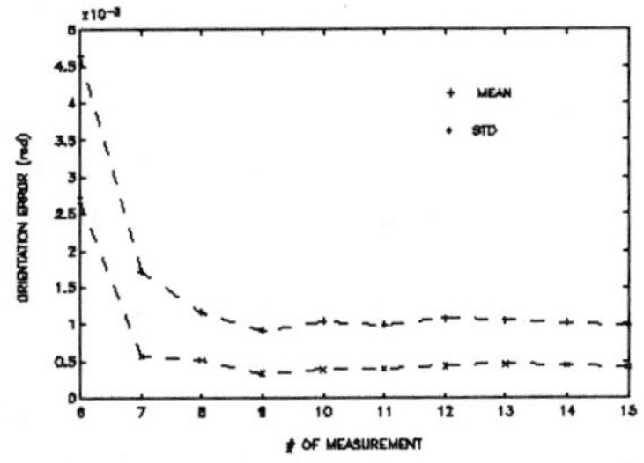

FIG. 3. Orientation errors after the entire robot is calibrated.

REFERENCES

1. L. J. Everett, M. Driels and B.W. Mooring, "Kinematic Modeling for Robot Calibration," *Proc. IEEE Int. Conf. on Robotics and Automation*, 1987, pp.183-189.
2. S. Hayati, "Robot Arm Geometric Parameter Estimation," in *22nd IEEE Int. Conf. on Decision Control*, 1983, pp.1477-1483.
3. K. S. Robert, "A New Representation for a Line," *Proc. Int. Conf. on Computer Vision & Pattern Recognition*, 1988, pp.635-640.
4. P. L. Broderick and R.J. Cipra, "A Method for Determining and Correcting Robot Position and Orientation Errors Due to Manufacturing," *J. of Mechanisms, Transmission and Automation in Design*, **110**,3-10(1988).
5. H. Zhuang, Z.S. Roth and F. Hamano, "A Complete and Parametrically Continuous Kinematic model," to appear in *IEEE Trans. on Robotics and Automation*, 1992.
6. C. H. Wu, "A Kinematic CAD Tool for the Design and Control of a Robot Manipulator," *Int. J. of Robotics Research*, **3**(1),58-67(1984).

Modelling and Observer Design of a Hydraulic Manipulator Actuator

Zhenfu Lin, Eric Jackson
International Submarine Engineering Ltd.
1734 Broadway St, Port Coquitlam, B.C. CANADA V3C 2M8

Mehrdad Saif
Simon Fraser University
Burnaby, B.C, CANADA V5A 1S6

ABSTRACT

Because of their stiffness and high force-to-weight ratio, hydraulic manipulator actuators are very useful for a variety of applications, especially for manipulating heavy objects. In addition, they are well suited for use in harsh environments such as underwater applications. These actuators are difficult to control, however, due to the lack of accurate models. Commercial hydraulic actuators exhibit a number of nonlinear and nonideal conditions such as expansion effects in the hydraulic lines, underlap, nonlinearity and hysteresis characteristics in the servovalve, and stiction.

This paper presents the investigation, modelling and force control of the hydraulic system. The aforementioned nonideal characteristics of the hydraulic drive were experimentally determined and mathematically modelled. Based on the experimental results, a nonlinear sixth-order model of the system was obtained. Next a linearized model was obtained and then validated against the nonlinear model. The linearized sixth-order model was then reduced using a balanced model reduction technique; the two least controllable/observable modes were removed. A state Observer was then designed and simulated based on a particular operating point.

INTRODUCTION

Hydraulic actuators are used in many applications because they generate large forces, have a high force-to-weight ratio and are very stiff. As a result, their applications are found in teleoperated manipulators, mobile equipment, lifting equipment -- especially those handling heavy weights -- and in the aerospace and machine tool industries. Hydraulic systems are difficult to control accurately, however, because of their nonlinear characteristics and the lack of practical mathematical models. This paper describes a project that involved modelling a hydraulic manipulator actuator and designing and simulating an observer for it. The observer is intended to be used in a force controller for the actuator

The manipulator under investigation is shown in Figure 1.1. The piston and the robotic arm are rigidly connected; thus, the robotic arm is driven by the actuator, whose force is generated as a result of the differential pressure across the piston. For this project, we consider only one degree of freedom. The physical arrangement of the hydraulic system is illustrated in figure 1.2. The pump energizes the fluid which is then forced into the cylinder through the servovalve orifice and eventually back into the reservoir. The fluid that goes through the servovalve forms a hydraulic flow that pushes the piston connected to the robotic arm. In this way, the actuator can move forward or backward depending on the direction of flow.

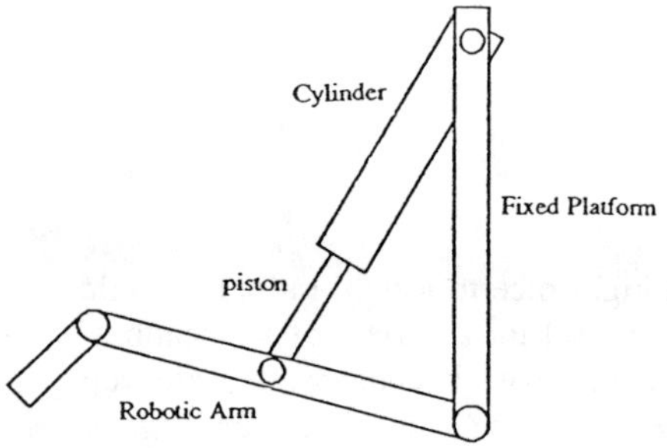

Figure 1.1 Hydraulic Manipulator Actuator

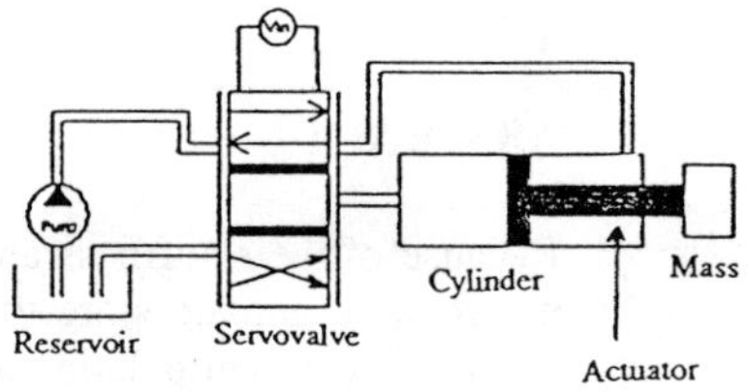

Figure 1.2 Typical Hydraulic System

The servovalve orifice area through which the fluid flows is controlled by a voltage source. Therefore, varying the servovalve voltage input controls the amount of flow.

The external electronic and computer interface that provides the means for data acquisition and control is depicted in figure 1.3. The voltage source for controlling the hydraulic system is the output of a D/A converter incorporated into the electronic and computer interface. As figure 1.3 shows, the computer is connected to three sensors through A/D converters: two pressure sensors and an angular position sensor.

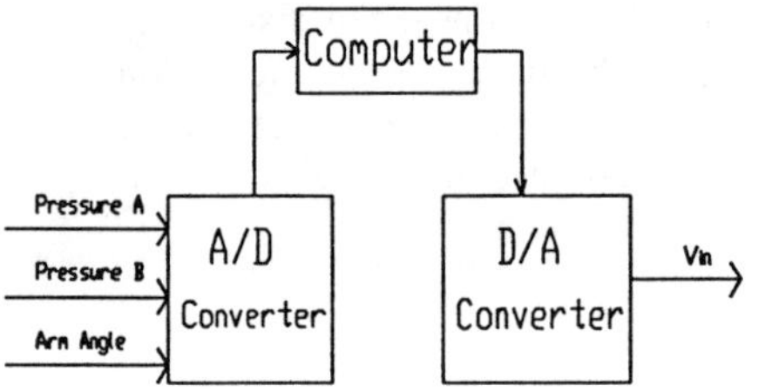

Figure 1.3 Computer Interface

Some researchers have spent considerable effort trying to find ways to control the force of a hydraulic manipulator. Based on the assumption that the piston position is fixed and has zero velocity, Conrad and Jensen [1] modelled the hydraulic system as a third order linearized system, and through computer simulation they designed and compared different controllers using various control strategies. They concluded that modern control techniques yielded far better performance than that of traditional controllers. For an earlier thesis project, Springle [2] also designed and implemented a controller similar to the one suggested by Conrad and Jensen.

The work done by Conrad and Jensen and by Springle provides a pathway for understanding the hydraulic system and for designing controllers for it. However, more work is needed in order to design a practical controller. While based on the results obtained by Conrad and Jensen and by Springle, this project investigates some aspects of hydraulic actuator not considered before.

This project examines a more realistic situation, considering many non-ideal conditions. The servovalve exhibits underlap characteristics and the hydraulic line expands and contracts when there is a change in pressure. The force and differential pressure are no longer equivalent as they were in the previous studies because the area on both sides of the piston is different. The sum of the pressures on both sides of the piston may also not be equal to the supply pressure due to the underlap effect in the servovalve. And, unlike the arrangement of fixed piston position in the previous work, the piston here is allowed to move. Taking these non-ideal characteristics into account, a sixth-order mathematical model was obtained in this work, as compared to the third-order one generated both by Conrad and Jensen and by Springle. This model is then used to design an observer, which can be used in a state feedback controller.

This paper presents a discussion of how the system parameters were determined, of the derivation of the nonlinear mathematical model, and of the linearized model of the hydraulic actuator. It also investigates the observability, controllability, balanced model reduction and empirical techniques used to simplify the model. Due to the nonlinear nature of the system, the actuator should be linearized about a number of operating points, with a controller designed for each, if it is to operate on a wide range of force conditions. In this paper, a state observer is designed for one such operating point.

MODELLING

Friction

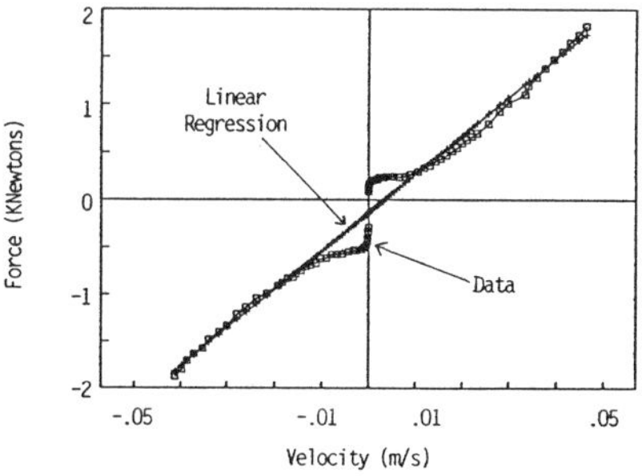

Figure 2.1 System Friction Profile

The friction inherent in the system consists of two types: Coulombic and viscous. Coulombic friction is caused by the rubbing action in the contact between the piston and the cylinder wall. On the other hand, viscous friction is a consequence of the fluid shear in the mating contact of the piston and the cylinder wall to resist motion and is a linear function of the fluid used [3]. Figure 2.1 is the friction profile for the system. The nonlinear region near the origin in the profile represents the Coulombic friction while the linear region describes the viscous friction.

The Coulombic friction is very difficult to control due to its nonlinear nature. Efforts were made during the project to reduce the Coulombic friction. For this paper, while not considering Coulombic friction effects, we take viscous friction into consideration. The slope in the friction profile represents the friction constant, k_{fr}.

Spring Effects

The spring effect of the hydraulic system primarily comes from two sources: fluid compression and hydraulic hose expansion. When there is an abrupt change in pressure, hydraulic fluid compresses to a certain extent depending on the type of fluid. The bulk modulus of elasticity, β, being a function of the type of fluid used, describes the extent to which the fluid is compresses or expands. This constant was obtained from the previous work by Springle [2].

The second source of spring effect is hydraulic hose expansion. The hydraulic line connecting the servovalve and the cylinder is expandable. When there is a pressure increase, the hose expands in its cross section area and contracts in length. Its relationship is, to the first-order approximation, a linear one, and the constant depends only on the specific hose with a specific volume. For this system, the data was supplied by the manufacturer.

Servovalve Characteristics

The servovalve can be envisioned to have four orifices as shown in figure 2.2. When the spool is shifted to the right relative to the center position, A_{b2} and A_{a2} are nonzero, and the flow goes from P_s to Return through A_{b2} and A_{a2}. Likewise, when the spool is shifted to the left, A_{b1} and A_{a1} are nonzero, and the flow goes from P_s to Return through A_{a1} and A_{b1}.

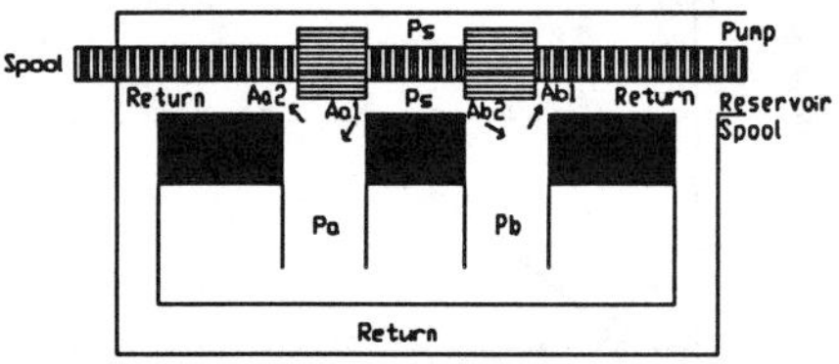

Figure 2.2 Servovalve Model

At the center position, if the servovalve is not underlapped, then all four orifice areas are zero. However, as demonstrated later, the servovalve of interest has about 1% underlap on the return but none on the supply pressure. Furthermore, the servovalve has hysteresis associated with the input current vs spool position relationship.

The dynamics of a typical servovalve's spool position in response to an input voltage is of fifth-order as demonstrated in Springle [2]. However, for most applications, a second order model is sufficient Springle [2].

The underlap of the servovalve was also confirmed by experimental data. The data was taken in an "open circuit" configuration in which flow only occurs in the form of internal leakage as a result of underlap. P_a and P_b were measured as the input voltage varied from -10 volts to +10 volts. The measurements are illustrated in Figure 2.3, which shows that the degree of underlap is about 1.5%.

The hysteresis of the servovalve comes from two sources: mechanical (due to static friction of the spool) and electromagnetic hysteresis. It is highly desirable to reduce or to eliminate hysteresis because it causes the system to be indeterministic in the sense that a single input point maps to a region of possible output points. The hysteresis of the servovalve was measured and is illustrated also in figure 2.3. The hysteresis of the hydraulic system comes from the servovalve alone because the rest of the system of hysteresis free. Figure 2.3 also shows the curve crossings near zero pressures are separated by about 0.3 V, representing about 1.5% hysteresis (20 V maximum swing.)

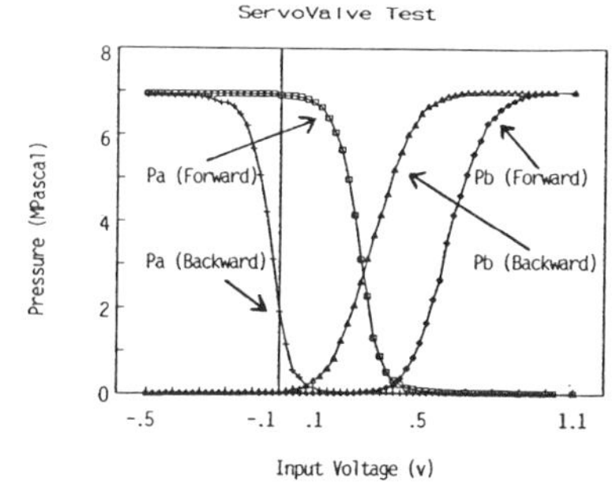

Figure 2.3 Open Loop Experimental Data

After the aforementioned characteristics of the system were studied and the constants were determined, a system of six nonlinear equation was generated. These equations, however, do not take into account the hysteresis characteristics. The nonlinear model was then validated against the experimental data, as illustrated in Figure 2.4 and Figure 2.5. The outputs are the two pressures on both sides of the piston (see Figure 1.2). In order to use linear theory for designing a controller, a linearized model was obtained. The response of the linear model is almost identical to the nonlinear; therefore, both the linear and nonlinear models are good approximations to the actual response.

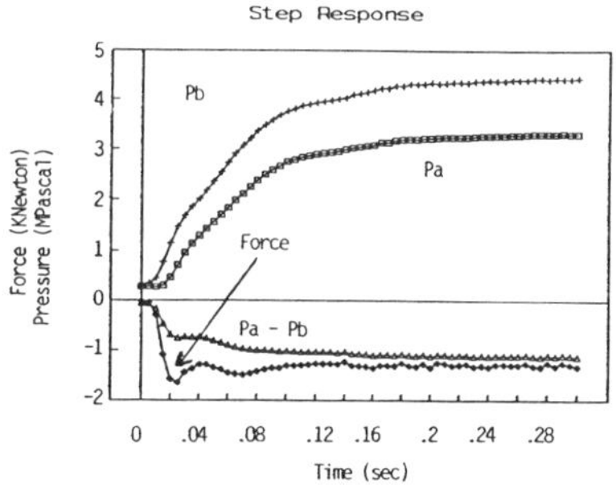

Figure 2.4 System Response (5 volts)

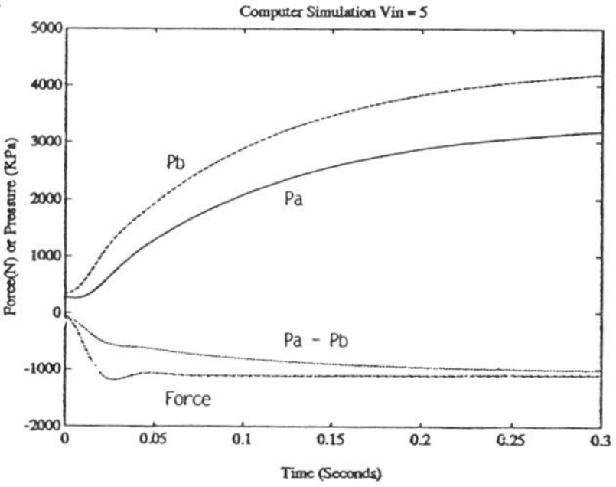

Figure 2.5 Nonlinear Model(5 V)

STATE OBSERVER DESIGN

A state observer was designed in order to provide complete state feedback for a controller, and to estimate the spool position in the presence of hysteresis. The linear model obtained has two modes that are nearly uncontrollable/unobservable; therefore, the balanced model reduction technique [4] was applied to reduce the model to a fourth-order one, with two modes eliminated. A full order observer was designed based on this reduced model. The observer was then simulated using PRO-MATLAB in a SparcStation. Because the largest pair of poles {-69±j123} are much larger than the other two poles {-25, -0.16}, to determine the pole locations for the observer required trial and error, and fine tuning. Figures 2.1 through 2.4 illustrate the convergence of the four states for a step input, with an arbitrary initial condition. (Solid curves represent the actual states; dashed are the estimated) Note that some initial conditions result in larger (sometimes much larger) overshoots than others.

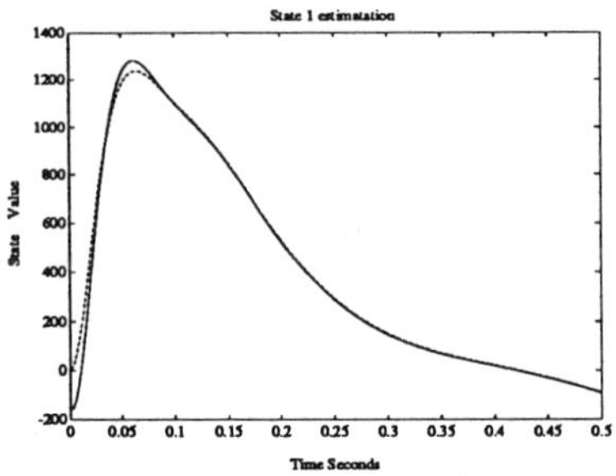

Figure 3.1 State 1 Estimation

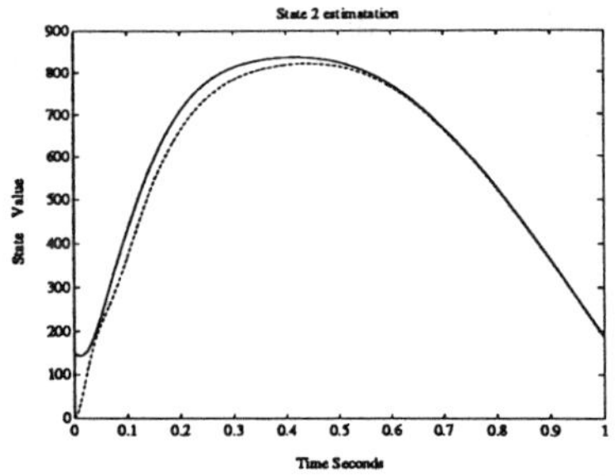

Figure 3.2 State 2 Estimation

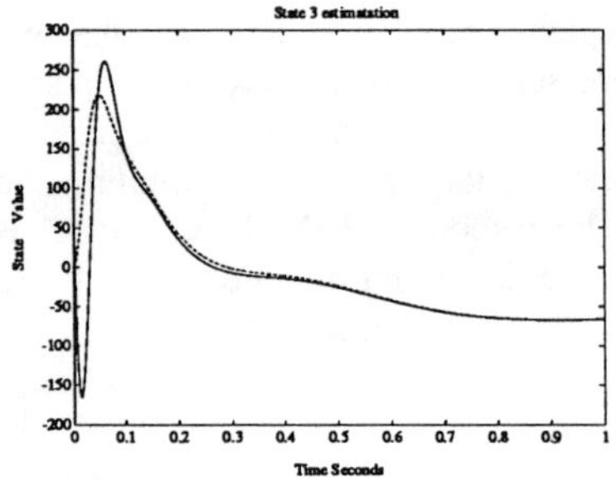

Figure 3.3 State 3 Estimation

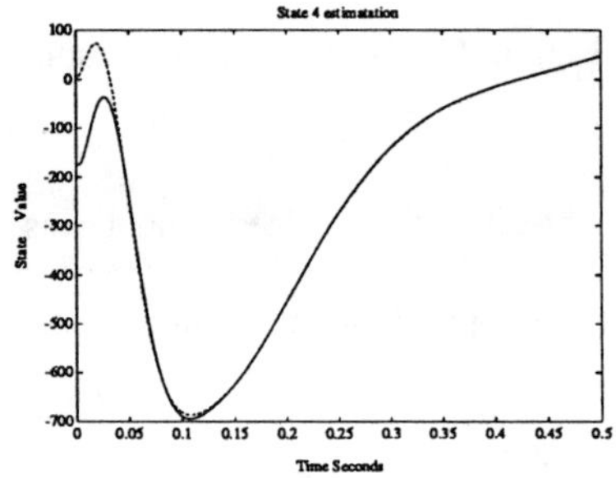

Figure 3.4 State 4 Estimation

CONCLUSION

Several parameters of the hydraulic system were experimentally determined, and a sixth-order nonlinear model and its linearized model were generated. Furthermore, each aspect of the system was studied; these include viscous and Coulombic friction, the fluid and hose compliance, and underlap and hysteresis characteristics. The linearized version was then reduced to a fourth order system using the balanced model reduction technique because two of the original modes were nearly uncontrollable/unobservable. A full order observer was designed and simulated. However, neither the observer nor a controller has been implemented at the present time.

There is still a great deal of work to be done in order to produce an effective force controller. We are also considering alternative control techniques such as adaptive controllers and those employing fuzzy logic and neural networks which are less model dependent. Because the system is nonlinear and has the aforementioned nonideal conditions, an adaptive or model independent controller is highly desirable.

REFERENCE

1. F. Conrad and C.J.D. Jensen (1987)., "Design of Hydraulic Force Control State Estimate Feedback", IFAC 10th Tiennial World Congress, Munich,FRG.

2. G. Springle (1989)., "A Hydraulic Pressure Controller for Robotic Manipulators", B.A.Sc. thesis, Simon Fraser University, Burnaby, B.C.

3. P. Dransfield (1981)., "Hydraulic Control Systems - Design and Analysis of Their Dynamics", Lectures Notes in Control and Information Sciences, Springler-Verlag

4. B.C. Moore (1981)., "Principal Component Analysis in Linear Systems: Controllability, Observability, and Model Reduction", IEEE Transactions on Automatic Control., AC-26, pp. 17-31.

5. K.J. Åström and B. Wittermark (1990)., *Computer Control Systems*,Eaglewood Cliff: Prentice Hall. pp.41-50,pp.224-278.

6. McCloy D. and H.R. Martin (1980)., Control of Fluid Power, Waterloo: Ellis Harwood Ltd. pp.133-200, pp.9-53.

7. K. Ogata (1970)., *Modern Control Engineering*, Eaglewood Cliff: Prentice Hall. pp.663-669.

THE SEMIGRAPHICAL SOLUTION OF THE DIRECT KINEMATICS OF GENERAL PLATFORM MANIPULATORS

J. ANGELES and K. E. ZANGANEH
McGill Research Centre for Intelligent Machines &
Department of Mechanical Engineering, McGill University
817 Sherbrooke Street West, Montreal H3A 2K6, Quebec, Canada

ABSTRACT

Presented in this paper is a semigraphical method for the computation of all real direct kinematic solutions of platform manipulators with general geometries. Within the proposed solution procedure we set up an upper bound on the maximum number of solutions. Then, the direct kinematic problem is reduced to two bivariate equations in the sines and cosines of two unknown angles. Each of these equations defines a contour in the plane of the two angles involved. The intersections of these contours provide actual and spurious possible solutions, a supplementary equation being then used to discriminate between actual and spurious solutions.

INTRODUCTION

Platform manipulators are multibody mechanical systems comprising an end-effector platform, henceforth abbrivated as EEP, connected to a base platform (BP) by several kinematic subchains leading to an architecture with multiple closed kinematic loops. The direct kinematic problem of these manipulators is, in general, more challenging than the inverse problem, mainly because of the nonlinearities involved. Solutions of the direct kinematics of different platform manipulators with simplified geometries have been presented in the literature [1–6]. Recently, Raghavan [7] presented a numerical solution to direct kinematics of the general Stewart platform based on the continuation method. Moreover, he considered Stewart platforms with complex-valued parameters and conjectured that the problem can admit up to 40 solutions.

In this paper we present a semigraphical method for solving the direct kinematics of general platform manipulators. By exploiting some properties of the underlying equations, an upper bound is first obtained on the maximum number of solutions. Then, the problem is reduced to two bivariate equations in the sines and cosines of two unknown angles. Each of these equations in turn, defines a contour in the plane of the unknown angles. The intersections of the two foregoing contours provide all possible solutions, actual and spurious. Then, using a supplementary equation the actual solutions are discriminated from the foregoing set.

Figure 1 depicts the general platform manipulator under study. It comprises a mechanism with six legs connecting a mobile EEP to a BP. The legs at each end are joined to the platforms by spherical joints, the centers of the latter being, in general, non-coplanar. Moreover, each leg is driven by a prismatic actuator.

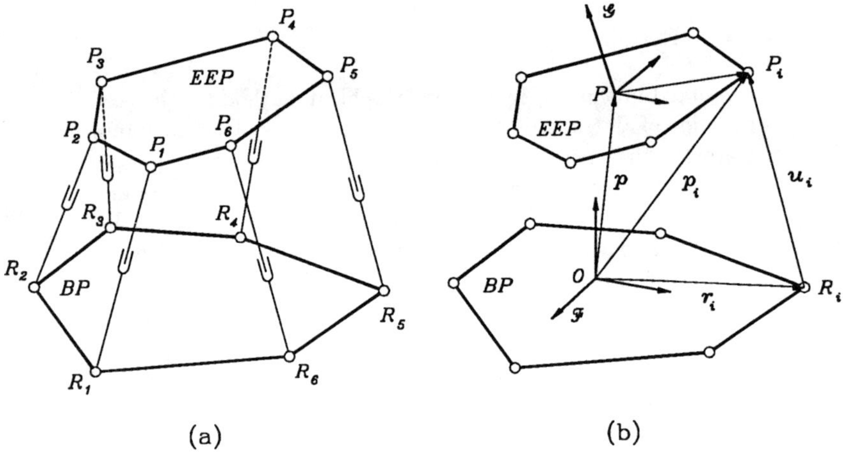

FIG. 1. General platform manipulator

The direct kinematic problem (DKP) of the manipulator under study is defined as: *Given the lengths of the six legs, determine the corresponding EEP pose, i.e., the orientation and position of the EEP.*

THE DIRECT KINEMATIC EQUATIONS

In view of the general platform manipulator shown in Figure 1, we assign a fixed coordinate frame $\mathcal{F}$ to the BP at point O, and a moving coordinate frame $\mathcal{G}$ to the EEP at point P. Moreover, the attachment points of the ith leg on the BP and EEP are denoted by R_i and P_i, respectively. Furthermore, the EEP is moved from the home configuration (HC) to the current configuration (CC) with the aid of the six actuators. Hence, if at the HC the vector directed from point P to P_i is denoted by $\boldsymbol{\sigma}_i$, and at the CC by $\boldsymbol{\pi}_i$, then we can write

$$\boldsymbol{\pi}_i = \boldsymbol{Q}\boldsymbol{\sigma}_i, \qquad \text{for } i = 1, \cdots, 6 \tag{1}$$

where $\boldsymbol{Q}$ is the rotation matrix relating the orientation of the EEP at HC to that at CC. Moreover, at CC we denote the positions of the vertices P_i and R_i by $\boldsymbol{p}_i$ and $\boldsymbol{r}_i$, respectively. Furthermore, vector $\boldsymbol{u}_i$ is defined as directed from R_i to P_i, for $i = 1, \cdots, 6$, and $\boldsymbol{p}$ is the position vector of P, as shown in Figure 1. Referring to this figure, the geometrical relations below can be readily derived:

$$\|\boldsymbol{u}_i\|^2 \equiv \|\boldsymbol{p}_i - \boldsymbol{r}_i\|^2 = \|\boldsymbol{p} + \boldsymbol{Q}\boldsymbol{\sigma}_i - \boldsymbol{r}_i\|^2, \qquad \text{for } i = 1, \cdots, 6 \tag{2}$$

where $\|\cdot\|$ denotes the Euclidean norm of $(\cdot)$. After expanding the rigth-hand-side of eq.(2), we obtain

$$\frac{1}{2}\boldsymbol{p}^2 + (\boldsymbol{Q}\boldsymbol{\sigma}_i - \boldsymbol{r}_i)^T\boldsymbol{p} - \boldsymbol{r}_i^T\boldsymbol{Q}\boldsymbol{\sigma}_i + s_i = 0, \qquad \text{for } i = 1, \cdots, 6 \tag{3}$$

where $s_i = \frac{1}{2}(\boldsymbol{\sigma}_i^2 + \boldsymbol{r}_i^2 - \boldsymbol{u}_i^2)$ and $(\cdot)^T$ represents the transpose of $(\cdot)$. The six equations in eq.(3) constitute the DKP of the general platform manipulator under study.

MAXIMUM NUMBER OF SOLUTIONS

In order to derive an upper bound on the maximum number of solutions, we rewrite eq.(3) in matrix form, namely,

$$\boldsymbol{A}\boldsymbol{x} = \boldsymbol{0} \tag{4}$$

where the 5-dimensional vector $\boldsymbol{x}$ is defined as $\boldsymbol{x} \equiv [\boldsymbol{p}^2 \quad \boldsymbol{p} \quad 1]^T$ and $\boldsymbol{A}$ is the 6×5 matrix shown below:

$$\boldsymbol{A} \equiv \begin{bmatrix} 1/2 & (\boldsymbol{Q}\boldsymbol{\sigma}_1 - \boldsymbol{r}_1)^T & -\boldsymbol{r}_1^T\boldsymbol{Q}\boldsymbol{\sigma}_1 + s_1 \\ \vdots & \ddots & \vdots \\ 1/2 & (\boldsymbol{Q}\boldsymbol{\sigma}_6 - \boldsymbol{r}_6)^T & -\boldsymbol{r}_6^T\boldsymbol{Q}\boldsymbol{\sigma}_6 + s_6 \end{bmatrix} \tag{5}$$

Now, for eq.(4) to admit a nontrivial solution, all determinants of the 5×5 submarices of $\boldsymbol{A}$ should vanish. This results in a set of six nonlinear equations in $\boldsymbol{Q}$. However, only five of them are independent. if we denote the determinant of the ith submatrix by Δ_i, then

$$\Delta_i = \sum_{\substack{j=1 \\ j \neq i}}^{6} (-1)^{i+j}\eta_j\Delta_{ij} \tag{6}$$

where $\eta_j \equiv \boldsymbol{r}_j^T\boldsymbol{Q}\boldsymbol{\sigma}_j - s_j$, $\Delta_{ij} \equiv \det(\boldsymbol{A}_{ij})$ and $\boldsymbol{A}_{ij}$ is the 4×4 submatrix of $\boldsymbol{A}$ obtained by deleting the ith and jth rows and the last column of $\boldsymbol{A}$. The expansion of Δ_{ij} yields in turn an expression of the form

$$\Delta_{ij} = \frac{1}{2} \sum_{\substack{k=1 \\ k \neq i,\, k \neq j}}^{6} (-1)^{i+j+k}\Delta_{ijk} \tag{7}$$

where $\Delta_{ijk} \equiv \det(\boldsymbol{A}_{ijk})$ and the 3×3 matrix $\boldsymbol{A}_{ijk}$ is also a submatrix of $\boldsymbol{A}$, namely, that obtained by deleting the first and last columns together with the ith, jth and kth rows of $\boldsymbol{A}$. Using the above notation, some symmetric properties can be exploited among the foregoing subdeterminants that ease their computations. We thus have

Lemma: Every determinant Δ_i, for $i = 1, \cdots, 6$, is biquadratic in $\boldsymbol{Q}$.

Proof: We first show that every subdeterminant Δ_{ijk} is linear in $\boldsymbol{Q}$. To do so, we choose an arbitrary subdeterminant Δ_{123} and expand it as

$$\Delta_{123} = (\boldsymbol{Q}\boldsymbol{\sigma}_6 - \boldsymbol{r}_6)^T[(\boldsymbol{Q}\boldsymbol{\sigma}_4 - \boldsymbol{r}_4) \times (\boldsymbol{Q}\boldsymbol{\sigma}_5 - \boldsymbol{r}_5)] \tag{8}$$

Using the identity relation $\boldsymbol{Q}\boldsymbol{Q}^T = \mathbf{1}$, with $\mathbf{1}$ being the 3×3 identity matrix, the right-hand side of eq.(8) is expanded as

$$\begin{aligned}\Delta_{123} = \boldsymbol{\sigma}_6^T[\boldsymbol{\sigma}_4 \times \boldsymbol{\sigma}_5 - \boldsymbol{\sigma}_4 \times \boldsymbol{Q}^T\boldsymbol{r}_5 + \boldsymbol{\sigma}_5 \times \boldsymbol{Q}^T\boldsymbol{r}_4 + \boldsymbol{Q}^T(\boldsymbol{r}_4 \times \boldsymbol{r}_5)] \\ -\boldsymbol{r}_6^T[\boldsymbol{Q}(\boldsymbol{\sigma}_4 \times \boldsymbol{\sigma}_5) - \boldsymbol{Q}\boldsymbol{\sigma}_4 \times \boldsymbol{r}_5 + \boldsymbol{Q}\boldsymbol{\sigma}_5 \times \boldsymbol{r}_4 + \boldsymbol{r}_4 \times \boldsymbol{r}_5]\end{aligned} \tag{9}$$

It is apparent from eq.(9) that Δ_{123} is linear in $\boldsymbol{Q}$. Therefore, from eq.(6) it can be concluded that Δ_i is biquadratic in $\boldsymbol{Q}$.

Now, every rotation matrix $\boldsymbol{Q}$ can be written in the form

$$\boldsymbol{Q} = [\,\boldsymbol{m} \quad \boldsymbol{n} \quad \boldsymbol{m} \times \boldsymbol{n}\,] \tag{10}$$

subjected to the conditions

$$\boldsymbol{m}^T\boldsymbol{n} = 0, \quad \boldsymbol{m}^T\boldsymbol{m} = 1, \quad \boldsymbol{n}^T\boldsymbol{n} = 1 \tag{11}$$

where, $\boldsymbol{m} \equiv [\,m_1 \quad m_2 \quad m_3\,]^T$ and $\boldsymbol{n} \equiv [\,n_1 \quad n_2 \quad n_3\,]^T$. Hence, if we adjoin eqs.(11) to any three subdeterminants Δ_i, a system of six independent equations is obtained that can be solved for the six unknown components of $\boldsymbol{m}$ and $\boldsymbol{n}$. On the other hand, each Δ_i is of order four, while eqs.(10) are of order two; hence, the whole system has a degree of $4 \times 4 \times 4 \times 2 \times 2 \times 2 = 512$. However, the system does not make any distinction between the legs in a given layout and in the mirror image of this layout. Hence, the foregoing number has to be divided by two, thereby deriving an upper bound of 256 on the maximum number of solutions.

ELIMINATION PROCEDURE AND CONTOUR EQUATIONS

In order to derive the bivariate equations of the two contours, we have to eliminate four of the unknowns. To do so, we pick up five equations among eq.(6) and four independent equations from the identity relation, $\boldsymbol{Q}^T\boldsymbol{Q} = \mathbf{1}$. The foregoing nine independent equations can be written as a system of nine hemogeneous equations in the form

$$\boldsymbol{N}\boldsymbol{y} = \mathbf{0} \tag{12}$$

where all entries of the 9×10 matrix $\boldsymbol{N}$ are functions of vector $\boldsymbol{n}$ only, while $\boldsymbol{y} \equiv [m_1^2 \quad m_1m_2 \quad m_1m_3 \quad m_1 \quad m_2^2 \quad m_2m_3 \quad m_2 \quad m_3^2 \quad m_3 \quad 1]$. Now, eq.(12) can be partitioned into two subsystems, each consisting of eigth equations, i.e.,

$$\boldsymbol{N}_1\boldsymbol{y} = \boldsymbol{0}, \qquad \boldsymbol{N}_2\boldsymbol{y} = \boldsymbol{0} \tag{13}$$

We now eliminate m_1, m_2 and m_3 dyalitically [8] from eqs.(13) by multiplying their two sides by m_1 and then combine each of the resultant system with its associated one in eq.(13). This yields two systems of homogeneous equations, each of them constituting a set of sixteen independent equations of the form

$$\boldsymbol{N}_1'\boldsymbol{y}' = \boldsymbol{0}, \qquad \boldsymbol{N}_2'\boldsymbol{y}' = \boldsymbol{0} \tag{14}$$

where $\boldsymbol{y}' \equiv \left[\boldsymbol{y}''^T \quad \boldsymbol{y}^T\right]^T$ and

$$\boldsymbol{y}'' \equiv [m_1^3 \quad m_1^2m_2 \quad m_1^2m_3 \quad m_1m_2^2 \quad m_1m_2m_3 \quad m_1m_3^2]^T$$

From the definitions of $\boldsymbol{y}'$ and $\boldsymbol{y}''$, and conditions (11), it is apparant that any solution of eqs.(14) should be nontrivial, and hence, the determinants of the coefficient matrices should vanish, i.e.,

$$\det(\boldsymbol{N}_1') = 0, \qquad \det(\boldsymbol{N}_2') = 0 \tag{15}$$

the latter two determinants being functions of $\boldsymbol{n}$ only. We express now the components of this vector in spherical coordinates in order to reduce the number of unknowns to only two, i.e.,

$$\boldsymbol{n} \equiv [\sin\theta_1\cos\theta_2 \quad \sin\theta_1\sin\theta_2 \quad \cos\theta_1]^T \tag{16}$$

By substituting eq.(16) into eqs.(15), two bivariate equations in θ_1 and θ_2 are derived, namely

$$f_1(\theta_1, \theta_2) \equiv \det(\boldsymbol{N}_1') = 0, \qquad f_2(\theta_1, \theta_2) \equiv \det(\boldsymbol{N}_2') = 0 \tag{17}$$

The two functions f_1 and f_2 define two contours $\mathcal{C}_1$ and $\mathcal{C}_2$ in the $\theta_1 - \theta_2$ plane. The intersections of these contours result in all possible solutions, actual and spurious. Below we describe how to discriminate between spurious and actual solutions. For each pair (θ_1, θ_2), we first solve eq.(12) in a nonhomogeneous form, by transferring the last column of $\boldsymbol{N}$ to the right-hand side of the equation, and compute the associated three components of $\boldsymbol{m}$. Then, we pick up four arbitrary equations among eqs.(4), and solve the resultant system for p^2 and $\boldsymbol{p}$. The actual solutions of the problem are only those which can satisfy the equation

$$p^2 - \boldsymbol{p}^T\boldsymbol{p} = 0 \tag{18}$$

NUMERICAL EXAMPLE

Using the above procedure, the contour intersections and the position and orientation of an arbitrary platform manipulator were obtained. The parameters of the platform manipulator were chosen as

$\boldsymbol{r}_1 = [-2,\, -2,\, -1]^T, \quad \boldsymbol{r}_2 = [-4,\, 0,\, -1]^T, \quad \boldsymbol{r}_3 = [-1,\, 2,\, 3/2]^T,$

$\boldsymbol{r}_4 = [2,\, 2,\, 3/2]^T, \quad \boldsymbol{r}_5 = [5,\, 0,\, -1/2]^T, \quad \boldsymbol{r}_6 = [3,\, -2,\, -1/2]^T$

$\boldsymbol{\sigma}_1 = [-8.258 \times 10^{-3},\, -1.294,\, 1.256]^T, \quad \boldsymbol{\sigma}_2 = [-0.75,\, -0.75,\, 2.806]^T,$

$\boldsymbol{\sigma}_3 = [-1.233,\, 1.337,\, 2.166]^T, \quad \boldsymbol{\sigma}_4 = [1.651 \times 10^{-2},\, 2.587,\, -2.511]^T,$

$\boldsymbol{\sigma}_5 = [1.0,\, 1.0,\, -3.742]^T, \quad \boldsymbol{\sigma}_6 = [0.617,\, -0.669,\, -1.083]^T$

$\boldsymbol{u}_1^2 = 71.522,\ \boldsymbol{u}_2^2 = 89.0,\ \boldsymbol{u}_3^2 = 69.321,\ \boldsymbol{u}_4^2 = 73.321,\ \boldsymbol{u}_5^2 = 76.25,\ \boldsymbol{u}_6^2 = 62.229$

The superimposed contours are shown in Figure 2, in which $\mathcal{C}_1$ is shown with solid line. In this figure, the contours overlap at some parts comprising of spurious and actual solutions.

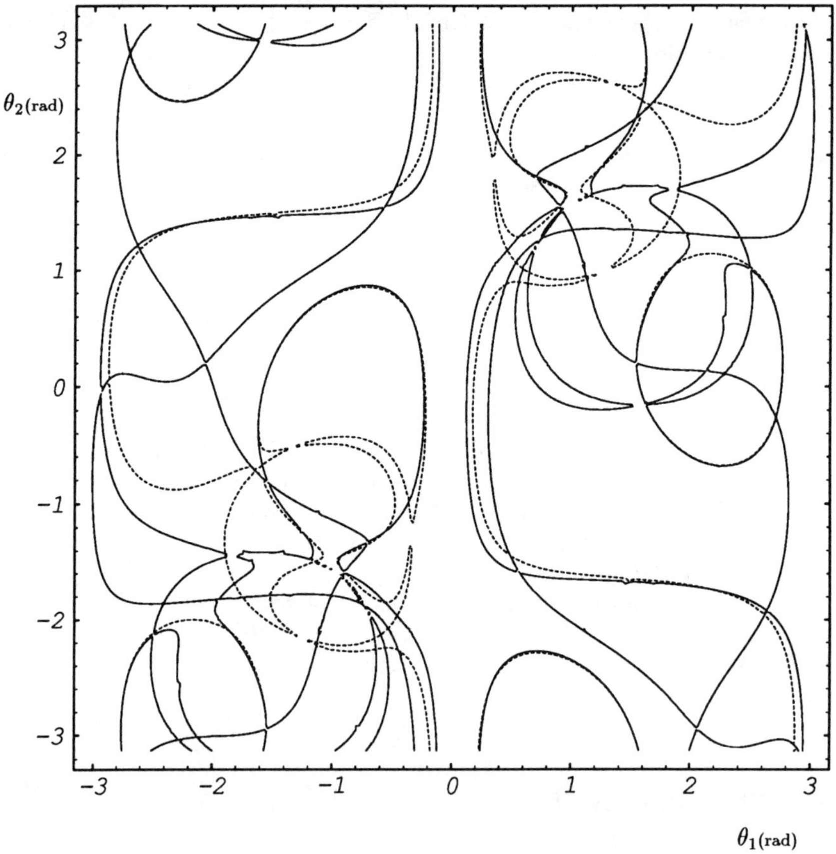

FIG. 2. Contours $\mathcal{C}_1$ and $\mathcal{C}_2$ of the platform manipulator

Using superimposed plots of the two contours and eq.(18), the actual solutions are found. One of these solutions is recorded below:

$$\theta_1 = 0.5236 \text{ rad}, \quad \theta_2 = 1.0472 \text{ rad}, \quad \boldsymbol{n} = [\,0.250 \quad 0.433 \quad 0.866\,]^T,$$

$$\boldsymbol{m} = [\,0.250 \quad -0.893 \quad 0.374\,]^T, \quad \boldsymbol{p} = [\,1.0 \quad -2.0 \quad 8.0\,]^T$$

CONCLUSIONS

In this paper, the equations for the direct kinematic problem of platform manipulators with general geometries were derived. An upper bound was obtained for the maximum number of solutions. Then, the equations were reduced to two bivariate equations, each defining a contour in the plane of the two unknown angles involved. The superimposed plots of contours yield all possible solutions, actual and spurious. By using a discriminating equation, the actual roots were identified in the aforementioned set.

ACKNOWLEDGEMENTS

The research work reported here was made possible under NSERC (Natural Sciences and Engineering Research Council, of Canada) Grants A4532, STRGP 205 and EQP00-92729. Support from IRIS, the Institute for Robotics and Intelligent Systems, a network of Canadian centres of excellence, is also acknowledged. The two authors completed this work while on leave, the first at the Institute B for Mechanics of the Technical University of Munich, under an Alexander von Humboldt Research Award, the second at the University of Nice at Sophia Antipolis, working in Prof. Peter Sander's group.

REFERENCES

1. K. M. Lee and K. Shah, "Kinematic Analysis of a Three-Degrees-of-Freedom In-Parallel Actuated Manipulator," IEEE *J. Robotics and Automation* **4**(3), 354–360 (June 1988).
2. S. Charentus and M. Renaud, "Modelling and Control of a Modular, Redundant Robotic Manipulator," in *Experimental Robotics, Vol. 1*, 508–527, edited by M. Thoma and A. Wyner (Springer-Verlag, New York, 1989).
3. K. J. Waldron, M. Raghavan and B. Roth, "Kinematics of a Hybrid Series-Parallel Manipulation System," *J. Dynamic Syst., Meas., and Contr.* **111**(2), 211–221 (June 1989).
4. C. Innocenti and V. Parenti-Castelli, "Direct Position Analysis of the Stewart Platform Mechanism," *Mech. Mach. Theory* **25**(6), 611–621 (1990).
5. V. Parenti-Castelli and C. Innocenti, "Forward Displacement Analysis of Parallel Mechanisms: Closed Form Solutions of PRR-3S and PPR-3S Structures," *J. Mech. Design* **114**(1), 68–73 (March 1992).
6. J. P. Merlet, "Direct Kinematics and Assembly Modes of Parallel Manipulators," *Int. J. Robotics Res.* **11**(2), 150–162 (April 1992).

7. M. Raghavan, "The Stewart Platform of General Geometry Has 40 Configurations," in *Advances in Design Automation, Vol. 2*, 397–402, edited by G. A. Gabriele (ASME Press, New York, 1991).
8. G. Salmon, *Lessons Introductory to the Modern Higher Algebra* (Chelsea Publishing Co., New York, 1964).

DYNAMIC ANALYSIS OF MANIPULATORS BY THE VECTOR-NETWORK METHOD

M.J. RICHARD, I. BINDZI and C.M. GOSSELIN

DEPARTEMENT DE GENIE MECANIQUE
UNIVERSITE LAVAL
QUEBEC (CANADA)
G1K 7P4

ABSTRACT

This paper presents the dynamic analysis of tree-structure robot manipulators based on the vector-network method. The vector-network technique arises from the application of fundamental theorems of dynamics along with the exploitation of principles from graph theory. The dynamical equations of motion of the manipulator, expressed in terms of Lagrangian generalized coordinates, are automatically derived by using a matrix form of the Newton-Euler equations of motion. The method appears to be a simple systematic procedure for formulating all of the equations of motion for any multi-rigid-body constrained system.

INTRODUCTION

Robot manipulators are complicated mechanical systems with a highly non-linear dynamic behavior. There has been in recent years steadingly increasing interest in self-formulating computer programs for the computer-aided design of robots, viewed as multi-rigid-body constrained mechanical systems (very often only open-chain systems are considered). Among those programs are the NEWEUL program of Kreuzer and Schiehlen [1], based on a Newton-Euler formulation for the generation of symbolic equations of motion using D'Alembert's principle, and the RESTRI program [2] originally devoted to three-dimensional constrained mechanical system.

Most of the existing models for the dynamic analysis of manipulators are based on recursive dynamic schemes using either Newton-Euler formulation [3,4], D'Alembert's principle [1], or Lagrangian dynamics [5,6]. Such developments are very tedious and the necessary insight in the given equations of motion is not apparent for the latter, written for each segment separately, are very complex even when they are written in symbolic form. Wittenburg [7] has applied, with some success, graph theory for the analysis of mechanical systems and his work has greatly been improved by Roberson and Schwertassek [8]. However their attempt to analyze the mechanical system as a whole have not fully exploited the capabilities of the different laws of graph theory since the discretization of the mechanical system is not complete (the mechanical system is replaced by vertices representing centers of mass of the individual bodies and by edges representing the joints between bodies). With such a discretization, there is a need for further analytical development in order to take into account the dimensions and inertia of the rigid bodies. The vector-network method [2,9] reduces the mechanical system to its most simplified form by including in a graph all the parameters involved in the motions of its individual bodies.

VECTOR-NETWORK TECHNIQUE

The fundamental bases of the vector-network technique have been presented in recent papers [10,11] eventhough the method utilized for the analysis presented here is slightly different from its original form fully detailed in [2]; these latter references are recommended for a good understanding of the vector-network technique. Figure 1 shows the discretisation of the Stanford manipulator with the vector-network method.

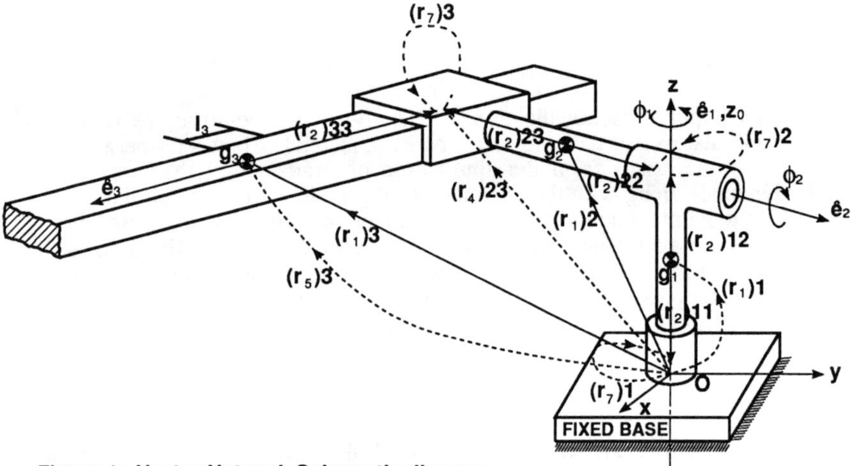

Figure 1: Vector Network Schematic diagram of the Standford manipulator

In its most general form, the principle of orthogonality may be stated as follows: "If the scalar products of the dynamic and kinematic variables associated with each edge of a system graph are summed over all of the edges in the graph the sum will be zero". [13] This principle arises primarily from the orthogonality of the dynamic and kinematic variables associated with any mechanical system, which can easily be established (see [13]). Consider the most general mechanical system, containing all types of elements, N_1 - rigid body elements, N_2 - rigid-arm elements which locate interconnecting points between a rigid body and the mechanical system, N_3 - specified points in the mechanical system whose motion is a known function of time, N_4 - moment relative to a given rigid body's centre of mass, N_5 - external forces and /or moments acting on a given rigid body, N_6 - spring (translational or rotational) damper system, N_7 - kinematic joint. Application of the orthogonality principle permits the derivation of the following equation:

$$\{\delta r_1\}^T \left(A_{15}\{F_5\} - [\underset{-}{m}]\{\ddot{r}_1\}\right) + \{\delta\tilde{\pi}_1\}^T \left([D]\,A_{15}\{T_5\} - [I]\{\dot{\omega}_1\} - [\{\underline{\underline{\omega}}_1\}]\,[I]\{\omega_1\}\right) + \delta W = 0 \tag{1a}$$

with: $$\delta W = \{\delta r_1\}^T A_{16} \{F_6\} + \{\delta \tilde{\pi}_1\}^T [D] A_{16} \{T_6\} + \{\delta r_7\}^T \{F_7^a\} + \{\delta \pi_7\}^T \{T_7^a\} \tag{1b}$$

The matrices A_{ij} are N_i x N_j matrices (i=1,2,3, and j= 4,5,6,7) representing the relations between elements of groups N_i and N_j.

Equation (1) is equivalent to the formulation of D'Alembert's principle for the mechanical system. The orthogonality principle could also be exploited to derive Jourdain's principle on virtual powers, using the virtual variations in the speed of the elements of the schematic diagram as kinematic variables.

MANIPULATOR EQUATIONS OF MOTION

First, to derive the equations of motion for a manipulator let

$$\sigma_i = \begin{cases} 1 & \text{for a pin joint} \\ 0 & \text{for a prismatic joint} \end{cases} \quad \text{and} \quad \bar{\sigma}_i = 1 - \sigma_i \tag{2}$$

For the entire manipulator, we thus have the relations:

$$\{r_7\} = [\bar{\sigma}] [\{\hat{e}_a'\}] \{q\} \quad \text{and} \quad \{\gamma_7\} = [\sigma] [\{\hat{e}_a'\}] \{\dot{q}\} \tag{3a; 3b}$$

$$([\bar{\sigma}] [\{\hat{e}_a'\}])^T \{F_7^c\} = \{0\} \quad \text{and} \quad ([\sigma] [\{\hat{e}_a'\}])^T \{T_7^c\} = \{0\} \tag{3c;3d}$$

The following relations may easily be demonstrated;

$$\{\dot{\gamma}_7\} = [\sigma] [\{\hat{e}_a'\}] \{\ddot{q}\} + [\{\underline{\underline{\gamma}}_1\}] \{\gamma_7\} \tag{4a}$$

$$\begin{aligned} \{\ddot{r}_7\} = & [\bar{\sigma}] [\{\hat{e}_a'\}] \{\ddot{q}\} - [\{\underline{\underline{r}}_7\}] (A_{17}^{-1})^T [\sigma] [\{\hat{e}_a'\}] \{\ddot{q}\} \\ & - [\{\underline{\underline{r}}_7\}] (A_{17}^{-1})^T [\{\underline{\underline{\gamma}}_1\}] \{\gamma_7\} \\ & + [\{\underline{\underline{r}}_7\}] (A_{37} A_{17}^{-1})^T \{\dot{\gamma}_3\} \\ & + [D]^T ([\{\underline{\underline{\omega}}_1\}] [\bar{\sigma}] [\{\hat{\hat{e}}_a'\}] \{\dot{q}\} \\ & + [\{\underline{\underline{\omega}}_1\}] [\{\underline{\underline{\omega}}_1\}] [\bar{\sigma}] [\{\hat{\hat{e}}_a'\}] \{q\}) \end{aligned} \tag{4b}$$

$$\begin{aligned} \{\dot{\omega}_1\} = & [D] (A_{17}^{-1})^T [\sigma] [\{\hat{e}_a'\}] \{\ddot{q}\} + [D] (A_{17}^{-1})^T [\{\underline{\underline{\gamma}}_1\}] \{\gamma_7\} \\ & - [D] (A_{37} A_{17}^{-1})^T \{\dot{\gamma}_3\} \end{aligned} \tag{5a}$$

$$
\begin{aligned}
\{\ddot{r}_1\} = {} & \left(A_{17}^{-1}\right)^T [\bar{\sigma}] \left[\{\hat{e}'_a\}\right] \{\ddot{q}\} - \left(A_{17}^{-1}\right)^T [\underline{\underline{X}}_2]^T \left(A_{17}^{-1}\right)^T [\sigma] \left[\{\hat{e}'_a\}\right] \{\ddot{q}\} \\
& - \left(A_{17}^{-1}\right)^T [\underline{\underline{X}}_2]^T \left(A_{17}^{-1}\right)^T \left(\left[\{\underline{\underline{\gamma}}_1\}\right] \{\gamma_7\} - A_{37}^T \{\dot{\gamma}_3\} \right) \\
& - \left(A_{27}\, A_{17}^{-1}\right)^T \left[\{\underline{\underline{\dot{r}}}_2\}\right]^T A_{14}^T [D]^T \{\omega_1\} \\
& + \left(A_{17}^{-1}\right)^T [D]^T \left(\left[\{\underline{\underline{\omega}}_1\}\right] [\bar{\sigma}] \left[\{\hat{\tilde{e}}'_a\}\right] \{\dot{q}\} \right. \\
& \left. + \left[\{\underline{\underline{\omega}}_1\}\right] \left[\{\underline{\underline{\omega}}_1\}\right] [\bar{\sigma}] \left[\{\hat{\tilde{e}}'_a\}\right] \{q\} \right) \\
& - \left(A_{37}\, A_{17}^{-1}\right)^T \{\ddot{r}_3\}
\end{aligned}
\tag{5b}
$$

let: $$[\underline{M}] = A_{17}^{-1} [\underline{m}] \left(A_{17}^{-1}\right)^T \quad \text{with} \; [\underline{M}]^T = [\underline{M}] \tag{6}$$

$$[K] = A_{17}^{-1} [J] \left(A_{17}^{-1}\right)^T \tag{7a}$$

and $$[J] = [D]^T [I] [D] + [\underline{\underline{X}}_2] [\underline{M}] [\underline{\underline{X}}_2]^T \tag{7b}$$

The "vectorial" matrix $[\underline{M}]$ (each component of this matrix is a three-dimensional unit matrix E_3) is called the "overall mass matrix". [J] is the inertia matrix, in the inertial reference frame, of individual "overall rigid bodies" (composed of one rigid body and of point masses). The similarities between equations (6) and (7a) allow to define matrix [K] as an "overall inertia matrix".

$$
\begin{aligned}
\{P_t\} = {} & \left([\bar{\sigma}] \left[\{\hat{e}'_a\}\right]\right)^T [\underline{M}] \left\{ [\underline{\underline{X}}_2]^T \left(A_{17}^{-1}\right)^T \left(\left[\{\underline{\underline{\gamma}}_1\}\right] \{\gamma_7\} - A_{37}^T \{\dot{\gamma}_3\} \right) \right. \\
& + A_{27}^T \left[\{\underline{\underline{\dot{r}}}_2\}\right]^T A_{14}^T [D]^T \{\omega_1\} - A_{37} \{\ddot{r}_3\} \\
& - [D]^T \left(\left[\{\underline{\underline{\omega}}_1\}\right] [\bar{\sigma}] \left[\{\hat{\tilde{e}}'_a\}\right] \{\dot{q}\} \right. \\
& \left. \left. + \left[\{\underline{\underline{\omega}}_1\}\right] \left[\{\underline{\underline{\omega}}_1\}\right] [\bar{\sigma}] \left[\{\hat{\tilde{e}}'_a\}\right] \{q\} \right) \right\} \\
& + \left([\bar{\sigma}] \left[\{\hat{e}'_a\}\right]\right)^T A_{17}^{-1} \left(A_{15} \{F_5\} + A_{16} \{F_6\}\right)
\end{aligned}
\tag{8a}
$$

$$
\begin{aligned}
\{P_r\} = {} & \left([\sigma] \left[\{\hat{e}'_a\}\right]\right)^T A_{17}^{-1} [\underline{\underline{X}}_2] \; [\underline{M}] \left\{ [\underline{\underline{X}}_2]^T \left(A_{17}^{-1}\right)^T \left(\left[\{\underline{\underline{\gamma}}_1\}\right] \{\gamma_7\} - A_{37}^T \{\dot{\gamma}_3\} \right) \right. \\
& + A_{27}^T \left[\{\underline{\underline{\dot{r}}}_2\}\right]^T A_{14}^T [D]^T \{\omega_1\} - A_{37}^T \{\ddot{r}_3\} \\
& - [D]^T \left(\left[\{\underline{\underline{\omega}}_1\}\right] [\bar{\sigma}] \left[\{\hat{\tilde{e}}'_a\}\right] \{\dot{q}\} \right. \\
& \left. \left. + \left[\{\underline{\underline{\omega}}_1\}\right] \left[\{\underline{\underline{\omega}}_1\}\right] [\bar{\sigma}] \left[\{\hat{\tilde{e}}'_a\}\right] \{q\} \right) \right\} \\
& + \left([\sigma] \left[\{\hat{e}'_a\}\right]\right)^T A_{17}^{-1} \left(A_{15} \{T_5\} + A_{16} \{T_6\} + A_{14} \left[\{\underline{\underline{r}}_2\}\right] A_{26} \{F_6\} \right. \\
& - [\underline{\underline{X}}_2]\, A_{17}^{-1} \left[A_{15} \{F_5\} + A_{16} \{F_6\}\right] \\
& \left. - [D]^T \left[\{\underline{\underline{\omega}}_1\}\right] [I] \{\omega_1\} \right)
\end{aligned}
\tag{8b}
$$

The column-vector $\{P_t\}$ (of dimension n) represents the effects of external and compliance forces together with those associated with the coriolis and relative translational accelerations. The column-matrix $\{P_r\}$ has the same representation as $\{P_t\}$, except that in this case, all the terms are rotational.

The dynamical equations of motion for the manipulator are thus:

$$\left[\left([\overline{\sigma}]\,[\{\hat{e}_a'\}]\right)^T \;\vdots\; \left([\sigma]\,[\{\hat{e}_a'\}]\right)^T\right] \begin{bmatrix} [\underset{-}{M}] & \vdots & [Y]^T \\ \cdots & \cdots & \cdots \\ [Y] & \vdots & [K] \end{bmatrix} \begin{bmatrix} [\overline{\sigma}] & [\{\hat{e}_a'\}] \\ & \\ [\sigma] & [\{\hat{e}_a'\}] \end{bmatrix} \{\ddot{q}\} = \{P\} + \{Q\} \tag{9}$$

with:

$$[Y] = -\,A_{17}^{-1}\,[\underset{=}{X}_2]\,[\underset{-}{M}] \tag{10a}$$

$$\{P\} = \{P_t\} + \{P_r\} \tag{10b}$$

$$\{Q\} = \{F_7^a\} + \{T_7^a\} \tag{10c}$$

The equations of motion of the manipulator (9) are thus written in the known form:

$$[W]\,\{\ddot{q}\} = \{P\} + \{Q\} \tag{11}$$

with [W] being a positive definite symmetric matrix analogous to an inertia tensor. The manipulator's equations of motions have been developed through a Newton-Euler formalism in order to bring some insight in the internal structure of the vector-network method. It has thus been shown that the dynamic analysis of a manipulator (multi-rigid body system) is made through the dynamical equilibrium equations of each joint separately.,

The constraint forces and torques at the joints are easily extracted (at each time increment for direct dynamic analysis) from corresponding cutset equations (see [2]. In the case of an inverse dynamic analysis, equation (11) is used by isolating the n x 1 column-matrix {Q}.

CONCLUSION

In the above developments it was shown how the vector-network technique performs the dynamic analysis of any kind of tree-structured manipulator. The vector-network algorithm is very versatile since it allows one to extract any kind of information pertinent to the manipulator at any moment during the course of the simulation (compliance forces or torques, length of springs $\{r_6\}$, ...). Also, the dynamical equations of motion being expressed in terms of relative (or natural) coordinates (which are also the Lagrangian generalized coordinates for the manipulator), the simulation of the manipulator's dynamic response should not be time consuming according to conclusion of a paper from Garcia and al. (12). The algorithm developed in this paper has been implemented into the initial RESTRI computer program [2].It is expected that this technique will provide new tools to launch pilot studies and stimulate further research in the field of robotics.

REFERENCES

1.Kreuzer, E.J., and W.D. Schiehlen: "Equations of Motion and Equations of Stress for Robots and Manipulators: An Application of the NEWEUL formalism". Proceedings of ROMANSY 84, the 5th CISM-IFTOMM Symposium; MIT Press (1985).

2.Richard, M.J.,: "Dynamic Simulation of Constrained Multi-Body Systems Using Vector-Network Techniques", Ph.D. Thesis, Queen's University, Kingston, Ontario (1985).

3.Pennock, G.R., and A.T. Yang: "Dynamic Analysis of a Multi-Rigid-Body Open-Chain System",Journal of Mech., Trans., and Auto. in Design, Vol. 105, March 1983.

4.Walker, M.W., and D.E. Orin: "Efficient Dynamic Computer Simulation of Robotic Mechanisms", Robot Motion. Planing and Control, p. 107, MIT Press (1982).

5.Hollerbach, J.M.: "A Recursive Lagrangian Formulation of Manipulator Dynamics and a Comparative Study", Robot Motion, Planing and Control, p. 73, MIT Press (1982)

6.Paul, R.P.: "Robot Manipulators. Math., Prog. and Control", MIT Press (1981).

7.Wittenburg, T.,: "Dynamics of Systems of Rigid Bodies", Teubner, Stuttgart, (1977).

8.Roberson, R.E., Schwertassek, R.,: "Dynamics of Multibody", Springer-Verlag, (1988).

9.Bindzi, I.: "Modélisation Dynamique de Systèmes Mécaniques de Corps Rigides par la Méthode du Réseau Vectoriel", M.Sc. Thesis, Laval University, Quebec (1990).

10.Andrews, G.C., Richard, M.J., Anderson, R.J.: "A General Vector-Network Formulation with Kinematic Constraints", Mech. Mach. Theory, Vol. 23, no 3, pp. 243-256 (1988)

11.Richard, M.J., Anderson, R., Andrews, G.C.: "Generalized Vector-Network Formulation for Multibody Systems", ASME Journal of Dyn. Systems, Measur. and Control, Vol. 108, p. 322 (1986).

12.Unda, J., Garcia de Jalon, J., Losantos, F. and Enparantza, R.: "A Comparative Study on Some Different Formulations of Constrained Mechanical Systems", ASME Journal of Mech., Trans., and Auto. in Design, Vol. 109, p. 466 (1987).

13.Andrews, G.C.: "Dynamics Using Vector-Network Techniques", Course-lecture Notes, Department of Mech.Eng., University of Waterloo, Waterloo (Ontario), 1975.

DYNAMIC SIMULATION OF MULTIPLE-CONTACT POWER GRASP FOR DEXTEROUS HANDS

Khalid Mirza, Mark D. Hanes, and David E. Orin

Department of Electrical Engineering
The Ohio State University
Columbus, Ohio 43210

ABSTRACT

Dynamic simulation of highly constrained systems using rigid-body dynamic equations encounters severe problems and limitations. This paper presents the development of a dynamic simulator which uses a compliant contact model to solve many of these problems. It is capable of simulating multiple contacts between an object and robot hand with friction as well as slipping and rolling effects and therefore can be used for analyzing power grasps. The simulator's grasp stability results are compared with theoretically computed ones using a model of the DIGITS Grasping System.

INTRODUCTION

Power grasps are characterized by multiple points of contact between the object grasped and the surfaces of the fingers and palm [1]. They maximize the load carrying capability of the robot hand and are highly stable in nature because of form closure, resulting from the large number of distributed contact points. In spite of the advantages, research in power grasp has been quite minimal, principally because of the analytical difficulties resulting from multiple contacts.

Dynamic simulation requires the solution of the forward dynamics problem: given the input torques and the initial state, solve for the joint/object positions, velocities and accelerations. In a highly-constrained system such as that of a power grasp, one of the major difficulties is to determine the internal forces. The difficulty arises in part because of the large matrix which must typically be inverted when dealing with a multiple closed-loop system. Another problem is to compute passive forces of constraint due to friction, including wedging effects. By simply adding Coulomb friction to the rigid-body dynamic model, one may find cases when no solution exists or multiple solutions occur [2]. In order to simulate realistic motion of the object grasped, slipping and rolling effects at contacts must also be modeled. Slipping is an integral part of power grasp as an object is maneuvered into a stable configuration, and rolling effects can have a significant impact on grasp stability, especially if the object has rotational motion. All of these problems are further complicated by the changing topological structure resulting from the making and breaking of contacts, and the motion of contact location on the surfaces.

One way to solve these problems is by introducing compliance in the system. A compliant contact model is developed for the dynamic simulator which uses local geometric information to compute contact forces. This approach effectively decouples the closed kinematic loops resulting from the multiple contacts. The resulting relatively simple and efficient equations can then be used to compute the motion of the finger links and the object. The simulator also incorporates the system dynamics, physical constraints within the grasping system, and rolling and sliding effects of the object at the contacts. The DIGITS Grasping System, a four-fingered dexterous grasping system currently under development at The Ohio State University [3], has been used as a target system for the simulator. (See Figure 1.) The following sections present details of the contact model and the algorithm using a planar model of the DIGITS System.

CONTACT MODEL

The compliant contact model developed consists of a pair of spring and damper units which are attached between the finger link and the object in normal and tangential directions to the finger link surface (Figure 2). This simulates the normal and tangential (frictional) forces at the contact.

First, based on the finger link and object positions, contact is detected if the object surface penetrates into the finger surface. Springs are then attached from the deepest penetration point on the object's surface to the perpendicular projection on the finger surface. This generates the initial impact force. Subsequently, the spring attachment points on the two surfaces are kept fixed and any relative motion between the finger link and the object results in an extension of the springs which generates the corresponding contact force. The spring displacement c and velocity $\dot{c}$ are computed using local geometric information. The resulting force f is given by

$$f = Kc + B\dot{c},$$

where K and B are spring stiffness and damping constants, respectively. The choice of K and B controls the behavior of the contact and can model different material's surface characteristics. Hence, this contact model introduces compliance in the system and simultaneously specifies all contact forces.

SLIPPING AND ROLLING EFFECTS

As described in the previous section, the spring attachment points are kept fixed as long as contact is maintained between the two surfaces. However, for the special case of slipping and rolling, the spring attachment points are allowed to move in order to simulate slipping and rolling effects.

Slip is detected when the computed tangential force F is larger than the threshold set by the normal force N at the contact:

$$|F| > |\mu N|,$$

where μ is the coefficient of friction. To impose the frictional constraint, F is restricted to μN. This change in the tangential force is made consistent with the corresponding spring configuration by moving its attachment point on the finger link surface.

The rolling effect can be analyzed by studying the vector system at the contact as shown in Figure 3. Vectors $\underline{s}$ and $\underline{t}$ are the contact vectors to the two spring attachment points, and $\underline{c}$ is the spring displacement vector. Initially, vector $\underline{t}$ is perpendicular to the finger surface. A rotation of vector $\underline{t}$ accompanied with translation of the object relative to the finger surface indicates a rolling effect. For this case, vector $\underline{t}$ is adjusted to align with the vector to the deepest penetration point, $\underline{r}$, which moves the contact location on the object surface to compensate for the object roll. Vector $\underline{s}$ is adjusted to keep the tangential force unchanged, which allows the contact location on the finger surface to track the rolling object.

Hence, the slipping and rolling effects are modeled at each contact independently just by analyzing the local geometric information. The adjusted contact forces, which are now consistent with all the given constraints, are subsequently used to solve the forward dynamics problem.

SIMULATION RESULTS AND CONCLUSIONS

The dynamic simulator is currently implemented on Sun workstations using X Windows as a graphical interface. (See Figure 4.) The output of the simulator can also be viewed in a three-dimensional shaded-color representation on a G.E. Graphicon workstation. Presently, the simulator uses a planar model of the DIGITS System grasping a circular object. However, it is flexible enough to be adapted to new systems and objects and thus has potential uses in a wide range of grasping research. All initial conditions may be completely specified by the user, including joint torque profiles and an external wrench on the object.

The simulator has been used in part to verify several theoretical results on power grasp stability, as discussed in [4, 5]. Linear programming was used to determine the maximum weight handling capability of DIGITS System for various size objects and friction levels. Previously, it was unknown as to whether these results were physically realizable. The simulator corroborated the predicted results and found grasp configurations capable of improved weight handling. It also demonstrated that to obtain certain grasp configurations, wedging is required.

The simulator has been found to be a significant tool for power grasp research. Using only local geometric information and simple dynamic equations, it allows simulation of multiple contacts with sliding, rolling, and wedging effects. Additional work is proceeding to extend the simulator to handle arbitrary convex objects and should prove beneficial in on-going research into the design of power grasp controllers.

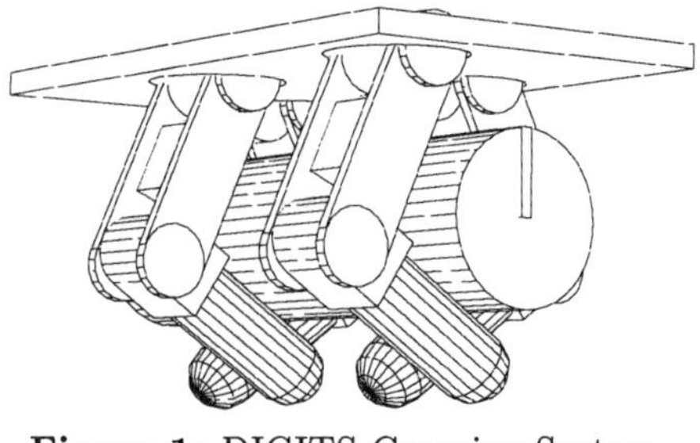

Figure 1: DIGITS Grasping System.

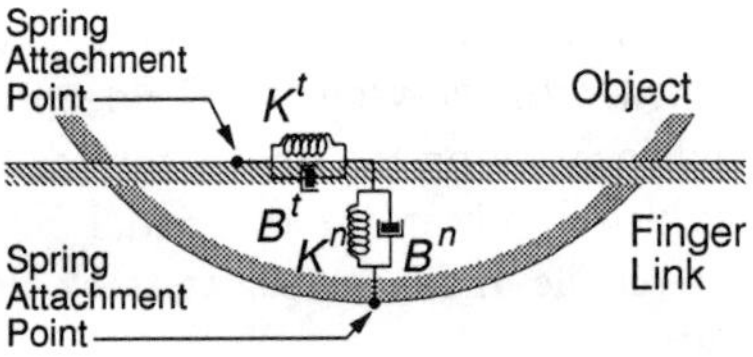

Figure 2: Contact model.

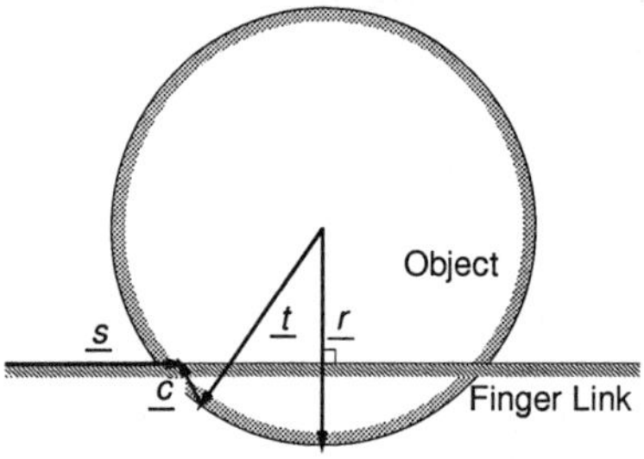

Figure 3: Contact vector model.

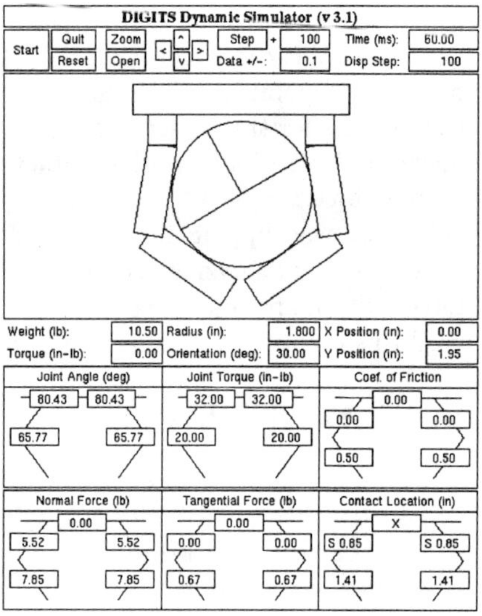

Figure 4: Simulator in X-Windows.

REFERENCES

[1] M. R. Cutkosky, "On grasp choice, grasp models, and the design of hands for manufacturing tasks," *IEEE Transactions on Robotics and Automation*, vol. 5, no. 3, pp. 269–279, June 1989.

[2] P. E. Dupont, "The effect of coulomb friction on the existence and uniqueness of the forward dynamics problem," in *Proc. of the 1992 IEEE International Conference on Robotics and Automation*, (Nice, France), pp. 1442–1447, May 1992.

[3] S. J. Coe and K. J. Waldron, "Mechanical design of the DIGITS System," in *Proc. of the 1989 1st National Applied Mechanisms and Robotics Conference*, (Cincinnati, Ohio), November 1989.

[4] K. Mirza and D. E. Orin, "Control of force distribution for power grasp in the DIGITS System," in *Proceedings of the 29th Conference on Decision and Control*, vol. 3, (Honolulu, Hawaii), pp. 1960–1965, December 1990.

[5] M. D. Hanes, S. C. Ahalt, K. Mirza, and D. E. Orin, "Power grasp force distribution control using artificial neural networks," *Journal of Robotic Systems*, vol. 9, no. 5, July 1992.

A new Technique of Eliminating Void in Workspace of a Manipulator with an Offset Wrist

K. S. Lee, J. H. Kim, K. W. Jeong
Research Scientists, Robotics Group
Dept. of Mech. Eng., RIST, Pohang, KOREA, 790-600

Y. Youm
Professor
Dept. of Mech. Eng., POSTECH, Pohang, KOREA, 790-600

ABSTRACT

Most conventional manipulators with an offset wrist have a problem that the end-effector cannot be freely oriented. The link offset in wrist prevents the manipulator from orienting the end-effector at a certain region. We defined this region as a void. In this paper, this void is analyzed and a new technique which can eliminate such voids in workspace is proposed.

1. Introduction

A serial manipulator has at least 6 degrees of freedom in order to position and orient the end-effector in space. Some manipulators are provided with a fixed offset between two of the wrist axes so that a stronger wrist configuration can be obtained to handle heavier payloads. These manipulators have different kinematic characteristics compared with the manipulators which have no link offset in wrist. There are differences in the orienting region of the wrist and the workspace of the manipulator between the offset wrist and non-offset wrist.

R.P.Paul[1] presented that the orienting region of the wrist of which the rotational axes intersect at one point is determined by the magnitude of twist angles between the axes in the wrist. When their twist angles,α_4 and α_5, are different, there always exist some region in which the manipulator cannot be oriented in workspace. When the twist angles are equal and 90° respectively, the orienting region is a sphere form. Almost all the wrists are designed according to the afore mentioned theory so that their kinematic parameters,α_4 and α_5, are equal and 90°. K.C.Gupta[2] introduced some basic concepts regarding the workspace of manipulators having a non-offset wrist. For such a system, the workspace has a structure of the primary and secondary workspaces which are affected by the hand size.

For most manipulators with an offset wrist, there exist some region in which the end-effector cannot be oriented for the designinated position in cartesian work space. Even though their kinematic parameters,α_4 and α_5, are equal and 90°, there exist a void shown in Fig.1. These manipulators do not have primary workspace.

In this study, the reason why the manipulators having an offset wrist lose their orienting motion in some regions in workspace is investigated. A new design technique of an offset wrist is also proposed to solve this problem.

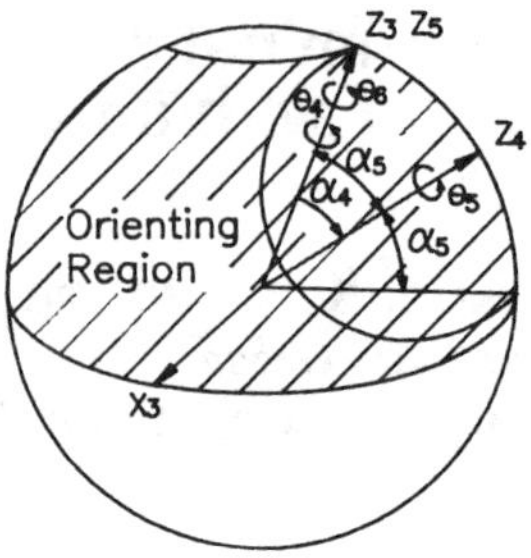

Fig.1 Orienting region of an offset wrist when $\alpha_4=\alpha_5=90°$

2. Orienting region of a manipulator with an offset wrist

2. 1 Model of a manipulator

We assumed a manipulator with an offset wrist as shown in Fig.2 in order to investigate the orienting region and the workspace. There is a fixed offset between joint 4 and 5 of the wrist. The twist angles between the Z axes are 90°. The zero position of the coordinates is denoted by using a standard notation shown in Fig. 3.

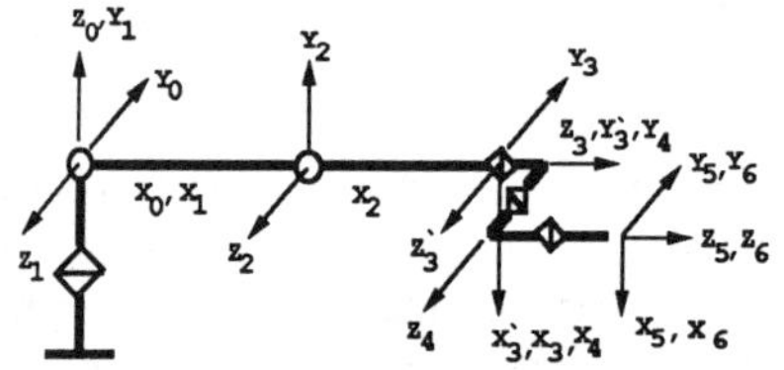

Fig. 2 Model of a manipulator with an offset wrist

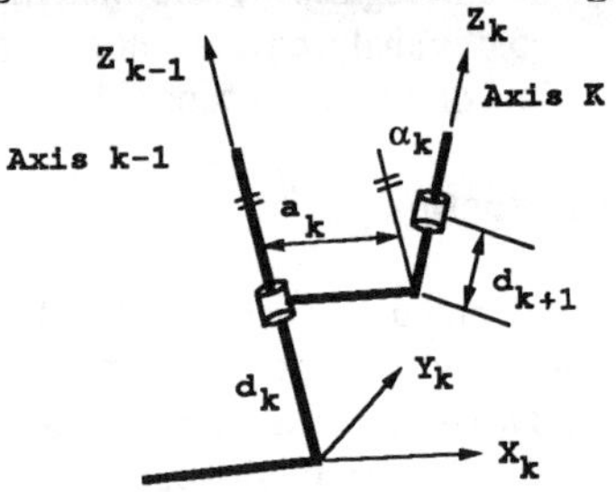

Fig. 3 Standard link notation

2.2 The orienting region of the manipulator

A manipulator with an offset wrist loses its orienting motion in some region at any positions in workspace. For example, if the manipulator is located at a position shown in Fig.4 and the desired direction of the end-effector at this point is $\vec{a}$, the manipulator cannot move along this direction, $\vec{a}$. If we draw all the direction at which the manipulator can direct at P_5, there exist a corn in which the Z_5 axis cannot reach, as shown in Fig.5. If the desired direction at this point is in this corn, the manipulator never position this direction.

The positioning point of the manipulator with an offset wrist must be the origin of frame 5 located at the tip of offset link. The manipulator can be reconstructed as shown in Fig.6. For this manipulator, the upper arm was defined virtually as the third side of the triangle composed of the original upper arm and offset link, and then the three axes of wrist come to intersect mutually at the tip of the offset link. In this reconstructed model, the twisted angles, α_4 and α_5, between axes are different. The kinematic characteristics of these two manipulators are almost identical. We can easily realize that the manipulator shown in Fig. 6 has some region in which the manipulator cannot orient the end-effector arbitrarily from the theory presented by R.P.Paul[1]. Therefore, the manipulator under being considered also has some void as shown in Fig.1.

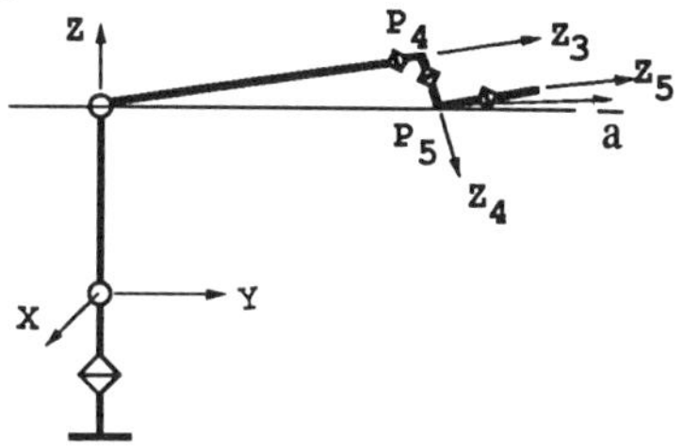

Fig.4 Wrist direction which the manipulator cannot orient

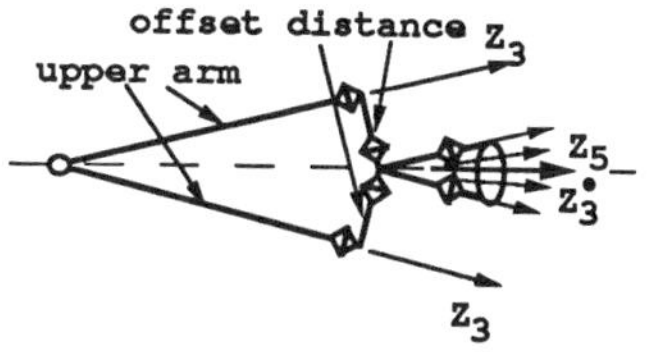

Fig. 5 A corn in which wrist cannot orient

Consequently, the arm direction of a manipulator with an offset wrist is not the Z_3-axis direction any more and there is another real arm direction $Z_3°$. The twist angle between $Z_3°$ and Z_4 is no more perpendicular, although α_5 remain 90° as shown in Fig.7. The difference between α's prevents the manipulator orienting any region forward the arm direction. We defined this region as a void in which the manipulator cannot orient the end-effector.

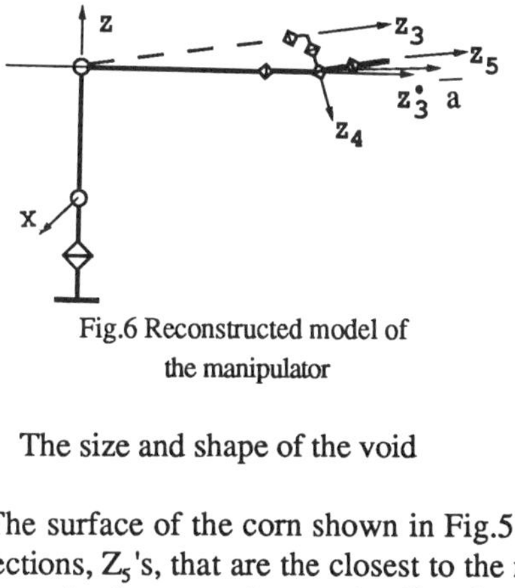

Fig.6 Reconstructed model of the manipulator

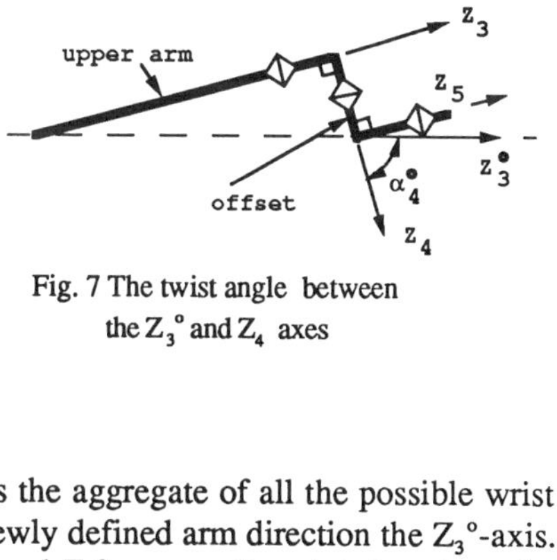

Fig. 7 The twist angle between the $Z_3°$ and Z_4 axes

2.3 The size and shape of the void

The surface of the corn shown in Fig.5 is the aggregate of all the possible wrist directions, Z_5's, that are the closest to the newly defined arm direction the $Z_3°$-axis. The angles between these wrist directions and $Z_3°$ are greatly related to the twist angles between the $Z_3°$ and Z_4 axes at each position as shown in Fig.7. The twist angle between the $Z_3°$ and Z_4 axes was defined as $\alpha_4°$. We investigated how different the twist angles are according to the value of joint 4 at the point and how different they are at other points in workspace. Some points which are thought to be positioned frequently in workspace are selected as shown in Fig.8 and the twist angles between the $Z_3°$ and Z_4 axes are examined at each point. The cross section of the base of the corns are drawn in Fig.9 as a result. As shown in Fig.9, the corn shrinks horizontally as the manipulator stretches its arm and expand horizentally as the manipulator shrinks. The base of the corn constitutes a circle at a position and an ellipse at the other positions. This kind of manipulator does not have the primary workspace where the manipulator can position and orient the end-effector arbitrarily.

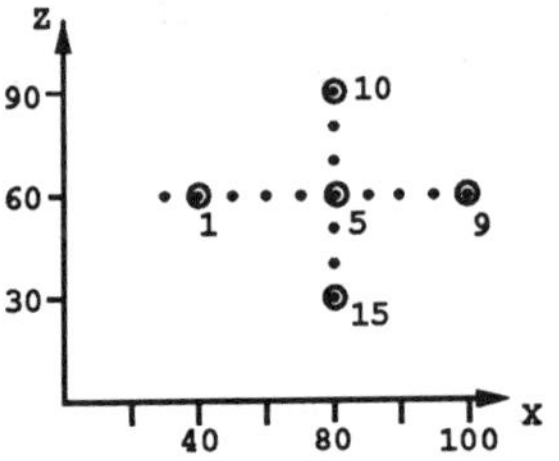

Fig.8 Selected positions for computer simulation

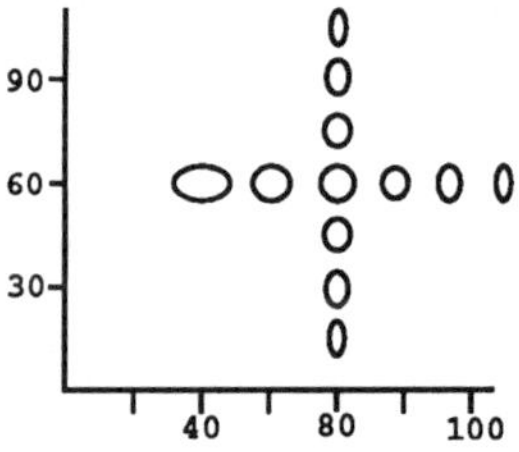

Fig.9 Variation of the corn according to the position

3. A new design technique of kinematic parameters of an offset wrist

As the results of the previous discussions, we assumed that if we design the kinematic parameters, α_4 and α_5, which are different initially but the twist angle $\alpha_4°$ is equal α_5 after positioning, we can eliminate the void. But the angle $\alpha_4°$ is different depending on the location of the manipulator and the value of the joint 4 , even at the same position. This means that we may not eliminate all voids in workspace, because there are various $\alpha_4°$'s. Therefore we must select only one of them when designing the manipulator.

In order to examine our assumption, we selected one position numbered 5 in Fig.8. At this position, the twist angle $\alpha_4°$ is unique regardless of joint 4. According to our assumption, we can design the kinematic parameters at least at that point. We calculated the twist angle $\alpha_4°$ at this position and designed α_5 to be the same with $\alpha_4°$ after the manipulator positionning. We investigated how the void varies using a computer simulation. From the Fig.7, the twist angle between the $Z_3°$ and Z_4 axes is

$$\alpha_4° = 90 - \Delta\alpha$$

where,

$$\Delta\alpha = \text{atan}\,(d_5/l_3).$$

Here $\Delta\alpha$ is 7.125° and $\alpha_4°$ is 82.875° for the manipulator under considered. The original value of α_5 is -90°, but it must be the same with $\alpha_4°$ to eliminate the void. Considering the direction of rotation, the α_5 is -82.875.

The result of computer simulation is shown in Fig.10 (a). The outmost circle in (a) shows the region in which the manipulator cannot orient when α_4=90°, α_5= -90° and the innermost circle represented as a dot shows the region when α_4 = 90°, α_5 = -82.875°. The middle circles show how the region varies. The void become smaller as α_5 varies from -90° to -82.875° and finally disappear. These results partly certify our assumption correct. As we mentioned above, the twist angle $\alpha_4°$ varies depending on the manipulator positions and the value of joint 4 even at the same position. It was difficiult to imagine any strategy which is assumed to be satisfied with all these conditions. Therefore, we applied the same strategy used above at other positions in workspace and investigated how the void changes. Fig. 10 (b),(c) and (d) show the variation of voids at several positions numbered to 1,10 and 15 in Fig. 8, respectively. Each case shows the same result. That is, the outmost ellipse shows the void when α_4=90°, α_5= -90° and the void becomes smaller as α_5 varies from -90° to -82.875° and finally disappears. The innermost ellipse represented as a line shows the void when α_4 = 90° and α_5 = -82.875°. These results represent an unexpected phenomenon. If we design their kinematic parameters by using the method mentioned above, all voids can be eliminated. The reason can be certified by the following explanation.

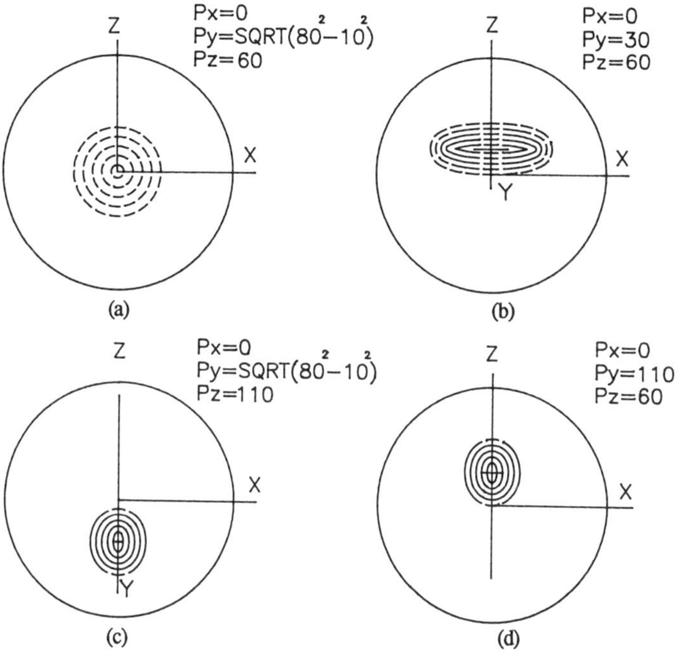

Fig. 10 Variation of void according to α_5

(a) At point 5 in Fig. 8 (b) At the other points in Fig. 8

Fig. 11 Trajectory of Z_5 vectors

The rotating axis of α_5 is X_4. When the manipulator rotates joint 4 to position, the frame 4 rotates around the Z_3-axis which is rotating around the new arm direction $Z_3{}^{\circ}$. The Z_5's are wrist directions and the X_4's are X directions of frame 4 when α_4 = 90° and α_5 = -90° in Fig.11. If the manipulator rotates joint 4, the Z_5 changes to $Z_5{}^1$, $Z_5{}^2$, $Z_5{}^3$, ...and similarly for the X_4-axis. Changing the α_5 from -90° to -82.875° induces a similar effect that each Z_5 turns 17.125° around each corresponding X_4. After compensating α_5, all Z_5's coincide to $Z_3{}^{\circ}$ like Fig.11(a) when the base of the corn forms a circle, and gather on a small curve like Fig.11 (b) when the bottom of the corn spapes an ellipse. If rotating joint 5 about each Z_5, all wrist directions constitute a complete sphere when Z_5's coincide to Z_3°. When Z_5's gather on a small curve, wrist directions form an ellipsoid. Thus the voids are eliminated in workspace.

Based on our investigation, a new design technique is proposed to solve the problem of an offset wrist. From the Fig. 12., the kinematic parameters, α_4 and α_5, must be designed using the following equations.

$$\alpha_5 = \mathrm{acos}\left(\frac{l_3'^2 + d_5^2 - l_3^2}{2\, l_3'\, d_5}\right)$$

where,

$$l_3'^2 = d_5^2 + l_3^2 - 2\, l_3\, d_5 \cos(180^{\circ} - \alpha_4).$$

Here l_3 is the length of upper arm and d_5 is the length of the offset link.

First, determine one of α's and then calculate the other from the above equation.

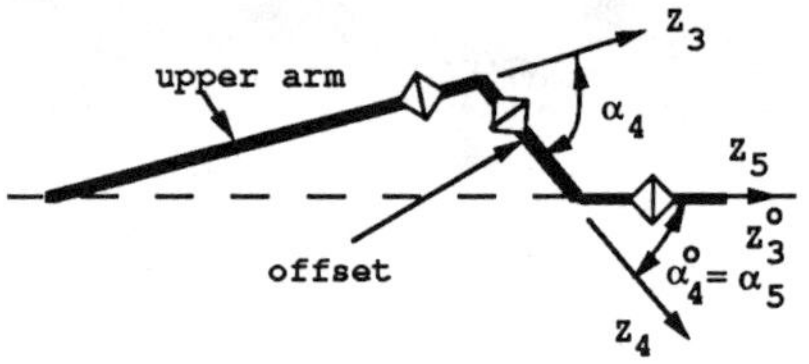

Fig.12 Kinematic parameters, α_4 and α_5

4. Conculsion

In this paper, we studied the kinematic characteristics of a manipulator with an offset wrist. It was found that there always exist some region in which the wrist cannot be oriented at the designated position in the workspace and this kind of manipulator does not have a primary workspace. Based on the analysis presented in this study, we proposed a new design technique of kinematic parameters of an offset wrist. The proposed method was shown to effectively solve this problem and thus to improve performance of the manipulator.

5. References

1. R. P. Paul, and C. N. Stevenson, "Kinematics of Robot Wrist," Journal of Robotics Research. Vol. 2, No. 1, (Spring 1983).
2. K. C. Gupta, and B. Roth, "Design Consideration for Manipulator Workspace," ASME Journal of Mechanical Design, Vol. 104, (October 1982).
3. R. Featherstone,"Position and Velocity Transformations Between Robot End-Effector Coordinates and Joint Angles," Journal of Robotics Research, Vol. 2, No. 2, (Summer 1983).
4. M. Takano, K. Yashima and S. Yada," Computer Simulation System of Kinematics and Dynamics of Robot," Journal of the Faculty of Engineering, The University of Tokyo, Vol. 36, No. 4.(1982).

DESIGN OF AN AUTONOMOUS ROBOT GRIPPER

P.GORCE - P.BIDAUD - M. PICARD
Laboratoire de Robotique de Paris
Université de Paris 6 - CNRS - ENSAM
Tour 66 - 4 place Jussieu
75252 PARIS Cedex 05

ABSTRACT

The Robotic Laboratory of Paris has developped a flexible gripper equipped with tactile sensors and a local system control. This system offers flexibility and integrability capabilities to answer the needs for assembly of multi-component mechanical systems. In this paper, advances in gripper mechanical design, sensor integration and intelligent local control from an hardware and a software point of view are discussed.

1. INTRODUCTION:

In industrial application, grippers are often used to manipulate a small number of parts with similar size and weight in repetitive operations which require a minimal dexterity [1]. Because of their lack of flexibility [2] and their bad adaptativity in front of uncertainties, these grippers cannot be used in complex applications.

To answer flexibility constraints in the grasp, many researches and developments have been undertaken in the past ten years. From a mechanical point of view, many works have been realized on the following area: - the analysis of the grasp and more especially the stability of this one [3], [2], [4], [5], - design of finger mechanisms [7], [8], [9]. Therefore, several grippers with tactile sensors [6], [7] and several multifingered hands [8], [9], [10] have been developped.

To implement a functions on such system complex sensory controlled, it is necessary to introduce decisional capabilities [11]. This suppose both a software and an hardware adapted organization.

In this paper, first we develop a set of mechanical considerations to answer the constraints of multi-component assembly. Which basically suppose: - the capabilities to realize a stable grasp of objects of various shapes, - the adaptation of the grasping forces (internal force) to the variation of the physical behaviour of the manipulated object, - the measurement of external forces. Then, we describe the gripper we built and the hardware integration of these differents functions. Finally, we present a software structure based on intelligent primitives which regulate informations with respect to tasks to realize.

2. DESIGN OF GRIPPER MECHANISMS:

2.1. Analysis of mechanical considerations:

The gripper mechanisms can be considered as composed of linkages (finger mechanisms) and a set of contact pairs (the "jaws"). The design of finger linkages has to integrate overall dimension constraints and force/motion transmission caracteristics determined by the "task". The kinematic caracteristics of the set of contact pairs have to be define such as the static equilibrum of the object is satisfy. This condition can be expressed in any frame by:

$$[G] . [f] = [F] \qquad (1)$$

where column of [G], called the grasp matrix, are the plucker's coordinates of contact wrenches, [f] is a n dimensionnal column vector of contact wrench magnitudes (n is being the number of contact wrenches) to substain an arbitrary external wrench (regrouping the load of the object and assembly forces).
Ignoring unidirectional and frictionnal constraints, the necessary condition to assure force closure is:

$$\text{rank}([G]) = 6 \qquad (2)$$

The conditions for no breaking contact and no sliding that the contact force magnitudes must satisfy are:

$$f_{zi} > 0 \qquad (3)$$

$$\mu^2 f_{zi}{}^2 \geq f_{xi}{}^2 + f_{yi}{}^2 \qquad (4)$$

where the contact normal f_{zi} is defined as inward direction from outer object face and the tangential components of each contact force f_{xi} and f_{yi} are defined in tangential plan at the contact.
The grasp matrix can be considered as a projector of the contact force magnitude vector into a generalized force vector [18]. The contact force which project to a particular generalized forces are not unique if the number of contact wrenches is greater than the rank of [G]. In this case, the possible contact force magnitudes can be expressed as:

$$[f] = [fp] + [fh] \qquad (5)$$

where [fp] is the particular solution or a manipulation force to realized static equilibrum of the grasp [5], [14], [16], [17], [18]. [fh] is the homogeneous solution, called force closure [13] or internal force [5], [14], [15], [16], [17], [18] it is a vector of contact magnitudes within the nullspace of [G].
Concerning the grasp others definitions have been given: [13] takes into account the stiffness of each finger in the study of the stability of one polyhedron, [12] defines that the grasp is stable with elastic fingers if, according to the grasping configuration, the potential function is minimized.

2.2. Analysis of hardware and software considerations:

Reactivity capabilities can be introduce by the use of:

- force sensors to measured the elementary components of forces developped by the fingers on the object and the external forces,
- position sensors to measure a kinematic variables of finger mechanisms,
- computing resources:
  . to prevent a sliding of the object by means of tactile informations associated to position informations,
  . to compute an adapted force transmission submit at an external forces,
  . to compute the finger coordinates,
- communications in order to facilitate the integration of the gripper in the global production structure.

3. THREE FINGERED HAND DESCRIPTION:

3.1. Mechanical description:

The gripper linkages are derived from the O.EVANS mechanism (figure 1).The finger mechanisms are connected to a mechanical part "the nut" and transforms the movement of the linear actuator into three concentric translations (figure 2).

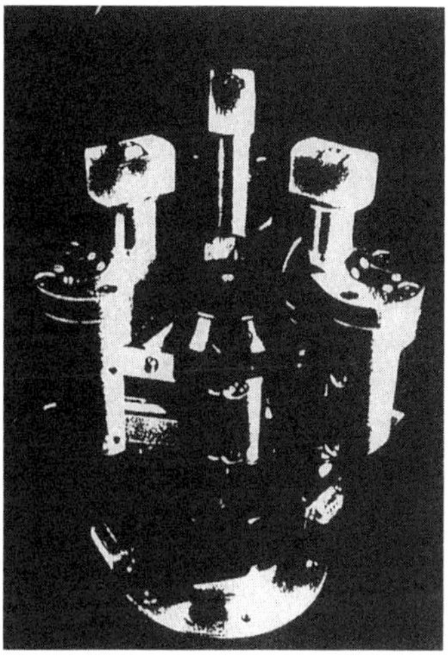

Figure 1: Robot hand description

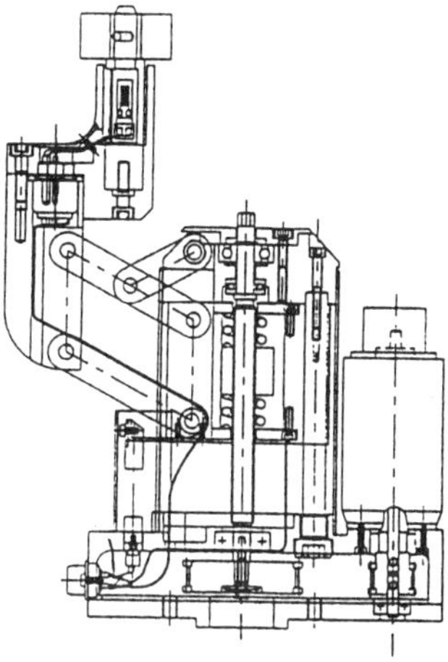

Figure 2: Mechanism description

The 2D model of the O.EVANS mechanism is given on figure 3:

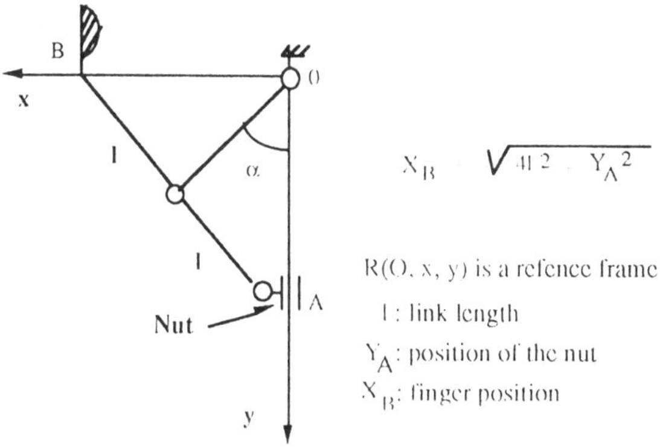

Figure 3: Plan model of the O.EVANS linkage

To obtain an good dynamics in grasping force application a spring (with stiffness K) was integrated between the nut and the finger mechanisms.
Each contact of the three-fingered are considered as ponctual with friction to satisfy the stability condition.

3.2. Robot gripper instrumentation:

Force and position sensors are integrated in the mechanical structure of the gripper.

. Position sensors allow to measure: angular position of the motor (incremental optical sensor), nut position (rectilinear potentiometer) and the reference position (Photologic slotred optical switches). The finger tip coordinate are expressed in reference frame R(0, x, y) as:

$$x = \sqrt{4l^2 - y^2} \qquad y = \frac{q}{(2p)} \cdot \frac{1}{n} \cdot q_r \qquad (6)$$

with: x finger position, y nut position, q_r angle of the motor axis and p pitch of nut.

Each finger was design to be a compliant mechanical part of which deformation are measured by extensometric technics. Each finger could receive one or three strain gauges to measure respectively the normal components or the resultant force (three components) at the contact. They can also receive strain gauges to measure the contact pressure (type FSR from Interlink).
The grasp force can be deduce as follow:

$$F_{finger} = \frac{F(nut)}{3tg\alpha} \qquad (8)$$

with: $$tg\alpha = \frac{2l-y}{2l-x} \quad \text{and} \quad F(nut) = K\,.\,(Ymes - Y) \qquad (9)$$

$$F_{finger} = \frac{2l - \sqrt{4l^2-y^2}}{3(2l-y)K(Ymes-Y)} \qquad (10)$$

3.3. Local controller and interface:

The autonomy of this gripper is obtained by the controller architecture and the programming method we use. The distributed organization of control provides an efficient solution for integration of intelligent primitives.

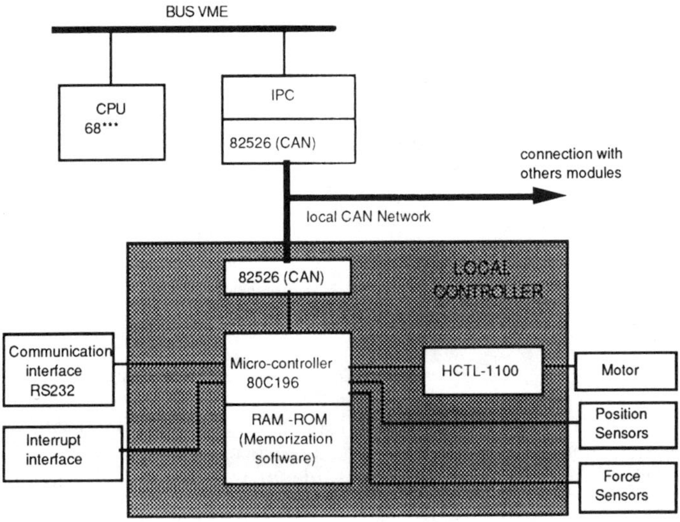

Figure 4: Local controller and its interface

The hardware of the gripper controller (see figure 4) involves a numerical controller HCTL1100 from HP, a power supplier chip L298, a microcontroller 80C196 from Intel. This local controller can communicates with a remote computer, based on VME bus or PC bus, over a networks field-bus CAN from Intel.
We can notice that the physical proximity of the hardware components of the controller from the sensors and the actuator contitutes an important factor for fiability and security of the grasp which call a great amount of data exchange and processing.

Because of communication protocole differencies between the CAN bus and the VME bus we have develop a software package to control

- completness of data,
- data flows,
- access time to data.

in data exchanges (figure 5).

The first function is realized by the DEBUG function which control the validity of the data. The second is insured by the treatement functions which allows to update the data of buffer. The last one is realized by a buffer which allows data exchange and a possibility for each processor to convert in its own format the data which are in the buffer. Is by this one, that the set of transfert is realize. It is composed of two parts:
- data: set of exchanged variable between the process by communication and data formating system,
- the state structure: this memory area is reserved to the gripper state description.

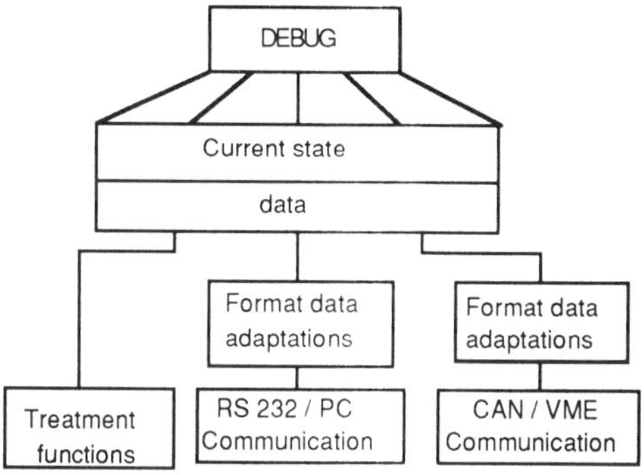

Figure 5: Sofware structure

Such a complex system requires decision-making capabilities to achieve an optimal grasp of objects with various shapes and errors recovery. Therefore, it seems preferable to use decentralized decision-making so that the effector can control its own tasks.

The implementation of the above principle naturally requires the use of declarative programming in order to facilitate the introduction of heuristic rules and meta-knowledge. The "real-time" constraints introduced by the robotic application requires, an efficient inference engine. For this purpose, we have developed (using "C" langage) a light (64K) inference engine of the Prolog kind.

4. FONCTIONALITIES:

Gripper task specification consists to define the sequences as well as the order of their execution. For that, we have decomposed the task in several basic functions which realize elementary actions. The basic primitives are for example:

- **Initialization**: primitive which initialize the incremental position sensor and the tactile sensors.
- **Position (X_c, V)**: the opening (or closing) of the gripper to a refence position with a given velocity,
- **Grasping (F_c)**: this procedure allows applied a given grasp force by the means of the finger positions,

- **Force reading (F_n or F_x, F_y, F_z)**: This function allows to measured the normal component or the resultant (normal and tangential components) forces at the contact finger/object.

These actions are conditioned by pre and post conditions linked to the existence of a situation. By an organization of these, the basic functions allow to define the operating sequences (or procedures) according to the considered application. For that, we have developped "expert" functions which evaluate the pre and post conditions.

5.CONCLUSION:

In this paper, we have discribed the realization an autonomous robot gripper. On this gripper tactile sensors and a local controller have been implemented. Today, this gripper is used in industrial applications.

BIBLIOGRAPHIE:

1. A.R.Mannaa, M.Akyurt, A.K.El Kalay, "Enhanced gripping mechanisms for industrial robots", International Journal of Robotics and Automation, Vol 6, n°3, 1991, p156-160,
2. M.R.Cutkosky, "On grasping choice, grasp models and design of hands for manufacturing tasks", *IEEE Transaction on Robotics and Automation*, Vol 8, n°3, 1989,
3. J.K.Salisbury, "Kinematic and force analysis or articulated hand", *Ph D dissertation, STANFORD University, Dept of Mechanical Eng*, May 1982,
4. J.W.Jameson, L.J.Leiffer, "Quasi-static analysis: a method for predicting grasp stability", *Proc IEEE International Conference on Robotic and Automation*, 1986, p876-883,
5. Z.Ji, B.Roth, "Direct computation of grasping force three-finger tip-prehension grasps", *Journal of Mechanisms Transmissions, and Automation in Design*, Vol 10, dec 88, p405-412,
6. C.Bao, H.Van Brussel, "A sensory controlled gripper system", *Robotics and Autonomous systems*, Vol 6, n°3, Jul 1990, p283-295,
7. D.T.Pham, E.Tacgin, "CAD of devices for gripping prismatic components", Robotica, Vol10, p229-237,
8. J.K.Salisbury, B.Roth, "Kinematic and force analysis of articulated mechanical hands", *Journal Mech. Trans. Auto. in Design*, Vol105, 1983,
9. S.C.Jacobsen and al, "Design of the UTAH/MIT dexterous hand", *Proc IEEE International Conference on Robotics and Automation*, 1986,
10. J.C.Guinot, P.Bidaud, S.Zeghloud, "Modelisation and simulation of force-position control for a manipulator gripper", *Robotics Reseach the 3th International Symposium*, 1986, p315-319,
11. J.Xiao, R.A.Voltz, "Design and motion constraints of parts mating planning in the presence of uncertainties", *Proc IEEE Conference on Robotics and Automation*, 1988,
12. H.Hanafusa, H.Asada, "Stable prehension by a robot hand with elastic fingers", *In Proc 7th ISIR Tokyo, Japan*, 1977, p 361-368,
13. V.Nguyen, "The synthesis of the stable grasp in the plan", *Proc IEEE International Conference on Robotics and Automation*, 1986, p884-889,
14. Y.C.Parck, G.P.Starr, "Finger force computation for manipulation of an object by a multifingered Robot hand", *Proc IEEE International Conference on Robotics and Automation*, 1989, p930-935,
15. J.Kerr, R.Roth "Analysis of multifingered hands", International Journal of Robotics Research, Vol 4, n°4, 1986, p3-17,
16. A.Bicchi, "Analysis and control of power grasping", IEEE/RSJ International Workshop on Intelligent Robots and Systems IROS'91, nov 3-5 1991, p691-697,
17. T.Yoshikawa, K.Nagai, "Evaluation and determination of grasping forces for multi-fingered hands", Proc IEEE International Conference on Robotics and Automation, 1988, p245-248,
18. R.P.Podhorodeski, A.A.Goldenberg, R.G.Fenton, "Analitycal bases for the contact forces leading to internal forces for point friction contact grasps", Conference on Apllied Mechanism and Robotics, Cincinnati, Nov 1989,

DESIGN OF SHAPE MEMORY ALLOY ACTUATORS FOR ROBOTICS

C. LIANG and C. A. ROGERS
Center for Intelligent Material Systems and Structures
Virginia Polytechnic Institute and State University
Blacksburg, Virginia 24061-0261
Tel. (703) 231-2900, Fax (703) 231-2903

ABSTRACT

Shape memory alloys (SMA) have been used as actuators in robotics applications. Because of the lack of proper design guides, the design of SMA actuators has been based on experimental trial and error. This paper presents a comprehensive investigation of the design of SMA actuators for robotics applications, including constitutive models of shape memory alloys and theoretical design tools for the design of SMA actuators and springs.

INTRODUCTION

The investigation of shape memory alloys, especially Nitinol, has been an area of interest and activity since the 1960's when Nitinol was first discovered. However, the constitutive modeling of SMA, i.e., the design tools, have not been very well developed, which has hindered the application of SMA. The design of SMA actuators has been based on experimental trial and error. Because of the length restriction, Detailed literature review cannot be provided. Interested readers may refer to (Funakubo, 1987) for more information on SMA properties and applications; and to (Liang and Rogers, 1990) for information regarding the modeling techniques of shape memory alloys.

This paper presents a comprehensive investigation on the design of SMA actuators for robotics applications, including constitutive models of shape memory alloys and theoretical design tools for the design of SMA actuators and springs. Because of space limitation, we cannot provide detailed discussion; readers can refer to the references given in each section for more information.

CONSTITUTIVE MODELING OF SMAs

Tanaka (1982) has presented a one-dimensional thermomechanical constitutive relation for shape memory materials based on the thermomechanical Helmholtz free energy concept and metallurgical phenomenology to give a qualitative assessment of the stress-strain relations of shape memory alloys. Liang and Rogers (1990) further developed this model and modified Tanaka's relations in order to quantitatively study the constitutive relations of shape memory alloys, including the shape memory effects, and provide design tools for engineering applications.

The basic governing equations for SMA can be expressed as

$$\dot{\bar{\sigma}} = D\dot{\bar{\epsilon}} + \Theta\dot{T} + \Omega\dot{\xi}, \tag{1}$$

where D, Θ and Ω are the Young's modulus, thermoelastic tensor, and phase transformation tensor, respectively. The variables, $\bar{\epsilon}$, $\bar{\sigma}$ and T, are the Green strain, the second Piola-Kirchhoff stress and temperature, respectively. A $(\dot{})$ represents a time derivative. The martensite fraction, ξ, an internal variable, is expressed phenomenologically by the following equations in the phase transformation range. The martensitic transformation $(A \rightarrow M)$ is given by

$$\xi = \frac{1-\xi_A}{2}\cos[a_M(T-M_f)+b_M\sigma]+\frac{1+\xi_A}{2}, \tag{2}$$

while the reverse transformation $(M \rightarrow A)$ is given by

$$\xi = \frac{\xi_M}{2}\{\cos[a_A(T-A_s)+b_A\sigma]+1\}, \tag{3}$$

where a_A, a_M, b_A, and b_M are material properties derived from the four transition temperatures, M_f (Martensitic finish), M_s (Martensitic start), A_s (Austenitic start), and A_f (Austenitic finish) and the stress influence coefficient which describes how the transition temperatures vary with stress. The quantities, ξ_A and ξ_M, are the initial martensite fractions for each transformation process, i.e., heating or cooling. The independent variables required for this constitutive model include the four transition temperatures, Young's modulus, thermoelasticity tensor, phase transformation tensor (or the maximum recoverable strain which is considered as constant in the interested temperature range), and the stress influence coefficient. A detailed explanation of the constitutive model and the associated constants is given in Liang and Rogers (1990).

Based on the one-dimensional model, the two-way shape memory effect (SME) of SMA has been modeled by Zhang and Rogers (1992). An internal stress term which represents the accumulation of dislocations around crystal boundaries during the two-way SME training process has been derived. The internal stress term is used as an initial condition to change the phase transformation hysteresis. This model can provide a qualitative description of the two-way SME.

The one-dimensional model of shape memory alloys has been experimentally verified. The detailed experimental process and verification can be found in Dye (1990) and Liang (1990). The model can provide quantitative prediction of the stress-strain relations of SMA, free shape recovery effect, and recovery stress-strain-temperature relations. The model over-predicts the recovery stress when the initial strain is more than 2% because the model assumes that the strain is only associated with the phase transformation and no plastic deformation is considered. However, since 2% initial strain is generally recommended for SMA actuators for higher repeatability because of the fatigue of SMA, this model may be used to design SMA actuators.

DESIGN OF SMA ACTUATORS

There are two basic types of SMA actuators. The actuator shown in Fig. 1 (a) is called a bias force actuator, which uses a spring to generate the restoring force. The SMA spring or wire is first elongated at a low temperature before installation. When the SMA spring is heated, the recovery force generated by the shape memory effect (SME) "pulls" the spring to store energy in the spring. When the SMA actuator is cooled, the potential energy stored in the bias spring is released to strain the SMA spring (or wire) back to its initial position, resulting in a deformation-reformation cycle. The actuator depicted in Fig. 1 (b) is called a differential SMA actuator which includes an opposing SMA element used to create an active bias force and a deformation SMA element with some initially-stored energy. A specially designed cooling and heating strategy for the two SMA elements can produce a differential motion path. This paper can only provide a brief introduction, interested readers please refer to (Liang and Rogers, 1992b) for more detailed discussion.

Design of Bias Spring SMA Actuators

The stress-strain-temperature relations for a one-dimensional SMA actuator is governed by Eq. (1). The general rate form of the stress-strain relation for the bias spring of the actuator may be expressed as

$$\dot{\bar{\sigma}} = -\frac{kL}{s}\dot{\bar{\epsilon}}, \tag{4}$$

where K is the spring constant of the spring, L the length of the SMA wire, and s the cross-sectional area of the SMA wire. Using the quasi-static assumption reduces the problem to a first-order differential equation and modifying the initial conditions will allow to incorporate the influence of the working load. Substituting Eq. (4) into Eq. (1) results in

$$(1+\frac{sD}{kL})\dot{\bar{\sigma}} = \Theta\dot{T} + \Omega\dot{\xi}, \tag{5}$$

where

$$\dot{\xi} = \frac{\partial\xi}{\partial\bar{\sigma}}\dot{\bar{\sigma}} + \frac{\partial\xi}{\partial T}\dot{T}. \tag{6}$$

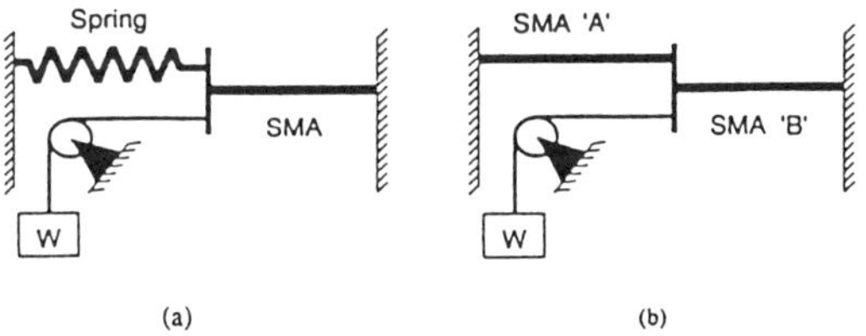

Figure 1: (a) Bias SMA actuator; (b) Differential SMA actuator

From Eqs. (1), (4) and (6), the following expression can be derived

$$(1+\frac{sD}{kL}-\Omega\frac{\partial\xi}{\partial\bar{\sigma}})d\bar{\sigma} + (-\Theta-\frac{\partial\xi}{\partial T})dT = 0. \tag{7}$$

Since $\frac{\partial^2\xi}{\partial\bar{\sigma}\partial T} = \frac{\partial^2\xi}{\partial T\partial\bar{\sigma}}$, a closed-form solution of Eq. (7) can be found as follows if the material properties (i.e., D, Θ, and Ω) are assumed to be constant:

$$(1+\frac{Ds}{kL})\bar{\sigma} - \Omega\xi - \Theta T = constant. \tag{8}$$

The initial conditions and boundary conditions for a bias SMA actuator may be modified to account for the external load 'W' as shown in Fig. 1 (a). After applying the initial condition to determine the *constant*, the following governing equation for the SMA element of a bias spring SMA actuator can be derived as

$$\sigma - \sigma_0 = \Omega'(\xi-\xi_0) + \Theta'(T-T_0), \tag{9}$$

where

$$\begin{cases} \Theta' = \frac{\Theta}{1+\frac{sD}{kL}} \\ \Omega' = \frac{\Omega}{1+\frac{sD}{kL}} \end{cases}, \tag{10}$$

In the design of a bias spring SMA actuator, the important design parameters are the properties of the SMA wire, i.e., the transition temperatures, D, Θ, and Ω; and the geometric parameters of the SMA spring (or wire) like the length and cross-sectional area. The spring constant is the basic design parameter for bias spring actuators. From Eqs. (9) and (10), all the undefined geometric variables are normalized as one term, $\frac{kL}{sD}$, which simplifies the design process. According to different working requirements such as the amount of stroke 'S', it is possible to choose the best $\frac{kL}{sD}$ and then the individual k, L, or s according to some other secondary design limitations such as space restrictions, etc.

Design of Differential SMA Force Actuators

The design parameters for this type of SMA actuators are the geometrical variables of the SMA element 'A' and 'B' and material constants. Both elements are assumed to be made of the same SMA material here in order to simplify the analysis. In this paper all the variables with superscript (A) are related to SMA element 'A' and superscript (B) related to element 'B'. The original length of 'A' and 'B' are L_A and L_B. The initial temperature for both element 'A' and 'B' is the same as the surrounding temperature T_0. The other initial conditions for element 'A' are zero, while the initial conditions for element 'B' are

$$\begin{cases} \epsilon_0^B = \epsilon_{res} \\ \xi_0^B = \frac{\epsilon_{res}}{\epsilon_L} \\ \sigma_0^B = \frac{W}{s} \end{cases}, \tag{11}$$

where ϵ_{res} is the residual martensitic strain (or the initial strain), ϵ_L the maximum recoverable strain, and W the load that models the working load. The governing rate form of the constitutive equations for this system are

$$\begin{cases} \dot{\sigma}^A = D\dot{\epsilon}^A + \Omega\dot{\xi}^A + \Theta\dot{T}^A \\ \dot{\sigma}^B = D\dot{\epsilon}^B + \Omega\dot{\xi}^B + \Theta\dot{T}^B \end{cases}. \tag{12}$$

The geometric compatibility of elements 'A' and 'B' and the equilibrium condition of the system is assured by

$$\begin{cases} L_A\dot{\epsilon}^A = -L_B\dot{\epsilon}^B \\ s^A\sigma^A = s^B(\sigma^B - \sigma_0^B) \end{cases}. \tag{13}$$

The governing equation and compatibility equation together yield:

$$L_A(\sigma^A - \Omega\xi^A - \Theta T^A) + L_B(\sigma^B - \Omega\xi^B - \Theta T^B) = constant. \tag{14}$$

The *constant* can be determined with the initial conditions as in the case of a bias spring SMA actuator.

Design of Flexible Non-Joint Robotics Fingers

(Chaudhry and Rogers, 1991) has investigated the large shape control with SMA actuators. Figure 2 shows the time-lapse picture of a beam deforming with an

SMA actuator wire which is surface-guided on the beam. This special configuration utilizes the large stroke capability of SMA and the nonlinear post-buckling characteristics of the beam to achieve large shape change. To achieve the deformation shown in the Fig. 2, less than 1% strain from the SMA is needed because the SMA actuator must follow the deformation of the beam which has a small longitudinal deformation. The deformation of the beam is determined by the force generated by the actuators. The rotation of the tip of the beam can reach more than 90° as long as the actuator provides enough force (it is always easy to increase the number of SMA actuators to the beam to generate more force). Based on this concept, a new SMA actuator configuration which may be used as non-joint robotics fingers is being investigated. This type of SMA actuator is essentially the same as the bias SMA actuators shown in Fig. 1 (a) except that the beam provides a nonlinear spring constant relation. Superelastic alloy (SMA with extremely low transition temperature) is now being used to make the beam to ensure a full recovery of the actuator. Detailed design of this type of SMA actuators will be presented in another paper.

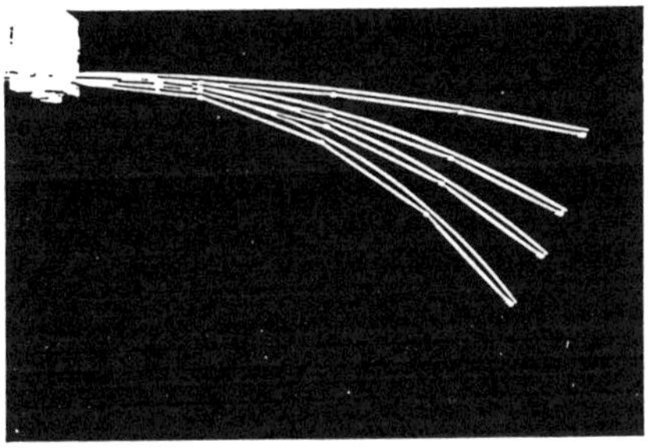

Figure 2. Time-lapse picture of a beam deformed by a surface-guided (passing through three points on the beam) SMA actuator (0.015 in. nitinol wire).

DESIGN OF SMA SPRINGS

Since most SMA actuators are in the form of springs or coils in order to achieve more stroke, the design of SMA springs has also been investigated (Liang and Rogers, 1992c). Based on the multi-dimensional constitutive relation (Liang and Rogers, 1992a), the one-dimensional shear stress and strain relation of shape memory alloys can be expressed as

$$\tau - \tau_0 = G(\gamma - \gamma_0) + \frac{\Omega}{\sqrt{3}}(\xi - \xi_0) + \frac{\Theta}{\sqrt{3}}(T - T_0), \tag{15}$$

where G is the elastic shear modulus.

The design of conventional springs is standard and can be found in most machine design textbooks. The normalized force-displacement relation of an SMA spring can be derived as

$$\frac{y}{y_e} = \frac{F}{F_e} - \frac{\Omega}{\sigma_e}\xi. \tag{16}$$

where $y_e = 2\pi N R^2 \sigma_e / \sqrt{3} r G$ and $F_e = \pi r^3 \sigma_e / 2\sqrt{3} R$. R is the mean spring radius, r the radius of the spring wire, N the number of effective coils, and σ_e the elastic stress limit of the SMA.

The force-displacement expression of Eq. (16) is very convenient in the design and application of shape memory alloy springs as actuators and other control elements. The phase transformation tensor is independent of temperature. Other parameters, such as y_e, F_e, and σ_e, are temperature-dependent. All the design variables of SMA springs, i.e., SMA radius (r), spring radius (R), and the number of active coils (N) appear only in the normalization factors, y_e and F_e.

The shape memory effect of SMA springs may also be derived in a similar manner to their force-displacement relations. Interested readers may refer to (Liang and Rogers, 1992c) for more detailed discussion and numerical examples.

SUMMARIES

This paper presents brief introductions to the constitutive modeling of SMA and theoretical design tools for the design of SMA actuators and springs. This research may be easily extended to the field of robotics.

ACKNOWLEDGMENT

The authors gratefully acknowledge the support of NSF Presidential Young Investigator Award, Grant No. MSS 9157086; Dr. Ken Chong Grant Monitor.

REFERENCES

1. Chaudhry, Z. and Rogers, C. A., 1991, "Bending and Shape Control of Beams Using SMA Actuators," *Journal of Intelligent Material Systems and Structures,* Vol. 2, No. 4.

2. Dye, E. T., 1990, Master thesis, An experimental Investigation of the Behavior of Nitinol, Department of Mechanical Engineering, Virginia Polytechnic Institute and State University, Blacksburg, Virginia.

3. Funakubo, H., 1987, Shape Memory Alloys, Gordon and Breach Science Publishers, New York.

4. Liang, C., 1990, The Constitutive Modeling of Shape Memory Alloys, PhD Dissertation, Virginia Polytechnic Institute and State University, Blacksburg, Virginia.

5. Liang, C. and Rogers, C. A., 1990, "One-Dimensional Thermomechanical Constitutive Relations for Shape Memory Materials", *Journal of Intelligent Material Systems and Structures*, 2(1).

6. Liang, C. and Rogers, C. A., 1992a, "The Multi-Dimensional Constitutive Relations of Shape Memory Alloys", *Journal of Engineering Mathematics*, in press.

7. Liang, C. and Rogers, C. A., 1992b, "Design of Shape Memory Alloy Actuators,", *Journal of Mechanical Design*, in press.

8. Liang and Rogers, 1992c "Design of SMA Springs with Their Application in Vibration Control," *Journal of Vibration and Acoustics*, in press.

9. Tanaka, Y and Nagaki, O., 1982, "A Thermomechanical Description of Materials with Internal Variables in the Process of Phase Transitions", *Ingenieur-Archiv*, 51:287–299.

10. Zhang, X. D., Liang, C., and Rogers, C. A., 1992, "Modeling of Two-Way Shape Memory Effects," *Journal of Philosophy*, in press.

PART 2 : SENSING AND SENSORS

Tactile-Assisted Robotic Object Manipulation

Pavan Sikka and Hong Zhang
Department of Computing Science
University of Alberta
Edmonton, Alberta, Canada T6G 2H1

ABSTRACT

The sense of touch plays an important role in robotic object manipulation. An important aspect of using tactile sensors for object manipulation (as opposed to object recognition) is that the forces acting on the object being manipulated distort the tactile image so that the tactile image is no longer a true representation of the contact. In this paper, we use tactile sensors to recover information about the status of the grip and also about the position and orientation of the held object during manipulation from such "distorted" tactile images. This information can be used to recover the transformation from the hand to the object being held, making it possible for the robot to reorient itself automatically in case of slippage. Thus, tactile sensors can be used to monitor a task in real-time. Experimental results are given to demonstrate the effectiveness of the proposed method.

INTRODUCTION

Although robots are able to perform tasks effectively in a structured environment, they are unable to function in unstructured environments. This can be attributed to their very limited sensing ability. A key requirement for robots to perform in a changing and uncertain environment is the ability to sense and track the changes taking place in the environment. Vision, ultrasound, and force are some of the modalities that have been studied extensively in this context.

In this paper, we consider an application of tactile sensing. Tactile sensors consist of an array of sensors each of which usually measures a quantity that is related to the deformation in the area of contact or the normal component of the forces acting within the area of contact. It is only recently that robotics researchers have started to explore the domain of tactile sensing.

Most of the research in tactile sensing for robots has concentrated in exploring the use of tactile sensors for object recognition and in the design of tactile sensors. In contrast, we show how tactile sensors can be used to monitor the progress of a task by obtaining the position and orientation of the object held in the sensorized robot hand. Although this is not central to the planning of the task, this function is extremely important in making robots more autonomous in their operation. It gives robots the ability to detect problems as the task is being performed and also to recover from some of those problems on their own. In the following sections, we will first present a general discussion on the use of tactile sensors in manipulation. We will then outline and discuss our method of monitoring task progress using tactile sensors. Following this, we will describe the hardware used to conduct our experiments and also the results obtained form these experiments. We finally conclude with some comments on possible future work.

TACTILE SENSORS IN MANIPULATION

In object recognition, the information derived from the tactile sensors is about the shape of the contact area. The forces acting on the object are known and can be controlled to obtain "good" images. On the other hand, in object manipulation, the forces acting on the object change constantly and cannot be fully controlled. The "images" obtained from the tactile sensors are distorted by these contact forces. For example, a force acting perpendicular to the contact area causes the contact force distribution to be skewed. A force acting in the plane of the area of contact causes tangential forces that may cause the object to slip. We need to be able to recover information about the "true" contact from these "distorted" images.

To monitor the progress of the given task, tactile sensors can be used to ensure that the object is properly gripped at all times during the task, and to detect slippage as it occurs. Since they provide information about the shape of the contact, they can also be used to recover the position and orientation of the grasped object with respect to the tactile sensors.

Slip Detection

Most of the work in slip detection involves tactile sensors of zero dimension, i.e., tactile sensors consisting of only one sensing element. Some examples of this type of sensors are the Skin Acceleration Sensor (SAS) described by Howe *et al* [4] and the Intrinsic Tactile Sensor (ITS) described by Bicchi *et al* [1, 2]. The SAS is essentially an accelerometer and it responds to vibrations generated by the making and breaking of contact. The ITS is a six-axis force/torque sensor built in the interior part of the fingertip of the robot end effector. Bicchi *et al* [1, 2] and Bicci [3] describe the use of this sensor in determining the grasping force and in detecting slip.

Wojcik [5] presents a number of different ways of detecting slip using a tactile sensor. The simplest method is based on the center of gravity of the image. A second method thresholds the images before calculating the center of gravity. A third method uses edge detection to find the object contour and then calculates the center of the object contour. A problem with the above approach is that slip caused by rotation cannot be detected by the above method. Another problem with this approach is caused by the effect of forces acting on an object during manipulation. If a large force acts on the object, then the image may change sufficiently for the algorithm to signal slip - thus detecting slip when it has not occured. The method described here takes care of these problems.

Recovering Information About Orientation

In this section, we describe a method that can be used to detect slip and to obtain information about the orientation of the object held by the robot arm. The procedure used for monitoring the task and recovering the information about object orientation is shown below.

1. Obtain two tactile images T_l and T_r, one from each tactile sensor mounted on the hand.
2. Filter and threshold T_l and T_r to obtain binary tactile images T_{lb} and T_{rb}.
3. Combine the two processed binary images T_{lb} and T_{rb} into one consolidated binary image T_c.
4. Compute the covariance matrix Λ_c associated with the tactile image T_c. Since the image is 2-dimensional, we need to compute σ_{xx}, σ_{yy}, and σ_{xy}.
5. Compute the center of contact. Also compute the orientation of the area of contact using the equation $\theta = \frac{1}{2}\tan^{-1}\frac{2\sigma_{xy}}{\sigma_{yy}-\sigma_{xx1}}$
6. If the change in the center of the image or the orientation of the image exceeds the specified threshold, then signal the controlling program that slip has occured.

Combining Tactile Images

If no forces are acting on the object held by the robot, then each tactile sensor yields a reasonably good image of the shape of the part of the object that is in contact with the tactile sensor. In this situation, reasonably reliable information about slip and orientation can be obtained independently from either sensor. When a force is acting on the object being held, it is not possible to obtain reliable information either about slip or about the orientation by using one sensor alone. However, we find that if we combine the information from both the sensors, we are once again able to get reasonably good information about slip and orientation. Although different ways of combining the two images are possible, we find that simply ORing the two images, which is very efficient, produces satisfactory results.

EXPERIMENTAL RESULTS

The experimental system consists of a PUMA 260 six-degree-of-freedom manipulator arm, a servo-gripper with adjustable gripping force (Schunk), a six-axes force-torque sensor mounted on the wrist of the PUMA 260, and two tactile sensors (each an array of 16x16 "tactels") mounted one each on the two gripper jaws (Interlink Electronics, Inc). Robot motion control is carried out on a 68020-based card (Heurikon) while tactile data acquisition and processing are performed on a 68000-based card. A Sun-3/160 system provides the user interface.

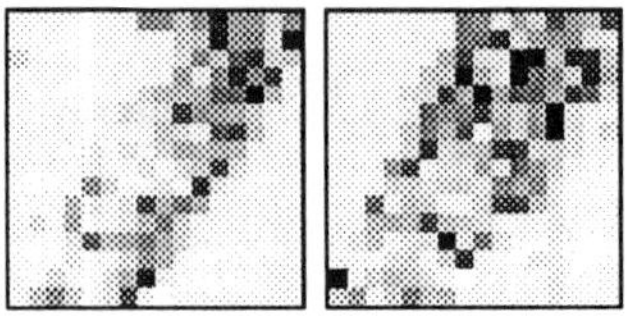

Figure 1: The "raw" images obtained from the two sensors.

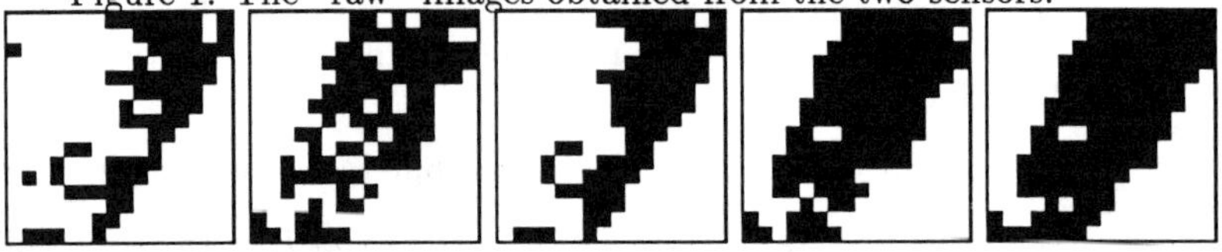

Figure 2: Case 1 : Object held in hand is oriented diagonally.

In this section, we describe the results obtained from applying our method to two different cases. In each case, the object held in the gripper was a rectangular bar. In the first case, the object was held in the hand at an inclination of 50 degrees. In the second case, the object was held at an inclination of 50 degrees and a contact force was applied to the object.

Case 1 : Inclination of 50 degrees without Contact Forces

The peg was held at an inclination of 50 degrees and centered at (7.5, 7.5). No contact forces were acting on the object. Figure 1 shows the grey-scale plots of the raw images obtained from the tactile sensors. The images are quite noisy.

The first two images in Figure 2 show the result of thresholding these raw images. The contact shape is now clearer. However, these figures do contain "mechanical" noise. There are some isolated tactels (0s surrounded by 1s or vice versa) that need to be filtered.

The next two images in Figure 2 show the result of filtering the first two images in Figure 2. The center of the image and the orientation θ obatined in this case are shown in Table I. Both images give us reasonably good estimates of the orientation. The last image in Figure 2 was obtained by combining the two filtered images.

Case 2 : Inclination of 50 degrees with Contact Forces

The figures mentioned so far corresponded to the situation where no external forces were acting on the held object. Figure 3 corresponds to the situation where the same peg is gripped at the same inclination as case 1 but with a force acting on it.

As before, the first two images in Figure 3 were obtained by thresholding the original images and the next two images in Figure 3 were obtained by filtering the first two images in this figure. The last image was obtained by combining

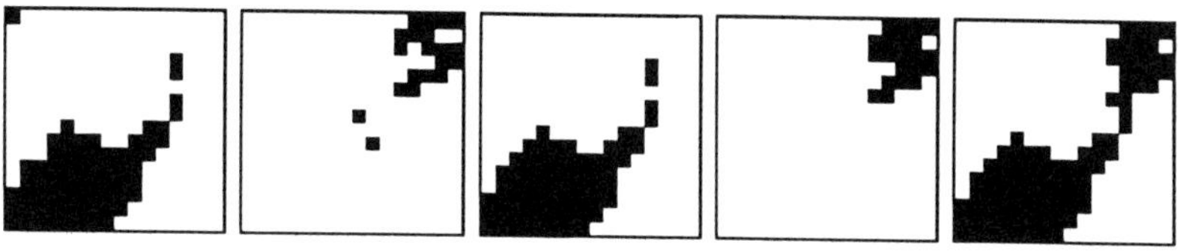

Figure 3: Case 2 : A force perpendicular to the plane of the sensors is applied to the tip of the object

Table I. Results obtained by our method.

	$\|\theta\|$ (degrees)	$\|\Delta\theta\|$ (degrees)	$\|x_c\|$ (tactels)	$\|\Delta x_c\|$ (tactels)	$\|y_c\|$ (tactels)	$\|\Delta y_c\|$ (tactels)
Case 1 : Inclination 50						
left	52.52	2.52	9.36	1.86	6.29	1.21
right	49.7	0.30	8.00	0.50	6.54	0.96
combined	50.12	0.12	8.00	0.50	6.82	0.68
Case 2 : Inclination 50, Contact Force						
left	37.41	12.59	5.50	2.00	11.47	3.97
right	74.21	24.21	13.00	5.50	2.14	5.36
combined	46.07	3.93	7.22	0.28	9.33	1.83

the filtered images.

This case illustrates a serious problem associated with most slip detection algorithms. Table I shows the centers and orientations obtained from each individual image. These estimates are poor and are useless for recovering the orientation. Furthermore, the centers of the two images also do not correspond to the real situation. An algorithm based on the change in center alone would incorrectly indicate slip in each of the two cases. In contrast, the orientation obtained from the combined image is again reasonably good.

The results are all summarized in Table I.

Remarks on the Results

From the results presented above, we can make the following observations.

- The change in the center of the image is only one indicator of slip. Using this criterion alone, incorrect decisions can be made (for example, in [5]). Slip may not be detected by this method if the object rotates around the center. In short, both changes in center as well as changes in orientation need to be considered to detect slip.
- If the decisions are made on the basis of individual images then a force acting on the object can distort the images enough so as to lead to incorrect decisions if both location and orientation are considered. This effect of a distorting force can be compensated by simultaneously considering images obtained from tactile sensors mounted on the two opposite sides of the gripper.

- The algorithm is quite robust with respect to small distortions in the image. For example, we found that in the worst case a missing tactel (or an extra tactel) leads to an error of about 2 degrees in the orientation. However, the many different sources of noise make it difficult to obtain very accurate results from the algorithm. Random noise, poor contact, and mechanical crosstalk all distort the images to limit the accuracy of the procedure.

CONCLUSION

In this paper we have first presented a brief survey about the use of tactile sensors in object recognition and in object manipulation. We then presented a method of obtaining the orientation of the object held in the hand and of monitoring the execution of a manipulation task. We presented experimental results to demonstrate that our method is robust (it works even when the forces acting on the object are not known - typical in most object manipulation situations).

There are a number of issues that need to be tackled. First, there is a need to come up with alternative methods of combining two images and characterizing them in terms of applicability to particular situations. Second, there is the related issue of the need to calibrate the tactile sensors.

A more general issue that underlies this work and needs to be explored extensively is the different ways in which tactile sensors can be used in object manipulation tasks. A related issue is the possibility of using the force distributions directly in manipulation tasks as opposed to using binary images (as is usually done in object recognition).

Acknowledgements

This work was supported in part by the National Science and Engineering Research Council of Canada under Grant OGP0042194 and by FANUC of Japan.

References

[1] A. Bicchi and P. Dario. Intrinsic tactile sensing for artificial hands. In R. Bolles and B. Roth, editors, *Robotics Research*. MIT Press, 1987.

[2] A. Bicchi, J. K. Salisbury, and P. Dario. Augmentation of grasp robustness using intrinsic tactile sensing, 1989.

[3] A. Bicci, M. Bergamasco, P. Dario, and A. Fiorillo. Integrated tactile sensing for gripper fingers, 1988.

[4] R. D. Howe, N. Popp, P. Akella, I. Kao, and M. R. Cutkosky. Grasping, manipulation, and control with tactile sensing, 1990.

[5] Piotr J. Wojcik. Application of tactile sensors for slippage detection of a robot gripper. In *IFIP International Conference, Roma, Italy*, 1990.

RESTORATION OF IMAGES CORRUPTED BY ADDITIVE AND MULTIPLICATIVE NOISE, BLUR AND NON-LINEARITY

M. Ahmadi, M. El-Gabali and M. Shridhar
Department of Electrical Engineering
University of Windsor
Windsor, Ontario, N9B 3P4, Canada

ABSTRACT
This paper discusses the development of a new technique for the restoration of images degraded by multiple noise sources. The degradation is caused by additive, multiplicative, focus blur and/or nonlinearities. The image is modelled as a Markov Random Field, while the noise sources are assumed to be Gaussian processes with known means and variances. The blur is assumed to be caused primarily due to 'out of focus' conditions and a model for blur is used in the restoration algorithm. Test results with a variety of images indicate that the algorithm yields satisfactory restoration even in the presence of high levels of noise and nonlinearities.

INTRODUCTION

Restoration of images degraded by noise has been the subject of extensive investigations over the past several years. In these proposed restoration techniques many researchers modelled the image as a Markov Random Field (MRF) while the additive noise assumed to be a Gaussian process with known mean and variance [1]-[5]. In a recent work [5], the authors used the MRF image model to develop a restoration algorithm for gray-scale images degraded by high levels of additive noise. This algorithm, besides yielding optimal restoration, allowed for a parallel implementation of the technique. The authors have since extended this technique to images degraded by either additive noise and blur [6] or multiplicative noise and blur as well as nonlinear distortion [7]. By using an approximation scheme originally developed by Lee [8], the authors were able to develop a restoration algorithm that can be applied to images degraded by both forms of noise (additive and multiplicative) as well as blur and non-linearity distortion. The derivation of the algorithm and the results obtained with test images are presented in this paper.

PROBLEM STATEMENT

In this section the assumptions about the statistical characterization of the scene, the corrupting noise and the basic problem statement will be presented.

We will restrict our definitions to two-dimensional discrete random fields defined over a finite lattice Lt of pixels. Capital letters will be used to represent random variables and random fields while lower case letters will be used to represent specific realizations.

ASSUMPTIONS:

1. The model of the degraded image is given by:

$$Y = C\,[H * F(X)]^t . W_m + W_a \qquad (1)$$

At any given location (i,j) the noisy image is described by:

$$y_{ij} = C \left\{ \sum_{kl} H(k,l)\; F[x(i+k,\; j+l)] \right\}^t w_{m_{ij}} + w_{a_{ij}} \qquad (2)$$

for all $(i,j) \in Lt$

Where:

W_a	is a representation for additive noise.
W_m	is a representation for multiplicative noise.
C,t	are the nonlinearity constants.
H	is the blurring window of angle ϕ.

$$H(k,l) = \begin{cases} \cos^2\phi & (k,l) = (0,0) \\ \dfrac{\sin^2\phi}{n_w} & (k,l) \in \zeta_{ij} \end{cases} \tag{3}$$

ζ_{ij} represents all pixels neighbouring (i,j) inside a window centered at (i,j). The figure below shows an example of 3 by 5 window.

$$\begin{bmatrix} o & o & o & o & o \\ o & o & (i,j) & o & o \\ o & o & o & o & o \end{bmatrix}$$

In this representation
$l_w = 2,\ l_k = 1,\ W = 2\,l_w + 1,\ h = 2\,l_h + 1,\ n_w = W.h - 1$
n_w represents the number of pixels inside the blurring window excluding the center pixel.

2. X is an homogeneous and isotropic MRF with {X = x} as a specific realization.
Note that:

{X = x} stands for $\{X_{ij} = x_{ij},\ (i,j) \in Lt\}$
$x_{ij} \in [-1,1]$ is a real number representing a 'label' at location (i,j).
$F(x_{ij}) \geq 0$ is a nonnegative integer representing the corresponding grey level.
The state space of X contains N elements, i.e. the number of grey levels in X is finite = N.

The probability density function of X is given by [11]:

$$P\ (X = x) = \frac{e^{U(x)}}{Z} \tag{4}$$

Where:

Z	is the normalizing constant
U(x)	is the energy function of the form:

$$U(x) = \sum_{ij} \left\{ \alpha\, x_{ij} + \beta \sum_{kl \in \eta_{ij}} \theta_{kl}\, R(x_{ij}, x_{kl}) \right\} \tag{5}$$

where:

(i) θ represents positive weights used to reward the interactions between the center pixel and its neighbours based on their relative distances with respect to the center pixel.
(ii) α and β are model parameters.
(iii) $R(x_{ij}, x_{kl})$ is given by:

$$R(x_{i_{ij}}, x_{kl}) = 1 - \frac{(x_{ij} - x_{kl})^2}{2} \tag{6}$$

Clearly: $-1 \leq R(x_{ij}, x_{kl}) \leq 1$

3. F is a linear mapping and is given by:

$$F(x_{ij}) = f_m + f_d\, x_{ij} \tag{7}$$

Arranging the state space of X in increasing order we can write:

$$x_{ij} \in \{x_1, x_2, \ldots, x_N\} \tag{8}$$

Where: $x_1 = -1$, $x_N = 1$, $x_r > x_t$ for $r > t$ $1 \le t, r \le N$

It can be verified that:

$$f_m = \frac{F(1) + F(-1)}{2} \tag{9}$$

$$f_d = \frac{F(1) - F(-1)}{2} \tag{10}$$

4. Given X the RVs Y_{ij}, $\{(i,j) \in Lt\}$ are conditionally independent and depend only on the X values inside the blurring window centered at (i,j).

5. The noisy image y consists of M grey levels ($M \ge N$)

6. The corrupting noise sources W_a and W_m are Gaussian independent and identically distributed (i.i.d.) with zero and unity means (respectively) and known (estimable) variances
$W_a = N(0, \sigma^2{}_a)$, $W_m = N(1, \sigma^2{}_m)$

MATHEMATICAL FOUNDATION

The restoration algorithm is iterative and at each iteration the state of the current pixel is to be updated based on the available data. The available data include the original record and the current realization elsewhere. In the formulation to be given below we will maximize the posterior marginal probability at each pixel. Let {K} represent the clock cycle (timing) of the current realization, then, {K + 1} represents the timing of the updated state.

The maximization problem can be written in the form:

$$\max_{x_{ij}} P(X_{ij} \mid Y, \{X \backslash X_{ij}\})$$

Using Baye's theorem, after neglecting the effect of the constant terms, we get:

$$P(X_{ij}^{\{K+1\}} \mid Y, (X\backslash X_{ij})^{\{K\}}) \;\alpha\; P(Y \mid X_{ij}^{\{K+1\}}, X_{\zeta_{ij}}^{\{K\}})\; P_{ij}(X_{ij}^{\{K+1\}} \mid X_{\eta_{ij}}^{\{k\}}) \tag{11}$$

Where:

$X\backslash X_{ij}$	means excluding the random variable X_{ij} from the random field X.
$X_{\zeta_{ij}}$, $X_{\eta_{ij}}$	represent a set of random variables belonging to X and indexed by ζ_{ij} and η_{ij} respectively.
$P_{ij}(X_{ij} \mid X_{\eta_{ij}})$	represents the contribution of the neighboring pixels to the conditional probability of the center pixel.

For ease of notation Eqn. (11) is simplified as:

$$P(X_{ij} \mid Y, (X\backslash X_{ij})) \;\alpha\; P(Y_{ij} \mid X_{\zeta_{ij}})\; P_{ij}(X_{ij} \mid X_{\eta_{ij}}) \tag{12}$$

Where $P_{ij}(X_{ij} \mid X_{\eta_{ij}})$ is given by:

$$P_{ij}(X_{ij}|X_{\eta_{ij}}) \;\alpha\; \exp\left\{ \alpha\, x_{ij} + \beta \sum_{kl} \theta_{kl}\, R(x_{ij}, x_{kl}) \right\} \tag{13}$$

and

$$P(Y_{ij}|X_{\zeta_{ij}}) = P\left[Y_{ij} = C\, \{K_0(\phi)+K_1(\phi)x_{ij}\}^t \boxtimes n_{ij} | X_{\zeta_{ij}} \right] \tag{14}$$

$$k_o(\phi) = \left\{ f_m + \frac{f_m \sin^2\phi}{n_w} \sum_{k\ell \in \zeta_{ij}} x_{k\ell} \right\} \geq o$$

$$k_1(\phi) = f_d \cos^2\phi \geq o$$

THE PROPOSED RESTORATION TECHNIQUE

In this section, the authors will develop a unified approach for estimating the clean image when the corrupting noise is known to be additive and/or multiplicative. Under those conditions, it is possible to derive robust and computationally efficient algorithms for estimating the clean image by using the linearizing approximation proposed by Lee [8]. It is worth noting that when the corrupting noise is known to be only multiplicative or only additive, more efficient algorithms have been proposed [5], [6] and [7].

This approach will be illustrated by considering only the multiplicative noise. Then the noisy image is given by

$$Y_{ij} = C\,\{ K_0(\phi) + K_1(\phi)\, x_{ij} \}^t\, w_{m_{ij}} = g(x_{ij})\, w_{m_{ij}} \tag{15}$$

where $g(x_{ij})$ is defined as

$$g(x_{ij}) = C\,\{ K_0(\phi) + K_1(\phi)\, x_{ij} \}^t \tag{16}$$

One can now derive the following linear approximation for Y_{ij}:

$$Y_{ij} \approx A_1\, w_{m_{ij}} + A_2\, g(x_{ij}) + A_3 \tag{17}$$

where

$$A_3 = \bar{w}_{m_{ij}}\, \bar{g}_{ij}(x_{ij}) - A_1\, \bar{w}_{m_{ij}} - A_2\, \bar{g}(x_{ij})$$

$$A_1 = \bar{g}(x_{ij}) \quad \text{and} \quad A_2 = \bar{w}_{m_{ij}}$$

The specific results lead to,

$$Y_{ij} \approx \bar{w}_{m_{ij}}\, g(x_{ij}) + \bar{g}(x_{ij})\,(w_{m_{ij}} - \bar{w}_{m_{ij}}) \tag{18}$$

where $\bar{w}_{m_{ij}}$ and $\bar{g}(x_{ij})$ are computed as averages within a suitably chosen window. From equation (18) one can write

$$w_{m_{ij}} = \frac{y_{ij} - \bar{w}_{m_{ij}}\, g(x_{ij}) + \bar{g}(x_{ij})\, \bar{w}_{m_{ij}}}{\bar{g}(x_{ij})}$$

The significance of equation (18) is brought out by rewriting the expression as

$$y_{ij} \cong \bar{w}_{m_{ij}}\, g\,(x_{ij}) - \bar{g}\,(x_{ij})\, \bar{w}_{m_{ij}} + \bar{g}\,(x_{ij})\, w_{m_{ij}}$$

$$\cong h\,(x_{ij}) + \bar{g}\,(x_{ij})\, w_{m_{ij}} \tag{19}$$

where $h(x_{ij}) = \bar{w}_{m_{ij}}\, g\,(x_{ij}) + \bar{g}\,(x_{ij})\, \bar{w}_{m_{ij}}$

Thus using this approximation, the model for images corrupted by multiplicative noise is reduced to the additive case for which very efficient algorithms have already been derived [5].

When the image is corrupted by both additive and multiplicative noise sources a similar approach leads to the following:

$$y_{ij} \cong \bar{w}_{m_{ij}}\, g\,(x_{ij}) - \bar{g}\,(x_{ij})\, \bar{w}_{m_{ij}} + \bar{g}\,(x_{ij})\, (w_{m_{ij}} + w_{a_{ij}}) \tag{20}$$

By defining

$$h\,(x_{ij}) \stackrel{\Delta}{=} \bar{w}_{m_{ij}}\, g\,(x_{ij}) - \bar{g}\,(x_{ij})\, \bar{w}_{m_{ij}} \tag{21}$$

and

$$v_{ij} \stackrel{\Delta}{=} \bar{g}\,(x_{ij})\, w_{m_{ij}} + w_{a_{ij}}$$

one gets a similar expression for y_{ij}:

$$y_{ij} \cong h\,(x_{ij}) + v_{ij} \tag{22}$$

Equation (22) reveals that with this approximation the restoration algorithms developed for additive noise can be effectively used for image corrupted by multiple noise sources using the following Eqns.,

$$\hat{x}_{ij} = \frac{\alpha + \beta \sum_{k\ell} \theta_{k\ell}\, x_{k\ell} + \dfrac{\bar{y}_{ij}\, k_1\,(\phi)}{\sigma_a^{\,2}}}{\beta \sum_{k\ell} \theta_{k\ell} + \left[\dfrac{k_1(\phi)}{\sigma_a}\right]^2} \tag{23}$$

where

$$\bar{y}_{ij} = y_{ij} - k_o\,(\phi) \quad (i,j) \in L_t$$

The main advantage of this approach lies in the observation that one algorithm can be utilized for the restoration of images corrupted by additive and/or multiplicative noise with and without blur and non-linearity.

RESULTS AND DISCUSSION

The algorithm developed in the previous section was applied to several images degraded by high levels of additive/multiplicative noise and satisfactory restoration was achieved in most cases.

Figure (1a) shows a three level image which is generated synthetically by the authors and corrupted by compound noise sources with equal additive and multiplicative noise components at a SNR of 1 (Fig. 1b).

The restored image is shown in Fig. (1c). It is clear that satisfactory restoration has been achieved in this case.

CONCLUSION

The authors have clearly shown that an effective restoration algorithm for images corrupted by compound noise sources has been realized using a linear approximation scheme. Such an approach allows one to use efficient restoration algorithms that have been developed for images degraded by additive noise sources.

REFERENCES

[1] S. Geman and D. Geman "Stochastic Relaxation, Gibbs Distribution and Bayesian Restoration of Images", IEEE Trans. on PAMI, Vol. 6, No. 6, pp. 721-741, 1984.

[2] H. Derin, H. Elliot, R. Cristi and D. Geman, "Bayes Smoothing Algorithms for Segmentation of Binary Images Modelled by Markov Random Fields", IEEE Trans. on PAMI, Vol. 6, No. 6, pp. 707-720, 1984.

[3] J. Besag, "On the Statisical Analysis of Dirty Pictures", J.R. Statis, Soc. B (1986) 48, No. 3, pp. 259-302.

[4] R. Cristi, "Application of Markov Random Fields to Smoothing and Segmentation of Noisy Pictures", Proc. IEEE ICASSP, 1988, pp. 1144-1147.

[5] M. Shridhar, M. Ahmadi and M. El-Gablai, "Restoration of Noisy Images Modelled by Markov Random Fields With Gibbs Distribution", IEEE Trans. on CAS, Vol. 36, No. 6, 1989, pp. 884-890.

[6] M. El-Gabali, M. Shridhar, M. Ahmadi and A.J.H. Kekre, "Restoration of Motion Blurred Images", Proc. of 26th Asiloman Conf. on Signals, Systems and Computers, Nov. 1990, pp. 549-552.

[7] M. Ahmadi, M. Shridhar, G.A. Jullien and W.C. Miller, "Restoration of Images Corrupted by Multiplicative Noise, Blur and Nonlinearity", Proc. of 9th Kobe Intern. Symp. on Electronics and Information Sciences, June 1991, pp. 31-1 - 31-7.

[8] Jon-Sen Lee, "Digital Image Enchancement and Noise Filtering by the Use of Local Statistics", IEEE Trans. on PAMI, Vol. 2, No. 2, March 1980, pp. 165-168.

Fig. (1 A) Original Image

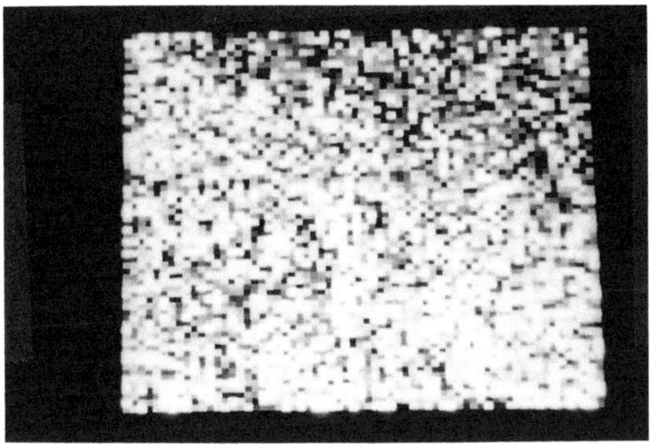

Fig. (1 B) Degraded Image with addition and multiplicative noise, and non-linearity S/N = 1

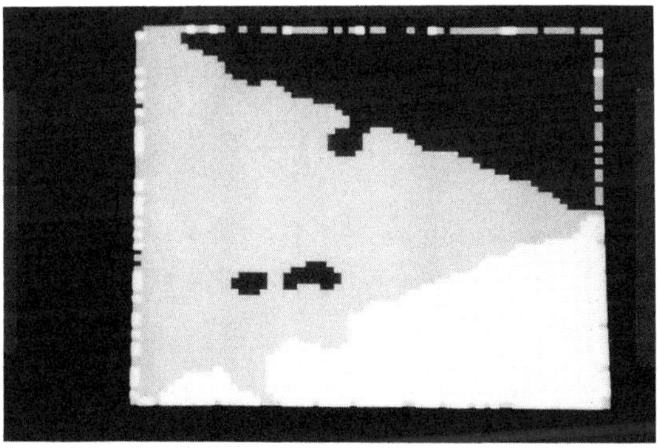

Fig. (1 C) Restored Image

A Simplified RAC-Based Camera Calibration Algorithm

Hanqi Zhuang, Zvi S. Roth, Xuan Xu and Kuanchih Wang
Robotics Center and Department of Electrical Engineering
Florida Atlantic University
Boca Raton, FL 33431

Abstract

A simplified Radial Alignment Constraint (RAC)-based method, which is a modification of Tsai's camera calibration algorithm, is presented in this paper. The simplified algorithm reduces the computational complexity since only three 2x2 matrix inversions are required to calibrate a camera.

1. Introduction

This paper presents a simplified camera calibration algorithm based on Tsai's calibration technique [1]. Camera calibration has received wide attention in the fields of photogrammetry, computer vision and robotics ever since stereo cameras were introduced to perform 3-D measurements with advantages of noninvasiveness, fully automated operation, modest accuracy and low cost. Among the various camera calibration techniques, the Radial Alignment Constraint (RAC)-based calibration method proposed in [1] has been recognized as an excellent method for its use of off-the-shelf TV cameras and lenses, compensation of lens distortion, and its capability of high accuracy and high speed.

The simplified RAC-based camera calibration described in this paper accelerates computation speed. The idea behind the simplified calibration algorithm is that during the first stage of the calibration, by selecting only camera calibration points which lie on the x and y axes of the world coordinate system, the radial alignment constraint equations of the camera are simplified. Instead of using a linear least-squares algorithm to estimate five variables in Stage one and three variables in Stage two, as was done in [5], two variables linear least-squares estimations are repeated three times in the new method. As is well known, a two variables linear least squares problem has a simple closed-form solution, which results in a significant reduction of the computation cost. For this method to be applicable, it is required that the ratio of camera scale factors be given *a priori*.

The paper is organized as follows. Tsai's camera model is reviewed in Sections 2. The simplified RAC-based algorithm is given in Section 3. Finally, experimental results are presented.

2. A Camera Model

Fig. 2.1 illustrates the basic geometry of the camera model. $\{x_w, y_w, z_w\}$ denotes the world coordinate system. $\{x, y, z\}$ denotes the camera coordinate system, whose origin is at the optical center point O, and whose z axis coincides with the optical axis. $\{X, Y\}$ denotes the image coordinate system centered at O_I (the intersection of the optical axis z and the image plane). $\{X, Y\}$ lies on a plane parallel to the x and y axes. $\{X_f, Y_f\}$ (not shown in the figure) denotes the computer image coordinate system used in the computer memory. (X_u, Y_u) are the image coordinates of an object point P_u if a perfect pin hole model is used, whereas (X_d, Y_d) are the actual image coordinates of such a point due to lens distortion.

The camera model is made up of a sequence of four transformations from the world coordinate system to the computer image coordinate system.

Step 1. Transformation from the world coordinates (x_w, y_w, z_w) to the camera coordinates (x, y, z),

$$\begin{bmatrix} x \\ y \\ z \end{bmatrix} = R \begin{bmatrix} x_w \\ y_w \\ z_w \end{bmatrix} + t \tag{2.2.1}$$

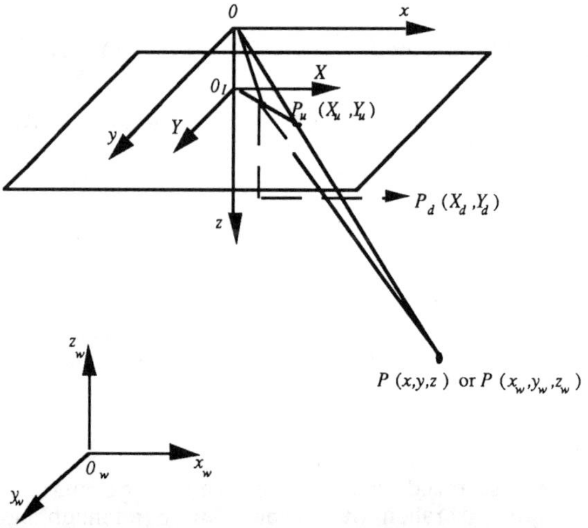

Fig. 2.1 Geometry of the camera model

where the rotation matrix R and translation vector $\boldsymbol{t}$ are written as

$$R = \begin{bmatrix} r_1 & r_2 & r_3 \\ r_4 & r_5 & r_6 \\ r_7 & r_8 & r_9 \end{bmatrix} \tag{2.2.2}$$

and $\boldsymbol{t} = [t_x \ t_y \ t_z]^T$ (2.2.3)

Step 2. Transformation from the camera coordinates (x, y, z) to the ideal image coordinates (X_u, Y_u) using perspective projection with the pin hole model,

$$X_u = f\frac{x}{z} \tag{2.2.4a}$$

$$Y_u = f\frac{y}{z} \tag{2.2.4b}$$

where f is the effective focal length.

Step 3. Transformation from the ideal image coordinates to actual image coordinates taking into account radial lens distortion. That is

$$X_d(1 - kr^2)^{-1} = X_u \tag{2.2.5a}$$

$$Y_d(1 - kr^2)^{-1} = Y_u \tag{2.2.5b}$$

where (X_d, Y_d) are true or distorted image coordinates, $k > 0$ is the distortion coefficient, and

$$r^2 \equiv X_d^2 + Y_d^2 \tag{2.2.6}$$

(2.2.5) is a simplified radial lens distortion model similar to that given in [5].

Step 4. Transformation from the actual image coordinates (X_d, Y_d) to the computer image coordinates (X_f, Y_f),

$$X_f = \mu_x X_d + C_x \tag{2.2.7a}$$

$$Y_f = \mu_y Y_d + C_y \tag{2.2.7b}$$

where C_x and C_y are the computer image coordinates of the origin in the image plane, μ_x and μ_y are scale factors in the x and y directions, respectively. Define an image coordinate system $\{X, Y\}$ by

$$X = X_f - C_x \tag{2.2.8a}$$

$$Y = Y_f - C_y \tag{2.2.8b}$$

Combining the four steps, the camera model, relating the world coordinate system $\{x_w, y_w, z_w\}$ to the image coordinate system $\{X, Y\}$, is

$$\frac{X}{1 - kr^2} = f\,\mu_x \frac{r_1 x_w + r_2 y_w + r_3 z_w + t_x}{r_7 x_w + r_8 y_w + r_9 z_w + t_z} \tag{2.2.9a}$$

$$\frac{Y}{1 - kr^2} = f\,\mu_y \frac{r_4 x_w + r_5 y_w + r_6 z_w + t_y}{r_7 x_w + r_8 y_w + r_9 z_w + t_z} \tag{2.2.9b}$$

μ_x and μ_y can often be obtained from the camera data sheet. If however μ_x and μ_y together with f are treated as unknowns, one should bear in mind that these parameters are not independent of one another. At most two parameters from the set $\{\mu_x, \mu_y, f\}$ can be identified. Define two new parameters as

$$f_x \equiv f\mu_x \tag{2.2.10a}$$

$$f_y \equiv f\mu_y \tag{2.2.10b}$$

Also define

$$\mu \equiv \frac{\mu_y}{\mu_x} = \frac{f_y}{f_x} \tag{2.2.11}$$

μ is termed the *ratio of scale factors*. Then from (2.2.5), (2.2.6) and (2.2.11), the distortion term can be written as

$$kr^2 = \alpha r'^2 \tag{2.2.12}$$

where $\alpha \equiv k\mu_y{}^2$ (2.2.13)

and $r'^2 \equiv \mu^2 X^2 + Y^2$ (2.2.14)

With the above modifications and by using (2.2.9), the new camera model, relating the world coordinates (x_w, y_w, z_w) to the image coordinates (X, Y), is

$$\frac{X}{1 - \alpha r'^2} = f_x \frac{r_1 x_w + r_2 y_w + r_3 z_w + t_x}{r_7 x_w + r_8 y_w + r_9 z_w + t_z} \tag{2.2.15a}$$

$$\frac{Y}{1 - \alpha r'^2} = f_y \frac{r_4 y_w + r_5 y_w + r_6 z_w + t_y}{r_7 x_w + r_8 y_w + r_9 z_w + t_z} \tag{2.2.15b}$$

In this model, the parameters to be calibrated are: the *extrinsic* parameters of R and $\mathbf{t}$, and the *intrinsic* parameters f_x, f_y, μ, α, C_x and C_y. Note that f_y is computed by multiplying f_x by μ. The focal length f can be computed from the modified intrinsic parameters if camera specifications are given by the camera manufacturer [1].

3. The Simplified RAC-Based Camera Calibration Algorithm

The simplified RAC-based camera calibration algorithm for the coplanar case consists of two stages. In the first stage, camera calibration points are restricted to lie on the x and y axes of the world coordinate system, by which the RAC equations (2.2.15) are simplified. This simplification makes it possible to use only 2x2 matrix inversions. As in [1], it is assumed that C_x, C_y and μ are given.

3.1 Stage 1 — Computation of the Rotation Matrix R, and the translation parameters t_x and t_y

In this stage, only calibration points which lie on the x and y axes of the world coordinate system are used; see Fig. 3.1. If the x and y axes of the world coordinate system are out of the field of view, these can be translated to within the field of view by a simple transformation.

A. *Computation of the image coordinates* (X_i, Y_i)

Let N be the number of calibration points. Then for i = 1, 2, ..., N,

$$X_i = X_{f,i} - C_x; \quad Y_i = Y_{f,i} - C_y$$

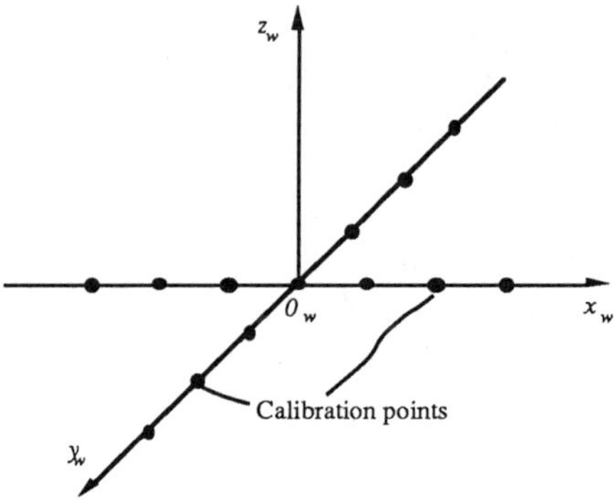

Fig. 3.1 Calibration points in the first stage

where $X_{f,i}$ and $Y_{f,i}$ are the computer image coordinates for the ith point.

B. Computation of the intermediate parameters $\{v_1, v_4\}$

Let $z_w = y_w = 0$, and $x_w \neq 0$, which corresponds to those calibration points on the x axis of the world system. For the ith calibration point on the x_w axis, from (2.15)

$$\begin{bmatrix} x_{w,i}Y_i & -\mu\, x_{w,i}\, X_i \end{bmatrix}\begin{bmatrix} v_1 \\ v_4 \end{bmatrix} = \mu\,(X_i - Y_i \frac{X_0}{Y_0}) \tag{3.1.1}$$

where (X_0, Y_0) are the image coordinates of the origin of the world coordinate system. When $N > 2$, an overdetermined system of linear equations for the unknowns $\{v_1, v_4\}$ can be established and solved by a linear least-squares algorithm in a closed form.

Remark: Clearly both t_y and Y_0 should not be simultaneously zero. To avoid such a case, the projection of the origin of the world coordinate system on the image plane should be away from the origin of the image coordinate system.

C. Computation of the intermediate parameters $\{v_2, v_5\}$

Let $z_w = x_w = 0$, and $y_w \neq 0$, which corresponds to those calibration points on the y axis of the world system. For the ith sample on the y_w axis,

$$\begin{bmatrix} y_{w,i}Y_i & -\mu\, y_{w,i}\, X_i \end{bmatrix}\begin{bmatrix} v_2 \\ v_5 \end{bmatrix} = \mu\,(X_i - Y_i \frac{X_0}{Y_0}) \tag{3.1.4}$$

When $N > 2$, an overdetermined system of linear equations for the unknowns $\{v_2, v_5\}$ can be established and solved by a linear least-squares algorithm or by a closed form solution.

The derivation of (3.1.4) is similar to the derivation of (3.1.1).

D. Computation of R, t_x ***and*** t_y

Step 1. Let C be a 2x2 matrix defined as

$$C = \begin{bmatrix} v_1 & v_2 \\ v_4 & v_5 \end{bmatrix}$$

If no row or column of C vanishes, then

$$t_y^2 = \frac{S_r - \sqrt{S_r^2 - 4\,(v_1v_5 - v_4v_2)^2}}{2\,(v_1v_5 - v_4v_2)^2} \tag{3.1.5a}$$

where $S_r = v_1^2 + v_2^2 + v_4^2 + v_5^2$. Otherwise,

$$t_y^2 = (v_i^2 + v_j^2)^{-1} \tag{3.1.5b}$$

where v_i and v_j are the non-zero elements in the row or column of C.

Step 2. Determine the sign of t_y.

Assume that $t_y > 0$, and calculate

$$r_i = v_i t_y \qquad i = 1, 2, ..., 5 \tag{3.1.6a}$$

$$t_x = v_3 t_y \tag{3.1.6b}$$

and for a particular calibration point compute

$$x = r_1 x_w + r_2 y_w + t_x \tag{3.1.6c}$$

$$y = r_4 x_w + r_5 y_w + t_y \tag{3.1.6d}$$

If both sign(x) = sign(X) and sign(y) = sign(Y), then sign(t_y) = 1. Otherwise sign(t_y) = -1.

Step 3. Computation of R

If sign(t_y) = 1, retain r_i, i = 1, 2, 4, 5, and t_x obtained from (3.1.6). Otherwise, change the sign of every r_i and t_x.

Let

$$r_3 = (1 - r_1^2 - r_2^2)^{1/2} \tag{3.1.7a}$$

$$r_6 = -\,\text{sign}(r_1 r_4 + r_2 r_5)(1 - r_4^2 - r_5^2)^{1/2} \tag{3.1.7b}$$

$$[r_7\ r_8\ r_9]^T = [r_1\ r_2\ r_3]^T \text{x} [r_4\ r_5\ r_6]^T \tag{3.1.7c}$$

where "x" denotes a vector cross product. Go to Stage 2 to compute f_x. If $f_x > 0$, retain R, otherwise the sign of f_x needs to be changed, and

$$r_3 = -(1 - r_1^2 - r_2^2)^{1/2} \tag{3.1.8a}$$

$$r_6 = \text{sign}(r_1 r_4 + r_2 r_5)(1 - r_4^2 - r_5^2)^{1/2} \tag{3.1.8b}$$

$$[r_7\ r_8\ r_9]^T = [r_1\ r_2\ r_3]^T \text{x} [r_4\ r_5\ r_6]^T \tag{3.1.8c}$$

3.2 Stage 2 — Computation of t_z, α, f_x and f_y

In this stage, the calibration points are not constrained to the x and y axes of the world coordinate system; see Fig. 3.2. The remaining parameters, t_z, α, f_x and f_y, can be estimated using a two-variable linear least-squares algorithm.

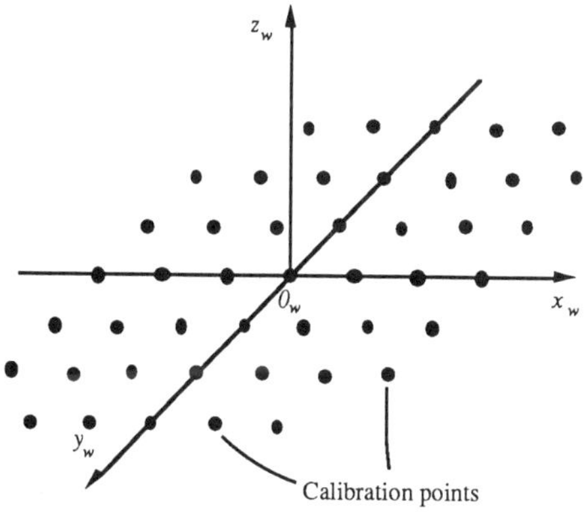

Fig. 3.2 Calibration points are not restricted on x_w and y_w in the second stage

Let $z_{w,i} = 0$. For the ith calibration point, from (2.15a)

$$\begin{bmatrix} X_i w_i & -X_i X_0^{-1} t_x r'^2_0 + x_i r'^2_i \end{bmatrix} \begin{bmatrix} f_x^{-1} \\ \alpha \end{bmatrix} = x_i - X_i X_0^{-1} t_x \tag{3.2.1}$$

where $x_i = r_1 x_{w,i} + r_2 y_{w,i} + t_x$ (3.2.2)

$$w_i = r_7 x_{w,i} + r_8 y_{w,i} \tag{3.2.3}$$

$$r'^2_0 = \mu^2 X_0^2 + Y_0^2 \tag{3.2.4}$$

When $N > 2$, an over-determined system of linear equations for the unknowns α and f_x can be established and solved by a linear least-squares algorithm or by a closed form solution. After f_x is found,

$$f_y = f_x \mu \tag{3.2.5}$$

$$t_z = (1 - \alpha r'^2_0) f_x t_x X_0^{-1} \tag{3.2.6}$$

4. Experimental Results

The experimental system consisted of two CCD cameras (Electrophysics, model number CCD1200), two 25 mm TV lenses, an IBM-AT personal computer, an ITEX video imaging board with its driver software, a camera calibration board, and a coordinate measuring machine (CMM). The camera calibration board is a glass plate painted with vapor deposited metallic chromium. It has 10x10 dot array points with center to center distance of 10 mm, and diameter (of each point) of 2 mm. The flatness of the calibration board is within ± 0.003 mm and the center to center accuracy of the calibration points is within ± 0.002 mm.

To estimate the image coordinates of a calibration point, an adaptive thresholding algorithm was devised. The centroid of the binary image of calibration point was selected as the estimate of its image coordinates. This procedure could yield image coordinates accuracy to within 1/5 of a pixel.

To estimate the ratios of the scale factors μ of the two cameras, an algorithm given in [2] was applied to compute μ. Since two cameras and lenses with the same models were used, identical values of μ for two cameras were obtained. The estimated μ is 1.2401506.

The algorithms given in [1] and Section 3 were employed to compute camera models with the estimated μ. For the original RAC-based algorithm, 16 points were used for the identification of the camera parameters and 8 points were used to estimate the 3D accuracy of these models. Tables 4.1 lists the estimated 3D accuracy of the identified camera models for a single camera with two views. The results obtained by using two cameras with single view were compatible to those listed in Table 4.1.

The simplified RAC-based algorithm, as expected, demonstrates a tradeoff between computational efficiency and accuracy. The original RAC-based Algorithm is superior in terms of accuracy. One reason is that only a limited number of calibration points (7 points) were used in the first stage computation of the simplified algorithm. Improvements can be made by increasing the number of calibration points contained within the field of view for the simplified RAC-based algorithm. On the other hand, the simplified algorithm is much faster. Computer simulation showed that the simplified RAC algorithm is about five times faster than the original RAC algorithm, not including the image preprocessing time.

Table 4.1 3D error with single camera (two views)
for field-of-view = 40x50 mm^2
Test points are on a plane 20 *mm* away from the calibration plane

Unit *mm*	Average Error			Maximum Error		
	x	y	z	x	y	z
RAC-Based Algorithm	0.0077	0.0045	0.0235	0.0290	0.0102	0.0459
Simplified RAC Algo.	0.0520	0.0245	0.0530	0.0661	0.0504	0.1019

References

[1] Tsai, R. Y., "A Versatile Camera Calibration Technique for High Accuracy 3D Machine Vision Metrology Using Off-the-Shelf TV Cameras and Lenses," *IEEE J. Robotics Automat.*, Vol. RA-3, No. 4, Aug. 1987 pp. 323-344.

[7] Xu, X., "Camera Calibration Techniques," M. S. Thesis, Florida Atlantic University, in preparation.

A LIQUID-CRYSTAL-BASED SURFACE-DEPTH TO GRAY-SCALE-IMAGE TRANSDUCER

J.J. Soltis and C.K. Yu
Department of Electrical Engineering
University of Windsor
Windsor, Ontario, N9B 3P4, Canada

ABSTRACT

This paper reports experimental result obtained using a novel liquid-crystal-based surface-depth to gray-scale-image transducer. True surface profiling of a manufactured metal part is possible by using the part as the ground electrode and converting the local surface electric field to the gray-scale image using the properties of liquid crystals.

INTRODUCTION

Consider metal parts which are primarily flat in which surface defects (e.g. pores or pits, protrusions, burs, cracks) need to be measured. The entire surface should be characterized in less than 1 second (e.g. surface of piston head).

Since the parts are metallic they can be charged to a voltage and used as a DC or AC electrode for the surface to be measured. Since twisted nematic liquid-crystal display devices are field-effect in nature, a non-contacting surface-depth measuring device as shown in Fig. 1 is possible. In operation, bipolar DC voltage pulses are applied between the single transparent electrode of the device and the part to be measured. The liquid crystal responds to the electric field which is related to the underlying surface undulations. With the voltage selected to provide mid-gray viewing for a smooth surface, protrusions are viewed as darker areas and surface pits as whiter areas. This image is then triggered-digitized in 16 msec and the resulting frame processed in the conventional fashion using the camera/computer image processing system. The entire process, with the total surface area analysed, can be extremely rapid. The device converts the surface depth information of a part to a corresponding gray-scale image. The liquid-crystal material itself has good innate contrast and resolution. Resolution is best when the distance of the part (under test) to the liquid-crystal layer is minimized, since this reduces the electric field spreading effects.

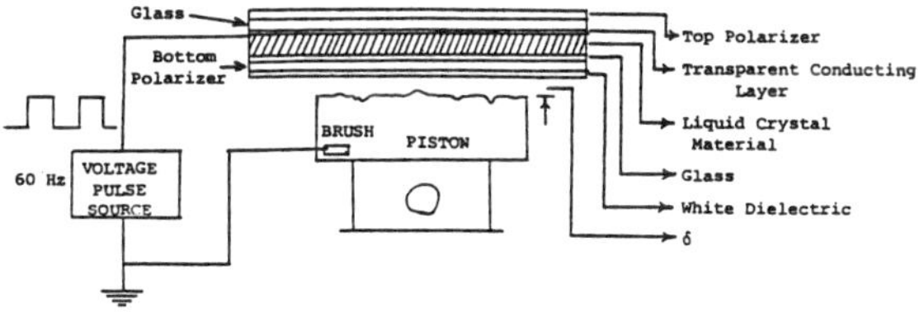

Figure 1 - Surface-Depth to Grey-Scale Image Transducer

EXPERIMENT DESCRIPTION

The experimental set-up is shown in Fig. 2. The sample is shown in Fig. 3.

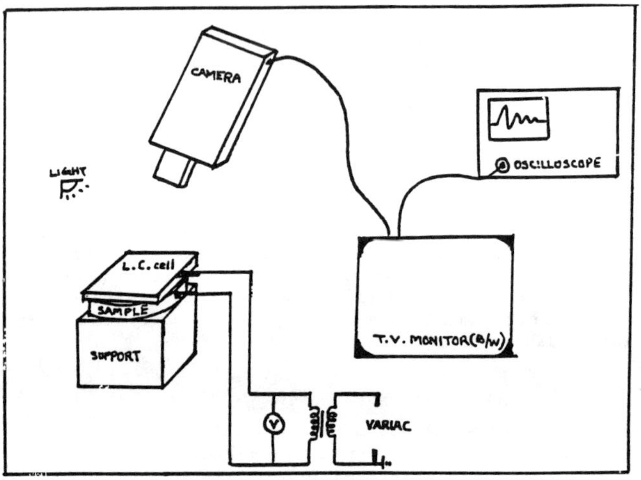

Figure 2 - Experimental Set-up

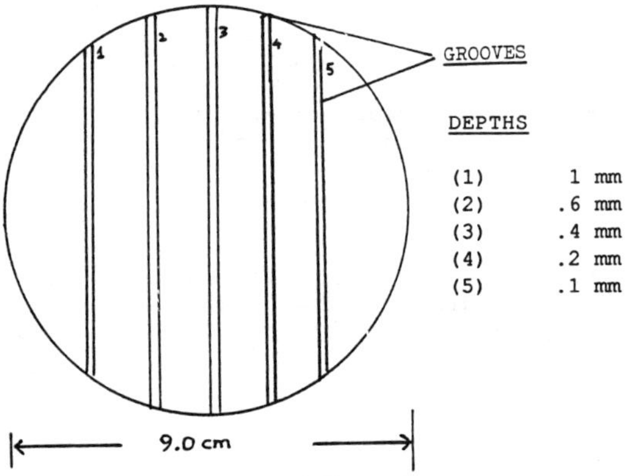

Figure 3 - Sample Steel Disk

The L.C. cell is placed on top of the sample with the side without the transparent electrode closest to the surface of the sample. It is important to ensure that the L.C. cell is lying flat on top of the sample.

The camera positioned over the L.C. cell is focussed so that the monitor is displaying only the L.C. cell and the position of the table lamp is adjusted so that the monitor displays the clearest image possible and it should also ensure that the L.C. surface is evenly illuminated. The time scale on the oscilloscope is set at 5 micro second per cm and the vertical scale is set at 0.1 V/cm in order to display signal from the T.V. monitor. The voltage is applied to both the connectors of the L.C. cell and the sample being used. It is important that the voltage be increased slowly and be aware of any breakdown or sparking in the L.C. cell. **NOTE:** In air, breakdown voltage is 4.7 KV at a spacing of 0.1 cm as measured with sphere gaps. For our case the smallest spacing between the electrodes is 1.1 mm and largest is 2 mm.

EXPERIMENTAL OBSERVATIONS

A fundamental problem with the present L.C. cell is that since (in conjunction with the part surface) it forms a capacitor, the opposite charges of the electrodes cause the upper glass plate to bend as depicted in Fig. 4. This undesirable distortion causes the centre portion to be darker than the surrounding area.

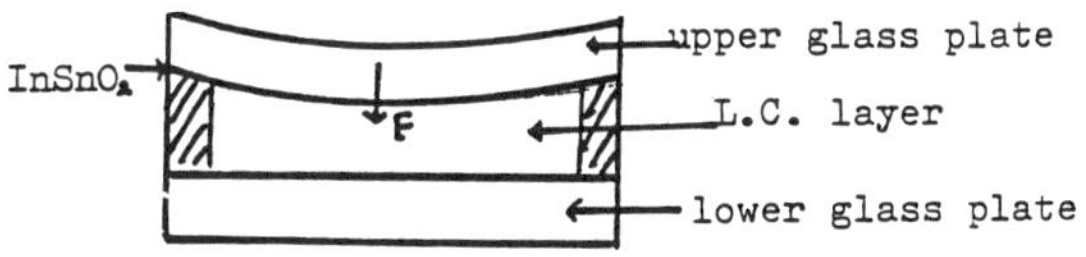

Figure 4 - Deflection of Glass Plate (Upper)
Due to Bending Moment

With the system described above the voltage is slowly increased. At about 400 V, the centre portion of the L.C. cell has turned dark as seen in Fig. 5. At the same time as noted on the monitor, only 4 out of the 5 grooves are seen. The smallest groove which is 0.1 mm in depth cannot be seen. The image is not good enough yet to show which line has a higher brightness. The signal from the monitor is seen on the oscilloscope as shown in Fig. 7. Due to the uneven blackness of the image produced on the monitor as in Fig. 6, the signal was not quite as expected. The part of signal representing the grooves are indicated by arrows.

The voltage is now increased to about 600 V. As seen in Fig. 8, the darker region has now spread throughout the whole L.C. cell. Only the two leftmost grooves remain clearer. This is best seen on the T.V. monitor as shown in Fig. 9. Note that the leftmost line (deepest grooves = 1 mm) is much brighter than the one to its left. The second line (from left) is in turn brighter than the third (0.6 mm and 0.4 mm. respectively). The corresponding signal from the monitor as observed on the oscilloscope is seen in Fig. 10. The signal also clearly shows the difference in depth of the grooves. Further increasing the voltage to about 800 V, the L.C. cell becomes dark throughout and no image is seen on the monitor.

The voltage is now returned to OV and a white sheet of paper is inserted between the L.C. cell and the sample. The voltage is then slowly increased and about 800 V, the L.C. cell turns dark throughout and 4 out of 5 grooves shown up very clearly as white lines of different brightness. This is illustrated in Fig. 11. Also noted from the figure, the deeper groove (from left) shows up as a brighter line. Further increasing the voltage causes the L.C. cell turns dark throughout.

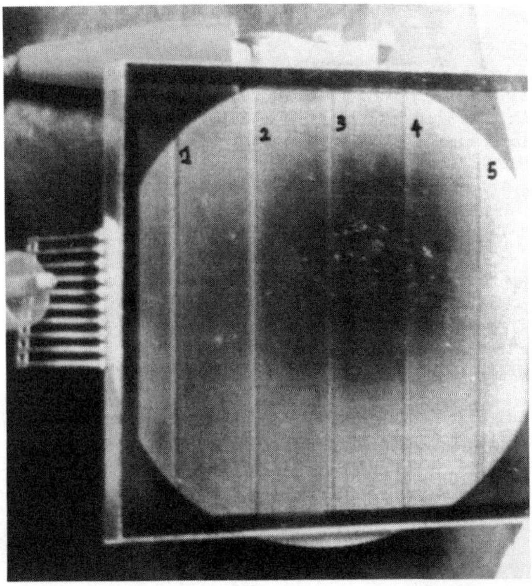

Figure 5 - L.C. Cell at 400 V

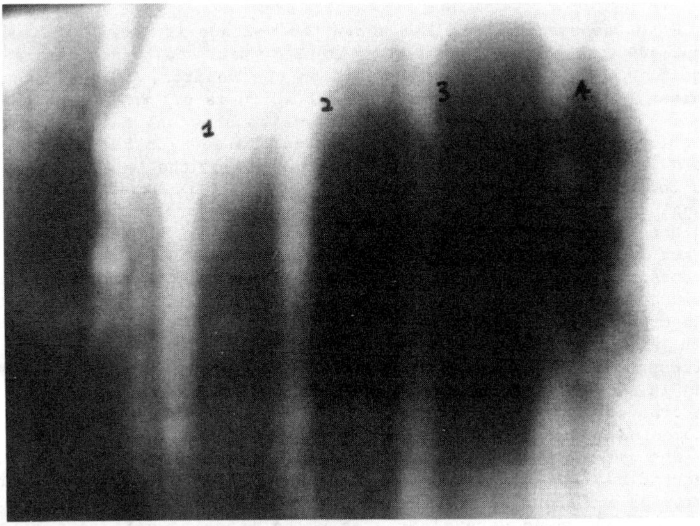

Figure 6 - L.C. Cell as Seen on Monitor at 400 V

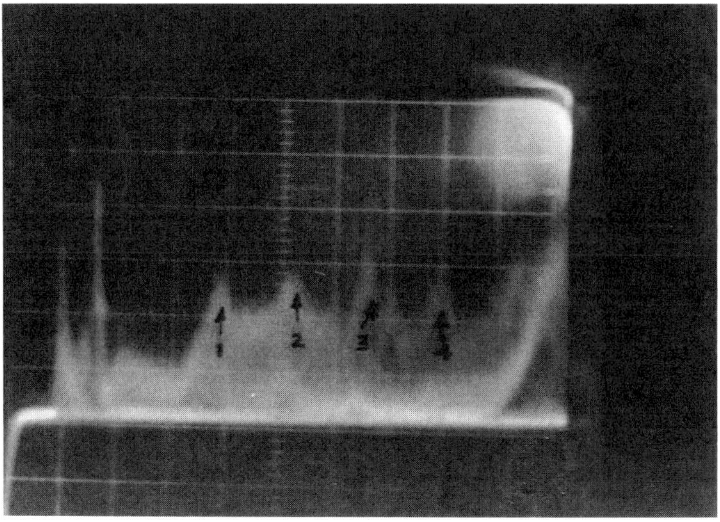

Figure 7 - Signal On Oscilloscope at 400 V

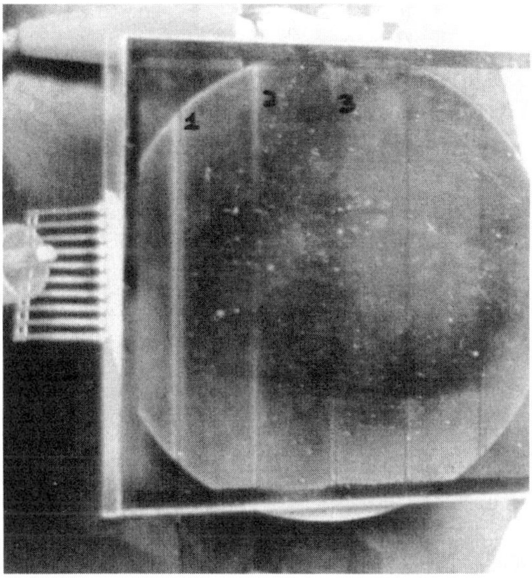

Figure 8 - L.C. Cell at 600 V

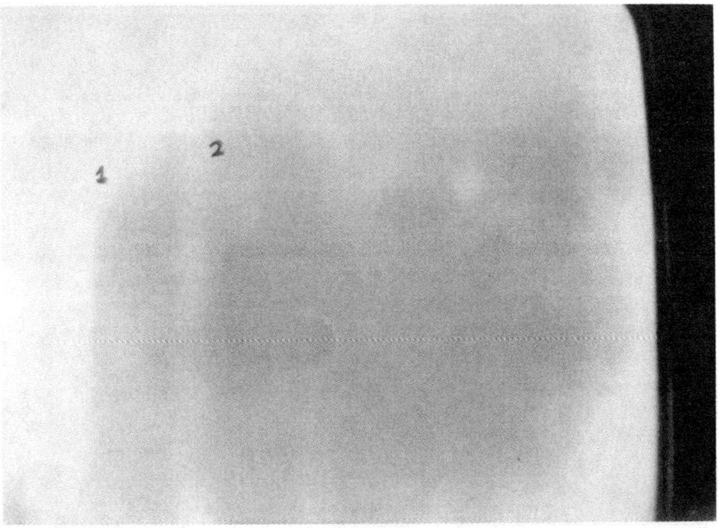

Figure 9 - L.C. Cell as Seen on Monitor at 600 V

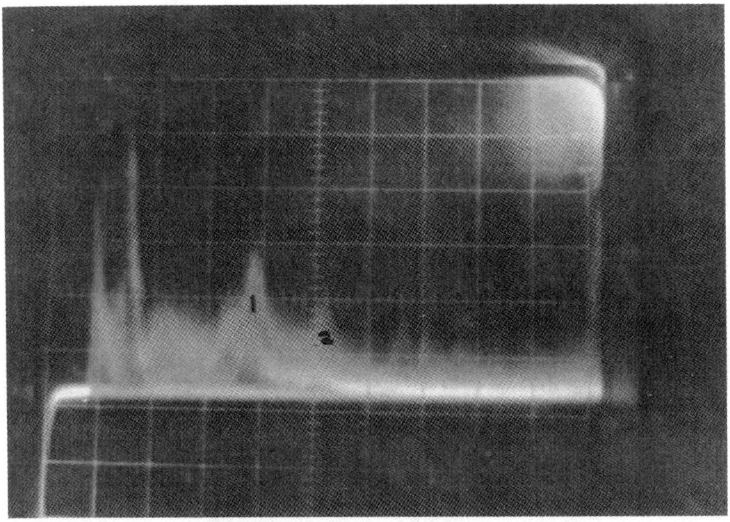

Figure 10 - Signal on Oscilloscope at 600 V

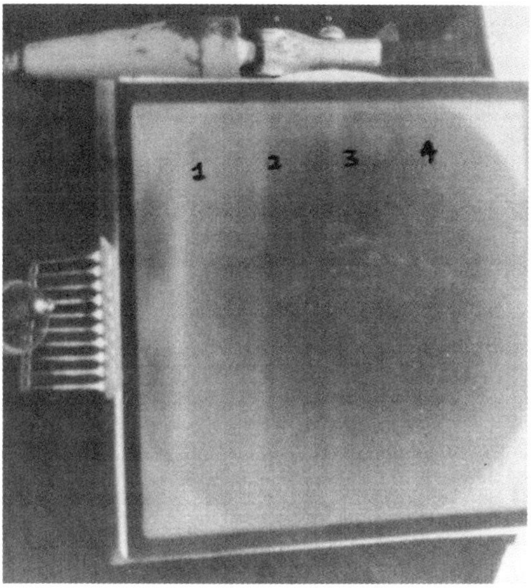

Figure 11 - L.C. Cell at 800 V With Paper in Between

DISCUSSION AND CONCLUSIONS

The present results verify the essential principles of the transducer. By incorporating a matrix support structure the distortion associated with plate bending can be avoided. Optimal results are expected when cells are fabricated utilizing nematic liquid crystal microdroplets distributed in a solid plastic matrix [1]. For the latter, the possibility of surface roughness measurement appears possible. This is currently being studied.

REFERENCES

[1] J.W. Doane, N.A. Vaz, B.G. Wa and S. Zumer, "Field controlled light scattering from nematic microdroplets", Appl. Phys. Lett. 48(4), pp. 269-271, January 1986.

A TWO-COMPONENT FORCE SENSOR FOR THE EXPLORATION OF UNKNOWN PLANAR SURFACES*

J.L. FABIAN
Columbia Gas of Ohio, Columbus, OH

H. HEMAMI
Department of Electrical Engineering
The Ohio State University, Columbus, OH

ABSTRACT

The two-component contact force sensor is developed for robotic probing of unknown surfaces. The two components are the vertical and tangential forces that a robot or end effector applies to a planar surface. An experimental testbed is constructed to study the sensor characteristics, to probe unknown surfaces, and to test the effectiveness of a simple control algorithm. The two-component sensor provides normal and tangential force information to a robot controller. As a simple control algorithm is developed to regulate normal force to prespecified limits while effectively advancing the sensor along the surface without loss of contact and/or manual intervention. Both smooth contours and sharp corners are explored. Contact force information is used off-line for shape recognition and relative surface friction measurement.

INTRODUCTION

Exploration of unknown surfaces by a robotic probe is facilitated by smooth gliding [1] or rolling [2] of the probe on the unknown surface [3,4]. One way to achieve effective gliding or rolling is by the detection of incremental motion between the probe and the unknown surface. If both the probe and the unknown surface are imbedded with appropriate sensors [5] the detection and estimation of relative motion and its control is feasible [6]. An alternative method is the measurement of both the normal and tangential forces at the point of contact. The normal component can be controlled to maintain contact, and the tangential component can be controlled to effect gliding or rolling. The control will also naturally depend on the relative friction properties of the probe and the surface to be explored. The measurement and control of the normal and tangential components of the contact force by a sensor for gliding on an unknown planar surface and the estimation of the shape of the surface are the subject of this paper [7].

The sensor is a three link arm restrained at the joints by springs. The end link is a wheel for making contact with unknown surfaces. A drawing of the sensor is shown in Figure 1. The unconstrained shape of the sensor is similar to a bent finger. This

---

* This work was supported, in part, by NIDRR Grant No. H133E 80017 and, in part, by the Department of Electrical Engineering, The Ohio State University.

shape allows the sensor to smoothly grope along surfaces with different curvatures. Three Hall effect I.C. sensors are mounted respectively on the three links to measure angular position of the joints.

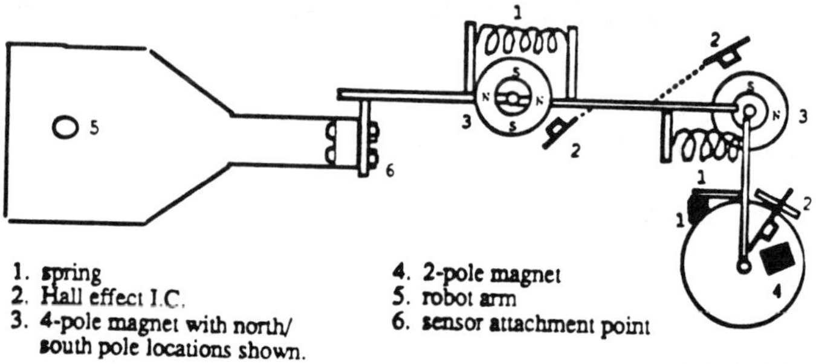

Fig. 1: Finger-like sensor with its major component.

The system extracts three voltage signals $V_{\theta 1}$, $V_{\theta 2}$, and $V_{\theta 3}$ from the Hall Effect I.C.'s. These signals are processed to specify the position (θ_1, θ_2, θ_3) and torques (u_1, u_2, u_3) at the finger joints. The above processing is by three lookup tables that are constructed from a preliminary calibration of the system. The three lookup tables are as follows. The first lookup table provides the three angular positions for a given set of three I.C. signals. The second lookup table provides gravity effects based on the orientation of the finger relative to earth. The third lookup table provides the three torques at the joints. The normal and tangential components of force are computed from the outputs of the three lookup tables under the assumptions that the movement of the probe is slow and consequently the angular velocities and acceleration in the equations of motion can be ignored.

The sensor is mounted on a Rhino XR series robot with three degrees of freedom — two translations and one rotation — all in a plane.

Different experiments are carried out to explore planar surfaces whose shapes are unknown to the system: large and small circles and concave and convex corners. The performance of the system is evaluated in terms of the sensors' dynamic response, the estimates of the radii of the circles, the length of time to arrive at the estimate, the maintenance of the vertical component of force within desired bounds, and a measure of the coefficient of dry Coulomb friction arrived at from the two components of force.

Feasibility and future evolvement of such sensors are discussed.

CONTROL

The end effector on which the sensor is mounted is a Rhino XR series robot. It has three degrees of freedom in the plane — two degrees of translation and one degree of rotation. The end effector has three stepper motors with optical encoders that transmit position information to the computer.

The movement strategy is based on the magnitude of the normal force and the point of contact on the third link of the sensor (Fig. 2).

EXPERIMENTS

Three kinds of surface were probed with the system: circular shapes of aluminum (or paper-covered aluminum), concave corners, and a rectangular post.

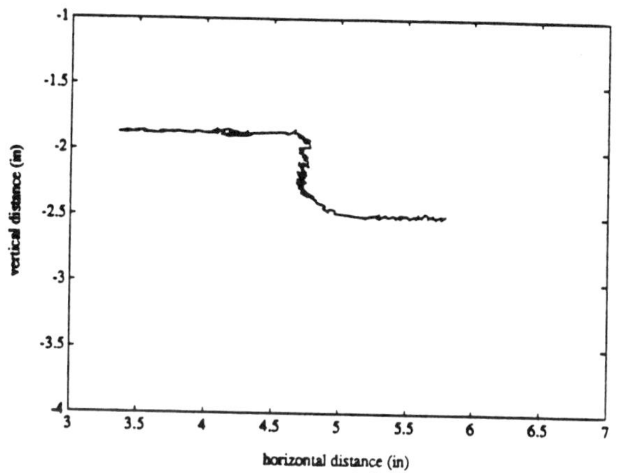

Fig. 3: The coordinates of the point of contact in the robot's x-y coordinate system while probing a concave corner are plotted.

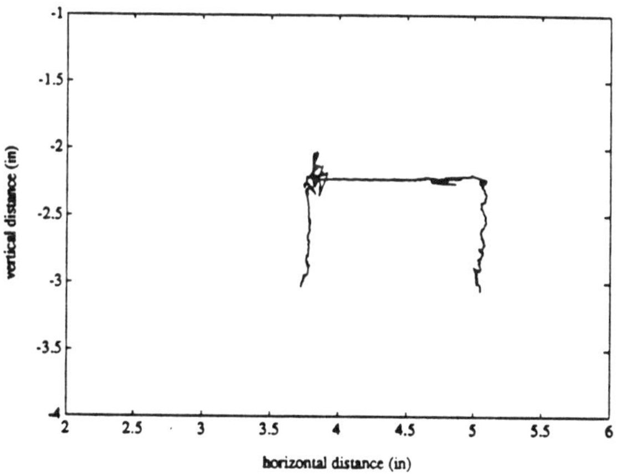

Fig. 4: The coordinates of the point of contact in the robot's x-y coordinate system while probing a rectangular post of width 1.25 in are plotted.

REFERENCES

1. J. Bay, "Constrained motion of a 3-D manipulator over unknown constraints: the robotic grouping problem," Ph.D. Dissertation, The Ohio State University, September 1988.

2. R.E. Goddard, "Nonholonomic dynamic control of robotic manipulators with applications to autonomous probing," Ph.D. Dissertation, The Ohio State University, August 1990.

3. P. Dario and D. DeRossi, "Tactile sensors and the gripping challenge," *IEEE Spectrum* , vol. 22, no. 5, pp. 46-52, August 1985.

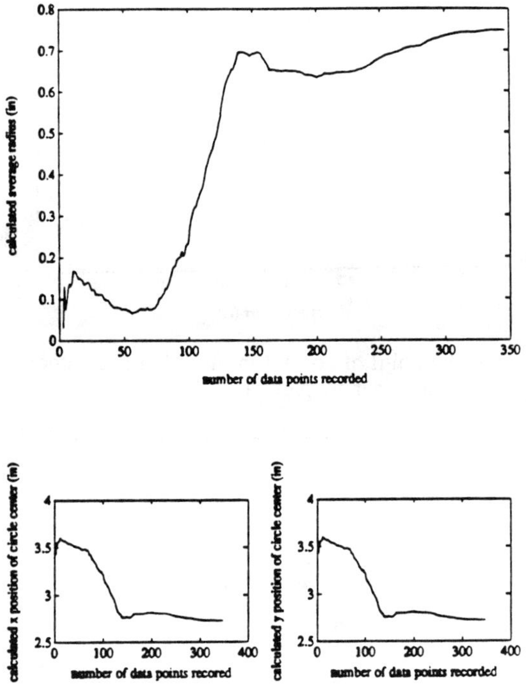

Fig. 5: Average radius and the coordinates of the center for small circle as the number of recorded contact points increases..

4. R.S. Johansson and A. Vallbo, "Tactile sensory coding in the glabrous skin of the hand," *Trends in Neuroscience*, vol. 6, no. 1, pp. 27-37, January 1983.

5. M. Taylor, B. Kawarizadeh, H. Hemami, and K. Boyer, "A preliminary experiment in dual channel tactile information flow," to appear in *Journal of Robotic Systems*, June 1993.

6. R.D. Howe and M.R. Cutkosky, "Sensing skin acceleration for slip and texture perception," *Proc. 1989 IEEE Intl Conference on Robotics and Automation*, Scotsdale, AZ, May 1989, vol. 1, pp. 145-150.

7. J.L. Fabian, "Unknown surface probing with a two-dimensional force sensor," M.S. Thesis, The Ohio State University, December 1991.

Noncontact Perception of an Unknown Surface by Using Multiple Laser Scanners

KEVIN B. SMITH and YUAN F. ZHENG
Department of Electrical Engineering
The Ohio State University
Columbus, Ohio 43210

ABSTRACT

Current laser digitizing systems use a single Laser Displacement Sensor (LDS) held in a fixed orientation to world coordinates while digitizing. Because the LDS's accuracy is dependent on the orientation between the LDS and the surface being digitized, the set of surfaces that can be accurately digitized is restricted to surfaces with limited tangential variations. This restriction is overcome by using an articulated machine to position the LDS. The articulated machine allows the LDS's position and orientation to be changed at each location digitized. However, determining what the position and orientation for the LDS should be when the surface is unknown requires extrapolating the surface from information acquired while digitizing. The authors propose gathering local information about the unknown surface by using multiple LDS. This local information is used in extrapolating the unknown surface at the next location to be digitized. This article overviews an articulated multiple LDS digitizing system built by the authors and the extrapolating approach used.

INTRODUCTION

Digitizing unknown surfaces quickly, reliably, and accurately is important to the manufacturing industry. In the past, coordinate measuring machines (CMM) using touch probes have been used for digitizing sculptured surfaces. Digitizing using a CMM with a touch probe is accurate and reliable; however, it is also slow. This has motivated several tool manufacturers to replace the touch probe with a Laser Displacement Sensor (LDS). In addition, 3-axis machining centers are now being used to position the LDS while digitizing.

The LDS can collect sculptured surface data faster than a touch probe. In addition, the LDS does not require mechanical parts to come in contact with the sculptured surface, thus protecting the surface from possible damage or movement. However, the LDS's accuracy depends on the position and orientation of the laser with respect to the sculptured surface. The LDS is the most accurate when its orientation is normal to the sculptured surface and the surface is within the LDS's effective stand-off range [1].

On 3-axis machines, the LDS is held in a fixed orientation while digitizing [2]. This limits the types of sculptured surfaces that can be digitized. Because the LDS's orientation must be within a specified range of sensor-to-surface perpendicularity, only sculptured surfaces that fall within that range can be digitized. Work has been done to increase the LDS's surface-to-sensor orientation range [3]. However, using an articulated machine for positioning and orienting the LDS, a more general sculptured surface (one with high tangential variations) can be digitized. The

LDS's position and orientation can then be continuously changed during the digitizing process. Thus, narrower variations in sensor-to-surface orientation can be maintained. Use of an articulated machine with multiple LDSs is suggested by the authors. The multiple LDS are used to gather local information about the unknown surface which will aid in extrapolating the surface's position and normal vector at the next digitized location.

To study the issues involved with maintaining a near normal orientation between an unknown sculptured surface and the LDS, a non-contact 3-D digitizing system with six degrees-of-freedom was constructed. This system uses two Keyence LB-72/LB-12 LDS positioned by a six degrees-of-freedom PUMA robot, see Fig. 1.

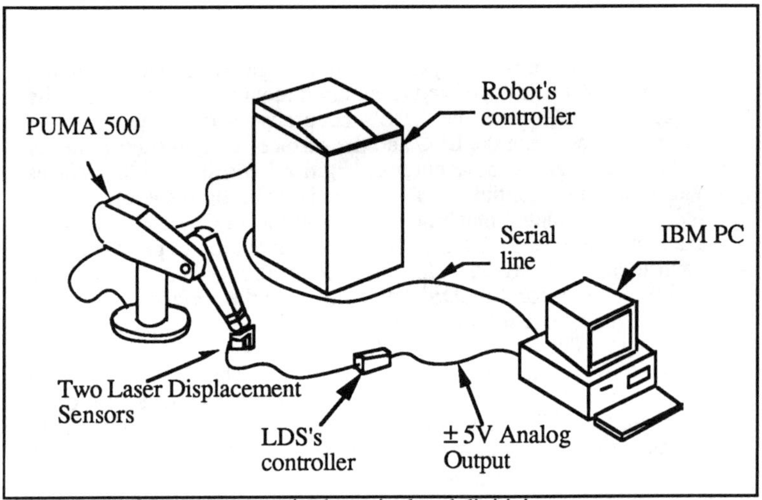

Figure 1 Major components in the articulated digitizing system

The output of the LDS is sent to an analog-to-digital converter in the IBM PC, and the positioning information is sent to the robot's controller via the PC's serial port. This system provides the means to implement and evaluate proposed digitizing and extrapolating approaches.

DIGITIZING APPROACH

In this digitizing system, the surface is digitized one row at a time. The rows are formed by the intersection of parallel planes with the unknown surface. The intersection of a plane with the surface forms a plane curve, see Fig. 2. The position and orientation of the LDS is known with respect to world coordinates from the robot's controller. From this information, a 4x4 homogeneous transformation matrix, T_l, is created establishing the relationship between the local coordinates of the LDS with respect to world coordinates. Thus,

$$P_o = T_l\, P_l, \text{ where } P_o = \begin{bmatrix} X_o \\ Y_o \\ Z_o \\ 1 \end{bmatrix} \text{ and } P_l = \begin{bmatrix} X_l \\ Y_l \\ Z_l \\ 1 \end{bmatrix} = \begin{bmatrix} 0 \\ 0 \\ -\rho \\ 1 \end{bmatrix}.$$

Using this transformation, the displacement value ρ is converted to the coordinates of a point on the unknown surface.

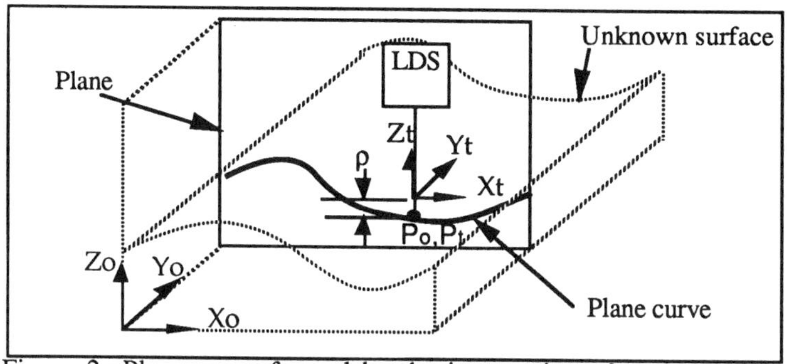

Figure 2 Plane curve formed by the intersection of a plane with the unknown surface. The LDS is kept within the plane while digitizing a row.

Digitizing with Multiple LDS

Digitizing with multiple LDSs consists of positioning and orienting a set of LDSs which are oriented parallel to each other. Once positioned, each LDS provides coordinates for a different point on the plane curve (see P1, P2, and P3 in Fig. 3). From these points, local information about the unknown surface can be determined, such as location, tangent, and curvature. The amount of information which can be determined depends on the number of LDSs in the set. One point is needed to determine position. Two points are needed to determine the tangent. And three points are needed to determine the curvature. These points can be digitized using only one LDS. However, the LDS must be mechanically moved to each point digitized. The LDS set, on the other hand, only has to be positioned once to measure multiple points which substantially increases the digitizing rate. The accuracy of this local information is a function of the density of the digitized points in relation to the curvature of the surface. Good approximations can be obtained provided the tangent or curvature is nearly constant over the set of digitized points.

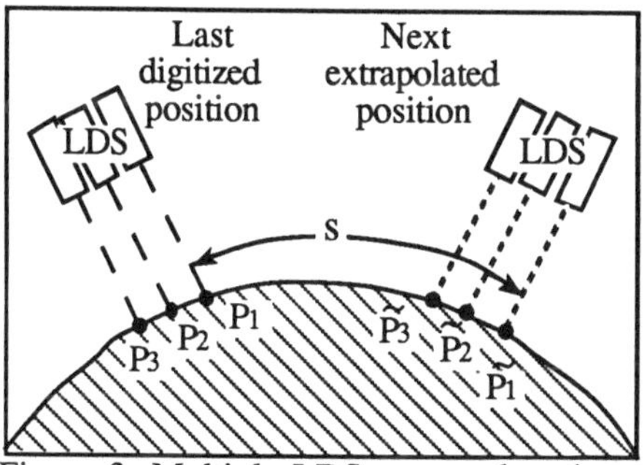

Figure 3 Multiple LDSs at two locations. The sampling arc length is represented by s and P1,P2, and P3 are digitized points.

Approximating Slope and Curvature

The slope and curvature at a point on a plane curve can be approximated by using local discrete points. The slope, m, of a plane curve, defined as $m = \frac{dy}{dx}$, at a point P_1 is approximated using the coordinates of P_1 and the coordinates of another local point P_2, see Fig. 4. The approximated slope, m_1, of the curve at P_1 is $m_1 \approx \frac{Y_1-Y_2}{X_1-X_2}$ and the approximated tangent at the point P_1 is the line that passes through the point P1 with the slope of m_1.

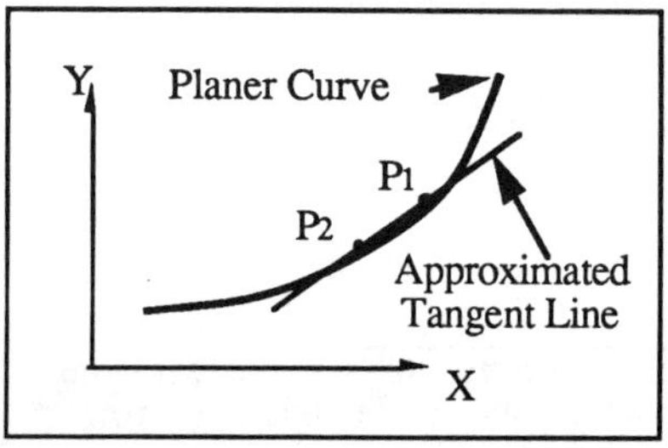

Figure 4 Approximating the Tangent

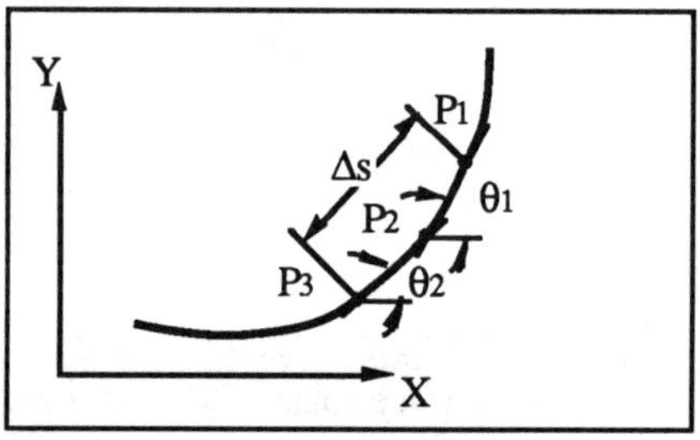

Figure 5 Approximating the Curvature

The curvature, κ, is defined as $\kappa = \left|\frac{d\theta}{ds}\right|$ where θ is the angle between the curve's tangent line and the horizontal axis and s is the curve's arc length. The curvature, κ_1 at the point P_1 can be approximated using P_1 and two other local points, P_2 and P_3 (see Fig. 5). With the coordinates of the points P_1, P_2, and P_3 being respectively (X_1,Y_1), (X_2,Y_2), and (X_3,Y_3), the curvature, κ_1, is approximated by

$$\kappa_1 = \left|\frac{\theta_1-\theta_2}{\Delta s}\right|, \tag{1}$$

where

$$\theta_1 = \arctan\left(\frac{Y_1-Y_2}{X_1-X_2}\right), \tag{2}$$

$$\theta_2 = \arctan\left(\frac{Y_2-Y_3}{X_2-X_3}\right), \tag{3}$$

and

$$\Delta s = \sqrt{(X_1-X_2)^2+(Y_1-Y_2)^2} + \sqrt{(X_3-X_2)^2+(Y_3-Y_2)^2}. \tag{4}$$

As the points P2, and P3 move closer to P1, the approximation of the slope and curvature improves. In the limit, the approximated values m_1 and k_1 become the actual slope and curvature values of the the curve at the point P1.

CONSTANT CURVATURE EXTRAPOLATION

For each location digitized, a position and orientation vector of the LDS set must be determined. Optimally, the LDS is positioned so the surface is in the middle of its measuring range, and the LDS is oriented normal to the surface. To determine these optimal vectors for the LDS, the position and tangent vectors of the surface must be known. Because the surface is assumed to be unknown, the desired surface vectors are approximated by extrapolating the information collected from the LDS. The extrapolation approach used assumes that the surface's plane curve of a digitized row has constant curvature along the sampling arc length. Therefore, either a circle or a line is used for the extrapolating equation.

A circle has constant curvature, κ, equal to the inverse of its radius, R, that is κ = 1/R. The extrapolating circle is found directly by solving for the circle which passes through the last digitized points, P1, P2, and P3, see Fig. 6.

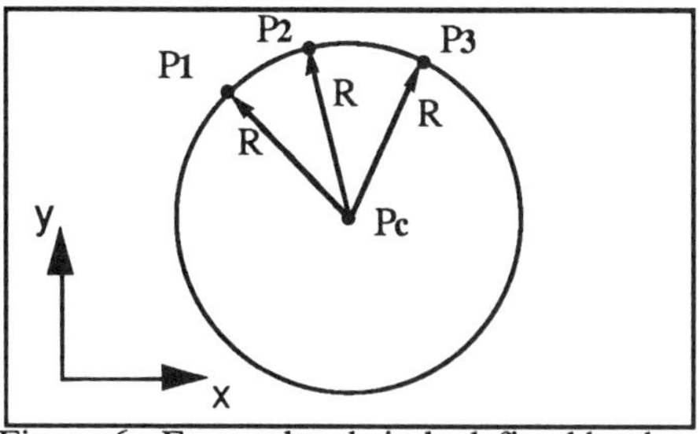

Figure 6 Extrapolated circle defined by three points

The center of the circle Pc is determined by taking the vector norm of the difference between Pc and the digitized points and setting them equal to each other. That is

$$R^2 = \|P_c - P_1\| = \|P_c - P_2\| = \|P_c - P_3\|. \tag{5}$$

Expanding (5) and writing the result in matrix notation gives

$$\begin{bmatrix} x_3^2 - x_1^2 + y_3^2 - y_1^2 \\ x_3^2 - x_2^2 + y_3^2 - y_2^2 \end{bmatrix} = \begin{bmatrix} x_3 - x_1 & y_3 - y_1 \\ x_3 - x_2 & y_3 - y_2 \end{bmatrix} \begin{bmatrix} x_c \\ y_c \end{bmatrix}. \tag{6}$$

Once the center point coordinates have been determined using (6), the radius, R, is determined using (5). The equation for the extrapolated circle is

$$(x - x_c)^2 + (y - y_c)^2 = R^2. \tag{7}$$

If only the tangent vector is to be approximated, then the curve is extrapolated by a line passing through the last two digitized points. The equation for the extrapolated line is

$$y = m_1(x - x_1) + y_1. \tag{8}$$

From the extrapolation equation used, the plane curve's position and tangent vectors at the next location to digitize can be readily determined given a desired sampling interval. The LDS is then moved to this position with an orientation perpendicular to the curve's tangent vector and within the digitizing plane.

EXPERIMENT

An experiment was conducted to illustrate the effect that orientation between the LDS and surface has on digitizing accuracy. In this experiment, a row of thirty points on a sheet of paper were repeatedly digitized. The paper was taped to a flat surface and the orientation between the LDS and the paper was held fixed while the row of points was digitized consecutively for ten times. The orientation of the LDS was then changed and the row was again digitized ten times. The average coordinate values of the digitized points for each LDS orientation were plotted in Fig. 7. Also, all ten measurements of each point were plotted for the 90° LDS orientation. The variation in measurements of the same point with the same LDS orientation is attributed to the robot's 0.004 inch repeatability and the LDS's 0.002 inch resolution.

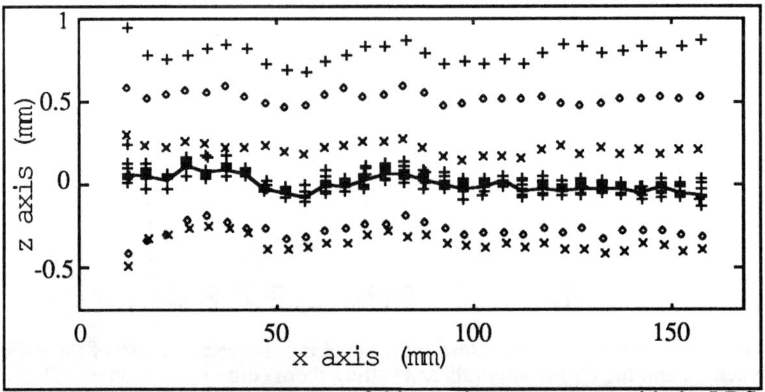

Figure 7 Average measurements of thirty points on a paper with the LDS at different orientations. From bottom to top, the LDS's orientation is {'x'=60°, 'o'=70°, '+'=90°, 'x'=100°, 'o'=110°, and '+'=120°}. The actual measurements are plotted for the 90° (perpendicular) case and a line drawn through the average value.

The difference in the average values with different sensor-to-surface orientations is due to the LDS's accuracy being a function of orientation and errors in the robot's calibration. This example illustrates that accuracy is affected by sensor-to-surface orientation.

CONCLUSIONS

An articulated digitizing system using multiple LDSs was built to study issues related to noncontact digitizing technology. Because a LDS's accuracy is dependent on sensor-to-surface orientation, an articulated machine was used. The application studied was digitizing unknown surfaces which requires surface extrapolation from locally acquired information. This article presented how the information is obtained when using Multiple LDSs to obtain the local information about the unknown surface.

ACKNOWLEDGEMENTS

This work was supported by the Engineering Research Center for Net Shape Manufacturing (ERC/NSM) at The Ohio State University.

REFERENCES

1. "Keyence LB-70,LB-72 Laser Displacement Sensor Instruction Manual," *Keyence Corporation,* 1991.
2. "Sharnoa Integrated Machining Systems," *Sharnoa Corporation,* 1991.
3. K. Saito, et al., "Noncontact 3-D Digitizing and Machining System for Free-Form Surfaces," *Annals of the CIRP,* vol. 40, p. 483-486,1991.

AUTONOMOUS SHAPE EXPLORATION USING SIMPLE SENSORS[1]

J. G. HOLLINGER and J. S. BAY
Bradley Department of Electrical Engineering
Virginia Polytechnic Institute and State University
Blacksburg, Virginia 24061-0111

ABSTRACT

Using the techniques discussed here, an unknown surface can be explored and estimated by a robotic manipulator without operator assistance. The information necessary to model the surface is obtained from the contact location between the manipulator and the surface and the sensed normal direction at that point. This information is then processed with a Kalman filter to iteratively estimate geometric parameters that describe the surface. The estimate's covariance matrix, updated by the Kalman filter each iteration, provides an indication of the uncertainty of the estimate. In addition, the covariance provides a mechanism by which the manipulator can be taught to explore the surface optimally (optimal in the sense that no other trajectory on the surface would cause the estimate to converge faster.)

Real-time, autonomous exploration and curvature determination experiments were conducted with a MERLIN 6540 Industrial Robot. The exploration results for spheres of known diameter show that the autonomously generated trajectories cause the surface estimate to converge faster than predetermined trajectories such as spirals.

INTRODUCTION

The anticipated use of robots in unstructured environments requires methods for determining the shape of objects encountered. These methods would be the first step in grasping and manipulation operations and obstacle avoidance. Research relating to this topic has been performed in computer vision using monocular and stereo images of objects [7]. Unfortunately, among its other drawbacks, this approach requires enormous amounts of computing power and expensive equipment such as cameras, frame grabbers and laser range finders.

An alternate approach to object recognition is through tactile sensing. Instead of generating the massive amounts of data in images, tactile exploration provides sequential data acquisition. The informational content of the data depends on the trajectory followed on the object's surface. Some trajectories will provide more useful information than others. Because the time it takes to complete an exploration trajectory is proportional to its length, it would be desirable to determine the optimal trajectory to minimize the exploration time.

---

[1] This work was supported by the Office of Naval Research under grant no. N00014-91-J-1621

A similar strategy in industrial applications uses component measuring machines [6]. A component measurement machine generates highly accurate measurements at a set of discrete points using high-precision sensors. In contrast, the manipulator described here generates a stream of noisy measurements from a variety of simple and inexpensive sensors. It is hoped that the lack of highly accurate measurements can be overcome by the use of statistical filters, surface models and optimal search algorithms.

AUTONOMOUS EXPLORATION

The goal of our autonomous exploration algorithm is to generate a geometric model of the surface being explored as quickly and efficiently as possible. The surface model has the implicit form:

$$S(x, P) = 0 \tag{1}$$

where P is a vector of constant geometric parameters that fully describe the surface model and x is a vector of Cartesian coordinates expressed in the inertial coordinate system. This form can easily be used to represent quadrics and superquadrics. The general equation for quadrics describe an ellipsoid and the associated parameter vector, P, would describe the length, rotation and translation of its three major axes.

The task of generating a description of an object becomes equivalent to estimating the geometric parameter vector, P. Because a sequence of surface measurements will be generated by the tactile probe as it travels over the object, a Kalman filter can be used to iteratively update an estimate of the parameter vector. For quadrics, sensor data consisting of contact location and local surface normal can be formulated as a linear combination of the geometric parameters and used as the measurement vector for the Kalman filter [3]. This estimator can easily be extended to superquadrics by employing an extended Kalman filter as described in [2].

The Kalman filter equations also provide an update for the covariance matrix of the estimate. If the covariance matrix is relatively large, the estimate is not yet reliable and the filter equations will place more emphasis on incoming sensor data. As the covariance decreases, the estimate is more reliable, so the incoming, noisy, sensor data are de-emphasized.

The autonomous exploration algorithm used here relies on the fact that the covariance matrix is an explicit mathematical function of the geometric coordinates of the manipulator (the location of contact with the surface and the local surface normal are measured in the inertial coordinate frame.)

To determine a search direction, a scalar cost function derived from the covariance matrix is formulated. Next, the rate of change of this cost function is found. This gives a measure of the time rate of change in the uncertainty of the estimate. We then compute the spatial (directional) derivative of the result. Examining the directional derivative for extrema, we can pinpoint the direction in which the manipulator should travel in order to maximize the decrease in the uncertainty of the estimate. The resulting search direction will cause the estimate to rapidly converge, thus reducing the exploration time and distance traveled. The mathematical details of the process are fully described in [1].

It has been shown [4] that surface curvature is a valuable invariant property of an object. Mean curvature, Gaussian curvature and principle curvature provide excellent features for qualitatively describing a surface. Therefore, the object's local radius of curvature will be determined during exploration. This will be in addition to estimating the geometric parameters of the surface model.

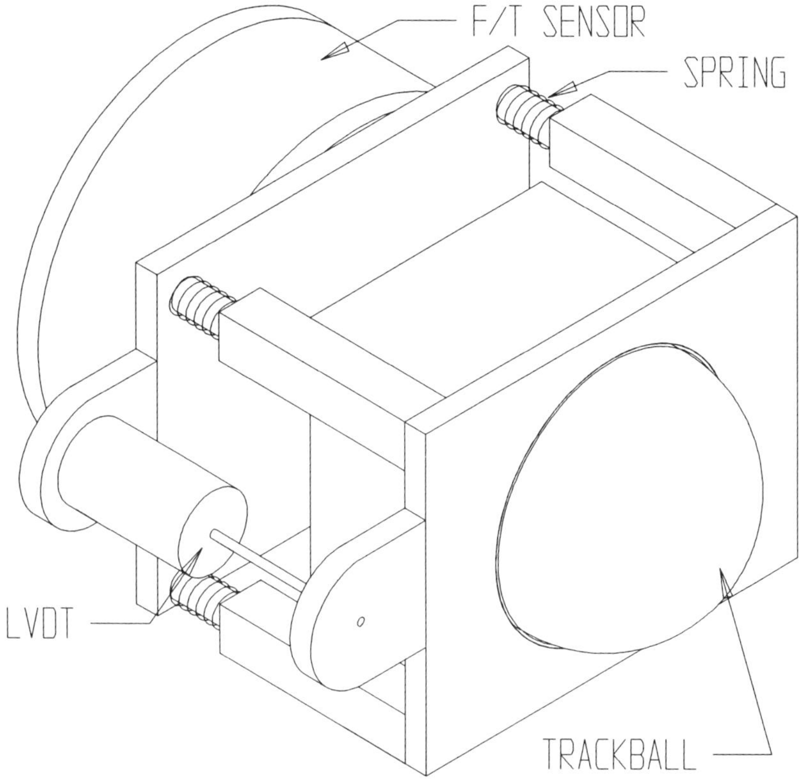

Figure 1. End-effector used for tactile exploration.

EXPERIMENTAL APPARATUS

To test the algorithm, a MERLIN 6540 industrial robot has been instrumented with a wrist mounted force/torque sensor and a specialized end-effector made from an industrial trackball (see figure 1.) The MERLIN is a six degree of freedom robot with a 40 inch reach, 0.001 inch repeatability and 50 pound payload capacity.

The industrial trackball serves two important purposes. First, it provides a near-frictionless contact between the manipulator and the surface being explored. This eliminates tangential forces. As a result, the only force measured by the force/torque sensor will be along the surface normal.

The second purpose of the trackball is to measure the surface's local radius of curvature. Optical encoders mounted inside the trackball housing measure the rotation of the ball. This can be used in conjunction with the known radius of the trackball to find the traversed arc length on its surface. Since the arc length on the trackball is equal to the arc length traversed on the surface, its local radius of curvature can be determined by:

$$\rho = \frac{d\ell}{d\theta} \tag{2}$$

where $d\ell$ is the incremental arc length traversed on the surface and $d\theta$ is the incremental angular change in the surface normal.

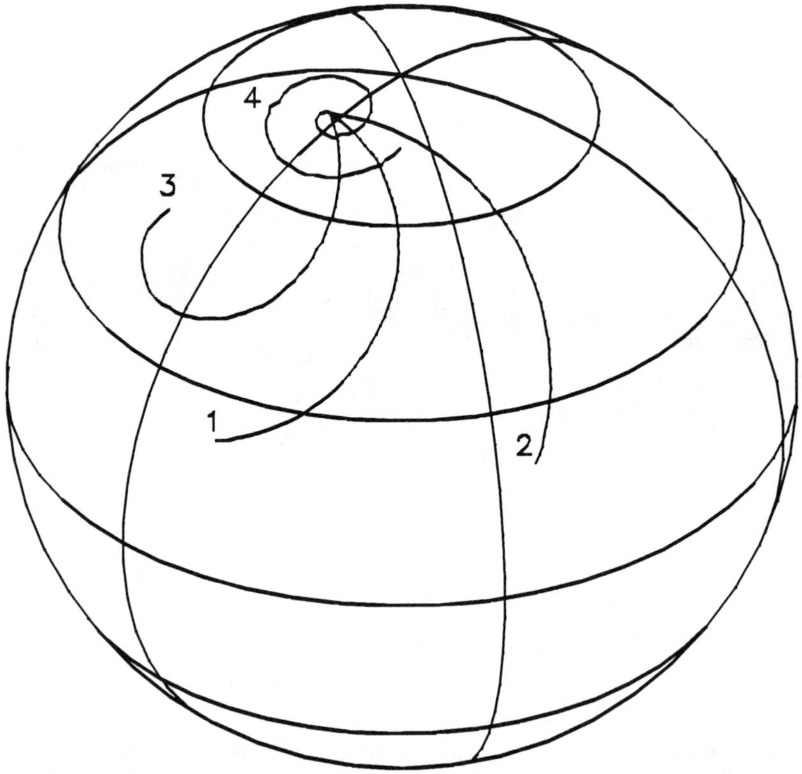

Figure 2. Exploration paths on plexiglass sphere. Path 1 is optimal. Paths 2 and 3 are 30° to the left and right of optimal and path 4 is an outward spiral.

The housing of the trackball is mounted on a sliding mechanism that is attached to the force/torque sensor. The mechanism consists of four posts attached to the trackball housing that slide through brass bushings imbedded in the force/torque sensor's mounting plate. Springs on each post allow for a compliant contact with the surface and enables a robust force controller to be implemented with a stepper motor robot. A fuzzy logic force controller similar to the one described in [5] was used to regulate the contact force. A linear variable displacement transducer mounted between the force/torque sensor and the trackball housing measures the displacement of the springs. This distance is incorporated into the kinematics of the robot and provides an accurate measurement of the contact location.

RESULTS

Experiments were performed to determine the effectiveness of the estimator and the autonomous exploration algorithm. These experiments were conducted on a plexiglass sphere with a 9 inch radius. To initiate shape estimation, the robot was manually positioned on the sphere. Four paths were then used to explore it (the first 50 steps of each path are shown in figure 2.) The first path was the optimal trajectory generated by the autonomous exploration algorithm. The second and third paths were shifted to the left and right of the optimal trajectory by ± 30 degrees and the fourth path was a preprogrammed outward spiral.

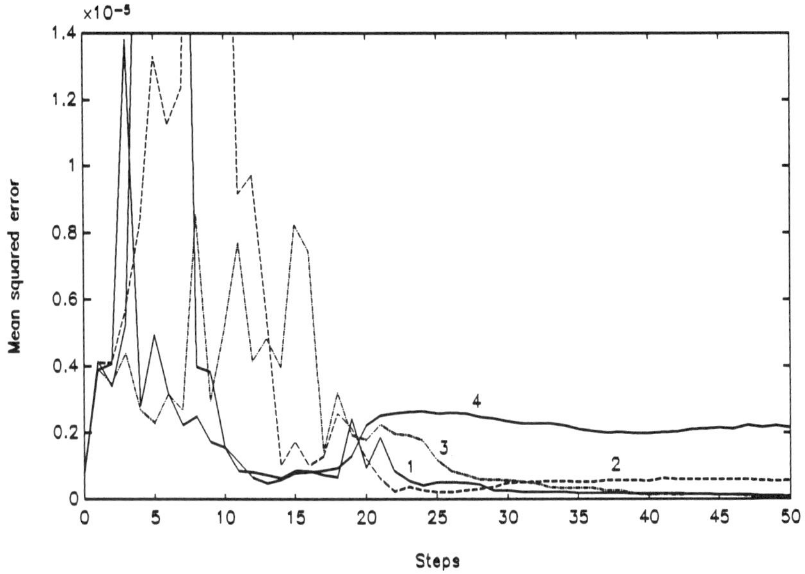

Figure 3. Mean squared error for the parameters related to the length of the estimated ellipsoids' axes. The four curves correspond to the four exploration trajectories.

The spheres were modeled as quadrics (ellipsoids). The mean squared error of the estimated geometric parameters representing the length of the ellipsoid's three axes are shown in figure 3. The optimal trajectory results in the fastest converging estimate, followed by the two trajectories shifted to the left and right of the optimal. The preprogrammed spiral trajectory is clearly suboptimal.

Unfortunately, the estimates had not yet converged after 50 steps. The first three exploration paths had to be terminated at this point because the configuration of the manipulator caused the wrist to over flex. However, significantly more data points were collected with the spiral trajectory. This allowed the estimate to converge. After 383 steps the estimator reported that the lengths of the estimated ellipsoid's axes were 9.5826, 9.3324 and 8.4524 inches. The true lengths were 9 inches each. Further experiments indicate that the first three estimates would have converged if the trajectories could have been extended.

When determining the local radius of curvature, the surface measurements were filtered by averaging the current and previous measurements. The incremental change in arc length and incremental change in surface normal were computed from the difference between the current, averaged, surface measurement and the averaged measurement occurring two steps earlier. The local radius of curvature was then computed with equation (2). The radius of curvature for the four paths on the sphere are summarized in table I.

TABLE I. Radius of curvature for four paths on 9 inch sphere.

Path	Average (inches)	Variance ($inches^2$)
1	9.3559	0.9222
2	9.4645	1.3090
3	8.4654	2.7270
4	8.6115	3.3349

CONCLUSIONS

A robot instrumented with simple and inexpensive sensors, such as a force/torque sensor and a trackball, has been successfully taught to explore and estimate objects. In particular, an autonomous exploration algorithm has been developed to optimize the rate of convergence of a geometric estimate of the object. Experimental results have verified the optimality of this exploration algorithm and its effectiveness at producing an accurate estimate. The object's radius of curvature has also been determined from the ratio of the incremental arc length traversed on its surface and the incremental change in surface normal.

These experimental results demonstrate that tactile exploration is a viable alternative to computer vision for estimating the geometric shape of an object.

REFERENCES

[1] Bay, J. S., "A fully autonomous active sensor-based exploration concept for shape sensing robots," *IEEE Transactions on Systems, Man, and Cybernetics*, vol. 21, no. 4, pp. 850-860, July/August 1991.

[2] Bay, J. S., "An estimator based exploration engine for autonomous learning systems," *Proc. IEEE International Conference on Systems, Man, and Cybernetics*, Charlottesville, VA, October 1991.

[3] Bay, J. S., "Tactile shape sensing via single- and multi-fingered hands," *Proc. IEEE International Conference on Robotics and Automation*, pp. 290-295, Scottsdale, AZ, 1990.

[4] Besl, P. and R. Jain, "Intrinsic and extrinsic surface characteristics," *Proc. IEEE Conference on Computer Vision and Pattern Recognition*, pp. 226-233, San Francisco, CA, 1985.

[5] Hollinger, J. G., R. A. Bergstrom, and J. S. Bay, "A fuzzy logic force controller for a stepper motor robot," *Proc. IEEE International Symposium on Intelligent Control*, Glasgow, Scotland, August 1992.

[6] Menq, C. H., H. T. Yau, and G. Y. Lai, "Automated precision measurement of surface profile in CAD-directed inspection," *IEEE Transactions on Robotics and Automation*, vol. 8, no. 2, pp. 268-278, April, 1992.

[7] Sclaroff, S. and A. Pentland, "Closed-form solutions for physically-based shape modeling and recognition," *Proc. IEEE Conference on Computer Vision and Pattern Recognition*, pp. 238-243, Lahaina, Maui, Hawaii, June 1991.

An Integrated Vision System Target Seeker

Rafael M. Iñigo, Begoña C. Arrúe
University of Virginia School of Engineering and Applied Science

ABSTRACT

The system described in this paper is the result of a project whose goal was to detect and track military targets. The system should work in real time for targets varying in size from a considerable portion of the image (multi-pixel case) to very small targets (single pixel case). The possibility of multiple objects in the image was also considered. Only the multi-pixel system is discussed in this paper.

INTRODUCTION

The system described in this paper consists of several subsystems. The first subsystem detects target motion and produces a segmented edge image of the various objects present in the image. Next, the centroid of each segmented object is obtained and the window centered on it. A logarithmic spiral mapping(LSM) is performed on each window. This makes the mapped image invariant to rotation and scaling. The images are classified by means of a higher order neural network with input from the computation plane of the log-spiral map. Classification is performed in parallel for all windows.

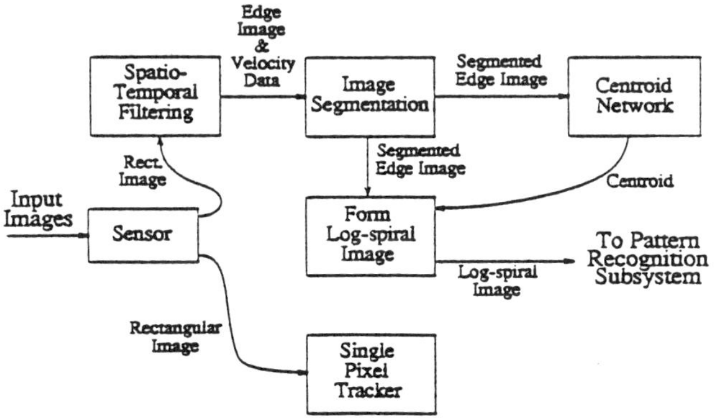

Figure 1. Block diagram of the system

Although the centroid calculator can be used for tracking, a separated *line correlator target tracker (LCTT)* is used as back-up. This completes the *Multi-pixel* branch. The single pixel branch is able to track a single pixel target in a high noise environment. A block diagram of the system is given in Fig.1. Below we describe each multi-pixel block in more detail.

EDGE AND MOTION DETECTION ALGORITHM

A modified spatio-temporal gradient scheme was developed to detect motion. It consists of two stages: spatio-temporal filtering and velocity estimation. Edge and motion information are produced concurrently.

The input image sequence **f(x,y,t)** is filtered by convoluting it with a spatio-temporal Gaussian function **G(x,y,t)** , producing a filtered image **T(x,y,t)** . The second partial derivative of **T** with respect to **s** (directional vector) is equal to the convolution of **f** and the second partial of **G** w.r.t. **s** . From this, a constraint equation in the **s** direction is obtained which provides object velocity and edge. By obtaining this function in at least two directions, the aperture problem is solved. A range of velocities can be detected by using several values of σ in **G** . Good results were obtained in simulations with sequences of synthetic and real images.

A problem with the algorithm is the enormous number of operations required. By means of several improvements such as hierarchical filtering, approximations for the Gaussian derivative filters, etc., the number of multiplications and additions can be reduced from about 1.2 billion to 4 million. Since many of the processes can be carried out in parallel, the latter is well within the capability of state-of the-art digital processors [1].

The algorithm can be implemented by means of a dedicated digital processor or using DSP chips.

SEGMENTATION ALGORITHM

The output of the motion/edge detection subsystem is a binary edge image containing the moving objects present in the image and, possibly, some noise pixels. It is necessary, then, to segment the image, i.e. to separate the objects into individual windows.

Using known digital picture processing techniques, it is determined what pixels are adjacent to each other, by means of connectivity analysis, and also whether these pixels have the same velocity. If both criteria are satisfied, the pixels belong to the same object. A rectangular window is then generated to enclose each object. Situations in which part of an object falls within the window of another are considered by eliminating edges that do not satisfy the connectivity criterion.

Simulations, starting with an image sequence, proceeding through the motion/edge detection algorithm and outputing the segmented windows have been very satisfactory. The output of this subsystem consists of as many windows as there are objects in the image. The location of these windows with respect to the image coordinates is known from this algorithm.

Implementation by a digital processor is straight forward.

CENTROID CALCULATION

Centroid calculation can be performed by standard DPP techniques. However, this is a part of the system that can be designed and implemented with artificial neural networks [2]. This will reduce processing time by orders of magnitude.

Centroid determination with analog networks requires a two step process:
a) Convolution of the data set **f(x,y)** with a filter set **h(x,y)** ,

$$z(x,y) = f(x,y)*h(x,y)$$

b) Peak detection of the resulting operation.

Due to convolution properties, the centroid of **z** is the sum of the centroid of **f** and the centroid of **h** . By proper choice of h(x,y) , the centroid of f(x,y) occurs at the peak value of **z** .

It is very easy to implement convolution by means of a network of operational amplifiers(op-amps) and resistors, with inputs from the sensor. The output of this network is the input to a Hopfield-Tank neural net peak detector. The output of the HT net is the centroid location. The peak detector network is also implementable with op-amps and resistors.

The complete centroid calculation subsystem has been simulated both in software and using SPICE with excellent results. A small, 5x5 network was also implemented in hardware and it performed as expected.

LOGARITHMIC SPIRAL MAPPING

The output of the *Centroid Calculation Network* , i.e. windowed binary edge images of each object centered on the centroid, is mapped by means of a LSM in order to take advantages of the properties of this mapping for object classification.

It is necessary, first, to transform the rectangular homogeneous tessellation to an exponential one, i.e. to rings of pixels with arc of ring shape and a ring radius that increases exponentially. For simulations this has been done in software, but there are two possibilities for hardware implementation: one uses dedicated digital circuitry, the other artificial neural nets. Once the image has been transformed to exponential tessellation, it is mapped to a computation plane by means of the transformation $w = \ln(z)$, where $z = x + jy$ and $w = u + jv$. Scalings about the optical axis (centroid) in image plane are mapped to shifts along the u axis in computation plane. Rotations about the optical axis are mapped to shifts along the v axis [3]. Using this mapping it is possible to classify images of an object with different sizes and orientations.

Notice that starting in block III (Centroid calculation), all the operations can be performed in parallel for all the windows.

MODIFY HIGH ORDER NEURAL NETWORK (MHONN)

As stated before, the centroid calculation eliminates object translation in the original image; the LSM converts the object rotation in the original image into a horizontal cyclic shift (CS) and the object scaling into a vertical ordinary shift (OS) in the log-spiral mapped image. Therefore, the task of the MHONN is to achieve pattern recognition independent of these two types of shifts.

Reid [4] proposed a high order neural network (HONN) with only the second-order terms to achieve ordinary translational invariant pattern recognition. The original second-order neural network developed by Reid consists of only two layers of neurons. Reid used one layer of adjustable weights and directly implemented translational invariance constraints by setting certain weight values to be the same. In his architecture, Reid only utilized these constraints to reduce the independent weights number, therefore speeded up the training process, but the actual number of weights was not reduced by the invariance constraints. The advantage of this setting is that the network is very simple. However, the disadvantage is the huge memory requirement for weight storage, or the equivalent number of weights for hardware implementation.

In our system, even though the MHONN has to handle not only the OS in the vertical direction, but also the CS in the horizontal direction, the nature of the problem is still of second order (i.e., it is necessary and sufficient for the classifier to utilize the correlation information between each pair of input signals to achieve the desired recognition). So, a second-order neural network is used as the pattern recognizer. However, we made several major modifications to the original second-order network to meet our special needs as well as to improve the network performance. The term **feature** is defined as a **vector** connecting two pixels in the input plane. The length and orientation of this vector specifies a unique spatial separation between a pair of pixels.

For our system, the input of the HONN is binary, so AND gates instead of multipliers are used to extract the correlation information between each pair of the input pixels. The first modification consists in adding a layer of summers in-between the AND layer and the output layer. The significance of this modification is that it reduces the number of weights by orders of magnitude, which means that the memory reduction in software implementation, and the number of components in hardware implementation is reduced proportionally. It must be noted that simulations were also performed with grey level images using "software multipliers".

In the modified architecture, the output signals of the AND gates corresponding to each unique type feature are first summed together, and then, the summed signal is fed into **one** adaptable weight.

The above seemingly simple modification can save a tremendous amount of memory space or hardware components. The analytical derivation is as follows: the general second-order neuron transfer function is given in Eqn (1):

$$y_i = \Theta\left[\sum_{j=1}^{N}\sum_{k=j+1}^{N} w^i{}_{jk}\, x_j\, x_k\right] \tag{1}$$

where x_j stands for the jth pixel located at row r_j and column c_j. After applying the translational invariant constraints onto the weight matrix $W_i = \left[w^i{}_{jk} \right]$, the weights $w^i{}_{jk}$'s will group into classes. Weights in each class will have the same value, no matter what the value might be. The above modification can be represented by rearranging the summations in the above equation according to the classes:

$$y_i = \Theta \left[\sum_{j=1}^{N} \sum_{k=j+1}^{N} w^i{}_{jk}\, x_j\, x_k \right] = \Theta \left[\sum_{m,n} w^i{}_{mn} \sum_{v\,(m,n)} x_j\, x_k \right] \tag{2}$$

where $m = r_k - r_j$; $n = c_k - c_j$; (r_k, c_k) represents the row and column location of pixel x_k; $v(m,n)$ is a set of ordered pairs $(x_j\,,\, x_k)$ in which the vector from x_j to x_k belongs to the same type of feature specified by $(\, m\,,\, n\,)$.

After this modification, the *delta learning rule* is used to train the network.

For an N × N input image, the number of weights in the original HONN is $C_{N^2}{}^{2} = O\,(\,N^4\,)$, but after the modification, the number of weights in the MHONN is reduce to $2\,(\,N^2 - N\,)$, and the trade-off is only to add $2\,(\,N^2 - N\,)$ summers. This modification reduces not only the memory requirement by a factor of $O(N^2)$, but also the complexity of the simulated HONN, and, of course, the number of components in hardware implementation.

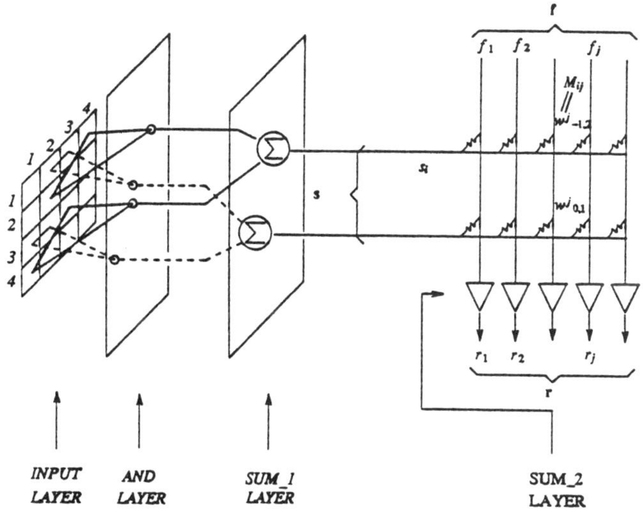

Figure 2. MHONN with associative recall learning mechanism

After feature detection by means of the modified HONN, classification is performed by means of an associative memory. Associative memories are relatively easy to implement in hardware if the number of patterns to be recognized is not too large. In our case we have a limited number of patterns (several views of an ICBM), hence an AM seems to be the best solution. Simulation results were very satisfactory, including displaced images (in image space) and noisy images. A diagram of the complete HONN-AM subsystem is given in Fig.2 .

TRACKING

Once an object has been classifed by the MHONN, its window is updated frame by frame by means of the centroid calculator which keeps the centroid at the center of the window. This provides information on object displacement in the x-y (image) plane, i.e. it gives the Δx and Δy values. However, centroid location is independent of object size and orientation, consequently, information on scaling and rotation on the optical axis is determined by means of the line correlator target tracker [5]. Δu and Δv values in the computation plane of the LSM correspond, respectively, to changes in scale and in orientation angle. Thus, applying the LCTT to the LSM computation plane of the chosen window, will provide the desired information.

The information provided by the centroid calculator and the LCTT becomes unreliable when parts of the object fall outside the image. Fortunately this will occur only when the sensor is very close to the target, at what point impact is imminent.

CONCLUSION

The IVSTS described in this paper was simulated as a whole. Its performance, using real ICBM images was very satisfactory under a variaty of situations. Parts of the system can be implemented using neural networks. Analog hardware implementation of these parts is feasible.

REFERENCES

1. Hsing, C., Iñigo, RM., and McVey, E.S., "Image Motion Detection and Estimation", *Joint SPIE-IECON Conf.*, Philadelphia, PA, November 5, 1989.
2. Himes, G.S., Iñigo, RM., "Centroid Calculation Using Neural Networks", *Journal of Electronic Imaging*, Vol. 1, No. 1, January 1992.
3. Sigda, J., "Log-Spiral Sensor Study", Quarterly report to USAF-SDC, Contract # DASG60-88-C-0030, September 1990.
4. Reid, M.B., Spirkovska, S., and Ochoa, E., "Simultaneous Position, Scale, and Rotation Invariant Pattern Classification Using Third-Order Neural Networks", *The International Journal of Neural Networks - Research and Applications*, Vol. 1, No. 3, pp. 154-159, July 1989.
5. Narathong, C., and Iñigo, RM., "Neuron Network Target Tracker", *SPIE Conference on Applications of Artificial Neural Networks Proceedings*, Orlando, Florida, April 18-20, 1990.

ACKNOWLEDGMENTS: The research on which this paper is based was supported, in part, by USAF-SDC, Contract # DASG60-88-C-0030.

REAL-TIME HOMOMORPHIC FILTERING OF HDTV IMAGES

H. J. Kaufman and M. A. Sid-Ahmed
University of Windsor
Dept. of Electrical Engineering
Windsor, Ontario, Canada, N9B 3P4

ABSTRACT

This paper describes the hardware implementation of a 2-D homomorphic filter which is capable of operating at the high real-time rates that are encountered in high-definition television (HDTV). Log pre-processing and antilog postprocessing circuits are utilized in conjunction with a high-frequency emphasis 2-D IIR filter to provide homomorphic filtering.

INTRODUCTION

An image $f(x,y)$ is defined as a two-dimensional light intensity function of spatial coordinates (x,y). In the illumination-reflectance model of an image, the illumination component $i(x,y)$ due to the amount of source light incident on the scene being viewed and the reflectance component $r(x,y)$ due to light reflected by objects in the scene are considered to vary independently, so that the image is formed multiplicatively, $f(x,y)$ being expressed by

$$f(x,y)=i(x,y)\ r(x,y)$$

Illumination is directly responsible for the dynamic range achieved by the

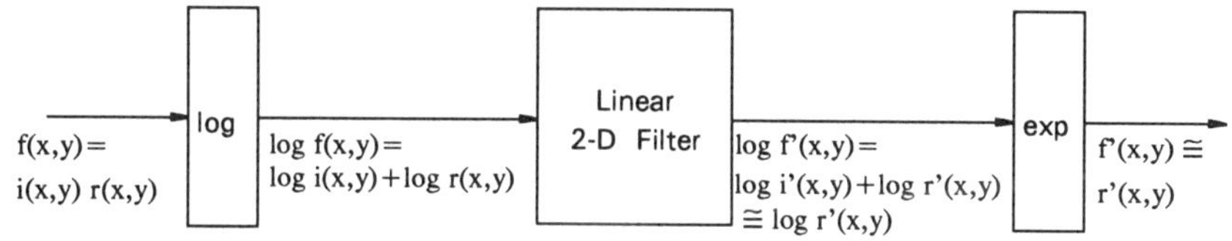

Fig. 1. Homomorphic Filter

pixels of an image while contrast is a function of the reflectance.
By means of homomorphic filtering, [2], as depicted in the block diagram of Fig. 1, it is possible to process the illumination and reflectance components of an image separately. In this scheme, input data is logarithmically converted prior to performing a linear filtering operation (usually high pass filtering) and the result is converted back to the linear scale. In the present work the linear operation selected is a 2-D high emphasis Butterworth (linear phase) recursive filter of order 2 x 2. To a first approximation we can assume as in [2] that the illumation component contains mostly low frequencies in its logarithm and that the logarithm of the reflectance component contains mostly high frequencies. The effect of the 2-D high emphasis filter will be to attenuate the low frequencies and amplify the high frequencies. This results in an enhancement of the reflectance components, which represent the objects in the scene, and an attenuation of the illumination component, which represents the light distribution. The net result is a clearer image.

REAL-TIME HARDWARE PROTOTYPE

A prototype of a 2-D real-time homomorphic filter based on the proposed method of realization given in [1] was constructed using off-the-shelf components.

Log and Antilog Circuits

Circuitry for log and antilog converters was developed based on the temperature compensated transdiode configuration [4]. Suitable high-speed op-amps and transistors (LM 318, 2N2920) are used for operation at video frequency. The input signal is assumed to be an analog baseband video signal containing only the luminance component.

2-D Semi-systolic Filter

In the hardware prototype of the homomorphic filter, the linear filter, ie. the middle block of Fig. 1, was realized as a 2-D semi-systolic filter [1]. This realization as originally proposed requires line delays of differing lengths. For practical implementation using standard commercially available components, this structure was modified to use two types of PE instead of only one type throughout, as shown in Fig. 2.

Line delays are based on CCD's, each storing one line of samples. These may be implemented in hardware as is done in [3].

The prototype filter is of order 2 x 2. Although a higher order structure could be used, the 2 x 2 structure provides an adequate approximation to the filter specifications in the case of a recursive filter. A 2 x 2 structure is economical in hardware and, as is apparent from Fig. 2, only three banks of PE's, are required, each bank containing 3 PE's and only four line delays. Type I PE's

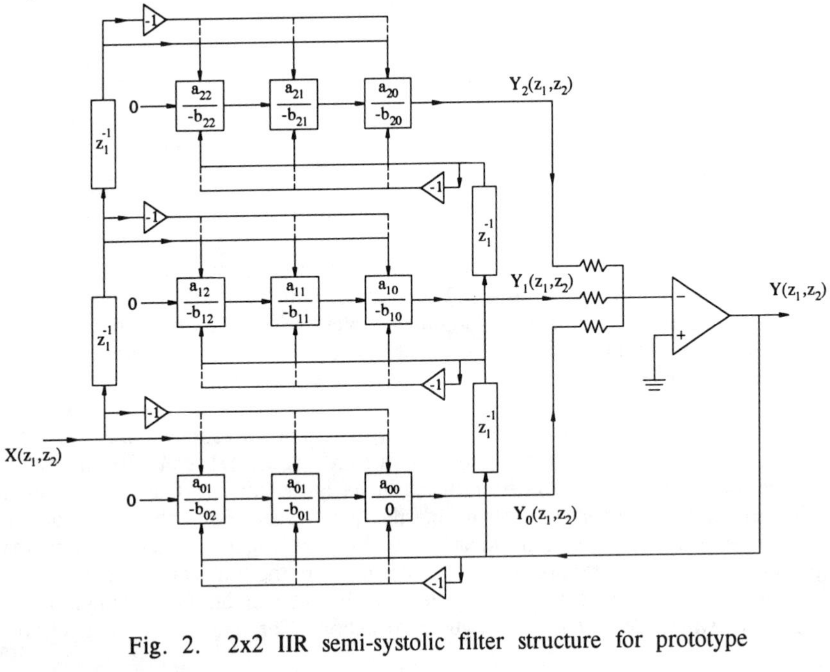

Fig. 2. 2x2 IIR semi-systolic filter structure for prototype

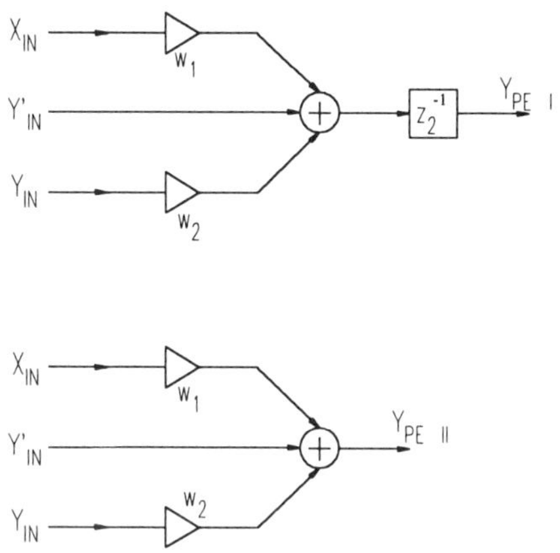

Fig. 3. (a) Type I PE
(b) Type II PE

are used everywhere except at the end of each bank of PE's, where a type II PE is used. Both PE types are shown in Fig. 3. Note that the type II PE is a modified version of the type I PE in which the pixel delay element denoted z_2^{-1} is excluded. Having two types of PE in the structure allows the use of standard length line delays. Since the end of stage PE's do not contain storage elements, there is no longer a latency of one pixel time in processing, so that the feedback connections to the lowermost bank of PE's (ie. the bank whose PE's contain the local variables b_{01}, b_{02}) are different.

Another change which was made for greater ease of construction with off-the-shelf components is the introduction of dual broadcast data lines, allowing either non-inverted or inverted data to be sent to each PE. (Note that in the literature on systolic structures, lines through which PE's receive the same data in parallel are called broadcast lines.) In this systolic structure the local variables of the PE's are the filter coefficients, that is, the set of numbers $\{\{a_{ij}\}\ \{b_{ij}\}\}$ in the transfer function $H(z_1,z_2)$ that approximate the desired filter response as determined by a design procedure. As will be seen when the actual hardware for each PE is described the dual broadcast lines provide a convenient method for assigning the correct sign to the data reaching each PE assuming that local variables are set within each PE without regard to sign..

Type I and type II PE's, as defined in Fig. 3, were constructed using high-speed op-amp circuits. This was done for the type I PE as shown in Fig. 4. The op-amp is configured as an inverting weighted summer and drives a passive delay line with resistive termination. The circuitry for a type II PE is the same as for a type I PE except that it consists solely of the inverted summer op-amp circuit, the delay line (with resistive termination) not being included.

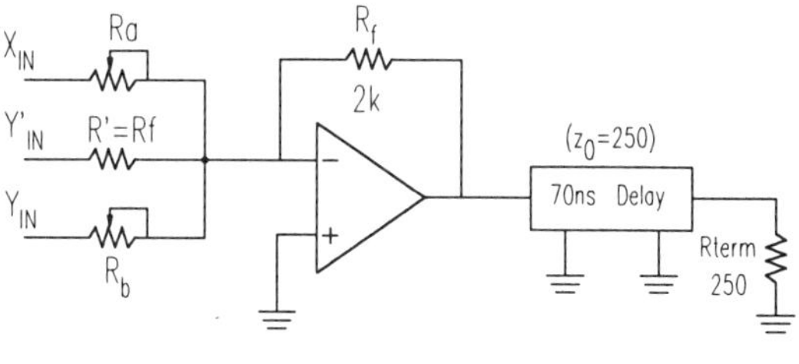

Fig. 4. Op-amp circuit for PE (type I shown)

With reference to Fig. 4 the summing amplifier configuration produces the weighted sum $(-R_f/R_a)X_{IN} + (-R_f/R_a)Y_{IN} + Y'_{IN}$. Comparing with Fig. 3, $w_1=(-R_f/Ra)$ and $w_2=(-R_f/R_b)$. For greater ease in setting coefficient values in the experimental prototype, R_a and R_b are variable resistors. Given that an inverting rather than non-inverting sum is obtained with the PE hardware and that the actual filter coefficients, which the weights represent, are signed numbers, the correct signs of weights in the PE can be established by appropriate connections to either the inverting or non-inverting broadcast data lines (Fig. 2).

The op-amp used in the PE's of the prototype is an EL2020C (Elantec). This is a high-speed monolithic (50MHz) current feedback type op-amp [5], specifically designed for video and other applications that require high slew rates, fast settling times and low power consumption. In contrast with conventional op-amps, those such as the EL2020C that use the current feedback topology provide nearly constant bandwidth and settling time over a wide range of closed-loop voltage gains. Higher performance (200 MHz) current feedback op-amps specifically designed for HDTV and other comparable high-speed applications such as the EL400 (Elantec) are available in an 8 pin mini-DIP package that is pinout compatible with the EL2020C. These higher performance op-amps could be used instead of the EL2020C in Fig. 2 for HDTV application.

A passive delay line is used to implement a pixel delay (denoted z_2^{-1} in the z-domain). In the time domain, a signal $x(m,n)$ is delayed by one pixel time resulting in a delayed version of the signal $x(m,n-1)$. The passive delay line used in the PE's of the prototype is a type 1514-70D (Data Delay Devices). Lumped constant delay lines of this type are widely used in a number of electronic applications. For instance, in color TV receivers the luminance signal is delayed to synchronize it with the chrominance signal. They are generally designed using low-pass LC filters as basic units cascaded in stages having linear phase characteristics overall. The type 1514-70D delay line has a delay time T_d=70ns and a characteristic impedance z_0=250Ω. Accordingly the delay is terminated in a 250Ω resistor. This delay line is available in a four pin SIP package that does not take up much circuit board space. Lumped constant delay lines with a clock input and sample-and-hold output are available in 8 pin DIP packages (eg. Reticon 5100 series) and can be used instead if desired.

REFERENCES

1. M.A. Sid-Ahmed, "Realization of 2-D IIR filters using sample-and-hold circuitry", *1991 IEEE International Symposium on Circuits and Systems*, Vol. 5, pp. 2467-2470, June 1991.

2. A.V. Oppenheim, R.W. Schafer and T.G. Stockham Jr., "Non-linear filtering of multiplied and convolved signals", *Proc. IEEE*, Vol. 56, No. 8, pp. 1264-1291, August 1968..

3. H.J. Kaufman and M.A. Sid-Ahmed, "2-D analog filters for real-time video signal processing", *IEEE Trans. on Consumer Electronics*, Vol. 36, No. 2 pp. 137-141, May 1990.

4. J.K. Roberge, "*Operational Amplifiers: Theory and Practice*", Wiley, 1975.

5. J.A. Dutra, "Current feedback op amps raise gain-bandwidth product", *Electronic Design* , July 14, 1988.

6. K. Endoh and M Lembke, "Apply delay lines in digital circuits", *Electronic Design,* pp. 85-88, June 22, 1989.

INSPECTION OF SURFACE TEXTURE OF MACHINED PARTS USING OPTICAL FOURIER PROCESSING

V.M. Huynh and S. Kurada
Department of Mechanical Engineering,
University of Windsor,
Windsor, Ont., N9B 3P4
Canada.

ABSTRACT

A two-dimensional optical Fourier transform technique based on a machine vision system was developed to analyze the surface texture of machined surfaces. Samples from different machining methods, namely, flat lapping, grinding, reaming and turning, were used in this study. A number of parameters, such as spectrum peak, RMS value and mean frequency, derived from the spectra, were used to characterize the surface. Correlation curves were obtained to establish the relationship between these parameters and those derived from the stylus method.

1. INTRODUCTION

Surface texture is one of the most important design specifications for machined parts as it influences the function and performance of the assembly. Accurate characterization of surface texture is vital in the design and production of mechanical parts since the texture influences corrosion resistance, uniformity in subsequent painting and plating operations and lubrication performance. Accurate determination of the surface texture and its functional properties is also of significant economic importance in production industries. This is because the part cost increases if the finished product either fails to meet the requirements or is over specified because of improper characterization or inaccurate measurement.

According to the American National Standards Institute [1], surface texture is defined as the repetitive or random deviations from the nominal surface which form the three-dimensional topography of the surface. It includes roughness, waviness, lay and flaws.

Various optical techniques have been devised to provide surface texture information and have significant advantages over the mechanical stylus method which is considered as a standard for roughness measurement. These techniques include laser speckle, interferometry, light scattering and the optical Fourier transform (OFT). Among these, the OFT technique is a versatile one for studying manufactured surfaces as it provides two dimensional texture information.

The OFT was first introduced by Pernick [2] to study the surface texture of optical surfaces which are at the low end of the roughness scale. This technique was later used by Thwaite [3] to apply to machined metal surfaces which have a larger range of roughness up to 0.6 μm. Huynh *et al.*[4] extended this technique to study the ground surfaces of different materials having roughness up to 1.2 μm. The objective of this work is to apply this technique to different surface textures which are created by different manufacturing processes such as flat lapping, grinding, milling, reaming and turning.

2. METHOD AND PROCEDURE

The scattering of light from a micro-rough surface which is illuminated by a monochromatic light source can be explained and predicted by the diffraction phenomena of light waves. The far field distribution of the scattering is the Fraunhofer diffraction pattern and is approximated by the Fourier transform of the phase of the surface roughness height function[5]. A lens or an optical system can be used to bring the pattern to the focal plane (Fourier plane) of the camera so that the image can be captured and processed. For roughness amplitudes in the order of a fraction of the illumination wavelength, the power spectrum can be directly related to the roughness. However, for larger roughness, this linear relationship does not hold. The objective of this work is to establish experimentally the relationship between the power spectrum of the diffraction pattern and the texture of surfaces of different manufacturing processes.

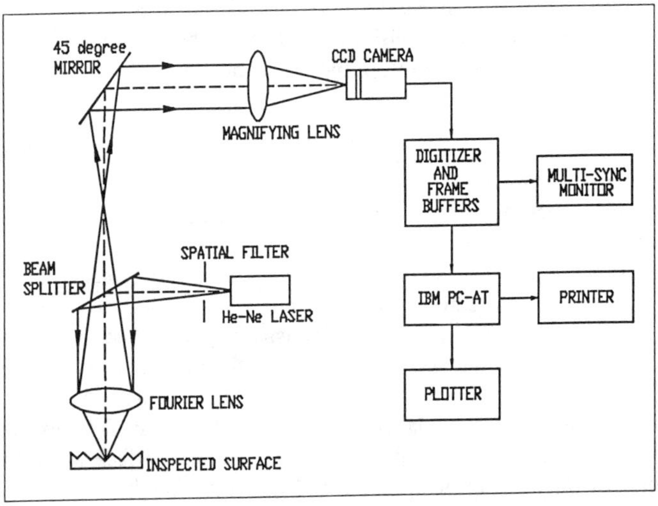

FIGURE 1: A block diagram of the experimental set-up

A schematic diagram of the set-up used in this work is shown in Figure 1. The system consists of a 2 mW He-Ne laser light source (with a wavelength = 0.632 μm) used in conjunction with a beam splitter to illuminate the surface through a 50 μm spatial filter. A transform lens is of 25 mm focal length and 0.25 numerical aperture. The second lens, having a 100 mm focal length, is for magnifying the image of the pattern onto the image plane of the CCD camera which has 480x525 pixel resolution. The analog signal from the camera is then transferred to a frame grabber which digitizes the image into 512x512x8 bit pixels. The digitized image is stored in one of the four frame buffers available on the frame grabber board for subsequent analysis by the IBM PC-AT. The image of the pattern is displayed on a multi-sync video monitor for viewing.

To obtain the Fourier spectrum, the optical pattern is digitized along its principal axes to derive the distribution of intensity of the pattern. The axial distance in the optical pattern is then transformed to a spatial frequency value via the relationships derived from the setup configuration and are given in Reference[4].

3. RESULTS AND DISCUSSION

a) Spectrum Analysis.

The characteristics of a typical flat lapping sample are shown in Figure 2. The machining marks extend in two directions perpendicular to each other (Figure 2a).

a)

b)

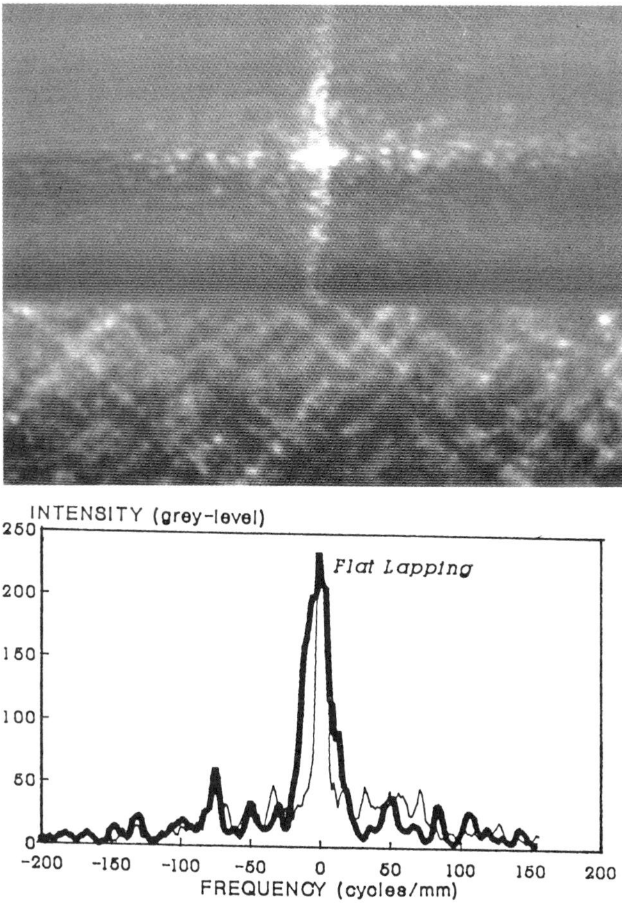

FIGURE 2: Characteristics of a flat lapping sample (Ra = 0.5 μm).a) Images of the surface and its Fourier pattern, b) The optical Fourier spectra.

The resulting Fourier pattern has the shape of a cross. The illumination plane in this case is parallel to one of the directions of the lay. The optical Fourier spectra were obtained for the two principal directions as shown in Figure 2(b). By examining the pattern, one can say that the surface roughness is close to a random surface even though there are some small discrete peaks in the mid frequency range of the spectrum.

b) Correlation of Spectra obtained by the two Methods

To assess the characteristics of the OFT, the spectra obtained by the optical method were correlated with those obtained from the mechanical stylus method in terms of spectrum peak, RMS values and mean frequencies.

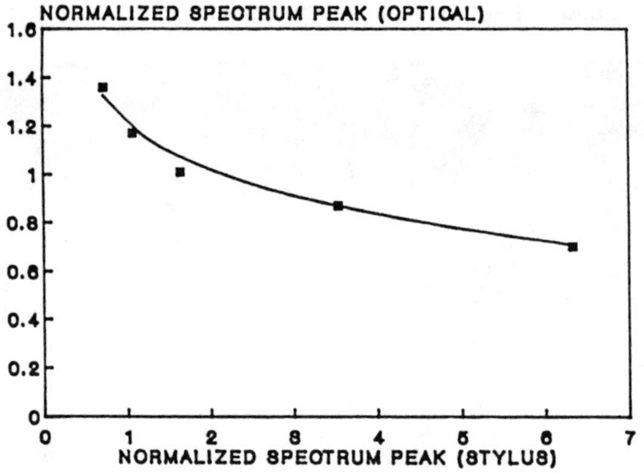

FIGURE 3: Correlation between the optical spectrum peak and that obtained by the stylus method for surface-ground samples.

There is a power-law relationship between the spectrum peaks obtained from the two methods for surface-ground samples as shown in Figure 3. Note that the data are normalized with respect to the standard calibrated sample having Ra = 0.5 μm. A similar relationship can be established between the spectrum RMS values obtained from the two methods for these ground samples.

The mean frequencies for samples from different machining processes were calculated using the optical spectra. These were then compared to those obtained by the stylus method. In general, an accuracy of better than 10% was observed. However, greater deviations (maximum 21%) were found for very rough surfaces (greater than 0.4 μm). This may due to the error in using the small sampling area which may not represent the characteristic of very rough surfaces.

c) Surface Roughness Measurement from the Optical Spectra

Since the peak and RMS of the OFT spectrum can be related to those obtained by the stylus method, one can use either one of these parameters to characterize the surface roughness. To achieve this, the spectrum peak derived from the optical Fourier spectrum was arbitrarily chosen to represent the optical surface roughness.

The peak values of surface-ground samples having different roughness were plotted against the average surface roughness Ra obtained from the stylus instrument as shown in Figure 4. Again, a power-law relationship exists between the two parameters. Similar data for different machining processes were obtained and plotted as shown in Figure 5. It is observed that there is a general trend for the calibration curves of all the processes, however, the scatter in the data is large. Therefore, it can be said that no unique curve for all surfaces of different textures.

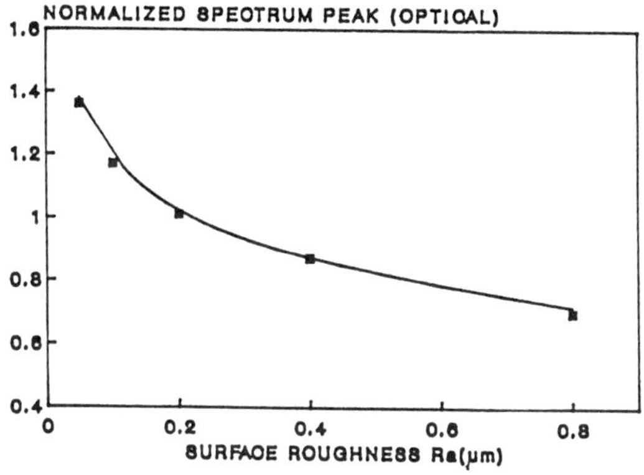

FIGURE 4: Correlation between optical spectrum peak and surface roughness Ra for surface-ground samples

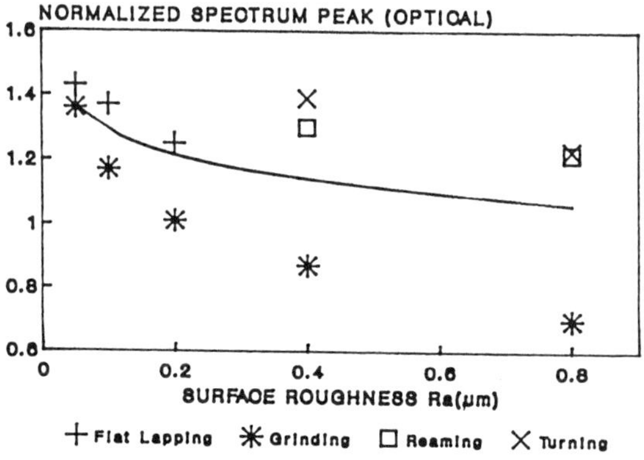

FIGURE 5 Correlation between optical spectrum peak and surface roughness Ra for samples machined by different processes.

4. CONCLUSIONS

A two-dimensional optical Fourier transform technique was developed for analyzing the surface texture of machined parts. Optical Fourier spectra were obtained for flat lapped, ground, reamed and turned samples. These were then correlated with the spectra obtained from the mechanical stylus method. Good correlation was found for the RMS value, the spectrum peak and the mean frequency between the two methods. Optical roughness calibration curves were obtained for samples produced by different machining processes.

The proposed optical method provides a convenient way to assess the surface texture of parts manufactured from different machining processes in a production environment. The texture information so obtained can be used not only for quality control but also for monitoring machine tool vibrations and tool wear during machining.

ACKNOWLEDGMENT

This research project is supported by the Natural Science and Engineering Research Council of Canada under grant number A5286.

REFERENCES

1. "Surface Texture", *American National Standards Institution,* (ANSI B46.1 - 1978, New York, 1978.)

2. Pernick, B.J., "Surface Roughness Measurements with an Optical Fourier Spectrum Analyzer", *Applied Optics,* **18,** 796, (1979.)

3. Thwaite, E.G., " A Quantitative Comparison of the Wavelength Spectrum of a Surface obtained by Optical Fourier Transformation with calculations from Profile Measurements", *Wear,* **83**, 181-187, (1982).

4. Huynh, V.M., Kurada, S. & North, W., "Texture Analysis of Rough Surfaces using Optical Fourier Transform", *Journal of Measurement Science and Technology,* **2,** 831-837, (1991).

5. Allardyce, K.L., "Diffraction of Rough Reflective Surfaces", *Applied Optics,* **26**, 2364-2375, (1987).

COMPUTATIONALLY EFFICIENT NONLINEAR PREFILTERING TECHNIQUE FOR EDGE DETECTION

Sung-Jea Ko

Department of Electrical and Computer Engineering
The University of Michigan
Dearborn, MI 48128

ABSTRACT

In this paper, we investigate the efficacy of the selective average (SA) prefiltering on the subsequent detection of edges in digital images. It is shown that the SA filter enhances blurred edges while attenuating noise. An efficient implementation method of the SA filter is introduced. Experimental results indicate that the SA filter is a desirable alternative to other edge enhancing filters which are currently used as the prefiltering techniques.

I. INTRODUCTION

In image processing applications, preserving edges in the image while filtering it is often important. In some situations, however, it is desirable to enhance the gradients or slopes of edges rather than to simply preserve them. The classical technique that counteracts blur without knowledge about the blur is essentially linear high freqency emphasis filtering, and thus it is very sensitive to noise [1],[2].

In this paper, a new class of edge enhancing filters, the *selective average* (SA) filter, is proposed and analyzed as an alternative to the high frequency emphasis filters. The SA filter consists of several subfilters, and at each pixel the input value is replaced by the output of the subfilter which is the closest in value to the input. Two types of subfilters, average and median filters, are employed in the proposed filters. Therefore, the SA filter is very simple to implement and computationally efficient. For the sake of simplicity, only one-dimensional (1-D) SA filters will be analyzed in this paper.

II. SELECTIVE AVERAGE (SA) FILTERS

Before proposing the SA filter, it is worthwhile to review the median filter. Let X_i and Y_i be the input and output of a filter, respectively. The median filter at a point is the median value of the input data inside the window centered at that point.

Definition 1: The output of the median filter with the window size $2N+1$ is defined by

$$Y_i = median\{X_{i-N}, ..., X_i, ..., X_{i+N}\} \tag{1}$$

The median filter preserves edge gradients but cannot enhance them.

Next we define the SA filter which can be effectively used for enhancing edges in images. This filter utilizes three subwindows containing samples to the left, right, and centered about the central sample X_i in the window.

Definition 2: The output of the SA filter with window sizes 2N+1 and P is then defined by

$$Y_i = \begin{cases} A_{i1}, & \text{if } |A_{i1} - M_i| \leq |A_{i2} - M_i| \\ A_{i2}, & \text{otherwise} \end{cases} \tag{2}$$

where $A_{i1} = average\{X_{i-N}, X_{i-N+1}, ..., X_{i-1}\}$, $A_{i2} = average\{X_{i+1}, X_{i+2}, ..., X_{i+N}\}$, and $M_i = median\{X_{i-P}, .., X_i, ..., X_{i+P}\}$.

This filter at each point selects the average which is closer in value to the sample median value M_i. The SA filter is effective in enhancing edges while suppressing noise. Furthermore, a repeated usage of the SA filter tends to convert blurred edges into ideal step edges. For example, Fig. 1 illustrates the performance of the median and SA filters when the input is a noisy blurred edge. Median filtering suppresses noise well, but it fails to enhance the edge. The SA filter, however, produces a step-type edge while reducing the noise reasonably well. A piecewise constant image can be obtained by repeated SA filtering. This property is particularly useful for contour or edge detection [3], [4]. Statistical edge enhancing properties of the SA filter will be discussed in the next section.

The average subfilters of the SA filter can be efficiently implemented using a recursive structure as in [5]. With the recursive averaging structure, the SA filter with $P = 0$ can be realized with one scaling multiplication, four additions, and one comparison at each point *independent* of the window length $2N + 1$.

III. STATISTICAL PROPERTIES

The noise suppression characteristics of SA filters are investigated when a constant signal is embedded in additive white noise. The edge enhancement properties are examined by considering blurred edges which are corrupted by additive white noise. In this section, attention will be placed on SA filters with $P = 0$, mainly because these are mathematically more tractable than SA filters with $P > 0$.

A. Noise Suppression

Property 1: For independently and identically distributed (i.i.d.) inputs, the output density function $g(y)$ of the SA filter with window size $2N + 1$ is given by

$$\begin{aligned} g(y) = 2f_A(y)\{&[1 - F_X(y)]F_A(y) + F_X(y)[1 - F_A(y)] \\ &+ \int_y^\infty f_X(x)[1 - F_A(2x - y)]dx + \int_{-\infty}^y f_X(x)F_A(2x - y)dx\} \end{aligned} \tag{3}$$

where $F_X(.)$ is the input distribution function, and $F_A(.)$ and $f_A(.)$, respectively, are the distribution and density functions of A_{i1} or A_{i2}.

From this density function it can be easily shown that $g(y)$ is symmetric about m, i.e., $g(m + y) = g(m - y)$ if $f(x)$ is symmetric around m. Thus, the SA filter is an unbiased estimator of the mean, and $E[X_i] = E[Y_i] = m$.

Using (3), the output variances of the SA filter were computed through numerical integration for i. i. d. Gaussian inputs with mean zero and variance one ($\eta(0,1)$). The results associated with different window sizes are tabulated in Table 1.

B. Edge Enhancement

We next consider the effects of SA filters in the neighborhood of 1-D noisy blurred edges. The 1-D input sequence representing a noisy blurred edge can be modeled as

$$X_i = \begin{cases} N_i, & i \leq 0 \\ h_i + N_i, & 0 < i \leq d \\ H + N_i, & i > d \end{cases} \tag{4}$$

where $\{N_i\}$ is a zero mean i. i. d. sequence, and $\{h_i\}$ is a deterministic sequence that determines the blurred edge shape. The positive constants H and d parameterize the degree of blurring. Here d is an integer. The $\{X_i\}$ represents an noisy step edge when $d = 0$, and a noisy ramp edge if $\{h_i\}$ is a linear trend with $h_0 = 0$ and $h_{d+1} = H$. The edge in (4) is called a concave/convex edge if $\{h_i\}$ is convex [i.e., $l_{i-1} \leq l_i$] for $1 \leq i \leq T$, and concave [i.e., $l_{i-1} > l_i$] for $T < i \leq d+1$ where T is an integer, $1 \leq T \leq d+1$, and $l_i = h_i - h_{i-1}$.

Property 2: The output density function $g_i(y)$ of the SA filter near the edge in (4) is given by

$$\begin{aligned} g_i(y) = {} & f_{A_{i1}}(y)\{F_{A_{i2}}(y)[1 - F_{X_i}(y)] + F_{X_i}(y)[1 - F_{A_{i2}}(y)] \\ & + \int_y^{\infty} f_{X_i}(x)[1 - F_{A_{i2}}(2x - y)]dx + \int_{-\infty}^{y} f_{X_i}(x)F_{A_{i2}}(2x - y)dx\} \\ & + f_{A_{i2}}(y)\{F_{A_{i1}}(y)[1 - F_{X_i}(y)] + F_{X_i}(y)[1 - F_{A_{i1}}(y)] \\ & + \int_y^{\infty} f_{X_i}(x)[1 - F_{A_{i1}}(2x - y)]dx + \int_{-\infty}^{y} f_{X_i}(x)F_{A_{i1}}(2x - y)dx \end{aligned} \tag{5}$$

where $F_{X_i}(.)$ is the input distribution function, and $F_{A_{i1}}(.)$ and $f_{A_{i1}}(.)$, respectively, are the distribution and density functions of A_{i1} or A_{i2}.

Note that the density functions f_{X_i}, $f_{A_{i1}}$, and $f_{A_{i2}}$ depend on the location of the window in the neighborhood of the edge in (4).

The edge enhancement characteristics of a filter for the noisy blurred edge can be examined using the expected value of the output. This quantity was computed though numerical integration by using the output density function $f_{Y_i}(y)$ in (5) along with the assumption that f_{N_i} is $\eta(0,1)$. The results associated with the noisy ramp and concave/convex edges with $H = 10$ and $d = 4$, respectively, are plotted in Fig. 2. The results indicate that the SA filters are essentially edge enhancing filters.

IV. EXPERIMENTAL RESULTS

Fig. 3(a) shows the Shepp-Logan phantom image with 128 x 128 pixels [6]. This image was blurred by the 5x5 average filter and then corrupted by

Gaussian noise of variance 36. The resulting image is shown in Fig. 3(b). Fig. 4 shows the results of the seperable median and SA filters on the noisy image when $N = 3$. The SA filters enhanced edges in the image. The median filter suppressed the noise well, but could not enhance edges. The filtered images in Fig. 4 were passed through the Sobel operator with threshold 75 and their results are also shown in Fig. 4. It is seen that the edge map for median filtering has lost many details of the original image. The results in this section indicate that the SA filter is an effective edge enhancing prefilter for edge detection. In summary, the SA filter is a desirable alternative to the median filter due to its edge enhancing characteristics and its simplicity in implementation.

REFERENCES

[1] W.K. Pratt, *Digital Image Proc.* 2nd ed. New York: Wiley, 1991.

[2] A. Rosenfeld and A.C. Kak, *Digital Picture Processing*, vol. 1, 2nd ed. New York: Academic, 1982.

[3] K. Minato, Y. -N. Tang, G.W. Bennett, and A.B. Brill, "Automatic contour detection using a fixed-point Hachimura-Kuwahara filter for SPECT attenuation correction," *IEEE Trans. Med. Imaging*, vol. MI-6, no 2, pp. 126-133, June 1987.

[4] Y.H. Lee, and A.T. Fam, "An edge gradient enhancing adaptive order statistic filter," *IEEE Trans. Acoust., Speech, Signal Processing*, vol. ASSP-35, pp. 680-695, May 1987.

[5] A. V. Oppenheim, and R. W. Schafer, *Discrete-Time Signal Processing*, Prentice Hall, 1989.

[6] L. A. Shepp and B. F. Logan, "The Fourier reconstruction of head section," *IEEE Trans. Nuclear Science*, vol. NS-21, pp. 21-43, June 1974.

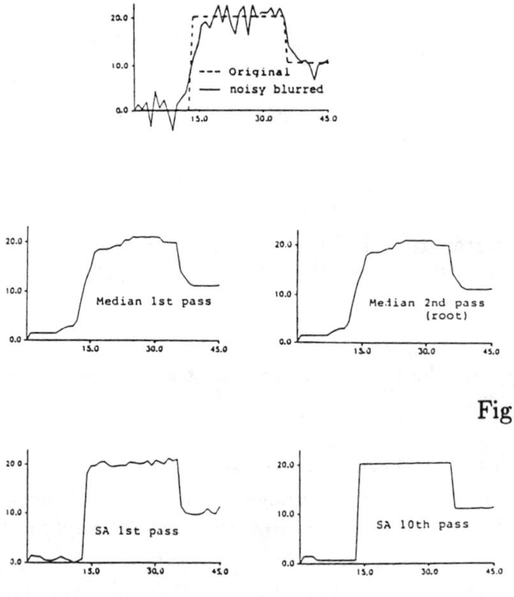

Fig. 1. 1-D simulation results.

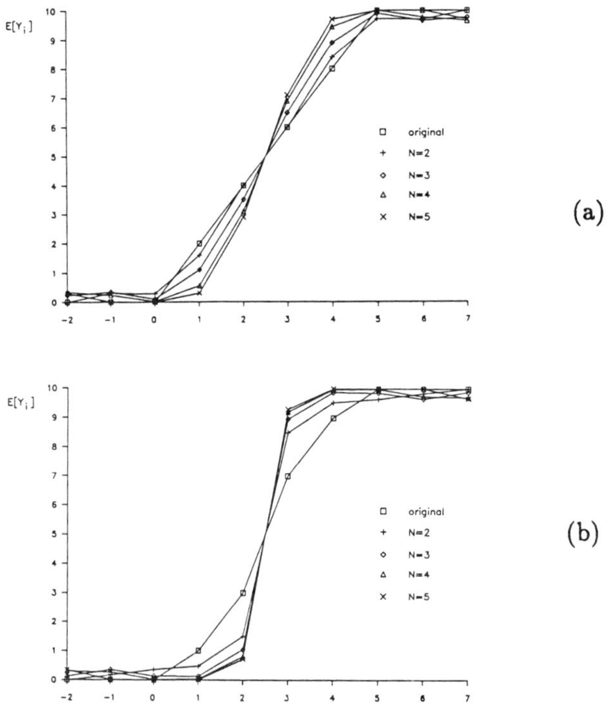

Fig. 2. Edge enhancing characteristics of SA filters with various windows; (a) Ramp edge, and (b) Concave/convex edge.

Table 1

N	Output Variance
2	0.368
3	0.254
4	0.198
5	0.160
6	0.137
7	0.121
8	0.107

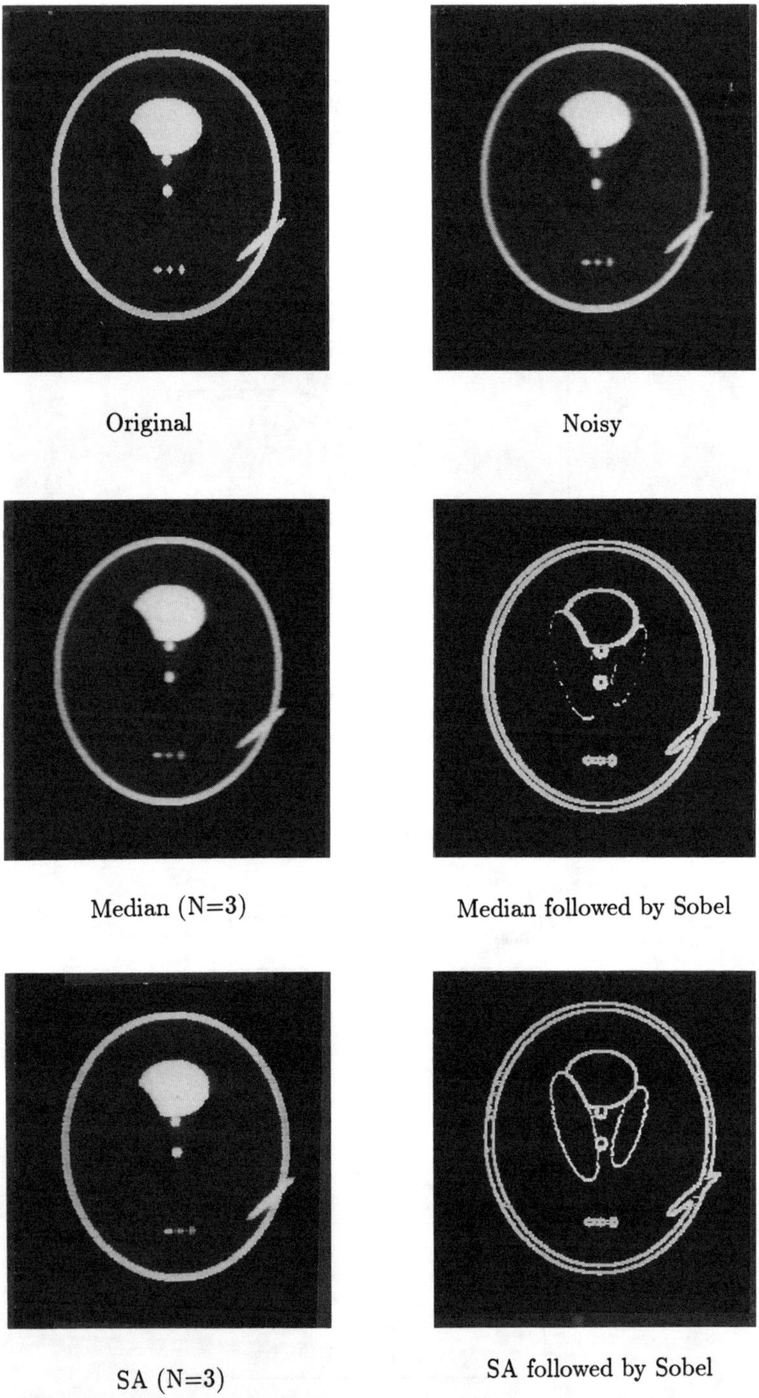

Fig. 3. 2-D experimental results.

A CASE-BASED APPROACH TO THE PERCEPTION-CONTROL-ACTION CYCLE

JIM ANTONISSE
antonisse@ai.mitre.org
Autonomous Systems Laboratory,
MITRE Corporation
McLean, VA 22102

HARRY WECHSLER
wechsler@gmuvax2.gmu.edu
Department of Computer Science
George Mason University,
Fairfax, VA 22030

ABSTRACT

This paper represents preliminary explorations in three intersecting areas in the intelligent control of perception and action in robotics. The discussion is set in the context of Perception-Control-Action architectures and involves, first of all, the learning and coordination of a robot's perceptual actions by the application of Case-based Reasoning (CBR) techniques. A second area involves the application of Distributed Associative Memory (DAM) techniques to the indexing problem in CBR to achieve reactive, flexible control over perceptual actions. Finally, this work finds its application in the context of automated depth perception using stereo vision and represents part of the emerging field of Active Vision.

INTRODUCTION

Stereo vision is a computationally intensive process that determines the range to a surface from the appearance of patches of the surface in the fields of view of a pair of sensors. Stereo is computationally hard in part because it induces a search among multiple competing hypotheses of what parts of the respective fields of view correspond to identical parts of the external scene. Some systems perform this search by changing the angle between the lines of sight of the sensors, called the vergence angle. Changing vergence is a key element in how an existing robot vision system, the PRISM-3 Realtime Stereo Vision System [1], automatically executes its ranging function.

Experience with PRISM shows that the automated vergence search is easily misled by ambiguous correspondences. Humans, however, can easily use keyboard control to specify a sequence of PRISM commands that bring identical parts of a surface into correct vergence. This paper represents our initial steps in efficiently encoding a similarly proficient vergence search in a robotic vision system. The approach we are exploring is to use Case-Based Reasoning (CBR) techniques to encode human command sequences, and to exploit Distributed Associative Memory (DAM) techniques to achieve robust active stereo vision routines.

THE PCA CYCLE

Stereo vision using vergence is an inherently active process. We are exploring this process in the context of the Perception-Control-Action (PCA) architecture [2], which makes the close interaction of action and sensing explicit. In PCA (Figure 1) all perception is mediated by schemata. These are the data structures that implement memory, e.g., a map data structure might implement part of spatial memory. Schemata reflect the interpretation of sensor and internal state information to the system. In our case, schemata will reflect both perceptual states and contextual action sequences. Control uses schemata to determine what actions to invoke. Actions include not only external agentive acts such as manipulation, sensing, and navigation, but also internal acts such as anticipation (e.g., to set up expectancies for planning) and learning (the modification of schemata). The latter effect schemata directly, while the former only change schemata by side-effecting the world in ways that may later be sensed.

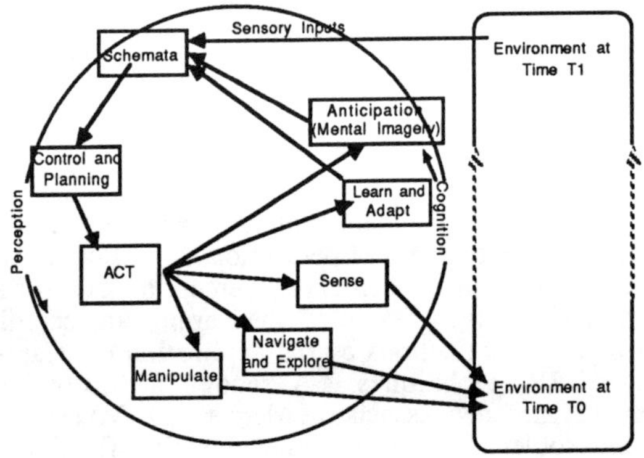

Figure 1. The PCA Cycle. Schemata are instantiated by sensory inputs (of both internal and external modalities) and by anticipation actions, and modified by learning actions. Both learning and anticipation are linked to external manipulation, navigation, and sensing actions.

HORIZONTAL ARCHITECTURES AND CASE BASED REASONING

The PCA cycle is consistent with what might be called the "classical" approach to robotics. This approach consists of a vertical division of labor among a set of capabilities that are envisioned as taking place sequentially, starting with sensing, continuing with interpretation and planning, and concluding with commitment to a set of actions. This approach encouraged explicit modeling of the objects of interpretation and their causal interrelationships as knowledge representations (e.g., frames, rules, CAD models).

The realtime requirements of robot control has highlighted some of the computational liabilities of explicit representations, and recent work has focussed on horizontal architectures for robotics implementations in which representation in discouraged. This development (led in particular by Brooks [3]) leads to robot control which is implemented holistically along complete competences with respect to a specific task. The approach has resulted in a number of demonstrations of autonomous robot capabilities, including small robots that learn to walk [4], find soda cans [5], and map a room [6].

Brooks' horizontalist ideas have been generalized to systems that share more traditional organizational structures, as in [7], where layered competences are compiled from task-level descriptions using the Rex/Gapps language [8]. The recent integration of the PCA principles in a layered design ([9]) represents an explicit synthesis of traditional vertical decomposition of tasks (e.g., first perceive, then act), and the horizontal synthesis more recently espoused. This paper extends these ideas by considering CBR cases as an explicitly represented horizontal processing thread.

If one has a traditional vertical model of a robot application domain, then one way of looking at the move to a horizontal approach is as an attempt to directly implement successful paths (or sets of paths) through the vertical model. The Case-Based

Reasoning approach to AI embodies this intuition [10]. In CBR one collects episodes of problem solving, stores these in memory, and retrieves them later in such a way that they are recalled situations similar but not necessarily identical to the original problem situation. Thus new situations “bring to mind” previous cases which are adapted to the present situation. The original problem solving episode may have been constructed in the context of a theory of a domain (see, e.g., [11]). In this case the theory provides a fallback position when no case case applies to a situation. However, most CBR systems use cases that are given directly so that novel situations must be accommodated by case adaptation mechanisms.

DISTRIBUTED ASSOCIATIVE MEMORIES

In real-world applications the events originally giving rise to a problem-solving episode cannot be expected to be exactly repeated. However similar occasions may often be expected to recur. The key issue in CBR is how to effectively index into the case base so that appropriate cases are recalled in new situations. When one considers CBR systems in which the event space is covered by the indexing scheme (i.e., in which the system will have *some* response for any situation), it becomes apparent that the indexing scheme induces a control regimen for switching among cases in a changing environment. Meanwhile, the key technical issue to be resolved for horizontal architectures to scale to systems with a rich repertoire of behaviors is how to manage switching among the behaviors. Such switches in behavior must take place in the presence of noisy environmental signals that only incompletely or partially indicate the relevant states of the world.

These requirements for the retrieval of a complete behavior on the basis of a partial or approximate cue for the behavior suggest the neural network techniques fo Distributed Associative Memory (DAM) for the retrieval task. DAM techniques exploit the well-known capacity for matrix representations and operations to serve as a content-addressable memory (see, e.g., [12]). A DAM works using a simple linear systems theory approach to retrieval. Given a set of vector inputs $\mathbf{x}_i$ of length N that is to be associated with another vector set $\mathbf{y}_i$ of length M, one may construct an NxM memory $\mathbf{M}$ by summing the outer products $\mathbf{y}_i^T \mathbf{x}_i$. The resulting matrix represents a distributed form of the original associations in the sense that, if the original inputs were orthogonal and if one "presents" the input to the system by matrix multiplying $\mathbf{Mx}_i$, the result or output of the system is $\mathbf{y}_i$ (for $\langle \mathbf{x}_i^T \mathbf{x}_i \rangle = 1$).

If the inputs are not orthogonal crosstalk between the stored representations occurs. This crosstalk is detrimental to exact replication of the $\mathbf{y}_i$ on presentation of the associated $\mathbf{x}_i$. However, under certain conditions the $\mathbf{y}_i'$ actually retrieved by *arbitrary* input $\mathbf{x}_i$ will be the "best fit" $\mathbf{y}_i$ in a principal components sense, where the principal components are derived from the statistics of the population of observed $(\mathbf{x}_i,\mathbf{y}_i)$ pairs. This implies that when the set of observed $(\mathbf{x}_i,\mathbf{y}_i)$ pairs is not normative, we may sometimes interpret "crosstalk" to be a form of error correction in the data or behavioral generalization in the system response.

APPLICATION: ROUTINES FOR STEREO VISION

The hypothesis we explore in this paper is that DAM techniques can serve as an effective mechanism for indexing into the appropriate intermediate-level stereo vision behaviors of a real-time robotic visual system. The behaviors are temporal sequences of actions that implement a vergence search. The original input-output pairs are behavior fragments of a certain length. These are to be learned, generalized, and applied to achieve effective from stereo vision routines.

Stereo Vision Geometry

Depth perception in stereo vision depends on the triangulation of points in the scene as they are respectively perceived in the right and left sensor fields. Figure 2 illustrates an example of stereo vision geometry using a symmetric sensor configuration similar to that employed in PRISM.

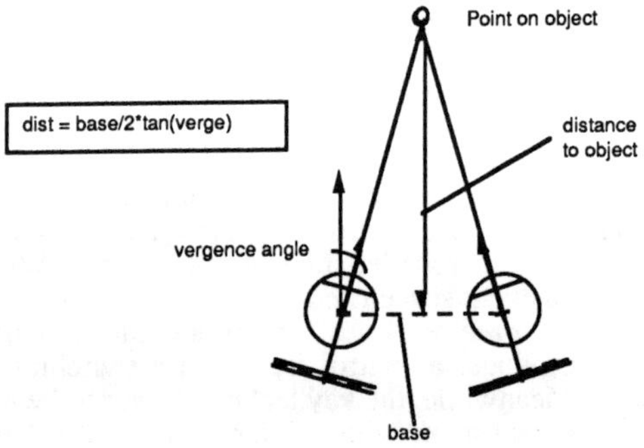

Figure 2: Stereo Geometry.

Stereo vision algorithms confront a world presented as a field of incident light (originating or reflected from a potentially irregular set of surfaces) sampled by the sensor's receptive elements. The resulting projection of portions of a scene on the sensor fields leads to a requirement for search among combinations of left and right sensor readings for correct correspondences. Figure 3 demonstrates the different search/surface configurations possible. Figure 3a corresponds to negative disparity, where the verge angle between the sensors should be reduced to improve the pixel-to-pixel correspondence of the left and right images, Figure 3b corresponds to zero disparity, where pixel-to-pixel correspondence is at its peak, and Figure 3c to positive disparity, where increased vergence would improve correspondence.

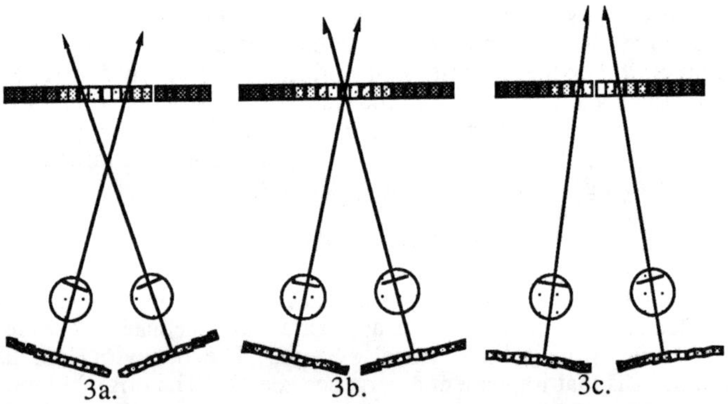

Figure 3: Disparity, Range, and the Correspondence Problem.

For a numerical example, let the intensity values of the 15 points on the surface in Figure 3 be: [6 14 1 11 11 11 2 11 4 6 13 15 12 12 1]. Let the left and right 1D input images of the 12 pixels recorded (accounting for inverse optics) be, respectively: [1 11 11 11 2 11 4 6 13 15 12 12 1] and [6 14 1 11 11 11 2 11 4 6 13 15 12]. Define, further, a four-element window for each field of view that may be placed over any of the eight contiguous blocks of four in the field of view. We may then compute correlation statistics, e.g., the covariance, over pairs of the left and right windows to test the hypothesis that the windows reflect the same part of the scene. Numbering the windows in a field of view from -4 at the far left to 4 on the far right, we may usefully generate correlation statistics on the following pairs of windows from the left and right: {(-4 4) (-4 3) (-3 3) (-2 3)... (3 -4) (4 -4)}. Using the example pixel values and the covariance statistic, this results in the 1D disparity map shown in Figure 4. This procedure was originally developed in a study of the human visual system [13] and is very similar to the process used by PRISM.

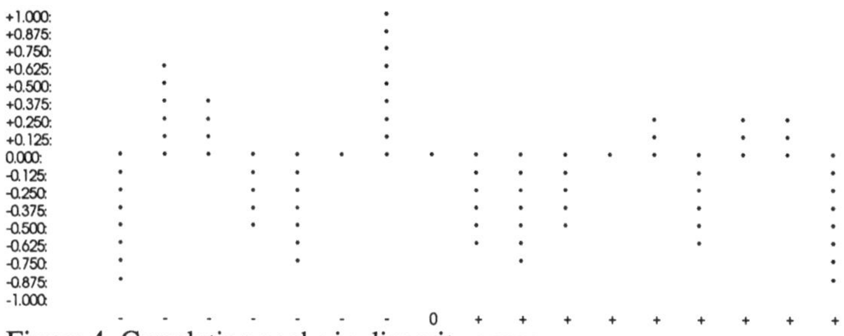

Figure 4. Correlation peaks in disparity space.

(Computing correspondence is an expensive proposition for wide field-of-view/high-resolution systems. Many natural and artificial vision systems, including both PRISM and our simulation, reduce computational expense by reducing the size of the fields over which such processes are applied. However, discussion of this facet of the work is beyond the scope of the present paper.)

Implementation of Perceptual Actions

We have implemented a simulation of the above visual search process as an interactive game in which we may specify three different kinds of visual acts. Verge actions change the angle of vergence at which disparity is computed. Search scale actions which change the increments at which the vergence angle is changed and visual scale actions which simultaneously change the width and resolution of the field of view are also implemented. Perceptual processing and display precedes each step and forms the context for deciding what action to take.

The form of this game follows the actual problem of finding the correct verge position in the PRISM system quite closely, hence represents a reasonable analog of an actual robotics problem. A problem solving episode in the game represents a sequence of these three types of actions interspersed by perceptions of the world. In this respect the game forms a PCA cycle in the small.

Viewing the game from a CBR perspective we may readily see sequences of game actions as cases. There is in general an issue in CBR of how to delineate a case. In the active vision context there is a similar choice in how cases are accumulated. For instance, we can acquire cases on the basis of the dimension of the DAM. We have chosen, instead, to introduce a marker which terminates an arbitrarily long case interactively. This forces us to view the DAM as a system for finding domain features in a continuous input stream. An autoassociative memory (i.e., $(\mathbf{x}_l,\mathbf{x}_l)$) is used for storing these cases since retrieval will determine what next set of actions to take given a partial sequence of past actions and a present perceptual state.

The approach we are taking is similar to that of Anderson [1], in which he used input feedback to build up a set of statistical features to best characterize associations in a domain. Since he used normalized, zero-mean input vectors in an autoassociative memory, the features he derived corresponded to the eigenvectors of (a scalar multiple of) the covariance matrix of the ($\mathbf{x}_i$,$\mathbf{y}_i$) associations. His result transfers to our application, since we normalize our input vectors, use autoassociation, and may assume that there would be no systematic bias in the opposed pairs of operators we have defined over large samples of action sequences.

SUMMARY AND STATUS

The goal of the vergence search is to find the verge angle that focuses the corresponding patches of the sensor fields of view on the same points in the scene. The success of this project will depend in large part upon having a case-base of successful searches from humans into which to index. However, a little reflection reveals that indexing cases only as action sequences is inadequate, since cases are context-invoked--we are so far ignoring an important motivator for applying CBR techniques, handling context brought about by multiple goals.

We now encode the perceptual context through measured of the correlation peak, its location in the disparity vector, and a PRISM goodness-of-peak measure. Notably absent is any direct representation of the contents or structure of the images in the right and left sensor fields (which we almost certainly use this in the PRISM context). Our hypothesis is that the stark representation of perceptual context will be adequate to this task. A natural metric for scoring the goodness of a case as a problem solving episode is just the ratio of this estimated distance and the actual distance (where distance may be calculated according to the relation in Figure 2.) Our first set of experiments will test this and (we believe) show the power law of practice in an increasing human performance capability, followed by experiments in the DAM usage of the acquired cases.

REFERENCES

1. H. K. Nishihara, "Real-Time Implementation of a Sign-Correlation Algorithm for Image-Matching", Unpublished manuscript describing PRISM-2.
2. H. Wechsler, *Computational Vision*, Academic Press, 1990.
3. R. A. Brooks, "A Robust Layered Control System For A Mobile Robot," *IEEE Journal of Robotics and Automation*, **RA-2**(1), March, 1986.
4. P. Maes, "Situated Agents Can Have Goals," in *Theory and Practice from Biology to Engineering and Back*. P. Maes (editor). Special Issue of the *Journal of Robotics and Autonomous Systems*. **6**(1&2).
5. J. H. Connell, " A Colony Architecture for an Artificial Creature," MIT Artificial Intelligence Lab Technical Report 1151, October 1989.
6. M. J. Mataric, "Behavioral Synergy Without Explicit Integration," AAAI Spring Symposium Series, *SIGART Bulletin*, ACM Press, **2**(4), August 1991, 130-133.
7. R. P. Bonasso, "Integrating Reaction Plans and Layered Competences Through Synchronous Control," *SPIE Conference Proceedings*, Boston, MA, November, 1990.
8. L. P. Kaelbling, "Goals as Parallel Program Specifications," *Proceedings of the National Conference on Artificial Intelligence*, St. Paul, MN, August, 1988.
9. H. J. Antonisse and H. Wechsler, "Integrating Perception, Action, and Learning," AAAI Spring Symposium Series, *SIGART Bulletin*, ACM Press, **2**(4), August 1991, 25-29.
10. R. C. Schank and D. B. Leake, "Creativity and Learning in a Case-Based Explainer," *AI Journal*, **40**, 1989.
11. C. Hammond, "Case-Based Planning: An Integrated Theory of Planning, Learning, and Memory", Yale PhD. Thesis, 1986.
12. Y.-H. Pao, *Adaptive Pattern Recognition and Neural Networks*, Addison-Wesley, 1989.
13. D. Marr, *Vision*, W. Freeman, 1982.
14. J. A. Anderson, J. W. Silverstein, S. A. Ritz, and R. S. Jones, "Distinctive Features, Categorical Perception, and Probability Learning: Some Applications of a Neural Model," *Psychological Review*, **84**:413-451.

Integrating New Sensors into a NASREM Implementation

Dr. Ronald Lumia
Robot Systems Division
National Institute of Standards and Technology
Bldg 220, Rm B127
Gaithersburg, Maryland 20899

Abstract

It is rare when the requirements of a system remain constant between the conception of the implementation and its delivery. Changes in specification occur and these modifications cannot be ignored in the methodology of developing systems. One way to approach this problem is to employ a functional architecture which is supple enough to support the inevitable changes without requiring significant system redesign. This paper discusses the integration of a new vision sensor, which supports active vision, into the existing robot system implementation at NIST. The lab implementation uses the NASA/NIST Standard Reference Model for Telerobot Control System Architecture (NASREM) as its functional architecture. The four degree of freedom active vision platform is described along with the process of its integration.

1. Introduction

The Intelligent Controls Group of the Robot Systems Division at NIST has been implementing the NASREM architecture for several years. NASREM is a hierarchical system with six levels where each level is divided into task decomposition, world modeling, and sensory processing. A more detailed description of NASREM can be found in [1]. To implement NASREM, interfaces for the modules were required. The interfaces for the lower two levels can be found in [2-7]. The approach has been to develop the architecture independently of the application by creating a robot control system testbed.

Since the NASREM architecture has been developed in a technology independent manner, there are two types of modifications which must be allowed. First, the architecture must be able to support different application areas. The NASREM architecture has been use for underwater vehicles [8], mining vehicles [9], autonomous land vehicles [10], as a basis for the Next Generation Controller, as well as for its originally stated space application. Second, the architecture must facilitate the integration of new sensors even if the application does not change so that the implementation may evolve with technology. This paper describes the evolution of the testbed when the capability of active vision is added.

The paper is organized as follows. First, a short description and rationale for active vision is provided. Then, the NASREM testbed is described both before and after the addition of the active vision capability. Then, a tracking capability using active vision is described. Finally, some application areas are suggested for future research.

2. Active Vision

Active vision is a new research area where controlling the motion of the sensor is just as important as the capabilities of sensor processing. This leads researchers away from the search for the "correct" image processing algorithm, e.g., optimal edge or corner extractors, toward the development of an understanding of the relationship between feature extraction and motion of the sensor.

In order to explore this direction the Intelligent Controls Group has constructed The Real-time Intelligently Controlled Optical Positioning System (TRICLOPS), four degree of freedom, three camera device, as shown in Figure 1. The first degree of freedom rotates the cameras in a manner analogous to that of a human neck. The second degree of freedom tilts all three cameras, allowing the device to look up and down. The central camera is rigidly attached to this degree of freedom. Since it is equipped with a wide angle lens, this camera provides images with a wide field of view but with low spatial resolution. The remaining two degrees of freedom are the independently controlled vergence cameras. The focus of this paper is the integration of TRICLOPS into the NASREM testbed.

3. NASREM Testbed

The layout of the NASREM testbed before the addition of TRICLOPS is shown in Figure 2 in the unshaded area. The system consists of two major parts: the hardware associated with the robot and the hardware related to the vision processing. The robot is the Robotics Research Corporation[1] (RRC) K-1607, a seven degree of freedom manipulator, which incorporates joint torque loops to minimize the effects of drive train nonlinearities, especially motor friction and harmonic drive compliance, to reduce the apparent motor inertia seen at the actuator output. The RRC controller provides all of the power electronics to drive each joint motor. The RRC controller provides an interface which allows an external computer system to issue joint torque commands to the manipulator every 2.5 ms.

The desired torques for each joint are calculated by the NIST developed controller in the Manipulation rack, a multi-processor system based on Motorola 680x0 processor boards in a VME backplane. Currently, there are seven processors in the Manipulation target system which execute the Primitive and Servo level code. The interface to the RRC controller is through an IEEE 488 parallel link. Code development is in ADA on a SUN-3 host, and the cross-compiled code is downloaded, along with a run-time kernel, to the target system. The SUNs are connected through an ethernet and a variety of software is available on the system to assist in the code development process.

The camera image is processed by the Pipelined Image Processing Engine (PIPE), a special purpose computing architecture designed specifically for image processing. PIPE is a preprocessor for iconic (spatially indexed) images. The output of PIPE is interfaced to the Perception rack through a dedicated 680x0 computer. Currently, the

1. Products named in this paper are listed for purposes of information only. There is no implied endorsement of any products or implication that they are the best available for the purpose.

Perception rack has four computers to process the output of PIPE and interface with the World Model. The Perception rack is interfaced with a SUN host in the same way as the Manipulation rack.

A connection between the Manipulation rack and the Perception rack must exist to use the two systems in concert. This connection is easily removed during the debugging phase of program development so as to reduce contention for the system hardware by programmers.

4. Integrating TRICLOPS into the NASREM Testbed

As stated in the introduction, the ability to integrate new sensors into the system is critical for the evolution of a system. The active vision sensor, however, is significantly more complex than most sensors. It is a system, complete with its own computers for control. In spite of this, the integration was rapid for several reasons. First, relatively little software had to change, and even where changes were required, the organization provided by the NASREM architecture structured the changes. Second, the complete computing infrastructure could be used unchanged. Third, the active vision system required the addition of only 20% of the software to adapt the current system for use as the active vision platform.

The addition of TRICLOPS into the NASREM testbed is shown in Figure 2 as the shaded region. The procedure used to integrate active vision into the system was:

Clone a VME rack of computers

Copy and modify the NASREM controller used for the RRC robot

Develop the interfaces which are unique to TRICLOPS

Develop the unique algorithms associated with TRICLOPS

Perform process to processor assignment to achieve design specifications

4.1. Clone a VME rack of computers

For flexibility in research, as opposed to economic incentives, it was decided to create another VME bucket of 680x0 processors to control TRICLOPS. The reason is that the contention for hardware is minimized during the development of the capabilities of the system. The three VME racks, as shown in figure 2, can have programmers independently developing software. When the entire system is to run in an integrated fashion, all of the backplanes must be connected together. All that this required, other than the purchase of the equipment, is to connect a bus extender card from the TRICLOPS VME rack to the Perception rack. From the standpoint of camera output, the only requirement is the obvious need to connect the TRICLOPS cameras to the inputs of PIPE.

4.2. Copy and modify the NASREM controller used for the RRC robot

The NASREM controller code for the RRC robot was simply copied for use as the controller for TRICLOPS. Most of the modifications are obvious. The world model must changed in order to have new forward and inverse kinematics, number of joints (4 for TRICLOPS, 7 for RRC robot), Jacobian, inertia matrix. These changes are relatively easy but more important, it is clear from the structure of the code imposed by the NASREM architecture precisely which processes must change.

4.3. Develop the interfaces which are unique to TRICLOPS

The information in the interface between TRICLOPS and its controller is the same. It consists of desired joint torques, with actual joint torque and positions as feedback. Therefore, the robot controller and TRICLOPS controller from that standpoint are identical. The physical mechanism associated with the interface, however, is different for the devices. Consequently, the connection between TRICLOPS and its controller, a standard parallel port, had to be implemented.

4.4. Develop the unique algorithms associated with TRICLOPS

Several unique algorithms were required for TRICLOPS. First, the vergence control algorithms, where the outer two cameras would verge on a single point in space were required. This was implemented as a new trajectory generation algorithm at the primitive level of the new TRICLOPS controller. Also, an algorithm which computes position from triangulation was required in the perception branch.

4.5. Perform process to processor assignment to achieve design specifications

In order to achieve human-like motion performance, a design specification was to have TRICLOPS' closed loop bandwidth to be 50 Hz. To achieve this, the sampling frequency for the motors associated with TRICLOPS is 2 KHz. This requirement is much more stringent than that of the robot controller. Consequently, the process to processor assignment, a process described in [4], needed to change to achieve this goal.

5. Conclusion

A functional architecture is helpful because it provides the basis for the organization of the hardware and software which implements it. When new requirements for a system dictate the addition of new sensors or even new systems, the functional architecture provides guidance for a smooth and rapid integration.

6. References

[1] Albus, J.S., McCain, H.G., Lumia, R., "NASA/NBS Standard Reference Model Telerobot Control System Architecture (NASREM)," NIST Tech. Note 1235, NIST, Gaithersburg, MD, July, 1987.

[2] Fiala, J., "Manipulator Servo Level Task Decomposition," NIST Technical Note 1255, NIST, Gaithersburg, MD, October, 1988.

[3] Kelmar, L. "Manipulator Servo Level World Modeling," NIST Tech. Note 1258, NIST, Gaithersburg, MD, December, 1989.

[4] Fiala, J., "Note on NASREM Implementation," NISTIR 89-4215, NIST, Gaithersburg, MD, December, 1989.

[5] Chaconas, K. J., Nashman, M., "Visual Perception Processing in a Hierarchical Control System: Level 1," NIST Tech. Note 1260, June, 1988.

[6] Wavering, A. "Manipulator Primitive Level Task Decomposition," NIST Technical Note 1256, NIST, Gaithersburg, MD, October, 1988.

[7] Kelmar, L. "Manipulator Primitive Level World Modeling," NIST Tech. Note 1273, NIST, Gaithersburg, MD, December, 1989.

[8] Albus, J.S., "System Description and Design Architecture for Multiple Autonomous Undersea Vehicles," NIST Technical Note 1251, NIST, Gaithersburg, MD, September, 1988.

[9] Huang, H.M., "Hierarchical Real-Time Control System for Use with Coal Mining Automation," Fourth Conference on the Use of Computers in the Coal Industry, Morgantown, WV, June 1990.

[10] Szabo, S., Scott, H.A., Murphy, K.N., Legowik, S.A., "Control System Architecture for a Remotely Operated Unmanned Land Vehicle," Proceedings of the 5th IEEE International Symposium on Intelligent Control, Philadelphia, PA, September, 1990.

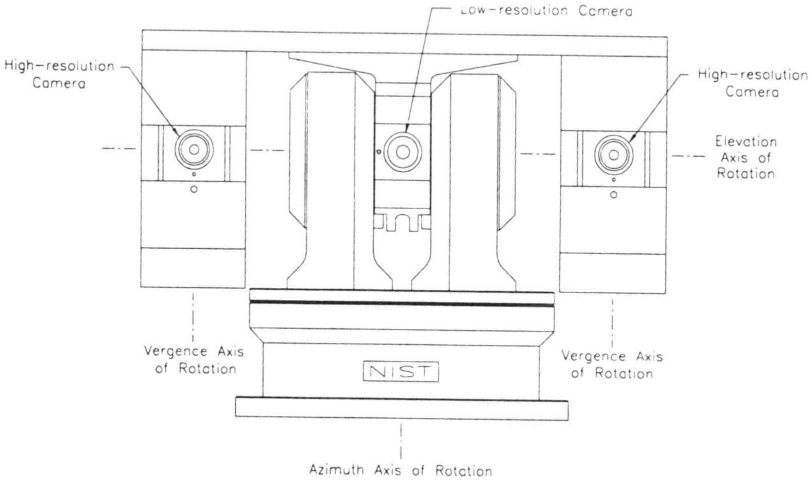

Figure 1. Front View of TRICLOPS

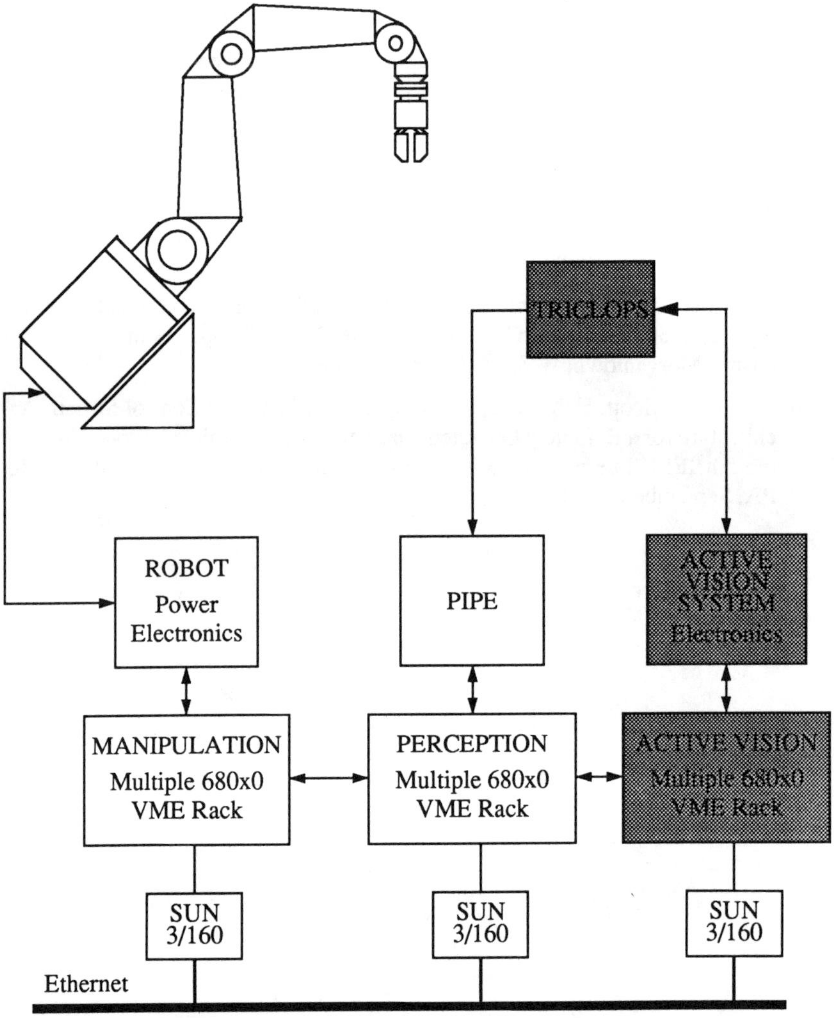

Figure 2. NASREM Testbed for Manipulation and Active Vision

Integrate Sensing and Situated Control for Robust Manipulation

Vincent S. Hwang
MITRE Corporation, 7525 Colshire Drive, McLean, Va 22102

ABSTRACT

Intelligent robots equipped with sophisticated sensors and on-line reasoning capabilities will play a major role in future Space operations. Such robots use their sensors to perceive the surroundings, reason about what action plan to do next, and execute the action plan. This paper describes the use of situated control and advanced sensing for robust manipulation in a semi-structured environment. We demonstrated such a system in the control of various robot arms equipped with wrist-mounted CCD camera and force-torque sensor in the placing of a real-sized Orbital Replacement Unit (ORU) mockup onto its base at Goddard Space Flight Center. Same approach can also be used to control mobile robots.

INTRODUCTION

Intelligent robots equipped with sophisticated sensors and on-line reasoning capabilities will play a major role in future Space operations. Such robots use their sensors to perceive the surroundings, reason about what action plan to do next, and execute the action plan. The operating environment is mostly predictable. However, exception and mistake can occur. The cost of any mistake occurring is high. These robots must operate in a fail-safe mode. Straightforward factory automation is not sufficient. In addition to using fault-tolerant and reconfigurable hardware components and architecture, robot designers must make sure that all actions are continuously monitored to ensure safe completion of the mission.

A conventional approach to robot control is to first generate all actions for the given goal and then execute these actions one by one. The goal is accomplished when all actions are completed. This simple approach is not sufficient when fail-safe operation is required. Instead, a more sophisticated plan-generate-monitor control model should be used [1]. In this approach, the robot checks the progress continuously. Upon the completion of an action, the robot ensures that the current action completes successfully and the next action is readily executable. Significant deviation from the expectation causes the robot to generate and execute corrective actions. These actions will either get the ongoing action back to normal operating conditions or terminate the ongoing mission. We call such style of robot control the situated control.

This paper reports on MITRE's research on applying situated control to fail-safe robust manipulation [2,3]. Given a mission, the actions, plans, and contingency plans are compiled. Normal operating envelope, expected initial and termination conditions are identified. During the execution, when everything proceeds as expected, the robot executes the action plans. When observation deviates from the expectation, the robot analyzes the difference, selects an appropriate contingency plan, and executes the plan to ensure a fail-safe operation.

A prototype is developed based on this approach. We demonstrated the utilities of this prototype at Goddard Space Flight Center on two robots including the Cincinnati Milacron T3 industrial robot and the Puma 762 robot. These robots were equipped with a wrist-mounted CCD camera and a force-torque sensor. The test scenario is the placement of a real-sized Orbital Replace Unit (ORU) mockup onto its base. The camera is used to guide the robot to position the ORU near the top of the base. Once there, a motion servoed force control algorithm is used to guide the robot to complete the attachment. The robot performs robustly under various operating conditions. Same approach can also be used to control mobile robots.

SITUATED ROBOT CONTROL

This section describes the use of situated control to robust manipulation control. Figure 1 shows the control flow. Robot actions are generated either by user or by some off-line planner. The actions are entered into a data structure called agenda. The position of objects are stored in a data structure called the World Model (WM). The Task Interpreter (TI) examines the agenda and the WM to select appropriate actions to process. Actions are processed by TI in the following ways: 1) dispatch robot and sensing commands to appropriate controllers, 2) activate monitors attached with the commands. The monitors are specialized programs that are used to ensure that the robot does not breach the normal operating envelope. The robot controller executes the robot commands and reports the completion status back to the system. The sensor controller executes the sensing commands, processes the sensor data, and returns the information to be stored in the WM. When the precondition of a monitor evaluates to true, the contingency plan of the monitor is entered into the agenda and is eligible to be processed in the next interpretation cycle. When an action is completed, all the monitors related to that action are deactivated. The interpretation cycle continues until the agenda becomes empty or the mission is accomplished.

The control of the system is intrinsically asynchronous. Multiple robot commands, sensing commands, and monitors can be executed in parallel. TI uses a simple priority scheme to prioritize the actions in the agenda and always selects the action with highest priority to process next. Once a command is dispatched, TI is ready to select the next action. Explicit synchronization between the actions is

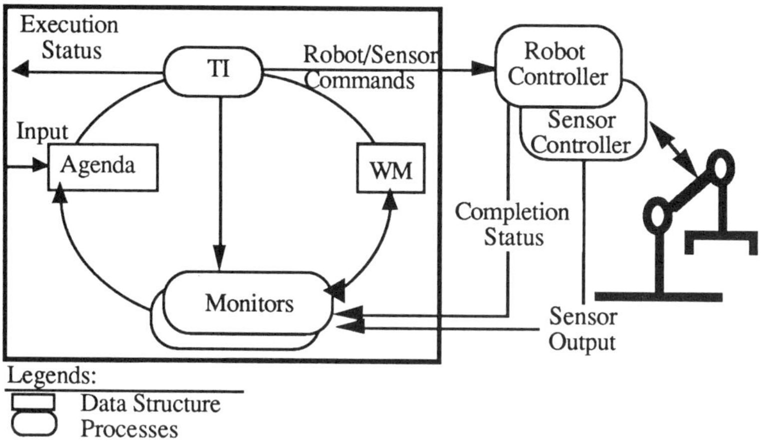

Figure 1: System Architecture

required. For example, sequential execution of a sequence of actions can be done by forcing the start of an action depends on the completion of the previous action.

Two robot motion primitives are used. They are gross-motion and compliant-motion. Gross motion primitive moves the robot wrist to a designated position and orientation. Attached to this primitive is a monitor that checks for excessive force on the arm. The monitor stops the arm motion when excessive force is detected. Compliant-motion primitive controls the fine motion of the arm. This primitive contains a monitor to watch for stagnation and breach of operating envelope. Reference-by-name mechanism is used. The values of variables are evaluated each time they are used. Such mechanism is particular useful for the representation of robot trajectory way points and robot motion destination.

COMPUTING POSITION AND ORIENTATION OF 3D OBJECTS

This section describes the perception module. This module is used to compute the position and orientation (3D pose) of objects [4]. It proceeds as follows: 1) digitize images, 2) extract compact homogeneous regions, 3) match these regions against the stored object model, and 4) compute the pose of the object. The computation cost of matching is quite high for general 3D objects. We used artificial markers to reduce the computation requirement. Marker consists of 5 coplanar circular dots are placed at known locations with respect to the ORU. The 3D pose of the ORU is known once the 3D pose of the marker is computed.

The perception module needs to operate under various lighting conditions. An adaptive method is used to extract compact homogeneous regions. First, parameters of compact homogeneous regions are defined. At run time, a list of possible intensity values is computed. For each selected value, a list of candidate regions is computed. A final candidate list is generated by eliminating duplications. This algorithm is implemented on a DataCube with a connected component labeling board. This algorithm works quite well for images taken under various lighting conditions.

Two approaches to 3D pose computing are investigated. The first approach is the so-called quadrangle method. This method computes the 3D coordinates of the vertices of a quadrangle relative to the camera frame from their image coordinates. The second approach uses an optimization approach. The pose of a point set is computed by finding the values that optimize a cost function measuring the distance between observed image features and predicted image features. The pose computed using the quadrangle method is reasonable for real images. The accuracy improves as the distance between the camera and the marker decreases. However, analysis indicates that the accuracy and the stability of pose computing is best when 3D markers and the optimization approach are used.

Two sensing primitives are used to encode sensing actions. They are compute-pose and find-object. Compute-pose primitive is used when object is in the camera's field-of-view. This primitive returns the 3D pose of the object or a failure status if the object is not found. Find-object primitive is used to actively search for objects. The find-object primitive combines the gross-motion and compute-pose primitives to direct the camera through a spiral motion to search for the object. The search stops when the object is found or when the robot breaches a given operating volume.

FINE MOTION CONTROL

The position servoed force control is implemented as a finite state machine [5]. Several states and privileged robot motions are defined. The privileged motions are used to move the robot from one state into another state.

Our fine motion algorithm is derived from the work of Laugier [6]. For each assembly and disassembly mission, a graph of disassembly paths is generated based on contact geometry of parts and tool. Each node of the graph constitutes a contact situation. For each contact situation, a set of privileged robot motion is generated. Through this motion, sub graphs are spawned and new contact situations are generated. The spawning process stops when no privileged motion can be generated. A path from the start node to a terminating node constitutes a potential solution path. Each candidate solution path is evaluated based on a heuristic function. This value reflects the likelihood that a solution path can be used to achieve the goal state (i.e. inserted assembly) when applied to the force controlled manipulator. The path with the highest value is always chosen first.

Following this approach, we first tested a hand coded fine motion strategy. The solution path was manually generated by analyzing the geometry of the ORU engagement mechanics. The resulting software module is robust and effective. Later, we implemented a contact force-torque simulator on IGRIP. Using this simulator, we are able to generate fine-motion strategy for objects consist of parallelepipeds and cylinders based on the algorithm described in the previous paragraph. Future work includes applying the solution graph to the control of real robot manipulator in assembly and disassembly.

EXPERIMENT & SUMMARY

Figure 2 shows the experiment setup at GSFC. A real-sized ORU is attached to the wrist of the robot. A SONY XC77 CCD camera with a VAFI (variable focal length and auto-iris) lens is mounted on a plate attached to the wrist. Mounted between the robot wrist and the ORU is a Lord force-torque sensor. A planar marker is used to aid the locating of ORU. Figure 3 shows the hardware setup of the experiment.

The approximate location of the ORU was given to the robot. The robot was first directed to go to 15" above the ORU before descending to 2" above the ORU. Several additional snapshots were taken to update and refine the estimated location of the ORU. Once there, the robot was directed to activate the fine motion module to attach the ORU onto its base. We performed the experiment under various lighting conditions. We also moved the base of the ORU during the descend. In all cases, the robot was able to detect the movement of the base, modify the descending trajectory, and robustly complete the attachment operation.

This paper addresses an approach of integrating situated control and advanced sensing in a dynamic and semi-structured environment. A prototype is implemented and demonstrated. Fail-safe operation is supported by monitors and situated control. The architecture of the prototype is domain-independent, and will be useful to other applications.

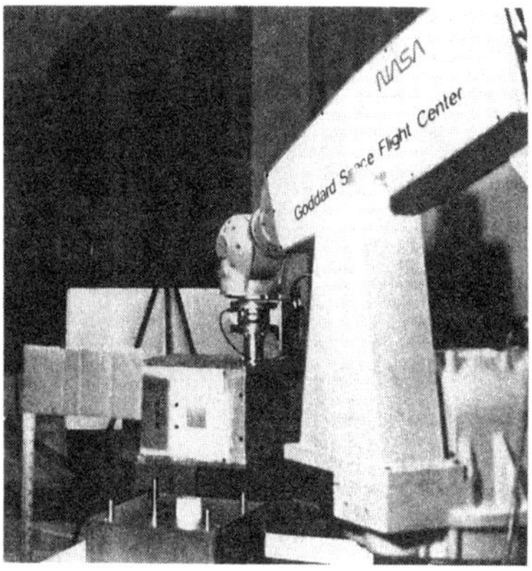

Puma Robot with ORU Markup, Base, and Wrist Assembly

(Right) A Closeup View: Wrist Camera, Force/Torque Sensor, ORU, and Base

Figure 2: Laboratory Setup at GSFC

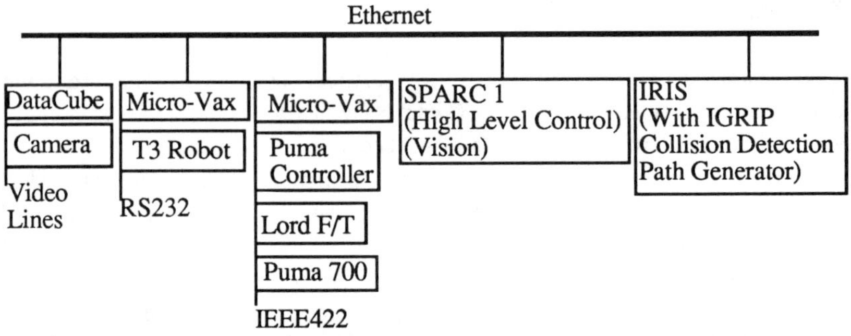

Figure 3: Hardware Setup at GSFC

REFERENCES

1 J. Sanborn and J. Hendler, A Model of Reaction for Planning in Dynamic Environments, International Journal of AI in Engineering (1988)

2 R. Bonasso, V. Hwang, J. Sanborn, W. Stoney, Investigating Robot Safety and Robustness in an Autonomous System Laboratory, *i-SAIRAS* (Japan 1990)

3 A. Castellano, V. Hwang, W. Stoney, Test of a Vision-Based Autonomous Space Station Robotic Task, Proceedings of SPIE Cooperative Intelligent Robots in Space II (Boston 1990)

4 V. Hwang, R. Keizer, T. Winkert, P. Spidaliere, On Pose Determination using Point Features, Proceedings of SPIE Cooperative Intelligent Robots in Space III (Boston 1990)

5 E. Huber, Automatic Fine Motion Assembly Control from a Contact Space Model, Internal MITRE document (1992)

6 C. Laugier, Planning Fine Motion Strategies by Reasoning in the Contact Space, *IEEE Journal of Robotics and Automation* (1989)

APPLICATION OF STEREO VISION TO OBJECT RECONSTRUCTION IN INDUSTRIAL ENVIRONMENTS

RAFIC BACHNAK
Dept. of Electronics Engineering Technology
Franklin University
Columbus, Ohio

MEHMET CELENK
Dept. of Electrical & Computer Engineering
Ohio University
Athens, Ohio

ABSTRACT

This paper describes a stereo-based multi-camera system for object reconstruction in an industrial environment. The system consists of N number of cameras arranged in an N/2 number of periodic stereo pair structure. The construction of the structural description is performed in two stages: (i) the 3-D data from different perspectives are inferred independently, and (ii) the extracted information from neighboring camera pairs are integrated to reconstruct the complete surface of the object.

INTRODUCTION

Developing three-dimensional (3-D) vision systems is of prime importance in many applications of computer vision [1-4]. Among several existing techniques, stereo has been one of the most used methods. It is based on the use of a pair of cameras displaced in position in such a way that they have a common view of the scene. The images of the camera pair are explored to extract the desired scene information. This involves three main procedures: camera modeling and calibration, image matching, and 3-D data calculations.

Although progress has made stereo vision a potentially practical technique for 3-D measurements [5-8], a system for complete object reconstruction is yet to be designed. Such a system is vital for applications involving object grasping, recognition and inspection. This paper proposes a stereo-based multi-camera system for complete 3-D measurements and surface reconstruction in industrial environments. A robot workspace is sensed by a set of N cameras arranged in an N/2 periodic stereo pair structure. The extracted 3-D information from individual camera pairs are then combined to generate the surface description of the object. The approximation to the surface is given by the minimal of the quadratic variations in the surface orientation with initial surface constructed using Lagrange polynomials.

OVERALL SET-UP OF THE SYSTEM

The overall arrangement of the system is shown in Fig. 1, where the subscript i refers to the ith camera pair number. All camera pairs are defined with respect to a global coordinate system at O. It is assumed that the camera pairs are identical and placed at a distance D_o from the origin O.

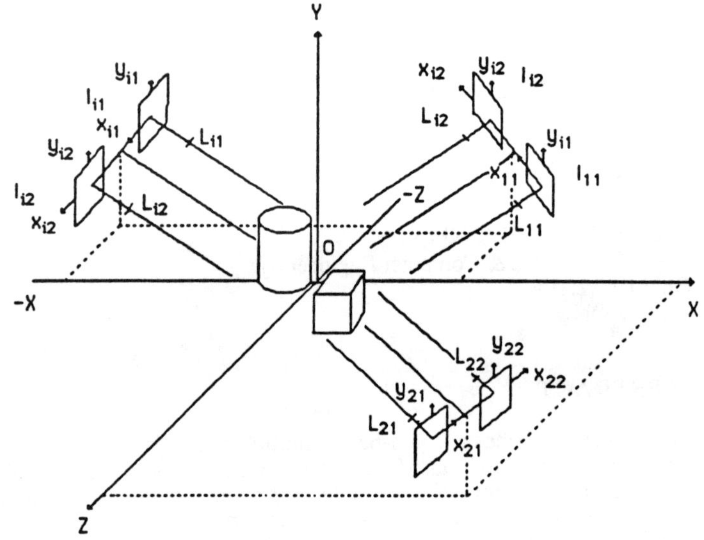

Fig. 1. Overall arrangement of the system.

The relationship between the image and the scene coordinates of a scene point are derived in [9]. They are given by

$$\begin{vmatrix} (R_{11}+a_{p1} R_{31}) & (R_{12}+a_{p1} R_{32}) & (R_{13}+a_{p1} R_{33}) \\ (R_{21}+a_{p2} R_{31}) & (R_{22}+a_{p2} R_{32}) & (R_{23}+a_{p2} R_{33}) \\ (R_{11}+a_{p3} R_{31}) & (R_{12}+a_{p3} R_{32}) & (R_{13}+a_{p3} R_{33}) \end{vmatrix} \begin{vmatrix} X_p \\ Y_p \\ Z_p \end{vmatrix} = \begin{vmatrix} A_{p1} \\ A_{p2} \\ A_{p3} \end{vmatrix} \qquad (1)$$

where

R_{ij} (i=1, 2, 3 and j=1, 2, 3) are the elements of the rotation matrix,
$a_{p1} = (x_{po1}/D_i)$, $a_{p2} = (y_{po1}/D_i)$, $a_{p3} = (x_{po2}/D_i)$,
$A_{p1} = B/2 - (D_o/D_i)\, x_{po1}$, $A_{p2} = (D_o/D_i)\, y_{po1}$, and
$A_{p3} = B/2 + (D_o/D_i)\, x_{po1}$

Here x_{poi} and y_{poi} (i=1, 2) refer to the output image components of the ith reference point. B is the baseline of the stereo pair and D_i is the effective focal length of the camera. a_{p1}, a_{p2}, a_{p3}, A_{p1}, A_{p2}, and A_{p3} are computed using the known parameters D_o and B, and calculated variables x_{po1}, y_{po2}, z_{po1}, and D_i.

To compute the 3-D data, it is necessary to determine the point pairs in the two images that correspond to the same object point. This is referred to as the correspondence problem. In order to solve the correspondence problem, we assume that the objects of interest are within the view of both cameras. In addition, two generally used constraints are employed: the uniqueness constraint and the detailed match constraint. The uniqueness constraint limits each image point to correspond to one point in the real world. The detailed match constraint considers candidate edges of similar orientation more likely to be valid matches. The matching procedure consists of three main processes and is described in [9].

DATA INTERPOLATION

A stereo system gives the depth information only at some object points. There are several methods that generate a description of the scene based on the known data [10-12]. The ultimate goal is to fit to the given data a surface that best approximates the actual object. Existing techniques can be classified under two major approaches: i) the constructed function may be derived to interpolate the given data, such that

$$Z_{ij}=f(X_i, Y_j); \quad i=0,\ 1, 2, \dots, N \quad j=0,\ 1, 2, \dots, M \tag{2}$$

Here, the surface passes through all original points. This approach is desirable when the data points are reliable and their values must be preserved. ii) the constructed function approximates the overall behavior of the known points. This approach alters the original data but gives reasonable results if the data points are not precise. The one-dimensional Lagrange polynomials are given, for a set of n+1 known depth points (Z_i, i=0,1, ... n) corresponding to distinct values (X_i; i=0,1,2, ... n), by

$$P_n(X) = \sum_{k=0}^{n} L_k(X) \tag{3}$$

where

$$L_k(X) = Z_k \prod_{\substack{i=0 \\ i\neq k}}^{n} (X-X_i)/(X_k-X_i)$$

The behavior of the polynomial depends on the position of the points used in its generation. To perform the initialization in the base plane of the space, Eq. (3) is applied to the rows and columns separately. The value of the constructed surfaces at a point (X,Y) is calculated locally based on the data at relatively nearby points. We consider here only single valued functions where there can be only one depth value for each point on the XY plane. This restriction can be relaxed if a model of the object is assumed. To ensure the continuity and smoothness from one scene point to the next, the last data point used in a particular polynomial is taken as the first point in the next. The data are generally sparse, therefore the horizontal and vertical operations give different results. The error is minimized in the least square sense by averaging both results. The operation may be performed in two-dimensional using

$$P_{n+m}(X, Y) = \sum_{k=0}^{n} \sum_{l=0}^{m} L_k(X)\, L_l(Y) \tag{4}$$

where $L_k(X)$ and $L_l(Y)$ are the one dimensional Lagrange polynomials associated with the interpolation points X_i and Y_j, respectively.

When the stereo data is not reliable for accuracy and consistency, a surface that passes through the known points is not the best representation of the actual object. In such a case, a surface approximating the known points may be desirable. Using Grimson's approach [10], the function to minimize is given by

$$\psi = \int\int [(\partial^2 f/\partial x^2)^2 + 2\,(\partial^2 f/\partial x\,\partial y)^2 + (\partial^2 f/\partial y^2)^2]\,dx\,dy + \sigma_1 \sum (f(x,y) - Z(x,y))^2 \qquad (5)$$

The first term of Eq. (5) gives the minimal solution to the quadratic variations in the surface orientation while the second term represents a least squares quantity, where the summation is taken over all points of known depth. The constant σ_1 is a scale parameter that reflects the desired fit.

EXPERIMENTAL RESULTS

The stereo system was tested on several real objects, with image acquisition and digitization performed through a VICOM image processor. The gray level images are of size 512x512 pixels with 256 gray levels. The host computer was a VAX 11/750 under the UNIX 4.3 BSD operating system. In the experiment of Fig. 2, the 3-D information of some points were computed. The average difference between the calculated values and actual measurements was 1.2 %.

Fig. 2. Objects used in an experiment for testing the system's accuracy.

To study the performance of the Lagrange polynomials, we first consider the second order function $Z=98 - 0.3X^2 - 0.4Y^2 - 1.3X - 0.8Y$. This function is used to generate some data points that are used to reconstruct second order Lagrangian polynomials from which the surface is reconstructed. The percent error between the reconstructed surface and the original function was 2.14. The reconstructed data points are given as input to the quadratic variations algorithm. As a result, the percent error is reduced to 1.39. Another example that uses synthetical data is that of Figs. 3 and 4. Lagrange polynomials are used with data overlapping for surface continuity and smoothness. The surface is reconstructed assuming that each point on the XY plane has one depth value Z. Fig. 3 shows the output of the Lagrangian functions while Fig. 4 shows the output of the quadratic variations method.

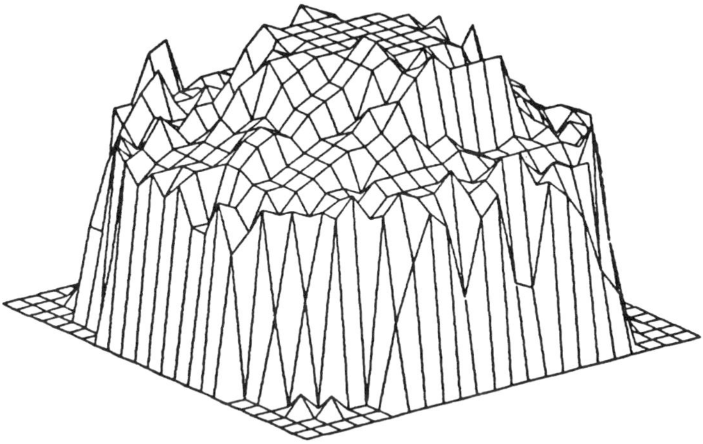

Fig. 3. Output of the Lagrange polynomials.

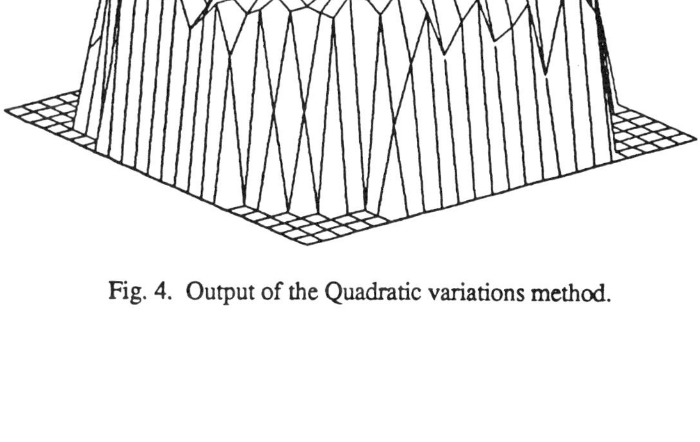

Fig. 4. Output of the Quadratic variations method.

CONCLUSIONS

The multi-camera system presented in this paper addresses the reconstruction of objects in industrial environments. It consists of N number of cameras arranged in an N/2 number of periodic stereo pair structure. We have studied the application of the Lagrange polynomials for fitting a function to a set of scattered 3-D points. The value of the constructed surface at a point Z(X, Y) is calculated locally in finite intervals based on the data at relatively nearby points. This produces a large number of polynomials, however, it requires less computational time than a global solution. Overlapping of data is used when constructing the polynomials to ensure the continuity and smoothness of the surfaces from one scene point to the next. Addressing the problem of common data between two adjacent camera pairs and the possibility of objects with multi-valued surface functions must be addressed before the system can be successfully employed for real applications.

REFERENCES

1. R. A. Jarvis, "A perspective on range finding techniques for computer vision," IEEE Trans. on Patt. Anal. and Machine Intell., vol. PAMI-5, no. 2, pp. 122-139, 1983.
2. E. W. Kent, and M. O. Shneier, "Eyes for automation" IEEE Spectrum, pp. 37-45, March 1988.
3. R. P. Kruger and W. B. Thompson, "A technical and economic assessment of computer vision for industrial inspection and robotic assembly," Proc. IEEE, vol. 69, no. 12, pp. 1524-1538,1981.
4. D. Nitzan, "Three-dimensional vision structure for robot applications," IEEE Trans. on Patt. Anal. and Machine Intell., vol. 10, no. 3, pp. 291-309, 1988.
5. Y. C. Kim and J. K. Aggarwal, "Positioning three-dimensional objects using stereo," IEEE Journal of Robotics and Automation, vol. RA-3, no. 4, pp. 361-373,1987.
6. T. Tsumiyama and T. Huang, "Motion stereo for navigation of autonomous vehicles in man-made environments," Pattern Recognition, vol. 20, no. 1, pp. 105-113, 1987.
7. Y. Yakimovsky and R. Cunningham, "A system for extracting three-dimensional measurements from a pair of T.V. cameras," Computer Graphics and Image Processing, vol. 7, pp. 195-210, 1978.
8. W. Hoff and N. Ahuja, "Surfaces from stereo: integrating feature matching, disparity estimation and contour detection," IEEE Trans. Pattern Anal. and Machine Intell., Vol. 11, no. 5, pp. 121-136, 1989.
9. R. Bachnak and M. Celenk, "Three-dimensional object reconstruction from multiple views of a stereo-based vision system," SPIE vol. 1293, Applications of Artificial Intelligence VIII (1990), pp. 128-139.
10. W.E.L. Grimson, "An implementation of a computational theory of visual surface interpolation," Comp. Vision, Graphics, and Image Processing, vol. 22, 39-69, 1983.
11. D. Terzopoulos and M. Vasilescu, "Adaptive surface reconstruction," Sensor Fusion III: 3-D perception and Recognition, Paul S. Schenker, Editor, Proc. SPIE 1383, pp. 257-264, 1990.
12. D. Lee, "Surface reconstruction with discontinuities," Sensor Fusion III: 3-D perception and Recognition, Paul S. Schenker, Editor, Proc. SPIE 1383, pp. 297-304, 1990.

PART 3 : ROBOT CONTROL

Multiple Finger Control

John T. Wen and Steve H. Murphy
Center for Intelligent Robotic Systems for Space Explorations
Rensselaer Polytechnic Institute
Troy, NY 12180

Abstract

This paper considers the motion and force control of multiple fingers holding a common payload. The main technical challenge posed by this problem is the relative motion between fingers and the payload at the grasp contact. We extend our previous work of the rigid grasp case by separately considering the motion and force control problems. Three motion control strategies are discussed: feedback linearization, arms–as–actuators, and energy Lyapunov function method. Several feedforward and feedback internal force control strategies are also discussed.

1 Introduction

With the miniaturization of actuators and sensors and advances in microprocessor, high performance finger control is increasingly becoming a reality. There have been many multi–finger hand designs, starting with the well–known MIT/Utah hand [1] and many others. There have also been many theoretical studies on kinematic, dynamic, and control issues related to multi–finger hands, for example, [2, 3] to just name a few. At the same time, there has been a great deal of interest in the multiple–arm control. The principal difference between the multi–finger control and multi–arm control is in the contact model; in the former, there can be relative motion at the grasp contact (based on some model), and in the latter, the contact is rigid, i.e., all force/torque can be directly transmitted through the contact. In this paper, we extend our earlier work in multiple–arm control to the multi–finger case. In particular, the move/squeeze decomposition proposed in [4] can be generalized with some simple modifications. As a result, the motion loop and force loop can be separately designed and analyzed.

In the rest of the paper, we will first present a dynamical model of the multi–finger grasp a common load in Section 2, and then discuss the motion and force control problems in Section 3.

2 Dynamic Model

Multiple arm system with contact motion can be modeled as

$$M\ddot{\theta} + C\dot{\theta} + k + J^T f = \tau \tag{1}$$

$$M_c \alpha_c = A^T f - b_c - k_c \tag{2}$$

$$Hf = 0 \tag{3}$$

$$v^+ = J\dot{\theta} \tag{4}$$

$$
\begin{aligned}
v^- &= A v_c = v^+ + H^T W &(5)\\
\alpha^+ &= J\ddot{\theta} + \dot{J}\dot{\phi} &(6)\\
\alpha^- &= A\alpha_c + a = \alpha^+ + H^T\dot{W} + \dot{H}^T W &(7)
\end{aligned}
$$

where H specifies the directions the contact is free to move (see Fig.1 for various examples) and W parameterizes the contact velocity. The first two equations model the arm dynamics and payload dynamics separately. The rest of the equations model the kinematic constraint relationships. It is assumed in this model that the payload is nominally rigid. Note that the stacked notation is used in these equations, e.g., θ denotes the stacked joint variables of all arms, f is the stacked contact force/torque, etc.

There are two key concepts in multiple finger grasping: grasp stability and grasp manipulability. First define an effective grasp matrix and Jacobian (by eliminating the directions where force cannot be exerted and velocity cannot be transmitted, respectively):

$$G = A^T \widetilde{H} \qquad J_h = \widetilde{H}^T J. \tag{8}$$

where $\tilde{H}$ is a full rank matrix whose range coincides with the null space of H. A grasp is stable if any $f_c = A^T f$ can be supported by an appropriate choice of τ. An equivalent analytical statement of this condition is that G is onto (i.e., G is full rank, since G is typically fat). A dual concept is grasp manipulability. A grasp is manipulable if every v_c can be achieved with some appropriately chosen $\dot{\theta}$. This condition can be equivalently stated as

$$R(J_h) \supset R(G^T) \tag{9}$$

where $R(\cdot)$ denotes the range space of a linear map.

A complete parameterization of f that satisfies (3) is

$$f = \widetilde{H}\rho. \tag{10}$$

3 Multi–Finger Motion and Force Control

Feedback Linearization and Arms-as-Actuators

We begin by eliminating α_c, $\ddot{\theta}$:

$$(G^T M_c^{-1} G + J_h M^{-1} J_h^T)\rho = J_h M^{-1}\tau - \eta_1 \tag{11}$$

where

$$\eta_1 = \widetilde{H}^T A(-b_c - k_c + a) + \widetilde{H}^T J M^{-1}(C\dot{\theta} + k).$$

If the complete kinematic and dynamic information is known and none of the fingers is in a singular configuration, then the following control law can be implemented:

$$\tau = J_h^T (J_h M^{-1} J_h^T)^{-1}(\eta_1 + (G^T M_c^{-1} G + J_h M^{-1} J_h^T)\rho_d) \tag{12}$$

where ρ_d is an effective control signal to be specified later. With this control law, we have

$$(G^T M_c^{-1} G + J_h M^{-1} J_h^T)(\rho - \rho_d) = 0. \tag{13}$$

For motion control, ρ and ρ_d are both in $R(G^T)$ (the component in $N(G)$ only contributes to the internal force). In this case, $\rho = \rho_d$. Now, ρ_d can be chosen to either

Line Contact

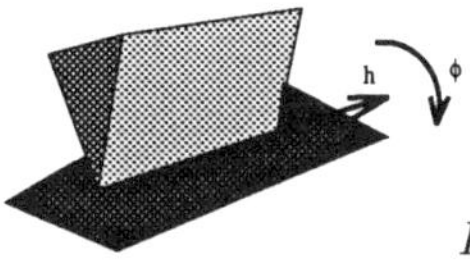

$$Hf = [h^T \quad 0]\begin{bmatrix} N \\ F \end{bmatrix} = 0 \quad H^T W = \begin{bmatrix} h \\ 0 \end{bmatrix}\dot{\phi}$$

Sliding Contact

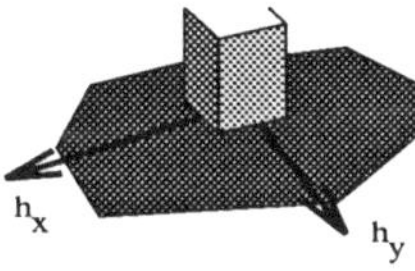

$$Hf = \begin{bmatrix} 0 & h_x^T \\ 0 & h_y^T \end{bmatrix}\begin{bmatrix} N \\ F \end{bmatrix} = 0 \quad H^T W = \begin{bmatrix} 0 & 0 \\ h_x & h_y \end{bmatrix}\begin{bmatrix} v_x \\ v_y \end{bmatrix}$$

Point Contact

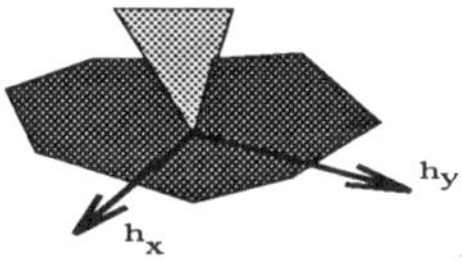

$$Hf = \begin{bmatrix} I & 0 \\ 0 & h_x^T \\ 0 & h_y^T \end{bmatrix}\begin{bmatrix} N \\ F \end{bmatrix} = 0 \quad H^T W = \begin{bmatrix} I & 0 & 0 \\ 0 & h_x & h_y \end{bmatrix}\begin{bmatrix} \omega_z \\ v_x \\ v_y \end{bmatrix}$$

Point Contact with Friction

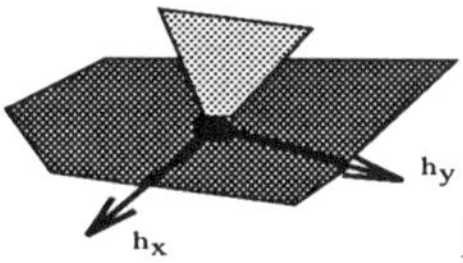

$$Hf = [I \quad 0]\begin{bmatrix} N \\ F \end{bmatrix} = 0 \quad H^T W = \begin{bmatrix} I \\ 0 \end{bmatrix}\omega$$

Soft Finger

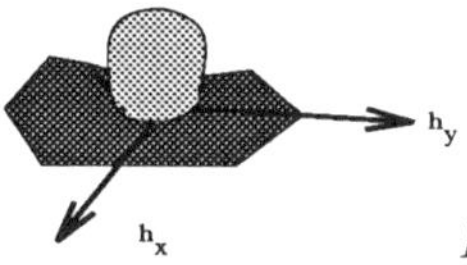

$$Hf = \begin{bmatrix} h_x^T & 0 \\ h_y^T & 0 \end{bmatrix}\begin{bmatrix} N \\ F \end{bmatrix} = 0 \quad H^T W = \begin{bmatrix} h_x & h_y \\ 0 & 0 \end{bmatrix}\begin{bmatrix} \omega_x \\ \omega_y \end{bmatrix}$$

Figure 1: Some Typical Contact Models

obtain decoupled double integrator (as in feedback linearization) or a required f_c (as in arms–as–actuators). In the former case,

$$\rho_d = (G^T M_c^{-1} G)^{-1} G^T (M_c^{-1}(b_c + k_c) + u). \tag{14}$$

Then, in the closed loop

$$G^T(\alpha_c - u) = 0. \tag{15}$$

If the grasp is stable, then $N(G^T) = \{0\}$. This implies

$$\alpha_c = u \tag{16}$$

which is a standard linear control problem.

In the arms–as–actuator case, f_c is regarded as the effective control variable. Again assuming a stable grasp, then any control strategy involving f_c can be obtained by solving

$$G\rho_d = f_c. \tag{17}$$

Note that grasp manipulability is assumed in either of the above schemes.

Energy Lyapunov Function Approach

The approach presented above has the appealing feature of reducing the original complex nonlinear problem to a much simpler problem. However, exact kinematic and dynamic model of all the fingers and the payload are required for its implementation. For a single manipulator or multiple manipulators with rigid grasp, the inherent Hamiltonian structure of the system has been exploited to motivate a class of controllers that are simple in structure [5, 6, 4]. We call this approach the energy Lyapunov Function approach since an energy motivated Lyapunov Function is used to show the stability of the closed loop system. We now extend this approach to our multi–finger model. The overall formulation is very similar to the rigid grasp case in [4, 7], therefore, some of the details are omitted for the sake of space.

The basic idea is to choose the joint torque in the following form:

$$\tau = \tau_m + \tau_s + \tau_{ff}. \tag{18}$$

The first term, τ_m, is the motion control which we will choose shortly based on a Lyapunov analysis. The second term, τ_s, is for the internal force control. In order to avoid coupling between the motion and force loops, τ_s is chosen to be of the following form:

$$\tau_s = J_h^T \tilde{G} \Xi \tag{19}$$

so that it will not affect the motion. The last term, τ_{ff}, compensates for all the gravity load and is chosen as

$$\tau_{ff} = k - J^T \widetilde{H} \rho_c$$

where ρ_c is any solution of

$$G\rho_c = k_c.$$

If the grasp is stable, then the gravity equation above is always solvable.

To see the stability of the closed loop motion control system, consider the following energy–motivated Lyapunov function candidate:

$$V = \frac{1}{2} v_c^T M_c v_c + \frac{1}{2} \dot{\theta}^T M \dot{\theta} + U^* \tag{20}$$

where U^* is the desired potential energy. As in the rigid grasp case, U^* may be chosen as a function of the joint variables, finger–tip variables, or payload configuration variables. In the first two cases, only the steady state convergence can be shown, and a phenomenon called *Jamming* [4] can occur. For simplicity, we shall consider only the payload configuration feedback case here. In this case, the motion control is of the following form:

$$\tau_m = -J_h^T F_P - J_h^T F_D - D\dot{\theta} \tag{21}$$

where

$$GF_P = K_p \Delta x_c \qquad GF_D = K_v v_c \tag{22}$$

and K_p, K_v, D are positive definite gains. Notice that if the grasp is stable, then (22) is always solvable.

After differentiating V along the solution, and using the fact that $\frac{1}{2}\dot{M} - C$ is skew symmetric, we obtain (with some simple algebra omitted)

$$\dot{V} = -v_c^T K_v v_c - \dot{\theta}^T D\dot{\theta}. \tag{23}$$

By the Invariance Principle, the system would reach a steady state. If J_h is *asymptotically nonsingular* (i.e., nonsingular as $t \to \infty$), then at the steady state, we have

$$GF_P = 0$$

which implies $\Delta x_c = 0$ as desired.

Internal Force Control

Through an orthogonal decomposition, internal force control can be separately considered from the motion control. In fact, any stable motion control law can be used in conjunction the force control. It has recently been pointed out in the force control literature [8, 7, 9] that orthogonal decomposition in the force/torque space (and, as a dual, in the velocity/angular velocity space) can lead to unit inconsistency problem. Some remedy was proposed in [7, 9]. We will not address the subtlety here, except to note that the precaution taken in [7, 9] may be needed here as well. The detail will be communicated in the future. In the ensuing analysis, we shall assume fixed units have been chosen for all physical variables.

For force control, we first rewrite (11) as (note that the force control is of the form (19))

$$(G^T M_c^{-1} G + J_h M^{-1} J_h^T)\rho = J_h M^{-1} J_h^T \tilde{G}\Xi - \eta_2 \tag{24}$$

where η_2 includes the motion control and all inertial forces. Since gravity is compensated for, η_2 tends to zero asymptotically for an asymptotically stable motion control law (or a constant if only steady state is reached). Note that Ξ does not affect η_2 since it does not cause any motion. If we decompose ρ as (with all units fixed)

$$\rho = \tilde{G}\xi + G^T\zeta \tag{25}$$

and regard ξ as a parameterization of the internal force, then, under the assumption of Jacobian nonsingularity (actually, we only need $J_h^T \tilde{G}$ to be full rank), we have

$$\xi = \Xi + \eta \tag{26}$$

where η lumps together all inertial forces, does not depend on Ξ or ξ, and tends either to zero or a constant. Now, the same force control analysis as in the multi–arm (rigid grasp) case can be applied. In the absence of force measurements, Ξ can be chosen as an open loop control. If force measurement (i.e., ρ) is available, then the algorithm of choice is integral force feedback. Frequently, only a partial contact force can be measured (e.g., contact normal force via tactile sensors). Then a combination of integral feedback and feedforward can be used.

4 Conclusion

The motion and force control problem of a multi–finger grasp is analyzed in this paper. By defining an effective Jacobian and grasp matrix, the move/squeeze decomposition concept for the rigid grasp case can be applied. Several motion control and force control strategies are discussed.

Acknowledgment

This work is supported by the Center for Intelligent Robotic Systems for Space Explorations at the Rensselaer Polytechnic Institute funded by the National Aeronautics and Space Administration under grant NAGW 1333.

References

[1] S.C. Jacobsen, J.E. Wood, D.F. Knutti, and K.B. Biggers. The Utah/MIt hand: Work in progress. In M. Brady & R. Paul, editor, *Robotics Research.* MIT Press, 1984.

[2] J. Salisbury. *Kinematic and Force Analysis of Articulated Hands.* PhD thesis, Stanford University, Palo Alto, CA., 1982.

[3] M.R. Cutkosky. *Robotic Grasping and Fine Manipulation.* Kluwer Academic Publishers, Hingham, MA, 1985.

[4] J.T. Wen and K.Kreutz. Stability analysis of multiple rigid robot manipulators holding a common rigid object. In *Proc. 27th IEEE Conference on Decision and Control*, Austin, TX, 1988.

[5] M. Takegaki and S. Arimoto. A new feedback method for dynamic control of manipulators. *ASME J. Dynamic Systems, Measurement and Control*, 102, June 1981.

[6] J.T. Wen and D.S. Bayard. Simple robust control laws for robotic manipulators, part I: Non–adaptive case. In *JPL/NASA Telerobotics Workshop*, January 1987.

[7] S. Murphy and J. Wen. Force decomposition in robot force control. In *Proc. 1991 American Control Conference*, Boston, MA, June 1991.

[8] J. Duffy. The fallacy of modern hybrid control theory that is based on orthogonal complements of twist and wrench spaces. *Journal of Robotic Systems*, 7(2):139–144, Feb. 1990.

[9] J.T. Wen and S.H. Murphy. Position/force control in multiple–manipulator systems. In *Proc. AIAA Space Programs and Technologies Conference*, Huntsville, AL, March 1992.

A DISCRETE-TIME VIRTUAL PASSIVE CONTROL ALGORITHM FOR LINEAR DYNAMIC SYSTEMS

GORDON K. F. LEE AND ERIC CLARK*

Mars Mission Research Center, North Carolina State University, Raleigh, NC

ABSTRACT

The virtual passive control approach has recently been shown to be an effective technique for robust compensation of linear as well as nonlinear systems, when the system is open-loop stable and behaves as a block second-order plant. Mechanical systems as well as current space structure designs process this property whereby inertia, damping and stiffness parameters characterize the dynamics.

Implementation of such controllers using digital systems may lead to instability however unless discretization issues are addressed. This paper presents a discrete-time representation of the Virtual Passive Approach and develops the robust controller structure for linear systems. The algorithm is then applied to a model of an actual experimental testbed; the test article is a ten-bay truss structure located at NASA Langley's Spacecraft Dynamics Branch. Results illustrate the robustness properties of the algorithm and expose issues which need to be addressed in future research activities.

I. INTRODUCTION

Many of the current space structure designs require large dimensional lightweight components; for example Space Station is made up of many struts and joints interlaced in a grid structure. Thus typical operations such as rendezvous and docking between structures or the reorientation of a single structure may cause unwanted vibration along the system due to its high degree of flexibility.

Researchers have addressed this problem from many different points of view. For example, in [1,2], authors have looked at smart composite materials which may employ distributed actuators and sensors to handle vibration suppression. Others [3,4, for example], have looked at robust controllers which damp the vibration through active or passive compensation.

The approach developed here is based upon the virtual passive control methodology. This strategy is a robust compensator that may be applied to nonlinear as well as linear systems and systems with unknown parameters or even unknown structure. The approach considers the system as one which can equivalently be represented by a mechanical system (spring-mass-dashput structure). By designing passive second-order block controllers, it can be shown that the overall system can not be destabilized, provided the open-loop "behaves" like a mechanical system.

For completeness sake, a brief summary of the continuous-time linear virtual passive control strategy is presented here [3].

---

* This work was supported by NASA Grant No. NAGW-1331 to the Mars Mission Research Center.

Consider a pseudo-model as:

$$M\ddot{x} + D\dot{x} + K\dot{x} = Bu$$
$$y = H_a\ddot{x} + H_v\dot{x} + H_d x \qquad (1)$$

representing a plant with mechanical parameters M (mass matrix), D (damping matrix) and K (spring constant matrix).

A controller may be selected with second order dynamic blocks as:

$$M_c\ddot{x}_c + D_c\dot{x}_c + K_c x_c = B_c u_c$$
$$y_c = H_{ac}\ddot{x}_c + H_{vc}\dot{x}_c + H_{dc} x_c \qquad (2)$$

Here x_c represents the controller states of dimension n_c, $\{M_c, D_c, K_c\}$ are the controller mass, damping and stiffness matrices, respectively, which are usually selected as positive definite symmetric, B_c is the force distribution matrix, and $\{H_{ac}, H_{vc}, H_{dc}\}$ are the acceleration, velocity and displacement output matrices, respectively. The vector u_c is the controller input which is actually the plant output y. The controller output y_c is actually the plant input u, assuming no reference input is added. Note (2) is really a pseudo controller model as the actual controller to be implemented can be modeled by a transfer matrix with gains tuned on-line, if desired. Model (2) is a baseline structure for control gain selection.

Then augmenting the dynamics yields

$$M_t\ddot{x}_t + D_t\dot{x}_t + K_t x_t = 0 \qquad (3)$$

where

$$M_t = \begin{bmatrix} M & -B H_{ac} \\ -B_c H_a & M_c \end{bmatrix}, \qquad D_t = \begin{bmatrix} D & -B H_{vc} \\ -B_c H_v & D_c \end{bmatrix},$$

$$K_t = \begin{bmatrix} K & -B H_{dc} \\ -B_c H_d & K_c \end{bmatrix}, \qquad \underline{x}_t = \begin{bmatrix} x \\ x_c \end{bmatrix} \qquad (4)$$

The idea is to force $\{M_t, D_t, K_t\}$ to be positive definite resulting in (4) being asymptotically stable, by proper selection of $\{M_c, D_c, K_c\}$ and $\{H_{dc}, H_{vc}\}$. Note H_{ac} is usually constrained in actual mechanical system designs.

In [3], various continuous-time controller designs are developed using the virtual passive control structure. With the increase flexibility of computer implementation, a discrete-time version of (2) must be developed and analyzed in terms of sampling time requirements and digitization effects. This is the objective of this paper.

II. THE DISCRETE-TIME VIRTUAL PASSIVE CONTROLLER ALGORITHM

There are several approaches for developing a discrete-time virtual passive control (VPC) algorithm in order to implement the method on a computer. In the most general case, one can consider a discrete-time model for the plant in terms of discretized mechanical parameters as:

$$M\,x(k+2) + D\,x(k+1) + K\,x(k) = B_2\,u(k+2) + B_1\,u(k+1) + B_o\,u(k)$$
$$y(k) = H_1\,y(k-1) + H_o(k-2) + H_a\,x(k) + H_v\,x(k-1) + H_d\,x(k-2) \qquad (5)$$

This general model can be rationalized by performing a simple bilinear-z transformation on (1) and observing the corresponding time difference representation.

Similarly a VPC in discrete-time form can be constructed in parallel to (5). Let the subscript c denote the control dynamics. Then

$$u(k) = F_1\, u(k-1) + F_o\, u(k-2) + H_{ac}\, x_c(k) + H_{vc}\, x_c(k-1) + H_{dc}\, x_c(k-2) \quad (6)$$

Augmenting (5) and its control dynamics counterpart, and using (6), a time-domain equivalent to (4) becomes

$$\begin{bmatrix} (Mq^2+Dq+K) & -(B_2q^2+B_1q+B_o)(q^2I-F_1q-F_o)^{-1}(H_{ac}q^2+H_{vc}q+H_{dc}) \\ -(B_{2c}q^2+B_{1c}q+B_{oc})(q^2I-H_1-H_o)(H_aq^2+H_vq+H_d) & (M_cq^2+D_cq+K_c) \end{bmatrix} \begin{bmatrix} x(k) \\ x_ck \end{bmatrix} = 0 \quad (7)$$

with q the advanced time operator. The discrete-time plant model is such that $(Mq^2 + Dq + K)^{-1}\,(B_2q^2 + B_1q + B_o)$ is a stable transfer matrix by proper selection of the sampling period T. The overall system matrix of (7) must be kept stable by proper controller parameter selection.

As (7) provides many degrees of freedom in the control parameters, a simplified form of (5) is selected for this preliminary investigation. Consider a discrete-time plant model:

$$\begin{aligned} Mx(k+2) + Dx(k+1) + Kx(k) &= B(u(k+2) + 2u(k+1) + u(k)) \\ y(k+2) + 2y(k+1) + y(k) &= H_a\, x(k+2) + H_v\, x(k+1) + H_d x(k) \end{aligned} \quad (8)$$

The controller is of the same form as (8) but with subscript c.

Then augmenting the dynamics yields:

$$M_t\, x_t(k+2) + D_t\, x_t(k+1) + K_t\, x_t(k) = 0 \quad (9)$$

where M_t, D_t and K_t are now in the same form of (4).

Now if the controller parameter can be chosen such that the eigenvalues of

$$A_t = \begin{bmatrix} 0 & I \\ -M_t^{-1} K_t & -M_t^{-1} D_t \end{bmatrix}$$

are stable, the overall system of (7) is stable. It was found that stability of (7) can usually be assumed if the following conditions are met:

1. M_t is positive definite
2. $M_t + D_t + K_t$ is positive definite
3. $M_t - D_t + K_t$ is positive definite
4. Each eigenvalue $\{\lambda_i\}$ of M_t is greater in magnitude than the corresponding eigenvalue $\{\lambda_i\}$ of K_t, when the eigenvalues are sequenced in same order.

Here M and K are assumed stable by proper selection of the sampling period. A formal proof of this theorem is currently being constructed.

In order for these conditions to be met, the off-diagonal terms of M_t, D_t, and K_t must be symmetric. That is:

$$BH_{ac} = H_a^T B_c^T, \qquad BH_{vc} = H_v^T B_c^T, \qquad BH_{dc} = H^{vd} B_c^T$$

To guarantee M_t positive definite, we add a direct feedback term:

$$B\, H_{ac}\, M_c^{-1}\, B_c\, H_a,$$

which effectively "neutralizes" the off-diagonal matrices in M_t.

Further conditions include:

$$M + K > D$$

$$M_c + K_c > D_c$$

To insure condition (4) above, it is suggested that the off-diagonal terms of M_t and K_t be set equal, and that M_c be set equal to K_c.

It was found that this discrete-time VPC algorithm works well in keeping the closed-loop system stable. Proper tuning of the control parameters will improve overall performance.

III. THE EXPERIMENTAL TESTBED

In order to evaluate the discrete-time VPC, a numerical model of a test article located at NASA Langley Research Center was selected.

The experimental testbed under study is shown in Figure 1. It consists of 10 bays, each of which is a cube of 10 inches in dimension. Fourteen rods complete the bay configuration -- three sets of four rods for the six faces of the cube and single laced diagonals that cross with the diagonal in the opposite face of the bay. The lacing pattern has repeating double bays for structural integrity. The 10-bay mock-up is hung in a vertical direction from the ceiling, extending over 8 feet. A detail description of the testbed is provided in [4].

In order to perform initial benchmarks on the characteristics of the virtual control algorithm for the 10-bay testbed, a finite element model was generated using MSC/NASTRAN. An additional mass load of 20 pounds is placed at the bottom bay. The load is used to constrain the significant frequency range at under 10Hz (typical for other experimental testbeds at NASA Langley Research Center).

The 44 nodes or truss joints are modeled as masses with magnitude assigned according to the number of struts joining at the node. Actuators and sensors are also modelled as lumped masses.

Using MSC/NASTRAN, the systems nodes were generated. A reduced-order state-space model was produced as a preliminary model for the 10-bay mock-up. The states are displacements and vibrations, inputs are thrusters and outputs are displacements from the strain gauges. Displacements are in inches, velocities are in in/sec, inputs are in lbs, outputs are in inches.

IV. RESULTS

The ten bay testbed was modelled in state-space form resulting in 22-pairs of damping ratio, frequency of oscillation plant parameters. First the open-loop plant was excited with an impulse; output one's response is shown in Figure 2. For complete-ness sake, a continuous-time VPC was designed according to (4) and a typical result shown in Figure 3. Note the improvement in reduced settling time and in the magnitude of the overshoot.

A bilinear-z, analog approximation discrete-time VPC was then constructed in order to see the effects of digitization. Figure 4 illustrates the results. By proper selection of the sampling period, the local controller approximation performs adequately.

Finally the direct digital VPC using the algorithm discussed in Section II was implemented. Several control parameter sets were tried and tuning such parameter stability constraints is an area of future research. These results are discussed in more detail at the conference. In all the simulation results involving digitization, a zero-order hold sampler was modelled for the D/A and ideal sampler for the A/D.

Results indicate that the direct digital VPC is a viable approach in implementing the virtual passive controller design directly (without first employing a continuous-time design and applying an analog-invariance approximation method).

V. CONCLUSIONS

This paper introduced the virtual passive control methodology and presented a direct digital algorithm which may be used for computer implementation of this technique. In numerical simulations, the discrete-time VPC seems to be a feasible control methodology for robust compensation, even under uncertainty.

As this paper presents preliminary results, several issues are still being addressed. First a formal proof of the stability of the digital VPC algorithm is in order. Currently a Lyapunov formulation is being considered to refine this algorithm.

Secondly, as with the continuous-time case, the VPC provides a guideline in selecting the robust controller parameters. However tuning such parameters to improve performance is still a design issue and some formalism as to the relationship between the sets of controller parameters one can choose and the performance measure one desires (including uncertainty bounds) would make the approach even more appealing.

Finally this approach is being investigated for real-time implementation on several experimental testbeds, including the ten-bay structure at NASA Langley Research Center. These investigations may provide additional insight into the characteristics of the virtual passive control methodology.

VI. REFERENCES

1. B. Wada, "Adaptive Structures", 30th Structures, Structural Dynamics and Materials Conference, Mobile, pp. 1-11 (1989).
2. F. Kuwao, S. Yoshihara, S. Motohashi, K. Takahara, and M. Natori, "Vibration Characteristics and Shape Control of Adaptive Planar Truss Structures, 30th Structures, Structural Dynamics and Materials Conference, Mobile, pp. 1136-1144. (1989).
3. J. N. Juang and M. Phan, "Robust Controller Designs for Second-Order Dynamic Systems: A Virtual Passive Approach", NASA Tech. Mem. TML-102666 (1990).
4. E. Clark and G. K. F. Lee, "Modeling and Control Issues for a 10-Bay Mock-up of Flexible Space Structures," Proc. of the ISMM Int'l Conf. on Microcomputer Applications, Long Beach (1990).

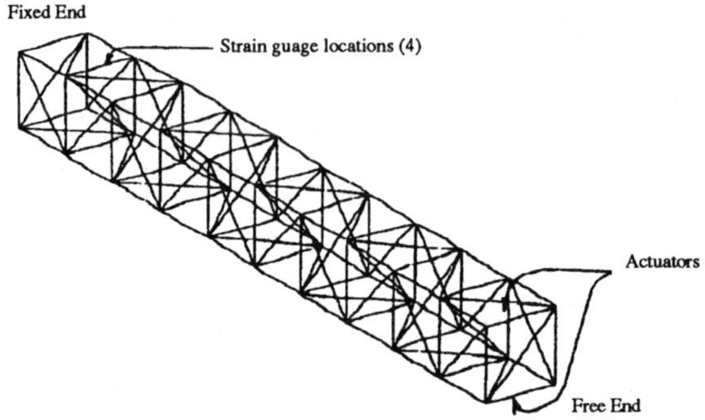

Figure 1: A diagram of the 10-bay mock-up experimental testbed

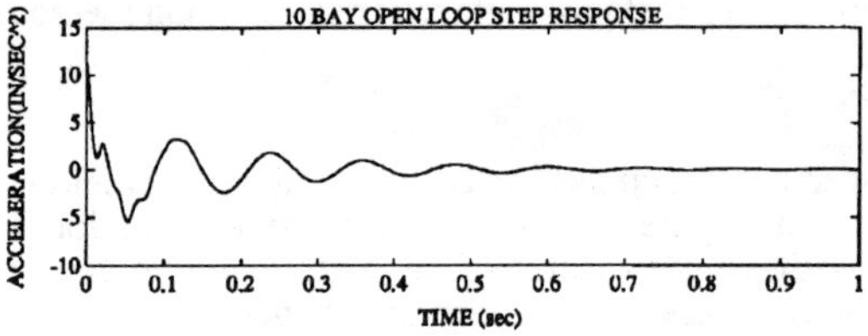

Figure 2: Open-loop response of output one

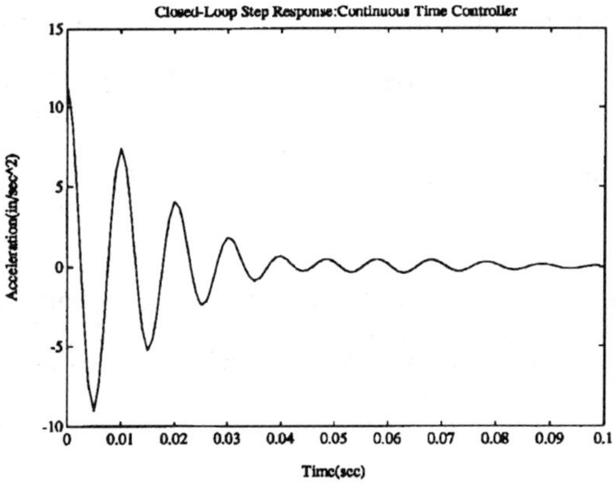

Figure 3: Closed-looped response of output one - continuous-time case

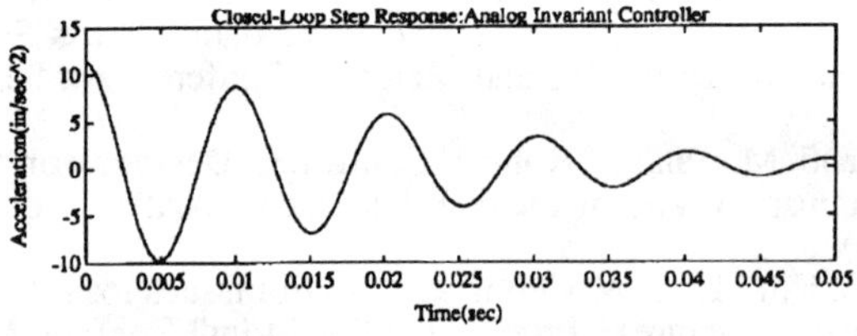

Figure 4: Closed-loop response - analog approximate method

FORCE CONTROL OF PLANAR POWER GRASP IN THE DIGITS SYSTEM

Chi H. Chen, Khalid Mirza, and David E. Orin
Department of Electrical Engineering
The Ohio State University
Columbus, OH 43210

ABSTRACT

The implementation and experimental results of planar power grasp in the DIGITS grasping system are presented in this paper. A system model is developed for the force distribution analysis, and it is used to solve for the internal forces for a planar power grasp by including constraints of force balance, maximum joint torque limits, and friction effects. With open-loop force control only, a maximum weight of 30 lbs was stably grasped by a two-finger system.

INTRODUCTION

Over the past decade, a number of multifinger dexterous hands have been designed and studied to provide the flexibility desired in grasping and manipulating objects [1]. However, one of their limitations is their inability to apply enough force at the finger tips to handle heavy objects, due to their delicate mechanical structure. To achieve the same objective, power grasp is an important alternative to finger-tip or precision grasp [2]. With power grasp, an object is enveloped by the fingers, with multiple points of contact between the surfaces of the fingers and the object. With the enveloping characteristic providing form closure, power grasp provides better grasp stability [3] and may be effectively used to handle heavier objects [4].

To provide a stable grasp without slippage or crushing, the force distribution between the grasping system and the object is a key area of investigation in achieving power grasps. In our earlier paper [4], an algorithm was developed to find the force distribution for power grasps in the DIGITS grasping system, under development at The Ohio State University [5]. We estimated the maximum weight that could be theoretically handled for a given object radius and surface friction as well as specified joint torque limits for the DIGITS system. For an object with a one inch radius, it was estimated that a load of 32 lbs could be stably handled by a two-finger system.

In this paper, the implementation of the power grasp operation with force distribution control is presented for a two-finger DIGITS system. To effectively realize power grasp, four different stages are defined and developed here — Preshape, Compliant Touch, Grip and Release. Experimental results are also presented for force control in the Grip stage.

THE DIGITS GRASPING SYSTEM

The DIGITS (Dexterous Integrated Grasping with Intrinsic Tactile Sensing) grasping system is a joint project of the Mechanical Engineering and the Electrical Engineering Departments at The Ohio State University [5]. It is designed to perform both precision grasps and power grasps. Figure 1 shows a graphical model of the DIGITS. Currently only the two-finger version is available to investigate the planar power grasp case with the base joint at the palm (axis perpendicular to it) fixed.

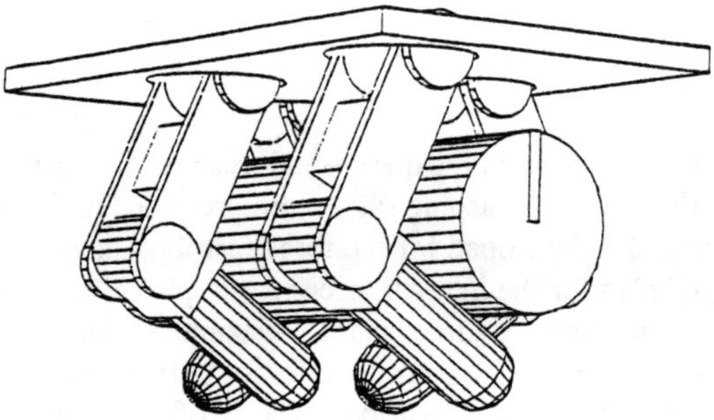

Figure 1: A graphical model of DIGITS grasping system with four fingers

POWER GRASP CONTROL STAGES

Realizing the planar power grasp process in the DIGITS system is the focus of this paper. To provide a smooth operation for power grasp under supervisory control, four stages have been developed and implemented: Preshape, Compliant Touch, Grip, and Release. Table 1 lists all the grasp stages and their main functions.

Table 1: List of the grasp process stages and their main functions

Stage	Function(s)
Preshape	• Configures the fingers to an appropiate grasp posture. • Synchronizes all finger joint motions.
Compliant Touch	• Closes each finger to achieve a light touch on to the object without applying large impact forces.
Grip	• Applies the desired clinch level specified by the user. • Uses force control algorithm.
Release	• Returns the fingers to a predefined release position. • Synchronizes all finger joint motions.

DIGITS MODEL

To realize planar power grasp in the DIGITS system, the system model shown in Figure 2 is developed and analyzed. This model is similar to our earlier model [4], except that the width of the fingers are added in order to be more realistic. The model is developed for the planar case with a point of contact on each link including the palm. For any shape (including irregular shapes), an object can be inscribed within a cylindrical volume whose nominal radius is determined from the circumference of the object. As a result, the object is modeled to be cylindrical. Such a simple model is acceptable in most cases since force control allows the fingers to adapt to any irregularities in shape. Furthermore, a cylindrical model also simplifies the derivation of the inverse kinematic equations [6]. All contacts between the surfaces of the fingers and the object are modeled as point contacts with friction. Joint torques are also included in this model and are used in our force distribution analysis.

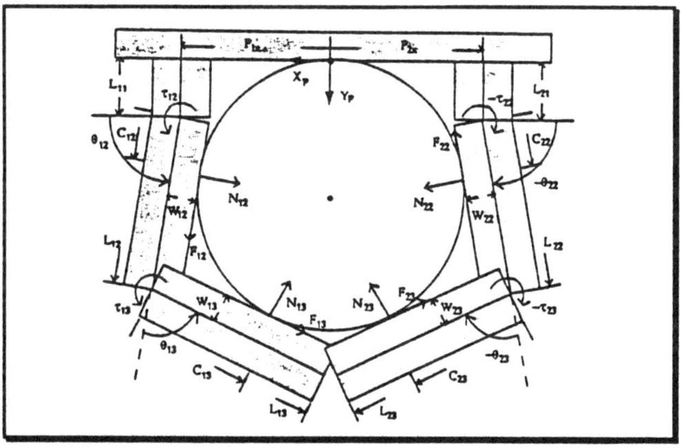

Figure 2: The DIGITS planar model with all parameters and variables shown

FORCE DISTRIBUTION ALGORITHM

A key area of investigation to effectively realize power grasp is the force distribution problem. It is a constrained optimization problem in that the internal forces are to be computed to balance the forces on the object while limiting them to be within the actuators' capabilities and to avoid slippage. The problem is underspecified so that the load may be maximized or balanced among the fingers, or some other objective may be optimized. In our earlier paper [4], an algorithm was developed to find the force distribution for power grasp and was based on linear programming. Using the algorithm, the theoretical maximum weight which can be handled by DIGITS was computed for a given object size. Here the same algorithm, modified for the system model presented earlier, is used to compute the required joint torques. Profiles for these are shown in Figure 3.

In Figure 3, two three-dimensional plots, generated using linear programming, are given for the symmetrical case for the joint torque profiles for joints 2 and 3. The results are for an object with a radius of 1.125 in and for joint torque limits of 16 in-lb. The clinch level is provided by the human operator and is defined as the desired normal force to be applied to the object at the contacts. It ranges from 0 to 10 lbs with a resolution of 0.1 lb. The estimated weight of the object is also supplied, and it ranges from 0.5 to 10 lbs with a resolution of 0.5 lb. A friction coefficient of 0.5 was used to generate the profiles.

Joint Torque Profile for Joint 2 **Joint Torque Profile for Joint 3**

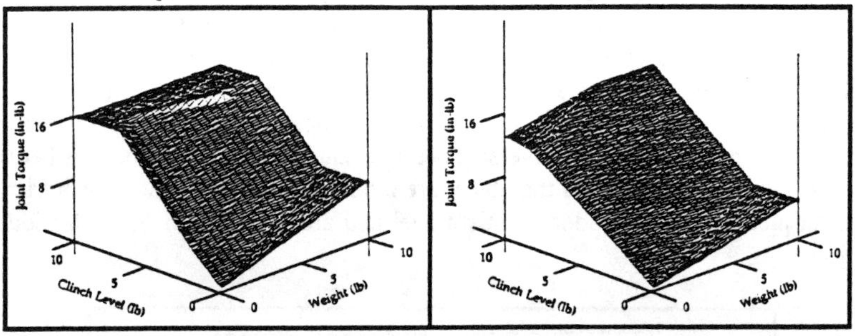

Figure 3: Torque profiles for joints 2 and 3 for an object with radius of 1.125 in.

EXPERIMENTAL RESULTS

Grip control provides open-loop control of force with a closed-loop current controller. Its main function was to follow the joint torque profiles shown in Figure 3 for a varying clinch level furnished by the operator. In this section, an experiment was performed for one particular case. An object with a specified radius and weight was chosen, and the results of grip control are presented. The actual motor current of Joint 2 of Finger 1 is compared to the commanded motor current as the clinch level is varied.

For the experiment, a steel pipe with a radius of 1.125 in and weight of 1.7 lbs was chosen as the object grasped. Then, a look-up table containing the joint torque profiles for the chosen object was generated by using the force distribution algorithm. The values in the look-up table were converted from joint torque to commanded motor current using a relationship derived empirically. Currently, the DIGITS system allows only one set of joint torque profiles, for a specific object. The clinch level in the table ranges from 0 to 10 lbs with a resolution of 0.1 lb.

To perform the experiment, a stepped ramp of clinch level, shown in Figure 4(a), was sent to the grip control where the commanded motor currents were obtained from a look-up table. Figures 4(b) and 4(c) show the comparison between the commanded motor current and the actual motor current of Joint 2 of Finger 1. The results show that the actual current closely follows the commanded value.

Although no force feedback signal is currently available, the operator can often estimate the clinch level through visual observation of the object grasped. With joystick control, the operator can effectively control the force applied to the object. Also, in all of the experiments of grip control, the open-loop force controller successfully stabilized the system even in the presence of external disturbances.

One additional experiment was completed during the Grip stage. The maximum weight which could be handled by two DIGITS fingers was investigated. The long

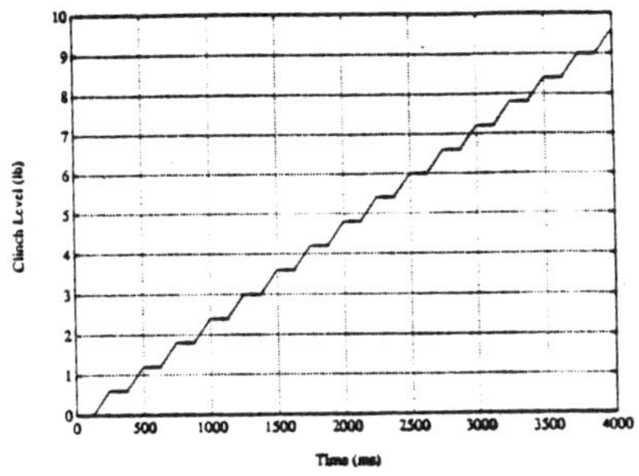

(a) Commanded clinch level

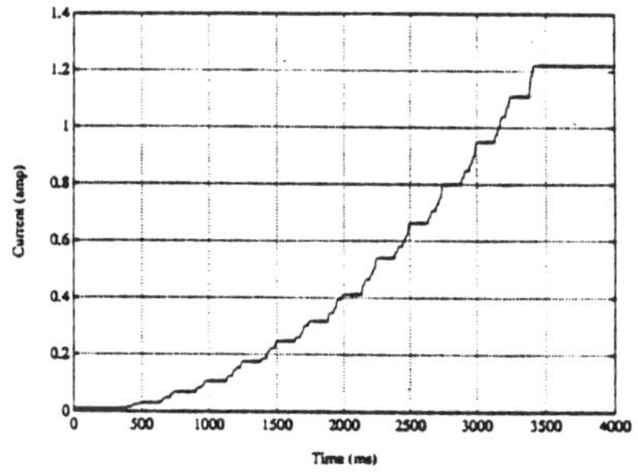

(b) Commanded motor current for Joint 2 of Finger 1

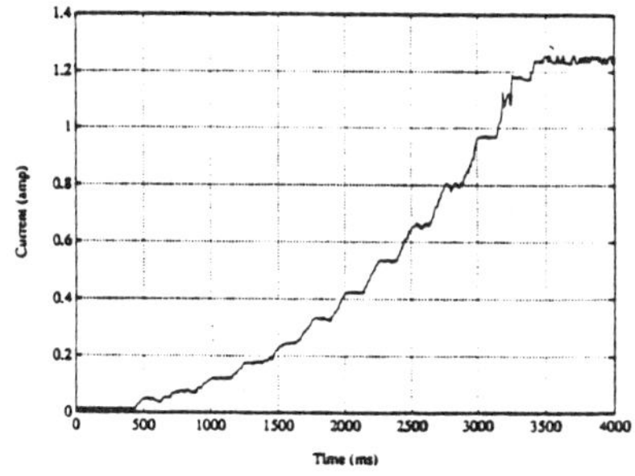

(c) Actual motor current for Joint 2 of Finger 1

Figure 4: Experimental results of grip control

steel pipe with a radius of 1.125 in was grasped by DIGITS. With constant joint torques for a clinch level of 10 lbs, the weight of the pipe was increased by adding weights to the two ends of the pipe in a balanced fashion. Before the pipe dropped from the DIGITS grasp, a maximum weight of 30 lbs was observed. The theoretical maximum weight which can be handled with DIGITS for a similar setup, obtained from our earlier analysis [4] , is also close to 30 lbs.

SUMMARY AND CONCLUSIONS

This paper has presented a brief description of the DIGITS system and the development of force control for the planar power grasp process. Four stages of power grasp — Preshape, Compliant Touch, Grip and Release — were developed to provide a smooth operation for power grasp.

The force distribution algorithm derived from our system model allows an operator to specify the size and weight of the object, and it generates a set of optimized joint torques for a desired clinch level. In this work, the ability to stably grasp an object in power grasp has been demonstrated. With open-loop force control only, the two-finger DIGITS grasping system can handle a 1.125 in radius pipe with a maximum weight of 30 lbs.

In this paper, only the static case in power grasp is investigated. To extend beyond the open-loop force control to manage object slippage and twisting, an impedance control scheme is under investigation to take the dynamic effects of the DIGITS system into account. A dynamic control algorithm is being developed to compensate for any external disturbances to the object to insure a stable grasp.

Although a two-finger DIGITS system is used in this work, a three-dimensional configuration for the DIGITS grasping system is being explored to provide a greater flexibility in grasping. Several graphical models have been developed by Mirza and Orin [4] and Coe and Waldron [5], with experimentation for a four-finger system to be forthcoming.

REFERENCES

[1] Gregory P. Starr, "Experiments in Assembly Using a Dexterous Hand," *IEEE Transactions on Robotics and Automation,* vol. 6, pp. 342-347, June 1990.

[2] J. Napier, "The Prehensile Movements of the Human Hand," J. Bone Joint Surg., vol. 38B, pp. 902-913, November 1956.

[3] M. R. Cutkosky, "On Grasp Choice, Grasp Models, and the Design of Hands for Manufacturing Tasks," *IEEE Transactions on Robotics and Automation,* vol. 5, pp. 269-279, June 1989.

[4] Khalid Mirza and David E. Orin, "Control of Force Distribution for Power Grasp in the DIGITS System," *Proc. 29th IEEE Conf. on Decision and Control*, pp. 1960-1965, Honolulu, Hawaii, December 1990.

[5] S. J. Coe and K. J. Waldron, "Mechanical Design of the DIGITS System," in *Proc. of the 1989 1st. National Applied Mechanisms and Robotics Conference,* Cincinnati, Ohio, November 1989.

[6] Chi H. Chen, "Multiprocessor Force Control of Power Grasp in the DIGITS System," M.S. Thesis, The Ohio State University, Department of Electrical Engineering, Columbus, Ohio, 1991.

SIMULATION & CONTROL ENVIRONMENT: AN INTEGRATED ENVIRONMENT FOR FAULT TOLERANT TESTING OF SPACE ROBOTS

Nagy M. Shehad and Daniel T. Wick
The MITRE CORPORATION, Houston

ABSTRACT

Space is a particularly hazardous environment in which to maintain and repair robots. Thus, fault tolerance is a major concern in the design and testing of space robots. The following paper describes a software support environment that will support the design and testing of fault tolerant space robots.

INTRODUCTION

NASA plans to construct the Space Station *Freedom* (SSF) in one of the most hazardous environments known to humankind - space. It is of utmost importance that the mechanical attachment devices of the SSF function as planned. In addition, it is equally important that systems and procedures to assemble the SSF are both safe and effective. To ensure that the systems and procedures will meet all safety requirements, NASA is developing the Space Systems Automated and Integration and Assembly Facility (SSAIAF). This facility will test the *integration* of flight systems, the telerobotic systems, and humans that will be involved in the assembly and maintenance of the SSF. Both mechanical and dynamic performance of these elements in a micro gravity environment will be tested using realtime simulations.

Traditionally, the approach to building such hardware-in-the-loop, human-in-the-loop simulations has involved development of unique software to support each scenario. This scenario-specific approach results in continuous software development and maintenance, which consumes large portions of a project's budget and can adversely affect a project's schedule. A proposed SSAIAF software architecture, Simulation and Control Environment (SCE), will support the development of generic software components, called modules, that can be reused in numerous simulation scenarios. This approach will reduce the requirements for new simulation development.

This paper presents a brief overview of the SSAIAF. This is followed by a more in-depth description of the SCE proposed to be used by SSAIAF in developing the real-time simulations. The paper concludes with a description of how the SSAIAF can support the development of fault tolerant space robots.

SSAIAF SYSTEMS

SSAIAF systems will provide the resources for all testing and development activities undertaken in the facility. The SSAIAF systems are shown graphically in figure 1.

Computing resources. Computing systems in this environment include hardware with sufficient processors, memory and I/O capacity to support two simultaneous tests on SSAIAF systems. The currently envisioned system will include three host computers, peripherals, and computer control consoles.

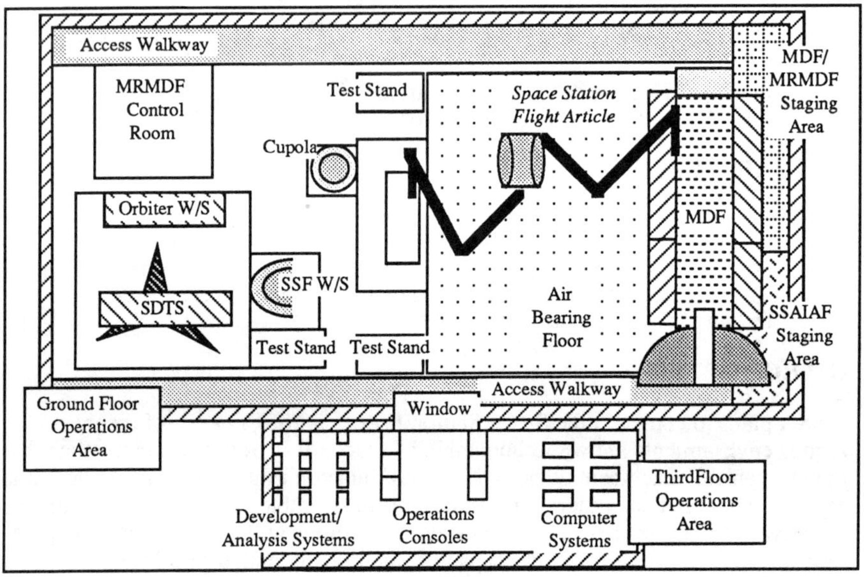

FIG. 1 SSAIAF systems.

Manipulator Development Facility. The Manipulator Development Facility (MDF) is a stand-alone system that provides a realistic test and training environment for astronauts who will use the Shuttle arm. The MDF permits astronauts or test engineers to lift structural components out of a mockup of the Orbiter cargo bay and position them for deployment or erection.

Mobile Remote Manipulator Development Facility. The Mobile Remote Manipulator Development Facility (MRMDF) is a stand-alone system that provides a realistic test and training environment for astronauts who will use the Space Station arm. The MRMDF permits astronauts or test engineers to grapple and position a station payload, a platform, and another object for deployment or erection on the Space Station.

Air -Bearing Floor. The Air-Bearing Floor (ABF) is a large, smooth surface area (70' by 90') upon which pallets can float on a cushion of air. Each pallet can carry over a thousand pounds of weight. The cushion of air permits the pallets to move in two dimensions as if there were no friction. Thus, the air-bearing floor simulates weightlessness in two dimensions.

Six-Degree-of-Freedom Dynamic Testing System. The Six-Degree-of-Freedom Dynamic Testing System (SDTS) is a computer-controlled hydraulic table (Stewart Platform) that can move in six planes of motion at one time (right/left, back/forward, up/down, pitch, yaw, and roll). The SDTS can lift large objects weighing up to 10,000 lbs. and move them to a position, as directed, within a tolerance of 0.1 inch. The orientation of the SDTS is such that it usually moves objects vertically.

Simulation and Control Environment. The Simulation and Control Environment (SCE) is a software system that integrates all elements required for the development of SSAIAF realtime simulations in a comprehensive and cost effective application generation environment. The modular approach allows the reusability of software developed for previous simulations. The SSAIAF simulations developed using the SCE will execute on the test computing systems. SCE development activities utilizing a user-friendly, icon-driven user interface will be accomplished on development/maintenance workstations. The SCE architecture and usage is further discussed below.

THE SIMULATION AND CONTROL ENVIRONMENT

SCE Architecture

The proposed software architecture for SCE will include both Commercial-Off-the-Shelf (COTS) and custom components. SCE uses a layered architecture (figure 2) that supports modular and reusable software modules as well as application generation techniques that provide fast development and ease of change.

The SCE has been designed in a modular architecture for three reasons. First, this modular approach supports the maintenance of the entire system. This permits system maintainers and developers to revise and update any particular module without affecting the entire SCE. This modular design is also POSIX-compliant so that it can be easily migrated to other platforms as hardware continues to evolve.

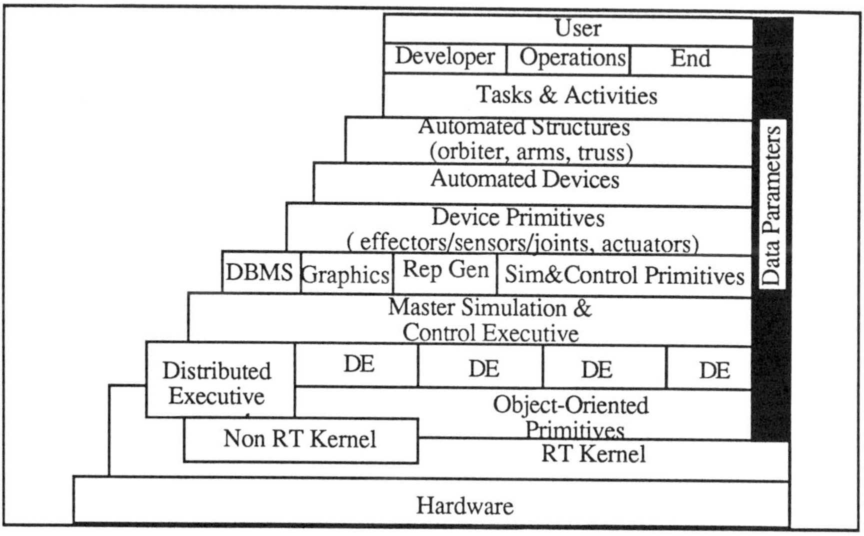

FIG. 2 SCE Modular Design.

The second reason for this architecture is to support an automated approach to the development of simulations. The SCE is designed to maximize the reuse of previous development efforts. To this end, a developer can select a highly developed software module or portions of previous simulations and incorporate those in a new effort. For example, a developer could select arm models from one simulation and propulsion models from another and combine those in a new simulation. Thus, a developer has the ability to reuse software at any level that is most appropriate for the development effort being undertaken.

The third reason for this architecture is to facilitate simulation generation by automating many attributes traditionally determined by the simulation designer. For example, a simulation executive/scheduler within the SCE will automatically allocate software modules (processes) to one or more CPUs communicating through shared memory, in order to maximize the efficiency of the simulation. In the traditional approach these decisions would be made by a designer in an arbitrary way and evaluated in a trial-and-error manner.

Major SCE Elements

The SCE is comprised of four major elements (figure 3). First, the math models are implemented in the form of modules and can be considered as building blocks for the simulation developer. The modules themselves are built using lower level primitives. The second element is an icon-based user interface that allows the simulation developer to define a particular simulation using a flow diagram-like notation that expresses the data-dependencies between the various modules, the simulation rates, delays, etc. The data representing a particular simulation are then stored as a set of Simulation Definition File(s). The third element is an application generator that translates the data created by the simulation definition element into another set of files called the Simulation Data File(s) and Executive Input File(s). The Executive Input File(s), which are generated following test computer dependent directives, are then shipped to the test environment, in addition to the simulation initialization data. The

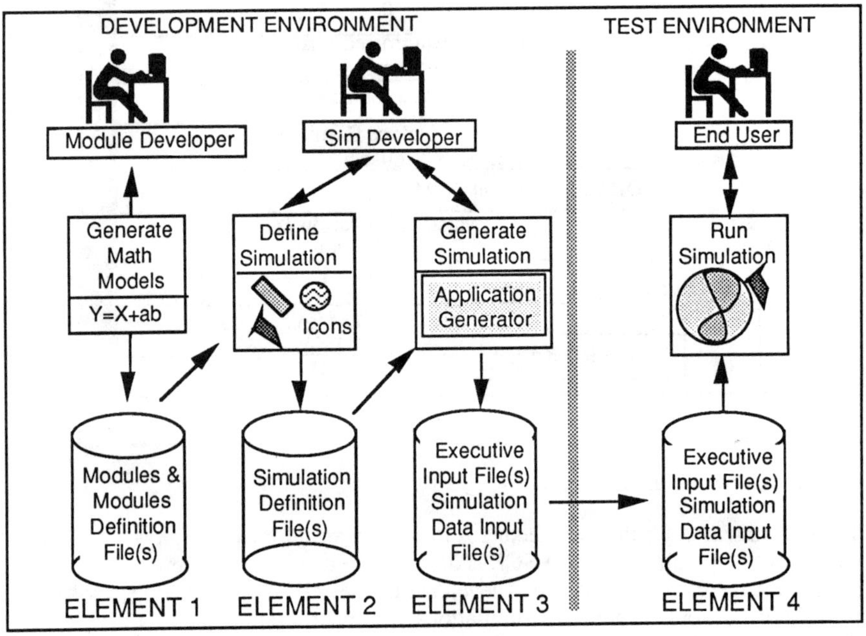

FIG. 3 Major SCE Elements.

fourth element of the SCE, which is resident on the test computer complex, uses the Executive Input File(s) to setup the execution environment at run time and actually schedules and controls the simulation and automatically allocates parallel processes (when available) to multiple CPUs.

Simulation Life Cycle Using SCE

The use of the SCE in the simulation life cycle is displayed in figure 4. The life cycle begins with the generation of simulation requirements. If existing SCE capabilities cannot support these requirements, SCE or Module developers will develop additional capabilities, e.g. a robot model, to support the simulation requirements.

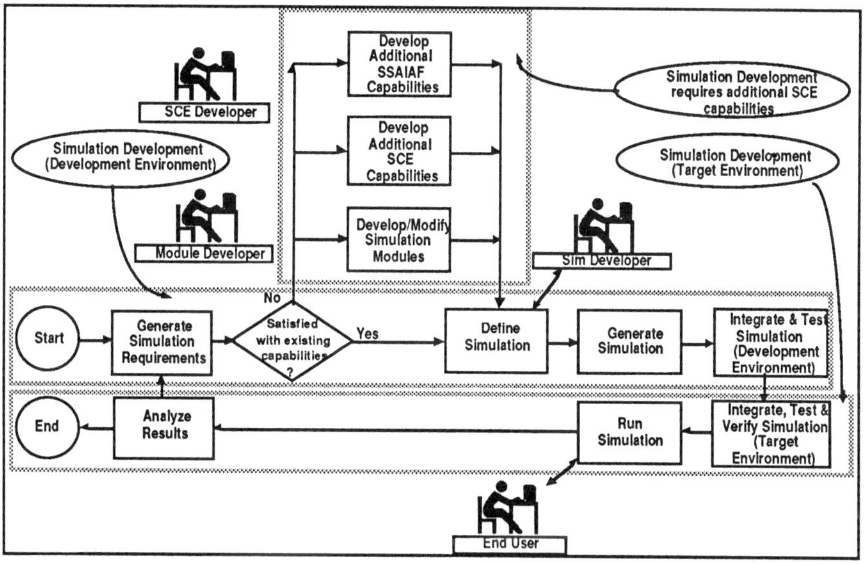

FIG. 4 Simulation Life Cycle Using SCE.

Once, the necessary capabilities are available, a simulation is defined and developed. This might include combining a Space Station Freedom model, a robot model, and ephemeris modules (gravity, solar, etc.) into a simulation. The simulation is generated using pre-developed modules and tested. Once the simulation is verified, the end user will run the simulation and evaluate the results.

Developing Fault Tolerant Robots

Rapid prototyping. The SCE will support rapid prototyping of fault tolerant space robots by permitting the user to select modules ranging from simple to complex, and to combine the modules in a simulation to investigate the action of these systems in space. The system will contain robot and environment (gravity and solar) modules and will automatically combine these complex modules into a testable simulation. These will be displayed in the developer's workstation screen. The developer will then run a simulation using various scenarios.

Test and verification The SSAIAF will be designed to move robot and flight-like articles as they would move in space to provide form-and-fit verification testing over a range of possible conditions. The SCE will support the evaluation of the complex dynamics and forces that cannot be tested elsewhere. The fault tolerance of robots can be tested within the utilizing SCE and SSAIAF test systems. Repeated and long term testing is possible, as well as off-nominal operations (eg., collisions).

Human-in-the-loop-testing. Just as important as system verification, the tasks requiring human-robot teams must also be verified to ensure that they can be accomplished in a safe and efficient manner. In other words, simple hardware check-out is not sufficient for these systems. Often, a human will have to work in conjunction with a robot. Thus, actual or simulated testing on human-robot teams must be performed. The SSAIAF will provide capabilities which support verification of systems and the tasks in space.

Problem resolution Experience in space operations has shown that anomalous situations occur which require realtime resolution (e.g., utilizing a robot in an untested scenario). The SCE could support rapid prototyping of possible solutions and testing of untested robotic operations. Interfaces with Johnson Space Center operations could include data and voice links with Shuttle and Space Station astronauts as they try to resolve on-orbit problems.

A repository of simulation objects. Finally, the SCE will serve as a repository of space systems models that can support the testing of future space systems. Future space robots will be designed and tested. The SCE will support the archiving of simulation modules and the results of prior testing for the benefit of future robot designers.

CONCLUSION

Space is one of the most hazardous environments in which humans must work. Because of the difficulty of maintaining and repairing robots, fault tolerance is a major concern of NASA in the design and testing of space robots. Testing space robots prior to launch requires a simulated environment. As stated above, the traditional approach to building, such as human-in-the-loop simulations, has involved development of unique software to support each scenario. However, in the SSAIAF the SCE approach represents a departure from tradition. The modular design maximizes the reuse of previously developed simulations. An entire simulation need no longer be developed from "scratch." Instead, developers can combine components of previous efforts and concentrate their efforts on the unique elements of the new simulation. In addition, the modular design of the SCE supports its own maintenance and permits migration to new platforms as hardware continues to evolve. The result of this approach is an adaptable software environment that supports the rapid development and test of space robots in a cost-effective manner.

ADAPTIVE MOTION-FORCE CONTROL OF ROBOTS WITH UNCERTAIN CONSTRAINTS

RICARDO CARELLI and VICENTE MUT
Instituto de Automática
Universidad Nacional de San Juan
San Martín 1109 (o), 5400 San Juan, Argentina

ABSTRACT
This paper addresses the problem of motion-force control of constrained robots in the presence of uncertainties both in dynamic robot parameters and in the constraint function. The controller design is based on singular model robot representation. In a previous paper, an adaptive solution is proposed in relation to dynamic robot uncertainties. In this work, it is additionally considered the uncertainties in the constraint function, related to which two design solutions are presented. One solution is the explicit identification of the constraint function and the other one is the direct estimation of the Jacobian matrix of a transform function. Simulation results show a good performance of the proposed adaptive controller under uncertain constraints.

Keywords: robot control, force control, nonlinear control, adaptive control.

I. INTRODUCTION

Constrained motion robot control is concerned with the control of a robot whose end-effector mechanically interacts with the environment. Several approaches have been suggested, such as impedance control, force control and hybrid motion/force control. The *hybrid position/force* control method was proposed by Raibert and Craig [5]. In a paper by McClamroch and Wang [4], a theoretical framework is presented for the study of stability of robots under constrained tasks. Singular model robot representation is used with the constraint description and the effects of constraint forces explicitly included, and global conditions for feedback stabilization and tracking are presented.

Most high performance model-based robot control strategies rely on the exact cancellation of the robot nonlinear dynamics. The high uncertainty in some robot parameters, such as link inertia and payload, has led many researchers in the last ten years to consider the design of globally stable *adaptive controllers* for robots. Adaptive controllers for constrained robots based on a singular model representation have been recently proposed in [1], [3], [6]. However, in all this papers it is assumed an exact knowledge of the constraint equation. It is of course desirable to relax this non realistic assumption. In this work we present an adaptive controller designed to achieve the control objective under uncertainties on both, the robot dynamics and the constraint function. The design is based on the singular model representation by McClamroch and Wang [4]. We believe that the controller presented in this paper is a quite general solution for the adaptive control problem of constrained robots.

II. CONSTRAINED ROBOT MODEL AND MAIN PROPERTIES

In the absence of friction and other disturbances, the joint-space (robot coordinates) *dynamics* of an n-link constrained rigid robot manipulator can be written as,

$$M(q)\ddot{q} + C(q,\dot{q})\dot{q} + g(q) = f + u \tag{1}$$

where $f \in \mathbb{R}^n$ the vector of generalized constraint forces, and $u \in \mathbb{R}^n$ the vector of generalized control input forces. Following [4], consider that the environmental constraints satisfy the *constraint equation*,

$$\phi(q) = 0 \text{ with } \phi(q):\mathbb{R}^n \to \mathbb{R}^m \tag{2}$$

Also, the generalized constraint forces in robot coordinates are given by $f = J'(q)\,\lambda$, where $\lambda \in \mathbb{R}^m$ is the vector of generalized multipliers and $J(q)=\partial\phi(q)/\partial q$ a Jacobian matrix. Now, still following [4], let's assume that there exists a function $\Omega(q_2)$: $\mathbb{R}^{n-m} \to \mathbb{R}^m$ such that $\phi(\Omega(q_2),q_2) = 0$, for all $q_2 \in \mathbb{R}^{n-m}$. Consider the *nonlinear coordinate transformation* $X:\mathbb{R}^n \to \mathbb{R}^n$,

$$x = X(q) = \begin{vmatrix} q_1 - \Omega(q_2) \\ q_2 \end{vmatrix} \tag{3}$$

and its inverse transformation $Q:\mathbb{R}^n \to \mathbb{R}^n$, where vector partitions $q'=(q_1',q_2')$ and $x'=(x_1',x_2')$ are used. Also, let's define the Jacobian matrix $T(x)=\partial Q(x)/\partial x$, which is nonsingular,

$$T(x) = \begin{vmatrix} I_m & \partial\Omega(x_2)/\partial x_2 \\ 0 & I_{n-m} \end{vmatrix} . \tag{4}$$

Then, the dynamic equation (1) can be expressed in terms of the vector x as,

$$\overline{M}(x)\ddot{x} + \overline{C}(x,\dot{x})\dot{x} + \overline{g}(x) = T'(x)f + T'(x)u \tag{5}$$

where, $\overline{M}=T'M\,T$, $\overline{C}=T'C\,T+T'M\,\dot{T}(x)$ and $\overline{g}=T'g$.

Introducing the partitioning of the identity matrix $I_n=[E_1' \mid E_2']$, where E_1 is mxn and E_2 is (n-m)xn, the following *reduced form equations* are obtained,

$$E_1\overline{M}(x_2)E_2'\,\ddot{x}_2+E_1\overline{C}(x_2,\dot{x}_2)E_2'\,\dot{x}_2+E_1\overline{g}(x_2)=E_1T'f+E_1T'u \tag{6}$$

$$E_2\overline{M}(x_2)E_2'\,\ddot{x}_2+E_2\overline{C}(x_2,\dot{x}_2)E_2'\,\dot{x}_2+E_2\overline{g}(x_2)=E_2T'u. \tag{7}$$

Some fundamental properties of the dynamic motion equation (5) (and reduced form (6-7)) are given in [1].

Equations (6) and (7) can be written as,

$$E_1\overline{M}(x_2)E_2'\,\ddot{x}_2+E_1\overline{C}(x_2,\dot{x}_2)E_2'\,\dot{x}_2+E_1\overline{g}(x_2)=E_1\xi(x_2,\dot{x}_2,\ddot{x})\theta \tag{8}$$

$$E_2\overline{M}(x_2)E_2'\,\ddot{x}_2+E_2\overline{C}(x_2,\dot{x}_2)E_2'\,\dot{x}_2+E_2\overline{g}(x_2)=E_2\xi(x_2,\dot{x}_2,\ddot{x}_2)\theta \tag{9}$$

III. UNCERTAINTY IN ROBOT DYNAMICS: ADAPTATION

Consider the robot model (6-7) and (8-9). The dynamic parameter vector θ of the manipulator and payload is considered constant but unknown.

Using a computed torque structure, we consider the following *adaptive impedance/force control law*,

$$E_1T'(x_2)u = E_1\hat{\overline{M}}(x_2)E_2'\,w + E_1\hat{\overline{C}}(x_2,\dot{x}_2)E_2'\,\dot{x}_2 + E_1\hat{\overline{g}}(x_2) - E_1T'(x_2)J'(x_2)\lambda_d+E_1E_1'G_fE_1T'(x_2)J'(x_2)[\hat{\lambda}-\lambda_d] \tag{10}$$

$$E_2T'(x_2)u = E_2\hat{\overline{M}}(x_2)E_2'\,w + E_2\hat{\overline{C}}(x_2,\dot{x}_2)E_2'\,z + E_2\hat{\overline{g}}(x_2) \tag{11}$$

where $\hat{\overline{M}}$, $\hat{\overline{C}}$ and $\hat{\overline{g}}$ are estimates of $\overline{M}$, $\overline{C}$ and $\overline{g}$ with $\hat{\theta}$. The signals w, z are defined as,

$$w = \ddot{x}_{2d} - G_v\dot{e}_2 - G_pe_2; \qquad w \in \mathbb{R}^{n-m} \tag{12}$$

$$z = \dot{x}_2 - \nu; \qquad z \in \mathbb{R}^{n-m} \tag{13}$$

$$\nu=(1/(p+\rho))[\ddot{e}_2+G_v\dot{e}_2+G_pe_2);\ p=d/dt;\ \rho>0;\ \nu \in \mathbb{R}^{n-m} \tag{14}$$

with $e_2(t)=x_2(t)-x_{2d}(t)$ and $x_{2d}(t)=E_2X(q_d(t))$ by using the corresponding coordinate transformation. G_f is an mxm constant nonnegative definite feedback gain matrix.

Remark 1. Due to properties given in [1], a part of control law (10-11) can be expressed as,

$$E_1\hat{\bar{M}}(x_2)E_2'w+E_1\hat{\bar{C}}(x_2,\dot{x}_2)E_2'\ \dot{x}_2+E_1\hat{\bar{g}}=E_1\varphi(x_2,\dot{x}_2,w,y)\hat{\theta} \tag{15}$$

$$E_2\hat{\bar{M}}(x_2)E_2'\ w+E_2\hat{\bar{C}}(x_2,\dot{x}_2)E_2'\ z+E_2\hat{\bar{g}}=E_2\varphi(x_2,\dot{x}_2,w,y)\hat{\theta} \tag{16}$$

with $\varphi \in \mathbb{R}^{n\times p}$, $y=\dot{x}_2$ in (15), y=z in (16). ▽▽▽

To update the parameter vector $\hat{\theta}$ we consider an integral *update law*,

$$\dot{\hat{\theta}} = -\Gamma\ [E_2\varphi(x_2,\dot{x}_2,w,z)]'\nu \tag{17}$$

where $\Gamma \in \mathbb{R}^{p\times p}$ is symmetric positive definite.

Proposition 1 (see proof in [1]). For the above controller, the following holds: a) $\tilde{\theta} = (\hat{\theta} - \theta) \in L_\infty^p$, b) $\nu \in L_2^{n-m}\cap L_\infty^{n-m}$, c) e_2 and $\dot{e}_2 \in L_2^{n-m} \cap L_2^{n-m}$, d) $e_2(t) \to 0$ as $t \to \infty$, $q(t) \to q_d(t)$ as $t \to \infty$, e) Steady-state force error is bounded and inversely proportional to the force gain matrix G_f norm. Zero error force tracking is achieved under persistent excitation of matrix φ.

IV. UNCERTAINTY IN CONSTRAINT FUNCTION: CONSTRAINT ESTIMATION

A. Identification of the constraint function

The first approach we propose is the explicit identification of the parameters in a previously defined structure of the constraint function. Particularly, we consider constraint functions which admit a linear parameterization of the form,

$$\phi(q)= \varphi(q)\ \theta \tag{18}$$

where θ are the parameters to be identified and $\varphi(q)$ is a regressor matrix. For example, the following quadratic function structure is representative of a common family of constraints,

$$\phi_l(q)=a_0+\sum_1^n a_i q_{(i)}+\sum_{i=1}^{n}\sum_{j=i+1}^{n} b_{i,j}q_{(i)}q_{(j)}+\sum_1^n c_i q^2_{(i)}=0 \tag{19}$$

where $\phi_l(q)$ is a generic element of $\phi(q)$: $\mathbb{R}^n \to \mathbb{R}^m$ and,

$$\theta' = (a_0,\ldots a_i\ldots,\ldots b_{i,j}\ldots,\ldots c_i\ldots)$$
$$\varphi(q)=(\ldots,q_{(i)},\ldots,\ldots,q^2_{(i)},\ldots,\ldots q_{(i)}q_{(j)},\ldots).$$

Parameter vector θ can be estimated by means of the gradient method in the discrete version,

$$\hat{\theta}(k+1) = \hat{\theta}(k) + \Gamma\, \varphi'(q)\, \varepsilon \qquad (20)$$
$$\varepsilon = \hat{\phi}(q) - \phi(q) = -\phi(q)$$

with k the discrete instant, Γ is a constant gain matrix and $\phi(q)=\varphi(q)\theta$.

Another typical identification method is standard least squares [2]. When applied to (18) we obtain,

$$\hat{\theta}(k+1) = \hat{\theta}(k) - K(k)\, \varphi(k+1)\, \hat{\theta}(k) \qquad (21)$$

where, $K(k) = \{P(k)\, \varphi'(k+1)\, [\lambda + \varphi(k+1)\, P(k)\, \varphi'(k+1)]^{-1}\}$ with $P(k+1)=[I - K(k)\, \varphi(k+1)]\, P(k)/\lambda$.

Compared with the gradient method, least squares extracts more information about the parameters being estimated. Estimated parameters $\hat{\theta}$ are used as follows. The Jacobian $T(x)$, equations (4) can be written as a function of parameters θ. An estimation of T is obtained after replacing θ by $\hat{\theta}$. T estimated is now used in controller calculation. Similarly, the desired force f_d is defined in aggreement to the estimated constrained function $\hat{\phi}(q)$ as $f_d = J'(x_2)\lambda$. The identification approach requires that robot motion q (and consequently q_d) be sufficiently exciting to assure convergence of estimated parameters to those of the true constraint function.

B. Direct estimation of the Jacobian $T(x_2)$.

The second approach we consider in relation to uncertainty in constraint function is the direct estimation of the Jacobian $T(x_2)$. From equation (3), the partition $\partial\Omega(x_2)/\partial x_2$ in (4) can be expressed as $\partial q_1/\partial q_2$. Now $T(x_2)$ can be estimated after calculation of,

$$(\partial\Omega(x_2)/\partial x_2)_k = \text{matrix}[\Delta q_{1,i}/\Delta q_{2,j}]_k = \text{matrix}[p_{i,j}(k)], \qquad (22)$$
$$i=1,\ldots m;\; j=1,\ldots n-m$$

where $\Delta\cdot(k) = \cdot(k) - \cdot(k-1)$. This estimation assumes a sampling period T_s. It is also based on only two subsequent measurements of vector q. A predictor - estimator algorithm which is based on three measurements is the following one,

$$\widehat{(\partial\Omega(x_2)/\partial x_2)}_k = \text{matrix}[\hat{p}_{i,j}(k)], \qquad (23)$$

where, $\hat{p}_{i,j}(k) = \eta(k)\, p_{i,j}(k+1/k) + (1-\eta(k))\, p_{i,j}(k)$
with, $p_{i,j}(k+1/k) = p_{i,j}(k) + (p_{i,j}(k)-p_{i,j}(k-1))$.

$\eta(k)$ is a prediction confidence factor with values

between 0 and 1,

$$\eta(k) = [1-|tgh(\alpha\ \varepsilon(k))|]$$
$$\varepsilon(k) = [p_{i,j}(k/k-1) - p_{i,j}(k)]/p_{i,j}(k)$$

with $\varepsilon(k)$ the relative prediction error and α a scaling constant.

This predictor-estimator is computationally more involved, but can achieve better controller performance.

The desired force $f_d(k)$ must be consistent with the constraint.In this approach, it is defined by means of a desired module and a direction calculated from $p_{i,j}(k)$ terms. It is worth-noting that the estimators requiere a supervision to avoid dividing by zero for null variations of q_2.

VI. CONCLUSIONS

In this work we have presented a solution to the adaptive control of constrained robots in the presence of uncertainties both in the robot model parameters and the constraint function. The controller design is based on a singular systems model representation. Simulations show a good control performance even in the worst case of simultaneous uncertainties in dynamics and constraint. The main contribution of this paper is the consideration of constraint function uncertainty, which is an assumption of practical relevance.

REFERENCES

1. Carelli, R., Kelly R. "Adaptive control of constrained robots modeled by singular systems". Proc. of the IEEE Conference on Decision and Control, 1989.
2. Goodwin G., Sin K. S. "Adaptive filtering prediction and control". Prentice-Hall, 1984.
3. Jong-Hann J., Li-Chen Fu. "Efficient adaptive hybrid control strategies for robots in constrained manipulation". Proc. IEEE Inter. Conference on Robotics and Automation, 1991.
4. McClamroch N.H., Wang D. "Feedback stabilization of constrained robots", IEEE Trans. on Automatic Control, vol.33, No.5, pp.419-426, May 1988.
5. Raibert M., Craig J.J. "Hybrid position/force control of manipulators", ASME Journal of Dynamic Systems, Measurement and Control, vol.102, pp.126-133, 1981.
6. Su C., Leung T. P., Zhou Q. "Adaptive control of robot manipulators under constrained motion". Proc. of the Conference on Decision and Control, 1990.

ACTIVE VEHICLE SUSPENSION WITH ADAPTIVE PREVIEW CONTROL

E. ESMAILZADEH, Professor and H. BATENI, Ph.D. Candidate
Department of Mechanical Engineering
Sharif University of Technology
Tehran, IRAN

ABSTRACT

Most vehicle suspensions are composed of passive spring and damper elements, although improved suspension performance is achieved if an active system is employed to control forces and relative velocities. An adaptive control technique for active vehicle suspension is discussed. This method could automatically self-tune the active suspension such that the disturbances of dynamic parameters have no effect on the dynamic characteristics of the system. The concept of preview control is shown to be interesting and may well be applied to the guidance of vehicles over statistically described surfaces by incorporating a state simulator.

INTRODUCTION

The purpose of suspension systems for ground transport vehicles is to isolate passenger and freight compartments from disturbances due to road irregularities. The input forces which are transmitted to a vehicle through its suspension should be reduced in order to ensure a comfortable ride. This must be done with a minimum loss of wheel-to-road contact on rough roads and terrains.

Active suspension system is employed as the mean of overcoming many of the performance limitations inherent with the use of a passive suspension system, [1]. Servo-valve controlled pneumatic suspensions of the body feedback type is investigated by Esmailzadeh [2] which consists of a pneumatic actuator parallel with the suspension spring and a servo-valve to control the flow of the air as shown schematically in Figure 1.

Several researchers [3]-[6] have described different approaches to study semi-active and active suspension systems employed for an on-road or off-road vehicles. Linear full-state feedback control of commercial vehicle suspensions was examined by Chalasani, [4]. Optimal controllers, however, minimize a defined performance index and could not have the capability of adapting to significant system parameter changes. On the other hand, an adaptive control scheme initially diagnoses the input disturbances, parameter variations and/or structural variances and then redetermine the controller parameters and/or configuration via an on-line adaptive algorithm. Current hardware technology and knowledge of modern adaptive control theory allow us to employ a sophisticated electronically controlled active suspension system in a vehicle at a reasonable cost.

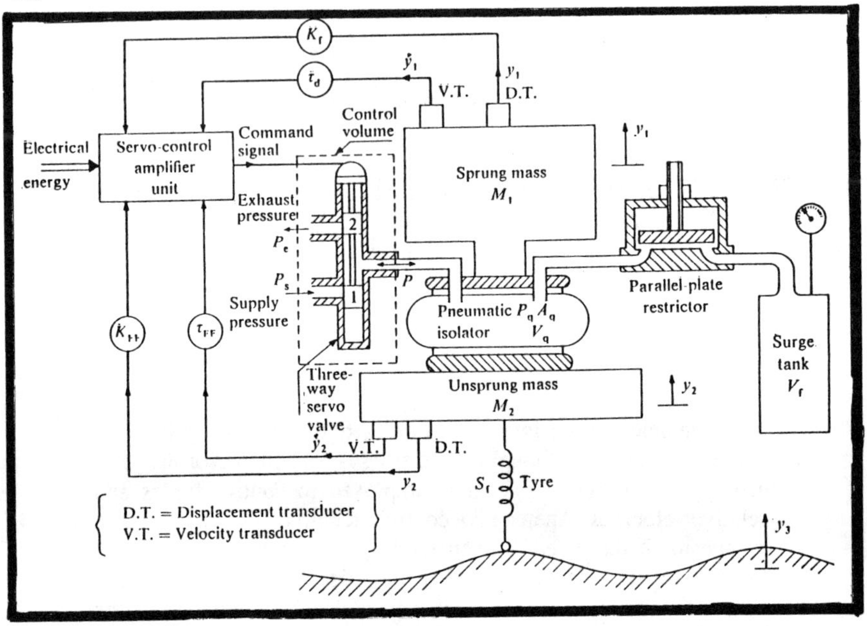

Figure 1. Servo-valve controlled pneumatic vehicle suspension.

The use of state vector feedback in active suspension systems has been investigated by some researchers, [4] and [7]. There are two major drawbacks to be reported:

(i) The system is only optimal for specific values of the dynamic parameters, namely, the tire stiffness, sprung and unsprung masses. By varying any of these parameters the feedback should be re-optimized for the present condition, accordingly.

(ii) It is assumed that all the components of the system are ideal and, therefore, have zero value for the time delay. On the contrary, for such a system at its best condition, the time delay is about 0.01 second.

In this paper a method will be proposed to overcome these difficulties. With a thorough investigation of presently available systems, a study of the active suspension system with adaptive preview control is carried out.

ADAPTIVE SYSTEM WITH LOAD CHANGES

Variation of the value of the sprung mass in a vehicle is completely normal and probable. Having different numbers of passenger in a car, the sprung mass may increase or decrease accordingly. The adaptive system must be capable of adapting itself with these changes. The new value of the sprung mass should be measured and fed back into the system in the form of a control signal in order to adjust the feedback coefficient of the state variables for the present condition. It is well understood that such a control signal does not regulate the dynamic characteristics of the system directly, but it only adapts the system with respect to the new condition by means of changing the parameters of the controller.

The road irregularities normally produce disturbances in the interacting forces between sprung and unsprung masses. These irregularities may be considered as *noise* in the performance of the adaptive system.

It is interesting to note that, whenever the payload is increased, the mean value of the amplitude of the sensed vibratory signal changes, too. One sensible method of eliminating the noises from the main signal is to increase the reaction time in the adaptation system, or simply, that of the sensing device.

TIRE STIFFNESS ADAPTABILITY

The stiffness of a tire is directly proportional to the pressure of the air trapped inside the tire, and therefore,the stiffness varies whenever the pressure is higher or lower than its regular set value. In this situation the stiffness should be used as a control signal in order to change the parameters of the controller.

The air pressure of the tire, after the first stage of pre-processing, would be converted to a signal which is directly proportional to the tire stiffness. The pre-processor, being an adaptive control mechanism, is chosen to be a simulator with the same dynamic characteristics as that of the tire. The tire stiffness signal will then be fed into the decision system.The schematic representation for the tire stiffness and the sprung mass adaptive control system is shown in Figure 2.

In order to eliminate the effect of the undesirable noise caused by the road disturbances, which in turn would change the tire pressure, on the dynamic characteristics of the decision system one may select a rather larger value for the time constant of the pressure transducer.

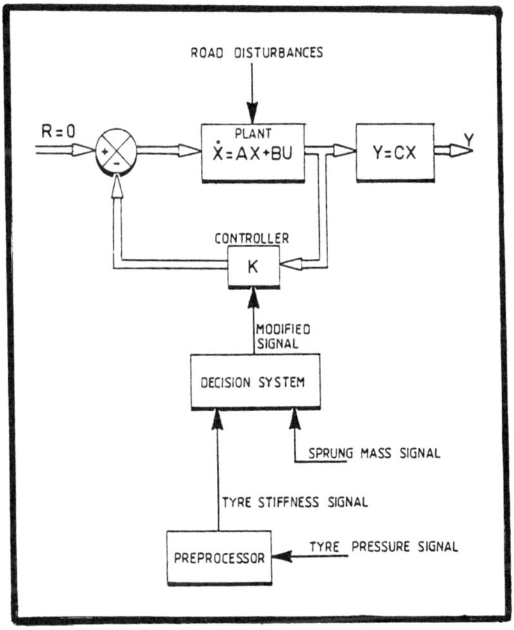

Figure 2. Schematic diagram for adaptive control system.

PREVIEW CONTROL SYSTEM

There is a strong similarity between the function of a human operator to steer a vehicle along a winding road and the function of a suspension system to guide a vehicle over the vertical part of a roadway profile.The accuracy with which a driver can keep his vehicle at the center of a lane for a given speed depends, among other things, on his visibility of the roadway ahead of the car. In the active suspension system discussed previously it was assumed that future disturbances are unpredictable and unknown. A consequence of this assumption is a control scheme in which control action is decided by operating upon instantaneous command signals. It is reasonable to expect that the performance may be improved if future information about command signals is available. This type of control problem is referred to as *Preview control.*

It must be noted that in an actual system, in spite of state variables feedback, the reaction time constants of the plant,namely the actuator, and the controller are rather large. This indicates that the actuator will react whenever the wheel has passed the road elevation, and hence, the suspension system can not eliminate the undesired disturbances. Due to this drawback the road surface profile should be viewed well in advance and even before the wheel has reached to that point.

ADAPTIVE PREVIEW CONTROL

It has already been established that the state variables are greatly affected by both the road surface profile and the dynamic characteristics of the sprung and unsprung masses. Moreover, all the state variables should be previewed together and even before they are fed back into the system. In other words, the system should be capable of not only previewing the road surface profile, but also to simulate the suspension system in a simulator unit being in the form of a state estimator.

The measured state variables with various time constants and delays are input to the main controller unit in order to obtain a suitable actuation signal, and hence, giving the actuator enough time to react well before the wheel reaches the road elevation under consideration.

Figure 3 shows schematically a full scale adaptive suspension system with the preview control and the simulator unit. The irregularities of the road surface are sensed well in advance by a preview sensor. Normally it is directly installed to the suspension mechanism and, therefore, the observation of the preview system is a combination of the *in-future* road disturbance and the *in-present* condition of the unsprung mass. By comparing the two signals in the first pre-processor, an output signal proportional to the actual road surface profile is produced. The simulator unit, using the output signal together with the load condition and the tire stiffness signals, simulates the suspension system and predicts the state variables, accordingly.

In general various parameters, such as distance between the view point and wheel (L), vehicle speed (V), time lag of main controller (τ_3), and time constant of the plant (τ_2) have been fully considered in these state variables. The loop indicated by (*) in Figure 3 provides the estimated state variables explained above. The output signals from this section should have a time precedence exactly equal to those time constants in order to give the system enough time to react accordingly.

The remaining parts of the system are somewhat similar to that of an optimal full state vector feedback used for an active suspension system.

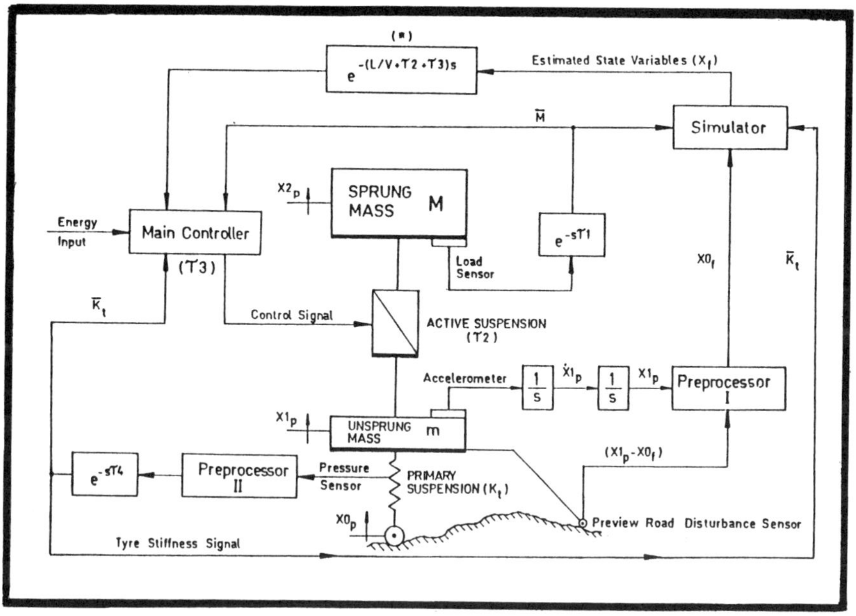

Figure 3. Adaptive suspension system with preview control and simulator unit.

CONCLUSIONS

An active suspension for a vehicle represents a class of controls in which the input may be easily known before it reaches the plant, in this case, the vehicle. Moreover, it could well respond to preview information or adjust its dynamics in response to a number of sensed aspects of the vehicle motion and disturbances. The most important advantage of an adaptive vehicle suspension system is its capability of adapting with the loading condition, tire stiffness and road profile irregularities. This system could provide a better ride comfort, improved road holding and overall stability with a continuous wheel-to-road contact. The concept of preview control, i.e., a control scheme in which an input is sensed before it reaches the controlled plant is discussed. It can be concluded that the reduction in vibration and clearance space that a preview control system may bring about is quite considerable when compared with an optimized system that does not use preview information, [5]. An example for the implementation of the proposed system could be seen in the case of the high speed trains and guided vehicles. The previewed data are the information obtained from a small *warning* wheel which travels in front of the vehicle and senses the road surface irregularities. In the case of trains with steel wheel the complexity of the tire stiffness adaptive system is completely eliminated. It has been proposed that a small wagon with a specific suspension system may be employed as the simulator unit and the state variables of the wagon could be used to predict the state variables for the main suspension system.

REFERENCES

1. Esmailzadeh, E., "Design Synthesis of a Vehicle Suspension System using Multi-parameter Optimization", *Vehicle System Dynamics*, **7**(2), 83-96 (1978).
2. Esmailzadeh, E., "Servo valve Controlled Pneumatic Suspensions", *J. of Mechanical Engineering Science*, **21**(1), 7-18 (1979).
3. Margolis, D.L. "The Response of Active and Semi-active Suspensions to Realistic Feedback Signals", *Vehicle System Dynamics*, **11**(6), 267-282 (1982).
4. Chalasani, R.M., "Ride Performance Potential of Active Suspension Systems-Part I.", *ASME Monograph*, AMD **80** DSC(2) 187-235 (1986).
5. Esmailzadeh, E. and Bateni, H., "Optimum Vehicle Active Suspensions with Full State Feedback Control" *Technical Report,* Sharif Univ. of Tech., Mechanical Engineering Department, Tehran (1991).
6. Kawagoe, K. and Iguchi, M., "Semi-active Control and Optimum Preview Control Application to Vehicle Suspension", *JSAE Rev.* 24-31 (1985).
7. Crolla, D.A. , et.al., "Active Suspension Control for an Off-road Vehicle", *Proc., Instn. Mech. Engrs.*, **201**(D1), 1-10 (1987).
8. Thompson, A.G., "Optimal and Suboptimal Linear Active Suspensions for Road Vehicles", *Vehicle System Dynamics,* **13,** 61-72 (1984).
9. Chalasani, R.M., "Ride Performance Potential of Active Suspension Systems-Part II", *Symposium on Simulation and Control of Ground Vehicles and Transportation Systems*, ASME Monograph, AMD, **80,** DSC, **2**, 205-234,(1986).
10. Hall, B.B, and Gill, K.F., "Performance Evaluation of Motor Vehicle Active Suspension Systems", *Proc. Instn. Mech. Engrs.,* **201** (D2), 135-148 (1987).

Edge Finishing Using Hybrid position/force Control of an XY Table[1]

Gregory P. Starr[2]

Department of Mechanical Engineering
The University of New Mexico
Albuquerque NM 87131

Clifford S. Loucks[3]

Sandia National Laboratories
Albuquerque NM 87185-5800

Abstract

Edge finishing and deburring of parts is a tedious operation that should be automated. This paper describes the use of a direct-drive XY table in the automated edge finishing of machined parts. The table is faster and more accurate than an articulated robot, thus where planar motion in a small work area is sufficient it is preferable. Hybrid position/force control is used to guide the table (with mounted workpiece) past the tool and maintain the contact force at the desired level. A six-axis force sensor on the tool spindle is used to measure contact force. We identified a dynamic model for the table from experimental measurements and used this model to design a position/force controller for the table. An example application of the table in the deburring of an actual jet engine turbine hub is presented.

1 INTRODUCTION

The one area of manufacturing which has stubbornly resisted efforts toward automation and computer assistance is the edge and surface finishing of machined parts[1]. Although many parts are created through computer-aided design, then produced on CNC machines, when they come off the machining line they still require labor-intensive hand finishing.

The 1970s might be classified as the golden years of deburring and surface finishing because of the wealth of innovations and classification of world-wide efforts[2]. During the 1980s work emphasized robotic deburring and finishing. In the 1990s, environmental concerns will play a greater role in manufacturing, and more automation will be introduced to exclude workers from hazardous materials. This must be true of finishing operations as well[2].

Sandia National Laboratories has established the Robotic Edge Finishing Laboratory (REFL)[3] to investigate the technology required for automated deburr-

1. This work is performed at Sandia National Laboratories and is supported by the U.S. Department of Energy under Contract DE-AC04-76DP00789.

2. Currently on sabbatical with the Intelligent Machine Principles Division, Sandia National Laboratories.
3. Intelligent Machine Principles Division.

ing, chamfering, and blending of machined edges. Successful performance of these tasks requires monitoring and control of the contact force between the edge-finishing tool and the workpiece.

The quality of the finished edge is dependent on the maintenance of a constant contact force. Thus, if control of the tool-workpiece contact force can be improved, quality of the finished edge will improve. The direct-drive XY table used in our work has faster dynamic response than a typical articulated robot, and can therefore achieve higher quality edge-normal force control.

In addition, the simplicity of an XY table is attractive when compared to a typical articulated robot. For the edge finishing of small planar parts, the greater dexterity offered by an articulated robot is unnecessary, and its greater complexity and expense tends to discourage potential users.

2 SYSTEM COMPONENTS

The components of interest are the XY table, the force sensor, and the control computer. A sketch of the system is shown in Figure 1. The spindle is stationary, while the table carries the workpiece.

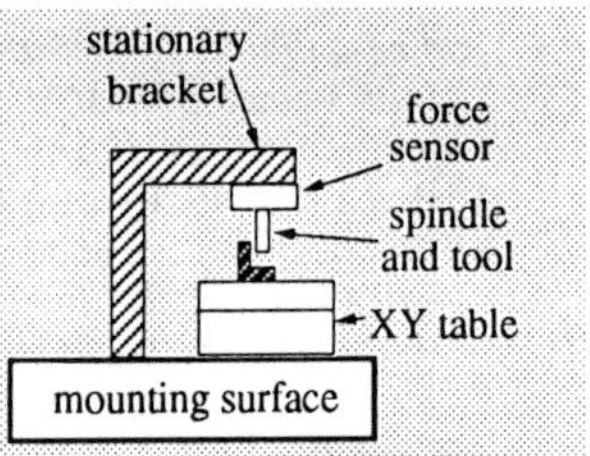

Figure 1. Table-Based Edge Finishing System.

The XY table is powered by three-phase linear actuators driven by PWM DC amplifiers. Both table axes are equipped with incremental encoders with a resolution of 12700 counts/inch.

A JR3 force sensor is used to sense the tool-workpiece contact force. The maximum force without saturation is 25 lb, with a resolution of 0.2 oz.

The control computer for the edge finishing system is built around a Motorola 68020 processor and a VME backplane, using the VxWorks real-time operating system.

3 CONTROLLER DESIGN

This section describes the development of a hybrid position/force controller for the XY table. By expressing table commands along to the normal and tangential axes of a task-relevant constraint frame, this controller allows the table both to move the workpiece past the deburring tool and to control the contact force.

Hybrid Position/Force Control

This technique, first proposed by Raibert and Craig[4], identifies those directions along which position must be controlled, and the orthogonal directions along which force must be controlled. For a planar edge finishing task, position must be controlled in the edge-tangential direction, while force must be controlled in the edge-normal direction.

There are three coordinate frames of interest, the table frame {T}, the force sensor frame {F}, and the workpiece edge frame {E}. These frames are shown in

Figure 2. The tangential position controller uses a position tracking error formed in the {E} tangential direction to actuate the table, while the normal force controller uses a force error formed in the {E} normal direction to actuate the table. Both table commands are summed.

Figure 3 shows a block diagram of the hybrid position/force control system. The **R** matrices convert between the three coordinate frames, while the diagonal selection matrices **S** select either the normal or tangential vector component.

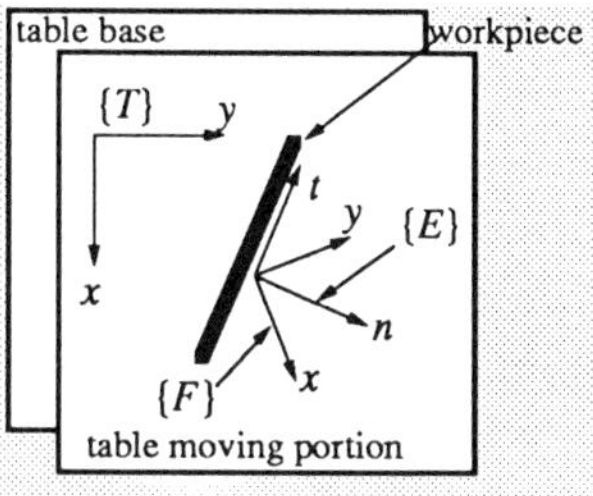

Figure 2. Coordinate Frames.

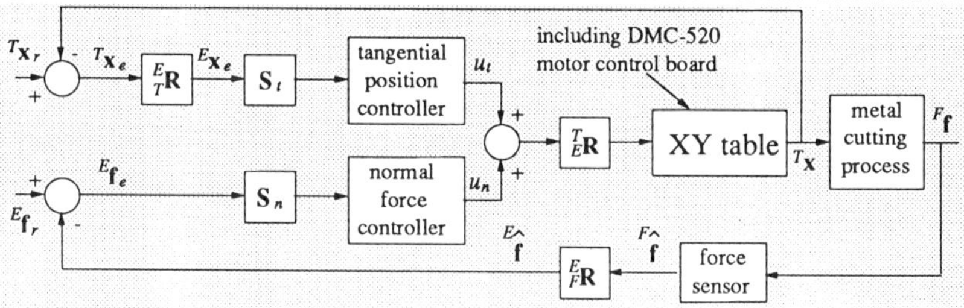

Figure 3. Block diagram of hybrid position/force control system.

Note that for a curved edge the relationship between {E} and fixed frames {T} and {F} will change with time, and the relationship between these frames must be updated online.

Models for both the table and the table+cutting process were identified to be:

$$G(z) = \frac{X(z)}{X_r(z)} = \frac{0.0284z^{-2} + 1.1346z^{-3} + 0.9128z^{-4}}{1 - 0.2089z^{-1} - 0.5426z^{-2} - 0.2158z^{-3} - 0.0326z^{-4}} \frac{\text{count}}{\text{count}} \tag{1}$$

$$G(z) = \frac{F(z)}{X_r(z)} = \frac{3.2241z^{-7} + 6.4190z^{-8} + 2.4456z^{-9}}{1 - 0.7654z^{-1} - 0.0363z^{-2} - 0.1632z^{-3}} \frac{\text{mlb}}{\text{count}} \tag{2}$$

There is a pure integration in the table model, but not in the table + cutting process. This is due to the metal removal, which erodes the surface against which the reaction force is generated. If a non-cutting tool were used, one would expect the integration to return.

Tangential Position Control

As shown in Figure 3, the tangential controller receives the tangential component of position trajectory tracking error and generates the tangential-direction table command. The plant transfer function for this controller is that of equation (2). For edge finishing, the reference position trajectory will nominally be a ramp, with constant velocity (feedrate). A Type II position controller, with its zero steady-state position error in velocity-following, is therefore desirable. The dominant poles are located at $z = 0.66$, with a natural frequency of over 8 Hz. The PID controller is

$$D(z) = \frac{0.1975 - 0.1876z^{-1}}{1 - z^{-1}} \tag{3}$$

The response of both the model and the actual system to a ramp input of speed 10 ipm is shown in Figure 4. The actual position tracking error has a peak about 30% higher than the model, but the dynamics of the error rejection are comparable, and the error does approach zero, as expected.

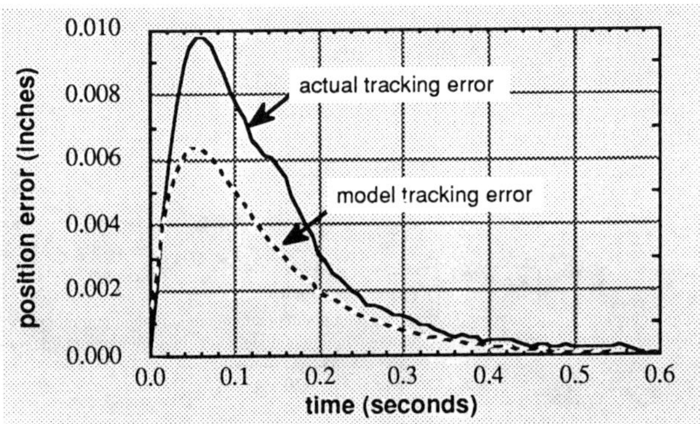

Figure 4. Table velocity tracking error, 10 ipm.

Normal Force Control

As previously indicated, both X and Y table axes are position-controlled by a commercial motor controller board. To achieve control of force using a motion-controlled device, one must command motion in response to force. As shown in Figure 3, the normal force controller accepts force error and generates position commands. The dynamic relationship of force to motion is called *impedance*. If the motion command is *position*, the device will behave as a generalized spring. If the motion command is *velocity*, we obtain a generalized damper, and likewise if *acceleration* is commanded proportional to force, the device will behave as a pure inertia. A combination of the three yields a generalized mechanical impedance.

The plant here is the table + metal cutting process, with model given by equation (3). Using this plant transfer function, a PD force controller was designed using the root locus method. The resulting normal force PD controller is

$$D(z) = 0.036 - 0.022z^{-1} \tag{4}$$

The dominant poles have a natural frequency of over 6 Hz.

A comparison of the model and the actual unit step force responses are shown in Figure 5. The overshoot of the actual system is about 12% higher than the model, and the response is a little faster.

Note that the final value in Figure 5 is not quite unity. This arises from the lack of an integration in the plant. Rather than add an integrator to the controller, we prefer the faster response of the PD controller. While not infinite, the low frequency gain is still high, and should yield adequate steady state force accuracy.

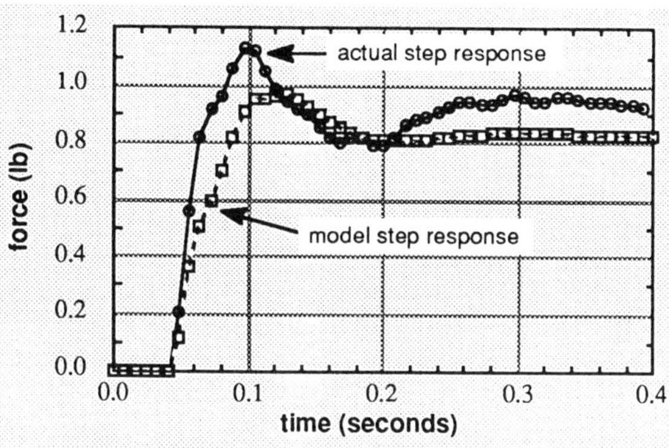

Figure 5. Actual and model unit step force responses.

4 REFERENCE TRAJECTORY DETERMINATION.

Referring to the block diagram of Figure 3, the reference trajectory ${}^{T}\mathbf{x}_r$, or work-piece outline, must be known in advance in $\{T\}$ coordinates. If a CAD model of the part is available, this may be used, with appropriate cutter radius compensation. If an engineering drawing is available, this may be digitized and used. However, a simple method of obtaining the reference trajectory is to use the table itself to acquire the part outline.

The force sensor allows the table to perform "guarded moves," which are moves in a *direction* that terminate upon receipt of a preset threshold force. With this capability, the table can repeatedly contact the part, storing its location at contact. The locus of these stored points defines the part outline or reference trajectory in $\{T\}$, inherently including cutter radius compensation. The table must be positioned near the start of the trajectory, and must know the edge orientation at the start. It can then perform successive guarded moves at a preset stepsize to "learn" the part. Stepping from one point to the next along the edge tangent is done using the last-computed edge orientation.

5 EDGE FINISHING A JET ENGINE TURBINE HUB

This section presents the procedure and results of using the XY table system to perform edge finishing on a jet engine turbine hub. The hub is circular, 9.470 inches in diameter, with 86 "fir tree" shaped slots in the periphery, which receive the turbine blades. During manufacturing, these slots are broached, then a "pre-break" operation is performed to chamfer the edges before further smoothing/stress-relief processing. This pre-break operation requires a chamfer of 0.020±0.010" on the edges of the slots. The operation is currently done manually, is very tedious, and takes more than one day per part. The trajectory of the "fir tree" was obtained using the learning procedure described in Section 4. Figure 6 shows the trajectory. Note that this is a rather "intricate" contour, with small radii of curvature. In learning this trajectory, position data were taken at a spacing of 0.1 mm.

The turbine hub is made of high temperature steel alloy, thus a carbide deburring cutter was used. A feedrate of 3 ipm and a reference contact force of 1.5 lb gave acceptable material removal. The normal contact force during the operation is shown in Figure 7. The completion time for one fir tree was 16 seconds, for a total hub finishing time (both sides) of approximately 45 minutes. This is an order of magnitude decrease in processing time compared to the current manual operation.

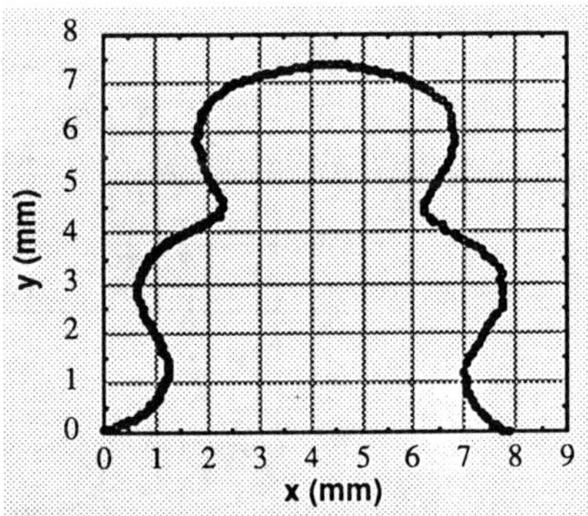

Figure 6. Learned Trajectory of Jet Engine Turbine Hub.

6 CONCLUSIONS

Although there is still work to be done in experimenting with several parameters of this XY table-based system; notably feedrate, contact force, and tool; the results to date have been very encouraging. Edge finishing of an intricate contour was successful, and the simplicity of a two axis table *vs* an articulated robot is attractive for industrial applications. The computational requirements are not high, and may be fulfilled by a PC, a more realistic choice for industry.

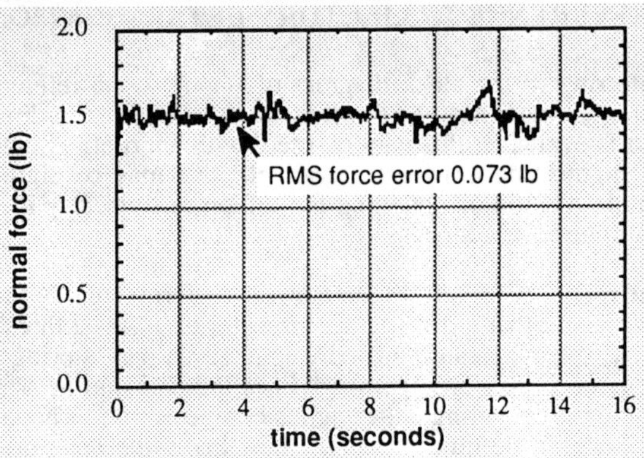

Figure 7. Normal contact force during turbine hub edge finishing.

REFERENCES

[1] B. L. Dawson, "Automated Surface Finishing," *Aerospace Engineering*, 9 (8), pp. 25-28, 1989.

[2] L. K. Gillespie, "Deburring and Surface Finishing: The Past Ten Years and Projections for the Next Ten Years," Allied-Signal Aerospace Company, Kansas City Division, KCP-613-4437, September 1990.

[3] C. B. Selleck and C. S. Loucks, "A System for Automated Edge Finishing," Proc. 1990 IEEE International Conference on Systems Engineering, Pittsburgh, PA, August 1990.

[4] M. Raibert and J. Craig, "Hybrid Position/Force Control of Manipulators," *ASME Journal of Dynamic Systems, Measurement, and Control,* June 1981.

Force and Fine-Position Control of Multiple Planar Robotic Mechanisms

Shahram Payandeh   Mehrdad Saif

School of Engineering Science
Simon Fraser University
Burnaby, British Columbia
V5A 1S6
CANADA

Abstract

One of the approaches to increase the dexterity of a single robot manipulating system is a design philosophy which consists of multiple robotic mechanisms. Although applications of such design philosophy are endless, there are many problems still remain to be addressed. One of these problems is the control of the contact forces (grasping forces) between the mechanisms and the object and the position of the grasped object. This paper addresses this problem. First, a model of the mechanisms in contact with the grasped object is postulated; second the problem of controlling the grasping forces and positions of the mechanisms is formulated in the linear multi-input/multi-output system and finally, a centralized optimal controller is proposed.

1 Introduction

Although robot manipulators have been successfully applied to various tasks, their versatility are limited. For example, when a manipulator is assigned to pick objects with different sizes and attributes, it is usually necessary to change its end-effector with another in order for the manipulator to establish a secure grip. The changing of the end-effectors becomes inevitable when in addition to pick-and-place tasks, the manipulator has to perform a tooling operation (drilling a hole).

One approach which can increase the versatility of a single manipulating mechanism is the utilization of a design philosophy which incorporates a collection of mechanisms. There are numerous problems involved in implementing such systems. One of the problems is the control of contacting forces between the manipulators and the object, [1] , [2] , [3]. Especially, in the case of dextrous end-effectors, it is sometime required to displaced the grasped object so that, for example, it is possible to insert the object into the other, [4] , [5], [6], [7] .

The objectives of this paper are: 1) to present a model of a planar grasp between the collection of robotic mechanisms and the object and 2) to formulate the problem in the state-space domain and to propose a linear optimal centralized controller where the outputs of the system can follow desired values.

The paper is organized into following sections: in section (2) a model of the mechanisms in contact with the grasped object is presented; section (3) gives grasping force and position controller for multiple robotic mechanisms; section (4) demonstrate the results of this paper using two 2DOF planar mechanisms (fingers) in contact with the object and finally section (5) presents discussions and outlines the future work.

2 A Model of Mechanisms in Contact with the Object

In this paper, the contact between the mechanisms and the object is modeled as soft contact where the source of compliance arises from the soft rubber type material attached to their end-points.

Let the linearized model of each planar finger in contact with the object expressed with respect to its configuration parameters (θ_i) be written as: (the plane being perpendicular to the direction of gravity)

$$\mathbf{M}_{i_\theta}\ddot{\theta}_i + \mathbf{C}_{i_\theta}\dot{\theta}_i = \tau_{act_i} - \mathbf{J}_i^T f_{ext_i} \tag{1}$$

where, ($\mathbf{M}_{i_\theta}$) is the inertia matrix of the ith finger, ($\mathbf{C}_{i_\theta}$) is the linearized velocity dependent terms of the dynamic model; (τ_{act_i}) is the actuating torque vector and (f_{ext_i}) is a vector of force acting on the ith finger end-point (grasping force vector) which is expressed with respect to its end-point reference coordinate frame. ($\mathbf{J}_i$) is a linear mapping (Jacobian) from the space of joint velocities to the end-point velocities and ($\theta_i = (\theta_1, \theta_2, \theta_3)_i^T$).

Equation (1) can be expressed in the coordinate frame located at the end-point of the manipulators as:

$$\mathbf{M}_i\ddot{x}_i + \mathbf{C}_i\dot{x} = f_{act_i} - f_{ext_i} \tag{2}$$

where ($\mathbf{M}_i$) is the inertia matrix expressed in the (i^{th}) manipulator end-point frame and ($\mathbf{C}_i$) is the velocity dependent term.

The dynamic model of the object expressed with respect to its coordinate frame can also be written as(assuming that the grasped object can only translates):

$$\sum_{i=1}^{n} f_{ext_i} = \mathbf{M}_o\ddot{x}_o \tag{3}$$

in the above equation (f_{ext_i}) is the grasping forces of the (i^{th}) mechanism where its components are expressed with respect to the reference coordinate frame of the object.

For the planar grasp only forces can get transmitted to and from the object. These forces are functions of both the mechanisms and object displacements, and the known model of the compliant-tip material ($\mathbf{K}_{c_i}, \mathbf{C}_{c_i}$), or:

$$f_{ext_i} = f(x_i, \dot{x}_i, x_o, \dot{x}_o, \mathbf{K}_{c_i}, \mathbf{C}_{c_i})$$

Rewriting equations (2) and (3) we have:

$$\begin{cases} \mathbf{M}_1\ddot{x}_1 + (\mathbf{C}_1 + \mathbf{C}_{c_1})\dot{x}_1 + \mathbf{K}_{c_1}x_1 - \mathbf{C}_{c_1}\dot{x}_o - \mathbf{K}_{c,1}x_o = f_{act_1} \\ \vdots \\ \mathbf{M}_i\ddot{x}_i + (\mathbf{C}_i + \mathbf{C}_{c_i})\dot{x}_i + \mathbf{K}_{c_i}x_i - \mathbf{C}_{c_i}\dot{x}_o - \mathbf{K}_{c,i}x_o = f_{act_i} \\ \vdots \\ \mathbf{M}_o\ddot{x}_o + (\mathbf{C}_{c_1} + \mathbf{C}_{c_2} + \cdots)\dot{x}_o - \mathbf{C}_{c_1}\dot{x}_1 - \mathbf{C}_{c_2}\dot{x}_2 - \cdots \\ (\mathbf{K}_{c_1} + \mathbf{K}_{c_2} + \cdots)x_o - \mathbf{K}_{c_1}x_1 - \mathbf{K}_{c_2}x_2 - \cdots = 0 \end{cases} \tag{4}$$

In the above equation, (x_i) is the position and orientation vector of the (i^{th}) finger end-point; $(f_{act_i} = (f_{act_x}, f_{act_y}, f_{act,\theta})^T)$ is the vector of actuating force for a 3DOF planar finger and (x_o) is the vector of the linear displacement of the object.

3 Contact Force and Fine-Position Controller

The above equations can be transformed into the standard state-space model as:

$$\dot{x} = \mathbf{A}x + \mathbf{B}u \tag{5}$$

where $(x = (x_1, x_2, \cdots, x_i, x_o, \dot{x}_1, \dot{x}_2, \cdots, \dot{x}_i, \dot{x}_o)^T) \in \mathbf{R}^n$ is the state vector of the compounded finger/object system and $(u = (f^T_{act_1}, f^T_{act_2}, \cdots, f^T_{act_i})^T) \in \mathbf{R}^r$ is the actuating force vector.

The output vector of the system can be written as:

$$y = \mathbf{C}x \tag{6}$$

where (y) is a vector composed of grasping forces and the positions of the fingers and object.

Given the dynamic model of the fingers and object, the objective of the controller is for the outputs y_1 of the system to track some desired reference constant [8] (or periodically changing) input vector (y_r) containing the desired grasping forces and the position of object. Defining:

$$\dot{z} = y_r - y_1 = y_r - \mathbf{C}_1 x \tag{7}$$

and augmenting equation (6) with (7) results in (n+p) dimensional system:

$$\begin{aligned} \dot{\bar{x}} &= \bar{\mathbf{A}}\bar{x} + \bar{\mathbf{B}}u + \mathbf{D}y_r \\ \bar{y} &= \bar{\mathbf{C}}\bar{x} \end{aligned} \tag{8}$$

where

$$\bar{x} = \begin{bmatrix} x \\ z \end{bmatrix}, \ \bar{\mathbf{A}} = \begin{bmatrix} \mathbf{A} & \mathbf{0} \\ -\mathbf{C}_1 & \mathbf{0} \end{bmatrix}, \ \bar{\mathbf{B}} = \begin{bmatrix} \mathbf{B} \\ \mathbf{0} \end{bmatrix}, \ \mathbf{D} = \begin{bmatrix} \mathbf{0} \\ \mathbf{I} \end{bmatrix}, \ \bar{\mathbf{C}} = \begin{bmatrix} \mathbf{C} & \mathbf{0} \\ \mathbf{0} & \mathbf{I} \end{bmatrix}$$

It is now desired to obtain the proportional-integral (PI) optimal control law (u) of the form

$$u = \mathbf{K}_1 x + \mathbf{K}_2 \int \dot{z} dt \tag{9}$$

such that the following performance measure is minimized subject to (8)

$$J = \frac{1}{2}\int_0^{\infty} (\|\bar{x}\|^2{}_{\mathbf{Q}} + \|u\|^2{}_{\mathbf{R}})dt \tag{10}$$

and the closed-loop eigenspectrum of the system are assigned to desired locations. Using the sequential approach of ([9]), it can be shown that for a fixed $\mathbf{R}$, if $\mathbf{Q}$ in (10) is appropriately selected, the optimal control law that would achieve the above objectives would be given by:

$$u = -\mathbf{R}^{-1}\bar{\mathbf{B}}^T\mathbf{P}\bar{x} = \mathbf{K}\bar{x} \tag{11}$$

It can also be shown [10] that the pair $(\bar{\mathbf{A}}, \bar{\mathbf{B}})$ is completely controllable if and only if $\{\mathbf{A}, \mathbf{B}\}$ is controllable and,

$$\boldsymbol{\Omega} = \begin{bmatrix} \mathbf{B} & \mathbf{A} \\ \mathbf{0} & -\mathbf{C} \end{bmatrix}$$

has rank $(n+p)$, where (p) is the number of outputs that need to regulated.

The summary of the optimal modal servo controller design in a sequential manner is as follows: at the ith round, the model of the mechanisms in contact with the grasped object is aggregated to a first or second order system depending on whether a real or a complex conjugate pair of poles are being placed. An appropriate weighting matrix is found such that the pole(s) is(are) placed at desired location(s). The calculated weighting matrix and the optimal control law are then transformed to the original higher dimensional space. In this space, the eigenstructure of the aggregated system is preserved in an invariant subspace of the original higher dimensional closed loop system. At this point the ith round of the sequential process is complete and the sequential algorithm is continued in this manner until all the poles are assigned. Finally, the overall desired state weighting matrix and the optimal feedback gains are calculated by simply adding up the weights which are obtained at each round of the process.

4 Example

This section presents the performance of the optimal controller position and force controller using two 2DOF fingers and a rigid object.

Each finger is modelled to have two rotary joint with the dimension of each link given as ($l_1 = 0.153m$ and $l_2 = 0.123m$) and mass of ($m_1 = 0.053Kg$ and $m_2 = 0.02Kg$) [11]. The initial configuration parameters of the fingers are $\theta_{1_1} = 135°, \theta_{1_2} = 45°, \theta_{2_1} = -90°, \theta_{2_2} = 90°$. The compliant material of both finger-tips are modeled as linear springs having stiffness of ($1000N/m$). The mass of the grasped object $m_0 = 0.2Kg$. The objective of the simulation is for each finger to exert forces equal to 5 and then 2 Newton on the grasped object and moving the object 1 and then 2 cm from its initial grasp configuration.

Since the proposed controller of equation (9) requires full state feedback of the finger/object system, an observer is designed [12] for estimating the state of the object(i.e. it is assumed that the position and velocity of the grasped

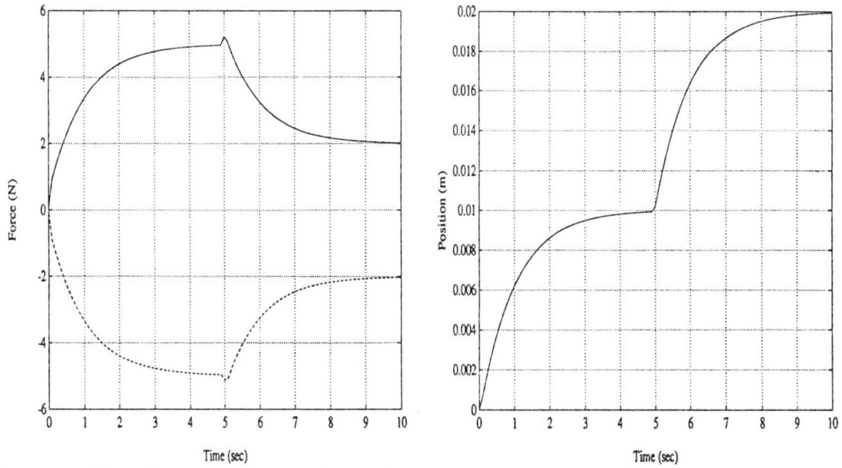

Figure 1: The force responses of the fingers to 5 and 2 Newton Grasping force and the position response of the grasped object.

object are not measurable). Following the procedure for designing the optimal controller outlined in the previous section, gains of the controller were selected such that the closed-loop response has minimum overshoot. Figure 1 shows the response of the controller for step changes in the desired grasping forces and the position of the grasped object.

5 Discussions and Future Work

This paper presents a model of robotics mechanisms (fingers) in contact with the grasped object. In this model compliant material is attached between the end-points of the mechanisms and the object. Using this model, a centralized force and position controller is proposed. The feature of this controller is that it minimizes a performance index while it also places the closed loop poles of the system at desired locations. This feature of the controller is important in multiple mechanical fingers grasping a fragile object where the objective is to shape the transient response of the system. The results of this paper is demonstrated using two 2DOF fingers.

A number of issues remain that needs to be addressed: 1) extension of the controller to the spatial contact configurations; 2) investigation into existence of a decentralized optimal force controller and 3) suboptimal modal projective control strategies are already under investigation for force and position control in multiple robotics applications.

Acknowledgment This research was supported by Natural Science and Engineering Research Council (NSERC) of Canada.

References

[1] Yun, X. , *Coordination of Two-Arm Pushing*, Proceedings of International Conference on Robotics and Automation, 1991, pp. 182-187

[2] Akella, P., Siegwart, R. and Cutkosky, M., *Manipulation with Soft Finger: Contact Force Control*, Proceedings of International Conference on Robotics and Automation, 1991, pp. 652-657

[3] Mason, M. T. and Salisbury, J. K., *Robot Hands and the Mechanics of Manipulation*, The MIT Press, 1984

[4] Schneider, S. A. and Canon, R. H. , *Object Impedance Control for Cooperative Manipulation*, Proceedings of IEEE International Conference on Robotics and Automation, 1989, pp. 1076-1087

[5] Murray, R. M. and Sastry, S. S. , *Control Experiments in Planar Manipulation and Grasping*, Proceedings of IEEE International Conference on Robotics and Automation, 1989, pp. 624-629

[6] Hsu, P., *Control of Multi-Manipulator Systems- Trajectory Tracking, Load Distribution, Internal Force Control and Decentralized Architecture*, Proceedings of IEEE International Conference on Robotics and Automation, 1989, pp. 1234-1239

[7] Payandeh, S. and Goldenberg, A., *Grasp Impedance: Examples of Fingers' Targeted Impedance*, Proceedings of IEEE Intelligent Control Conference, 1988

[8] Payandeh, S. and Goldenberg, A., *A Robust Force Controller: Theory and Experiments*, Proceedings of International Conference on Robotics and Automation, 1991, pp. 36-41

[9] Saif, M., , *Optimal Linear Regulator Pole Placement by Weight Selection*, International Journal of Control, Vol. 50, 1989, pp. 399-414

[10] Davison, E. J. and Smith, H. W., *Pole Assignment in Linear Time-Invariant Multivariable Systems with Constant Disturbances*, Automatics, Vol. 7, 1971, pp. 489-498

[11] Murray, R. M., *Control Experiments in Planar Manipulation and Grasping*, Electronics Research Laboratory, College of Engineering, University of California, Berkeley, Report No. UCB/ERL M89/3

[12] O'Reilly, J., *Observers in Linear Systems*, Academic Press, London, 1983

ROBUST ADAPTIVE CONTROL OF ROBOT MANIPULATORS

Jong-Hwan Kim, Keun-Mo Koo, and Hyun-Sik Shim
Dept. of Electrical Engineering,
Korea Advanced Institute of Science and Technology,
373-1 Kusung-dong, Yusung-gu, Taejon 305-701, KOREA
e-mail : johkim@kit.kaist.ac.kr

ABSTRACT

This paper proposes a robust algorithm for continuous-time direct adaptive control of robot manipulators based upon Lyapunov direct method. The upper bound of the disturbances is estimated as well as the robot parameters, and the feedback gain is adjusted by using the estimated upper bound such that the predetermined tracking error precision is achieved under the disturbances.
Key Words : adaptive control, robust control, robot control, Lyapunov stability, feedback gain.

INTRODUCTION

Recently, some researches have been done on direct adaptive control to improve the accuracy in robot manipulator control. These show the globally convergent property of the overall system in the disturbance free case. However, under the unknown disturbances, the desired tracking error precision can not be obtained if the feedback gains of the system are not properly selected. If we have some knowlegde about the disturbances, we may select the feedback gain to achieve the desired tracking error precision, but it is hard to know a prior knowledge about the disturbances.

In this paper, a robust algorithm for continuous-time direct adaptive control of robot manipulators is proposed based upon Lyapunov direct method. In this scheme, adaptive estimation laws containing deadzones are proposed to estimate the upper bound of the disturbances in addition to the robot parameters. The feedback gain is updated to get the desired tracking accuracy by using the estimated upper bound of the disturbances. Hence, in our scheme the prior knowledge about the disturbances is not required.

ROBUST ADAPTIVE CONTROL

The dynamics of the manipulator with n degrees of freedom is described by the following equation :

$$T = M(q)\ddot{q} + H(q, \dot{q}) \qquad (1)$$

where

$$\begin{array}{rcl} q: & \text{Joint angle ,} & q \in R^{n\times 1} \\ T: & \text{Joint torque ,} & T \in R^{n\times 1} \\ M(q): & \text{Inertia matrix ,} & M \in R^{n\times n} \\ H(q,\dot{q}): & \text{Coriolis, centrifugal, frictional} & \\ & \text{and gravitational forces ,} & H \in R^{n\times 1}. \end{array}$$

The desired trajectory is represented by q_d. Define position error as

$$\tilde{q} = q - q_d.$$

Let $s \triangleq \dot{\tilde{q}} + \lambda\tilde{q}$ and consider the following control law :

$$\begin{aligned} T &= M(q)\ddot{q}_r + H(q,\dot{q}) - \frac{1}{2}\dot{M}(q,\dot{q})s - k(t)s \\ &= W(q,\dot{q},\ddot{q}_r,s)\Theta - k(t)s \end{aligned} \tag{2}$$

where $\ddot{q}_r$ is defined as $\ddot{q}_d - \lambda\dot{\tilde{q}}$. W and Θ are defined from the following relation:

$$M(q)\ddot{q}_r + H(q,\dot{q}) - \frac{1}{2}\dot{M}(q,\dot{q})s = W(q,\dot{q},\ddot{q}_r,s)\Theta \tag{3}$$

where

$$\begin{array}{rll} \Theta: & \text{physical parameter vector ,} & \Theta \in R^{p\times 1} \\ W: & \text{matrix of known function ,} & W \in R^{n\times p}. \end{array}$$

In practical application, there may exist bounded disturbances in control input T. We represent these disturbances by vector function $\nu(t) \in R^{n\times 1}$. Let us assume that there exist bounded positive constants α_1, α_2, and α_3 which satisfy the following relations :

$$\begin{aligned} \|\nu(t)\| &\leq \alpha_1 + \alpha_2\|e\| + \alpha_3\|\dot{e}\| \\ &= \alpha^T\phi \end{aligned} \tag{4}$$

where $\alpha = (\alpha_1\ \alpha_2\ \alpha_3)^T$ and $\phi = (1\ \|e\|\ \|\dot{e}\|)^T$. The unknown parameter vector α is estimated by a parameter adaptation algorithm. The estimates of α and Θ are defined as $\hat{\alpha}$ and $\hat{\Theta}$, respectively. By using the estimate of Θ, (2) can be represented as

$$\begin{aligned} T &= \hat{M}(q)\ddot{q}_r + \hat{H}(q,\dot{q}) - \frac{1}{2}\dot{\hat{M}}(q,\dot{q})s - k(t)s \\ &= W(q,\dot{q},\ddot{q}_r,s)\hat{\Theta} - k(t)s \end{aligned} \tag{5}$$

where $\hat{M}$ and $\hat{H}$ are the estimates of M and H, respectively. Considering the effect of $\nu(t)$ and applying (5) for T to (1) yield

$$M(q)\dot{s} = \tilde{M}(q)\ddot{q}_r + \tilde{H}(q,\dot{q}) - \frac{1}{2}\dot{\hat{M}}(q,\dot{q})s - k(t)s + \nu(t) \tag{6}$$

where $\tilde{M} = \hat{M} - M$ and $\tilde{H} = \hat{H} - H$.

Lemma 1 *Consider the manipulator described by (1) with the control law (5) and the adaptation algorithms which are given as follows:*

$$\dot{\hat{\Theta}} = \begin{cases} -\Gamma_1 W(q,\dot{q},\ddot{q}_r,s)^T s & \text{if } \|s\| > \lambda\epsilon_1 \\ 0 & \text{if } \|s\| \leq \lambda\epsilon_1 \end{cases} \tag{7}$$

$$\dot{\hat{\alpha}} = \begin{cases} \Gamma_2 \|s\| \phi & \text{if } \|s\| > \lambda\epsilon_1 \\ 0 & \text{if } \|s\| \leq \lambda\epsilon_1 \end{cases} \tag{8}$$

$$k(t) = \frac{\hat{\alpha}^T \phi}{\lambda\epsilon_1} \tag{9}$$

where λ, ϵ, and w are given positive constants, and Γ_1 and Γ_2 are $p \times p$ and 3×3 positive definite matricies, respectively. The norm $\|s\|$ is defined as $(s^T s)^{\frac{1}{2}}$. Then the following holds along the system (6):

$$\dot{V} < 0 \text{ for all } \|s\| > \lambda\epsilon_1$$

where $V(t)$ is a Lyapunov function :

$$V(t) = \frac{1}{2} s^T M(q) s + \frac{1}{2}\tilde{\Theta}^T \Gamma_1^{-1} \tilde{\Theta} + \frac{1}{2}\tilde{\alpha}^T \Gamma_2^{-1} \tilde{\alpha} \tag{10}$$

where $\tilde{\Theta} = \hat{\Theta} - \Theta$ and $\tilde{\alpha} = \hat{\alpha} - \alpha$.

Proof : The time derivative of (10) is

$$\dot{V}(t) = s^T M(q)\dot{s} + \frac{1}{2} s^T \dot{M}(q,\dot{q}) s + \tilde{\Theta}^T \Gamma_1^{-1} \dot{\hat{\Theta}} + \tilde{\alpha}^T \Gamma_2^{-1} \dot{\hat{\alpha}}. \tag{11}$$

Substituting (6) for $M(q)\dot{s}$ in (11) and using the relation (3) yield

$$\begin{aligned} \dot{V}(t) &= s^T(W\tilde{\Theta} - k(t)s + \nu(t)) + \tilde{\Theta}^T \Gamma_1^{-1} \dot{\hat{\Theta}} + \tilde{\alpha}^T \Gamma_2^{-1} \dot{\hat{\alpha}} \\ &\leq -s^T k(t) s + \tilde{\Theta}^T (W^T s + \Gamma_1^{-1} \dot{\hat{\Theta}}) + \|s\| \|\nu(t)\| + \tilde{\alpha}^T \Gamma_2^{-1} \dot{\hat{\alpha}}. \end{aligned}$$

From (4), (7), and (8), the above inequality can be expressed as

$$\dot{V}(t) \leq -s^T k(t) s + \hat{\alpha}^T \phi \|s\|.$$

From (9), it can be easily verified that

$$\dot{V}(t) < 0 \text{ when } \|s\| > \lambda\epsilon_1. \qquad \blacksquare$$

Then we can obtain the following theorem as a result.

Theorem 1 *Consider the manipulator given by (1) with the control law (5) and the adaptation algorithms (7), (8), (9). Then, there exists a finite time t_f such that*

$$\|\tilde{q}(t)\| < \epsilon \,, \quad \|\dot{\tilde{q}}(t)\| < 2\lambda\epsilon \quad \text{for all} \quad t > t_f$$

where ϵ is a constant satisfying $\epsilon > \epsilon_1$.

Proof : Since $s(t)$ is continuous in time t and has a bounded time derivative, the following inequality is hold for a finite time t_f.

$$\|s(t)\| < \lambda\epsilon \quad \text{for all} \quad t \geq t_f \tag{12}$$

The tracking error vector $\tilde{q}(t)$ can be expressed as

$$\tilde{q}(t) = \int_0^t e^{-\lambda(t-\tau)} s(\tau) d\tau.$$

From (12) we thus get

$$\|\tilde{q}(t)\| < \lambda\epsilon \int_0^t e^{-\lambda(t-\tau)} d\tau < \epsilon \quad \text{for all} \quad t > t_f.$$

The boundedness of $\dot{\tilde{q}}(t)$ can be easily shown as follows.

From the definition of the vector $s(t)$, we get

$$\|s(t)\| = \|\dot{\tilde{q}}(t) + \lambda\tilde{q}(t)\| < \lambda\epsilon.$$

Thus, we can obtain

$$\|\dot{\tilde{q}}(t)\| < \lambda\epsilon + \lambda\|\tilde{q}(t)\| = 2\lambda\epsilon \quad \text{for all} \quad t > t_f. \qquad \blacksquare$$

CONCLUSION

A robust direct adaptive control scheme for robot manipulators which can achieve the desired tracking error precision under the unknown disturbances has been proposed based upon Lyapunov direct method.

In this scheme, the boundedness of all the signals of the robot system is guaranteed by the adaptation algorithms containing deadzones. In addition to the robot parameters the upper bound of the unknown disturbances is estimated to maintain the tracking error within the desired accuracy by adjusting the feedback gain properly.

REFERENCES

[1] J. J. Craig, P. Hsu, and S. S. Sastry, "Adaptive control of mechanical manipulators," *Int. J. Robotics Res.*, vol.6, pp.16-28, 1987.

[2] J. J. E. Slotine and W. Li, "Composite adaptive control of robot manipulators," *Automatica*, vol.25, no.4, pp.509-519, 1989.

[3] R. Johansson, "Adaptive control of robot manipulator motion," *IEEE Trans. Robotics and Automat.*, vol.6, no.4, pp.483-490, 1990.

[4] B. B. Peterson and K. S. Narendra, "Bounded error adaptive control," *IEEE Trans. Automat. Contr.*, vol.27, no.6, Dec. 1982.

[5] J. Imura, T. Sugie, and T. Yoshikawa, "Adaptive robust control of robot manipulators," *in Proc. of Int. Symposium on the M.T.N.S.*, Kobe, Japan, June 1991.

[6] M. J. Corless and G. Leitmann, "Continuous state feedback guaranteeing uniform ultimate boundedness for uncertain dynamic systems," *IEEE Trans. Automat. Contr.*, vol.26, no.5, pp.1139-1144, 1981.

[7] J. J. Craig, *Introduction to robotics : mechanics and control.*, Addison-Wesley, 1986.

[8] S. Sastry and M. Bodson, *Adaptive control : stability, convergence, and robustness*, Prentice-Hall International, 1989.

SIMULATION AND A COMPARATIVE ANALYSIS OF POSITION & FORCE CONTROL TECHNIQUES IN ROBOTICS

Ali Meghdari, Associate Professor
School of Mechanical Engineering
Sharif University of Technology
Tehran, Islamic Republic of Iran

Sassan Raghibizadeh, M.Sc.
Faculty of Petroleum Engineering
University of Petroleum Industry
Ahwaz, Islamic Republic of Iran

ABSTRACT

This article contains a comparative and qualitative analysis of a number of Position and Force control techniques applied to a 3-DOF planar manipulator. Simulated position control methods in this paper are namely the PID, Computer Torque Method (CTM), Craig's, and Seraji's Adaptive Methods. Force/Hybrid Control techniques such as Active Stiffness method, Khatib's Hybrid Method, and Seraji's Adaptive Method for Hybrid (Position-Force) control have been also studied and simulated.

INTRODUCTION

In recent years extensive research have been focused on the issues of position and force control in robotics and a number of promising techniques have evolved [1-10]. To gain a basic understanding of these issues, some of these methods have been studied, simulated, and compared. They were compared in the sense of the rate of overshoot in step response, sensitivity to a change in controller parameters, accuracy in following a prescribed trajectory, stability, and consumed energy. The Force/Hybrid control methods are also simulated for the case where robot end-effector is following a trajectory tangent to a wall and applying a force in the normal direction.

MANIPULATOR DYNAMICS

Dynamics of an n-degree of freedom manipulator is generally defined by a set of highly non-linear strongly coupled second order differential equations as[1,2,3]:

$$M(\Theta)\ddot{\Theta} + V(\Theta,\dot{\Theta}) + G(\Theta) + F(\dot{\Theta}) + \tau_d = \tau \qquad (1)$$

where $M(\Theta)$ is the (nxn) generalized inertia matrix; and $V(\Theta,\dot{\Theta})$, $G(\Theta)$, $F(\dot{\Theta})$ are the (nx1) vectors representing the velocity, gravity, and the friction terms, respectively. Parameters Θ, $\dot{\Theta}$, $\ddot{\Theta}$ are the (nx1) joint position, velocity, and acceleration vectors. τ is the (nx1) vector of joint torques or forces applied by actuators, and τ_d is the (nx1) vector of joint torques or forces due to external disturbances.

POSITION CONTROL TECHNIQUES

In robotics, the position control problem is stated as: for the obtained instantaneous values of joint positions and velocities Θ, $\dot{\Theta}$ from sensors, and for the desired values of θ_d, $\dot{\theta}_d$, $\ddot{\theta}_d$ from the trajectory generator, one should compute the torque vector $\tau(t)$ such that $\theta(t)$ reaches the desired values of $\theta_d(t)$ with the least error. Figure 1 shows a simple diagram representing the position control problem from the Input-Output point of view. The position control techniques that are simulated and compared here are namely the PID, Computer Torque Method, Craïg's, and Seraji's Adaptive Methods.

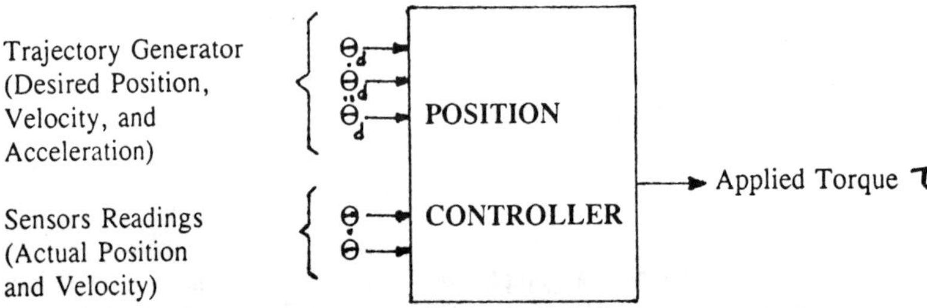

Fig. 1. General Diagram of Position Control Methods.

Most of the current robots utilize Proportional-Integral-Derivative (PID) controllers, which are simple in design and each degrees of freedom of a robot is controlled separately and coupling effects are generally ignored [4].

To obtain greater accuracy, it is necessary to utilize the robot dynamic equations in control. In Computed-Torque-Method (CTM), the controller utilizes the dynamic equations in a portion of the feedforward in order to instantaneously compute and apply the required joint torques. The greatest problem of this method is the large amount of computations involved, and the possibility of existing errors in the robot dynamic model and parameters. To reduce the first problem Nigam and Lee [5] have applied the recursive and serial nature of the Newton-Euler dynamics method, and designed a system of seven microprocessors of the kind MC68020 which computes the required joint torques for a 6-DOF robot in about 0.6 milliseconds. And to diminish the second problem Fu et. al. [1] have employed the adaptive method in linear systems in order to determine the involved parameters in the linearized model of the dynamic equations about the desired trajectory. This method uses a feedforward component computed from the dynamic equations in order to first direct the robot around the desired trajectory, and then by applying a feedback component it will eliminate the feasible errors.

By linearizing the dynamic equations about a hypothetical state, one can replace a non-linear system with a time varying linearized system [6,7]. Seraji [6] employs the Model-Reference-Adaptive-Control method in order to control the obtained linearized system. His method is presented in Cartesian space, but since the dynamic equations are not directly used, therefore we have employed the technique in joint space.

Craig [8] presents an adaptive method for identification and control of the unknown system parameters without linearizing the dynamic equations. He proves the stability of such a system and its robustness for bounded disturbances. The control law in this method is very similar to that of the Computer-Torque-Method. The difference is in the on-line identification of the system's unknown dynamic parameters which are used in controlling the robot.

It is generally assumed that the robot dynamic structure is completely known in adaptive systems, and the only unknowns are the dynamic parameters utilized. This assumption is not very correct for the case of friction forces and errors due to joints and links flexibility and loose connections at the joints. To overcome these effects that are generally difficult to model, but are repeated alike in repeating a trajectory, Learning Control Methods are presented [8]. In addition, for cases where repeating a trajectory is not possible, several methods based on Time-Delay-Control [9,10] are presented. These methods employ information received from the previous instant (t-l) to identify and control the robot at the desired instant (t).

In addition to the above techniques that are generally applicable in joint-space, other methods known as Resolved-Motion-Control have been presented for robot control in Cartesian space [1]. They are named as Resolved-Motion-Control due to the fact that for generating a desired motion in Cartesian space, one has to resolve the Cartesian motion simultaneously to a coordinated joints motion.

Simulations and a Comparative Look

A three degree-of-freedom planar robot with link lengths as $l_1 = l_2 = 0.5$ m, and $l_3 = 0.25$ m is considered for simulating the techniques overviewed in previous section. Their performance have been analyzed for two different trajectories. To evaluate the step response of the methods consider the first trajectory shown in Figure 2 for which the robot changes its configuration by a step input from $\Theta_1 = [0,0,0]^T$ at t=0 sec. to $\Theta_2 = [0,0,45°]^T$ and stops for 2 seconds at that position. Then at t = 2 sec. it moves to $\Theta_3 = [0,45°,45°]^T$ by a step and stops for 2 sec. at the new position. Finally at t=4 sec. it moves to $\Theta_4 = [45°,45°,45°]^T$ by a step and stops for 2 sec. at the this configuration.

In the second trajectory shown in Figure 3, robot changes its configuration from $X_1 = [-0.9,0,180°]^T$ to $X_2 = [-0.9,0.6,135°]^T$ in about 1 second. Then it moves to the configuration $X_3 = [0.9,0.6,45°]^T$ in 3 seconds, and later to $X_4 = [0.9,-0.3,-45°]^T$ in 1.5 seconds, and finally to configuration $X_5 = [0.9,0.3,0°]^T$ in one second, then stops there for 2 seconds. The total motion time is 8.5 seconds. Note that vector $X = [x,y,\varphi]^T$ is a vector representing robot hand's configuration in base frame at the center O.

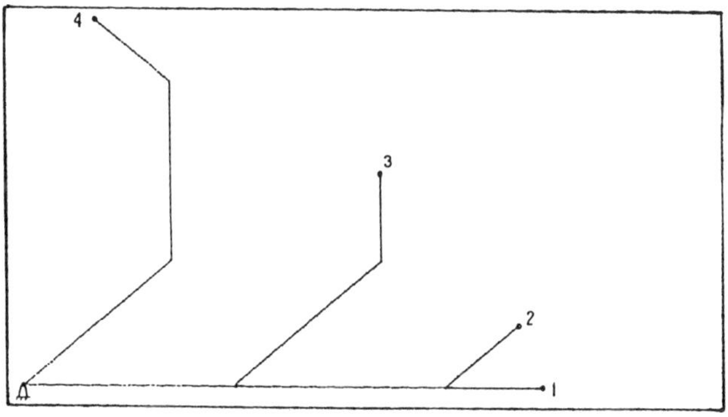

Fig. 2. Robot Motion Trajectory for Position Control.

Control methods may be compared by various means. We have qualitatively compared above methods for 8 different cases. The first trajectory shown in Figure 2 is used for test #1, where as the second trajectory shown in Figure 3 is used for tests #2 through 8. Table 1 describes these tests and their objectives.

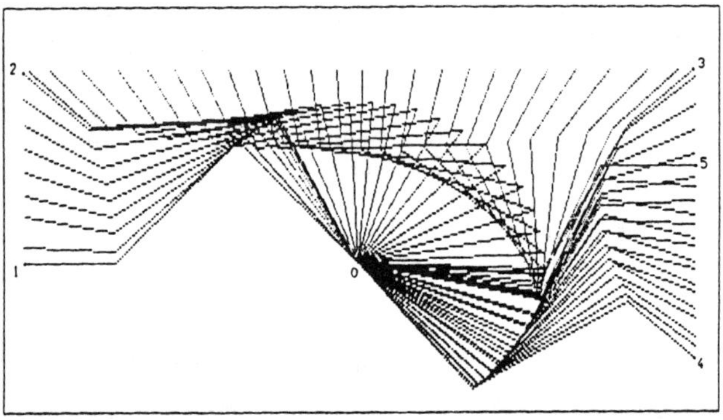

Fig. 3. Robot Motion Trajectory for Position Control.

Table 1. Various Tests to Evaluate PC-Methods Performance.

Test-1	Test-2	Test-3	Test-4	Test-5	Test-6	Test-7	Test-8
Step Response (Normal Conditions)	Trajectory Tracking (Normal Conditions)	Increasing Sample Time by 50 %	Sudden Change in Loading Conditions	Increasing Controller Coefficients by 50 %	Decreasing Controller Coefficients by 50 %	Doubling the Velocity	Increasing Velocity by 4 times the Norm. Cond.

In test-7 the total motion time is one-half (4.25 sec.) and in test-8 it is one-forth (2.12 sec.) of the normal case. The sampling times for CTM, PID, Craig's, and Seraji's methods were chosen to be 0.004 sec., 0.002 sec., 0.004 sec., and 0.001 sec., respectively. For each test, the maximum error, maximum consumed power, sum of the squares of error, steady state error, the total consumed energy for each joint, the maximum end-effector error in x and y directions and sum of their squares are computed and compared. Figure 4 shows the errors in all joints obtained from test-1 of each control method described above [4].

In simulations, it was tried to select higher values for controllers coefficients. But note that in reality these values depends on the parameters such as robots natural frequency, noise in sensors, motor power, data taking frequency, and systems stability. Obtained results indicate that the PID method maintains the least level of accuracy. Table 2 displays the maximum error and the square root of the sum of the squares of errors of end-effector (SSQR error).

Table 2. Error Comparison Among Position Control methods.

METHODS	PID	CTM	Seraji	Craig
SSQR error	107	28	21	1
MAX error	68	11	18	1

It is observed that the maximum error exists in the PID method being 68 times greater than the Craig's method. And the maximum error in CTM is lower than Seraji, but Seraji's method presents a greater accuracy for the overall trajectory. The steady state error in Seraji's method is lower than others, and is followed by Craig's, PID, and

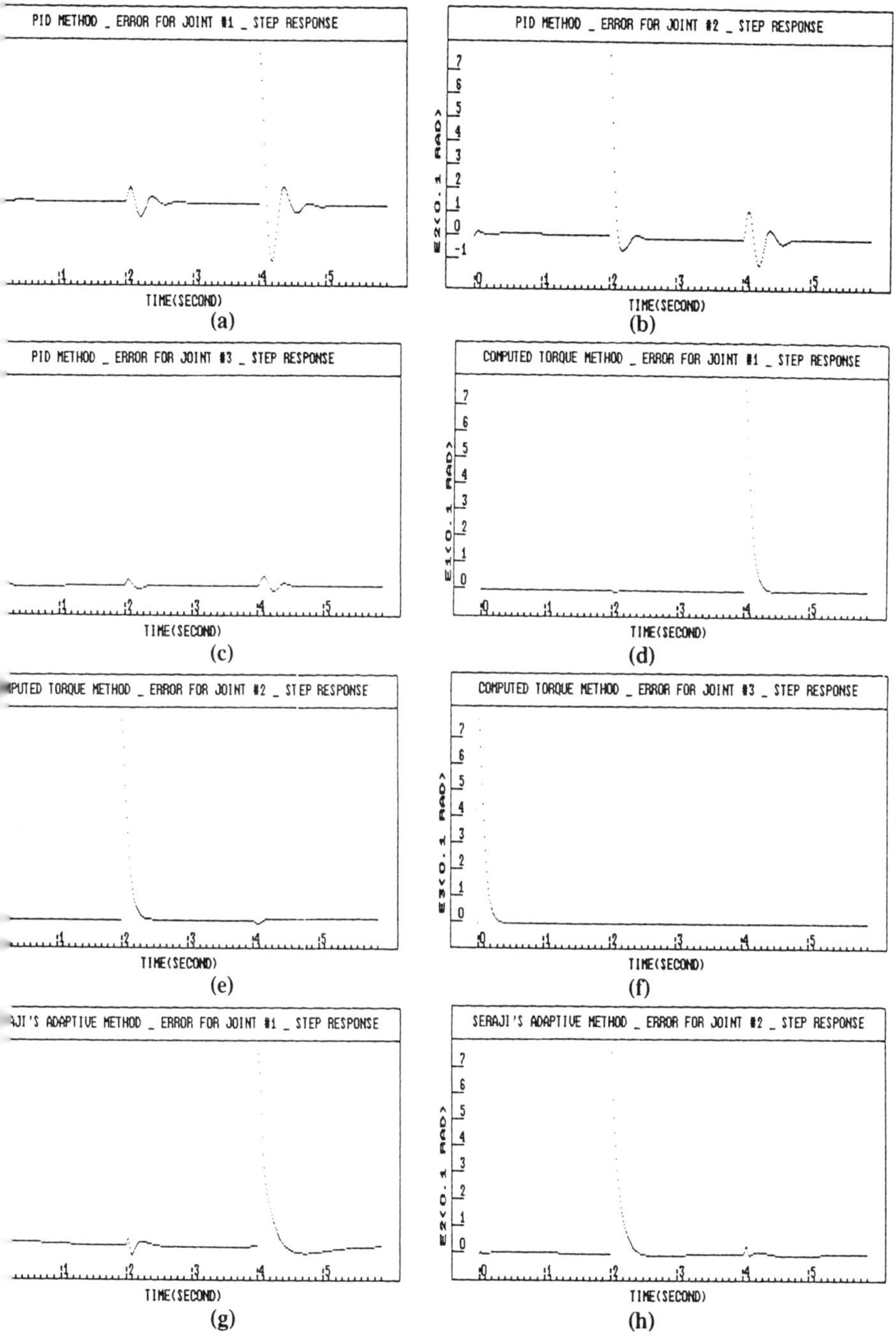

Fig. 4. Step-Response Errors of the Position Control Methods.

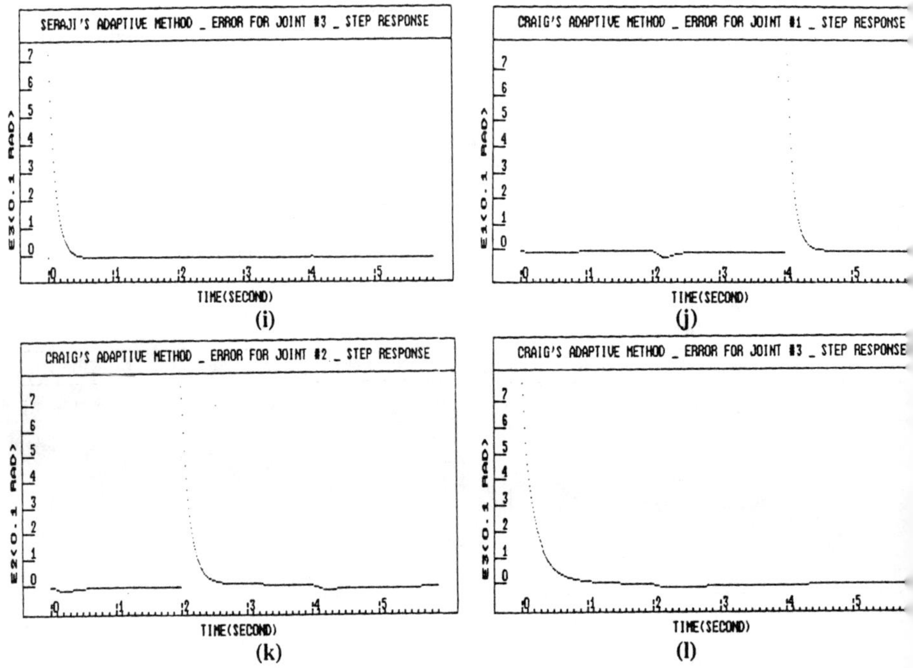

Fig. 4. Step-Response Errors of the Position Control Methods (continued).

CTM accordingly. In normal speed for every joint, Craig's method requires the lowest maximum consumed power followed by CTM, Seraji, and PID, accordingly. Increasing speed of motion results in a rapid increase in maximum consumed power in Seraji's method which may result in a non-linear behavior and saturation of motors.

Stability and robustness of the Craig's method is proven in [8]. And also CTM method is shown to be robust for small errors in parameter values [8]. It is generally difficult to prove stability for PID method, and this method is generally suitable for cases where coupling and nonlinearity effects in robot dynamics are minimized. Stability of Seraji's method is only guaranteed at low speeds due to the assumptions made in proving its stability. Simulation results obtained from test-8 indicate that the Seraji's method becomes unstable at higher speeds. And a small sampling time must be chosen for this method. All methods provide greater accuracy when their controller coefficients are increased. Variations of sampling times around the figured values have little effect on accuracy. Finally, PID and Seraji's methods are the cheapest due to low computations.

FORCE CONTROL TECHNIQUES

When a robot hand is not moving freely in space, and is in contact with a surface, position control in normal direction is no longer possible and robot looses some degrees of freedom. In this case, on may use force control in directions where motion is limited (force subspace), and apply position control in free directions (position subspace). To control a robot in such conditions, Hybrid (position & force) control techniques are generally used. Since description of contact forces and robot hand motion is basically done in Cartesian space, most force control techniques are formulated in Cartesian space. Whitney [11] briefly describes some of these techniques and the important issues involved. For a number of assembly cases having a passive compliance is sufficient. For example, for peg-in-hole insertion one can utilize the Remote-Center-Compliance (RCC) inorder to create the required active flexibility in the robot hand [2,3].

This section briefly describes three of the force control methods namely, Active Stiffness Method, Khatib's Hybrid Method, and Seraji's Adaptive Method for Hybrid Control. Their performance in controlling a 3-DOF revolute planar manipulator have been simulated and the results are presented.

Active Stiffness method is actually a simple position control method of the Proportional Derivative (PD) kind, in which the proportional controller coefficients are set such that the desired active stiffness matrix is created for the robot hand [2,10].

Seraji's Adaptive Method for Hybrid Control utilizes equations of motion written in a Constraint Frame [13]. This frame is defined in accordance to the geometry of the contacted surface such that some of its axis represent the force-control direction(s), and others the position-control direction(s). In other words, the robot's hand workspace $\{X\}$ which is of order n for an n-DOF robot is divided into two independent subspaces of force control $\{Z\}$ with an order of p, and position control $\{Y\}$ with an order of q, such that $p + q = n$. An advantage in employing constraint frame is that subspaces $\{Z\}$ and $\{Y\}$ are independent of each other and one could describe the performance of position and force controllers independently of each other [13].

In contrast with most position-force control methods that are formulated in constraint frame, Khatib's Hybrid control method provides the ability to describe the equations in any arbitrary suitable frame. In this method, in a portion of the feedforward similar to CTM, inertia, friction, Coriollis, and Centrifugal effects are eliminated [12].

Simulations and a Comparative Look

Figure 5 displays the performance of the above methods in following a trajectory. Robot end-effector is moving against a wall in vertical direction from the point $X_1 = [0.6,-0.5,0]^T$ to the point $X_2 = [0.6, 0.5, 0]^T$. It is clear that it is only possible to apply a force in a normal direction to the wall ($\hat{X}_0$-direction, or the force control subspace). But it is possible to position control the robot hand in vertical and angular directions ($\hat{Y}_0$ and φ directions, or the position control subspace with two dimensions). At first robot hand is positioned at X_1 and does not apply any force to the wall. By starting the motion it stays at position X_1 for one second and applies a desired force of 10 newton to the wall. Then it moves from configuration X_1 to X_2 and applies the same force throughout the motion. Finally it stops at configuration X_2 for 2 seconds applying the same force.

Sampling times considered for the Active Stiffness Method, Seraji's Method, and Khatib's Method are 0.002 sec., 0.001 sec., and 0.004 sec., respectively. Simulation results indicate that the Active Stiffness Method does not have the ability for accurate and simultaneous control of force and motion and is only preferred for simpler tasks such as applying a force with no motion involved. Its main advantages are simplicity, cost, force-feedback requirements, and ability to run on the existing controllers.

Figure 6 displays the position and force errors in both Hybrid control methods by Khatib and Seraji in following the trajectory shown in Figure 5. The results indicate that overall Khatib's method provides a greater accuracy. And the steady state error in

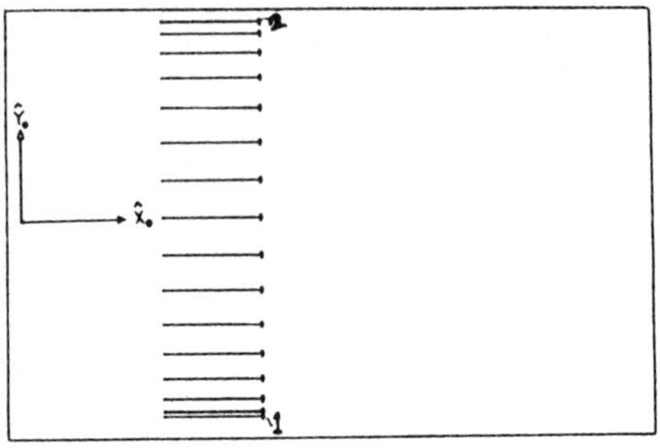

Fig. 5. The Desired Trajectory for Robot Position & Force Control.

Seraji's method is much less than the Khatib's method. From many points of view, Khatib's method is very similar to the Computed-Torque-Method, and Seraji's Hybrid control is very similar to its Position-Control-Method discussed in previous sections. A main advantage in Seraji's method, is that it requires much less computations and it does not demand the solution to the complex dynamic equations. In Seraji's method all computations are made in constraint frame, and changing the geometry of contact may change this frame and the structure of equations.

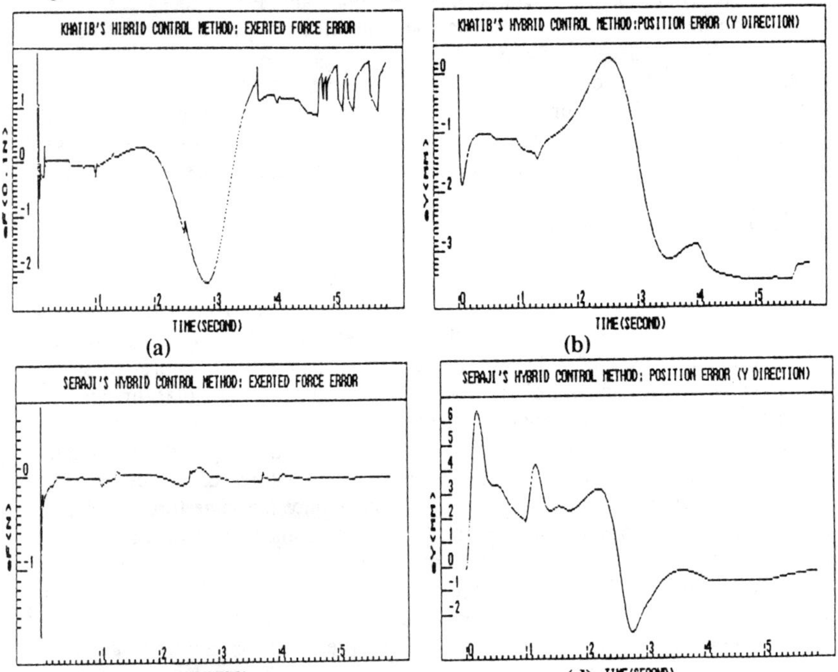

(a) (b) (c) (d)

Fig. 6. Positin and Force Errors in Khatib's and Seraji's Methods.

CONCLUSIONS

Simulation results for position control methods like the PID, CTM, Seraji's

Adaptive Method, and Craig's Adaptive Method are presented. In addition force control methods such as Active Stiffness, Seraji's Adaptive Method for Hybrid Control, Khatib's Hybrid Control were also simulated and the results are presented. The results indicate that for position control with high accuracy and stability at high speeds, Craig's method is very suitable. In the case where dynamic parameters are constant and are accurately known, CTM method provides a similar results to that of Craig's. Both methods require large amount of computations. If the stability in Seraji's method is ensured, it provides high accuracy (specifically at the steady state) and requires much less computations. This method is not very suitable for high speeds. PID method is the cheapest and its accuracy is limited.

Active stiffness method in force control does not provide a suitable accuracy, but requires a low cost, and is able to run on most current industrial robots. If the stability in Seraji's Hybrid method is at higher speeds is ensured, it provides a suitable accuracy in steady state and requires much less computations. If position and force control at high speed is desired, Khatib's method is very suitable, specially if the dynamic parameters are accurately known. Its controller structure is adaptable to any geometry of the surface contact, but it requires a large amount of computations.

REFERENCES

1. K.S. Fu, R.C. Gonzalez, and C.S.G. Lee, Robotics; Control, Sensing, Vision, and Intelligence, (Mac-Graw Hill, 1987).
2. J. J. Craig, Introduction to Robotics: Mechanics and Control, (Addison Wesley,1989).
3. M. Shahinpoor, A Robot Engineering Text Book, (Harper & Row, 1987)
4. S. Raghibizadeh, "Kinematics, Dynamics, Trajectory Generation, Control, and Simulation of a 3-DOF Robot, and Design & Fabrication of an I/O Interface Board for Control Applications", M.Sc. Thesis, (June 1991), Sharif University of Tech., Iran.
5. R. Nigam, C. Lee,"A Multiprocessor-based Controller for the Control of Mechanical Manipulators", IEEE J. of Robotics & Automation, Vol. RA-1, No. 4,(December 1985).
6. H. Seraji, "Direct Adaptive Control of Manipulators in Cartesian Space", J. of Robotic Systems, (1987).
7. H. Seraji, M. Jamshidi, and M. Shahinpoor, "Linear Multivariable Control of Two-Link Robots", J. of Robotics Systems, (1986).
8. J.J. Craig, Adaptive Control of Mechanical Manipulators, (Addison Wesley, 1988).
9. K.Y. Toumi, and O. Ito, "A Time Delay Controller for Systems with Unknown Dynamics", J. of Dyn. Sys., Measurement & Control, (March, 1990).
10. T.C. Steve Hsia, T.A. Lasku, and Z. Guo, "Robust Independent Joint Controller Design for Industrial Robot Manipulators", IEEE Trans. on Industrial Electronics, Vol. 38, No. 1, (February, 1991).
11. D.E. Whitney, "Historical Perspective and State of the Art in Robot Force Control", The Int. J. of Robotics Research, Vol. 6., No. 1, (Spring 1987).
12. O. Khatib, "A Unified Approach for Motion and Force Control of Robot Manipulators: The Operational Space Formulation", IEEE J. of Robotics and Control, Vol. RA-3, No. 1, (February, 1987).
13. H. Seraji, "Adaptive Force and Position Control of Manipulators", J. of Robotics Systems, (1987).
15. C.Y. Kuo, and S.P.T. Wang, "Nonlinear Robust Hybrid Control of Robot Manipulators", Trans. of the ASME, Dyn. Sys. & Control, (March 1990).

A FORCE COMMANDED IMPEDANCE CONTROL SCHEME FOR ROBOTS WITH NONLINEARITIES

Julio J. González[†], Luciano Chirinos[†], Bruce S. Widmann[‡] and Glenn R. Widmann[‡]

† Colorado State University
Department of Electrical Engineering
Institute of Robotic Studies
Fort Collins, CO 80523

‡ Hughes Aircraft Company
Missile Systems Group
Automotive Controls Engineering
Canoga Park, CA 91304-7925

ABSTRACT. This paper will demonstrate that the blending of Hybrid Control and Impedance Control into a Hybrid Impedance Control Strategy possesses the advantage of optimizing the robustness-performance control tradeoff for the case of a realistic nonlinear robot. The theoretical conclusions will be verified through both simulations and actual experiments performed on the PUMA 560 robot, using a testbed facility developed at Colorado State University.

1. INTRODUCTION

Hybrid Impedance Control (HIC) was introduced in the literature by Anderson and Spong [1]. This scheme was proposed in an attempt to unify two presumedly conflicting control implementations of compliant motion control, namely Hybrid Control (HC) and Impedance Control (IC). Their work was primarily of a philosophical nature and dealt with a simplified linearized ideal robot. As such, it did not address the robustness issue.

Recently, some researchers studied the application of HIC to a realistic robotic manipulator. Lasky and Hsia [2] investigated the force aspect of Hybrid Impedance Control. Their theoretical work was only substantiated by simulations. Their model did not consider the issue of hard nonlinearity, such as slip-stick friction. Liu and Goldenberg [3] addressed both the force and position aspects of Hybrid Impedance Control and presented experimental results. However, their investigation specifically centered on the effects of unmodeled dynamics on the system performance and the effects of hard nonlinearities were not considered.

The central idea of this paper is that HIC combines the properties of high robustness of IC and high performance of HC. This paper focuses the investigation in improving the equilibrium level between force control robustness and performance in the inevitable presence of commonly ocurring hard nonlinearities, such as discretization, actuator saturation, and slip-stick friction. A Force Commanded Impedance Control (FCIC) scheme (i.e., the force aspect of HIC) will be introduced. It will be demonstrated, both theoretically and experimentally, that the proposed FCIC strategy will provide improved performance over a Force Commanded Control (FCC) implementation (i.e., the force aspect of HC) for a similar degree of robustness.

2. PLANT MODEL FOR FORCE CONTROL

The most general dynamics expression for an environmentally constrained n degree of freedom (DOF) robotic manipulator with a gear ratio $\boldsymbol{r} \in \mathcal{R}^n$, relates the joint displacements $\boldsymbol{q} \in \mathcal{R}^n$, and the environmental contact forces $\boldsymbol{f}_{env} \in \mathcal{R}^n$, to the applied actuator voltages $\boldsymbol{u} \in \mathcal{R}^n$, as presented by [4]:

$$\boldsymbol{u} = \boldsymbol{M}(\boldsymbol{q})\ddot{q} + \boldsymbol{V}(\boldsymbol{q},\dot{\boldsymbol{q}})\dot{q} + \boldsymbol{J}_u^T(\boldsymbol{q})\boldsymbol{f}_{env} + g_u(\boldsymbol{q}) + \boldsymbol{u}_{SS} \quad (1)$$

with the terms for the plant dynamics, as related to voltages, defined as: $\boldsymbol{M}(\boldsymbol{q})$:

symmetric mass matrix of link and motor terms, $\boldsymbol{V}(\boldsymbol{q},\dot{\boldsymbol{q}})$: matrix of gyroscopic related and motor viscous friction terms, $\boldsymbol{J}_u(\boldsymbol{q})$: plant Jacobian matrix, $\boldsymbol{g}_u(\boldsymbol{q})$: vector of gravitational loading terms, $\boldsymbol{u}_{SS}$: vector of terms describing slip-stick friction.

2.1. One DOF Force Control Model

For the simplicity of analysis and experimental verification, only a one DOF case will be investigated. Consider the system configuration, displayed in Figure 1, in which a control voltage u applied to the PUMA 560 shoulder joint results in the end effector exerting a vertical force f_{env} to a purely elastic stationary environment with a rectilinear stiffness coefficient K_e. Under all the aforementioned simplification conditions, (1) reduces to:

$$u = m\ddot{q} + v\dot{q} + \gamma l f_{env} + g_u + u_{SS} \tag{2}$$

where m and v are the appropriate elements of the matrices $\boldsymbol{M}(\cdot)$ and $\boldsymbol{V}(\cdot,\cdot)$, l represents the length of the shoulder link, γ is a conversion factor from torque to voltage, and u_{SS} is characterized by the stiction and Coulomb components U_S and U_C, respectively. The characteristics of the environmental force becomes: $f_{env} = K_e x$, where x is the deformed environment position as measured from the undeformed environment stationary position x_o. Introducing f_{env} and the rectilinear/rotational relationship $\dot{x} = l\dot{q}$ into (2), we arrive at:

$$u = \frac{m}{K_e l}\ddot{f}_{env} + \frac{v}{K_e l}\dot{f}_{env} + \gamma l f_{env} + g_u + u_{SS}\ . \tag{3}$$

The linear portion of the constrained dynamics (3) is characterized by the type-0, second-order transfer function:

$$\frac{K_p\omega_n^2}{s^2 + 2\xi\omega_n s + \omega_n^2} \triangleq H_p(s) \text{ where: } \left\{ \begin{array}{lcl} K_p & = & 1/(\gamma l) \\ \omega_n & = & l\{(\gamma K_e)/m\}^{1/2} \\ \xi & = & \frac{1}{2}v/\left\{l(\gamma K_e m)^{1/2}\right\}\ . \end{array} \right. \tag{4}$$

Examination of (4) reveals that large K_e produce high ω_n and small ξ, which leads to poor relative stability. The experimentally determined plant parameters are shown in Table 1.

3. FCC SCHEME

A single-loop FCC law $U(s) = G_f(s)[F_d(s) - F_{env}(s)] + \hat{g}_u/s$ acting on plant (3) produces the closed-loop response:

$$F_{env}(s) = \frac{G_f(s)H_p(s)}{1 + G_f(s)H_p(s)}\left\{F_d(s) - \frac{U_{SS}(s)}{G_f(s)}\right\}\ . \tag{5}$$

The multi-loop FCC strategy (i.e., with inner position loop) is not presented because its discrete-time implementation results in unstable performance for practical sampling period values [4].

4. FORMULATION OF A SINGLE-LOOP FCIC SCHEME

The plant dynamics (2), expressed in the s domain, becomes:

$$\begin{aligned} U(s) &= (ms + v)sQ(s) + \gamma l F_{env}(s) + g_u/s + U_{SS}(s) && (6) \\ &= \gamma Z_r(s)sX(s) + \gamma l F_{env}(s) + g_u/s + U_{SS}(s). && (7) \end{aligned}$$

where $sX(s)$ is the rectilinear speed, and $Z_r(s) \triangleq (ms + v)/(\gamma l)$ is the manipulator input impedance.

The concept of output impedance can be used to relate the desired and applied forces [1]. This relationship can be shown with the aid of a circuit-like electrical graph architecture shown in Figure 2, where $F_d(s)$ represents the Thèvenin desired force, $F_{env}(s)$ the output applied force, $Z_f(s)$ the desired robot Thèvenin impedance, and $Z_e(s)$ is the environment impedance. For an elastic environment, $Z_e(s)$ is a purely capacitive load (i.e., $Z_e = K_e/s$). By Ohm's law, the "current" is given by: $sX(s) = [F_d(s) - F_{env}(s)]/Z_f(s)$. Substituting $sX(s)$ into (7), the necessary control law to adjust the desired manipulator output impedance $Z_f(s)$ results:

$$U(s) = U_z(s) + \gamma l F_{env}(s) + \frac{g_u}{s} + U_{SS}(s) \tag{8}$$

where:

$$U_z(s) \triangleq \gamma \frac{Z_r(s)}{Z_f(s)} [F_d(s) - F_{env}(s)] \tag{9}$$

is the impedance control law.

According to (8), the control law (9) should compensate for the reflected environmental force $\gamma l F_{env}$, and the nonlinear term (i.e., gravity and friction). The former task is practically accomplishable, since γ, l and f_{env} (the applied force) are fairly easy to quantify through direct measurements. As for the nonlinear term, only gravitational loading can be reasonably compensated for, since it is a well defined geometrical nonlinearity. However, this is not true for slip-stick friction, due to the non repeatability in the value of the stiction friction and the uncertainty in the true value of the Coulombic friction.

Consequently, the control signal, with incomplete compensation, will be of the form:

$$U_z(s) = U(s) - \hat{\gamma}\hat{l}\hat{F}_{env}(s) - \hat{g}_u/s \tag{10}$$

where $\hat{\gamma}$, $\hat{l}$, $\hat{F}_{env}$ and $\hat{g}$ are the estimates ot the true characteristics.

Assuming all estimates are very close to the true values, substituting (6) into (10), and using $F_{env}(s) = K_e l Q(s)$, the compensated plant dynamics becomes:

$$U_z(s) = s(ms + v)Q(s) + U_{SS}(s) = \frac{s(ms + v)}{K_e l} F_{env}(s) + U_{SS}(s)\,. \tag{11}$$

Thus, environmental force compensation transforms the linear plant dynamics $H_p(s)$ (4), into $H_{pc}(s) \triangleq K_e l/s(ms+v)$ (i.e., second-order, type-1 system). Combining (11) and (9), the closed-loop system response results:

$$F_{env}(s) = \frac{G_f(s)H_{pc}(s)}{1 + G_f(s)H_{pc}(s)} \left\{ F_d(s) - \frac{U_{SS}(s)}{G_f(s)} \right\} \tag{12}$$

where the controller transfer function $G_f(s) \triangleq \gamma \frac{Z_r(s)}{Z_f(s)}$ is all dictated by impedances.

Environmental force compensation has conflicting effects on the design of a robust single-loop FCIC discrete-time controller. For example, it moves uncompensated plant poles from the highly underdamped location $s_{1,2} = -\xi\omega_n \pm \omega_n(1-\xi^2)^{1/2}$ to the overdamped location $s_1 = 0$, $s_2 = -v/m$. On the other hand, it incorporates the high K_e into the DC plant gain. This drawback makes the discrete-time implementation of the single-loop FCIC unstable [4]. However, this drawback is eliminated if an inner position loop is considered, since then the resulting compensated position transfer function $1/s(ms+v) \triangleq H^q_{pc}(s)$ is independent of K_e. Consequently, since K_e is not present, it is possible to use a high gain inner-loop proportional controller (i.e., $G_q(s) \equiv K_q$). The resulting open-loop position transfer function $K_q/s(ms+v)$ tolerates large values of K_q before the discretely-implemented inner position loop becomes unstable [4]. Additionally, this also has the advantage of rejecting the friction disturbance. This analysis justifies the attempt to implement a multi-loop FCIC strategy.

5. FORMULATION OF A MULTI-LOOP FCIC SCHEME

The analysis performed for a single-loop architecture is readily extended to the multi-loop architecture (Figure 3), by just considering the latter as a special case of the former in which the inner loop dynamics assumes the role of the "plant".

The compensated dynamics (11) can be written as: $U_q(s) = Q(s)/H^q_{pc}(s) + U_{SS}(s)$, where the notation $U_z(s)$ has been changed to $U_q(s)$ to explicitly refer to the inner position loop. Using the inner loop controller $U_q(s) = G_q(s)[Q_d(s) - Q(s)]$, the inner-loop dynamics can be expressed by:

$$Q_d(s) = \frac{Q(s)}{H_I(s)} + \frac{U_{SS}(s)}{G_q(s)} \tag{13}$$

where $H_I(s) \triangleq H^q_{pc}(s)G_q(s)/[1 + H^q_{pc}(s)G_q(s)]$ is the inner-loop "plant" dynamics. The dynamics (13) can be also expressed in terms of an inner-loop input impedance $Z_I(s)$:

$$Q_d(s) = \frac{Z_I(s)}{K_e l} sX(s) + \frac{U_{SS}(s)}{G_q(s)} \tag{14}$$

where $Z_I(s) \triangleq K_e/[sH_I(s)]$. Proceeding as in Section 4, the impedance control law results:

$$U_z(s) \equiv Q_d(s) \triangleq \frac{1}{K_e l}\frac{Z_I(s)}{Z_f(s)}[F_d(s) - F_{env}(s)] \; . \tag{15}$$

Combining (13) and (15), using $F_{env}(s) = K_e l Q(s)$, gives the system output response:

$$F_{env}(s) = \frac{K_e l G_f(s) H_I(s)}{1 + K_e l G_f(s) H_I(s)} \left\{ F_d(s) - \frac{U_{SS}(s)}{G_f(s) G_q(s)} \right\} \; ; \; G_f(s) \triangleq \frac{1}{K_e l}\frac{Z_I(s)}{Z_f(s)} \; . \tag{16}$$

6. FCC VS. FCIC ROBUSTNESS-PERFORMANCE COMPARISON

The same lag-phase controller $G_f(s) = K_f(sT_{lead}+1)/(sT_{lag}+1)$ will be used to compare the FCC and FCIC strategies. The reason for an lag-phase controller is the elimination of the distortion due to control signal saturation [4]. Therefore, only the slip-stick friction nonlinearity will be considered when discussing system performance. The FCC and FCIC strategies are compared with the aid of Table 2.

Robustness related equations (17) and (18) represent dominant open-loop transfer functions obtained by setting $T_{lag} \gg (\omega_n \xi)^{-1} = 2m/v$, and then using standard techniques employed for reducing the order of the system dynamics. Performance related equations (19) and (20) are the steady state response of (5) and (16), to an step of amplitude F_d.

For the purpose of performance comparison, a similar degree of robustness is set in both cases. Stability of the resulting dominant first order system depends on the open-loop gain and on the sampling period T_f, as compared to the open-loop time constant T_{lag}. Since $T_f = 15$ msec and $T_{lag} = 4$ sec are selected in both cases, then similar degrees of robustness can be obtained by selecting the same open-loop gain. To obtain an open-loop gain of 10, a value of $K_f = 0.25$ (V/N) is needed in (17) and a value of $K_f = 0.3$ (mrad/N) in (18). Using the parameters of Table 1, together with an inner-loop position gain of $G_q(s) \equiv K_q = 2,000$ (V/rad), then the steady state disturbance forces become $U_C/K_f = 2$ N in (19) and only $U_C/K_f K_q = 0.833$ N in (20). This performance improvement will be verified experimentally.

7. EXPERIMENTAL RESULTS

The curves in Figure 4 represent closed-loop step responses to a step of amplitude 10 N. Curves (a) (simulated) and (b) (experimentally measured) correspond to the FCC strategy. Curves (c) (simulated) and (d) (experimentally measured) correspond to the FCIC implementation of Figure 3. The observed horizontal portions of curves (b) and (d) where f_{env} remains periodically constant is theorized to be the result of the slip-stick characteristics of the force sensor not considered in the presented model. The simulated responses are clearly representative of the true system responses. The FCIC strategy indeed shows a performance improvement over the FCC strategy for the same degree of robustness.

8. CONCLUSIONS

The main feature characterizing the FCIC implementation is environmental force compensation. This compensation slows the plant dynamics, allowing for a high-gain stable inner position loop. On the other hand, the FCC strategy does not possess the slowering action provided by environmental force compensation, which results in an unstable discretely-implemented inner position loop. The high gain inner-loop of the FCIC strategy helps to reject the slip-stick friction nonlinearity, thus providing improved performance over the FCC strategy, for a similar degree of robustness.

References

[1] R.J. Anderson and M.W Spong, "Hybrid Impedance Control of Robotic Manipulators," *IEEE Journal of Robotics and Automation,* Vol. 4, N° 5, pp. 549-556, October 1988.

[2] T.A. Lasky and T.C. Hsia, "On Force-Tracking Impedance Control of Robot Manipulators," *Proc. IEEE Int. Conf. on Robotics and Automation,* Sacramento, CA, pp. 274-280, April 1991.

[3] G.J. Liu and A.A. Goldenberg, "Robust Hybrid Impedance Control of Robot Manipulators," *Proc. IEEE Int. Conf. on Robotics and Automation,* Sacramento, CA, pp. 287-292, April 1991.

[4] J.J. González and G.R. Widmann, "Investigation of Nonlinearities in the Force Control of Real Robots," *IEEE Trans. on Systems, Man, and Cybernetics.* (To appear).

Table 1: Puma 560 Experimental Parameter Values.

Linear Parameters							Nonlinear Parameters		
m	(V-s^2)	0.1		K_p	(N/V)	40	g_u	(V)	−2.4
v	(V-s)	1.8	$K_E =$	ω_n	(s^{-1})	90	U_C	(V)	0.5
l	(m)	0.42	77,200 N/m	ξ		0.1	U_S	(V)	0.6

Table 2: Comparison of Force Control Strategies.

Robustness-Performance	Type of Force Control: FCC	Type of Force Control: FCIC
Robustness Related Equation	$G_f(s)H_p(s) \simeq \frac{K_pK_f}{sT_{lag}+1}$ (17)	$K_e l G_f(s)H_I(s) \simeq \frac{K_e l K_f}{sT_{lag}+1}$ (18)
Performance Related Equation	$f_{ss}^{step} = \frac{K_fK_p}{1+K_fK_p}\left\{F_d - \frac{U_c}{K_f}\right\}$ (19)	$f_{ss}^{step} = \frac{K_e l K_f}{K_e l K_f+1}\left\{F_d - \frac{U_C}{K_fK_q}\right\}$ (20)

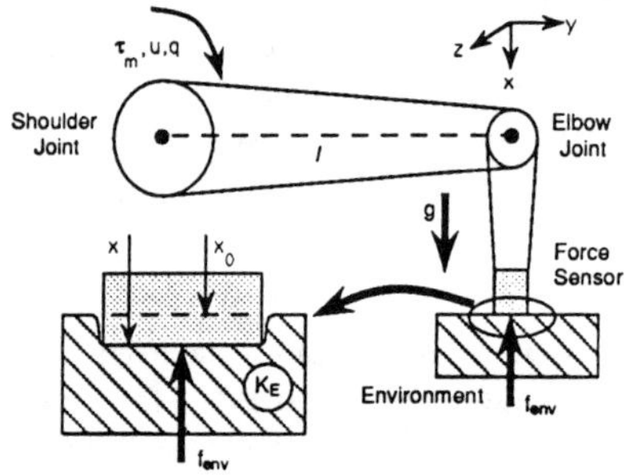

Figure 1: Manipulator Configuration.

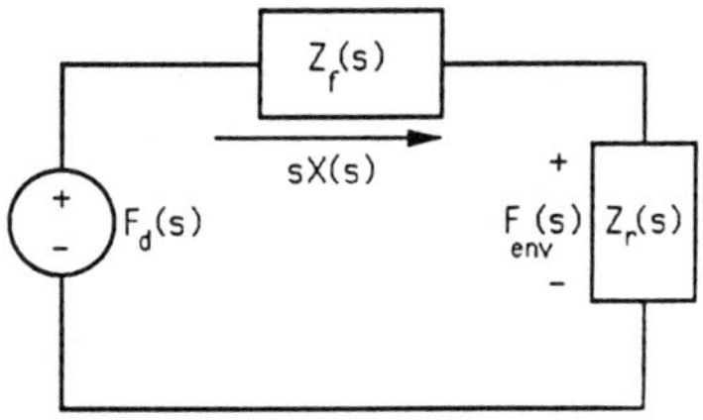

Figure 2: FCIC Electrical Graph.

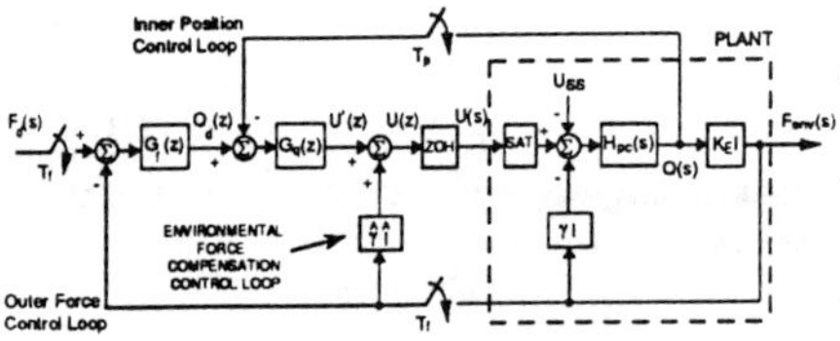

Figure 3: Discrete-Time FCIC Implementation.

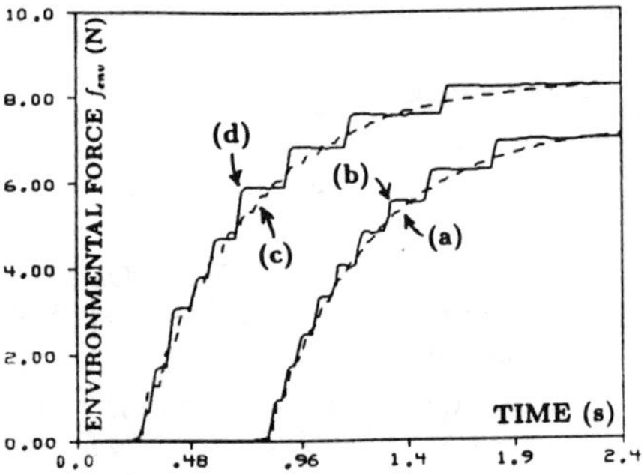

Figure 4: FCC and FCIC Step Responses.

A NEW APPROACH TO ADAPTIVE MANIPULATOR CONTROL PART I: THEORY

R. Colbaugh* **H. Seraji†** **K. Glass***

* New Mexico State University, Las Cruces, NM 88001
†Jet Propulsion Laboratory, California Institute of Technology, Pasadena, CA 91109

Abstract

This two-part paper presents a new approach to adaptive motion control of manipulators. The control schemes designed using this approach are very general and computationally efficient since they do not require knowledge of either the mathematical model or the parameter values of the robot dynamics, and are implemented without calculation of the robot inverse kinematic transformation. All of the proposed control strategies are globally stable in the presence of bounded disturbances, and in all cases the size of the residual tracking errors can be made arbitrarily small. In this paper, which is the first of the two parts, the proposed approach to adaptive manipulator control is summarized and several new direct adaptive schemes are given. The second of the two parts presents computer simulation and experimental results which demonstrate that accurate and robust trajectory tracking can be achieved by using the proposed controllers.

1. Introduction

Increasing the applicability and versatility of robot manipulators requires the development of control systems capable of providing performance superior to that obtainable with conventional robot controllers. Of particular interest is designing control strategies that ensure accurate end-effector trajectory tracking despite uncertainties in the robot dynamics and payload, external disturbances, and sensor noise. Many researchers have investigated this problem in recent years, and as a result of these studies two broad approaches have emerged for achieving accurate trajectory tracking in the presence of uncertainties and disturbances. The first approach, usually called *robust control*, utilizes (assumed known) upper bounds on the various uncertainties to design controllers that compensate for the lack of complete and accurate information regarding the system [e.g., 1]. Alternatively, in *adaptive control*, an attempt is made to compensate for uncertainties by adjusting the controller parameters on-line based on the observed system performance [e.g., 2]. An advantage of the adaptive approach is therefore the potential to continually improve performance while the robot is executing the task.

This two-part paper examines the adaptive motion control problem for rigid link manipulators. Over the past decade this topic has attracted considerable attention, and no attempt will be made here to deal comprehensively with the vast literature in this area. Instead, adaptive motion control will be discussed from the perspective of the representation of the robot dynamics utilized in the development of the control strategy. Within this framework, two broad classes of control schemes can be identified. The first utilizes the following representation of the robot dynamic model:

$$\mathbf{T} = W(\theta, \dot{\theta}, \ddot{\theta})\mathbf{p} \tag{1}$$

where $\mathbf{T} \in \Re^n$ is the joint torque vector, $\theta \in \Re^n$ is the joint coordinate vector, $\mathbf{p} \in \Re^p$ is the vector of unknown constant dynamic model parameters, and $W \in \Re^{n \times p}$ is the "regressor" matrix that embodies the structure of the robot dynamics and is assumed to be known. This "linear-in-the-unknown-parameters" model was first used in robot parameter estimation algorithms, and subsequently formed the basis for developing a class of globally stable algorithms for adaptive motion control [e.g., 3]. Following [4], we shall call these schemes *indirect adaptive controllers* because the adaptive laws provide an explicit estimate of $\mathbf{p}$, which is then used in the control law. Although these controllers have performed well both in computer simulations and experiments, there are some potential difficulties associated with this approach. For example, implementation of these schemes with robots possessing more than two or three joints can be computationally intensive, particularly if it is desired to implement them directly in Cartesian-space. More seriously, their design requires precise knowledge of the structure of the entire robot dynamic model, including friction and transmission effects, which is often an unrealistic requirement in practice. Additionally, recent studies have indicated that these indirect adaptive controllers can lack robustness to unmodelled dynamics, sensor noise, and other disturbances [5].

The second approach to adaptive motion control considered here will be termed *direct adaptive control* [4] because the adaptive laws adjust the controller gains directly without parameter estimation. These schemes typically use the following dynamic model representation in their development:

$$\mathbf{T} = H_\theta(\theta)\ddot{\theta} + \mathbf{V}_\theta(\theta, \dot{\theta}) + \mathbf{G}_\theta(\theta) \tag{2}$$

where $H_\theta \in \Re^{n \times n}$ is the symmetric positive-definite inertia matrix, $\mathbf{V}_\theta \in \Re^n$ represents the torques due to Coriolis and centripedal acceleration as well as all friction effects, and $\mathbf{G}_\theta \in \Re^n$ is the gravity torque vector. Ordinarily the development of direct adaptive controllers has proceeded by assuming that very little information is available concerning either the structure or the parameter values of the terms in (2); indeed, it is often only required to know that H in (2) is symmetric and positive-definite to develop the controller design [e.g, 6]. As a consequence, these methods are equally applicable to trajectory tracking in joint-space or task-space. However, the control algorithm derivations rely on the assumption that the adaptive elements in the controller can vary significantly more rapidly that the elements in the dynamics (2), which has caused some concern regarding the stability of the resulting control systems in fast motions. Nevertheless, it must be noted that these direct adaptive controllers are extremely simple, computationally efficient, require very little model information, and have demonstrated good tracking performance and robustness characteristics in simulations and experiments with industrial six degree-of-freedom robots.

The goal of this two-part paper is to introduce a new class of adaptive robot controllers that enjoys the simplicity, computational efficiency, and robustness of the direct adaptive controllers [e.g., 6] and at the same time possesses the attractive characteristic of unconditional global stability of the indirect adaptive controllers [e.g., 3]. Very recently there has been progress in this direction, with Stepanenko and Yuan [7] proposing a joint-space trajectory tracking strategy and Colbaugh, Seraji, and Glass [8] presenting a Cartesian-space impedance controller that provides motion tracking as a special case. In each case the control scheme possesses a direct adaptive structure and global stability is demonstrated without making assumptions regarding the rate of change of the robot dynamics. This paper, which is the first of two parts, presents a new class of direct adaptive controllers for robot trajectory tracking. The control laws are derived directly in Cartesian-space, do not require knowledge of either the mathematical model or the parameter values of the robot dynamics, and are implemented without calculation of the robot inverse kinematic transformation. As a consequence, the schemes are very general and computationally efficient, and can be implemented with either nonredundant or redundant manipulators. All of the proposed control strategies are globally stable in the presence of bounded disturbances, and in all cases the size of the tracking errors can be made arbitrarily small.

2. Problem Formulation

Let $\mathbf{x}$ define the position and orientation of the robot end-effector relative to a user-defined reference frame and note that, in the most general case, the elements of $\mathbf{x}$ are local coordinates for some smooth manifold $\mathcal{M}$ of dimension m. For example, for arbitrary placement and orientation of the end-effector in a three-dimensional workspace, $\mathbf{x}$ is chosen as some parametrization of the non-trivial six-dimensional manifold $\Re^3 \times SO(3)$, where $SO(3)$ denotes the group of rigid body rotations. The forward kinematic and differential kinematic relationships between the robot joint coordinates θ and the Cartesian end-effector coordinates $\mathbf{x}$ can be written as

$$\mathbf{x} = \mathbf{f}(\theta), \qquad \dot{\mathbf{x}} = J(\theta)\dot{\theta} \tag{3}$$

where $\mathbf{f} : \Re^n \to \mathcal{M}$ and $J \in \Re^{m \times n}$ is the end-effector Jacobian matrix. It is assumed that $m \leq n$ so that both nonredundant and redundant robots can be considered together. Consider the robot dynamic model in task-space:

$$\mathbf{F} = H(\mathbf{x})\ddot{\mathbf{x}} + V_{cc}(\mathbf{x}, \dot{\mathbf{x}})\dot{\mathbf{x}} + \mathbf{G}(\mathbf{x}) + \mathbf{d}(t) \tag{4}$$

where the robot is assumed to be in a nonsingular configuration (i.e., J is of full row rank), $\mathbf{F} \in \Re^m$ is the Cartesian-space control input, $H \in \Re^{m \times m}$ is the symmetric, positive-definite inertia matrix, $V_{cc} \in \Re^{m \times m}$ quantifies Coriolis and centripedal acceleration effects and is related to H via $\dot{H} = V_{cc} + V_{cc}^T$, $\mathbf{G} \in \Re^m$ is the vector of gravity torques, and $\mathbf{d} \in \Re^m$ is a vector of bounded but otherwise arbitrary disturbances. It is well known that the dynamics (4) represents a Hamiltonian system and as a consequence possesses a well-understood structure [e.g., 9]. For example, in addition to the properties already mentioned, it can be shown that $V_{cc}(\mathbf{x}, \dot{\mathbf{x}})\dot{\mathbf{x}} = D(\mathbf{x})[\dot{\mathbf{x}}\dot{\mathbf{x}}]$, where $D \in \Re^{m \times m^2}$ is the matrix of Christoffel symbols associated with H and $[\dot{\mathbf{x}}\dot{\mathbf{x}}] = [\dot{x}_1\dot{x}_1, ..., \dot{x}_1\dot{x}_m, \dot{x}_2\dot{x}_1, ..., \dot{x}_m\dot{x}_m]^T \in \Re^{m^2}$. Furthermore, it is known that the terms $H, \mathbf{G}, D$ are bounded functions of $\mathbf{x}$ whose time derivatives $\dot{H}, \dot{\mathbf{G}}, \dot{D}$ are also bounded in $\mathbf{x}$ and depend linearly on $\dot{\mathbf{x}}$, while the matrix V_{cc} is bounded in $\mathbf{x}$ and depends linearly on $\dot{\mathbf{x}}$. All of these properties are established in [8], and will prove very useful in the development of the proposed adaptive controllers.

The control schemes proposed in this paper consist of two subsystems: a class of adaptive schemes which provide the Cartesian-space control input $\mathbf{F}$ required to ensure that the system (4) evolves from its initial state to the desired final state along some specified trajectory $\mathbf{x}_d(t)$ (where $\mathbf{x}_d$ is bounded with bounded derivatives), and an algorithm for mapping the control input $\mathbf{F}$ to a physically realizable joint-space control torque $\mathbf{T}$. The $\mathbf{F} \to \mathbf{T}$ map is simply $\mathbf{T} = J^T\mathbf{F}$ for nonredundant robots, but proper construction of this map for redundant robots provides a powerful approach to effective redundancy utilization [10-13]. For example, this approach provides a very natural means of using any available redundancy to globally minimize the robot kinetic energy. This desirable objective is easily achievable with the following very simple $\mathbf{F} \rightarrow \mathbf{T}$ map [10]:

$$\mathbf{T} = J^T\mathbf{F} \tag{5}$$

Because this transformation also gives the (unique) $\mathbf{F} \rightarrow \mathbf{T}$ map for nonredundant robots, we shall for simplicity use (5) for both nonredundant and redundant robots in this paper. It is stressed, however, that considerable flexibility exists in selecting an appropriate $\mathbf{F} \rightarrow \mathbf{T}$ map for a redundant robot [10-13]. The development of the proposed class of adaptive Cartesian-space control strategies is the main focus of this two-part paper, with the theory being summarized in the present part and simulation and experimental results being presented in the second part.

3. Summary of Adaptive Control Schemes

The proposed class of adaptive Cartesian-space control strategies is summarized in this section of the paper. Let us rewrite the task-space dynamic model as

$$\mathbf{F} = H(\mathbf{x})\ddot{\mathbf{x}} + D(\mathbf{x})[\dot{\mathbf{x}}\dot{\mathbf{x}}] + \mathbf{G}(\mathbf{x}) + \mathbf{d}(t) \tag{6}$$

where the identity $V_{cc}(\mathbf{x}, \dot{\mathbf{x}})\dot{\mathbf{x}} = D(\mathbf{x})[\dot{\mathbf{x}}\dot{\mathbf{x}}]$ was used. Examination of (6) suggests the following control law structure:

$$\mathbf{F} = A(t)\mathbf{z} + B(t)[\dot{\mathbf{x}}\mathbf{y}] + \mathbf{f}(t) + K_p(t)\mathbf{e} + K_v(t)\dot{\mathbf{e}} \tag{7}$$

where $\mathbf{e} = \mathbf{x}_d - \mathbf{x}$, the quantities $A(t) \in \Re^{m\times m}$, $B(t) \in \Re^{m\times m^2}$, $\mathbf{f}(t) \in \Re^m$, $K_p(t) \in \Re^{m\times m}$, $K_v(t) \in \Re^{m\times m}$ are adaptive gains whose update laws are to be determined, and $\mathbf{y}, \mathbf{z} \in \Re^m$ are terms that depend on the desired end-effector trajectory $\mathbf{x}_d$ and whose precise form is to be established. In this control law, the first three terms on the right side represent feedforward components intended to functionally match elements in the robot dynamics (6), while the last two terms are adaptive feedback elements. Observe that, although $A, B, \mathbf{f}$ are clearly the adaptive counterparts to $H, D, \mathbf{G}$ in (6), nothing is assumed regarding the structure of these dynamic model terms except for the dimension of the matrices.

Applying the control law (7) to the manipulator dynamics (6) yields the tracking error dynamics:

$$H(\mathbf{z} - \ddot{\mathbf{x}}) + D[\dot{\mathbf{x}}(\mathbf{y} - \dot{\mathbf{x}})] + K_v\dot{\mathbf{e}} + K_p\mathbf{e} + \mathbf{\Phi}_G + \Phi_H\mathbf{z} + \Phi_D[\dot{\mathbf{x}}\mathbf{y}] - \mathbf{d} = \mathbf{0} \tag{8}$$

where $\mathbf{\Phi}_G = \mathbf{f} - \mathbf{G}, \Phi_H = A - H$, and $\Phi_D = B - D$. The structure of (8) suggests the following approach to designing direct adaptive schemes for trajectory tracking:

1. Select $\mathbf{y}$ and $\mathbf{z}$ so that (8) has the structure of a dissipative Hamiltonian system with generalized velocity $\mathbf{y} - \dot{\mathbf{x}}$ [9], and so that small $\mathbf{y} - \dot{\mathbf{x}}$ ensures accurate tracking.
2. Specify adaptation laws for $A, B, \mathbf{f}, K_p, K_v$ so that these gains are bounded and their adjustment tends to decrease $\mathbf{y} - \dot{\mathbf{x}}$.

This approach can be utilized to derive a broad class of adaptive schemes with control laws of the general form (7) and adaptation laws possessing the following structure:

$$\begin{aligned}
\mathbf{f}(t) &= \mathbf{f}(0) - \gamma_1 \int_0^t \| \dot{\mathbf{e}} \| \mathbf{f} dt + \beta_1 \int_0^t \mathbf{q} dt \\
A(t) &= A(0) - \gamma_2 \int_0^t \| \dot{\mathbf{e}} \| A dt + \beta_2 \int_0^t \mathbf{q}\mathbf{z}^T dt \\
B(t) &= B(0) - \gamma_3 \int_0^t \| \dot{\mathbf{e}} \| B dt + \beta_3 \int_0^t \mathbf{q}[\dot{\mathbf{x}}\mathbf{y}]^T dt \qquad (9) \\
K_p(t) &= K_p(0) - \gamma_4 \int_0^t \| \dot{\mathbf{e}} \| K_{pt} dt + \beta_4 \int_0^t \mathbf{q}\mathbf{e}^T dt \\
K_v(t) &= K_v(0) - \gamma_5 \int_0^t \| \dot{\mathbf{e}} \| K_{vt} dt + \beta_5 \int_0^t \mathbf{q}\dot{\mathbf{e}}^T dt
\end{aligned}$$

where $\mathbf{q}(t) \in \Re^m$ is a weighted position-velocity tracking error whose precise form depends on the choice of $\mathbf{y}$ and $\mathbf{z}$, and the γ_i and β_i are positive scalar constants. Thus the proposed class of control laws can be parameterized by $\mathbf{y}$, $\mathbf{z}$, and $\mathbf{q}$.

As an illustration of this approach to controller design, several triples $\mathbf{y},\mathbf{z},\mathbf{q}$ that satisfy the above criteria and yield globally stable control strategies are listed below. In all cases the tracking errors can be made arbitrarily small through proper selection of the control law parameters. Additionally, if there are no external disturbances and the final desired end-effector position/orientation $\mathbf{x}_d$ is a constant, then asymptotic stability is achieved. Detailed stability proofs and analyses of convergence properties are presented in [14]. It is stressed that the controllers listed below are representatives of the proposed class and that the list is not exhaustive. For example, it is very natural using this approach to exploit any knowledge regarding the mathematical model or parameter values of the manipulator dynamics by incorporating this information into the definitions of $\mathbf{y}$ and $\mathbf{z}$ [14]. Finally, it is noted that the list of controllers is one of increasing computational complexity, with the simplest schemes requiring slightly more robot model information for their implementation.

control scheme 1: $\mathbf{y} = \mathbf{0}$, $\mathbf{z} = \mathbf{0}$, $\mathbf{q} = \dot{\mathbf{e}} + \epsilon\mathbf{e}/(1+ \| \mathbf{e} \|^2)$ (10a)
control scheme 2: $\mathbf{y} = \dot{\mathbf{x}}_d$, $\mathbf{z} = \ddot{\mathbf{x}}_d$, $\mathbf{q} = \dot{\mathbf{e}} + \epsilon\mathbf{e}/(1+ \| \mathbf{e} \|^2)$ (10b)
control scheme 3: $\mathbf{y} = \dot{\mathbf{x}}_d$, $\mathbf{z} = \ddot{\mathbf{x}}_d + \lambda\dot{\mathbf{e}}$, $\mathbf{q} = \dot{\mathbf{e}} + \epsilon\mathbf{e}/(1+ \| \mathbf{e} \|^2)$ (10c)
control scheme 4: $\mathbf{y} = \dot{\mathbf{x}}_d + K\mathbf{e}$, $\mathbf{z} = \ddot{\mathbf{x}}_d + K\dot{\mathbf{e}}$, $\mathbf{q} = \dot{\mathbf{e}} + K\mathbf{e}$ (10d)
control scheme 5: $\mathbf{y} = \dot{\mathbf{x}}_d + K\mathbf{e}$, $\mathbf{z} = \ddot{\mathbf{x}}_d + K\dot{\mathbf{e}} + \lambda\mathbf{q}$, $\mathbf{q} = \dot{\mathbf{e}} + K\mathbf{e}$ (10e)

4. Conclusions

This two-part paper presents a new class of direct adaptive controllers for robot trajectory tracking. The control laws are implementable either in joint-space or in Cartesian-space for redundant or nonredundant robots, and do not require knowledge of the mathematical model or the parameter values of the manipulator dynamics. It can be shown that the control strategies are globally stable in the presence of bounded disturbances, and that the size of the tracking errors can be made arbitrarily small [14]. In the present paper, which is the first of two parts, the proposed approach is summarized and several new direct adaptive schemes are given. Additionally, comments are made regarding the relative advantages and disadvantages of the different controllers. The second of the two parts presents computer simulation and experimental results which demonstrate that accurate and robust trajectory tracking can be achieved by using the proposed controllers.

5. Acknowledgements

The research described in this paper was supported in part through grants from the Department of Energy (WERC) and the General Motors Corporation and in part through contracts with the Jet Propulsion Laboratory and the National Aeronautics and Space Adminstration.

6. References

1. Slotine, J-J., "The Robust Control of Robot Manipulator", *International Journal of Robotics Research*, Vol. 4, No. 2, 1985, pp. 49-63
2. Dubowsky, S., and D. DesForges, "The Application of Model-Referenced Adaptive Control to Robotic Manipulators", *ASME Journal of Dynamic Systems, Measurement, and Control*, Vol. 101, 1979, pp. 193-200
3. Slotine, J-J. and W. Li, "On the Adaptive Control of Robot Manipulators", *International Journal of Robotics Research*, Vol. 6, No. 3, 1987, pp. 49-59
4. Narendra, K. and A. Annaswamy, *Stable Adaptive Systems*, Prentice Hall, Englewood Cliffs, NJ, 1989
5. Reed, J. and P. Ioannou, "Instability Analysis and Robust Adaptive Control of Robotic Manipulators", *IEEE Transactions on Robotics and Automation*, Vol. 5, No. 3, 1989, pp. 381-386
6. Seraji, H., "A New Approach to Adaptive Control of Manipulators", *ASME Journal of Dynamic Systems, Measurement, and Control*, Vol. 109, No. 3, 1987, pp. 193-202
7. Stepanenko, Y. and J. Yuan, "Robust Adaptive Control of a Class of Nonlinear Mechanical Systems with Unbounded and Fast-Varying Uncertainties", *Automatica*, Vol. 28, No. 2, 1992, pp. 265-276
8. Colbaugh, R., H. Seraji, and K. Glass, "Direct Adaptive Impedance Control of Robot Manipulators", *Journal of Robotic Systems*, Vol. 10, No. 3, 1993 (in press)
9. Schutz, B. F., *Geometrical Methods of Mathematical Physics*, Cambridge University Press, Cambridge, 1980
10. Colbaugh, R. and K. Glass, "Cartesian Control of Redundant Manipulators", *Journal of Robotic Systems*, Vol. 6, No. 4, 1989, pp. 427-459
11. Seraji, H., "Configuration Control of Redundant Manipulators: Theory and Implementation", *IEEE Transactions on Robotics and Automation*, Vol. 5, No. 4, 1989, pp. 472-490
12. Seraji, H. and R. Colbaugh, "Improved Configuration Control for Redundant Robots", *Journal of Robotic Systems*, Vol. 7, No. 6, 1990, pp. 897-928
13. Colbaugh, R. and K. Glass, "On Controlling Robots with Redundancy", *Robotics and Computer-Integrated Manufacturing: An International Journal*, Vol. 9, No. 2, 1992, pp. 121-135
14. Colbaugh, R., H. Seraji, and K. Glass, "A New Class of Adaptive Controllers for Robot Trajectory Tracking" (to appear)

A NEW APPROACH TO ADAPTIVE MANIPULATOR CONTROL PART II: EXPERIMENTS

R. Colbaugh*   H. Seraji†   K. Glass*

* New Mexico State University, Las Cruces, NM 88001
†Jet Propulsion Laboratory, California Institute of Technology, Pasadena, CA 91109

Abstract

A new approach to adaptive motion control of manipulators is introduced in the first part of this two-part paper. In the present paper, which is the second of the two parts, the performance of the proposed controllers is illustrated through computer simulations and experiments with industrial robots. The simulation study utilizes complete and faithful kinematic and dynamic models of a PUMA 762 manipulator, and demonstrates that accurate and robust trajectory tracking can be achieved by using the proposed schemes. Experimental results are presented for a PUMA 560 manipulator and confirm that the control strategies provide a simple and effective means of obtaining high performance trajectory tracking.

1. Introduction

There is considerable interest in developing manipulator control strategies capable of providing accurate end-effector trajectory tracking despite uncertainties in the robot dynamics and payload and in the presence of external disturbances and sensor noise. Adaptive controllers, which can compensate for uncertainties and disturbances by adjusting the controller parameters online based on the observed system performance, provide an attractive potential solution to this control problem. In the first part of this two-part paper a new approach to adaptive manipulator control is introduced. The control schemes designed using this approach are very general and computationally efficient, and can be shown to be globally stable in the presence of bounded disturbances. Additionally, these controllers are capable of achieving accurate task-space trajectory tracking with no knowledge of either the mathematical model or the parameter values of the robot dynamics, and without requiring the inverse kinematic transformation.

This paper, which is the second of two parts, illustrates the performance of the proposed controllers through computer simulations and experiments with industrial robots. The computer simulation study examines both the nominal performance and the robustness of the new control schemes when applied to a six degree-of-freedom (DOF) PUMA 762 manipulator. The simulations are conducted using a graphics-based robot simulation environment that incorporates models of all important dynamic subsystems such as joint stiction, sensor noise, and transmission effects [1]. The hardware experiments are performed in the JPL Robotics Research Laboratory and utilize a PUMA 560 arm and controller and a DEC MicroVAX II computer [2]. These experiments parallel the computer simulation study in that both the nominal performance and the robustness of the proposed control strategies are examined.

2. Simulation Results

The class of adaptive robot control schemes described in the first part of this two-part paper is now applied to an industrial robot in a series of computer simulation studies. The robot chosen for this simulation study is a six DOF PUMA 762 manipulator. Considerable effort was devoted to develop complete and accurate kinematic and dynamic models for this robot, and to integrate these models into a graphics-based robot simulation environment. The simulation software is written in 'C' and is hosted on a Silicon Graphics Personal IRIS workstation. A complete description of this flexible and modular simulation environment is beyond the scope of this paper; the interested reader is referred to [1]. It is noted, however, that the simulation package incorporates models of all important dynamic subsystems and phenomena, such as joint stiction, sensor noise, and transmission effects, and therefore provides the basis for a meaningful evaluation of controller performance. Additionally, it should be mentioned that the development of this simulation environment drew extensively from the work reported in [3,4].

While a full description of the PUMA 762 manipulator is presented in [3] and is not repeated here, it is illuminating to briefly characterize this robot. The PUMA 762 possesses a conventional design with six revolute joints configured in a "waist-shoulder-elbow-wrist" arrangement. Precise values for all kinematic and dynamic model parameters for this robot are given in [3]; here it is simply noted that the overall reach of the arm is approximately 50in and the total weight of the arm is almost 850lb.

The simulation study consists of two parts: a *nominal performance study* in which no external disturbances are applied to the robot dynamics, and a *robustness study* where a payload-related disturbance is introduced. It is stressed that the robot models in both the nominal and robustness studies include all rigid body dynamic terms as well as realistic models of joint friction, sensor noise, and transmission effects. The results presented here are samples selected from a comprehensive computer simulation study which was carried out to test the performance of the proposed class of controllers. These examples are chosen for presentation because they are typical of the larger study and not because they represent the best performance attainable with the control schemes. Indeed, no effort was made to "tune" the controller gains to obtain exceptional performance. In this study, the control strategies are applied to the mathematical model of the PUMA 762 manipulator through computer simulation with a sampling period of two milliseconds. All integrations required by the controller are implemented using a simple trapezoidal integration rule with a time-step of two milliseconds. All information concerning the dynamic model, parameter values, and simulation strategy for this study can be found in [1,3,4]. It must be emphasized that this information is used only to simulate the robot behavior and is *not* used in the control law computation. Finally, it is noted that throughout the computer simulation studies, the unit of length is inch, the unit of time is second, the unit of angle is degree, and the unit of force is pound, unless stated otherwise.

The first simulation in this study demonstrates the capability of the control scheme (7),(9),(10b) to provide accurate task-space trajectory tracking under nominal conditions, in which no external disturbances are present. The PUMA arm is initially at rest with joint-space position $\theta(0) = [0, 90, -180, 0, -90, 0]^T$. The robot is commanded to smoothly move its end-effector 20in in the x-direction, 20in in the y-direction, and 20in in the z-direction in 2 seconds according to the following temporal trajectories: $x(t) = x_i + 10(1 - \cos(\pi/2)t)$, $y(t) = y_i + 10(1 - \cos(\pi/2)t)$, $z(t) = z_i + 10(1 - \cos(\pi/2)t)$ for $t \subset [0, 2]$, where $[x_i, y_i, z_i]$ are the initial coordinates of the end-effector and all motions are specified relative to the base coordinate frame. Additionally, the robot is required to smoothly change its end-effector orientation from an initial orientation matrix R_{init} to a final orientation matrix R_{final}, where

$$R_{init} = \begin{bmatrix} -1 & 0 & 0 \\ 0 & -1 & 0 \\ 0 & 0 & 1 \end{bmatrix} \quad , \quad R_{final} = \begin{bmatrix} \sqrt{3}/2 & -1/2 & 0 \\ 1/2 & \sqrt{3}/2 & 0 \\ 0 & 0 & 1 \end{bmatrix}$$

and where each rotation matrix R is specified relative to the base frame of the robot. The temporal trajectory for this orientation change is specified using the familiar "equivalent-angle-axis" parameterization of $SO(3)$ given by $\phi\hat{\mathbf{k}} \in \Re^3$, where $\hat{\mathbf{k}}$ is a unit vector aligned with the axis of rotation from the current to the desired orientation and ϕ is the angle between these orientations measured about this axis [5]. In this simulation the desired orientation trajectory is $\phi_d = 75°(1 - \cos(\pi/2)t)$ and $\hat{\mathbf{k}}_d = [0,0,1]^T$ for $t \subset [0,2]$. Thus the robot is commanded to translate its end-effector approximately 35in and rotate its end-effector through 150° in 2 seconds.

The control strategy (7),(9),(10b) is utilized to achieve the specified trajectory tracking. The adaptation gains $A, B, \mathbf{f}$ are set to zero initially, while K_p, K_v are initialized to $K_{p0} = diag(100 \;\; 100 \;\; 100 \;\; 500 \;\; 500 \;\; 500)$ and $K_{v0} = diag(50 \;\; 50 \;\; 50 \;\; 200 \;\; 200 \;\; 200)$. The remaining controller parameters are set as follows: $\gamma_i = 0.001$ and $\beta_i = 2$ for $i = 1, 2, ..., 5$ and $\epsilon = 2$. Note that the unit of angle is radian in all controller parameters, and this accounts for the large numerical values of K_{p0} and K_{v0}. The results of this simulation are given in Figures 1a and 1b, and indicate that accurate task-space trajectory tracking is achieved.

The next simulation in this study investigates the robustness of the proposed controllers to external disturbances. Note that if the manipulator payload mass m is time-varying, the skew-symmetry of the matrix $\dot{H} - 2V_{cc}$ is disrupted; this fact was noted in [6] and is demonstated in [7] using a simple example. This result is important since most adaptive trajectory tracking schemes (including the ones proposed in the first part of this paper) exploit this skew-symmetry in their development. As a consequence, if the payload varies with time and skew-symmetry is lost, then the "extra" terms in the manipulator dynamic model which arise from the time-variation of the payload represent an external disturbance to the nominal model. To examine the effect of this disturbance, the controller (7),(9),(10e) is utilized to repeat the task specified in the first simulation, but here a time-varying payload is introduced with mass given by $m(t) = 25(1 + cos\frac{\pi}{2}t)$. Thus the payload smoothly decreases from 50lb to 0lb during the trajectory. In this simulation the adaptation gains $A, B, \mathbf{f}, K_p, K_v$ are set to zero initially, and the remaining controller parameters are set as follows: $K = diag(2), \lambda = 10$, and $\gamma_i = 0.001$, $\beta_i = 2$ for $i = 1, 2, ..., 5$. The results of this simulation are given in Figures 2a and 2b, and indicate that accurate task-space trajectory tracking is achieved despite the gross and continuous variation in payload mass.

3. Experimental Results

The class of robot control schemes described above is now applied to an industrial robot in a series of experiments. The experimental facility utilized for this study is the JPL Robotics Research Laboratory and consists of a Unimation PUMA 560 arm and controller, and a DEC MicroVAX II computer. The

MicroVAX II hosts the RCCL (Robot Control "C" Library) software, which was originally developed at Purdue University [8] and subsequently modified and implemented at JPL. During the operation of the arm, a hardware clock constantly interrupts the I/O program resident in the Unimation controller at a preselected sampling period T_s, which can be chosen as 7, 14, 28, or 54 milliseconds; for all experiments reported here $T_s = 7$ms. At every interrupt, the I/O program gathers information about the state of the arm (such as joint encoder readings), and interrupts the control program in the MicroVAX II to transmit this data. The I/O program then waits for the control program to issue a new set of control signals, and then dispatches these signals to the appropriate joint motors. Therefore, the MicroVAX II acts as a digital controller for the PUMA arm and the Unimation controller is effectively bypassed and is utilized merely as an I/O device to interface the MicroVAX II to the joint motors. A complete description of this testbed is beyond the scope of the paper; the interested reader is referred to [2].

The simple control scheme (7),(9),(10a) was selected for implementation in the present experimental study. This choice was motivated by the extreme computational efficiency of the control law (7),(9),(10a), and also the desire to investigate the tracking accuracy obtainable with a scheme with no feedforward elements. Note that computational efficiency is particularly important in view of the modest computing power available and the relatively high communication overhead in the testbed [2]. The experimental study consists of a nominal performance study and a robustness study, and in this way parallels the computer simulation investigation. In the nominal performance study, the scheme (7),(9),(10a) is implemented for trajectory tracking control of the PUMA arm waist joint θ_1 while the other joints are held steady using the Unimation controller. In contrast, in the robustness study the strategy (7),(9),(10a) is commanded to track a trajectory similar to the one specified in the nominal study, but now the other joints are given a large motion using the Unimation controller. Observe that because the controller (7),(9),(10a) is implemented for controlling the waist joint only and therefore is provided with sensor data for this joint alone, the motion of the joints distal to the waist joint have the effect of a time-varying payload. Recall from the previous section that this results in the application of an external disturbance to the nominal robot dynamics.

The first experiment considered is the nominal performance test summarized above. The PUMA arm is initially at rest with joint-space position $\theta(0) = \mathbf{0}$. The waist joint angle is commanded to change smoothly from the initial position $\theta_1 = 0$ to the goal position $\theta_1 = \pi/2$ in 2 seconds using the cycloidal trajectory $\theta_{d_1} = (\pi t - sin\pi t)/4$ for $t \subset [0, 2]$. The control law (7),(9),(10a) is utilized to achieve the desired trajectory tracking. The adaptation gains $\mathbf{f}, K_p, K_v$ are initialized to zero and the remaining controller parameters are set to the following values: $\epsilon = 1.5, \beta_1 = 4, \beta_2 = 8, \beta_3 = 64$, and $\gamma_i = 0$ for $i = 1, 2, 3$. While the arm is in motion, the reading of the waist joint encoder at each sampling instant is recorded directly from the arm, converted into degrees and stored in a data file. Figure 3 shows the desired and actual trajectories of the waist joint angle. It is seen that the joint angle $\theta_1(t)$ tracks the desired trajectory $\theta_{d_1}(t)$ very closely; indeed, the peak value of the tracking error $e(t) = \theta_{d_1}(t) - \theta_1(t)$ is 1.40°. The initial lag in the θ_1 response is due to the large stiction (static friction) present in the waist joint. Figure 4 shows the tracking performance of the waist joint for the same motion using the Unimation controller, which is operating with a sampling period of 1 millisecond. It is noted that the peak joint tracking error in Figure 4 is 5.36°, which produces 4 centimeters peak position error at the end-effector. By comparing Figures 3 and 4, it is evident that the tracking performance of the adaptive controller is noticeably superior to that of the Unimation controller, despite the fact that the Unimation control loop is seven times faster than the adaptive control loop.

The next experiment investigates the performance of the proposed controller when additional robustness is required for successful completion of the desired task. In this experiment, the control law (7),(9),(10a) is utilized to track a desired waist joint trajectory similar to the one specified in the first experiment, and the arm is initially in the same configuration. In this case, however, the Unimation controller is used to smoothly move the shoulder joint θ_2 and elbow joint θ_3 from the initial zero positions to $\theta_2 = -90°$ and $\theta_3 = 90°$ during the execution of the task. Observe that this is a gross and rapid variation in the arm configuration, which produces a large and fast variation in the arm inertia as seen at the waist joint. As indicated above, this variation presents a large external disturbance to the nominal waist joint dynamics. The desired waist joint trajectory for this experiment is $\theta_{d_1} = (2\pi t/3 - sin(2\pi t/3))/6$ for $t \subset [0, 3]$, and all adaptive gains and controller parameters are set to the values used in the first experiment. The results of this experiment are shown in Figure 5, which indicates that accurate trajectory tracking is achieved despite the large external disturbance.

4. Conclusions

This two-part paper presents a new approach to direct adaptive motion control of manipulators. Theoretical issues regarding this approach are summarized in the first part of the paper. In the present paper, which is the second of the two parts, the capabilities of the proposed control schemes are illustrated through a computer simulation study and an experimental investigation. The results of these studies demonstrate that the controllers provide a simple and effective means of obtaining high performance trajectory tracking. Future work will focus on the application of the proposed approach of controller development to the problem of robot compliant motion control.

5. Acknowledgements

The research described in this paper was supported in part through grants from the Department of Energy (WERC) and the General Motors Corporation and in part through contracts with the Jet Propulsion Laboratory and the National Aeronautics and Space Adminstration.

6. References

1. Glass, K. and R. Colbaugh, "A Computer Simulation Environment for Manipulator Controller Development" (to appear)
2. Seraji, H., T. Lee, and M. Delpech, "Experimental Study on Direct Adaptive Control of a PUMA 560 Industrial Robot", *Journal of Robotic Systems*, Vol. 7, No. 1, 1990, pp. 81-105
3. Bedewi, N., E. Aronne, and P. Chang, "Robot Mathematical Modeling, Simulation, and Parameter Indentification", Final Report for NASA Contract NAS5-29412, Task 24, March 1988
4. Bedewi, N., "RDSIM: Robot Dynamics Simulation", Final Report for NASA Contract NAS5-29412, Task 34, March 1989
5. Craig, J., *Introduction to Robotics: Mechanics and Control*, Addison-

Wesley, New York, 1989, Second Edition
6. Song, Y. and R. Middleton, "Dealing with the Time-Varying Parameter Problem of Robot Manipulators Performing Path Tracking Tasks", *Proc. 29th IEEE Conference on Decision and Control*, Honolulu, HA, December 1990
7. Colbaugh, R., H. Seraji, and K. Glass, "A New Class of Adaptive Controllers for Robot Trajectory Tracking" (to appear)
8. Hayward, V. and R. Paul, "Introduction to RCCL: A Robot Control 'C' Library", *Proc. IEEE International Conference on Robotics*, Atlanta, GA, May 1984

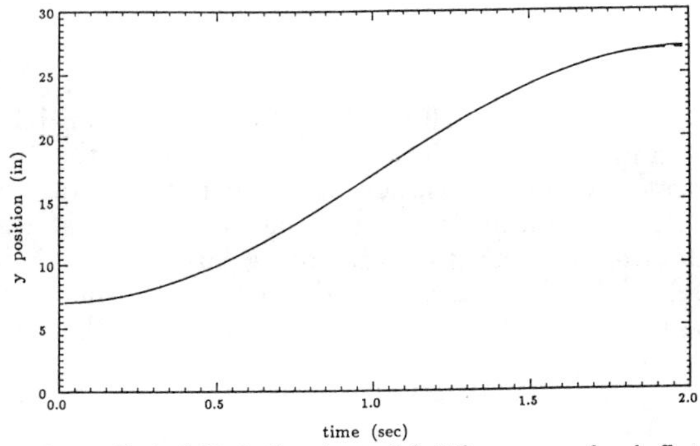

Figure 1a: Desired (dashed) and actual (solid) response of end-effector coordinate y in PUMA Arm Simulation (nominal case)

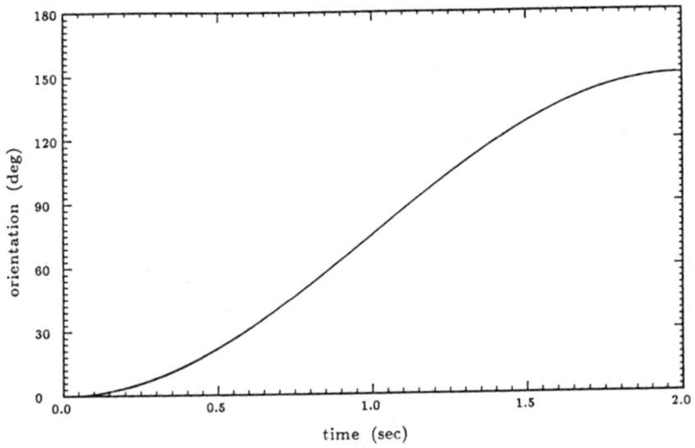

Figure 1b: Desired (dashed) and actual (solid) response of end-effector orientation angle ϕ in PUMA Arm Simulation (nominal case)

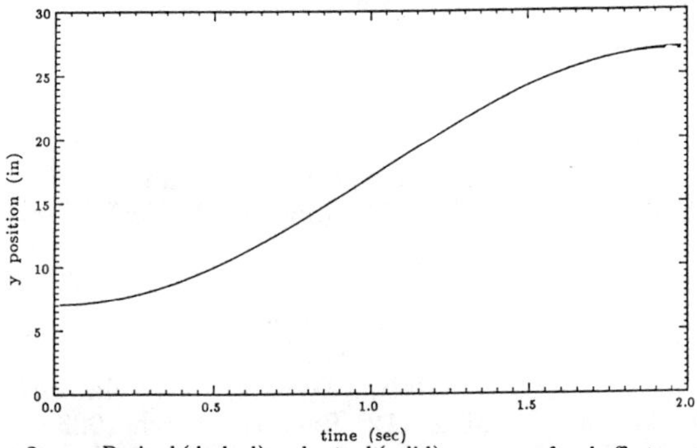

Figure 2a: Desired (dashed) and actual (solid) response of end-effector coordinate y in PUMA Arm Simulation (time-varying payload)

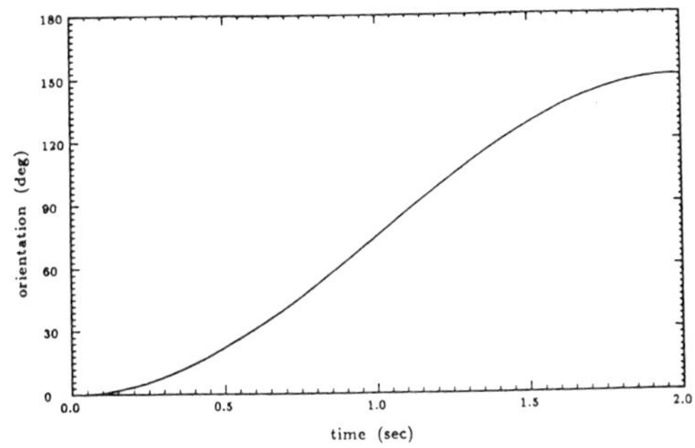

Figure 2b: Desired (dashed) and actual (solid) response of end-effector orientation angle ϕ in PUMA Arm Simulation (time-varying payload)

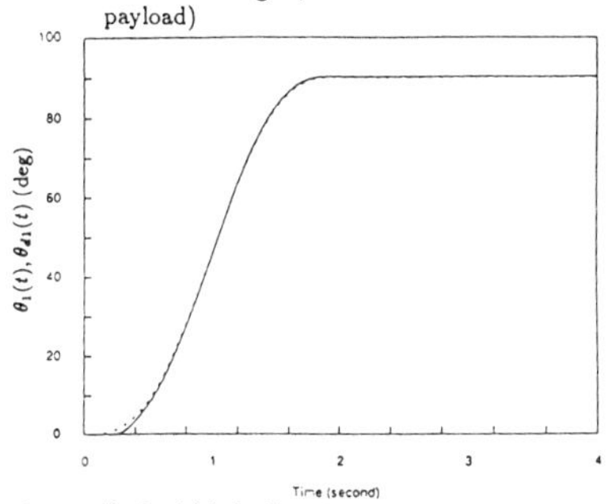

Figure 3: Desired (dashed) and actual (solid) response of PUMA waist angles under adaptive controller

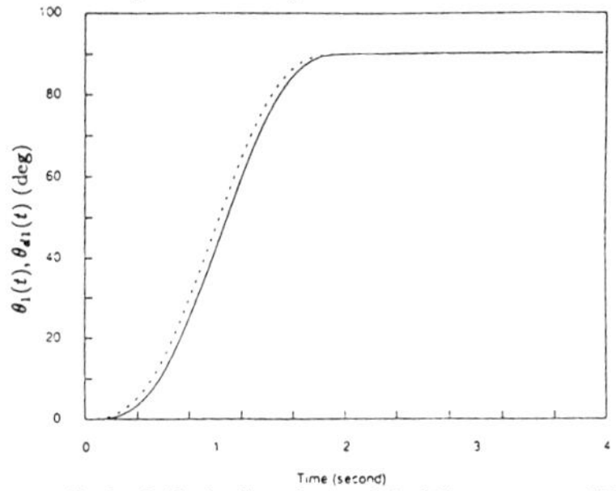

Figure 4: Desired (dashed) and actual (solid) response of PUMA waist angles under Unimation controller

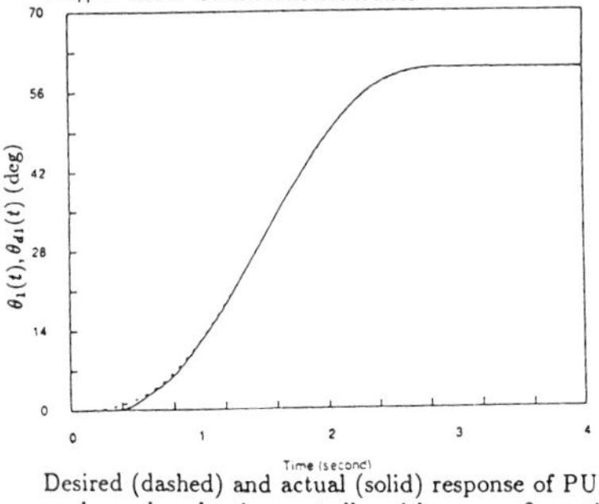

Figure 5: Desired (dashed) and actual (solid) response of PUMA waist angles under adaptive controller with arm configuration change

PART 4 : PARALLEL AND REDUNDANT ROBOTS

Repeatable Inverses for Kinematically Redundant Manipulators and Their Associated Algorithmic Singularities†

Rodney G. Roberts Anthony A. Maciejewski

School of Electrical Engineering
Purdue University
West Lafayette, Indiana 47907

ABSTRACT

One approach to resolving kinematic redundancy in a robotic system and guaranteeing repeatability is to augment the manipulator Jacobian with the appropriate number of gradient rows. A variety of such techniques have been proposed including configurational control and the extended Jacobian. Unfortunately such augmenting schemes often introduce artificial singularities, called algorithmic singularities, which can seriously restrict the region of the workspace in which the manipulator can operate. This paper will examine the algorithmic singularities of two particular classes of optimal repeatable control strategies and the influence certain considerations have on the location and presence of algorithmic singularities in the joint space.

I. INTRODUCTION

There are many techniques in the current robotics literature for resolving kinematic redundancy, most of which are performed at the kinematic level. Due to the extra freedom associated with redundancy, a particular criterion is often optimized during redundancy resolution. A popular criterion that has been investigated in the literature is the minimum joint velocity norm criterion, which is achieved under pseudoinverse control. This is an important criterion when using Jacobian control since the Jacobian can only be used to locally approximate changes in the end effector position. Weighted-pseudoinverses can be applied when it is desired to minimize a weighted-least squares criterion of the joint velocities. Unfortunately, pseudoinverse control will, in most cases, result in a control that drifts in the sense that the manipulator does not return to its initial configuration when performing a cyclic end effector motion [2]. Thus, when a cyclic end effector motion is specified, the resulting joint motion is not, in general, cyclic.

For obvious reasons, joint space drift is undesirable, particularly when there is a concern that joint limits may be exceeded, kinematic singularities may be encountered, or static collisions can occur. Thus, designing repeatable control strategies is a topic that has received recent interest. One way to accomplish this goal is to augment the Jacobian with a gradient row. Various criteria can be used for choosing this row vector. Possible methods for determining an appropriate augmenting vector include minimizing an objective function of the manipulator's configuration as in the case of the extended Jacobian [1], requiring the manipulator to satisfy a configurational constraint [5], or choosing

† This work was supported in part by Sandia National Laboratories under contract number 18-4379B. Additional support was provided by the NEC Corporation and the TRW Foundation.

a repeatable strategy which is nearest to a desired nonrepeatable strategy [3]. This paper will be primarily concerned with the latter method. Unfortunately, resolving redundancy using any of the above methods can result in artificial singularities which can seriously hinder the operation of the manipulator, particularly if it is required to move in large regions of the workspace.

This paper will examine the algorithmic singularities of two types of repeatable control strategies. Both methods are optimal in a given region of the joint space and are chosen from a given set of inverses. This work will focus on how the presense and location of the algorithmic singularities of the optimal inverses are influenced by the region of joint space interest and the set of allowable augmenting vectors. The paper is outlined in the following manner: Section II presents some preliminary material including a discussion of augmented Jacobians and their associated algorithmic singularities. A short discussion on the distance between the repeatable inverse and the pseudoinverse is given since this distance can be used to identify configurations which are near algorithmic singularities. A brief discussion of two methods for obtaining repeatable control strategies which are nearest to a desired nonrepeatable control strategy is presented in Section III. Section IV illustrates through two examples how algorithmic singularities can occur and how the choice of the basis of allowable augmenting vectors and the region of interest affect the occurrence of algorithmic singularities in the optimal inverse. Finally, conclusions appear in Section V.

II. PRELIMINARIES

In this paper the manipulator's configuration will be denoted by the n-vector θ, while the manipulator's end effector position and orientation will be given by the m-vector $\mathbf{x}$. For kinematically redundant manipulators, $m < n$. The Jacobian equation is given by

$$\dot{\mathbf{x}} = J\dot{\theta} \tag{1}$$

where $\dot{\mathbf{x}}$ represents the end effector velocity, $\dot{\theta}$ represents the joint velocity, and J is the manipulator Jacobian. In order to control the manipulator, it is necessary to invert (1). In this paper only control strategies of the form

$$\dot{\theta} = G\dot{\mathbf{x}} \tag{2}$$

will be considered where G is a generalized inverse of J. Furthermore, only manipulators with a single degree of redundancy will be considered so that $m = n - 1$. In this case, any generalized inverse can be written as

$$G = J^{+} + \hat{\mathbf{n}}_J \mathbf{w}^T \tag{3}$$

where $^{+}$ denotes the pseudoinverse, $\hat{\mathbf{n}}_J$ is a unit length null vector of the Jacobian, and the m-vector $\mathbf{w}$ uniquely determines G. Note that the distance of the generalized inverse G from the pseudoinverse is given by $\|G - J^{+}\|_2 = \|\mathbf{w}\|_2$.

One way of obtaining a generalized inverse G is to simply augment the Jacobian with the appropriate number of rows and then perform a matrix inversion, provided that the augmented Jacobian is nonsingular. For the case of a single degree of redundancy, augmenting the Jacobian with $\mathbf{v}^T$ results in an inverse whose $\mathbf{w}$ vector is given by

$$\mathbf{w}^T = -\frac{\mathbf{v}^T J^{+}}{\hat{\mathbf{n}}_J \cdot \mathbf{v}}. \tag{4}$$

The term that presents a problem at an algorithmic singularity is the quantity $\hat{\mathbf{n}}_J \cdot \mathbf{v}$. At the algorithmic singularity this term becomes zero causing $\mathbf{w}$ to take on an infinite value. Therefore $\|G - J^+\|_2$ becomes excessively large as the configuration approaches an algorithmic singularity and can thus be used as a measure of how close the manipulator is to an algorithmic singularity.

III. NEAREST REPEATABLE CONTROL STRATEGIES

This section will briefly describe two methods for obtaining repeatable control strategies. These two methods design repeatable strategies which are nearest to a desired nonrepeatable inverse G_d. The first method minimizes the criterion

$$\|G_r - G_d\|_\Omega^2 = \int_\Omega \|G_r - G_d\|_2^2 \, d\theta = \int_\Omega \|\mathbf{w}_r - \mathbf{w}_d\|_2^2 \, d\theta \tag{5}$$

where Ω is a simply-connected, singularity-free subset of the joint space. Equation (5) provides a measure of the closeness of two inverses on some important subset Ω of the joint space. The subset Ω may be chosen based on some optimal configuration at which one would like the manipulator to operate.

For very simple manipulators one can solve a set of partial differential equations to determine the repeatable strategies [3]. For more complicated manipulators the optimization must be performed numerically using a basis of allowable augmenting vectors consisting of gradient vector functions. Suppose that a set of N linearly independent gradients $\{\mathbf{v}_1, \mathbf{v}_2, \ldots, \mathbf{v}_N\}$ has been chosen. The span of this set is a space of gradients of the form

$$\mathbf{v} = \sum_{i=1}^{N} a_i \mathbf{v}_i \tag{6}$$

where the a_i are real scalar constants. Note that any scaled version of the augmenting vector results in the same inverse so that a normalization of the coefficients a_i is in order. The problem then becomes an $(N-1)$-dimensional optimization. One final consideration relates to the possible existence of algorithmic singularities for certain combinations of a_i in the domain Ω. Such combinations of coefficients should be avoided when determining an optimal solution.

The above optimization can be relatively difficult since the criterion function is, in general, highly nonlinear. Even when a minimum is obtained it is difficult to determine whether it is in fact a global minimum. Rather than minimizing the distance between the repeatable inverse and the desired inverse, it is possible to minimize the distance between their associated null spaces. The distance between the two null vectors $\mathbf{v}_r$ and $\mathbf{v}_d$ will be given by

$$\|\mathbf{v}_r - \mathbf{v}_d\|_\Omega = \left[\int_\Omega \|\mathbf{v}_r - \mathbf{v}_d\|_2^2 \, d\theta \right]^{\frac{1}{2}} \tag{7}$$

where once again Ω is a region of interest in the joint space which will typically have the form $I_1 \times I_2 \times \cdots \times I_n$ where each I_i is an interval which will depend on the manipulator's joint limits and the specified task at hand. Let $\mathcal{V}$ denote a space of allowable augmenting vectors which will typically be given as the span of a finite set of orthonormal gradient vectors where orthogonality is given by the unique inner product associated with (7), i.e., the inner product given by $< \mathbf{u}, \mathbf{v} >_\Omega = \int_\Omega \mathbf{u} \cdot \mathbf{v} \, d\theta$. Let this set of gradient vector fınctions be given by $\{\phi_1, \phi_2, \cdots, \phi_N\}$. By definition then, $< \phi_i, \phi_j >_\Omega = \delta_{ij}$ where $< \cdot, \cdot >_\Omega$ is the inner product on $\mathcal{L}_2(\Omega)$ and δ_{ij} is simply the Kronecker delta. The space of allowable augmenting vectors $\mathcal{V}_N$ is then just the span $\{\phi_1, \phi_2, \cdots, \phi_N\}$. Corresponding to this basis is a Gramian matrix M defined by

$$M_{ij} = \int_{\Omega} (\phi_i \cdot \hat{\mathbf{n}}_G)(\phi_j \cdot \hat{\mathbf{n}}_G)\, d\theta \tag{8}$$

where $\hat{\mathbf{n}}_G$ is the normalized null vector of G_d^T. It has been shown in [4] that the optimal coefficients can be calculated from the singular value decomposition of M and are given as the singular vector associated with the largest singular value of M. It has also been shown that the largest singular value of M is a measure of how well the optimal augmenting vector matches the associated null space of G_d.

IV. ALGORITHMIC SINGULARITIES

In order to illustrate how the choice of allowable augmenting vectors affects the presence of algorithmic singularities for the two methods described in the previous section, a specific example will be presented. Consider a planar manipulator consisting of two orthogonal prismatic joints and a third revolute joint of unit length (a PPR manipulator). This manipulator would have as its Jacobian

$$J = \begin{bmatrix} 1 & 0 & -\sin\theta_3 \\ 0 & 1 & \cos\theta_3 \end{bmatrix} \tag{9}$$

and a unit length null vector $\hat{\mathbf{n}}_J = \frac{1}{\sqrt{2}}\left[\sin\theta_3 \quad -\cos\theta_3 \quad 1\right]^T$. It is desired to find a repeatable inverse as a function of θ_3 which is close to the pseudoinverse in the sense of equation (5) where the region of interest is the θ_3 interval $[-\pi/4, \pi/4]$.

The manipulator in this example is simple enough to analytically calculate the nearest repeatable inverses for infinite dimensional augmenting spaces. In particular, for the set of all repeatable inverses which are functions of θ_3 only, it has been shown [3] that the nearest optimal repeatable inverse has the form

$$\mathbf{w} = \frac{k\cos\theta_3 + \sin\theta_3}{\sqrt{2}(k^2+1)}\left[1 \quad -k\right]^T \tag{10}$$

where $\mathbf{w}$ is defined in (3). This solution is parameterized by the scalar k which is determined by the limits of integration. For the region of interest being considered here, i.e. $[-\pi/4, \pi/4]$, k is identically zero. Note that at the center of this region, the inverse matches the pseudoinverse exactly. The vector $\mathbf{w}$ is bounded by $1/\sqrt{2}$, so that no algorithmic singularities result from this control. Thus the optimal inverse for this method is very well behaved.

In general, it is not possible to analytically calculate the nearest repeatable control strategy. However, one can use a direct method by considering control strategies which are obtained by augmenting the Jacobian with a gradient row. Thus the goal at hand will be accomplished by considering augmenting rows which are gradients and functions of θ_3 only. Two sets of augmenting vectors will be considered, with one being a subset of the other. Both optimizations will once again be taken over the θ_3 interval $[-\pi/4, \pi/4]$.

First, consider a very small basis of augmenting vectors given by the span of $\mathbf{e}_1$ and $\mathbf{e}_2$. For this particular example, the optimal coefficients are $a_1 = 0$ and $a_2 = 1$ so that the augmenting row for the optimal repeatable method is given by $[0 \quad 1 \quad 0]$. In order to apply the null space approximation method, it is necessary to normalize the basis elements in the norm $\|\cdot\|_\Omega$. The orthonormal basis for the null space approximation method is then given by $\phi_i = \sqrt{2/\pi}\mathbf{e}_i$. The corresponding Gramian matrix M is calculated using (8) and is found to be

$$M = \begin{bmatrix} 0.0908 & 0 \\ 0 & 0.4092 \end{bmatrix}. \tag{11}$$

The largest singular value is easily seen to be 0.4092 and the vector of optimal coefficients is $[0 \quad 1]^T$ so that the optimal augmenting vector is ϕ_2. The optimal augmenting row then is $[0 \quad 1 \quad 0]$, which is identical to that obtained from the first method. Note that, as will be seen shortly, this is not the case in general. Note also that this augmenting row results in an algorithmic singularity when $\cos\theta_3$ is zero, which is, however, outside of the selected design region Ω. In fact, all augmenting rows obtained from the given basis will result in an algorithmic singularity somewhere in the joint space. This motivates choosing a larger basis in the hope that algorithmic singularities can be avoided.

Next consider the optimal repeatable inverse over a set of inverses determined by the span of $\mathbf{e}_1$, $\mathbf{e}_2$, and $\mathbf{e}_3$. The θ_3 interval of interest will once again be taken to be $[-\pi/4, \pi/4]$. For this basis, the augmenting row for the nearest optimal repeatable method is given by $[0 \quad -0.6428 \quad 0.7660]$. However, when applying the null space approximation method, the matrix M is given by

$$M = \begin{bmatrix} 0.0908 & 0 & 0 \\ 0 & 0.4092 & -0.4502 \\ 0 & -0.4502 & 0.5000 \end{bmatrix} \tag{12}$$

which results in a choice of $[0 \quad -0.6707 \quad 0.7418]$ as an optimal augmenting row. While these augmenting rows are slightly different, they both represent a significant improvement over the previous basis. A quantitative measure of this improvement is given by the largest singular value of M which for this basis is 0.9071. Note that each of these augmenting vectors is close to the vector $[0 \quad -0.7071 \quad 0.7071]$, which is exactly the unit length null vector $\hat{\mathbf{n}}_J$ at the center of the region of interest. Using this augmenting vector would result in a strategy that exactly matches the desired pseudoinverse behavior at configurations with $\theta_3 = 0$; however, algorithmic singularities will occur precisely when $\cos\theta_3 = -1$, so that while this particular augmenting row provides a perfect match to the pseudoinverse at configurations where $\theta_3 = 0$, it does not perform as well as either the nearest optimal repeatable or null space approximation methods.

In order to illustrate how the choice of the region Ω affects the existence of algorithmic singularities, consider a planar PRR manipulator which, for simplicity, is described in absolute angles. The Jacobian is given by

$$J = \begin{bmatrix} 1 & -\sin\psi_2 & -\sin\psi_3 \\ 0 & \cos\psi_2 & \cos\psi_3 \end{bmatrix} \tag{13}$$

and its null vector is given by $\mathbf{n}_J = [\sin(\psi_3 - \psi_2) \quad -\cos\psi_3 \quad \cos\psi_2]^T$ where ψ denotes the absolute angle formulation of the revolute joints. Unlike the previous example, this manipulator has kinematic singularities which occur when $\cos\psi_2 = \cos\psi_3 = 0$. It will thus be necessary to take this into consideration when choosing a region of joint space interest. Once again consider the augmenting basis determined by the span of $\mathbf{e}_1$, $\mathbf{e}_2$, and $\mathbf{e}_3$. Let

$\Omega = [3\pi/8, 5\pi/8]^3$ so that no kinematic singularities are present in the design region. Applying the null space approximation method results in the augmenting row $[0.8129 \quad 0.4118 \quad 0.4118]$ which has algorithmic singularities when $\psi_3 = \psi_2$ so that such singularities are present in the region of interest. In fact, it turns out that any augmenting row corresponding to the span of the standard basis results in an algorithmic singularity. Thus one is not able to calculate the nearest optimal repeatable control strategy given this particular set of augmenting vectors. Note that this is a difference between the two methods, namely that it is possible to calculate an optimal augmenting row for the null space approximation method even if the resulting control strategy has an algorithmic singularity in Ω. Thus, when only applying the null space approximation method, one should check that an algorithmic singularity does not occur in the design region.

Next, consider a design region given by $\Omega = [63\pi/256, 65\pi/256]^3$. Since this is a smaller region of interest, it is possible to choose an allowable augmenting vector which does not result in an algorithmic singularity in the region Ω. Thus the nearest optimal repeatable method has a solution which is given by the augmenting row $[0 \quad -0.7660 \quad 0.6428]$. The null space approximation method solution is given by $[0 \quad -0.7071 \quad 0.7071]$. Both augmenting rows result in algorithmic singularities, but these singularities occur outside of Ω.

V. CONCLUSIONS

This paper has examined the algorithmic singularities of two classes of optimal repeatable control strategies. Each method calculates the inverse from a set of repeatable strategies which is nearest to a desired nonrepeatable inverse in a given region of the joint space. It was shown that both the existence and the location of algorithmic singularities is a function of the set of allowable augmenting vectors and the design region. It was also seen that the null space approximation method, unlike the nearest optimal repeatable method, did not distinguish algorithmic singularities in the design region. Finally, from the examples presented, it would appear that the methods result in similar augmenting rows.

REFERENCES

[1] J. Baillieul, "Kinematic programming alternatives for redundant manipulators," in *Proc. IEEE Int. Conf. Robotics Automat.*, (St. Louis, MO), pp. 722-728, March 1985.

[2] C. A. Klein and C. H. Huang, "Review of pseudoinverse control for use with kinematically redundant manipulators," *IEEE Trans. on Syst., Man, Cybern.*, vol. SMC-13, no. 3, pp. 245–250, March/April 1983.

[3] R. G. Roberts and A. A. Maciejewski, "Nearest optimal repeatable control strategies for kinematically redundant manipulators," *IEEE Trans. Robotics Automat.*, vol. 8, no. 3, pp. 327-337, June 1992.

[4] R. G. Roberts and A. A. Maciejewski, "Repeatable generalized inverse control strategies for kinematically redundant manipulators," in *Proc. 30th IEEE Conf. Decision Contr.*, (Brighton, United Kingdom), pp. 2428-2434, December 11–13, 1991.

[5] H. Seraji, "Configuration control of redundant manipulators: Theory and implementation," *IEEE Trans. Robotics Automat.*, vol. 5, no. 4, pp. 472–490, August 1989.

LEARNING INVERSE KINEMATICS OF REDUNDANT MANIPULATORS

David DeMers & Kenneth Kreutz–Delgado
University of California, San Diego

Abstract

The inverse kinematics problem for manipulators with redundant degrees of freedom is classically ill–posed. In practice, solution of such underdetermined problems is done numerically utilizing a differential method where the problem has been regularized, either by adding constraints equal to the number of excess degrees of freedom or by optimization of a side functional.

Topological results indicate that advantages may accrue by a direct computation of the inverse function; however, because of the ill–posedness of the problem, little research has been done in this direction. This paper indicates how the ill–posed inverse kinematics problem may be reformulated as a set of well–posed problems, and summarizes results of the author's previous work which shows how direct inverse functions for manipulators with redundant degrees of freedom may be constructed from samples of joint and position/orientation data alone. Expending off–line resources on learning the inverse kinematics directly can then result in fast and accurate on–line computation, with redundancies resolved at run–time.

1 Introduction

The forward kinematic function $f : \Theta^n \rightarrow \mathcal{W}^m \subset \mathcal{X}^m, \mathcal{W}^m = f(\Theta^n)$, maps a set of n joint parameters to the m-dimensional task space (typically $n \geq m$),

$$x = f(\theta),\ x \in \mathcal{W}^m$$

where f is continuous and differentiable. The inverse kinematics problem is to solve $f(\theta) = x$ for a desired target value $x = x_d$.

Computation of inverse kinematics of robot manipulators with redundant degrees of freedom is difficult, and is the subject of much current research. In general, there is no closed form solution. Therefore, in practice, "indirect" differential methods such as pseudo–inverse and gradient descent techniques are used. Computation of inverse kinematics directly is attractive in that a direct inverse kinematics solver can provide a fast solution at run time. Unfortunately, computation of the direct inverse kinematics is problematic due to two manifestations of ill–posedness. First, for any target in the workspace, there are a finite number of solution branches. The second cause of ill–posedness is due to the higher dimensionality of the configuration space. Thus each of the solution sets will generally contain an infinite number of solutions, all belonging to a so–called "self–motion manifold", a manifold of dimensionality equal to the number of redundant degrees of freedom. Consequently, no global inverse kinematics function exists, and a local inverse is one–to–many.

Therefore, in order to regularize the ill–posed problem to obtain a well–posed problem, two resolutions are needed. First, because of the multiplicity of solution sets, solution branch resolution will be necessary. That is, we must somehow choose between the finite number of manifolds which solve $f(\theta) = x_d$. And second, because the inverse solution branches have non–trivial dimensionality, redundancy resolution must then occur. Thus we must choose one of the members of a single solution branch of $\{\theta : f(\theta) = x_d\}$.

Fortunately, the configuration space may be partitioned into a finite number of disjoint regions over each of which inverse mappings *do* exist. Moreover, these regions can be identified by analyzing input–output (joint angle & position) samples of the forward mapping, *without* explicit knowledge even of the forward kinematics function.

This paper shows that the partitioning can be accomplished by the use of learning methods to regularize the ill–posed inverse kinematics problem. The proposed approach exploits the fact that the configuration space can be separated into regions bounded by the surfaces specified by the co–regular sets of the forward kinematics function. For the non–redundant manipulator, these regions are bounded by the singularities, and map one–to–one onto the workspace. For the manipulator with redundant degrees of freedom, these regions are bounded by more complicated surfaces, each containing a single self–motion manifold for each workspace location in the range of the kinematic mapping. The redundancy can be resolved by a parameterization of the self–motion manifolds. An augmented forward kinematic mapping from $\Theta^n \rightarrow \mathcal{W}^m \times t^{n-m}$ can be constructed, where t^{n-m} denotes locations on a self–motion manifold. This mapping is one–to–one and onto, and invertible. Function approximation schemes can be used to estimate the inverse function,

$$f_i^{-1} : \mathcal{W}^m \times t^{n-m} \mapsto \Theta^n$$

for the now–regularized problem. Globally, then, the direct inverse consists of a finite set of f_i^{-1}. At run–time, a choice of f_i^{-1} and a choice of t^{n-m} can be made, to select the desired solution branch, and to resolve the redundancy.

2 Topological Motivation

Given a value in the configuration space, θ, we would like to be able to compute two functions; first, a function which identifies the particular branch it belongs to, and second, a function which produces an appropriate parameter value along the pre–image (or "self–motion") manifold. That is, knowing that $f(\theta) = x_d$ for a specified x_d, we would like to have additional functions $\mathcal{B}(\theta)$ and $\mathcal{T}(\theta)$

$$\mathcal{B}(\theta) : T^n \rightarrow \{1, \ldots, k\}$$

such that $\mathcal{B}(\theta)$ identifies the particular branch that θ is on, and

$$\mathcal{T}(\theta) : T^n \rightarrow \mathcal{M}^{n-m}$$

where $\mathcal{T}(\theta)$ identifies θ with a parameterization of the redundancy on the self–motion manifold $\mathcal{M}^{n-m}$. $\mathcal{T}(\theta)$ locates θ on the solution branch $\mathcal{B}(\theta)$ associated with the given output value $x = f(\theta)$.

2.1 Topological Results

The following are some elementary topological results which serve to motivate the proposed learning approach. See [8] for proofs.

Definitions:

Let Θ be a compact manifold of dimensionality[1] n, and $\mathcal{W}$ be a compact manifold of dimensionality m. For f, a smooth map from Θ to $\mathcal{W}$, let the differential df_θ be the map from the tangent space of Θ at $\theta \in \Theta$ to the tangent space of $\mathcal{W}$ at $f(\theta)$. The set of points in Θ which map to $x \in \mathcal{W}$ is the *pre-image* of x, denoted by $f^{-1}(x)$. df_θ has a natural representation given by an m by n *Jacobian* matrix whose elements consist of the first partial derivatives of f w.r.t. a basis of Θ. We call the *critical points* of f those elements of the set $\mathcal{C}$, defined as the set of all $\theta \in \Theta$ such that df_θ has rank less than the dimensionality of $\mathcal{W}$. Elements of the image of $\mathcal{C}$, $f(\mathcal{C})$ are called the *critical values*. The critical values form manifolds in $\mathcal{W}$. The set $\mathcal{R} = \mathcal{W} \backslash \mathcal{C}$ are the *regular values* of f. For $\theta \in \Theta$, if $\exists \theta^* \in f^{-1}(f(\theta))$, $\theta^* \in \mathcal{C}$, we call θ a *co-regular point*[2] of f.

Lemma 1 (Pre–Image Theorem) *$\forall x \in \mathcal{R} = \mathcal{W} \backslash \mathcal{C}$, $f^{-1}(x)$ is a submanifold of Θ of dimensionality* dim(Θ) - dim($\mathcal{W}$).

Let $f : \Theta \rightarrow \mathcal{W}$ be a smooth mapping on compact manifolds Θ and $\mathcal{W}$, with $m = n$. Lemma 1 yields the fact that the pre-image of a regular value in the workspace of a non-redundant manipulator consists of 0-dimensional manifolds, i.e. points in the configuration space. If $n > m$, the pre-image of a regular value in the workspace of a redundant manipulator consists of manifolds of dimensionality $n - m$. The following theorems apply in the non-redundant case, thus let $m = n$.

Lemma 2 *$\forall x$, $x \in \mathcal{R} = \mathcal{W} \backslash \mathcal{C}$, the cardinality of $f^{-1}(x)$ is finite.*

Lemma 3 ("Stack of Records" Theorem) *$\forall x$, $x \in \mathcal{R}$, $\exists$ an open neighborhood of x, $\mathcal{U}$ such that $f^{-1}(\mathcal{U}) = \bigcup_{i=1}^{n} V_i$, where $V_i \cap V_j = \emptyset$ for $i \neq j$, and each V_i is an open neighborhood of an element of $f^{-1}(x)$. f maps each V_i diffeomorphically onto $\mathcal{U}$.*

Theorem 1 *$\forall \theta_1, \theta_1 \in \Theta$, $f(\theta_1) = f(\theta_2) \Rightarrow$ any continuous path between θ_1 and θ_2 contains $\hat{\theta}$ such that* $\det J(\hat{\theta}) = 0$ *(that is, $\hat{\theta} \in \mathcal{C}$).*

Let Θ be the n-torus, T^n, and $\mathcal{W}$ be $\mathbb{R}^n$ (the case of the non-redundant manipulator). The set of critical points of f form manifolds in T^n, some of which *separate* T^n into the disjoint regions V_i which map onto $\mathcal{W}$. See [3].

For manipulators with redundant degrees-of-freedom, the critical points do not form separating surfaces in the configuration space; however, the pre-image sets of the critical values do. These are *co-regular surfaces* composed of co-regular points, which appear as $m - 1$ dimensional torii (see [3]). Regions in the workspace which are surrounded by critical value manifolds have pre-images in the configuration space which consist of a finite number of disjoint regions separated by the co-regular surfaces. Lemma 3 generalizes (under appropriate conditions on f) in the redundant case such that the pre-image "neighborhoods" consist of manifolds of dimensionality $n - m$. Let $\mathcal{C}_i$ be the simply connnected regions of the configuration space bounded by co-regular surfaces. The image of $\mathcal{C}_i$ under f is a connected[3] region of the workspace, $\mathcal{W}_i$. Let f_i denote f restricted to the domain $\mathcal{C}_i$. Then ($\mathcal{C}_i$, $\mathcal{W}_i$, f_i, T^{n-m}) form a

[1] Hereafter, we will no longer write the explicit superscripts for dimensionality of Θ and $\mathcal{W}$; they are assumed to have dimensionality n and m, respectively.

[2] Following the terminology of [3].

[3] Though not necessarily *simply* connected; it may contain a workspace "hole".

locally trivial fiber bundle. For a redundant manipulator operating in the task space SE(3), there are a finite number of such bundles. An upper bound is known (see [3]), though the actual number depends on the particular manipulator, and is typically much smaller. For non–redundant manipulators, the number of regions is no greater than the number of inverse solutions.

3 Learning Inverse Kinematics

It can be seen immediately from the above that our goal of identifying the pre–image solution sets and parameterizing redundancy can be achieved by identifying the $\mathcal{C}_i$ and parameterizing the fibers (self–motions manifolds). Local inverse functions can then be approximated from samples of (θ, x) pairs, one function on each region. Therefore the problem of developing direct inverse kinematics functions, globally, requires solving two sub–problems; identification of the distinct $\mathcal{C}_i$ and $\mathcal{W}_i$, and then parameterizing the fibers of the resultant discrete set of locally trivial fibrations. These problems are equivalent to the problems of finding the functions $\mathcal{B}(\theta)$ and $\mathcal{T}(\theta)$, from above.

3.1 Configuration Branch Resolution

From a sufficiently large collection of input–output pairs (θ, x) we can sample the workspace at discrete locations and find all of the x points which lie within a specified neighborhood. The pre–image of this neighborhood contains the corresponding θ points, which can be grouped into disjoint pre–image regions using unsupervised learning (clustering). The distinct pre–image sets lie entirely inside of some solution branch region $\mathcal{C}_i$, and are separated by co–regular surfaces. The $\mathcal{W}_i$ can be identified as connected workspace regions. The $\mathcal{W}_i$ are typically *not* disjoint, but overlap wherever multiple solution branches exist. For each of the $\mathcal{W}_i$, all of its corresponding $\mathcal{C}_i$ can be located and identified as distinct regions.

A short summary of the procedure follows. A more detailed description of the approach can be found in [6], with application to a non–redundant manipulator (the Puma 560). Essentially, first a large number of samples of (θ, x) pairs are gathered. A local neighborhood around some point x_0 in the workspace is chosen. All θ points whose x counterpart map into the neighborhood are grouped by use of the technique of hierarchical clustering. x_1 is chosen nearby x_0, and the process repeated. The groups found are identified with the corresponding previously discovered groups. The x_i are chosen to "sweep" through the workspace (or the portion of interest). When completed, the θ points which map to a neighborhood of any of the x_i will be labeled as to group. These groups are identified with a particular solution branch, $\mathcal{C}_i$. A classifier can then easily be constructed from the labeled data, which can be used to identify the $\mathcal{C}_i$ in which any new θ point lies.

This procedure thus partitions the data into separate $\mathcal{C}_i$, solving the first sub-problem. The classifier approximates our desired function $\mathcal{B}(\theta)$.

3.2 Parameterization of Redundancy

The topology of the self–motion manifold of an all–revolute joint manipulator is known[4] *a priori*; it is T^{n-m}. There are a number of ways to parameterize torii from

[4]There are some complications, since the manifold is a submanifold of T^n, not $\mathbb{R}^n$, it may also have a non–trivial winding number corresponding to whether it "cuts" one or more generators of the

samples. One of the easiest approximations to use is to fit a deformable map of the same topology to the data. See [5] for application to the redundant three–link planar positioning manipulator. Using a topology–preserving elastic network (see [7]) we can construct an interpolation of the self–motion manifold. Nearby self–motion manifolds can have qualitatively similar parameterizations developed by using the previously learned elastic net as the initial conditions for the new manifold.

The second subproblem is solved, then, by construction of explicit parameterizations for the self–motion manifolds of a sampling of workspace locations. This paramaterization approximates our desired function $\mathcal{T}(\theta)$. $\mathcal{T}(\theta)$ is invertible; given a value of the redundancy parameters, θ can be computed.

3.3 Inverse Kinematics

Direct inverse functions can now be approximated. Given x_d, a workspace point for which joint angles are desired, the set of $\mathcal{W}_i$ in which it lies can quickly be determined. The inverse solutions for $f(\theta) = x_d$ consist of the self–motion manifolds in each of the $\mathcal{C}_i$. Explicit determination of *which* solution branch (which $\mathcal{C}_i$) to choose, and *where* on the self–motion manifold, can be made at run–time based on any arbitrary user–specified criteria. Thus for each of the $\mathcal{C}_i$ in principle we can construct the inverse function

$$f_i^{-1} : \mathcal{W} \times t^{n-m} \rightarrow \Theta$$

where f_i^{-1} consists of two parts,

$$g_i : \mathcal{W} \mapsto \mathcal{T}_i^{n-m}$$

and

$$\mathcal{T}_i^{-1} : t^{n-m} \mapsto \Theta$$

$g_i(x_d)$ identifies the self–motion manifold which projects to x_d in the workspace, and $\mathcal{T}_i^{-1}(t^{n-m})$ inverts the parameterization, and thus maps the values of the redundancy parameters to unique configuration space points.

The f_i^{-1} can be learned by application of the previously learned $\mathcal{B}(\theta)$ and $\mathcal{T}(\theta)$. The $(\theta,\, x)$ data can be partitioned into separate groups for each $\mathcal{C}_i$ using $\mathcal{B}(\theta)$; constructed e.g. as a neural network classifier, as described above. Within each group, the data can be augmented using $\mathcal{T}(\theta)$ into $(\theta,\, (x,\, t^{n-m}))$ pairs. These are samples from the invertible augmented forward kinematics function; thus function approximation methods can be applied to estimate the $(x; t^{n-m}) \rightarrow \theta$ mapping directly. In particular, feedforward neural networks can be used to approximate the direct inverse.

4 Conclusions

We have discussed how theoretical topological analysis of the robot kinematics function serves to motivate construction of approximate inverse functions over subsets of the reachable workspace, such that the entire workspace is covered (with the possible exception of boundary singular values; although the the domain of the inverse functions can be as close to the singularities as desired). We have shown how the two subproblems, global regularization of the multiple solution branches and local

torus. However, for manipulators with a single redundant degree–of–freedom, there are only a small, finite number of homotopy classes within which the self–motion manifold will fall. See [3].

regularization of the redundancy, can be approximately solved using only samples of the input–output behavior of the forward kinematics of the manipulator. Once these regularizations can be made, a *global* direct inverse function can be constructed by constructing a finite set of *local* inverse functions. At run–time, then, inverse kinematics can be computed in constant time, and the choice of solution branch and value of the redundancy parameter can be made as late as desired.

Acknowledgements

This research was partially supported by NSF Presidential Young Investigator Award IRI–9057631 and a Fellowship from the California Space Institute.

References

[1] Baker, Daniel & Charles Wampler, "On the Inverse Kinematics of Redundant Manipulators", *Intl. J. of Robotics Research* **7**, No. 2, 3–21, (March, 1988).

[2] Baker, Daniel, "Some Topological Problems in Robotics", *Math. Intelligencer* **12**, No. 1, 66–76 (1990).

[3] Burdick, Joel, "Kinematics and Design of Redundant Robot Manipulators", Stanford Ph.D. Thesis, Dept. of Mechanical Engineering, July 1988.

[4] Burdick, Joel, "On the Inverse Kinematics of the Redundant Manipulators: Characterization of the Self–Motion Manifolds", *Proc. 1989 IEEE Int. Conf. Robotics & Automation*, pp. 264–270.

[5] DeMers, David & Kenneth Kreutz–Delgado, "Learning Global Topological Properties of Robot Kinematic Mappings for Neural Network based Configuration Control", in Goldberg, Kenneth, ed. *Neural Networks for Robotics*, (to appear).

[6] DeMers, David & Kenneth Kreutz–Delgado, "Learning Global Direct Inverse Kinematics", in Moody, et al. (eds.), *Advances in Neural Information Processing Systems 4*, 589–594 (Morgan Kaufmann, 1992).

[7] Durbin, Richard & David Willshaw, "An Analogue Approach to the Traveling Salesman Problem using an Elastic Net Method", *Nature*, Vol. 326, p. 689, 1987.

[8] Guillemin, Victor & Alan Pollack, *Differential Topology*, (Prentice-Hall, 1974).

[9] Morozov, Vladimir A., *Methods for Solving Incorrectly Posed Problems* (Springer–Verlag, 1984).

[10] Ritter, Helge J., Thomas M. Martinetz, & Klaus J. Schulten, "Topology–Conserving Maps for Learning Visuo–Motor–Coordination", *Neural Networks*, Vol. 2, pp. 159–168 (1989).

Extended Impedance Control Using Real and Virtual Sensor-Based Redundancy Resolution

Mark K. Long and Paul G. Backes
Jet Propulsion Laboratory
California Institute of Technology
Pasadena, California

ABSTRACT

An approach to redundancy resolution for kinematically redundant robot manipulators is described which employs a fixed redundancy parameter and a complete task space impedance model with real and virtual sensory input. The approach can be applied to a wide variety of manipulators and specific examples are given for the seven degree-of-freedom Robotics Research Corporation K-1207 arm. The task space of the manipulator is generalized and impedance models are applied to each task space degree of freedom. Command and sensor based motion is achieved through utilizing inputs from a trajectory generator, operator input devices, force sensors, joint limit and singularity sensors, and other real and virtual sensors.

INTRODUCTION

This paper describes a method for redundancy resolution of kinematically redundant manipulators. This is one of the prototype technologies being developed in the JPL Supervisory Telerobotics (STELER) Laboratory for local/remote telerobotic operations. Telerobotics must provide reliable operations in partially structured and unstructured environments as well as provide a wide range of capabilities with modest computing power. Supervisory Telerobotic systems have a wide variety of applications including maintenance, inspection, experiment tending, and servicing. To achieve these goals the technologies of dextrous manipulation using kinematically redundant arms, active compliant control, and multi-sensor shared control have been brought together in a unified control framework. This single control structure will enable successful completion of tasks while being able to respond to unplanned scenarios, unstructured environments and anomalies.

Previous approaches to redundancy resolution include pseudoinverse and augmented (or composite) Jacobian methods. A wide variety of pseudoinverse approaches have utilized projection operators on the null space of the end-effector Jacobian to resolve the redundancy. These methods cannot in general guarantee cyclic or conservative motion and more importantly often cannot guarantee direct control of the entire mechanism. This often results in some unspecified internal motion of the mechanism. These methods also rely heavily on optimization functions to resolve the redundancy with only local results and often with objective functions of questionable utility to the task at hand. Augmented Jacobian methods focus on defining kinematic functions that provide forward and differential kinematic relationships to attempt to fully specify the motion of all degrees of freedom of the mechanism. These approaches have been fairly successful[1, 2, 3] but still have some difficulties. Often the forward kinematic relationships are difficult to compute and may not be defined over the entire workspace. Additionally the majority of these functions do not have closed form Jacobian relationships and require numerical techniques for computation. Many of the problems with artificial singularities of these new kinematic functions and discontinuities in switching from one function or subtask have been addressed using damped least squares for position controlled applications [4, 1, 2]. However, damped least squares with joint velocity weighting introduces tracking error throughout the workspace and requires the selection of weighting matrices which is often non-intuitive. Many of these issues have not been addressed for force or impedance control applications.

The approach to control of kinematically redundant manipulators described here uses a task space of the same dimension as the robot. This is desirable from the standpoint of directly controlling all degrees of freedom and eliminating uncontrolled internal motions which can pose safety, collision, and cyclicity problems. Additionally it is desirable to make use of all available degrees of freedom to complete kinematically challenging tasks in unstructured environments that may not be part of the planned suite of capabilities.

The task space parameterization does not change over time. Instead an impedance model is applied to the full dimension of the task space and shared control techniques are used to combine real and virtual sensory inputs to the impedance model. In this way, degree(s) of freedom normally referred to as redundant are unified with the task space and reference trajectories as well as sensor data cause motion of the entire mechanism. Additionally since the task space parameterization is fixed, issues of switching discontinuity do not arise and singular regions associated with the task space parameterization are fixed. The entire task space of the manipulator can be used to accomplish tasks as motion is mapped to the appropriate degrees of freedom. Thus online numerical techniques are not required to compute differential (Jacobian) relationships. When operating in these regions potential fields may be applied to the impedance model to inhibit movement in the singular or increasingly singular directions. Alternatively, if position based impedance control is being used, the damped least squares methods can be used with the tracking error producing velocity weighting only active in the singular regions. The parameterization of redundancy discussed here generally have some physical significance, however, there also exist a wide variety of kinematic arrangements that may not have a "natural" redundancy parameter that has physical significance or spans a significant portion of the workspace. In such cases one can rightly inquire into the motivation behind such a design from a task execution perspective and determine the minimum set of parameters that will completely describe the configuration of the manipulator.

This paper places emphasis on application of the techniques to the Robotics Research Corporation K-1207 manipulator, however many of the results are applicable to other kinematically redundant manipulators with various degrees of freedom. Previous work has described techniques for compliant motion control [5, 6, 7], shared control[8], and redundancy resolution [9, 1, 10, 2]. Here we bring together some of these techniques, extending the task space to 7 DOF and integrating redundancy into a shared control framework. The 7 DOF kinematics are handled using a composite Jacobian approach [9] according to the approach reported in[2]. The redundancy is parameterized by the "arm angle" [10]. The approach used for compliant control is a form of impedance control [5] most directly related to the work reported in [11]. The impedance model is extended to include an impedance model for the extra degree of freedom with a corresponding sensor input providing feedback to the model. The redundancy resolution and compliant control are integrated into the shared control framework [8] which allows the integration of real and virtual sensors into a single control framework.

EXTENDED IMPEDANCE CONTROL

First we review position based impedance control. Given a standard position based impedance trajectory filter:

$$M \cdot (\ddot{X}_d - \ddot{X}_c) + B \cdot (\dot{X}_d - \dot{X}_c) + C \cdot (X_d - X_c) = F_T \tag{1}$$

where M, B, and C are the desired impedance mass, damping, and stiffness respectively, X_d is the desired trajectory of a task space variable, X_c is the commanded task space trajectory which includes sensor input, and F_T is the total of real and virtual force sensor inputs [12]. The subset of real and virtual sensors is specified in the input parameter set, and could, for example, include end-effector forces, operator inputs, and joint limit potentials. F_T is discussed below.

In the current implementation, the joint commands are given in velocities. Given $\ddot{X}_d$, $\dot{X}_d$, X_d, and F_T from shared control(to be reviewed below), and M, C and B constants the above general impedance formulation results in the following trajectory filter:

$$\begin{aligned} \ddot{X}_c(N+1) &= \ddot{X}_d(N+1) + M^{-1} \cdot \{B \cdot [\dot{X}_d(N+1) - \dot{X}_c(N)] \\ &\quad + C \cdot [X_d(N+1) - X_c(N)] - F_T(N+1)\} \\ \dot{X}_c(N+1) &= \dot{X}_c(N) + \ddot{X}_c(N+1) \cdot \Delta t \end{aligned} \tag{2}$$

Equation 3 is a scalar equation and can be applied easily to the linear velocity of the end-effector as well other scalar task space parameters, such as additional task space parameters defined for kinematically redundant manipulators. For the Robotics Research arm, the additional task that parameterizes the redundancy is the "arm angle" (described below). Care must be taken in the implementation of Equation 3 for angular velocity. The representation of orientation, the differencing of orientation quantities, as well as the units of C must be chosen consistently. The current implementation uses quaternions for representation of orientation and the operations are defined accordingly [13].

REDUNDANCY RESOLUTION

While the classical problem of end-effector position and orientation control for a "spatial" manipulator can be handled by a six DOF robot arm many tasks require the use of all available degrees of freedom. For example to completely control all seven degrees-of-freedom of the Robotics Research arm, the task space of the manipulator is parameterized by the position and orientation of the end-effector and the "arm angle". The "arm angle", represented by ψ, is defined by the angle between the plane passing through the shoulder, elbow and wrist point of the manipulator and some reference plane[10]. A composite Jacobian is formed from the individual Jacobians that relate the rate of change of the joint angles to the rate of change of the task parameters. Here the composite Jacobian, J^C, is given by [2]:

$$J^C_{7\times7} = \begin{pmatrix} J^{\omega}_{3\times7} \\ J^{v}_{3\times7} \\ J^{\psi}_{1\times7} \end{pmatrix} \tag{3}$$

where J^{ω} is the angular velocity Jacobian, J^v is the linear velocity Jacobian, and J^{ψ} is the arm angle Jacobian. J^{ω} and J^v are readily available using [10]:

$$\begin{pmatrix} J^{\omega}_r \\ J^{v}_r \end{pmatrix} = \begin{pmatrix} \hat{\mathbf{z}}_1 & \hat{\mathbf{z}}_2 & \cdots & \hat{\mathbf{z}}_7 \\ \hat{\mathbf{z}}_1 \times \mathbf{P}_{1,r} & \hat{\mathbf{z}}_2 \times \mathbf{P}_{2,r} & \cdots & \hat{\mathbf{z}}_7 \times \mathbf{P}_{7,r} \end{pmatrix} \tag{4}$$

where r is the velocity reference point, $\hat{\mathbf{z}}_i$ is the z axis of joint i, and $\mathbf{P}_{i,r}$ is the position vector from the ith link frame to the velocity reference point r. See [10]. The arm angle Jacobian is available from [10, 2]:

$$J^{\psi} = \frac{(\hat{\mathbf{w}} \times \mathbf{p})^T}{\| \mathbf{p} \|^2}\mathbf{E} + \left\{ \frac{\hat{\mathbf{V}}^T \mathbf{w}}{\| \mathbf{h} \|^2}(\hat{\mathbf{w}} \times \mathbf{h})^T - \frac{\hat{\mathbf{w}}^T \mathbf{e}}{\| \mathbf{w} \| \| \mathbf{p} \|^2}(\hat{\mathbf{w}} \times \mathbf{p})^T \right\} \mathbf{W} \tag{5}$$

where $\mathbf{w} = \mathbf{P}_{0,7}$; $\mathbf{e} = \mathbf{P}_{0,4}$; $\mathbf{p} = \mathbf{e} - \hat{\mathbf{w}}(\hat{\mathbf{w}}^T \mathbf{e})$; $\mathbf{V}_T$ is the vector specifying the reference plane; $\mathbf{h} = (\mathbf{w} \times \hat{\mathbf{V}}) \times \mathbf{w}$; $\mathbf{E}$ is the elbow linear velocity Jacobian, and $\mathbf{W}$ is the wrist linear velocity Jacobian. Notice that most of the required data for J^C is available as a by product from a forward kinematic iteration [2]. The composite Jacobian approach extends the task space to completely specify all of the degrees of freedom of the manipulator. Thus the manipulator is no longer treated as redundant.

SHARED CONTROL

Shared Control [8] is a technique for combining sensor based motion, teleoperation and autonomous control inputs in a single control structure. Inputs from the operator and real and virtual sensors are merged to give a combined sensor input to the control system either through F_T or through modifying the desired trajectory X_d from the reference trajectory X_r which is supplied by the trajectory generator. A single control structure is used that is configured by an input parameter set to allow a full spectrum of control modes from teleoperation, supervised autonomy to full autonomous task execution.

Typically, the operator uses a hand-controller, joystick or other input device to specify manipulator task space trajectories or modifications to nominal trajectory currently being provided by the system trajectory generator or the operator inputs can be treated as sensor inputs which see the end-effector as an impedance. Compliant teleoperation is a special case where the reference trajectory has zero velocity and the trajectory is completely specified by the operator. Note that through the input parameter set the operator may select to supply input to all degrees of freedom (7 in the case of the RRC K-1207) or only a subset.

Real sensors such as force/torque sensors and proximity sensors are used to input forces to the impedance model, with the ultimate result of modifying the state commands to the arm. This is the normal mode of operation for active compliance control. A proximity sensor for the elbow or sensor skin can be used to detect obstacles and provide force input to the impedance model to have the arm comply with obstacles in the workspace. If so specified by the input parameter set this sensor data can be used to modify the nominal trajectory in the velocity domain.

Virtual sensors allow other operational conditions and events to be handled in a unified manner. The most common virtual sensor is the joint limit and kinematic singularity sensor. This sensor applies a virtual force to the full impedance model (7 DOF force) in a direction that moves the manipulator away from the joint limit or singularity. In practice, this method has been found the most desirable way of dealing with joint limits and singularities when the operator is directly commanding the arm. As the operator moves the arm near a joint limit or singular workspace region, instead of a fault occurring or the arm being stopped, the virtual force generated by the sensor keeps the joint from its limit. See [12]. Other virtual sensors include dithers that serve to alleviate friction problems in autonomous pin insertion and sensors associated the arm angle or other task parameters beyond the position and orientation of the end-effector. For example, a world model data-base can provide a potential field for obstacle avoidance that can be used as an input to the arm angle impedance model to utilize the kinematic redundancy to avoid workspace obstacles. Other virtual sensors can represent potential fields of virtual forces that position the arm angle in posture with desirable force or velocity transmission properties either locally or computed globally based on the arm destination and contact location.

Thus the terms F_T and $\dot{X}_d$ in the above control equations are a sum of operator, real sensor and virtual sensor inputs. With a unified control framework the control mode can be specified by an input parameter set to a single algorithm without the need to flight qualify multiple control algorithms and switch control software on the fly.

$$\dot{X}_d = \dot{X}_r + \sum \dot{X}_{Sensor} \quad \text{and} \quad F_T = (F_r - F_a) + \sum F_{Sensor} \tag{6}$$

where $\dot{X}_r$ and F_r are the reference velocity and force respectively and F_a is the measured actual end-effector force. $\dot{X}_{Sensor}$ and F_{Sensor} are the additions to the nominal force and velocity commands from all real and virtual sensor input.

INTEGRATION

This formulation of compliance control, redundant arm control, and shared control in a unified framework with real and virtual sensors provides the ability to execute tasks requiring any permutation of control modes available from the various sensors sources and task degrees of freedom. In addition to specifying the processing mode for each of the real and virtual sensors, the input parameter set has the capability to specify a unique coordinate frame in which each of the sensors act. For example, if the current task requires opening a door, while the gripper may be grasping the handle, through shared control a single degree of freedom of the operator's hand controller may be placed at the door hinge. Thus the motion of the operator turning about the hinge is transformed from the hinge to the door grasp point and the end-effector is commanded to travel in the appropriate arc. The control mode can be specified in a variety of ways. Since the control configuration parameter set can be quite large, often a "macro interface" is used to compactly represent the parameters for various tasks. See [14].

Now with the task space command vector $\dot{X}_c$ and the task space to joint map J^C the joint servo commands can be computed. Although Jacobian inverse routines could be used, a damped-least squares inverse is used to allow further task prioritization and singularity robustness. Once the control configuration is specified the input to the joint servos can be computed using Equations 6,3, and 3 and the damped-least squares [2, 4]:

$$\dot{\theta}_d = [J^T \cdot W_T \cdot J + W_V]^{-1} \cdot J_T \cdot W_T \cdot \dot{X}_c \tag{7}$$

where W_T is a diagonal task weighting matrix that relates the relative priorities of the tasks. W_V is a diagonal velocity weighting matrix which weights the norm of joint velocities. It is important to note that while a non-zero W_V matrix will provide robustness to singularities by limiting excess joint velocities at the expense of end-effector tracking, it will also induce tracking error. By setting W_T to identity and W_V to zero a standard inverse Jacobian result is retrieved with the same algorithm.

CONCLUSIONS

The compliant control scheme for kinematically redundant manipulators presented in this paper provides a framework for supervisory and shared control task execution.

Capabilities provided by this approach include complete utilization of the manipulator degrees of freedom, the use of real and virtual sensors for control, and flexibility to complete a wide range of planned and unplanned tasks in unstructured environments. The proposed scheme alleviates the need to flight qualify multiple controllers and switch control for various tasks through the use of a single control methodology that provides a rich parameters set for selecting control modes. This control scheme has been implemented in the laboratory and is currently undergoing test and evaluation.

ACKNOWLEDGMENTS

The work described in this paper was performed at the Jet Propulsion Laboratory, California Institute of Technology, under contract with the National Aeronautics and Space Administration for the Space Station Freedom Engineering Prototype Development Program.

References

[1] H. Seraji, Mark K. Long, and T. Lee. Motion control of 7 dof arms: The configuration control approach. In *IEEE Int. Conf. on Robotics and Automation*, 1991.

[2] Mark K. Long. Task directed inverse kinematics for redundant manipulators. *Journal of Intelligent and Robotic Systems*, (To Appear), 1992.

[3] Wyatt S. Newman and Mark E. Dohring. Augmented impedance control: An approach to compliant control of kinematically redundant manipulators. In *IEEE Int. Conf. on Robotics and Automation*, 1991.

[4] H. Seraji and R. Colbaugh. Improved configuration control for redundant robots. *Journal of Robotic Systems*, 7(6):897–928, 1990.

[5] N. Hogan. Impedance control: An approach to manipulation: Part i — theory. *ASME Journal of Dynamic Systems, Measurement, and Control*, 107:1–7, March 1985.

[6] Dale A. Lawrence and R. Michael Stoughton. Position-based impedance control: Achieving stability in practice. In *Proceedings AIAA Conference on Guidance, Navigation, and Control*, pages 221–226, Monterey, Ca, August 1987.

[7] Paul G. Backes. Generalized compliant motion with sensor fusion. In *Proceedings 1991 ICAR: Fifth International Conference on Advanced Robotics, Robots in Unstructured Environments*, pages 1281–1286, Pisa, Italy, June 19–22 1991.

[8] Paul G. Backes. Documentation of space station telerobotics system technology: Shared control. volume ii: Detailed technical description — generalized compliant motion with shared control. Technical Report D-8575 (internal document), Jet Propulsion Laboratory, April 1991.

[9] S.Y. Oh, D. Orin, and M. Bach. An inverse kinematic solution for kinematically redundant robot manipulators. *Journal of Robotics Systems*, 1(3):235–249, 1984.

[10] K. Kreutz-Delgado, M. Long, and H. Seraji. Kinematic analysis of 7 dof manipulators. *JPL Engineering Memorandum No.347-90-275*, January 1990.

[11] Mark K. Long and Paul G. Backes. Impedance based shared control of a redundant robot. In *1992 IASTED International Conference on Control and Robotics*, Vancouver, British Columbia, August 1992.

[12] Paul G. Backes. Multi-sensor based impedance control for task execution. In *IEEE Int. Conf. on Robotics and Automation*, Nice, France, May 1992.

[13] Mark K. Long. Trajectory generator and robot control for 1992 supervisory telerobotics laboratory. Technical Report 3473-92-001, Jet Propulsion Laboratory Engineering Memorandum, 1992.

[14] Paul G. Backes and Kam S. Tso. Umi: An interactive supervisory and shared control system for telerobotics. In *Proceedings IEEE International Conference on Robotics and Automation*, pages 1096–1101, Cincinnati, Ohio, May 1990.

Optimization of the Dynamic Performance of Redundant Robots in the Presence of Faults†

Christopher L. Lewis Anthony A. Maciejewski

School of Electrical Engineering
Purdue University
West Lafayette, Indiana 47907

ABSTRACT

Robots working in remote or hazardous environments require additional measures to ensure their usability upon the failure of an actuator. This paper uses the concept of worst-case dexterity to define kinematic and dynamic fault tolerance measures for redundant robots. These measures are then used to specify the operating configuration which is optimal in the sense that the robot's dexterity remains high even if one of the joints fails.

I. INTRODUCTION

The use of robots in remote and/or hazardous environments greatly increases the cost and in some cases eliminates the possibility of repairing failed actuators [1],[2],[6]. For this reason kinematically redundant robots have been proposed for these applications. The additional degrees of freedom permit a degree of dexterous motion of the end effector even after an actuator has failed. The extra degrees of freedom may also be used, prior to failure, to maintain configurations that minimize the loss of dexterity resulting from the failure of an actuator [3],[5]. To optimize this kind of fault tolerance for redundant robots, the notion of dexterity must be quantized.

Heuristically, the dexterity of a manipulator is described as the ease in which motion of the end effector can be accomplished. Since end effector velocity is controlled indirectly through the joint rates, the notion of kinematic dexterity is defined as a quantification of the relationship between the joint rates and the end effector's velocity. Taking this relationship one step further, the notion of dynamic dexterity is a quantification of the relationship between the actuator torques and the end effector's acceleration. This paper will take a look at the various measures of kinematic and dynamic dexterity. Then, to form measures of fault tolerance, these dexterity measures will be adapted to include the additional assumption that any one of the manipulator's actuators could fail. The minimum reduced kinematic and dynamic dexterity resulting from a single joint failure is defined to be the respective measure of kinematic and dynamic fault tolerance. A configuration wherein the robot retains a high dexterity for all possible single joint failures is said to be fault tolerant. Using these measures of fault tolerance a three-link planar mechanism is examined and comparisons are drawn between these two measures.

II. KINEMATIC AND DYNAMIC DEXTERITY

The mathematical relationship between the joint rates, $\dot{\theta}$, and the end effector velocity, $\dot{x}$, is given by the linear Jacobian equation

$$J\dot{\theta} = \dot{x}. \tag{1}$$

† This work was supported in part by Sandia National Laboratories under contract number 18-4379B. Additional support was provided by the National Science Foundation under grant CDR 8803017 to the Engineering Research Center for Intelligent Manufacturing Systems.

The Jacobian, $J \in R^{m \times n}$, like all matrices may be decomposed using the singular value decomposition (SVD) so that

$$J = U \Sigma V^T \tag{2}$$

where $U \in R^{m \times m}$ and $V \in R^{n \times n}$ are orthogonal matrices and $\Sigma \in R^{m \times n}$ is a diagonal matrix consisting of the singular values, denoted σ_i, which are typically given in descending order. The SVD representation of J is convenient due to the physical interpretation of the singular values and vectors. The singular values define the size of the possible motions of the end effector due to any given joint rate having a unit norm. The right singular vectors, v_i, describe the direction, or normalized linear combination of joint rates, which cause a motion of size σ_i in the unit u_i direction of Cartesian space.

The singular values have been used to form a variety of kinematic dexterity measures for manipulators. For instance, the ratio of the largest to the smallest singular value is the condition number of J [4]. Physically, it is clear that the condition number is a measure of the eccentricity of the velocity ellipsoid. Another popular measure is the product of all of the singular values, the determinant of J, which is proportional to the volume of the velocity ellipsoid [7]. The sum of the singular values, the trace of J, has also been used. If the Cartesian trajectory were known *a priori* then the distance to the edge of the ellipsoid in the direction of the known trajectory is a path specific dexterity measure. Finally, the smallest singular value σ_m of the Jacobian is a measure of the worst-case available velocity. Throughout the remainder of this work the worst-case kinematic dexterity, denoted by km and defined as

$$km = \sigma_m(J), \tag{3}$$

will be used as a measure of kinematic dexterity.

Unfortunately the relationship between actuator torques and end effector acceleration is not as simple. Following the development in [8], the dynamic equations of motion for robotic manipulators are generally given by

$$M(\theta)\ddot{\theta} + h(\theta, \dot{\theta}) + g(\theta) = \tau \tag{4}$$

where $M(\theta) \in R^{n \times n}$ is the moment of inertia matrix, $h(\theta, \dot{\theta}) \in R^n$ is a vector of torques caused by the centrifugal and Coriolis forces, $g(\theta) \in R^n$ is the vector of torques due to gravity and τ is the resulting vector of joint torques. It is well known that the inertia matrix, $M(\theta)$, for realistic systems is at all times invertible. The relationship between joint accelerations and the end effector acceleration is given by

$$J\ddot{\theta} + \dot{J}\dot{\theta} = \ddot{x} \tag{5}$$

which is obtained by differentiating (1). Since we are only concerned with optimal configurations, $\dot{\theta}$ may be set to zero. Thus by combining (4) and (5) one obtains

$$\ddot{x} = JM^{-1}(\tau - g) \tag{6}$$

which describes the mapping between joint torques and end effector acceleration. Noting that $JM^{-1}g$ is a uniquely determined vector at every point in the joint space, the applied torque to end effector acceleration mapping may be characterized in a manner that is analogous to the kinematic mapping. However, the acceleration ellipsoid is shifted so that it is centered at $JM^{-1}g$.If the manipulator is in motion, the Coriolis and centrifugal forces have a similar shifting effect on the acceleration ellipsoid. However, neither gravity nor

Coriolis forces change the shape of the ellipsoid. Therefore, the the worst-case dynamic dexterity, dm, is defined as the length of the smallest axis of the acceleration ellipsoid, and is given by

$$dm = \sigma_m(JM^{-1}). \tag{7}$$

III. A DEFINITION OF FAULT TOLERANCE

The kinematic and dynamic dexterity measures defined in the previous section will now be extended to form a measure of fault tolerance for redundant manipulators. It will be assumed that failed actuators can can be locked in their failed position. Fault tolerance may thus be quantified in terms of the velocity and acceleration ellipsoids that result from failing actuators. The minimum remaining dexterity over all possible joint failures is defined to be the measure of fault tolerance.

In the kinematic case, the Jacobian that relates the joint rates of the manipulator to the end effector velocity with the fth actuator locked, denoted by ${}^fJ$, is given by

$${}^fJ = [\, j_1 \quad \cdots \quad j_{f-1} \quad j_{f+1} \quad \cdots \quad j_n \,] \tag{8}$$

where the column j_f has been removed and where the remaining j_i's are identical to their values in J. The kinematic fault tolerance measure, denoted kfm, can now be given in terms of the failure Jacobians as

$$kfm = \min_{f=1\ldots n} \sigma_m({}^fJ). \tag{9}$$

The dynamic fault tolerance measure is quite similar in structure to the kinematic measure. It is the minimum of the dynamic measures of the system reduced by locking one of the joints. The dynamic equations of motion for a manipulator with the fth actuator locked have the form

$${}^fM\,{}^f\ddot{\theta} + {}^fh + {}^fg = {}^f\tau \tag{10}$$

where ${}^fM$ is the inertia tensor of the original system with the fth row and column removed, ${}^f\theta$ is the vector of the remaining joint variables, and ${}^fh$ is the Coriolis and centrifugal forces arising from the dependence of M on θ and is algebraically similar to h with $\dot{\theta}_f$ set to zero and the fth element eliminated. The gravitational force, ${}^fg$, is g with the fth element eliminated and ${}^f\tau$ is the vector of remaining joint torques. A dynamic fault tolerance measure, denoted dfm, may be defined from the dynamic dexterity of the reduced system as

$$dfm = \min_{f=1\ldots n} \sigma_m({}^fJ\,{}^fM^{-1}). \tag{11}$$

IV. A 3DOF PLANAR EXAMPLE

In this section, the fault tolerance of a three-link planar manipulator is studied. All three links are chosen to have length $l_i = \frac{1}{3}$ m and moments of inertias $I_i = \frac{1}{12}m_i l_i^2$. Two different mass ratios are compared. In the first case, the link masses are $m_i = \frac{1}{3}$kg. In the second case, $m_1 = .8$kg and $m_2 = m_3 = .1$kg. Note that variations in the mass ratio are used since dm is inversely related to the total mass. The kinematic and dynamic dexterity and fault tolerance measures were evaluated throughout the joint space. The results are given in the form of contour plots in Figs. 1-3. Note that only the variables θ_2 and θ_3 need to be considered since all of the measures are independent of θ_1.

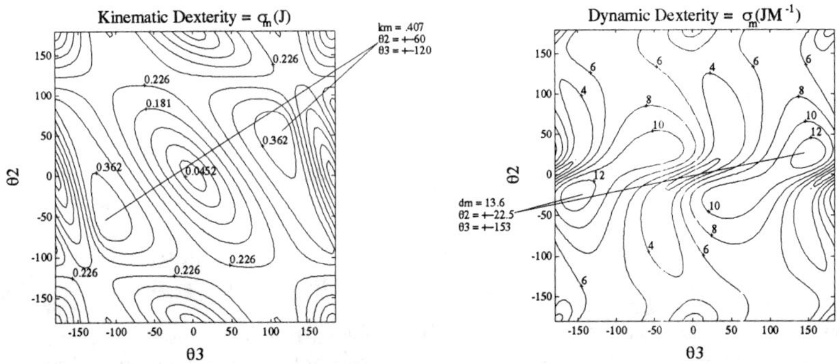

Fig. 1 The Worst-Case Dexterity of a 3 DOF Robot
$(l_1 = l_2 = l_3 = \frac{1}{3}m,\ m_1 = m_2 = m_3 = \frac{1}{3}kg)$

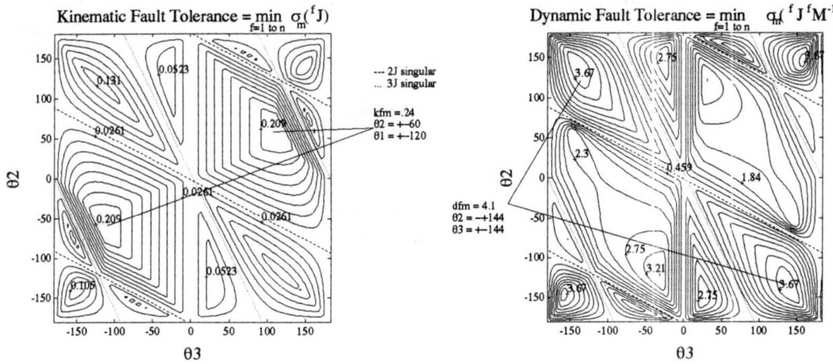

Fig. 2 Kinematic and Dynamic Fault Tolerance of a 3 DOF Robot
$(l_1 = l_2 = l_3 = \frac{1}{3}m,\ m_1 = m_2 = m_3 = \frac{1}{3}kg)$

The kinematic and dynamic dexterity measures for the first mass ratio case are shown in Fig. 1. The minima for both measures occur at the kinematic singularities which can be identified by $\sin\theta_2 = \sin\theta_3 = 0$. Clearly, the zeros of the dynamic dexterity measure must match those of the kinematic measure since they are both due to a rank deficiency in J. The effect of M^{-1} in the definition of dm is basically limited to skewing the contours around these fixed minima. It is interesting to note, however, that the maxima for both measures are also relatively close to each other. The robot configurations for these maxima are illustrated in Fig. 4 (a) and (c).

The kinematic and dynamic fault tolerance measures for this robot are shown in Fig. 2. By comparing Fig. 1 with Fig. 2, the most striking feature is that the point minima at the singularities have expanded to become several valleys of minima that partition the joint space. Physically, these minima represent all of the singularities of the resulting two-link robots that can occur after an arbitrary joint failure. Thus it is easy to geometrically describe the conditions that they represent. For example, the singularity which occurs on the line $\theta_3 = 0$ is due to a failure of joint 1, which puts the resulting two-link robot into a reach singularity since j_2 and j_3 are co-linear. Likewise if $\theta_3 = -2\theta_2$ then the end effector lies on the same line as the first link and j_1 and j_2 are co-linear, thus resulting in a singularity if joint 3 fails. A similar geometric argument describes the lines of singularities associated with the loss of joint 2.

The existence of so many minima that partition the joint space has a profound effect on the resulting dynamic fault tolerance. Since the zeros of these two measures must match, there is a far less pronounced skewing effect due to M^{-1} when calculating dfm. Thus for all practical purposes the robot is restricted to a small region of the joint space if it is to maintain some level of fault tolerance. It is interesting to note, however, that the globally optimal dynamic fault tolerant configuration is not in the same partition of joint space as the globally optimal kinematic fault tolerant measures (see Fig. 4 (b) and (d)). However, this is a minor point since the local maxima in each region are comparable. It is also interesting to note that despite the radical difference between the contours of km and kfm, both of their optimal configurations occur at the same point, which is isotopic [4] (see Fig. 4 (a) and (b)). This is fortunate since one would not like to sacrifice the current normal performance of the robot in anticipation of a future joint failure.

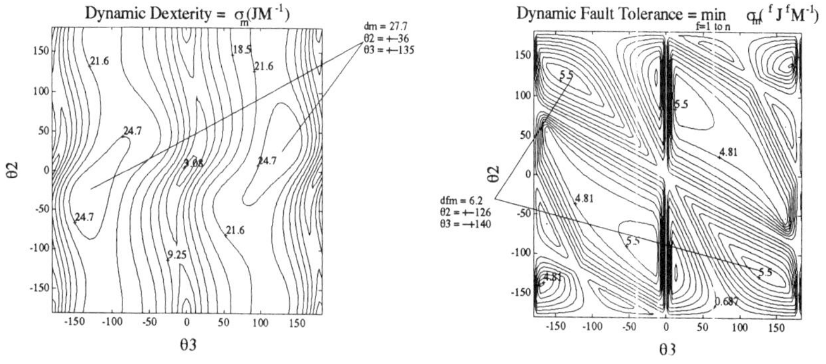

Fig. 3 Dynamic Dexterity and Fault Tolerance of a 3 DOF Robot

$$(l_1 = l_2 = l_3 = \tfrac{1}{3}m,\ m_1 = .8kg, m_2 = m_3 = .1kg)$$

Finally, Fig. 3 illustrates the effect of changing the distribution of mass on the dynamic measures. Since the link lengths are the same as in the previous case, the kinematic measures are identical to those in Figs. 1 and 2. The contours of dm are now quite different from the previous case, however, the optimal configuration has not substantially changed (see Fig. 4 (c) and (e)). Moving more of the mass towards the base of the robot tends to reduce the effect of the dynamics and therefore makes dm take on more of the characteristics of km. The same is true for the dynamic fault tolerance measure although the lines of singularities are still by far the most defining feature.

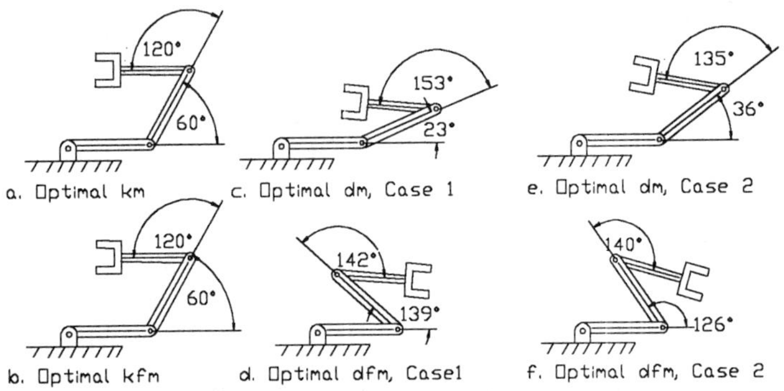

Fig. 4 Optimal km, dm, kfm, and dfm Configurations
(Case 1: $m_1 = m_2 = m_3 = \frac{1}{3}kg$, Case 2: $m_1 = .8kg, m_2 = m_3 = .1kg$)

V. CONCLUSIONS

This work has extended the notions of kinematic and dynamic dexterity to define measures of kinematic and dynamic fault tolerance for redundant robotic manipulators. These measures were then used to identify operating regions in which the robot remains usable even when one of the joints must be locked due to a failure. It was shown that the minima of the kinematic fault tolerant measure were the single most important factor in defining such regions.

REFERENCES

[1] H. Arai and S. Tachi, "Position Control of a Manipulator with Passive Joints Using Dynamic Coupling," *IEEE Trans. Robotics Automat.*, vol. 7, no. 4, pp. 528–534, 1991.

[2] B. Christensen, W. Drotning, and S. Thunborg, "Model-Based, Sensor-Directed Remediation of Underground Storage Tanks," *J. Robotic Sys.*, vol. 9, no. 2, pp. 145–159, 1992.

[3] R. Colbaugh and M. Jamshidi, "Robot Manipulator Control for Hazardous Waste-Handling Applications," *J. Robotic Sys.*, vol. 9, no. 2, pp. 215–250, 1992.

[4] C. A. Klein and T. A. Miklos, "Spatial Robotic Isotropy," *Int. J. Robotics Res.*, vol. 10, no. 4, pp. 426–437, 1991.

[5] A. A. Maciejewski, "The Design and Control of Fault Tolerant Robots for Use in Hazardous or Remote Environments," *Robotics and Remote Systems: Proc. Fourth American Nuclear Society Topical Meeting*, pp. 633–642, Albuquerque, NM, Feb. 24-28, 1991.

[6] E. Papadopoulos, and S. Dubowsky, "Failure Recovery Control for Space Robotic Systems," *Proc. 1991 American Control Conf.*, pp. 1485–1490, Boston MA, June 26-28, 1991.

[7] T. Yoshikawa, "Manipulability of Robotic Mechanisms," *Int. J. Robotics Res.*, vol. 4, no. 2, pp. 3–9, 1985.

[8] T. Yoshikawa, "Dynamic Manipulability of Robot Manipulators," *J. Robotic Sys.*, vol. 2, no. 1, pp. 113–124, 1985.

A PARALLEL REDUNDANT MANIPULATOR : SPEED-R-MAN AND ITS CONTROL

C. REBOULET, C. LAMBERT, N. NOMBRAIL

CERT-ONERA - Automatic Control Department
2 avenue Edouard Belin (BP 4025)
31055 - TOULOUSE Cedex (France)

ABSTRACT

Attention have focused for many years on parallel manipulators due to their intrinsic capabilities, structural stiffness and light weight of mobile parts. Relatively recent work, starting with [9] has demonstrated the possibility of carrying out purely translation movements with the help of new parallel structure morphologies. The results presented in this paper continue in this line and SPEED-R-MAN is the latest development on parallel manipulators in CERT. After a brief description of this new robot, organised around a kinematically redundant parallel structure, an analysis based on the dynamic manipulability index is presented. The basic idea is to split the problem into two sub-tasks: the principal manages the end-point kinematics and the second continuously chooses the best posture, that is, the one which produces the greatest acceleration possible taking into account the direction of the trajectory and the load transported.

Keywords : Parallel manipulator - Redundant Manipulator - Dynamic Manipulability index

1. INTRODUCTION

In recent years, a great deal of research has been carried out into means of improving performance in the field of trajectory speeds. Even though this problem is of capital importance for many applications, surprisingly it can be seen that there are very few robots on the market capable of manipulating light-weight objects at high speeds (three cycles/second) in a large work-space. Because of their rigidity and lightness, parallel structures are better adapted to the resolution of this type of problem.

This interest in parallel architectures is not new. Among earlier authors, [1] followed by [2], [3] and [4] may be mentioned. Much of this early work, however, made use of morphologies which were better adapted to acceleration and/or effort control. They were often used to carry out fine manipulation tasks calling for a certain dexterity [5], [6], [7], [8]. It is much later that applications requiring high accelerations have been considered.

Relatively recent work, starting with [9], has demonstrated the possibility of carrying out purely translation movements with the help of new parallel structure morphologies. Other works have since presented equally original architectures [10], [11].

The results presented in this paper continue in this line and describe a new robot manipulator morphology organised around a kinematically redundant parallel structure. Thus, for a given position of the end effector, there exists an infinity of possible attitudes.

Several performance indices have been proposed to solve the problem of redundancy [12], [13]. In this paper, the analysis described uses the dynamic manipulability index. Then, an original control structure is deduced, specially adapted to redundant kinematics. The basic idea is to split the problem into two sub-tasks : the principal manages the end-point kinematics and the second continuously chooses the best posture, that is, the one which produces the greatest acceleration possible taking into account the direction of the trajectory and the load transported.

2. PRESENTATION OF THE KINEMATIC STRUCTURE

Figure 1 shows the morphology of the Speed-R-Man robot manipulator. The property which permits purely translation movements is obtained thanks to an original combination of three steel band transmissions. This arrangement allows the end-effector to maintain a constant orientation. Moreover, only traction or compression efforts are transmitted to the end-effector. The three connecting rods which transmit these efforts are in carbon fibre and weigh less than 100 grammes.Thus, with the exception of the six actuators grouped on the fixed support, the mass in movement is very low and does not exceed 500 grammes when unloaded.

Finally, each sub-arm is composed of a connecting rod driven by a set of two pneumatic actuators. In fact, this robot manipulator can be considered as a kind of multi-arm system acting on a single end-effector.

Despite the presence of six actuators, this robot has only three external degrees of freedom ; it is, therefore, kinematically redundant. As will be seen, this particularity makes possible a notable gain in the possibilities of acceleration.

Figure 1 : Experimental set-up

3. DYNAMIC ANALYSIS

For the purposes of the mathematical analysis, a set of two actuators is considered to be equivalent to a rotary system (crank) with an adjustable length r_i. Each rotary system has an inertia of I_{V_i} and drives the end-effector, mass M_T, through a connecting rod.

Figure 2 represents the simplified skeleton of a sub-arm in the plane.
Notice that the centres of rotation of all the joints of the sub-arm (i) have been displaced by a quantity $\overrightarrow{C_iO}$ so as to facilitate the mathematical study without prejudicing the validity of the representation.

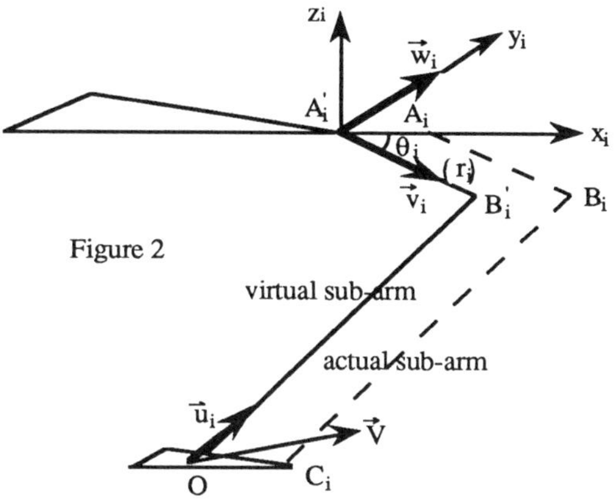

Figure 2

$\vec{u}_i$ unit vector of the connecting rod i

$\vec{v}_i$ unit vector of the crank i

$\vec{w}_i$ unit vector of the rotation axis of the crank i

$\dot{\theta}_i$ rotation speed of the revolute joint i

$\dot{r}_i$ translation speed of the prismatic joint i

As to the geometrical modelisation, it can be reduced to the case of the DELTA structure. Consequently, the results are similar to those presented in [15] or [10].

We shall now analyse the dynamic manipulability. To do this, the inertia matrix as well as the Jacobian matrix must be established.

3.1. Jacobian matrix.

If the velocities are projected according to the direction of the ith connecting rod, we obtain :

$$\vec{V} \bullet \vec{u}_i = \vec{V}_{B'_i} \bullet \vec{u}_i \tag{1}$$

where :

$$\vec{V}_{B'_i} = \dot{r}_i\, \vec{v}_i + r_i\, \dot{\theta}_i \left(\vec{w}_i \times \vec{v}_i\right) \tag{2}$$

Relation (1), projected onto a reference coordinates frame gives

$$\dot{p}_i = \mu_i\ \dot{r}_i + \lambda_i\, r_i\, \dot{\theta}_i \tag{3}$$

if

$$\begin{cases} \mu_i = \vec{v}_i \bullet \vec{u}_i & (4) \\ \lambda_i = \left(\vec{w}_i \times \vec{v}_i\right) \bullet \vec{u}_i & (5) \end{cases}$$

$\dot{p}_i$ represents the speed in the direction of connecting rod i which can be written

$$\dot{p}_i = u_i^T\ \dot{X} \tag{6}$$

where $\dot{X}$ is a vector formed by the components of the speed vector $\vec{V}$:

$$\dot{X}^T = [\dot{x}\ \ \dot{y}\ \ \dot{z}] \tag{7}$$

If U is the matrix formed by the unit vectors u_i :

$$U = [u_1\ \ u_2\ \ u_3]$$

then relation (6) is written

$$\dot{p} = U^T\, \dot{X} \tag{8}$$

Matrix formulation of relation (3) gives

$$\dot{p} = J\, \dot{q} \tag{9}$$

with
$$J = [\mu \ \ \nu]$$
$$\dot{q}^T = [\dot{r} \ \ \dot{\theta}]$$

where μ (resp. ν) is a diagonal matrix whose elements are μ_i(resp. ν_i), with :
$$\nu_i = \lambda_i \, r_i$$

The generalised pseudo-inverse [16] is written
$$\dot{q} = J^+ \dot{p} + (J^+ J - I)\,\kappa \qquad (10)$$
where
$$J^+ = J^T (JJ^T)^{-1}$$
and
$$k^T = [k1 \ \ k2] \ \text{ is any arbitrary vector.}$$

The second term of (10) represents the projection of k on the nul-space of J.

From (10), it is easy to deduce
$$\dot{r}_i = \frac{1}{\mu_i^2 + \nu_i^2}\left[\mu_i \, \dot{p}_i - \nu_i^2 \, \kappa_{1i} + \mu_i \, \nu_i \, \kappa_{2i}\right] \qquad (11)$$

$$\dot{\theta}_i = \frac{1}{\mu_i^2 + \nu_i^2}\left[\nu_i \, \dot{p}_i + \mu_i \, \nu_i \, \kappa_{1i} - \mu_i^2 \, \kappa_{2i}\right] \qquad (12)$$

Relations (8) and (9) immediately give the expression of the inverse Jacobian matrix in the particular case of $\dot{r} = 0$:
$$\dot{\theta} = J_0^{-1} \dot{X} \qquad (13)$$
with
$$J_0^{-1} = \nu^{-1} \, U^T \qquad (14)$$

The ith line of the inverse Jacobian matrix is thus written
$$\dot{\theta}_i = \frac{1}{\nu_i} u_i^T \dot{X} \qquad (15)$$

It is also easy to deduce from (3) the internal motions of the joints that do not produce an end-point motion. This gives :
$$\dot{\theta}_i = - \frac{\mu_i}{\nu_i} \dot{r}_i \qquad (16)$$

This relation is very important as it will be used to decouple two sub-tasks. It enables the posture of the manipulator to be modified without modifying the end point motion which is the essential task.

Now, we will create the inertia matrix in operational space.

3.2. Inertia matrix.

Let I_V be the diagonal matrix whose elements are the moments of inertia I_{Vi} about the axes y_i of the rotating elements. Let Γ be the torque vector.

If we neglect the Coriolis terms and the velocity and gravity related terms, which have little influence in the case of a very light parallel robot, the dynamic model in the operational space can be written :

$$J_o^{-T}\,\Gamma = M_x\,\ddot{X} \tag{17}$$

where the inertia matrix, M_X, is written :

$$M_x = M_T + J_o^{-T}\,I_v\,J_o^{-1} \tag{18}$$

where M_T is the diagonal matrix whose elements are the end effector mass.

3.3. Posture adaptation.

The unit sphere defined by :

$$\|\Gamma\|^2 \leq 1 \tag{19}$$

is mapped into an ellipsoid defined by :

$$\ddot{X}^T\,M_X^T\,J_0\,J_0^T\,M_x\,\ddot{X} \leq 1 \tag{20}$$

Apart from the fact that it uses the inertia matrix in Cartesian space, this ellipsoid is similar to that defined by [17]. It is called «dynamic manipulability ellipsoid ».

The basic idea, therefore, consists in varying the posture of each sub-arm so as to modify the form and the orientation of the ellipsoid and thus maximise the acceleration in the direction specified by the task.

Suppose the end effector is moving in the direction defined by the unit vector s ; we call α the distance from the center to the surface of the ellipsoid along the directional vector.

As $\ddot{X} = \alpha\,s$, we deduce from (20) : $\alpha = 1/\sqrt{s^T\,M_X^T\,J_0\,J_0^T\,M_X\,s}$

To optimize the acceleration in the direction s, we choose the three lengths r_i that maximise α.

Figure 3 and 4 show some results . To simplify the simulations, we consider the case where the end effector is in the vertical plane of the sub-arm 1. The figures represent the robot projected in this plane.

The first figure shows the evolution of the dynamic manipulability ellipsoid for three different postures. The choice of the posture has a great influence on the form of the ellipsoid and on the value for a given direction of the end effector.

On the others figures, the redundant structure is compared with a non-redundant one, similar to the DELTA structure where the radius of the arms are fixed. For the redundant robot, we have optimized α :

- in the horizontal direction with the end effector unloaded (figure 4a),
- in the vertical direction with the end effector loaded (figure 4b)

The gain on acceleration is very important : 50 % for the first case and 20 % for the second case. It is thus interesting to adapt the posture in function of the trajectory to be followed but also whether the end effector is loaded or not.

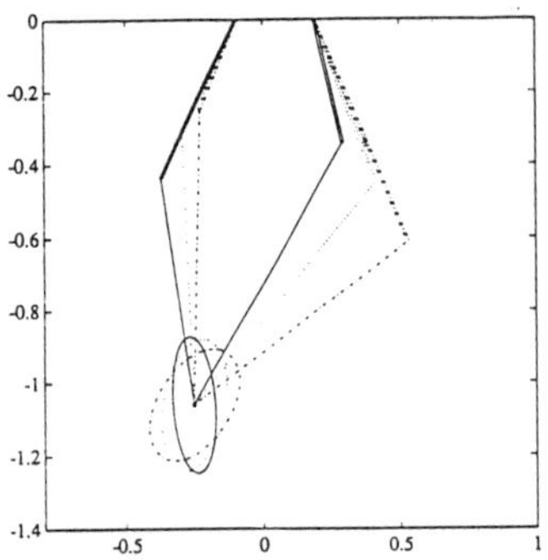

Fig.4(a)

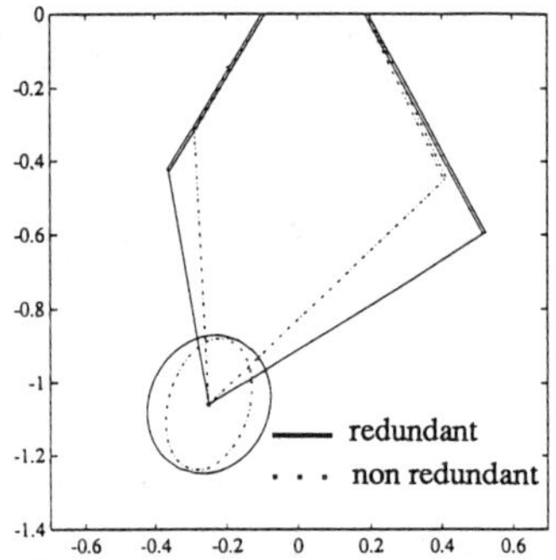

Figure 3
Deformation of the ellipsoid
with the posture

Fig.4(b)

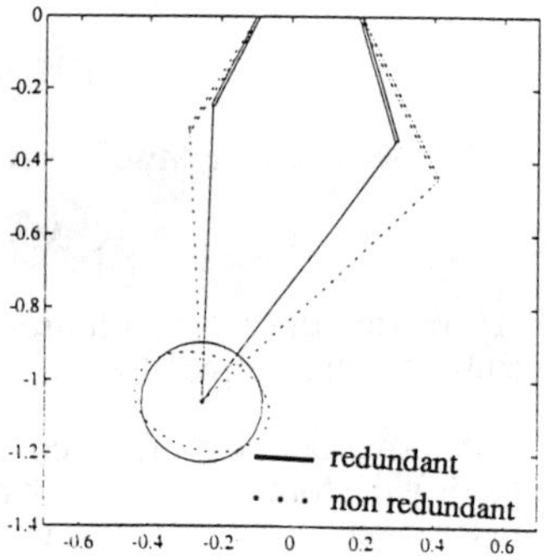

Figure 4
Comparisons between the redundant
robot and a non redundant one

The general control scheme will now be described.

4. CONTROL SCHEME

Control generally deals with two sequential problems : trajectory planning followed by trajectory tracking. For a redundant robot, a third problem has to be considered between the others : the choice of joint trajectories.

The part dealing with trajectory tracking has been structured on two levels. The first is intended to decouple the set of two actuators. The second decouple the sub-arms by taking into account the main dynamic interactions. This is done using the inertia matrix in joint space.

The problem of trajectory planning is far more complex. With a view of simplification and in order to allow real time computation, it is supposed that limits on cartesian velocity and acceleration are constant along the prescribed path. These limits, fixed a priori, must permit a good trajectory tracking even in the worst conditions. On an other hand, it is important to maximise these values in order to avoid under-using the possibilities of the robot. Thus, the posture is adapted in the most constraining situation to find the optimal limits on velocity and acceleration.

Then, the cartesian path is described and the joint coordinates must be determined. The control has in fact two aims :

- the first is, naturally, to satisfy the trajectory specified by the task
- the second is to modify the robot posture in order to optimise the dynamic manipulability ellipsoid.

Two sub-tasks thus work in concert to develop the reference vector in joint space. Certain components are intended to satisfy the end effector trajectory ; others act in the nul-space of the Jacobian. They correspond to internal joint movement which does not produce any displacement of the end effector

The general control scheme is shown in figure 5.

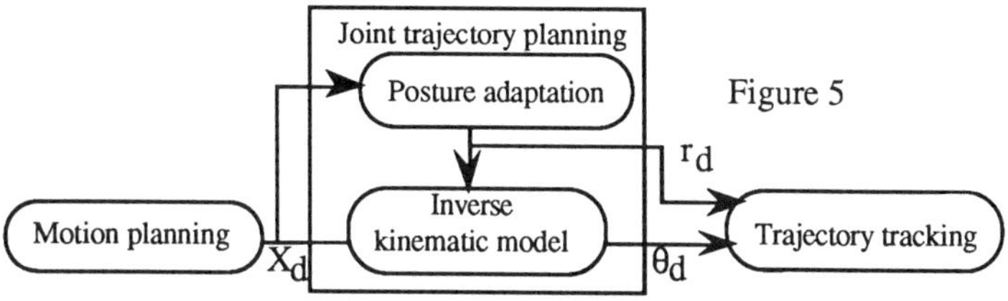

Figure 5

5 - CONCLUSION

One of the main aims of industrial robotization is an increase in productivity. One way of accomplishing this is to design extremely light-weight robots, capable of high rates of acceleration, and to develop trajectory generation algorithms which enable displacement times to be reduced. This present study meets both objectives. SPEED-R-MAN has been designed according to a kinematically redundant parallel architecture with three external degrees of freedom. A dynamic analysis has shown the interest of adapting the posture so as to have the highest possible rate of acceleration during the trajectory. A large part of the control structure is operational ; the algorithm permitting posture adaptation is in the process of being set up.

REFERENCES

[1] Stewart, A., (1966) "A platform with six degrees of freedeom" - Proc. of the Institution of Mechanical Engineers - Vol.180, Part I, n°15, pp 371-386.

[2] McCallion, H., Johnson, G.R. and Pham, D.T. (1979) "A compliant device for inserting a peg in a hole" - The Industrial Robot, pp 81-87.

[3] Inoue, H., Tsusaka, Y. ,Fukuizumi, T. (1986)"Parallel manipulator" - The Third Int. Symposium of Robotics Research, MIT Press, Cambridge, Ma, pp 322-327.

[4] Reboulet, C. , Robert, A. (1986) "Hybrid control a a manipulator equipped with an active compliant wrist" - The Third Int. Symposium of Robotics Research, MIT Press, Cambridge, Ma, pp 237-241.

[5] Merlet, J.P., (1992) "Direct kinematics and assembly modes of parallel manipulator" - Int. Journal of Robotics Research - Vol.11, n°2, pp 150-162.

[6] Amirat, Y., Artigue, F., Pontnau, J. (1992) "Six degrees of freedom parallel robots with C_5 links" - Robotica, vol. 10, pp 35-44

[7] Hayward, V. and Kurtz, R.(1989) "Preliminary study of a serial parallel redundant manipulator" NASA Conf. on Telerobotics, Pasadena, USA.

[8] Gosselin, C. and Angeles, J. (1989) "The optimum kinematic design of a spherical three-degree-of-freedom parallel manipulator" - Trans. of the ASME, Vol.111, pp 202-207 - June.

[9] Clavel, R. (1988) "DELTA a fast robot with parallel geometry" - Proc. of the Int. Symposium on Industrial Robots, pp 91-100, April.

[10] Pierrot, F., Dauchez, P., Fournier, A. (1991) "Hexa a fast six-dof fully parallel robot" - Proc. of the 5th Inter. Conf. on Advanced Robotics, Pisa (Italy), June.

[11] Hervé, J.M.and Sparacino, F. (1991) "Structural synthesis of parallel robots generating spatial translation" Proc. of the 5th Inter. Conf. on Advanced Robotics, Pisa (Italy), June.

[12] Yoshikawa, T. (1984) "Analysis and control of robot manipulators with redundancy" - The first Int. Symposium of Robotics Research, MIT Press, Cambridge (USA), pp 735-747.

[13] Chang, P.H. (1989) "Development of a dexterity measure for kinematically redundant manipulators" - Proc. American Control Conf;, pp 496-506, June.

[14] Pierrot, F., Reynaud, C. and Fournier, A. (1990) "DELTA : a simple and efficient parallel robot" Robotica, vol. 8, pp 105-109.

[15] Sternheim, F. (1987) "Computation of the direct and inverse kinematic models of the Delta 4 parallel robot" Robotersysteme 3, pp 199-203.

[16] Albert, A. (1972) "Regression and the Moore-Penrose pseudo-inverse", Academic Press.

[17] Tadokara, S., Kimura, I. and Takamori, T. (1991) "A measure for evaluation of dynamic dexterity based on a stochastic interpretation of manipulator design" Proc. IEEE Robotics and Automation, pp 509-514.

CONTACT AND IMPACT CONFIGURATIONS FOR KINEMATICALLY REDUNDANT ROBOTS

I. D. Walker
Department of Electrical and Computer Engineering
Rice University, Houston, Texas 77251-1892

ABSTRACT

Interaction of a robot with its environment (whether accidental, or intentional, as part of a planned task), inevitably involves contact or collision. This paper examines new methods to evaluate the effects of impact forces at the end effector on robot manipulators. Two new measures of the vulnerability of any arm to impacts in varying directions are given. Two impact ellipsoids corresponding to these measures are defined and analyzed. The effect of different configurations of redundant arms on (potentially damaging) impact forces at their end effectors during contact with the environment are reviewed and discussed. New methods for examining the optimal configurations of redundant manipulators under impact task constraints are discussed.

INTRODUCTION

Kinematic redundancy has been advocated as a useful tool in robotics and investigated by many researchers. The self-motion inherent in redundant robots can be used to evade obstacles, avoid singular configurations, and many other tasks. For a review of the properties and applications of kinematic redundancy, see [5, 6]. Most analysis has been performed for planning motions based on the kinematic equations. However, when considering dynamic interactions between robots and their environment, kinematic redundancy can also be used to alter the dynamic effects of contact or collision of a robot with its environment. In earlier work [7, 8], it has been shown how the configuration of the robot affects the impact forces produced at the moment of contact with the environment (for example when picking up a payload or during force control). For a single robot in contact with a solid surface, the magnitude of the impulsive contact force can be minimized (or maximized) by choosing the appropriate configuration (basically the robot "braces" itself for collision) [7]. For the case of grasping and regrasping an object held by multiple fingers or manipulators, the analysis can be extended to find "best" regrasp points and configuration [8]. Applications to force control were investigated in [11]. In this paper, we see how the contact model used in [7] can be used to define measures of "impact bracing" via appropriate ellipsoids in end effector space.

This work has applications for redundant robots in force control and assembly, such as the Space Station (SSRMS) redundant arms for assembly and/or docking. The results could also be important for applications in cluttered workspaces in hazardous environments, such as hazardous waste cleanup, where collisions with the environment could be extremely dangerous.

ROBOT MODEL

We begin by reviewing the models of robot and impact used for the contact scenario. We assume the kinematic relationship between the n joint variables $\underline{\theta}$ and the m end effector position variables of the robot $\underline{x}$ is given by

$$\underline{x} = f(\underline{\theta}) \tag{1}$$

Differentiation of this relationship leads to the expression

$$\underline{v} = \dot{\underline{x}} = [J(\underline{\theta})]\dot{\underline{\theta}} \tag{2}$$

where the $m \times n$ matrix $[J]$ is the manipulator Jacobian relating the joint velocities $\dot{\underline{\theta}}$ to the end effector velocities $\dot{\underline{x}}$. The dynamic equations are assumed given as

$$\underline{\tau} = [I(\underline{\theta})]\ddot{\underline{\theta}} + \dot{\underline{\theta}}^T[P(\underline{\theta})]\dot{\underline{\theta}} - [J(\underline{\theta})]^T\underline{F} \tag{3}$$

where $\underline{\tau}$ is the vector of joint torques, $[I(\underline{\theta})]$ is the $n \times n$ inertia matrix (symmetric, positive definite for all $\underline{\theta}$), $\dot{\underline{\theta}}^T[P(\underline{\theta})]\dot{\underline{\theta}}$ contains the nonlinear centripetal and coriolis effects, and $\underline{F}$ is the force applied at the end effector.

IMPULSIVE CONTACT MODELS

In the following, we consider the instantaneous (impulsive) effects of contact, modeled essentially as a collision. Such a model should provide the magnitude, direction, and point of application of impulsive forces resulting from contact or collision. This implies analysis of, for a full dynamic model of a general manipulator, the impulse forces and impulse couples that will be experienced at the impact (contact) point, and at all points throughout the internal manipulator structure. Notice that both sets of impulses are of interest - internal impulses may damage the manipulator, whereas the contact point impulses may result in damage to either the manipulator or the contacted point in the environment.

The derivation in this section follows that in [7], and is repeated here for completeness. The approach followed is that of applying the classical theory of dynamical systems [10]. Earlier work made restrictive assumptions on the types of contact allowed [9, 14, 15], or used simplified mass-spring-damper models for the manipulators. Here we include full manipulator dynamical models [4] for general contact types (both plastic and elastic contact).

We assume that the bodies in contact at the moment of collision are rigid, and that each is a series of connected rigid bodies. Then the mathematical theory of rigid body collisions [10] may be utilized as follows. Assume that there are two bodies in collision. If the velocities of bodies 1 and 2 immediately before collision are $\underline{v}_1$ and $\underline{v}_2$ respectively, then we have the following expression:

$$[(\underline{v}_1 + \Delta\underline{v}_1) - (\underline{v}_2 + \Delta\underline{v}_2)]^T\underline{n} = -e(\underline{v}_1 - \underline{v}_2)^T\underline{n} \tag{4}$$

where $\underline{n}$ is the normal vector to the plane of contact of the two bodies, $\Delta\underline{v}_1$ and $\Delta\underline{v}_2$ are the change in $\underline{v}_1$ and $\underline{v}_2$ immediately following collision, and $0 < e < 1$ is a constant denoting the type of collision taking place. The value of e is 0 for purely plastic collisions (the colliding bodies have zero relative velocity to each other at the point of contact immediately following collision) and 1 for purely elastic collisions (the impulsive interaction forces between the bodies are normal to the tangent plane at the collision point). Values of e between 0 and 1 indicate intermediate cases of the above two.

Now it is assumed that the impact occurs in a small period of time Δt and in the model the idealization $\Delta t \to 0$ is made. In this (infinitesimally small) period of time all velocities and angular velocities remain finite, and thus there is no change in positions or orientations of any bodies in the system. However, an impulsive force $\hat{\underline{F}}$ is created at the collision point by the impact. This is a force $\underline{F}(t)$ such that

$$\lim_{\Delta t \to 0} \int_t^{t+\Delta t} \underline{F}(s)ds = \hat{\underline{F}}$$

where $\hat{\underline{F}}$ is a finite quantity. This impulsive force induces finite changes in the velocities and angular velocities of the bodies, and assuming that the volume surrounding the colliding points is infinitesimally small, (4) holds. Next, to find the relationship between $\hat{\underline{F}}$ and the velocity increments in (4), consider the integral of (3) from t to $t + \Delta t$. This is given as

$$\int_t^{t+\Delta t} [I(\underline{\theta})]\ddot{\underline{\theta}}ds + \int_t^{t+\Delta t} \dot{\underline{\theta}}^T [P(\underline{\theta})]\dot{\underline{\theta}}ds - \int_t^{t+\Delta t} [J(\underline{\theta})]^T \underline{F}ds = \int_t^{t+\Delta t} \underline{\tau}ds$$

Since the positions and velocities are finite at all times as $\Delta t \to 0$, the integral involving $[P]$ becomes zero, as does that involving $\underline{\tau}$, and thus we are left with (performing the integral over the first term)

$$[I(\underline{\theta})][\dot{\underline{\theta}}(t + \Delta t) - \dot{\underline{\theta}}(t)] = [J(\underline{\theta})]^T \hat{\underline{F}}$$

Define $\Delta\dot{\underline{\theta}}$ to be $[\dot{\underline{\theta}}(t + \Delta t) - \dot{\underline{\theta}}(t)]$. Then we have, at the moment of collision

$$\Delta\dot{\underline{\theta}} = [I(\underline{\theta})]^{-1}[J(\underline{\theta})]^T \hat{\underline{F}} \tag{5}$$

This equation, together with (4), forms the basis for our model of impulsive contact phenomena. It expresses the relationship between the impulsive contact force and the instantaneous velocity increment corresponding to the induced object velocity increment at the point of contact. Notice that the relationship $\Delta\underline{v} = [J(\underline{\theta})]\Delta\dot{\underline{\theta}}$ still holds at the instant of collision for the arm. Therefore

$$\Delta\underline{v} = [J(\underline{\theta})][I(\underline{\theta})]^{-1}[J(\underline{\theta})]^T \hat{\underline{F}} \tag{6}$$

at the instant of impact. In the following section, we use the above relationships between the (instantaneous) changes in velocities and the impact forces to develop measures of evaluating the robots ability to withstand impacts in differing directions.

IMPACT MEASURES AND ELLIPSOIDS FOR ROBOTS

We have reviewed above the relationship between the impulsive forces of contact and their effects on the robot via (5). In the following we use (5) to define two measures, together with related ellipsoids, which qualify the ability of a given robot to withstand impacts.

The use of ellipsoids to model the ability of a robot to perform various tasks is well-established. Yoshikawa [13, 12] has defined a manipulability ellipsoid, a manipulating force ellipsoid, and a dynamic manipulability ellipsoid for single manipulators. These quantities have been used extensively in analysis of single robot motion tasks. Ellipsoids have also proved to be extremely useful in understanding task and motion-related properties of multiple manipulators, for example [1, 2, 3].

Recalling again (5), we define first the **dynamic impact measure**, which is a scalar value w_{di} given by

$$w_{di} = \sqrt{det([J^+]^T[I][I]^T[J^+])} = \sqrt{det([J^+]^T[I]^2[J^+])} \tag{7}$$

where J^+ is a pseudoinverse of J, given in general by $J^+ = AJ^T(JAJ^T)^{-1}$ for some positive definite $n \times n$ matrix A. If the singular value decomposition (SVD) of $[J^+]^T[I]$ is given by

$$[J^+]^T[I] = [U][\Sigma][V]^T$$

where U and V are orthogonal matrices, and Σ is given by $\Sigma = diag[\sigma_1......\sigma_n]$, with the singular values of $[J^+]^T[I]$ given by $(\sigma_1, ..., \sigma_n)$, then the measure w_{di} can be found as

$$w_{di} = \sigma_1\sigma_2.....\sigma_n \tag{8}$$

The dynamic impact measure models the ability of the robot to withstand impacts at the end effector (see (5), and compare with the well-known measure of manipulability w for (2) given by $w = \sqrt{det([J][J^T])}$ [13]).

The **dynamic impact ellipsoid** in $\Re^m$ corresponding to this measure is defined by

$$(\underline{u}\varepsilon\Re^m : \underline{u}^T[J][I]^{-2}[J]^T\underline{u} \leq 1) \tag{9}$$

Thus, from (5), we see that the dynamic impact ellipsoid is formed by those contact impulse forces $\underline{u}$ which correspond to (instantaneous) changes in joint velocities $\Delta\underline{\dot{\theta}}$ with unit norm or less, i.e. $(\Delta\underline{\dot{\theta}} : \|\Delta\underline{\dot{\theta}}\| = \Delta\underline{\dot{\theta}}^T\Delta\underline{\dot{\theta}} \leq 1)$. The dynamic impact ellipsoid provides an intuitive interpretation of the susceptibility of the robot in a given configuration $\underline{\theta}$ to impact impulse forces. This is easily seen via the singular values of $[J^+]^T[I]$ and the singular vectors (columns of U). The principal axes of the ellipsoid are given by the columns $\underline{u}_i$ and the corresponding singular values represent relative magnitudes of the impulsive forces required to produce the unit ball of changes in joint velocities of the robot. (see Figure 1. for an example planar example, i.e., $m = 2$). The volume of the ellipsoid is proportional to w_{di}.

The dynamic impact ellipsoid can be viewed as being dually related to the dynamic manipulability ellipsoid introduced by Yoshikawa [12], which quantified the degree of arbitraryness of changing the acceleration at the end effector given

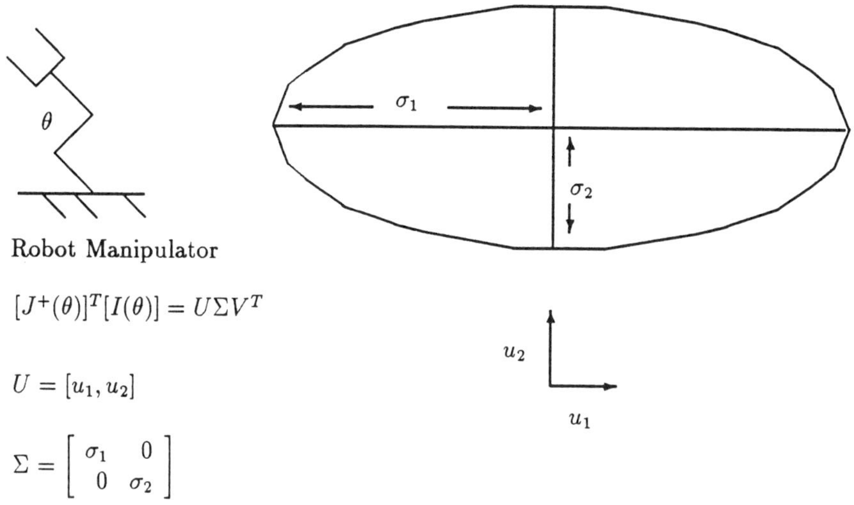

1: Planar Dynamic Impact Ellipsoid.

norm bounds on the joint torques. The dynamic impact ellipsoid in contrast highlights directions in which the robot is dynamically "stiff" (rather than dynamically maneuverable, as for the dynamic manipulability ellipsoid) to changes at the end effector. The duality notion is similar to that between the manipulability and manipulating force ellipsoids [13] for purely kinematic analysis.

The dynamic impact ellipsoid quantifies the arbitraryness of varying impulsive contact forces with bounded changes in joint velocities. Large (small) singular values indicate large (small) relative impact forces in the corresponding direction u_i for unit joint velocity changes. Different tasks will beg different shapes for the ellipsoid. Small singular values corresponding to directions which will produce large changes in velocities for relatively small impacts, are undesirable for tasks where damage to the robot is of primary concern. Alternatively, for tasks such as nuclear waste handling, where damage to the environment is most critical, large singular value directions are undesirable, since they correspond to directions in which the robot will resist even large impact forces. Therefore, given an application, there are critical directions in which impacts (planned or unplanned) are most and least desired. These critical directions are immediately clear once the ellipsoid is calculated. Therefore the dynamic impact ellipsoid is a good means to evaluate the ability of a robot to withstand impact in different directions at the end effector.

We next introduce another useful measure based on (5). The **generalized impact measure** is a scalar value w_{gi} given by

$$w_{gi} = \sqrt{det([J^+]^T[I][J]^+)} \qquad (10)$$

This measure can be calculated as

$$w_{di} = \bar{\sigma}_1 \bar{\sigma}_2 .....\bar{\sigma}_n \tag{11}$$

where the SVD of $[J^+]^T[I]^{1/2}$ is given by

$$[J^+]^T[I]^{1/2} = [\bar{U}][\bar{\Sigma}][\bar{V}]^T$$

and $\bar{\Sigma} = diag[\bar{\sigma}_1......\bar{\sigma}_n]$ The generalized impact measure can be seen to be a measure of the magnitude of the effects of collision using a weighted norm, as discussed below.

Finally, the **generalized impact ellipsoid** in $\Re^m$ is defined by

$$(\underline{v} \varepsilon \Re^m : \underline{v}^T [J][I]^{-1}[J]^T \underline{v} \leq 1) \tag{12}$$

The generalized impact ellipsoid is formed by those contact impulse forces $\underline{v}$ which produce (instantaneous) changes in joint velocities with unit norm or less, where the norm in this case is the 2-norm weighted by the inertia matrix ($\|\underline{x}\|_2 = \underline{x}^T[I]\underline{x}$). The volume of the generalized impact ellipsoid is proportional to w_{di}. Notice that the quantity $\underline{F}^T[J][I]^{-1}[J]^T\underline{F}$, for a given $\underline{F}$, is the change in kinetic energy $\Delta\dot{\underline{\theta}}^T[I]\Delta\dot{\underline{\theta}}$ of the robot due to the collision (from (5)). An intuitive interpretation of the size and shape of the generalized impact ellipsoid using the SVD of $[J^+]^T[I]^{1/2}$, may be made similar to that discussed above for the dynamic impact ellipsoid. The significance of this second measure, and its related ellipsoid, will become clearer when we consider self-motion of redundant manipulators in the following section. The generalized impact ellipsoid can be viewed as being dual to the generalize inertia ellipsoid introduced by Asada and discussed in [12].

The two new ellipsoids introduced above could be useful tools in the planning and analysis in robot tasks where contact with the environment is desired or likely. For given robot configurations (notice that both measures depend on the configuration via J and I) the analysis provides information about the relationship between the magnitudes and directions of the impact effects. In the following, we discuss how the analysis can be applied to the case of choosing self-motion for redundant robots to "brace" for collision.

REDUNDANT MANIPULATOR IN CONTACT WITH SOLID SURFACE

Consider the situation of a kinematically redundant arm in contact with a solid object, such as a wall. In this case, we have

$$\underline{v}_2 = \Delta\underline{v}_2 = \underline{0} \tag{13}$$

in (4). Substituting the above into (4), and noticing from (6) that

$$\Delta\underline{v}_1 = \Delta\underline{v} = [J(\underline{\theta})][I(\underline{\theta})]^{-1}[J(\underline{\theta})]^T\hat{\underline{F}} \tag{14}$$

for the manipulator, we can solve for the contact impulse force magnitude $\hat{F}$ as (using the fact that $\hat{\underline{F}} = \hat{F}\underline{n}$ [7]):

$$\hat{F} = \frac{-(1+e)(\underline{v})^T \underline{n}}{\underline{n}^T([C])\underline{n}} \qquad (15)$$

where $[C]$ is given by

$$[J(\underline{\theta})][I(\underline{\theta})]^{-1}[J(\underline{\theta})]^T$$

Notice that in the above expression for $\underline{\hat{F}}$ the dependence of the impulse force on the contact geometry $\underline{n}$, the impact velocity and $\underline{v}$, and the contact type denoted by e is clearly shown. Notice that the scalar $\underline{n}^T[C]\underline{n}$ is the value for the change in kinetic energy due to the unit impulse in direction $\underline{n}$. It is clear that reducing the impact velocity $\underline{v}$ will reduce the magnitude of the impact, as would be expected. For many cases however, the contact geometry and impact velocity will be limited and determined by the application. In general manipulation operations, time constraints will usually dictate a minimum acceptable speed of operation. The question arises as to how to reduce the impact effects given $\underline{n}$ without reducing the speed of the task operation represented by $\underline{v}$.

For kinematically redundant arms, the configuration can be altered using the self-motion. The issue of choice of "best" configuration for impact is discussed in [7]. However, conventional techniques to minimize $\underline{\hat{F}}$ are nonintuitive in general, and a means of visualizing the relative effects of different self-motions would be useful. The measures introduced in this paper can be used to conveniently visualize the effects of using the self motion, i.e. choosing different contact configurations. Notice that the shape and volume of the ellipsoids change with configuration. This feature can be used to evaluate different possible configurations given a required impact geometry and velocity for the end effector. Different applications will explicitly provide preferred directions for high and low resistance to impact forces, as discussed in the previous section. Given an application, the self-motion can be used to align the impact ellipsoid with the desired task force directions, in order to "brace" the arm appropriately for the task. In this way, the analysis of this paper can be used to more intuitively use the self-motion of redundant arms in contact and impact tasks.

CONCLUSIONS

The case of contact between a robot and its environment is considered. Given a model of impact relating the change in joint velocities to the impulsive forces generated by the collision, new measures for evaluating the effect of impacts are introduced. Two new ellipsoids are defined which model the impact scenario intuitively. Using these ellipsoids, the directions at the end effector which can best and least withstand impact are clearly seen. Application of these methods to use of the self motion of a redundant manipulator to minimize the magnitude of impact with a solid wall are given.

ACKNOWLEDGEMENTS

This work was supported in part by the National Science Foundation under grant MSS-9024391 and by DOE Sandia National Laboratory Contract #18-4379A.

REFERENCES

1. P. Chiacchio, S. Chiaverini, L. Sciavicco, and B. Siciliano. Global task space manipulability ellipsoids for multiple armed systems. *IEEE Transactions on Robotics and Automation*, RA-7(5):678–685, 1991.

2. S. Lee. Dual redundant arm configuration optimization with task-oriented dual arm manipulability. *IEEE Transactions on Robotics and Automation*, RA-5(1):78–97, 1989.

3. Z. Li and S.S. Sastry. Task-oriented optimal grasping by multifingered robot hands. *IEEE Transactions on Robotics and Automation*, RA-4(1):32–44, 1988.

4. J.K. Mills. Manipulator transition to and from contact tasks: A discontinuous control approach. In *Proceedings of 1990 IEEE Conference on Robotics and Automation*, pages 440–446, Cincinnati, OH, 1990.

5. D.N. Nenchev. Redundancy resolution through local optimization: A review. *Journal of Robotic Systems*, 6(6):769–798, 1989.

6. B. Siciliano. Kinematic control of redundant robot manipulators: A tutorial. *Journal of Intelligent and Robotic Systems*, 3(3):201–210, 1990.

7. I.D. Walker. The use of kinematic redundancy in reducing impact and contact effects in manipulation. In *Proceedings 1990 IEEE Conference on Robotics and Automation*, pages 434–439, Cincinnati, OH, 1990.

8. I.D. Walker. Dynamics and regrasping for multiple cooperating robot manipulators. In *Proceedings 1991 American Control Conference*, pages 2509–2514, Boston, MA, 1991.

9. Y. Wang and M.T. Mason. Modeling impact dynamics for robotic operations. In *Proceedings of 1987 IEEE Conference on Robotics and Automation*, pages 678–685, Raleigh, NC, 1987.

10. J. Wittenberg. *Dynamics of Systems of Rigid Bodies.* Stuttgart: B.G. Teubner, 1977.

11. B.J. Yi, I.D. Walker, R.A. Freeman, and D. Tesar. Geometric stability in force control. In *Proceedings of 1991 IEEE Conference on Robotics and Automation*, pages 281–286, Sacramento, CA, 1991.

12. T. Yoshikawa. Dynamic manipulability of robot manipulators. In *Proceedings of 1985 IEEE Conference on Robotics and Automation*, pages 1033–1038, St. Louis, MO, 1985.

13. T. Yoshikawa. Manipulability and redundancy control of redundant mechanisms. In *Proceedings of 1985 IEEE Conference on Robotics and Automation*, pages 1004–1009, St. Louis, MO, 1985.

14. Y.F. Zheng and H. Hemami. Mathematical modeling of a robot collision with its environment. *Journal of Robotic Systems*, 2(3):289–307, 1985.

15. Y.F. Zheng and F.R. Sias. Two robot arms in assembly. In *Proceedings of 1986 IEEE Conference on Robotics and Automation*, pages 1230–1235, San Francisco, CA, 1986.

ACTIVE SPRING DESIGN VIA REDUNDANTLY ACTUATED PARALLEL MECHANISMS

Robert A. Freeman and Byung-Ju Yi

Department of Mechanical Engineering

THE UNIVERSITY OF TEXAS AT AUSTIN

ABSTRACT

The design of programmable spring mechanisms is discussed. The structure of these devices is of the parallel mechanism type with the spring effect resulting from the employment of redundant, direct drive, actuation systems. The basic concept is illustrated by an uni-directional device. Previous work and potential applications of the active spring are addressed.

INTRODUCTION

The application areas of springs are numerous. Historically, the study of springs has been divided into five catagories according to the types of tasks the spring mechanisms are to perform [6].

The first category includes the well-known problem of static force balancing [3, 4, 5, 11]. The second category of spring use in mechanisms is for inertial load balancing [1, 7]. The use of spring mechanisms for energy storage (category 3) is very broad [5, 9]. Vibration resonance control and isolation follows as the fourth category [6, 10]. As the last (fifth) category, springs have been used to correct for positioning error of the end-point of a robot by allowing for controlled, force induced, displacement of the end-point. "Remote Center Compliance" (RCC) devices [17], which are passive spring mechanisms, have been conceived and built for this purpose. However, current technology is largely restricted to task specific designs which allow, at best, only minimal adjustment of stiffness due to the passive nature of the actuation system in these mechanisms. "Instrumented Remote Center Compliance" (IRCC) devices [8, 16], which consist of passive spring mechanisms along with force sensors for feedback adjustment, have also been conceived. However, they are sluggish in response.

As mentioned, previous spring mechanism design has been task-specific with the systems unable to actively control (adjust) the spring rate independent of position (deflection). These drawbacks prevent attainment of satisfactory performance in the categories mentioned above. Therefore, the development of actively adjustable and non-position dependent springs is desirable for general purpose spring applications. One potential solution to this problem employs input load control of redundantly actuated parallel mechanisms and is discussed in the following.

Redundantly actuated linkages have more force inputs (actuators) than they have kinematic (motion) freedoms. It is therefore possible to actively control (adjust) the internal load state (preload) of the mechanism. This internal loading, along with the nonlinear linkage geometry, creates an effective stiffness directly analogous to that of a wound metal spring. The required preload control cannot be accomplished without redundant force inputs, nor can it be arbitrarily adjusted using elemental springs as inputs. However, by proper coordination of redundant actuator efforts the desired effective stiffness can be created, actively adjusted, and planned in a feedforward fashion. In the work presented here, we introduce the basic spring mechanism concept in terms of a four bar mechanism and potential applications of active spring mechanisms are addressed.

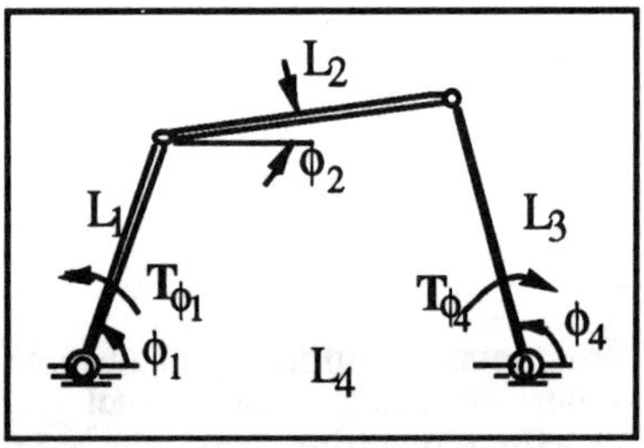

FIG. 1. Four Bar Mechanism with Redundant Actuation

FOUR-BAR EXAMPLE

Consider the four-bar mechanism of Figure 1. For clarity, in the discussion to follow, consider further that gravity acts normal to the plane of the four-bar. This mechanism has one degree-of-freedom (1 DOF) of motion, and hence, its kinematic state is completely defined by a single parameter (e.g., ϕ_1). However, if a second motor is placed in the structure, say at ϕ_4, the linkage can sustain internal loads while remaining in a state of static equilibrium. This is accomplished by setting the effective load $(\mathbf{T}_{\phi_1})^*$ at the selected "minimum" joint set ϕ_1 equal to zero. This requires that

$$(\mathbf{T}_{\phi_1})^* = \mathbf{T}_{\phi_1} + (g_1^4)\,\mathbf{T}_{\phi_4} = 0, \tag{1}$$

where the Jacobian which relates the joint ϕ_4 to the joint ϕ_1 is

$$g_1^4 = \frac{d\phi_4}{d\phi_1} = \frac{L_1 S_{1\text{-}2}}{L_3 S_{4\text{-}2}}, \tag{2}$$

and $\mathbf{T}_{\phi_1}$ and $\mathbf{T}_{\phi_4}$ are the efforts of the actuated joints. Here, $S_{1\text{-}2}$ and $S_{4\text{-}2}$ denote $\sin(\phi_1 - \phi_2)$ and $\sin(\phi_4 - \phi_2)$, respectively. Therefore, the effort at either joint (e.g., $\mathbf{T}_{\phi_4}$) can be arbitrarily selected and the system then put in equilibrium by supplying a balancing effort at the other joint (e.g., $\mathbf{T}_{\phi_1}$) according to Eq. (1) . Thus the internal load state of the linkage can be actively adjusted independent of its motion by controlling the redundant actuator input efforts.

The internal load, or preload, acts as an effective system stiffness to external disturbance, either force or displacement, as illustrated by the following development. Consider, that a set of equilibrium loads have been determined for an arbitrary desired equilibrium configuration of the four-bar of Fig. 1. Further, consider that the balanced input loads, $\mathbf{T}_{\phi_1}$ and $\mathbf{T}_{\phi_4}$, are held constant. Now, assume that a small displacement is imposed on the linkage changing its configuration. The effective load at the generalized input coordinate (eg., ϕ_1) is no longer zero, but is

$$\Delta(\mathbf{T}_{\phi_1})^* = \Delta(\mathbf{T}_{\phi_1} + (g_1^4)\,\mathbf{T}_{\phi_4})$$
$$= \mathbf{T}_{\phi_4}\,\Delta(g_1^4). \tag{3}$$

Therefore, the mechanism is no longer in equilibrium $((\mathbf{T}_{\phi_1})^* \neq 0)$, due to the displacement induced change in the Jacobian (g_1^4), and will respond according to the system dynamics as referenced to ϕ_1. This effect is directly analogous to the force induced by an implosed displacement in a wound metal spring.

Thus, the mechanism exhibits an effective stiffness to external disturbance due to the redundant actuation. This stiffness can be determined and expressed analytically by considering the change in the effective load at ϕ_1 due to a small change in configuration, as represented by a small change in ϕ_1. Recalling Eq. (3), the change in the effective load can be expressed as

$$\Delta(\mathbf{T}_{\phi_1})^* \sim \mathbf{T}_{\phi_4}\left(\frac{dg_1^4}{d\phi_1}\right)\Delta\phi_1 \tag{4}$$

since the linkage geometry can be considered to be dependent on the single parameter ϕ_1. Noting that the change in the effective load in a unidirectional spring (see Fig. 2) is

$$\Delta F^* = -\,K\,\Delta x, \tag{5}$$

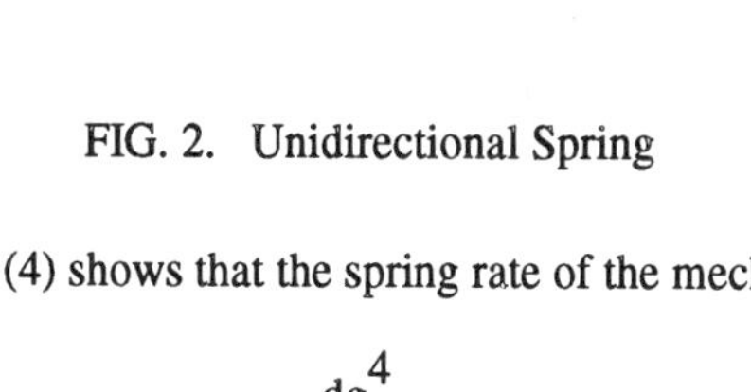

FIG. 2. Unidirectional Spring

comparison with Eq. (4) shows that the spring rate of the mechanism, referenced to ϕ_1, is

$$K_{\phi_1} = -\,\mathbf{T}_{\phi_4}\left(\frac{dg_1^4}{d\phi_1}\right) = -\,h_{11}^4\,\mathbf{T}_{\phi_4}, \tag{6}$$

where the constitutive relationship between effort and stiffness is given by

$$h_{11}^{4} = \frac{d^2\phi_4}{d\phi_1^2} = \frac{L_1C_{1-2} - L_3(g_1^4)^2C_{4-2} + L_2(g_1^2)^2}{L_3S_{4-2}} \tag{7}$$

with

$$g_1^2 = \frac{d\phi_2}{d\phi_1} = \frac{L_1S_{1-4}}{L_2S_{4-2}} \tag{8}$$

and is itself purely a function of position. The stiffness defined by Eq. (6) is a local, or kinematically instantaneous, property of the mechanism configuration and actuator efforts. The primary difference between the proposed spring mechanism concept and an actual elemental spring is that the spring rate is a non-linear function of displacement (if the efforts are held constant).

Thus, the operational concept of the proposed spring mechanism requires redundant actuation and can be explained as follows. First, an equilibrium configuration is specified. Then the desired spring rate is specified and the effort required at joint ϕ_4 is determined by solution of Eq. (6). Finally, the effort required at joint ϕ_1, is found from Eq. (1). Thus the proposed spring mechanism is capable of generating , and continuously adjusting, a desired stiffness independent of displacement by simple local servo effort control without recourse to external disturbance measurement, and feedback.

The one dimensional case can be directly extended to multi-dimensional cases. Parallel mechanisms are employed to design active spring mechanisms. The excess of potential input locations in parallel mechanisms allows the implementation of redundant actuation. Fig. 3 and Fig. 4 illustrate planar and spherical 3-DOF parallel mechanisms, respectively, which have been investigated with regard to stiffness generation in our previous work [13, 14].

In other previous work [12, 15], conditions for full stiffness generation in general closed-chain parallel mechanisms were derived from an analysis of the purely geometric constitutive relationship between effort and stiffness, and load distribution methodologies were developed which illustrated how redundantly actuated systems can independently generate motion and stiffness.

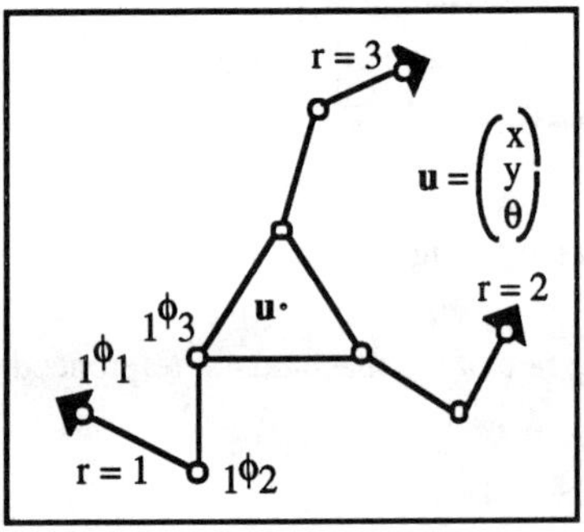

FIG. 3. 3-DOF Planar Mechanism

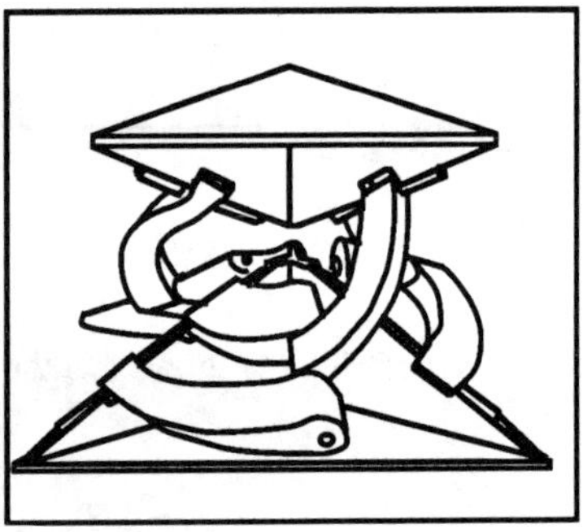

FIG. 4. 3-DOF Spherical Mechanism

APPLICATIONS

The novel concept of active spring generation via redundant actuation has illustrated considerable promise towards the various previously mentioned applications. Much research remains towards the generalization of active spring generation in several physical domains and its practical applications to various automated processes and machine designs. Potential research items are summarized below.

a. Generalization of antagonistic redundant actuation for active spring design : This will provide the framework for the future research with the goal being a unified methodology for active spring generation in general physical domains.

b. Applications : Initially, design parameters and specific task-based criteria for each application should be determined. The application areas can be divided into five major categories ;

•RCC device design [14] : architectural design of robot end-effector position correction device with an actively adjustable spring rate (AASR) and its performance analysis.

•Automatic suspension system design : design and performance analysis of multi-dimensional suspension systems using AASR.

•Bio-mechanics applications [13] : analysis of the active spring generation existing in general biomechaical systems with application to spring assistive mechanism design and human assistive devices.

•Load balancing technology : investigation of load balancing methodology in general machinery and robotic mechanisms using AASR.

•Variable natural frequency wave form generator design : design of variable wave generators which are useful in experimentation in which certain waves with complicated frequencies are required.

REFERENCES

1. Conley, W.J., "Integration of Nonlinear Springs into Machines by Analytical Synthesis," Master Thesis, University of Florida, Gainesville, 1975.
2. Cutkosky, M.R., and Wright, P.K., "Active Control of a Compliant Wrist in Manufacturing Tasks," ASME J. Eng. for Industry, Vol. 108, No. 1, pp. 36-43, 1986.
3. Griffis, M.W., "Kinestatic Control : A Novel Theory for Simultaneously Regulating Force and Displacement," Ph.D Dissertation, Department of Mechanical Engineering, University of Florida, Gainsville, 1991.
4. Hain, K., Applied Kinematics, McGraw Hill Book Company, New York, 1967.
5. Jobes, C.C., "Synthesis of a Controllable Circuit Breaker Mechanism," Ph.D Dissertation, West Virginia University, 1987.
6. Lakshminarayana, K. and Dizioglu, B., "Synthesis of Spring-Restrained Mechanisms for Amplitude-Independent Natural Frequency of Oscillation," Mechanism and Machine Theory, Vol. 7, pp. 167-190, 1972.

7. Matthew, G.K., "The Closed Form Synthesis of Spring Parameters to Balance Given Force Functions in Planar Mechanisms," Ph.D Dissertation, University of Florida, Gainsville, 1975.
8. Nevins, J.L. and Whitney, D.E., "The Force Vector Assembly Concept," Proc. First CISM-IFTOMM Symposium on Theory and Practice of Robots and Manipulators," Udine, Italy, 1973.
9. Oledzki, A., "Dynamics of Piano Mechanisms," Mechanism and Machine Theory, Vol. 7, pp. 373-385, 1972.
10. Srinath, K. and Karmakar, R., "Vibratory Conveying by Nonsinusoidal Excitation," Proc. the Institution of Mechanical Engineers, Vol. 202, pp. 405-408, 1988.
11. Streit, D.A. and Gilmore, B.J., "Perfect Spring Equilibrators for Rotatable Bodies," 1988 ASME 20th Biannual Mechanism Conf., Kissimmi, FL, pp. 487-496.
12. Yi, B.J. and Freeman, R.A., "Antagonistic Stiffness Characteristics Inherent in the Coordination of General Robotic Linkage Systems," Flexible Mechanism, Dynamics, and Robot Trajectories, 1990 ASME 21st Biennial Mechanism Conf., Chicago, Il, DE-Vol. 24, pp. 133-141.
13. Yi, B.J. and Freeman, R.A., "Modeling and Control of the Impedance Properties in Biomechanical Systems," Proc. ASME Winter Annual Meeting Conf., Bio-engineering Division, Atlanta, GA. Dec. 1991.
14. Yi, B.J. and Freeman, R.A., "Synthesis of Actively Adjustable Springs by Antagonistic Redundant Actuation," To Apear in ASME Journal of Dynamic Systems, Measurements, and Control, Sept. 1992.
15. Yi, B.J. "Analysis of Redundantly Actuated Mechanisms with Applications to Design and Control of Advanced Robotic Systems," Ph.D Dissertation, Department of Mechanical Engineering, The University of Texas Austin, Austin, TX, Dec. 1991.
16. Van Brussel, H. and Simons, J., "The Adaptive Compliance Concept and its Use for Automatic Assembly by Active Force Feedback Accomodations," Proc. 9th Symp. on Industrial Robots, SME, pp. 167-181, 1979.
17. Whitney, D.E. and Nevins, J.L., "What is the Remote Center Compliance (RCC) and What Can It Do ?," Internal Report P-728, The Charles Stark Draper Lab., Cambridge, MA., 1978.

NON-INVARIANCE DUE TO DYNAMICS FORMULATION IN THE FORCE OPTIMIZATION PROBLEM FOR REDUNDANTLY ACTUATED ROBOTIC SYSTEMS

MEYER NAHON
Department of Mechanical Engineering
University of Victoria
Victoria, British Columbia, Canada

ABSTRACT

Mechanical hands, legged vehicles and cooperating manipulators can be classified as *redundantly actuated robotic systems.* The redundant actuation comes about because these mechanisms are equipped with more actuators than they have degrees of freedom. As a result, many different sets of actuator torques can be used to produce a desired motion of the system—some of these causing the actuators to act antagonistically. In order to avoid this situation, optimization techniques are commonly proposed to find an optimal set of actuator torques. These are then used as force/torque setpoints for the actuators in a hybrid position/force control system.

When solving for an optimal set of actuator torques, the minimization of 'internal force' is commonly proposed as a desirable objective. This is achieved by minimizing the norm of the solution vector to the underconstrained inverse dynamics problem describing the motion of these systems. However, there exists many ways in which the inverse dynamics equations can be formulated, and each of these will result in a *different* minimum-norm solution. While other types of non-invariance of the solution to the underconstrained inverse dynamics problem have been noted by a number of authors, this paper treats the problem of non-invariance of the solution with variations in the dynamics formulation. A physical interpretation of this non-invariance is presented and an example is used to highlight the effect it can have on the solution produced.

INTRODUCTION

Controllers for robotic systems often require the use of an *inverse dynamics* model in order to calculate the actuator torques required in order to achieve a desired motion of the system. In the case of systems containing closed kinematic chains, as shown in Figure 1(a), this problem is usually formulated as a set of simultaneous linear equations which must be solved for actuator and constraint wrenches. For the systems under consideration, it is presumed that the desired motion of one *pole* is known—the grasped object in the case of a mechanical hand of cooperating manipulators, and the body of the machine in the case of a walking machine. Using this information, it is possible to calculate the motion of every other member in the system if it is assumed that the p fingers, manipulators or legs are not kinematically redundant. This consists of the independent solution of p inverse kinematics problems which yield all the joint coordinates, velocities and accelerations in the system. Given these inputs, there are a number of ways in which the dynamics equations of the system can be written, each of which reduces to a system of linear equalities of the form

$$\mathbf{A}_1\mathbf{x} = \mathbf{b}_1 \tag{1}$$

The work reported here was completed with the support of the Natural Sciences and Engineering Research Council of Canada (NSERC Research Grant #OGP 0105684).

where $\mathbf{A}_1$ is a matrix of dimension $m \times n$, $\mathbf{x}$ is a vector of dimension n which contains the wrenches acting in the system, while $\mathbf{b}_1$ is a vector of dimension m which represents the motion of the system. The numerical values of m and n depend on the method used to formulate the dynamics equations. The following sections discusses two ways in which this can be done—each differing in complexity, computational requirements, and amount of information generated. Irrespective of how the equations are formulated, if the system is redundantly actuated, n will always be greater than m by an amount r, the degree of actuation redundancy.

The fact that the linear system given by eq.(1) contains more unknowns than equations implies that there exist many solutions which will satisfy the equations. This allows us to formulate an optimization problem in order to find the 'best' solution in a pre-established sense. In this light, eq.(1) constitutes a set of *constraints* to an optimization problem in which we minimize a certain *objective function*, $f(\mathbf{x})$, of vector $\mathbf{x}$ of *design variables*. Often, the objective function to be minimized is chosen as $f(\mathbf{x}) = \mathbf{x}^T\mathbf{x}$, thereby minimizing the solution component in the nullspace of $\mathbf{A}_1$. Since that component satisfies $\mathbf{A}_1\mathbf{x} = \mathbf{0}$, it contributes nothing to the motion of the system represented by the right-hand-side of eq.(1), and is therefore associated with the 'internal forces' in the system. A number of authors have noted the non-invariance of this solution with changes in units and changes in frames of reference [1], [2], [3], [4]. The present work discusses the non-invariance in the minimum-norm solution arising from variations in the formulation of eq.(1).

DYNAMICS

Compact Formulation

The most compact and most commonly used method of formulating the dynamics equations—see e.g., [4], [5] [6]—considers only the motion of the grasped payload. This has the advantage that the size of the resulting set of linear equations is greatly reduced and permits a fast solution. However, it only includes the contact wrenches in the vector of design variables, $\mathbf{x}$. In the case of mechanical hands and cooperating manipulators, the single body for which the equations of motion are written is the grasped object, while in the case of walking machines, that body is the complete machine considered as a single rigid body—i.e., neglecting the internal joints of the machine.

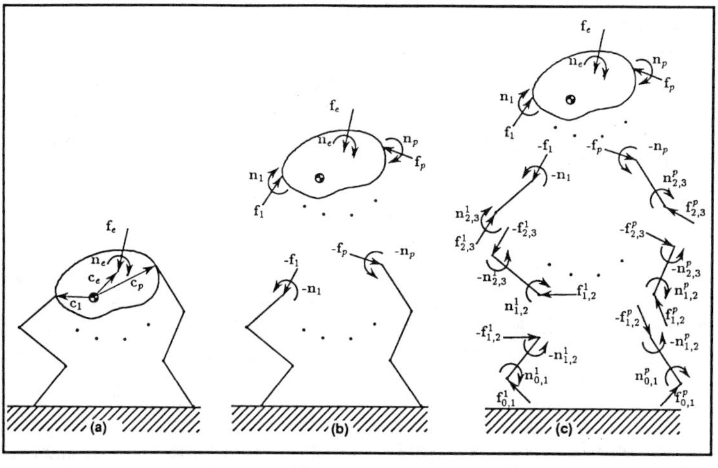

Figure 1 A Common Payload Supported by p Fingers, Legs or Manipulators

Referring to Figure 1(b), the vector equations of motion of the payload are given by

$$\sum_{i=1}^{p} \mathbf{f}_i = m_o(\mathbf{a}_o + \mathbf{g}) - \mathbf{f}_e \tag{2a}$$

$$\sum_{i=1}^{p} (\mathbf{n}_i + \mathbf{c}_i \times \mathbf{f}_i) = \mathbf{I}_o\dot{\boldsymbol{\omega}}_o + \boldsymbol{\omega}_o \times \mathbf{I}_o\boldsymbol{\omega}_o - (\mathbf{n}_e + \mathbf{c}_e \times \mathbf{f}_e) \tag{2b}$$

where m_o and $\mathbf{I}_o$ represent the payload's mass and centroid moment of inertia, $\boldsymbol{\omega}_o$ and $\mathbf{a}_o$ are its angular velocity and translational acceleration, and $\mathbf{c}_i$ is the vector from the payload's centroid to contact point i. Furthermore, $\mathbf{f}_e$ is the vector of the net external applied force on the payload, of dimension 3, $\mathbf{n}_e$ is the vector of net external applied torques, also of dimension 3 and $\mathbf{c}_e$ is the vector from the object's centroid to the point of application of the external force. These can be written more compactly in the form of eq.(1) with

$$\mathbf{A}_1 = \begin{bmatrix} \mathbf{0} & \mathbf{1} & \mathbf{0} & \mathbf{1} & \cdots & \mathbf{0} & \mathbf{1} \\ \mathbf{1} & \mathbf{C}_1 & \mathbf{1} & \mathbf{C}_2 & \cdots & \mathbf{1} & \mathbf{C}_p \end{bmatrix}, \qquad \mathbf{x} = \begin{bmatrix} \mathbf{n}_1 \\ \mathbf{f}_1 \\ \vdots \\ \mathbf{n}_p \\ \mathbf{f}_p \end{bmatrix} \tag{3a}$$

$$\mathbf{b}_1 = \begin{bmatrix} m_o(\mathbf{a}_o + \mathbf{g}) - \mathbf{f}_e \\ \mathbf{I}_o\dot{\boldsymbol{\omega}}_o + \boldsymbol{\omega}_o \times \mathbf{I}_o\boldsymbol{\omega}_o - (\mathbf{n}_e + \mathbf{c}_e \times \mathbf{f}_e) \end{bmatrix}, \qquad \mathbf{C}_k \equiv \frac{\partial(\mathbf{c}_k \times \mathbf{v})}{\partial \mathbf{v}} \tag{3b}$$

and $\mathbf{v}$ is an arbitrary vector of dimension 3. Using eqs.(3a) and (3b), eq.(1) now represents 6 scalar equations in, at most, $6p$ unknowns—i.e, p contact wrenches of 6 elements each. Matrix $\mathbf{A}_1$ is of dimension $6 \times 6p$ when a general grasping wrench of dimension 6 can be exerted, rendering the system underdetermined whenever $p > 1$. This will usually be the case for multiple manipulators handling a common payload. If contact torques cannot be exerted—i.e., the contact wrench is of dimension 3—, the matrix $\mathbf{A}_1$ is of $6 \times 3p$, and the system is underdetermined whenever $p > 2$. This situation normally arises in walking machines and mechanical hands since they cannot usually exert torques at the foot/ground or fingertip/object contact.

Complete Formulation

A more comprehensive method for writing the dynamics equations of the system explicitly considers all wrenches acting in the system. The force and moment balance equations are written for each link in the chain resulting in a large system of equations, where vector $\mathbf{x}$ contains the contact, actuator *and* constraint wrenches.

The system is first broken down into its individual rigid bodies as shown in Figure 1(c). The Newton-Euler equations for link j in parallel path i are written as:

$$\mathbf{f}^i_{j-1,j} - \mathbf{f}^i_{j,j+1} = m^i_j(\mathbf{a}^i_{cj} + \mathbf{g}) \tag{4a}$$

$$\mathbf{n}^i_{j-1,j} - \mathbf{n}^i_{j,j+1} + \hat{\mathbf{c}}^i_{j,j-1} \times \mathbf{f}^i_{j-1,j} - \hat{\mathbf{c}}^i_{j,j} \times \mathbf{f}^i_{j,j+1} = \mathbf{I}^i_j\dot{\boldsymbol{\omega}}^i_j + \boldsymbol{\omega}^i_j \times \mathbf{I}^i_j\boldsymbol{\omega}^i_j \tag{4b}$$

where $\mathbf{f}^i_{k,k+1}$ and $\mathbf{n}^i_{k,k+1}$ are, respectively, the force and moment applied by link k of path i to link $k+1$ in the same path, and $\hat{\mathbf{c}}^i_{k,l}$ represents a vector from the centroid of link k to the origin of frame l. The centroid moment of inertia and mass of link j are denoted by $\mathbf{I}^i_j$ and m^i_j, respectively. Finally, $\mathbf{a}^i_{cj}$ is the translational acceleration of the centroid of link j, while $\boldsymbol{\omega}^i_j$ is the angular velocity of that link.

The wrench acting at the distal end of the last link of each path is equal and opposite to the wrench applied by that link on the payload. Therefore, we can write

$$\mathbf{f}_i = \mathbf{f}^i_{b_i,b_i+1}, \qquad \mathbf{n}_i = \mathbf{n}^i_{b_i,b_i+1}, \qquad i = 1,\ldots,p \tag{5}$$

where b_i denotes the number of links in th i-th finger, leg or manipulator. Equations (4a) and (4b) are written for each link in each path of the system, while eqs.(2a) and (2b) are written for the moving pole. Once these vector equations are expressed in a single frame, they can be assembled into a single system of simultaneous equations of the form given by eq.(1) with

$$\mathbf{A}_1 = \begin{bmatrix} \mathbf{0} & \mathbf{1} & \cdots & \mathbf{0} & \mathbf{1} & \mathbf{0} & \mathbf{0} & \mathbf{0} & \mathbf{0} & \cdots & \mathbf{0} & \mathbf{0} \\ \mathbf{1} & \mathbf{C}_1 & \cdots & \mathbf{1} & \mathbf{C}_p & \mathbf{0} & \mathbf{0} & \mathbf{0} & \mathbf{0} & \cdots & \mathbf{0} & \mathbf{0} \\ \mathbf{0} & \mathbf{0} & \cdots & \mathbf{0} & \mathbf{0} & \mathbf{0} & \mathbf{1} & \mathbf{0} & -\mathbf{1} & \cdots & \mathbf{0} & \mathbf{0} \\ \mathbf{0} & \mathbf{0} & \cdots & \mathbf{0} & \mathbf{0} & \mathbf{1} & \widehat{\mathbf{C}}^1_{1,0} & -\mathbf{1} & -\widehat{\mathbf{C}}^1_{1,1} & \cdots & \mathbf{0} & \mathbf{0} \\ \vdots & & \ddots & & \vdots & \vdots & & & & \ddots & & \vdots \\ \mathbf{0} & \mathbf{0} & \cdots & \mathbf{0} & -\mathbf{1} & \mathbf{0} & \mathbf{0} & \mathbf{0} & \mathbf{0} & \cdots & \mathbf{0} & \mathbf{1} \\ \mathbf{0} & \mathbf{0} & \cdots & -\mathbf{1} & -\widehat{\mathbf{C}}^p_{b_{pp},b_{pp}} & \mathbf{0} & \mathbf{0} & \mathbf{0} & \mathbf{0} & \cdots & \mathbf{1} & \widehat{\mathbf{C}}^p_{b_{pp},b_{pp}-1} \end{bmatrix} \tag{6a}$$

$$\mathbf{x} = \begin{bmatrix} \mathbf{n}_1 \\ \mathbf{f}_1 \\ \vdots \\ \mathbf{n}_p \\ \mathbf{f}_p \\ \mathbf{n}^1_{0,1} \\ \mathbf{f}^1_{0,1} \\ \mathbf{n}^1_{1,2} \\ \mathbf{f}^1_{1,2} \\ \vdots \\ \mathbf{n}^p_{b_{pp}-1,b_{pp}} \\ \mathbf{f}^p_{b_{pp}-1,b_{pp}} \end{bmatrix}, \qquad \mathbf{b}_1 = \begin{bmatrix} m_o(\mathbf{a}_o + \mathbf{g}) \\ \mathbf{I}_o\dot{\boldsymbol{\omega}}_o + \boldsymbol{\omega}_o \times \mathbf{I}_o\boldsymbol{\omega}_o \\ m^1_1(\mathbf{a}^1_{c1} + \mathbf{g}) \\ \mathbf{I}^1_1\dot{\boldsymbol{\omega}}^1_1 + \boldsymbol{\omega}^1_1 \times \mathbf{I}^1_1\boldsymbol{\omega}^1_1 \\ \vdots \\ m^p_{b_{pp}}(\mathbf{a}^p_{cb_{pp}} + \mathbf{g}) \\ \mathbf{I}^p_{b_{pp}}\dot{\boldsymbol{\omega}}^p_{b_{pp}} + \boldsymbol{\omega}^p_{b_{pp}} \times \mathbf{I}^p_{b_{pp}}\boldsymbol{\omega}^p_{b_{pp}} \end{bmatrix}, \qquad \widehat{\mathbf{C}}^i_{k,l} \equiv \frac{\partial(\widehat{\mathbf{c}}^i_{k,l} \times \mathbf{v})}{\partial \mathbf{v}} \tag{6b}$$

for any 3-dimensional vector $\mathbf{v}$.

DISCUSSION

As mentioned earlier, optimization techniques can now be applied to minimize the norm of the vector $\mathbf{x}$. However, as is apparent from eqs.(3a) and (6b), the elements of that vector differ depending on the formulation used for the dynamics equations. We can therefore expect that the minimization of $\mathbf{x}^T\mathbf{x}$ will yield different results for these two formulations.

Since the minimum-norm solution minimizes the component of the vector $\mathbf{x}$ which contributes nothing to the motion represented by the right-hand-side of eq.(1), we can interpret the two solutions as follows:

1. The minimum-norm solution to eqs.(1), (3a), and (3b) minimizes the forces which contribute nothing to the motion of the payload, since this is the motion represented by the right-hand-side in this case.

2. The minimum-norm solution to eqs.(1), (6a), and (6b) minimizes the forces which contribute nothing to the motion of the payload and the links of the manipulators, since this is the motion represented by the right-hand-side in this case.

NUMERICAL EXAMPLE

The two formulations of the dynamics equations are now applied to the two-fingered planar mechanical hand shown in Figure 2 in order to highlight their differences. The finger links have a mass of 1 kg each, while the grasped object object has a mass of 2 kg. The object, initially with its center 150 mm below the palm is raised by 100 mm in 0.1 s. The object has zero velocity and acceleration at the beginning and end of its trajectory.

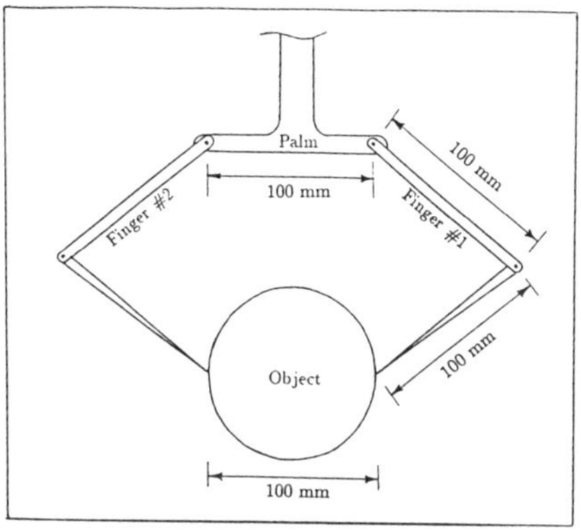

Figure 2 A Planar Two-Fingered Hand Grasping an Object

Using the compact formulation, the vector $\mathbf{x}$ contains four variables: the x and y components of the two fingertip/object contact forces. There are three scalar equations governing the motion of the object in the plane, and so $r = 1$ (the system has one more actuator than degrees of freedom). When the complete formulation is used, the vector $\mathbf{x}$ contains 16 elements, and the system is governed by 15 scalar equations (3 equations for each of 5 bodies in the system), and again $r = 1$. It should be noted that no inequality constraints have been included to account for the limitations of the passive fingertip/object contacts and the present formulation effectively considers that the fingertips are actively grasping the object. Reference [7] outlines how these limitations may be included in the formulation of the optimization problem.

The results obtained when minimizing $\mathbf{x}^T\mathbf{x}$ for each formulation are shown in Figure 3. The actuator torques τ_1 and τ_2 acting in Finger #1 are shown, as are the fingertip/object contact forces in the x and y directions. It is apparent that the minimum-norm solutions of these two systems differs substantially. Because the motion of the object is purely vertical, the first formulation produces a solution with no horizontal x-component of the contact force. By contrast, the complete formulation results in a solution in which this component is non-zero, and is opposed by an equal and opposite horizontal force at the tip of Finger #2 (not shown). The vertical y-component of the contact force is identical for both formulations. Finally, because the contact forces differ for the two formulations, the actuator torques also differ.

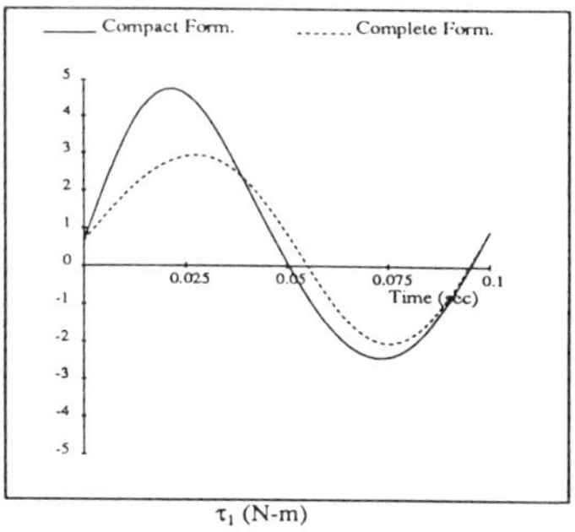

τ_1 (N-m)

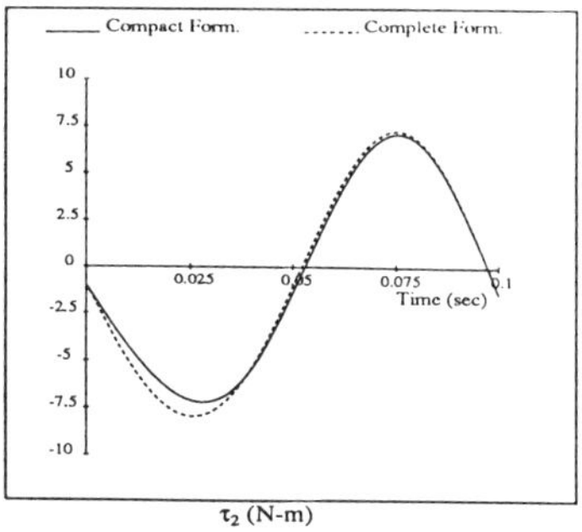

τ_2 (N-m)

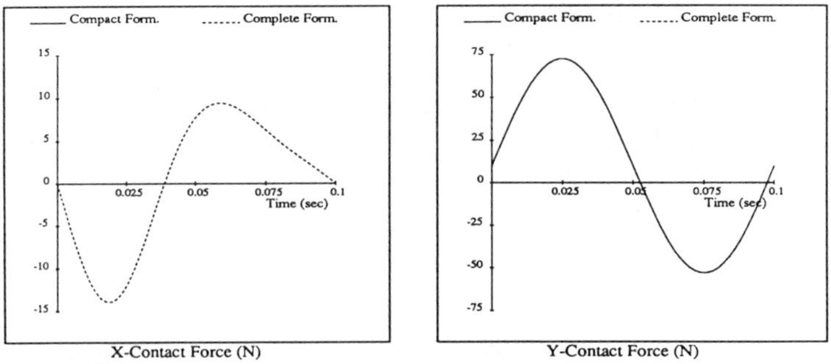

Figure 3 Actuator Torques and Contact Forces for Finger # 1

CONCLUSIONS

The method used to formulate the inverse dynamics equations for redundantly-actuated robotic systems has a significant effect on the solution obtained when the minimum-norm solution to these equations is chosen. This occurs because the vector whose norm is being minimized contains components which vary with the formulation used. Two formulations of the dynamics equations were presented here. In the first case, the vector **x** contains only the contact wrenches acting between the common payload and the tips of the fingers, legs or manipulators in the system. In the second case, the vector **x** contains those wrenches, as well as all the actuator and constraint wrenches acting at the joints of the figers, legs or manipulators.

In both cases, the minimum-norm solution minimizes the solution component which contributes nothing to the motion of the system considered. In the first case, only the motion of the payload is considered, and the solution obtained therefore contributes only to the motion of the payload. In the second case, the motion of the payload *and* the fingers, legs or manipulators is considered, and so the solution contributes to the motion of the complete system. A numerical example was presented to show that these solutions differ substantially from each other.

References

[1] Lipkin, H., and Duffy, J., 1988, "Hybrid Twist and Wrench Control for a Robotic Manipulator", *Transactions of the ASME, Journal of Mechanisms, Transmissions, and Automation in Design*, Vol. 110, pp. 138–144.

[2] Bruyninckx, H., 1991, "Some Invariance Problems in Robotics", Katholieke Universiteit Leuven, Dept. of Mech. Eng., Division PMA, Report 91R4.

[3] Murphy, S. H., and Wen, J. T., 1991, "Force Decomposition in Robot Force Control", *Proceedings of the 1991 American Control Conference*, Boston, Mass., pp. 2533–2538.

[4] Nahon, M., and Angeles, J., 1991, "Minimization of Power Losses in Cooperating Manipulators", *Proceedings of the 1991 American Control Conference*, Boston, Mass., pp. 3044–3049.

[5] Alberts, T. E., and Soloway, D. I., 1988, "Force Control of a Multi-Arm Robot System", *Proceedings of the 1988 IEEE International Conference on Robotics and Automation*, Philadelphia, pp. 1490–1496.

[6] Nakamura, Y., Nagai, K., and Yoshikawa, T., 1987, 'Mechanics of Coordinative Manipulation by Multiple Robotic Mechanisms', *Proceedings of the 1987 IEEE International Conference on Robotics and Automation*, Raleigh, N.C., pp. 991–998.

[7] Nahon, M., and Angeles, J., 1991, "Optimization of Dynamic Forces in Mechanical Hands", *Transactions of the ASME, Journal of Mechanical Design*, Vol. 113, pp. 176–173.

ON THE MANIPULABILITY OF TWO-ARM ROBOTS

Yann BOUFFARD and Pierre DAUCHEZ
LIRMM - UMR C 9928- Université Montpellier II
34095 Montpellier Cedex 5 - FRANCE

ABSTRACT

The goal of this paper is to compare various ways of defining a force manipulability for two-arm robots. This analysis is done in the framework of the symmetric hybrid position/force task description for two-arm robots, proposed by Uchiyama and Dauchez a few years ago. More particularly, the assembly of two objects with two arms is considered. Several force ellipsoids are defined in this case and their usefulness in achieving the task is pointed out. Simple examples are used all along the paper to clarify the results.

INTRODUCTION

It is well known that multi-arm systems offer enhanced capabilities over single-arm mechanisms. For example, manipulation of heavy or flexible objects cannot be properly done with one arm. We happen to be interested in using two cooperative arms for assembling two objects, with each being held by one arm [1]. The contact forces between the objects are controlled in our approach. We have experimentally observed (on our real setup built around two Puma's 560) that some configurations of the arms are more convenient than others for achieving such an assembly task, although some other criteria like the assembly strategies can also play an important role in the success rate [2]. This is due to the fact that, depending upon the arms configuration, the contact forces can or cannot be properly controlled. Therefore, the key issue is to find a good location of the arms for achieving the task.

This problem has been treated by several researchers in the case of single-arm robots and the concept of manipulability has then been introduced [3, 4]. In this case, this manipulability can be represented by ellipsoids, as recalled in this paper. However, the "classical" ellipsoids do not take into account the possible differences between the robot actuators capabilities. If one does, weighted ellipsoids can be obtained. We compare them to the classical ones in the case of a simple 2-degree-of-freedom (dof) planar robot. Polytopes are also considered in this comparison. Our idea is then to consider two cooperative arms as a single robot (with two arms), such that the results obtained in the single-arm case can easily be adapted to our assembly problem with two arms. To do that, we use the definitions of external and internal forces initially proposed in [5] for the manipulation of a single rigid object, and we present an adaptation of these definitions to our particular problem. It allows us to define various ellipsoids in the two-arm case: several external force ellipsoids, several internal force ellipsoids and one generalized force ellipsoid. The information they contain is analyzed and their utilization in our assembly task is clarified. Finally, conclusions and future research plans are given.

MANIPULABILITY OF A SINGLE-ARM ROBOT

Definitions

The *manipulability* of a robot is a way of measuring its manipulating ability. According to that concept, a force manipulability ellipsoid and a velocity manipulability ellipsoid can be defined for a single arm. In the rest of this paper, we will only be concerned with the force manipulability concept. With this assumption, let us consider a n-dof arm and a m-dimensional taskspace. The well-known static relationship between the task force vector **h** and the joint torque vector τ implies the use of the transpose of the robot jacobian matrix **J** as follows:

$$\tau = \mathbf{J}^T \mathbf{h} \tag{1}$$

Then, the unit sphere in the joint torque space (normalized torques):

$$\tau^T \tau = 1 \tag{2}$$

maps into the task force space ellipsoid (for a given robot configuration):

$$\mathbf{h}^T [\mathbf{J}\ \mathbf{J}^T]\ \mathbf{h} = 1 \tag{3}$$

which is called *force manipulability ellipsoid.* It of course depends on the robot configuration, via ($\mathbf{J}\ \mathbf{J}^T$). It is usually said that this ellipsoid represents the maximum force the robot can exert in all the task directions, for a given configuration. It can therefore be used to define the best robot location to achieve a task for which the force to be exerted or controlled is well defined. Actually, if we consider a subspace of the joint torque space limited by the actual maximum amplitudes of the robot torques (i.e. actuators capabilities), we obtain a so-called *torque polytope*, approximated by a sphere in eq.2. This polytope maps into a *force polytope* by transformation (1). The force ellipsoid is again only an approximation of this force polytope. It can be numerically shown that this approximation is less and less accurate when n increases. However, it is still interesting to use it, because of the analytical form (eq.3) that the polytope does not have.

In the case of a 2-dof arm with revolute joints and unit length links evolving in a (x,y) plane, Figure 1 presents the torque polytope (dotted surface) and the torque sphere (τ_1 and τ_2 are the two components of τ) and Figure 2 presents their mapping in the task space, for a configuration where the robot joint angles are $q_1 = +127$ degrees and $q_2 = -127$ degrees. The approximation is quite obvious, and is even worse if the actuators do not have the same maximum torques, as will be seen now.

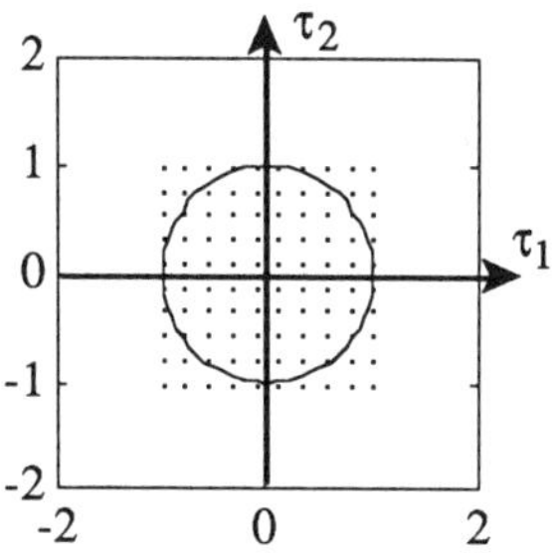

FIG. 1. Torque space

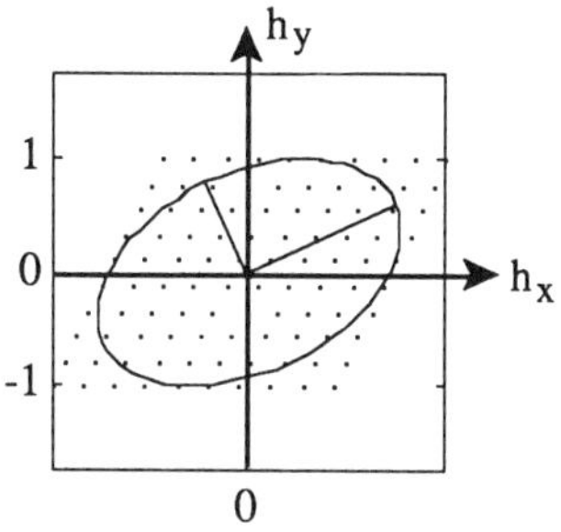

FIG. 2. Task space

Weighted ellipsoids

For the same example as before, if $\tau_{2max} = 1.5\ \tau_{1max}$, Figure 3 shows the new torque polytope, of which the unit sphere becomes a terrible approximation. A better solution is then to define a *weighted ellipsoid* instead of a unit sphere by:

$$\tau^T\ [\mathbf{W}\ \mathbf{W}^T]\ \tau = 1 \qquad (4)$$

where **W** is a weighting diagonal matrix, the ith component of which is $1/\tau_{imax}$. This torque ellipsoid (Figure 3) maps into a weighted force manipulability ellipsoid defined by:

$$\mathbf{h}^T\ [\mathbf{J}\ \mathbf{W}\ \mathbf{W}^T\ \mathbf{J}^T]\ \mathbf{h} = 1 \qquad (5)$$

which is a much better approximation of the force polytope than the non-weighted force ellipsoid, as can be seen in Figure 4. In addition, one can notice that the principal axes of this new ellipsoid are different from that of the "classical" one (Figure 2). This means that the non-weighted force ellipsoid generally gives wrong results on the possible force magnitude in every task space direction. Finally, a very important point is that we have computationally proved (but no space is available in this paper to detail the algorithms) that we can extract from the force polytope (with a least-square method) an ellipsoid which is identical to the proposed weighted ellipsoid. This is very interesting to compute some force ellipsoids in the case of two-arm robots, as will be seen later in this paper.

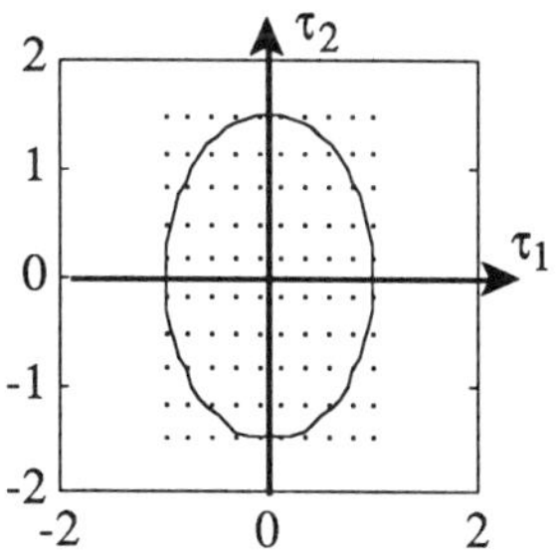

FIG. 3. Weighted torque ellipsoid

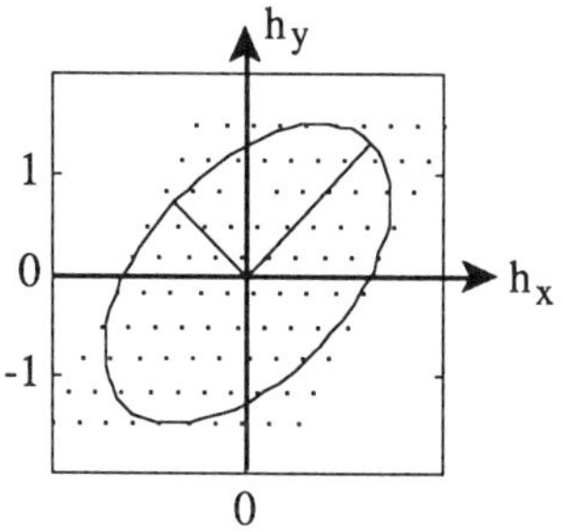

FIG. 4. Weighted force ellipsoid

STATICS OF A TWO-ARM ROBOT

In order to extend the concept of force manipulability to two-arm robots, one must find a definition of a task force vector **h** for this new mechanism. Since we are mainly interested in controlling the contact forces between two objects, a definition of such forces could be sufficient. Nevertheless, we prefer to adapt to our assembly problem the vector **h** defined in [5] for the coordinated manipulation of a single rigid object. In this approach, **h** includes a so-called external force/moment vector $\mathbf{f}_a$ (producing the motion of the object) and a so-called internal force/moment vector $\mathbf{f}_r$ (creating constraints in the object). The definitions of these vectors is as follows:

$$\mathbf{f}_a = \mathbf{f}_{b1} + \mathbf{f}_{b2} \tag{6}$$

$$\mathbf{f}_r = 1/2\,(\mathbf{f}_{b1} - \mathbf{f}_{b2})$$

where $\mathbf{f}_{bi}$ represents the force/moment vector exerted by arm i on the object at the tip of *the virtual stick i*. The virtual stick i is defined as an imaginary rigid body joining the end effector i to a user-defined point on the object. Therefore, the tips of the virtual sticks coincide (and more strictly two reference frames attached to these tips coincide), which yields the extremely simple expressions reported in eq.6.

For the assembly of two objects being held by two arms, we propose to regard the two objects in contact as a single object, to define two virtual sticks between the end effectors and the contact point and to use $\mathbf{f}_r$ in eq.6 to describe the "contact forces". Of course $\mathbf{f}_a$ can also be used, in the sense that this external force vector produces the global motion of the set {arm 1-objects being assembled-arm 2}. This global motion can be used to go to a location with a better internal force manipulability.

A problem is that during the insertion, the relative motion of the objects will yield a large discrepancy between the tips of the virtual sticks, so that eq.6 does not hold anymore. To cope with this difficulty, a solution is to make the sticks vary in function of time, so that their tips always coincide [1].

For the arm i, there exists a jacobian matrix $\mathbf{J}_i$ such that:

$$\tau_i = \mathbf{J}_i^T \mathbf{f}_{bi} \tag{7}$$

where τ_i represents the joint torque vector of arm i. Therefore, from the definitions of eq.6, we obtain the following equation:

$$\tau = \begin{bmatrix} \tau_1 \\ \tau_2 \end{bmatrix} = \begin{bmatrix} 1/2\mathbf{J}_1^T & \mathbf{J}_1^T \\ 1/2\mathbf{J}_2^T & -\mathbf{J}_2^T \end{bmatrix} \begin{bmatrix} \mathbf{f}_a \\ \mathbf{f}_r \end{bmatrix} = \mathbf{J}^T \mathbf{h} \tag{8}$$

which represents the mapping between the torque space and the force space for the two-arm robot.

FORCE ELLIPSOIDS FOR A TWO-ARM ROBOT

Assume, for the sake of simplicity, that all the actuators (of the two arms) have the same torque capabilities. From eq.8, we can define a force ellipsoid whose equation is given by eq.3, with the definitions of **h** and **J** of eq.8. If the actuators capabilities are different, it is not difficult to define a weighted force ellipsoid for the two-arm robot, as presented before for the single arm case. We call it a *generalized force ellipsoid*, because **h** is a generalized force/moment vector which includes external and internal forces and moments. One must notice that this ellipsoid is defined in the **h** task space, which is the cartesian product of the "classical" force task space (the one appearing in eq.3 for example) by itself.

However, other ellipsoids can be defined. As a matter of fact, eq.7 allows us to define a force polytope for arm i, which is a mapping of the torque polytope in the virtual stick force task space. These two force polytopes (one per arm) can be combined according to eq.6 to get an external force polytope and an internal force polytope for the two-arm system, from which we can extract an *external force ellipsoid* and an *internal force ellipsoid* by a least square method. As mentioned before, these two ellipsoids are identical to those which are analytically defined by:

$$\mathbf{f_a}^T [\mathbf{J_a} \mathbf{J_a}^T] \mathbf{f_a} = 1 \qquad (9)$$

$$\mathbf{f_r}^T [\mathbf{J_r} \mathbf{J_r}^T] \mathbf{f_r} = 1$$

where $\mathbf{J_a}^T$ and $\mathbf{J_r}^T$ represent the matrices which relate the two-arm robot torque vector τ of eq.8 to $\mathbf{f_a}$ and $\mathbf{f_r}$, respectively. Therefore, from eq.8, it can be seen that:

$$\mathbf{J_a}^T = \begin{bmatrix} 1/2\mathbf{J_1}^T \\ 1/2\mathbf{J_2}^T \end{bmatrix} \text{ and } \mathbf{J_r}^T = \begin{bmatrix} \mathbf{J_1}^T \\ -\mathbf{J_2}^T \end{bmatrix} \qquad (10)$$

As an example, let us consider two prismatic 1 dof arms, the tips of which coincide (the virtual sticks are null), and moving along an x axis. In that case, $f_{bi} = \tau_i$ (which are now scalars: f_{bi} is the only component of $\mathbf{f_{bi}}$, along the x axis). Therefore $\mathbf{J_i}^T = 1$, $\mathbf{J_a} = [1/2 \;\; 1/2]$ and $\mathbf{J_r} = [1 \;\; -1]$. The equation of the generalized force ellipsoid in the (f_a, f_r) plane is then:

$$1/2\, f_a^2 + 2\, f_r^2 = 1 \qquad (11)$$

It shows a maximum magnitude of f_a equal to $\sqrt{2}$ and a maximum magnitude of f_r equal to $\sqrt{2}/2$. In addition, the equations of the external and internal force ellipsoids defined by eq.9 are respectively:

$$1/2\, f_a^2 = 1 \text{ and } 2\, f_r^2 = 1 \qquad (12)$$

which are segments on the x axis. They show the same maximum magnitudes for f_a and f_r than those given by eq.11. Obviously, from eq.6, the real maximum magnitude of f_a is 2 (with normalized torques) and that of f_r is 1. This difference is again due to the fact that the ellipsoids are approximations of the polytopes.

From these calculations, one can see that the external (respectively internal) force ellipsoid defined by eq.9 can be obtained from the generalized force ellipsoid by setting $f_r = 0$ (respectively $f_a = 0$).

But other cuts can be made in the generalized ellipsoid, which yield the creation of an infinity of external and internal force ellipoids. For instance, we may want to exert an external force corresponding to the weight of the manipulated object. We will then get the internal force ellipsoid corresponding to that external force constraint. Notice that this ellipsoid is defined in the "classical" force/moment space, while the generalized one is defined in the cartesian product of this space by itself. Therefore, we can conclude that there is somehow less information in the external and internal force ellipsoids than in the generalized one. However, they are easier to use once f_a (or possibly f_r for other applications than assemblies) is given.

CONCLUSION

Starting with the definition of the force manipulability ellipsoid for single-arm robots, we have shown the limitations of this concept and have then proposed its extension to the two-arm robot case. This extension has been made possible by considering the two arms as a single robot and by using a very simple static model of such a robot, based upon an ad hoc formulation of the task (virtual sticks). We have demonstrated that several force ellipsoids could be defined for a two-arm system. The results presented through very simple examples have shown that these ellipsoids do not all contain the same information but can all contribute to solving the problem of finding the best location to achieve a given task (assembly in our case).

However, for real 6 dof robots, a classical geometrical representation of these ellipsoids is not possible anymore (their dimensions are 6 or 12) and therefore their usefulness may not be obvious. Consequently, we are currently investigating a way of representing the manipulability which would be really user-friendly but would not loose the direction information contained in the ellipsoids (some scalar representations of the manipulability, such as the surface of the ellipsoid, do exist, are easy to use, but do not carry much information).

REFERENCES

1. P. DAUCHEZ and X. DELEBARRE, "Force-Controlled Assembly of Two Objects with a Two-Arm Robot", *Robotica*, volume 9, 1991, pp. 299-306.
2. Y. BOUFFARD-VERCELLI and E. DEGOULANGE, "Experimental Analysis of Assembly Strategies with a Force-Controlled Two-Arm Robot", *IEE Colloquium on Multi-Arm Robotics*, London, 17 June 1992, pp. 4/1-4/3.
3. T. YOSHIKAWA, "Manipulability of Robotic Mechanisms", *2nd Int. Symp. of Robotic Research*, Kyoto, August 1984, pp. 91-98.
4. S. CHIU, "Task Compatibility of Manipulator Postures", *Int. Journal of Robotics Research*, volume 7, n° 5, 1988, pp. 13-21.
5. M. UCHIYAMA and P. DAUCHEZ, "A Symmetric Hybrid Position/Force Control Scheme for the Coordination of Two Robots", *IEEE Int. Conf. on Robotics and Automation*, Philadelphia, 24-29 April 1988, pp. 350-356.

ROBUST CONTROL EXPERIMENTS WITH A FAST PARALLEL ROBOT

Philippe FRAISSE and François PIERROT
LIRMM - UMR C 9928- Université Montpellier II
34095 Montpellier Cedex 5 - FRANCE

ABSTRACT

In order to perform fast motions with parallel robots, some kind of robust control is definitely necessary. As a first step towards such a control scheme, we mainly present here the results we have obtained on a fast 3 DOF DELTA-type parallel robot. To avoid the chattering phenomenon encountered with classical **V**ariable **S**tructure **C**ontrol laws, we present here a modified algorithm called **VSC-HF** (High Frequency). Then, the chattering phenomenon disappears and one can build a robust control for fast motions. Several experimental results are presented to prove the efficiency and the robustness of our control algorithm.

INTRODUCTION

Many studies have been dedicated to the analysis, the design or the kinematic optimization of parallel robots and recently, a lot of new parallel structures have been proposed. All these very interesting works are mainly focused on different ways to solve the direct kinematic problem **[1]**, or to derive the dynamic models of closed-loop mechanical systems **[2]**. But one can notice that only few papers were dedicated to the control of these manipulators, even if some of them present very good features for high-speed motions of for heavy-payload carrying. For instance, the well known DELTA robot **[3]** is able to perform fast motions **[4]**, but it is often used with only very small objects. This may become a problem if one is interested in using industrial grippers or force sensors, or if one plans to move "heavy" objects during fast motions...

Obviously, the use of control schemes including the dynamic effects could be used, but the problems encountered during the identification process (static friction model ?) and the additionnal computation power required, lead us to propose other control schemes which do not need the precise knowledge of the robot dynamics. Moreover, we think that robust independent joint controllers could be an interesting feature for parallel robots: we might then servo robots only in joint space ...

In these conditions, control of complex parallel robots has to face the problem of working with a non-linear and ill-known system. Robust control, such as **V**ariable **S**tructure **C**ontrol **[5]**, allows us to avoid this difficulty. In the domain where this technique is valid, the system is forced to dynamically behave according to a differential equation defined *a priori* . This equation can be represented by a hypersurface in the phase space, called "sliding surface". In theory, with such an approach, one can expect to get performances independent of the payload and the acceleration effects ...

In this paper, we first recall the problem inherent to VSC technique: the chattering phenomenon. We present a modified algorithm of the VSC called VSC-HF. The last part of the paper is intended to prove the robustness of our new control scheme working on a DELTA robot prototype.

VSC-HF APPROACH

As usually done with the classical VSC approach, let us consider the linear model of a robot actuator such that:

$$\dot{\mathbf{X}} = \mathbf{A}\ \mathbf{X} + \mathbf{B}\ \mathrm{U} + \gamma \tag{1}$$

where $\mathbf{X}=[q\ \dot{q}\ ]^t$ is the state vector, q is the joint variable, **A** is the (2x2) state matrix, γ is the (2x1) external disturbance vector wich represents the resistant torque (this includes all the dynamics of the robot...), U is the control variable and **B** is a (2x1) control vector. Notice that the control variable U is defined as a discontinuous function **[6]** such that:

$$\left.\begin{array}{l} \mathrm{U}(\mathbf{X},t) = \alpha \ \text{ if } S(\mathbf{X},t) > 0 \\ \\ \mathrm{U}(\mathbf{X},t) = -\alpha \ \text{ if } S(\mathbf{X},t) < 0 \end{array} \quad (\alpha > 0) \right| \tag{2}$$

where $S(\mathbf{X},t)$ represents the so-called sliding surface.

To get a really simple robust controller we choose as α the physical maximum input voltage of the controlled actuators. Then the trajectory described in the phase plane will stay on the sliding surface until the maximun capabilities of the actuators are reached. In the case of a time-independent surface $S(\mathbf{X})$, the less restrictive existence condition to get the sliding mode is:

$$S.\dot{S} < 0 \tag{3}$$

If the condition expressed in Eq. **(3)** is always true, then the control system is governed by the differential equation of the sliding surface $\mathbf{S}(\mathbf{X})=0$. In the general case we choose:

$$S(\mathbf{X}) = \mathbf{C}\ .\ \tilde{\mathbf{X}} \tag{4}$$

where **C** is a (1x2) vector, $\tilde{\mathbf{X}}$ is the tracking error vector $(\mathbf{X_d}-\mathbf{X})$, and $\mathbf{X_d}$ is the desired state vector. Equation **(3)** allows us to calculate vector **C** which gives the sliding mode.

However this control scheme cannot be implemented because of a chattering phenomenon **[7]**: the use of the discontinuous term produces an oscillation on the state variables. We propose to introduce an additionnal term Y in the sliding surface to increase drastically the oscillation frequency far over the bandwidth of the controlled system and so to reduce the influence of the oscillation. We obtain the following equation:

$$S = \mathbf{C} \cdot \tilde{\mathbf{X}} + Y \tag{5}$$

$$\left.\begin{aligned} \dot{\mathbf{W}} &= \mathbf{A_{hf}} \cdot \mathbf{W} + \mathbf{B_{hf}} \cdot U \\ Y &= \mathbf{C_{hf}} \cdot \mathbf{W} \end{aligned}\right| \tag{6}$$

where $\mathbf{A_{hf}}$ is a (2x2) state matrix, $\mathbf{B_{hf}}$ is a (2x1) control vector, and $\mathbf{C_{hf}} = [1\ 0]$. We choose $\mathbf{A_{hf}}$ so that the system **(6)** represents a low-pass filter; **W** is the (2x1) state vector of this filter.
If we consider a perfectly discontinuous control, the low-pass filter order must be greater than 2. Since it is impossible to get a real system that works exactly according to Eq. **(2)** (hysteresis, bandwidth...), the filter order may be equal to 2.

Let us present now the numerical application for the 3-DOF fully-parallel DELTA robot and the implementation on our experimental setup. The travelling plate is connected to the base plate by three kinematic chains actuated by PARVEX brushless motors controlled in torque. Each motor is equipped with a digital encoder whose resolution is 10000 pulses per revolution.

We have first performed a rough identification process which consists in approximating the arm connected to the motor *via* a belt transmission (ratio: 1/10) to a second order system such that (for every actuator):

$$\dot{\mathbf{X}} = \begin{bmatrix} 0 & 1 \\ 0 & -1/T \end{bmatrix} \cdot \mathbf{X} + \begin{bmatrix} 0 \\ K_m K_{da} K_c/T \end{bmatrix} \cdot U + \begin{bmatrix} 0 \\ K_m K_{da} K_c \cdot \gamma_2/T \end{bmatrix} \tag{7}$$

where $\mathbf{X}=[q\ \dot{q}]^t$, K_m is the Voltage to Velocity static gain, K_c is the optical encoder gain, K_{da} is the Digital to Analog Converter gain, T is the time constant, q is the actual position and U is the input voltage and γ_2 is the external disturbance. We have identified the following values: K_m=837.7 rad/s/Volt, K_c=1592 pulses/rd, K_{da}=1/204.8 V/pulses and T= 0.4 sec. From Eq. **(4)**, the sliding surface is defined as:

$$S = [\,C_1 \ \ 1\,] \cdot (\mathbf{X_d} - \mathbf{X}) \tag{8}$$

where $\mathbf{X_d}$ is the desired state vector and C_1 is a positive constant.The input vector U is defined by (Eq. 2) with $\alpha = V_{max}$ (V_{max} = 10 V is the maximum input voltage for the servo-motor). The necessary sliding condition (Eq. **(3)**) gives the following result: $0 < C_1 < 200\ s^{-1}$. To overcome the chattering phenomenon, we use the additional term Y such that :

$$A_{hf} = \begin{bmatrix} 0 & 1 \\ -\omega_o^2 & -2.\omega_o \end{bmatrix} \qquad B_{hf} = \begin{bmatrix} 0 \\ -\omega_o^2.K_{hf} \end{bmatrix} \qquad \textbf{(9)}$$

where ω_o is the bandwidth of this second order low-pass filter defined by the state equation (Eq. **(9)**) and K_{hf} must be adjust such that the oscillation frequency be forced at the proper value.

EXPERIMENTATIONS ON A FAST PARALLEL ROBOT

We are going to present the DELTA robot controller and the experimental results that prove the efficiency and the robustness of the VSC-HF algorithm *versus* a classical PID (which is the classical independent joint controller...). We have used a multi-transputer system based on the INMOS T805 whose boards have been developed in our laboratory. Since the oscillation frequency f_0 ($f_0 = \omega_o/2\ \pi$) required by the new VSC-HF algorithm is, in the case of the DELTA robot, about 100 kHz, we built the non-linear block (Eq. **(2)**) and the low-pass filters (Eq. **(6)**) with analog circuits.
We first studied our approach with step responses tests (performed without payload). In a step response test, the desired state vector is: $\mathbf{X_d} = [\ 1\ 0\ ]^t$. If the system works in sliding mode then S=0. Thus, without the additional term Y, the system behaviour is described by:

$$\frac{1}{C_1}\dot{q} + q = 1 \qquad \textbf{(10)}$$

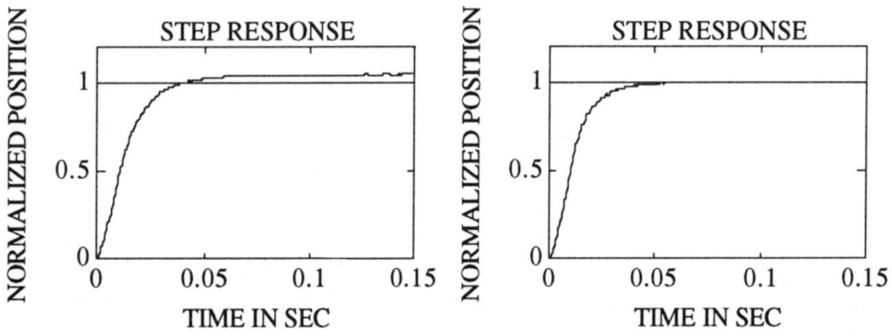

(a) FIG. 1. (b)

On FIG.1. (a), we show the system behaviour with a first-order sliding surface (that can be represented by a line in the phase space) defined by Eq. **(5)** ($C_1 = 70s^{-1}$). As expected, the response is defined by Eq. **(10)**, but Y introduces a steady state error (which depends on the resistant torque $\gamma_2(t)$ and on K_{hf}) whose value is:

$$\lim_{t \to \infty} (q_d - q)(t) = \frac{K_{hf}\,\gamma_2(t)}{C_1}$$

To eliminate the steady state error, we propose to add an integral term in the sliding surface. If the sliding condition is respected (Eq. **(3)**),we obtain the response shown on FIG. 1. (b).
With such a controller, we have carried out many tests to analyze the robustness of this control law. We compare the VSC-HF control law with the classical independent-joint PID controller. The performances criterion I_p is defined as:

$$I_p = \sum_{i=1}^{3} \int (q_{di} - q_i)^2 \, dt \qquad \textbf{(11)}$$

We have choosen 4 different payloads (no travelling plate, with travelling plate, 0.2 kg, 0.5 kg), and performed for each payload 3 motions with different velocities and accelerations (25 m/s^2, 37 m/s^2, 50 m/s^2). On FIG. 2, the trials number 1 to 3 are related to motions without the travelling plate (i.e. no coupling effect !), the trials number 4 to 7 are related to motions with the rods connected on the travelling plate, and so on. These 12 trials were carried out with a cartesian-space trajectory generation.

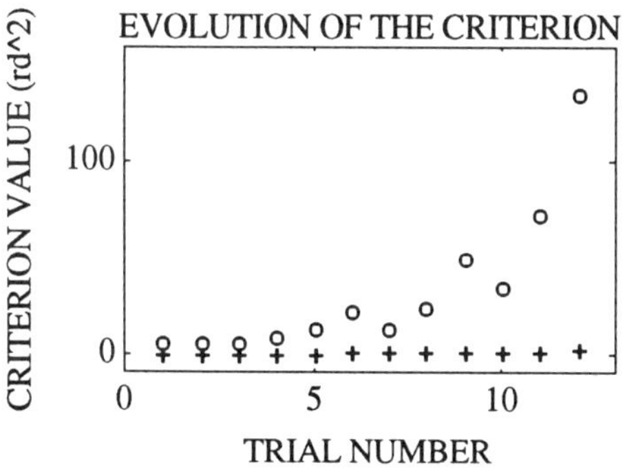

FIG. 2.

FIG. 2 shows how robust our control scheme is, in comparison with classical PID scheme: the I_p criterion is quite constant with VSC-HF (+), when it goes from 5 to 140 rd^2 with PID (o).

CONCLUSION

We think that fast parallel robots might be controlled in joint space with robust controllers, on one hand, to reject the disturbances coming from hanging objects, and, on the other hand, to avoid servoing them in cartesian space with a non-analytical direct model. In this paper we have studied one solution: the VSC approach. To avoid the classical chattering phenomenon problem, we have proposed a new control algorithm called VSC-HF that is different from the one presented previously by Slotine with the boundary layer **[7]**. The new algorithm consists in adding in the sliding surface an additional term the time constant of which is much smaller than the smallest time constant of the previous system. After a rough identification (much more simpler than the identification required to obtain the dynamic parameters), we realized an analog circuit to perform that function. We have performed several experimentations on a real setup, in which we show the efficiency and the robustness of the VSC-HF algorithm versus a classical PID algorithm. For the future, we plan to take into account some force information in a robust-hybrid control scheme **[8]**.

REFERENCES

1. J.-P. MERLET, "Parallel manipulators, part I", Rapport de Recherche INRIA N° 646, March 1987.
2. J. ANGELES and S. K. LEE, "The formulation of dynamical equations of holonomic mechanical systems using a natural orthogonal complement", Journal of Applied Mechanics, Vol. 55, pp. 243-244, March 1988.
3. R. CLAVEL, "DELTA, a fast robot with parallel geometry", Proc. of ISIR, pp. 91-100, Lausanne, Switzerland, April 1988.
4. F. PIERROT and al., "High speed control of a parallel robot", Proc. of the IEEE IROS, pp. 949-954, Tsuchiura, Japan, July 1990.
5. H. HASHIMOTO and al., "Practical realization of VSS controller for robotic arms", Proc. IEEE IECON 86, pp. 34-40, 1986.
6. V. UTKIN, "Discontinuous control systems: state of the art in theory and applications", Proc. IFAC World Congress, pp. 75-93, Munich, 1987.
7. J.J. SLOTINE, "The robust control of robot manipulators", The Internationnal Journal of Robotics Research, Vol 4 (2), pp. 49-64, 1985.
8. P. FRAISSE and al., "Towards robust hybrid control for two-arm robots", Proc. of the IEEE IROS '91, pp. 331-336, Osaka, 2-5 November 1991.

Trajectory Planning for Cooperating Robots Using Maximum Payload Under Time-Optimal Control

Pierre Larochelle J.M. McCarthy

Department of Mechanical Engineering
University of California, Irvine
Irvine, CA 92717

Abstract

Minimum time movements characteristically require that a set of motors in the robotic system be driven at their maximum torque throughout the motion. Here we increase the payload for a given movement until a minimum time solution is no longer feasible. This defines the maximum payload for the system on the associated path. The path is then optimized in order to find the trajectory which affords the greatest payload for the robotic system. To illustrate the algorithm a detailed analysis of two cooperating robotic arms developed at Odetics Inc. is presented. The result is that while 20 (lbs) is the maximum payload along the linear trajectory, the optimum path affords a maximum payload of 240 (lbs).

1 Introduction

Time-optimal control of robotic systems determines the torque histories that minimize the transit time required for the system to move from an initial to a final configuration. Performance evaluation of such systems determines the maximum payload of the system for a given path. Here, the focus is on determining the path upon which the robotic system is capable of moving its maximum payload while being driven to minimize time.

In this work we use the software package, TORUS, for Time-Optimal Robot System, which embeds Bobrow's time-optimal control strategy into a payload maximization algorithm. The program determines, for given path endpoints, the maximum payload that the system can carry, and the minimum time in which the system can move that maximum payload on the optimum path.

2 Literature Survey

The work of Bobrow, Dubowsky, and Gibson (1985), first presented the true minimum time solution along a prescribed path . This was the first work to consider the complete, non-linear, dynamics and torque constraints for robotic systems. With the solution to the problem of time-optimal control for robotic manipulators Bobrow (1988) was able to derive a path planning technique which produces time-optimal manipulator motions in a workspace containing obstacles . In this work time-optimal control is used to find the optimum path to avoid the obstacles as well as to minimize the time to perform the motion along that optimum path thereby demonstrating the ability of time-optimal control to be used as a tool to couple the path planning problem with the dynamics problem in robotics. Further theoretical work by McCarthy and Bobrow (1990) yielded a theorem which determines the minimum number of saturated actuators for a general robotic system under time-optimal control . This theorem demonstrates how time-optimal control forces robotic systems to their actuator limits.

The notion of measuring the performance of a robotic system was expanded to include dynamics by Wang and Ravani, (1988) . Their method results in either load-optimal motion with a time constraint, or with near time-optimal motion with load constraint. However, computationally their methods are very difficult to implement. In Larochelle and McCarthy, (1991), a method of determining the maximum payload for a robotic system along a completely specified path was presented. Here, we extend that work and present numerical results of determining a maximum payload path.

3 Time-Optimal Control

Here we describe how the *time-optimal trajectory* is found. The equations of motion for the system are written in terms of the path parameter s and its first two time derivatives, (Bobrow et al., 1985). For each s and $\dot{s}$ these equations become a set of $n+p+m+1$ linear equations in $2(p+m+1)$ unknowns; where n is the number of coordinates used to describe the system, p is the number of actuators, and m is the number of constraint equations. The equations of motion and the bounds on $\boldsymbol{\tau}$ and $\boldsymbol{\lambda}$ combine to form a linear programming problem;

$$\begin{aligned}
&\text{Maximize/minimize } \ddot{s} \text{ subject to:}\\
&\mathbf{a}(s)\ddot{s} - [B]\boldsymbol{\tau} - [C]^T\boldsymbol{\lambda} = \mathbf{d}(s,\dot{s})\\
&\boldsymbol{\tau}^{min} \leq \boldsymbol{\tau} \leq \boldsymbol{\tau}^{max}\\
&\boldsymbol{\lambda}^{min} \leq \boldsymbol{\lambda} \leq \boldsymbol{\lambda}^{max}
\end{aligned} \tag{1}$$

The second step is to construct the *maximum velocity curve.* The maximum velocity curve is found by numerically determining $\max(\dot{s})$ for each s along the path.

Once the maximum velocity curve has been constructed the time-optimal trajectory can be found. First, using Eq. 1 solve for the minimum acceleration $\ddot{s}$ at the end point of the path and integrate backwards in time to construct the *minimum acceleration curve.* This integration is performed until the curve intersects the maximum velocity curve or one of the axis in the $\dot{s}$-s phase plane. Similarly, find

Link	Length (m)	Mass (kg)
K	0.6136	29.5
L	0.5128	6.81
H	0.7620	varied

Table 1: Kinematic Parameters

the *maximum acceleration curve* by integrating forward in time from the beginning of the path. Connecting these curves in the $\dot{s}$ - s plane are *switching points.* These points were found using the *ladder search algorithm*, (see Chu, 1990).

4 Maximizing the Payload

In this section we present a discussion of payload maximization using the time-optimal trajectory in the $\dot{s}$ - s phase plane.

The simplest equation of motion for a rigid body is: $F = m\ddot{x}$; where, F is the total force on the body, m is the mass of the body, and $\ddot{x}$ is the acceleration of the body. If the total force on the body is fixed and m is increased then $\ddot{x}$ must decrease. In the context of general robotic systems, which have bounds on their actuators, as the payload for the path is increased the maximum and minimum accelerations $\ddot{s}$ that the system can apply to the workpiece are decreased. Therefore, as the payload is increased the time-optimal trajectory falls in the $\dot{s}$-s plane. Continued increase of the payload will result in the time-optimal trajectory intersecting the s axis.

We now outline the algorithm. In order to find the maximum payload the system can carry along a path the mass of the workpiece is incrementally increased. As the maximum payload is approached the time-optimal trajectory in the $\dot{s}$-s plane is brought down toward the s axis. When the payload has exceeded the capacity of the system there is some point s^* on the path at which there is no valid solution to Eq. 1. Physically, at s^* the motor limits, or the reaction force limits, would have to be exceeded for the robot system to continue along the path. Thereby determining the maximum payload.

In order to find the optimum trajectory the given path is varied using a non-linear optimization routine to find the path which maximizes the payload for the system to go from its start position to its goal position. This is accomplished by manipulating the control points of the path, determining the maximum payload for that path, and then iterating until we have converged upon the optimum path.

5 Case Study

To illustrate the application of the algorithm described in this paper two cooperating robots were used. The robots are positioned such that their base joints are 0.5 (m) apart, see Fig. 1. The robots are described by their geometry, mass, and peak continuous motor torques as listed in Tbl. 1 and Tbl. 2. The path studied is shown below, see Fig. 2. The start and goal positions are shown, as well as the initial and final configurations of the robot system. This path is linear in that

Motor	Max. Torque (Nm)	Min. Torque (Nm)
θ	452.0	-452.0
ϕ	226.0	-226.0
ψ	67.8	-67.8

Table 2: Dynamic Parameters

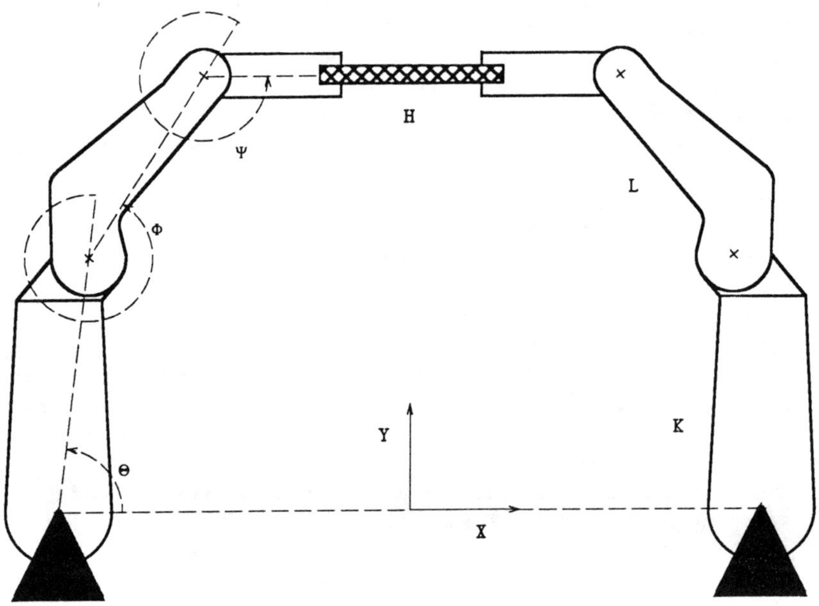

Figure 1: Two Cooperating Odetics Manipulators

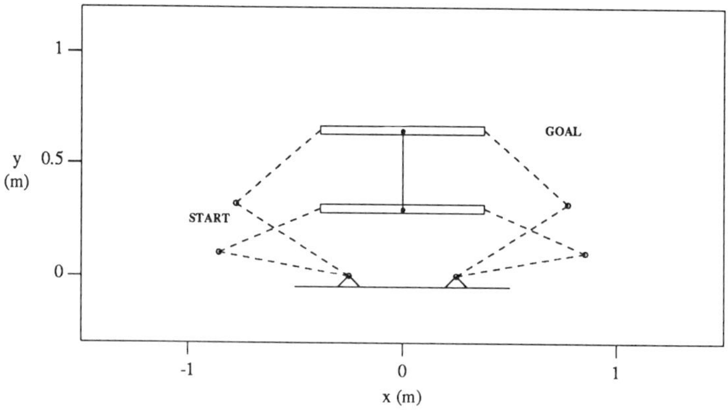

Figure 2: Linear Vertical Path

all of the path variables, $\{x, y, \alpha\}$, vary linearly with respect to s. The maximum payload for this path was found to be 20 (lbs). The time-optimal control solution is presented in Fig. 3. Note that the slope of the time-optimal trajectory is near zero and that all of the actuators are saturated throughout the motion. The optimum path is shown in Fig. 4. An equivalent path may be obtained by reflecting this path about the y-axis. The time-optimal control solution is presented in Fig. 5. Note that the optimum path reduces the wrist torques. This path moves the payload in such a manner as to avoid placing a large load on the wrists. Thereby affording more effective use of the robot's shoulder and elbow actuators. The maximum payload was found to be 240 (lbs). This is a dramatic payload increase of 220 (lbs) over the maximum payload of the linear path.

6 Conclusion

In this paper we analyzed a cooperating robot system using movements planned for maximizing the payload while under time-optimal control. We studied the maximum load that can be carried by the robots working on a linear path from the desired start position to goal position and found the maximum payload to be 20 (lbs). We then found the optimum trajectory which affords the maximum payload of 240 (lbs). The dramatic increase indicates that the computational intensity involved in computing optimum trajectories for robotic systems is warranted.

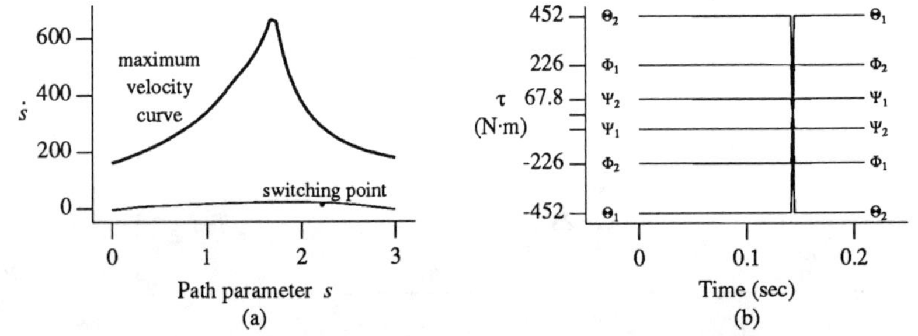

Time-optimal control for the closed chain robot system on a linear path. a) time-optimal trajectory, b) joint torques.

Elapsed time: .21 sec. Payload: 20 lb.

Figure 3: Linear Path with Maximum Payload.

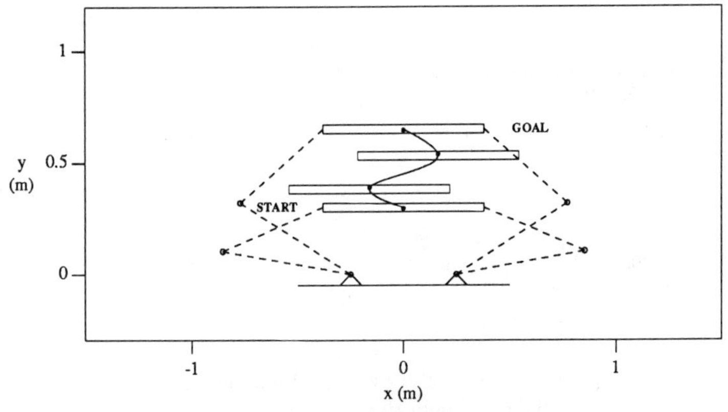

Figure 4: Optimum Path

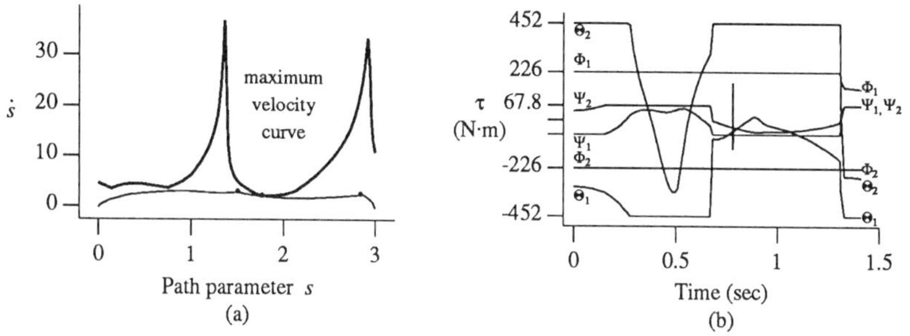

Time-optimal control for the cooperating robots on the maximum payload path. a) time-optimal trajectory, b) joint torques.

Elapsed time: 1.41 sec. Payload: 240 lb.

Figure 5: Optimum Path with Maximum Payload.

7 Acknowledgements

The support of the National Science Foundation, grant DDM-9102567, is gratefully acknowledged. In addition, this research was supported in part by the University of California, Irvine through an allocation of computer time on the Convex-C240.

References

[1] J. E. Bobrow. Optimal Robot Path Planning using the Minimum-Time Criterion. *IEEE Journal of Robotics and Automation*, 4(4):443–450, August 1988.

[2] Bobrow, J., Dubowsky, S. and Gibson, J. Time-Optimal Control of Robotic Manipulators along Specified Paths. *International Journal of Robotics Research* , Fall 1985.

[3] Bobrow, J., and McCarthy, J.M. The Number of Saturated Actuators and Constraint Forces during Time-Optimal Movement of a General Robotic System. *IEEE Transactions on Robotics and Automation*, April 1992.

[4] Chu, V. *Master's Thesis*, Univ. of Calif., Irvine, 1990.

[5] P. Larochelle and J. M. McCarthy. Performance Evaluation of Cooperating Robot Movements using Maximum Load under Time-Optimal Control. *IEEE Transactions on Robotics and Automation*, 3, April 1991.

[6] L. Wang and B. Ravani. Dynamic Load Carrying Capacity of Mechanical Manipulators *ASME Journal Dynamic System, Measurement and Control*, 110:46–61, March 1988.

GRAPHICAL REPRESENTATION OF THE SINGULARITY LOCI OF PLANAR PARALLEL MANIPULATORS

Clément M. Gosselin and Jaouad Sefrioui
Département de Génie Mécanique
Université Laval
Ste-Foy, Québec, Canada, G1K 7P4

ABSTRACT

In this paper, analytical expressions describing the singularity loci of planar three-degree-of-freedom parallel manipulators are derived. They are obtained using the analytical form of the Jacobian matrix, which is known to be singular when the manipulator is in a degenerate configuration. These expressions are then used to superimpose graphical representations of the manipulators' singularity loci and workspace. It is shown that, for a given orientation of the end-effector, the singularity locus in the (x, y) plane is a quadratic form which can be either an ellipse, a hyperbola or a parabola. Examples are given to illustrate the results. The graphical representation of the singulatity loci is a very useful tool and it has been included in a CAD package developed for parallel manipulators.

1 Introduction

The optimum kinematic design of manipulators has attracted the attention of researchers for some years. More specifically, most of the work that has been done on the subject has been directed towards the optimization of dexterity or manipulability indices which are known to be related to the kinematic accuracy [1–3]. In the context of parallel manipulators, kinematic accuracy is a very important issue. Indeed, it is well known that a special type of local degeneracy can occur in the motion of these manipulators. Therefore, it is very important to be able to identify singular configurations at the design stage in order to improve the system's performance.

In this paper, analytical expressions describing the singularity loci of planar three-degree-of-freedom parallel manipulators are derived. It is shown that, for a given orientation of the end-effector, the singularity locus in the (x, y) plane is a quadratic form which can be either an ellipse, a hyperbola or a parabola. These expressions are then used to superimpose graphical representations of the manipulators' singularity loci and workspace. Examples are given to illustrate these results. The aim of the graphical tool developed here is to help the designer by allowing him to determine the exact locus —in the Cartesian space— of the singular configurations of a particular design and hence to be able to make decisions on whether or not the singularities encountered are acceptable.

2 Singularity analysis

Let $\boldsymbol{\theta}$ and $\mathbf{x}$ be the vector of joint coordinates and the vector of Cartesian coordinates of a n degree-of-freedom parallel manipulator, respectively. These vectors are related

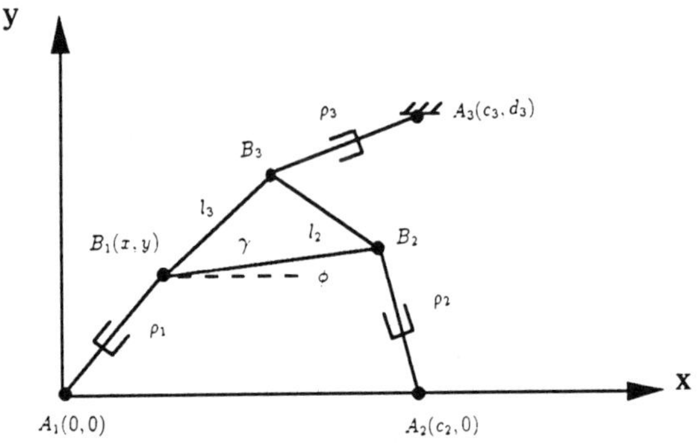

Figure 1: Planar three-degree-of-freedom parallel manipulator.

by nonlinear kinematic constraints which, when derived with respect to time, yield a velocity equation of the form:

$$\mathbf{B}\dot{\boldsymbol{\theta}} + \mathbf{A}\dot{\mathbf{x}} = \mathbf{0} \tag{1}$$

where $\mathbf{0}$ is the n-dimensional zero vector and where $\mathbf{A}$ and $\mathbf{B}$ are n-dimensional square matrices which are configuration dependent. These matrices are called Jacobian matrices.

In [4], Gosselin and Angeles have suggested a classification of the different types of singularities encountered in closed kinematic chains by considering the Jacobian matrices $\mathbf{A}$ and $\mathbf{B}$ defined above. They have identified three types of singularities, namely: *i*) singularities of type *I* which occur when matrix $\mathbf{B}$ is singular, *ii*) singularities of type *II* which occur when matrix $\mathbf{A}$ is singular, and *iii*) singularities of type *III* which occur when the nonlinear kinematic constraints degenerate.

In this paper, the singularity loci associated with the singularities of type *I* and *II* of the planar three-degree-of-freedom parallel manipulator are studied. The third type of singularity will be assumed to be avoided by a proper choice of the kinematic parameters.

3 Planar three-degree-of-freedom parallel manipulators

A planar three-degree-of-freedom parallel manipulator is represented in Fig. 1. The three prismatic actuators are mounted on unactuated revolute joints and allow the positioning and the orientation of a platform in the plane. This type of manipulator has been studied in the past (see for instance [5]) and its potential applications include part handling, welding, deburring and milling. It can also be build with revolute actuators, as shown in [5]. The kinematic architecture presented in Fig. 1 is completely general since no symmetry or special geometric parameters are assumed.

3.1 Computation of the Jacobian matrices

By inspection of Fig. 1, the inverse kinematic problem is readily solved. Indeed, one can write the length of each of the legs of the platform as:

$$\rho_1^2 = x^2 + y^2 \tag{2}$$

$$\rho_2^2 = (x + l_2\cos\phi - c_2)^2 + (y + l_2\sin\phi)^2 \tag{3}$$

$$\rho_3^2 = [x + l_3\cos(\phi+\gamma) - c_3]^2 + [y + l_3\sin(\phi+\gamma) - d_3]^2 \tag{4}$$

where c_2, c_3 and d_3 are coordinates of the position of the fixed unactuated revolute joints, $\rho_i, i = 1, 2, 3$ is the length of the ith leg (the ith joint coordinate), l_1, l_2 and γ

are related to the geometry of the platform and x, y, ϕ are the Cartesian coordinates of the platform, i.e., the position of one point and the orientation with respect to a fixed coordinate frame. We can then define $\mathbf{x} = [x, y, \phi]^T$, as the vector of Cartesian coordinates and $\boldsymbol{\theta} = [\rho_1, \rho_2, \rho_3]^T$, as the vector of joint coordinates. The differentiation of eqs.(2–4) with respect to time will lead to an equation of the form of eq.(1) where **A** and **B** are the 3×3 Jacobian matrices. These matrices are written as:

$$\mathbf{A} = \begin{bmatrix} a_{11} & a_{12} & a_{13} \\ a_{21} & a_{22} & a_{23} \\ a_{31} & a_{32} & a_{33} \end{bmatrix} \tag{5}$$

and

$$\mathbf{B} = \mathrm{diag}(\rho_1, \rho_2, \rho_3) \tag{6}$$

with

$$a_{11} = x \tag{7}$$
$$a_{12} = y \tag{8}$$
$$a_{13} = 0 \tag{9}$$
$$a_{21} = (x - c_2) + l_2 \cos\phi \tag{10}$$
$$a_{22} = y + l_2 \sin\phi \tag{11}$$
$$a_{23} = l_2(y\cos\phi - (x - c_2)\sin\phi) \tag{12}$$
$$a_{31} = (x - c_3) + l_3\cos(\phi + \gamma) \tag{13}$$
$$a_{32} = (y - d_3) + l_3\sin(\phi + \gamma) \tag{14}$$
$$a_{33} = l_3[(y - d_3)\cos(\phi + \gamma) - (x - c_3)\sin(\phi + \gamma)] \tag{15}$$

3.2 Singularities of type *I*

As stated above, these singularities occur when the determinant of matrix **B** is equal to zero. Mathematically, this condition leads to: ($\rho_i = 0$, $i = 1 \text{ or } 2 \text{ or } 3$). However, since the prismatic joints do not have an infinite range of motion, this singularity will occur whenever one of the joints reaches one of its limits, i.e., when $\rho_i = \rho_{i,min}$ or $\rho_i = \rho_{i,max}$ for $i = 1, 2$ or 3. This is in fact the boundary of the Cartesian workspace and it is easily obtained [5].

3.3 Singularities of type *II*

Singularities of type *II* occur when matrix **A** is singular, i.e., when $\det(\mathbf{A}) = 0$, where **A** is given in eq.(5). From eq.(5) and from the latter condition, the analytical expression for the singularity locus can be written as:

$$E_1 x^2 + E_2 y^2 + E_3 xy + E_4 x + E_5 y = 0 \tag{16}$$

where the coefficients, E_i, $i = 1, 2, ...5$, are functions of the geometric parameters of the robot and of the orientation of the platform.

The nature of the curve represented by eq.(16) depends on the following quantity [6]:

$$\Delta = E_1 E_2 - \frac{E_3^2}{4} \tag{17}$$

which can be computed from the expressions of the E_i's. This leads to

$$\Delta = A_2 \sin^2\phi + A_1 \sin\phi\cos\phi + A_0 \tag{18}$$

with

$$A_0 = -(d_3 l_2 - c_2 l_3 \sin\gamma)^2 \quad (19)$$
$$A_1 = 2(c_3 l_2 - c_2 l_3 \cos\gamma)(d_3 l_2 + c_2 l_3 \sin\gamma) \quad (20)$$
$$A_2 = -[(c_3 l_2 - d_3 l_2 - c_2 l_3 \cos\gamma - c_2 l_3 \sin\gamma)(c_3 l_2 + d_3 l_2 - c_2 l_3 \cos\gamma + c_2 l_3 \sin\gamma)] \quad (21)$$

From [6], we have the following result: if Δ is positive the singularity curve is an ellipse, if Δ is negative the singularity curve is a hyberbola and if Δ is equal to zero, the curve is a parabola. Therefore, since Δ itself is a function of the orientation of the platform, the nature of the curve can change, for a given manipulator, with the orientation of the platform.

3.4 Manipulators of general architecture

In this case, the sign of Δ will change with angle ϕ. Because of the form of the function $\Delta(\phi)$, there will always be 4 values of ϕ which will lead to parabolic singularity loci. In fact, these points are nothing but the roots of $\Delta(\phi) = 0$ and the latter equation can be written as a quartic equation in $\tan(\frac{\phi}{2})$.

3.5 A class of manipulators with a special architecture

It is interesting to study a special class of manipulators for which the three points on the platform B_1, B_2 and B_3 are collinear. This simplification of the architecture is mathematically expressed by setting angle γ to zero. The coefficients of eq.(16) are then simplified and the expression of Δ becomes

$$\Delta = -\frac{E_3^2}{4} \quad (22)$$

which is a negative semi-definite quantity. Therefore, the singularity curve for this type of manipulator is always a hyperbola. The expressions for the two branches of the hyperbola and the expression of the asymptotes are easily obtained. In this case, three-dimensional representations of the singularity loci in the (x, y, ϕ) space have been plotted but are not shown here because of space limitation.

4 Superposition of the singularity locus and the workspace of the manipulator

One of the main objectives of the derivations presented in this paper was to include the expressions describing the singularity loci in a package aimed at the computer aided design of parallel manipulators. Such a package — called SIMPA, which stands for *Simulation Interactive de Manipulateurs Parallèles* — has been developed at Laval [7] and the analytical expressions of the singularity loci obtained above have been included in it in order to be able to plot the singularity curves and to superimpose them on the workspace of the manipulator. This feature is of great help to the designer since it allows him to get a very clear picture of the location of the singular configurations of type *II* in the workspace of the manipulator. (It is recalled that the singularities of type *I* are simply located at the boundary of the Cartesian workspace.)

A few examples of the graphical results obtained with our computer package, SIMPA, are given in Figs. 2 and 3. Fig. 2 shows plots obtained for a manipulator of general architecture while Fig. 3 presents results obtained with a manipulator of the special class discussed above.

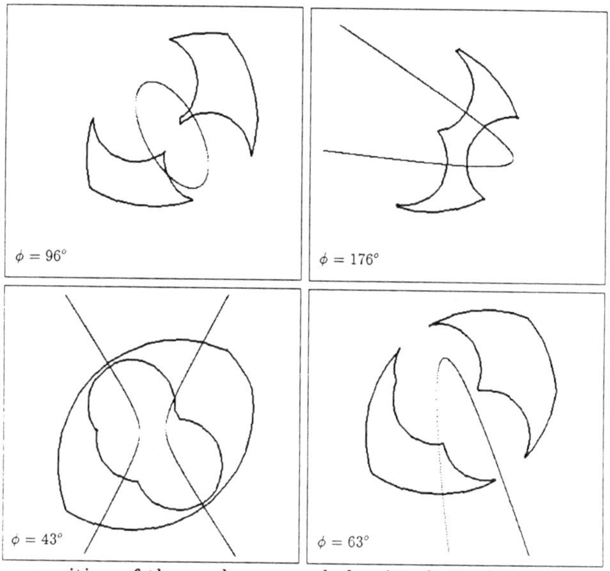

Figure 2: Superposition of the workspace and the singularity locus of a planar 3-dof manipulator with l_2=4, l_3=3, $\gamma = 30°$, c_2=7, c_3=0, d_3=6, for different values of angle ϕ.

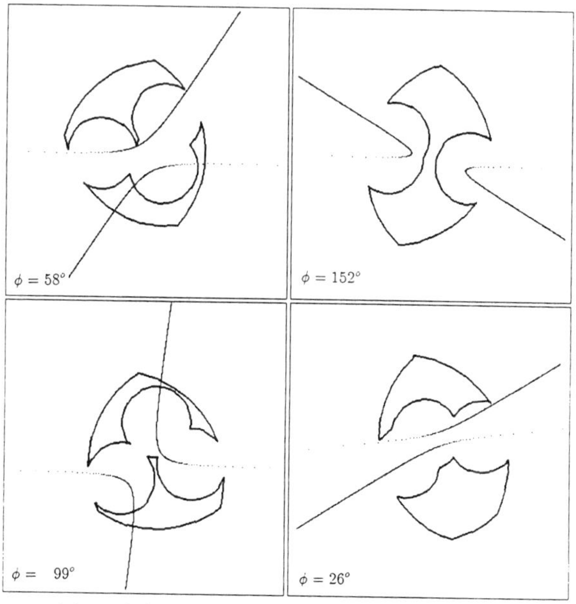

Figure 3: Superposition of the workspace and the singularity locus of a planar 3-dof manipulator with l_2=4, l_3=1, $\gamma = 0$, c_2=5, c_3=2, d_3=0, for different values of angle ϕ.

5 Conclusion

Analytical expressions describing the singularity loci of planar three-degree-of-freedom parallel manipulators have been obtained in this paper. These expressions have been used to obtain plots of the singularity curves superimposed on the workspace of the manipulator. Fisrt of all, the two types of local degeneracies that can occur with parallel manipulators have been recalled. As shown elsewhere, the first type of degeneracy is located at the boundary of the Cartesian workspace and can be obtained easily. Therefore, the derivations presented here have focused on the second type of degeneracy. The equations have been obtained by imposing the vanishing of the determinant of one of the Jacobian matrices of the manipulator. It has been shown that, in general, the curve obtained in the xy plane, i.e., for a given value of the orientation of the platform, is either a hyperbola, an ellipse or a parabola. A class of manipulators for which the singularity locus is always a hyperbola has also been identified. The graphical superposition of the singularity locus and the Cartesian workspace, which has been implemented in a CAD package dedicated to parallel manipulators, is of great interest in the context of design since it allows the prediction of singular configurations and hence leads to a better optimization of the performance of the system.

6 Acknowledgements

This work was completed under research grants from the Natural Sciences and Engineering Research Council of Canada (NSERC) and le Fonds pour la Formation des Chercheurs et l'Aide à la Recherche (FCAR) of Québec. The authors also wish to acknowledge the help of Simon Laverdière who wrote and implemented the software allowing the graphical representation of the singularity loci.

References

[1] J.K. Salisbury and J.J. Craig, 'Articulated hands: force control and kinematic issues', *The International Journal of Robotics Research*, Vol. 1, No. 1, pp. 4–17, (1982).

[2] T. Yoshikawa, 'Manipulability of robotic mechanisms', *The International Journal of Robotics Research*, Vol. 4, No. 2, pp. 3–9, (1985).

[3] C.A. Klein and B.E. Blaho, 'Dexterity measures for the design and control of kinematically redundant manipulators', *The International Journal of Robotics Research*, Vol. 6, No. 2, pp. 72–83, (1987).

[4] C. Gosselin and J. Angeles, 'Singularity analysis of closed-loop kinematic chains', *IEEE Transactions on Robotics and Automation*, Vol. 6, No. 3, pp. 281–290, (1990).

[5] C. Gosselin and J. Angeles, 'The optimum kinematic design of a planar three-degree-of-freedom parallel manipulator', *ASME Journal of Mechanisms, Transmissions, and Automation in Design*, Vol. 110, No. 1, pp. 35–41, (1988).

[6] A.C. Jones, *Introduction to Algebraical Geometry*, OXFORD University Press, (1912).

[7] C. Gosselin, S. Laverdière and J. Côté, 'SIMPA: A graphic simulator for the CAD of parallel manipulators', Proceedings of the *ASME Computers in Engineering Conference*, San Francisco, August 2–6, (1992).

Dual-Arm Impedance Based Generalized Compliant Motion

Paul G. Backes
Jet Propulsion Laboratory
California Institute of Technology
Pasadena, California

ABSTRACT

A unified algorithm for multi-sensor based dual-arm cooperative control of manipulators is presented which allows the commonly held object to act as an impedance relative to real and virtual sensor inputs. Each sensor control is provided with a unique task frame before merging all resultant control in a common frame. The algorithm allows a wide range of supervisory and shared control task execution within a constrained computation environment such as may be expected in space telerobotics. Move-squeeze decomposition is used to compute the forces at the two grasp points which cause internal body forces and external move forces. Forces in the two subspaces are then controlled independently. Squeeze forces are controlled with active damping and the move forces, trajectory, and sensor based motion are controlled with position based impedance control.

I. INTRODUCTION

Dual-arm robot control will be valuable in many application domains from military to undersea to hazardous waste to space. The research described in this paper was motivated by the needs of space telerobotics. Space telerobotics has many constraints, many of which also apply to terrestrial applications. These constraints include limited computational resources, unstructured environments, spatial and temporal separation of operator and robot control, a wide range of tasks and safety concerns.

Cooperative dual-arm control provides the ability for two manipulators to manipulate a commonly grasped object to execute a task. This may be valuable when handling heavy or extended objects or objects which are difficult to grasp securely with one arm. Integration of multiple sensor based control increases the span of tasks which may be executed. This paper describes a unified control algorithm integrating cooperative dual-arm control with multi-sensor based impedance control. Parameterization of the algorithm provides a means for specifying the desired behavior of motion from a general dual-arm task execution capability. An earlier version of the algorithm [1] has been implemented in the JPL Supervisory Telerobotics (STELER) laboratory where it has been used to execute a variety of supervisory and shared control tasks such as satellite capture and fluid coupler mating. In the earlier method, impedance parameters were not used as the control parameters describing the interaction between the commonly held object and its environment, as is done here.

II. MOVE-SQUEEZE DECOMPOSITION

Multiple arms holding an object can apply both forces[1] which cause motion of the object and forces which cause internal forces to build up within the object without motion. The former are called external or move forces and the later are called internal or squeeze forces [2, 3]. Execution of a task requires the control of both of these force subspaces. Description of the desired contact interaction between the commonly held object and its environment requires description of the desired move subspace forces. Description of the desired internal object forces requires description of the squeeze subspace forces. Decomposition of the forces applied by multiple robots on an object into move and squeeze subspace components was originally described in [2] and subsequently by various authors [4, 5, 3, 6, 7, 8]. It is assumed that the arms, the commonly held object, and the grasps are all rigid.

A commonly held object is grasped by one robot at location G_1 and a second robot grasps the object at location G_2. A third point on the object, C, is the location of potential contact with the environment. The two arms exert forces and torques on the object, f_1, n_1 and f_2, n_2, which are combined in the vectors F_1 and F_2, i.e., $F_1 = [n_1, f_1]^T$, $F_2 = [n_2, f_2]^T$. The vector of forces at C is then

$$F_c = \Phi_{c,1} \cdot F_1 + \Phi_{c,2} \cdot F_2 = [\Phi_{c,1}, \Phi_{c,2}] \cdot \begin{bmatrix} F_1 \\ F_2 \end{bmatrix} = A^T \cdot F_{12} \tag{1}$$

The components of the force vector F_{12} which are in the nullspace of A^T comprise the squeeze forces, F_{12s}, and the components of F_{12} which are in the vector space of A^T comprise the move forces, F_{12m}, [2] i.e.,

$$F_{12} = F_{12s} + F_{12m} \tag{2}$$

The move forces are computed using the Moore-Penrose inverse of A^T, i.e.,

$$F_{12m} = A^{+^T} \cdot F_c = A \cdot (A^T \cdot A)^{-1} \cdot A^T \cdot F_{12} \tag{3}$$

The force vector F_{12} is known by measurement with wrist force-torque sensors. The squeeze forces are then computed with

$$F_{12s} = F_{12} - F_{12m} \tag{4}$$

The move and squeeze forces (and torques) for the two manipulators are then extracted from F_{12m} and F_{12s}

$$\begin{aligned} F_{12m} &= [n_{1m}, f_{1m}, n_{2m}, f_{2m}]^T \\ F_{12s} &= [n_{1s}, f_{1s}, n_{2s}, f_{2s}]^T \end{aligned} \tag{5}$$

The actual move and squeeze forces are then available for use in control. Given the mass properties of the common object, the gravity forces of the composite object beyond the force sensors are computed and subtracted from the measured F_{12}. Each robot is assigned to support half of the mass properties of the common object.

III. KINEMATIC RELATIONSHIPS

The kinematic relationships between the two manipulators, the common object, and the various motion sources are shown in figure 1. An additional sensor would be

[1]Throughout this paper, "position" will refer to both location and orientation, and "force" will refer to both force and moment.

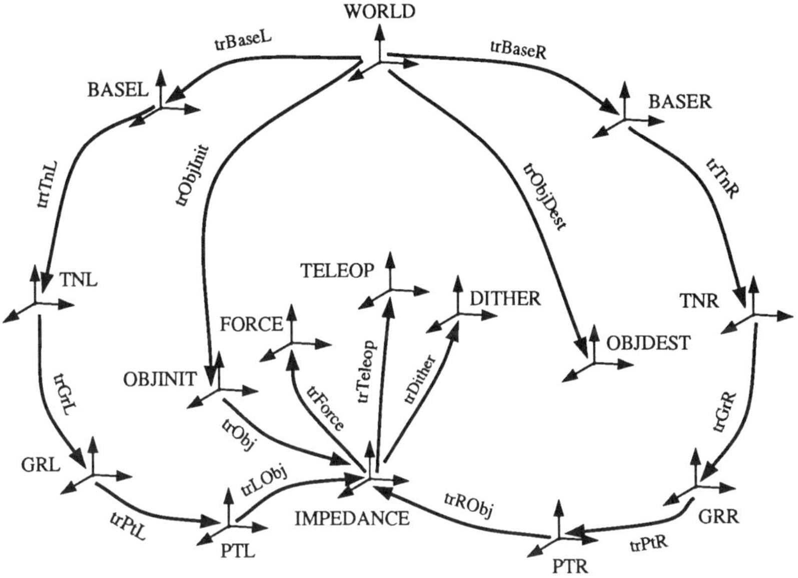

Figure 1: Kinematic diagram for dual-arm generalized compliant motion

added similarly to the force, teleop, and dither inputs. The figure shows the various coordinate frames and the transformations between them. The transform product $trObjInit \cdot trObj$ represents the currently commanded position of the common object. The motion of the common object is computed independent of the motion of the manipulators using position based impedance control as described in Section V below. The kinematic ring equations used to describe the motion of the two arms constrained to the motion of the common object are

$$trBaseL \cdot trTnL \cdot trGrL \cdot trPtL \cdot trLObj = trObjInit \cdot trObj \tag{6}$$

$$trBaseR \cdot trTnR \cdot trGrR \cdot trPtR \cdot trRObj = trObjInit \cdot trObj \tag{7}$$

The motion of the left and right robots are treated equally, each dependent on the motion of the common object and the input parameters. In Equation (6), the trBaseL, trGrL and trLObj transforms are constant and given in the input parameters. Each sample interval the $trPtL$ transform is computed by squeeze control (see Section IV below), and then $trTnL$ is computed using Equation (6). Computations for the right robot are equivalent to the left except for variations due to different input parameters. Once $trTnL$ and $trTnR$ are computed each sample interval, inverse kinematics for each arm are used to compute the equivalent joint angles for the robots. Joint position servos then drive the joints to the new computed angles.

IV. SQUEEZE CONTROL

Squeeze control is used to compute $trPtR$ and $trPtL$ of Equations (6) and (7). Each sample interval move-squeeze decomposition, as explained in Section II, is used to compute the squeeze forces in the GRL and GRR coordinate frames. Either or both of the robots can be controlled to track the desired squeeze force with the choice specified by the input parameter set. Squeeze control will be described for only the left arm here although it is equivalent for the right arm. Each sample interval, the vector of six perturbations along and about the axes of the GRL frame is computed with

$$\delta X_{lsq} = K_{lsq} \cdot (F_{lsqr} - F_{lsqa}) \tag{8}$$

where K_{lsq} is the diagonal matrix of gains, F_{lsqr} are the desired (reference) squeeze forces and F_{lsqa} are the measured actual squeeze forces. The $trPtL$ transform is then updated with the new perturbations with

$$trPtL = trDel_{lsq} \cdot trPtL \tag{9}$$

where $trDel_{lsq}$ is the perturbation transform generated from δX_{lsq} [1].

V. IMPEDANCE BASED GENERALIZED COMPLIANT MOTION

Impedanced based generalized compliant motion is used here to control the motion of the common object based upon the trajectory generator, force control of the move forces from move-squeeze decomposition, and motion based upon several other sensors. Generalized compliant motion is a control algorithm which allows multiple sensor based control for task execution [9, 10, 1]. Generalized compliant motion, as described in [9, 10, 1], does not directly control the manipulator (single arm case) or the common object (dual arm case) as an impedance relative to its environment. The formulation of generalized compliant motion described here does treat the common object as an impedance relative to its environment. The flexibility of generalized compliant motion for task execution is maintained, but extended for the impedance control framework. The sensors can be either real, e.g., a force-torque sensor, or virtual, e.g., a computed distance to collision. Each sensor is provided an individual task space for control.

In impedance control the desired relationship between motion and forces in each task frame degree of freedom (DOF) is specified recognizing the physical interaction between the robot and its environment [11, 12, 13]. Impedance control can be implemented either as torque-based impedance control [14] or position-based impedance control [15, 16]. The scheme described here utilizes position-based impedance control but may be expanded to torque-based in future implementations. The target impedance of the common object relative to its environment has inertia matrix M, damping matrix B and stiffness matrix K. The Cartesian impedance control equation is given by

$$M \cdot (\ddot{X}_c - \ddot{X}_r) + B \cdot (\dot{X}_c - \dot{X}_r) + K \cdot (X_c - X_r) \;=\; F_r - F_a + F_t + F_d + F_l + F_j \tag{10}$$

where X_r is the reference trajectory, X_c is the commanded position, F_r is the reference (desired) force, F_a is the actual (measured) force, F_t is a virtual force representing the teleoperation inputs, F_d is a virtual force representing a dither input, F_l is a virtual force representing proximity sensor [17] distance and tilt control, and F_j is a virtual force representing joint space joint limit and and joint singularity avoidance [18, 1].

Equation (10) can be implemented with

$$
\begin{aligned}
{}^I\ddot{X}_c^{n+1} = {}^I\ddot{X}_r^{n+1} + M^{-1} \cdot [\, {}^IF_r^{n+1} - {}^IF_a^n \\
+ {}^IF_t^n + {}^IF_d^n + {}^IF_l^n + {}^IF_j^n \\
-B \cdot (\, {}^I\dot{X}_c^n - {}^I\dot{X}_r^{n+1}) - K \cdot (\, {}^IX_c^n - {}^IX_r^{n+1})]
\end{aligned} \quad (11)
$$

$$ {}^I\dot{X}_c^{n+1} = {}^I\dot{X}_c^n + {}^I\ddot{X}_c^{n+1}\Delta\tau \quad (12) $$

$$ {}^IX_c^{n+1} = {}^IX_c^n + {}^I\dot{X}_c^{n+1}\Delta\tau \quad (13) $$

where $\Delta\tau$ is the sample interval, the preceding superscript indicates the coordinate frame (I in the equation), and the post superscript indicates the sample interval, i.e., n is the present sample interval and $n+1$ is the next sample interval. In the actual implementation, the position ${}^IX_c$ represents the commanded position of the common IMPEDANCE frame. The transformation equivalent to ${}^IX_c$ is computed and used in Equations (6) and (7) in place of $trObjInit \cdot trObj$.

VI. CONCLUSIONS

The dual-arm impedance based generalized compliant motion scheme provides general dual-arm task execution capability. Position based impedance control is used for the interaction between the commonly held object and its environment. Specifying desired control behavior via impedance parameters should allow a natural specification of parameterization for task execution.

ACKNOWLEDGEMENTS

The work described in this paper was performed at the Jet Propulsion Laboratory, California Institute of Technology, under contract with the National Aeronautics and Space Administration.

REFERENCES

[1] Paul G. Backes. Dual-arm supervisory and shared control space servicing task experiments. In *Proceedings AIAA Space Programs and Technologies Conference*, Huntsville, AL, March 24-27 1992. AIAA paper No. 92-1677.

[2] P. Dauchez and M. Uchiyama. Kinematic formulation for two force-controlled cooperating robots. In *Proceedings Third International Conference on Advanced Robotics (ICAR)*, pages 457–467, Versailles, France, October 1987.

[3] John T. Wen and Kenneth Kreutz. Stability analysis of multiple rigid robot manipulators holding a common rigid object. In *Proceedings of the 27th Conference on Decision and Control*, Austin, Texas, December 1988.

[4] Yoshihiko Nakamura, Kiyoshi Nagai, and Tsuneo Yoshikawa. Mechanics of coordinative manipulation by multiple robotic mechanisms. In *Proceedings IEEE International Conference on Robotics and Automation*, 1987.

[5] Masaru Uchiyama, Naotoshi Iwasawa, and Kyojiro Hakomori. Hybrid position/force control for coordination of a two-arm robot. In *Proceedings IEEE International Conference on Robotics and Automation*, pages 1242–1247, 1987.

[6] Masaru Uchiyama and Pierre Dauchez. A symmetric hybrid position/force control scheme for the coordination of two robots. In *Proceedings IEEE International Conference on Robotics and Automation*, 1988.

[7] K. Kreutz and A. Lokshin. Load balancing and closed-chain multiple arm control. In *Proceedings American Control Conference*, pages 2148–2154, Atlanta, Georgia., June 1988.

[8] Ian Walker and Rober Freeman. Internal object loading for multiple cooperating robot manipulators. In *Proceedings IEEE International Conference on Robotics and Automation*, pages 606–611, 1989.

[9] Paul G. Backes. Generalized compliant motion task description and execution within a complete telerobotic system. In *Proceedings IEEE International Conference on Systems Engineering*, August 9–11 1990.

[10] Paul G. Backes. Generalized compliant motion with sensor fusion. In *Proceedings 1991 ICAR: Fifth International Conference on Advanced Robotics, Robots in Unstructured Environments*, pages 1281–1286, Pisa, Italy, June 19–22 1991.

[11] N. Hogan. Impedance control: An approach to manipulation: Part i — theory. *ASME Journal of Dynamic Systems, Measurement, and Control*, 107:1–7, March 1985.

[12] N. Hogan. Impedance control: An approach to manipulation: Part ii — implementation. *ASME Journal of Dynamic Systems, Measurement, and Control*, 107:8–16, March 1985.

[13] N. Hogan. Impedance control: An approach to manipulation: Part iii — applications. *ASME Journal of Dynamic Systems, Measurement, and Control*, 107:17–24, March 1985.

[14] N. Hogan. Stable execution of contact tasks using impedance control. In *Proceedings IEEE International Conference on Robotics and Automation*, pages 1047–1054, 1987.

[15] Dale A. Lawrence and R. Michael Stoughton. Position-based impedance control: Achieving stability in practice. In *Proceedings AIAA Conference on Guidance, Navigation, and Control*, pages 221–226, Monterey, Ca, August 1987.

[16] Dale A. Lawrence. Impedance control stability properties in common implementation. In *Proceedings IEEE International Conference on Robotics and Automation*, pages 1185–1190, 1988.

[17] Hamamatsu Photonics K.K. Optical displacement sensor with tilt sensing (h3065, c3359-00), specifications and instructions manual. 1126 ICHINO-CHO, HAMAMATSU CITY, JAPAN.

[18] Paul G. Backes. Multi-sensor based impedance control for task execution. In *Proceedings IEEE International Conference on Robotics and Automation*, Nice, France, May 1992.

Force Regulation in Multiple Manipulator Systems

Mike J. Ryan, Steve H. Murphy, John T. Wen
Center for Intelligent Robotic Systems for Space Explorations
Rensselaer Polytechnic Institute
Troy, NY 12180

Abstract

An important advantage of multiple manipulator systems is the possibility to regulate the internal force in the payload. Common force control strategies include impedance control, position accommodation force control and Jacobian force control. This paper mainly considers the latter two schemes for which detailed dynamic models for the arms and the environment are not required. Comparison between the these two methods are made and some experimental results on decentralized position accommodation force control are discussed.

1 Introduction

Multiple–manipulator systems significantly extend the capabilities of a single manipulator. With two or more manipulators a great number of significant tasks become feasible that would be extremely difficult, if not impossible with a single manipulator. This paper focuses on an important aspect of the multiple arm system control: regulation of the forces exerted by the manipulators on the payload. Successful force control will provide the capabilities necessary for many of the multi–arm tasks.

There have been various force control strategies, for example, for rigid grasping arms in [1, 2], and multiple fingers in [3, 4]. These strategies fall into the following categories:

1. Impedance control: The end effector is made to mimic a desired impedance through feedback linearization of the arm dynamics and direct force cancellation.

2. Position accommodation: The end effector is commanded to follow the arm trajectories corresponding to a desired impedance relation.

3. Jacobian force control: The arm is servoed directly based on the force error.

Direct force cancellation in impedance control can lead to instability due to the delay in force measurement [5]. Hence, we shall consider only the latter two cases. This paper will focus only on the position accommodation control as it is the first one that we have implemented experimentally. We will contrast its features to the Jacobian control whenever appropriate. In Section 2, we will describe the position accommodation and Jacobian force control methods and their comparison. In Section 3, some experimental results related to the decentralized force control are presented. A discussion on the force resolution issue in position accommodation control is given in Section 4.

2 Force Control Strategies

As explained in the introduction, we will focus on two force control strategies: position accommodation control and Jacobian force control. We shall first consider these methods in the context of a single arm force control (in contact with the environment) which is the basis of the decentralized multiple arm force control. Then the application to the multiple arm force control problem will be addressed.

Position Accommodation Control

The position accommodation control is similar in spirit to the impedance control where the end effector frame of the arm is commanded to mimic a desired mechanical impedance driven by the sensed force:

$$M_{des}\alpha_{des} + D_{des}v_{des} + K_{des}(x_{des} - x_{ref}) = -B(f - f_{des}), \tag{1}$$

However, in impedance control feedback linearization and force cancellation are used to achieve this relationship exactly (at least in theory), but in position accommodation, any position controller can be used to follow the desired trajectory obtained from (1).

When this approach is used for force control, i.e., f_{des} is regarded as a force setpoint, K_{des} is usually set to zero in the constraint direction, so the desired setpoint will move beyond the constraint surface until the constraint force reaches the required setpoint. For the sake of illustration, consider a joint level proportion–derivative and gravity compensation (PDG) controller:

$$\tau = -K_p(\theta - \theta_{des}) - K_v\dot{\theta} + k(\theta). \tag{2}$$

To see the effect of this controller together with the desired impedance equation (1), consider the special case that the arm is in contact with an infinite rigid environment and arm is in a motion equilibrium. In this case, we have the force balance equation $\tau = J^T f$. After linearization and substitution of the impedance equation and inverse kinematics, we have a second order equation

$$M_{des}JJ^T\ddot{\eta} + D_{des}JJ^T\dot{\eta} + B\eta = v \tag{3}$$

where η is the integral of $\Delta f = f - f_{des}$ and v is a constant vector. If the parameters in the impedance equation are chosen so that (3) is a stable system, then η would tend to a constant (assuming that the arm never breaks off contact) and Δf would tend to zero as desired. However, the inclusion of M_{des} actually complicates matter. Depending on whether the left hand side of (3) is underdamped or overdamped, there may be oscillation in the force profile. Furthermore, this behavior depends on the robot configuration when contact occurs, due to the Jacobian terms in the equation. We recommend setting M_{des} to zero, so η would converge monotonically. In this case, the joint torque controller is of the form

$$\tau = -K_p(\theta - \theta_{ref}) - K_v\dot{\theta} - J^{-1}D_{des}^{-1}B\int_0^t \Delta f(s)\,ds. \tag{4}$$

which is an integral force feedback controller with fixed spring and damper. As discussed in [6], integral force feedback possesses many desirable features (which agrees with the conclusion of an experimental investigation in [5]), especially in terms of robustness with respect to force measurement time delay when the environment is stiff.

Jacobian Based Direct Force Control

The position accommodation control converts the force error into a position error and performs joint servoing based on the position error. It is also possible to directly servo the joints based on the force error itself. The general form of the controller is

$$\tau = \tau_m + k(\theta) + J^T F \tag{5}$$

where τ_m is a position control law (say, PD or PID) and F is the force controller. As discussed in [7], F can be chosen to be in either the feedforward or feedback form (or combination thereof). As mentioned before, integral force feedback is a particularly desirable choice, for both single arm and multiple arm force control.

The force control strategies described above can be applied to multiple arm control in two ways:

1. Decentralized: Each arm is controlled independently according to its own force control objective. In the position accommodation case, each arm has an impedance equation to follow:
$$M_{des_i}\ddot{\delta x}_i + D_{des_i}\dot{\delta x}_i + K_{des_i}\delta x_i = A_i^T(f_i - f_{bias_i}) \tag{6}$$
where A_i^T is the grasp Jacobian of the ith arm and $\delta x_i = x_i - x_{ref_i}$. This approach is the basis for the multiple–manipulator operating system KALI [8]. In general, x_{ref_i} and f_{des_i} should be chosen to satisfy the kinematic constraint. In the Jacobian force control case, a force setpoint f_{des_i} is specified for each arm.

2. Centralized: The arms are cooperatively controlled to allow for separate consideration of external and internal force control. In this case, all tip forces are propagated to a common frame C through the individual grasp Jacobian: $f_c = A^T f$. This force is used for the external force compliance.

   For the internal force control, we extract an internal force variable ξ from $f = \widetilde{A^T}\xi$ where $\widetilde{A^T}$ is a full rank matrix whose range coincides with the null space of A^T. As explained in [9], the internal force can be parameterized by ξ, and strategies presented before can be used for its control.

While the decentralized force control system is inherently simple, it has distinct disadvantages: The decoupling of motion and force control cannot be achieved. Internal force control and external compliance cannot be independently specified. The system will compensate for load gravity force as if it were an external force, requiring additional load compensation. The centralized force control system solves these difficulties at the expense of added calculations.

In general, position accommodation offers several advantages for multiple–manipulator systems:

- The basic controller for each manipulator remains unchanged. This is an advantage where existing manipulator controllers must be used and cannot be changed.
- Switching between controllers for cooperative modes and independent modes is not necessary.
- The calculation requirements are minimal.
- If the inverse kinematics function is sufficiently intelligent, then one manipulator may move through a singularity without necessarily adversely affecting the control of internal forces.
- The system will accommodate to external forces in the same manner as internal forces.

The disadvantages of such an approach are:

- Compliance is achieved through the controllers. Different controllers, PD vs PID for example, and different controller gains will require different impedance functions for equivalent performance.
- There is a potential for unstable interactions between the joint position control loop and the force compliance loop. If the joint controllers and force compliance calculations are digital and have the same bandwidth, then the interactions will be high, if the joint controllers are analog with much higher bandwidth, as is typical of most robot controllers, then the interactions are greatly diminished.

Overall this type of multiple–manipulator control represents one of the most straightforward implementations. As it can be implemented on many robotic systems it is a fundamental baseline to which other methods may be compared.

3 Experimental Results on Position Accommodation Control

We have been conducting multiple arm control experiments in the Center for Intelligent Robotic Systems for Space Explorations (CIRSSE) Dual Arm Laboratory at Rensselaer Polytechnic Institute. There are two PUMA560 type manipulators, each equipped with a Lord 15/50 force/torque sensor. The computer platform is a VME bus based system with five 68020 single board computers. The control software utilizes the CIRSSE–developed CTOS/MCS real time robot operating system. Only decentralized position accommodation has been experimented at this stage.

A number of experimental results is shown in this section. The following aspects of the decentralized position accommodation control are examined:

1. The effect of underlying controller (PD vs. PID).
2. Compression and tension tests.

Effect of Underlying Controller

A major drawback of the position accommodation force control is the dependence of the force control performance on the underlying position controller. For example, a good position controller may not perform well when it is used in the force control context. Fig.1 shows a force control experiment where the same impedance is specified, but the underlying controller is PD in one case and PID in the other. The result show a stable response for the PD case and an oscillatory response in the PID case, even though PID is the controller of choice for position control due to its better steady state response.

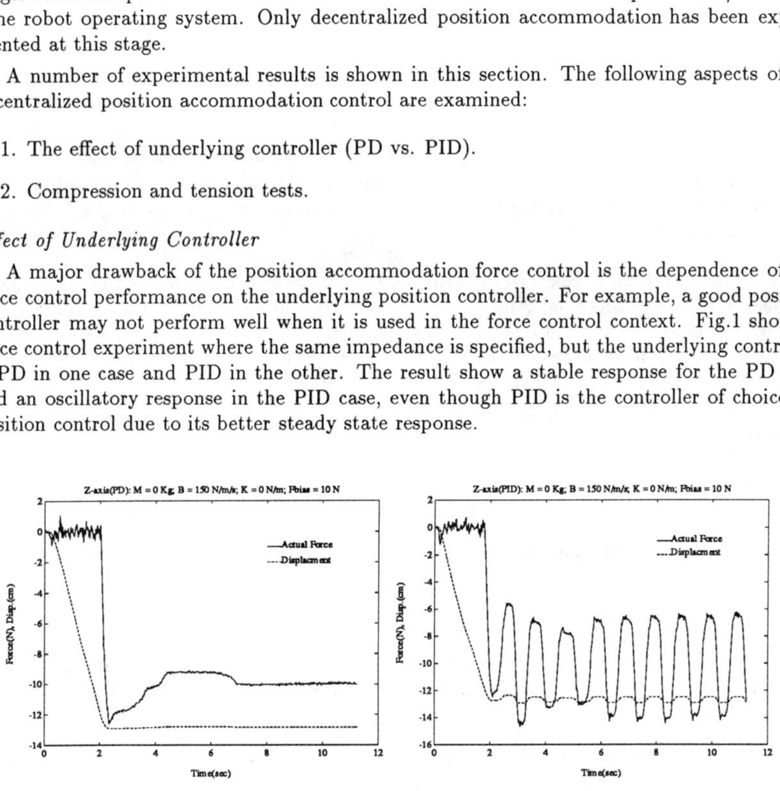

Figure 1: Comparison of PD and PID controllers

Compression and Tension Tests

Fig.2 shows the results of the arm pushing and pulling against each other (each with 5N force) along the longitudal axis of the held strut. The dash line shows the compliance in the closed kinematic chain and the variation in the force corresponds to the variation in the compression displacement variable. Note also the two arms show a net drift since there is no spring programmed in this case.

4 Force Resolution Issue

For most robotic systems, the feedback of joint positions is done via optical encoders. The resolution of these encoders determines the absolute *positioning resolution* of the manipulator. This positioning resolution will correspondingly determine a *force resolution* related to

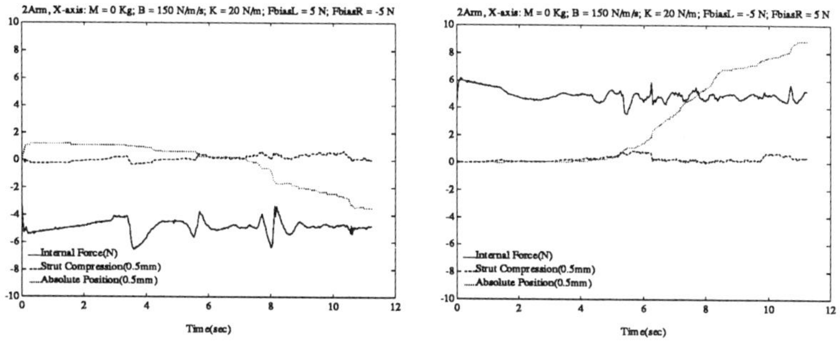

Figure 2: Compression and Tension Tests

Table 1: Approximate Force Resolution: Z-axis

Parameter	PD Pos. Controller	PID Pos. Controller	Units
Lumped-Stiffness, k_l	99	9090	N/m
Force Resolution, Δf_l	0.01	0.9	N

the force mechanisms. The significance of the force resolution in relation to position accommodation control is discussed in this section.

The linear force resolution, Δf_l, of the manipulator can be approximated from the linear position resolution of the manipulator, Δp_l, and the lumped linear stiffness of the manipulator's position-controller, force sensor, and gripper, k_l, as $\Delta f_l = \Delta p_l \, k_l$ Estimating the linear positioning resolution of the PUMA arm to be $\Delta p_l \approx 0.1$[mm] [10], the approximate linear force resolution for both PD and PID position-controllers can be found, and is shown in Table 1 along with the estimated lumped-stiffness.

As can be seen in Table 1, there is a significant difference in the lumped-stiffness and corresponding force resolution between the PD and PID position-controllers. It can be inferred that this would greatly affect the performance of the position accommodation force control and was demonstrated to do so in the experiments shown in the previous section. The experimental runs with the PID controller showed sustained oscillations when contacting a rigid environment. These oscillations have been seen to persist even when heavily damped impedance parameters have been used, thus leading to the conclusion that the oscillations may have been caused *not* by too much forward-gain, but rather represented a *limit-cycle* behavior due to the coarseness of the force resolution.

To avoid the oscillations seen in Sec. 3, a PD position controller is used for most tasks. This brings out the quandary of position-based force control: for good positioning accuracy it is desired to have a very stiff manipulation mechanism(i.e. PID control), but this will correspondingly produce coarser force resolution. In practice, it may be necessary to switch controllers between free motion and contact tasks.

5 Conclusion

Two common multiple–arm force control strategies: position accommodation force control and Jacobian force control are considered in this paper. Experimental results showed the limitation of the position accommodation decentralized force control law: dry friction, effect of underlying position control law, and coupling between the position and force loops. We are currently conducting experimental investigation of centralized controller and Jacobian force control.

Acknowledgment

This work is supported by the Center for Intelligent Robotic Systems for Space Explorations at the Rensselaer Polytechnic Institute funded by the National Aeronautics and Space Administration under grant NAGW 1333.

References

[1] J.T. Wen and K.Kreutz. Stability analysis of multiple rigid robot manipulators holding a common rigid object. In *Proc. 27th IEEE Conference on Decision and Control*, Austin, TX, 1988.

[2] P. Backes. Dual–arm supervisory and shared control space servicing task experiments. In *Proc. AIAA Space Programs and Technologies Conference*, Huntsville, AL, March 1992.

[3] Y. Nakamura, K. Nagai, and T. Yoshikawa. Mechanics of coordinative manipulation by multiple robotic mechanisms. In *Proc. 1987 IEEE Robotics and Automation Conference*, pages 991–998, Raleigh, NC, March 1987.

[4] M. Uchiyama and P. Dauchez. A symmetric hybrid position/force control scheme for the coordination of two robots. In *Proc. 1988 IEEE Robotics and Automation Conference*, pages 344–349, Philadelphia, PA, May 1988.

[5] R. Volpe. *Real and Artificial Forces in the Control of Manipulators: Theory and Experiments*. PhD thesis, Carnegie Mellon University, Pittsburgh, PA., September 1990.

[6] J.T. Wen and S.H. Murphy. Position and force control of robot arms. *IEEE Transactions on Automatic Control*, 36(3):365–374, March 1991.

[7] J.T. Wen and K. Kreutz-Delgado. Motion and force control of multiple robotic manipulators. *Automatica*, to appear in 1992.

[8] V. Hayward and A. Topper. Kali a control system for multiple arm robots, 1989.

[9] S. Murphy and J. Wen. Force decomposition in robot force control. In *Proc. 1991 American Control Conference*, Boston, MA, June 1991.

[10] M. J. Ryan. Implementation of robotic force control with position accommodation. Master's thesis, Rensselaer Polytechnic Institute, Troy, NY., June 1992.

THREE BRANCH HYBRID-CHAIN MANIPULATORS: DESIGN CONSIDERATIONS

Ron P. Podhorodeski, Kenneth H. Pittens
Dept. of Mechanical Engineering, University of Victoria
P.O. Bow 3055, Victoria, B.C., CANADA, V8W 3P6

ABSTRACT

This work examines design concerns related to hybrid-chain manipulators. Based on capability symmetry, work volume and kinematic & dynamic concerns, three branch symmetrically actuated hybrid-chains with passive spherical branch ends are proposed as a preferred structure for hybrid-chain manipulators. A kinematically simple (KS) class of branches suitable for hybrid-chain devices is examined. The characteristics of potential hybrid-chains based on KS revolute-only branches are considered in further detail. The requirement of a unique solution to the forward displacement problem and the existence of uncertainty configurations within the workspace leads to the recommendation of utilizing redundancy in joint encoding and actuation.

1. INTRODUCTION

A hybrid-chain manipulator consists of serial branches acting in parallel on a common end effector (e.e.). Each of the serial branches consists of both actuated and passive joint degrees of freedom (DOF). Fig. 1 illustrates an example three branch hybrid-chain. A serial-chain manipulator (e.g., common industrial robots) can be considered to be a limiting case of a hybrid-chain, i.e., this extreme consists of one branch with all joints actuated. Similarly, fully parallel-chains such as the Stewart Platform[1] can be considered the other extreme, i.e., they have only one joint actuated per branch. Between these two extremes there exists a wide spectrum of possible numbers of branch and actuated joints resulting in potentially useful manipulator structures.

This work examines design concerns related to hybrid-chain manipulators. Based on this examination a preferred structure for hybrid-chain manipulators is proposed. Furthermore, a class of KS branches suitable for hybrid-chains is defined (Section 2). Three branch hybrid-chains based on the revolute joint only versions of the KS branches are examined in further detail. The requirement of unique solutions to basic kinematic and force problems leads to the recommendation of utilizing redundancy in joint actuation and encoding (displacement sensing). Design considerations allowing the effective implementation of such redundancy are given in Section 3 and conclusions in Section 4.

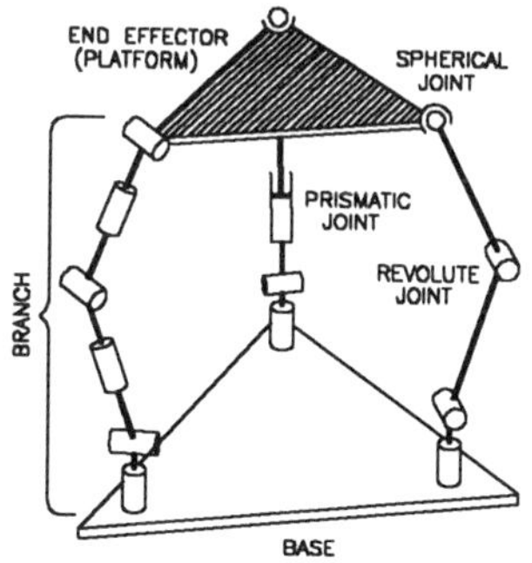

Fig. 1 – A Hybrid-Chain Manipulator

2. HYBRID-CHAIN MANIPULATORS: DESIGN CONSIDERATIONS

2.1 Potential Applications

In general, serial-chain based manipulators have the advantage of providing a large work space of reasonable dexterity compared to parallel-chain manipulators. Parallel-chain manipulators provide better relative payload capacity as well as higher stiffness and accuracy. Parallel-chains can also be dynamically advantageous since passive base distal joints reduce manipulated masses. However, with a fully-parallel actuated chain these advantages come at the expense of available work volume, often due to branch interference.

Serial-chain devices are utilized in relatively low payload to manipulator weight ratio tasks (e.g., 1/25) requiring low to medium accuracy and a large work volume. Fully parallel-chain devices have been recommended for applications such as: assembly [2,3] (high speed and accuracy); master-arms[4,5] (compactness, stiffness and work volume symmetric characteristics); and flight simulation[1] (high load). With proper design, hybrid-chain manipulators exploiting the advantages of both fully parallel and serial-chain devices can potentially be developed. These characteristics include: reasonable work volume; high stiffness, accuracy, payload to manipulator weight ratio and dynamic performance; and work volume symmetric performance characteristics. In addition, with joint actuation redundancy hybrid-chains allow actuator load optimization.

To achieve the above advantageous characteristics a hybrid-chain should have: a minimal number of branches to avoid branch interference (work volume concern); passive end joints to allow reduction of base distal masses; and the ability to accommodate actuation redundancy for actuator load optimization capabilities. In order to have work volume symmetric performance characteristics the branches of a hybrid should be identical in structure & actuation and should be symmetric in their placement around the common payload (e.e.).

2.2 Forward and Inverse Kinematic and Force Problems

Dependent on application implementation of a manipulation device will require solution of basic forward and inverse kinematic and force problems. [1] For example: a motion generator may only require an inverse displacement solution; a displacement input device (master) requires a forward displacement solution; a basic robot manipulator may require forward displacement, inverse displacement and inverse velocity solutions; and force reflective master-arms or "force-controlled" manipulators require forward and inverse displacement, velocity and force solutions. The ability and ease of solution of these problems for hybrid-chain manipulators is a basic design concern of structure and actuation.

Inverse Displacement, ID - From the known e.e. pose the position of the ends of all branches can be found. Serial-chain techniques allow resolution of joint displacement values. For KS branches direct ID solutions can be found using techniques such as [6,7]. Spherical branch end joints are particularly attractive for achieving closed form ID solutions.

Forward Displacement, FD - Unlike serial-chains, a closed form FD solution is not available for non-redundantly joint encoded (i.e., six joints for spatial manipulation) hybrid-chains. Iterative approaches for general chains have been proposed[8] and 16th order polynomials of a single variable have been used for special chain structures[9]. The difficulty in implementation of such solutions is that the FD solution is not unique, e.g., up to 16 solutions exist for the cases of [9]. With appropriate use of redundant encoding the FD solution for hybrid-chains

[1] Forward problems involve resolving e.e. quantities knowning joint values and inverse problems resolving joint quantities given e.e. values.

can be rendered unique and in closed form.

Inverse Velocity, IV - IV solutions are required for finding required rates at the actuated joints. For both nonredundant and redundant actuation velocity solutions can be easily found using concepts of screw reciprocity[10,11]. The e.e. is constrained to instantaneous motions belonging to the screw system defined by the intersection of the screw systems defined by the joints of each branch. Hence, if any of the branches are in a degenerate configuration an available DOF(s) at the e.e. will be lost. Having a fewer number of branches reduces the total number of these branch associated degeneracies.

Forward Velocity, FV - For nonredundant encoding the FV solution requires consideration of a linear system of equations[10,11]. With appropriate redundancy in joint encoding direct forward velocity solutions are possible.

Inverse Force, IF - The IF solution requires consideration of a linear system of six equations in terms of as many unknowns as actuated joints. The unknowns are wrench intensities associated with the reciprocal screws related to the actuated joints[11]. Uncertainty configurations[12] of the hybrid-chain correspond to configurations where the hybrid-chain is unable to sustain arbitrary loading and unconstrained instantaneous twisting of the e.e. becomes possible. Appropriate redundancy in joint actuation allows elimination of uncertainty configurations. With redundant actuation the solution of the wrench intensities is underdetermined and joint forces (torques) can be optimized for load objectives.

Forward Force, FF - Direct FF solutions for KS branches can be found.

2.3 Preferred Basic Structures

A Preferred Hybrid-Chain Structure: From the previous discussion a hybrid-chain allowing: i) symmetrical (and redundant) actuation; ii) passive spherical ends; iii) a least number of branches to avoid branch interference and branch degeneracy; and iv) base proximal actuator mounting is desirable.

Non-redundant symmetrically actuated hybrid-chains capable of six DOF include "6-1,1,1,1,1,1" (fully parallel), "3-2,2,2", "2-3,3" and "1,6" (strictly serial), "number of branches-actuated joints/branch" configurations. Of these, the "6-1,1,1,1,1,1" is prone to branch interference, and the "2-3,3" and "1-6" cannot have passive spherical ends. [2] Redundant symmetricly actuated hybrid-chains include "6-2,2,2,2,2,2" "5-2,2,2,2,2", "4-2,2,2,2", "3-3,3,3" and "2-4,4". Branch interference eliminates "6-2,2,2,2,2,2" and "5-2,2,2,2,2". "2-4,4" does not have passive ends leaving the "4-2,2,2,2" and "3-3,3,3" configurations. The "3-3,3,3" configuration is less prone to encounter branch degeneracies and interference.

Therefore, a hybrid-chain formed from three branches with passive spherical ends can be considered a preferred basic hybrid-chain structure.

A Class of Kinematically Simple Branches: Construction and kinematic modeling of the branches is simplified if unnecessary offsets of the branch links are eliminated and if the joint directions are chosen successively parallel or perpendicular. Therefore, the definition of the KS branches becomes: i) all branches must have a passive spherical end joint group; ii) the main-arms consist of three joints; iii) all unnecessary link offsets are avoided and iv) joint directions are successively parallel or perpendicular.

If all possible main-arm structures formed from combinations of revolute and prismatic joints are considered, and fundamentally degenerate and kinematically equivalent structures are eliminated, the above class can be shown to consist of: (a) five structures characterizing all potential KS revolute-only main-arms (Fig. 2a); (b) nine structures characterizing all potential KS two revolute, one

[2] The "2-3,3" structure with passive spherical ends cannot control rotation about a line passing through the passive spherical joints.

prismatic main-arms (Fig. 2b); (c) two structures characterizing all potential KS one revolute, two prismatic joint based main-arms (Fig. 2c) and (d) one structure characterizing all potential KS three prismatic joint based main-arms (Fig. 2d). In the context of three branch hybrid-chains, any three of the above KS structures could be used to support a common payload platform. In the remainder of this work, three branch hybrid-chains based on the revolute-only KS branches are considered in further detail.

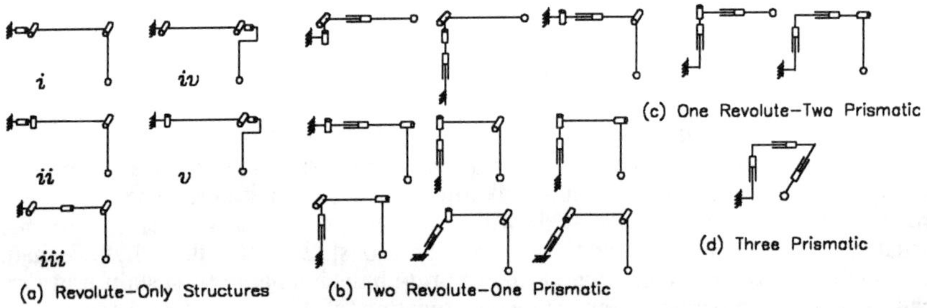

Fig. 2 – Kinematically Simple Potential Branch Structures

3. REVOLUTE-ONLY KS BRANCH HYBRID-CHAINS

3.1 Revolute-Only Hybrid-Chain Characteristics

Each of the five unique revolute-only KS branches (Fig. 2a) can be classified as one of two basic structure types. These structure types are I) branches with a revolute elbow mounted on a pointer base (*ai* and *aii*) and II) branches with a pointer elbow mounted on a revolute base (*aiii*, *aiv* and *av*). The work volume of *ai* is pseudo-spherical, that of *aii* is spherical and those of *aiii*-*av* are tori formed from revolution of a sphere about a revolute joint. The pointer-revolute structures have larger individual branch work volumes than the revolute-pointers. For the assembled three branch hybrid-chain this work volume advantage will still exist but is lessened due to loss of available work volume from branch interference and maximal reach of the individual branches. With appropriate design and assembly, lower arm caused interference for revolute-pointer structures can be eliminated.

In terms of dynamic performance the revolute-pointer structures may be advantageous since their lower arms are constrained to motion in respective planes. The planar motion of the lower limbs of revolute pointers is also advantageous in terms of compactness of the hybrid-chain. Base proximal mounting of actuators is simpler for pointer-revolute structures. However, high precision toothed belts and gears make base proximal mounting feasible for most joints of the revolute-pointers (see Section 3.3).

Having no degeneracies within the branch workspace is highly desirable since branch degeneracy leads to the loss of available instantaneous motion at the e.e. (Section 2.2). [3] The passive spherical end decouples the degeneracies into main-arm and spherical end joint group caused degeneracies. Pointer-revolute branch *ai* is degenerate when the first joint axes intersects the passive spherical end ("over-shoulder" degeneracy). This degeneracy cannot be avoided by design and is work volume internal. Similarly, revolute-pointer branch *av* is in a degenerate configuration within its work volume when the second joint axis intersects

[3] All of the revolute-only branches are degenerate in "straight-arm" configurations. However, these are maximal reach (work volume boundary) configurations of the hybrid-chain.

the passive spherical end. If the upper-arm is designed shorter than the lower arm the branches *aii*, *aii* and *aiv* have no work volume internal degeneracies. All degeneracies for these branch main-arms are work volume boundary positions making the branches strong candidates for hybrid-chain application.

3.2 Solution of the Forward and Inverse Problems

ID & IV: The "translational decoupling" provided by the spherical end joints makes derivation of analytical ID solutions very simple. Similarly analytical IV solutions can be found. Space limitations prevent their presentation here.

FD: If the main-arm joints are nonredundantly encoded (two joints/branch) the FD solutions for hybrid-chains formed from three KS revolute-only branches can be described by a 16th order polynomial of one variable. The FD solution in this case is not unique (up to 16 solutions). With redundant encoding of the main-arm joints (three joints/branch), analytical expressions for the location of the three branch ends can be easily found. Knowing the platform and e.e. geometry it is simple to resolve the e.e. pose (a unique solution).

FV: Redundantly encoded main-arm joints allows the translational velocity of the three branch ends to be found. This yields velocities of three points on the e.e. allowing the translational and angular velocity of the e.e. to be easily resolved.

IF: The associated reciprocal screw quantities involved in the resolution of instantaneous kinematics and force problems for all KS branches are zero pitch screws passing through their respective spherical end joints. Depicted in Fig. 3 is a three *aiv* branch hybrid-chain. Also shown is a branch schematic with associated reciprocal screws indicated. If the chain is nonredundantly actuated two of these reciprocal screws (e.g., $\mathbf{W}_{1_k}$ and $\mathbf{W}_{3_k}$ if joints 1 & 3 are actuated) for each branch and therefore six in total for the hybrid-chain) are involved in the solutions. The hybrid-chain becomes uncertain when these six screws do not span the full six-system of all screw quantities. Workspace internal uncertainty configurations exist for all nonredundantly actuated three branch hybrid-chains formed from the KS branches. If all nine main-arm joints are actuated there becomes nine associated reciprocal screw quantities ($\mathbf{W}_{1_k}$, $\mathbf{W}_{2_k}$ and $\mathbf{W}_{3_k}$, $k = 1, 2, 3$) available to span the six-system of all screws. This use of redundant actuation effectively eliminates uncertainties. Furthermore, this redundant actuation makes an order of ∞^3 IF solutions available for load sharing optimization.

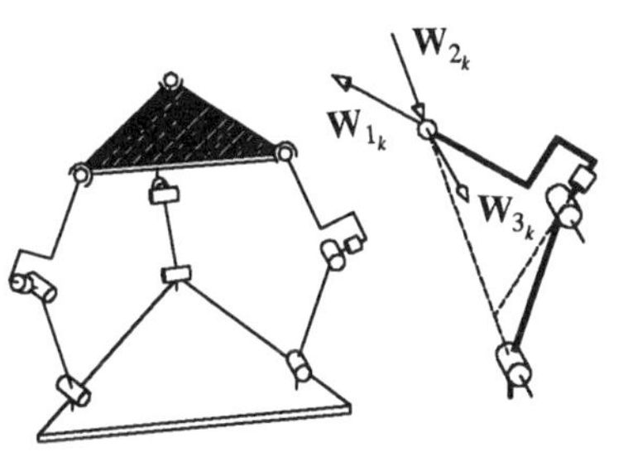

Fig. 3 – Hybrid-Chain & Reciprocal Screws

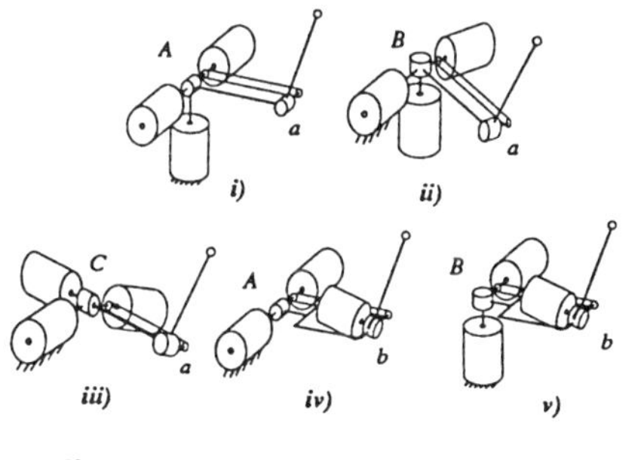

Fig. 4 – Main-arm Actuation Schematic

3.3 Achieving Actuation Redundancy

Assembly of all of the unique KS revolute-only branches can be achieved using three different types of shoulder and two different types of elbow joint assemblies. The use of precision toothed belts and gears can allow all of the actuators to be base proximal mounted in most cases. The distribution of joint assemblies (marked A, B, & C for shoulder and a & b for elbow assemblies) and base proximal actuator mounting is shown in the schematic of Fig. 4.

3.4 Passive Joint Design

The end joints of the branches are to be passive and spherical. The use of ball and socket joints is possible but in general will not provide a sufficient motion ability. A "Cardan-type" construction from three successively perpendicular revolute joints provides a greater range of motion. However, this arrangement introduces a source of work-volume internal degeneracies (wrist degeneracy). The use of mechanical stops on the branches can prevent the encounter of such degeneracies but will lead to a loss of available e.e. work volume. Double-pointer based wrist constructions[13] would allow the elimination of wrist degeneracies in the work volume and may be ideal for hybrid-chain applications.

4.0 CONCLUSIONS

Hybrid-chain manipulators have the potential of providing a combination of the advantages of both strictly serial-chain and fully parallel-chain devices. Three symmetrically actuated branches with passive spherical ends can be considered a preferred basic hybrid-chain structure. This conclusion is based on capability symmetry, work volume, base proximal actuation, and kniematic & force problem solvability concerns. The appropriate use of encoding redundancy of the branch main-arm joints for such a chain yields simple and unique forward kninematic solutions. As well, the appropriate use of actuation redundancy of the branch main-arm joints eliminates work volume internal uncertainty configurations and allows actuator load sharing optimization.

REFERENCES

[1] Stewart, D., "A Platform with Six Degrees of Freedom", *Proc. Inst. Mech. Eng. (London)*, Vol. 180, Part 1, No. 15, pp. 371-386, 1965.

[2] McCallion, H., Truong, P.D., "The Analysis of a Six Degree-Of-Freedom Workstation for Mechanised Assembly", *5th World Congress on Theory of Mach. and Mech.*, pp. 611-616, 1979.

[3] Pierrot, F., Fournier, A., Dauchez, P., "Towards a Fully-Parallel 6 DOF Robot for High-Speed Applications", *1991 IEEE Int. Conf. on Rob. & Auto.*, Sacramento, CA, pp. 1288-1293, 1991.

[4] Fischer, P., Daniel, R., Siva, K.V., "Specification and Design of Input Devices for Teleoperation", *1990 IEEE Int. Conf. Rob. & Aut.*, Cincinatti, OH, pp. 540-545, 1990.

[5] Kim, W.K., Tesar, D., "Study on Structural Design of Force-Reflecting Manual Controllers", *21th ASME Mech. Conf.*, DE-Vol. 24, Chicago, IL, pp. 481-488, 1990.

[6] Pieper, D., "The Kinematics of Manipulators Under Computer Control", *Ph.D. Thesis*, Stanford University, 1968.

[7] Paul, R.P., Robot Manipulators: Mathematics, Programming, and Control, *MIT Press*, 1981.

[8] Podhorodeski, R.P., "A Screw Theory Based Forward Displacement Solution for Hybrid Manipulators", *2nd Nat. Appl. Mech. & Rob. Conf.*, Cincinnati, OH, Nov. 1991, Pap. No. IIIC.2.

[9] Innocenti, C., Parenti-Castelli, V., "Direct Position Analysis of the Stewart Platform Mechanism", *Mech. and Mach. Theory*, Vol. 25, No. 6, pp. 611-621, 1990.

[10] Kumar, V., "Instantaneous Kinematics of Parallel-Chain Robotic Manipulators", *ASME 21st Mech. Conf.*, Chicago, IL, pp. 279-285, 1990.

[11] Podhorodeski, R.P., Pittens, K.H., "Kinematic Design and Analysis of Hybrid-Chain Manipulators", appearing *IASTED Int. Conf. Contr. & Auto.*, Vancouver, CANADA, August 1992.

[12] Hunt, K.H., Kinematic Geometry of Mechanisms, *Claridon Press*, Oxford, 1978.

[13] Stanisic, M.M., Duta, O., " Symmetrically Actuated Double Pointing Systems: The Basis of Singularity-Free Robot Wrists", *IEEE Trans. Rob. & Auto.*, Vol. 6, No. 5, Oct. 1990.

VIRTUAL FORCE CONTROL OF A SIX-DEGREE-OF-FREEDOM PARALLEL MANIPULATOR*

C.C. Nguyen† S.C. Antrazi‡ C.E. Campbell, Jr.§
Catholic University of America
NASA/Goddard Space Flight Center

Abstract

The novel concept of virtual force control is introduced in this paper. This concept avoids sudden step transition of position control to contact force control resulting in contact force vibration when the robot end-effector makes contact with the environment. A virtual force/position control scheme consists of two loops, the force control loop and the position control loop. While the position control loop regulates the free motion, the force control loop regulates the contact force after making contact with the environment and the virtual force measured by a range sensor called capaciflector in the virtual environment. Experimental results of a one-degree-of-freedom virtual force control scheme applied in berthing an orbital replaceable unit are reported and compared with those of conventional pure contact force control cases.

1 Introduction

A robotic assembly task such as mating the male and female parts of an electrical connector or inserting an orbital replacement unit (ORU) into its fixture is generally carried out in two modes, the non-compliant mode and the compliant mode. In the non-compliant mode, since the robot end-effector is moving in free space and is not in contact with its working environment, only pure position control is necessary. On the other hand, in the compliant mode, since the robot end-effector is in contact with the environment, not only position but also contact force control is employed in order to avoid damages to the end-effector and the workpiece. Numerous control schemes such as accommodation, joint compliance, impedance control, and hybrid position/force control have been developed for the compliant mode [1,2]. Most developed force control schemes assume that some degrees of freedom (DOF) are lost when the non-compliant mode is transferred to the compliant mode. And through proper modification of a selection matrix [2], certain DOFs are selected to be force-controlled and the remaining DOFs to be position-controlled. As a result, at the instance the robot end-effector makes contact with the environment, a sudden step transition from position control to force control occurs in those DOFs selected to be force-controlled. The contact force impulse caused by this sudden control mode transition excites the robot structural resonances. These in turn produce unwanted contact chattering and consequently damages to the contact surface especially when the end-effector touches a very stiff object and is controlled by a high-gain force control system [3]. One obvious solution to the above problem is to perform the assembly task at a very slow speed so that the impact of force at contact is minimized. A main drawback of this method is the unnecessarily long operation time which is very critical for some tasks especially when performed in space. The above drawback of conventional force control has motivated researchers to search for ways of making a smooth transition of position into force control. This paper introduces new contact force control concept called *virtual force control* which enables the force controller to be activated even before the robot end-effector workpiece touches the environment. In particular, the virtual force control is developed for a six DOF parallel manipulator whose mechanism have attracted many researchers because of the ability to perform fine and precise motion [8,9]. The paper is organized as follows. It first briefly

*Research funded by Goddard Space Flight Center under Grant NAG 5-780
†Professor of Electrical Engineering, Catholic University of America
‡Graduate Research Assistant
§NASA/Goddard Space Flight Center

reviews the properties and operation principle of the capaciflector. Then using a special case of one DOF berthing of an ORU, the paper introduces the concept and a control scheme of the virtual force control. After that, it develops three supporting processes of the virtual force control scheme, the prefiltering, the off-line calibration and modeling and the on-line distance determination. Finally results of an experiment in which the proposed control scheme is used to berth an ORU to its fixture are presented and compared to those of two conventional pure contact force control cases.

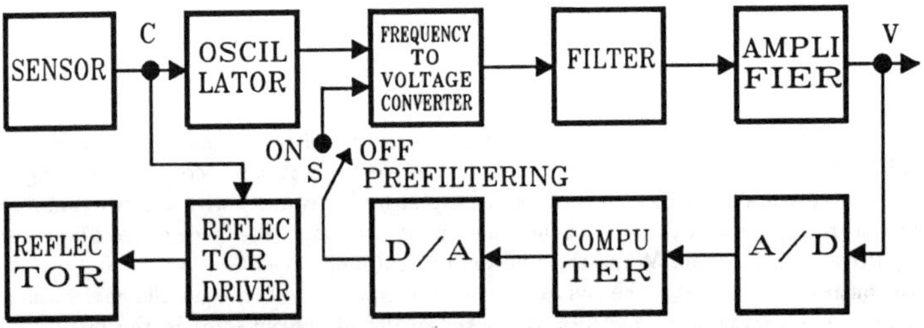

Figure 1 Block diagram of the Capaciflector circuit

2 The Capaciflector

A *capacitive sensor* is a sensor possessing the ability to sense the capacitance of the capacitor between the sensor plate and a grounded object. The total capacitance c_t that the sensor senses is the sum of the capacitance of the capacitor between the sensor and ground c_{sg} and the capacitance of the capacitor between the sensor and the object c_{so}. Therefore the range and the sensitivity of the sensor can be increased by decreasing c_{sg}. This can be done by using insulation materials to increase the *stand off* distance between the sensor plate and ground. However the added insulation introduces many problems such as causing the robot arm on which the sensor is mounted to be bulkier than necessary or impeding the heat flow of the robot arm to outside environment. Consequently, in order to resolve the above problem, the *capaciflector* (*capaci*tive re*flector*) which is a capacitive sensor backed a reflector element (shield) was developed [4]. The reflector prevents field lines from the sensor plate to return directly to ground and thereby increases the number field lines from the sensor plate to the object. In order to make the capaciflector useful such as for control applications, a circuit should be designed to convert the measured capacitance into the corresponding DC voltage. Figure 1 shows the block diagram of a circuit currently used for the capaciflector. As seen from the figure, the measured capacitance serves as the input capacitance of an oscillator whose frequency depends on the capacitance. Via a converter, the frequency of the oscillator is then converted to a corresponding DC voltage which then filtered and amplified. Finally from the fact that the distance between the sensor plate and the object is inversely proportional to the capacitance of the capacitor between the plate and the object, the output of the above capaciflector circuit is a measure of the distance. As a result, the capaciflector can be used to measure the distance from a grounded object to the sensor plate. Since its development, the capaciflector has been found in several applications such as docking [5] and imaging [6].

3 Virtual Force Control

Figure 3 depicts an experiment arrangement which is used to explain the concept of virtual force control. As shown in the figure, a six DOF parallel manipulator consisting of a base platform, a payload platform and six linear actuators is used to berth an ORU into its fixture. Contact is measured by a six DOF force sensor mounted between the ORU and the manipulator. To enable the reader to follow the explanation easily, one DOF berthing is considered. Assuming that the ORU is perfectly aligned with its fixture, if a conventional force control scheme is employed, the manipulator is controlled by a pure position control scheme to move the ORU down onto the fixture at a desired speed and to stop the motion when a contact is established by a force sensor reading along the vertical axis. At the instance of making contact, a sudden stepwise transition of position into force control takes place. The force control scheme then takes over and controls the manipulator to push the ORU onto its fixture until a desired contact force is applied. The sudden stepwise transition of position into force control often creates contact force disturbances which may damage the contact surface.

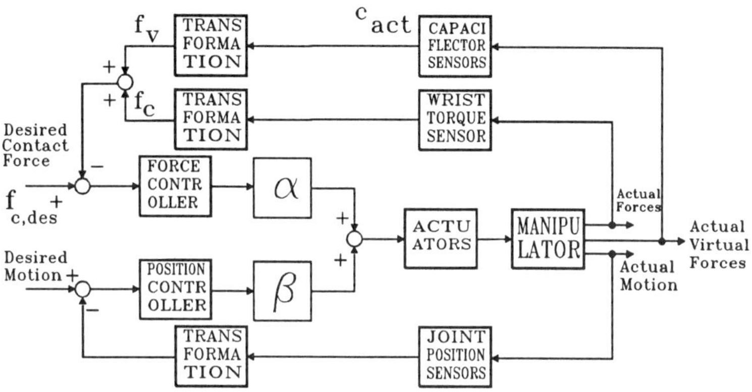

Figure 2 The virtual force/position control scheme

If a capaciflector is mounted at the bottom of the ORU and used to measure the capacitance between the ORU bottom plate and the fixture upper plate, then a *virtual environment* can be defined as the zone bounded by $d_v \leq d \leq 0$. The corresponding capacitance range for the virtual environment is given by $c_v \leq c \leq \infty$. Since the capaciflector circuit is driven by a power supply with finite voltage, its output voltage will reach a maximum value which is usually the power supply voltage when the capacitance reaches ∞. Consequently the output voltage of the capaciflector circuit is equal to its maximum value when the ORU bottom plate touches the fixture upper plate. In addition, the capaciflector capability of continuously monitoring the capacitance change as the ORU approaches the fixture enables a velocity control system to regulate the ORU velocity to be zero at contact. Through proper transformation, the capacitance can be transformed into a *virtual force* which pulls or pushes the ORU as it enters the virtual environment.

Figure 2 illustrates a general force/position control system composed of two loops, the lower position loop controlling the ORU motion outside the virtual environment and the upper force loop controlling the virtual force inside the virtual environment and the contact force $f_{c,des}$. In the position loop, actual manipulator joint displacements measured by joint position sensors are transformed into the corresponding Cartesian motion via the manipulator forward kinematic transformation. The difference between the actual and desired motion is then used to drive a pure position controller whose operation is controlled by a switching variable β. The force loop has two subloops, the virtual force subloop utilizing a capaciflector to regulate the motion of the ORU inside the virtual environment in order to achieve a desired trajectory of the virtual force and the contact force subloop utilizing a wrist force sensor to apply a desired force on the fixture after the ORU makes contact with the fixture. The combination of the actual virtual force f_v and the actual contact force f_c is then compared with the desired contact force. The difference between the combined force and the desired contact force is then used to drive a force controller whose operation is controlled by a switching variable α.

For this particular one DOF berthing case, the switching variables α and β assume the following properties:

$$\alpha = \begin{cases} 0 & \text{if } d > d_v \\ 1 & \text{if } d \leq d_v \end{cases} \quad \beta = \begin{cases} 1 & \text{if } d > d_v \\ 0 & \text{if } d \leq d_v \end{cases} \tag{1}$$

where d denotes the distance of the ORU bottom plate to the upper plate of the fixture and d_v the upper boundary of the virtual environment. The transformation converting actual capacitance c_{act} to virtual force f_v is given by $f_v = c_{act} - c_{max}$. According to (1), the position control loop is the only loop which is activated for $d > d_v$ and is turned off as soon as the ORU enters the virtual environment. At the same instance, the force control loop is activated. In the virtual environment, since no contact has been made resulting in no contact force measured by the wrist force sensor, only the virtual force subloop is activated. Inside the virtual environment the virtual force is negative and is acting like a force pulling the ORU into the fixture upper plate. The subtraction of the desired contact force from the virtual force is bigger than the desired force. Thus the force controller should control the ORU to move into the direction of the pulling virtual force (toward the fixture upper plate) so that the resulting virtual contact force is reduced. As the pulling virtual force decreases to zero, the ORU velocity should decrease proportionally. When the ORU makes contact with the fixture, the virtual force is equal to zero yielding full control to the contact force subloop. In other words, the virtual force subloop is fully activated in the virtual environment and is switched off as soon as the ORU bottom plate touches the fixture upper plate. If the ORU velocity is controlled so that it is zero when the virtual force is zero then a soft landing of the ORU bottom plate on the fixture upper plate is secured and no contact force disturbance is produced.

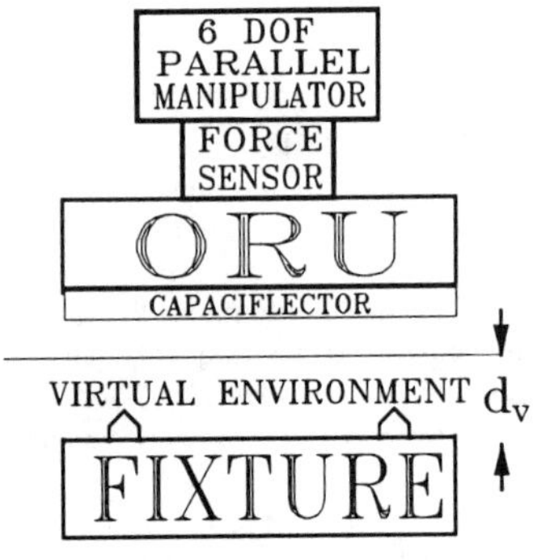

Figure 3 The one DOF ORU berthing

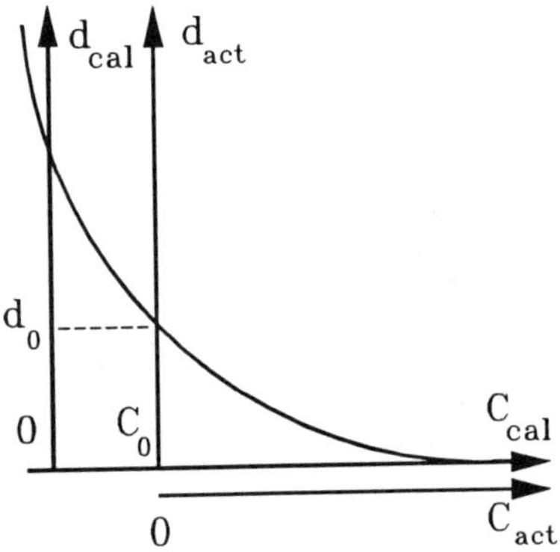

Figure 4 The calibration curve

4 Virtual Force Control Supporting Processes

This section presents the developments of three processes required in the implementation of the proposed control scheme, namely the prefiltering, the off-line calibration and modeling and the on-line distance determination.

4.1 The Capaciflector Prefiltering

According to Figure 1, while Switch S in the ON position, the capaciflector is being prefiltered. Capaciflector prefiltering is to set the capacitance value to zero at a vertical position. This is done by feeding back the capacitance value to a computer via an Analog/Digital converter board, processing it and sending an off-set voltage out to the frequency-to-voltage converter via a Digital/Analog converter board until the capaciflector output voltage is equal to zero. Prefiltering can be performed *off-line* for calibration and modeling and *on-line* for distance determination.

4.2 The Off-Line Calibration and Modeling

This section presents a method of obtaining *off-line* a modeling equation of the distance d_{cal} between the ORU bottom plate and the fixture upper plate as a function of the capacitance c_{cal}. First, the capaciflector is prefiltered at a vertical location about 10 inches above the fixture upper plate. Then the parallel manipulator is controlled to move the ORU down onto the fixture with an increment of 0.05 inch and the capacitance value is measured and recorded at each increment. Using the recorded data, orthogonal functions are used to derive the coefficients a_0, a_1,...,a_6 of the following 6-th order polynomial

$$d_{cal} = f_1(c_{cal}) = a_0 + a_1 c_{cal} + a_2 c_{cal}^2 + a_3 c_{cal}^3 + a_4 c_{cal}^4 + a_5 c_{cal}^5 + a_6 c_{cal}^6 \tag{2}$$

A typical plot of the above polynomial is given in Figure 4.

4.3 The On-Line Distance Determination

Before moving the boundary of the virtual environment, the distance between the current vertical location of the ORU to the fixture must be determined. Recalling that the modeling function is obtained for the region 10 inches $\leq d \leq 0$, prefiltering at an unknown vertical location $d_0 \leq 10$inches is equivalent to shifting the vertical axis in the positive direction of c_{cal} an amount of c_0 (see Figure 4). In other words, if the capacitance value c_0 at d_0 can be determined, then the calibration function $d_{cal} = f_1(c_{cal})$ can be modified to obtain a function f_2 relating c_{act} to d_{act}, namely $d_{act} = f_2(c_{act})$. Thus substituting $d_{cal} = d_{act}$ and $c_{cal} = c_{act} + c_0$ into $d_{cal} = f_1(c_{cal})$ yields

$$d_{act} = f_1(c_{cal}) = f_1(c_{act} + c_0) = f_2(c_{act}) = a_0 + a_1(c_{act} + c_0) + \ldots + a_6(c_{act} + c_0)^6 \tag{3}$$

Consequently in order to obtain c_0, first the capaciflector is prefiltered at the unknown vertical position d_0 and the ORU is controlled to move a small known distance Δ_d and the capacitance change Δ_c caused by the movement is measured and recorded. The slope $\frac{\Delta_d}{\Delta_c}$ is then computed and substituted into the derivative of $f_2(c_{act})$ with respect to c_{act} and evaluated at $c_{act} = 0$, to obtain

$$\frac{\Delta_d}{\Delta_c} = a_1 + 2a_2 c_0 + 3a_3 c_0^2 + 4a_4 c_0^3 + 5a_5 c_0^4 + 6a_6 c_0^5 \tag{4}$$

The roots of (4) are then solved using a computer subroutine and c_0 is selected to be the smallest positive root. Finally c_0 is substituted into $d_{cal} = f_1(c_{cal})$ to calculate d_0.

5 Experimental Study of a Special Case

This section is devoted to report results of an experiment in which the one DOF virtual force control scheme is applied to berth the ORU onto its fixture. The experiment is carried out in four phases: the distance determination phase, the pure position control phase, the virtual force control phase and the contact force control phase, which are described below:

- **The Distance Determination Phase:** Starting at an unknown vertical location $d_0 \leq 10$ in, using the approach in Section 4.3, the distance of the ORU bottom plate to the fixture upper plate is determined.
- **The Pure Position Control Phase:** Based upon the distance d_0 determined in the above phase, the parallel manipulator is controlled to move the ORU with full speed down to the fixture and the motion is stopped as the ORU reaches a distance d_v of about 0.5 in. above the fixture upper plate.

- **The Virtual Force Control Phase:** As soon as the ORU reaches d_v, the virtual force control scheme is fully activated and the pure position control is switched off via the variables α and β, respectively. The capacitance c_{act} which is measured by an X-shaped capaciflector mounted in the center of the ORU bottom plate and is in the range from 0 Volt (at d_0) to 10 Volts (at contact) is converted to the corresponding virtual force by the transformation $f_v = c_{act} - 10$. Thus the virtual force has a value of -10 lb (at d_0) and a value of 0 lb (at contact). Being pulled by the negative virtual force to the fixture plate, the ORU is controlled by the virtual force control scheme so that its motion causes the virtual force to decrease from -10lb to 0lb smoothly without any overshoot, ensuring a soft landing of the ORU bottom plate on the fixture upper plate with almost zero contact force.

- **The Contact Force Control Phase:** As soon as the contact between the ORU and the fixture is established by the capaciflector reading of zero virtual force, the contact force control scheme is activated and the virtual control scheme is switched off. A constant gain force controller is employed to keep pushing the ORU on the fixture plate until a desired contact force $f_{c,des}$ is exerted.

Figure 5 shows the time histories of the virtual force and the contact force of the above experiment. To compare the performance of the virtual force control with pure contact force control, two additional experiments with a pure contact force control scheme are conducted, one with a velocity of 0.3 in/sec and another with a velocity of 0.062 in/sec starting at the same distance d_0 as in the virtual force control scheme. Figure 6 shows the time history of the contact force of the three experiments and Table 1 summarizes their performance in terms of maximum overshoot, time to contact, and settling time. As seen from the table, the pure contact force control scheme with high velocity has short operation time but suffers from high maximum overshoot and one with low velocity has low maximum overshoot but long operation time. The virtual force control scheme achieves both short operation time and negligible maximum overshoot.

Control Scheme	Velocity	Max. Overshoot	Time to Contact	Settling Time
Pure Contact	0.3 in/sec	77 lbs	9 sec	15 sec
Pure Contact	0.062 in/sec	7 lbs	43 sec	50 sec
Virtual	0.3 in/sec	0.5 lbs	9 sec	28 sec

Table 1 Experiment results of ORU berthing

6 Conclusion

This paper has presented the novel concept of virtual force control whose development was based on a range sensor, called capaciflector. The basic properties of the capaciflector were discussed. Results experiments conducted to evaluate the performance of the proposed virtual force control scheme for berthing an ORU with its fixture confirmed that the virtual force control scheme provides superior performance as compares to pure contact force control scheme. Current activities focus on developing a three DOF virtual force control scheme which will be integrated with a capaciflector-based scanner system for pose determination [6]

References

[1] **Khatib, O.** "A Unified Approach for Motion and Force Control of Robot Manipulators: The Operational Space Formulation," *IEEE Journal of Robotics and Automation*, Vol. RA-3, No. 1, pp. 43-53, 1987.

[2] **Raibert, M.H. and Craig, J.J.**, "Hybrid Position/Force Control of Manipulators," *J. Dynamic Systems, Measurement, Control*, 102, pp. 126-133, 1981.

[3] **McKerrow, P.J.**, Introduction to Robotics, *Addison Wesley*, 1990.

[4] **Vranish, J.M., R.L. McConnell, and Mahalingam, S.** "Capaciflector Collision Avoidance Sensors for Robots," *Computers and Electrical Engineering: An International Journal*, Vol. 17, Number 3, pp. 173-180, 1991.

[5] **Cheung, E., Manzano, M., and McConnell, R.**, "Docking Orbital Replacement Units with Capaciflectors," *Proceedings of the Fourth International Symposium on Robotics and Manufacturing*, Santa Fe, New Mexico, November 1992.

[6] **Campbell, Jr., C.E.**, "Robotic Berthing Using Capaciflector Imaging," *Proceedings of the Fourth International Symposium on Robotics and Manufacturing*, Santa Fe, New Mexico, November 1992.

[7] **Catalano, D.A.**, "WP-04 Orbit Replaceable Units Robotics Maintenance Final Test Report," *Analex Corporation*, Goddard Space Flight Center, November 1990.

[8] **Nguyen, C.C., Antrazi, S., Zhou, Z-L, and Campbell, Jr., C.E.**, "Analysis and Implementation of a 6 DOF Stewart-Platform-Based Robotic Wrist," *Computers and Electrical Engineering: An International Journal*, Vol. 17, Number 3, pp. 191-204, 1991.

[9] **Nguyen, C.C., Antrazi, S., Zhou, Z-L, and Campbell, Jr., C.E.**, "Adaptive Control of a Stewart Platform-Based Manipulator," Invited Paper, to be published in a special issue of *Journal of Robotic Systems*, 1992.

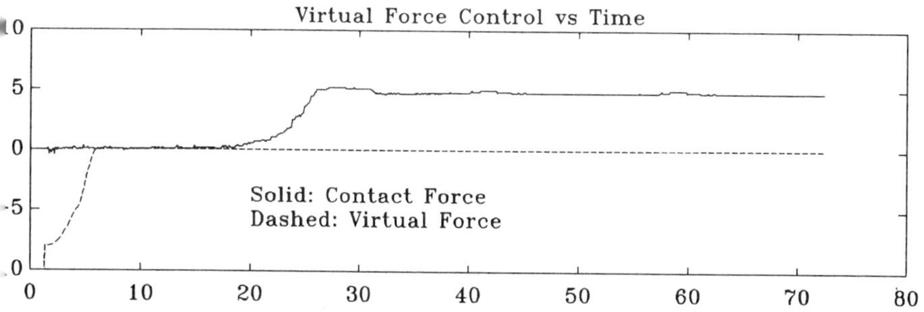

Figure 5 Force time history of the virtual force control scheme
(vertical axis=force in lb.; horizontal axis=time in sec.)

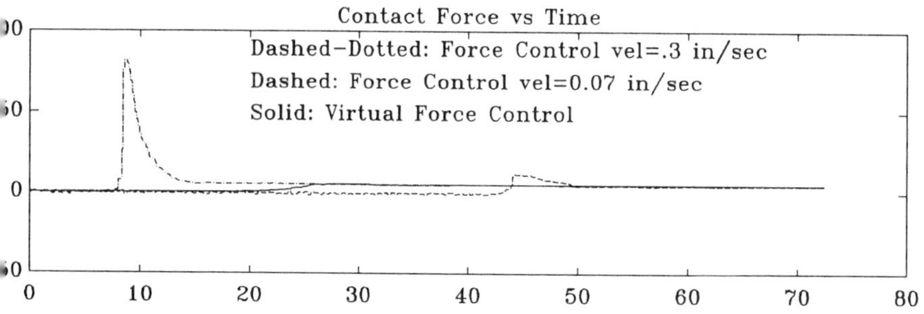

Figure 6 Contact force time history of the three control schemes
(vertical axis=force in lb.; horizontal axis=time in sec.)

Multiple Arm Manipulation of Loosely Held Objects

Kenneth Kreutz-Delgado
University of California, San Diego

Abhi Jain
Jet Propulsion Laboratory/California Institute of Technology

Daryush Agahi
University of California, San Diego

Abstract

In this paper, some illustrative examples are given that demonstrate the use of the Spatial Operator Algebra for dyamical modeling and analysis of a system of multiple parallel manipulators manipulating an articulated object. In general, the contacts between the manipulators and the held object are "loose", meaning that some contact degrees–of–freedom are not constrained by the contact. "Articulated" means that the held object has internal degree–of–freedom to be controlled by the grasping manipulators (as would be required when the object is, say, a pair of scissors).

A partitioning is performed into a *primary* system of manipulators and a *secondary* system of task objects and environment, together with *closure constraints* that describe the effect of mutual interactions between manipulators, task objects, and the environment. With this partitioning, a large part of the dynamics remains independent of the specified nature of the task object to be manipulated. This allows for optimization of the time–invariant parts of control and path–planning algorithms and for efficient modification of such algorithms at run–time to accomodate a variety of different held objects, grasps, and environmental contraints.

1. Constrained Equations of Motion

Details of the use of the Spatial Operator Algebra for the modeling and analysis of complex multibody systems can be found in references [1,3,4] . The multiple–arm system model discussed here is presented in detail in the earlier paper [2] and is only briefly touched upon here. In essence, both the primary and secondary systems are symmetrically described as tree–structures, with possibly mobile bases, subject to contact and environmental constraints. The environmental constraints describe interactions between the systems and the environment as well as closed loop closure conditions that may exist within each subsystem. With β denoting the generalized speeds for the joints, the equations of motion for a single manipulator are given in operator form by

$$\begin{aligned} V &= \phi^* H^* \beta, \qquad \alpha = \phi^*(H^*\dot{\beta} + a) \\ f &= \phi(M\alpha + b) = \phi M \phi^* H^* \dot{\beta} + \phi(M\phi^* a + b + Bf(0)) \\ T &= Hf = H\phi M\phi^* H^* \dot{\beta} + H\phi(M\phi^* a + b + Bf(0)) \\ &= \mathcal{M}\dot{\beta} + \mathcal{C} + Bf(0) \end{aligned} \tag{1.1}$$

where $\mathcal{M} = H\phi M\phi^* H^*$ is the *mass matrix* of the serial chain, and $\mathcal{C} = H\phi(M\phi^* a + b)$ consists of the coriolis and centrifugal joint forces. The Jacobian matrix J, which maps the joint rates β to the contact points velocity, $\hat{V} = J\beta$, is given by $J = B^*\phi^* H^*$.

The primary system here is defined as a collection of serial manipulators mounted on a common base. For the jth manipulator, we have the quantities V_j, M_j, ϕ_j, H_j etc. Define

$$\phi_P \triangleq \mathrm{diag}(\phi_j), \quad H_P \triangleq \mathrm{diag}(H_j), \quad M_P \triangleq \mathrm{diag}(M_j) \quad \text{etc.}$$

The dynamics of the multiarm system are then given by (1.1) with the quantities therein being replaced by the corresponding quantities with the subscript "P". The collective Jacobians of all the manipulators is given by $J_P = \mathrm{diag}(J_j) = B_P^*\phi_P^* H_P^*$. Let $\hat{V}_P$ denote the vector of spatial velocities at the environmental and grasp contact points for the primary system. Thus, $\hat{V}_P = J_P\beta_P$.

For the task object, referred to as the secondary system, we use the same notation as for the primary system except for the use of the subscript S.

A "loose" grasp of the tool/task-object by the primary system permits relative motion between the the task object and the manipulators, leading to grasp constraint equations of the form

$$\hat{Q}_P\hat{V}_P + \hat{Q}_S\hat{V}_S = 0 \,. \tag{1.2}$$

Interactions between the primary system and the environment and the task–object and the environment lead to constraint equations of the form

$$Q_P\hat{V}_P = 0 \quad \text{and} \quad Q_S\hat{V}_S = 0 \,. \tag{1.3}$$

Defining the overall constraint matrix A as

$$A \triangleq [A_P \;\; A_S], \quad \text{where} \quad A_P \triangleq \begin{pmatrix} \hat{Q}_P \\ 0 \\ Q_P \end{pmatrix}, \quad \text{and} \quad A_S \triangleq \begin{pmatrix} \hat{Q}_S \\ Q_S \\ 0 \end{pmatrix}$$

the closure constraints can be collectively expressed in the form:

$$A\begin{pmatrix} \hat{V}_P \\ \hat{V}_S \end{pmatrix} = [A_P \;\; A_S]\begin{pmatrix} J_P & 0 \\ 0 & J_S \end{pmatrix}\begin{pmatrix} \beta_P \\ \beta_S \end{pmatrix} = 0 \,. \tag{1.4}$$

It is assumed that steps have been taken to ensure that $[A_P J_P \quad A_S J_S]$ is of full row rank. Note that changes in the grasp/environmental constraints require only the updating of the A_P and A_S matrices.

Given the above set–up, it is now possible to effectively simulate, analyze, or control the combined primary–secondary system [1]–[6].

2. Constraint Conditions

Assume for simplicity that all constraints are on the *relative* spatial velocity and there are no environmental constraints so that $\hat{Q}_P = -\hat{Q}_S \triangleq Q$. Listed below are examples of commonly encountered constraints and the associated partitioned constraint matrix:

$$Q = \begin{pmatrix} B_{\omega\omega} & B_{\omega v} \\ B_{v\omega} & B_{vv} \end{pmatrix}$$

We use the triad of vectors $[x, y, z]$ to denote the constraint joint frame, and define $\tilde{x}$ and x^* by $\tilde{x}y \equiv x \times y$ and $x^*y \equiv x.y$, where "×" and "." denote the vector cross product and dot product respectively.

(i) **0 dof joint:** Rigid grasp. $B_{\omega\omega} = B_{vv} = I_3$ and $B_{\omega v} = B_{v\omega} = 0$.

(ii) **1 dof rotary joint about x:** (a) Interbody hinge joints, (b) grasp of a cylindrical handle. $B_{\omega\omega} = \tilde{x}$, $B_{\omega v} = B_{v\omega} = 0$, and $B_{vv} = I_3$.

(iii) **1 dof rolling joint about x:** Rolling without slipping of wheels/paint roller (with vector from axis to contact point r) on ground/wall. $B_{\omega\omega} = \tilde{x}$, $B_{\omega v} = 0$, $B_{v\omega} = -\tilde{r}$, and $B_{vv} = I_3$.

(iv) **2 dof rotary joint about x and y:** "soft finger" grasps. $B_{\omega\omega} = z^*$, $B_{\omega v} = B_{v\omega} = 0$, and $B_{vv} = I_3$.

(v) **3 dof spherical joint:** (a) Rolling without slipping of a spherical caster on ground (with vector from axis to contact point r),, (b) an interbody ball and socket joint ($r = 0$). $B_{\omega\omega} = 0$, $B_{\omega v} = 0$, $B_{v\omega} = -\tilde{r}$, and $B_{vv} = I_3$.

(vi) **1 dof helical joint about x with pitch p:** Threading a bolt. $B_{\omega\omega} = \tilde{x}$, $B_{\omega v} = 0$, $B_{v\omega} = -pI_3$, and $B_{vv} = I_3$.

(vii) **2 dof cylindrical joint along and about x:** (a) Insertion of a peg in a hole, (b) Internal joint for a plunger/syringe. $B_{\omega\omega} = \tilde{x}$, $B_{\omega v} = B_{v\omega} = 0$, and $B_{vv} = \tilde{x}$.

(viii) **2 dof prismatic joint in x, y plane** Polishing/sanding a surface. $B_{\omega\omega} = I_3$, $B_{\omega v} = B_{v\omega} = 0$, and $B_{vv} = z^*$.

(ix) **6 dof joint** (a) Fully mobile base, (b) loss of grasp contact. $B_{\omega\omega} = B_{\omega v} = B_{v\omega} = B_{vv} = 0$.

Some of the Q matrices above are not of full rank and they should be converted to full rank matrices before use in the algorithms. Furthermore, note that the constraints in (iii), (iv) and (v) are nonholonomic. For a given system, the overall A constraint matrix in (1.4) is formed by assembling the Q matrices from each individual grasp and environmental constraint on the system.

3. Application Examples

In this section we look at an example of a primary multi–manipulator system interacting with articulated task objects and focus primarily on the modeling of the articulated objects and the associated grasp and environmental constraints. In the first example, the task–object is a pair of scissors/shears being held by two manipulators. The second example is that of a roller being pushed along the floor by a single manipulator. Since the modeling of the manipulators and their dynamics is well understood, we will focus on the task–objects and the constraints on them, and those between them and the manipulators.

The scissors/shears are shown in Figure 1 and the roller is shown in Figure 2. Each of them has a single internal 1 dof rotary joint. In fact, a closer look at the figures reveals that when regarded as free bodies they have the same kinematical structure, and they each can be regarded as a two link object with a 1 dof internal rotary joint. With the two objects regarded as free objects, we attach a 6 dof joint between link 2 and the inertial frame, which is equivalent to making link 2 completely mobile with respect to the inertial frame. Keeping in mind the isomorphism between the two objects, we proceed to give some details about modelling the kinematical constraints between the primary and secondary systems. When regarded as free objects, the task–objects have 7 dofs. C_1 and C_2 denote the point of contact where the constraint conditions apply. Also $l(x, y)$ indicates the vector from x to y, and I_m denotes the $m \times m$ identity matrix.

Constraint Matrix: The constraints for the shear example consist of the rigid grasp constraints at C_1 and C_2 as in (i) above. For the case of the roller example, the constraint at C_1 is due to a rigid grasp as in (i), while the constraint due to the contact at C_2 is as in (iii).

Link Kinematical and Dynamical Parameters:

Link 1

$$\theta_S(1) \in \mathcal{R}^1; \quad \beta_S(1) = \dot{\theta}_S(1); \quad H_S^*(1) = \begin{pmatrix} h(1) \\ 0 \end{pmatrix} \in \mathcal{R}^6; \quad h(1) = \begin{pmatrix} 0 \\ 0 \\ 1 \end{pmatrix} \quad \text{in link 1 fram}$$

$$M_S(1) \in \mathcal{R}^{6\times6}; \quad T_S(1) \in \mathcal{R}^1; \quad \phi_S(2,1) = I_6; \quad \phi_S(1, C_1) = \phi_S(l(O_1, C_1)) = \begin{pmatrix} I_3 & \hat{l}(O_1, C \\ 0 & I_3 \end{pmatrix}$$

Link 2

$$\theta_S(2) \in SE(3); \quad \beta_S(2) = \begin{pmatrix} \omega_S(2) \\ v_S(2) \end{pmatrix} \in \mathcal{R}^6; \quad H_S^*(2) = I_6; \quad M_S(1) \in \mathcal{R}^{6\times6}$$

$$T_S(2) = \begin{pmatrix} N_S(2) \\ F_S(2) \end{pmatrix} \in \mathcal{R}^6; \quad \phi_S(3,2) = I_6; \quad \phi_S(2, C_2) = \phi_S(l(O_2, C_2) = \begin{pmatrix} I_3 & \hat{l}(O_2, C_2) \\ 0 & I_3 \end{pmatrix}$$

Object Kinematics:

$$V_S = \phi_S^* \beta_S; \quad \phi_S^* = \begin{pmatrix} I_6 & I_6 \\ 0 & I_6 \end{pmatrix}; \quad H_S^* = \begin{pmatrix} H_S^*(1) & 0 \\ 0 & I_6 \end{pmatrix}; \quad \beta_S = \begin{pmatrix} \dot{\theta}_S(1) \\ \beta_S(2) \end{pmatrix}$$

$$\Rightarrow \quad V_S = \begin{pmatrix} V_S(1) \\ V_S(2) \end{pmatrix} = \begin{pmatrix} H_S^*(1)\beta_S(1) + \beta_S(2) \\ \beta_S(2) \end{pmatrix}$$

Note that , $V_{C_1} = \phi_S^*(1, C_1) V_S(1)$, $\quad V_{C_2} = \phi_S^*(2, C_2) V_S(2)$. Thus,

$$B_S = \begin{pmatrix} \phi_S(1, C_1) & 0 \\ 0 & \phi_S(2, C_2) \end{pmatrix}, \quad \text{and} \quad \hat{V}_S = \begin{pmatrix} V_{C_1} \\ V_{C_2} \end{pmatrix} = B_S^* V_S$$

$$\Rightarrow \quad \hat{V}_S = J_S \beta_S, \quad \text{where} \quad J_S = \begin{pmatrix} \phi_S^*(1, C_1) H_S^*(1) & \phi_S^*(1, C_1) \\ 0 & \phi_S^*(2, C_2) \end{pmatrix} \in \mathcal{R}^{12\times7}$$

Note that J_S^* is always onto, i.e., J_S is always full rank and $\text{Ker} J_S^* \equiv \mathcal{R}^5$. It is easy to verify that

$$\mathcal{M}_S = \begin{pmatrix} H_S(1) M_S(1) H_S^*(1) & H_S(1) M_S(1) \\ M_S(1) H_S^*(1) & M_S(2) \end{pmatrix}$$

The shear has no environmental constraints on it and both constraints are due to the two manipulators grasping at the points C_1 and C_2. On the other hand, the roller has one constraint on it due to the manipulator grasping at the point C_1, and a second one due to its contact with the environment (the "ground") at C_2, where its roller can only roll and not slip on the ground.

The examples above illustrate how very different objects and their interactions with the manipulators and the environment can be modelled in a very simple, concise and systematic manner. Some of the issues involved in kinematic control of the combined primary/secondary system, and in particular the control of the secondary system by the primary system, is discussed in the companion paper [6].

4. Acknowledgement

The research described in this paper was performed in part at the Jet Propulsion Laboratory, California Institute of Technology, under contract with the National Aeronautics and Space Administration. K. Kreutz–Delgado and D. Agahi are partially supported by NSF PYI grant No. IRI–9057631.

References

[1] G. Rodriguez, K. Kreutz-Delgado and A. Jain, "A Spatial Operator Algebra for Manipulator Modeling and Control", *The International Journal of Robotics Research*, 1990.

[2] A. Jain, K. Kreutz-Delgado, and G. Rodriguez, "Recursive Dynamics of Multiarm Robotic Systems in Loose Grasp of Articulated Task Objects", in *3rd International Symposium on Robotics and Manufacturing*, (Vancouver, Canada), June 1990.

[3] G. Rodriguez and K. Kreutz-Delgado, "Recursive Mass matrix Factorization and Inversion: an Operator Approach to Open and Closed-Chain Multibody Dynamics", JPL Publication 88-11, Jet Propulsion Laboratory, Pasadena, CA, 1988.

[4] G. Rodriguez, A. Jain and K. Kreutz-Delgado, "Spatial Operator Algebra for Multibody Dynamics", *Journal of Astronautical Sciences*, 1992 Jan–Mar, Vol. 40 No. 1, pages 27–50.

[5] J. Wen and K.Kreutz-Delgado, "Motion and Force Control of Multiple Robotic Manipulators", *Automatica*, accepted for publication.

[6] K. Kreutz-Delgado, D. Agahi, and A. Jain, "Generalized Velocities and Coordinates of Multiple Arm Systems", *Fourth International Symposium on Robotics and Manufacturing*, Nov 1992.

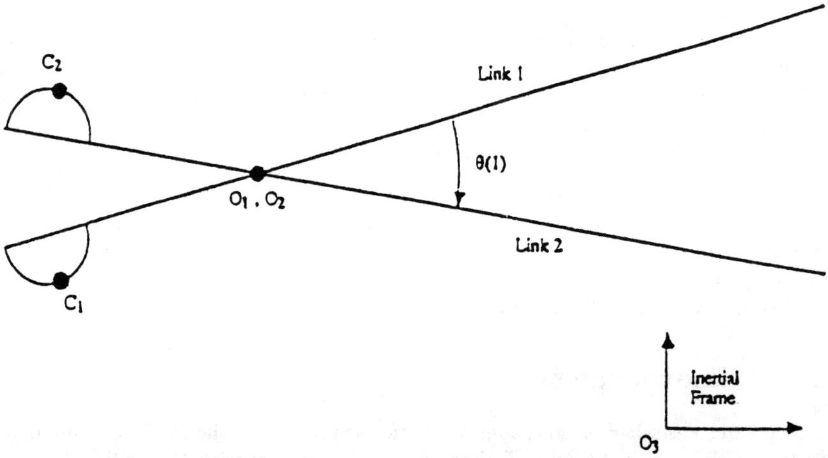

Figure 1. Shears/Scissors

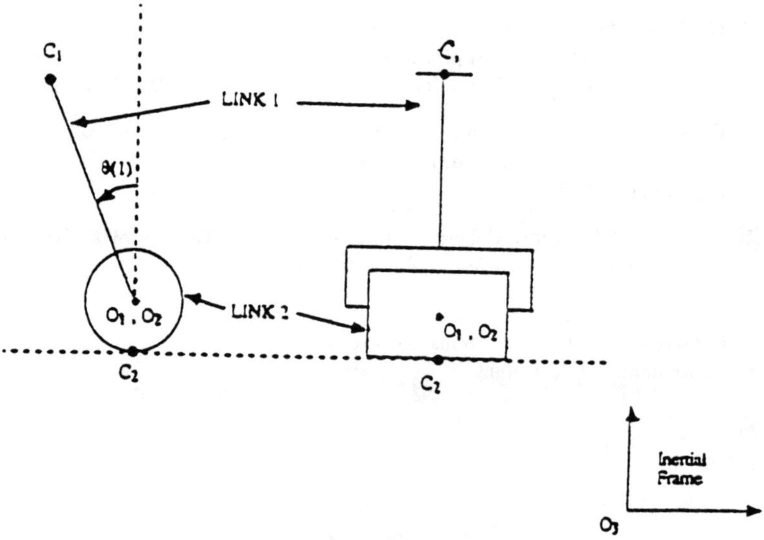

Figure 2. Roller

PAYLOAD KINEMATIC CALIBRATION USING A DUAL MANIPULATOR SYSTEM

W. F. Dellinger
Swales & Associates, Inc.
5050 Powder Mill Road
Beltsville, MD 20705

J. N. Anderson
Center for Manufacturing Research
Box 5077, Tennessee Technological University
Cookeville, TN 38505

ABSTRACT

An algorithm is presented that determines the kinematics of a payload with multiple degrees-of-freedom. Using a dual manipulator system equipped with wrist force/torque sensors, vectors normal to the constraint surface dictated by the payload are found. These vectors are used to determine the payload's kinematic structure. Once the structure is known, the kinematic parameters are determined using vector geometry.

INTRODUCTION

The kinematic calibration of a payload is necessary when two manipulators pick up an object whose kinematic structure and parameters are unknown. In this case the directions in which the object will allow movement are unknown. Using a force controller, calibration motions can be performed. The resulting data can then be analyzed to determine the kinematic structure and parameters of the payload. These parameters provide the manipulators with the information required to effectively handle the payload without damage.

This paper describes an algorithm that was designed for this purpose. The case considered herein is that of an unknown payload with two non-parallel revolute joints. The payload's links are assumed to be rigid, the joints need not be frictionless, and the manipulators used in the calibration have themselves been previously calibrated such that their kinematics are known. Other payloads involving one and two revolute and prismatic joint combinations are analyzed in [1].

Organization of this paper is as follows. The next section describes the process of determining the payload's structure. This involves collection and analysis of force/torque, position, and orientation data to locate the payload's joint axes. Once these axes are known, the payload's kinematic parameters can be calculated using the method outlined in the third section. Finally, conclusions are drawn from this work.

STRUCTURAL ANALYSIS

To generate data from which the kinematic information is extracted, the robot system is commanded to perform maneuvers which cause motion along the constraint surface. These are called calibration maneuvers and are performed with the left manipulator held fixed while the right manipulator, henceforth called the test arm, is commanded independently. The test arm is controlled by adding small bias torques τ_b to the joint torque signals. The τ_b are calculated as

$$\tau_b = K_v J^{-1}(q_m)\frac{\|V^*\| - \|V_m\|}{\|V_m\|}V_m, \tag{1}$$

where K_v is a positive definite "velocity error" gain matrix, $J^{-1}(q_m)$ is the inverse Jacobian of the test arm, $\|V^*\|$ is a target magnitude of the Cartesian velocity, and V_m is the instantaneous (measured) Cartesian velocity given by

$$V_m = \frac{\Phi(q_m[(k+1)T]) - \Phi(q_m[kT])}{T}. \tag{2}$$

Here $\Phi(q_m[kT])$ represents the forward kinematics of the test arm at time kT based on the measured joint angles q_m and T is the sample period. Use of V_m speeds the motions between test points.

After moving the test arm along the constraint surface, motion is stopped at a test point p_{t_i}. With the test arm operating under position control, it is commanded by adding a zero mean, white noise process to each of the test arm's commanded joint angles. The test arm is thus excited such that motions are attempted around p_{t_i} and position data are collected. The data points are used pairwise to calculate differential position vectors, δX_{ij}, which are used to define the tangent plane, $\mathcal{P}_{T_i}$. The vector normal to the constraint surface at p_{t_i}, n_i, can be found as the vector normal to $\mathcal{P}_{T_i}$.

The n_i provide information about the payload's structure. A payload with two revolute joints has as a constraint surface a generalized torus when the two joint axes are not parallel. (The torus will have zero volume if the two axes are perpendicular and intersecting.) Planar motion results when the two joint axes are parallel. For a torus, the n_i will typically have different directions such that the orientation change between any pair of p_{t_i} will not be about the same axis as that between any other pair. Thus by calculating the equivalent axis of rotation between pairs of test points the case of a toroidal constraint surface (indicating two non-parallel revolute joints) can be distinguished from the surfaces of other joint combinations. These other cases are considered in [1].

Determination of Payload Axes

Once the structure of the payload has been determined, the locations and directions of the payload's joint axes are found. Figure 1 shows the frames associated with a two-jointed payload. The origins of the left and right end-effector frames ($\{l\}$ and $\{r\}$), payload joint frames ($\{1\}$ and $\{2\}$), and an intermediate frame ($\{3\}$) relative to the base frame are given by p_l, p_r, p_{org1}, p_{org2}, and p_3, respectively. Axes z_1 and z_2 are the joint axes of the payload.

The first axis determined in the proposed algorithm is z_2. This axis runs through the center of the torus and can be at any angle relative to the base frame. All of the n_i intersect z_2, but not at a common point. However, the projections of the n_i (n_{P_i}) onto a plane perpendicular to z_2 ($\mathcal{P}_{\perp 2}$) will intersect z_2 at a common point. Figure 2 shows how the x_b-y_b plane can be rotated by α_b (about z_b) and β_b (about $-x_b$) to obtain $\mathcal{P}_{\perp 2}$. In the proposed algorithm, the values of α_b and β_b are iterated until the pairwise intersection points of the n_{P_i}

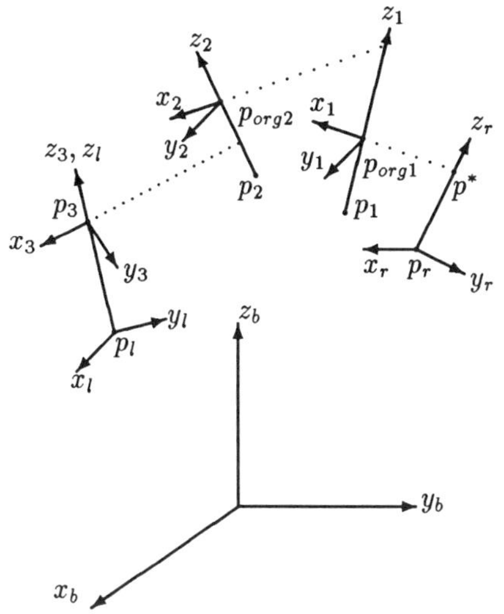

Figure 1: A Two-Jointed Payload

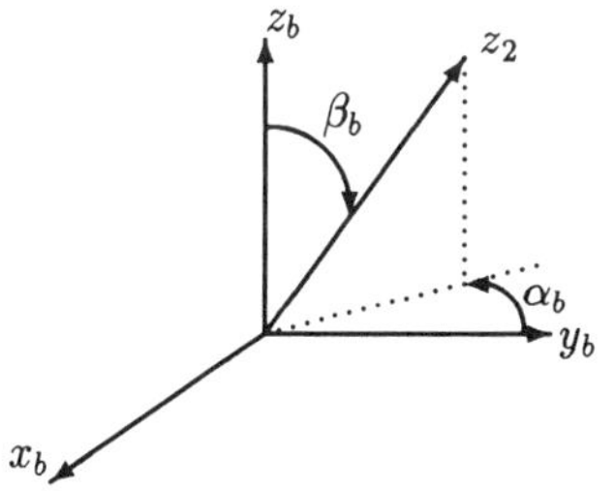

Figure 2: Determination of $\mathcal{P}_{\perp 2}$

converge to a common point. To do this, a cost function is defined as $J = \frac{1}{2}e^T e$, where e contains the errors between pairs of intersection points. That is, for three projected vectors,

$$e = \begin{bmatrix} P_{12} - P_{13} \\ P_{12} - P_{23} \\ P_{13} - P_{23} \end{bmatrix}. \tag{3}$$

The vector containing the two coordinates of the intersection of n_{P_i} and n_{P_j}, P_{ij}, is found by solving the equation of two intersecting lines,

$$P_{ij} = p_{P_i} + \lambda_i n_{P_i} = p_{P_j} + \lambda_j n_{P_j}, \tag{4}$$

for λ_i and λ_j. In (4), p_{P_i} is the projection of p_{t_i} onto $\mathcal{P}_{\perp 2}$.

Defining $\theta_k = [\alpha_{bk} \ \beta_{bk}]^T$, Newton's method [2] can be applied to minimize J using the update

$$\theta_{k+1} = \theta_k - \left(\frac{\partial^2 J}{\partial \theta_k^2}\right)^{-1} \mu \left(\frac{\partial J}{\partial \theta_k}\right), \tag{5}$$

where μ is chosen for speed of convergence. After convergence, the direction of z_2 is given by the third column of the rotation matrix defined by α_b and β_b. The point p_2 is chosen to be the intersection of z_2 with a plane parallel to $\mathcal{P}_{\perp 2}$ and passing through one of the p_{t_i} [1].

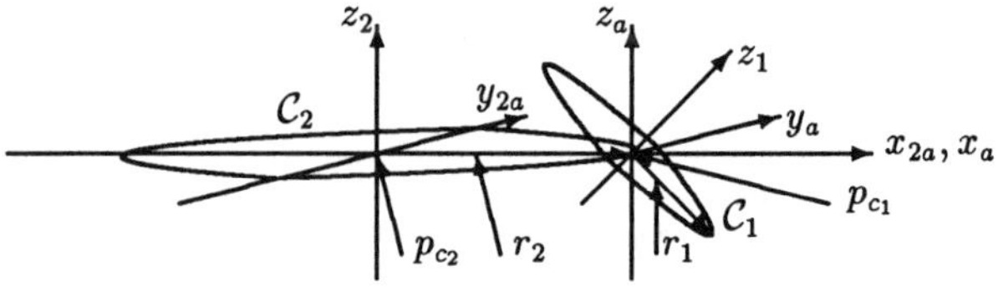

Figure 3: Structure of the Payload

The process used to identify the direction of z_1 and a point on z_1, p_1, is described next with the aid of Figure 3. When the motion of the payload is confined to rotation about z_1 only, the origin of frame $\{r\}$ traces out the circle $\mathcal{C}_1$ with a radius of r_1. Note that the plane containing $\mathcal{C}_1$ is perpendicular to z_1, and z_1 passes thru the center of $\mathcal{C}_1$ at p_{c_1}. The locus of points defined by rotating the center of $\mathcal{C}_1$ about z_2 defines the circle $\mathcal{C}_2$ centered on z_2 at p_{c_2} with a radius of r_2. The toroidal constraint surface is the surface defined by $\mathcal{C}_1$ as it is rotated about z_2. All of the p_{t_i} lie on this surface.

The frames shown in Figure 3 are defined such that the origin of frame $\{a\}$, p_{c_1}, lies in any plane, $\mathcal{P}$, that contains z_2. The direction vector of x_a (and x_{2a}) is perpendicular to z_2 and in $\mathcal{P}$ (pointing away from z_2). The direction vector of z_a is that of z_2. The orientation and location of z_1 (and $\mathcal{C}_1$) are fixed in frame $\{a\}$. The circle $\mathcal{C}_1$ can be modeled as a circle in the x_a-y_a plane, centered at the origin and having the radius r_1, that has been rotated about z_a by α_a and about x_a by β_a. This is equivalent to rotating z_a such that it coincides with z_1.

Given the above definitions, the relationship of frame $\{a\}$ with the base frame is established by r_2 and the signed distance of p_2 to p_{c_2}, γ. The locus of points obtained by rotating the p_{t_i} to $\mathcal{P}$ is used to define an area $\mathcal{A}$ of $\mathcal{P}$ that must contain p_{c_1}. This imposes bounds on r_2 and γ. The relationship of the base frame to $\mathcal{C}_1$ (and also z_1) is then established by α_a, β_a and r_1. The iterative algorithm that determines these parameters uses the circles defined by rotating the test points about z_2, $\mathcal{C}_{t_i}$. This two stage minimization procedure is as follows:

1. The values $r_2(i)$, $\gamma(i) \in \mathcal{A}$ are used to locate the base of a unit normal, n_{z1}, of a plane, $\mathcal{P}_{\perp 1}$, at a point which is taken to be p_{c_1}.

2. The orientation of n_{z1} is set to $\alpha_a(j)$ and $\beta_a(j)$ relative to x_a and y_a.

3. The n_i are rotated to the points defined by the intersection of the $\mathcal{C}_{t_i}$ and $\mathcal{P}_{\perp 1}$ and then projected onto $\mathcal{P}_{\perp 1}$.

4. If the cost function J (defined as before, i.e., J models the summed squares of the pairwise distances between intersection points of the projected vectors on $\mathcal{P}_{\perp 1}$) exceeds a threshold, the steepest descent direction is determined, the updates $\alpha_a(j+1)$ and $\beta_a(j+1)$ are calculated, j is set to $j+1$ and the algorithm returns to step 2. Otherwise convergence to a common intersection point, p_{int}, has occurred and the algorithm proceeds to the next step.

5. If $\epsilon = p_{\text{int}} - p_{c_1}$ is not zero, p_{int} is rotated about z_2 until it lies in $\mathcal{A}$, if it is not already in $\mathcal{A}$. This point in $\mathcal{A}$ defines the values $r_2(i+1)$ and $\gamma(i+1)$. The assignments $j = 1$ and $i = i + 1$ are made and the algorithm returns

to step 1. The values used for $\alpha_a(1)$ and $\beta_a(1)$ are the last values obtained from step 4. If $\epsilon = 0$, a circle is fit to the intersection points from step 3 yielding r_1.

The results of this algorithm clearly define the direction of z_1 and a point on z_1 (p_{c_1}).

KINEMATIC CALIBRATION

One of the more common sets of parameters used to describe the kinematics of a mechanism is the Denavit-Hartenberg set [4]. This set includes link length, offset, joint angle, and twist, labeled a_i, d_i, θ_i, and α_i, respectively, where the subscript i denotes the joint frame number. In the case of a manipulator, the frames are assigned to each joint, starting at the base and working out to the end-effector. In this analysis, frames must be assigned to the payload joints.

The left and right manipulator end-effector frames, frames $\{l\}$ and $\{r\}$ in Figure 1, have known positions (p_r and p_l) and orientations (R_r and R_l) with respect to the base frame. The procedure for assigning the first payload coordinate frame is as follows. The axis of rotation, z_1, is available from the structural analysis of the previous section. The x_1 axis lies along the common normal between z_r and z_1, which can be obtained as follows [5].

The equations of the lines that lie along the z_r and z_1 axes can be written as

$$l_r(b) = bz_r + p_r \tag{6}$$

and

$$l_1(a) = az_1 + p_1, \tag{7}$$

respectively. The variables a and b, which determine point locations on the lines, must be assigned such that the line connecting points $l_r(b)$ and $l_1(a)$, $v = l_1(a) - l_r(b)$, is the common normal. Since v must be perpendicular to z_r and z_1, $z_r \cdot v = 0$ and $z_1 \cdot v = 0$. Solving these equations for a and b yields

$$b = \frac{z_r \cdot (p_1 - p_r) - [z_1 \cdot (p_1 - p_r)][z_1 \cdot z_r]}{1 - [z_1 \cdot z_r]^2} \tag{8}$$

and

$$a = z_1 \cdot (bz_r - p_1 + p_r). \tag{9}$$

When $\|z_1 \cdot z_r\| = 1$ (parallel axes), the denominator of (8) equals zero. In this case a is chosen to be zero and the value of b becomes $b = z_r \cdot (p_r - p_1)$.

Using these values for a and b, $p^* = l_r(b)$ and $p_{org1} = l_1(a)$, as shown in Figure 1. The x_1 axis is given by

$$x_1 = \frac{p_{org1} - p^*}{\|p_{org1} - p^*\|}. \tag{10}$$

The angle between x_r and x_1, θ_1, can be easily found by several methods.

The link offset, d_1, is the signed distance between p^* and p_r. In view of (6), the sign of d_1 is the same as that of b. Thus

$$d_1 = \frac{b}{|b|}\|p^* - p_r\|. \tag{11}$$

Given the direction of x_1, as calculated by (10), the link length, a_1 (the distance between p_{org1} and p^*), is positive. Thus

$$a_1 = \|p^* - p_{org1}\|. \tag{12}$$

The link twist, α_1 (the angle between z_r and z_1 taken about x_1), is determined in the same manner as θ_1. The four parameters which describe frame {2} relative to frame {1}, θ_2, d_2, a_2, and α_2, are determined by the same process used to find θ_1, d_1, a_1, and α_1.

In general, the Denavit-Hartenberg convention cannot be used to describe frame $\{l\}$ in terms of frame {2} because both frames have already been assigned. However the use of an intermediate frame (frame {3}) solves this problem. This results in an additional set of four kinematic parameters, given by θ_3, d_3, a_3, and α_3, which describe frame {3} relative to frame {2}.

The final set of parameters is used to describe frame $\{l\}$ relative to frame {3}. As can be seen from Figure 1, only two of the four parameters are non-zero; the angle between x_3 and x_l, θ_4, and the signed distance from p_3 to p_l, d_4. These parameters are determined in the same manner as before.

With the calculation of these ten parameters, the kinematics of the payload are known. All calculations were performed with relatively little information generated by calibration maneuvers.

CONCLUSIONS

This paper presents an algorithm which determines the kinematic structure and parameters of an unknown payload. The case considered here is that of a payload having two revolute joints, each with a single degree-of-freedom. Other combinations of joints are considered in [1]. Simulations of the method using MATLAB demonstrate accurate results.

REFERENCES

1. Dellinger, W. F., *Kinematic Calibration of Unknown Payloads With Multiple Degrees-of-Freedom*, Ph.D. Dissertation, Tennessee Technological University, 1992.
2. Gill, P. E., W. Murray, and M. H. Wright, *Practical Optimization*, Academic Press, 1981.
3. Paul, Richard P., *Robot Manipulatros: Mathemtaics, Programming, and Control*, MIT Press, 1981.
4. Walker, M. W., and D. E. Orin, "Efficient Dynamic Computer Simulation of Robotic Mechanisms," ASME Journal of Dynamic Systems, Measurement and Control, Vol. 104, pp. 205–211, Sept., 1982.
5. Jeger, M., and B. Eckman *Vector Geometry and Linear Algebra*, John Wiley & Sons, Ltd., 1967.

FORWARD POSITION ANALYSIS OF PARALLEL MECHANISMS WITH THREE GENERAL OPEN-SUBCHAINS BASED ON SELECTION OF INDEPENDENT JOINTS

Chang-De Zhang, Shin-Min Song
Department of Mechanical Engineering
University of Illinois at Chicago
P.O. Box 4348
Chicago, Illinois 60680

Abstract

This paper studies the closed-form solutions of the forward position analysis of parallel mechanisms with three general open-subchains. Each of the three subchains has three single degree-of-freedom joints and is attached to the base link through either revolute joint or prismatic joint and connected to the output link through spherical joint. Since the output link has six degrees of freedom, only six joints among the nine are independent. There are 84 possible ways to select six independent joints from the nine joints. These 84 combinations are classified into the following three categories: *Category 2:2:2* (the three numbers represent the number of independent joints at the three subchains respectively), *Category 3:2:1*, and *Category 3:3:0*. It is shown that the highest order of the resultant polynomials is 16 for the 27 combinations in the first category and is 4 for the 54 combinations in the second category. There is no solution for the 3 combinations in the last category.

1 Introduction

During the past decade, a remarkable attention has been directed to study the parallel manipulators since they possess the following advantages compared to their serial counterparts: high stiffness, high load capacity, quick dynamic response, and high position accuracy. Particularly, there is a significant number of publications [1-10] addressing forward position analysis of parallel manipulators since the forward position analysis of a parallel manipulator is necessary when 1) it is controlled by using Cartesian control scheme; 2) it is used as a position or force/torque sensor; 3) it is applied as a master manipulator in a teleoperation system. In general, there are two approachs for forward position analysis: numerical approach [4-5] and closed-form solution approach [1-3,6-9]. The first approach applies numerical algorithems to compute the unknowns from a system of position constraint equations. The second approach utilizes various elimination methods to derive a single polynomial in one unknown from a system position constraint equations. If the order of the polynomial is less than or equal to four, the solution has analytical closed-form. The advantages of the closed-form solution approach was clearly stated in [2].

In this paper, we will present the closed-form solutions of forward position analysis of the parallel mechanisms with three general open-subchains (see Figure 1), based on selection of independent joints. Each general open-subchain has three single d.o.f. joints and one spherical joint through which the subchain connected to the moving platform. Since the moving platform has six d.o.f., only six out of the nine single d.o.f. joints are independent. Hence, there are 84 ways to select six independent joints. According to the number of independent joints which each subchain has, the 84 combinations can be classified into three categories: *Category 2:2:2*, *Category 3:2:1*, and *Category 3:3:0*. The three numbers represent the numbers of independent joints in three subchains. The numbers of combinations in the three categories are 27, 54, and 3 respectively.

In the following, the basic equations for the forward position analysis will be derived first. Then, the forward position analysis of the three categories will be conducted respectively.

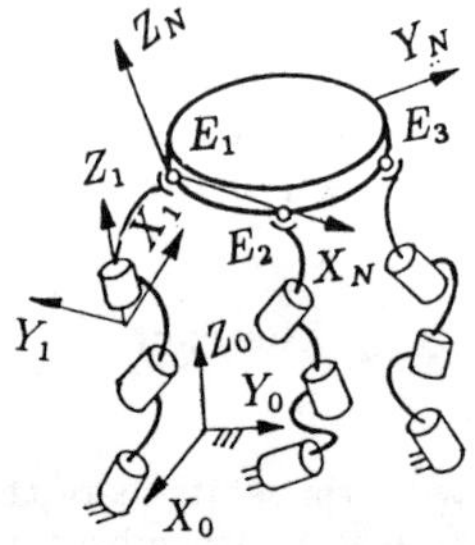

Figure 1: A Parallel Mechanism with Three General Open-Subchains

2 Basic Equations

Figure 1 shows the parallel mechanism with three general open-subchains. The type of the three single d.o.f. joints of each subchain is either revolute or prismatic. The coordinate system X_0-Y_0-Z_0 (or S_0) is attached to the base platform, while the coordinate system X_N-Y_N-Z_N (or S_N) is fixed at the moving platform with its origin coincident with E_1, X_N-axis passing through E_2 and Y_N-axis lying in the moving platform. The coordinate system X_i-Y_i-Z_i is fixed at the last link of each subchain. The transformation matrix ${}^0\boldsymbol{T}_N$ from S_0 to S_N is described as:

$$
{}^0\boldsymbol{T}_N = \begin{bmatrix} l_x & m_x & n_x & x \\ l_y & m_y & n_y & y \\ l_z & m_z & n_z & z \\ 0 & 0 & 0 & 1 \end{bmatrix} \tag{1}
$$

where the nine rotational elements are related by the following equations:

$$l_x^2 + l_y^2 + l_z^2 = 1 \tag{2}$$
$$m_x^2 + m_y^2 + m_z^2 = 1 \tag{3}$$
$$l_x m_x + l_y m_y + l_z m_z = 0 \tag{4}$$
$$n_x = l_y m_z - l_z m_y \tag{5}$$
$$n_y = l_z m_x - l_x m_z \tag{6}$$
$$n_z = l_x m_y - l_y m_x \tag{7}$$

Mathematically speaking, the forward position analysis of the parallel mechanisms with three general open-subchains is to determine all elements of ${}^0\boldsymbol{T}_N$, and the three dependent joints when six independent joint variables are given. There are totally fifteen unknowns. Since E_is are located on the plane X_N-O_N-Z_N, n_x, n_y, n_z will not be involved in the following analysis. Hence, we need only to determine twelve unknowns. Besides the three equations in Equations (2-4), nine more equations are required to be derived in order to completely determine all the twelve unknowns.

Since the position of E_i with respect to S_0 obtained through coordinate transformation from S_N to S_0 should be identical to that obtained through coordinate transformation from S_i to S_0, the following constraint equations for each subchain i can be derived:

$$^0\boldsymbol{T}_i \begin{bmatrix} r_i \\ s_i \\ t_i \\ 1 \end{bmatrix} = \begin{bmatrix} p_i l_x + q_i m_x + x \\ p_i l_y + q_i m_y + y \\ p_i l_z + q_i m_z + z \\ 1 \end{bmatrix} \quad i = 1, 2, 3 \tag{8}$$

where $[r_i\ , s_i\ , t_i]^t$ represents the position of E_i with respect to S_i, $[p_i\ , q_i\ , 0]^t$ represents the position of E_i with respect to S_N, and $^0\boldsymbol{T}_i$ is the transformation matrix from S_i to S_0.

Equation (8) includes nine equations that form the basic equations for the forward position analysis, along with the three equations in Equations (2-4). Based on the basic equations, the closed-form solutions of the forward position analysis of *Category 2:2:2*, *Category 3:2:1* and *Category 3:3:0* are derived, respectively in the following.

3 Category 2:2:2

In this category, each of the three subchains has two independent joints and there are 27 combinations of the six independent joints.

From the kinematic point of view, any two adjacent links can be equivalently replaced by one general link if their common joint variable is known. That is, each general subchain with two known independent joints can be replaced by a one-link subchain. In this way, the parallel mechanism shown in Figure 1 becomes a structure with three subchains as shown in Figure 2. Depending on whether the three dependent joints are revolute and/or prismatic joints, Figure 2 may represet four different types of structures: *RRR-3S*, *PRR-3S*, *PPR-3S* and *PPP-3S*. Since the forward position analysis of these structures has been conducted in [7-9], it is not repeated here. The results are a 16-th order polynomial for each of the first two structures, a 12-th order, and a 8-th order polynomials for the last two structures, respectively. This completes the forward position analysis of *Category 2:2:2*.

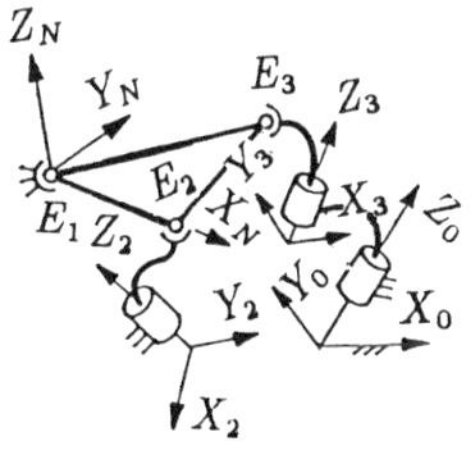

Figure 2: A Structure with Three Subchains

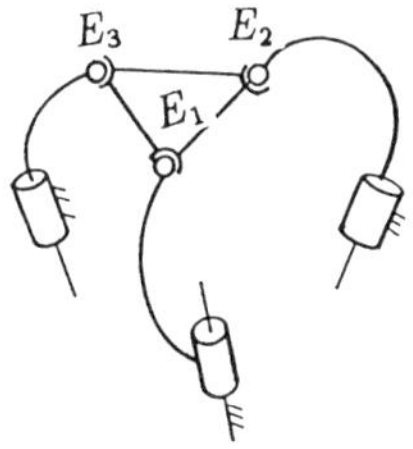

Figure 3: A Structure with Two Subchains

4 Category 3:2:1

This category contains 54 combinations to select the six independent joints. It is assumed that subchains 1, 2 and 3 have three, two and one independent joints, respectively. Under this assumption there are nine combinations. It is sufficient for us to study these nine combinations since the results can be applied to the remaining combinations by switching the subchain numbers.

It has been shown that a stewart platform has analytical closed-form solutions of forward position analysis if it satisfies the following geometrical condition: One rotational d.o.f. of the output (or the base) link is decoupled from the other five d.o.f. [3]. We will show that all the combinations in *Category 2:2:2* satisfy this geometrical condition.

Referring to Figure 1, a positional unit vector along the X_N axis has five d.o.f. and it can be fully determined by the five independent joints in subchains 1 and 2. The rotational d.o.f. about this positional vector is determined by the independent joint in subchain 3. Hence, one rotational d.o.f. of the output link is decoupled from its other five, and the above mentioned geometrical condition is satisfied. This means that analytical closed-form solutions are available.

4.1 Determination of x, y, and z

Since subchain 1 has three independent joints which are known, the position of E_1 with respect to S_0, i.e., x, y, and z, can be uniquely determined through performing the forward position analysis of subchain 1. Once the position of E_1 has been determined, subchain 1 can kinematically be replaced by a spherical joint with its center at E_1. This is shown in Figure 3.

4.2 Determination of l_x, l_y, l_z and d_2 or θ_2

Subchain 2 has two independent joints and can be replaced by a one-link subchain, as shown in Figure 3. The coordinate system X_2-Y_2-Z_2 (S_2) is fixed at the last link of the equivalent subchain 2. The transformation matrix from S_2 to S_0 can be described according to the Denavit-Hartenberg method [10]:

$$ {}^0\boldsymbol{T}_2 = \begin{bmatrix} c\theta_2 & -s\theta_2 & 0 & a_2 \\ s\theta_2 c\alpha_2 & c\theta_2 c\alpha_2 & -s\alpha_2 & -s\alpha_2 d_2 \\ s\theta_2 s\alpha_2 & c\theta_2 s\alpha_2 & c\alpha_2 & c\alpha_2 d_2 \\ 0 & 0 & 0 & 1 \end{bmatrix} \tag{9} $$

It should be pointed out that the indices of the link length and link twist are modified for the convenience of expression. According to Equation (8), the following three equations can be derived:

$$ r_2\, c\theta_2 - s_2\, s\theta_2 + a_2 = p_2\, l_x + x \tag{10} $$
$$ r_2\, c\alpha_2\, s\theta_2 + s_2\, c\alpha_2\, c\theta_2 - t_2 s\alpha_2 - s\alpha_2 d_2 = p_2\, l_y + y \tag{11} $$
$$ r_2\, s\alpha_2\, s\theta_2 + s_2\, s\alpha_2\, c\theta_2 + t_2 c\alpha_2 + c\alpha_2 d_2 = p_2\, l_z + z \tag{12} $$

It should be mentioned that x, y, and z in the above equations are knowns since they have been determined according to the independent joints in subchain 1. Based on Equations (2,10-12), the analytical closed-form solutions of l_x, l_y, l_z and dependent joint variable d_2 or θ_2 can be derived according to the joint type of the dependent joint.

Type of the Dependent Joint: P

Since the dependent joint is a prismatic joint, d_2 is unknown. Eliminating d_2 through multiplying Equation (11) by $c\alpha_2$ and Equation (12) by $s\alpha_2$ and adding them together yield:

$$ r_2 s\theta_2 + s_2 c\theta_2 = (p_2 l_y + y)c\alpha_2 + (p_2 l_z + z)s\alpha_2 \tag{13} $$

Now, l_x can be calculated from Equation (10), l_y and l_z from Equations (2, 13) by solving a second order polynomial, and then d_2 from Equation (11) or (12).

Type of the Dependent Joint: R

Since the dependent joint is a revolute joint, θ_2 is unknown. Equations (10-11) can be rearranged as:

$$(r_2\, c\theta_2 - s_2\, s\theta_2)\, c\alpha_2 = (p_2\, l_x + x - a_2)\, c\alpha_2 \tag{14}$$
$$(r_2\, s\theta_2 + s_2\, c\theta_2)\, c\alpha_2 = p_2\, l_y + y + t_2\, s\alpha_2 + s\alpha_2\, d_2 \tag{15}$$

Equations (10,12) can be rewritten as:

$$(r_2\, c\theta_2 - s_2\, s\theta_2)\, s\alpha_2 = (p_2\, l_x + x - a_2)\, s\alpha_2 \tag{16}$$
$$(r_2\, s\theta_2 + s_2\, c\theta_2)\, s\alpha_2 = p_2\, l_z + z - t_2\, c\alpha_2 - c\alpha_2\, d_2 \tag{17}$$

Squaring both sides of Equations (14-17) and adding them together yield a linear equation in l_x, l_y and l_z:

$$A_1 l_x + A_2 l_y + A_3 l_z + A_4 = 0 \tag{18}$$

where A_is are constants. They are not provided due to space limitation and so are other coefficients in the following derivation. Squaring both sides of Equations (14-15) and adding them together lead to:

$$B_1 l_x^2 + B_2 l_y^2 + B_3 l_x + B_4 l_y + B_5 = 0 \tag{19}$$

Eliminating l_z from Equations (2,18) gives:

$$C_1 l_x^2 + C_2 l_y^2 + C_3 l_x l_y + C_4 l_x + C_5 l_y + C_6 = 0 \tag{20}$$

Eliminating l_y from Equations (19-20) yields:

$$D_1 l_x^4 + D_2 l_x^3 + D_4 l_x^2 + D_5 l_x + D_6 = 0 \tag{21}$$

The resultant equation is a 4-th order polynomial and its roots can be determined analytically. Once l_x has been calculated from Equation (21), l_y can be computed from Equation (19), and l_x from Equation (18). Then, θ_2 can be computed from Equations (10-11) through determining $s\theta_2$ and $c\theta_2$.

4.3 Determination of m_x, m_y, m_z and θ_3-θ_4, θ_3-d_4, d_3-θ_4, or d_3-d_4

Since subchain 3 has one independent joint, it can be replaced by a two-link subchain, as shown in Figure 3. The coordinate system X_3-Y_3-Z_3 (S_3) is fixed at the last link of the equivalent subchain 3. The transformation matrix of S_3 with respect to S_0 can be determined according to the D-H notation [10] as:

$$ {}^0\boldsymbol{T}_3 = \begin{bmatrix} c\theta_3 & -s\theta_3 & 0 & 0 \\ s\theta_3 & c\theta_3 & 0 & 0 \\ 0 & 0 & 1 & d_3 \\ 0 & 0 & 0 & 1 \end{bmatrix} \begin{bmatrix} c\theta_4 & -s\theta_4 & 0 & a_4 \\ s\theta_4 c\alpha_4 & c\theta_4 c\alpha_4 & -s\alpha_4 & -s\alpha_4 d_4 \\ s\theta_4 s\alpha_4 & c\theta_4 s\alpha_4 & c\alpha_4 & c\alpha_4 d_4 \\ 0 & 0 & 0 & 1 \end{bmatrix} \tag{22}$$

According to Equation (8), the following three equations can be derived:

$$\begin{aligned} &r_3(c\theta_3\, c\theta_4 - c\alpha_4\, s\theta_3\, s\theta_4) - s_3(c\theta_3\, s\theta_4 + c\alpha_4\, s\theta_3\, c\theta_4) + t_3 s\alpha_4\, s\theta_3 \\ &\qquad + a_4\, c\theta_3 + s\alpha_4\, s\theta_3\, d_4 = p_3\, l_x + q_3\, m_x + x \end{aligned} \tag{23}$$

$$\begin{aligned} &r_3(s\theta_3\, c\theta_4 + c\alpha_4\, c\theta_3\, s\theta_4) - s_3(s\theta_3\, s\theta_4 - c\alpha_4\, c\theta_3\, c\theta_4) - t_3 s\alpha_4\, c\theta_3 \\ &\qquad + a_4\, s\theta_3 - s\alpha_4\, c\theta_3\, d_4 = p_3\, l_y + q_3\, m_y + y \end{aligned} \tag{24}$$

$$\begin{aligned} &r_3\, s\alpha_4\, s\theta_4 + s_3\, s\alpha_4\, c\theta_4 + t_3\, c\alpha_4 + c\alpha_4\, d_4 + d_3 \\ &\qquad = p_3\, l_z + q_3\, m_z + z \end{aligned} \tag{25}$$

It should be noticed that x, y, z, l_x, l_y and l_z are knowns since they have been determined according to the independent joints in subchains 1 and 2.

Before we calculate the remaining unknowns, let us first simplify Equations (23-24). Multiplying Equation (23) by $c\theta_3$ and Equation (24) by $s\theta_3$, and adding them together result in:

$$r_3\, c\theta_4 - s_3\, s\theta_4 + a_4 = (x + p_3\, l_x + q_3\, m_x)c\theta_3 + (y + p_3\, l_y + q_3\, m_y)s\theta_3 \tag{26}$$

Similarly, multiplying Equation (23) by $s\theta_3$ and Equation (24) by $c\theta_3$, and subtracting them yield:

$$\begin{aligned} &-r_3\, c\alpha_4\, s\theta_4 - s_3\, c\alpha_4\, c\theta_4 + t_3\, s\alpha_4 + s\alpha_4\, d_4 \\ &\quad = (x + p_3\, l_x + q_3\, m_x)s\theta_3 - (y + p_3\, l_y + q_3\, m_y)c\theta_3 \end{aligned} \tag{27}$$

In the following, the closed-form solutions of m_x, m_y, m_z, and the two dependent joint variables are derived successively, based on the five equations in Equations (3-4,25-27). Since the two dependent joints may be prismatic and/or revolute, there are four combinations of dependent joints.

Types of the Dependent Joints: R-R

Since both the dependent joints are revolute, θ_3 and θ_4 are unknown, and d_3 and d_4 are considered as given. Let us first eliminate θ_3 and θ_4. Squaring both sides of Equations (25-27) and adding them together yield:

$$E_1\, s\theta_4 + E_2\, c\theta_4 = E_3\, m_x + E_4\, m_y + E_y\, m_z + E_6 \tag{28}$$

Equation (25) can be rewritten as:

$$F_1 s\theta_4 + F_2 c\theta_4 = F_3 m_z + F_4 \tag{29}$$

Based on Equations (28-29), $s\theta_4$ and $c\theta_4$ can be expressed as:

$$s\theta_4 = G_1\, m_x + G_2\, m_y + G_3\, m_z + G_4 \tag{30}$$
$$c\theta_4 = H_1\, m_x + H_2\, m_y + H_3\, m_z + H_4 \tag{31}$$

Since $s^2\theta_4 + c^2\theta_4 = 1$, the below equation can be obtained:

$$I_1 m_z^2 + (I_2 m_y + I_3 m_x + I_4)m_z + I_5 m_y^2 + (I_6 m_x + I_7)m_y + I_8 m_x^2 + I_9 m_x + I_{10} = 0 \tag{32}$$

Eliminating m_z from Equations (4,32) gives:

$$J_1 m_y^2 + (J_2 m_x + J_3)m_y + J_4 m_x^2 + J_5 m_x + J_6 = 0 \tag{33}$$

Eliminating m_z from Equations (3-4) results in:

$$J_7 m_y^2 + J_8 m_x m_y + J_9 m_x^2 + J_{10} = 0 \tag{34}$$

Eliminating m_y from Equations (33-34) yields:

$$J_{11} m_x^4 + J_{12} m_x^3 + J_{13} m_x^2 + J_{14} m_x + J_{15} = 0 \tag{35}$$

Hence, the resultant equation is a 4-th order polynomial in m_x. Once m_x has been calculated from Equation (35), then m_x can be derived from Equation (34), and m_z from Equation (4) or Equation (32) in case that $l_z = 0$. Afterwards, θ_4 can be determined from Equation (30-31). Finally θ_3 can be determined from Equations (23-24).

Types of the Dependent Joints: R-P

Since the dependent joints are R and P joints respectively, θ_3 and d_4 are unknowns and d_3 and θ_4 are knowns. In order to eliminate d_4, let us rewrite Equations (28,25) as:

$$d_4^2 + K_1 d_4 + K_2 m_x + K_3 m_y + K_4 m_z + K_5 = 0 \tag{36}$$

$$K_6 d_4 + K_7 m_z + K_8 = 0 \tag{37}$$

Eliminating d_4 from Equations (36-37) gives:

$$K_9 m_z^2 + K_{10} m_z + K_{11} m_x + K_{12} m_y + K_{13} = 0 \tag{38}$$

Eliminating m_z from Equations (4,38) leads to:

$$K_{14} m_y^2 + (K_{15} m_x + K_{16}) m_y + K_{17} m_x^2 + K_{18} m_x + K_{19} = 0 \tag{39}$$

It should be noticed that the above equation has the same structure as Equation (33). Eliminating m_y from Equations (34,39) gives a 4-th order polynomial in m_x with similar form as Equation (35), and the coefficients can be obtained from J_{11}-J_{15} by replacing J_i $(i = 1, 2, \cdots, 6)$ with K_i $(i = 14, 15, \cdots, 19)$. After m_x has been determined through solving the 4-th order polynomial, m_y can be determined from Equation (34) and m_z from Equation (4) or Equation (39) for the case that $l_z = 0$. Then, d_4 and θ_3 can be calculated from Equation (37) and Equations(23-24), respectively.

Types of the Dependent Joints: P-R

In this case, d_3 and θ_4 are unknowns. we can rearrange Equations (26-27) in the forms of:

$$\begin{aligned} c\alpha_4(r_3\, c\theta_4 - s_3\, s\theta_4) = {} & [(x + p_3\, l_x + q_3 m_x) c\theta_3 \\ & + (y + p_3\, l_y + q_3\, m_y) s\theta_3 - a_4] c\alpha_4 \end{aligned} \tag{40}$$

$$\begin{aligned} -c\alpha_4(r_3\, s\theta_4 + s_3\, c\theta_4) = {} & (x + p_3\, l_x + q_3 m_x) s\theta_3 \\ & -(y + p_3\, l_y + q_3\, m_y) c\theta_3 - t_3 s\alpha_4 - s\alpha_4\, d_4 \end{aligned} \tag{41}$$

Eliminating θ_4 from Equations (40-41) gives:

$$L_1 m_y^2 + (L_2 m_x + L_3) m_y + L_4 m_x^2 + L_5 m_x + L_6 = 0 \tag{42}$$

Similarly to the above case, a 4-th order polynomial in m_x can be derived through eliminating m_y from Equations (34,42). Once m_x has been computed from the resultant polynomial, m_y can be determined from Equation (34) and m_z from Equation (4) or Equation (42) for the case that $l_z = 0$. Afterwards, d_3 and θ_4 can be derived from Equation (25) and Equations (30-31), respectively.

Types of the Dependent Joints: P-P

For this case, d_3 and d_4 are unknowns. Since Equation (26) does not contain d_3 and d_4, it can be used to determine m_x, m_y and m_z along with Equations (3-4). Equation (26) can be rewritten as:

$$M_1 m_x + M_2 m_y + M_3 = 0 \tag{43}$$

Eliminating m_y from Equations (34,43) leads to a 2-nd order polynomial in m_x. After m_x has been determined from the resultant polynomial, m_y and m_z can be derived from Equation (43) and Equation (4), respectively.

Hence, we have completed the derivation of the analytical closed-form solutions of *Category 3:2:1.*

5 Category 3:3:0

This category contains three selections. Let us assume that the two subchains with three independent joints are subchains 1 and 2. Hence, x, y, and z can be uniquely determined according to the three constraint equations of subchain 1. Similarly, l_x, l_y, and l_z can be determined from the constraint equations of subchain 2. However, l_x, l_y and l_z may not satisfy Equation (2). Furthermore, m_x, m_y, and m_z can not be determined. Hence, the selected six joint variables are actually not independent and these selections should be avoided.

6 Conclusion

In this paper, the closed-form solutions of the forward position analysis of parallel mechanisms with three general open-subchains were derived, based on selection of six independent joints from nine single d.o.f. joints. The 84 selections were divided into three categories: *Category 2:2:2*, *Category 3:2:1* and *Category 2:2:2*. It was shown that the highest order of the resultant equations is 16 for the 27 combinations in the first category and is 4 for the 54 combinations in the second category. There is no solution for the 4 combinations in the last category.

7 Acknowledgement

The financial support of the National Science Foundation through a Presidential Young Investigator Award (No. NSF DDM-8657920) to the second author is acknowledged.

References

[1] Nanua, P., Waldron, K.J., and Murthy V., "Direct Kinematic Solution of a Stewart Platform", *IEEE Transactions on Robotics and Automation,* Vol. 6, No. 4, pp 438-444, 1990.

[2] Griffis, M. and Duffy, J., "A Forward Displacement Analysis of a Class of Steward Platforms." *Journal of Robotic Systems*, Vol. 6, No. 6, pp. 703-720, 1989.

[3] Zhang, C.D. and Song, S.M., "Forward Kinematics of a Class of Parallel (Stewart) Platforms with Closed Form Solutions", *Journal of Robotic Systems*, Vol. 9, No. 1, pp. 93-112, Jan., 1992.

[4] Innocenti, C. and Parenti-Castelli, V., "A Novel Numerical Approach to the Closure of the 6-6 Stewart Platform Mechanism", *Proceedings of the IEEE International Conference on Robotics and Automation*, Sacramento, California, pp. 851-855, April, 1991

[5] Raghavan, M., "The Stewart Platform of General Geometry Has 40 Configurations", *Proceedings of the 1991 ASME Design Engineering Conference, DE-Vol. 32-2, Advances in Design Automation*, Vol. 2, pp. 397-401, Miami, 1991

[6] Zhang, C.D. and Song, S.M., "Forward Position Analysis of Nearly General Stewart Platforms", appear in the *proceedings of ASME Mechanisms Conference*, 1992.

[7] Innocenti, C. and Parenti-Castelli, V., "Direct Position Analysis of the Stewart Platform Mechanism", *Mechanism and Machine Theory*, Vol. 25, No. 6, pp. 611-621, 1990.

[8] Parenti-Castelli, V., and Innocenti, C., "Direct Displacement Analysis for Some Classes of Parallel Mechanisms", *Eighth CISM-IFToMM Symposium on Theory and Practice of Robots and Manipulators*, July 2-6, 1990, Cracaow, Poland.

[9] Parenti-Castelli, V., and Innocenti, C., "Forward Displacement Analysis of Parallel Mechanisms. Closed-Form Solution of PRR-3S and PPR-3S Structures", *Proceedings of the 21st Biennial Mechanisms Conference: Mechanism Synthesis and Analysis*, ASME, DE-Vol. 25, pp. 111-116, Chicago, Sept., 1990.

[10] Craig, J. J., *Introduction to Robotics: Mechanics and Control* 2th Edition, Addison-Wesley, 1989.

A ROBOTIC CRANE SYSTEM UTILIZING THE STEWART PLATFORM CONFIGURATION

ROGER BOSTELMAN, NICHOLAS DAGALAKIS, and JAMES ALBUS
Robot Systems Division,
National Institute of Standards and Technology
Gaithersburg, Maryland 20899 U.S.A.
June 1992

ABSTRACT

The Robot Systems Division of the National Institute of Standards and Technology (NIST) has been experimenting for several years with new concepts for robot cranes. These concepts utilize the basic idea of the Stewart Platform parallel link manipulator. The unique feature of the NIST approach is to use cables as the parallel links and to use winches as the actuators. So long as the cables are all in tension, the load is kinematically constrained, and the cables resist perturbing forces and moments with equal stiffness to both positive and negative loads. The result is that the suspended load is constrained with a mechanical stiffness determined by the elasticity of the cables, the suspended weight, and the geometry of the mechanism. Based on these concepts, a revolutionary new type of robot crane, the NIST SPIDER (Stewart Platform Instrumented Drive Environmental Robot) has been developed that can control the position, velocity, and force of tools and heavy machinery in all six degrees of freedom (x, y, z, roll, pitch, and yaw).

Depending on what is suspended from its work platform, the SPIDER can perform a variety of tasks. Examples are: cutting, excavating and grading, shaping and finishing, lifting and positioning.

A 6-meter version of the SPIDER has been built and critical performance characteristics and control methods analyzed.

1. INTRODUCTION

A new crane design utilizing six cables to suspend a load platform was first developed by the National Institute of Standards and Technology (NIST) in the early 1980's. A Defense Advanced Research Projects Agency (DARPA) sponsored program on robot crane technology produced the design, development and testing of three different sized prototypes to determine the performance characteristics of this proposed robot crane design. A description of the overall DARPA program and the results of this research are presented in [1]. Initial testing of these prototypes showed that this design results in a stiff load platform [2, 3]. This platform (see figure 1) can be used in typical crane operations, or as a robot base, or a combination of both. Applications of this new crane design in the construction industry are illustrated [4].

The objective of this paper is to briefly describe the SPIDER (Stewart Platform Instrumented Drive Environmental Robot), it's functional capabilities, and it's operational procedures. The primary function of the SPIDER is to lift, maneuver, and position large loads with precise control of position and force in all six degrees of freedom. The SPIDER has the following potential functional capabilities: cutting, excavating and grading, shaping and finishing, lifting and positioning, flexible fixturing, and transporting manipulators. We have already developed several applications with SPIDER. These capabilities can be targeted for a variety of specific applications such as: fighting oil well fires or hazardous waste site inspection and clean-up.

In response to the oil fires set in Kuwait during the Persian Gulf War, NIST adapted its robot crane technology to assist in extinguishing oil well fires. A system called the NIST Oil Well Fire Fighting Robot (NOWFFR) was designed and constructed. Early in 1991, the NOWFFR design was modified for applications related to nuclear and toxic waste site cleanup. The new robot design is called the "SPIDER". The SPIDER is light weight and easily assembled. The SPIDER can be made mobile by attaching vehicles to the feet so that it can be driven over rough terrain.

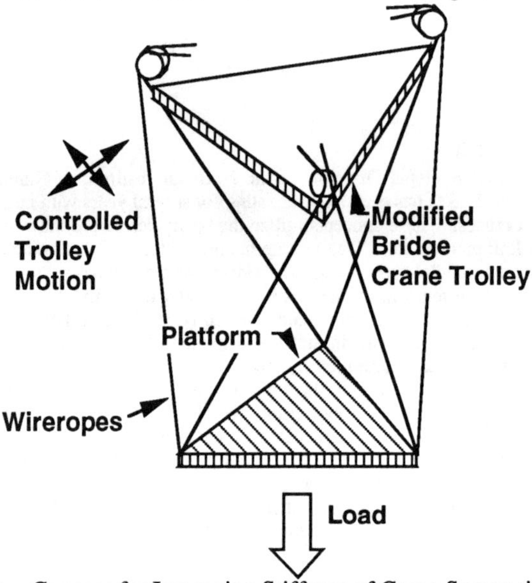

Figure 1 - Design Concept for Improving Stiffness of Crane Suspension Mechanisms

A number of advantages of the SPIDER over current technology are: rigid support and precise maneuverability of large loads, remote positioning of tools and equipment, executing precise motions with tools/equipment to accomplish complex tasks, high lift-to-weight ratio, resistance to environmental perturbations, accurate control of loads by a novice operator , reduced crew size.

The next section describes the system design, showing both the SPIDER design and a comparison to conventional cranes, followed by a section describing the different control modes of the SPIDER. A summary and conclusion section follows, as well as a list of references for the paper.

2. SYSTEM ARCHITECTURE

The SPIDER design evolved from previous Stewart Platform work with actuators and cables/winches in place of actuators. When all six cables are in tension, the stable platform is kinematically constrained, and there exists a known mathematical relationship between the lengths of the six cables and the position and orientation of the platform. The theory of this mathematical relationship has been known for many years and was first embodied in a Stewart platform for testing tires in the 1950's [5], and applied to aircraft flight simulators during the 1960's and '70's. It was applied to cable driven manipulators by Landsberger [6,7] and to cranes by NIST in the 1980's [1].

2.1 SPIDER Design

The framework of the SPIDER is a six legged structure resting on three support points. The legs are 6 meter aluminum tubes that are ten centimeters in diameter and constructed in an octahedron geometry. These can be seen in Figure 2, which is a photograph of the NOWFFR taken before it was modified to become the SPIDER.

Figure 2 - Photograph of the NOWFFR before it was modified to become the SPIDER. Centered is a heat shield/chimney and a chain saw attached to the platform and ready to cut an oak log.

The top of the SPIDER structure consists of a triangle. Each vertex of the triangle supports two cables. Together, the six cables support a lower work platform. The connecting joints at the vertices consist of ball and socket joints. Due to its octahedron geometry, all forces are directed through points at the vertices. As a result there are no bending or twisting moments generated by the load. Each member of the structure is always in pure compression or tension except for supporting its own

Figure 3 - One of three pairs of winches used to maneuver the platform.

The SPIDER structure thus provides near maximum strength and stiffness possible for any given mass of structural material.

Six cables, made of 5mm diameter braided steel cable and supporting the lower work platform, are arranged to kinematically constrain the work platform such that its stiffness is determined by the tensile elasticity of the cables. The length of each cable is controlled by a winch (see figure 3) having a 455 kg load rating. By controlling the six winches simultaneously, an operator can maneuver the lower work platform in six degrees of freedom. The work platform is made of aluminum I-beams. It can be a variety of sizes. Studies at NIST have shown that a 2 to 1 ratio of the size of the upper triangle to the lower platform is the most stiff for lateral force loading [2]. A mathematical model of the robot and a preliminary investigation of its workspace limits are discussed in [4].

The three support points of the octahedron are carried by three vehicles for mobility. To accomplish steering and velocity commands provided by an operator, computer coordination of vehicle track motions is necessary.

2.2 Comparison to Conventional Cranes

Existing cranes of many different types from many manufacturers, are able to lift comparable loads, but cannot stabilize the loads in rotation or sway and have no means of controlling forces or torques on the load. Under ideal conditions, a highly skilled crane operator can provide some measure of oscillation damping. However, for precise orientation, a crew of riggers is needed to manually stabilize the load from rotating and swinging and to manually guide the load into its final desired position with adverse conditions such as wind. An expert operator cannot prevent perturbations such as wind from causing the load to sway, in some cases by more than a meter. Novice operators of conventional cranes may have difficulty in preventing heavy loads from colliding with objects in the environment.

The principal advantage of the SPIDER is that it provides sufficient control to allow even a novice operator to position a load without sway to within a few millimeters in x, y, and z, and to control orientation without oscillation to within one degree in roll, pitch, and yaw. Force sensors on the SPIDER winch mechanisms could also allow the operator (with computer assistance) to control forces and torques on a load after it comes into contact with the environment. The control provided by the SPIDER could thus reduce the size of the crew needed to manually position loads from three or four, to zero or one. An additional advantage of the SPIDER is its high lift-to-weight ratio. Due to its octahedron geometry, the SPIDER requires no counter weight and experiences no twisting or bending moments. As a result, it can lift at least five times its own weight. This is significantly more than any robot or crane in current use.

3. CONTROL MODES

On the SPIDER, the lengths of the six cables are controlled by six winches. These can be controlled and coordinated by a computer. Input commands from a six-axis joystick enable an operator to control the motion of the work platform in all six degrees of freedom (x, y, z, roll, pitch, and yaw). The operator can thus manuever the stable platform, and whatever load or tool that is attached to it, over a large working volume. The SPIDER has three control modes: Manual mode, Computed Manual Mode with a Stewart Platform Joystick and Computed Manual Mode with a Spaceball Joystick.

3.1 Manual Mode

The joystick used to control the SPIDER in the Manual mode is a Stewart Platform with linear potentiometers. By orienting this joystick so that the potentiometers are roughly parallel to the SPIDER cables, the pose of the platform can be controlled under master-slave rate control. In this mode, the operation of each winch is controlled by the displacement of its corresponding potentiometer. Over a limited range wherein the cables and potentiometers remain roughly parallel, the SPIDER motion can be controlled directly by the potentiometers without a computer. The joystick potentiometers generate analog voltages for amplifiers to control the winches.

3.2 Computed-Manual Mode with the Stewart Platform Joystick

The SPIDER also has a Computed-Manual mode. In this mode the joystick signals are switched to send analog voltages from the joystick to a computer. Joystick potentiometers signals are fed into an analog-to-digital (A/D) board embedded in the computer. From these signals the computer calculates the joystick input position and orientation and translates it to SPIDER command cable lengths. Feedback is received from cable length and force sensors located near the winches. This enables closed loop platform pose control. Cable travel encoders generate phase quadrature signals for an encoder input board embedded in the computer. Cable tension sensors are input into an A/D board in the computer. Command signals are output from the computer via a digital to analog (D/A) board and sent to the winch amplifiers. Control algorithms being developed for the SPIDER include forward and reverse kinematics that are necessary to allow control of the platform.

3.3 Computed-Manual Mode with the Spaceball Joystick

This is similar to the previously described control mode except that in this case a spaceball joystick is used in order to input the platform position and orientation rate control commands. The controller of this joystick measures the Cartesian axes

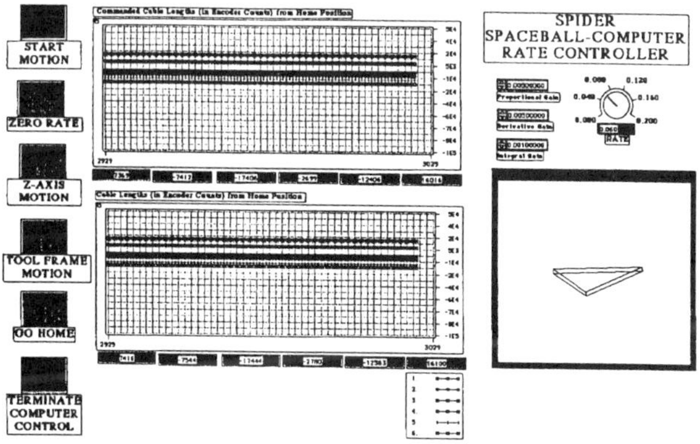

Figure 4 - Computer Control Panel used as Operator Interface to the SPIDER

components of the force and moment applied to a ball and sends them to the computer it is interfaced with, every time they exceed a threshold. The computer then interprets them to translation and rotation motion commands respectively. It also supports nine True/False Boolean buttons which allow the selection of a large number of control options. Currently only a few of these buttons are used. Three are used in order to increment, decrement or set equal to zero the commanded motion rate. A few more are used in order to constrain the platform directions of motion.

The computer display front panel was designed to be simple, intuitive to understand, and real time interactive. Figure 4 shows the current version of the front panel. In the default state, motion takes place with respect to the base frame attached to the suspension plane of the moving platform. Pressing the "Z-AXIS MOTION" button forces the computer to accept the Z axis motion commands only. Pressing the "TOOL FRAME MOTION" button forces the computer to command motion, with respect to a frame attached to the moving platform. The function of the rest of the buttons is obvious from the label they carry underneath them. An animation window will allow, in its final form, the three dimensional space display of the moving platform and its workspace. The issued motion commands will be executed and displayed in that window. An impact of the platform with objects in its workspace will be detected ahead of time and proper action can be initiated. The operator can set the rate of motion to zero and then try different motion senaria in the animation window, save the commands and play back any desired sequence. Currently the computer uses a PID servo motion controller. The gains of the controller can be set interactively from the screen and keyboard.

4. SUMMARY AND CONCLUSIONS

Two models of the SPIDER have been designed and constructed. A two meter model has been used to test mobility issues. A six meter model has been used to test lifting and load positioning perameters and to analyze the size and shape of the work volume. Experiments were done on the SPIDER to measure payload, work volume, and platform-movement precision. A weight of 455 kg. was maneuvered by the SPIDER platform while in manual mode. The load was carried to the limits of the work volume until cables began to go slack.

Two conditions under which two of the cables always become loose for any value of the weight W have been identified. The first condition occurs when the vector of the weight W crosses a vertical place through one of the suspension lines. The second condition occurs when the weight vector W crosses an imaginary line created by the intersection of two planes through two pairs of active cables. These two conditions are mathematically explained [4].

An experiment to show platform-movement precision and stability was done with a chain saw attached to the work platform. A depth of cut to within 1 mm could be made in a solid oak log. Deep cuts could also be made with ease even while driving the tip of the chain saw blade directly into the oak log.

Future research on SPIDER will include integrating more advanced sensing capability, such as machine vision, and additional mechanical analysis and testing. The long range goal is to build a full-scale working prototype jointly with an industrial partner.

5. REFERENCES

[1] Dagalakis, N. G., Albus, J. S., Goodwin, K. R., Lee, J. D., Tsai, T., Abrishamian, H., and Bostelman, R. V., "Robot Crane Technology Program -- Final Report", NIST Technical Note 1267, National Institute of Standards and Technology, Gaithersburg, MD, July 1989.

[2] Dagalakis, N. G., Albus, J. S., Wang, B. L., Unger, J., and Lee, J. D., "Stiffness Study of a Parallel Link Robot Crane for Shipbuilding Applications", Transactions of the ASME Journal of Offshore Mechanics and Artic Engineering, Vol. 111, No. 3, pp.183-193, 1989.

[3] Unger, J., and Dagalakis, N. G., "Optimum Stiffness Study for a Parallel Link Robot Crane Under Horizontal Force", Proceedings of the Second International Symposium on Robotics and Manufacturing Research, Education and Applications, Albuquerque, NM, November 16-18, 1988.

[4] Albus, J. S., Bostelman, R. V., Dagalakis, N. G., "The NIST SPIDER, A Robot Crane", to be published in the Journal of Robotic Systems, July 1992.

[5] Stewart, D., "A Platform with Six Degrees of Freedom", Proc. of the Inst. of Mechanical Eng. Volume 180 (15), Part I: 371-386, 1965-1966.

[6] Landsberger, S.E., "Design and Construction of a Cable-Controlled Parallel Link Manipulator", Master's Thesis, Dept. of Mechnical Engineering, Massachusetts Institute of Technology, September 1984.

[7] Landsberger, S.E., Sheridan, T.B., "A New Design for Parallel Link Manipulators," Proc. Systems Man and Cybernetics Conf., 812-814, Tucson, AZ, November 1985.

A STEWART PLATFORM-BASED FORCE REFLECTING HAND CONTROLLER

M.D. Bryfogle* C.C. Nguyen† S.C. Antrazi‡ P.C. Chiou
Science Applications International Corporation
Catholic University of America

Abstract

Force feedback can enhance the efficiency of a teleoperation system by providing the operator with a sense of *feel* of the forces and torques arising from the interaction of the slave manipulator with the remote environment. This paper addresses the kinematic analysis and control of a Parallel FOrce Reflecting Hand Controller (PFORHC) whose design and implementation are based on a fully parallel Stewart Platform mechanism. Kinematic analysis for the PFORHC is performed and a fixed-gain PD control scheme is developed for force feedback control. Experimental results show that the force control scheme utilizing a hand grip force sensor provides smaller steady-state errors as compared to the case utilizing no handgrip force sensor.

1 Introduction

Research in teleoperation, in which an operator controls the motion of a *slave manipulator* located at a remote environment by handling a *master mechanism*, has attracted tremendous attention during the last decade because of its potential applications in hostile and hazardous environments, such as nuclear and chemical waste sites, space and undersea [1]-[6]. A master mechanism which provides the operator with the above capability is classified as a force-reflecting hand controller (FRHC). This paper deals with a force-reflecting hand controller whose design was based on a fully parallel Stewart Platform mechanism [8]. This device referred hereto as the **P**arallel **FO**rce **R**eflecting **H**and **C**ontroller (PFORHC) was developed under a Small Business Innovative Research (SBIR) program funded by NASA Resident Office at the Jet Propulsion Laboratory for teleoperation in space.

2 The PFORHC

Figure 1 provides a system block diagram of the subject manipulator teleoperation system. A human operator introduces a pose command to the system by clenching the handgrip of the PFORHC master mechanism and moving his hand. Position sensors within the mechanism allow for the computatioı of the handgrip pose (In this paper, the word *pose* is used to denote both position and orientation of a rigid body and the word *wrench* is used to denote both force and torque.) This information is transmitted to the slave manipulator as a pose command where it is resolved into joint position commands through the slave manipulator inverse kinematics. A wrench sensing cell in the slave manipulator measures the interaction wrench of the slave manipulator with its environment. This information is transmitted back to the master mechanism where it is resolved into actuator commands to represent the slave wrenches by exertions against the operator hand. Figure 2 depicts the PFORHC mechanism which comprises a rigid body handgrip, six linkages termed legs, and a base or frame. Each of the six legs makes a separate and direct, distal connection to the handgrip by means of a four degrees of freedom (DOF) spherical joint comprising four revolute axes intersecting at a point. The proximal end of each leg makes a separate and direct connection to the base frame by means of a two DOF Hooke joint comprising two revolute axes intersecting at a point. The leg

*Science Applications International Corporation, McLean, Virginia 22102
†Professor of Electrical Engineering, Catholic University of America, Washington DC 20064
‡Graduate Research Assistant

embodies a prismatic joint between the proximal and distal joints with the distal or upper leg portion translating relative to proximal or lower leg portion. As a result of the configuration of legs and joints, the handgrip possesses six DOFs. An actuator is mounted within the structure of each of the proximal joints to develop forces within each leg and thereby drive the prismatic joint. Within the handgrip workspace and actuator limitations, the leg system can exert an arbitrary wrench against the handgrip and operator. The actuator configuration consists of two frameless, direct current electric motors mounted inside deep groove ball bearings on opposite ends of a motor drive shaft. The drive comprises a simple cable wrapping the motor shaft between the two motor rotors. One cable end terminates directly on the upper, sliding leg for leg retraction while the other end passes over a turning shaft mounted distally on the lower leg and terminates on the sliding leg to provide extension. The motors selected are permanent magnet, samarium cobalt, brush type, rotary motors from Inland Motors, Model # QT-1401-D. The poles of the two motors are angularly displaced on the motor shaft to be completely out of phase to effectively achieve a doubling of the number of poles. The cable configuration for the power transmission effectively implements direct drive with no backlash. A stainless steel 0.040 *in.* diameter wire rope forms the cable and makes 4.5 turns about a "V" groove on the motor shaft at a radius of 0.2 *in*. The combination of motors and cable action radius allows for the generation of 35 *lbf* tension in the cable. The actuators are backdriveable with an average of 1.25 *lbf* applied to the sliding portion of the leg. At full force, a single actuator consumes 434 Watts.

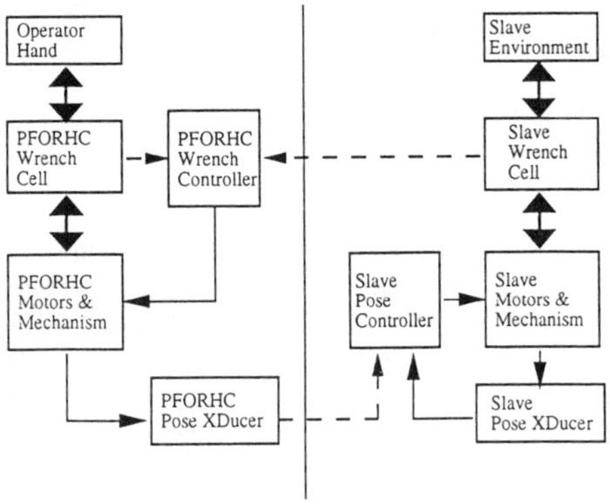

Figure 1 Block diagram of the teleoperation system

Figure 2 The PFORHC

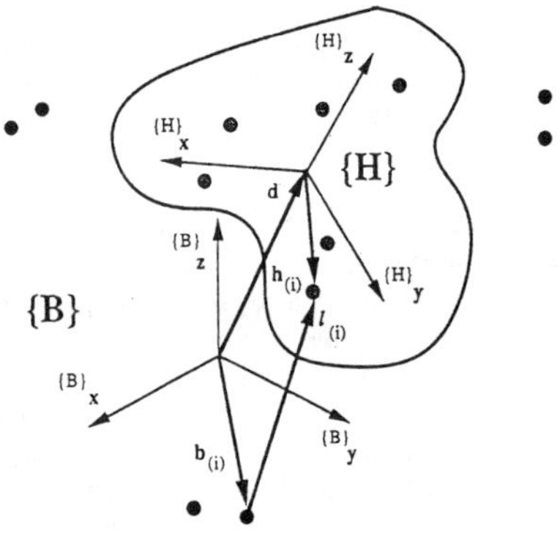

Figure 3 Frame assigment in the PFORHC

Linear position transducers are placed within each leg. The transducer selected is the Lucas-Schaevitz Induced Voltage Position Sensor (IVPS) model #D1051-0001-IVPS-400. A 50 pound range, six DOF wrench sensing cell, supplied by JR^3 Corporation is mounted in the handgrip between the operator hand interface and the handgrip structure. The cell provides a feedback means whereby the friction and inertia forces within the mechanism could attempt to be nullified.

Figure 3 depicts a general vector configuration used to describe the fully parallel mechanism. The global coordinate frame, $\{\mathbf{B}\}$, is shown fixed in the base. The vectors, $\mathbf{b}_i$, for $i = 1, 2, \ldots 6$ describe the position of the attachment points of the proximal leg joints in the base relative to the base coordinate origin. The handgrip is depicted as an arbitrary rigid body with a coordinate frame $\{\mathbf{H}\}$ assigned. The rotation matrix expressing coordinate axes of the frame $\{\mathbf{H}\}$ in frame $\{\mathbf{B}\}$ will be denoted as $\Re$. The vectors, $\mathbf{h}_i$ for $i = 1, 2, \ldots 6$, describe the position of the attachment points of the distal joints in the handgrip relative to the handgrip coordinate origin. The vectors, $\mathbf{l}_i$, describe the position of the $\mathbf{h}_i$ relative to the $\mathbf{b}_i$. The vector $\mathbf{d}$ describes the position of the origin of frame $\{\mathbf{H}\}$ relative to the origin of frame $\{\mathbf{B}\}$. All these vectors are expressed in frame $\{\mathbf{B}\}$. The $\mathbf{h}_i$ may sometimes be expressed in the handgrip frame and when so expressed are written as ${}^{\{\mathbf{H}\}}\mathbf{h}_i$.

Many selections of generalized coordinates are possible for the fully parallel mechanism. The most convenient set is the length between terminal joints for each leg, denoted herein as the scalar θ_i for $i = 1, 2, \ldots 6$ representing $||\mathbf{l}||$. This selection is made primarily on the basis of ease transducer placement within the leg, actuation, and mathematical formulation. The θ_i are written as a column vector in R^6 as $\boldsymbol{\theta}$. The time derivative of θ_i, $\dot{\theta}_i$, represents the leg speed (time rate of leg length change) and not the total time derivative of the vector $\mathbf{l}$.

3 The PFORHC Kinematic Analysis

The inverse kinematic relationship specifies the θ_i as a function of the handgrip displacement, $\mathbf{d}$, and handgrip orientation, $\Re$. With $\mathbf{d}$ and $\Re$ given, the position of the handgrip joints relative to the corresponding base joints can be determined as:

$$\mathbf{l}_i = (\mathbf{d} + \mathbf{h}_i) - \mathbf{b}_i = (\mathbf{d} + [\Re]\, {}^{\{\mathbf{H}\}}\mathbf{h}_i) - \mathbf{b}_i \tag{1}$$

The magnitude of the $\mathbf{l}_i$ vector yields θ_i as follows:

$$\theta_i = \left(\left[[\mathbf{d} + [\Re]\, {}^{\{\mathbf{H}\}}\mathbf{h}_i] - \mathbf{b}_i \right]^T \left[[\mathbf{d} + [\Re]\, {}^{\{\mathbf{H}\}}\mathbf{h}_i] - \mathbf{b}_i \right] \right)^{\frac{1}{2}} \tag{2}$$

Equation (2) represents a closed-form solution to the PFORHC inverse kinematics in the sense that if the position expressed by $\mathbf{d}$ and the orientation expressed by $\Re$ of frame $\{\mathbf{H}\}$ are given, the θ_i can be determined using (2).

In the context of robotic systems, a Jacobian matrix, $\mathcal{J}$, is defined as a linear mapping of the first differential of the mechanism generalized coordinate vector (mechanism actuator speeds) to the translational and angular velocity vectors of a specified body within the mechanism (handgrip). We write $[\mathbf{v} \quad \boldsymbol{\omega}]^T = \mathcal{J}\dot{\boldsymbol{\theta}}$ where $\mathbf{v}$ denotes the translational velocity vector in R^3 of the handgrip expressed in frame $\{\mathbf{B}\}$, $\boldsymbol{\omega}$ denotes the angular velocity vector (in the sense of an instantaneous rotation about an instantaneous axis of rotation with speed components projected on an orthogonal set of Cartesian frame axes) in R^3 of the handgrip expressed in frame $\{\mathbf{B}\}$, and $\dot{\boldsymbol{\theta}}$ denotes vector in R^6 of the leg speeds. The derivation of $\mathcal{J}$ for the PFORHC follows from the formulation of the existence of an inverse Jacobian. From above we have $\dot{\boldsymbol{\theta}} = \mathcal{J}^{-1}[\mathbf{v} \quad \boldsymbol{\omega}]^T$.

We observe that for each leg, the translational speed vector is the projection of the total translational velocity of the upper joint onto the leg axis unit vector. The total translational velocity of each handgrip joint is taken as the sum of the handgrip origin translational velocity and the translational velocity induced by the angular velocity of the handgrip. Thus

$$\dot{\theta}_i\, \mathbf{s}_i = \left(\left[\mathbf{v} + \boldsymbol{\omega} \times [\Re]\, {}^{\{\mathbf{H}\}}\mathbf{h}_i \right]^T \mathbf{s}_i \right) \mathbf{s}_i = \left([\mathbf{v} + \boldsymbol{\omega} \times \mathbf{h}_i]^T\, \mathbf{s}_i \right) \mathbf{s}_i \tag{3}$$

where $\mathbf{s}_i = \mathbf{l}_i / \| \mathbf{l}_i \|$, denotes the unit vector describing the leg direction, and $\times$ denotes the vector cross product operation. The inner product in Equation (3) can be distributed and the resulting equation factored to form:

$$\dot{\theta}_i \, \mathbf{s}_i = [\, \mathbf{s}_i{}^T \quad (\mathbf{h}_i \times \mathbf{s}_i\,)^T \,] \begin{bmatrix} \mathbf{v} \\ \\ \omega \end{bmatrix} \mathbf{s}_i \tag{4}$$

Forming the inner product of both sides of the above vector equation with $\mathbf{s}_i$ yields the scalar relation:

$$\dot{\theta}_i = [\, \mathbf{s}_i{}^T \quad (\mathbf{h}_i \times \mathbf{s}_i\,)^T \,] \begin{bmatrix} \mathbf{v} \\ \\ \omega \end{bmatrix} \tag{5}$$

Equation (5) can be written simultaneously for each of the six legs to yield an expression in vector matrix form as

$$\dot{\theta} = \begin{bmatrix} \mathbf{s}_1{}^T & (\mathbf{h}_1 \times \mathbf{s}_1\,)^T \\ \mathbf{s}_2{}^T & (\mathbf{h}_2 \times \mathbf{s}_2\,)^T \\ \mathbf{s}_3{}^T & (\mathbf{h}_3 \times \mathbf{s}_3\,)^T \\ \mathbf{s}_4{}^T & (\mathbf{h}_4 \times \mathbf{s}_4\,)^T \\ \mathbf{s}_5{}^T & (\mathbf{h}_5 \times \mathbf{s}_5\,)^T \\ \mathbf{s}_6{}^T & (\mathbf{h}_6 \times \mathbf{s}_6\,)^T \end{bmatrix} \begin{bmatrix} \mathbf{v} \\ \\ \omega \end{bmatrix} \tag{6}$$

Using (6) the Jacobian matrix can be obtained as:

$$\mathcal{J} = \begin{bmatrix} \mathbf{s}_1{}^T & (\mathbf{h}_1 \times \mathbf{s}_1\,)^T \\ \mathbf{s}_2{}^T & (\mathbf{h}_2 \times \mathbf{s}_2\,)^T \\ \mathbf{s}_3{}^T & (\mathbf{h}_3 \times \mathbf{s}_3\,)^T \\ \mathbf{s}_4{}^T & (\mathbf{h}_4 \times \mathbf{s}_4\,)^T \\ \mathbf{s}_5{}^T & (\mathbf{h}_5 \times \mathbf{s}_5\,)^T \\ \mathbf{s}_6{}^T & (\mathbf{h}_6 \times \mathbf{s}_6\,)^T \end{bmatrix}^{-1} . \tag{7}$$

The forward kinematics determines the handgrip coordinate origin position vector, $\mathbf{d}$, and handgrip orientation, $\Re$, as a function of the vector of leg lengths, θ. A numerical solution is required for the forward kinematics since it involves solving six nonlinear equations in six unknowns. A Newton-Raphson approach developed by Nguyen et al [9] is employed here to solve the forward kinematics for the PFORHC.

4 The Wrench Feedback Control System

The design philosophy of the PFORHC has been that the proper goal of wrench feedback teleoperation system research is to make the entire teleoperation system transparent to the operator - to make the operator feel as if he were actually on site and an integral part of the slave robot. Certainly, the operator should not be cognizant of any dynamics in the hand controller mechanism. In all likelihood, the operator should be made cognizant of the dynamics of the slave robot. Simply stated, it is the slave which performs the work and not the human, therefore it is the slave dynamics which should be experienced by the operator.

The human has limitations which no amount of "software massaging" can overcome. In particular, the human has a very limited bandwidth in tracking oscillations by either providing counterbalancing forces through his arm or by position tracking. This is intuitive and has been shown experimentally [5,6]. Certainly the actuation force of human muscle has limitations. It is coupled to a mechanical link with mass, the bone. Such a system has bandwidth limitations. In direct interaction with a vibrating member, the human is usually thrown back. In a force feedback system, if the human is thrown back, a new position command is generated to the slave robot which will usually set up a limit cycle oscillation. It is completely fitting to low pass filter the force information throughout the system to remove frequencies beyond the human bandwidth. Beyond that, the human should experience the force as realistically as possible. Computational hardware will eventually and probably very soon, have the capability to compute the complete dynamics of a slave manipulator to account for the entire mass content of the manipulator and its grasped object at rates which are suitable for teleoperation. If a very dynamically sound hand controller mechanism exists, then it will surely be useful to provide the dynamics of the slave to the operator through the mechanism. Success will depend upon the integrity of the hand controller.

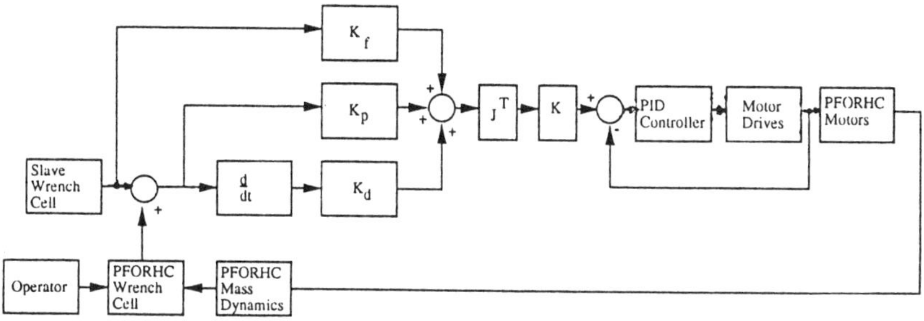

Figure 4 Block diagram of the wrench feedback control system

As shown in Figure 4, the control law for the wrench feedback within the PFORHC relies upon an inner and outer control loop. The inner loop comprises a current control loop around each of the actuators. This loop was introduced basically to compensate for electrical dynamics of the actuators. The outer loop attempts to compensate for the friction and mass content of the PFORHC mechanism. The inner control loop is particularly simple to implement. The desired current command from the outer loop is fed as a voltage signal to a local current controller and amplifier package supplied by Inland Motor. The current controller implements a fixed proportional, integral, and derivative (PID) control law. The bandwidth of the controller and motor combination is in the kilohertz range. The control module provides a port for an externally mounted current limiting resistor. This can be used in addition to software clamps to provide a measure of safety against excessive forces. The outer control loop provides a direct feedforward term based on a theoretical force and a compensating term to attenuate the inertial and friction forces. The approach to the compensator design is intuitive as opposed to analytic for this highly nonlinear system. A fixed gain, proportional plus derivative (PD) control action acting on the wrench error measured at the handgrip is implemented. The proportional term provides a large proportional factor to speed response and reduce steady state error, while the derivative term is introduced for stabilization of the large proportional gain. The determination of the gain values proceeds empirically from laboratory testing.

5 Experimental Study

For the purpose of experimentally evaluating the PFORHC subsystem comprising the mechanism and wrench feedback controller, the subsystem was excited with computer-simulated feedback wrenches from the slave robot, while the mechanism was mechanically constrained to move only along the base frame **y** axis. The force response of the PFORHC to a 10 lbf step input in the positive **z** direction with and without the outer PD control loop is shown in Figures 5 and 6, respectively. These data show both the command step and force response. During the test, the handgrip was moved along the **y** axis to create a disturbance. Observing the force trajectories, we found that the exclusion of the external PD controller causes the force trajectory to have a steady-state error of about 1 lbf. On the other hand, although the inclusion of the external PD controller causes the transient response to have larger overshoot as compared to the previous case, it decreases the steady state error to zero. After the bench tests, the PFORHC subsystem is successfully coupled to a GE P60 manipulator at the Center for Intelligent Machines and Robotics at the University of Florida. The qualitative testing demonstrates system level closed loop control with wrench feedback from the slave. While coupled to the slave, a total closed loop computational update rate of 83 Hz. is measured.

6 Conclusions and Recommendations

The research of this paper embraced the thesis that a superior force reflecting hand controller could be constructed from the fully parallel mechanism and that the implementation of a friction and inertia canceling control loop about the entire mechanism based upon wrench sensing in the mechanism handgrip would be beneficial. The feasibility from a kinematic and mechanical viewpoint is demonstrated by the design and construction of the PFORHC mechanism. The wrench cell in the handgrip has shown a measure of improvement in the steady-state tracking capability. It is anticipated that the outer loop fixed gain controller performance can be improved by further development to specifically include experimental system identification with associated gain tuning. Adaptive gain controllers are likely candidates for further development. The fundamental nonlinear and time varying system characteristics favor adaptive controllers. Additionally, the computational update rate of 83 Hz which is experimentally established indicates the feasibility of on-line gain computations for the adaptive controller with the current hardware configuration.

Acknowledgements

The research presented in this paper was funded by the NASA resident office at the Jet Propulsion Laboratory under Small Business Innovation Research Contract (**SBIR**) contract #NAS7-1069. The authors wish to express their gratitude to NASA, the Jet Propulsion Laboratory, and Dr. Neville Marzwell, the technical monitor.

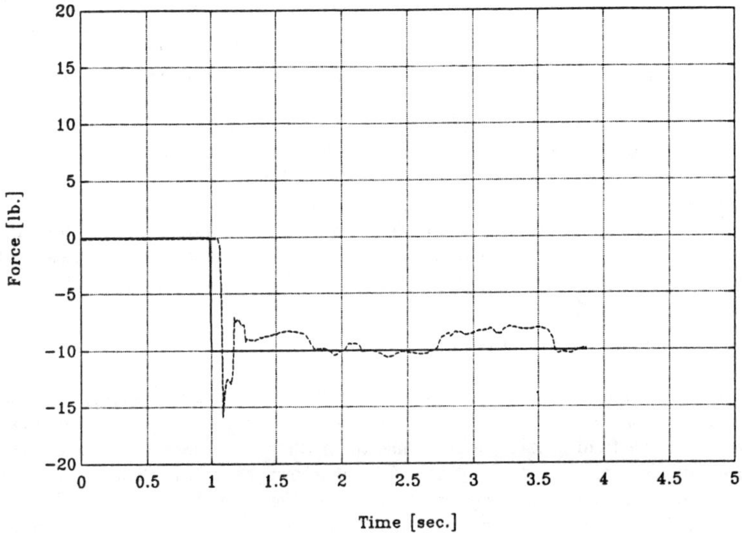

Figure 5 Time responses of the Z-axis forces with external PD controller

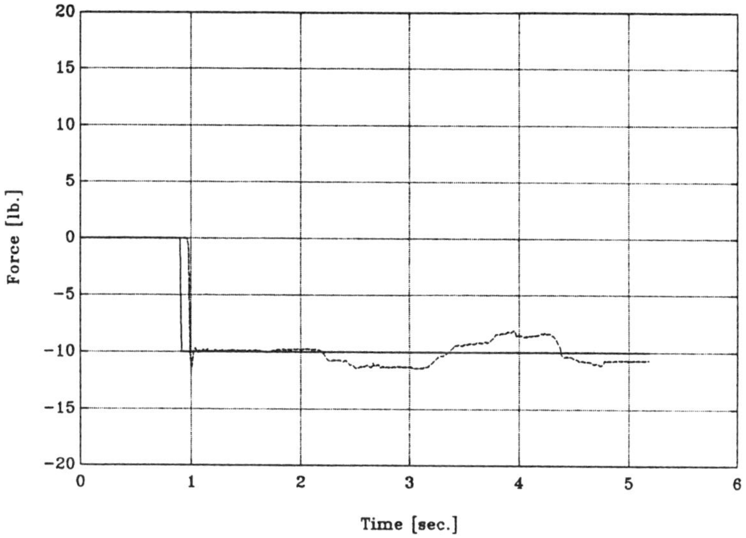

Figure 6 Time responses of the Z-axis forces without external PD controller

References

[1] **Goertz, R.C. and Thompson, W.M.**, "Electronically Controlled Manipulator," *Nucleonics*, pp. 46-47, November 1954.

[2] **Bejczy, A.K., and Salisbury, J.K.**, "Controlling Remote Manipulators Through Kinesthetic Coupling," *Computers in Mechanical Engineering*, Vol. 2, July 1983, pp. 48-60.

[3] **Handlykken and Turner, T.**, "Control System Synthesis for a Six Degree of Freedom Universal Force Reflecting Hand Controller", *Proceedings of the 19th IEEE Conference on Decision and Control*, Albuquerque, New Mexico, Dec. 10-12, 1980.

[4] **Martin, H.L. and Hammel, W.R.**, "Joining Teleoperation with Robotics for Remote Manipulation in Hostile Environments," *Conference Proceedings Robots 8, Robotics International of Society of Manufacturing Engineers*, Vol. 1, Dearborn, Michigan, June 1984.

[5] **McRuer, D.T. and Krendel, E.S.**, "The Human Operator as a Servo System Element," Part I, *Journal of the Franklin Institute*, Vol. 267, May 1959, pp. 381-404.

[6] **McRuer, D.T. and Krendel, E.S.**, "The Human Operator as a Servo System Element," Part II, *Journal of the Franklin Institute*, Vol. 267, June 1959, pp. 511-536.

[7] **Noll, A.M.**, "Man-Machine Tactile Communication," *Ph.D. Dissertation*, Polytechnic Institute of Brooklyn, June 1971.

[8] **Stewart, D.**, "A Platform with Six Degrees of Freedom," *Proc. Institute of Mechanical Engineering*, Vol. 180, part 1, No. 5, pp. 371-386, 1965-1966.

[9] **Nguyen, C.C., Antrazi, S.S., Park, J-Y., and Z-L. Zhou**, "Analysis and Experimentation of a Stewart Platform-Based Force/Torque Sensor," accepted for publication in *International Journal of Robotics and Automation*, 1992.

CONTROL OF REDUNDANT DIRECT-DRIVE ROBOTS

W. J. Chung
Automation Research Center
Pohang Institute of Science and Technology
Pohang P. O. Box 125, 790-600, KOREA

W. K. Chung and Y. Youm
Department of Mechanical Engineering
Pohang Institute of Science and Technology
Pohang P. O. Box 125, 790-600, KOREA

J. H. Eun
Research Institute of Science and Technology
Pohang P. O. Box 135, 790-600, KOREA

ABSTRACT

This paper presents the kinematic and dynamic control of redundant direct-drive robots. The POSTECH 7-DOF Direct Drive Robot is chosen as an example of redundant direct-drive robot since it includes a planar 3-DOF subarm with one redundancy. For the kinematic control of the robot, the virtual link concept is used to divide the inverse kinematics of this robot into that of planar 3-DOF shoulder-elbow subarm and that of spatial 4-DOF elbow-wrist subarm. For the dynamic control of the planar 3-DOF subarm, the $\boldsymbol{J}$-Minor based Dynamic Control (JMDC) is used to guarantee the stability of joint torque by preserving the signs of full row-rank minors of a manipulator Jacobian.

INTRODUCTION

A robotic manipulator is called (kinematically) "redundant" if it possesses more degrees of freedom (DOF) than are required to complete a specified task. Redundancy in a manipulator structure yields increased dexterity and versatility for performing a task due to the infinite number of joint motions called the "self-motion". The POSTECH Direct Drive (DD) Robot was designed to have 7 degrees of freedom (DOF) with one additional redundancy [1]. Considering the limitations of conventional 6-DOF manipulators in singularity avoidance, obstacle avoidance, and flexibility in motion, we developed the 7-DOF robot including a planar 3-DOF subarm. The robot is a SCARA type, and has the direct drive mechanism to eliminate mechanical uncertainty in the driving mechanism and to enable faster motion [1].

Recently, Seraji *et al.* [2] presented a kinematic analysis of an anthropomorphic 7-DOF serial link spatial manipulator having a spherical-revolute-spherical joint arrangement (i.e., 3R-1R-3R). They parameterized the redundancy by a scalar variable which corresponds to the angle between the arm plane and a reference frame. Seraji's approach, however, cannot be applied to SCARA type spatial manipulators including planar redundant subarms. In this paper, a new approach to the inverse kinematics of SCARA type spatial redundant manipulator such as the POSTECH 7-DOF DD Robot is presented using the *virtual link* concept [3].

The dynamic control of redundant manipulators has been mainly focused on minimizing torque loadings at the joints. A dramatic difficulty with using conventional methods of local torque optimization [4–6] is the instability that these methods can induce physically unachievable joint torques. In order to overcome this instability problem, this paper also presents a new real-time dynamic control law which is based on the kinematic property of *full row-rank minors* of a manipulator Jacobian, and thereby can guarantee the stability of joint torques. The effectiveness of the new approach for torque reduction is evaluated by computer simulation in comparison with the conventional methods.

KINEMATIC CONTROL

Virtual Link Concept

A virtual link, as referred to in [3], is defined as a fictitious link connecting the conjunction joints between subarms. In a serially connected planar n-link manipulator, the number of virtual links, n_{vir}, should be in a range of $n/3 \leq n_{\text{vir}} \leq n/2$ because the subarms belong to either two or three link modules. The subarm module is defined as the simplest form to create a virtual link, which can be either 2-link or 3-link system. Thus the subarm module is considered as a tractable virtual link system to solve the inverse kinematic problem for a planar manipulator with large redundancy. The determination of the subarm module in a virtual link system is described in [3].

Inverse Kinematics

The POSTECH DD Robot is a 7-DOF manipulator with planar 3R(Shoulder)-1R (Elbow)-3R(Wrist), as shown in the schematics of Fig. 1. In order to separate the inverse kinematic problem of the planar 3-DOF subarm from the overall inverse kinematic problem of POSTECH 7-DOF DD Robot, the position of the wrist and the elbow angle should be previously determined. The position of wrist, (p_x^W, p_y^W, p_z^W), and the elbow angle, θ_4, can be easily obtained for a given end-effector position and orientation.

For the POSTECH 7-DOF DD Robot, the planar 3-DOF subarm (ℓ_1-ℓ_2-ℓ_3) is substituted with the virtual link $\ell_{1,\text{vir}}$ as shown in Fig. 2. The intermediate variables, $\ell'_{1,\text{vir}}$ and $\theta'_{1,\text{vir}}$, are introduced for further simplification. The relationships among $\ell_{1,\text{vir}}$, $\ell'_{1,\text{vir}}$, $\theta_{1,\text{vir}}$, and $\theta'_{1,\text{vir}}$ are given by

$$\ell_{1,\text{vir}} = \sqrt{(\ell'_{1,\text{vir}})^2 - d_4^2}\,, \tag{1}$$

$$\theta_{1,\text{vir}} = \theta'_{1,\text{vir}}\,, + \text{atan2}(d_4, \ell_{1,\text{vir}})\,. \tag{2}$$

The $\ell'_{1,\text{vir}}$ and $\theta'_{1,\text{vir}}$ are easily obtained from the geometry shown in Fig. 2:

$$\ell'_{1,\text{vir}} = \sqrt{(p_x^W)^2 + (p_y^W)^2} - d_5 s_4\,, \tag{3}$$

$$\theta'_{1,\text{vir}} = \text{atan2}(\sin\theta'_{1,\text{vir}}, \cos\theta'_{1,\text{vir}})\,, \tag{4}$$

where

$$\cos\theta'_{1,\text{vir}} = \frac{p_x^W}{\ell_{1,\text{vir}} + d_5 s_4}\,, \tag{5}$$

$$\sin\theta'_{1,\text{vir}} = \frac{p_y^W}{\ell_{1,\text{vir}} + d_5 s_4}\,, \tag{6}$$

and $s_4 \triangleq \sin\theta_4$. Therefore $\ell_{1,\mathrm{vir}}$ and $\theta_{1,\mathrm{vir}}$ can be obtained from Eqs. (1) and (2), and thus the position of the elbow, (p_x^E, p_y^E, p_z^E), is determined as follows:

$$p_x^E = \ell_{1,\mathrm{vir}} \cos\theta_{1,\mathrm{vir}}\,, \tag{7}$$

$$p_y^E = \ell_{1,\mathrm{vir}} \sin\theta_{1,\mathrm{vir}}\,, \tag{8}$$

$$p_z^E = 0\,. \tag{9}$$

Based on the elbow position obtained in the above, we have only to solve the inverse kinematic problem for the planar 3-DOF subarm shown. Here, the extended Jacobian method [7] is adopted for the inverse kinematics of the planar 3-DOF subarm because this subarm has only one redundancy. This method exploits the orthogonality between the null space vector ($\boldsymbol{n}_J$) of the Jacobian matrix and the gradient vector (∇H) of a performance function ($H(\boldsymbol{\theta})$), which is expressed by

$$\boldsymbol{n}_J \cdot \nabla H = 0\,. \tag{10}$$

For known p_x^E and p_y^E, Eq. (10) as well as the kinematic equations given by

$$\ell_1 c_1 + \ell_2 c_{12} + \ell_3 c_{123} = p_x^E \tag{11}$$

$$\ell_1 s_1 + \ell_2 s_{12} + \ell_3 s_{123} = p_y^E\,, \tag{12}$$

where $c_2 \triangleq \cos\theta_2$, $c_3 \triangleq \cos\theta_3$, $c_{23} \triangleq \cos(\theta_2+\theta_3)$, $s_2 \triangleq \sin\theta_2$, $s_3 \triangleq \sin\theta_3$, and $s_{23} \triangleq \sin(\theta_2+\theta_3)$, constitute three non-linear algebraic equations. Having obtained the values of θ_1, θ_2, θ_3, and θ_4, it now remains to find the wrist angles, i.e., θ_5, θ_6, and θ_7. These angles can be obtained by using homogeneous transformation matrices, which is omitted due to space limitation.

DYNAMIC CONTROL

J-Minor based Dynamic Control

It was illustrated in [8] that the small values of full row-rank minors of a manipulator Jacobian matrix can result in a remarkable increase of the torque required to track the desired end-effector trajectory of a redundant manipulator. According to Borrel and Liégeois [9], the full row-rank minors have close relation to the *aspects* which are functions of manipulator's configurations. Based on the concept of aspects defined in [9], the $\boldsymbol{J}$-Minor based Dynamic Control, which is hereinafter abbreviated "JMDC," was proposed in [8].

The JMDC method for the dynamic control of redundant manipulators aims at preserving an aspect by using the performance function $H(\boldsymbol{\theta})$ which has a direct control over each full row-rank minor. In this method, the command joint velocity is generated by

$$\dot{\boldsymbol{\theta}}_d^{i+1} = \boldsymbol{J}^+(\boldsymbol{\theta}^i)\,\dot{\boldsymbol{x}}_d^i + \kappa\left\{\boldsymbol{I} - \boldsymbol{J}^+(\boldsymbol{\theta}^i)\,\boldsymbol{J}(\boldsymbol{\theta}^i)\right\}\nabla H(\boldsymbol{\theta}^i) \tag{13}$$

where i denotes the current state at time $t = i \cdot \Delta t$ (Δt: sampling time); $\boldsymbol{\theta}^i$ and $\dot{\boldsymbol{\theta}}^i$ are the measured values of joint angle and velocity at time t, respectively. A good candidate for $H(\boldsymbol{\theta})$ is the minor index [10] which is defined by

$$H(\boldsymbol{\theta}) = \prod_{i=1}^{p} \delta_i\,, \tag{14}$$

where δ_i's for $i = 1, 2, \ldots, p$ are the full row-rank minors (or minors of rank m) of the Jacobian matrix $\boldsymbol{J} \in \Re^{m \times n}$; with $p = {}_nC_m$, the combination of m taken out of n.

By using the values of command joint velocity, $\dot{\boldsymbol{\theta}}_d^i$ and $\dot{\boldsymbol{\theta}}_d^{i-1}$, the command joint angle $\boldsymbol{\theta}_d$ and the command joint accceleration $\ddot{\boldsymbol{\theta}}_d$ are generated by simple numerical integration and differentiation, respectively. The command torque induced by the JMDC is given in the form of computed-torque control as follows:

$$\boldsymbol{\tau} = \boldsymbol{M}(\boldsymbol{\theta}^i)\left\{\ddot{\boldsymbol{\theta}}_d^i + \boldsymbol{K}_v(\dot{\boldsymbol{\theta}}_d^i - \dot{\boldsymbol{\theta}}^i) + \boldsymbol{K}_p(\boldsymbol{\theta}_d^i - \boldsymbol{\theta}^i)\right\} + \boldsymbol{N}(\boldsymbol{\theta}^i, \dot{\boldsymbol{\theta}}^i), \tag{15}$$

where $\boldsymbol{M}(\boldsymbol{\theta}) \in \Re^{n \times n}$ is a symmetric, positive definite inertia matrix, and $\boldsymbol{N}(\boldsymbol{\theta}, \dot{\boldsymbol{\theta}}) \in \Re^n$ is a vector containing terms such as Coriolis, centripetal, and gravitational torques; $\boldsymbol{K}_p \in \Re^{n \times n}$ and $\boldsymbol{K}_v \in \Re^{n \times n}$ are position and velocity feedback gain matrices, respectively.

Computer Simulation

The comparative evaluation of the JMDC method against conventional methods of local torque optimization was performed for the planar 3-DOF subarm model of the POSTECH 7-DOF DD Robot using computer simulation. The presumed movement of the end-effector is a straight-line Cartesian trajectory starting and ending with zero velocity, with constant acceleration and deceleration over the first and the last half of the movement, respectively. The initial configuration is given by $\boldsymbol{\theta}^0 = [-60° \;\; 140° \;\; -140°]^T$.

For the simulation of the JMDC method, the minor index is used as a performance function, which is expressed for the planar 3-DOF manipulator model as:

$$H = (\ell_1\ell_2 s_2 + \ell_1\ell_3 s_{23})\,\ell_2\ell_3 s_3\,(-\ell_2\ell_3 s_3 - \ell_1\ell_3 s_{23}). \tag{16}$$

As shown in Fig. 3, the arm motion for this method is uniform and stays within one kind of aspects. As anticipated from this arm motion, the joint torques are drastically held down at low values. This fact is illustrated in Fig. 4 which shows the torque norm profiles for the four methods, namely inertia-weighted pseudoinverse method [4], null space vector method [5], redundancy decomposition control method [6], and NTDC method.

CONCLUSION

When a spatial redundant manipulator has planar links with serial joints, these links can be handled by the virtual link concept. In this paper, this concept was applied for dividing the inverse kinematics of the POSTECH 7-DOF DD Robot into that of planar mechanism (shoulder-elbow) and that of spatial mechanism (elbow-wrist). The proposed approach can be applied to the SCARA type spatial redundant manipulators whose the degrees of redundancy are included in their planar mechanisms.

This paper also applied the dynamic control method, named "$\boldsymbol{J}$-Minor based Dynamic Control (JMDC), " to the planar 3-DOF subarm model of the POSTECH 7-DOF DD Robot. The key idea of JMDC method is at the kinematic level to preserve the aspect to which an initial configuration belongs, and thereby at the dynamic level to guarantee the stability of joint torque. It was proved by computer simulations that the JMDC method can generate reasonably low values of joint torques in a global sense when compared with the conventional methods of local torque optimization.

REFERENCES

1. K. W. Jeong, W. K. Chung, and Y. Youm, "Development of POSTECH 7-DOF direct drive robot," in *Robotics and Manufacturing, Vol. 3*, edited by M. Jamshidi and M. Saif (ASME Press, New York, 1990).

2. K. Kreutz-Delgado, M. Long, and H. Seraji, "Kinematic analysis of 7 DOF anthropomorphic arms," in *Proc. IEEE Int. Conf. on Robotics and Automation* (Cincinnati, OH), May 1990, pp. 824–830.

3. W. J. Chung, W. K. Chung, and Y. Youm, "Inverse kinematics of planar redundant manipulator using virtual link and displacement distribution schemes," *Proc. IEEE Int. Conf. on Robotics and Automation* (Sacramento, CA), April 1991, pp. 926–932.

4. O. Khatib, "Dynamic control of manipulators in operational space," in *Proc. 6th IFToMM World Conf. Theory of Machines and Mechanisms*, New Delhi, India, Dec. 1983, pp. 15–20.

5. J. M. Hollerbach and K. C. Suh, "Redundancy resolution of manipulators through torque optimization," *IEEE J. Robotics and Automation*, **3** (4), 308–316, 1987.

6. S. Hirose and S. Ma, "Redundancy decomposition control for multi-joint manipulator," in *Proc. IEEE Int. Conf. Robotics and Automation* (Scottsdale, AZ), May 1990, pp. 119–124.

7. J. Baillieul, "Kinematic programming alternatives for redundant manipulators," in *Proc. IEEE Int. Conf. on Robotics and Automation* (St. Louis, MO), March 1985, pp. 722–728.

8. W. J. Chung, Y. Youm, and W. K. Chung, "*J*-Minor based Dynamic Control (**JMDC**) for kinematically redundant manipulators," in *Proc. IEEE/RSJ Int. Workshop on Intelligent Robots and Systems (IROS'91)* (Osaka, Japan), Nov. 1991, pp. 143–148.

9. P. Borrel and A. Liégeois, "A study of multiple manipulator inverse kinematic applications to trajectory planning and workspace determination," in *Proc. IEEE Int. Conf. Robotics and Automation* (San Francisco, CA), April 1986, pp. 1180–1185.

10. W. J. Chung, W. K. Chung, and Y. Youm, "Null Torque based Dynamic Control (NTDC) for Kinematically Redundant Manipulators," in *Proc. American Control Conference* (Chicago, Illinois), June 1992, pp. 1212–1216.

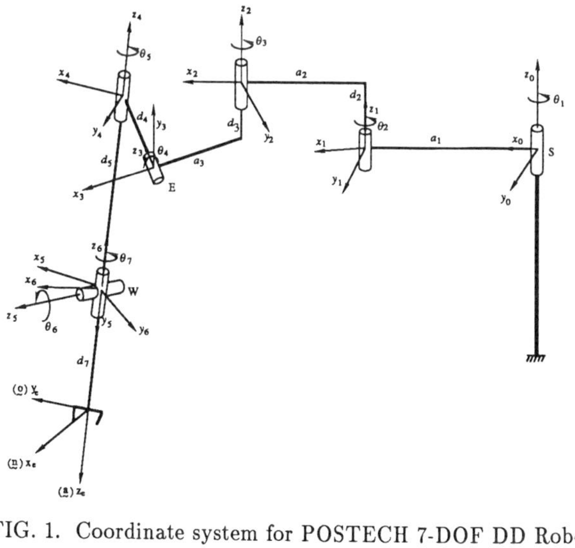

FIG. 1. Coordinate system for POSTECH 7-DOF DD Robot.

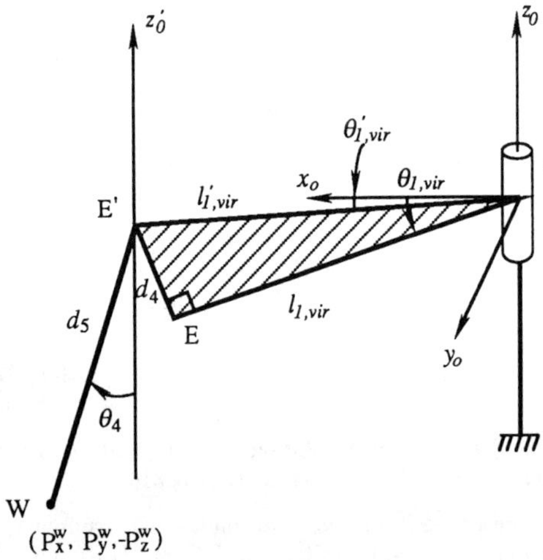

FIG. 2. Virtual link for the planar 3-DOF subarm.

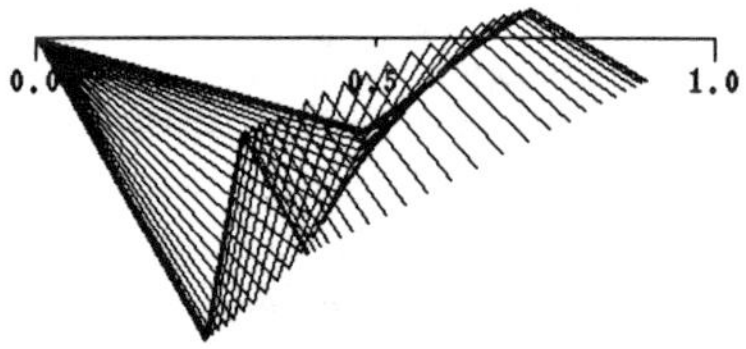

FIG. 3. Arm Motion for the JMDC

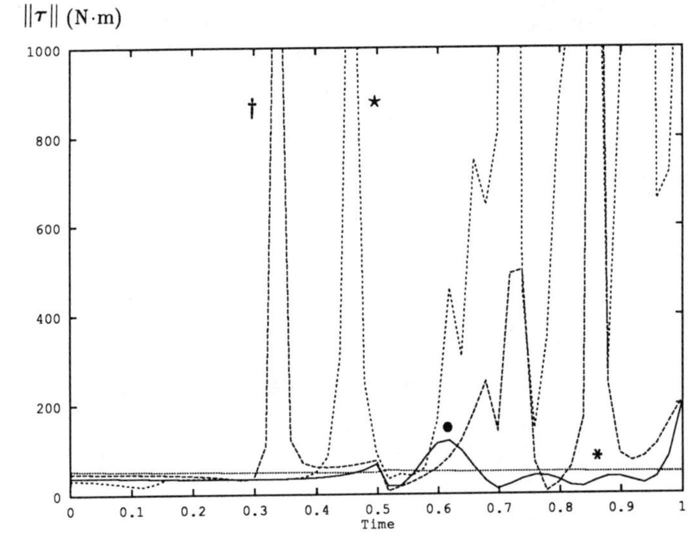

● : Inertia-Weighted Pseudoinverse Method
† : Null Space Vector Method
⋆ : Redundancy Decomposition Control Method
∗ : JMDC Method

FIG. 4. Profiles of torque norm $\|\tau\|$ for the four methods.

Generalized Velocities and Coordinates of Multiple–Arm Systems

Kenneth Kreutz-Delgado

University of California, San Diego

Abhi Jain

Jet Propulsion Laboratory/California Institute of Technology

Daryush Agahi

University of California, San Diego

Abstract

A multiple–arm robotic system will undergo a variety of discrete degree–of–freedom changes as it grasps different objects in different ways. A Spatial Operator Algebra of multiple body system dynamics is used to generate a minimal set of velocities "on the fly" via an $O(N)$ recursion as a multiple–arm system handles a variety of task objects over time. These velocities are either directly integrable to a minimal set of true generalized coordinates for the system, or can be mapped to a minimal set of integrable velocities.

1. Introduction

Multiple–arm (or finger) grasp of an object, or even simple single arm contact with the physical environment, results in a discontinuous change in the number of configuration degrees–of–freedom that describe allowable motions. A multi–armed robotic system will undergo a variety of such discrete changes as it handles different objects, in different ways, with each object put to a variety of different uses. Each combination of object–type, grasp–type, and object–use–type can correspond to a distinct, and different, number of degrees–of–freedom and set of true generalized coordinates.

Typically, the *unconstrained* configuration variables and their velocities are kept as the motion control variables (or integration variables when integrating the equations of motion for simulation purposes), even when closed–chain grasp or contact constraints are in effect. Generally, then, with respect to the unconstrained configuration variables and velocities a multiple–arm robotic system is overconstrained. This fact can result in numerical difficulties (since the unconstrained configuration variables are not typically independent and must evolve on some constraint surface) and unnecessary computational cost (since a larger set of configuration variables is being used than is needed and since numerically robust ODE–solution algorithms must typically be used).

This paper discusses algorithms, developed using a Spatial Operator Algebra of multibody system dynamics, that enable the generation of a minimal set of velocities "on the fly" as a multiple arm system handles a variety of objects in a variety of ways. These velocities are either directly integrable to a minimal set of true generalized coordinates for the system, or can be mapped to a minimal set of integrable velocities. As a result, a minimal parameterization of the configuration of the closed–chain system can be determined, even as the system changes its interconnection topology over time. With a set of true generalized velocities and coordinates at hand, the numerical instabilities associated with integrating overconstrained configuration variables can be avoided.

2. Generalized Velocities

The model of the combined multi–manipulator/task–object system used here is also used in the companion paper [4] and is discussed in greater detail in [5]. In the absence of grasp constraints and environmental constraints, the interconnection topologies of the primary and secondary systems are both given by tree structures. Henceforth, we will assume there are no environmental constraints on either system. In the absence of the grasp constraints, the tree–structured multi–manipulator system, the primary system, has joint generalized velocities β_P, while the tree–structured task–object, the secondary system, has internal configuration generalized velocities β_S. β_P and β_S are assumed to be physically meaningful velocities that are integrable or related to integrable quantities. For example, $\beta_P = \dot{\theta}$ for an all revoluted joint primary system. A grasp of the secondary system by the primary system occurs at a finite set of grasp contact points. The velocity of the contact points located on the primary system is denoted by $\hat{V}_P$, while the velocity of the contact points on the secondary system is denoted by $\hat{V}_S$. The Jacobian relationships between $\hat{V}_P$ and β_P, and $\hat{V}_S$ and β_S are given by ([4]–[6])

$$V_P = J_P\beta_P = B_P^*\phi_P^*H_P^*\beta_P$$

$$V_S = J_S\beta_S = B_S^*\phi_S^*H_S^*\beta_S$$

The kinematic constraints due to grasp constraint conditions at the grasp contact points are given by (see equations (1.2) and (1.4) of the companion paper [4])

$$A_P\hat{V}_P + A_S\hat{V}_S = 0 \tag{2.1}$$

which can be put into the form

$$\begin{pmatrix} A_P & A_S \end{pmatrix}\begin{pmatrix} J_P & 0 \\ 0 & J_S \end{pmatrix}\begin{pmatrix} \beta_P \\ \beta_S \end{pmatrix} = 0 \tag{2.2}$$

or

$$A\mathcal{J}\beta = 0 \tag{2.3}$$

where

$$A \triangleq \begin{pmatrix} A_P & A_S \end{pmatrix}, \quad \mathcal{J} \triangleq \begin{pmatrix} J_P & 0 \\ 0 & J_S \end{pmatrix}, \quad \text{and} \quad \beta \triangleq \begin{pmatrix} \beta_P \\ \beta_S \end{pmatrix}.$$

Because it is assumed that there are no environmental constraints acting on either system, $A_P = \hat{Q}_P$ and $A_S = \hat{Q}_S$ in terms of the notation used in the companion paper [4]. Note that with the addition of the grasp constraint conditions (2.3), the overall combined primary/secondary system is, in general, now a closed–chain system.

Constraint (2.3) is satisfied if and only if $\mathcal{J}\beta \in \operatorname{Ker} A$, or equivalently, if and only if

$$\mathcal{J}\beta = \tilde{A}\lambda, \quad \text{where} \quad \operatorname{Im}\tilde{A} = \operatorname{Ker} A \quad (\text{and thus } A\tilde{A} = 0) \tag{2.4}$$

for some $\lambda \in \mathcal{R}^{\nu_o}$ where $\nu_o = \dim \ker A$. The parameter λ is to be used as a partial set of generalized velocities that describe overall closed–chain system motions. In section 4, we will discuss a way to choose λ such that it is physically meaningful.

Note that once a value for λ has been determined, then the joint velocities β are at least partially specified. For example, if the Moore–Penrose pseudoinverse $\mathcal{J}^+$ is available for $\mathcal{J}$, we can compute the Moore–Penrose solution by $\beta^+ = \mathcal{J}^+\lambda$, yielding joint velocities, β^+, that will produce the desired velocity λ. At the block–component level, this gives

$$\mathcal{J}\beta = \tilde{A}\lambda \quad \Longrightarrow \begin{pmatrix} J_P & 0 \\ 0 & J_S \end{pmatrix} \begin{pmatrix} \beta_p \\ \beta_S \end{pmatrix} = \begin{pmatrix} \tilde{A}_P \\ \tilde{A}_S \end{pmatrix} \lambda$$

Thus

$$J_P\beta_P = \tilde{A}_P\lambda \quad \Longrightarrow \quad \beta_P^+ = J_P^+\tilde{A}_P\lambda \tag{2.5}$$

$$J_S\beta_S = \tilde{A}_S\lambda \quad \Longrightarrow \quad \beta_S^+ = J_S^+\tilde{A}_S\lambda \tag{2.6}$$

It is obvious from (2.5) and (2.6) that additional velocity–like elements $\xi_P \in \mathcal{R}^{\nu_p}$ and $\xi_S \in \mathcal{R}^{\nu_s}$ are provided by a minimal parametrization of the null–spaces of J_P and J_S,

$$\beta_P^o = \tilde{J}_P\xi_P \in \text{Ker}\, J_P \quad \text{and} \quad \beta_S^o = \tilde{J}_S\xi_S \in \text{Ker}\, J_S \tag{2.7}$$

where $\text{Im}\, \tilde{J}_P = \text{Ker}\, J_P$, $\text{Im}\, \tilde{J}_S = \text{Ker}\, J_S$, $\nu_p = \dim \text{Ker}\, J_P$, and $\nu_s = \dim \text{Ker}\, J_S$.

In principle, then, a minimal and complete set of velocities that describe the closed–chain primary/secondary system is given by

$$V_G = \text{col}(\lambda, \xi_P, \xi_S) \in \mathcal{R}^{\nu}, \quad \nu = \nu_o + \nu_p + \nu_s \ .$$

The question to be addressed is how to choose λ, ξ_P, and ξ_S such that V_G is a physically meaningful set of generalized velocities.

3. The Jacobian Null–Space Components

It is shown in [1] that the Moore–Penrose solutions (2.5) and (2.6) can be computed via an $O(N)$ link–to–link recursion, where N is the total number of links in the combined systems in this case. It is also shown in [1] how to obtain alternative solutions $\beta_P^g = J_P^g\tilde{A}_P\lambda$ and $\beta_S^g = J_S^g\tilde{A}_S\lambda$ via the use of a simpler $O(N)$ recursion, where, unlike the Moore–Penrose solutions β_P^+ and β_S^+, β_P^g and β_S^g may have components in the nullspaces of J_P and J_S respectively. In either case, it is therefore possible to compute joint velocities that will provide the desired velocity λ using an $O(N)$ algorithm that can adjust to real–time changes in the interconnection topologies of the primary or secondary systems.

The development of the algorithms for the generalized inverses J_P^g and J_S^g done in [1] also result in efficient $O(N)$ algorithms capable of computing null–space parametrizations ξ_P and ξ_S that have a simple physical interpretation. Specifically, the components of ξ_P give the values of the joint velocities for those joints that are in excess of those needed to position and orient the tips of the manipulators at the contact points [7]. Thus, integrating the velocity ξ_P gives the values of the joint variables for the excess (i.e., redundant) joints. Similar considerations hold for ξ_S. Having found ξ_P and ξ_S via the use of the $O(N)$ algorithms given in [1], an $O(N)$ algorithm for computing the null–space vectors β_P^o and β_S^o (see eq. (2.7)) is given by

$$\beta_P^o = (I - \mathcal{L}_P)^* C_P^* \xi_P, \quad \beta_S^o = (I - \mathcal{L}_S)^* C_S^* \xi_S \tag{3.1}$$

where (3.1) corresponds to an iterative algorithm also presented in [1].

4. A Task–Space Based Generalized Velocity

We will now consider an important special case where λ has physical meaning as the generalized velocities of the unconstrained secondary task–object system, $\lambda = \beta_S$.

Suppose A_P is invertible, then (2.1) yields

$$A_P J_P \beta_P + A_S J_S \beta_S = 0 \quad \Longrightarrow \quad J_P \beta_P + A_P^{-1} A_S J_S \beta_S = 0 \ . \tag{4.1}$$

Thus, with no loss of generality, we can assume that $A_P = I$ in this case. Hence

$$A = (I \ \ A_S), \quad \text{and} \quad \tilde{A} = \begin{pmatrix} -A_S \\ I \end{pmatrix}$$

Therefore, (2.3) and (2.4) yield

$$J_S \beta_S = J_S \lambda \quad \text{and} \quad J_P \beta_P = -A_S J_S \beta_S$$

and we have achieved $\lambda = \beta_S$ (assuming that $\operatorname{Ker} J_S = \{0\}$ for simplicity), which is physically meaningful. By assumption, β_s can be integrated to produce generalized coordinates for the held task–object.

Note that a necessary and sufficient condition for the secondary system to be controllable by the primary system, is that the equation

$$J_P \beta_P = -A_S J_S \beta_S \tag{4.2}$$

be consistent and that $\operatorname{Ker} A_S J_S = \{0\}$. These two conditions hold for all β_S if and only if

$$\operatorname{Im} A_S J_S \subset \operatorname{Im} J_P \quad \text{and} \quad \operatorname{Ker} A J_S = \{0\} \ . \tag{4.3}$$

A sufficient condition for controllability is obviously given by having no primary system manipulators being kinematically singular (i.e., J_P has full row rank) and $\operatorname{Ker} A = \{0\}$ (since $\operatorname{Ker} J_S = \{0\}$ by assumption).

5. Example

Consider the case of multiple–arm rigid grasp of a rigid body (see [2] or [3] for more details). Assuming single degree of freedom joints for simplicity, we have $\beta_P = \dot{\theta}_P$. Also $\beta_S = V_C = \operatorname{col}(\omega_c, v_c)$ gives the angular and linear velocity of the held object. Then

$$V_P = J_P \dot{\theta}_P, \quad V_S = V_C \quad \Longrightarrow \quad \beta = \begin{pmatrix} \dot{\theta}_P \\ V_C \end{pmatrix}$$

where C is an arbitrary reference point which is fixed with respect to the held object (e.g., the center of mass point).

The Jacobian and constraint matrix are respectively

$$\mathcal{J} = \begin{pmatrix} J_P & 0 \\ 0 & I \end{pmatrix}, \quad \text{and} \quad A = \begin{pmatrix} I & -\mathcal{A} \end{pmatrix}.$$

with

$$\mathcal{A}^T = \mathcal{A}^T(C) \triangleq [\phi(C, O_1), ..., \phi(C, O_l)] \in \mathcal{R}^{6 \times 6l}$$

assuming l rigid grasp contact points at tip positions O_1 through O_l [3].
We see that the special case considered in the previous section holds with $A_P = -\mathcal{A}$. Let

$$\tilde{A} = \begin{pmatrix} \mathcal{A} \\ I \end{pmatrix} \quad (A\tilde{A} = 0)$$

then

$$\mathcal{J}\beta = \tilde{A}\lambda$$

or

$$\begin{pmatrix} J_P & 0 \\ 0 & I \end{pmatrix} \begin{pmatrix} \dot{\theta}_P \\ v_C \end{pmatrix} = \begin{pmatrix} \mathcal{A} \\ I \end{pmatrix} \lambda$$

$$\Longrightarrow \quad \lambda = V_C \quad \text{and} \quad J_P\dot{\theta}_P = \mathcal{A}V_C \ .$$

λ is seen to be the physically meaningful spatial velocity of the held object. As described in section 3, the secondary generalized velocities are given by ξ, where a general primary system null velocity is of the form $\dot{\theta}^o = (I-\mathcal{L})^*C^*\xi$, ξ being the values of the excess joint velocities $\dot{\theta}'$, $\dot{\theta}' = \xi$. Thus, in this case, $V_G = \mathrm{col}(V_C, \dot{\theta}')$. Since $\mathrm{Ker}\,\mathcal{A} = \{0\}$, a sufficient condition for kinematic controllability of the held object is that no primary system manipulator be kinematically singular.

6. Conclusion

A general framework has been presented which allows physically meaningful and integrable generalized velocities to be computed for a closed–chain system composed of a primary multi–manipulator system grasping a secondary task–object system. These velocities, and joint space control velocities, can be computed using $O(N)$ recursive algorithms. Sufficient conditions for kinematic control of the held object by the primary multi–arm system using $O(N)$ recursive algorithms (to compute β^+, or β^g, and β^o given V_G) are also given.

7. Acknowledgements

The research described in this paper was performed in part at the Jet Propulsion Laboratory, California Institute of Technology, under contract with the National Aeronautics and Space Administration. K. Kreutz–Delgado and D. Agahi gratefully acknowledge the support of NSF PYI grant No. IRI–9057631.

References

[1] K. Kreutz-Delgado, D. Agahi, "An Operator Approach to Recursive Jacobian Generalized Inversion", preprint ECE Dept., UCSD, submitted to *IEEE Trans. Robotics and Automation*, March 1992.

[2] G. Rodriguez and K. Kreutz-Delgado, "Recursive Mass matrix Factorization and Inversion: an Operator Approach to Open and Closed-Chain Multibody Dynamics", JPL Publication 88-11, Jet Propulsion Laboratory, Pasadena, CA, 1988.

[3] K. Kreutz-Delgado and A. Lokshin, "Load balancing and Closed Chain Multiple Arm Control", in *1988 American Control Conference*, page 2148. The full version

of this report, with corrections, is available as JPL Engineering Memorandum 347-88-233, Jet Propulsion Laboratory, Pasadena, CA, 1988.

[4] K. Kreutz-Delgado, A. Jain, and D.Agahi, "Multiple Arm Manipulation of Loosely Held Objects", *Fourth International Symposium on Robotics and Manufacturing*, Nov 1992, Sante Fe, NM.

[5] A. Jain, K. Kreutz-Delgado, and G. Rodriguez, "Recursive Dynamics of Multiarm Robotic Systems in Loose Grasp of Articulated Task Objects", in *3rd International Symposium on Robotics and Manufacturing*, (Vancouver, Canada), June 1990.

[6] G. Rodriguez, A. Jain and K. Kreutz-Delgado, "Spatial Operator Algebra for Multibody Dynamics", *Journal of Astronautical Sciences*, 1992 Jan–Mar, Vol. 40 No. 1, pp. 27–50.

[7] K. Kreutz-Delgado, and D. Agahi, unpublished notes.

FORWARD KINEMATICS IN ANALYTICAL FORM OF THE TOPOLOGICALLY-SYMMETRIC 4-4 FULLY-PARALLEL MECHANISM

C. Innocenti
Dipartimento di Ingegneria delle costruzioni meccaniche
Facoltà di Ingegneria - Università di Bologna
Viale Risorgimento, 2 - 40136 Bologna - Italia

ABSTRACT

The paper presents the direct position analysis of the general-geometry 4-4 fully-parallel mechanism featuring two N-shaped sets of legs. The geometry of both base and platform is general. The analysis provides 16 closure configurations in the complex field. A numerical example is reported.

INTRODUCTION

The currently growing interest in the application of parallel manipulators to manufacturing environments demands an answer to a number of questions concerning, first of all, selection of the most suitable mechanical arrangement and the formulation of its kinematic model. While from a hardware standpoint it is not difficult to devise a series of different fully-parallel mechanisms, their kinematic characterization can be problematic, especially with regard to solution of the direct position analysis.

As is well known, the direct position analysis - also referred to as forward kinematics - aims to find all possible rigid body positions of the output controlled link (platform) with respect to the frame (base) for any given set of actuator displacements. For a fully-parallel mechanism this means finding all the closure configurations of the mechanism. Since the actual number of all allowable platform positions is of interest, analytical methods must be adopted in order to reduce the problem to the solution of an algebraic equation with one unknown. The degree of such equation directly provides the number of closures of the mechanism in the complex field.

This paper presents forward kinematics in analytical form of the fully-parallel mechanism shown in Fig. 1. The six variable-length legs, which provide the platform with six degrees of freedom relative to the base, are arranged in such a way as to form two N-shaped sets of three legs each (the mechanism itself will henceforth be referred to as N-N mechanism). The four connection points on both base and platform, centers of simple or double spherical pairs, make the manipulator fall within the category of the 4-4 fully-parallel mechanisms.

Other 4-4 fully-parallel schemes have been devised which differ in the arrangement of the legs [1,2]. However, the N-N arrangement is notable in that it features topological symmetry with respect to two section surfaces (see lines a-a and b-b in Fig. 1). The N-N arrangement was first presented in [1] where, in the special case of base and platform both planar, the analytical-form forward kinematics was solved, and 16 closures found. However, if an absolutely general geometry is considered - and in particular the four connection points on both base and platform are no longer constrained to lie on a plane - the kinematic model adopted in [1] is no longer applicable, and a more general one is needed.

In this paper the kinematic model of the general-geometry N-N mechanism is first introduced, allowing five closure equations with four unknowns to be written. By a specifically-developed elimination procedure, a final algebraic equation of degree 16 is obtained and, through linear back substitution, 16 closure configurations are determined in the complex field.

It is worth noting that relaxation of the geometric constraints required in [1] do not increase the number of possible closure configurations. However, the proposed procedure is well-suited also for the N-N mechanism considered in [1], which is only a geometric-peculiar (as opposed to structurally-peculiar) N-N arrangement.

A numerical example is reported that confirms the new theoretical result.

Finally, the equivalence between the forward kinematics of the N-N manipulator and the displacement analysis of well-known classes of spatial mechanisms is shown. Accordingly, the proposed procedure represents an alternative analytical approach to the displacement analysis of other classes of open and closed kinematic chains.

KINEMATIC MODEL

As far as the direct position analysis is concerned, every actuated kinematic pair can be thought of as frozen. Hence the N-N mechanism becomes a structure, and the analysis reduces to finding all assembly configurations of the structure.

Figure 2 represents the structure, called N-N structure, deriving from the N-N mechanism of Fig. 1 when the leg lengths are frozen. It is worth noting that the kinematic model represented by the N-N structure does not apply to the forward kinematics of the N-N mechanism only, but also to every mechanism that, when all inputs are frozen, reduces to the N-N structure.

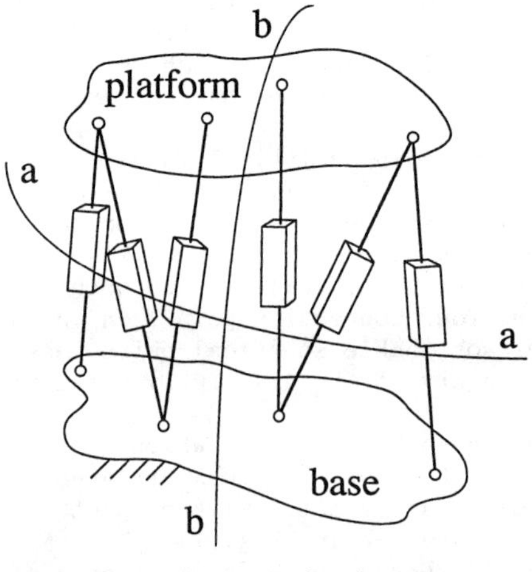

FIG. 1. The N-N mechanisms.

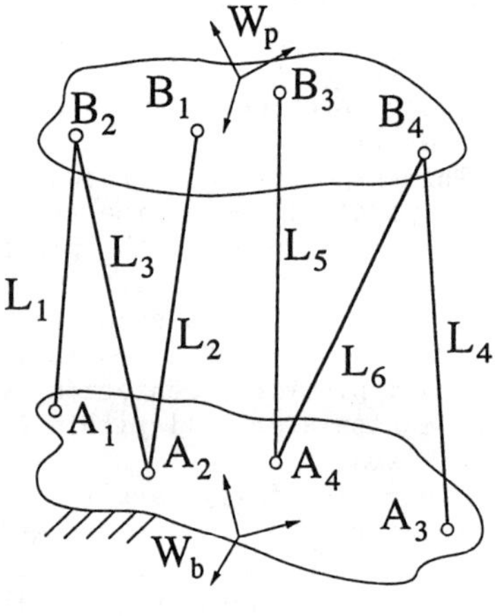

FIG. 2. The N-N structure.

With reference to Fig. 2, let W_b and W_p be two arbitrary reference systems fixed to base and platform respectively. The geometry of the N-N structure is described by the coordinates in W_b of points A_j, $j=1,..,4$, the coordinates in W_p of points B_j, $j=1,..,4$, and the leg lengths L_j, $j=1,..,6$.

Structure disassembling

In order to parametrize all possible closure configurations of the N-N structure, the structure itself is disassembled as shown in Fig. 3, where both base and platform are allowed to carry two V-shaped groups of two legs each.

Due to legs $A_1B_2^*$ and $A_2B_2^*$, point B_2^* can describe a circular trajectory fixed to W_b with center Q_1 and radius H_1. Similarly, point B_4^* is confined to a circle fixed to W_b with center in Q_2 and radius H_2. Moreover, points A_2^* and A_4^* must lie on circles fixed to W_p, with centers in Q_3 and Q_4, and radii H_3 and H_4 respectively.

Position of points B_2^* and B_4^* in W_b, and points A_2^* and A_4^* in W_p, can be expressed through:

$$(B_2^*-Q_1) = H_1\cdot(C_1\cdot\underline{a}_1 + S_1\cdot\underline{b}_1) \tag{1}$$
$$(B_4^*-Q_2) = H_2\cdot(C_2\cdot\underline{a}_2 + S_2\cdot\underline{b}_2) \tag{2}$$
$$(A_2^*-Q_3) = H_3\cdot(C_3\cdot\underline{a}_3 + S_3\cdot\underline{b}_3) \tag{3}$$
$$(A_4^*-Q_4) = H_4\cdot(C_4\cdot\underline{a}_4 + S_4\cdot\underline{b}_4) \tag{4}$$

where $C_j=\cos(\theta_j)$ and $S_j=\sin(\theta_j)$, $j=1,..,4$; $\underline{a}_j$ and $\underline{b}_j$ are mutually orthogonal unit vectors parallel to the plane of the circular trajectory involved, and θ_j is the angular position of the moving point relative to $\underline{a}_j$.

The key phase of the disassembling procedure lies in considering the two groups of points $G_b=\{A_2, B_2^*, A_4, B_4^*\}$ and $G_p=\{A_2^*, B_2, A_4^*, B_4\}$

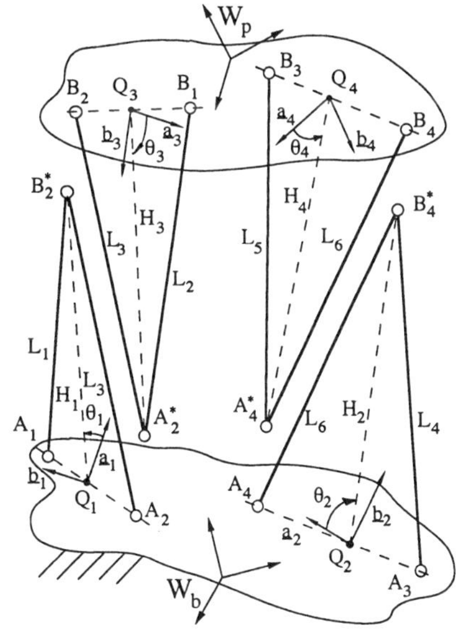

FIG. 3. The disassembled structure.

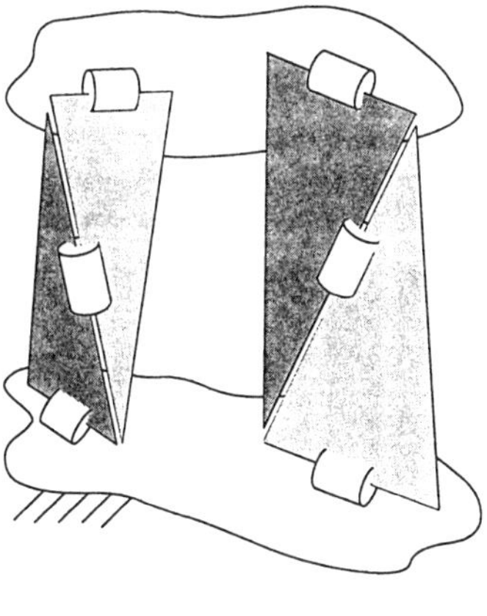

FIG. 4. The 6R mechanism.

as the vertices of two tetrahedrons. If a set of values θ_j, j=1,..,4, is found such that tetrahedrons G_b and G_p are congruent, the lower and upper subchains of Fig. 3 can be joined together by superimposition of point A_2 on A_2^*, A_4 on A_4^*, B_2 on B_2^*, and B_4 on B_4^*. Hence a closure configuration of the N-N structure results.

Concluding, the finding of the closure configurations of the N-N structure is equivalent to determining the set of values for θ_j, j=1,..,4, such that tetrahedrons G_b and G_p can superimpose.

Closure Equations

Two congruent tetrahedrons have corresponding edges of equal length. Since edges $A_2B_2^*$ and $A_4B_4^*$ of tetrahedron G_b are respectively equal to edges $A_2^*B_2$ and $A_4^*B_4$ of tetrahedron G_p, only four significant equations can be written:

$$(B_4^*-B_2^*)^2 = (B_4-B_2)^2 \tag{5}$$
$$(A_4-A_2)^2 = (A_4^*-A_2^*)^2 \tag{6}$$
$$(B_2^*-A_4)^2 = (B_2-A_4^*)^2 \tag{7}$$
$$(B_4^*-A_2)^2 = (B_4-A_2^*)^2 \tag{8}$$

However, two sets of four points might not superimpose if only conditions on mutual distances are satisfied: indeed no distinction would be possible - by equating only edge lengths - between a tetrahedron and its reflection with respect to one of its faces. Thus a further closure equation is needed:

$$[(B_2^*-A_2)\times(B_4^*-A_4)]\cdot(A_4-A_2) = [(B_2-A_2^*)\times(B_4-A_4^*)]\cdot(B_4-B_2) \tag{9}$$

which specifies that point B_4^* lies, with respect to the plane through $A_2B_2^*A_4$, on the same side as point B_4 with respect to plane $A_2^*B_2A_4^*$.

The equation set (5)-(9) represents all and only the conditions for the N-N structure to be assembled. By substituting in (5)-(9) the expressions (1)-(4), the following relations can be obtained:

$$\sum_{i,j=0,2} e_{ij}\cdot C_1^{u(i)}\cdot S_1^{v(i)}\cdot C_2^{u(j)}\cdot S_2^{v(j)} = 0 \tag{10}$$

$$\sum_{i,j=0,2} f_{ij}\cdot C_3^{u(i)}\cdot S_3^{v(i)}\cdot C_4^{u(j)}\cdot S_4^{v(j)} = 0 \tag{11}$$

$$m_0 + m_1\cdot S_1 + m_2\cdot C_1 + m_3\cdot S_4 + m_4\cdot C_4 = 0 \tag{12}$$

$$n_0 + n_1\cdot S_2 + n_2\cdot C_2 + n_3\cdot S_3 + n_4\cdot C_3 = 0 \tag{13}$$

$$\sum_{\substack{i,j=0,2\\ i+j\neq 0}} p_{ij}\cdot C_1^{u(i)}\cdot S_1^{v(i)}\cdot C_2^{u(j)}\cdot S_2^{v(j)} + \sum_{i,j=0,2} q_{ij}\cdot C_3^{u(i)}\cdot S_3^{v(i)}\cdot C_4^{u(j)}\cdot S_4^{v(j)} = 0 \tag{14}$$

where coefficients e_{ij}, f_{ij}, m_j, n_j, p_{ij}, and q_{ij} depend only on the geometry of the N-N structure. Moreover, the integer functions u(k) and v(k) that appear in (10), (11) and (14) are defined as follows:

$$u(k) = [k-\mathrm{mod}(k,2)]/2; \qquad v(k) = \mathrm{mod}(k,2) \tag{15}$$

where mod(k,2) is the remainder of the division of integer k by 2.

SOLUTION OF THE EQUATION SET

Equations (10) and (13) can be linearly solved to provide S_2 and C_2. Indeed they can be put in the form:

$$\begin{bmatrix} e_{01}+e_{11}\cdot S_1+e_{21}\cdot C_1 & e_{02}+e_{12}\cdot S_1+e_{22}\cdot C_1 \\ n_1 & n_2 \end{bmatrix} \begin{bmatrix} S_2 \\ C_2 \end{bmatrix} = \begin{bmatrix} -e_{00}-e_{10}\cdot S_1-e_{20}\cdot C_1 \\ -n_0-n_3\cdot S_3-n_4\cdot C_3 \end{bmatrix} \quad (16)$$

and, by Cramer's rule, the following expressions are obtained:

$$S_2 = \alpha_1/\alpha_0; \qquad C_2 = \alpha_2/\alpha_0 \quad (17)$$

where:

$$\alpha_0 = \sum_{i=0,2} a_{i0}\cdot C_1^{u(i)}\cdot S_1^{v(i)}; \qquad \alpha_1 = \sum_{i,j=0,2} a_{ij1}\cdot C_1^{u(i)}\cdot S_1^{v(i)}\cdot C_3^{u(j)}\cdot S_3^{v(j)}$$

$$\alpha_2 = \sum_{i,j=0,2} a_{ij2}\cdot C_1^{u(i)}\cdot S_1^{v(i)}\cdot C_3^{u(j)}\cdot S_3^{v(j)} \quad (18)$$

In relations (18), the coefficients a_{i0}, a_{ij1}, and a_{ij2} (i,j=0,..,2) are constant quantities.

Now the congruence condition $S_2^2+C_2^2=1$ is imposed. After rationalization has been performed, from (17) and (18) it follows that

$$\sum_{i,j=0,4} E_{ij}\cdot C_1^{u(i)}\cdot S_1^{v(i)}\cdot C_3^{u(j)}\cdot S_3^{v(j)} = 0 \quad (19)$$

If expressions (17) for S_2 and C_2 are substituted in (14), equation (19) can replace equations (10) and (13) in the closure equation set, which will come out as composed by equations (19), (11), (12), and (14), and does not contain the unknown θ_2.

The elimination of θ_4 can be accomplished in a similar way. From equations (11) and (12), S_4 and C_4 are obtained as:

$$S_4 = \beta_1/\beta_0; \qquad C_4 = \beta_2/\beta_0 \quad (20)$$

where:

$$\beta_0 = \sum_{i=0,2} b_{i0}\cdot C_3^{u(i)}\cdot S_3^{v(i)}; \qquad \beta_1 = \sum_{i,j=0,2} b_{ij1}\cdot C_1^{u(i)}\cdot S_1^{v(i)}\cdot C_3^{u(j)}\cdot S_3^{v(j)}$$

$$\beta_2 = \sum_{i,j=0,2} b_{ij2}\cdot C_1^{u(i)}\cdot S_1^{v(i)}\cdot C_3^{u(j)}\cdot S_3^{v(j)} \quad (21)$$

Then condition $S_4^2+C_4^2=1$ leads to:

$$\sum_{i,j=0,4} F_{ij}\cdot C_1^{u(i)}\cdot S_1^{v(i)}\cdot C_3^{u(j)}\cdot S_3^{v(j)} = 0 \quad (22)$$

Moreover, by substituting (20) into (14), the following equation is obtained:

$$\sum_{i,j=0,4} G_{ij} \cdot C_1^{u(i)} \cdot S_1^{v(i)} \cdot C_3^{u(j)} \cdot S_3^{v(j)} \quad (23)$$

Equations (19), (22), and (23) are the result of eliminating θ_2 and θ_4 from the original set of five closure equations.

Elimination of θ_3

If the well-known identities:

$$S_j = 2 \cdot t_j/(1+t_j^2); \qquad C_j = (1-t_j^2)/(1+t_j^2) \qquad (j=1,3) \quad (24)$$

are substituted in (19), (22) and (23), and rationalization is performed, the three-equation closure set can be put in the form:

$$\sum_{j=0,4} L_{ji} \cdot t_3^j = 0 \qquad (i=1,2,3) \quad (25)$$

where terms L_{ij} ($j=0,..,4$; $i=1,2,3$) are fourth-order polynomials in t_1. Let us now consider the following 8x7 matrix:

$$D = \begin{bmatrix} L_{01} & L_{11} & L_{21} & L_{31} & L_{41} & 0 & 0 \\ 0 & L_{01} & L_{11} & L_{21} & L_{31} & L_{41} & 0 \\ 0 & 0 & L_{01} & L_{11} & L_{21} & L_{31} & L_{41} \\ L_{02} & L_{12} & L_{22} & L_{32} & L_{42} & 0 & 0 \\ 0 & L_{02} & L_{12} & L_{22} & L_{32} & L_{42} & 0 \\ 0 & 0 & L_{02} & L_{12} & L_{22} & L_{32} & L_{42} \\ L_{03} & L_{13} & L_{23} & L_{33} & L_{43} & 0 & 0 \\ 0 & L_{03} & L_{13} & L_{23} & L_{33} & L_{43} & 0 \end{bmatrix} \quad (26)$$

and its 7x7 minors D_1 and D_2 which are obtained by suppressing the 7th and 8th rows respectively. It can be proved that equations (25) are satisfied for the same value of t_3 if and only if the following conditions simultaneously hold:

$$\det(D_1) = 0; \qquad \det(D_2) = 0 \quad (27)$$

Clearly, the common roots of equations (27) are all and only the roots of the G.C.D. of their left hand sides. The left hand sides of equations (27) are 28th-order polynomial in t_1, while their G.C.D. has proved to be a 16th-order polynomial in t_1. Hence the following algebraic closure equation results:

$$\sum_{j=0,16} h_j \cdot t_1^j = 0 \quad (28)$$

which provides 16 values for t_1 in the complex field.

Back substitution

Let t_{1j} ($j=1,..,16$) be the generic root of equation (28). For $t_1=t_{1j}$ ($j=1,..,16$) the first two equations of set (25) admit a first-order G.C.D. whose root provides the value t_{3j} for t_3. Finally, from relations (17) and (20) via (18) and (21), the values θ_{2j} and θ_{4j} for θ_2 and θ_4 can be obtained. Hence 16 closure configurations of the N-N structure are possible in the complex field.

TABLE I. Closure configurations in terms of (x,y,z) coordinates in reference system W_b of points B_j, $j=1,..,4$

x	y	z
configuration 1		
(1.608763E+01, 0.000000E+00)	(-1.852584E+00, 0.000000E+00)	(-1.843367E+00, 0.000000E+00)
(1.726984E+01, 0.000000E+00)	(1.382764E+00, 0.000000E+00)	(3.912926E+00, 0.000000E+00)
(1.974174E+01, 0.000000E+00)	(-8.492805E+00, 0.000000E+00)	(-4.768250E+00, 0.000000E+00)
(8.239902E+00, 0.000000E+00)	(9.372715E-01, 0.000000E+00)	(1.848502E+00, 0.000000E+00)
configuration 2		
(1.067692E+01, 0.000000E+00)	(-1.385060E+00, 0.000000E+00)	(8.868608E+00, 0.000000E+00)
(1.609930E+01, 0.000000E+00)	(1.138127E-01, 0.000000E+00)	(5.214681E+00, 0.000000E+00)
(9.111545E+00, 0.000000E+00)	(4.852607E-01, 0.000000E+00)	(1.661790E+01, 0.000000E+00)
(8.222179E+00, 0.000000E+00)	(7.783908E-01, 0.000000E+00)	(3.660347E-01, 0.000000E+00)
configuration 3		
(1.372346E+01, 0.000000E+00)	(-2.264431E+00, 0.000000E+00)	(5.913042E+00, 0.000000E+00)
(7.390128E+00, 0.000000E+00)	(-3.788609E+00, 0.000000E+00)	(7.514855E+00, 0.000000E+00)
(2.044209E+01, 0.000000E+00)	(-6.690478E+00, 0.000000E+00)	(7.040030E+00, 0.000000E+00)
(8.065935E+00, 0.000000E+00)	(5.845494E-01, 0.000000E+00)	(-6.349159E-01, 0.000000E+00)
configuration 4		
(9.642472E-01, 0.000000E+00)	(1.000152E+00, 0.000000E+00)	(8.972837E+00, 0.000000E+00)
(4.975394E+00, 0.000000E+00)	(-3.996742E+00, 0.000000E+00)	(6.987454E+00, 0.000000E+00)
(1.958589E+00, 0.000000E+00)	(9.000626E+00, 0.000000E+00)	(7.970982E+00, 0.000000E+00)
(-4.019515E+00, 0.000000E+00)	(-6.006980E+00, 0.000000E+00)	(5.962486E+00, 0.000000E+00)
configuration 5		
(-3.911487E+00, 0.000000E+00)	(1.425535E+00, 0.000000E+00)	(-1.050506E+00, 0.000000E+00)
(-2.349299E+00, 0.000000E+00)	(4.511394E+00, 0.000000E+00)	(-6.798292E+00, 0.000000E+00)
(-1.101665E+01, 0.000000E+00)	(1.339793E+00, 0.000000E+00)	(2.887673E+00, 0.000000E+00)
(1.531839E+00, 0.000000E+00)	(-3.879708E+00, 0.000000E+00)	(-6.072915E+00, 0.000000E+00)
configuration 6		
(-3.406220E+00, 0.000000E+00)	(9.251524E-01, 0.000000E+00)	(2.836743E+00, 0.000000E+00)
(1.070730E+00, 0.000000E+00)	(-3.540067E+00, 0.000000E+00)	(5.076999E+00, 0.000000E+00)
(-2.489106E+00, 0.000000E+00)	(8.927391E+00, 0.000000E+00)	(1.776986E+00, 0.000000E+00)
(-4.138021E+00, 0.000000E+00)	(-6.858423E+00, 0.000000E+00)	(-1.840908E+00, 0.000000E+00)
configuration 7		
(1.093958E+00, 0.000000E+00)	(1.069512E+00, 0.000000E+00)	(9.089050E+00, 0.000000E+00)
(-2.831550E-01, 0.000000E+00)	(-3.058947E+00, 0.000000E+00)	(3.984210E+00, 0.000000E+00)
(-5.768888E-01, 0.000000E+00)	(1.044392E-01, 0.000000E+00)	(1.698062E+01, 0.000000E+00)
(8.040719E+00, 0.000000E+00)	(9.525048E-01, 0.000000E+00)	(3.195934E+00, 0.000000E+00)
configuration 8		
(1.568153E+00, 0.000000E+00)	(3.706291E-01, 0.000000E+00)	(-9.055654E+00, 0.000000E+00)
(-3.175528E+00, 0.000000E+00)	(3.075776E+00, 0.000000E+00)	(-5.159544E+00, 0.000000E+00)
(8.925859E+00, 0.000000E+00)	(3.304164E+00, 0.000000E+00)	(-1.086080E+01, 0.000000E+00)
(-2.521441E+00, 0.000000E+00)	(-6.118335E+00, 0.000000E+00)	(-4.139501E+00, 0.000000E+00)
configurations 9 and 10		
(-4.056042E+00, 4.530990E+00)	(-1.628426E+00,-2.391371E+00)	(-5.561472E+00,-6.157323E+00)
(-1.626067E-01,-3.176660E+00)	(-3.737845E+00, 1.145171E+00)	(4.929591E+00,-2.585781E+00)
(-1.708110E+01, 7.612648E+00)	(2.108477E+00,-5.448956E+00)	(-1.025334E+01,-1.714754E+01)
(8.472479E+00, 5.894656E-01)	(8.118243E-01, 6.209365E-01)	(-8.014292E-01, 2.672571E+00)
configurations 11 and 12		
(-3.898381E+00, 4.980967E-01)	(7.813074E-01,-4.792217E-01)	(1.475541E+00, 2.308820E+00)
(-3.500045E+00,-6.010743E-01)	(2.925492E+00,-1.564861E+00)	(-5.067338E+00, 1.886123E+00)
(-8.617747E+00, 3.048053E+00)	(3.914110E+00, 1.780197E+00)	(8.261828E+00, 3.039096E+00)
(-3.517661E+00,-3.118726E+00)	(-6.793018E+00,-1.536492E+00)	(-4.909018E+00, 3.347437E+00)
configurations 13 and 14		
(-6.644785E+01,-5.686988E+01)	(2.473570E+01, 2.373291E+00)	(5.645220E+01,-7.482098E+01)
(1.634090E+00,-4.708228E+00)	(-4.550205E+00, 8.106169E-01)	(6.611636E+00,-2.650232E+00)
(-1.653177E+02,-7.232936E+01)	(6.115523E+01,-1.397973E+01)	(6.550860E+01,-1.778318E+02)
(-1.157150E+01,-1.665248E+00)	(-1.067306E+01,-2.316921E+00)	(4.613547E+00,-1.317772E+01)
configurations 15 and 16		
(1.827234E+01, 7.170257E-01)	(-2.169805E+00, 1.287640E+00)	(-1.745193E-01,-6.587047E+00)
(1.894774E+01,-3.572407E-01)	(5.618599E+00, 1.888485E+00)	(-1.175553E+00,-2.637060E+00)
(2.314578E+01, 9.957343E-01)	(-7.592532E+00,-1.364595E-01)	(4.211375E+00,-8.657496E+00)
(9.873087E+00,-5.188505E-02)	(1.833769E+00,-5.980503E-01)	(1.014363E+00,-5.669192E+00)

NUMERICAL EXAMPLE

The proposed procedure has been applied to an N-N structure whose dimensions, in arbitrary length unit, are given in the following. With reference to Fig. 2, the coordinates in W_b of points A_j, $j=1,..,4$, are: $A_1=(8,1,-3)$, $A_2=(7,5,0)$, $A_3=(-2,3,1)$, $A_4=(4,-7,2)$. The coordinates in W_p of points B_j, $j=1,..,4$, are: $B_1=(1,1,9)$, $B_2=(5,-4,7)$, $B_3=(2,9,8)$, $B_4=(-4,-6,6)$. The leg lengths are:

L_1=11.57, L_2=11.53, L_3=11.57, L_4=10.48, L_5=17.2, L_6=9. Sixteen closures are obtained, eight of which are real, the remaining complex. All closures are reported in Table I in terms of coordinates in W_b of points B_j, j=1,..,4. Only one closure is reported out of every couple of complex conjugate ones. All closure configurations have been successfully verified by inverse position analysis.

REMARKS

With reference to Fig. 2, it is worth noting that each couple of legs sharing one extremity can be regarded as a rigid triangle. Accordingly, four triangles can be spotted that are hinged both to one another and to the base or platform (see Fig. 4). Hence the N-N structure is equivalent, from a kinematic point of view, to the single-loop 6R structure represented in Fig. 4.

It is well known that the search for all closures of a 6R structure is equivalent to either performing the displacement analysis of a 7R single-loop spatial mechanism, or the inverse displacement analysis of a 6R serial manipulator. Both issues admit 16 solutions [3,4], which agrees with the result found here for the N-N structure.

The algorithm presented in this paper, which requires a reduced number of closure equations if compared to existing procedures [3,4], can be efficiently employed to analyze a rather general variety of 7R single loop mechanisms and 6R serial chains.

CONCLUSIONS

In this paper a method for solving the forward kinematics in analytical form of a type of 4-4 fully-parallel mechanism, has been reported. Unlike available results in the literature, base and platform are not restricted to being planar. By an original procedure that eliminates three unwanted unknowns from a five-equation set, the method results in a 16th-order algebraic equation with one unknown. Hence 16 closure configurations are possible in the complex field. A numerical example has confirmed the new result.

Acknowledgments

The financial support of MURST by funds 40%-60% is acknowledged.

REFERENCES

1. W. Lin, J. Duffy, M. Griffis, "Forward Displacement Analyses of the 4-4 Stewart Platforms", Proc. of the 21st Biennial Mechanisms Conference, Chicago, IL, Sept. 16-19, 1990, pp. 263-269.
2. Innocenti C., Parenti-Castelli V., "Analytical Form Solution of the Direct Kinematics of a 4-4 Fully In-Parallel Actuated Six Degree-of-Freedom Mechanism", Proc. of the Ro.man.sy '92, Udine, Italy, Sept. 1-4, 1992.
3. H.J. Lee, C.G. Liang, "Displacement Analysis of the General Spatial 7-Link 7R Mechanism", Mechanism and Machine Theory, Vol. 23, No. 3, 1988, pp. 219-226.
4. M. Raghavan, B. Roth, "Inverse Kinematics of the General 6R Manipulator and Related Linkages", Proc. of the 21st Biennial Mechanisms Conference, Chicago, IL, Sept. 16-19, 1990, pp.59-65.

A Stewart Platform as an Adaptive Structure --- Its Application to Active Vibration Control

Zheng Geng and **Leonard S. Haynes**
Intelligent Automation, Inc., 1370 Piccard Drive, Rockville, MD 20850

James D. Lee
Department of Mechanical Engineering, George Washington University, Washington, D.C. 20052

ABSTRACT
A study is performed of a six degree-of-freedom active vibration isolation system based on a Stewart platform mechanism to be used for precision control of a wide range of space-based structures as well as earth-base systems. This paper presents a part of the study results, which includes a new Terfenol-D actuator design and analysis, a design of Stewart platform as a vibration isolation device, and a local force feedback algorithms for active vibration control and simulation results.

§1 Introduction

Future space structure will be made of lightweight and lightly damped materials. Typically, truss-type structures are considered for most space applications since they are efficient and amendable to in-space assembly or development. However, to meet the stringent accuracy and performance requirements, the flexible structural vibrations inherent to the light weighted truss will have to be "damped" or controlled either actively or passively. A typical passive damping technique, which is effective in damping out higher frequency vibration, usually adopts viscoelastic materials. The better performance with less weight can be expected by using active controls.

The requirements for vibration control in space applications can be classified into two levels: vibration isolation at a component level and structural vibration suppression at a system level. In component level isolation, a vibration isolation and articulation device is to provide articulation while reducing the vibration transmission between the vibration source and the desired vibration-free parts of a system, from which the vibration disturbance should be minimized. In a system level vibration control, the active vibration control devices or active elements are to exert active control force such that the structure vibration modes are suppressed in a specific frequency range. For simplicity reason, both types of applications mentioned above are refereed as active vibration control(AVC) in this paper.

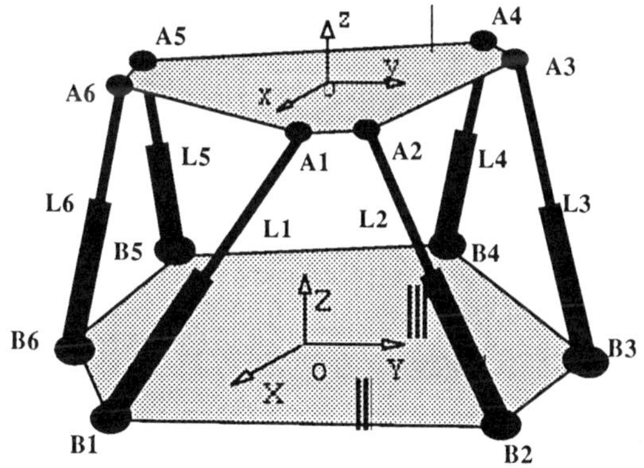

Figure 1. A Stewart Platform

To achieve active vibration control, adaptive structure concept was proposed[5]. An adaptive structure is defined as a structural system whose geometrical configuration and inherent structural characteristics can be changed in order to adapt to environmental changes or to meet the mission requirements. An adaptive structure can be formed through the use of electronically controlled active elements consisting of actuators and sensors, so they function as normal load carrying members and simultaneously provide structural adaptation capabilities.

In a generic scenario of active vibration isolation applications, the vibration in a structure should be characterized as multiple multiple degree of freedom(DOF) phenomenon. For instance, it may have both translation components (along X, Y, Z axes) as well as rotational components (Roll, Pitch, Yaw). Therefore the issue of how to arrange active elements of an adaptive structure in an "optimal" manner to perform multiple DOF vibration isolation tasks needed to be addressed, both at system level and at component level. Due to the complexity of the multiple DOF system dynamics and control design, very few research activities of multiple DOF active vibration isolation systems have been reported.

We have adopted the concept of Stewart platform to deal with multiple DOF vibration isolation problems of precision control for space-based structures as well as earth-base systems. A typical Stewart platform (Figure 1) consists of six variable-length actuators connecting a mobile plate to a base plate. As the lengths of actuators change, the mobile plate of the platform is able to move in all six DOF with respect to the base[1][2]. The 6-DOF positioning capability of a Stewart platform provides a germane mechanism for handling multiple DOF vibration control problems.

Stewart platforms have numerous features which make them particularly attractive for active vibration control. Stewart platform systems use the minimum number of linear actuators possible to provide 6 DOF motion control, which is widely used for flight simulators and 6 DOF robotic manipulators. One of the major advantages of Stewart platforms for space applications is the inherent capability for providing articulation between subsystems as well as vibration isolation with the same mechanical system. Consequently there are significant savings in system complexity and weight, as well as a reduction in the requirements for associated modeling efforts. Stewart platforms can be designed to carry large loads and remain stable in the unpowered configuration. These features are useful in solving problems related to testing a space structure in a gravity environment. The other important property of a Stewart platform for vibration control applications is that, neglecting the gravity and inertial load of actuators, all the forces transmitted between the mobile plate and the base plate of a Stewart platform must be totally axial forces in each of the six actuators. This means that if the axial forces can be measured and eliminated, then all the forces and hence all the vibration can be eliminated. This is certainly not true of a conventional truss structure where the forces at the connection nodes include components in all six DOF.

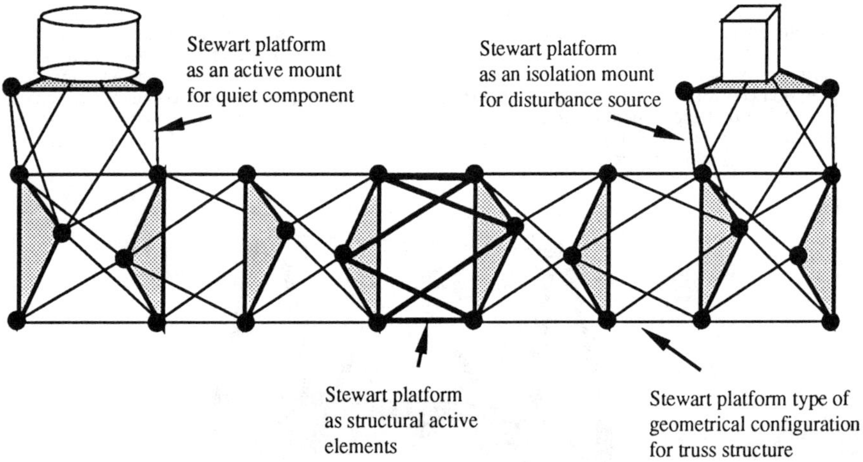

Figure 2 Stewart Truss Structure

We have proposed a concept of "Stewart truss structure" (Figure 2) as a candidate for active member arrangement. In a Stewart truss structure, each bay of the truss is configured as a Stewart platform. If all six elements on one bay of the truss is replaced by active element equipped with sensor and actuator, this bay of truss become an actively controlled Stewart platform, which can be used to prevent the structural vibration generated by the structure connected to one side of the platform from propagating to the structure connected on the other side of the platform. At a component level, a Stewart platform can be controlled as an active mount for quiet components and as an isolation device for disturbance sources. This paper studies the design and control problems of a class of multiple DOF vibration isolation system based on the concept of Stewart truss and the Stewart platform mechanism.

The vibration control application of a Stewart platform is quite different from those of using Stewart platform as a flight simulator[1] or a multiple DOF parallel link manipulator[2][3]. The stroke required for damping structural vibration is of the order of magnitude of tens of micron, and the frequency response performance should achieve the range of kHz. Force requirements for vibration control devices vary from applications. For the application we aimed at, the force generation capability of each actuator is within the range of hundreds of pound. In the past, the actuators used in building Stewart platforms have been hydraulic, electrical, or pneumatic actuators, which usually have low bandwidth. Consequently, their response times were too slow for adequate vibration isolation at higher frequencies. In section 2, we will discuss our efforts to build magnetostrictive linear actuators using Terfenol-D material. The mechanical designs and modelling of the actuator are detailed. Our 6-DOF active vibration damping system is built based on the Stewart platform mechanism, which incorporates six Terfenol-D actuators. Section 3 is devoted to local force feedback algorithm for active vibration control. Simulation results are given in section 4.

§ 2 Design and Implementation of Long Stroke, High Force and Wide Bandwidth Terfenol-D Linear Actuators

In order to implement active vibration control, effective actuation techniques should be studied. Due to the special requirements of space applications, the vibration control actuators should be simple, reliable, and have a high strength-to-weight ratio as well as a large amplitude of motion. Existing electromagnetic, pneumatic, or hydraulic actuators and their associated driving and supporting structures are heavy or bulky compare to the amount of work to be performed. While incremental improvements are continually made to such actuators, the technologies themselves are mature, and it is not realistic to expect significant improvements which depend upon the same basic electromagnetic, pneumatic, or hydraulic principles.

Consequently, researchers have attempted to develop improved actuators by exploiting new materials. Such materials include shape memory alloys and polymers, piezoelectric ceramics and polymers, magnetostrictive materials, and various mechanochemical gels and polymers.

Piezoelectric actuators have attracted considerable interest in the active vibration control community, because they are simple, reliable, compact and light weight. Piezoelectric actuators have a very broad useful bandwidth, and their power consumption is not large. Their mechanical properties can be tailored to particular applications. However the low stroke capability of these materials is a concern.

A promising actuator employs the magnetostrictive alloy Terfenol-D. We developed a new form of a Terfenol-D actuator and have fabricated and used six of the designed actuators in our study for active vibration control. Terfenol-D can achieve accurate linear or oscillatory motion in response to a magnetic field generated by a low voltage electric current. In addition to achieving the largest strain and stroke of any commercially available material, Terfenol-D driven actuators offer relatively high force and microsecond response times.

2.1. Terfenol-D Actuator

The magnetostrictive properties of nickel and nickel alloys were discovered in the 1840's. In the 1970's, researchers at the Naval Surface Weapons Center developed magnetostrictive alloys of terbium(Tb), dysprosium(Dy), and iron(Fe), which exhibited strains 50 times as large as nickel alloys and 10 times as large as piezoelectric ceramics. They named this material Terfenol-D. The Young's Modules of Terfenol-D is moderately high ($2.5 - 3.5 \times 10^{10}$ N/m^2), which allows for high force generation and the high energy density, which makes Terfenol-D devices quite compact[4].

Several versions of Terfenol-D actuators have been built by a few companies, but the maximum linear displacement has been limited to ± 63.6 microns (±2.5 thousandths of an inch). We decided, at the outset of this effort, to attempt to build a Terfenol-D actuator which had greater displacement. Our goal was ± 127 microns (± 5 thousandths of an inch). Another goal was to implement an actuator which had both a flux monitoring capability and a direct strain measuring capability. This was not available from any known commercial source of Terfenol actuators. Our Terfenol-D actuator incorporates an instrumented Terfenol-D rod, an excitation coil to provide the magnetic driving field, a permanent magnet assembly to provide a magnetic bias field, and a mechanical preload mechanism.

2.2. Instrumented Terfenol Rod

The Terfenol-D rod is dimensioned for particular applications. The length of the rod determines the stroke, while the cross sectional area relates the force output. The power output is proportional to the volume of the Terfenol-D material. The stiffness of the rod, together with the attached mass of actuator, determines the resonant frequency of the actuator.

The instrumented Terfenol-D element we used has a Terfenol-D rod with 0.00635 meter (1/4 inch) in diameter and approximately 0.1524 meter (6 inches) long. The instrumentation we developed consists of a 50 turn flux pickup coil and a pair of strain gauges. The strain gauges are wired in a bridge network, but only one of the gauges are attached rigidly to the Terfenol. The other gauge provides temperature and external field balance. If only one strain gauge had been used, it would have been much more sensitive to those noise sources. The flux pickup coil can be used to directly measure the flux change in the Terfenol, and the strain gauge bridge directly measures the elongation of the rod.

2.3. Signal Coil

The Terfenol rod elongates in response to a magnetic field generated by the signal coil. The wire size is selected to match the actuator impedance to the drive electronics. The coil we used was hand-wound and has a 12 ohm resistance. It can generate 500 Oersteds per ampere. The coil is hollow so that the Terfenol rod can be inserted into the center of the coil.

2.4. Permanent Magnetic Bias

With no magnetic bias, Terfenol will only expand in the presence of a magnetic field, independent of the direction of the field. In order to make a working device from Terfenol-D, magnetic bias is required to set a magnetostriction cause by coils in the center of its linear operating range so that an alternating current applied to the signal coil will cause the actuator to alternately extent and contract. In other words, with the magnetic bias, the Terfenol will expand or contract depending on the direction of the control field generated by the signal coil.

Permanent magnets are used in our actuators for the pre-bias. In the magnet assembly used in our actuators, there are actually three magnetic cylinders, and two spacers. We permanently glued the magnetic components together and had the entire configuration re-magnetised as a unit. After this modification, the actuators worked well. We have achieved a travel of ± 127 microns (± 5 thousandths of an inch).

2.5. End Caps and Mechanical Preload

The performance of Terfenol-D is improved by compressive preload. A mechanical pre-stress device is incorporated into the actuator. The actuator is assembled as follows: First insert the lower magnetic circuit cap into the outer housing so that it seats on the bottom of the outer housing. Then insert the permanent bias magnets. Then insert the coil inside of the permanent magnets. Then insert the upper preload cap with the spring facing outward. Finally the outer cap screws into the housing, depressing the spring to achieve the mechanical preload. The brass bushing in the outer cap is designed to provide smooth motion of the actuator stem as the Terfenol expands and contracts, but we believe there is significant binding which results in some non-linearity. We will design a more appropriate bearing configuration in subsequent designs.

2.6. Model of a Terfenol-D Actuator

The stroke-current characteristics of Terfenol-D actuators are non-linear and exhibit hysteresis. However, in a small range of motion, the characteristics can be approximated as a linear system:

$$\delta = \overline{K}_u u \tag{2.1}$$

where

δ is axial extension caused by magnetic field change; u is applied current to generate magnetic field; K_u is the Terfenol-D characteristic constant.

The force p generated by a Terfenol-D rod due to the magnetic field is given by:

$$p = E A \frac{\delta}{L} = K_d \delta \tag{2.2}$$

where

A is cross section area in m^2; E is Young's Modules, $3.5*10^{10}$ N/m^2; L is rod length ; δ is the change in rod length; δ/L is dimensionless strain. K_d is a proportional coefficient relating the deformation and the force of Terfenal_D.

A mathematic model for the actuator is then given by:

$$f = K_d [(x - z) + \overline{K}_u u] \tag{2.3}$$

where x and z are the axial coordinates of both ends of the actuator, aligned in the same direction as the force f exerts. Defining $K_u = K_d \overline{K}_u$, we can rewrite the model as:

$$f = K_d (x - z) + K_u u \tag{2.4}$$

where K_d is the characteristic constant determined by material properties of the specific Terfenol-D device, as well as by the structure of the coils. This model is reasonably accurate at the frequencies below the first resonance of the actuator assembly, which in the case of our actuators, is on the order of kHz.

§3 Force Feedback Control Law Design

There are a number of approaches proposed in recent years to active vibration isolation. The control designs of the most of these approaches rely on the model information of the structural dynamic response. In reality, a truth model of a lightly damped space structure has an infinite number of lightly damped modes; Consequently, higher order modes are truncated from the model in order to perform effective control design. This type of model-based control design, such as LQG optimal control, pole-zero cancellation, etc., will work well as long as the modes included in the model are accurately compensated, and the unmodeled modes are sufficiently damped. In the case where there are closely spaced higher order unmodelled modes, the performance of such control design will severely deteriorate, even resulting instability.

One of the emphasis of our study is to develop a robust control system using Terfenol-D actuators for the purpose of vibration isolation in the presence of inaccurate and unmodelled structural dynamics. Consider the pole-zero pattern of a system with flexible structures connected by active control elements, such as actuators and sensors. Structural control designers normally work with sensor-actuator pairs that are collocated. The advantage of a collocated sensor-actuator pair is that the undamped system transfer function is assured to have an alternating pole-zero pattern on the imaginary axis of s-plane, and the phase changes of the system transfer function are bounded to a region no wider that 180°. This is important from structural control perspective since other sensor-actuator configuration typically result in non-minimum phase zeros, or non-alternating pole-zero patterns, both of which will lead to difficult control situations. Based on the analysis of poles and zeros characteristics of flexible structures, a local integral force feedback control designs will be discussed in this section. This method is general and robust in the sense that they are based solely on the generic features of flexible structures without specifying model dynamics.

It is well known in the Evans root-locus method that as the gain changes, the poles of a closed loop system move from the poles to the zeros of the open loop system, and the loci will depart from a pole p_j at the angle θ_d , and arrive at a zero z_j at the angle θ_a, where

$$\theta_d = \sum_{i} \theta_{zi} - \sum_{\substack{i \\ i\neq j}} \theta_{pi} + r(180°) \quad , \quad \theta_a = \sum_{i} \theta_{pi} - \sum_{\substack{i \\ i\neq j}} \theta_{zi} + r(180°) \tag{3.1}$$

and r is a non-zero integer. With the alternating pole-zero pattern of a typical flexible structure, one can find that if proportional feedback gain is used, the closed loop poles will likely locate on the straight line connecting pairs of open loop poles and zeros, since the departure and arrival angles of the root loci are either 90° or -90° according to the above formula. Therefore the damping ratio is approximately the same as that of an open loop system. On the other hand, if a differential feedback or an integral feedback control law is used, both the departure and arrival angles of the loci will change by 90°, according to the above formula. The feedback law adds an additional pole or zero into system, and can bring the undamped open loop poles deeper into the left hand side of the s-plane, thereby increasing the system damping ratio.

The integral feedback law can be expressed as:

$$u = G_I \int f\,dt \tag{3.2}$$

where f is the measured force from the force sensor and u is the control current signal to the actuator. G_I is the integral gain. In a digital implementation, integral control can be formulated in an extremely simple fashion:

$$u(k+1) = u(k) + G_I\, f(k)\, \Delta \tag{3.3}$$

where Δ is the integral interval. Notice that pure integral feedback adds a pole at the origin of the s-plane which may sometimes cause stability problems. A modified version of an integral feedback control law is given as:

$$u(k+1) = \alpha\, u(k) + G_I\, f(k)\, \Delta \tag{3.4}$$

where the "forgetting factor" $a < 1$. This amounts to shifting the pole at the origin of the s-plane slightly to the left along the real axis, which should not significantly effect the loci of the flexible poles.

It should be mentioned that the vibration isolation performance of the integral force feedback system depends on the exact dynamics of the open loop system, however integral feedback can always increase a system's stability margin and damping ratio for a general type of systems, and therefore it is robust to system changes. The other advantage of integral feedback control is its simplicity in implementation.

§4. Simulation Results and Analysis

In order to verify the local force integral feedback control law for our Stewart platform-based vibration isolator, extensive simulations were performed. The general dynamic model of a Stewart Platform[3] was combined with the analytic model of the Terfenol-D actuators. A set of physical dimension parameters such as actuator length, weight, etc. which match our prototype cubic Stewart

Platform system were specified for the simulation. All simulation programs were written in C language on IBM 386 and Macintosh workstation environments. The simulation step size was chosen to be 0.0001 second.

In the results presented as follows, we simulated a typical case of multidimensional vibration isolation using our designed Stewart platform with Terfenol-D actuators as an active control device. Multidimensional vibration disturbances were assumed on the base plate of the platform. To be specific, a set of six random magnitude and phase sinusoidal forces with components along the X, Y, and Z directions were generated to excite each attachment point of the actuator. Since the magnitude and phase of these exciting forces were randomly selected these disturbances generated not only displacements along the X, Y, and Z directions, but also rotational vibrations. Our vibration disturbance pattern for the simulations was, therefore, quite general. Under this complicated vibration excitation pattern, the internal structure modes were activated. We then scanned the frequency range from 0.5 Hz to 250 Hz and found that the maximum resonant peak response of the mobile plate, in terms of vibration magnitude, could be obtained at the frequency of 39.8Hz. We then selected this frequency as the worst case of system vibration and used that frequency in our active control simulation. The simulation results at this frequency can be found in the Figures entitled "Mobile plate Motion V.S. Base Disturbance (Freq=40Hz)" (X, Y,Z,Roll,Pitch Yaw). The measured forces between the actuators and the mobile plate are shown in the Figures entitled "Measured Force of Link Number", and demonstrate large magnitudes which suggests a large energy transmission from the base plate to the mobile plate. The lengths of each actuator were also monitored through the simulation.

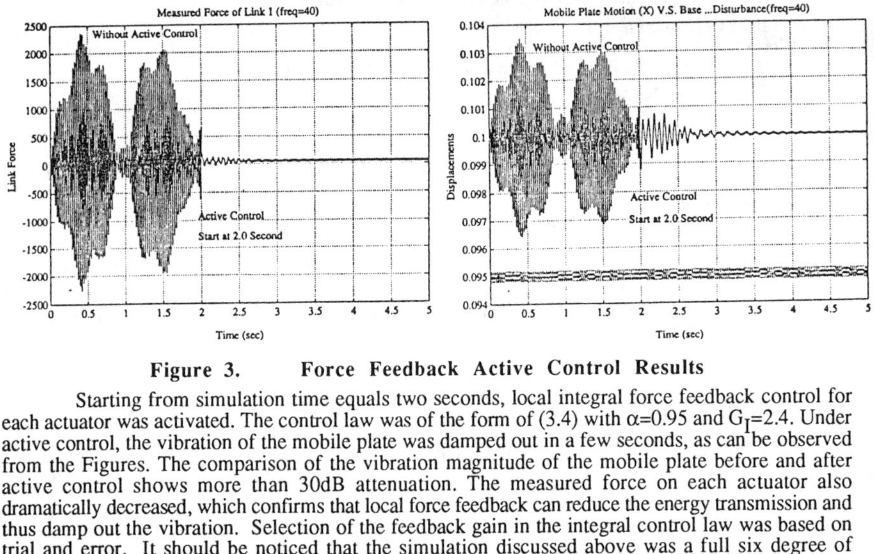

Figure 3. Force Feedback Active Control Results

Starting from simulation time equals two seconds, local integral force feedback control for each actuator was activated. The control law was of the form of (3.4) with α=0.95 and G_I=2.4. Under active control, the vibration of the mobile plate was damped out in a few seconds, as can be observed from the Figures. The comparison of the vibration magnitude of the mobile plate before and after active control shows more than 30dB attenuation. The measured force on each actuator also dramatically decreased, which confirms that local force feedback can reduce the energy transmission and thus damp out the vibration. Selection of the feedback gain in the integral control law was based on trial and error. It should be noticed that the simulation discussed above was a full six degree of freedom simulation including both the characteristics of the Terfenol-D actuators and the dynamics of the cubic configuration of our Stewart Platform vibration isolation system. .

Acknowledgements

Authors would like to extend their thanks to the supports of this work from NASA under the grant #NAS7-1134 and from US Army Strategic Defense Command under grant #DASG90-91-C-0041.

References

[1] D. Stewart, "A platform with six degree of freedom", _Proc. Inst. Mech Eng._, London, Vol.180, 1965, p 371

[2] E.F.Fichter, "A Stewart Platform - Based Manipulator : General Theory and Practical Construction", _Int. J. of Robotics Research_, Vol.5, No.2, 1986, p157

[3] Z.Geng,L.Haynes,J.Lee, and R.Carroll,"On the dynamic model and kinematics analysis of a class of Stewart platform", _Journal of Robotics and Autonomous Systems_, Vol. 9, 1992

[4] Mel Goodfriend, K.Shoop, "Adaptive characteristics of the magnetostrictive alloy Terfenol-D for active vibration control", _Proc. of Recent Advances in Active Control of Sound and Vibration_, Blacksburg, VA, April 1991, p 199

[5] B.Weda, J.Fanson, J.Garba, and G.Chen,"Adaptive structure to meet requirements for future space missions", _AIAA SDM Issues of international space station._ Williansburge, VA, April, 1988

Design, Implementation, and Experiments with a Thirty-Degree-of-Freedom 'Hyper-Redundant' Robot

Gregory S. Chirikjian † Joel W. Burdick

School of Engineering and Applied Science
California Institute of Technology
Pasadena, CA 91125

Abstract

"Hyper-redundant" robots have a very large degree of kinematic redundancy. These robots are analogous to "snakes," "elephant trunks," and "tentacles." Potential benefits of hyper-redundancy include superior obstacle avoidance ability, novel forms of locomotion, and robust tentacle-like grasping. In previous work, the authors have developed a novel and efficient framework for hyper-redundant robot kinematic analysis and algorithm development. This paper reviews the implementation of 30 degree-of-freedom planar hyper-redundant robot which demonstrates and validates this earlier work. Experiments in manipulation, obstacle avoidance, locomotion, grasping, and grasped object reorientation are summarized.

1. Introduction

Hyper-redundant robots have previously been called 'highly articulated,' 'tentacle,' 'snake-like,' 'tensor-arm,' 'elephant-like,' and 'spine.' The authors have coined the word 'hyper-redundant' to capture the essence of these related concepts. To date, hyper-redundant robots have remained largely a laboratory curiosity because: (1) previous kinematic modeling techniques have not been efficient or well suited to hyper-redundant robot task modeling; (2) the mechanical design and implementation of hyper-redundant robots has been perceived as unnecessarily complex; and (3) hyper-redundant robots are not anthropomorphic and thus require different programming paradigms. The authors have undertaken a broadly based investigation which we hope will eventually overcome many of the obstacles to practical hyper-redundant robot deployment. In previous papers, we have presented novel techniques for unified kinematic analysis of hyper-redundant robots [1-5]. These techniques have also been used as the basis of novel obstacle avoidance [2], locomotion [4], and grasping schemes [4] (these papers should be consulted for references to other approaches to hyper-redundant mechanism analysis and previous developments of hyper-redundant robot prototypes).

This paper briefly reviews our previous work and presents the design and implementation of a thirty degree-of-freedom (DOF) planar hyper-redundant robot which was developed to verify and validate our analytical approach. We also present experiments in hyper-redundant manipulation, obstacle avoidance, locomotion, grasping, object manipulation, and dual arm hyper-redundant manipulation. Some of these operating modes, such as grasped object reorientation, have not previously been considered or demonstrated in the literature.

2. Review of Planar Hyper-Redundant Robot Kinematics

Our kinematic analysis method is based on a *backbone curve* which captures the hyper-redundant robot's important macroscopic features. Kinematic analysis and

† Currently at the Dept. of Mechanical Engineering, Johns Hopkins University

task planning operations reduce to the determination of the proper time varying behavior of the backbone curve. Continuous backbone curves shapes which result from planning algorithms can be used to directly control the geometry of continuous morphology robots. For discretely segmented morphologies, such as the variable geometry truss (VGT) robot described in this paper, the backbone curve nominally captures the robot geometry closely, though not exactly. *Fitting algorithms* (Section 2.3) are used to determine the discrete actuator displacements which will cause the robot to adhere as closely as possible to the continuous backbone curve solution. In this paper, we focus on planar hyper-redundant kinematics, as our actual robot is planar.

2.1. Parametrization of the Backbone Curve

Hyper-redundant robot backbone curves, can be *nonextensible* (fixed length), or *extensible* (variable length). An intrinsic backbone curve parametrization consists of both the parametrization of points along the curve and the choice of a reference frame at each curve point. Points on both extensible and nonextensible planar and spatial curves can be parametrized in the form:

$$\overline{x}(s,t) = \int_0^s l(\sigma,t)\overline{u}(\sigma,t)d\sigma \tag{1}$$

where $s \in [0,1]$ measures distance along the backbone curve. $\overline{x}(s,t)$ is a vector from the robot base to the point s on the backbone curve. $\overline{u}(s,t)$ is the unit vector tangent to the curve at s. $l(s,t)$ is the length of the curve tangent and takes the form:

$$l(s,t) = 1 + \epsilon(s,t) > 0 \tag{2}$$

where $\epsilon(s,t)$ is the local extensibility of the robot at s. $\epsilon(s,t) > 0$ indicates local extension, while $\epsilon(s,t) < 0$ indicates contraction. In the planar case, let $\overline{x}(s,t) = [x(s,t), y(s,t)]^T$ denote the position of point s with respect to the backbone curve base:

$$x(s,t) = \int_0^s l(\sigma,t)\sin\theta(\sigma,t)d\sigma; \qquad y(s,t) = \int_0^s l(\sigma,t)\cos\theta(\sigma,t)d\sigma. \tag{3}$$

$\theta(s,t)$ is the clockwise measured angle which the tangent to the curve at s makes with the y-axis:

$$\theta(s,t) = \int_0^s \kappa(\sigma,t)l(\sigma,t)d\sigma \tag{4}$$

where $\kappa(s,t)$ is the classical *curvature function* of the backbone curve. $l(s,t)$ and $\theta(s,t)$ (or alternatively, $l(s,t)$ and $\kappa(s,t)$) are termed the *intrinsic shape functions* of the backbone curve, as they completely describe the backbone curve shape.

A *backbone reference frame* at s is a set of right-handed orthonormal vectors with frame origin coincident with point s in the curve. In the planar case, the reference frame basis vectors are the unit backbone curve tangent and normal vectors:

$$\mathbf{Q}(s,t) = \begin{bmatrix} \sin\theta(s,t) & \cos\theta(s,t) \\ \cos\theta(s,t) & -\sin\theta(s,t) \end{bmatrix} \tag{5}$$

2.2. Inverse Kinematic Schemes

For given shape functions, the forward kinematics can be found by integration of (1). However, the inverse kinematic problem, which corresponds to finding a backbone curve which satisfies tasks constraints, is more difficult. There are an infinite number of such curves. Previous work has focused on two methods for resolving hyper-redundancy: a "modal" approach, and one based on the calculus of variations. In the modal approach, the shape functions are restricted to the 'modal' form:

$$\theta(s,t) = \sum_{i=1}^{N_\theta} a_i(t)\Phi_i(s) \qquad l(s,t) = \sum_{i=N_\theta+1}^{N} a_i(t)\Phi_i(s) \tag{6}$$

where Φ_i is a *mode function*, and a_i is a *modal participation factor*. N equals or exceeds the total number of task constraints. Hyper-redundancy is resolved in the modal approach by constraining the backbone curve to N effective DOF. In the planar case, these DOF are distributed between N_θ 'bending modes' and $(N - N_\theta)$ 'extension' modes. The $\{\Phi_i\}$ are predetermined functions chosen by the programmer. Thus, the backbone curve geometry becomes solely a function of the $\{a_i\}$. The inverse kinematics problem reduces to finding the $\{a_i\}$ which satisfy task constraints.

The restriction to modal form is practically useful, as it often leads to closed or nearly closed form inverse kinematic solutions. Examples of closed form inverse kinematics can be found in [1,3,4]. It is not necessary to have closed form inverse kinematic solutions to use the modal method. Look-up tables and interpolation can store the mapping between participation factors and end-effector coordinates for any given set of modes. Alternatively, a numerical inverse kinematic solution method based on a *modal Jacobian* and a scheme analogous to resolved rate trajectory planning can be used [1,3].

Alternatively, the calculus of variations can be used to generate necessary conditions for the optimal backbone curve shape with respect to task constraints and a user-defined optimality criterion [5]. [5] considers optimality criteria based on a weighted measure of backbone bending. In the planar case, the minimization of the integral:

$$I = \frac{1}{2}\int_0^1 \alpha(s)\kappa^2(s)ds = \frac{1}{2}\int_0^1 \alpha(s)\dot{\theta}^2(s)ds \tag{7}$$

will minize the bending of a nonextensible backbone curve. $\alpha(s)$ is a scalar weighting function, which may be chosen to reflect practical constraints [5]. In this approach, we seek to find the intrinsic function $\theta(s)$ which minimizes (7) while satisfying end-effector constraints (3) and the boundary conditions $\theta(s=0) = 0$ and $\theta(s=1) = \theta_{ee}$, where θ_{ee} is the desired orientation of the backbone curve tip, or end-effector. [5] should be consulted for the solution to his problem and related examples.

2.2. Backbone Curve "Fitting" Procedures

For a VGT morphology (and other discretely segmented mechanisms), a "fitting" procedure is used to determine the individual actuator displacements which cause the VGT to adhere as closely as possible to the continuous backbone curve solution. Figure 1 shows one module of a planar VGT robot which consists of a serial chain of n such modules. Attached to each module is an intrinsic 'input' and 'output' frame. Fitting is done by constraining the module input/output frames to coincide exactly with the intrinsic frames of the backbone curve at $(n+1)$ points, $\{s_i\}$ $i = 0, \cdots, n$. Without loss of generality, these points are chosen uniformly, e.g. $s_i = i/n$. [5] outlines a procedure for optimally reparametrizing the length distribution function so that the physical locating of the fitting points may be nonuniform. The intrinsic frames at s_{i-1} and s_i define an inverse kinematic problem for the ith truss module.

This module inverse kinematic problem can be solved as follows. Let $\overline{r}^i_{i-1}$ be a vector from $\overline{x}(s_{i-1},t)$ to $\overline{x}(s_i,t)$:

$$\overline{r}^i_{i-1}(t) = \mathbf{Q}^T(s_{i-1},t)[\overline{x}(s_i,t) - \overline{x}(s_{i-1},t)] \tag{8}$$

The three vectors collinear with the three prismatic joint axes (Figure 1) can be computed as:

$$\begin{aligned} \overline{v}^i_j &= \overline{r}^i_{i-1} - \overline{n}^i_j + \mathrm{ROT}(-\overline{e}_3, \delta\theta^i)\overline{n}^i_j \qquad j = 1,2 \\ \overline{c}^i &= \overline{r}^i_{i-1} - \overline{n}^i_1 + \mathrm{ROT}(-\overline{e}_3, \delta\theta^i)\overline{n}^i_2 \end{aligned} \tag{9}$$

where $\delta\theta^i = \theta(s_i,t) - \theta(s_{i-1},t)$ and $\overline{n}^i_j$ are the vectors to the jth vertex of the ith platform in the frame affixed to that platform. For this example, $\overline{n}^i_1 = [-w_0/2, 0]^T$ and $\overline{n}^i_2 = [w_0/2, 0]^T$ where w_0 is the width of each horizontal face of the truss. The controlled degrees of freedom are the lengths of the prismatic actuators:

$$\lambda^i_j = ||\overline{v}^i_j||; \qquad \lambda^i_3 = ||\overline{c}^i|| \tag{10}$$

for $i = 1, ..., n$, and $j = 1, 2$. Thus, (9) and (10) provide the inverse kinematics solution for this module geometry based on the backbone curve geometry.

3. Manipulator Description

A design study of possible morphologies which would enable us to demonstrate the algorithms developed in our previous work concluded that a variable geometry truss morphology would meet all of our requirements. Our final design is a 30 DOF VGT structure consisting of 10 identical modules which are planar 3 DOF parallel manipulators (with the same kinematic structure of Figure 1). The prismatic actuators are implemented with D.C. servo motors and lead screw drives. The actuators can vary in length from a minimum of 12 inches to a maximum of 18 inches. The geometry of the module can be simply modified to trade off between module dexterity and mechanical advantage.

For manipulation and grasping experiments, castors are attached to the fixed elements of the modules in order to allow low friction motion over the laboratory floor, which is the plane of manipulation. For locomotion experiments, the manipulator is detached from its base, and turned on its side. Each module is equipped with rubber faced ratchet wheels which permit rolling in one direction only. This is required for standing wave locomotion [4]. For traveling wave locomotion, the wheels are locked in place, and the wheels' rubber surfaces are use as passive feet. These rubber surfaces are also used as the contact points for planar grasping operations. Because of its modularity, the robot can be separated into two 15 DOF robots for cooperative manipulation experiments [1].

The robot system is controlled by an a multiprocessing computer, which consists of Heurikon 68030 processing boards in a VME bus, which is in turn connected to a Sun Microsystems 4/260 computer. The WindRiver Systems VxWorks software system provides operating system and system management functions. The 68030 processors compute the kinematic, path planning, sensor processing, and task level coordination functions. They also implement the low level actuator position feedback control. The Sun computer is used primarily for user interface and data storage.

4. Experiments

It is difficult to summarize these experiments in this format, and interested readers can obtain copies of a video-tape of these experiments from the authors. Figure 2 shows the fully curled robot configuration, indicating its range of dexterity. Figure 3 shows one snap-shot of an obstacle avoidance maneuver which employs the the theory developed in [2]. Figure 4 shows two snap-shots of one 15 DOF component of the robot displaying locomotion based on the standing wave idea developed in [4]. The photo in Figure 5 shows how a hyper-redundant manipulator can be used as a grasping device, which can be used to maintain a 'power grasp' while still being able to incrementally rotate the object within the grasp, as formulated in [1,4].

5. Summary

This paper summarized the kinematics and implementation of a 30 DOF planar hyper-redundant robot which was developed to validate the authors' previous analytical research in hyper-redundant robotic systems. In previous work, we developed modeling techniques and algorithms which were highly efficient and sufficiently general to enable real-time inverse kinematics and motion planning for this complex class of robots. Further, this basic framework has been used to develop novel obstacle avoidance, locomotion, and grasping algorithms. The experiments detailed in this paper confirm the validity and efficiency of our previous work. Our hope is that our analytical and experimental investigations will elevate hyper-redundancy to a greater level of practicality.

Aknowledgements: This work was sponsored in part by a NASA Graduate Student Researchers Program fellowship (for the first author), NSF grants MSS-901779 and MSS-9157843, and an office of Naval Research Young Investigator Award.

6. References

[1] Chirikjian, G.S., "Theory and Applications of Hyper-Redundant Robotic Manipulators," Ph.D Thesis, Dept. of Applied Mechanics, California Institute of Technology, June 1992.

[2] Chirikjian, G.S., Burdick, J.W., "An Obstacle Avoidance Algorithm for Hyper-Redundant Manipulators," *Proc. IEEE Int. Conf. on Robotics and Automation* (ICRA), Cincinnati, OH, May 13-18, 1990.

[3] Chirikjian, G.S., Burdick, J.W., " Parallel Formulation of the Inverse Kinematics of Modular Hyper-Redundant Manipulators," *Proc. ICRA*, Sacramento, CA, April, 1991.

[4] Chirikjian, G.S., Burdick, J.W., "Kinematics of Hyper-Redundant Locomotion with Applications to Grasping," *Proc. ICRA*, Sacramento, CA, April, 1991.

[5] Chirikjian, G.S. and Burdick, J.W., "Kinematically Optimal Hyper-Redundant Manipulator Configurations," *Proc. ICRA*, Nice, France, May, 1992.

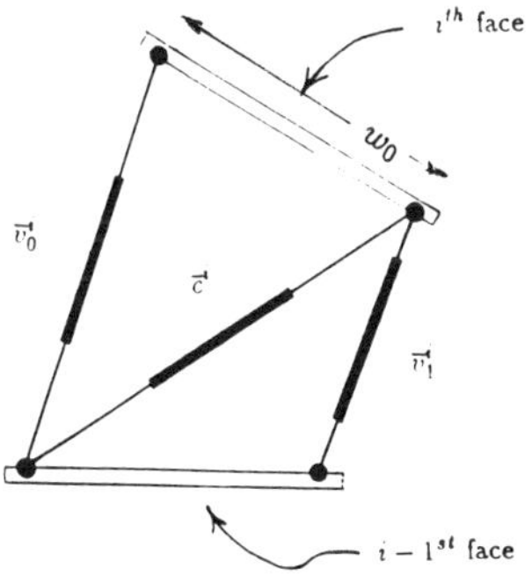

Figure 1: Schematic of one variable geometry truss module

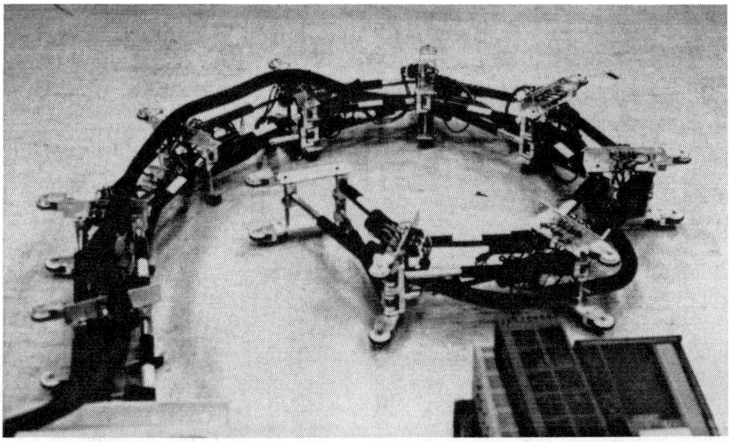

Figure 2: Photograph of Fully Curled Robot

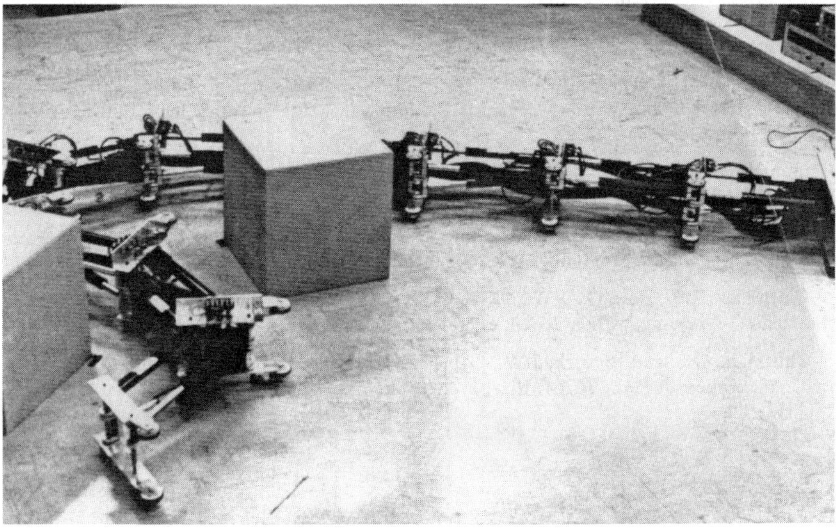

Figure 3: Obstacle Avoidance Demonstration

Figure 4: Standing Wave Locomotion Demonstration

Figure 5: Tentacle-Like Grasp and Object Reorientation Demonstration

DESIGN OF A CONTROLLER FOR ROBUST GAIN SCHEDULING

L. Chirinos†, J. J. Gonzalez†, M. W. Thompson†, and G. R. Widmann‡

† Colorado State University
Department of Electrical Engineering
Institute of Robotic Studies
Fort Collins, CO 80523

‡ Hughes Aircraft Company
Missile Systems Group
Automotive Controls Engineering
Canoga Park, CA 91304-7924

ABSTRACT

A technique to design a gain scheduling type of robust controller is introduced. This method uses the stability hypersphere concept [1] and [5] to define a stability hypervolume that encloses a bounded set of plant parameters.

1. INTRODUCTION

In most practical situations, robotics system parameters are only known within a limited degree. Variations are mainly due to component aging, varying payload, variability from robot to robot due to the particular manufacturing process (i.e. design controller for nominal robot must work for every robot manufactured). Therefore, a robust controller design is a practical necessity when dealing with realistic robotics systems.

The problem we undertook was to see if we could generate a set of controller parameters capable of guaranteeing the stability of a plant with bounded uncertainty. The set of controller parameters is chosen by assuring that the set of plant parameters were always mapped inside the stability hyperspheres. This was accomplished by intersecting the stability hyperspheres [1] corresponding to each element in the set of controllers, in such a way that the generated volume contains all the elements in the set of plants.

Early works in this direction [1] and [3] have shown that is possible to develop a controller capable of enlarging the hypersphere stability at its maximum. This is achieved by iterating over the controller parameters for a nominal plant. However, there is a limit on the maximum uncertainty that can be tolerated for a given controller order and also the performance for limit cases might be severely degraded.

In using our technique, we are able to introduce some performance specifications by providing an adequate desired characteristic polynomial. The selection of this polynomial will be a compromise between acceptable performance and size of the set of controllers needed to guarantee global stability. The price we have to pay is that a simple identification of the changing conditions is required in order to select the appropriate controller parameters from a look-up table.

An example using data from a single link flexible arm manipulator is given to demonstrate this design methodology.

2. PRELIMINARY DEFINITIONS

Consider a Single-Input Single-Output general feedback structure such as the one shown in Figure 1. Here the nth order s-domain transfer function of the nominal actual physical plant is given as

$$G(s) = \frac{B(s)}{A(s)} = \frac{b_n s^n + b_{n-1}s^{n-1} + \cdots + b_0}{a_n s^q + a_{n-1}s^{n-1} + \cdots + a_0} \tag{1}$$

from which the plant vector $v \in \Re^{2n+2}$ can be defined as

$$v \triangleq [b_0 \quad a_0 \quad \cdots \quad b_n \quad a_n]^t \tag{2}$$

Let us define $L(s)$ and $P(s)$ as the mth order controller polynomials as follows

$$\begin{aligned} P(s) &= p_m s^m + p_{m-1}s^{m-1} + \cdots + p_0 & (3)\\ L(s) &= l_m s^m + l_{m-1}s^{m-1} + \cdots + l_0 & (4) \end{aligned}$$

where the controller parameter vector $x \in \Re^{m+1}$ is given by

$$x \triangleq [p_m \quad l_m \quad \cdots \quad p_0 \quad l_0]^t \tag{5}$$

Then, the qth order closed-loop characteristic equation can be obtained as

$$K(s) = P(s)A(s) + L(s)B(s) = k_q s^q + k_{q-1}s^{q-1} + \cdots + k_0 s$$

where $q = m + n$. Let $k \in \Re^{q+1}$ be the characteristic equation vector as follows

$$k \triangleq [k_q \quad k_{q-1} \quad \cdots \quad k_0]^t \tag{6}$$

Now, if $A(s)$ and $B(s)$ are relatively prime and the controller order $m = n - 1$, then for a given desired characteristic vector $k \in \Re^{2n}$ the controller parameters vector x can be obtained from

$$Mx = k \tag{7}$$

where the matrix $M \in \Re^{2n \times 2n}$ is given by

$$M = \begin{bmatrix} & & & & & & b_0 & a_0 \\ & & b_0 & a_0 & \cdot & \cdot & \cdot & \cdot \\ b_0 & a_0 & \cdot & \cdot & \cdot & \cdot & \cdot & \cdot \\ \cdot & \cdot & \cdot & \cdot & \cdot & \cdot & \cdot & \cdot \\ \cdot & \cdot & \cdot & \cdot & \cdot & \cdot & b_n & a_n \\ \cdot & \cdot & b_n & a_n & & & & \\ b_n & a_n & & & & & & \end{bmatrix}$$

Equation (7) can also be rewritten as

$$Xv = k \tag{8}$$

where matrix $X \in \Re^{[2n] \times [2n+2]}$ is given by

$$X = \begin{bmatrix} & & & & & & p_n & l_n \\ & & p_n & l_n & \cdot & \cdot & \cdot & \cdot \\ p_n & l_n & \cdot & \cdot & \cdot & \cdot & \cdot & \cdot \\ \cdot & \cdot & \cdot & \cdot & \cdot & \cdot & \cdot & \cdot \\ \cdot & \cdot & \cdot & \cdot & \cdot & \cdot & p_0 & l_0 \\ \cdot & \cdot & p_0 & l_0 & & & & \\ p_0 & l_0 & & & & & & \end{bmatrix}$$

from which X can be seen as a linear transformation matrix that maps the plant vector $v \in \Re^{2n+2}$ into the coefficient vector $k \in \Re^{2n}$.

Since the plant vector v will not be exactly known, let us define the plant perturbations by considering that the plant parameters vector v corresponds to a set of vectors

$$S_v = \left\{ v \in \Re^{2n+2} : \ v^- \leq v \leq v^+ \right\} \tag{9}$$

and from (8) the characteristic vector k becomes the set of vectors

$$S_k \in \left\{ k \in \Re^{2n} : \ k^- \leq k \leq k^+ \right\} \tag{10}$$

where $v^- \leq v \leq v^+ \Longrightarrow v_i^- \leq v_i \leq v_i^+$ and $k^- \leq k \leq k^+ \Longrightarrow k_i^- \leq k_i \leq k_i^+$.

Theorem 1: Let (3) and (4) define a stabilizing controller x for the nominal plant v^0. Then, the radius of the largest stability hypersphere in the parameter space of v, centered at v^0, is given by

$$\rho(v^0, x) = \min\{r_0(v^0, x),\ r_n(v^0, x),\ r(v^0, x)\} \tag{11}$$

Proof: Bhattacharyya [1].

Figure 2 illustrates Theorem 4. Here, it is shown that a perturbation of magnitude $\Delta(v)$ of v^0 can not destabilize the closed-loop system unless $\Delta(v)$ intersect Π_0, Π_n or $\Pi(\omega)$ for some ω. Where Π_0, Π_n and $\Pi(\omega)$ correspond to stability hypersurfaces in the plant parameter space.

3. CONTROLLER DESIGN

In this section a design methodology for a robust gain scheduling type of controller is presented. The design is based on Theorem 1 and is somehow similar to the one proposed by Bhattacharyya [1]. The main difference is that our design uses the polynomial pole assignment technique to find a finite set of controller parameters. This set is computed for a finite set of plant parameters, selected from a suitable partition of the bounded uncertainty region. Once the controller parameter set that maps all the plant into the stability region is found, a look-up table can be built and the appropriate controller is selected whenever a particular plant model falls into its operation space. A satisfactory performance for a given set of controller parameters can be obtained by choosing an appropriate desired closed-loop characteristic equation. This idea is illustrated in Figure 3, where one can see that the controllers designed for plants V_2 and V_4 are enough to stabilize all the other plants. Here V_1 and V_4 are the perturbation outer limits.

The design procedure can be summarized as follows:

STEP 1. Select a finite set of plants such that (9) becomes

$$S_{v^*} = \{v_i : \ i = 1, \cdots, nv\} \subset S_v \tag{12}$$

where nv is the number of plant vectors chosen in the partition.

STEP 2. Find the finite set of controller parameters x_i

$$S_x = \{x_i : \ X_i v_i = k_i : \ i = 1, \cdots, nv\} \tag{13}$$

where k_i is the desired characteristic vector for the ith plant vector.

STEP 3. Find the finite set of stability hypersphere radius for every pair $\{ v_i, x_i \}$

$$S_\rho = \{\rho(v_i, x_i) : \ i = 1, \cdots, nv\} \tag{14}$$

STEP 4. Choose the minimum number of x_i's needed to cover the whole perturbation range.

STEP 5. If the whole range can not be covered, then reduce the spacing between some or all of the v_i's and $v_{i\pm 1}$'s and go to STEP 2. Otherwise STOP.

4. DESIGN EXAMPLE

This section describes an example to illustrate the design procedure discussed in Section 3. The CSU experimental flexible manipulator testbed in Figure 4 was used to provide the plant model used in this example.

After some manipulations it is possible to show [4] that the collocated control transfer function between the hub sensor output and the torque input can be approximated by

$$\frac{\theta_H(s)}{T(s)} = \frac{1}{I_T s(s+a)} \prod_{i=1}^{\infty} \frac{\omega_i^2}{\Omega_i^2} \frac{s^2 + 2\zeta_i' \Omega_{n_i} s + \Omega_{n_i}^2}{s^2 + 2\zeta_i \omega_{n_i} s + \omega_{n_i}} \tag{15}$$

where each antiresonance frequency Ω_i is separated by two consecutive resonance frequencies ω_i such that $\omega_{i-1} < \Omega_i < \omega_i$, for $i = 1, \cdots, \infty$. Also, the non-collocated control transfer function between the tip sensor output and the torque input by

$$\frac{\overline{y}_t(s)}{T(s)} = \frac{1}{I_T s(s+a)} \prod_{i=1}^{\infty} \frac{\omega_i^2}{\alpha_i^2} \frac{\alpha_i^2 - s^2}{s^2 + 2\zeta_i \omega_{n_i} s + \omega_{n_i}} \tag{16}$$

Although a true flexible arm model is inherently of infinite order, it has been shown that three flexible modes can produce an acceptable representation of its dynamics. In order to simplify the design procedure, a fourth order model will instead be used (this will account for one flexible mode).

The design objective is to generate a controller (or a set of controllers) able to cover the model perturbations that arise from different payloads at the tip of the flexible arm. The closed-loop response is desired to be underdamped.

STEP 1: The nominal plant v^0 corresponds to the no payload plant and the range of variation Δv arises from a total payload variation $\Delta m = +200.0$ gr.

The following parameter values were used to obtain the discrete set of plant vectors v.

$$a = \zeta = \zeta' = 0.20$$

Table 1 illustrates the variation in the plant parameters (15) and (16) as a function of payload variation.

STEP 2: The desired characteristic polynomial was chosen to be

$$K(s) = (s+1)(s+2)(s+3)(s+4)(s+5)(s+6)(s+7) \tag{17}$$

for all the cases. Then, solving for every x_i in (7), we obtain S_x.

STEPS 3 & 4: By finding S_ρ we are able to determine the controller which provides the greatest stability radius. This controller corresponds to the one designed for a payload of 120.0 gr and is defined by the polynomials

$$\begin{aligned} P(s) &= 1.8597s^3 - 30.6547s^2 + 37.1756s + 21.1661 \\ L(s) &= s^3 + 23.6393s^2 + 45.3512s + 159.2539 \end{aligned} \tag{18}$$

for collocated control and

$$\begin{aligned} P(s) &= -7.3432s^3 + 283.6272s^2 + 101.9238s + 21.1661 \\ L(s) &= s^3 + 23.6392s^2 + 95.9095 - 517.6406 \end{aligned} \quad (19)$$

for non-collocated control.

STEP 5: We stop here since this controller alone is capable of maping S_v to the s-domain left half plane.

Figures 5 and 6 show the collocated and non-collocated step responses of a single link flexible arm under different payloads. These simulations were done for payloads of 0 gr, 120.0 gr, and 200.0 gr when the unique controllers defined by (18) and (19) are used.

5. CONCLUSIONS

This paper presented preliminary results on the investigation of the design of a gain scheduling type of robust controller. The simulation results show that this methodology can be applied to some particular cases without big difficulties. In our particular case, this scheme lead us to a relatively simple controller design for a flexible arm. Here we took advantage of the smooth flexible mode variation under different masses on the tip of the flexible arm. Since the weight is the only information required to determine the actual plant parameters, a bigger payload range could be handled with a simple tip mass detector to switch between different controllers in a look-up table

Although the design example was given for a stability hypersphere, the methodology can be readily applied to the results given by Chieh [2] for a generalized stability hypersphere, which includes some performance conditions.

References

[1] S. P. Bhattacharyya, *"Robust Stabilization Against Structured Perturbations"*, Lecture Notes in Control and Information Sciences, No. 99, Springer-Verlag Berlin, Heidelberg 1987.

[2] Chieh-Li Chen and Neil Munro, *"Calculation of the Largest Generalized Stability Problem for the Maximum Setting-Time and Minimum Damping-Ratio Cases"*, IEEE AC-36, NO. 6, pp 756-759, June 1991.

[3] L. H. Keel, S. P. Bhattacharyya, and Jo W. Howze, *"Robust Control with Structured Perturbations"*, IEEE AC-33, No. 1, pp. 68-77, January 1988.

[4] E.M. Nebot, *"Modeling and Control of a Single Link Flexible Manipulator"*, Ph.D. Thesis, Dept. of EE, Colorado State University, Fort Collins, CO, Summer 1988.

[5] C. B. Soh, C. S. Berger, and K. P. Dabke, *"On the Stability Properties of Polynomials with Perturbed Coefficients"*, IEEE AC-30, No. 10, pp. 1033-1036, October 1985.

Tip Mass (gr)	It (Kg-m^2)	Ω_1 (rad/sec)	α_1 (rad/sec)	ω_1 (rad/sec)
000.0	0.4400	3.51602	11.18664	12.17551
020.0	0.4404	3.30427	11.18664	11.69594
040.0	0.4416	3.12576	11.18664	11.32082
060.0	0.4436	2.97280	11.18664	11.02085
080.0	0.4464	2.83999	11.18664	10.77615
100.0	0.4500	2.72318	11.18664	10.57305
120.0	0.4544	2.61957	11.18664	10.40194
140.0	0.4596	2.52683	11.18664	10.25588
160.0	0.4656	2.44321	11.18664	10.12981
180.0	0.4724	2.36730	11.18664	10.01991
200.0	0.4800	2.29801	11.18664	09.92377

Table 1: First mode model parameters for different payloads.

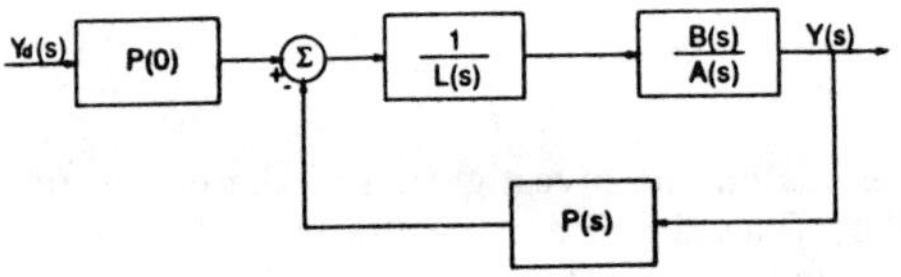

Figure 1: General Feedback Controller Structure.

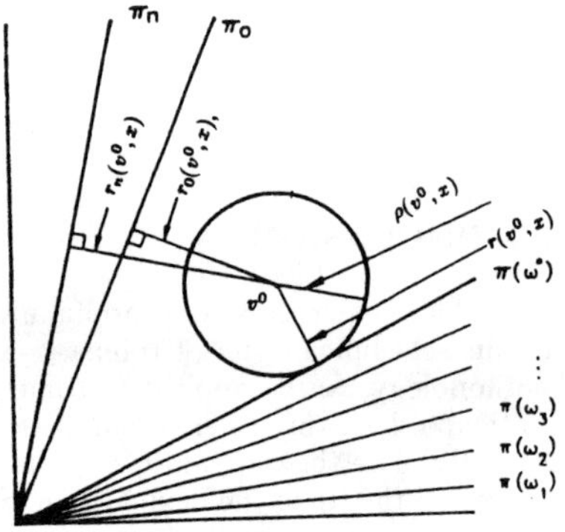

Figure 2: Stability Region in the Plant Parameter Space.

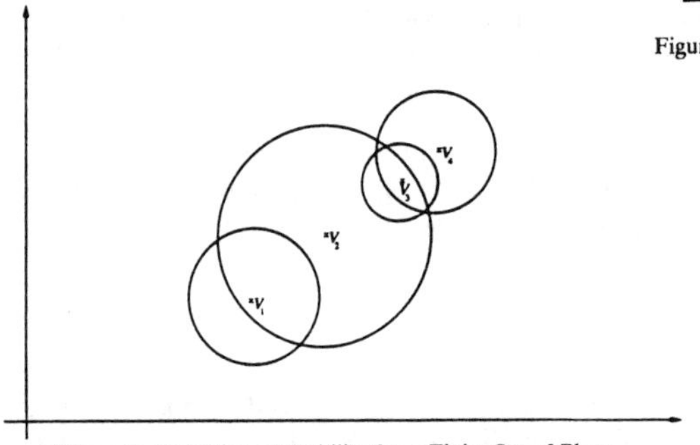

Figure 3: Definition of Stability for a Finite Set of Plant-Controller Parameter Vectors.

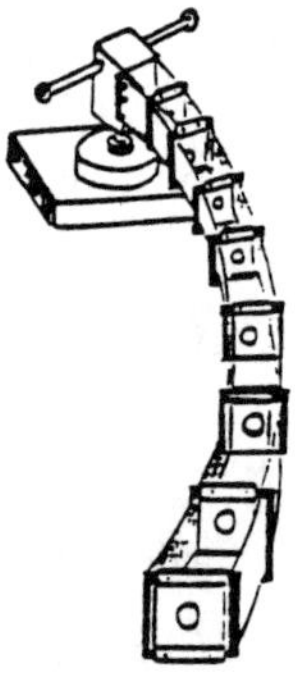

Figure 4: CSU Single Link Flexible Arm Manipulator Testbed.

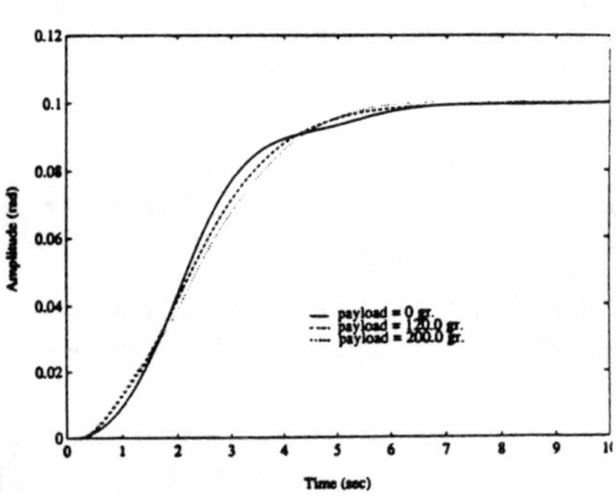

Figure 5: Step Responses for the Collocated Controller defined by (18).

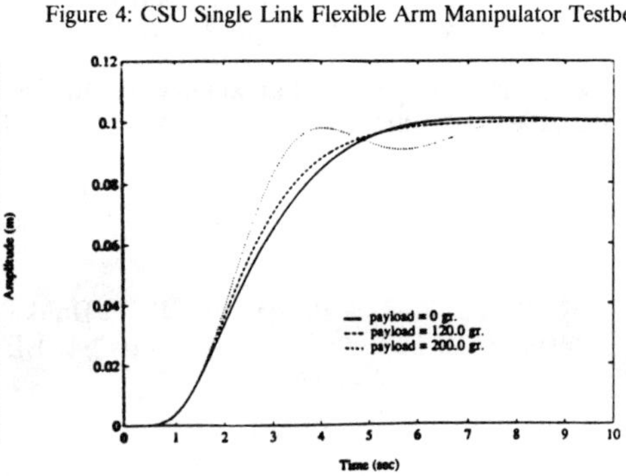

Figure 6: Step Responses for the Non-Collocated Controller defined by (19).

PART 5 : TELEROBOTICS AND SPACE APPLICATIONS OF ROBOTS

A Fail-Safe Tele-Autonomous Robotic System for Nuclear Facilities

Kam Sing Tso
SoHaR Incorporated
Beverly Hills, CA 90211

Paul G. Backes
Jet Propulsion Laboratory
California Institute of Technology
Pasadena, CA 91109

Abstract

A fail-safe tele-autonomous robotic system is proposed for use in advanced nuclear facilities. The design exploits technologies developed for space telerobotics. The target system consists of a graphical user interface for an operator to specify and execute robotic tasks, hand controllers for teleoperation, a 3-dimensional graphical simulator, and robot control software to drive both the graphical simulation and dual 6 degree-of-freedom robots to perform tasks using autonomous, teleoperated, and shared control modes. A preliminary design for a safety monitoring system for fail-safe operations is also described.

1 Introduction

Use of robotics in nuclear facilities can minimize personnel radiation exposure and reduce plant outage times. While existing commercial robot technology may be useful for some tasks, most maintenance and repair tasks require more advanced robotics technology. Our approach to achieving advanced robotics in future nuclear facilities is to provide robust, sophisticated robotic control and a well integrated user interface for performing operations. The control should include *fully autonomous* operation for performing precise and repetitive process steps, *teleoperation* for unusual maintenance and off-normal operations, and *shared control* for processes that are too complex to be handled in either fully autonomous or pure teleoperated operation. At present, few robotic systems have the capability to operate in all of these control modes. We propose to exploit recent advances from the results of NASA sponsored research at the California Institute of Technology Jet Propulsion Laboratory (JPL) for space telerobot servicing and apply it to advanced nuclear metallic fuel reprocessing operations.

The JPL/NASA Telerobotics Program was initiated in 1985 to explore the feasibility of using robots in space for on-orbit assembly, maintenance, and repair operations. The program has resulted in a fully operational robotic system with the following features: separation of the computing facilities into local and remote sites, autonomous motion generation in joint or Cartesian coordinates, force reflecting teleoperation with voice interaction between the operator and the robots, shared control between the autonomously generated motions and operator controlled teleoperation, and dual-arm coordinated trajectory generation [1]. The system has been used to carry out realistic experiments such as the exchange of an orbital replacement unit, bolt turning, surface cleaning, and satellite grappling, using a mixture of autonomous actions and teleoperation, with either a single arm or two cooperating arms [2].

The space telerobot technologies are suited for nuclear facilities for the following reasons: 1) The ground-remote operations for space telerobotics can be similarly applied to local-remote operations for nuclear facilities where operators remotely operate the robots in the high radiation environment of the shielding-cell. 2) The robot task primitives developed for space servicing are directly applicable to the nuclear reprocessing tasks.

2 System Architecture

The proposed architecture of the system is shown in Figure 1. The system is physically

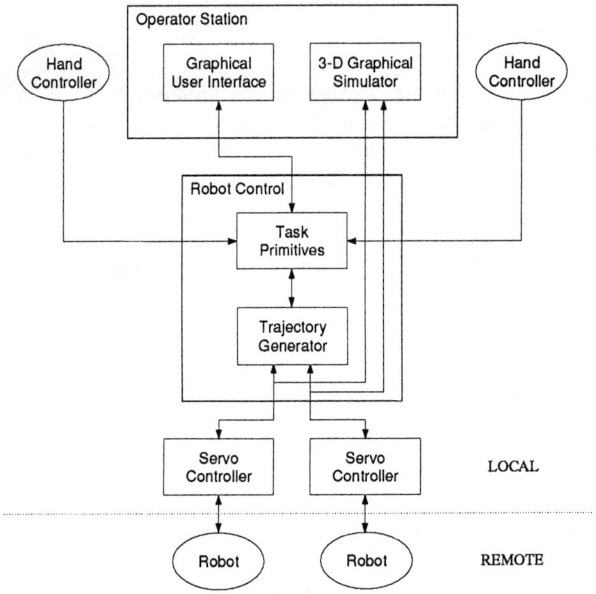

Figure 1: Overall Architecture of the Robotic System

divided into the *local* site, where the operator sits, and the *remote* site, where the manipulators work. The remote site components are minimized because of the high radiation environment in the shielding-cell. For most commercial robots, that means only the manipulators are to be placed in the remote site, with their servo controllers being placed outside the cell.

The system is designed as a hierarchy of levels:

- *Operator Level*: The top level of the hierarchy provides a user interface for the operator to perform robotic tasks. The operator can either send a command to execute the task autonomously, operate the hand controller to control the robot in teleoperation, or a combination of both in traded or shared control.

- *Task Level*: This level accepts task commands from the operator and provides task primitives to perform the tasks. Experience at JPL has shown that very few parameterized primitives are needed to accomplish most robotic servicing tasks.

- *Trajectory Level*: This level provides a trajectory generator which computes the manipulator setpoints required to execute the motions requested by the task primitives.
- *Servo Level*: This lowest level provides servo control of the manipulator joint positions. It implements the PID control loop and normally runs at a rate between 200 and 1000 Hz.

The following subsections discuss each level in detail except the Servo Level which is provided by the robot manufacturer.

2.1 Trajectory Level Control

The trajectory level control makes use of the Robot Control "C" Library (RCCL) [3]. RCCL provides routines to specify and queue motion requests, a trajectory generator which computes the manipulator setpoints required to execute the motions, and interface to the robot controller. The features which make using RCCL attractive include:

- *Multiple robot control.* RCCL can control multiple robots in independent and coordinated operations. Coordinated motion is accomplished by maintaining the relative positions of the robots while moving the object along a desired path.
- *Sensor Integration.* RCCL allows user-defined functions to be executed synchronously at the control level in every sampling period. The functions can make use of the sensor data read back from the hand controllers and wrist force sensors to implement teleoperation and compliant force control. The capabilities have been combined to implement shared control.
- *Robot simulation.* RCCL provides a simulation feature under which the control level, instead of connecting to a set of real manipulators, connects to a Unix process that simulates the actions of the arms.
- *Ease of application development.* RCCL robot applications are written as ordinary "C" programs. As a result, learning, development, and testing are similar to common software projects. Debugging can be handled with a standard debugger and, with the help of the simulator, can be done in a non-realtime fashion.

2.2 Task Level Control

Task execution is provided by means of a set of parameterized *primitives* which accept input from the operator. Experience at JPL has demonstrated that a small set of primitives is required to perform most robot servicing tasks [4]. The primitives can be divided into four categories:

1. *Guarded Motion Primitives.* The guarded motion primitives are Cartesian-Guarded-Motion, Joint-Guarded-Motion, and Move-to-Contact. The Cartesian- and Joint-Guarded-Motion move through *via points* specified as Cartesian frames

and joint angles respectively. During the motion, they actively check the values of the force/torque sensor and stop upon a sensed collision. The Move-to-Contact primitive moves the arm until contact and backs off toward the initial position until the contact force is reduced.

2. *Generalized Compliant Motion Primitive.* A Generalized-Compliant-Motion primitive has been implemented for use in general compliant motion tasks such as door opening, bolt turning, pin insertion, and leveling by specifying values of a parameter set such as the destination frame, compliance frame, tool frame, and compliance gains. It can also be used for shared control tasks by specifying which degrees of freedom are controlled autonomously and which are controlled by teleoperation input.

3. *Grasp Primitive.* The Gripper-Motion primitive provides both free and compliant grasps using either a pneumatic or servoed gripper.

4. *Teleoperation Primitive.* The teleoperation primitive uses hand controller inputs to move the arm, with or without force reflection.

The trading of control between autonomous and teleoperation is achieved by executing different primitives, and shared control is accomplished in the Generalized-Compliant-Motion primitive by modifying some degrees of freedom of the motion according to the input of the hand controllers.

2.3 Operator Level Control

The operator interface will include: 1) a graphical user interface to control the robotic system in tele-autonomous operation and monitor the status of the system, 2) hand controllers to manipulate the robots in teleoperation, and 3) a 3-D graphical system to simulate the robots.

The graphical user interface consists of the Operations, Task Primitives, Operation Building, and Status subwindows. Single task execution can be initiated by selecting a task primitive, which will open its parameter input subwindow, allowing the operator to select and input the specific parameters for that task. Primitives can be threaded together into operations which require a number of primitives to accomplish. The interface allows the operator to select either robots for task execution, and to select whether the execution is performed by the real or simulated robots. Real time robot states will be displayed on the screen for the operator to monitor the execution. At any time the operator can press the STOP button to abort the execution of the primitives. An uninterruptible hardware kill switch should also be connected to the robots in order to turn off manipulator power and apply brakes to the joints for emergency shutdowns.

The 3-D graphical display which provides robot motion simulation allows the operator to first preview the task execution before he/she sends the commands to the actual manipulators. This feature is especially useful in high risk environments such as nuclear facilities.

3 Fail-Safe Operations

In order to allow robots to operate critical tasks that are potentially hazardous, provisions for fail-safe operations are needed to be integrated into the system. Fail-safe operation means that when a loss of control occurs, there is no damage to either the robot or its immediate task environment as a result of the failure, and that the failure does not pose a threat to the safety of plant personnel or the external environment.

3.1 Fault Tree Analysis

Fault tree analysis [5] involves the construction of a logic diagram showing credible event sequences that could lead to a specified hazard. A partial fault tree for damage to the filter cartridge during replacement is shown in Figure 2.

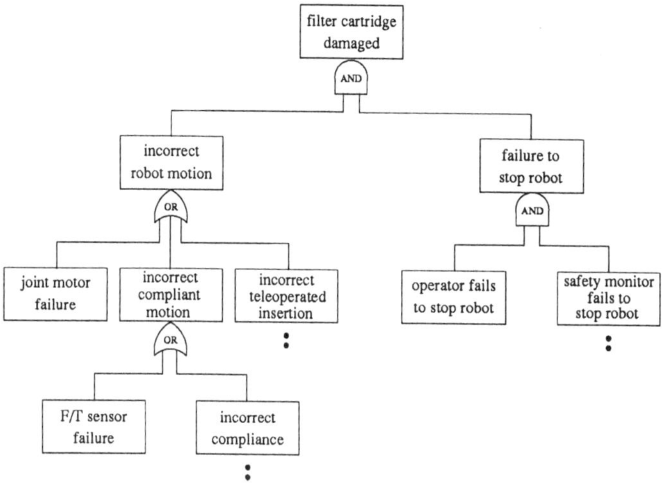

Figure 2: Top Level Fault Tree for Filter Cartridge Damage

In the fault tree diagram, when an OR gate links multiple events, all of them must be covered, but when an AND gate joins the failures, it is only necessary that one event be covered. Fault trees are valuable for 1) documenting all assumptions on anticipated inputs, outputs, and operations, 2) identifying missing classes of potential failures and providing a structure for reviews, 3) justifying the deliberate omission of certain classes of failures because they are derived from compound events and are thus "extremely unlikely."

3.2 Design of Safety Monitoring Provisions

The safety monitoring provisions reflects the previously described control hierarchy of the robotic system.

- *Servo Level*: The servo level should have a hardware kill switch with which the operator can turn off motor power and apply brakes to the joints for emergency robot motion shutdown.
- *Trajectory Level*: The servo level kill switch is useful only if the incorrect motion is slow enough for the operator to react. At this level, the software should monitor the joint velocities and force/torque values and stop the robot whenever any of the thresholds is exceeded.
- *Task Level*: The Trajectory Level safety monitor is limited because the thresholds are set independently from the tasks. If the thresholds are set too high,

damages to delicate objects may still occur. If they are set too low, false alarms may occur very often. The problem can be remedied at the Task Level by setting the thresholds according to task needs.

- *Operator Level*: The safety monitors are effective only if the incorrect actions cause a threshold to be exceeded. For other incorrect actions the vigilance of the operator is relied on to stop the motion. One way to alleviate the demand on the operator is to store the execution profile of the common operations. During execution, the software can detect failure by actively comparing the execution with the stored profile.

From the fault tree in Figure 2, we can see that if we can stop the arm from carrying out incorrect motion, we can have safe operation regardless of hardware and software failures.

4 Conclusions

The fail-safe tele-autonomous robotic system for nuclear reprocessing operations described in the paper exploits the technologies developed for space telerobotics. The system provides local-remote dual-arm sensored-based operations in three control modes: autonomous, teleoperation, and shared control. Also, safety provisions have been integrated to provide fail-safe operations. This system will provide greater capability than previous systems, particularly in task specification and control.

5 Acknowledgments

The research described in the paper was partly performed at the Jet Propulsion Laboratory, California Institute of Technology, under contract with the National Aeronautics and Space Administration.

References

[1] S. Hayati, T. S. Lee, K. S. Tso, P. G. Backes, and J. Lloyd, "A unified teleoperated-autonomous dual-arm robotic system," *IEEE Control Systems*, vol. 11, pp. 3–8, Feb. 1991.

[2] P. G. Backes, "Dual-arm supervisory and shared control space servicing task experiments," in *Proceedings AIAA Space Programs and Technologies Conference*, (Huntsville, AL), Mar. 1992.

[3] J. Lloyd, M. Parker, and R. McClain, "Extending the RCCL programming environment to multiple robots and processors," in *Proceedings of the 1988 IEEE International Conference on Robotics and Automation*, (Philadelphia, PA), pp. 465–469, Apr. 1988.

[4] P. G. Backes and K. S. Tso, "UMI: An interactive supervisory and shared control system for telerobotics," in *Proceedings of the 1990 IEEE International Conference on Robotics and Automation*, (Cincinnati, OH), pp. 1095–1101, May 1990.

[5] W. Vesely, F. Goldberg, N. Roberts, and D. Haasl, "Fault tree handbook," technical report NUREG-0492, U.S. Nuclear Regulatory Commission, Jan. 1981.

AUTONOMOUS BERTHING OF A WORK ATTACHMENT MECHANISM/ WORK ATTACHMENT FIXTURE*

C.C. Nguyen† S.C. Antrazi‡ A. Sciannella, Jr.‡
Robotics and Control Laboratory
Catholic University of America
Washington DC, 20064

C.E. Campbell, Jr. J.M. Vranish
Intelligent Robotics Laboratory
NASA/Goddard Space Flight Center
Greenbelt, Maryland 20771

Abstract

This paper deals with autonomous berthing of a Work Attachment Mechanism/Work Attachment Fixture (WAM/WAF) developed by NASA for berthing and docking applications in space. The WAM/WAF system enables fast and reliable berthing/unberthing of space hardware. A successful operation of the WAM/WAF requires that the WAM motor velocity be precisely controlled. The paper first describes the operating principle and design of the WAM/WAF and then presents the development of a control system used to regulate the WAM motor velocity. Finally it reports the results of an experiment in which the WAM/WAF is used to handle an orbital replaceable unit.

1 Introduction

Handling of space hardware such as Orbital Replaceable Units (ORU) [1] can be done by using robot fingers and finger interfaces. The fingers are mounted to the gripper of the robot manipulator and the interfaces to the ORU. Picking up the ORU is achieved by moving the robot manipulator to a location in which the fingers align with the interface, and closing the fingers to grasp the interface. The above strategy suffers from several drawbacks such as unsuccessful grasping of the fingers with the interface due to misalignment, requirement of highly accurate position and force control schemes causing on-line computational burden, and low payload capability due to insufficient holding force of the mating surfaces between the fingers and the interface. The above disadvantages of conventional handling strategy have motivated Goddard Space Flight Center (GSFC) to develop a berthing mechanism, called Work Attachment Mechanism/Work Attachment Fixture (WAM/WAF) [2] which enables fast and reliable berthing and unberthing of space hardware. Successful operation of the WAM/WAF requires that the velocity of the WAM motor be precisely controlled. This paper deals with the problem of autonomous berthing/unberthing of the WAM/WAF and is structured as follows. Section 2 describes the operating principle

*Research funded by Goddard Space Flight Center, Grants NAG 5-780 and NAG 5-1145
†Director and Professor of Electrical Engineering
‡Graduate Research Assistant

and design of the WAM/WAF and Section 3 presents the development of a control system designed to control the WAM motor velocity. Section 4 reports the results of an experiment in which the WAM/WAF is used to handle an ORU, and also discusses associated problems such as passive compliance and alignment. Section 5 reviews the paper and outlines some future activities.

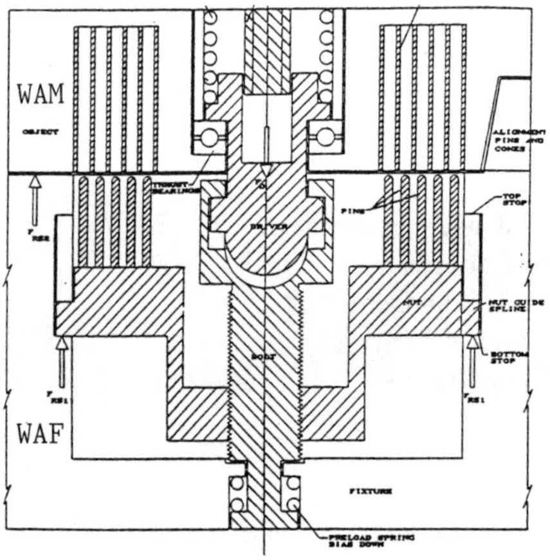

Figure 1: The WAM/WAF

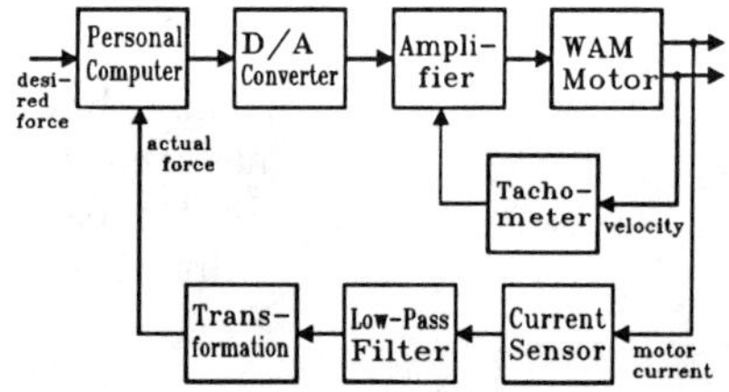

Figure 2: The WAM control system

2 The WAM/WAF

The development of the WAM/WAF was motivated by several drawbacks from which the original NASA fastener and its several improved versions have suffered. The first version of the NASA fastener was equipped with screws having low pitch machine threads which can be easily cross-thread by astronauts wearing gloves and space suits. Screws having high pitch ACME threads were used in a later version of the NASA fastener to resolve the above problem. However, spacecraft vibration can cause this type of high pitch thread screw to back out, according to a vibration simulation study. To overcome the latest problem, a following version of the NASA fastener incorporated a taper interface added on the top of the screw. However, the addition of the taper requires that large torques of the order of 100 ft-lb be applied to unscrew the fastener. Most of the above problems have been resolved in the design of the WAM/WAF, the latest version of the NASA fastener. This device was initially designed to fasten the Flight Telerobot Servicer (FTS) to the space station structure and can be used for other space applications such as serving as a leg interface device for robots walking in space to calibrate instruments for Earth Observing Systems (EOS). Figure 1 illustrates the main components of the WAM/WAF whose designed was based on the concept of *spline-locking screw*. As shown in the figure, the WAM consists of a driver, a motor driving the driver and an object. The WAF is composed of a fixture, a driver, a spline interface and a bolt with fine pitch. The object can be berthed to the fixture by a sequence of actions. First the driver is moved to rest on top of the bolt and then is rotated clockwise. A preload spring is installed under the bolt to reduce the friction between the driver head and the bolt to prevent them from moving together. As the splines of the driver and the bolt line up, the driver bias spring pushes the driver into the bolt. Now having enough torque to overcome the friction caused by the bolt preload spring, the driver and the bolt rotate together and the bolt moves downwards. The bolt continues to move and stops its motion when a desired *locking force* is exerted between the object and the fixture. The unberthing process is in the reverse order. Starting with the object locked to the fixture via the locking force, a torque with sufficient magnitude is applied to rotate the bolt and the driver together in a counter-clockwise direction to release the locking force. As the rotation continues, the preload spring keeps the driver in the downward position while the bolt moves upwards to push the object away from the fixture. Now the driver and the bolt rotate together until the bolt hits a stop and the driver splines and the bolt splines are automatically aligned. An upward motion of the WAM finally detaches the object from the fixture.

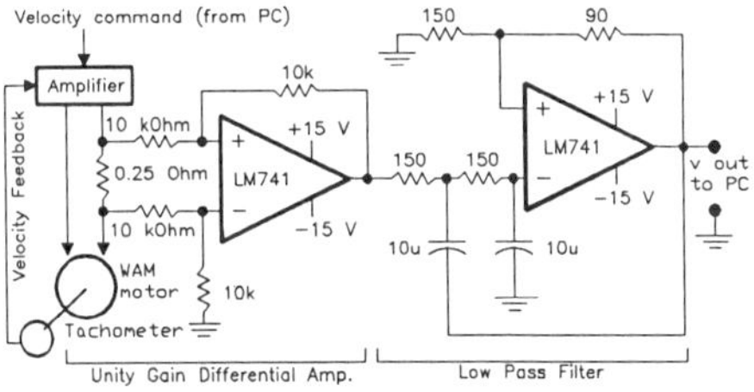

Figure 3: The current sensing circuit

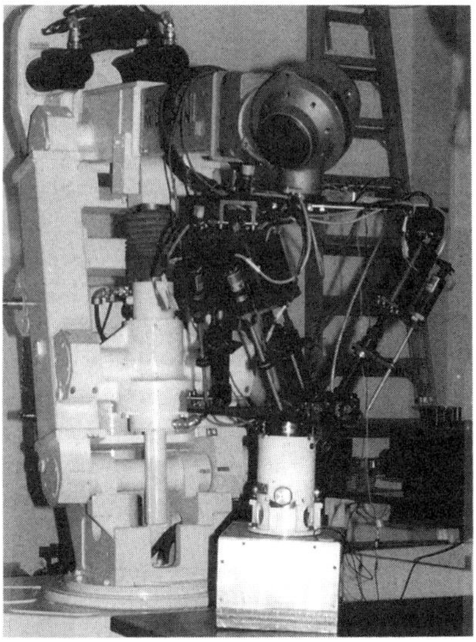

Figure 4: The GSFC testbed

3 The WAM Control System

As discussed in the previous section, a successful operation of the WAM/WAF relies heavily on the correct timing of the WAM motor velocity. In other words, for the berthing process, the WAM motor speed should be maintained at a constant value during the berthing and reduced at the end of the berthing so that the applied locking force is not too big for the motor to overcome at the beginning of the unberthing process. In addition, the motor should be stopped as soon as a desired locking force is reached. On the other hand, a sufficient torque should be applied at the beginning of the unberthing process to release the locking force and the motor should be stopped as soon as the bolt comes to a stop. Consequently, a control system which can sense applying torques and regulate the WAM motor velocity accordingly should be considered. The WAM motor is a permanent magnet motor driven by a pulse-width modulated servo amplifier that can be operated in both *velocity mode* and *torque (current) mode* with or without a tachometer. Figure 2 presents a control scheme consisting of two loops, the inner velocity control loop and the outer current sensing loop. The velocity control loop implemented in the velocity mode with a tachometer is used to maintain a constant motor velocity at selected time frames and the current sensing loop is used to monitor the motor current which is proportional to the motor torque due the permanent magnet characteristics of the motor. The mechanical stops at the end of the berthing and unberthing processes are detected by a surge of motor currents. Based upon the motor current value, a personal computer (PC) sends proper signals to the amplifier via a digital/analog board to rotate the motor in a selected direction (clockwise for berthing and counter-clockwise for unberthing) or to stop the motor. Figure 3 shows the current sensing network in which a high power resistor with a very low resistance value is placed in series with the motor to sense its current. The voltage dropping across the resistor, which is amplified by a differential amplifier represents a value proportional to the motor current to be sampled by the PC. Since the PC is generally sampling the current at a very low rate as compared to the switching frequency of the amplifier, a low-pass filter is connected in series with the differential amplifier to pass only the low frequency current to the PC, replicating the motor response to the amplifier output. Since the switching frequency of the amplifier is 20 kHz and the PC sampling frequency is approximately 50 Hz, a second order low-pass filter is designed with a cutoff frequency of 106 Hz. In addition, controlling software is written for the PC so that it can distinguish transient current spikes produced by initial motor acceleration, final motor deceleration, and unexpected transient frictional disturbances from steady-state currents caused by mechanical stops at the end of the berthing and unberthing processes. One problem which results from the above control scheme is the locking forces produced at the end of the berthing and unberthing processes. To minimize the berthing/unberthing time, the motor is controlled to maintain a relatively high velocity during the processes and stopped as soon as a steady-state current surge is detected. At the end of each process, the kinetic energy produced by the motor velocity will be transformed into the potential energy stored in the locking force. Both berthing and unberthing processes will need an additional torque to overcome the locking forces to break the object away from the fixture (at the beginning of the unberthing) and to break the spline interface away from the bolt (at the beginning of the berthing). However, since velocity mode is selected in the above control scheme, once a current limit is set using a designated potentiometer located on the amplifier for one process, the same current limit must be used for the reverse process unless the user decides to manually adjust the current limit before starting any reverse process. If the current limit stays the same for the reverse process, the amplifier is unable to supply the additional torque to overcome the preload locking force. One way to overcome this problem would be to run the motor at very low speeds, which result into extremely long berthing/unberthing processes. Instead, the motor is controlled to run at a high velocity during most of the berthing/unberthing processes, and then to slow down as the end of the processes is reached. Consequently the entire berthing/unberthing process is first timed and then the time can be used by the PC software program to determine when to slow down the motor. According to experimental results reported later, this method is quite reliable since the motor operating in the velocity mode can be controlled to maintain a repeatable velocity profile during the berthing and unberthing processes and is much simpler than the alternative method of employing switched resistor networks to change the amplifier current limit during different stages of the processes.

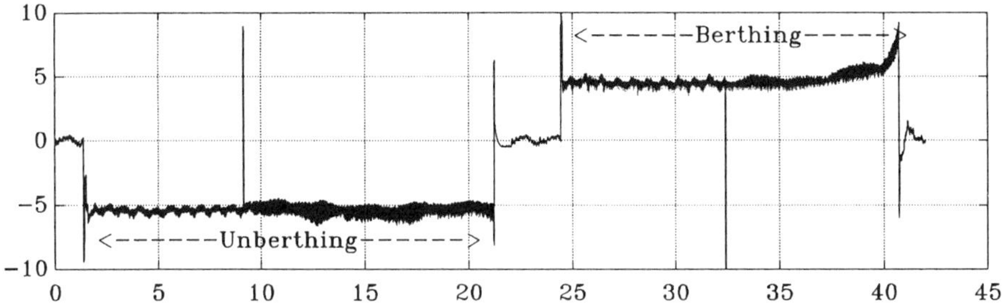

Figure 5: Motor current response *(hori. axis=time [sec]; vert. axis=current [A])*

4 Experimental Evaluation

The results of an ORU handling experiment are reported in this section. A testbed of the Intelligent Robotics Laboratory at GSFC is shown in Figure 4 where a *full size* WAM/WAF whose length is about a foot, is mounted to an ORU and the WAM to a compliance mechanism (CM) via a six degree-of-freedom (DOF) force/torque sensor. The base platform of the CM is mounted to the last link of a Cincinnati Milacron T3 robot. The design of the CM was based on the Stewart Platform and a complete analysis of its kinematics can be found in [3]. Arbitrary passive compliance (stiffness) of the CM can be achieved by selecting proper proportional gains for the position controllers of the CM legs. We now discuss the tasks to be carried out in the experiment. With the ORU laying flat on a table, the WAM is first manually[1] berthed to the WAF. Then the T3 robot is commanded to slightly push the ORU on the table and the forces/torques measured by the force/torque sensor are recorded as *fine alignment force/torque pattern*. After the T3 robot records the current ORU pose, the PC sends a signal to the amplifier to start the unberthing process by rotating the WAM motor counterclockwise. As soon as the unberthing process is completed, which is detected by a steady-state current surge sensed by the current sensor, the T3 robot is commanded to perform an upward vertical motion to move the WAM out of the WAF. Since the WAF was intentionally placed on a sloppy table, the ORU pose is slightly disturbed as the WAM leaves the WAF. Now the T3 is commanded to move the WAM back to a pose of about two inches above the ORU. The T3 robot then moves the WAM slowly down onto the WAF and stops when a contact between the WAM and the WAF is established through a reading of the force/torque sensor. The WAM is now situated in a rough alignment with the WAF and the berthing should not start until a fine alignment between the WAM and the WAF is obtained. In order to align the WAM with the WAF, an alignment scheme is activated to command the T3 robot to modify the ORU pose in very small increments until the fine alignment force/torque pattern is reproduced resulting in a fine alignment between the WAM and the WAF. Then the PC sends a signal to the amplifier to start the berthing process. When the WAM control scheme finishes the berthing process which is detected by a steady-state current surge, the T3 robot is commanded to lift the ORU off the table and move to a pre-recorded pose. Figure 5 presents the

[1]In manual mode, the WAM motor velocity profile is adjusted step-by-step using the PC keyboard

transient response of the motor current recorded during the above experiment. As illustrated by the figure, the first current spike of the unberthing process occurs when the motor is accelerated to a high velocity. After that, the motor is controlled to stay at the same velocity until it is decelerated (confirmed by the second current spike) to a low velocity. The motor continues to run with the low velocity until hitting a hard stop which is detected by a steady-state current surge and is decelerated (manifested by the third current spike) sharply to a complete stop (confirmed by zero current). Similarly in the berthing process, the motor is accelerated (confirmed by the fourth current spike) to a high velocity and stays at this velocity until it is decelerated (the fifth current spike) to a low velocity. Keeping at this low velocity, the motor slowly turns and is decelerated to a complete stop (manifested by the sixth current spike) as a desired locking force exerted between the bolt and the nut is detected by the corresponding motor current. Unlike the unberthing process in which the current increases sharply as the bolt hits the hard stop at the end of the process, the current in this case builds up exponentially due to the continuous build up of the locking force as the bolt is preloading the nut.

5 Conclusion

This paper has dealt with autonomous berthing/unberthing of a NASA fastener, called WAM/WAF. The operating principle and design of the WAM/WAF were described and an operational amplifier circuit which senses and filters the WAM motor currents was developed. The paper then presented a control scheme regulating the motor velocity based upon the motor current responses. Results of an ORU handling experiment in which fine alignment was obtained via a force-based alignment scheme and passive compliance was provided by the CM showed that the WAM/WAF can be autonomously berthed/unberthed using the developed control scheme. Recently, a *mini* WAM/WAF whose size is about one fifth of that of the full-size WAM/WAF was designed and built at GSFC. Autonomous berthing control schemes are being developed for this WAM/WAF and numerous potential WAM/WAF applications are being considered for maintenance and service of space hardware.

References

[1] **Catalano, D.A.**, "WP-04 Orbit Repleable Units Robotics Maintenance Final Test Report," *Analex Corporation*, Goddard Space Flight Center, November 1990.

[2] **Vranish, J.M.**, "Spline-Locking Screw Fastening Strategy (SLSFS)," *Proceedings of the Fourth World Conference of Robotics Research*, Pittsburgh, Pennsylvania, September 17-19, 1991.

[3] **Nguyen, C.C., Antrazi, S., Zhou, Z-L, and Campbell, Jr., C.E.**, "Analysis and Implementation of a 6 DOF Stewart-Platform-Based Robotic Wrist," *Computers and Electrical Engineering: An International Journal*, Vol. 17, Number 3, pp. 191-204, 1991.

[4] **Nguyen, C.C., Antrazi, S., Zhou, Z-L, and Campbell, Jr., C.E.**, "Adaptive Control of a Stewart Platform-Based Manipulator," Invited Paper, to be published in a special issue of *Journal of Robotic Systems*, 1992.

Robotic Berthing Via Capaciflector Imaging

Intelligent Robotics Laboratory
Goddard Space Flight Center, NASA

Charles E. Campbell, Jr., PhD.
Jacqueline Le Moigne, PhD.

Abstract

The capacitance between a *capaciflector* sensor and nearby objects can be used to determine the range to and the shape of those objects. The capacitance measured is a function of geometry, permittivity, and relative position and orientation (pose) of nearby objects to the sensor. The Intelligent Robotics Laboratory (IRL) at Goddard Space Flight Center has used a new sensor called the *capaciflector* to develop capacitance maps (images) of a platform with four pins.

An *X-shaped* sensor on the bottom of an ORU (Orbital Replacement Unit) box was moved in a raster pattern over a test pad by a 6 degree of freedom robot. The capacitance was measured while the robot was moving, and so one axis was sampled at high resolution and the other at low resolution.

Matched filtering, threshholding, and erosion was used to localize pins in the image. The algorithm in [3] was applied to determine the relative pose of the pins to the center of the image.

Introduction

The capaciflector, invented by John Vranish of Goddard Space Flight Center [7], has a capacitance-based object-detection range of up to 18 inches which is an order of magnitude better than that provided by standard capacitive sensors. Nearby objects can be usefully imaged within 6 inches, and the resulting images can be used to support robotic berthing and docking activities. This new sensor is composed of two strips of metal called the *shield* and the *sensor* (see Fig. 1). The shield uses a voltage follower circuit slaved to a low frequency signal ($\approx$ 10KHz - 30KHz) which drives the sensor. As objects come closer to the sensor, its accuracy improves. Figure 1 illustrates the capaciflector principle.

Capaciflectors use a shield to dramatically improve the detection range [7] as compared to typical capacitive sensors. The shield uses a voltage follower slaved to the oscillator driving the sensor strip. Since the sensor and the shield are in close proximity and are identically excited, the field lines emanating from the capaciflector are nearly the same as those that would emanate from

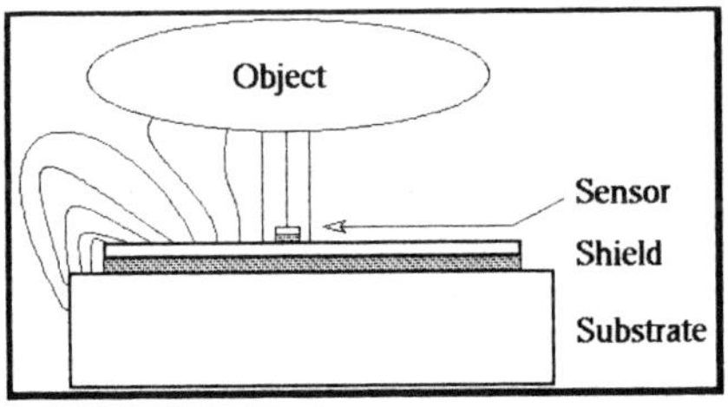

Figure 1: The Capaciflector Concept

the shield by itself. However, the central portion of the shield, facing away from the substrate, has field lines which remain reasonably parallel and orthogonal to the shield for the longest distance. These field lines are the ones which interact most with nearby objects. Hence, the sensor strip placed over the center of the shield generates nearly the same field lines that the bare shield would have in that region, but the sensor's field lines are the "best" (farthest reaching) field lines, interact with the substrate the least, and are therefore responsible for most of the capacitance to ground via nearby objects.

Capaciflector Sensor Processing

The capaciflector circuitry produces an analog signal and uses an input reference voltage to set it to zero output. The analog sensor output is converted to a 12-bit integer. The nominal range of the sensor is ±10V; however, the A/D board has a computer-controlled amplifier which serves to improve resolution when the capaciflector voltage is near zero (in steps of 2x, 4x, and 8x). Noise is suppressed by run-length averaging and a 3-pole digital filter.

The capaciflector's voltage reference is initialized during a "prefiltering" process. The algorithm has two main stages: the stepper, which serves to set up an interval which brackets the best voltage reference, and the binary searcher, which halves the interval until the best voltage reference is obtained. These algorithms were selected for their guaranteed convergence properties. Algorithm 1 details the method.

Algorithm 1 *Prefiltering Algorithm*

1. Start reference voltage control to −5V (which yields output voltage −10V). Delay 5 seconds to insure settling.
2. Stepper Phase: Increase reference voltage control by .5V steps until the capaciflector output goes positive. For each increase:
   (a) Take small block averages (6 samples each) of capaciflector output; label each block as BA_i for the ith block
   (b) Reject capaciflector output until $|BA_{i+1} - BA_i| <$ RejectThreshhold, where the *Rejection Threshhold* was selected to be .1V.

   (c) Use a large block average of 80 samples as the capaciflector output.

3. Binary Search Phase: Record last two voltage reference settings as $vref_-$ and $vref_+$.
   (a) If $|vref_+ - vref_-| <$ D/A converter resolution, stop.
   (b) Set the voltage reference $vref$ to $(vref_- + vref_+)/2$.
   (c) Perform settling time rejection as for prefiltering and a large block average of 80 samples as before.
   (d) If capaciflector output is ≈ 0, stop.
   (e) Else if capaciflector output is < 0: Set $vref_-$ to $vref$.
   (f) Else if capaciflector output is > 0: Set $vref_+$ to $vref$.

Capaciflector noise is suppressed by a run-length averaging filter whose output is pipelined through a three-pole digital filter. The run-length averaging filter nominally takes 80 samples, and the three-pole digital filter is composed of three pipelined single pole filters. The fg (filter gain) was selected as .9 and Δt was typically 50$m$seconds.

$$\overline{\text{capout}_i} = \text{capout}_i + fg\Delta t(\text{capout}_{i-1} - \text{capout}_i) \tag{1}$$

Capacitance Mapping

A Cincinnati Milacron T3 robot held an ORU box with four holes which mates with a four-pin berthing pad. On the bottom of the ORU was an "X" shaped sensor on a 4"x4" shield. This sensor was moved in a raster pattern over the targets given in Figures 4 and 5. The system constantly received information on where the robot was and what the capaciflector was currently reading. The capaciflector data was interpolated between robot pose sampling points to main regular sampling along the motion axis. The motion axis can be finely sampled with no time penalty, and both axis's sampling rates are user selectable.

In both Figures 4 and 5, the four pin berthing pad is imaged. The ORU pad is 15"x13" and the pins are 3/4" diameter, 2.5" high, tapered, and set on a 10"x10" square. Figure 4 shows the ORU via an imaging plane 4.5" inches above the ORU and parallel to the ORU. The second ORU picture, Figure 5, is taken with a rotation about the z (vertical) axis of 12° degrees and about the y axis of 4° degrees.

Capaciflector Image Processing

Once the range image has been acquired by the Capaciflector, and assuming that metal objects are being examined through air or vacuum (ie. permittivities are known), the image can be interpreted as a range image convolved with the Capaciflector's impulse response.

Various approaches have been taken to determine the pose of a robot [3] [5] [1]. The most common approach is to utilize interest points in the scene. In particular, it has been shown [3] [1] that by using four coplanar noncollinear points, there is a unique solution to the pose problem: this is known as the quadrangle method. In the case of the ORU imagery, the four pins are natural interest points from which the pose may be computed. The quadrangle method uses the spatial relationships of the four pins which are known in an absolute coordinate system. The locations of the four points are extracted and their 3-D coordinates are computed in the sensor's coordinate system. The relative pose is then determined by the 3-D affine transformation (three rotations, one translation) that takes the known absolute coordinates into the sensor coordinates.

Computing the pose from the Capaciflector data is then performed by following Algorithm 2.

Algorithm 2 *Pose Computation*

1. Extract from imagery all "dots": circular shaped, relatively uniform capacitance values
2. find the group of four "dots" which satisfy the known spatial relationships of the ORU pins
3. Compute the associated 3-D affine transformation

Extraction of "Potential Pins"

The *Image Processing step* processes an intensity image where the intensities represent range values. "Potential pins" are small circular regions in the image with equidistant points from the sensor, that means relatively uniform image intensities. These presumed pins are matched against a circular template of uniform intensity using correlation and a Fourier Transform. The correlation operation's peaks are extracted by a series of thresholding and erosion techniques, thus providing the centroids and the areas of all the potential pins. Figures 2 and 3 show the results of this image processing step on two ORU data sets, on axis and off axis. On the MasPar, a massively parallel computer, the time for the Image Processing is about 0.5 seconds.

Extraction and Labeling of the ORU Pattern

Extracting and labeling the ORU pattern is equivalent to matching each "potential pin" to one of the "model pins" and to choosing the proper combination of four matches that best fits the 4-pin pattern. Th matching process is represented by an interpretation tree [2]. Each node of the tree represents an individual match between a potential pin and a model pin. A path through four nodes of the tree (from root to leaf) represents four successive matches, ie. one global match to each 4-pin pattern. Each leaf of the tree is thus associated with a global match and for each leaf a confidence measure is computed. This measure describes how the spatial relationships between the four potential pins fit the given model. The match with the best confidence is then chosen. For more detail on step (2), see [4] and [6].

Computation of the 3-D Pose

Knowing the absolute coordinates of the quadrangle model and its sensor coordinates, we can compute the coefficients of the affine transformation: the corresponding equations are given by Hung and Yeh in [1]. Figures 2 and 3 summarize the results obtained. From the results of Table 1, the translation error is 0.86" and the rotation error is 0.78°; from the results of Table 2, the translation error is 1.5" and the rotation error is 2.5°.

These are preliminary results and they will be improved. The above errors can be explained by the low resolution of the data; this implies that one pixel error in locating the pins may correspond to a large error in range value which leads to large errors in the pose computation. One solution to this problem is to interpolate the data, which will reduce the errors in range values.

For the above examples, the combined time for steps (2) and (3) on the MasPar is 0.03 seconds, which is negligible compared to the Image Processing time. It is possible that more potential pins (clutter) may be extracted from the imagery, which would increase the size of the search tree and consequently the computation time: see [6] for a detailed study of these timings. This study shows that for 10 or 20 potential pins, the MasPar processing time is respectively 0.12 and 0.15 seconds.

References

1. M. Fishler and R. Bolles. Random sample consensus: A paradigm for model fitting with applications to image analysis and automated cartography. *Communications ACM*, 24(6):381 395, 1981.

2. W. Grimson. Object recognition by constrained search. In P. H. Winston and S. A. Shellard, editors, *Artificial Intelligence at MIT, Expanding Frontiers 2*, pages 289 313. MIT Press, 1990.

3. Y. Hung, P.-S. Yeh, and D. Harwood. Passive ranging to known planar point sets. In *IEEE Conference on Robotics and Automation*, pages 180 185, Mar 1985.

4. V. Hwang, T. Winkert, P. Spidaliere, and R. Keizer. On-point based pose computation under extreme lighting conditions. In *SPIE Cooperative Intelligent Robotics Space II Conference*, Boston, 1991.

5. Y. Lamdan, J. Schwartz, and H. Wolfson. Affine invariant model-based object recognition. *IEEE Transactions on Robotics and Automation*, 6(5):578 589, 1990.

6. J. L. Moigne, V. Hwang, and J. Dorband. Parallel tree interpretation for pose determination. In *Mannheim SuperComputer'91 Conference*, Mannheim, Germany, June 1991.

7. J. Vranish, R. McConnell, and S. Mahalingam. "capacíflector" collision avoidance sensors for robots. *Computers Elect. Engrng*, 17:173 179, Nov 1991.

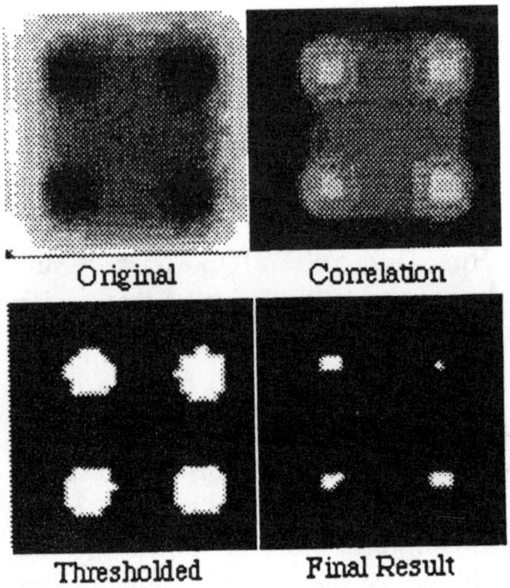

	Actual Pose	Computed Pose
TX(inch)	0.000	-0.4922
TY(inch)	0.000	0.1641
TZ(inch)	0.000	0.8584
RX(deg)	0.000	0.7769
RY(deg)	0.000	-0.2205
RZ(deg)	0.000	0.0000

Figure 2: On Axis Capaciflector Data

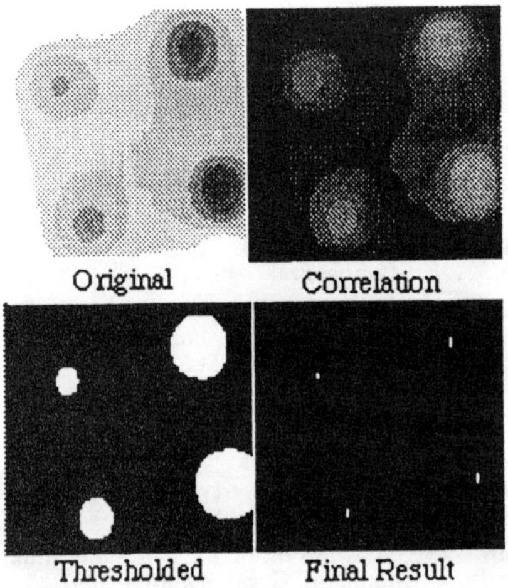

	Actual Pose	Computed Pose
TX(inch)	1.1859	1.0781
TY(inch)	0.0000	0.3750
TZ(inch)	3.9586	2.4121
RX(deg)	0.0000	-0.9649
RY(deg)	4.0000	6.1975
RZ(deg)	12.0000	14.5742

Figure 3: Off Axis Capaciflector Data

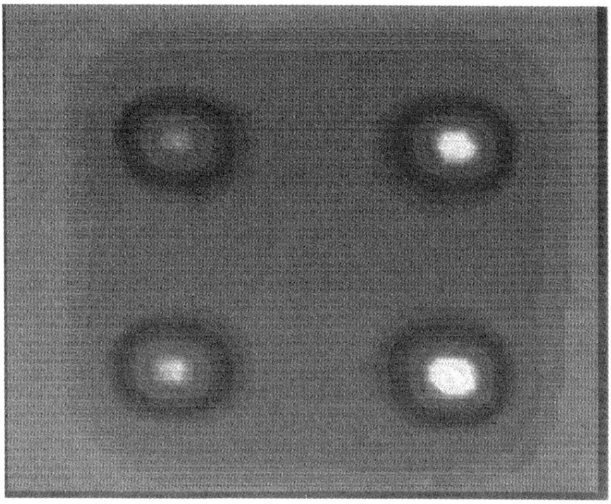

Figure 4: ORU Berthing Pad Image

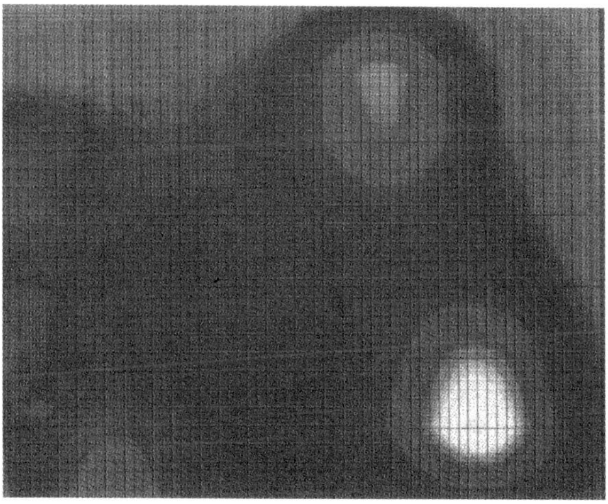

Figure 5: Tilted ORU Berthing Pad Image

A Dual and Redundant Arm Telerobot Task Execution System

Paul G. Backes, Robert Steele, Mark K. Long
Jet Propulsion Laboratory
California Institute of Technology
Pasadena, California

ABSTRACT

A telerobot task execution system is described which provides autonomous and shared control of single and dual-arm robotic systems for both redundant and non-redundant manipulators. The system, Modular Telerobot Task Execution System (MOTES), is capable of nominal task execution as well as monitoring and reflex motion. The system is designed to be the remote site execution system in a local-remote architecture space telerobot system. The system is written in Ada and runs in a VME environment on 68020 processors and initially controls both a Robotics Research 1207 7 degree of freedom manipulator and a PUMA 560 6 degree of freedom manipulator.

I. INTRODUCTION

A space based robot control system must satisfy various constraints including flexibility, robustness, safety and limited computational resources. Also, the robotic system may be remote from the operator, e.g, Earth based control of space robots. The architecture is then separated into the local site where the operator resides and the remote site is where the physical robots are. The robot control system described in this paper, Modular Telerobot Task Execution System (MOTES), has been designed to conform to these various constraints. A command interpreter is used to receive commands from the local site operator. The commands specify the permutations of control module behaviors for task execution. This method allows a fixed software system to provide a wide range of robot control behavior. The command interpreter approach has been previously used on unmanned robotic spacecraft such as Galileo [1]. A command to a module is a parameter set describing the desired behavior for that module. The architecture for trajectory generation and control is fixed but designed to provide a sufficient control capability.

This paper describes the design and implementation of the MOTES remote site telerobotics system for supervisory, shared, and teleoperation control of space robots which addresses the expected constraints of a flight system. MOTES is being implemented in the JPL Supervisory Telerobotics (STELER) laboratory initially for control of a 7 DOF redundant arm and a 6 DOF PUMA arm both for single arm control and cooperative dual-arm control. An earlier description of MOTES can be found in [2]. The paper is organized as follows. The system architecture and module functionalities are described in Section II. More detailed description of the modules are given in Section III, command types are given in Section IV and the implementation is described in Section V. Conclusions from the paper are given in Section VI.

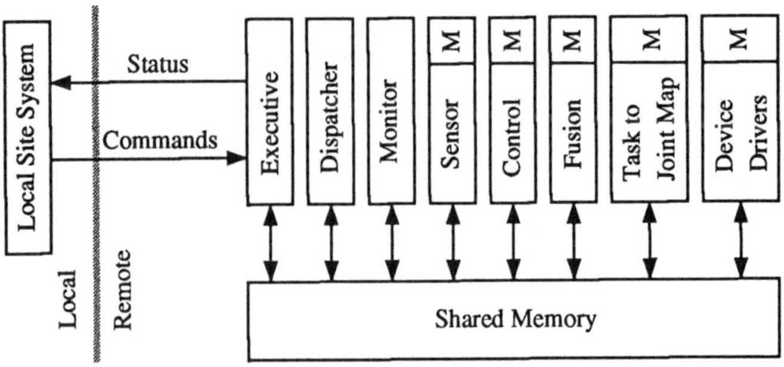

Figure 1: MOTES functional diagram

II. MOTES ARCHITECTURE

The MOTES system architecture is shown in figure 1. MOTES generates dynamic motion from a static description of the desired behavior. MOTES provides all task level control and task to actuator space mapping. The module types described below were selected because they represent different functionalities within the control system. There may be multiple modules of the same type, for example force, teleoperation, and collision avoidance control modules.

The **Shared Memory Module** provides access to all command parameters and system status information. The **Executive Module** handles communication with the local site system. It places new commands into the Task Command Queue and returns status and data. The **Dispatcher Module** controls the transition between execution states by checking the status of the various modules and specifying the appropriate commands and parameters to the various modules via shared memory. The **Monitor Modules** provide monitoring of the status of execution for both intended termination conditions and unintended error conditions. The **Sensor Modules** provide sensor data processing. The Sensor modules can represent both real and virtual sensors. The **Control Modules** provide the control associated with the various real and virtual motion sources. Each Control module generates a task space motion command. The **Fusion Module** merges the motion commands of the various Control modules into task space motion commands for the manipulators and other physical devices. The **Task to Joint Map Module** maps the task space command of the Fusion Module to the actuator space of the physical devices. The **Device Drivers Modules** communicate with the physical devices to send the actuator space commands and receive status data. It is assumed that the physical devices have their own low level control to implement the actuator space commands. The M in the various boxes of figure 1 indicates monitoring as described below in Section III.

III. MOTES MODULES

Each module interfaces to the rest of the system through shared memory with specified input and output parameters and functionality. This allows each module to be developed, tested and evolved independently. The modules operate asynchronously with the Dispatcher responsible for synchronizing the various modules via modification of command and state parameters in shared memory.

The Shared Memory is divided into the various sections described below. The Task Command Queue is the queue of commands sent from the local site. The Active Command Block contains the current command parameters resulting from local site commands or reflex action. The Reflex Command Queues are command queues for reflex actions to perform upon specific monitor events. The Reflex Table is a table which maps a monitor event to a reflex action. The System State Parameters are the system state as generated by system execution. Shared memory communication is achieved using procedures for reading and writing unique data records to and from shared memory. Two queuing mechanisms are used in the shared memory communication: first-in-first-out (FIFO) for commands and last-in-first-out (LIFO) for asynchronous communication between modules.

A command in the Task Command or Reflex Command Queues has two parts, the issue time and the functional command, a format similar to that used in the Galileo spacecraft Virtual Machine Language [1, 3]. The issue time can be absolute time, time relative to the end of the previous command, or immediate. Absolute time is useful when coordinating multiple manipulators. Absolute or relative time will usually be used in the first command of a new task so that the previous task is allowed to complete. The subsequent commands of the task will have immediate time. The functional command has two parts, the command identifier header which specifies the command number, the destination command queue (Task Command Queue or Reflex Command Queues), and which module the command is for, and the parameters of the command.

The Executive Module handles communication with the local site system. It places new commands into the Task Command Queue or Reflex Command Queues and returns status and data. The Executive communicates with the Dispatcher as necessary via shared memory. Normal task commands received from the local site are placed in a Command Queue so no specific communication with the Dispatcher is necessary. But if a command is received which needs immediate attention, then the Executive indicates via shared memory that there is a new command in the Task Command Queue requiring immediate attention. The Executive periodically sends the system execution status back to the local site. The Executive can also send blocks of collected data to the local site as commanded.

The Dispatcher controls the transition between commands by changing the Active Command Block parameters as specified by a monitor event. This may occur for various reasons including the successful and unsuccessful completion of a command. The Dispatcher acts as a command interpreter when issuing commands from the Task Command or Reflex Command Queues. The Dispatcher will issue the next command from the Task Command or Reflex Command Queues as specified by the issue time of the next command. If the issue time is not immediate, then the Dispatcher will wait for a termination condition or the arrival of an Interrupt command. If the issue time is immediate, then the next command is immediately issued. The status of execution is specified by monitor flags set by the various monitor modules. If a monitor flag indicates a monitor event, then the Dispatcher compares the flag state with the termination conditions as specified in the current command parameters. If the event is an acceptable termination condition, e.g., trajectory completed, then the next command of the queue is issued, if one exists. If there are no commands to be issued, then the Dispatcher issues the command sequence in the Default Reflex Command Queue, which is one of the Reflex Command Queues, and will continue until a new command from the local site is received or a monitor event occurs. If the monitor event is not an acceptable termination condition, e.g., force threshold exceeded, then the Dispatcher finds which Reflex Command Queue to use as specified by the entry in the Reflex Table which is associated with the event. The Dispatcher then executes the commands of the specified Reflex Command Queue, clears the Task Command Queue, and indicates to the Executive to report to the local site that the reflex action is occurring.

The Monitor Modules monitor the behavior of the system and set flags to indicate to the dispatcher when a monitor event has occurred. A monitor event is defined as a condition detected by a monitor module. Different monitor modules test for different conditions with testing parameters specified in the Active Command Block. Some examples of events sensed by the monitor modules are force-torque limits (collision), joint limits, and singularities. Monitoring occurs both in individual monitor modules and in the other modules. The M in the boxes of figure 1 represents the monitoring contained in the other modules. The reason is that monitoring occurs at the level where the information is generated and the context can be known. The Monitor Module is for monitoring states which have scope beyond the individual modules.

The Sensor Modules generate either real or virtual sensor data. Real sensor data is information that is directly related to an actual sensor while virtual sensor data is indirectly related to sensed information. Real sensors may include resolvers, potentiometers, force-torque sensors, proximity sensors, hand controllers, and vision systems. Virtual sensors may include distances to joint limits or distances to collision between objects based upon geometric models. Sensor information may also originate external to MOTES such as the position of an astronaut.

The Control Modules perform control for each real or virtual sensor as well as generating setpoints. The resulting command from each control module is placed in shared memory for the Fusion Module to read. Each Control module has input parameters in the Active Command Block which specify the desired behavior for the module. The Control modules perform control in task space for specific sensor data types. Each control module has a uniquely specified Cartesian control frame for control of the six Cartesian DOFs. The complete motion command specified by each control module is transformed to a common task space before being placed in shared memory. The common task space consists of the common Cartesian frame and the arm angle. The position trajectory generator is the control module which computes the desired task space position trajectory, including Cartesian frame and arm angle motion. Cooperative dual-arm control is achieved by utilizing task space control of the commonly held object [4]. Move-squeeze decomposition decomposes the forces from the two robots at the grasp points into internal, or squeeze, components and external, or move, components. A sensor module computes the move-squeeze decomposition. Control modules utilize the squeeze components to compute position perturbations for each arm to control the object internal forces. The move components are used for task space control of the common object.

The Fusion Module combines the commands from the various control modules as specified in the task parameters. For sensor based motion, the appropriate impedance model in each degree of freedom of the task space is used. The resulting motion command is then merged with the position trajectory generator motion to compute the command setpoint. The Task to Joint Map Module maps the task space command of the Fusion Module to the actuator space of the physical devices. The task space to joint space mapping is done using a composite Jacobian approach [5] using the method described in [6] for each arm. The Device Drivers Module communicates commands and status with the system hardware as well as perform computations which are hardware specific.

IV. COMMAND TYPES

There are various types of commands that the Executive can receive from the local site system. As described above, the commands have two parts: issue time and functional command. The functional identifier header within the functional command indicates the command type, which may include Interrupt and Module commands for each module. Additional command types, e.g., Cancel, may be added in the future. The commands are described below.

Module commands are placed in the command queue specified by the destination queue parameter in the command identifier header, e.g., the Task Command Queue or one of the Reflex Command Queues. Each module, e.g., joint limit monitor module or force-torque control module, receives data in an individual command. Each module command describes a module behavior or monitor condition. A sequence of Module commands is used to generate a collective command behavior. Each Module command has the previous command number as a parameter. If it does not agree with the actual previous command number, then the appropriate monitor event is set.

The Interrupt command causes the Executive to clear the Task Command Queue and set a monitor event specified in the command parameters. The Dispatcher will then read the monitor event and act on it as specified by the reflex table. The event might just tell the Dispatcher to read the next command of the Task Command Queue (which was placed behind the Interrupt command by the Executive as specified by the local site). The Interrupt command might also cause the Dispatcher to begin issuing commands from one of the Reflex Command Queues.

V. IMPLEMENTATION

The initial version of the MOTES system has been implemented in the JPL Supervisory Telerobotics (STELER) laboratory and controls a 7 DOF redundant manipulator. MOTES is written in the Ada programming language and runs in a VME environment on 68020 processors. Initial development of MOTES Ada code is done on a Sun-4/260 and compiled with a Sun native compiler. A Verdix VADSworks Ada cross compiler is used to produce the target code. The target code runs on top of the VADSworks operating system in the Heurikon 68020 environment. The hardware at the remote site includes a Robotics Research model K1207 7 DOF manipulator, a PUMA 560 6 DOF manipulator, a 6 DOF LORD wrist force-torque sensor, TRI servoed grippers, and the MOTES VME chassis and associated cards. Communication between the Robotics Research arm controller and the MOTES VME chassis is through a pair of Bit-3 memory interface cards. The direct servo control of the arm joint angles is provided by the manufacturer supplied controller. The local site 6 DOF hand controller [7] has a dedicated VME chassis for its control. The hand controller VME chassis and LORD force-torque sensor both communicate with the MOTES VME chassis via parallel interface. Communication with the TRI servoed gripper is via serial interface. The hand controller provides seven DOFs of input to control the Robotics Research Arm; six DOFs are the motion of the hand grip and one DOF is the trigger position with trigger polarity specified with a button on the grip.

VI. CONCLUSIONS

A telerobot task execution system is described which provides dual and redundant arm task execution for the remote site of a local-remote space telerobot system. Multiple real and virtual sensors can provide commanded motion simultaneously so that permutations of behaviors from control based on the various sensors can provide a wide range of capabilities with the system.

ACKNOWLEDGEMENTS

The research described in this paper was carried out by the Jet Propulsion Laboratory, California Institute of Technology, under a contract with the National Aeronautics and Space Administration. The authors would also like to thank Olen Adams for valuable conversations on Galileo spacecraft command sequencing.

REFERENCES

[1] Galileo Project. Galileo program description document - command and data subsystem, phase 9.1. Technical Report 625-355-06000, D-535 Rev. G, Jet Propulsion Laboratory, May 1989.

[2] Paul G. Backes, Mark K. Long, and Robert D. Steele. Designing minimal space telerobotics systems for maximum performance. In *Proceedings AIAA Aerospace Design Conference*, Irvine, CA, February 3-6 1992.

[3] Galileo Project. Galileo flight operations plan - galileo command dictionary. Technical Report PD 625-505, D-234, Jet Propulsion Laboratory, September 1989.

[4] Paul G. Backes. Dual-arm impedance based generalized compliant motion. In *Proceedings International Symposium on Robotics and Manufacturing*, Santa Fe, November 1992.

[5] S.Y. Oh, D. Orin, and M. Bach. An inverse kinematic solution for kinematically redundant robot manipulators. *Journal of Robotic Systems*, 1(3):235–249, 1984.

[6] Paul G. Backes and Mark K. Long. Redundant arm control in a supervisory and shared control system. In *Proceedings AIAA Space Programs and Technologies Conference*, Huntsville, AL, March 24-27 1992. AIAA paper No. 92-1578.

[7] D. McAffee and T. Ohm. Teleoperator subsystem/telerobot demonstrator: Force reflecting hand controller equipment manual. Technical Report D-5172 (internal document), Jet Propulsion Laboratory, January 1988.

SMART: A Modular Architecture for Robotics and Teleoperation

Robert J. Anderson

Sandia National Laboratories

Albuquerque, NM 87185[1]

Abstract

*This paper introduces SMART: Sandia National Laboratory's **M**odular **A**rchitecture for **R**obotics and **T**eleoperation. SMART is designed to integrate the different slave devices (e.g., large hydraulic arms, mobile manipulators, gantry robots), sensors (e.g., ultra-sonic sensors, force sensors), and input devices, (e.g., track ball, force-reflecting master, autonomous trajectory generators) required for waste management and environmental restoration tasks. The modular architecture allows for rapid synthesis of complex telerobotic systems. This paper introduces some sample modules and illustrates how the modules can be connected to achieve telerobotic behaviors. Examples include autonomous control, impedance control, and enhanced bilateral teleoperation.*

1.0 Introduction.

Sandia National Laboratories is developing technologies for remediating hazardous waste in remote environments. Operations involve a wide number of devices and sensors, utilize different levels of world modeling and require different levels of autonomous control and human intervention. In addition, one must insure that parts aren't damaged, container walls aren't breached, and that tasks are performed correctly.

In this paper we introduce SMART: (Sandia's Modular Architecture for Robotics and Teleoperation) a real-time control architecture for constructing telerobotic systems consisting of input devices, sensors and output devices. The architecture is generic. It can be used for any robot that can accept external position set points and with any sensor that can be interfaced to a VME bus. The architecture is both open and modular. Distinct modules are defined for each sensor, actuator, input device, and dynamic element. The hardware consists of commercially available cpus and processor boards running on a VME bus with a UNIX based operating system (VxWorks). The system is connected to a host processor and a graphics workstation using Ethernet. The software consists of libraries of C-routines. The architecture is parallel and real-time. Modules can run asynchronously, distributed arbitrarily across multiple cpus and attached processors. SMART uses dynamic interactions

[1]This work was performed at Sandia National Laboratories and supported by the U.S. Department of Energy under contract DE-AC04-76DP00789.

between modules to obtain desired behaviors. Servo loops are closed at fixed rates ranging from 100 Hz to 1KHz.

Using SMART it is possible to build flexible telerobotic workcells incorporating different input devices and robots. Sensors and world models work together to prevent the slave robot from damaging either itself or the environment.

1.1 Previous Research.

The work in this paper relates to many topics: open architecture control systems, telerobotic control systems, passivity based control systems, sensor integration, graphical control and impedance control. In this section we discuss some of the current research in these areas and how it relates to the SMART control architecture.

A number of researchers have introduced open-architecture control systems for robots based on UNIX and C. Some have involved replacing the robot controllers with generic robot controllers [10, 15, 16, 17]. Others have developed high-level interfaces to existing controllers [13]. These systems have been designed primarily for robotic control, and thus focus on issues such as task planning, programmability, and sensor interfacing. The SMART system does not plan tasks nor optimize the control architecture for particular sensors. Instead it provides an intuitive reconfigurable environment for telerobotics, with ample hooks for a high-level task planner.

In the area of telerobotic control architectures, a number of open architectures have been proposed for controlling unilateral systems [5, 6, 8, 12] , i.e., where the operator is not dynamically coupled to the robot. SMART works at a lower level then the unilateral control systems, providing a framework for developing sophisticated real-time dynamic behavior in a bilateral telerobotic system.

At NASA's Jet Propulsion Laboratory substantial work has been done in building high-performance bilateral teleoperation systems using special purpose hardware [3, 4, 9]. SMART works at a higher level than these bilateral systems, allowing greater discretion in choosing robots and sensors, and looser coupling between master and slave.

Finally, the underlying control equations for SMART are based on passive two-port elements. Passive system behaviors have long been used as a model for controlling autonomous systems [11] and have recently been applied to bilateral teleoperation systems [1, 7, 14]. A passive telerobotic controller can be proved stable in spite of the non-linear behavior of the operator, the robot, and the environment [2].

2. Building Teleoperator Control Systems.

SMART is based on two-port network theory. Each module in the SMART system has a network equivalent, either a one-port or two-port network. When joined together the modules form a complete circuit.

Two-port networks are a common tool for describing bilateral teleoperator systems. They give a compact representation, directly illustrate the passive nature of the underlying control laws, and are intuitively understood by a large audience. In this paper we extend this modeling tool to handle problems in sensor integration, world modeling and modular design as well.

To illustrate the SMART architecture we will build up a complex system a few modules at time. To build a working system we start with an initial one-port, connect it to any number of two-port elements, and then terminate the network with another one-port. For a valid connection the degrees of freedom (e.g., 1, 6 or N) and the operational space (i.e., world space (**W**) or joint space (**J**)) at the interface of both connected ports must match. Icons will be used to represent the connections. Each icon shows both a network and a pictorial representation of the module.

2.1 Building an autonomous robot controller

First we'll build a standard autonomous robot controller (Figure 1). A standard robot controller has three primary units, a module for generating trajectories, a module for computing the kinematics of the device, and a module for driving the joint servos. A termination one-port completes the system. These modules are introduced below.

The **TRAJECTORY MODULE** is a two-port which generates trapezoidal velocity profiles either in joint space or world space. In joint space the motion is interpolated, i.e., each joint motion starts and stops at the same time. In world space the motion is straight-line, i.e., following a straight-path in Cartesian space, and tracking the shortest rotational arc.

The **PUMA KIN MODULE** is a two-port which converts forces and velocities in world space to forces and velocities in joint space. This module requires the Jacobian and inverse kinematics routines for the manipulator, but doesn't require any information about the hardware connections.

The **PUMA JOINTS MODULE** is a one-port which maintains a direct connection to the PUMA robot hardware. It drives the PUMA to any position commanded by SMART. Because this module knows the true position of the slave robot, it can also be used to align or "sync" the SMART system to an initial configuration.

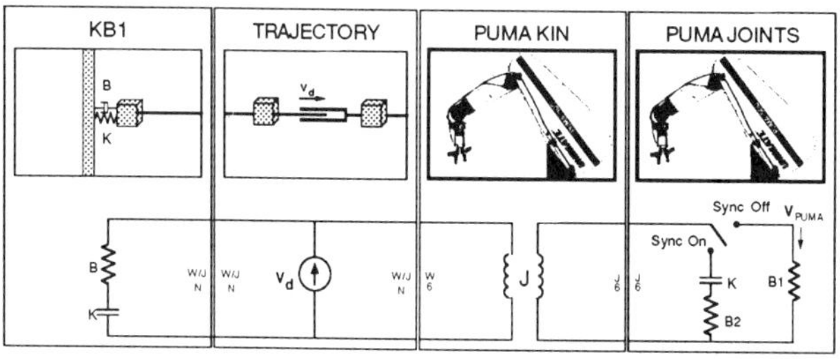

Figure 1: World space controller for PUMA

The **KB1 MODULE** is a one-port which terminates the left side of the network. It simulates the dynamics of a spring-damper system with stiffness, *K*, and damping, *B*. By providing a capacitive termination, flow is directed toward the right-side of the network. By modifying the stiffness and damping parameters, impedances ranging from very compliant to very stiff can be realized.

Combining these four modules results in the autonomous robot controller system shown above (Figure 1). In addition, to realize a joint-space controller for the PUMA the PUMA KIN module can simply be removed.

2.2 Previewing/monitoring motion graphically.

The previous example created a system for commanding straight-line motion of a robot—and provided nothing more than a standard commercial robot controller would have provided. In the next few sections we illustrate how the SMART architecture allows us to rapidly configure sophisticated telerobotic systems that exceed the capability of commercial systems.

For example, in order to allow previewing and monitoring of robot motion we have developed the VISUAL MODULE (Figure 2). The VISUAL module first initiates communication with a remote graphics host, and then continually updates joint information in the robot display. Placing this module adjacent to an input module allows visualization of operator intent. Placing it adjacent to a slave output module allows visualization of the actual slave motion. Figure 2 illustrates one example of a motion previewing and monitoring system.

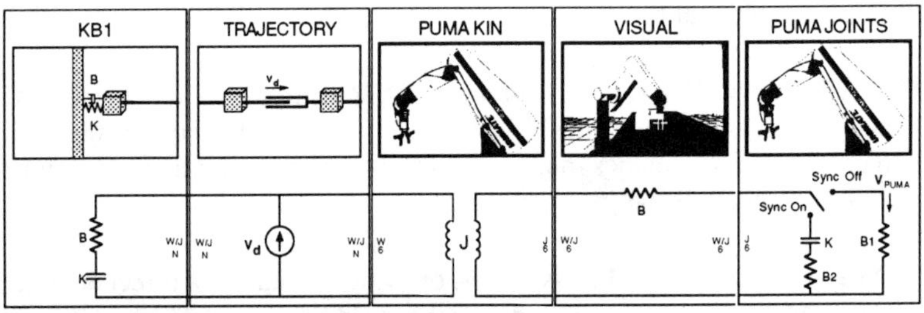

Figure 2: Motion monitoring/previewing system for PUMA 560.

2.3 Adding sensor information.

SMART is designed to work with external sensor input. The open software and hardware architecture makes it easy to attach devices through analog, digital and serial I/O channels. Two examples are given below.

The **PUMA JR3 JOINTS MODULE** (Figure 3) incorporates force feedback into the system using measurements from a JR3 force sensor. The force feedback signal enables contact operations such as part insertion to be achieved using a commercial robot.

The **SONIC MODULE** provides variable damping based on information obtained from a Sandia developed ultrasonic sensor array. It tells the operator if collision of the slave with a sensed object is imminent and will automatically slow the slave's velocity.

The **KB2 MODULE** and the **TASK_KB2 MODULE** both introduce additional compliance into system by simulating a spring, K, and dashpot, B. The compliance determines how the system reacts to sensor feedback. The KB2 Module works in either world space or joint space, while the TASK_KB2 Module works in task space by using the tool rotation matrix, R. Figure 3 illustrates a SMART system using both force sensor feedback and proximity based damping with the compliance aligned in task space. This ability to align the impedance terms in task space is commonly called impedance control [11].

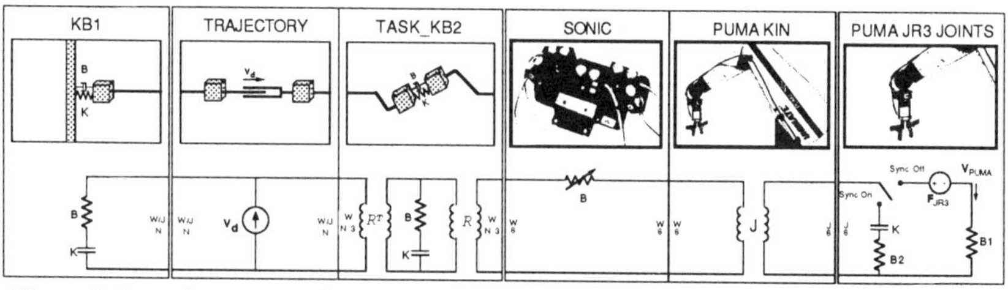

Figure 3: Impedance control system with force compliance and ultrasonic based damping.

2.4 Including operator inputs.

So far we have illustrated how SMART can be used for developing autonomous robot systems. In the next two sections, we illustrate how real-time operator input can be incorporated. In this section we focus on non-force-reflecting input devices.

There are many types of input devices that an operator can use to control the motion of a robot. Joy sticks, teach pendants and wands all allow some form of velocity or position input to the system. In this section we introduce a module for one such device, the **SPACE_BALL MODULE** (Figure 4). The SPACE BALL module takes velocity inputs from a CIS Dimension 6 space ball, and adds them to the other velocities in the system.

In addition, in order to record operator generated motions for future playback we have developed the **REPLAY MODULE**. The REPLAY module first records velocities without modification. Then upon request, it will play back the motion at a faster or slower speed. During playback the operator can still perturb the motion using other attached modules. As with all other non-force-reflecting input modules, the REPLAY module utilizes a force threshold. Motions which would result in this threshold being exceeded are automatically prevented. This capability represents a crucial safety feature of the SMART system.

Figure 4 illustrates a system in which the operator's space ball generated motion is recorded for playback. During playback, motion from the space ball is added to the overall velocity profile to fine tune the trajectory.

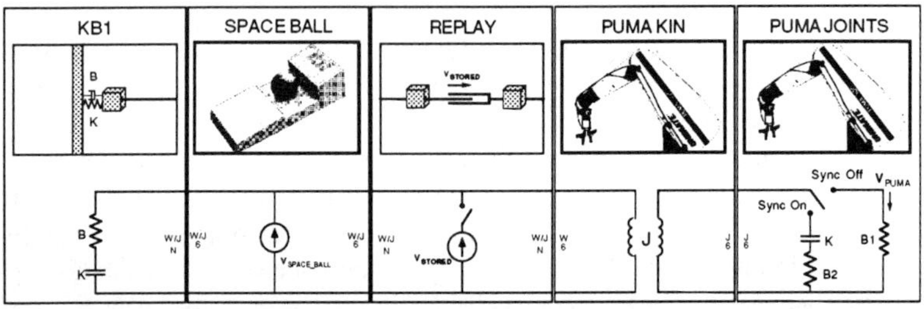

Figure 4: Space ball control of robot with replay .

2.5 Including force reflection.

Problems inherent in force reflection motivated the development of the SMART architecture. How can an operator be coupled into a generic robot system in a stable fashion? What forces should the operator feel? How do we incorporate world modeling and sensing into a force reflecting environment? Although SMART provides an excellent system integration mechanism for both autonomous, and non-force-reflecting telerobotic systems, its true strength lies in its ability to provide meaningful force information to the operator kinesthetically.

To implement bilateral force-reflection we introduce two modules, the **KRAFT MODULE** and the **KRAFT_KIN MODULE** (Figure 5). These modules take care of both the joint level hardware interface and the kinematics for a Kraft force-reflecting master device, respectively.

In addition, we introduce the **INDEX MODULE**. The INDEX MODULE allows the operator to index the motion of a slave device. Furthermore, this module incorporates both axis motion restriction, so that the slave robot can be constrained to move along a line, a plane, or a particular arc, and a force/position scale factor so that arbitrary power and impedance scaling is possible. Figure 5 illustrates how a Kraft master can be used for generic bilateral force-reflection with a PUMA manipulator.

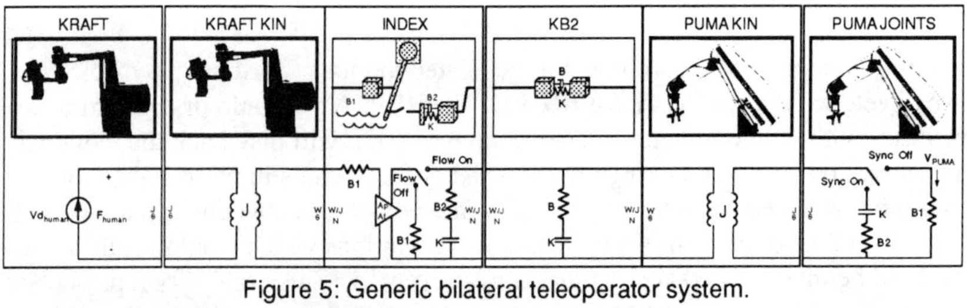

Figure 5: Generic bilateral teleoperator system.

2.6 Adding world model information.

There are two problems with using sensor information for detecting objects in a world environment. First, because of sensor delays, noise and non-collocation problems, it is difficult to get a high-fidelity response directly from sensors. Second, as in the case of force sensors, once contact is sensed it may be too late to respond—the damage is done. The solution is world modeling.

We have to date developed two modules for incorporating world models, the **LIMITS MODULE** and the **BOX MODULE**. The LIMITS module is a two-port module operating either in world space or joint space. It simulates a spring-damper whenever a position threshold is exceeded. It can be used to prevent the operator from exceeding the joint limits of the manipulator, or from straying outside a Cartesian workcell.

The BOX module is a two-port module which simulates a spring-damper whenever the slave device enters a pre-defined box region. In so doing, it simulates contact forces, preventing the operator from proceeding. By using simulated boxes that are larger than physical objects in the environment, collisions with the environment can be avoided entirely.

Figure 6 shows an enhanced telerobotic system with kinematics for master and slave, force and position scaling in world space, indexing with axis exclusion, space-ball superposition, world modeling, record/playback, graphics display, joint limits and force feedback.

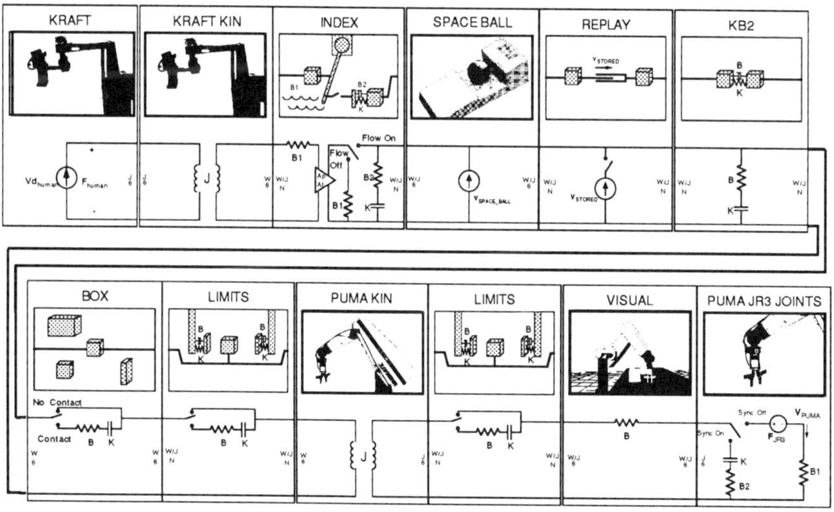

Figure 6: Enhanced bilateral telerobotic system.

4. Conclusions.

In this paper we have introduced SMART (Sandia's Modular Architecture for Robotics and Teleoperation). Sample SMART modules were introduced and examples of module connections were given to illustrate the rapid synthesis of telerobotic control systems. It was shown how autonomous control, impedance control, monitoring and previewing, and force-reflection could all be implemented within the SMART architecture. Current work involves improving the underlying computing power for the system and adding different modules to the system.

References.

[1] R. J. Anderson and M. W. Spong, "Bilateral Control of Teleoperators with Time Delay", *IEEE Transactions on Automatic Control*, vol. 34, pp. 494-501, 1989.

[2] R. J. Anderson and M. W. Spong, "Asymptotic Stability for Force Reflecting Teleoperators with Time Delay", *The International Journal of Robotics Research*, vol. 11, pp. 135-149, 1992.

[3] P. G. Backes and K. S. Tso, "UMI: An Interactive Supervisory and Shared Control System for Telerobotics", *IEEE International Conference on Robotics and Automation*, Cincinnati, Ohio, pp. 1096-1101, 1990.

[4] A. Bejczy and Z. Szakaly, "Universal Computer Control System (UCCS) for Space Telerobots", *IEEE International Conference on Robotics and Automation*, Rayleigh, NC, pp. 318-324, 1987.

[5] P. T. Boissiere and R. W. Harrigan, "An Alternative Approach to Telerobotics: Using a World Model and Sensory Feedback", *Third Topical Meeting on Robotics and Remote Systems*, Charleston, SC, pp. 1-9, 1989.

[6] R. A. Brooks, "A Robust Layered Control System for a Mobile Robot", *IEEE Journal of Robotics and Automation*, vol. RA-2, pp. 14-23, 1986.

[7] J. E. Colgate, "Power and Impedance Scaling in Bilateral Manipulation", *IEEE International Conference on Robotics and Teleoperation*, Sacramento, CA, pp. 2292-2297, 1991.

[8] S. Graves, L. Ciscon and J. D. Wise, "A Modular Software System for Distributed Telerobotics", *IEEE International Conference on Robotics and Automation*, vol. pp. 2783-2785, 1992.

[9] S. Hayati, T. Lee, K. Tso and P. Backes, "A Testbed for a Unified Teleoperated-Autonomous Dual-Arm Robotic System", *IEEE International Conference on Robotics and Automation*, Cincinnati, Ohio, pp. 1090-1095, 1990.

[10] V. Hayward and R. Paul, "Robot Manipulator Control Under UNIX RCCL: A Robot Control "C" Library", *International Journal of Robotics Research*, vol. 5, pp. 94-111, 1986.

[11] N. Hogan, "Impedance Control: An Approach to Manipulation", *ASME Journal of Dynamic Systems and Control*, vol. 107, pp. 1-7, 1985.

[12] R. Lumia, J. Fiala and A. Wavering, "The NASREM Robot Control System Standard", *Robotics & Computer-Integrated Manufacturing*, vol. 6, pp. 303-308, 1989.

[13] D. J. Miller and R. C. Lennox, "An Object-Oriented Environment for Robot System Architectures", *IEEE International Conference on Robotics and Automation*, Sacramento, CA, pp. 352-361, 1990.

[14] G. Raju, G. C. Verghese and T. B. Sheridan, "Design Issues in 2-port Network Models of Bilateral Remote Manipulation", *IEEE International Conference on Robotics and Automation*, Scottsdale, AZ, pp. 1316-1321, 1989.

[15] L. Salkind, "The SAGE Operating System", *IEEE International Conference on Robotics and Automation*, Scottsdale, AZ, pp. 860-865, 1989.

[16] E. M. Sollbach and A. A. Goldenberg, "Real-Time Control of Robots: Strategies for Hardware and Software Development", *Robotics & Computer-Integrated Manufacturing*, vol. 6, pp. 323-329, 1989.

[17] D. B. Stewart, D. E. Schmitz and P. K. Khosla, "Implementing Real-Time Robotic Systems Using CHIMERA II", *IEEE International Conference on Robotics and Automation*, Sacramento, CA, pp. 598-603, 1990.

Berthing Orbital Replacement Units with Capaciflector Sensors

Edward Cheung
NASA Goddard Space Flight Center / Jackson & Tull

Abstract

On-orbit servicing of spacecraft will require the changeout of their components in space. These components are usually configured into Orbital Replacement Units (ORUs) so that they can be handled by astronauts performing Extra-Vehicular Activity (EVA). Current research in robotics concerning supplementing or replacing EVA use vision and force feedback. Unfortunately, these sensor modalities do not offer a simple way of sensing proximity and avoiding collisions between the ORU and the environment. We propose to use a capacitive sensor for the detection of proximity of the ORU to objects. This sensor provides position and orientation feedback, which can be useful in the positioning and alignment of the ORU with respect to the docking fixture. In this paper we describe an algorithm that has successfully docked and undocked an ORU box in a space allowing approximately 0.5" clearance without touching. The resulting algorithm is simple to implement, has modest computational requirements, and is able to continuously prevent collisions of the ORU with the environment.

1. Introduction

Long term science missions proposed for Space Station Freedom (SSF) and Earth Observation System (EOS), have provided a major stimulus for developing subsystems that can be serviced on-orbit. These modules are usually packaged into Orbital Replacement Units (ORUs) that can be handled by astronauts performing Extra-Vehicular Activity (EVA). However, a study performed by NASA indicated that EVA effort alone will not be sufficient to support all ORU replacement activity required for the external maintenance of SSF [1]. In addition, other spacecraft such as EOS and Explorer Platform (EP) will be unmanned. Therefore, the utilization of robotics will be essential in the on-orbit servicing of spacecraft.

Our emphasis in research is the investigation of the role of a capacitive sensor [2] in the task of robotic changeout of ORUs. Traditionally, information on the task site is provided by vision and force feedback [3]. Unfortunately, these modes do not allow a simple means of sensing proximity and avoiding collisions between the ORU and the environment. In addition, we feel that proximity sensing capability will provide valuable data on the position and orientation of the ORU with respect to the fixture we wish to dock into. The sensor being used in our testbed relies on the capacitive effect to detect objects. The sensor output provides analog range information and it has a range of approximately 18". The sensor characteristics are described below in further detail.

2. Experimental setup

A simple setup was chosen as our initial test bed for ORU docking. All contacts between the ORU and the environment are avoided since our current hardware setup does not have any compliance to relieve contact forces that can build up. The task consists of docking the ORU into a fixture that is fastened to an aluminum sheet,

Figure 1. Here, docking is defined as positioning the ORU into the fixture while maintaining equal clearance on all sides. We desire to perform this task using only four capacitive sensors mounted on the bottom corners of the ORU and to use no prior knowledge of the position and orientation of the fixture. It is assumed that the operator will have some limited view of the task site in order to direct the overall motion of the ORU. However, collision avoidance and alignment of the ORU in the fixture is automatically handled by the docking algorithm using the four sensors.

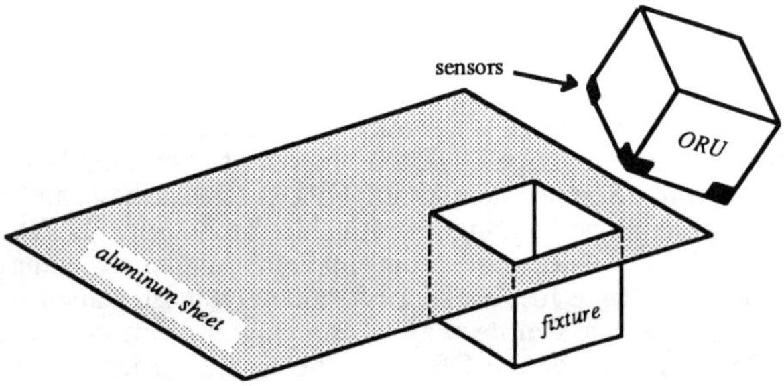

Figure 1. Sketch of the experimental setup.

The ORU is constructed out of aluminum and measures 13"x13.5"x9", while the fixture is constructed out of fiberglass covered by a layer of aluminum foil and measures 15"x15"x10". The overall clearance on the sides of the ORU while it is inserted into the fixture is thus between 3/4" and 1". The aluminum sheet measures 32"x50".

In principle, we can imagine the aluminum sheet to be part of the space craft, or the top face of other ORUs that are docked; their sides make up the sides of the fixture. In addition, there may be multiple open fixtures ready to accept ORUs. During the docking maneuver, there will be opportunity for the operator to specify which fixture he intends to use.

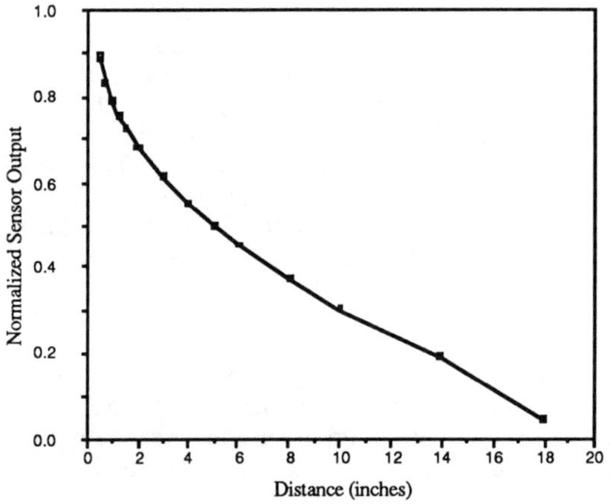

Figure 2. Plot of the normalized sensor output versus distance to test object.

The capacitive sensors were developed at the NASA Goddard Space Flight Center [2]. They are sensitive to all conductive and dielectric objects including metal objects and humans. As the plot in Figure 2 shows, their range is about 18". Four capacitive sensors are mounted on the bottom in the corners of the ORU. They are placed so that in addition to being sensitive to objects on the bottom of the ORU, they are also able to detect objects that are near the sides of the ORU. In this manner, some feedback on the position of the ORU can be gained while it is being inserted into the fixture. A plot of the data from one of the sensors (normalized to a scale of 0 to 1) versus the distance to a test object is shown in Figure 2. The sensors undergo a calibration procedure to ensure that they possess identical responses to objects. The data is then made more linear by using the logarithmic function, and then normalized to a scale of 0 to 1, corresponding to situations with no object, and close contact respectively.

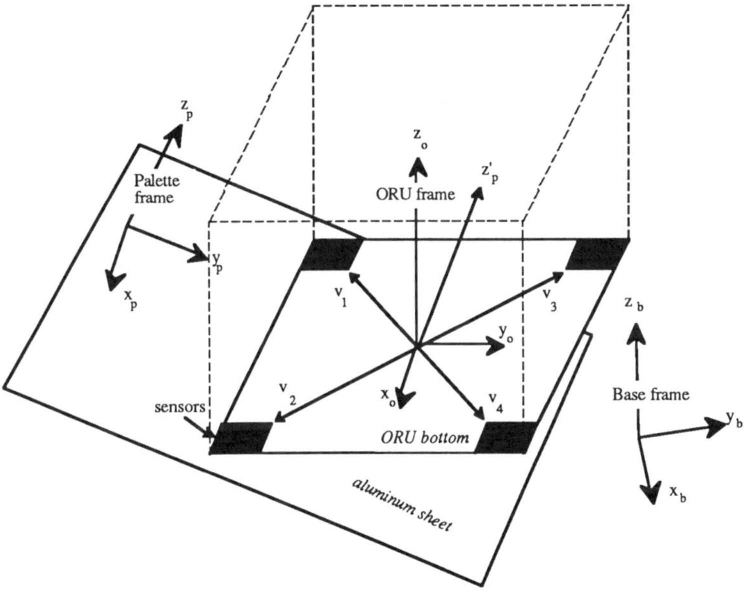

Figure 3. Vectors and frames used in the docking algorithm. The dark rectangles are the sensors, vector v_i points to sensor i.

3. The docking algorithm

The algorithm consists of five major steps, performed in the following sequence:

a) Leveling, the ORU is brought near the aluminum sheet until it is detected by at least one of the sensors. The bottom face of the ORU, which contains the sensors, is then automatically leveled against the aluminum sheet, and brought to a distance of 1" to the sheet.

b) Jogging, the operator uses keystrokes (or a joystick) to bring the ORU to the desired fixture until at least one of the sensors is near the edge of the fixture's hole. The motion of the ORU in this step is parallel along the face of the aluminum sheet. The distance to the sheet is adjusted continuously to maintain the 1".

c) Centering, the ORU is then automatically centered above the fixture's hole if at least one sensor detects the hole.

d) Rotating, the ORU is rotated above the fixture, aligning the ORU such that the sensor outputs are minimized.

e) Insertion, lastly the ORU is inserted into the fixture, while maintaining maximum clearance on all sides.

At all times the ORU's distance to the nearest object is monitored and adjusted. If the aluminum sheet moves away, the ORU follows; on the other hand, if there is motion towards the ORU by the aluminum sheet, or if there is an impending contact, the ORU's position is controlled to automatically and continually avoid collisions. Individual parts of the docking algorithm will be briefly discussed below.

3.1 Leveling

The desired orientation and position of the ORU after it has been leveled is such that the bottom face (with the sensors) is parallel and at a predetermined distance to the aluminum sheet, see Figure 3. The orientation of this latter object is represented by its normal vector z_p, which estimate z'_p is found using the data from the four capacitive sensors:

$$z'_p = z_o + K_s (s_1 v_4 + s_2 v_3 + s_3 v_2 + s_4 v_1) \tag{3.1}$$

where: s_i - output from the ith sensor.

v_i - vector describing the location of the ith sensor in the ORU frame.

z_o - normal of the ORU bottom in the ORU frame.

z'_p - estimated normal of the aluminum sheet in the ORU frame.

K_s - sensor gain.

To orient the ORU such that z_o becomes parallel to z'_p, we need to rotate over some axis perpendicular to both z'_p and z_o, by an amount equal to the angle between z'_p and z_o, these are found by:

$$a = z'_p \times z_o \tag{3.2}$$

$$alpha = K_o \operatorname{acos} (z'_p \bullet z_o)$$

where: a - the axis of rotation for the ORU.

x - the vector cross product.

alpha - amount of rotation.

K_o - orientation gain.

acos - inverse cosine.

• - vector dot product.

In addition to the rotation of the ORU, its position is also adjusted so that its distance to the aluminum plate is 2". This is done by moving the ORU along z'_p in response to the sensor data:

$$b = K_p (s_m - s_d)\ z'_p \tag{3.3}$$

where: b - ORU cartesian motion.

K_p - position gain.

s_m - output of sensor with highest reading.

s_d - desired sensor output : 0.67 for a distance of 2".

The ORU is positioned and oriented repeatedly until the angle between z'_p and z_o goes to zero and it is located a certain distance from the aluminum sheet. The ORU is then said to be leveled. Note that the dynamical behavior and convergence characteristics of this procedure is controlled by adjusting K_o, K_s, and K_p. Larger values for these gains will cause faster motion, at a risk of overshoot and other unstable behavior.

3.2 Jogging

Once the ORU is leveled, the operator can specify into which of the available fixtures he wishes to dock the ORU. This is done by jogging the ORU parallel to the aluminum sheet until one of the sensors detects the fixture's hole. Note that equation (3.3) is still used to maintain a constant distance between the aluminum sheet and the ORU. In this manner, collisions can be avoided if the aluminum sheet shifts. In addition, if there is a step in the aluminum sheet, due to perhaps a protrusion, the ORU can automatically hover over the obstruction.

$$b = K_p(s_m - s_d)\, z_o + o \tag{3.4}$$

where : b - ORU cartesian motion
o - Operator motion command in ORU frame
K_p - position gain

3.3 Centering

After the ORU is positioned near a fixture, the data from the sensors is used to generate an error signal to center the ORU in the hole. The key element in each of the major steps during docking is the goal to equalize the data from all four sensors. During the 'leveling' portion of the algorithm, the combined data is used to orient the ORU using only rotation to decrease the difference between the output of the sensors. In the case of 'centering', the combined data is used to move the ORU parallel to the aluminum sheet in order to reduce any differences between the sensor outputs. When all sensors are equal, the ORU is centered over the hole, with all four sensors being positioned an equal distance to the walls of the fixture. The control function used during centering is:

$$b = K_p(s_m - s_d)\, z_o + K_c\,(s_1 v_4 + s_2 v_3 + s_3 v_2 + s_4 v_1) \tag{3.5}$$

where : b - ORU cartesian motion
K_c - centering gain
K_p - position gain
s_d - desired sensor output : 0.67 for centering

The term $K_p(s_m - s_d)\, z_o$ on the right hand side of equation (3.5) causes the ORU to 'hover' with a constant sensor reading over the aluminum sheet. The quantity K_p controls the convergence characteristics and speed of the cartesian motion.

The remaining term on the right hand side of equation (3.5) moves the ORU away from the sensors that are detecting the aluminum plate towards those that aren't. In this manner, the ORU will be halted when it is centered over the fixture's hole.

3.4 Rotating

Once the ORU is centered over the hole, its sides can be aligned with the fixture using this step of the docking algorithm. It is important to note that when the data from all four sensors is averaged, we obtain a minimum when the sides of the ORU are parallel with the sides of the fixture. This point is known as a point of maximum clearance. There are four such locations as the ORU is rotated 360°. Data is sampled as the ORU is rotated in a search pattern to find one of the minima [4]. Once found, the ORU is then rotated around its z_o-axis until it can be inserted into the fixture.

3.5 Insertion

Once the ORU has undergone the previous other steps of the algorithm, it will be ready for insertion into the fixture. During this maneuver, for reasons of safety and optimal alignment, we wish to continue centering the ORU. Therefore, we can simply use equation (3.4), used during the 'centering' part of the algorithm. The only modification needed will be to raise the desired sensor level so that the ORU will continue to move in the direction of the z_o-axis until it detects the bottom of the fixture.

Note that safety is preserved even if there is insufficient clearance in the fixture due to it being too small, or because of a failure in the 'rotating' portion of the docking algorithm. If the command to 'insert' is given in the case of insufficient clearance, the ORU will be moved closer and closer to the aluminum plate, but will be halted once the desired sensor level has been achieved, preventing a collision.

4. Implementation and experimental results

The manipulator used to position the ORU is a Robotics Research K-1607 industrial robot arm. It is controlled by a custom Multibus based multiprocessor system. The sensor data processing and docking algorithm discussed above runs on an IBM PC-AT. Robot commands are sent from the PC via serial link to the Multibus system.

Table 1 summarizes typical execution times for steps of the docking algorithm, followed by a more detailed explanation.

Step	Time required (min.:sec)
Leveling	0:10
Jogging	0:30
Centering	0:20
Rotating	0:30
Insertion	0:10
Total	1:40

Table 1. Execution times for portions of the algorithm.

At the start of the experiment, refer to Figure 1, the ORU is positioned 12" away from the aluminum plate, and the angle between the two normals z_p and z_o is about 30°. The ORU is first leveled with the aluminum plate to 0.68° of error, and then brought to a distance of 0.9" from the plate. The ORU is then jogged a total of 26" to the edge of the fixture's hole until one of the sensors detect its presence. At that time the centering portion is invoked to automatically position the ORU in the middle of the fixture's hole, where it is then rotated for maximum clearance. Note that it is not necessary to allow the program to settle to zero error during centering since the rotation will still succeed. In addition, the centering algorithm is used again to further reduce errors during insertion. In the experiment, the rotation portion of the algorithm is able to align to 0.09° the sides of the ORU with the sides of the fixture, from an initial misalignment of about 26°. No misalignment between the ORU and the fixture is visible with the eye after the rotation portion has been completed. The ORU is then inserted approximately 9" into the fixture until the sensors detect its bottom.

Acknowledgement

The author expresses his gratitude to Stan Ollendorf and John Vranish (NASA Goddard Space Flight Center) for their support for this work.

References

[1] W. Fisher, C. R. Price, Space Station Freedom External Maintenance Task Team, Final Report, NASA JSC, 1990.

[2] J. Vranish, R. McConnell, S. Mahalingam, "Capaciflector" Collision-Avoidance Sensors for Robots, to be published.

[3] W Kim, P. Backes, S. Hayati, E. Bokor, Orbital Replacement Unit Changeout Experiments with a Telerobot Testbed System, Proceedings 1991 IEEE International Conference on Robotics and Automation, Sacramento, CA - April 1991.

[4] E. Cheung, R. McConnell, M. Manzo, Capaciflector Based ORU Docking, Proceedings 1991 Conference on Intelligent Robotic Systems for Space Exploration, Troy, NY - November, 1991.

An Operator Interface for Interactive Telerobot Task Description and Execution

Paul G. Backes, John Beahan, Bruce Bon
Jet Propulsion Laboratory
California Institute of Technology
Pasadena, California

ABSTRACT

An operator interface for interactive telerobot task description, simulation and execution is described. The interface is the local site of the local-remote telerobot system in the JPL Supervisory Telerobotics (STELER) laboratory. The remote site system provides control of a seven degree of freedom (DOF) redundant manipulator and a six DOF manipulator for supervised autonomous control task execution and a six DOF manipulator which carries cameras. The interface includes graphical and textual input with 3-D graphical simulation overlayed on video images of the scene. A hierarchical menu system allows the operator to interactively build task commands which can be simulated, edited, or executed on the real arms.

INTRODUCTION

Several studies have indicated that extra-vehicular activity (EVA) time for maintenance and repair of Space Station Freedom will be insufficient [1, 2]. Also, there will be a multi-year man-tended phase where astronauts will reside on the station only part time. One solution to reduce the servicing burden is to control space station robotic systems from Earth. A critical factor in specifying the control mode for the ground-remote system is the expected approximately 9 second round trip time delay between the Earth based local site operator control station and the Space Station based remote robot system [3]. The operator interface described in this paper provides supervised autonomous control to safely control the remote robotic system with the expected time delay. An earlier system is described in [4]. Figure 1 shows the local (Earth) and remote (Space Station) components of a ground-remote telerobot control system.

The local site operator interface includes a Sun workstation for the graphical user interface and a Silicon Graphics (SGI) graphics workstation for stereo graphics overlay on stereo video. The initial design of the operator control station called for a single monitor for video, graphics, and menu interface. Later it was learned that the windowing software for the menu interface available on the SGI workstation does not support stereo mode. Therefore a separate Sun workstation is used for interactive task description. In the single-workstation SGI design,

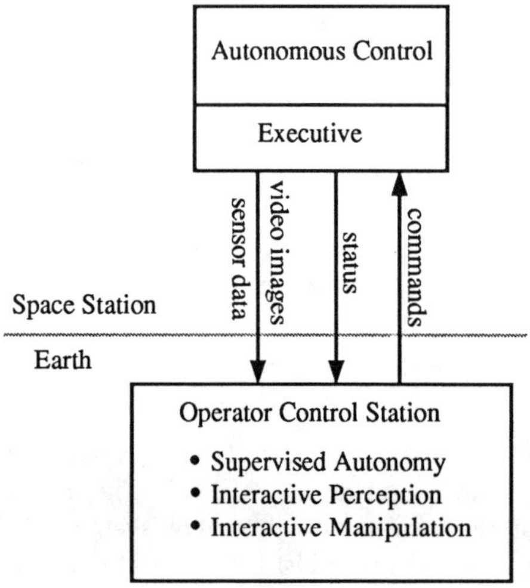

Figure 1: Ground and space elements of telerobot system

the video/graphics display is hidden when the operator is interacting with the window/text display, and vice versa. A spaceball is used to manipulate the graphical model on the SGI or to select destinations. TCP/IP is used for communication with the remote site system. The remote site system includes a six degree of freedom (DOF) manipulator and a seven DOF manipulator for task execution and a third manipulator carrying a camera platform containing four video cameras. An Ada language system is used for communication and control at the remote site [5].

The operator interface of the local operator control station (OCS) has two primary parts: perception and manipulation. Perception provides an interactive means for modeling the remote site scene. Manipulation provides interactive task description, simulation, editing, and execution. Central to the operator interface is the knowledge base which holds information on the state of the local and remote site systems and manipulation and perception data. The methodology of the local-remote system is to build and simulate manipulation and sensing commands on the ground, using a model of the robot and its environment stored in the knowledge base, which has been updated and validated with feedback sensory data. Simulations detect errors such as kinematic and geometric constraint violations. Successfully simulated commands may be sent to the remote site for execution.

INTERACTIVE PERCEPTION

The interactive perception module uses a combination of operator input and machine vision to create and refine the model of the task environment that resides in the knowledge base. Interactive perception utilizes computer power for precision measurement where computational requirements are heavy, and human perception for recognition, scene segmentation, and rough location designation, where reliable and efficient computer algorithms are unavailable. To aid human perception, the system provides views from multiple video cameras, including a stereo view for depth perception. Also, by displaying only the graphical model, the scene can be viewed from arbitrary viewpoints. The three primary functions initially planned are object localization, object model editing, and camera localization. Each of these utilizes machine vision to refine geometric information — least-squares fit to detect feature locations is used.

In *object localization*, the operator uses the spaceball to translate and rotate a graphics-overlay representing a geometric object model until reasonable registration has been achieved with video images of the object from multiple viewpoints. Then machine vision is used to accurately measure object position and orientation (pose). For *object model editing*, the operator uses the spaceball to move a 3-D cursor in order to designate the 3-D positions of features of interest. In the initial version of the OCS, the operator can add faces to a polyhedral object model by designating positions of vertices in sequence of each face. Vertex locations can be modified and faces can be deleted. Object model editing will therefore allow both creation of new models and editing of existing models. After the operator has finished editing an object model, machine vision may be used to refine the object model geometry to match detected features in the scene.

In *camera localization*, the operator moves a 3-D cursor with the spaceball to select known objects in the scene whose poses are well determined. After a set of known objects are selected, the operator uses the spaceball to alter the position and orientation of the cameras in the knowledge base (without actually moving the cameras), thus moving the ensemble of object overlays until reasonable registration of the known objects has been achieved. Then machine vision is used to accurately measure camera position and orientation. This is especially useful for cameras mounted on movable platforms, such as robot arms or pan-tilt heads, where small errors in camera orientation produce large errors in machine vision results.

INTERACTIVE MANIPULATION

The operator builds, simulates, edits, or executes a task command using interactive manipulation. A graphical menu interface provides the operator with manipulation options and suggested parameterization and the operator performs the high level cognitive reasoning to specify the specific task and task sequence. Supervised autonomous control is the basic control mode used. Here, individual autonomous commands are specified at the local site and sent for execution at the remote system. A task command can be simulated with local site graphical simulation and reedited and then resimulated until safety is assured for execution at the remote site. Individual commands can be concatenated into a task sequence. Task commands and sequences are saved in a command library.

An individual task command is initially generated by specifying the underlying macro for the command. Examples of basic macros are Cartesian guarded motion, move-to-touch, grasp, hinge, slide, insert, level, and push. Each macro has a graphical menu interface where the operator specifies the parameters for the macro for the specific task. Most of the parameters are automatically specified depending on the state of the system, e.g, which is the object being manipulated. The operator can also choose to leave certain parameters unspecified. If this is the case, then when the command is later selected for execution, the system will detect the missing parameters and automatically bring up the menu so the operator can specify the parameters. Along with the execution parameters are parameters for safety and termination condition monitoring. Besides manipulation commands, there are also sensing commands and reflex commands. Reflex commands specify to the remote system what type of reflex behavior to execute for the different types of monitor events which can be detected. The perception functions of the interface are used to specify objects and destinations for task commands.

PRIMARY SOFTWARE COMPONENTS

The perception and manipulation modules are called by the graphical user interface (GUI) in response to operator menu selections. The GUI is a window-based system utilizing X and Motif to provide button and pull-down menus, file-selection dialog boxes, textual and graphic status displays, and on-line help for operator interaction.

The video-graphics module uses the SGI Graphics Library (GL), and allows display of video images with accurate graphic overlays representing the current world model. It controls a window which covers the entire screen and provides for "viewing ports" within the larger window, each of which contains a single video image or a pair of images, viewable in stereo using special glasses. Using camera models produced by an existing camera calibration program[6, 7], it computes the necessary parameters for graphics and image display such that the graphics overlay video images properly and the operator can comfortably fuse the stereo pair. It displays video images received via the network communications module and provides a display of the database of objects with graphics overlay on the video or with any specified viewpoint without video.

The interactive perception module interacts with the operator primarily through the video/graphics and spaceball input modules. Operator interaction is used to manipulate object pose, object model geometry or camera pose, as the first phase of object localization, object model editing or camera localization, respectively. The machine vision module is called to refine the a priori data thus obtained, and the knowledge base module is called to update the world model with the new data. The machine vision module performs iterative least-squares fits as commanded by the operator-assisted perception module for object localization, object model editing, and camera localization operations.

The interactive manipulation module interacts with the operator using both the

GUI and the video/graphics and spaceball input modules. It calls the operator-assisted perception module in order to update the knowledge base as needed and also uses video/graphics for designation of 3-D locations and to display simulation results to the operator. The operator controls manipulation activities through control panels on the GUI.

The knowledge base provides representation of the robots and environment, including geometric models, mass properties and grasp points for manipulation, and information about the various operations in which each object can participate. The knowledge base represents the world as a directed, acyclic graph, whose nodes are objects related by parameterized coordinate transforms, and whose properties describe the physical object and how the object behaves under various operations, e.g., move, grasp, insert. It also contains all object information necessary for graphic display and machine vision.

Before being transmitted for execution, each physical action is simulated by a software replica of the robot and its environment. This replica is intended to detect those classes of errors which can be detected in the absence of precise geometric and mass information, and which are computationally feasible to simulate. These checks include kinematic constraints such as joint stops and singularities, and may include geometric constraints such as collision between the arm or payload and an obstacle in the environment.

CURRENT DEVELOPMENT STATUS

An initial version of the operator control station has been implemented and is used to send commands to a remote control system [5] for control of a 7 DOF manipulator. Also, stereo graphics are overlayed on static stereo video images which have been transmitted from the remote site. All modules are under further development.

CONCLUSIONS

The tools provided with the operator interface described in this paper will allow effective supervised autonomous control of a remote robotic system. Interactive task description allows the operator to specify the desired tasks, and preview simulation provides a priori safety checking of the execution before transmitting the commands to the remote robotic system. The operator control station described here represents the ground site of a prototype ground-remote telerobot control system being developed by NASA.

ACKNOWLEDGEMENTS

The research described in this paper was carried out by the Jet Propulsion Laboratory, California Institute of Technology, under contract with the National Aeronautics and Space Administration.

REFERENCES

[1] W. Fisher and C. Price. Space station freedom external maintenance task team final report. NASA Johnson Space Center, July 1990.

[2] N. Augustine. Report of the advisory committee on the future of the u.s. space program. December 1990.

[3] R. Aster, J.M. de Pitahaya, and G. Deshpande. Analysis of end-to-end information system latency for space station freedom. Jet Propulsion Laboratory, Internal Document D-8650, May 1991.

[4] Paul G. Backes. Ground-remote control for space station telerobotics with time delay. In *Proceedings AAS Guidance and Control Conference*, Keystone, CO, February 8-12 1992. AAS paper No. 92-052.

[5] Paul G. Backes, Mark K. Long, and Robert D. Steele. Designing minimal space telerobotics systems for maximum performance. In *Proceedings AIAA Aerospace Design Conference*, Irvine, CA, February 3-6 1992.

[6] Y. Yakimovsky and R. Cunningham. A system for extracting three-dimensional measurements from a stereo pair of tv cameras. *Computer Graphics and Image Processing*, 7;195–210, 1978.

[7] D. Gennery, T. Litwin, B. Wilcox, and B. Bon. Sensing and perception research for space telerobotics at jpl. In *Proceedings IEEE International Conference on Robotics and Automation*, pages 311–317, 1987.

ESA'S CONTROL DEVELOPMENT METHODOLOGY FOR SPACE A&R SYSTEMS

P. PUTZ

Department of Space Robotics, Dornier GmbH, Friedrichshafen, Germany

A. ELFVING

Robotics Section, ESA / ESTEC, Noordwijk, The Netherlands

ABSTRACT

In an effort to support the development of flexible space Automation and Robotics (A&R) systems, the European Space Agency ESA has begun to establish an A&R Control Development Methodology (CDM). The CDM encompasses guidelines for the design of control systems to accomplish inter-project unification and intra-project traceability . The CDM calls for a phased analysis and development approach to progress from user requirements to hardware and software implementations. A main focus is on a systematic structure of the requirements in order to clearly separate functional, application, operational, and implementational influences.

INTRODUCTION

Several space programs currently planned and under development define the need for quite complex and sophisticated Automation and Robotics (A&R) capabilities (e.g. internal automation and servicing of orbiting laboratories, platform external servicing, surface mobility and sample acquisition for planetary exploration). Design driving requirements for the resulting A&R systems are to account for evolving needs and technological capabilities for life times of up to 30 years, to meet very high safety and reliability demands, and to be realized within tight budgetary and time constraints. This implies high operational flexibility and built-in growth potential with controlled cost and risk.

There is wide agreement on the central importance of the A&R control subsystem to enable such quality and flexibility. The control subsystem involves several disciplines (control engineering, dynamics, simulation, computer hardware, software, human operations, etc.) which per se are highly mature in terms of widely accepted methods and tools, yet whose systematic coordination in a system engineering sense has remained somewhat unsatisfactory. Moreover, there has been concern from the Space Agencies that too little transfer of results among parallel A&R projects or from one generation of a system to the next has been taking place. In a scenario of shrinking budgets, this necessarily becomes an issue of prime interest.

In recognition of these facts, the European Space Agency ESA has initiated the development of guidelines and principles for the design of A&R systems in general and control systems in particular [1]. The main output of the respective program [2], [3] is the so called A&R Control Development Methodology (CDM) as a concise yet comprehensive framework for guiding the design creativity to the common benefits of developers and users. The CDM shall provide principles and rules for designing open and extendable A&R control implementation architectures within a simple overall structure, using unified and unambiguous terminology, and above all maintaining traceability between solution independent requirements and final realizations.

OVERVIEW OF THE CDM

The CDM above all defines a life cycle model for the A&R control subsystem, breaking the overall system analysis and design process into clearly defined individual steps or phases which are delimited by verifiable milestones and which are compatible with the phases proposed in other relevant standards [4].

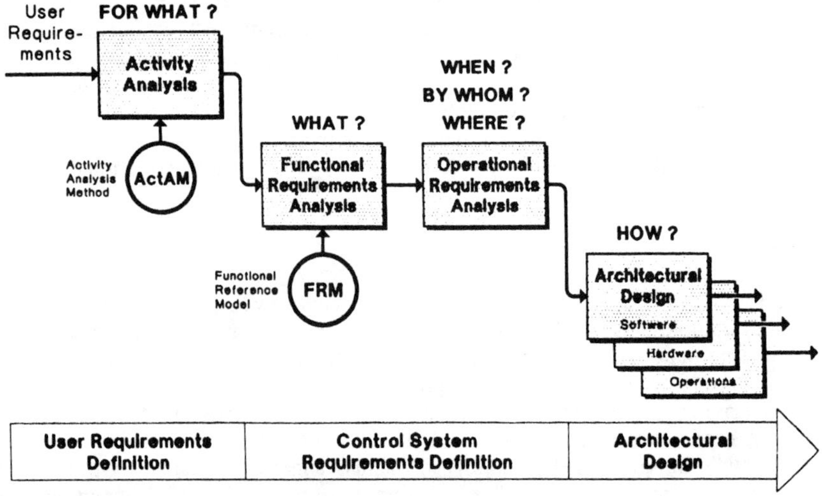

Fig.1. Main Steps of the CDM

Fig. 1 illustrates the main steps of the CDM in a simplified fashion. The first step is called Activity Analysis. It is part of the User Requirements Definition phase, where the overall A&R problem is formulated. Its input is the user provided top level A&R system requirements specification. A first key message of the CDM is that already the user requirements need to be refined in a systematic (hierarchical) structure. The proposed output from this step is a so called Activity Script, a semi-formal and fairly detailed prescription of the process which has to be automated by the A&R system (key question: FOR WHAT do I need automation and control ?).

The next phase is the refined analysis of the (control) subsystem requirements. Another key CDM concept is that this should be separated into two individual steps: Functional Requirements Analysis (purely stating WHAT control functions have to be performed) and Operational Requirements Analysis (specifying for each desired mode of operation WHEN, BY WHOM, and WHERE these control functions need to be executed).

The last step identified in Fig. 1 is the Implementation Architecture Design which, for the software part, coincides with the well known S/W Architectural Design phase. The point of the CDM is that also control hardware and control operations shall be "designed" to the same level in this step. The big difference to the preceding requirements analysis phase is that only design starts to say HOW the required functions are going to be realized. This involves the specific processors, bus systems, real time operating system services, algorithms, software communication schems, operational procedures, etc.

The remaining phases after Implementation Architectural Design are not shown in Fig. 1 any more, since they are considered the "conventional" steps of Detailed Design, Production, Integration, Testing, and finally Delivery and Maintenance. This appears to be so well known that it is not discussed within the CDM context.

The CDM is not limited to just prescribing a phased development, but for each of the shown steps it offers non-trivial support in the form of generic architecture schemes (reference models), guidelines on their use, and even computer based tool assistance. Two key CDM contributions are shown in Fig. 1: the Activity Analysis Method (ActAM) and the Functional Reference Model (FRM). They will be introduced in the following.

HIERARCHICAL ACTIVITY ANALYSIS

The Activity Analysis Method (ActAM) of the CDM shall produce refined user requirements suitable for further system and subsystem analysis. In order to formulate user requirements on the A&R system, one shall first analyze the process to be automated. This analysis shall result in a hierarchical description of A&R system activities to be performed. We propose a three layer hierarchy of activities: the very top level A&R Missions represent the highest level of activity for which an A&R system is responsible (e.g. "SERVICE a life Science Experiment" , "REPAIR a satellite"). Each A&R Mission can be decomposed into Tasks defined as the highest level activity performed on a single subject (e.g. "OPEN a door", "INSTALL a sample in a processing furnace"). Finally, each task shall be decomposeable into Actions (e.g. "GRIP a sample container", "MOVE the container to the freezer"). A&R missions, tasks, and actions can be formulated as commands with associated attributes which specify the particular circumstances.

The hierarchical activity analysis procedure outlined above can easily be seen to serve more than the purpose of user requirements formulation. We also consider it an important means for the development and optimization of operational scenarios, a common language for A&R system users, operations planners, developers, and managers. Thus, in order to account for the desired unambiguity, we have defined a semi-formal Activity Scripting Language (AcSL). AcSL shall not be "yet another robot programming language", but a guideline on typical A&R Missions, Tasks, and Actions. Indeed, AcSL is a scripting language and not an executable language .

FUNCTIONAL REQUIREMENTS ANALYSIS

The second major step of the CDM is the analysis of functional requirements for the A&R control system based upon the refined user requirements (Activity Script). The key concept at this point is to start with a purely functional analysis of the requirements. A function only says WHAT has to be done, strictly independent of HOW and BY WHOM. Typical control functions are servo control, inverse kinematic transformation, path interpolation, path planning, force/torque sensor data processing, etc. Even the most diverse applications use the same fundamental control functions, but in possibly quite different realizations.

In order to maintain the desired traceability to the activities identified in the user requirements, the control functions will be cast in the same overall structure, namely a three layer hierarchy responsible for achieving A&R missions, tasks, and actions. One of the central contributions of the CDM is to provide a general framework (logical model) for all A&R control functions. Since this shall serve as a reference for all possible applications, it has been called Functional Reference Model (FRM). The FRM is meant to be the essential functional and information architecture of general A&R control systems, independent of particular applications, operational scenarios, and implementations.

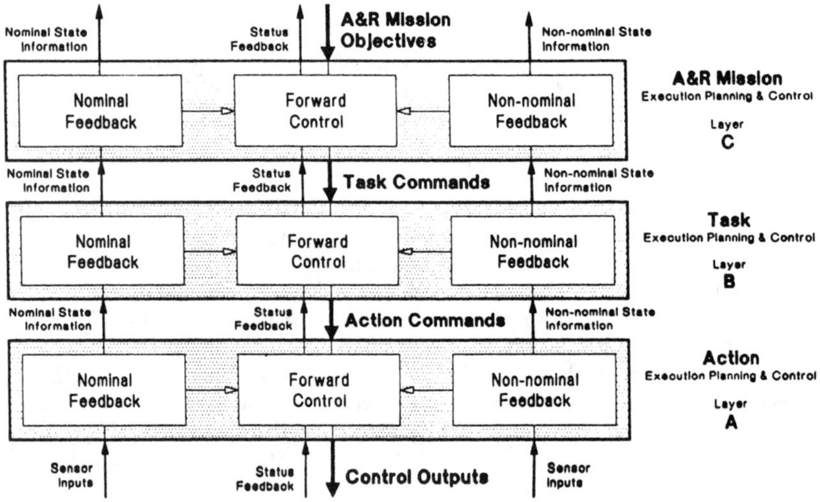

Fig. 2. Functional Reference Model (FRM) for A&R Control Systems

The top level view of the FRM is shown in Fig. 2. Just like the Activity Analysis scheme, the FRM consists of three hierarchical layers of control functions. The overall A&R control system is responsible for reaching a global goal, the A&R Mission Objectives. These are successively broken down into sub-goals: the tasks and actions, each handled by control functions on the appropriate layer, until the elementary "control outputs" can be issued to the A&R devices (e.g. currents to the joint motor drives).

The second structuring scheme of the FRM is a clear separation into "vertical" branches of control based on the concept of feedback. The center branch is called

Forward Control (FC). FC functions are responsible for activity decomposition, execution planning, and control by taking the most appropriate a-priori information known to each layer into account. FC functions could potentially exist without feedback, amounting to "blind" feedforward control action without regard to the actual effects on the controlled system.

The left branch of the FRM consists of Nominal Feedback (NF) functions for refinement and update of a-priori knowledge ("world models") based on the actual, but essentially expected, evolution of the process and consequently formulation of controlled adjustments of the FC. By FC and NF, "cascaded" control loops are closed on action, task, and A&R mission layers and are equipped with everything necessary for the "nominal" course of events. Besides FC and NF, the FRM foresees a third branch of "Non-Nominal Feedback" (NNF) functions responsible for the monitoring of discrepancies between actual and allowable states in both the FC and NF functions, diagnosis of their origins, and generation of directives (including recovery strategies and constraints) for FC. The NNF functions are thus reserved for "contingency" or "rescue" cases that are only triggered by irregularities not accounted for in the nominal FC / NF loop. A much more refined definition of the FRM is given in [3]. It shall also be acknowledged that the FRM concept and structure have been heavily and beneficially influenced by the NASREM architecture developed at the US NIST [5].

For given user requirements (activity analysis) of a control application, the functional requirements shall be derived in a hierarchical fashion according to the FRM framework. The resulting functional architecture (logical model) is called the Application Architecture (AA) since it is specific for the given problem.

OPERATIONAL REQUIREMENTS ANALYSIS

The third major step of the CDM according to Fig. 1 is the analysis and incorporation of the operational requirements on the A&R control subsystem. Operational requirements are understood as defining the desired modes of operation, namely the circumstantial fashion (especially w.r.t. the extent of human involvement) in which the control functions are to be performed. The underlying idea is that even in modes as different as "telemanipulation" and "automatic task program execution with online planning", the same fundamental control functions (from the FRM !) are involved, but allocated in quite different ways to humans and machines (software).

In other words, operational requirements are defined precisely by specifying the allocation of the control functions to the following domains: preparation or utilization (WHEN are the control functions performed?), human or machine (BY WHOM are the control functions performed?), and (for space applications) on ground or on board (WHERE are the control functions performed?). This brings in the key questions answered at this step (see Fig. 1). As a guideline, [3] shows examples of the allocation patterns for the most common operational modes.

The Operational Requirements Analysis step consists of identifying the functional allocations. The result is the final control subsystem requirements definition including the functional requirements from Step 2 and the operational requirements from Step 3. Because of the allocation, they are already grouped into requirements for control software, control hardware, and human control operations.

DESIGN AND PRODUCTION

The fourth and last step of the CDM in Fig. 1 is the Implementation Architecture Design. This is shown separate for control hardware, control software, and human control operations and is under the responsibilities of the respective specialists within a project. For the software part this corresponds to the Architectural Design phase [4] for which well established methods and tools exist. This is where the actual implementation decisions are taken on HOW the control functions are going to be realized. The result is the control software architecture, a physical model of the control subsystem. The main point is that, because of the tight connection to the preceding steps, this still exhibits the underlying FRM structure and thus provides the desired modularity, openness, extensibility, and interchangeability.

CONCLUSIONS

We have outlined the major steps of a newly developed A&R Control Development Methodology. The main feature of the CDM is a disciplined system engineering approach with clear phases. The phases exhibit built-in traceability due to the separation and distinct accommodation of independent influences. A main focus is on solution independent requirements analysis resulting in a functional architecture or logical model of the control system. For this purpose, the CDM offers an Activity Scripting Language (AcSL) and a Functional Reference Model (FRM) structured into implementation independent hierarchical layers. Operational requirements are treated as conditions on the allocation of the identified control functions. Powerful CASE tools can be employed to support the analysis and design steps.

First demonstration applications of the CDM have been begun. A prototype control system for a planned Spacelab A&R mission has been designed. Further plans call for getting even more experience by building implementations. When the guidelines in the CDM are mature enough, one can think of "automating" some of them by providing specially tailored computer tools which take over much of the routine work in analysis and design. The main ideas of the CDM are not specific to A&R control, and there will be efforts to extend them for even wider applications.

REFERENCES

1. A. Elfving and U. Kirchhoff, "Design Methodology for Space Automation and Robotics Systems", *ESA Journal*, 15, 149 - 164 (1991).
2. Dornier, "Evaluation of Standards for Robot Control System Architectures (Control Techniques)", *ESA Contract 8009/88/NL/JG Final Report (1990).*
3. Dornier, "Baseline A&R Control Development Methodology Definition Report", *ESA Contract 9292/90/NL/JG Report* (1992).
4. BSSC (ESA Board for Software Standardisation and Control), "ESA Software Engineering Standards", *Doc. Nr. ESA-PSS-05-0* (1991).
5. J.S. Albus, H.G. McCain, and R. Lumia, "NASA/NIST Standard Reference Model for Telerobot Control System Architecture (NASREM), *NIST Technical Note 1235* (1989).

Integrated Tools for Teleoperation

H. Das and P. Fiorini*
Jet Propulsion Laboratory,
4800 Oak Grove Dr., M/S 198-219,
Pasadena, CA 91109.

ABSTRACT

Tools and tool handling techniques designed for demonstrating the use of the Advanced Teleoperation System at JPL in satellite repair operations are described in this paper. Our model for the repair demonstration is the Main Electronics Box (MEB) replacement on the Solar Maximum Satellite Repair (SMSR) procedure developed by NASA and successfully performed by the crew of Space Shuttle STS-13 in 1984. A summary of the repair procedure consists of uncovering thermal protection blankets on the satellite, removing screws from the MEB panel, opening the panel, then removing electrical connectors from their sockets, cutting cords that hold electrical cables to the MEB panel, replacing the panel, and finally reversing the previous procedures. Methods for handling tools used in our SMSR procedure, along with details of the tool designs are discussed in this paper.

1 INTRODUCTION

In 1984 the crew of the Space Shuttle STS-13 successfully performed repair operations on failed components of the orbiting Solar Maximum Satellite. Using the Solar Maximum Satellite Repair (SMSR) procedure as a model, the Advanced Teleoperation (ATOP) Laboratory at JPL is demonstrating its capability for performing the satellite repair with a re-configurable computer controlled ATOP system [1] [4] [7] [8] designed as a prototype for space-based remote manipulation.

Specialized end effectors are necessary for the performance of many tasks in robotics. It is economically preferable to design inter-changeable effectors for particular operations than a single effector capable of performing all possible operations required of the robot. Many options are available in design for tooling and tool manipulation, each with its advantages and disadvantages. This paper describes the tools and tool handling procedures developed at the ATOP Laboratory for demonstration of satellite repair operations.

There have been a number of alternative methods proposed for robot-task manipulation. A tool operation strategy for the JPL Smart Hands [2] was proposed [6] in which tools are held in the grippers and power is delivered by electrical contacts that automatically engage upon pick-up. In an overview report of end effector use for space-based systems [5], the range of possibilities from simple 1 degree of freedom (dof) effectors to anthropomorphic hands are examined for extra-vehicular activity (EVA) tasks in space and it is concluded that anthropomorphic hands are the best solution.

The difference between

- a highly dexterous, versatile end effector capable of performing all tasks, and
- specialized tools used for particular tasks

is that it is difficult to design an end effector that can satisfy all tasks requirements. We propose that the best solution for end effector design for current technology is a mounting platform at the end of the robot with a set of interchangeable effectors designed for specialized tasks that can be attached to the platform. The action most often performed by the robot in the ATOP laboratory is simple gripping of parts for which a 2-jaw hand is sufficient. We believe that a simple 1 dof gripper is a suitable platform for a large number of repair tasks. A desired feature of this design is also to have force/torque sensing on the platform [6] so that the operator of the master-slave system can feel interaction forces and torques with all end effectors. With these 2 considerations we have chosen to use the JPL Smart Hand with a 2-jaw gripper and high bandwidth force/torque sensing permanently mounted on the robot end link and to have the ability to pick up specialized tools for particular operations. These *tools* are picked up as the need for them arises and they are returned to a tool holder when not needed.

*This work was carried out at the Jet Propulsion Laboratory, California Institute of Technology, under contract with the National Aeronautics and Space Administration. The authors are in the Robotic Systems and Advanced Computer Technology Section, JPL, California Institute of Technology, Pasadena, CA 91109.

2 TOOL STORAGE AND HANDLING

2.1 Tool Storage

The requirement on the tool storage system is that tools be held firmly in place during storage but be easily released when they are to be pick-up. When not in use, tools are held on a tool caddy beside the mock-up of the Solar Max Satellite. Velcro loop tape on mounting pads on the respective tools and corresponding Velcro hook tape on the tool caddy base keep the tools secured to the caddy. The advantages of this tool storage method are:

- minimal forces and torques are required to pick up tools,
- the tools are held securely when not in use,
- the compliance of the Velcro connection aids in reducing forces and torques when the robot makes contact with the tool. A rigid holding mechanism generates large forces and torques on tool pick-up due to intrinsic positioning errors, unless end effector compliance control is implemented.

2.2 Tool Pick-up

Two methods are currently employed for tool pick up at the ATOP system:

- Method 1 uses the gripping action of JPL Smart Hands to pick up tools from the tool caddy. The human operator squeezes the trigger on the hand controller grip to close the jaws of the JPL Smart Hand in order to pick up tools. Tools designed for this pick-up method have handles designed to fit into V-shaped grooves on the inside faces of the end effector jaws currently in place of the JPL Smart Hand. For picking up tools, the jaws are closed on to the tool handle for a secure grip then the tool is lifted off the tool caddy. The process is reversed for tool stowage. The procedure for tool pick-up with this method is illustrated on Figure 1.

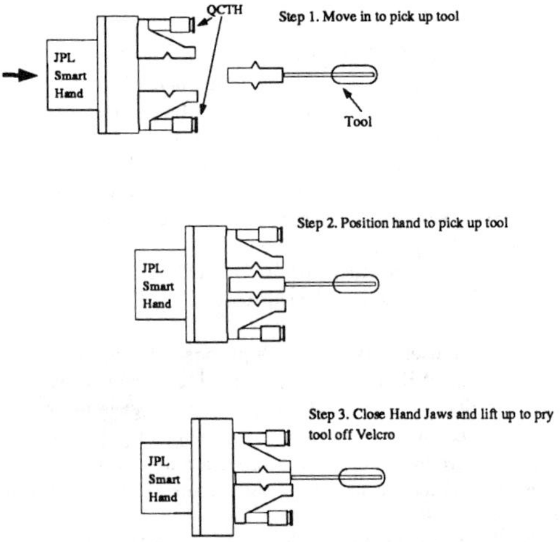

Figure 1: Method 1 for picking up a tool

- Method 2 uses a modified commercial quick connect/disconnect tool holder used in CNC machines called a Quick Change Tool Holder (QCTH) mounted on the JPL Smart Hand. The QCTH and a tool pick-up device were designed to allow the robot to pick up tools without external actuation but by following a set of movements to engage and release a spring-loaded lock. The procedure for tool pick-up with this method is illustrated on Figure 2. The first step is to move the robot hand so that a collar on the QCTH is squeezed when the QCTH contacts a spring-loaded plate mounted on a rail. Squeezing the collar releases 3 steel balls that are used to clamp the tool shank in the QCTH. The QCTH is ready to receive the tool shank when the collar is fully depressed and this is achieved by continuing the motion of the robot hand in the direction of the tool. Once the tool shank is within the QCTH, the robot hand is moved vertically up to pry the tool off the Velcro tape. The hand is then pulled back to release the QCTH collar. This action locks the tool in the QCTH and the tool is ready for use. Stowage of the tool is achieved by reversing the procedure.

The advantage of the first pick-up method is that it is simple and natural. Its disadvantage is that it uses the actuation of the Smart Hand and therefore prevents the motion of the grippers to be

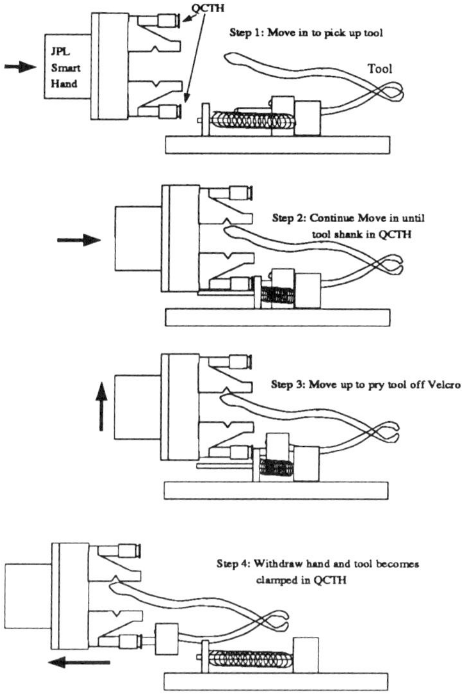

Figure 2: Method 2 for picking up a tool

used for actuation of the tool in subsequent task phases. The second pick-up method overcomes that disadvantage but is less natural and requires more involved manipulation.

The reason for the use of 2 methods is historical; the tools initially used on the ATOP system did not require additional mechanical actuation. The jaws on the JPL Smart Hand were used to hold the tool and move it to complete the task. The simplest method, Method 1, was therefore chosen for picking up the tools. The need for new tools that required mechanical actuation for performing operations resulted in the design of Method 2 for tool pick-up. Method 1 has been retained both for compatibility with the use of the original set of tools and because it is simple to use with tools not requiring mechanical actuation.

3 TOOL OPERATION

The tools currently in use at the ATOP and their applications are discussed in the following sub-sections. The criteria for design were that:

- the tools perform their respective operations,
- the tools be easily picked up and stowed away,
- the tools be light-weight and compact,
- tool actuation be intuitive and natural,
- the design and fabrication cost of the tools be minimized, and
- the tools operate both in a space and terrestrial environment.

These requirements have been met for all tools described in this memorandum.

3.1 Thermal Protection Blanket Tape Cutting

This operation is to cut tape at seams of the thermal protection blanket (TPB) used to shield the satellite from damaging cosmic radiation. The TPB used on the Solar Max Satellite mock-up consisted of gold Mylar film sandwiching a foam sheet. Overlapping panels of TPB were taped to cover the satellite. During the cutting operation, the tape at the seams of the TPB covering the Main Electronics Box (MEB) of the satellite is punctured and slit so that the TPB can be removed to expose the MEB. A knife shown on Figure 3 was designed for this operation. The handle fits into grooves in the grippers of the JPL Smart Hand while the knife blade is curved with a pointed tip to puncture the tape at

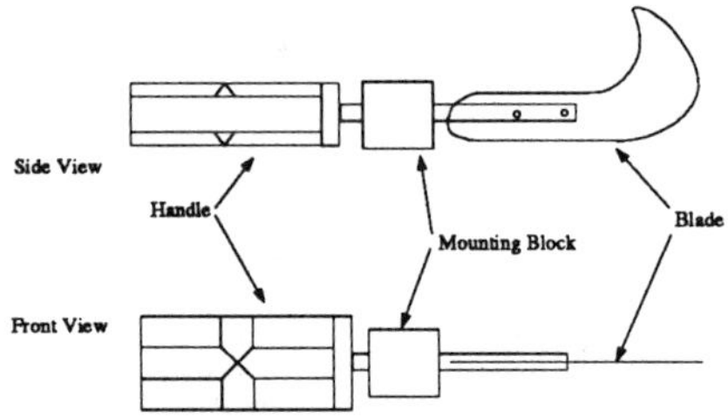

Figure 3: Thermal protection blanket tape cutting tool

the initiation of a tape cut. During tape cutting, the shape of the blade forces the cut to be aligned well by tensioning the tape at and beyond the cutting point. This tool has been used extensively in experiments for over a year and has proved to be reliable.

3.2 Screw Removal and Insertion

Fourteen 10-32 socket head screws hold the MEB panel on to the satellite and these need to be removed before the panel can be opened. In addition, on replacement of the MEB the screws have to be reinserted. A modified cordless screwdriver is used for this operation. A mounting block was added to the tool to allow attachment to the tool caddy. A tool handle was also designed for ease of tool handling using the gripping tool pick-up method. The tool is drawn on Figure 4. A 10-32

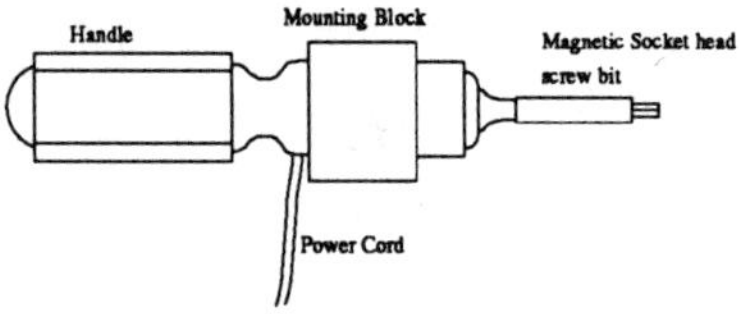

Figure 4: MEB panel screw removal and reinsertion tool

socket head magnetic bit is used on the tool so that the screw remains on the bit after removal. The screw can be bolted into another threaded hole either for storage or, after replacement of the MEB, to fasten the MEB to the satellite frame. The electrical wiring of the commercial tool was replaced to allow operation from the control room of the ATOP. Foot-operated switches are used to actuate the tool; one switch for unbolting and another for bolting in the screw.

3.3 Electrical Connector Screw Removal and Insertion

Eleven electrical connectors (50-pin, 15-pin and 9-pin) must be removed from the back of the MEB before it can be replaced. The procedure for removing the connectors is to first unbolt the captive screws (two for each connector) that attach the connectors to their sockets. This is done with a screw removal tool similar to the MEB panel screw unbolt/bolt tool. The difference between the two tools is that the connector screw removal tool has a #8 flat blade screwdriver as its bit. The tool is also used to bolt in the connector screws to re-mate the connectors to their respective sockets after installation of the new MEB.

3.4 Electrical Connector Removal and Re-mating

After the unbolting of the electrical connector screws, the connectors have to be pulled out of their sockets. The jaws of the JPL Smart Hand are too large to access individual connectors on the bank of connectors at the back of the MEB panel for this operation because the connectors are arranged close to each other. To overcome the problem, modified gripper attachments were designed for removing the electrical connectors from their sockets. An even more difficult procedure is re-mating the connectors and this requires the use of grippers that hold the connectors firmly and are small enough to insert the connectors into their respective sockets without interfering with neighboring connectors after the replacement of the MEB.

A pair of alternative grippers was designed to be picked up as tools using Method 2 for tool pick-up. The grippers are drawn on Figure 5. When held in the QCTHs, the new grippers extend length of the fingers by about 3 inches and reduce their size considerably. Grooves were designed on the new grippers to fit the back shells of the electrical connectors for ease of gripping. To improve the firmness of grasping and holding the connectors, the new grippers are bonded to a soft plastic coating which increases their surface friction coefficient. The compliance of the coating also aids in the stability of the slave robot on contact with hard environmental surfaces with the modified grippers.

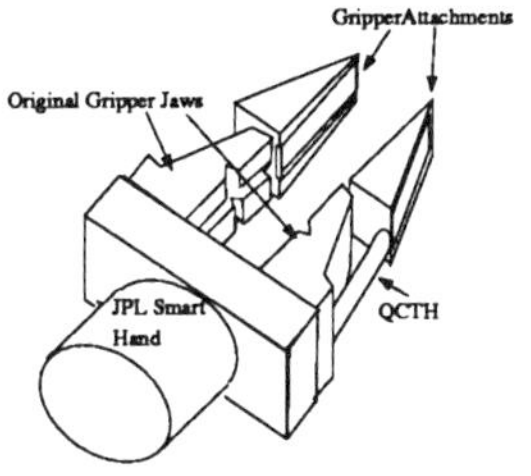

Figure 5: Modified grippers for electrical connector removal and insertion

3.5 Ty-Rap Cord Cutting

After the electrical connectors are unplugged from their sockets, the bundle of electrical cables held with Ty-Rap to the MEB has to be detached. The tool used for cutting the Ty-Rap cords that hold the bundled electrical cables to the MEB is shown on Figure 6. It is a modified diagonal cutter fitted with a mounting block for attachment to the tool caddy and a handle for pick-up by the QCTH. When held in the Smart Hand, the tool can be actuated by closing the jaws of the hand similar to the

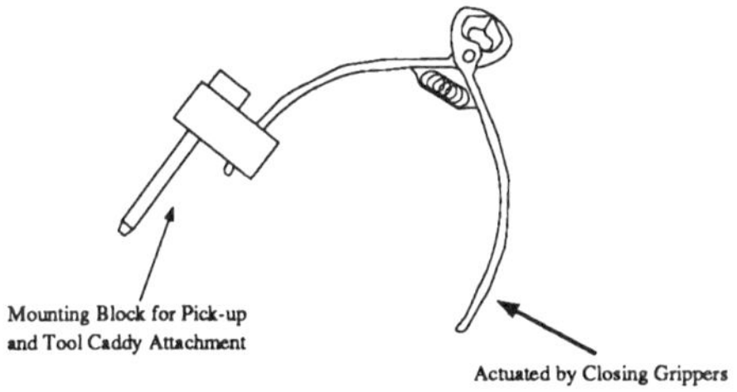

Figure 6: Ty-Rap Cord Cutting Tool

way fingers are used to squeeze the handle of the diagonal cutter when it is held in a human hand. Springs keep the cutter in the open position when not actuated. Cutting Ty-Rap cords is performed by the teleoperation system as it would if the tool were held in human hands; the tool blades are moved to the location of the cord and positioned to cut the cord, then the jaws of the robot hand are closed to actuate the tool and cut the cord. Rubber pads glued on the inside of the blades of the cutter retain the Ty-Rap cord after cutting and the cord may then be placed is a waste recovery bag for disposal. The reason for this additional feature is that debris must be kept minimal in space manipulation because of potential damage.

4 FORCE AND TORQUE OFFSETS

Force reflection is used on the ATOP system as aid to the operator and is often necessary for successful completion of the tasks performed. Previous work has shown that force reflection improves operator performance in teleoperated tasks [3]. A concern with the alternative tool mounting methods we use is the difference in forces and torques felt by the operator for the same environment interaction forces and torques because of the offset of the tool position from the force/torque sensor (F/TS).

In the example illustrated by Figure 7, the two mounting methods are shown with the tool of weight vector W_t held on the hand. The task interaction force is denoted by the vector F_e and torque M_e for both mounting cases.

The vector position of the tool tip from the F/TS is r_{t1} for tool pick-up method 1 and its center of mass vector position, C_1, is r_{g1} while the corresponding variables for tool pick-up method 2 are r_{t2} and r_{g2}. The force sensed at the F/TS is thus $F_e + W_t$ for both mounting methods but the torque is $M_e + r_{g1} \times W_t + r_{t1} \times F_e$ for method 1 and $M_e + r_{g2} \times W_t + r_{t2} \times F_e$ for method 2.

Since F_e, W_t, r_{t1}, r_{t2}, r_{g1} and r_{g2} are known, the torque readings can be biased in software to compensate for the different offsets of the tool center of mass and end position and it is possible to place the reference frame for interaction forces at any position. Ideally the reference point would be at a distance from the tool tip such that the operator feels torques as if the tool were held in the operator's hand. The torques fed back to the operator should therefore be biased so that the torque felt at the Force Reflecting Hand Controller (FRHC) is that at the handle of the tool.

Problems may arise in this compensation method if the tool has multiple contact points. An example is the cutting tool where the points of contact are along its blade. Also, if contact is made

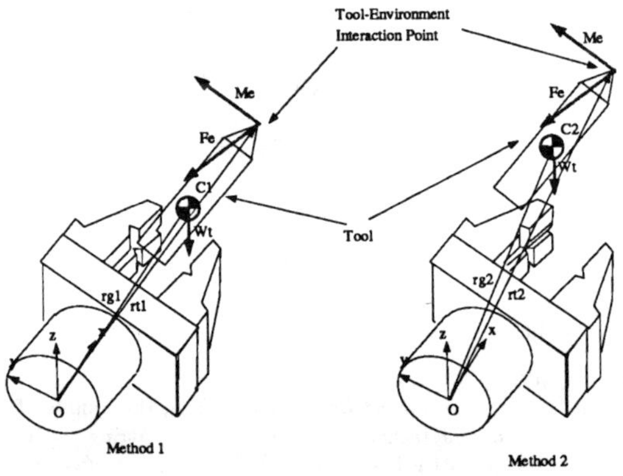

Figure 7: Effects of force and torque offsets

with the environment at points on the hand or on the tool other than the specified contact point r_t torques will be generated that cannot be corrected for unless the contact point is known.

In a space environment the above analysis still applies but with the tool weight vector $W_t = 0$. The preceding analysis has assumed that all forces and torques generated by motion of the tool are negligibly small. It is possible to have dynamic compensation, i.e. when forces and torques are not small, if instantaneous position, velocity and acceleration of the tool are known. In the ATOP system, acceleration of the end effector is neither measured nor computed and so dynamic compensation is not currently possible. Gravity force/torque compensation has been implemented on the ATOP system to compensate for the weight of the grippers and tools of arbitrary weight and therefore preventing the manipulator from yielding under the tool weight when in compliance control mode.

5 CONCLUSIONS

The tool operation method described in this paper is in use at the ATOP laboratory for demonstration of our SMSR procedures. Selected sub-tasks of the SMSR procedure have been used as candidate tasks in experiments evaluating human operator performance with various options available on the ATOP system [3]. The tools and tool procedures have performed reliably in the experiments and we are continuing the efforts of designing new tools for future operations.

References

[1] A. K. Bejczy, Z. F. Szakaly, and W. S. Kim. A laboratory breadboard system for dual-arm teleoperation. In *3rd Annual Workshop on Space Operations, Automation and Robotics (SOAR '89)*, NASA Johnson Space Center, Houston, TX, July 1989.

[2] A.K. Bejczy. Smart hand: Manipulator control through sensory feedback. Internal Document D-107, Jet Propulsion Laboratory, Pasadena, CA, January 1983.

[3] H. Das, H. Zak, W. S. Kim, A. K. Bejczy, and P. S. Schenker. Performance experiments with alternative teleoperator control modes for a simulated solar maximum satellite repair. In *5th Annual Symposium on Space Operations, Applications, and Research, (SOAR '91)*, NASA Johnson Space Center, Houston, TX, July 9-11 1991.

[4] P. Fiorini, H. Das, H. Zak, and A. Bejczy. Teleoperated repair in space: a system for baseline testing. In *International Advanced Robotics Programme (IARP), 1st Workshop on Robotics in Space*, Pisa, Italy, June 17-18 1991.

[5] A. H. Mishkin and B. M Jau. Space-based multifunctional end effector systems: Functional requirements and proposed designs, April 1988. JPL Publication 88-16.

[6] T. R. Ohm. End effector change options for the Model C Smart Hand. In *End Effector Change Mechanism Workshop*, NASA Goddard Space Flight Center, GreenBelt, MD, May 1990.

[7] P. S. Schenker, A. K. Bejczy, W. S. Kim, and S-K. Lee. Advanced man-machine interfaces and control architecture for dexterous teleoperations. In *IEEE Oceans '91*, Honolulu, HI, October 1991.

[8] Z. F. Szakaly and A. K. Bejczy. Performance capabilities of a JPL dual arm advanced teleoperation system. In *SOAR '90 Workshop*, Albuquerque, NM, July 1990.

ARTICULATED CALIBRATION EXPERIMENT (ACE) FOR THE EARTH OBSERVING SYSTEM

Stanford Ollendorf*
Robotics Branch
NASA/Goddard Space Flight Center
Greenbelt, Maryland 20071

Abstract

This paper presents the results of a study called the Articulated Calibration Experiment (ACE) whose objective is to investigate the feasibility of instrument calibration on the Earth Observing System (EOS) using a resident robots. After presenting the overview and the objectives of the study, the paper lists a set of robot design requirements derived from a set of task requirements. Preliminary results of dynamical analysis are also presented.

1 Introduction

In recent years, the servicing of spacecraft has significantly extended the life of NASA's orbiting assets. Beginning with the Solar Max Mission, where both instrument and subsystem modules were replaced, up to the current plans for the Hubble Space Telescope, spacecraft have increasingly been designed with servicing in mind. This is especially true for Shuttle-compatible payloads.

The cost benefits of servicing, however, depend on the scale of servicing required. For spacecraft such as the Earth Observing System (EOS) [1], a polar orbiting platform which is out of the reach of the Shuttle, the cost benefit of servicing through the use of large, expendable launch vehicles is lower than the cost benefit of servicing using small launch vehicles, when considering the size of the subsystem modules. Servicing at the instrument level, which can be accomplished by a *resident robot* uses small launch vehicles and is a cost effective approach when considering a reflight alternative.

The EOS Project Office instructed the Engineering Directorate at Goddard Space Flight Center to perform a Phase-A Study, called Articulated Calibration Experiment (ACE) [2,3] to prove the feasibility of instrument calibration by a resident robot such as one illustrated in Figures 1-3 that could evolve into more extensive instrument servicing and platform inspection and to assess the accommodation impact of a robot on the EOS-A platform. The EOS-ACE Study Project team, a team of civil servants and support contractors headed by the Office of Technical Management, was established to conduct the study.

The EOS/ACE Study Project team concluded that the use of a resident robot to perform accurate instrument calibration on the EOS-A platform was feasible. Of the types of robot concepts evaluated, a multi-arm robot was shown as being less complex and more maneuverable than the walking beam or other configurations. This robot would operate autonomously, with ground supervision as a contingency in the event

*Branch Head

of an anomaly. Furthermore, the robot would be designed to have minimal impact on spacecraft interfaces and spacecraft operations and to have no effect on the scientific instruments' performance. Its *non-interference* operation would minimize the cost impact resulting from the presence of the robot on the EOS-A platform and its use of platform support resources such as the power supply and data handling system.

Figure 1: The ACE robot

The major perturbation to the spacecraft that surfaced was the effect of robot motion and impact, while walking, on the attitude control system. To stay within allowable limits, low robot mass coupled with slow speed and active sensing would be required.

The study emphasized that the use of a robot on the EOS-A platform would be classed as an experiment so that the degree of its success would not affect the overall success of the mission. Maximum use was made of Flight Telerobotic Servicer-derived hardware and software for sizing and cost estimating purposes.

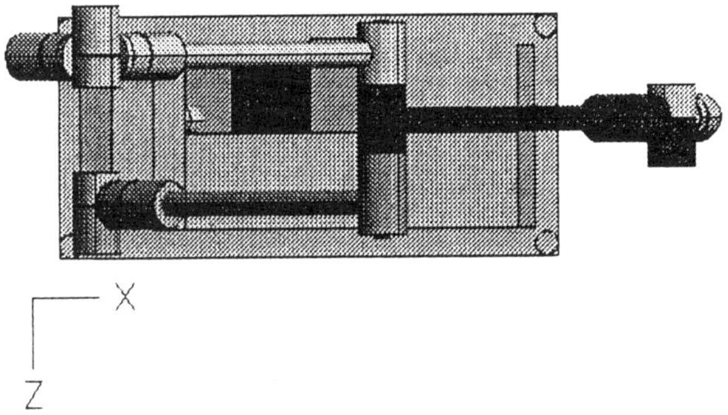

Figure 2: The ACE robot (top view)

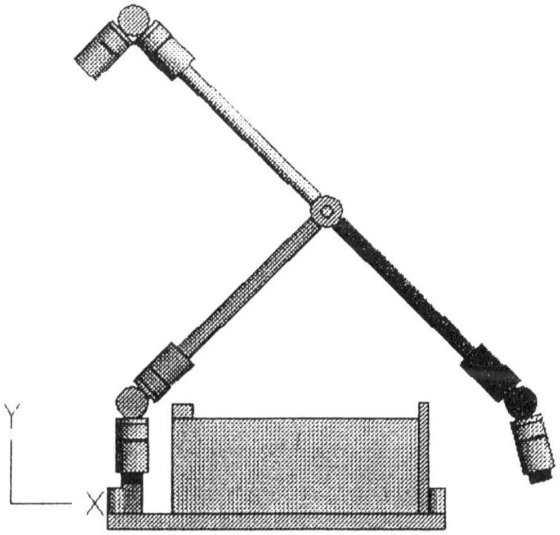

Figure 3: The ACE robot (side view)

2 Study Overview

An overview of the ACE system is given below:

- ACE system consisting of the robot and its calibration devices would be launched as part of the EOS-A1 payload
  - Would use a payload mounting plate or be directly mounted to the spacecraft structure
- The ACE would be classed as an *experiment*

  - EOS mission success would not depend on the ACE

- ACE is a semi-autonomous, dextrous manipulator that could perform multiple tasks:
  - Primary task is to calibrate and cross-calibrate instruments
  - Could inspect the EOS-A1 platform, instruments and itself for anomalies using video imaging
  - Could evolve to perform servicing tasks

- ACE would operate in concert with the EOS-A1 platform and ground support elements
  - Size, weight, power and data rate to be within EOS-A1 capabilities.

- ACE dynamics will be deliberate and slow so as not to cause attitude disturbances

3 Study Objectives

The following were objectives of the overall study effort:

- Determine calibration task requirements
- Establish robotic functional requirements and constraints
- Develop concept(s) for robot design
- Define support equipment, i.e., power supply, computers, etc.
- Establish spacecraft interface requirements
- Verify design concepts through analysis and simulation
- Develop ground and flight operational concepts
- Determine costs and perform trades
- Define robotic lab demonstration program

4 Task Requirements

The requirements for calibration tasks performed on the EOS are given below:

- Provides cross-calibration between instruments using same set of calibration targets
- Quantifies solar glint reflected from spacecraft and other instruments that degrades the accuracy of the solar diffuse calibration method.
- Provides a means for instruments to view the moon as a calibration source.
- Provides views of sun, moon and cold space to those instruments where such view are blocked.
- Provides blackbody calibration target with lower temperature comparable to that of coldest features on earth (225k).
- Provides higher level of redundancy for calibration function.

5 Robot Design Requirements

Based upon the task requirements as presented in Section 4, the requirements on the design of the ACE robot are derived and presented below:

- Robot access approach selected for ACE must:
  - Provide access to nadir side instruments requiring calibration
  - Provide access, as feasible, to perform spacecraft inspection
- The robot design selected must adhere to the following design requirements:
  - Require no changes to instrument payload
  - Minimize impact to instrument operations (fields of view)
  - Minimize impact to EOS spacecraft design and operations (including thermal management)
  - Meet requirements and constraints of general instrument interface specification
- Robot design must minimize the size, weight and power impacts to the spacecraft:
  - All are directly proportional to the required length and load-carrying capability and accuracy of the manipulator

6 Dynamic Analysis Results

The preliminary results of dynamical analysis of the ACE are summarized as follows:

- Moving at a rate of 200 sec or greater will keep the resultant torques to less than 1 n-m and accelerations less than .002 to .02 g's
- The impact force is directly proportional to the impact velocity, but the size of the impact velocity has no bearing on whether contact is maintained.
- Stiffer environments cause larger impact forces and increase the risk of bouncing upon impact
- Higher damping ratios have no effect on the peak impact force but decrease the chances of bouncing upon impact up to a point
- Larger masses increase the peak impact force but decrease the chances of bouncing upon impact

7 Conclusions

The study yielded the following results:

- The application of a robot to perform the instrument calibration task on the EOS platform proved to be feasible.

- A multi-arm robot concept moving about the instrument platform proved to be less complex and more dextrous than the walking beam and was operationally verified through graphic simulation.
- The robot weight was 171 pounds with a total system weight of 570 pounds including the calibration task set. Average power from the spacecraft was 236 watts, at a peak of 384 watts.
- Experiment will provide value-added capability to the calibration of EOS scientific instruments.

References

[1] **Perrygo, C.**, "ACE Instrument Calibration Requirements Document," *IDEA Corporation*, July 1991.

[2] **GSFC**, "Articulated Calibration Experiment (ACE)," Final Report, *Goddard Space Flight Center, NASA*, PEI # 1008.3, January 1992.

[3] **Routson, M.**, "EOS/ACE Kinematic Studies," Final Report, *Advanced Technology and Research Corporation*, November 1991.

Fault Tolerant Joint Development for the Space Shuttle Remote Manipulator System: Analysis and Experiment

Eugene Wu James Hwang
Lockheed Engineering & Sciences Company

John Chladek
NASA / Johnson Space Center

Abstract

A space-based fault-tolerant robot joint design with a dual-motor, single-output differential-based mechanical drive system was investigated. The mathematic model was developed and simulations were performed for various joint failure conditions. A scaled-down differential testbed was subsequently designed and built to validate the analytical results. Simulation and test results demonstrated that the design is capable of sustaining a single failure and absorbing the failure disturbance, and continuing to be operational with the remaining single drive mode.

1. Introduction

The Shuttle Remote Manipulator System (SRMS) is the 6-DOF robotic arm on board the Shuttle. It is used to deploy and retrieve payloads from or into the Orbiter cargo bay. The system does have a backup control mode and some redundant drive electronics which can be used to stow the arm away in the event of certain failures. It is merely fail-safe, but the operation can not be continued after the shutdown. Johnson Space Center/NASA has been investigating new manipulator systems and component fault tolerance designs [1,2]. The objective is to enhance the system performance with fail-operational capability, so that the operation of the SRMS may continue after a single failure. This project investigates fault tolerance designs of the manipulator joint. A dual input, one output differential has been proposed as the fault tolerant joint mechanism candidate [3,4].

In this paper, the differential gear train mathematic modeling is presented. The focus of this paper is on the failure transient responses of the differential system. Two most common failures, motor losing power and motor running away, are examined. It appears that satisfactory failure transient responses for both loaded and unloaded cases can be achieved. A scaled-down hardware testbed was constructed to validate simulation results. Test data were displayed and correlated to the analytical results. The consistency between test data and simulation results concluded that dual drive differential actuator is a good candidate for the fault tolerant joint design.

2. Modeling of a dual drive differential joint system

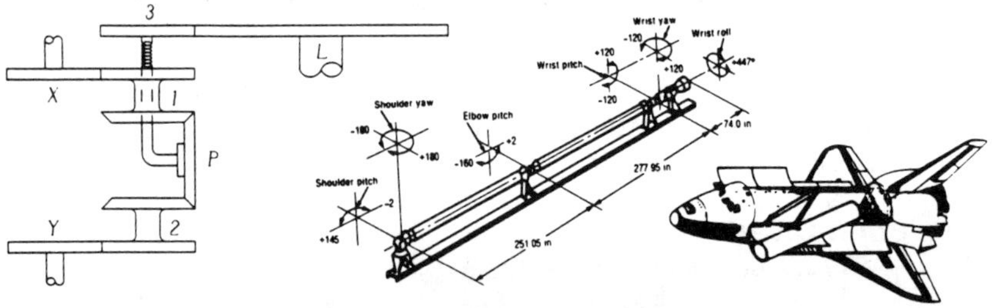

Figure 1. Dual drive(X & Y), single output(L), differential joint

A differential is a bevel planetary train with equal sun gears, constituting a velocity summing mechanism (Figure 1). Here, two driving motor shafts are directly coupled to the input gears, which are meshed with the sun gears of the differential. The arrangement allows the two input motors to run at different speeds, and yields the following kinematic equation:

$$\dot{\theta}_p = \frac{\dot{\theta}_1 + \dot{\theta}_2}{2}, \quad \frac{\theta_x}{\theta_1} = \frac{\dot{\theta}_x}{\dot{\theta}_1} = \frac{\ddot{\theta}_x}{\ddot{\theta}_1} = 1, \quad \frac{\theta_y}{\theta_2} = \frac{\dot{\theta}_y}{\dot{\theta}_2} = \frac{\ddot{\theta}_y}{\ddot{\theta}_2} = 1 \tag{1}$$

The speed reduction gearbox with gear ratio NL could be placed after the differential, yielding the equation:

$$\frac{\theta_3}{\theta_L} = \frac{\dot{\theta}_3}{\dot{\theta}_L} = \frac{\ddot{\theta}_3}{\ddot{\theta}_L} = N_L \tag{2}$$

For SRMS, most of the arm flexibility arises from the high ratio gearbox, therefore the arm flexibility is modeled as a gearbox spring and is placed after the differential in this case. An idealized inertial, viscous friction, gear stiffness model with a 100% transmission efficiency leads to the following torque equation:

$$N_L \left[K(\theta_p - \theta_3) - (J_3\ddot{\theta}_3 + B_3\dot{\theta}_3) \right] = J_L\ddot{\theta}_L + B_L\dot{\theta}_L \tag{3}$$

The transmission efficiency, η, defined as the ratio of the output power to the input power through two meshed gears, is very complicated in theoretical computation. Sliding losses along teeth, pressure angle of teeth, windage losses because of lubricant and even the speed-dependent losses all make the efficiency unpredictable. Above all, the difference in forward and backdrive efficiencies is found extremely important in the dual driving mechanism and is considered here. A forward-backdrive logic of the differential train is expressed as following:

$$K(\theta_p-\theta_3) + (J_1\ddot{\theta}_1+B_1\dot{\theta}_1) + (J_2\ddot{\theta}_2+B_2\dot{\theta}_2) = \eta_x[T_x-(J_x\ddot{\theta}_x+B_x\dot{\theta}_x)] + \eta_y[T_y-(J_y\ddot{\theta}_y+B_y\dot{\theta}_y)] \tag{4}$$

$$if\ \{[T_x-(J_x\ddot{\theta}_x+B_x\dot{\theta}_x)] - [T_y-(J_y\ddot{\theta}_y+B_y\dot{\theta}_y)] + K(\theta_p-\theta_3) + (J_1\ddot{\theta}_1+B_1\dot{\theta}_1) + (J_2\ddot{\theta}_2+B_2\dot{\theta}_2)\} \cdot \dot{\theta}_x$$

$$> 0 \rightarrow \text{motor } x \text{ forward driving}, \quad \eta_x = \eta_{F_x}$$

$$\leq 0 \rightarrow \text{motor } x \text{ backdriving}, \quad \eta_x = \frac{1}{\eta_{B_x}}$$

$$if\ \{[T_y-(J_y\ddot{\theta}_y+B_y\dot{\theta}_y)] - [T_x-(J_x\ddot{\theta}_x+B_x\dot{\theta}_x)] + K(\theta_p-\theta_3) + (J_1\ddot{\theta}_1+B_1\dot{\theta}_1) + (J_2\ddot{\theta}_2+B_2\dot{\theta}_2)\} \cdot \dot{\theta}_y$$

$$> 0 \rightarrow \text{motor } y \text{ forward driving}, \quad \eta_y = \eta_{F_y}$$

$$\leq 0 \rightarrow \text{motor } y \text{ backdriving}, \quad \eta_y = \frac{1}{\eta_{B_y}}$$

At last, with the energy being dissipated in friction, the power equation is given by:

$$\eta_x[T_x-(J_x\dot{\theta}_x+B_x\dot{\theta}_x)]\,\dot{\theta}_x+\eta_y[T_y-(J_y\dot{\theta}_y+B_y\dot{\theta}_y)]\,\dot{\theta}_y \tag{5}$$
$$=J_1\dot{\theta}_1\dot{\theta}_1+J_2\dot{\theta}_2\dot{\theta}_2+J_L\dot{\theta}_L\dot{\theta}_L+J_3\dot{\theta}_3\dot{\theta}_3+B_1\dot{\theta}_1^{\,2}+B_2\dot{\theta}_2^{\,2}+B_L\dot{\theta}_L^{\,2}+B_3\dot{\theta}_3^{\,2}+K(\theta_p-\theta_3)(\dot{\theta}_p-\dot{\theta}_3)$$

Surprisingly, even though the equation(5) involves non-linear terms, these dynamic equations for the system can be reduced to linear state equations through variable substitutions without approximation (the derivation is not shown here). It is expressed in the following state matrix form:

$$\begin{bmatrix}\dot{\theta}_p-\dot{\theta}_3\\ \dot{\theta}_x\\ \dot{\theta}_x\\ \dot{\theta}_y\\ \dot{\theta}_y\\ \dot{\theta}_L\\ \dot{\theta}_L\end{bmatrix}=\begin{bmatrix}0 & \frac{1}{2} & 0 & \frac{1}{2} & 0 & -N_L & 0\\ \frac{-K}{2(J_1+J_x\eta_x)} & \frac{-(B_1+B_x\eta_x)}{J_1+J_x\eta_x} & 0 & 0 & 0 & 0 & 0\\ 0 & 1 & 0 & 0 & 0 & 0 & 0\\ \frac{-K}{2(J_2+J_y\eta_y)} & 0 & 0 & \frac{-(B_2+B_y\eta_y)}{J_2+J_y\eta_y} & 0 & 0 & 0\\ 0 & 0 & 0 & 1 & 0 & 0 & 0\\ \frac{N_L K}{J_L+N_L^2 J_3} & 0 & 0 & 0 & 0 & \frac{-(B_L+N_L^2 B_3)}{J_L+N_L^2 J_3} & 0\\ 0 & 0 & 0 & 0 & 0 & 1 & 0\end{bmatrix}\begin{bmatrix}\theta_p-\theta_3\\ \dot{\theta}_x\\ \theta_x\\ \dot{\theta}_y\\ \theta_y\\ \dot{\theta}_L\\ \theta_L\end{bmatrix}+\begin{bmatrix}0 & 0\\ \frac{\eta_x}{J_1+J_x\eta_x} & 0\\ 0 & 0\\ 0 & \frac{\eta_y}{J_2+J_y\eta_y}\\ 0 & 0\\ 0 & 0\\ 0 & 0\end{bmatrix}\begin{bmatrix}T_x\\ T_y\end{bmatrix} \tag{6}$$

3. Rate controlled system and with the consideration of non-linear phenomena

The current SRMS could be operated in either position hold mode or joint rate mode. The position hold mode is only activated to close around the position servo to maintain the current joint angle setting. For the operation under rate mode, the invert Jacobians algorithm converts the hand controller Cartesian commands into resolved joint rate command for each joint, so that the end effector follows the commanded trajectory. Operations under joint rate mode when the manipulator is in motion represent the more demanding cases. The failure transient response under joint rate operation is more critical and examined here.

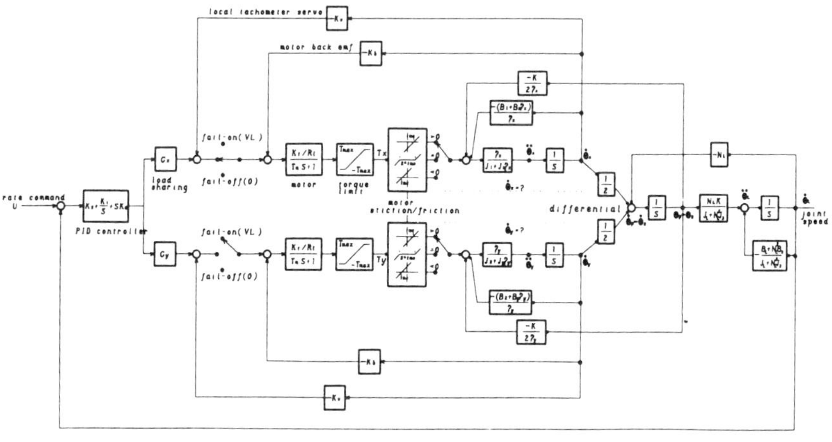

Figure 2. Dual drive fault tolerant joint closed loop system

The rate control mode utilizes rate feedback to close the loop(see Figure 2), with a PID-lag controller on the forward path. Due to the lightly damped open loop characteristics, extra feedback compensation is needed to be incorporated into the servo controller to improve the system damping especially when the arm is unloaded. The tachometer feedbacks from the individual input shafts are thus added to form the inner loop. To make the analysis complete, the motor transfer function and non-linear phenomena in the system, such as motor friction/stiction, torque limit due to amplifier current saturation are all included in the system. A dual drive differential system permits unequal load sharing between two actuators. However, an equilibrium load sharing is preferred during normal operations since unequal load sharing with a high load inertia often results in adversarial turning and effort canceling backdriving.

4. Failure transient responses

Representative manipulator joint-level failures include motor winding failure, servo controller power failure, sensor failure, bearing failure, gearbox failure, brake failure, cabling problem, connector problem, etc. They fall into three categories: fail-off with zero torque input, fail-on with maximum torque input, and the intermediate. The transient responses due to a fail-on type failure and a fail-off type failure are two typical situations and their mathematical simulations are examined here.

Using the existing SRMS shoulder yaw joint assembly as a guideline, reasonable physical parameters for a proposed differential fault tolerant joint are chosen here to be analyzed. For the current SRMS, the maximum rotational joint coarse rate for shoulder yaw is limited to 0.004 rad/sec when a heavy payload is attached. For an unloaded or lightly loaded arm, a fast coarse rate is possible. As a matter of fact, the failure could happen any time during a varying speed operation, or even before or after the action. It is assumed here that the joint failure occurs when the joint reaches a constant coarse rate of 0.004 rad/sec just for the sake of comparison.

Fail-off type failure is first examined, and the failure transient responses are shown in Figure 3. The loaded system perfectly absorbs the internal failure disturbance. The faulty motor y loses the speed and is finally stopped by motor friction, whiles, motor x sustains the whole load almost spontaneously such that failure transient is barely seen at the joint output. An unloaded arm reveals similar failure transient responses. The only difference is that joint output momentarily loses the speed due to a comparatively smaller payload angular momentum. The joint speed is regained in a very short period of time and results in a only 0.0182 radius off the targeted rotational position. By cascading a lag compensator, the position offset could be even reduced to 0.0018 radius.

Fail-on type failure is next examined and shown in Figure 4. Assuming that the system is first running at a commanded 0.004 rad/sec joint speed, and then suddenly the power amplifier or motor winding fails and puts out a maximum torque on y module. In both loaded and unloaded cases, this unexpected power input bypasses the high load impedance and generates a huge reactive torque to backdrive motor x. The reactive torque is much larger than the applied torque from motor x, such that motor x will turn in the opposite direction. Again, a loaded arm reveals a much less disturbance on the joint speed than an unloaded arm.

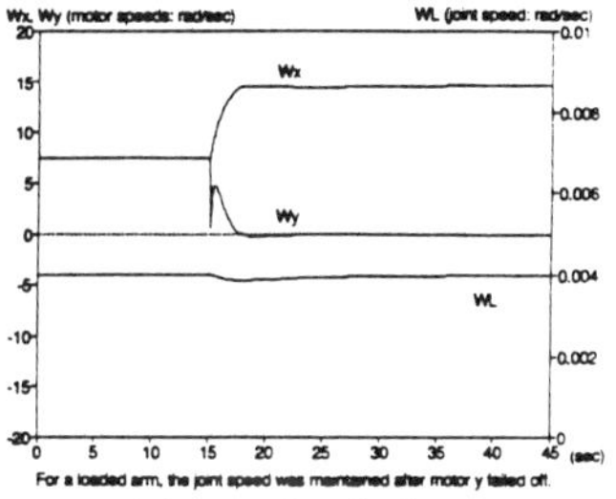

(a) arm loaded.

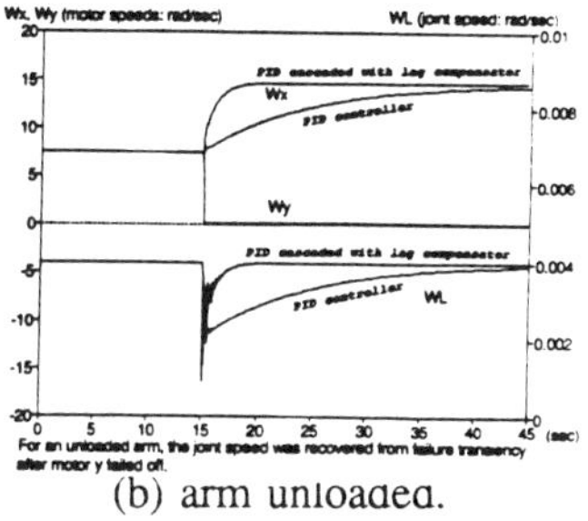

(b) arm unloaded.

Figure 3. Simulated fail-off failure transient responses

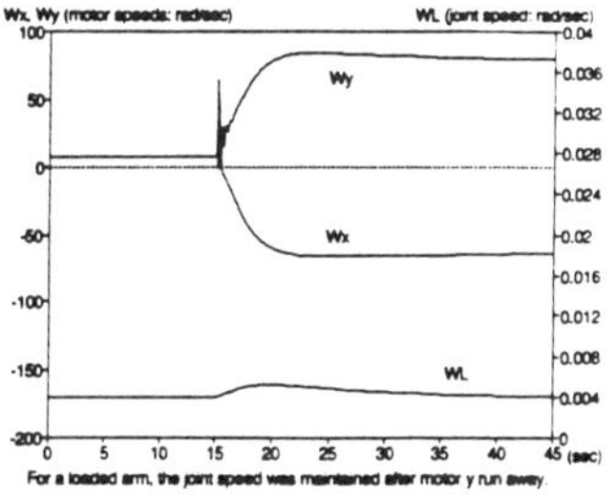

(a) arm loaded.

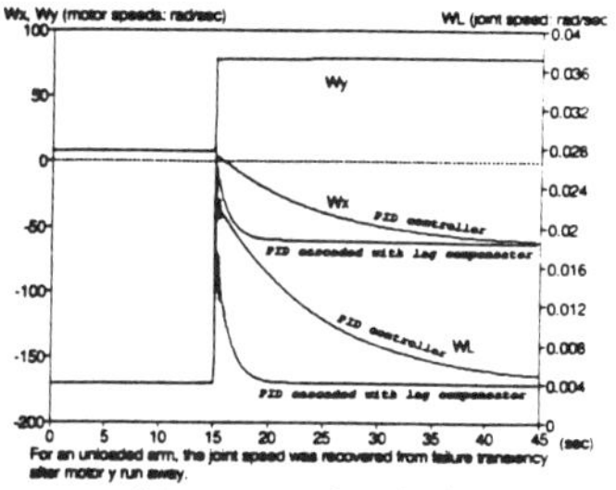

(b) arm unloaded.

Figure 4. Simulated fail-on failure transient responses

5. Testbed and Experimental data

To validate system simulation results, a hardware implementation was designed and the testbed was fabricated (Figure 5). This scaled-down testbed consists of motors with tachometer sensing, gearboxes, brakes, torque sensors, differential, and resolver. The flexible shaft couplers are used to connect all the components. They are used not only to solve the misalignment problems, but also to simulate SRMS soft joint stiffness. Employing the concept of effective torque, a pure inertial load is placed on the output of a gear-up gearbox to construct the simulated heavy inertial load being maneuvered in the 0-gravity environment. Two sets of such inertial load with different gear-ratio were made for comparison.

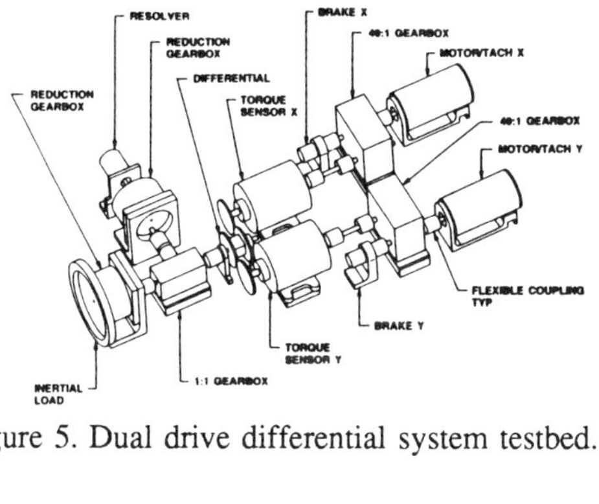

Figure 5. Dual drive differential system testbed.

The 1842/1 gear-ratio from SRMS shoulder yaw was scaled down to 40/1 on the testbed. The baseline joint speed was changed to 0.8 rad/sec accordingly. Figure 6 displays the fail-off failure transient responses of both heavy load and light load cases. The test data shows the consistency with simulation results in Figure 3. As expected, the system has a better fault tolerance capability when a heavier payload is attached.

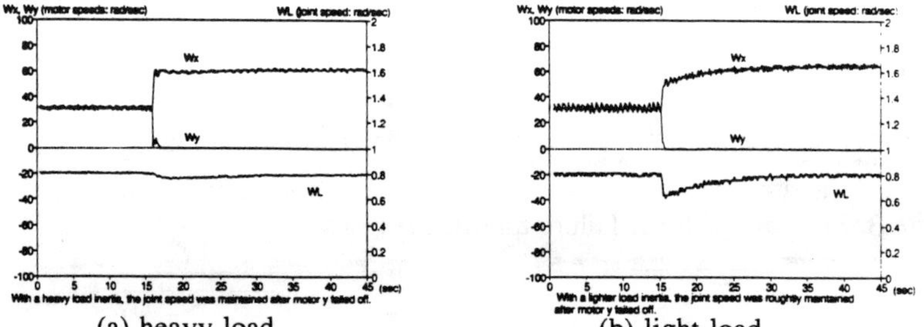

(a) heavy load. (b) light load.

Figure 6. Fail-off failure transient responses - experimental data

6. Conclusion

Due to the general requirements of heavy payload and low power consumption, SRMS' joints have unique characteristics, which are high gear reduction ratio and tremendous effective output inertia to motor inertia ratio. These unique dynamic characteristics combined with the differential's velocity summing feature enables a differential drive to absorb a failure transient and to remain operational in a single drive mode under the 0-gravity space-based environment.

The design of a fault tolerant joint by using differential mechanism and integrating control system design to minimize or absorb operational disturbances during the occurrence of a failure was addressed. The performance evaluation through analysis simulation and testbed experiment showed the consistency and promising results. Based on this preliminary study, design modifications that will enhance SRMS performance with fail-operational feature seem to be very feasible. This potential benefit could be even extended to future Space based manipulator design!

References

1. Charles R. Price, "Telerobotic Activities at Johnson Space Center", Proceedings of the NASA Conference on Space Telerobotics, JPL, Pasadena, California, USA, January 1989.
2. John T. Chladek, "Fault Tolerance for Space Based Manipulator Mechanisms and Control Systems", Proceedings of the First International Symposium on Measurement and Control in Robotics, Houston, Texas, USA, June 1990.
3. Eugene Wu, Myron Diftler, James Hwang, John Chladek, "A Fault Tolerant Joint Drive System for the Space Shuttle Remote Manipulator System", Proceedings of IEEE International Conference on Robotics and Automation, Sacramento, California, USA, April 1991.
4. Eugene Wu, James Hwang, John Chladek, "A Failure Tolerant Joint Design for the Space Shuttle Remote Manipulator System: Analysis and Experiment", Proceedings of IEEE Conference on Control Applications, Dayton, Ohio, USA, September 1992

PART 6 : MOBILE AND FLEXIBLE ROBOTS

DYNAMIC MODEL OF MULTIFINGERED ROBOT HANDS WITH ELASTIC JOINTS

Gongliang Guo, Jichuan Zhang, and William A. Gruver
Center for Robotics and Manufacturing Systems
University of Kentucky, Lexington, KY 40506-0108

ABSTRACT

Multifingered robot hands generally use two types of actuation systems: motor-long-range tendon actuation and motor-on-joint direct actuation. Due to the availability of small size servomotors, motor-on-joint direct actuation has been extensively applied to multifingered robot hands. The dynamic model of multifingered robot hands equipped with motor-on-joint direct actuation involves many factors, particularly joint elasticity, that must be considered in the design of systems for grasping and fine manipulation.

We propose an approach for simplifying the dynamic model of the elastic joints of a multifingered robot hand to reduce the number of generalized elastic coordinates (elastic degrees of freedom). Based on this simplified model, we obtain the equations of motion of a planar two-rigid-degree-of-freedom finger mechanism with elastic joints using a Lagrangian formulation, considering both the joint elastic motion and rigid motion.

INTRODUCTION

The dynamics and control of multifingered robot hands have been the topic of extensive research [1, 2]. These devices generally use two types of actuation schemes: motor-long-range tendon actuation (tendon-actuation) and motor-on-joint direct actuation (joint-actuation). Although mechanisms, kinematics, dynamics, and control of multifingered robot hands with tendon-actuation have been studied by many investigators [3-7], the dynamic model and control of multifingered robot hands with joint-actuation have been a formidable challenge. Research in this area involves many factors, particularly joint elasticity, that must be considered in the design of systems for grasping and fine manipulation at high speeds.

Through experiments using a single-bar robot mechanism with a single joint, Rivin et al. [8] showed that the elasticity of a robot mechanism results primarily from its joint elasticity. Ying and Huston [9] formulated the dynamics of an elastic beam-like mechanism using a finite element method. This approach required a large amount of computation for solving the elastic equation of motion. Based on D'Alembert's Principle, the dynamics of robot mechanisms with elastic joints were derived by Nikolic [10]; however, he did not simplify the elastic-joint model, so the derivation was complex. Mills [11] and Buffinton [12] analyzed the dynamics and stability of elastic-joint robots. In this research, we propose a new approach for simplifying the elastic joint model of a multifingered robot hand to reduce the number of elastic degrees of freedom. Based on this simplified model, we obtain the dynamic equation of a planar two-rigid-degree-of-freedom finger mechanism with elastic joints.

MODEL SIMPLIFICATION OF A ROBOTIC ELASTIC JOINT

In the joint-actuation system of a multifingered robot hand, a high speed servomotor and large ratio microreducer (e.g. harmonic transmission) are frequently used. It is evident that such an actuation system has elastic deformation. To improve the dynamic characteristics and capability for control of multifingered robot hands, the joint elasticity of the joint-actuation system is included in the

The equivalent torsional stiffnesses are

$$k_{E1} = k_1 N_1^2 N_2^2 \qquad k_{E2} = (k_2 + k_r) N_2^2 \qquad k_{E3} = k_3 \tag{3}$$

where k_1 is the stiffness of output shaft 1, k_2 and k_r are the stiffnesses of transmission shaft 2 and reducer, respectively, and k_3 is the stiffness of input shaft 3.
The revolute angles of the four circular plates are represented by

$$\theta_1 = \theta_{j(rigid)} = \theta_m / (N_1 N_2), \quad \theta_2 = \theta_{ro} / N_2, \quad \theta_3 = \theta_{Z1} / N_2, \quad \theta_4 = \theta_{j(rigid+elastic)} = \theta_j \tag{4}$$

where θ_m is the revolute angle of motor output shaft 1, θ_{ro} is the revolute angle of transmission shaft 2 of the reducer, θ_{Z1} is the revolute angle of gear 1, and θ_j is the revolute angle of link j.

Since the velocities and accelerations of a multifingered robot hand generally are not very large, the frequencies of its excitation forces are limited. Moreover, because the effect of higher order natural frequencies and main vibration modes on the motion of multifingered robot hands is small, we only consider here the lower order natural frequencies and vibration modes. The following subsection results a simplification of the four DOF joint model, shown in Fig. 2, into one rigid DOF and one elastic DOF.

Firstly, we introduce the transformation conditions for dynamic equivalence between the two basic models (shown in Fig. 3) composed of circular plates with torsional massless shafts.

$$N_a = \frac{k_i + k_{i+1}}{J_i} \qquad N_b = \frac{k'_{i+1}}{J'_i + J'_{i+1}}$$

(a) (b)

Fig. 3 Two basic models

If $p^2 << N_a^2$, Model (a) can be replaced by Model (b) which has the parameters:

$$J'_i = \frac{k_i J_i}{k_i + k_{i+1}} \qquad J'_{i+1} = \frac{k_{i+1} J_i}{k_i + k_{i+1}} \qquad k'_{i+1} = \frac{k_i k_{i+1}}{k_i + k_{i+1}} \tag{5}$$

where p is the natural frequency of the elastic-joint system.

If $p^2 << N_b^2$, Model (b) can be replaced by Model (a) which has the parameters

$$J_i = J'_i + J'_{i+1} \qquad k_i = \frac{k'_{i+1} (J'_i + J'_{i+1})}{J'_{i+1}} \qquad k_{i+1} = \frac{k'_{i+1} (J'_i + J'_{i+1})}{J'_i} \tag{6}$$

Secondly, the 4-DOF model of the elastic-joint system can be simplified into a 3-DOF model (shown in Fig. 4a) by replacing Model (b) by Model (a). Then the 3-DOF model again can be simplified to a 2-DOF model (shown in Fig. 4b) by replacing Model (a) by Model (b).

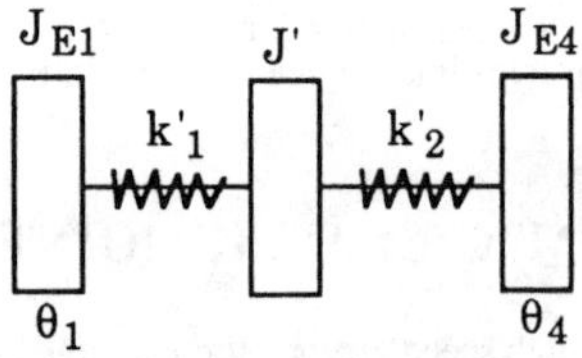

Fig. 4a Three-DOF model

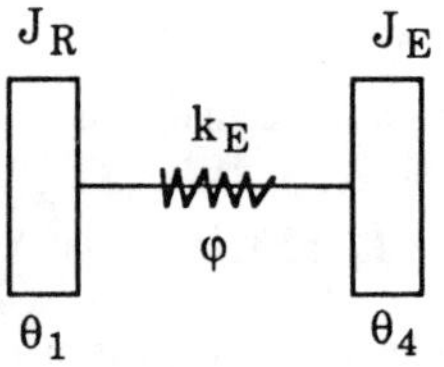

Fig. 4b Two-DOF model

equations of motion. A typical joint-actuation system of a rotary joint with one rigid degree of freedom (DOF) is shown in Fig. 1, including a motor, a reducer, and a gear box.

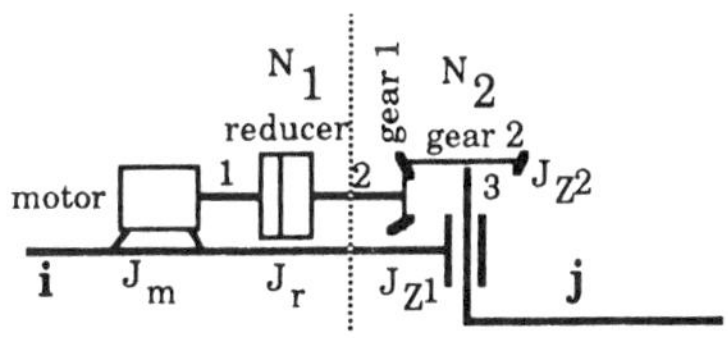

Fig. 1 Single motor-on-joint actuation system

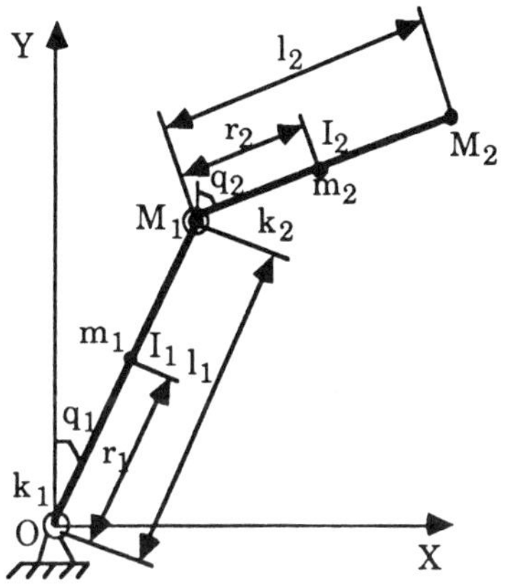

Fig. 5 Rigid two-DOF finger system

Joint elasticity is present in motor output shaft 1, transmission shaft 2, input shaft 3 of link j, and the reducer. Because the elastic deformation of the joint-actuation system in the direction of rotary motion (rotary deformation) is larger than that in other directions, we only consider rotary deformation (torsional vibration) of this system. Other deformations and damping are ignored in this treatment. To simplify the formulation, we assume that the joint-actuation system is composed of circular plates connected with torsional massless shafts. Using the equivalent principle of kinetic and potential energy for before and after transformation of the mechanical system, the single joint-actuation system can be represented as the equivalent elastic vibration model with four degrees of freedom (one rigid DOF and three elastic DOF) at input shaft 3 of link j as shown in Fig. 2.

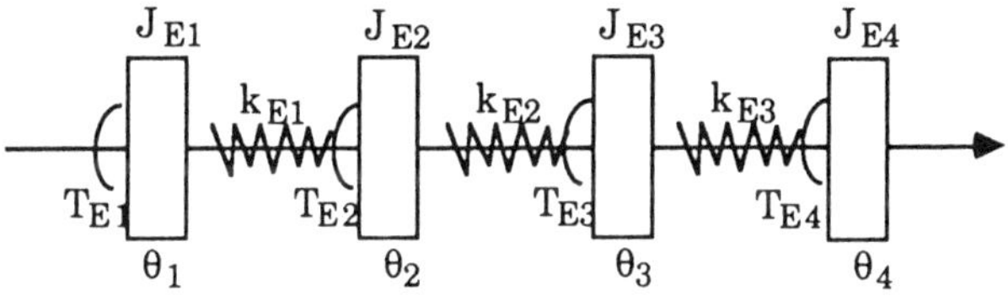

Fig. 2 Equivalent model of a robotic joint with elasticity

The equivalent moments of inertia are represented by

$$J_{E1} = J_m N_1^2 N_2^2 \qquad J_{E2} = J_r N_2^2 \qquad J_{E3} = J_{Z1} N_2^2 + J_{Z2} \qquad J_{E4} = J_3 \tag{1}$$

where J_m is the moment of inertia of the motor at the output shaft 1, J_r is the moment of inertia of the reducer at the transmission shaft 2, J_{Z1} is the moment of inertia of the gear 1 at the the transmission shaft 2, J_{Z2} is the moment of inertia of the gear 2 at the input shaft 3, J_3 is the moment of inertia of input shaft 3 of link j, N_1 and N_2 are the transmission ratios of the reducer and gear box, respectively.

The equivalent torques are

$$T_{E1} = T_m N_1 N_2 \qquad T_{E2} = T_{ro} N_2 \qquad T_{E3} = T_{Z1i} N_2 \qquad T_{E4} = T_{3o} \tag{2}$$

where T_m is the motor output torque, T_{ro} is the reducer output torque, T_{Z1i} is the input torque of gear 1, and T_{3o} is the output torque of input shaft 3.

The 3-DOF model has the following parameters:

$$J' = J_{E2} + J_{E3} \qquad k_1' = \frac{k_{E1} k_1''}{k_{E1} + k_1''} \qquad k_2' = \frac{k_{E3} k_2''}{k_{E3} + k_2''} \tag{7}$$

where $k''_1 = \frac{J_{E2} + J_{E3}}{J_{E3}} k_{E2}$ and $k''_2 = \frac{J_{E2} + J_{E3}}{J_{E2}} k_{E2}$

The 2-DOF model has the following parameters:

$$J_R = J_{E1} + \frac{k'_1}{k'_1 + k'_2} J', \qquad J_E = J_{E4} + \frac{k'_2}{k'_1 + k_2'} J', \qquad k_E = \frac{k'_1 k'_2}{k'_1 + k'_2}, \qquad \theta_4 = \theta_1 + \varphi \tag{8}$$

where θ_1 is the rigid degree of freedom of this joint, and φ is its elastic degree of freedom.

DYNAMICS OF A PLANAR TWO-RIGID-DOF FINGER

As an example of the method, we formulate the dynamic equations of a finger system consisting of a planar two-rigid-DOF plus two-elastic-DOF as shown in Fig. 5.

The revolute angular displacements θ_{m1} and θ_{m2} of the motor 1 and 2 are chosen as generalized coordinates of the rigid subsystem, and the elastic angular displacements φ_1 and φ_2 of joints 1 and 2 as the generalized coordinates of the elastic subsystem.

$$q_1 = \frac{\theta_{m1}}{N_{11}} + \varphi_1 \quad \text{and} \quad q_2 = q_1 + \frac{\theta_{m2}}{N_{22}} + \varphi_2 \tag{9}$$

where N_{11} and N_{22} are the total transmission ratios of the joint 1 and 2, respectively.

The dynamic equations for this system can be obtained from Lagrangian mechanics:

$$\frac{d}{dt}\left(\frac{\partial K}{\partial \dot{\theta}_{mj}}\right) - \frac{\partial K}{\partial \theta_{mj}} + \frac{\partial V}{\partial \theta_{mj}} = \tau_j \qquad j=1,2$$

$$\frac{d}{dt}\left(\frac{\partial K}{\partial \dot{\varphi}_i}\right) - \frac{\partial K}{\partial \varphi_i} + \frac{\partial V}{\partial \varphi_i} = 0 \qquad i=1,2 \tag{10}$$

where K and V are the kinetic energy and the potential energy, respectively.

We will perform this derivation in the following four steps. We first derive the position expressions of the mass center m_1 and m_2 of links 1 and 2, and the loads M_1 and M_2 as shown in the Appendix, and then compute the kinetic energy. We then develop the potential energy. Finally, we obtain the dynamic equations by differentiation.

Kinetic Energy

$$K = \frac{1}{2} m_1 (\dot{x}_{m1}^2 + \dot{y}_{m1}^2) + \frac{1}{2} M_1 (\dot{x}_{M1}^2 + \dot{y}_{M1}^2) + \frac{1}{2} m_2 (\dot{x}_{m2}^2 + \dot{y}_{m2}^2)$$
$$+ \frac{1}{2} M_2 (\dot{x}_{M2}^2 + \dot{y}_{M2}^2) + \frac{1}{2} J_{m1} \left(\frac{\dot{\theta}_{m1}}{N_{11}}\right)^2 + \frac{1}{2} J_{m2} \left(\frac{\dot{\theta}_{m2}}{N_{22}}\right)^2 + \frac{1}{2} (I_1 + J_{\theta 1}) \dot{q}_1^2 + \frac{1}{2} (I_2 + J_{\theta 2}) \dot{q}_2^2 \tag{11}$$

Potential Energy

$$V = [m_1 r_1 + (M_1 + m_2 + M_2) l_1] g \cos q_1 + \frac{1}{2} k_{11} \varphi_1^2 + \frac{1}{2} k_{22} \varphi_2^2 + (m_2 r_2 + M_2 l_2) g \cos q_2 \tag{12}$$

where J_{m1} and $J_{\theta 1}$ are the J_R and J_E of the first joint, and J_{m2} and $J_{\theta 2}$ are the J_R and J_E of the second joint, k_{11} and k_{22} are the equivalent stiffnesses k_E of joints 1 and 2, and g is the gravitational acceleration, τ_1 and τ_2 are the output torques of motors 1 and 2, I_1 and I_2 are the moments of inertia of links 1 and 2, respectively, about their mass centers.

Dynamic Equations

Substituting Eqs. (A1), (11), and (12) into Eq. (10), we obtain

$$M_{11}\ddot{\varphi}_1 + M_{12}\ddot{\varphi}_2 + M'_{11}\ddot{\theta}_{m1} + M'_{12}\ddot{\theta}_{m2} + C_{1ff} + C_{1fr} + C_{1rr} + k_{11}\varphi_1 + D_1 = 0 \quad (13)$$

$$M_{21}\ddot{\varphi}_1 + M_{22}\ddot{\varphi}_2 + M'_{21}\ddot{\theta}_{m1} + M'_{22}\ddot{\theta}_{m2} + C_{2ff} + C_{2fr} + C_{2rr} + k_{22}\varphi_2 + D_2 = 0 \quad (14)$$

$$M_{31}\ddot{\varphi}_1 + M_{32}\ddot{\varphi}_2 + M'_{31}\ddot{\theta}_{m1} + M'_{32}\ddot{\theta}_{m2} + C_{3ff} + C_{3fr} + C_{3rr} + D_3 = \tau_1 \quad (15)$$

$$M_{41}\ddot{\varphi}_1 + M_{42}\ddot{\varphi}_2 + M'_{41}\ddot{\theta}_{m1} + M'_{42}\ddot{\theta}_{m2} + C_{4ff} + C_{4fr} + C_{4rr} + D_4 = \tau_2 \quad (16)$$

The parameters are derived in the Appendix.

In the above model, Eqs. (13) and (14) are the Lagrangian equations for the generalized elastic coordinates; Eqs. (15) and (16) are the equations for the generalized rigid coordinates. To obtain the elastic motion of the robotic finger mechanism, we need only to solve Eqs. (13) and (14). Because the elastic displacements are small compared with the rigid displacements in the robot's motion, they do not affect the rigid motion of the joints. Thus, for small φ, $\sin(\theta+\varphi) \approx \sin\theta$ and $\cos(\theta+\varphi) \approx \cos\theta$, so that coupling C_{iff} between the elastic motions can be omitted.

Therefore, the elastic vibration equations (13) and (14) become

$$m_{11}\ddot{\varphi}_1 + m_{12}\ddot{\varphi}_2 + c_{11}\dot{\varphi}_1 + c_{12}\dot{\varphi}_2 + k_{11}\varphi_1 = m'_{11}\ddot{\theta}_{m1} + m'_{12}\ddot{\theta}_{m2} + c_{11}' + d_1' \quad (17)$$

$$m_{21}\ddot{\varphi}_1 + m_{22}\ddot{\varphi}_2 + c_{21}\dot{\varphi}_1 + c_{22}\dot{\varphi}_2 + k_{22}\varphi_2 = m'_{21}\ddot{\theta}_{m1} + m'_{22}\ddot{\theta}_{m2} + c_{22}' + d_2' \quad (18)$$

where the parameters are shown in the appendix.

Eqs. (17) and (18) can be rewritten in matrix form as:

$$[M(\theta_m)]\{\ddot{\varphi}\} + [c(\theta_m,\dot{\theta}_m)]\{\dot{\varphi}\} + [K]\{\varphi\} = \{p\} \quad (19)$$

where $[M]$, $[c]$, and $\{p\}$ are automatically generated during the rigid motion analysis, and the stiffness matrix $[K]$ is constant.

For multifingered robot hands, the elastic displacements of their joints are generally small, so that they can be considered to be small vibrations based on rigid displacements. The elastic displacements $\{\varphi\}$ of Eq. (19) can be computed by numerical methods.

CONCLUSIONS AND COMMENTS

We established the elastic-joint model with one-rigid-DOF and three-elastic-DOF. Using the equivalent principle of kinetic and potential energy for the transformation of the mechanical system, we transformed a 4-DOF elastic-joint model into a 2-DOF model. Using this simplification and a Lagrangian formulation, we derived the dynamic model of a planar, two-rigid-DOF finger mechanism with two elastic joints. Similarly, the equations of motion and a method for control of multijointed, multifingered robot hands with elastic joints can be obtained. By considering both the elastic and rigid motion of the robotic joints, the dynamic characteristics and performances of multifingered robot hands can be greatly improved. Work is currently in progress to validate this approach through simulations and experiments.

REFERENCES

[1] Cutkosky, M. R. and Kao, I., "Computing and Controlling the Compliance of a Robotic Hand," *IEEE Trans. on Robotics and Automation*, Vol. 5, No. 2, pp151-165, 1989.

[2] Grupen, R. A. and Weiss, R. S., "Force Domain Models for Multifingered Grasp Control," *Proceedings of the 1991 Int. Conf. on Robotics and Automation*, pp418-423, 1991.

[3] Salisbury, J. K. and Craig, J. J., "Articulated Hands: Force Control and Kinematic Issues," *The International Journal of Robotics Research*, Vol. 1, No. 1, 1982, pp4-17.

[4] Jacobsen, S. C. et al.,"The Utah/MIT Dextrous Hand: Work in Progress," *The International Journal of Robotics Research*, Vol. 3, No. 4, 1984, pp21-49.

[5] Nakamura, Y. et al.,"Dynamics and Stability in Coordination of Multiple Robotic Mechanisms," *The International Journal of Robotics Research*, Vol. 8, No. 4, 1989, pp44-61.

[6] Guo, G., Gruver, W. A., and Jin, K., "Control of Dynamic Grasping Systems using Neural Network Approximation," *Proceeding of the IEEE International Symposium on Intelligent Control*, August 13-15, 1991, Arlington, VA, pp196-202.

[7] Guo, G., Gruver, W. A., and Qian, X., "A Robotic Hand Mechanism with Rotating Fingertips and Motor-Tendon Actuation," *Proceeding of the IEEE International Conference on Systems, Man, and Cybernetics*, Oct. 13-16, 1991, pp1023-1028.

[8] Rivin, E. I., et al. "Analysis of Structural Compliance for Robot Manipulators," *Proceeding of the Robotics and Factories of the Future*, Springe-Verlag, Berlin, 1986, pp262-269.

[9] Ying, W. and Huston, R. L., "Dynamic Analysis of Elastic Beam-like Mechanism Systems," *J. of Mechanical, Transmission, and Automation in Design*, 1989, Vol. 111, pp626-629.

[10] Nikolic, I. "Determination of Elastodynamic Errors in Joints of Industrial Robots," *Robotica*, 1988, Vol. 6, pp213-219.

[11] Mills, J. K., "Stability and Control of Elastic-Joint Robotic Manipulators During Constrained-Motion Tasks," *IEEE Transactions on Robotics and Automation*, Vol. 8, No. 1, 1992, pp119-126.

[12] Buffinton, K. W., "Dynamics of Elastic Manipulators with Prismatic Joints," *ASME Transactions on Dynamic Systems, Measurement, and Control*, Vol. 114, No. 1, 1992, pp41-49.

Appendixes:

A: Position Expressions of Mass Center m_1 and m_2 and Load M_1 and M_2.

B: Parameters for. Eqs. 13, 14, 15, and 16.

C: Parameters for Eqs. 18 and 19.

They are omitted in this paper.

Recursive Dynamics of Flexible-Link Manipulators with Geared and Flexible Joints

Lorne Churchill & **Inna Sharf**

Department of Mechanical Engineering
University of Victoria
Victoria, British Columbia, Canada

Introduction

Dynamics simulation of robotic manipulators has been studied by many researchers and various formulations and solution algorithms have been proposed for this problem. In a large number of these works, the manipulator is modelled as an open chain of interconnected rigid or flexible bodies which represent its *links*. These are driven by *ideal actuators* that are often assumed to be point torque sources and, therefore, lack any dynamical characteristics of their own. More recently, as it became recognized that the actuators driving the links can have a significant effect on the dynamical characteristics of the system, a number of formulations have been proposed to include their dynamics in the model [1-4]. In achieving this goal, many researchers have focused on one or several dynamical characteristics of the actuators [1, 5], often in the context of a particular type of actuator, such as hydraulic [6].

In this paper we propose a novel formulation for including the dynamics of actuators in the model of a robotic manipulator. The distinct feature of the procedure is that the actuators are treated as constituent bodies of the chain, analogously to the links. Thus, our methodology considers a general open multibody chain made up of "actuator" bodies and "link" bodies, both of which may have general inertial and stiffness properties. To allow for gearing in the drive system, they are interconnected with *amplifiers*. The resulting formulation and the recursive solution algorithm are demonstrated with a series of simulation results showing the effects of gearing and joint flexibility on the response of a manipulator modelled after the Space Shuttle Arm.

Mathematical Model of the Link/Actuator Chain

The addition of finite sized actuators with amplifiers entails the specification of certain details of the structure of the bodies which make up the chain. In our analysis, we consider a general body to be made up of two parts: a *housing* and an *arm*. For link bodies, the housing vanishes and the arm represents what is traditionally referred to as the link. For actuator bodies, the housing is assumed to be rigid. However the arm, which subsumes the rotor and/or the shaft of a real actuator, may be flexible. It possesses one or more degrees of freedom with respect to the housing and these are the joint degrees of freedom.

The amplifier is rigid and without mass or inertia. It is made up of four parts: the *casing*, the *input shaft*, the *amplifier-mechanism* and the *output shaft*, each element distinguished for the purpose of identifying the amplifier's force and velocity transmission characteristics. The amplifier's casing is mounted on the housing of the previous body and its input shaft is attached to that body's arm (see Fig. 1). The amplifier-mechanism is supported by the casing and connects to the input and output shafts. The latter also attaches to the housing of the succeeding body in the chain.

The amplifier works by transmitting to the output shaft the summed motion of, first, the casing and, second, the relative motion of the input shaft *with respect to the casing*, amplified by the amplifier-mechanism. Forces applied to the output shaft are transmitted through the amplifier mechanism to the input shaft, directly if they are reactions forces or amplified if they are "actuated" forces. The amplifier also transmits the difference between the forces on the input and output shafts, through the amplifier casing to the actuator housing. Note that we allow the amplifier to redirect transmitted forces and velocities.

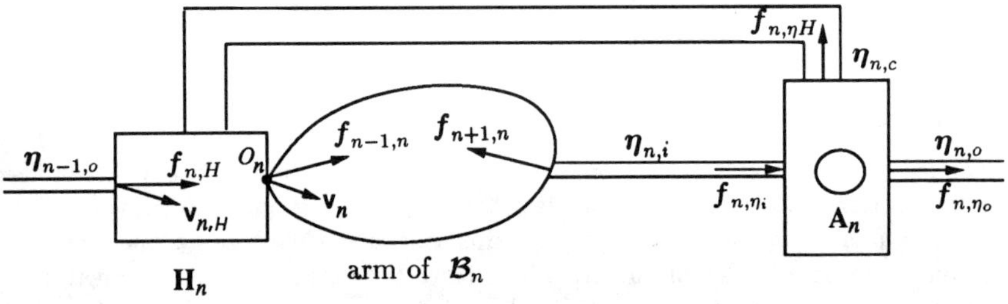

Figure 1: *Interbody Connections, Forces and Velocities*

All bodies in the chain are interconnected by amplifiers. This generality allows us to accomodate complex amplified joints by simply "stacking" actuators as successive bodies. When a link is connected to an actuator, velocities and forces are transmitted with unit amplification and the amplifier is said to be *identical.* As well, in view of physical limitations in the construction of amplifiers, we require that actuators with more than one rotational degree of freedom possess identical amplifiers.

Mathematically, we model the effects of an amplifier $\mathbf{A}_n$ via an *amplification matrix* of the form

$$\mathcal{A}_n \triangleq \Lambda_n(\mathcal{P}_n \boldsymbol{a}_n \mathcal{P}_n^T + \mathcal{Q}_n \mathcal{Q}_n^T) \tag{1}$$

where $\mathcal{P}_n$ is a projection matrix onto the actuated directions [7], $\mathcal{Q}_n$ is its orthogonal complement, $\boldsymbol{a}_n$ is a diagonal matrix of gearing ratios, and Λ_n is an (orthonormal) *redirection matrix* whereby we include the capability of amplifiers to change the direction of the motions they amplify. Accordingly, we define the redirected projection matrices:

$$\hat{\mathcal{P}}_n \triangleq \Lambda_n \mathcal{P}_n, \qquad \hat{\mathcal{Q}}_n \triangleq \Lambda_n \mathcal{Q}_n \tag{2}$$

We also note that in the absence of interdependent prismatic and rotational motion (screwlike joints,) $\mathcal{A}_n$ has a block diagonal form.

Equations of Motion

Our formulation is based on the Newton-Euler approach to the solution of the dynamics problem. The notation closely follows that of Sincarsin & Hughes [8] and hence all equations are written in terms of generalized (spatial) quantities. The presentation in this and the following two subsections is limited to rigid links and actuators.

The equations of motion of the housing and arm of body $\mathcal{B}_n$ follow from the well-established equations for a rigid body. They are:

$$\mathcal{M}_{n,H}\dot{\boldsymbol{v}}_{n,H} = \boldsymbol{f}_{nT,H} + \boldsymbol{f}_{nI,H} \tag{3}$$

$$\mathcal{M}_n \dot{\boldsymbol{v}}_n = \boldsymbol{f}_{nT} + \boldsymbol{f}_{nI} \tag{4}$$

where $\boldsymbol{v}_{n,H}$ and $\boldsymbol{v}_n$ are the generalized velocities of the housing and the arm, respectively, and $\boldsymbol{f}_{nT}$ and $\boldsymbol{f}_{nI}$ represent the generalized total and nonlinear forces. For most purposes it is sufficient to ignore the inertial effects of the motor housings, in which case we simply let $\mathcal{M}_{n,H} = \mathbf{O}$.

Interbody Forces

Figure 1 illustrates the forces transmitted through the various connections. The equations relating the forces acting on each part of this assembly are as follows:[1]

$$\begin{array}{rcll}
\text{housing } \mathbf{H}_n : & \boldsymbol{f}_{n,H} + \mathcal{R}_{n,n+1}^T \boldsymbol{f}_{n,\eta H} - \boldsymbol{f}_{n-1,n} + \boldsymbol{f}_{n,ext,H} & = \boldsymbol{f}_{nT,H} & (5) \\
\text{arm of } \mathcal{B}_n : & \boldsymbol{f}_{n-1,n} + \mathcal{R}_{n,n+1}^T \boldsymbol{f}_{n+1,n} + \boldsymbol{f}_{n,ext} & = \boldsymbol{f}_{nT} & (6) \\
\text{input shaft } \eta_{n,i} : & -\boldsymbol{f}_{n+1,n} - \boldsymbol{f}_{n,\eta_i} & = \mathbf{0} & (7) \\
\text{casing } \eta_{n,c} : & \boldsymbol{f}_{n,\eta_i} - \boldsymbol{f}_{n,\eta H} - \Lambda_n^T \boldsymbol{f}_{n,\eta_o} & = \mathbf{0} & (8) \\
\text{amplifier-mechanism } \mathbf{A}_n : & \mathcal{A}_n \boldsymbol{f}_{n,\eta_i} - \boldsymbol{f}_{n,\eta_o} & = \mathbf{0} & (9) \\
\text{output shaft } \eta_{n,o} : & \boldsymbol{f}_{n,\eta_o} - \mathcal{C}_{n+1,n}^T \boldsymbol{f}_{n+1,H} & = \mathbf{0} & (10)
\end{array}$$

where

$$\mathcal{C}_{n+1,n} = \begin{bmatrix} \mathbf{C}_{n+1,n} & \mathbf{O} \\ \mathbf{O} & \mathbf{C}_{n+1,n} \end{bmatrix} \qquad \mathcal{R}_{n,n+1} = \begin{bmatrix} \mathbf{1} & -\mathbf{r}_{n,n+1}^{\times} \\ \mathbf{O} & \mathbf{1} \end{bmatrix} \tag{11}$$

Here $\mathbf{C}_{n+1,n}$ is the rotation matrix from frame $\mathcal{F}_n$ to $\mathcal{F}_{n+1}$ and $\mathbf{r}_{n,n+1}$ is the vector connecting the origins of $\mathcal{F}_n$ and $\mathcal{F}_{n+1}$ (we assume only the arm has finite dimensions.) By including $\boldsymbol{f}_{n,ext}$ in equation (6) we have accounted for external forces that may act on $\mathcal{B}_n$. We also point out that, because of the presence of the amplifier, $\boldsymbol{f}_{n,n+1}$ does not represent the same force as $\boldsymbol{f}_{n+1,n}$.

Solving equations (7–10) for $\boldsymbol{f}_{n,\eta H}$ and $\boldsymbol{f}_{n+1,n}$ yields,

$$\boldsymbol{f}_{n,H} - \mathcal{T}_{n+1,n}^T \boldsymbol{f}_{n+1,H} + \boldsymbol{f}_{n,ext,H} + \boldsymbol{f}_{n,ext} \;=\; \boldsymbol{f}_{nT,H} + \boldsymbol{f}_{nT} \tag{12}$$

$$\boldsymbol{f}_{n-1,n} - \Gamma_{n+1,n}^T \boldsymbol{f}_{n+1,H} + \boldsymbol{f}_{n,ext} \;=\; \boldsymbol{f}_{nT} \tag{13}$$

where

$$\mathcal{T}_{n+1,n} \;\triangleq\; \mathcal{C}_{n+1,n} \Lambda_n \mathcal{R}_{n,n+1} \tag{14}$$

$$\Gamma_{n+1,n} \;\triangleq\; \mathcal{C}_{n+1,n} \mathcal{A}_n^{-T} \mathcal{R}_{n,n+1} \tag{15}$$

Equation (12) can also be viewed as an expansion of the total force acting on $\mathcal{B}_n$ in terms of the interbody forces.

Interbody Geometrical Constraints

In addition to interbody forces, we must also account for the geometric constraints implied by the interbody connections. The amplifier transmits velocities in accordance with the relation,

$$\boldsymbol{v}_{n,\eta_o} = \mathcal{A}_n^{-T}(\boldsymbol{v}_{n,\eta_i} - \boldsymbol{v}_{n,\eta_c}) + \Lambda_n \boldsymbol{v}_{n,\eta_c} \tag{16}$$

from which we deduce the velocity of a body's housing in terms of the velocities characterizing the previous body

$$\boldsymbol{v}_{n+1,H} = \mathcal{T}_{n+1,n} \boldsymbol{v}_{n,H} + \Gamma_{n+1,n} \boldsymbol{v}_{n,free} \tag{17}$$

[1] Quantities with a single numerical subscript are evaluated in the associated reference frame. The quantities $\boldsymbol{f}_{n-1,n}$ and $\boldsymbol{f}_{n+1,n}$ are evaluated in frame $\mathcal{F}_n$.

In the above $\boldsymbol{v}_{n,free}$, the velocity of the arm with respect to the housing may also be expressed in terms of the joint rates $\boldsymbol{v}_{n\gamma}$ and the projection matrix: $\boldsymbol{v}_{n,free} = \mathcal{P}_n \boldsymbol{v}_{n\gamma}$. The velocity of the arm at O_n is simply given by

$$\boldsymbol{v}_n = \boldsymbol{v}_{n,H} + \boldsymbol{v}_{n,free} \tag{18}$$

Note that only the actuator bodies possess $\boldsymbol{v}_{n,free}$.

Summary of Equations and Extension to Flexible Bodies

The governing equations outlined previously comprise all equations necessary to solve for the motion of the chain. Since the actuators and links are treated equally, the dynamics equations and interbody force relations have the same form for both types of bodies. Furthermore, the kinematics of the system is described by a recursive expression, also applicable to both links and actuators. The additional complexity due to gearing is accomodated by the amplification matrix in the corresponding expressions.

By way of summarizing we will extend the governing equations to the case of flexible links and actuators. We generalize the six dimensional state space for $\mathcal{B}_n$ to $(6 + s_n)$ dimensions, where s_n is the number of elastic degrees of freedom. The corresponding equations become:

- equation of motion:

$$\mathcal{M}_n \ddot{\boldsymbol{q}}_n + \mathcal{K}_n \boldsymbol{q}_n = \boldsymbol{f}_{nT} + \boldsymbol{f}_{nI} \tag{19}$$

- equations describing velocity kinematics:

$$\begin{aligned} \dot{\boldsymbol{q}}_{n+1,H} &= \mathcal{T}_{n+1,n} \dot{\boldsymbol{q}}_{n,H} + \mathcal{U}_{n+1,n} \dot{\boldsymbol{q}}_{n,free} \\ \dot{\boldsymbol{q}}_n &= \dot{\boldsymbol{q}}_{n,H} + \dot{\boldsymbol{q}}_{n,free} \end{aligned} \tag{20}$$

where

$$\begin{aligned} \mathcal{T}_{n+1,n} &\triangleq \begin{bmatrix} \mathcal{T}_{n+1,n,r} & \mathbf{O} \\ \mathbf{O} & \mathbf{O} \end{bmatrix} & \mathcal{U}_{n+1,n} &\triangleq \begin{bmatrix} \Gamma_{n+1,n} & \mathcal{S}_{n+1,n} \\ \mathbf{O} & \mathbf{O} \end{bmatrix} \\ \mathcal{S}_{n+1,n} &\triangleq \mathcal{C}_{n+1,n} \mathcal{A}_n^{-T} \Xi_{n+1,n} & \Xi_{n+1,n} &\triangleq \begin{bmatrix} \Psi_n(\mathbf{r}_{n,n+1}) \\ \frac{1}{2}\nabla^{\times} \Psi_n(\mathbf{r}_{n,n+1}) \end{bmatrix} \end{aligned} \tag{21}$$

and Ψ_n is the matrix of basis functions employed in discretizing the elastic deformation of the arm;

- interbody force relations:

$$\boldsymbol{f}_{n,H} - \mathcal{T}^T_{n+1,n,r} \boldsymbol{f}_{n+1,H} + \boldsymbol{f}_{n,ext,H} + \boldsymbol{f}_{n,ext,r} = \boldsymbol{f}_{nT,H} + \boldsymbol{f}_{nT,r} \tag{22}$$

$$\boldsymbol{f}_{n,int} - \mathcal{U}^T_{n+1,n} \boldsymbol{f}_{n+1,H} + \boldsymbol{f}_{n,ext} = \boldsymbol{f}_{nT} \tag{23}$$

where we note that, since the housing is assumed rigid, the first of these equations includes no elastic components.

We also partition the interaction between housing and the arm into actuated and constraint parts with:

$$\boldsymbol{f}_{n,int} = \mathcal{P}_n \boldsymbol{f}_{n,CF} + \mathcal{Q}_n \boldsymbol{f}_{n,\Box} \tag{24}$$

where $\boldsymbol{f}_{n,CF}$ represents the sum of control and frictional forces acting on the arm, and $\boldsymbol{f}_{n,\Box}$ is the joint constraint force. For a more detailed presentation of the governing equations for a chain of flexible links (without actuators) see reference [8].

Recursive Solution Algorithm

The Newton-Euler approach leads naturally to a recursive method of solution for the forward dynamics problem. The main results follow from the proof of a general theorem, that for a given body there exist $\boldsymbol{\Phi}_{n,H}$ and $\phi_{n,H}$ such that $\boldsymbol{f}_{n,H} = \boldsymbol{\Phi}_{n,H}\ddot{\boldsymbol{q}}_{n,H} + \phi_{n,H}$. With the following definitions

$$
\begin{aligned}
\widehat{\mathcal{M}}_n &\triangleq \mathcal{M}_n + \mathcal{U}^T_{n+1,n}\boldsymbol{\Phi}_{n+1,H}\mathcal{U}_{n+1,n} \\
\widetilde{\mathcal{M}}_n &\triangleq \mathcal{M}_n + \mathcal{U}^T_{n+1,n}\boldsymbol{\Phi}_{n+1,H}\mathcal{T}_{n+1,n} \\
\overline{\mathcal{M}}_n &\triangleq \mathcal{M}_n + \mathcal{T}^T_{n+1,n}\boldsymbol{\Phi}_{n+1,H}\mathcal{T}_{n+1,n} \\
\boldsymbol{f}_{nX} &\triangleq \boldsymbol{f}_{n,ext} + \boldsymbol{f}_{nI} - \mathcal{K}_n\boldsymbol{q}_n - \mathcal{U}^T_{n+1,n}\Big[\boldsymbol{\Phi}_{n+1,H}(\dot{\mathcal{T}}_{n+1,n}\dot{\boldsymbol{q}}_{n,H} \\
&\quad + \dot{\mathcal{U}}_{n+1,n}\dot{\boldsymbol{q}}_{n,free}) + \phi_{n+1,H}\Big] - \widehat{\mathcal{M}}_n\dot{\mathcal{P}}_n\dot{\boldsymbol{q}}_{n,\gamma} \\
\boldsymbol{f}_{nX,H} &\triangleq \left(\boldsymbol{f}_{n,ext} + \boldsymbol{f}_{nI} - \widehat{\mathcal{M}}_n\dot{\mathcal{P}}_n\dot{\boldsymbol{q}}_{n,\gamma}\right)_r - \mathcal{T}^T_{n+1,n,r}\Big[\boldsymbol{\Phi}_{n+1,H}(\dot{\mathcal{T}}_{n+1,n}\dot{\boldsymbol{q}}_{n,H} \\
&\quad + \dot{\mathcal{U}}_{n+1,n}\dot{\boldsymbol{q}}_{n,free}) + \phi_{n+1,H}\Big] + \boldsymbol{f}_{n,ext,H} + \boldsymbol{f}_{nI,H}
\end{aligned} \tag{25}
$$

we may derive an expression for the accelerations,

$$
\ddot{\boldsymbol{q}}_{n\gamma} = \widehat{\mathcal{M}}^{-1}_{n,PP}\mathcal{P}^T_n(\mathcal{P}_n\boldsymbol{f}_{nCF} + \boldsymbol{f}_{nX} - \widetilde{\mathcal{M}}_n\ddot{\boldsymbol{q}}_{n,H}) \tag{26}
$$

where the shorthand notation $\widehat{\mathcal{M}}_{n,PP}$ stands for $\mathcal{P}^T_n\widehat{\mathcal{M}}_n\mathcal{P}_n$. The proof is constructive, resulting in the following definitions for the unknown quantities:

$$
\begin{aligned}
\boldsymbol{\Phi}_{n,H} &\triangleq \mathcal{M}_{n,H} + \overline{\mathcal{M}}_n - \widetilde{\mathcal{M}}^T_n\mathcal{P}_n\widehat{\mathcal{M}}^{-1}_{n,PP}\mathcal{P}^T_n\widetilde{\mathcal{M}}_n \\
\phi_{n,H} &\triangleq \widetilde{\mathcal{M}}^T_n\mathcal{P}_n\widehat{\mathcal{M}}^{-1}_{n,PP}\mathcal{P}^T_n(\mathcal{P}_n\boldsymbol{f}_{n,CF} + \boldsymbol{f}_{nX}) - \boldsymbol{f}_{nX,H}
\end{aligned} \tag{27}
$$

The basic structure of the method presented here is the same as that of existing recursive formulations for manipulator dynamics [9]. In particular, the algorithm involves three main steps: an outward recursion for $\dot{\boldsymbol{q}}_{n,H}$ and $\dot{\boldsymbol{q}}_n$ with equations (20), a backward recursion for $\widehat{\mathcal{M}}_n$, $\widetilde{\mathcal{M}}_n$, $\overline{\mathcal{M}}_n$, $\boldsymbol{\Phi}_{n,H}$, $\boldsymbol{f}_{nX}$, $\boldsymbol{f}_{nX,H}$ and $\phi_{n,H}$ using equations (25) and (27), and a final outward recursion for the joint and elastic accelerations $\ddot{\boldsymbol{q}}_{n\gamma}$ with equation (26). The procedure retains an $O(N)$ computational complexity inherent to recursive methods, where N is the total number of bodies—links and actuators.

Results and Conclusions

The recursive solution algorithm presented here has been implemented in a computer simulation based on the simulation developed previously for the analysis of flexible-link manipulators [11]. Thus, to complete the paper we include simulation results demonstrating the effects of gearing and joint flexibility on the response of a manipulator modelled after the Space Shuttle Arm. The model consists of fourteen bodies—a base (the shuttle), six links, six joints and a payload—where two of the links, the booms of the Arm, are flexible. The properties of the links may be found in [10]. For the particular example presented here, we restrict ourselves to modelling the twist deformation in the rotors of the motors, while the two flexible links are approximated by Euler-Bernoulli beams.

Figure 2 contains the joint angle profiles for joints 4 and 5. These results were obtained by simulating a maneuver with prescribed, geared-down actuator angles given by $\theta_n(t) = 0.02(t - \frac{10}{\pi}\sin\frac{\pi}{10}t)$ rads for three models of the arm. Model NA (no actuators) includes links only; model RA uses rigid actuators with masses of 6.6 kg and gear ratios of 1842, 1260 and 740 for the actuators describing the shoulder (first two), elbow (third) and wrist (last three) joints. Finally, model FA contains flexible actuators with torsional stiffnesses of 5×10^6, 3×10^6 and 2×10^6 N-m^2 respectively. The results demonstrate that actuator dynamics has a significant effect on system response.

References

[1] T.J. Tarn, A.K. Bejczy, X. Yun & Z. Li, "Effect of motor dynamics on nonlinear feedback robot arm control," *IEEE Transactions on Robotics and Automation*, **7**(1), pp. 114-121, 1991.

[2] S.H. Murphy & J.T. Wen, "Dynamic modeling of geared and flexibly jointed manipulators," *Proc. IEEE Int. Conf. on Robotics and Automation*, pp. 1765-1770, 1991.

[3] E. Wu, M. Diftler, J. Hwang & J. Chladek, "A fault tolerant joint drive system for the space shuttle remote manipulator system," *Proc. IEEE Int. Conf. on Robotics and Automation*, pp. 2504-2509, 1991.

[4] T.-C. Lin & K.H. Yae, "The effects of harmonic drive gears on robot dynamics," *Advances in Design Automation*, DE-Vol. 32-2, pp. 515-522, 1991.

[5] M.C. Good, L.M. Sweet & K.L. Strobel, "Dynamics models for control system design of integrated robot and drive systems," *J. Dynamic Systems, Measurement, and Control*, **107**, pp. 53-59, 1985.

[6] N. Sepehri, F. Sassani & P.D. Lawrence, "An alternative approach for modelling a class of hydraulically-powered manipulators," *Robotics and Manufacturing, ISRAM'90*, **3**, pp. 57-64, 1990.

[7] P.C. Hughes, "Multibody dynamics for space station manipulators: dynamics of a chain of elastic bodies," Dynacon Report SS-2, 1985.

[8] G.B. Sincarsin & P.C. Hughes, "Dynamics of elastic multibody chains: part A—body motion equations," *Dynamics and Stability of Systems*, **4**(3,4), pp. 209-226, 1989.

[9] G.M.T. D'Eleuterio, "Dynamics of elastic multibody chains: part C—recursive dynamics," *Dynamics and Stability of Systems*, 1992.

[10] I. Sharf & C.J. Damaren, "Simulation of flexible-link manipulators: basis functionsl and nonlinear terms in the motion equations," *Proc. IEEE Int. Conf. on Robotics and Automation*, Nice, France, pp. 1956-1963, May 1992.

[11] I. Sharf & G.M.T. D'Eleuterio, "Computer simulation of elastic chains using a recursive formulation," *Proc. IEEE Int. Conf. on Robotics and Automation*, Philadelphia, Pennsylvania, pp. 1539–1547, 1988.

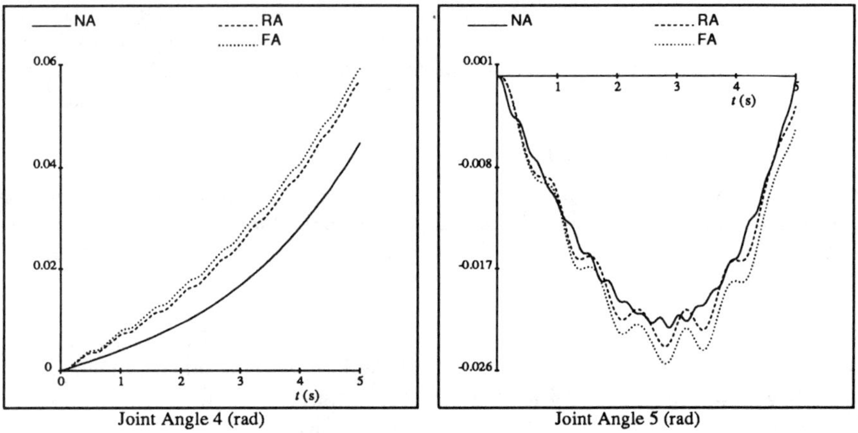

Figure 2: *Angle Profiles for Different Actuator Models*

A ROBOT CELL SIMULATOR FOR FMS DESIGN

ADEL S. ELMAGHRABY, MARCEL TAYSSOUN and WILLIAM SWAIN OTTMAN

Engineering Mathematics And Computer Science
Speed Scientific School
University of Louisville
Louisville, KY 40292

ABSTRACT

The paper presents a framework for simulation of alternative robot cell designs in the context of designing a Flexible Manufacturing System (FMS). A flexible manufacturing system simulator with full animation has been developed. Also, a 3-dimensional graphic robot cell simulator has also been developed allowing multiple robots within one cell. The suggested hierarchical framework allows for embedding a robot cell simulation as a lower level of detail in the overall flexible manufacturing floor simulator.

INTRODUCTION

Advances in personal workstation increased the demand for simulation packages for various fields of applications. These systems use simulation languages such as SIMSCRIPT and GPSS or general purpose languages such as C and FORTRAN [1]. The use of a robot system simulator embedded in a general simulation of a flexible manufacturing systems can be an important tool in making the simulation model more representative at various levels of detail and is the main purpose of this paper.

The abundance of these packages does not truly simplify simulation of flexible manufacturing systems. It is more important for the engineers to be able to specify the systems that are to be designed [2]. The job becomes more difficult with the fact that it takes a person an average of two years to become a good simulationist. This time period is important for the person to acquire both a knowledge in programming and statistics. Thus with the help of packages that do most of the work for a person, this task becomes available to more people from different fields.

Graphical User Interfaces (GUI) exemplify the ease of use provided by modern simulation packages. The idea behind the GUI is to minimize the effort and complexity required to describe or create a model. The intent therefore in the package presented in this paper was to eliminate code writing, thus allowing the user to create or edit a model by simply moving a mouse cursor and by interactions with menu options.

Discrete model simulation requires a user to define a working model that encompasses all aspects and properties of the system being modeled. This important point in simulation gives rise to the problem of detail. How detailed should the description of the model be? Should all components be defined and modeled precisely, or could one get away with simply approximating the behavior of some of the functions of the original system. This leads us to the topic of "Close-Enough" models [3]. This problem usually arises when the modeler has limited knowledge about the system and the way it functions. Although this method is not perfectly sound, it offers an alternative to having no model at all.

Integrating simulation models using a hierarchical approach allows us to overcome the shortcomings of the "Close-Enough" model by having different levels of detail. The main idea here is to break the simulation process into a series of smaller simulations or submodels. The example presented in this paper deals with the idea of Flexible Manufacturing Systems. Since most FMS systems deal with modeling factory floors it becomes inevitable to omit some of the details of individual machines in the factory model. A flexible manufacturing system simulator was developed with a lot of emphasis on the flow of control on a factory floor, but with less emphasis on the actual workings of the machines on the floor [4]. To counter balance this deficiency another simulator was designed to model in detail the workings of a certain type of machine namely a robot [5]. The aim here is to be able to go down in the hierarchy of the system model and emphasize the workings of a certain machine such as a robot. Since many FMS do not deal exclusively with single machines, the system allows grouping several robots in one cell. The following sections will give an overview of the two main components of the system.

THE FACTORY SIMULATOR

The creation of a flexible Manufacturing system is always preceded by a definition of the logical system. This system is represented in the software by a graph-like tree made up of nodes and lines [6]. The nodes represent the logical tasks required to produce or run a certain process or job. The lines represent the direction flow of the logical system. In order to create a logical diagram a user can simply use the friendly user interface to position the mouse at the desired location and proceed to click the left button upon which a node will appear on the screen denoting a certain task. After placing the desired number of nodes on the screen, the user can then proceed to connect these nodes using the connect option available in the software. The creation of the logical diagrams comes with full graph editing capabilities, such as deleting certain nodes or lines. An important characteristic that was defined in this project was the use of macros. A Macro is defined here to be a task that actually represents a number of tasks connected together. The reason behind this option is that in Flexible

Manufacturing System there are some operations which might appear more than once or might actually be made up of more than one task. In providing the macro option a user can create a certain logical diagram made up of nodes and lines with their corresponding attributes and save the whole thing under a file having a certain name. This saved file might represent the task diagram of a frequently used operation. Since the diagram has been defined and saved as a file, the user can now access it at any time in the future by simply referring to the name of the filename it was saved under preceded by the "@" sign. The significance of this is that the user can now insert this diagram into future drawings without having to redraw anything. The use of the "@" helps differentiate between nodes that are units and other nodes that actually represent a whole graph. The program then evaluates the operation name and upon sensing that a certain task has a name preceded by the letter "@", the system automatically displays a miniature logical diagram representing the Macro which is a lower level in the model hierarchy.

In the course of defining a Flexible Manufacturing System, the user must always be ready to supply the software with a resource file. Upon evaluating a logical diagram it is necessary for the software to proceed and try to assign each task to an actual physical machine. The process of doing this is referred to as Resource to Task Allocation. In order to do this it is therefore necessary to either create a new database or load an existing one. The attributes of the database can all be individually altered and modified by simply pointing the mouse to the desired field and changing it. The advantage of having the database in the design environment is that any changes or modifications that are applied to the database can immediately be used by the whole design. The attributes in the database are those of each individual machine. A machine is a physical device capable of carrying a logical operation. This machine can be defined vaguely by this simulator, or it can be defined precisely by the robot simulator that makes up the second simulation software.

After designing a logical diagram and loading or creating a resource file, the next most important step is to go through the resource to task allocation step. The object of this step is to assign a physical machine to each and every logical task. The assignment of a machine to each logical task is carried out according to an assigning algorithm. First a scan is made of all dedicated machines, these are machines that are capable of doing only one job. It is always necessary to use these machines and leave the more flexible machines that are capable of performing more than one job to be assigned later. The above process is accomplished by pointing to the Build function. If all resources are available, the system simply exits the current menu; otherwise an error indicating a shortage of resources is displayed. It is important to always check that the spelling of tasks conforms between the logical diagram and the database. The reason being that the algorithm uses a 'string compare' to search for available machines. Another cause of a shortage error could be that the number of machines needed to perform a process might be larger than the number of machines available.

After invoking the allocation algorithm the program proceeds to place the machines on the physical floor. The algorithm used in this part of the program always ensures that no machine blocks or hinders the path of another machine. The algorithm places the machines on the floor equidistantly regardless of the shape of the logical graph. It should be noted that the logical diagram does not need to conform with the two dimensional physical layout. The reason being that the physical layout has a direction of motion which goes from left to right and the algorithm always tries to find the best linear representation for a logical diagram, by trying as much as possible to place the machines in a straight line. In going through the resource to task allocation the algorithm searches and detects tasks which are macros. This is important because a logical diagram with only one node might actually translate to a physical layout with six different machines. These tasks which are macros are then loaded and inserted in the physical layout in their respected positions.

The final stage in the design of a Flexible manufacturing System, is the simulation of the system. This is achieved by getting the Simulate option. The main algorithm in the simulation design was the determination of the paths. It was necessary to assign the paths between the machines as they would actually appear on the factory floor. The main objective in the algorithm is to produce square paths. These are paths that change directions at right angles and never in a diagonal manner. The main idea behind this strategy is to avoid situations where paths might actually overrun the physical position of a machine on the floor. The other important aspect of the path algorithm is its ability to assign different paths to the outputs of machines, which is the case where some machines may send parts to more than one machine.

The simulation is executed based on the generated paths using a sharing technique. That is, every time a part is produced it is sent on an alternating path guaranteeing that no receiving machine is rendered idle more than the other. With all these things on hand, the simulation algorithm is now ready to simulate the factory floor shown in Figure 1.

During the simulation with animation of the factory floor it is possible for the user to point to any machine with the mouse and click the left button to view the most current statistical report on that machine. The report includes a three dimensional bar graph indicating the percentage of the time that the machine was idle and busy, plus the present status of the machine. Also available are statistic about the number of parts produced by that machine and the number of parts waiting on queue.

With all this comes the feature of interactively disabling a machine and continuing the simulation. This can be done by simply pointing to the Down option and clicking the left mouse button upon which the choices Yes and No are chosen to indicate whether a machine is down or not. Animation and three dimensional graphics are used to help the user monitor the flow of the simulation. This way the user can actually see the parts moving around the factory floor, while at the same time gather important statistical information about every machine on the floor.

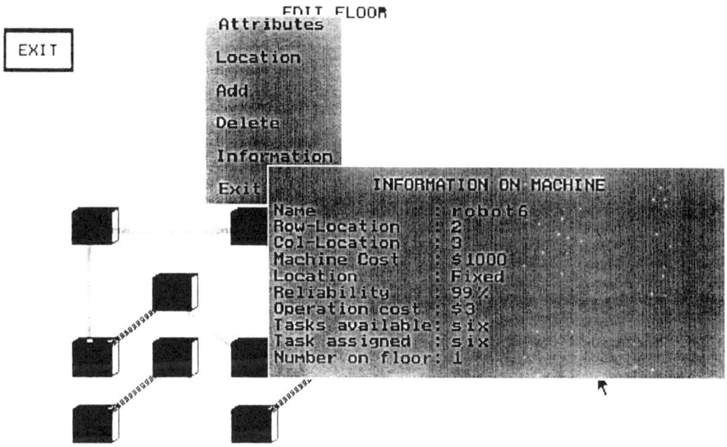

FIG. 1. Two Dimensional Representation of Floor Plan

THE ROBOT SIMULATOR

The second simulator is called ROBOMOD and provides the facilities for the modeling and simulation of robotic workcells. These defined models can then be integrated into the FMS simulator to create a more detailed model. The geometries of the robots and workcells are represented by wire-frame drawings. Three dimensional perspective computer graphics are employed to display these graphical models.

The only robot configuration capable of being modeled with the current implementation is one with a twisting base and rotational links. The base and all links must lie in the same vertical plane (i.e., no offset links are allowed). A maximum of nine links and a base are allowed. The shape of the links is box-like with varying dimensions, but the base must be square in the horizontal plane. The user defines the home position and rotation limits of each link.

The robot models are used to build a multiple robot workcell model. After defining the dimensions of the cell, up to five robot models may be inserted into it. The user specifies the position and orientation of each robot within the workcell.

The workcell simulator permits the user to interactively control the movements of any robot within the cell. For a better view of the robots, the user can change his/her perspective of the displayed workcell. This movement about the workcell allows the user to visually check for any collisions between robots. The bumping of the workcell walls by the robot can also be detected using this method.

The Robot Editor provides the facilities for the creation and editing of individual robot models. A model is built interactively by defining the physical dimensions and characteristics of a robot. Model building begins by first defining the dimensions of the base. Links are then stacked one atop the other as they are defined. The geometry of the robot is displayed and constantly updated during model construction. The editor is able to modify and delete the base or links of the robot. The size of any attached link can be changed by identifying the link and defining its new dimensions.

The Workcell Editor provides the facilities for the creation and editing of workcell models. A model is built interactively by defining the physical size and contents of a workcell. Model building begins by first defining the dimensions of the workcell. The geometry of the workcell is displayed and constantly updated during model construction and editing. This part of the software is very similar to the preceding part except that the user is allowed to view and edit a whole group of machines which might be used in conjunction with the FMS simulator.

The Workcell Simulator option allows the user to simulate the created workcells in three dimensions. Among the things displayed are the desired workcell, the current settings, and a list of the simulator commands. The current version of the simulator can only operate on one robot at a time in workcells containing multiple robots. The activation of a robot is toggled from one robot to the next. The links of the active robot are interactively controlled by the user. Viewpoint controls of the workcell are available to help the user zoom in and out of a workcell.

CONCLUSIONS

The paper explains the use of a two level hierarchical simulation for FMS. A robot cell simulator is used in conjunction with a manufacturing simulator. With both of these simulators currently working independently, it becomes important to have the robot cell simulator embedded into the FMS simulator thus giving the first one a more detailed description of the model. Future enhancements include porting the system to a UNIX system.

REFERENCES

1. N. A. Kheir, *Systems Modeling and Computer Simulation*, (Marcel Dekker, INC New York, 1988).
2. J. B. Schroer, "A Simulation Assistant for Modeling Manufacturing Systems," *Simulation*, (November 1989).
3. Lawrence Widman, "Semi-Quantitative “Close-Enough” Systems Dynamics Models: An Alternative to Qualitative Simulation," in *Artificial Intelligence, Simulation & Modeling* (John Wiley & Sons, New York, 1989).
4. Marcel Tayssoun, "Software System for Design and Simulation of Flexible Manufacturing Systems," (University of Louisville, M.S. Thesis, August 1990)
5. William Swain Ottman, "ROBOMOD An Off-Line 3D Graphic Robot Simulation System," (University of Louisville, M.S. Thesis, May 1991)
6. Adel S. Elmaghraby, R. S. Demeo, and J. Berry, "Testing an Expert System for Manufacturing", in *Artificial Intelligence and Simulation*, edited by W. Holmes, 62-64 (Society for Computer Simulation 1985).

Cooperation of Mobile Robots for Accident Scene Inspection

Raymond H. Byrne and John Harrington
Sandia National Labs
Albuquerque, NM 87185

ABSTRACT

A telerobotic system demonstration was developed for the Department of Energy's Accident Response group to highlight the applications of telerobotic vehicles to accident site inspection. The proof-of-principle system employs two mobile robots, Dixie and RAYBOT, to inspect a simulated accident site. Both robots are controlled serially from a single driving station, allowing an operator to take advantage of having multiple robots at the scene. The telerobotic system is described and some of the advantages of having more than one robot present are discussed. Future plans for the system are also presented.

INTRODUCTION

Mobile robots, ranging from telerobotic to fully autonomous, have been used to replace or assist humans in hazardous environments. Applications have included underwater exploration and construction, nuclear power plants, weapons research, explosive ordinance disposal, and battlefield weapons [1-3]. A significant amount of research has been conducted on robotics for nuclear power plants [3-9]. Robots have been developed to inspect pipes, perform routine maintenance, and help with decontamination and decommissioning. Robots were also used in the Three Mile Island and Chernobyl accidents [8,9].

This paper describes an accident response capability demonstrated to the Department of Energy's (DOE's) Accident Response Group (ARG). The mission of the DOE Accident Response Group is to efficiently manage and resolve accidents and significant incidents involving nuclear explosives that are in DOE custody at the time of the accident. ARG also provides support to the Department of Defense (DoD) in managing and resolving accidents and significant incidents involving nuclear materials in DoD custody at the time of the accident. The capability described in this paper involves the application of two telerobotic vehicles, Dixie and RAYBOT, to inspect a simulated accident site. Although the demonstration was developed for ARG, the principles involved are applicable in any unstructured environment where a preliminary assessment can be conducted by mobile robots.

Telerobotic systems can be divided into three main sub-systems: the operator control station, the communications link, and the remote robotic system. In the following discussion, the first section discusses the robotic vehicles, Dixie and RAYBOT. The operator console and communications link for the Dixie-RAYBOT system are described in the second section, leading into a brief presentation on the accident response demonstration. A summary section outlines plans for future research.

Dixie - RAYBOT

Both robots were designed around a Honda TRX-125 four-wheel all-terrain vehicle. The vehicles are fully actuated so steering, brake, throttle, and gear selection can be remotely controlled by the operator. Sensors on board the robots measure heading, roll angle, pitch angle, engine temperature, vehicle speed, battery voltage, and engine RPM. This information, as well as audio and video, is presented to the operator at the remote driving station.

Functionally, Dixie and RAYBOT are very similar. The main difference is that Dixie is equipped with a telescoping mast for the driving camera while RAYBOT carries a telescoping boom with an additional camera. Dixie is shown in Figure 1 with the mast extended and RAYBOT is shown in Figure 2. The control systems on the two vehicles are also quite different. The Dixie control system is built around a Motorola 6805 micro controller programmed in assembly language. Dixie's electronics were also designed for low power consumption so that the robot can stay in the field for extended periods of time.

The RAYBOT control system was designed to demonstrate a low cost, economical, and modular control system for telerobotic land vehicles. This was achieved by using a multiple processor architecture where dedicated processors handle low level functions such as servo control and data acquisition while a high level processor handles vehicle "intelligence". The high level processor communicates with the low level processors through dual-ported RAM. The RAYBOT control system is further described in [10].

RAYBOT is equipped with a camera mounted on a 5 degree-of-freedom boom. This gives the operator some flexibility in positioning the boom camera to observe an object. RAYBOT also carries a remotely controlled video multiplexer, which can select one of 4 video inputs. Two are used for on board cameras - the driving camera and the boom camera, leaving two open for other applications. These extra video inputs can be used by personnel at the scene to send back video from hand-held or tripod-mounted cameras.

OPERATOR CONSOLE and COMMUNICATIONS LINK

The telerobotic system used in the ARG demonstration consists of two vehicles, Dixie and RAYBOT, that can be controlled sequentially from a single driving station. The driving station is shown in Figure 3. Foot pedals (not shown) are used to control the brake and throttle while a steering yoke is used to control the vehicle steering. Switches select the following functions: parking brake; engine run/stop; gear shift (up/down); mast up/down (Dixie only); camera zoom and focus; camera on/off; and camera mode (rate control or position control). A momentary push-button switch is

Figure 1, Dixie with Mast Fully Extended

Figure 2, RAYBOT

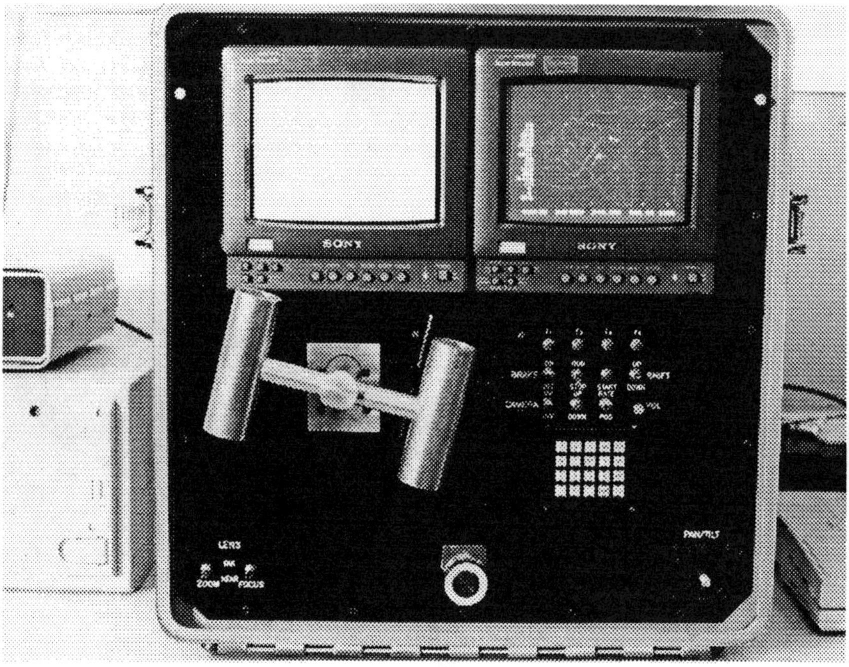

Figure 3, Remote Driving Station

used to control the starter. Four "soft-keys" and an alpha-numeric keypad with 20 keys allow the operator to switch menus and enter data. A joystick is used to control the driving camera pan/tilt unit on both vehicles and also the boom on RAYBOT. A volume control is provided for the audio feedback.

The driving station employs two monitors. One displays the video from the robot while the other can show maps, vehicle data, or other menus that help select different functions. Map displays aid in navigation, and are generated, stored, and retrieved as the robots are driven. Dead reckoning errors are reduced by triangulation off known landmarks.

Only one robot can be controlled at a time. To switch control from one vehicle to the other, the engine must be stopped and the parking brake applied. This ensures that the vehicle is in a safe state before control is relinquished. Switching control is then accomplished by pushing one key on the keypad. This allows the operator to switch control rapidly from one vehicle to the other.

Two separate RF links are used for communications between the robots and the driving station. The first is a 1200 baud full-duplex modem (410 MHz) for driving station - vehicle communications. Packets containing vehicle commands are sent from the driving station to the robot at 15 Hz. Sensor information is sent from the robot back to the driving station at approximately 7.5 Hz. A second RF transmitter

(900 MHz) on the robot sends video and audio to the operator console. The video transmitter is limited to line-of-sight operation which is usually the limiting factor in the range of the overall system.

SYSTEM DEMONSTRATION

A proof-of-principle demonstration was conducted for an ARG training class in August of 1990. Dixie and RAYBOT were used to explore a simulated plane crash, shown in Figure 4. The robots were controlled from an auditorium several miles from the simulated accident site. A microwave repeater on a tower was used to overcome line-of-sight problems with the video signal. The range of the RF modem was extended by having a transmitter/receiver close to the robots (~ 1 mile) and then using phone lines to complete the modem link to the driving station.

Figure 4, Dixie and RAYBOT Inspecting a Simulated Accident Site

Tasks that were demonstrated include a preliminary assessment of the accident site. Both robots were used to examine debris near the wreckage. Closer inspections were performed with the telescoping boom on RAYBOT using both robots to cooperatively position the boom camera. Approximate coordinates of different parts of debris could be determined using the map display at the driving station.

Once a field team arrived on the scene, the auxiliary video inputs on RAYBOT allowed the field team to send back real-time video from hand-held and tripod-mounted cameras. While the field team was performing their duties, the observers back at the auditorium used the driving camera pan/tilt units on Dixie and RAYBOT to watch the field team's activities closely.

SUMMARY & FUTURE RESEARCH

A proof-of-principle telerobotic accident scene inspection system was developed for DOE's Accident Response Group. Since the August, 1990 demonstration, several upgrades have been made to the system. The telescoping boom on RAYBOT has been replaced with a 5 degree of freedom manipulator which allows even greater flexibility in positioning the arm camera. Dixie has also been modified. The single driving camera has been replaced with a pair of cameras, allowing either stereo or normal video to be viewed at the remote driving station. Even with the improvements, a fieldable system would be very different from the proof-of-principle system currently being researched. One difference would be in the communications link between the driving station and the robots. The demo system uses an RF link which is impractical for this application. A fiber-optic communication link would most likely be used on a real system.

A second generation proof-of-principle system is currently under development that will use a fiber optic link. The new system will employ two robotic vehicles, a Surrogate Teleoperated Vehicle (STV) developed for the Army and a remotely operated High Mobility Multipurpose Wheeled Vehicle (HMMWV). Two Schilling Titan 7F hydraulic manipulators have been purchased and are being integrated onto the HMMWV. The Titan 7F is capable of lifting a 1200 lb. payload in the optimum configuration, which will allow an operator to remotely move large pieces of wreckage.

Overall, the ability to remove an operator from a potentially hazardous environment is one of the advantages of telerobotic systems. Although there are many disadvantages, such as cost and complexity, the combination of multiple cooperating robots and high strength dexterous manipulation should yield an effective accident response capability.

REFERENCES

1. James M. McNair, "Designing Robots for Demilitarization of Obsolete Chemical Munitions", American Nuclear Society Robotics & Remote Handling in Hostile Environments Conference, April 1984, pp.335-337.

2. Raymond H. Byrne and Paul Klarer," Sandia National Laboratories Military Robotic Technologies", Unmanned Systems, vol. 10, no. 2, Spring 1992.

3. "Undersea Vehicles Directory 1990-91", 4th Edition, Busby Associates, Inc., Arlington, Va., February 1990.

4. B. G. Kniazewycz and Ali Darvish, "Experience With the Surveyor^{TM} Mobile Robot in Radioactive Work Environments", Proceedings of the 34th Conference on Remote Systems Technology, November 1986, pp. 109-114, Washington, D.C.

5. K. Takehara, "Nuclear Power Plant Facility Inspection Robot", Advanced Robotics, vol. 3, no. 4, 1989, 321-331.

6. K. J. McDermott et. al., "Tomorrow's Mobile Robots in Today's Nuclear Power Plants", Proceedings of the 34th Conference on Remote Systems Technology, November 1986, pp.148-157, Washington, D.C.

7. Michael D. Pavelek et. al., "Operations and Achievements of Remote Equipment at Three Mile Island Unit 2", Nuclear Technology, vol. 87, no. 4, Dec. 1989, pp.1122-1133.

8. H. B. Meieran, "Robotic and Teleoperation Activities at the Chernobyl Atomic Power Station; Performance Successes and Limitations", IFS Teleoperation and Control International Symposium, July 12-15, 1988, pp.25-32, Bristol, UK.

9. Paul R. Bengel et. al., "Robotics Development at TMI-2", Proceedings of the 34th Conference on Remote Systems Technology, November 1986, pp. 139-143, Washington, D.C.

10. Raymond H. Byrne, "A Practical Implementation of a Hierarchical Control System for Telerobotic Land Vehicles", Proceedings of the National Telesystems Conference, May 19-20, 1992, Washington, D.C.

CONTROL OF MOBILE ROBOT WITH ONBOARD MANIPULATOR

KAI LIU and FRANK L. LEWIS
Automation & Robotics Research Institute
The University of Texas at Arlington
7300 Jack Newell Blvd., S.
Fort Worth, Texas 76118

ABSTRACT

This paper deals with the modeling and control of a wheeled mobile robot with an onboard manipulator. The overall system is thought of as a composite of two subsystems, the vehicle and the arm. A decentralized robust controller is designed for trajectory following by the arm end-effector. Bounded tracking errors are guaranteed even when the interactive forces between the arm and the vehicle are not measurable.

INTRODUCTION

With the extensive research into control theory for robotic systems and the widespread applications of industrial robots, more and more researchers have been recognizing the importance of locomotion in robotic systems [1, 3, 4]. The mobile robot system, which is a manipulator mounted on wheels, is becoming a main research topic. However, so far, most of the research interest is concentrated on path-planning problems such as collision avoidance, minimum-time navigation, and so on. Not much attention has been paid to control problems.

In this paper, the manipulator will be referred to as the *robot arm* and the wheeled frame will be referred to as the *vehicle*. In general, the motion, as well as the control, of the robot arm is referred to a *fixed* reference frame, for example to a base frame which is fixed on a floor. However, for the mobile robot systems, the base of the robot arm is never at a standstill; it should move with the vehicle. In this case, the motion, as well as the control, of the robot arm should be referred to a *moving* reference frame.

According to Newton's theory of motion, if the vehicle, as well as the base frame, keeps moving along a straight line with a uniform velocity, then, from the control point of view, there is no difference between the fixed base and the moving base. But if the vehicle moves along a curved path, or if the vehicle moves with a variable velocity, then the effect of the moving base cannot be ignored. Thus, we can see that the appearance of the movable base not only increases the dimension of the overall system but also increases the complexity of the control scheme.

This paper is concerned with the modeling and control of the wheeled mobile robot (WMR). The objective is to make the arm end-effector follow a specified trajectory. Intrinsically, WMR is a complex, multivariable dynamic system. On the other hand, WMR can be thought of as a composite of two subsystems; one is the vehicle with m degrees of freedom, and the other is the robot arm with n degrees of freedom. They form a composite system, in which the interconnections are interactive forces.

With this decomposition, the trajectory stability analysis of the overall system can then be accomplished in terms of the two simpler subsystems and in terms of the properties of the interactive forces. Based on the decomposition and stability analysis, a local controller can be designed for each subsystem that yields suitable performance of the overall system.

LOCATION AND VELOCITY MEASUREMENT FOR THE VEHICLE

In navigating the mobile robot on a horizontal x-y plane, an important problem is how to determine the location and the velocity of the vehicle. Almost all control strategies are based on the assumption that the correct information about the location and velocity is available. If the controller does not know the vehicle location and speed, talking about "trajectory following" is nonsensical.

In two dimensional space, the location of a mobile robot can be represented by a triple (x, y, θ) where x, y, and θ are, respectively, the 2-D position and orientation of the vehicle. Thus to locate the vehicle is the same thing as to obtain the state of $x(t)$, $y(t)$, and $\theta(t)$ at all times.

The location and velocity measurements of the vehicle fall into two categories -- relative measurements and absolute measurements.

Consider a differential-drive vehicle as shown in Figure 1. There are two driving wheels which are driven separately by two independent motors. The other two passive wheels which can rotate freely in any direction are used to keep the balance of the vehicle. Two sets of encoders and tachometers are attached to each motor such that the information about the rotation angle and rotation speed of each wheel is available. The rotation angle and rotation speed can be transferred to linear speed elements of each wheel in x- and y-direction ($v_x(t)$ and $v_y(t)$). Suppose that the mass-center of the vehicle is at point O, which is the middle point of the interconnecting line between the two wheels, then to locate the vehicle is equivalent to locate the point O. Let $v_{rx}(t)$ and $v_{ry}(t)$ ($v_{lx}(t)$ and $v_{ly}(t)$) indicate the linear velocity elements of the right-wheel (the left-wheel) in x- and y-direction. Using some fundamental mechanical principles, the linear velocity elements of the vehicle (point O) in x- and y-direction, $v_x(t)$ and $v_y(t)$, and the angular velocity, $\omega(t)$, are given by

$$v_x(t) = \frac{1}{2}(v_{rx}(t) + v_{lx}(t)), \quad v_y(t) = \frac{1}{2}(v_{ry}(t) + v_{ly}(t)) \tag{1}$$

$$\omega(t) = \frac{1}{2l}\left[\sqrt{v_{rx}^2(t) + v_{ry}^2(t)} - \sqrt{v_{lx}^2(t) + v_{ly}^2(t)}\right] \tag{2}$$

By integrating the above equations from t_0 to t along the path, the position information, $x(t)$, $y(t)$, and $\theta(t)$, can be calculated.

Since the location measured by the above technique is related to the initial states of the vehicle, and eventually this kind of measurement is an "incremental" process, there is a serious "error-accumulation" problem. The current measurement is accomplished based on the result of the previous measurements. Even if the measurement error at each step is very small, the accumulated error may be very big if the mission lasts a long time. To overcome this drawback, the whole mission may be divided into several small submissions. At the beginning of each submission, a so-

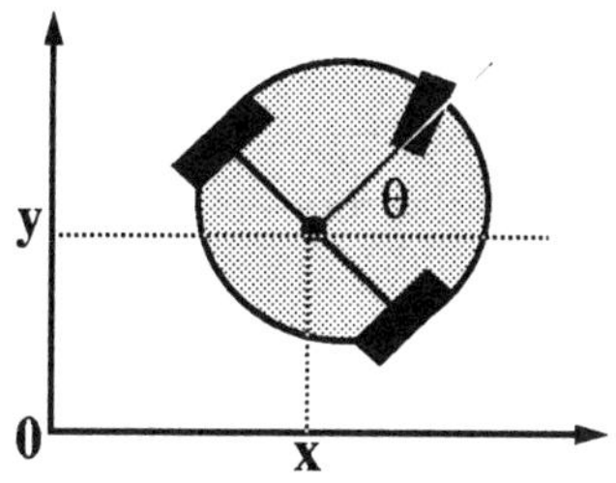

Figure 1. Differential-drive vehicle (Bottom view)

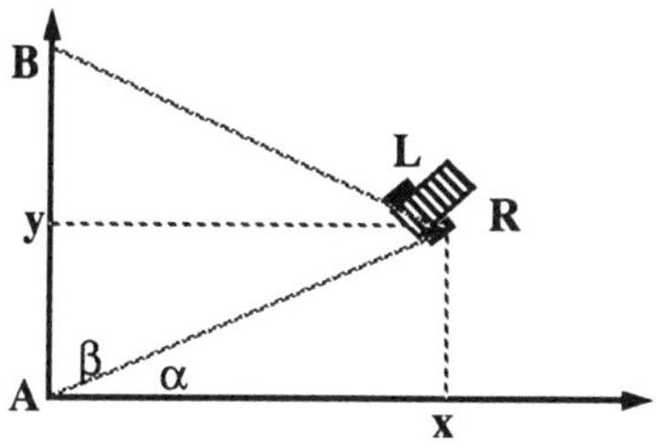

Figure 2. Absolute measurement

called "self-location" task should be done to let the vehicle return to its "home" position. Thus, the accumulated measurement errors are limited within each submission.

Using "self-location" to eliminate the accumulated measurement error has a limitation; that is, the range for maneuvering cannot be too large. For outdoor use, which needs extensive navigation, an absolute measurement technique would be applied. For example, suppose there are two laser rangefinders laid at points A and B (see Figure 2). The length of AB is known as $\overline{AB}$. The ranges between A and R, and between B and R can be measured by rangefinders as $\overline{AR}$ and $\overline{BR}$, respectively. Let (x_R, y_R) mark the coordinate of point R, we have

$$x_R = \frac{1}{2\overline{AB}}\sqrt{2(\overline{AB}^2\overline{AR}^2 + \overline{AB}^2\overline{BR}^2 + \overline{AR}^2\overline{BR}^2) - (\overline{AB}^4 + \overline{AR}^4 + \overline{BR}^4)} \tag{3}$$

$$y_R = \frac{1}{2\overline{AB}}(\overline{AB}^2 + \overline{AR}^2 - \overline{BR}^2) \tag{4}$$

Similar results can be obtained for the coordinate of point L. Thus, the location and the orientation of the vehicle can be obtained as

$$x = 0.5(x_R + x_L),\quad y = 0.5(y_R + y_L),\quad \theta = \operatorname{atan}\left(\frac{x_R - x_L}{y_R - y_L}\right) \tag{5}$$

To measure the velocity of the vehicle, a Doppler radar can be put at point A to measure the speed of point R along the line AR (v_R) and along the line AL (v_L). Since the angles α_R and α_L can be calculated during the position measurement, the velocity of the vehicle would be

$$v_x = \frac{1}{2}\left(\frac{v_R}{\cos(\theta - \alpha_R)} + \frac{v_L}{\cos(\theta - \alpha_L)}\right)\cos\theta \tag{6}$$

$$v_y = \frac{1}{2}\left(\frac{v_R}{\cos(\theta - \alpha_R)} + \frac{v_L}{\cos(\theta - \alpha_L)}\right)\sin\theta \tag{7}$$

$$\omega = \frac{1}{2\overline{LR}}\left(\frac{v_R}{\cos(\theta - \alpha_R)} - \frac{v_L}{\cos(\theta - \alpha_L)}\right) \tag{8}$$

Since the above mentioned measurement processes are independent of each other, the previous measurements have no effect on the current measurement. There is no "error-accumulation" problem. In other words, the measurement error is absolute error, its magnitude depends only on the accuracy of the measurement devices (radar, rangefinder, and so on).

CONTROL OF THE VEHICLE WITHOUT ROBOT ARM

To study the dynamics and the control of the mobile base, we assume that there is no slip at any time between the driving wheels and the ground. Let us take a look at the freebody diagram for the vehicle as shown in Figure 3. G_f represents the frictional force between the wheel and the ground parallel to the axles of the wheel. F_c represents the Centrifugal force due to the turning. Under no-slip condition, F_c is always equal to the sum of G_f. F_f represents the rolling frictional force due to the passive movement of the castors. Usually its magnitude is very small which can be ignored in our analysis. F_{lf} and F_{rf} represent, respectively, the frictional forces between the driving wheels and the ground. They are variable forces with constant upper bounds. Their magnitudes depend on the driving torques which are applied to the wheels by the driving motors.

Figure 4 shows the freebody diagram for the right-wheel. A control torque τ_r is applied to the wheel which is equivalent to a pair of forces F_r and $-F_r$, where $F_r=\tau_r/\alpha$. Note that $F_r \equiv F_{rf}$ as long as the frictional force F_{rf} is below its upper bound F_{mf}.

Selecting the angular position of the driving wheels $[\theta_r, \theta_l]$ as the generalized state variables, and the driving torques $[\tau_r, \tau_l]$ as the generalized forces, it is easy to show the dynamic equations of the vehicle

$$\begin{bmatrix} \frac{M\alpha^2}{2}+\frac{2I\alpha^2}{T^2} & \frac{M\alpha^2}{2}-\frac{2I\alpha^2}{T^2} \\ \frac{M\alpha^2}{2}-\frac{2I\alpha^2}{T^2} & \frac{M\alpha^2}{2}+\frac{2I\alpha^2}{T^2} \end{bmatrix} \begin{bmatrix} \ddot{\theta}_r \\ \ddot{\theta}_l \end{bmatrix} = \begin{bmatrix} \tau_r \\ \tau_l \end{bmatrix} \tag{9}$$

where M and I represent the mass and the moment of inertia of the vehicle, respectively; T is the vehicle track width. Note that equation (9) is expressed in "joint" (driving motor) space. To complete the trajectory following problem in world (Cartesian) space, two transformations must be done, they are: a) linear transformation between joint space and vehicle space

$$\begin{bmatrix} v(t) \\ \omega(t) \end{bmatrix} = \begin{bmatrix} \frac{\alpha}{2} & \frac{\alpha}{2} \\ \frac{\alpha}{T} & -\frac{\alpha}{T} \end{bmatrix} \begin{bmatrix} \dot{\theta}_r(t) \\ \dot{\theta}_l(t) \end{bmatrix}, \quad \begin{bmatrix} \dot{\theta}_r(t) \\ \dot{\theta}_l(t) \end{bmatrix} = \begin{bmatrix} \frac{1}{\alpha} & \frac{T}{2\alpha} \\ \frac{1}{\alpha} & -\frac{T}{2\alpha} \end{bmatrix} \begin{bmatrix} v(t) \\ \omega(t) \end{bmatrix} \tag{10}$$

$$s(t) = \int_{path} v(\sigma)\, d\sigma, \quad \theta(t) = \int_0^t \omega(\sigma)\, d\sigma \tag{11}$$

and b) nonlinear transformation between vehicle space and world space

$$\begin{cases} \dot{x}(t) = v(t)\cos(\theta(t)) \\ \dot{y}(t) = v(t)\sin(\theta(t)) \end{cases}, \quad \begin{cases} v(t) = \sqrt{\dot{x}^2+\dot{y}^2} \\ \theta(t) = \operatorname{atan}\{\dot{y}/\dot{x}\} \end{cases} \tag{12}$$

$$\begin{cases} dx = \cos(\theta(t))\, ds \\ dy = \sin(\theta(t))\, ds \end{cases}, \quad \begin{cases} ds = \sqrt{dx^2+dy^2} \\ \theta = \operatorname{atan}\{(dy)/(dx)\} \end{cases} \tag{13}$$

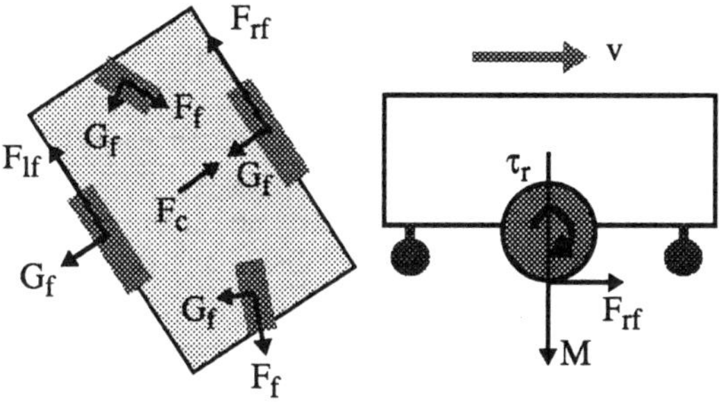

Figure 3. Freebody diagram for the vehicle

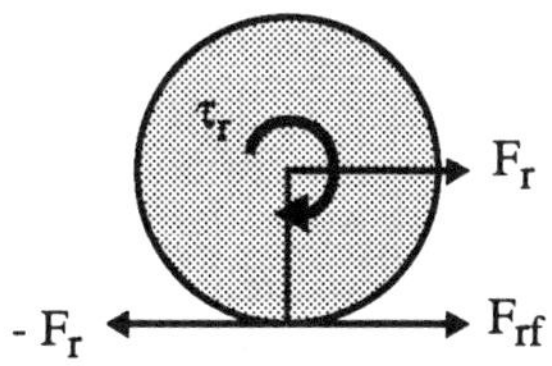

Figure 4. Freebody diagram for the right wheel

Figure 5 shows the computer simulation result when the vehicle moves starting from the origin of the world space and navigates longer than 40 miles in 50 minutes before it comes back to the origin. The position errors are less than 0.7 meters and the orientation error is less than 0.01° after the initial adjustments.

Although the result is good, there are still some problems in the control scheme.

(1) Even the tracking errors in joint space and in vehicle space may remain within a very small range (in this example, 10^{-9}-10^{-7}), it cannot guarantee that the tracking errors in world space remain within a small range. If there are external disturbances applied to the system, for instance strong wind results in a big error in joint space so that the vehicle temporarily deviates from its desired path, after a short period of adjustment, the errors in joint space should go to zero but the vehicle may still deviate from its desired path.

(2) If a robot arm is mounted on the vehicle to form a mobile robot system, then the motion of the robot arm may introduce external disturbances into the vehicle subsystem, especially if the vehicle and the robot arm moving with non-zero accelerations, then the interactive forces occur between them may destroy the stability of the whole system. In other words, when considering the whole mobile robot system, the interactive forces must be taken into account.

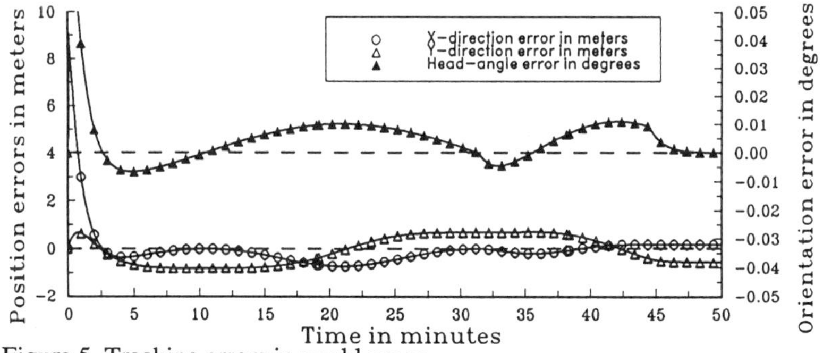

Figure 5. Tracking errors in world space

COMPOSITE SYSTEM MODEL FOR A MOBILE ROBOT

It is well known that Newton's laws, upon which classical mechanics is based, are valid only for the so-called *inertial* or *Newtonian* reference frame which is considered to be "fixed" in space.

For the ordinary industrial robot arm, the fixed base frame is generally chosen as an inertial frame. Referring to the inertial frame, one can derive Lagrange's equations from Newton's laws of motion by using the concept of virtual work.

However, for the mobile robot, the base of the robot arm is no longer fixed. It coincides with the vehicle frame which may move with linear and/or angular accelerations relative to the inertial reference frame (world space). To make Newton's laws valid, the fictitious *inertia force* and *inertia couple* must be applied through the mass center of the robot arm in the direction opposite to the linear acceleration and in the rotational sense opposite to the angular acceleration of the vehicle, respectively [6],[8]. Taking into account the effects of the moving vehicle on the motion of the robot arm, the dynamic equations can be written as [2]

$$M_R(q_R)\ddot{q}_R + N_R(q_R,\dot{q}_R) = T_R + J_R^T(q_R)f_V \tag{14}$$

where $q_R \in R^n$ is the generalized variable of the robot arm referring to its base; $M_R \in R^{n\times n}$ represents the symmetric, positive definite inertial matrix; $N_R \in R^n$ is the vector representing the centrifugal and Coriolis forces, the torques due to gravity, the unknown friction forces and external disturbances; $T_R \in R^n$ is the vector of joint torques supplied by the actuators; $f_V \in R^{3K\times 1}$ is the inertia forces applied by the vehicle where K is the number of the finite discrete points on which the mass of the robot is assumed to be concentrated. $J_R(q_R) \in R^{3K\times n}$ is a transformation which maps the 3K dimensional inertia forces into n dimensional joint space [6].

For the vehicle subsystem, assume that the mass of the vehicle is large enough so that the contact between the wheels and the ground is complete at any time. Taking into account the contact forces caused by the robot arm, the dynamic equations (9) can be rewritten as

$$\begin{bmatrix} \frac{M\alpha^2}{2} + \frac{2I\alpha^2}{T^2} & \frac{M\alpha^2}{2} - \frac{2I\alpha^2}{T^2} \\ \frac{M\alpha^2}{2} - \frac{2I\alpha^2}{T^2} & \frac{M\alpha^2}{2} + \frac{2I\alpha^2}{T^2} \end{bmatrix} \begin{bmatrix} \ddot{\theta}_r \\ \ddot{\theta}_l \end{bmatrix} = \begin{bmatrix} \tau_r \\ \tau_l \end{bmatrix} + J_V^T \begin{bmatrix} f_x \\ f_y \\ \tau_z \end{bmatrix} \tag{15}$$

where $f_R=[f_x, f_y, \tau_z]^T$ represent the contact forces owing to the reaction of inertia forces applied onto the robot arm and the accelerative action of the robot arm.

Note that f_R and f_V are not real forces in the sense of engineering, eventually they are fictitious forces. It is impossible to measure them by using standard force sensors. They are the functions of the acceleration of the vehicle and the manipulators. In practice, any physical system has an upper limit for its acceleration. These acceleration limitations are dependent on the particular physical structure of the systems as well as the power of the driving system. Even if the physical systems are out of control, their accelerations cannot exceed the maximum values. Thus, one can estimate the upper bounds for f_R and f_V in terms of the maximum acceleration of the vehicle and manipulators. This observation makes it possible to design some decentralized robust control schemes to separately control the vehicle and the robot subsystems subject to the interactive forces.

DECENTRALIZED ROBUST CONTROL OF A MOBILE ROBOT

Rewrite equations (14) and (15) in state-space form

$$M_R(q_R)\ddot{q}_R + N_R(q_R,\dot{q}_R) = T_R + F_V \tag{16-a}$$

$$M_V(q_V)\ddot{q}_V + N_V(q_V,\dot{q}_V) = T_V + F_R \tag{16-b}$$

where F_V and F_R represent uncertainties -- either the immeasurable interactive forces or the unmodeled dynamics. Note that even the exact magnitudes of F_V and F_R are unknown, we assume that (a) the nonlinear terms in F_V and F_R, if any, are confined to be sinusoidal functions and exponential functions with finite exponents, and (b) their norms are bounded by some known functions of the states, that is,

$$\|F_R\| \le \mathcal{F}_{RM}, \qquad \|F_V\| \le \mathcal{F}_{VM}\,. \tag{17}$$

Note these assumptions hold for most practice applications, at least so far as the mobile robot systems are concerned.

To implement the robust control scheme, we will first construct filtered tracking errors. Define the tracking errors for robot arm and vehicle as $e_i=q_{id}-q_i$, in the sequel the subscript *i* will be used to represent either *V* or *R*. Let the tracking errors pass through a stable low-pass filter, the filtered tracking errors are formed as

$$E_i = \dot{e}_i + \beta e_i, \qquad \beta > 0,\ i = V \text{ or } R \tag{18}$$

This is the standard filtered error used, for instance in [9]. It can be proven [7] that as long as the filtered errors E_i are bounded by some values, then the tracking errors are guaranteed to be bounded within a small ball whose radius can be reduced by selecting an appropriate design parameter β.

Define a continuous, *quasi*-saturation function $S_e(E_i,\gamma)$ as

$$S_e(E_i,\gamma) = \begin{cases} 1-e^{-\gamma E_i} & \text{if } E_i \ge 0 \\ -(1-e^{\gamma E_i}) & \text{if } E_i \le 0 \end{cases} \tag{19}$$

The function and its first-order derivative are continuous in E_i. When $|E_i|>5/\gamma$, the value of S_e goes to a constant (± 1); and when $|E_i|$ goes to zero, S_e goes to zero, too. This property makes the function an ideal candidate for the robust term in the control law.

The whole control law is a standard "computed-torque control" [2] with the estimated version of M and N, plus a robust term

$$T_i = \hat{M}_i(\ddot{q}_{id} - \beta\dot{e}_i) + \hat{N}_i - S_e(E_i,\gamma)\,[\Lambda_i + \Omega_i\beta|\dot{e}_i|] - \kappa E_i \tag{20}$$

where $\Omega_i>0$ and $\mathcal{N}_i>0$ are the maximum estimate errors for M and N, respectively;

$$\Lambda_i = \Omega_i|\ddot{q}_{id}| + \mathcal{N}_i + \mathcal{F}_{iM} > 0$$

κ is a positive design parameter. It is not hard to verify that with the controller (20), the derivative of a certain energy function will remain negative outside the neighborhood of the origin of the error space. According to *Lagrange Stability Theorem* [5], the boundedness of tracking errors is guaranteed and the maximum value of the error can be reduced by appropriately selecting the design parameters.

Figure 6 shows the computer simulation for a two-link mobile robot system. Two

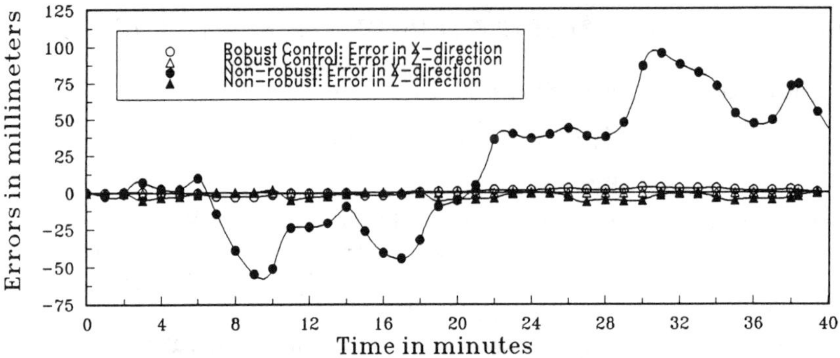

Figure 6. Position errors in Cartesian space under robust control

controller are used in this simulation study. The first one is a standard computed-torque control. The second is the robust controller (20). The difference is evident.

CONCLUSION

We consider the trajectory-following control of a robot arm mounted on a wheeled mobile bases. It shows that as long as the interactive forces between the robot arm and the vehicle are bounded by known functions of the states, there exists a robust control law which is continuous and guarantees that the whole tracking errors enter a neighborhood of the zero state in finite time and thereafter remain within that neighborhood.

REFERENCES

1. J. Borenstein, Y. Koren, "Motion control analysis of a mobile robot," *ASME J. of Dynamic Systems, Meas., and Control,* Vol.109, pp.73 - 79 (1987)
2. John J. Craig, *Adaptive Control of Mechanical Manipulators* (Addison-Wesley, 1989)
3. A. Hörmann and U. Rembold, "Development of an advanced robot for autonomous assembly," *Proc. IEEE International Conference on Robotics and Automation* (1991)
4. J.P. Huissoon, "Velocity profile generation for differential drive vehicles," *International Journal of robotics and Automation,* Vol.6, No.3, pp.140-148 (1991)
5. J.P. LaSalle, "Some extensions of Lyapunov's second method", *IRE Transactions on Circuit Theory,* Vol.CT-7, pp.520-527 (1960)
6. K. Liu and F.L. Lewis, "Stability Analysis of Decentralized Controller for Mobile Robots", *Advanced Robotics* (1992)
7. K. Liu and F.L. Lewis, "A Robust Control for Uncertain Multivariable Dynamical Systems with Applications in Mobile Robots", *Journal of Robotic Systems,* Vol 9(7) (1992)
8. J.L. Meriam, *Dynamics*, (John Wiley & Sons, Inc. New York, 1971)
9. J.E. Slotine and W. Li, "Adaptive manipulator control: a case study", *IEEE Transactions on Automatic Control,* Vol.AC-33 (1988)

Intelligent Task Planning and Environment Learning of a Mobile Robot System

Y. P. Chien, Q. Xue
Department of Electrical Engineering
Indiana University and Purdue University at Indianapolis

M. Palakal
Department of Computer and Information Science
Indiana University and Purdue University at Indianapolis

S. Chittajallu
Department of Mechanical Engineering
Indiana University and Purdue University at Indianapolis

Abstract

The intelligence of a mobile robot system is reflected by the ability to reason under different situations, the ability to adapt the changing work environments, and the ability to learn. In order to design an intelligent mobile robot system, the robot task and the work environment must be described intelligently. From the cognitive science study, we understand that human beings usually analyze a work sequence in terms of objects and their features. Therefore, we designed an object oriented knowledge base to describe the working environment and to plan the tasks of the robot. To increase the speed of task planning, special properties of a class of conjunctive planning systems are utilized.

1. Introduction

A new intelligent household robot, named VOLUNTEER, is being developed at Purdue University, Schools of Science, and Engineering and Technology at Indianapolis. In this project, many challenging problems such as sensor integration, environment modelling, task and path planning, and self learning need to be solved.

The problem in environment modelling is that the work environment of a household robot cannot be assumed known since the furniture arrangement in different houses are usually different. Even in the same house, the furniture arrangement can change at different time (eg. the position of a chair can change). On the other hand, the environment cannot be assumed completely unknown either since task and path planning in an unknown environment are very time consuming [1]. Therefore, an essential ability of an intelligent robot is that it can gradually learn the environment. In section 2, we describe an object oriented environment modelling method that facilitates environment learning.

The problem in task planning is that the automatic task planning must be done in real-time. Robot task planning is a conjunctive planning problem and have been studied for many years [2-4]. One major problem in conjunctive planning is goal interaction, i.e., the proposition which are processed later may negate the result of the proposition being processed. Some nonlinear planners such as RSTRIPS [3], DCOMP and NOAH [4] were proposed to solve the goal interaction problem. These task planner first plan the tasks without considering goal interaction, and then go back to change the plan if a goal interaction occurs. Consequently, these planners can only be used off-line. In section 3, we describe a fast task planner for VOLUNTEER.

2. Object oriented environment learning

Since VOLUNTEER is designed to be used in houses, its work environment is assumed unknown. However, this unknown environment is somewhat different from under sea or out space environments. In a household, most of the terrain remains the same all the time and only a small portion of the environment changes. When the robot initially enters a house, everything is unknown to the robot. Since locations of the walls and most large furniture will not change often, once the wall information is obtained by the robot through sensor detection or teaching, the information may not change in the future. Some furniture such as chairs or small tables may be moved within the house very often and the people in the house are moving all the time. These moving objects in the robot work space pose unknown obstacles to the robot motion. Therefore, the environment can be considered partially known. In a partially known environment, the robot still need to use sensors constantly for observing the workspace. Updating the environment model involves not only adding new information but also deleting the out-of-date information from the model. When the work environment is complicated, it is very time consuming to maintain and to update the environment data. Therefore, the environment description should be selected in such a way that updating environment model is simple and processing of query from the task commands can be done quickly.

An object oriented database offers such features. The object oriented database can provide a hierarchical structure to store and retrieve the environmental information. Moreover, it can be easily used for task decomposition in task planning. Therefore, it is used in VOLUNTEER for describing robot workspace. The structure of the object oriented database for VOLUNTEER is depicted in Figure 1. In this object oriented database, a building is defined as an object. All features (such as entrances, rooms, and corridors) associated with the building are defined as the attributes of the building object. Some of the attributes such as rooms and corridors can be considered as sub-objects. The features of these sub-objects are also defined as the attributes of the sub-objects. These attributes include the doors of the room, the geometric description of the room, and furniture inside the room. By the same token, attributes in the room object can also be considered as objects (a noun object). The defined level and the number of objects depends on the required tasks to be performed by the robot.

Each object is described by its features such as shape, location, color, relation to the neighbor object, and etc. The features of objects are obtained from sensory input. At present, we use ultrasonic sensors to obtain the positional features of the obstacles surrounding the robot. It is assumed that the robot can receive line segments obtained from the processed ultrasonic sensory information. The positional feature of an obstacle is represented by the polygonal boundaries of the obstacle.

Initially, each line segment is an object. The line objects are combined when they are connected. A method is associated to the object to combine the scattered environmental information into predefined and taught objects. If an object forms a closed chain, its positional features are completely known. Otherwise, its positional features are partially known. Another method associated with objects is to split an object into two objects when essential information is obtained. An object is splitted into two objects only when its additional features are identified by an operator or by pattern recognition rules. For example, the ultrasonic sensor cannot distinguish a table from a wall when the table is placed beside the wall, when the other feature of the table is available, the table can be identified. Then the table object need to be splitted from wall object. If part of an object is expected but does not exist from sensory information, further investigation to that object can be conducted. With the ability of combining and splitting objects according to the available information, this environment model can accept the scattered information and provide the information in terms of objects and there associated features.

In the object oriented database, all the information about an object are stored close to each other. Therefore, ,we can make the object recognition a method associated to the object. Objects for which the same method is used are grouped as a class. For example, the objects within which only the robot manipulator can operate comprise a class, and only the path planning methods for robot manipulators are associated with this class. The number of classes depends on the number of different path planning methods for different conditions implemented by the system. Since the robot work environment can range from totally known to completely unknown, depending upon the information of objects available and operation modes. Various path planning algorithms can be implemented as methods associated with classes of objects. For example, when the robot is in an environment-learning mode, the algorithm developed by Rao [5] can be adopted. When the room object belongs to a partially known object class, Lumelsky's algorithm can be adopted to find a feasible path. During the execution of a robot command, if an environment change is noticed but the change does not directly affect the planned motion based on the known information, the object oriented database is updated and the environment is still considered known. Lumelsky's algorithm is involved only when the preplanned path is not executable, and until the preplanned path is again recognized. Therefore, an object can be dynamically assigned to different classes based on the commands and environment conditions. It is assumed that the robot keeps updating the environment information. The robot is instructed to investigate the partially known objects whenever no specific task is assigned.

This object oriented database can be searched and modified locally so that the database maintenance and updating is relatively easy. Also path planing, task planning, and environment learning are tightly connected through this object oriented data base. The sensory information for path planning, task planning, and environment learning are shared between these different processes.

3. Goal Interaction Free Fast Task Planning

It can be observed that, in some robot applications, the goal (or subgoal) to be achieved by the robots can be decomposed into a set of subgoals which have explicit priority relationship according to the time sequence or the spatial order. If this property can be utilized in task planner, the speed of task planning can be increased greatly.

In STRIPS, only a minimum number of subgoals in a goal are clearly described and other subgoals that can be derived from these subgoals are not explicitly described. Although the description of subgoals is simple, the priorities of the subgoals cannot be assigned correctly since implicit subgoals are not shown. Therefore, a lot of effort in task planning is used to discover the implicit subgoals. Consequently, we conclude that a complete description of the subgoals and their priority is required in order to eliminate the process of finding the implicit subgoals and hence to increase the speed of task planning. It is found that the goal and subgoals of VOLUNTEER do have priorities and can be explicitly described.

The input to the task planner for VOLUNTEER is a user command. The output of the planner is a sequence of robot operations. In the knowledge-base, a robot operation is represented by four attributes: the name of the operation, the precondition, the deletion, and the postcondition. Each attributes is represented in predicate calculus. The preconditions of these robot operations are the subgoals to achieve the postconditions. The priorities of the preconditions can be assigned according to the following criteria:

(1) The time order among the subgoals, i.e., the subgoal happening earlier is assigned a higher priority number. A specific time instance is used as a reference.

(2) The spatial relationship among subgoals, i.e., the subgoal in which an object is locàted at a lower position is given a higher priority number. The position of a fixed object is used as a reference.

The priority numbers assigned to subgoals indicates the order that each subgoal should be achieved. Although a general conjunctive planning problem does not have the priority in their problem description, in most robot operations, priorities among the preconditions of an operation is usually obvious. For example, operation "putdown x onto y" is represented as:

putdown(X,Y)

P: $HOLD(X)_2 \wedge ATR(Y)_1$

D: HOLD(X),ATR(Y)

X: ON(X,Y)

where putdown(X,Y) is the name of the operation, P, D, and X stand for the precondition, deletion, and postcondition of the operation, respectively. Predicate ATR(Y) denotes that the robot is at position of Y. HOLD(X) means that the robot holds object X and ON(X,Y) means that the object X is on Y. Subscripts of the predicate represent the priority number of the predicate. A smaller number represents a lower priority.

It has been found that the assignment of the priority numbers for the preconditions of an operation is independent to the assignment of the priorities for the conjunctive robot tasks given by user although they follow the same criteria based on the time sequence and the spatial order. The priorities of the conjunctive tasks are given by the user or an interface program, while the priority numbers assigned to each subgoal for the operations are permanently written in the knowledge-base.

With the priority numbers attached to goals and the preconditions of operations (rules), a depth-first, goal-interaction-free, dynamic backward reasoning scheme can be naturally adopted to plan the task for VOLUNTEER. In the planner, the tasks given by the user is decomposed into a set of subgoals, each of which have a unique priority number based on the time or spatial relations. Furthermore, each subgoal can still be decomposed to lower level subgoals which also have unique priorities. Based on the priority numbers assigned to the subgoals, the planner knows exactly which one should be achieved first when several subgoals need to be achieved. It is proved that if the assignment of the priority to the conjunctive robot task and that of the precondition of all the robot operations is correct, the goal interaction will never occur by following the depth first search algorithm described above.

The task planner utilizes the priority assignment is greatly faster than the traditional task planners. Assume each rule have two preconditions and there are priorities among these preconditions. If the priority numbers are not assigned to these preconditions in the traditional task planner, the chance that a goal interaction free subgoal is selected is 0.5. In average case, only half the time a subgoal with high priority number is selected and goal interaction may occur. Therefore, the average number of subgoals searched for each operation is 3. Then the search tree is equivalent a trinary tree with depth n. The cost of task planning in the unit of robot operations is $O(3^n)$. In the best case, the subgoal with higher priority is always selected and no goal interaction will occur. The performance of this best case is essentially the same as that of the task planner using priority assignment. The search tree in the best case is a binary tree. Assume the binary tree is a complete tree with depth n, the cost of task planning is $O(2^n)$ in the unit of robot operations. Hence, the speed gain of the task planner utilizing the priority assignment over the traditional task planner is 1.5^n. If n=20, and a traditional task planner takes one hour to plan a particular task, the new task planner will take 1.08 seconds to plan this task (the speed gain is $1{,}5^{20}$=3325). Since the average number of preconditions of robot operations is usually larger than 2, the actual speed gain is even higher.

To further increase the speed of task planning, the internal representation of facts and operations are coded by binary numbers. For example, one 16 bit binary number is used to represent the code of a fact and an operation, and an additional 16-bit binary word is used to represent each parameter associated with the fact and operation. This representation increase the speed of planning by a large constant factor. A combination of hardware circuits and software program is used to implement the task planner. In this planner, instead of sequentially searching feasible robot operations to generate a subgoal, all the rules in the knowledge-base are compared with the subgoal parallelly by using hardware comparator.

4. Conclusion

We described an object oriented environment description and a fast task planning algorithm for a household robot. The advantage of this environment description is that the information is stored hierarchically so that the information storage and the retrieval is simple and less time consuming for complicated environment. The advantage of the proposed task planner is that it utilized the explicit priority assignment in task descriptions. This priority description can greatly increase the speed of task planning. This task planning method can be used for a fairly large subset of conjunctive problems that have priority among its subproblems. The time sequence and spatial order have been used to decide the priority numbers of the subproblems.

References

1. V. J. Lumelsky and Alexander A. Stepanov, "Dynamic Path Planning for a Mobile Automaton with Limited Information on the Environment," *IEEE Transactions on Automation Control,* AC-31,No.11, November 1986.
2. K.E. Olin and D.Y. Tseng, "Autonomous Cross-Country Navigation", *IEEE Experts,* August 1991, pp.16-30.
3. Fikes, R., and Nilson, N.J., "STRIPS: A New Approach to the Application of Theorem Proving to Problem Solving", *Artificial Intelligence,* 1971, pp 189-208.
4. Sacerdoti, E. D., "Non-linear Nature of Plans", *IJCAI-4,* pp.206-214.
5. N. S. V. Rao, "Algorithmic Framework for Learned Robot Navigation in Unknown Terrains," Computer, Vol. 22, No. 6, June, 1989, pp. 37-43.

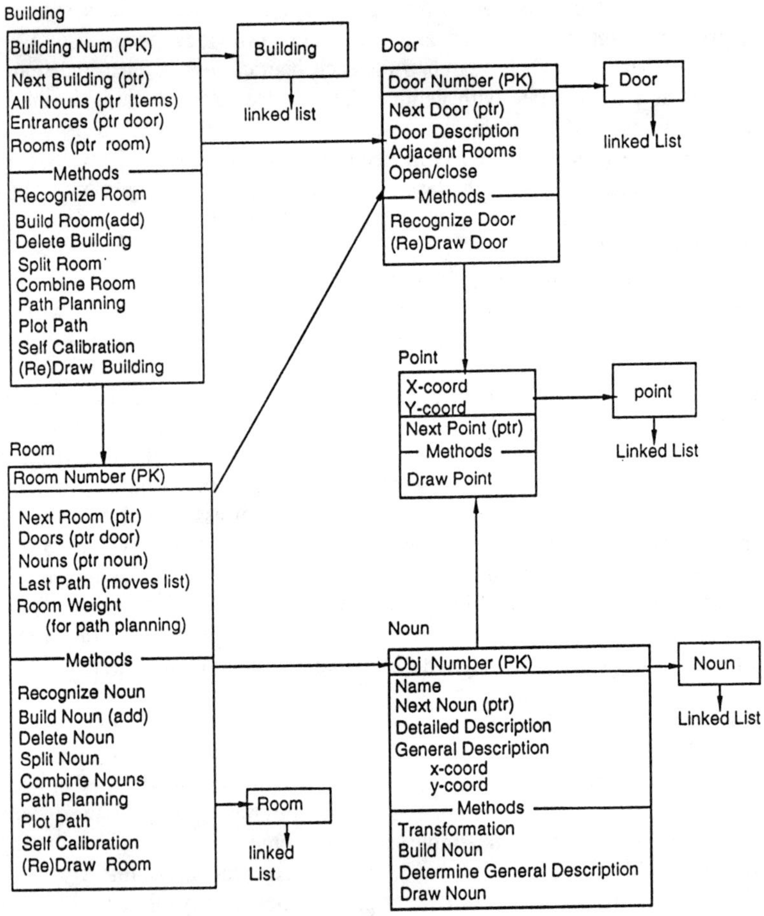

Figure 1. Object oriented environment model.

DESIGN AND TESTING OF A GRIPPER FOR THE AUTOMATIC AND FLEXIBLE ASSEMBLY OF A FAMILY OF PRECISE MECHANICAL COMPONENTS

J. El-Chaar, G.Galli, E.Imperio
IMU-Consiglio Nazionale delle Ricerche, Milano

C.R.Boër
Politecnico di Milano, Milano

Abstract

Our study is oriented to the design and realisation of a multiple gripper for automatic and flexible assembly of precise mechanical components in small/medium lots.
This study which leads to the realisation of a multifinger gripper according to traditional design technology, is interesting for the methodological approach used in the definition of the gripper functional characteristics. These are strictly connected to the product type to be handled, to the chosen assembly sequence, to the pallet type and to the used robotized system within a flexible automation domain.
In the Institute, a working group is evolving towards a new method to think the flexible integration of the assembly operation that could find the best application right for the precise mechanical workpieces.

Introduction

The present study undertaken at IMU Institute for Industrial Technologies and Automation of the National Research Council of Italy, is a part of the scientific and technological cooperation activities within a specific Research Project under the umbrella project called EUREKA-FAMOS.
The work is a contribution for the feasibility study of the design and the realisation of an automatic and flexible pilot plant for the manufacturing and assembly of mechanical group families, with high precision and tight tolerance characteristics.
The families can be considered different, but with comparable functionalities; that is each family differ in number, form of the components and the assembly modalities, but it is possible to apply the some technological processes.

The product

The present work is addressed to a family of product, named with substitution variant, that is, products having similar functionality but are different for the form and the dimension of same components, but maintain the same physical interfaces with the remaining of the product [1].
Specifically, a prototype of a multiple gripper has been designed and produced; this gripper can assemble a family of three lathe tool holders; each can be assembled according to different configurations slightly different bringing to 10 the total number

of tool holders, with a maximum weight of 960 g. Each tool holder has between 9 and 11 components, for a total number of different components to handle equal to 25. The tool holder family and the components show a large variability of characteristics: gripping morphology (flat internal and external surfaces, cylindrical external surfaces, threaded surfaces), dimensions (diameters between 3 and 14 mm, thickness between 1 and 84 mm), weight (between 2 and 690 g), assembly modes (positioning, insertion, forcing, screwing).

Assembly sequence

In relationship to such a typology of components and sub-assembly to be handled, the selection of a multiple gripper with parallel gripping surfaces has been thought early in the beginning of the study.
For the modelling of the multifinger gripper, that is to assess the functional parameters of the gripper, our approach started from the determination of the assembly sequence. Infact, the operation sequence that is used to assemble the components is considerably conditioning the gripper modelling.
Considering the physical and geometrical characteristics of the *contacts* (surface matching plane-plane, cylindrical-cylindrical, polyhedrical-polyhedrical, helical-helical) and *connections* (forced reversible or irreversible, glued or welded, screwed) it is possible to obtain a graphical representation (fig. 1) called *relational model* [2].
From such a model, giving the interdependent constraints of *"geometrical feasibility"*, *"mechanical feasibility"* and *"stability"*, it is possible to obtain a graphical representation of all possible assembly sequences [2], [3].

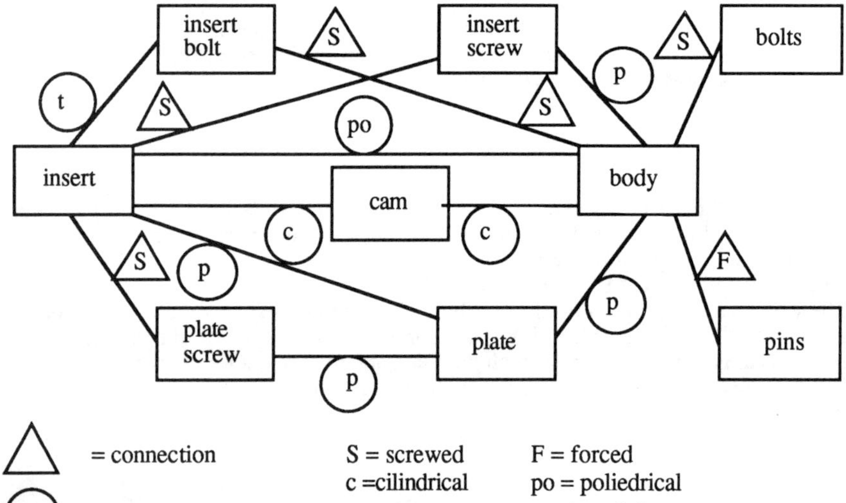

FIG. 1. Relational model

The "geometrical feasibility" is satisfied if a trajectory without collision exists to put in contact two subassemblies. The "mechanical feasibility" is satisfied if it is possible to obtain the contacts and the connections between the different components

respecting the product functional requirements. The "stability" is satisfied if the parts of each subassembly keep their relative position and are not spontaneously separating.
The choice of the best assembly sequence has been done "on the field" by detection of those compatible with the system product-manipulation system-feeding system and fixing as optimisation criteria the assembly sequence that gives the minimum number of change-over.
To determine the assembly sequences it was used the top-down method, that is by determination of disassembly sequences, assuming the following conditions:
- disassembly is a process in which each part can be removed from the assembly structure one by one, without collapsing the sub-assembly structure;
- the assembly sequence is the reverse sequence of disassembly.

Gripper modelling

Having defined a system of orthogonal axis in two dimensions, with the origin G coinciding with the center of gravity of the gripper (an axis is directed along the longitudinal development of the finger and the other axis is the half distance between the gripping surfaces) it is possible to model the gripper [4].
Starting from the knowledge of the assembly sequence and from the definition of all possible gripping surfaces of each component, it is possible to determine the preliminary dimensions of the fingers, as well as the minimum number of gripping couple taking into account the collision constraints between the components and, even if only partially, between the components and the gripper.

Gripper design

The real design phase then follows (cross section, gripping forces, inertia moments, etc.) and, in this case, we can have further collision constraints and, therefore, an iterative verification of the modelling.
The design of the multiple gripper has been oriented to the manipulation of a large number of components and sub-assemblies, such a design phase came after the analysis of the components and of their physical, geometrical and manipulation characteristics, these being independent from the gripper.
Afterwards, an iterative verification of the physical, geometrical and functional constraints of the mechanical components has been performed, in relation to the gripping mode, to the type of gripper (electromagnetic, articulated, by suction, etc.), the mode of operation (pneumatic, hydraulic, electromechanical, etc.), the auxiliary gripping fixtures and the robotic system.
The final design is a gripper with two symmetrical multiform fingers without sensors and with a system of pneumatic actuator. This result has been obtained by finding the gripping surfaces which optimise the kinematic and dynamic conditions of each pair gripper-part (fig. 2),[5].
Starting from the determination of the best position between each gripping surface and the components or sub-assembly, a prototype of feeding pallet has been designed which has also the function of working base for the assembly (fig.3), [5].
A graphical method has been used for the determination of the minimal number of references planes necessary to the manipulation of components and sub-assemblies
Such method allows to graphically represent for each gripping surface-component or sub-assembly pair the minimal and maximal position for a successful assembly operation. Two distinct intervals of possible distance exist for the reference planes which force to work on at least two planes selected in one of the intervals.
The results are satisfactory because it was possible to realize a gripper for the correct assembly of all the component typology except two parts that, if analysed according

to the *"redesign for assembly"* techniques, could be handled by the gripper too. The multiple gripper solution has been verified in a laboratory environment mounting it on a wrist of a SCARA type robot without sensors. The present solution is an interesting contribution to the automation of the assembly cycle and allows the reduction of the fixturing costs, of the time and of the number of gripper changes.

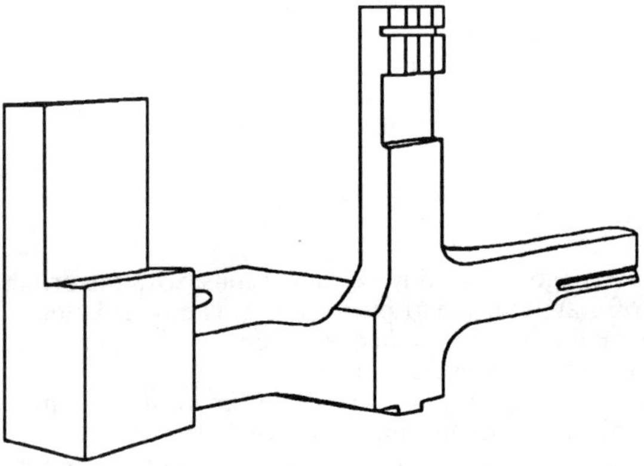

FIG. 2. Finger drawing

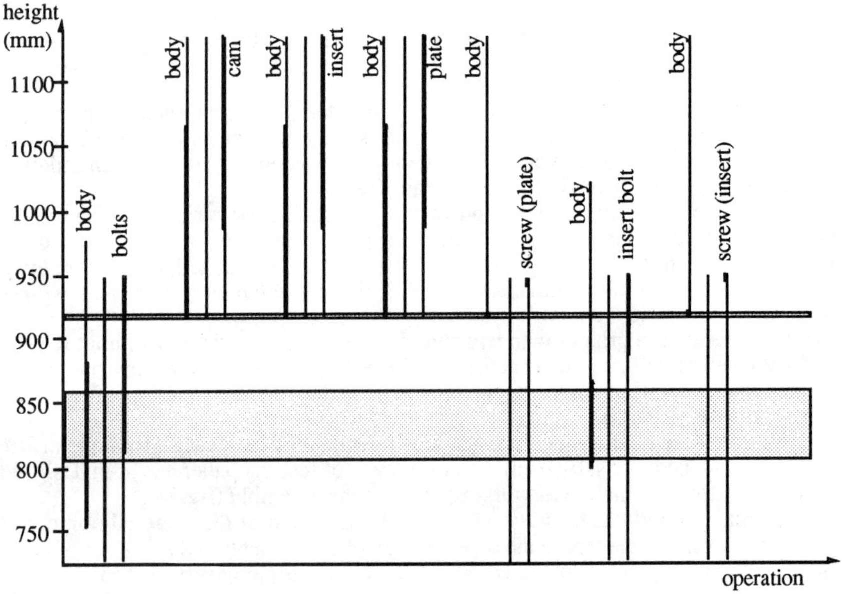

FIG. 3. Graphical determination of references planes

Conclusions

The CIM concept drives us towards a computer-automated, optimized and integrated manufacturing environment for producing our products, but all this could prove to be sub-optimal if the products and the processes themselves are not optimized and efficient in manufacture and assembly. Integrated product development provides a mean of matching the products to the manufacturing system. The challenge today is to solve the problem that regards the automated assembly for the industries that have middle and low output production and integrate it in the production process, toward the CIM technology. Following the above ideas, it is shown that a new concept for a *Flexible Assembly Center* (FAC) has to be developed which take into account the CIM methodology. The FAC is radically different from the known industrial robot technology and will be particularly useful in the assembly of precise mechanical groups.

We think that there will be an *universal assembly machine* for mechanical parts in different size and for mainly rotational or mainly prismatic groups. It will have also standard physical and information interfaces to allow its integration in group technology cell and systems and in a CIM environment.

This machine tool development should bring also benefits to the FMS technology because it will favor the integration of the different shop-floor components, but also the CIM technology will be affected by such a development [6].

CAD, CAM, CAPP will have to integrate the assembly process with the other manufacturing processes.

References

1. F. Lhote, M Meunier, R. Chossat. Products families and flexible assembly. Automation Congress 1988:What automation is needed in manufacturing industry, Paris, AFCET Publisher, pages 37-48.

2. L.S. Homem de Mello, A.C. Sanderson. A Basic algorithm for the generation of mechanical assembly sequences. Computer Aided Mechanical Assembly Planning, Kluwer Academic Publishers, pages 163-190, 1991.

3. A. Bourjault, C. Perrard. Design methodology for flexible assembly systems applied to a car pedal-board. Proceedings of the 20th International Symposium on Industrial Robot, pages 495-502, October 1989.

4. J.D. Wolter, R.A. Volz, A.C. Woo. Automatic generation of gripping positions. IEEE Transaction on Systems, Man and Cybernetics, Vol. SMC-15, No. 2, March/April 1985.

5. C. Brix, G. Galli. Contributo alla progettazione ed alla realizzazione di una cella flessibile per il montaggio automatico di una famiglia di portautensili per tornio: pinza di manipolazione, Tesi di laurea, Politecnico di Milano, Anno Accademico 1990/91.

6. C.R. Boër, J El-Chaar, E. Imperio. What is missing for a truly computer integrated assembly and manufacturing. International Symposium on Computer Integrated Manufacturing Systems-ISCIMS '92, Peking, June 1992.

The KAISER-II Quadruped Walking Robot:
Its Control System and Gait Algorithms

Zeungnam Bien, Yun Jung Lee, and Byungeui Min
Department of Electrical Engineering, KAIST
Daeduck Science Town, Taejoen, 305-701, Korea

ABSTRACT

This paper describes the control system architecture and the gait algorithms of an experimental quadruped walking robot, named as KAISER-II. The control system with multiprocessor-based hierachical structure is implemented for various functions, including supervision of total system, motion planning, 3-D visual information processing and actuator driving. For efficient autonomous locomotion, each module performs cooperative task. Especially, three kinds of experimental results are included to show the validity of the proposed system as well as the adaptive walking capability.

INTRODUCTION

To develop a walking robot system, various technologies in the fields of mechanical design, control, and computer science should be combined. We have constructed an experimental quadruped walking robot, named as KAISER-II, for the past 3 years.

In order to enhance the capability of autonomous locomotion, we have studied various problems including optimal design of an energy-efficient mechanism[1], optimal gait planning[2]-[3], 3-D self-positioning in an uncertain environment[4], and 3-D graphic simulation for remote monitoring as well as verification of walking motion[5].

This paper describes the control system architecture and the gait algorithms with particular focusing on integrated features for autonomous, adaptive walking. Three kinds of experimental results are given to show validity of the integrated control system.

DESCRIPTION OF THE KAISER-II

KAISER-II is a quadruped walking robot which has a total weight of 75 kg, a normal height of 820 mm, and a body of 416 mm by 430 mm (Fig. 1). The main frames and four legs are made of composite material called carbon fiber epoxy for lightening weight as well as increasing strength. Each 3-d.o.f. leg consists of a pantograph mechanism for motion in the vertical plane and a 4-bar, straight-line generator for sideway motion. This mechanism allows kinematically decoupled linear motion in each direction so that energy consumption can be reduced[7], and kinematic relations are simplified.

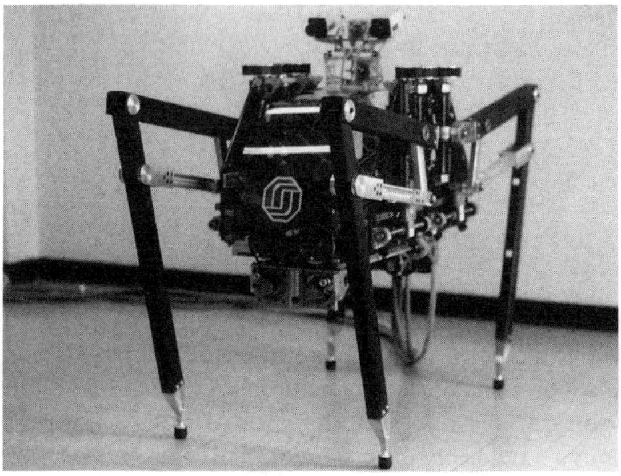

Fig.1 KAISER-II Walking Robot.

The 2-D pantograph mechanism is driven by two, 100 W, dc servo motors with ball screws, and a 60 W dc servo motor is used for the 4-bar mechanism.

In the center, at the top of the body, two cameras are mounted and they can change their sight of view in vertical and horizontal directions by means of step motors.

By 81 shielded twisted pairs and 38 power cables, the control signals for 14 motors and various sensors signals are transmitted between the KAISER-II and its control system.

CONTROL SYSTEM ARCHITECTURE

To endow a walking robot with the capability of autonomous locomotion, its control system should have some essential functions such as planning, sensing, and execution of motion[6]. We realize these functions with multiprocessor-based hierarchical structure which is divided into four modules; supervision of total system, motion planning, 3-D visual information processing, and servo control.

The supervisor coordinates each module so that the total system may accomplish the tasks assigned by human operator, and provide a menu-driven operator interface. Each menu item is closely related to each task. The menus can be categorized into three groups of operation modes. The first group consists of *jog* mode and *diagnosis* mode for off-line experiment. The operator can control each axis of legs as well as the body for up-down, fore-aft, and sideways motion in the jog mode, and can check the system status in the diagnosis mode. In the second group, *home* mode, *park* mode, and *walk* mode are provided. Since the KAISER-II rests on a chair, the home mode is used for standing up the robot and for moving legs to each initial position. The park mode is used for reverse motion of homing. In the walk mode, the robot executes walking with a periodic gait pattern according to various parameters such as speed, stride, duty factor, goal position, and gait type - crab, wave, and turning gait. The third group consists of higher priority modes to stop the robot at emergency and to reset the control system.

Due to the complexity of retracting useful information from natural scene, guide-mark patterns (GMPs) is used for practical application.

Using a pair of cameras with a proposed GMP, the vision system is capable of not only measuring the distance between obstacles and the robot, but also identifying the robot's 3-D position and orientation. The proposed GMP is a simple right-angled triangle pattern which has three feature points at each corner and has a coordinate frame. From these properties of the proposed GMP, a closed form of 3-D self-positioning algorithm is proposed[4].

In vision processing, first, a pair of images converted into binary images by thresholding, and coded by a chain coding technique. To reduce time of search for the guide mark, a window-based processing is employed.

The motion planner, which is an inherent part of the walking robot, generates appropriate gaits for a given environment. As an intermediate module between the supervisor and the servo controller, the motion planner also transfer commands and status between them. The detailed features of the motion planner will be described in next section.

According to a set of desired trajectories generated by the motion planner, the servo controller controls each leg's positions and speeds by means of 12 position controllers together with motors and servo amplifiers.

As for hardware configuration of the overall control system (Fig.2), a multiprocessor structure of a time-shared, common-bus type is adopted. A MVME 147 system is used for a host computer, and three 32-bit processor boards(MVME 133A) are assigned to the supervisor, the motion planner, and the vision system. To localize the servo control part having many hardware units, the servo controller is implemented to another 32-bit processor board (MVME 135-1) which has VSB (VME Subsystem Bus) for local interface.

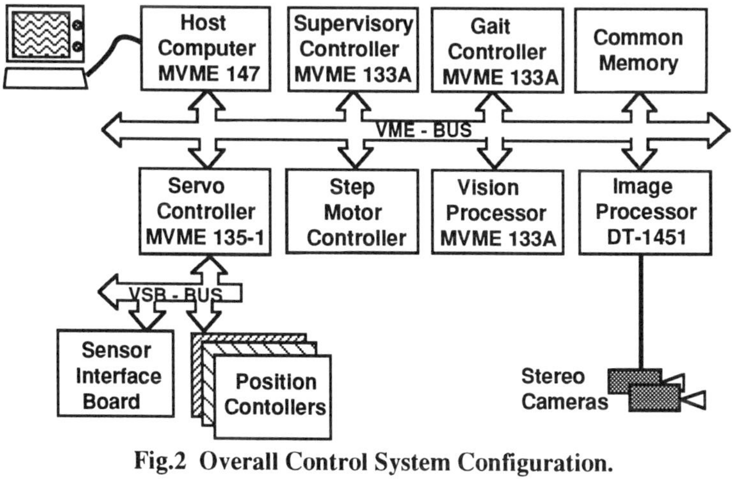

Fig.2 Overall Control System Configuration.

The communication between modules is accomplished by a predetermined handshaking protocol, via a dedicated space in common memory.

Each module has several jobs to concurrently process such as periodic polling of the communication area and executing the given jobs. To meet this multi-tasking requirement, the VMEexec real-time operating system is employed for each module.

GENERALIZATION OF GAIT ALGORITHM

For an efficient mediation between the robot and the environment, we have explored two intrinsic strategies: One is to determine the position and the orientation of the robot in the global navigation space in a real time fashion. The other is to structure the computational scheme of the robot for an easy manipulation of visual information in local navigation space. For this strategy, we have analyzed and developed several quadruped gaits by generalizing a typical walking algorithm to be independent of the environment and the type of gait.

To realize this concept, a conceptual model is proposed as shown in Fig.3 where the gait module is divided into three sub-modules; gait diagram generator, c.o.g. path generator and foothold generator. The output of each module is applied to the input of the servo command generator where actuator commands are calculated by solving the following inverse kinematics:

$$\begin{bmatrix} \mathbf{P}_i^{hi}(t) \\ 1 \end{bmatrix} = \begin{bmatrix} \left(\mathbf{R}_w^b(t) \cdot \mathbf{R}_b^{hi}\right)' & -\left(\mathbf{R}_w^b(t) \cdot \mathbf{R}_b^{hi}\right)' \cdot \left(\mathbf{R}_w^b(t) \cdot \mathbf{D}_b^{hi} + \mathbf{D}_w^b(t)\right) \\ \mathbf{0} & 1 \end{bmatrix} \begin{bmatrix} \mathbf{P}_i^{w}(t) \\ 1 \end{bmatrix}$$

Here the 3x3 rotational matrix $\mathbf{R}_b^a$ and the 3x1 displacement vector $\mathbf{D}_b^a$ are the components of the 4x4 matrix denoting the homogeneous transformation from frame $\{A\}$ to frame $\{B\}$ and the letters b, w, and hi denote the body, world and the hip coordinate frame, respectively.

Many previous works in gait analysis employ a moving body coordinate frame as a reference frame for specifying the position of foothold. But the foothold of a leg has only two discrete values in one locomotion cycle, which are present foothold and next foothold. Further, as can be seen in the above equation, to drive actuators in real experiments, all components except time-invariant matrices should be described in a world coordinate frame. This simple observation allows us to use a *pseudo world coordinate frame* as a reference frame. The origin of the pseudo world coordinate frame is set at the starting position of the vehicle's center in each locomotion cycle.

To calculate actuator commands, we need the informations of $\mathbf{R}_w^b(t)$ and $\mathbf{D}_w^b(t)$ which stand for the orientation and position of the vehicle body at time t, with respect to the pseudo world coordinate frame and these informations are given by the c.o.g. trajectory generator. The design of vehicle's trajectory is a very important process for obstacle crossing gaits, since it is impossible to cross over the terrain exceeding a certain degree of irregularity without postural modification of the body due to the geometrically limited reachable volumes of the machine. For this process, we developed an algorithm to decide the optimal body attitude and altitude on the slope which optimizes the static stability and also a proposed heuristic algorithm to cross over the stairs and the ditches by overcoming the geometric limitations of the reachable volumes.

The foothold generator in the conceptual model of Fig.3 is a major process in designing gaits over rough terrain that contains unsuitable regions for foot positioning.

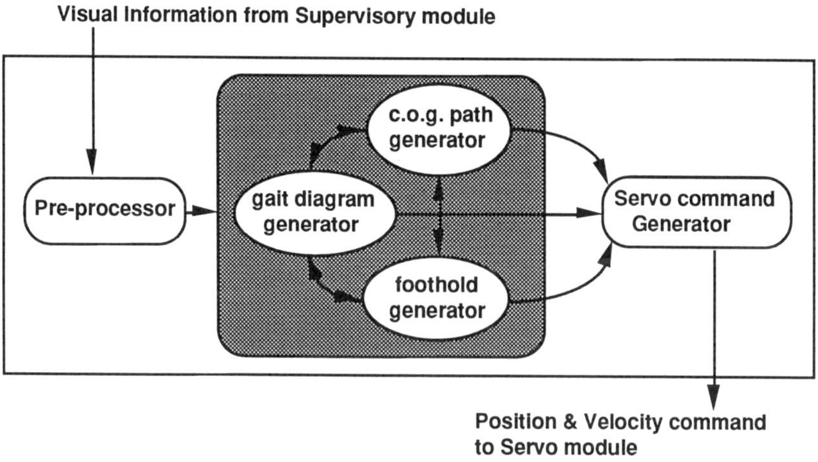

Fig.3 A Conceptual Model for Generalizing the Gait Module.

With the proposed gait algorithm integrated into the total control system, the following experiments are excuted (Fig. 4);

- walking across the horizental bar after the vision system measures the distance between the robot and the horizontal bar using a guide mark pattern.
- walking up and down the stairs of 40% grade.
- walking through the tunnel-typed obstacle which has lower height than the robot's normal one with continuous pitching motion of ±5 degrees.

CONCLUDING REMARKS

In this paper, a control system architecture and gait algorithms of a quadruped walking robot was described. To combine the modules of planning, sensing, and execution of motion for autonomous locomotion, a multiprocessor-based hierarchical structure of the control system and a generalized gait algorithm were proposed, and verified through experiments in structured environment.

To enhance the capability of autonomous navigation, several difficult problems still remain. In further research, we will concentrate on the path planning and the adaptive gait control for the quadruped walking robot.

REFERENCES

1. Yoon, Y. S. and Hong, H. J., "A Mechanical Design of a Quadruped Walking Machine for the Fastest Speed in Static Level Walking," *J. of the SME*, in press.

2. B. Min and Z. Bien, "On the Basic Characteristics and the Stability Properties of Quadruped Crab Gaits," Submitted to *Int. J.Robotics Research.*

3. Z. Bien, M. G. Chun, and H. S. Son, "An Optimal Turning Gait for a Quadruped Walking Robot," *Proc. IEEE/RSJ Int. Workshop* on IROS 91,Osaka, Japan, pp. 1511-1517, 1991.

4. Z. Bien, H. Y. Kwon, J. Youn, and I. H. Suh, "A closed form 3D self-positioning algorithm for a mobile robot using vision and guide-marks," *Robotica*, Vol.9, pp. 265-274, 1991.

5. B. W. Choi, M. J. Chung, and Z. Bien, "Development of Software Environment for the Kinematic Analysis of a Quadruped Robot," *Proc. IEEE/RSJ Int. Workshop* on IROS 89, Tsukuba, Japan, pp. 193-198, 1989.

6. D. E. Orin, "Supervisory Control of a Multilegged Robot," *Int. J. Robotics Research*, Vol.1, No.1, 1982.

7. S. Hirose, "A Study of Design and Control of a Quadruped Walking Vehicle," *Int. J. Robotics Research*, Vol.3, No.2, 1984.

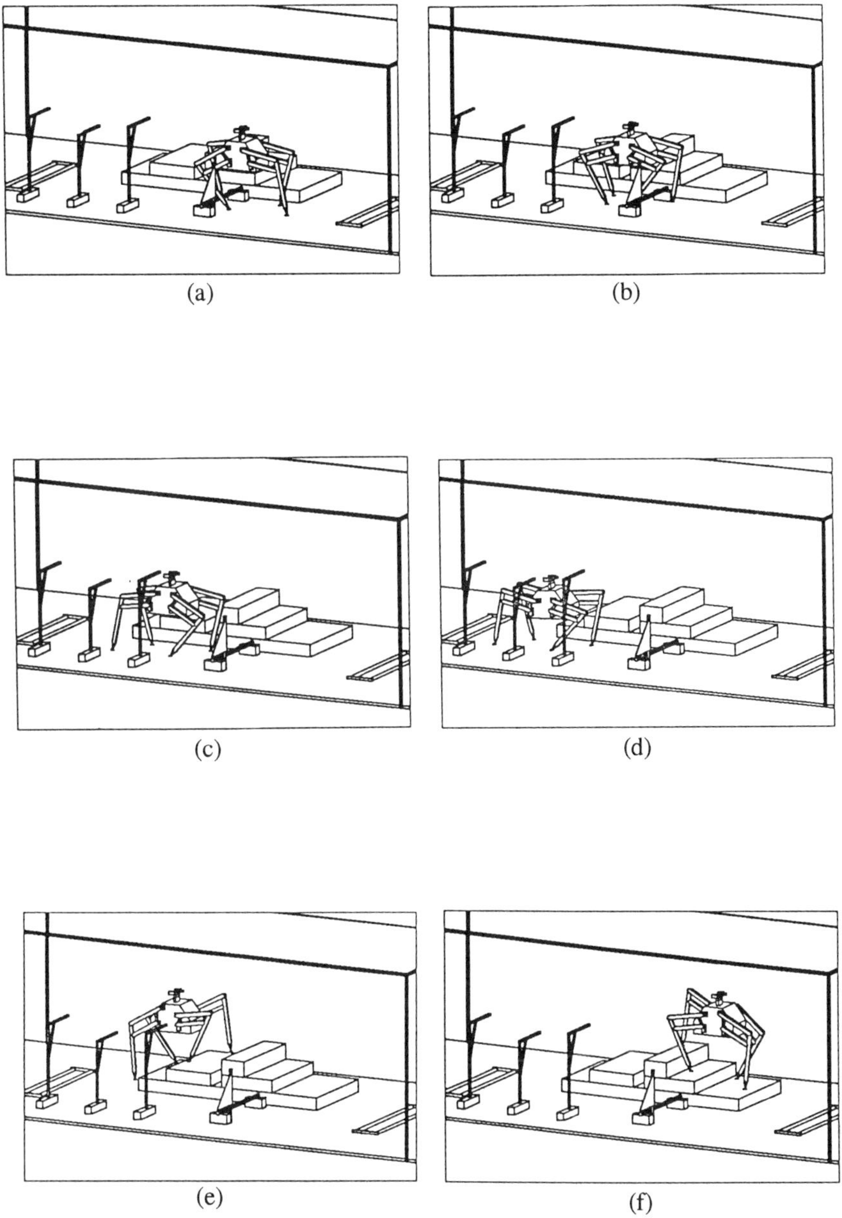

Fig.4 Structured obstacles and the KAISER-II emulated by 3-D graphic simulator and walking motions; (a)-(b) across the horizontal bar, (c)-(d) through the tunnel-typed obstacle, (e)-(f) up and down the stairs

CONTROL OF A FLEXIBLE-LINK MANIPULATOR*

H. Geniele, R.V. Patel and K. Khorasani

Dept. of Electrical & Computer Engineering
Concordia University
Montreal, Quebec
CANADA H3G 1M8

ABSTRACT

This work focuses on the tip position control of a flexible link which rotates in the horizontal plane. The dynamic model is derived using a Lagrangian assumed modes method based on Euler-Bernoulli beam theory. An output feedback control strategy is used. It consists essentially of two parts: an inner (stabilizing) control loop that moves the system's poles to appropriate positions in the left-half plane, and a feedback servo loop that allows tracking of the desired trajectory.

1. INTRODUCTION

Current robots are over-engineered in that their heavy construction is motivated mainly by a desire to simplify the control problem by reducing link flexibility. Lighter weight robots have the advantage of less costly frames. Their smaller and cheaper drive systems consume less energy. The link flexibility associated with such light weight robots, however, requires more sophisticated control techniques.

Several characteristics of the flexible link complicate the design of an effective control. A partial differential equation is needed to model this distributed parameter system. If expressed in terms of an ordinary differential equation (ODE), an infinite number of such equations are needed to represent the model exactly. Furthermore, this system of ODE's is highly nonlinear. Even if the flexible link is simplified by linearization, the resulting model is non-minimum phase. Straightforward application of output feedback will then cause migration of some of the poles to the right-half plane zeros resulting in instability of the closed-loop system.

This paper addresses the non-minimum phase behaviour by using an inner (stabilizing) control loop that ensures that the poles of the linearized system lie well within the left-half plane. An outer (servo) control loop yields the desired tracking of the reference input.

*This research was supported by the Institute for Robotics and Intelligent Systems (IRIS), and by NSERC (Canada) grants OGP0001345 and OGP0042515.

In Section 2 we present the Euler-Bernoulli based linearized model of the flexible link. The control is developed in Section 3. Section 4 contains simulation results. Section 5 concludes the paper with a discussion and suggestions for further work.

2. THE MODEL

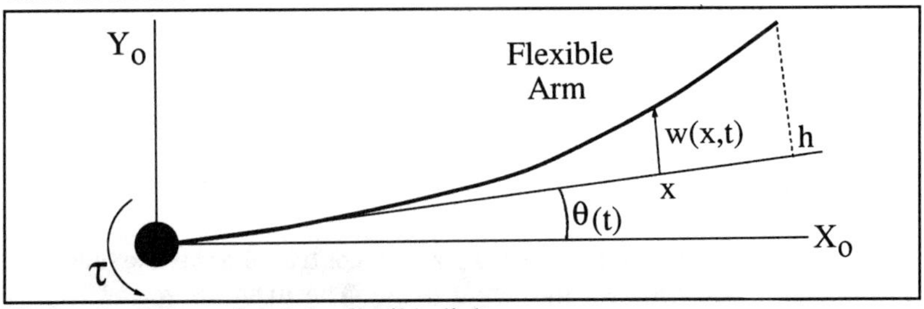

Figure 1. Schematic of the flexible link.

Figure 1 defines the position of any point along the link as $y(x,t)=x\theta(t)+w(x,t)$. We ignore the effects of rotary inertia and shear deformation by assuming the cross-sectional area of the arm to be small in comparison with its length h. Euler-Bernoulli beam theory and the assumed modes method allow us to express the deflection of a point located at a distance x along the arm as

$$w(x,t) = \sum_{i=1}^{n} \alpha_i(x)\, q_i(t)$$

where $q_i(t)$ is the generalized coordinate of the ith mode. As defined in [1], $\alpha_i(t)$ is the normalized, clamped-free eigenfunction of the ith mode. Following the derivation in [1] and by applying the Euler-Lagrange equations, we can write the equations of motion as

$$\ddot{\theta}\left(I_h + \frac{\gamma h^3}{3} + \gamma \sum_{i=1}^{n} q_i^2\right) + \gamma \sum_{i=1}^{n} \ddot{q}_i \int_0^h \alpha_i dx + 2\gamma\dot{\theta} \sum_{i=1}^{n} q_i \dot{q}_i + b\dot{\theta} = \tau \qquad (1)$$

$$\gamma\ddot{\theta} \int_0^h \alpha_j x dx + \gamma \ddot{q}_j + c_j \dot{q}_j + \gamma q_j \left(\omega_j^2 - \dot{\theta}^2\right) = 0$$

where I_h is the hub inertia, ω_j is the resonance frequency of the jth clamped-free eigenfunction, γ is the mass per unit length, and $b\dot{\theta}$ and $c_j\dot{q}_j$ have been added to account for viscous friction at the hub and structural damping of the arm, respectively. Equation (1) is linearized about an operating point and expressed in the perturbed state-space format

$$\begin{pmatrix} \delta\dot{z}_1 \\ \delta\dot{z}_2 \end{pmatrix} = \begin{pmatrix} [0]_{n+1} & [I]_{n+1} \\ M_k & M_c \end{pmatrix} \begin{pmatrix} \delta z_1 \\ \delta z_2 \end{pmatrix} + \begin{pmatrix} 0_{n+1} \\ \Gamma \end{pmatrix} \delta\tau \qquad (2)$$

$$\delta y = \left[h\ \alpha_1(h)\ \alpha_2(h) \ldots \alpha_n(h)\ 0_{n+1}^T\right] \left(\delta z_1^T\ \delta z_2^T\right)^T$$

where $z_1 = \left[\theta\ q_1\ q_2\ q_3\ .\ .\ .\ q_n\right]^T$ and $z_2 = \dot{z}_1$; δz_1 and δz_2 denote small perturbations in z_1 and z_2, respectively, about the operating point; $M_k = -M^{-1}K$,

$M_c = -M^{-1} C$ and $\Gamma = M^{-1}\left[1 \;\; 0_n^T\right]^T$ where

$$M = \begin{pmatrix} I_h + \dfrac{\gamma h^3}{3} & \gamma\int_0^h x\alpha_1 dx & \gamma\int_0^h x\alpha_2 dx & \dots & \gamma\int_0^h x\alpha_n dx \\ \gamma\int_0^h x\alpha_1 dx & \gamma & 0 & \dots & 0 \\ \gamma\int_0^h x\alpha_2 dx & 0 & \gamma & \dots & 0 \\ . & . & . & \dots & . \\ \gamma\int_0^h x\alpha_n dx & 0 & 0 & \dots & \gamma \end{pmatrix}$$

$$C = diag\left\{b, \left[c_i, i = 1,2,3,\dots, n\right]\right\}$$

and

$$K = diag\left\{0, \left[\gamma\omega_i^2, i = 1, 2, 3, \dots n\right]\right\}.$$

3. CONTROL

It has been shown in [2] and [3] that the tip transfer function *G(s)* of the system described in equation (2) is non-minimum phase, making the control design difficult. As explained in [3], when proportional plus derivative control is applied to G(s), the closed loop system becomes unstable with small increases of the controller gain. This difficulty can be overcome by using a feedforward compensator *T(s)* to place the zeros of the augmented plant $\hat{G}(s)=G(s)+T(s)$ in the left-half plane [4] (see Figure 2). When $\hat{Y}(s)$ is fed back with gain K_2, the resulting transfer function is $G_o(s) = G(s)[1 - K_2\hat{G}(s)]^{-1}$. The poles of *G(s)*, which, under conventional output feedback would move to the right-half plane, are now shifted toward the zeros of $\hat{G}(s)$, through the application of K_2, resulting in a stable $G_o(s)$.

Now, consider the outer loop with servo compensator *S(s)* as illustrated in Figure 2. This part of the control achieves tracking of the reference signal Y_{ref}. Stability of the overall closed loop system is assured for a range of values of the gain K_1 through the stabilizing action of the block $G_o(s)$.

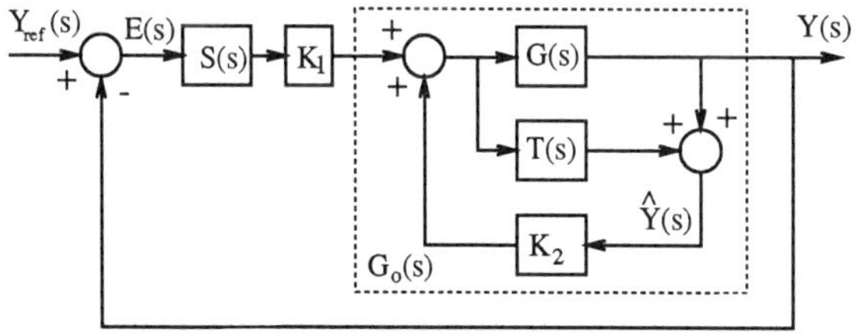

Figure 2. Block diagram of the control system.

It can be shown that the Laplace transform of the error signal is

$$E(s) = \left\{ 1 - G(s)\left[1 - K_2 \hat{G}(s) + K_1 S(s) G(s) \right]^{-1} K_1 S(s) \right\} Y_{ref}(s) \ .$$

As in the robust servomechanism problem ([5], [6]), if $y_{ref}(t)$ is a step signal, we set $S(s) = 1/s$ to ensure asymptotic tracking. The controller gains K_1 and K_2 ensure that the overall closed loop system is asymptotically stable. For the current application, we choose K_1 and K_2 such that the roots of the denominator polynomial of $E(s)$ are at acceptable locations in the left-half plane. The tip position error $e(t)$ then approaches zero as t approaches infinity with desired transient performance.

4. APPLICATIONS AND RESULTS

The nonlinear model (1) was simulated on a SUN 4/370 using MATLAB and the controller gains were computed for the configuration shown in Figure 2. Four elastic modes and one rigid-body mode were used to model the manipulator for the simulation. The parameters used for the simulation (see Table I) are those of an actual experimental manipulator constructed in our laboratory.

Table I. Model parameters.

h	1.2 m.
γ	1.13 kg./m.
I_h	0.27 kg.m.^{2}
b	0.59 Nt.m./rad.s^{-1}
EI	2.21 Nt.m.^{2}
c_1	0.6701 Nt.m./rad.s^{-1}
c_2	0.9297 Nt.m./rad.s^{-1}
c_3	0.9122 Nt.m./rad.s^{-1}
c_4	0.9129 Nt.m./rad.s^{-1}
ω_1	3.4 rad./s
ω_2	21.6 rad./s
ω_3	60.5 rad./s
ω_4	118.6 rad./s

The transfer function $T(s)$ was chosen such that the zeros of $\hat{G}(s)$ are sufficiently within the left-half plane, thus ensuring that the poles of $G_o(s)$, under the application of K_2, are adequately damped. The value of K_2 should be large enough to cause the poles of $G_o(s)$ to migrate toward the zeros of $\hat{G}(s)$.

Since the reference signal $y_{ref}(t)$ is taken to be a step function, choosing $S(s) = 1/s$ guarantees zero steady state error. The gains K_1 and K_2 are selected/adjusted so that the overall closed loop system is slightly underdamped yielding a quick response without excessive overshoot. Table II shows the pole-zero locations of transfer functions $G(s)$, $\hat{G}(s)$, $G_o(s)$ and the overall closed-loop transfer function for K_1 = 23132 and K_2 = -40.

Table II. Pole zero locations.

	ZEROS					POLES					
G(s)	67.8 ±80.1i	-49.9 ±43.1i	20.6 ±8.9i	-125.6	-7.1	-3.5 ±117.3i	-2.7 ±60.0i	-1.7 ±21.9i	-2.0 ±5.7i	-0.66	0
$\hat{G}$(s)	-3.7 ±117.3i	-2.8 ±60.0i	-7.5 ±20.7i	-8.0 ±15.0i	-12.0 ±2.0i	-3.5 ±117.3i	-2.7 ±60.0i	-1.7 ±21.9i	-2.0 ±5.7i	-0.66	0
G_o(s)	67.8 ±80.1i	-49.9 ±43.1i	20.6 ±8.9i	-125.6	-7.1	-3.7 ±117.3i	-2.8 ±60i	-7.6 ±20.8i	-7.8 ±15.0i	-11.9±2.2i	
Closed Loop	67.8 ±80.1i	-49.9 ±43.1i	20.6 ±8.9i	-125.6	-7.1	-3.7 ±117.3i	-2.9 ±59.9i	-6.5 ±23.1i	-15.0 ±10.8i	-3.7 ±7.2i	-3.93

Simulation

Figures 3 through 6 show the response of the manipulator under closed-loop control to a commanded step tip position of 20 cm. Also included is the response under LQ optimal state feedback control with output weighting (see dashed plots).

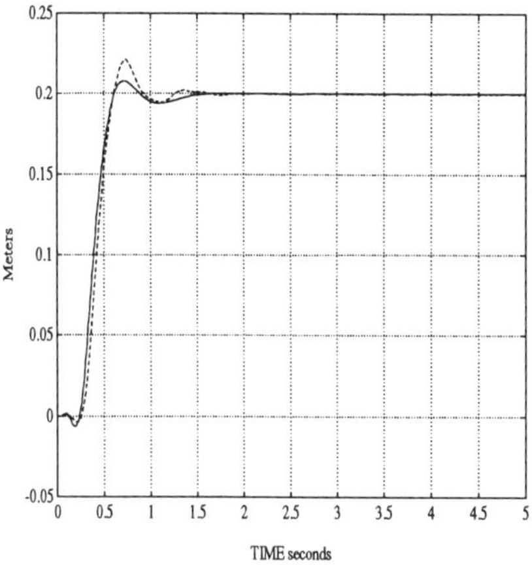

Figure 3. Net tip position *y(t)*.

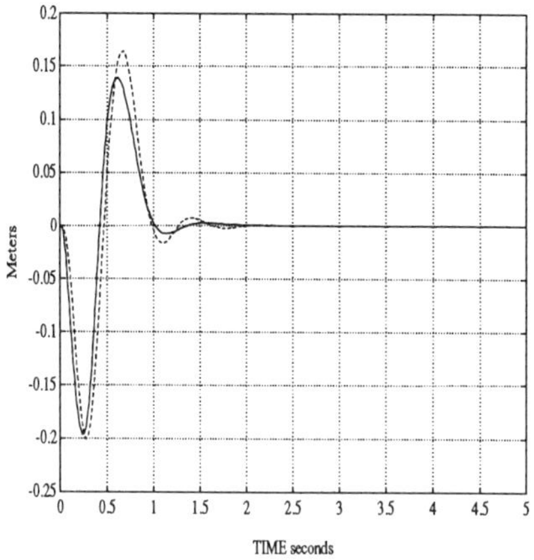

Figure 4. Tip deflection *w(x,t)*.

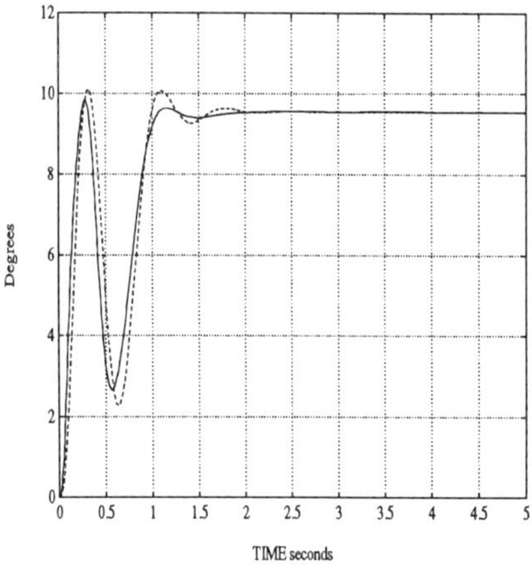

Figure 5. Hub angle $\theta(t)$.

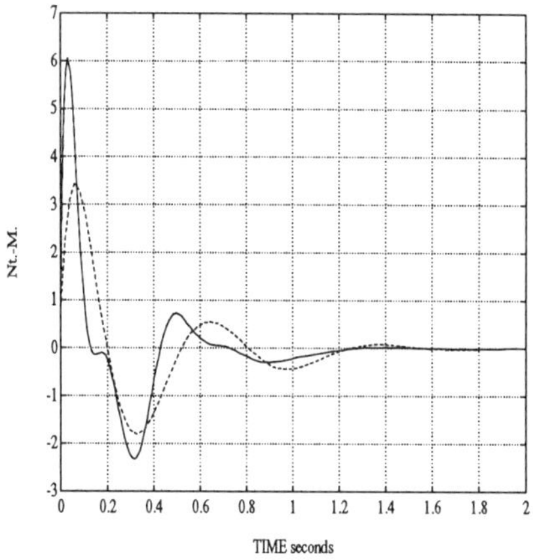

Figure 6. Control torque *T(t)*.

The flexibility of the manipulator was deliberately made large since we wish to test the ability of the controller to deal with this property. This high degree of flexibility, however, results in the significant tip deflection exhibited in Figure 4. Subjecting the system to a commanded step tip position substantially greater than 20 cm. greatly increases the deflection and violates the premise of the Euler-Bernoulli model. This problem will not occur for a practical manipulator possessing greater stiffness. In fact, the closed-loop poles for the stiffer practical manipulator could be placed even further within the left-half plane to yield an even faster response. An upper limit on the response speed would then be dictated by not only the tip deflection, but also the Coriolis and centrifugal forces. These nonlinearities are not very evident in the present simulation, but increase as the square of the hub velocity and would create problems for the linear controller.

5. CONCLUDING REMARKS

Figures 3 through 6 show that good closed-loop system response is obtained with a simple output feedback control strategy. In other words, we have succeeded in overcoming the problem caused by the non-minimum phase characteristic of the flexible link. The response compares favourably with that of the LQ controller, with the obvious advantage of not having to measure the system states. Further improvements in the control strategy can be made by using dynamic controllers in place of K_1 and K_2, eg. as obtained with H_∞ (robust) controller design techniques.

The problem of modifying the control to deal with model nonlinearities that become significant with high speed trajectories requires analyzing the zero dynamics of the nonlinear manipulator model and developing a nonlinear counterpart of $T(s)$ to handle the non-minimum phase characteristic of the system. The current work can also be used as the framework for an adaptive controller that would identify model parameters and achieve asymptotic tracking. These and other ideas are currently being investigated and will be reported in future publications.

REFERENCES

1. D. Wang and M. Vidyasagar, *"Transfer Functions for a Single Flexible Link,"* Int. J. Rob. Res., Vol. 10. No. 5, October 1991.

2. R. H. Cannon, Jr. and Eric Schmitz, *"Initial Experiments on the End-Point Control of a Flexible One-Link Robot,"* Int. J. Rob. Res., Vol. 3 No. 3, Fall 1984.

3. E.M. Nebot, G.K.F. Lee, and T. Brubaker, *"Experimental Test Bed for a Single Link Flexible Manipulator,"* American Control Conference, Vol. 3, p. 2096, 1989.

4. R.V. Patel and P. Misra, *"Transmission Zero Assignment in Linear Multivariable Systems, Part II: The General Case,"* Proc. Amer. Cont. Conf., Chicago, June 1992.

5. R.V. Patel and N. Munro, *"Multivariable System Analysis and Design,"* Pergamon Press, Oxford, 1982.

6. E.J. Davison and A. Goldenberg, *"Robust Control of a General Servomechanism Problem: The Servo Compensator,"* Automatica, Vol. 11, pp. 461-471, 1975.

AN OVERVIEW OF ROBOTIC STAGING AND INVENTORY FOR NUCLEAR MATERIALS IN MODIFIED RICHMOND IGLOOS

William D. Morse
Sandia National Laboratories

ABSTRACT

An overview is given for a robotic system that performs the automated placement, retrieval and inventory of nuclear materials in modified Richmond igloos.

INTRODUCTION

The need has arisen to maximize the staging capacity for nuclear materials in existing modified Richmond igloos [1]. The nuclear material is package in container drums that are dimensionally 20 inches in diameter and 30 inches high. In addition to maximizing staging capacity it has been paramount to simultaneously reduce personnel radiation exposures to the lowest possible level. There are two fundamental operations associated with staging of nuclear materials, namely, (1) placement & retrieval, and (2) control & accountability [2]. Both operations are accomplished robotically to eliminate personnel radiation exposure while enhancing safeguards and security.

STAGING CONFIGURATION

The modified Richmond igloo shown in Fig. 1 is a concrete structure having two sides that are approximately 40 feet deep, 12 feet 9 inches wide and 9 feet 11 inches high. The entrance doors are approximately 10 feet wide and 7 feet 9 inches high. The igloo has no environmental conditioning or internal electrical power. Floor conditions are moderate to poor with cracks and slight sloping.

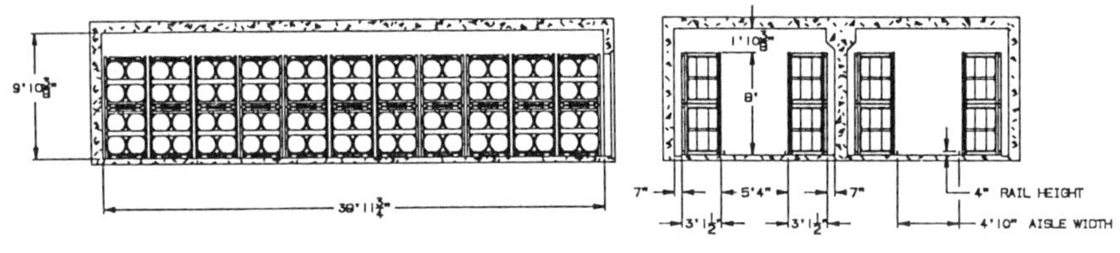

Figure 1. Modified Richmond Igloo

The drums are palletized horizontally as shown in Fig. 1 to provide direct access to each container while maximizing staging densities. The drums are palletized in groups of four or six. This configuration yields a maximum of five drums high and twenty-two deep.

AUTOMATED GUIDED VEHICLE

A narrow aisle Automated Guided Vehicle (AGV) turret lift truck executes placement and retrieval of palletized drums and also serves as a positioning platform for inventory sensor equipment [4]. Fig. 2 illustrates a man down turret lift truck capable of 180 degree fork rotation and picking up 3000 lb. loads perpendicular to the vehicles center line with a 65 inch free lift.[3].

Figure 2. Man Down Narrow Isle Turret Lift Truck

The AGV transverses a 62 inch center isle in the Richmond igloos using wire guidance in conjunction with rail guide physical barriers. The AGV is transported between igloos by a tractor and trailer. The AGV receives palletized drums at a pre-positioned pick up and delivery station adjacent to the igloo entrance. The AGV/pallet configuration provides the capability to accurately orient and position the pallets to obtain a depth of eleven pallets. Commands to the AGV are received from the Operational Control Unit (OCU) over an RF data link. The AGV has three possible modes of operation: (1) fully autonomous, (2) telerobotic, and (3) manual. The AGV control architecture shown in Fig. 3 consists of four basic segments, namely, communications, base vehicle control, fork control, and onboard intelligence. The AGV's onboard intelligence includes inherent safety features to preclude violation of any safety constraints.

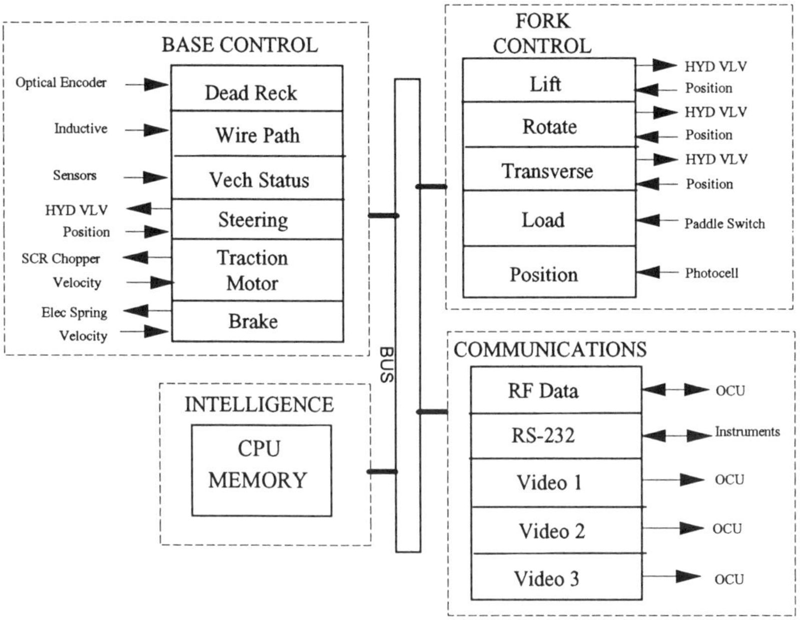

Figure 3. Automated Guided Vehicle Control Architecture

PHYSICAL INVENTORIES

Physical inventories conducted monthly require both an integrity inspection of the container drum and the collection of nuclear material identification data. To accomplish this inventory an automated sensor suite containing bar code readers and machine vision imaging equipment is integrated onboard the AGV.

Automated bar code reading is accomplished using fixed mount visible light sweep raster scanners. The scanned material identification data in conjunction with explicit location information is reconciled in real time with inventory data base records

Machine vision is used to compare reference images in real time to detect changes or anomalies. AGV accurate and consistent repositioning of imaging equipment is necessitated for pixel to pixel change detection. The AGV utilizes imaging position registration feedback for precision locating. Fig. 4 illustrates the major components and interfaces for the AGV robotic physical inventory configuration.

MEASUREMENT INVENTORIES

Gamma signature evaluation measurement inventories are conducted monthly on a random sample basis. A specially designed tungsten alloy collimator capable of shielding radiation from adjacent and surrounding materials is used to collect a gamma flux signature [5]. The collimator is integrated with the AGV for autonomous positioning and gamma signature collection. The measured gamma signature is compared in real time to the reference signature data base. The measurement equipment is integrated with the physical inventory equipment as shown in Fig. 4 allowing independent or simultaneous physical and measurement inventories.

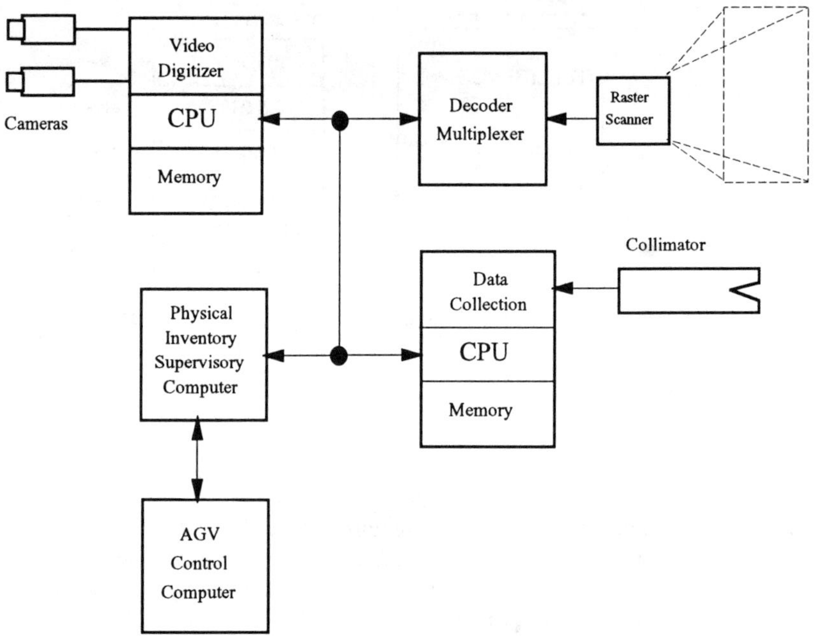

Figure 4. Physical & Measurement Inventory Configuration

REFERENCES

1. F. Martin, and B. C. Caskey, "Robotics Application for the Staging of Weapon Components," to be presented at the '92 INMM conference, Sandia National Laboratories, May 1992.
2. "Control and Accountability of Nuclear Materials," U.S. Department of Energy, Washington, DC., DOE 5633.3, June 1990.
3. Hyster Man-Down Turret Truck Specification Sheet; Model N30AH. Portland, OR: 1992.
4. P. Jackson, "Supervisory Control of AGVs In Materials Handling Computer Systems," *Proceedings of an Executive Briefing on Automated Guided Vehicles*, pp. 59-64, Nov. 1986.
5. D. J. Mitchell, H. L. Scott, K. W. Marlow, and P. E. Havey, "Radiation Signatures For Use In an Igloo Remote Inventory System Briefing," Sandia National Laboratories, Albuquerque, NM, July 1992.

AUTONOMOUS KINEMATIC CALIBRATION OF INDUSTRIAL HYDRAULIC MANIPULATORS

M. KHOSHZABAN and F. SASSANI
Department of Mechanical Engineering
The University of British Columbia, Canada

P. D. LAWRENCE
Department of Electrical Engineering
The University of British Columbia, Canada

ABSTRACT

This paper presents a new algorithm and instrumentation for kinematic parameter identification of hydraulic manipulators by forming a mobile closed-chain. Such manipulators, unlike conventional robots, may not have joint sensors to exploit in the calibration procedure. The new approach employs an extra sensing linkage, called "the calibrator". One end of the calibrator is attached to the manipulator's endpoint through a simple revolute joint while the other end has a passive ball joint attached to the ground, thus forming a closed mobile kinematic chain. The calibrator, by itself, consists of three links with three revolute joints. The three sensors attached to these joints read the relative angles of the links. Since hydraulic manipulators are usually heavy and powered, and therefore cannot be passively articulated into different configurations, a hierarchical identification algorithm using the recursive least-squares technique has been developed based on link-by-link movement of the manipulator, starting from the endpoint. The task kinematics and the offset angles of the sensors can be identified by this algorithm as well. The proposed method is fast, inexpensive, and easy to apply on site.

INTRODUCTION

Hydraulic manipulators are now being used as heavy-duty industrial robots and teleoperators in many application areas stretching from highly controlled working environments such as auto/aerospace industries and nuclear power reactors, to uncontrolled, hostile environmental conditions such as mining and forestry. Excavators and forest harvesters are two examples of such hydraulic machines being used in the latter area.

Many attempts have already been made to apply advanced control strategies to hydraulic manipulators to bring their operation under human supervision in the case of hydraulic teleoperators and/or under computer control in the case of heavy-duty industrial robots. One of the most powerful and effective control methods that may be used in this field is model-based control. However, the degree of success of this type of control strategy depends largely on a precise model for the actual system under control, including its kinematics, dynamics, and hydraulics. Even for the most sophisticated robots being used in research laboratories, one should not rely totally on nominal design specifications supplied by the manufacturers. In the case of the kinematic parameters, the differences between the actual and nominal values are mainly because of machine modifications, manufacturing tolerances, mounting errors, link/joint flexibility, backlash, temperature changes, and wear. Also, hydraulic manipulators are prone to severe damage, and their links and related parts need to be replaced frequently, which in turn, demands recalibration.

Investigations have shown that about 90% of the end effector position and orientation errors are because of the lack of knowledge about actual geometric values of the kinematic parameters (see e.g., [9]). Therefore, our primary interest in this paper is to present an efficient cost-effective method to calibrate these geometrical parameters for hydraulic manipulators. Here, we have used a robot model with modified Denavit-Hartenberg parameters to represent the actual manipulator, and a recursive least squares technique to identify the true kinematic parameters.

PROBLEM STATEMENT

Previous work in the area of kinematic calibration done by researchers mainly concentrates on moving the arms and sensing the joint angles as well as the end effector positions and orientations [1], [2], [3], and [4]. However, this method requires two different types of sensors and related equipment. The endpoint position sensors are not usually as accurate as joint angle sensors (e.g., stereo camera systems), or if they are, their prices are relatively high and their workspace is limited (e.g., laser tracking systems). Also, such sensors require precalibration (e.g., IREDs), a controlled environment (preferably, a laboratory), and some significant technical experties. They are not usually built to be used in outdoor conditions. Finally, since the principle behind their method of operation is different, the end effector position and orientation sensors are inconsistent with those of the angular positions. As a result, their random noise signature as well as some other deterministic error characteristics (e.g., caused by temperature changes, etc.) may be quite dissimilar.

The calibration method presented here, in contrast, is based on some recent work done by Bennett and Hollerbach [5]. *"If a manipulator is formed into a mobile closed kinematic chain, then its joint angle readings alone are enough to identify the kinematic parameters"* [5] providing that it has sufficient degrees of freedom (DOF's). This method appears to eliminate many of the above-mentioned problems if only we have a redundant manipulator with respect to the task's DOF's.

Many conventional hydraulic manipulators, such as excavators, usually have less than 6 DOF's. As an example, for a 4–DOF manipulator, we can have a point contact at its endpoint to close the chain. In fact, if n is the number of DOF's for the manipulator, and t is the number of task DOF's, then in order to form a mobile closed kinematic chain, we should have $n+t>6$, or in this case: $4+3=7>6$. In general, the degree of mobility in the chain is defined by $n+r-6$, which determines the number of powered joints in the closed-chain system.

Unlike open-chains, the joint movements in a mobile closed-chain are restricted; therefore, all other joints, except the powered ones in the closed-chain, should be set free to move arbitrarily. Unfortunately, this cannot be easily done in the case of a typical hydraulic manipulator even if its closed-chain has more than one degree of freedom. Furthermore, many conventional hydraulic manipulators may not have any joint angle sensors at all, or at least, at the time of calibration. The reason is that many of them have already been built for simple manipulation being controlled manually by a human being, or may utilize some other types of sensors (like a model-based vision system proposed in [6]). In every case, we have to modify the whole configuration of the system as well as the identification procedure in order to be able to use the closed kinematic chain approach.

The contribution of our work in the area of kinematic calibration is composed of two main points:

1. A new measurement device has been introduced to be attached to the endpoint of an existing hydraulic manipulator (an excavator-based log-loader); this device is an extra kinematic linkage with revolute joints having simple resolvers attached to its joints. We refer to it as *the calibrator*. The other end of our calibrator is attached to a passive spherical joint fixed to the ground or to the machine's chassis. The calibrator has two functions. First, to add a number of extra DOF's to the existing configuration in order to facilitate the movement of each hydraulic link while all other original joints are naturally fixed. Second, to provide the *necessary and sufficient* sensory data for identification purposes.
2. A new hierarchical calibration algorithm consistent with our calibrator characteristics has been developed based on link-by-link movement of the manipulator. In fact, at each set of movements, we have a new four-bar linkage to identify. The calibration procedure starts from the last link of the original manipulator (end effector), including the identification of task and calibrator kinematics, then proceeds to the next lower link, down to the base link.

In this way, we could solve the problem of hydraulic manipulator identification through a simple, fast, and economical method while providing sufficiently accurate results for model based control purposes.

KINEMATIC MODEL

The combined hydraulic manipulator and calibrator now forms a conventional robot with the exception that its endpoint (the ball joint) now is fixed to the ground. A modified version of Denavit-Hartenberg parameter representation, FIG. 1, has been chosen as the basis for our calibration procedure. The kinematics to be identified consist of four classical D-H parameters, θ_i, α_i, a_i, and d_i, plus the link twist angle β_i around the y_i axis [7]. From the nongeometric parameters, we only model the joint offsets of the sensors, θ^{off}.

The twist angle β_i is introduced because, for example, for two successive rotational parallel joints, small geometrical errors lead to large variations of the parameters. Without the twist angle β_i, we had to modify the parameters a_i and d_i for compensation of any nonparallelism between two successive joints even if these two parameters were initially correct. It is worthwhile to note that we use β_i only when we have two nominally parallel joints. In this case, we assume that the parameter d_i is error free, therefore, we use its nominal value. In other cases, β_i is simply set to zero.

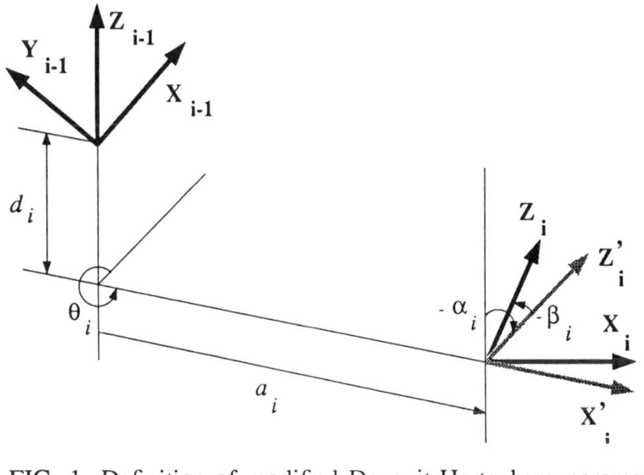

FIG. 1. Definition of modified Denavit-Hartenberg parameters

Having defined the kinematic parameters according to FIG. 1, we can construct a homogeneous transformation matrix as

$$
{}^{i-1}A_i = \begin{bmatrix} 1 & 0 & 0 & 0 \\ 0 & 1 & 0 & 0 \\ 0 & 0 & 1 & d_i \\ 0 & 0 & 0 & 1 \end{bmatrix} \begin{bmatrix} \cos\theta_i & -\sin\theta_i & 0 & 0 \\ \sin\theta_i & \cos\theta_i & 0 & 0 \\ 0 & 0 & 1 & 0 \\ 0 & 0 & 0 & 1 \end{bmatrix} \begin{bmatrix} 1 & 0 & 0 & a_i \\ 0 & 1 & 0 & 0 \\ 0 & 0 & 1 & 0 \\ 0 & 0 & 0 & 1 \end{bmatrix} \times
$$

$$
\begin{bmatrix} 1 & 0 & 0 & 0 \\ 0 & \cos\alpha_i & -\sin\alpha_i & 0 \\ 0 & \sin\alpha_i & \cos\alpha_i & 0 \\ 0 & 0 & 0 & 1 \end{bmatrix} \begin{bmatrix} \cos\beta_i & 0 & \sin\beta_i & 0 \\ 0 & 1 & 0 & 0 \\ -\sin\beta_i & 0 & \cos\beta_i & 0 \\ 0 & 0 & 0 & 1 \end{bmatrix} \tag{1}
$$

MEASUREMENT PROCEDURE

Suppose that there exist n powered but unsensed joints in a hydraulic manipulator. By moving only one joint at a time while freezing the others, we will get 1 DOF for the manipulator. For a passive point-task, the position is fixed and the orientation is unknown; therefore, $t=3$. Now, in order to form a one or more DOF's for the closed-chain, we need some extra linkage, with r DOF's, attached to the manipulator and the point-task. r must still satisfy the mobility equation $(1+r)+t>6$, or $r \geq 3$. The new configuration, on the whole, will consist of $n+r$ links with the $(n+r)$th link fixed to the ground.

Figure 2 shows a typical hydraulic manipulator with a three-link calibrator attached to its endpoint (bucket). The calibrator itself has three joint-angle sensors with no other sensor attached to the original manipulator. The measurement procedure starts from the outermost link, the bucket in FIG. 2. By moving the bucket and fixing all other links, different configurations for the calibrator and the bucket will be obtained and m_n different angles through the sensors can be measured accordingly. The geometric parameters for that link, along with the offset angles of the sensors and the task kinematics, may then be identified. Since the endpoint movements at this stage always remain on a single plane, we are not supplied with sufficient information to identify all kinematic parameters for the calibrator. However, this is not really a problem because the calibrator, itself, may be precalibrated very accurately in the laboratory.

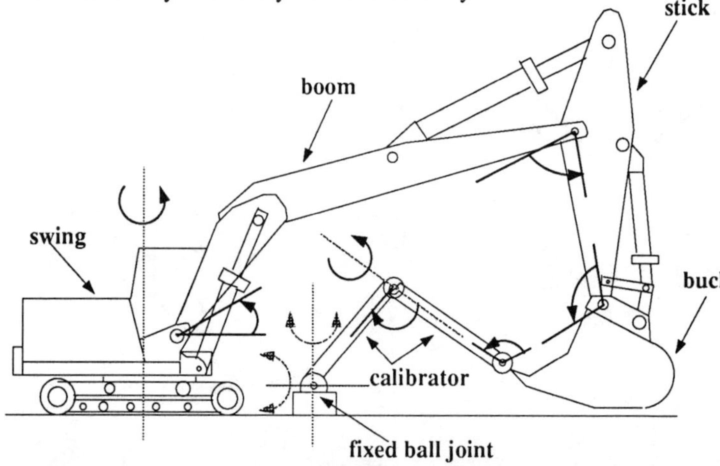

FIG. 2. Schematic of a typical hydraulic manipulator with the attached "calibrator".

The data acquisition procedure continues by moving the successive inner links one by one while fixing all other links. For each link i (where $i=1.....n$), m_i measurement points would be obtained in this way.

IDENTIFICATION TECHNIQUE

The direct kinematic model gives the task's position and orientation vectors with respect to the coordinate frame i-1 which is attached to the ith joint; i.e., ${}^{i-1}\mathbf{p}_{n+r}$ and ${}^{i-1}\mathbf{o}_{n+r}$ respectively. But, the position of the task (ball-joint) is fixed and its orientation is unknown (unsensed); therefore, only the three position equations can be exploited in the identification process. At a certain measurement point j_i, for the mobile closed-subchain formed from link i to link $n+r$, we may write

$$\left[{}^{i-1}\mathrm{p}_{n+r},\ 1\right]^T = {}^{i-1}A_i\left[{}^{i}\mathrm{p}_{n+r},\ 1\right]^T \tag{2}$$

where ${}^{i-1}A_i$ is a 4×4 homogeneous transformation matrix defined in eq. (1). Now, if we expand the position vector ${}^{i}\mathbf{p}_{n+r}$ in terms of the homogeneous matrices A's, we will have

$$\left[{}^{i-1}\mathrm{p}_{n+r},\ 1\right]^T = {}^{i-1}A_i\left(\prod_{k=i+1}^{n+r} {}^{k-1}A_k\right)[0,\ 0,\ 0,\ 1]^T \tag{3}$$

${}^{i-1}A_i$ contains the unknown modified D-H kinematic parameters for the ith link which is currently moving, while ${}^{k-1}A_k$ matrices have some known and unknown parameters from the very last link $n+r$ to link $i+1$.

For example, if the process begins from the machine's outermost link ($i=n$), then we will have α_n, a_n, d_n (or β_n) in ${}^{n-1}A_n$ matrix and also $\theta^{off}{}_{n+1}$ to $\theta^{off}{}_{n+r}$ in ${}^{k-1}A_k$ matrices to identify. But the joint angles for the calibrator, θ_{n+1} to θ_{n+r}, are all measured through the sensors while θ_i in matrix ${}^{i-1}A_i$ is still unmeasured (unsensed). However, for the jth measurement point, we realize that it only appears in the x and y components of the position vector ${}^{i-1}\mathbf{p}_{n+r}$. Now, maintaining the first two homogeneous matrices in the right hand side of eq. (1) and multiplying the remaining terms of eq. (3) to obtain the general expressions q_x, q_y, and q_z, we will get the three components of the position vector $\mathbf{p}$ in terms of θ_i as

$${}^{i-1}p_{x_{n+r}} = q_x.\cos\theta_i - q_y.\sin\theta_i \tag{4}$$

$${}^{i-1}p_{y_{n+r}} = q_x.\sin\theta_i + q_y.\cos\theta_i \tag{5}$$

$${}^{i-1}p_{z_{n+r}} = q_z + d_i \tag{6}$$

To eliminate θ_i, we square and add eqs. (4) and (5) to obtain a simple relation for the j_ith measurement point, as

$${}^{i-1}p^2_{x_{n+r}} + {}^{i-1}p^2_{y_{n+r}} = q_x^2 + q_y^2 \tag{7}$$

Equations (6) and (7) may be used in the identification procedure. In eq. (6), d_i is not independent of p_z; only their difference can be identified. Also, from eq. (7), only the norm of the position vector ${}^{i-1}\mathbf{p}_{n+r}$ in xy plane may be identified at the current ith identification stage.

For the next inner link ($i=n-1$), we have α_i, a_i,, and, in the case of having two successive parallel joints, β_n to identify. It is not necessary to recalibrate the sensors' offset angles since they have already been identified. The angle of joint i (i.e., θ_i) which connects link i-1 to link i is now fixed but unknown. Also offset length parameter d_i for link i is still to be identified. This procedure can be continued down to the first link. After calibrating the whole kinematics of the manipulator, we may identify the task kinematics with respect to the manipulator's base frame 0.

To have better estimates of the vector of the unknown D-H parameters, which we call it ϕ, especially for removing the influence of random noise from ϕ, it is better to have as many data points as possible, preferably from the whole workspace of the manipulator. Then, to solve for the unknown parameters in (6) and (7), an iterative least-squares technique may be employed since we are dealing with a nonlinear estimation problem. We do not intend to recall those techniques here again. The interested reader may find a number of good papers exist in the literature (e.g., [2], [3], [4], [5]). Rewriting eqs. (6) and (7) in a general vector form as

$$x = f\left(\theta, \theta^{off}, \phi, p\right) = 0 \tag{8}$$

and taking the differential of x with respect to the variations in link, calibrator, and task parameters, we obtain

$$\Delta x = \left[\frac{\partial f}{\partial \theta} \; \frac{\partial f}{\partial \theta^{off}} \; \frac{\partial f}{\partial \phi} \; \frac{\partial f}{\partial p}\right] \left[\Delta\theta \;\; \Delta\theta^{off} \;\; \Delta\phi \;\; \Delta p\right]^T = J\,\Delta\Phi \tag{9}$$

where J is the general Jacobian matrix. To evaluate J, one may use a symbolic differentiation program or simply apply a numerical routine using a finite difference method. The other option is to use simple vector relationships like the one proposed in [8]. Then, using the least squares, we come up with an estimate of parameter errors

$$\Delta\Phi = \left(J^T J\right)^{-1} J^T\,\Delta x \tag{10}$$

The parameter estimates are continuously updated as $\Phi_{new}=\Phi_{old}+\Delta\Phi$, and the iteration continues until $\Delta x \rightarrow 0$.

SIMULATION RESULTS AND DISCUSSION

We have applied the proposed identification algorithm to the simulated model of a Caterpillar 215B excavator attached to a three-link calibrator. The base coordinate frame 0 is attached to the first link's, i.e., to the swing's in Figure 2; therefore, θ_1^{off} and d_1 are taken as arbitrary constants, say 0. Including the calibrator kinematics, we have seven links to identify as well as the position vector of the ball-joint with respect to the origin of the base frame. As was mentioned earlier, we cannot identify all the kinematic parameters of the calibrator because we do not have sufficient excitation directions when moving only one link at a time and keeping the other links fixed. In fact, the movement of each link is limited to a single plane. However, the calibrator can be built and precalibrated with sufficient accuracy. In every case, the orientation of the last link, i.e., α_{n+r} is arbitrarily chosen because it is not attached to any further link. Finally, we need to specify at least one length parameter to scale the size of the chain [5], e.g., a_{n+r} or d_{n+r-1}, either of them belongs to the calibrator and is known in advance.

Table I gives the identified versus actual D-H parameters for the combined excavator and calibrator configuration. The arbitrary constant parameters are marked with an asterisk. The upper number in each cell is the actual parameter while the number in the denominator is the identified parameter value. For the calibrator, θ_5, θ_6, and $\theta_7$are actually the sensor offset angles from which, only θ_5 (or θ_5^{off}) has been attempted to be identified here since the other offset angles as well as the D-H parameters for the calibrator can be assumed to be known in advance.

For this simulation, it is assumed that the sensors are noisy with the noise level of ±0.05 degrees; yet, the results obtained are in good agreement with the actual values although the calibration errors propagate from the outermost link to the innermost one. d_2 is realized to have the maximum relative error of 6% among the parameters.

TABLE I: The Actual versus Identified D-H parameters

	manipulator				calibrator		
link #	1	2	3	4	5	6	7
θ (deg)	0.0* / 0.0	-30.0 / -29.8	90.0 / 89.7	30.0 / 30.1	90.0‡ / 90.2	0.0† / 0.0	90.0† / 90.0
α (deg)	-90.0 / -90.0	0.0 / 0.0	0.0 / -0.1	45.0 / 45.0	90.0† / 90.0	90.0† / 90.0	0.0* / 0.0
β (deg)	0.0† / 0.0	0.0 / 0.2	0.0 / 0.1	0.0 / 0.1	0.0† / 0.0	0.0† / 0.0	0.0* / 0.0
a (m)	0.508 / 0.519	5.190 / 5.178	1.800 / 1.792	1.143 / 1.142	0.0† / 0.0	0.0† / 0.0	3.500† / 3.500
d (m)	0.0* / 0.0	-.127 / -.135	0.0† / 0.0	0.0† / 0.0	0.0† / 0.0	3.000† / 3.000	0.083† / 0.083

*: arbitrary parameters. †: not identified. ‡: offset angles of the sensors.

In setting up the table, a number of different initial conditions were tried, but we always ended up with the same results shown in the table, indicating that a global minimum has been reached.

CONCLUSIONS AND FUTURE WORK

We have presented an off-line calibration algorithm to identify the geometric parameters of a hydraulic manipulator by attaching its endpoint to an external linkage and forming a mobile closed kinematic chain. The sensory data all comes from the attachment (calibrator).

One inherent drawback for the presented hierarchical identification technique is that the error found in the identification of outer links propagates towards the inner links. Therefore, as the identification proceeds to the next links, the parameters found become less accurate. One way to reduce these errors is using more accurate and less noisy sensors for the calibrator.

We have also simulated the technique for a number of various calibrator configurations as well as different task positions in the machine's physical workspace. We have found that for obtaining the most accurate results, there exists a limited subspace in which the ranges of the sensor readings would be optimum. Finding this subspace might be a challenging issue.

We are building a suitable calibrator and are going to conduct a number of experiments on a hydraulic machine (CAT 215B) at UBC to verify the proposed method.

ACKNOWLEDGMENTS

The authors would like to acknowledge the support of NSERC Center of Excellence funding (IRIS-PRECARN Project C6) for this work.

REFERENCES

1. S.A. Hayati, "Robot Arm Geometric Link Parameter Estimation," in *Proc. 22nd IEEE Conf. Decision Cont.*, (San Antonio, TX, Dec. 1983).
2. W.K. Veitschegger and C.H. Wu, "A Method for Calibrating and Compensating Robot Kinematic Error," in *Proc. IEEE Int. Conf. Robotics Automat.*, (Raleigh, NC, 1987).
3. C.H. An, C.G. Atkeson, and J.M. Hollerbach, *Model-Based Control of a Robot Manipulator* (MIT Press, Cambridge, MA, 1988).
4. J.M. Render, E. Rossignol, and M. Becquet, "Kinematic Calibration and Geometrical Parameter Identification for Robots," *IEEE Tran. Robotics Automat.*, **7**(6), 721–732, (Dec. 1991).
5. D.J. Bennett and J.M. Hollerbach, "Autonomous Calibration of Single-Loop Closed Kinematic Chains Formed by Manipulators with Passive Endpoint Constraints," *IEEE Trans. Robotics Automat.*, **7**(5), 597–606, (Oct. 1991).
6. I.J. Mulligan, A.K. Mackworth, and P.D. Lawrence, "A Model-Based Vision System for Manipulator Position Sensing," in *Proc. workshop on Interpretation of 3D Scenes*, (Austin, TX, Nov. 1989).
7. W.K. Veitschegger and C.H. Wu, "Robot Accuracy Analysis Based on Kinematics," *IEEE J. Robotics Automat.*, Vol. RA-2, (Sep. 1986).
8. D.E. Whitney, "The Mathematics of Coordinated Control of Prosthetic Arms and Manipulators," *ASME J. Dynamic Systems, Meas., Control*, 303–309, (1972).
9. D.E. Whitney, C.A. Lozinski, and J.M. Rourke, "Industrial Robot Forward Calibration Method and Results," *ASME J. Dynamic Systems, Meas., Control*, **108**, 1–8, (Mar. 1986).

HAZBOT: A HAZARDOUS MATERIALS EMERGENCY RESPONSE MOBILE ROBOT

H. W. Stone and G. Edmonds

Jet Propulsion Laboratory/California Institute of Technology
4800 Oak Grove Drive
Pasadena, CA 91109

ABSTRACT

This paper describes the progress that has been made towards the development of a mobile robot that can be used by hazardous materials emergency response teams to perform a variety of tasks including incident localization and characterization, hazardous material identification/classification, site surveillance and monitoring, and ultimately incident mitigation. In this application the use of robotic technologies, and in particular telerobotics, offers the opportunity to greatly reduce the risks of injury and loss of life by providing an alternative means for initially entering a hazardous site.

1 INTRODUCTION

This paper presents an overview of the design and implementation of a mobile robot system for responding to emergencies involving hazardous materials spills/releases The vehicle called HAZBOT II is the focus of a multi-year project which also involves members of the JPL Fire Department HAZMAT Team. HAZBOT's primary function will be to localize, characterize, identify, and mitigate hazardous chemical spills under emergency conditions.

The development of HAZBOT II has been motivated by:

- The dangers involved in responding to unknown or partially unknown chemical hazards and, in particular, the significantly increased risks of injury and loss of life to personnel involved in dealing with such incidents.
- The fact that remotely-controlled robots have the inherent ability to perform operations despite the presence of hazardous materials.
- The significant time required for human entry teams to prepare and suitup for initial entry into potentially hazardous sites. Preparation often takes 45 minutes.
- The limitations upon entry times by humans using self contained breathing apparatus (SCBA) and cumbersome protective equipment (e.g., flash suits, chemical suits, turn coats, etc.). Entry times are often limited to 10-20 minutes.
- The opportunity for remotely-operated vehicles to perform limited hazard abatement and remediation during initial entry.

A goal of this project will be to develop a working system within a few short years and to use it in actual HAZMAT incidents at JPL. To accomplish this two Andros [1] mobile robots were initially procured. One is being maintained as a working system while the other undergoes upgrades in response to system needs and requirements. Examples of other robotic systems which have been developed for dealing with hazardous materials are described in [?, ?, ?, ?, ?].

The paper is organized as follows. In Section 2, we describe the HAZBOT II system components and their function, making reference to the capabilities inherited from the original Andros system. Section ?? lists the system requirements derived from user inputs as well as results from experiments performed with JPL Fire Department HAZMAT team operators.

[1]Manufactured by Remotec Corporation, Oak Ridge, TN, USA.

Future plans are presented in Section ??. Section ?? summarizing our earlier accomplishments.

2 SYSTEM DESCRIPTION

HAZBOT II, depicted in Figure 1, consists of three basic elements:

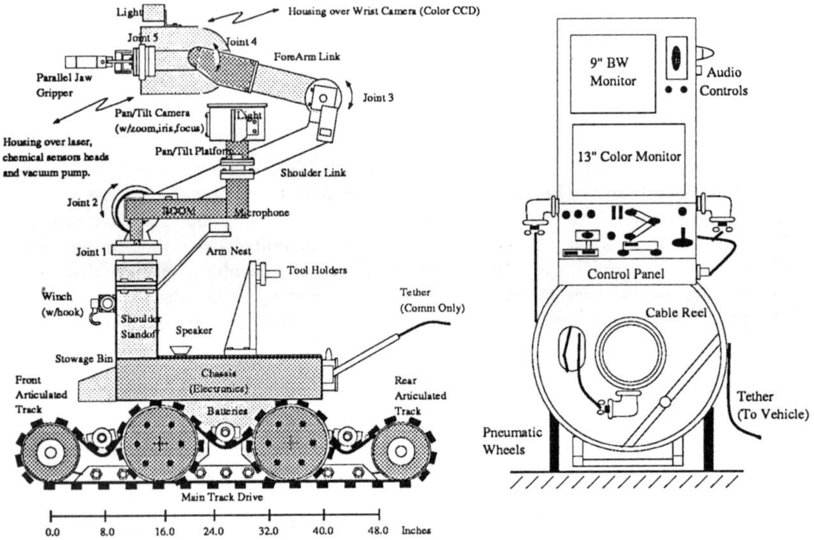

Figure 1: Schematic of HAZBOT II vehicle and control station

- An articulated, tracked mobile platform (i.e., chassis).
- A five degree-of-freedom robotic manipulator with a parallel jaw, two fingered gripper.
- A mobile control station with a detachable control panel, video monitors, and tether deployment reel.

2.1 Mobile Platform

HAZBOT II's mobile platform is driven by two sets of tracks, one on either side of the chassis. The track drive system is unique in that it also has a pair of front and rear articulated tracks which can be raised and lowered to vary the vehicle's "footprint" enabling it to climb stairs, over obstacles, and across ditches. By positioning the articulated tracks in their extreme position, vertically downward, the vehicle "stands on its toes" thereby greatly increasing the manipulator's vertical reach.

We have found the articulated tracks very useful in developing strategies for opening locked doors and other tasks. The tracks used in combination with the manipulator arm permit relatively dexterous manipulation tasks to be performed. To further this feature the original control system was modified to allow the operator to control the speed at which the front and rear tracks are raised/lowered.

The vehicle chassis contains most of the system's onboard electronics, including a microprocessor, pulse width modulated (PWM) amplifiers, video amplifiers, relay banks, audio amplifier, and the electronics portion of the combustible gas indicator and oxygen sensor. The chassis also contains the batteries which supply power for the entire vehicle, including manipulator and sensors. The chassis, which serves as an equipment platform, can carry approximately 100 lbs..

2.2 Manipulator Subsystem

The manipulator subsystem consists of a five degree-of-freedom manipulator and a parallel arc, two fingered gripper. The manipulator has a maximum reach of approximately 6 ft and is mounted on the front portion of the chassis with a 12 in. high standoff link. The standoff, not found on the original vehicle, increases the manipulator's vertical reach; improves the viewing angle obtained from the pan/tilt camera; provides space on the chassis for stowing tools; and, provides internal space to mount sensor electronics. The manipulator can reach and grasp objects on the floor in front of and to the sides of the chassis. It can also "fold" over backwards to retrieve tools from the top of its chassis or manipulate its tether. With the standoff and the ability to stand on its toes, the manipulator can reach 9 ft., enabling it to inspect and operate ceiling valves and similarly suspended equipment. Finally, the manipulator is capable of lifting approximately 100 lbs. at 3 ft. from the manipulator's base and 30 lbs. at 6 ft..

The manipulator subsystem components and controls are relatively simple. The individual manipulator joints are each driven by a permanent magnet, DC motor with gear head. The motors are rate controlled via the use of PWM amplifiers and switching circuits. The PWM amplifiers receive analog command signals from the onboard computer and custom I/O hardware. The mode of cnontrol is open-loop since the actuators/joints are currently void of position and/or velocity encoders.

The only significant drawback to the lack of joint encoders is that the operator has no means, other than visual feedback, for determining the manipulators configuration. Consequently, the operator must often use the onboard cameras to look at the manipulator as opposed to hazards. Except for this, open-loop rate control has been adequate.

The final component of the manipulator subsystem is a parallel jaw, two fingered gripper. The gripper can produce a squeeze force of up to 50 lbs. with a maximum finger separation is 8.5 in..

2.3 Control Station

The operator control station contains several elements mounted on a wheeled hand cart so that it can be easily deployed. The control station contains a 13 in. color video monitor; a 9 in. B/W video monitor; a 380 ft. tether; a removable control panel; and, a microphone/speaker for two way audio communications.

The control panel represents a complete redesign of the original commercially-supplied control panel. The original panel was found to be very confusing during experimentation with JPL HAZMAT Team members and often lead to undesirable and/or unintended motions of the manipulator and articulated tracks. A photograph of the current control panel are shown in Figure 2.

There are two basic differences between the original and current control panel layout/operation. First, a graphical side view image of the entire vehicle has been silk screened onto the control panel and the input devices (i.e., joysticks) for each of the vehicle actuators have been physically placed at the corresponding position of the actual joint location on the image. Using this very simple layout technique, it is obvious which joystick controls which joint. Second, the original switches used to "turn-on" a joint and specify the direction of motion have been replaced by spring loaded joysticks providing a signal proportional to the desired speed.

A combination of rotary joysticks and single axis pivot style joysticks have been used in the control panel. The style of the joystick used for a particular joint was a function of the orientation of the physical joint axis as depicted in the graphical image on the panel. For example, the shoulder joint is actuated via a rotary joystick (i.e., spring loaded rotary knob) whose axis of rotation is perpendicular to the control panel. This matches physically with the actual shoulder joint as depicted in the panel image.

To implement the above, a custom designed electronics board was designed to interface signals from the joysticks and switches to the existing control station I/O board and computer. (A limited budget precluded us from replacing the existing electronics and computer

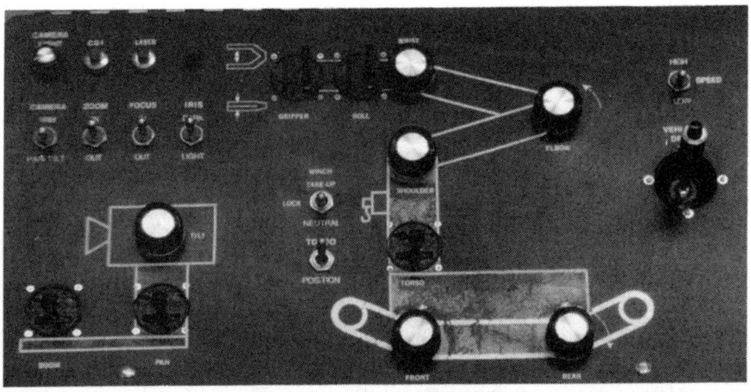

Figure 2: Layout of the current control panel.

at the time). Since the original system was capable of transmitting only one digitized analog speed setting, the new interface board was responsible for detecting which joystick had been activated, converting it into a binary directional signal for the corresponding joint, and multiplexing the analog joystick signals. It also served to map the new control panel functions into the wiring and logic of the original functions.

While the changes to the control panel are clearly not monumental in terms of electronic technology it reflects the fact that the users in this application are likely to have little if any prior experience or training in the operation of teleoperated multi-degree-of-freedom systems. During testing, operators from the JPL Fire Department HAZMAT team found the new control panel layout to be far superior to the original control panel. To them a panel layout of this type is easy to understand and much more intuitive.

The control panel also includes a two-axis joystick used to provide drive input commands (a button on the joystick serves to release the drive system brakes); switches to control the pan/tilt camera's zoom lens, focus, and iris; a switch to control the monitors, a switch to control the onboard lights, a switch to engage/disengage/lock a small winch; a switch to activate a wrist mounted laser (used to provide visual depth cues); and, a switch to activate the on-board chemical sensors.

Inputs from all of the above devices are collected and processed by custom hardware and input to a microprocessor, all of which is housed in a detachable control panel box. This microprocessor is responsible for packetizing the input data and shipping it to the vehicle's onboard microprocessor over a 1200 baud serial link in the tether.

2.4 Vision and Sensing

HAZBOT II is equipped with several sensors including two color CCD cameras, a two way audio system, a combustible gas indicator, and an oxygen level detector.

The CCD cameras are the primary means by which the operator can view the surrounding environment, locate objects, navigate, and position the end-effector. The first of the two cameras is equipped with a remotely-controlled zoom lens, focus, and iris. This camera is mounted on a pan/tilt platform attached to the torso via a motor driven boom. The boom can be positioned to optimally place the camera during critical tasks. This ability to independently position this camera has signficantly enhanced the operator's ability to control task execution and has made tasks like opening locked doors much easier and more reliable.

A second color CCD camera is mounted on the manipulator's wrist just behind the gripper's roll axis. This camera primarily supports manipulation but is also useful in navigating. The main limitation of this camera is that it looks almost directly down the back of the fingers. Consequently, it is often very difficult to judge the distance to an object. For operating in dimly lit environments, both cameras are equipped with halogen lamps whose illumination level is controllable via the operator.

To increase depth perception, a small 3.0 milliwatt, silicon-based "bullet" laser was mounted below the wrist link (see Figure 1) to provide a passive depth cue. The laser is mounted such that it projects a beam of light from a point below the wrist through the line which joins the tips of the two gripper fingers as illustrated in Figure 3. The laser is also equipped with a lens

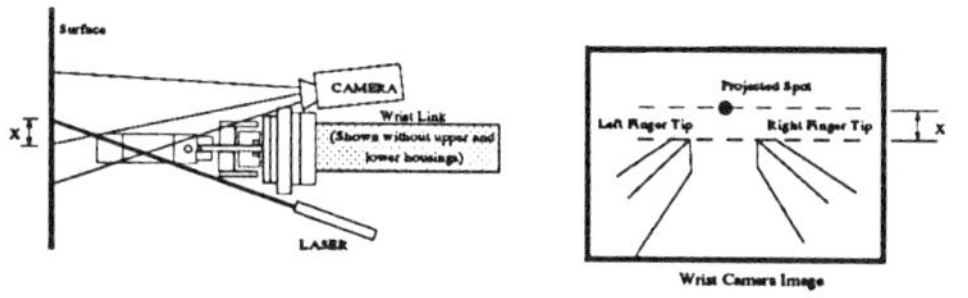

Figure 3: Geometry of the wrist camera's optical axis, the gripper fingers, and the laser beam.

for adjusting beam dispersion.

This setup provides a very simple form of structured lighting. When the gripper approaches an object, for example, a wall the laser will project a dot onto the wall which will appear to be above the fingers when viewed by the wrist camera. As the gripper moves toward the wall, the projected dot will move towards the centerline of the fingers. At the point at which the fingers are touching the wall, the dot will appear to intersect with the centerline between the finger tips. The movement of the dot relative to the fingers thus allows the operator to judge the distance to an object. During tests, operators from the JPL Fire Department HAZMAT team have found the laser to be extremely helpful. In particular, it increased the reliability and accuracy with which the operator can position the finger tips, which contain the chemical sensor's sampling probe, within close proximity (e.g., 1.0 inch) to a door jamb without hitting it.

The vehicle also has a two way audio system that allows the operator to communicate with potential victims. Audio signals from the combustible gas indicator and oxygen detector are also transmitted to the operator via this system. Audio feedback can also be used to determine whether the articulated tracks have touched the ground; to indicate whether or not the manipulator is touching an object; to listen for gas leaks; and to detect and locate explosions or violent chemical reactions. The use of audio alarms is very common among HAZMAT detection equipment. While we have not yet performed in-depth experiments into performance with and without audio feedback, our experiences indicate that it has much to offer in terms of aiding the operator during task execution.

A unique feature of HAZBOT II is its ability to detect the presence of combustible vapors and the level of oxygen in the surrounding atmosphere. A commercially-available combustible gas indicator with an oxygen sensor, integrated into the vehicle, forms the basis of this capability. To acquire air samples, the sensor's probe is machined directly into one of the gripper fingers. The oxygen cell and the combustible sensor head, along with the vacuum pump, are placed in a protective housing along with the laser, mounted to the underside of Link 4 (see Figure 1). A control panel on the front of manipulator standoff allows the operator to calibrate the sensors prior to vehicle deployment. The sensors have two types of output. One is an analog meter which reads in parts per million for combustible gases or % O_2. The other is an audio alarm which signals when preset thresholds are exceeded. The audio signals are automatically fed back to the operator via the onboard microphone. The analog reading can be viewed from the pan/tilt camera.

There are several important features about the scheme used to integrate these sensors. First, since the sensor probe is built into the finger tip, it detects vapors at any operator specified location. Heavy vapors near the ground can be detected just as readily as vapors rising to the ceiling or escaping through a door jamb. Second, the probe's permanent placement on the finger tips increases efficiency by eliminating the need repeatedly grasp the probe. Other useful tasks can also be performed with the gripper while simultaneously monitoring the atmosphere. This arrangement also makes use of a custom designed vacuum slip ring assembly which enables air samples to be drawn across the gripper's roll joint. The slip ring eliminates the need for externally mounted tubes and allows the gripper to rotate continuously.

3 SYSTEM REQUIREMENTS

The following is a list of fundamental system requirements obtained through numerous discussions with users and data from experiments conducted with the JPL Fire Department HAZMAT team. (A comprehensive discussion of these experiments is beyond the scope of this paper.) The requirements are:

- The ability to safely enter confined spaces containing toxic, explosive, and/or flammable vapors.
- The ability to navigate through environments containing stairways, ramps, etc.
- The ability to activate, deploy, and transmit data from chemical sensors.
- An interface which is intuitive to the operator.
- Limited need for the operator to perform complex spatial reasoning. (HAZMAT personnel typically have little, if any, prior experience in remotely operating complex multi degree-of-freedom articulated systems.)
- An effective means for perceiving depth.
- The ability to perform basic manipulation tasks including, but not limited to: the closure of butterfly valves, "K" bottle valves, and chain and yoke valves; stem and yoke valves, the opening and unlocking of doors; the opening and closing of fume hoods; and the manipulation of glassware (e.g., beakers, bottles, etc.).

The incorporation of technologies to meet the above requirements is the overall goal of this project. Several of these requirements have already been addressed, although not necessarily completely solved, by modifications that have been made to the vehicle to-date. In the following section we highlight the overall project plans.

4 FUTURE PLANS

The project is currently divided into two phases. Phase I will focus upon the development of technologies/capabilities to support teleoperations. In Phase II, will focus on the integration of specific sensor-based autonomous capabilities to assist the operator in coordinating system functions such as the automatic retrieval of tools. Other autonomous capabilities may also be required to control functions which the operator is inherently incapable of controlling.

At the end of each year of the project a complete end-to-end emergency response mission will be performed to demonstrate the system's evolving capabilities. Target mission scenarios include:

1992 Safe entry into a facility containing flammable/explosive vapors and use of the on-board manipulator to close the valve on a leaking compressed gas cylinder.

1993 Entry into a laboratory, opening of a fume hood, and detailed identification and characterization of the chemical hazard contained therein.

1994 Response to a hazardous materials incident in a high rise building in which the vehicle must autonomously ride an elevator.

1995 Travel to an incident site, identify a chemical hazard (e.g., glass container with damaged seal), grasp the container, place it into a secondary container, and remove it for subsequent disposal.

These scenarios will be performed under realistic conditions void of setups aimed at simplifying the task.

5 SUMMARY

We have begun developing a mobile robot that can be used by HAZMAT Teams to safely identify, locate, and eventually mitigate incidents involving hazardous chemical releases with the ultimate goal of reducing the risks of injury and loss of life. The project is unique in that it directly involves users in the specification of system requirements and the evaluation of system performance. We suspect that many of the capabilities currently being incorporated into HAZBOT II are also directly applicable to operations involving the storage and retrieval of hazardous materials, and chemical waste site cleanup.

In late September of 1991, the HAZBOT II vehicle performed its first end-to-end demonstration involving a scenario in which the vehicle:

- Navigated to the incident location from a distant (150-200 ft.) deployment site.
- Entered a building through a door with thumb latch style handle and door closer.
- Located and navigated to the suspected incident location (a chemical storeroom).
- Unlocked and opened the storeroom's door.
- Climbed over the storeroom's 12 in. high threshold to enter the storeroom.
- Located and identified a broken container of benzene.

ACKNOWLEDGEMENTS

We would like to thank the members of the JPL Fire Department, the JPL Fire Department HAZMAT Team and Occupational Safety office for their time and support. The research described in this paper was carried out by the Jet Propulsion Laboratory, California Institute of Technology, under a contract with the National Aeronautics and Space Administration.

References

[1] K. R. Brittain, S. O. Evans, and R. F. Sturkey. A compendium of robotic equipment used in hazardous environments. Technical Report EPRI NP-6697, Electric Power Research Institute, February 1990.

[2] H. B. Meieran. Frequent incidents increase robot potential. *HAZMAT World*, 2(9):pp. 70–73, September 1989.

[3] H. B. Meieran. Mobile robots responding to toxic chemical accidents. In *Proceedings of the Fourth American Nuclear Society Topical Meeting on Robotics and Remote Systems*, pages pp. 383–391, Albuquerque, NM, February 24-28 1991.

[4] H. B. Meieran and G. Mosci. The ansaldo s.m.t.-1 mobile robot. In *Proceedings of the Fourth American Nuclear Society Topical Meeting on Robotics and Remote Systems*, pages pp. 375–382, Albuquerque, NM, February 24-28 1991.

[5] K. A. Roy. Mobile robots audition for emergency response roles. *HAZMAT World*, 2(9):pp. 64–68, September 1989.

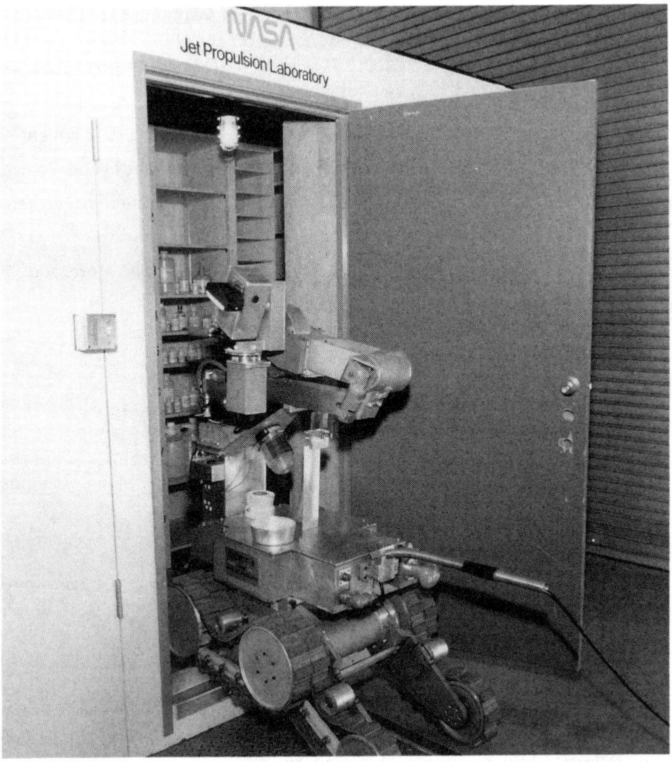

Figure 4: HAZBOT II climbing into a chemical store after having unlocked and opened its door with a specially designed tool.

MODEL REDUCTION OF FLEXIBLE MANIPULATORS

Jie Yang*† Yangsheng Xu† C.S. Chen*

† The Robotics Institute
Carnegie Mellon University
Pittsburgh, Pennsylvania 15213, USA

*Department of Electrical Engineering
The University of Akron
Akron, Ohio 44325, USA

Abstract

In this paper, we present a method of reducing the order of a marginally stable system based on the fact that translation transformations in the frequency domain preserve input-output properties of the system. We address the successful application of the method to model reduction of flexible manipulators with infinite-dimensional or finite-dimensional model. The method is also applicable for any other marginally stable model, such as elastic space trusswork and multi-dimensional space vehicle structure.

1 Introduction

Flexible manipulators have been increasingly interesting to many researches for light-weight, energy-efficiency, and less harmfulness concerns. Modeling and controlling such manipulators, however, is a challenge because conceptually the system must be characterized by a distributed parameter systems (DPS) with infinite number of modes, while controlling DPS requires a finite-dimensional model. A various approaches of methods can be employed to reduce the order of a model with either an infinite-dimensional or a finite-dimensional structure. The balanced truncation method [1] is based on the realization theory that for zero initial condition, the input-output behavior of the model is completely specified by the mutually controllable and observable subspace R^n. Model reduction by balancing techniques is applicable only for stable systems [1]–[5]. This is because, for the balancing techniques, the computation of the controllability and observability Gramians requires that the system be asymptotically stable. Some researches [6]–[9] have been directed to applying balancing method to flexible structure by assuming lightly damped and widely separated modes. This assumption is normally not valid in practice. For a flexible manipulator, as an elastic system, we have to face the fact that the system is marginally stable. This paper presents a method to reduce the model order of flexible manipulators using balancing techniques. The approach is based on the fact that translation transformations in the frequency domain preserve input-output properties of the system.

2 Modelling of Flexible Manipulators

In this section, we will discuss the modeling of flexible manipulators with infinite-dimensional and finite-dimensional structures. These two models will be used to demonstrate the procedure of model reduction in section 4.

2.1 Infinite-dimensional model

A one-link flexible manipulator can be modeled [10] by the Bernoulli-Euler beam equation:

$$EI\frac{\partial^4 y(x,t)}{\partial x^4} + \rho a\frac{\partial^2 y(x,t)}{\partial t^2} = 0 \tag{1}$$

where $0 < x < l$ and the equation has associated boundary conditions.

The transfer function for general Bernoulli-Euler beam equation can be found in [11] and the complete dynamics can be described by

$$\frac{\partial^2 Q(x,t)}{\partial t^2} + a^2\frac{\partial^4 Q(x,t)}{\partial x^4} = f(x,t) \tag{2}$$

$$Q(x,0)=Q_0(x),\quad \frac{\partial Q}{\partial t}(x,0)=Q_1(x),\quad \frac{\partial^2 Q}{\partial x^2}(0,t)=g_1(t)$$
$$\frac{\partial^3 Q}{\partial x^3}(0,t)=g_2(t),\quad \frac{\partial^2 Q}{\partial x^2}(l,t)=g_3(t),\quad \frac{\partial^3 Q}{\partial x^3}(l,t)=g_4(t)$$
$$0\le x\ge l,\quad t\ge 0,\quad a>0$$

To solve the above problem, we may define a *standardizing function*:

$$\begin{aligned} w(x,t) &= f(x,t)+Q_0(x)\delta'(t)+Q_1\delta(t)- \\ &-a^2\delta'(x)g_1(t)+a^2\delta(x)g_2(t)- \\ &-a^2\delta(l-x)g_3(t)-a^2\delta(l-x)g_4(t) \end{aligned} \tag{3}$$

and we convert the original problem into a zero initial and boundary condition problem for which the Green's function is

$$G(x,\xi,t)=\frac{4}{a}\sum_{n=1}^{\infty}\frac{\varphi_n(x)\varphi_n(\xi)}{\mu_n^2\varphi_n^2(l)}sin\, a\mu_n^2 t \tag{4}$$

where

$$\begin{aligned} \varphi_n(x) &= (sinh\,\mu_n l - sin\,\mu_n l)(cosh\,\mu_n x+cos\,\mu_n x)- \\ &-(cosh\,\mu_n l-cos\,\mu_n l)(sinh\,\mu_n x+sin\,\mu_n x) \end{aligned}$$

and the μ_n are the non-negative roots of the equation $cosh\,\mu l cos\,\mu l = 1$.
The transfer function is given by:

$$W(x,\xi,s)=\frac{4}{a}\sum_{n=1}^{\infty}\frac{\varphi_n(x)\varphi_n(\xi)}{\varphi_n^2(l)}\frac{1}{s^2+a^2\mu_n^4} \tag{5}$$

which has infinite poles:

$$s=\pm ja\mu_n^2,\qquad n=1,2,\ldots$$

2.2 Finite-dimensional model

For a multi-degree of freedom flexible manipulator, Lagrangian dynamics is an appropriate approach to model the system. As an example, the Self-Mobile Space Manipulator developed in the Robotics Institute of Carnegie Mellon University [12][13] is considered here. By a standard procedure based on Lagrangian dynamics, the explicit dynamics equation can be obtained.

$$M(q)\ddot{q}+K(q)q+C(q,\dot{q})=f \tag{6}$$

where M is the inertia matrix, K is the stiffness matrix, and C is the centrifugal and Coriolis torques.

The dynamic equation above is nonlinear due to the translational kinetic energy part in Equation (7). To linearize the model with reasonable accuracy, we assume that the speed of the robot is low so that centrifugal and Coriolis torques are negligible and the effect of link deformation on the variation of the inertia matrix is small so that the inertia matrix is a function of only joint angles. In this way the dynamic model can be linearized.

$$M\ddot{q}+Kq=f \tag{7}$$

It is obvious that the poles of the model lie on the $j\omega$ axis.

The model can be decoupled into two independent parts, tangential model and radial model. The tangential model represents the motion associated with joint 1 driving the tip mass in the circular direction, while the radial model describes the motion associated with joints 2 and 3 driving the tip mass in the plane perpendicular to the circular direction. It must be noted that the mass and stiffness matrices for both models are functions of the mixed joint angles. In this way, we can obtain two linearized models in the tangential and radial directions

$$\bar{M}_t\bar{q}_t''+\bar{K}_t\bar{q}_t=\bar{\tau}_t \tag{8}$$

$$\bar{M}_r\bar{q}_r''+\bar{K}_r\bar{q}_r=\bar{\tau}_r \tag{9}$$

3 Model Reduction for a Marginally Stable System

From the discussion above, it has been known that the model of flexible manipulator is marginally stable, i.e., the poles of the system lie on imaginary axis. In this section, we present a method of model reduction for such a system.

Consider the linear time invariant system

$$\dot{x}(t) = Ax(t) + Bu(t)$$

$$y(t) = Cx(t) \tag{10}$$

where $x(t) \in R^n$ is the state vector, $u(t) \in R^m$ is the input or the control vector, $y(t) \in R^p$ is the output vector, and A, B, C are matrices of appropriate sizes.

The transfer function of the system (1) is given by

$$H(s) = C(sI - A)^{-1}B \tag{11}$$

Definition 1: The symbol (A, B, C) represents a state space realization of the transfer function of the system.

Definition 2: The realization (A, B, C) is said to be minimal if it is completely controllable and observable.

Definition 3: (A, B, C) is said to be asymptotically stable if Re ($\lambda \mid A \mid) < 0$.

Definition 4: Let (A, B, C) be a minimal and asymptotically stable realization. (A, B, C) is said to be balanced if A, B and C satisfy the matrix Lyapunov equations with equal and diagonal controllability and observability Gramians, that is

$$AW_c + W_cA^T = -BB^T \tag{12}$$

$$A^TW_o + W_oA = -C^TC, \tag{13}$$

where $W_c = W_o = \Sigma = diag[\sigma_1\ \sigma_2\ \cdots \sigma_n]$ and Hankel singular values $\sigma_1 \geq \sigma_2 \geq \sigma_3 \geq \ldots \geq 0$.

Definition 5: A minimal realization (A, B, C) is said to be a J form realization if A is a Jordan matrix J and (J, B, C) has following structure:

$$J = diag[J_1\ J_2\ \cdots J_\alpha], \quad B = [B_{i1}\ B_{i2} \cdots B_{i\alpha}], \quad C = [C_{i1}\ C_{i2} \cdots C_{i\alpha}] \tag{14}$$

with

$$J_{ij} = diag[J_{ij_1}\ J_{ij_2}\ \cdots J_{ij_{r_{ij}}}], \quad B_{ij} = [B_{ij_1}\ B_{ij_2}\ \cdots B_{ij_{r_{ij}}}], \quad C_{ij} = [C_{ij_1}\ C_{ij_2} \ldots C_{ij_{r_{ij}}}].$$

If $(A, \tilde{B}, \tilde{C})$ is a minimal realization, there exists a non-singular matrix T such that

$$J = T^{-1}AT, \quad B = T^{-1}\tilde{B}, \quad C = \tilde{C}T \tag{15}$$

$$H(s) = C(sI - J)^{-1}B \tag{16}$$

In the rest of this section, we present an approach to model reduction of marginally stable system using balancing technique. We at first state the following theorems.

Theorem 1: The controllability and observability of a minimal realization J-form (J, B, C) depends only on the structures of J, B and C.

The proof of Theorem 1 can be obtained directly from the definitions of the controllability and observability.

Theorem 2: For minimal marginally stable realization $H(s)$, there exist translation transformations in frequency domain, $s' = s + a$, such that the poles of $H(s + a)$ lie strictly on the left half s-plane and the controllability and observability of $H(s + a)$ are identical to those of $H(s)$.

Proof: If $H(s)$ is a minimal realization, we have

$$H(s) = C(sI - J)^{-1}B$$

and

$$H(t) = L^{-1}[H(s)]$$

Let λ_{max} be the largest eigenvalue of the matrix J. For the marginally stable system $Re(\lambda_{max}) = 0$. we can choose

$$a = \varepsilon$$

where $\varepsilon > 0$.
Let $\hat{H}(t) = H(t)e^{-at}$, then the Laplace transform of $\hat{H}(t)$ can be determined by

$$\hat{H}(s) = H(s+a) = C[(s+a)I - J]^{-1}B = C(sI - \hat{J})^{-1}B$$

Thus

$$Re(\lambda \mid \hat{J} \mid) < 0$$

Since rank(J) = rank$(\hat{J})$, $(\hat{J}, B, C)$ is a minimal realization and J and $\hat{J}$ have same structure. From Theorem 1, $(\hat{J}, B, C)$ and (J, B, C) have the same controllable and observable subspace.

In the balanced realization, the *ith* Hankel singular value characterizes how controllable and observable the *ith* state is. Therefore the model can be reduced by neglecting the states corresponding to "small" Hankel singular values. This implies that the criterion for balanced model reduction depends only on the input-output information of the system. Thus it is possible to obtain a reduced order model by balancing method for an asymptotically stable system if the input-output map of the system can be determined. More specifically, we transform the marginally stable system into the asymptotically stable system by the translation transformation in the frequency domain and balance the resulting stable system. This procedure leads to the following algorithm.

Algorithm

1. Transform $H(s)$ into $\hat{H}(s) = H(s+a)$ so that $\hat{H}(s)$ has no poles on the right half s-plane nor on the $j\omega$ axis.
2. Find a minimal state space realization (A, B, C) of $\hat{H}(s)$.
3. Compute the balanced realization (A_b, B_b, C_b) from (A, B, C). That is

$$A_b = T^{-1}AT, \quad B_b = T^{-1}B, \quad C_b = CT,$$

   where T is the transformation matrix which makes the controllability and observability Gramians equal and diagonal. T can be obtained using any of methods presented in [1]–[5].
4. Determine the reduced order model (A_r, B_r, C_r) from (A_b, B_b, C_b). Recall that for the balanced realization, $W_c = W_o = \Sigma = diag[\sigma_1 \; \sigma_2 \; \cdots \sigma_n]$ and Hankel singular values $\sigma_1 \geq \sigma_2 \geq \sigma_3 \geq \ldots \geq 0$. If $\sigma_k \gg \sigma_{k+1}$, then Σ can be partitioned as

$$\Sigma = \begin{bmatrix} \Sigma_1 & 0 \\ 0 & \Sigma_2 \end{bmatrix}$$

   where $\Sigma_1 = diag[\sigma_1 \; \sigma_2 \; \cdots \sigma_k]$ and $\Sigma_2 = diag[\sigma_{k+1} \; \sigma_{k+2} \; \cdots \sigma_n]$.
   (A_b, B_b, C_b) can be partitioned accordingly

$$A_b = \begin{bmatrix} A_{b_{11}} & A_{b_{12}} \\ A_{b_{21}} & A_{b_{22}} \end{bmatrix}, \quad B_b = \begin{bmatrix} B_{b_1} \\ B_{b_2} \end{bmatrix}, C_b = [C_{b_1} \;\; C_{b_2}]$$

   to yield the reduced order model

$$A_r = A_{b_{11}}, \quad B_r = B_{b_1}, \quad C_r = C_{b_1}.$$

5. Obtain $\hat{H}_r(s)$ from (A_r, B_r, C_r) as

$$\hat{H}_r(s) = C_r(sI - A_r)^{-1}B_r.$$

6. The reduced order model of the original system is given by

$$H_r(s) = \hat{H}(s-a).$$

4 Model Reduction of Flexible manipulators

As discussed in section 2, the flexible manipulator can be modeled as either infinite-dimensional model or finite-dimensional system. In either case, it is generally needed to reduce the order of the model so that any model-based control scheme can be implemented in practice. In this section, we discuss the application of the method proposed in the previous section to the two models developed in section 2.

Example 1: Infinite-dimensional model

We list five modes of the infinite-dimensional model based on the parameters suggested in [10] in Table 1.

Table 1: Poles and zeros of the flexible manipulator

Pair	Poles	Zeros
1	0.00	±11.18
2	±15.42j	±60.45
3	±49.97j	±149.27
4	±104.25j	±277.58
5	±178.27j	±445.35

This is a tenth order model and is difficult to be implemented in real-time control. Using our method, we can reduce the model order and the resultant second order model is

$$H_r(s) = \frac{s^2 + 1.6013 \times 10^4 s - 7.6801 \times 10^5}{s^2 + 3.3671 \times 10^{-3} s - 3.3552 \times 10^{-4}}$$

The step responses of the tenth order model and the corresponding second order model are compared in Figure 1. It has been shown that the second order model is nearly identical to the tenth order model.

Example 2: Finite-dimensional model

The second example of flexible manipulators is the radial direction model developed in section 2 which has two inputs and four outputs and is manipulator configuration dependent. At $\theta_2 = \theta_3 = 60$ degree, the transfer function is given by

$$NUM_{11} = s^2(2.6823e+01s^4 + 3.7052e+03s^2 + 9.1806e+03)$$
$$NUM_{12} = s^2(1.7131e+03s^2 - 1.3589e+04)$$
$$NUM_{21} = s^4(-2.6823e+01s^2 - 3.5937e+03)$$
$$NUM_{22} = -1.9382e+03s^4$$
$$NUM_{31} = s^2(1.7131e+03s^2 - 1.3589e+04)$$
$$NUM_{32} = s^2(7.4510e+01s^4 + 7.6274e+03s^2 + 3.0999e+04)$$
$$NUM_{41} = s^4(-2.6823e+01s^2 - 5.4970e+03)$$
$$NUM_{42} = s^4(-7.4510e+01s^2 - 9.1002e+03)$$
$$DEN = s^4(s^4 + 2.3010e+02s^2 + 1.1032e+04)$$

where NUM_{ij}/DEN represents the transfer function from the torque τ_i to the angle θ_j. This is a 8th order MIMO system and using our method the reduced to 7th order is as follows.

$NR_{11} = 3.4686e-01s^6 + 2.6394e+01s^5 + 7.6040e+01s^4 + 3.6210e+03s^3 + 3.3186e+03s^2 + 7.0192e+03s - 9.9127e+02$

$NR_{12} = 1.6638e-01s^6 - 3.4956e-01s^5 + 3.8582e+01s^4 + 1.6517e+03s^3 + 1.6239e+03s^2 - 1.4633e+04s + 2.0666e+03)$

$NR_{21} = s^2(-4.6450e-04s^4 - 2.6773e+01s^3 + 3.7313s^2 - 3.5851D+03s + 5.0513e+02)$

$NR_{22} = s^2(-2.2280e-04s^4 + 2.4323e-02s^3 - 6.9621e-02s^2 - 1.9340e+03s + 2.7257e+02)$

$NR_{31} = 1.6667e-01s^6 - 2.5671e-01s^5 + 3.8477e+01s^4 + 1.6660e+03s^3 + 1.6091e+03s^2 - 1.4628e+04s2.0660e+03)$

$NR_{32} = 7.9945e-02s^6 + 7.4317e+01s^5 + 7.9596s^4 + 7.5946e+03s^3 - 1.7942e+02s^2 + 3.0496e+04s - 4.3073e+03)$

$NR_{41} = s^2(-4.1424e-04s^4 - 2.6699e+01s^3 + 3.6426s^2 - 5.4776e+03s + 7.7092e+02)$

$NR_{42} = s^2(-1.9869e-04s^4 - 7.4450e+01s^3 + 1.0504e+01s^2 - 9.0908e+03s + 1.2826e+03)$

$DENR = s^2(s^5 - 1.4289e-01s^4 + 2.3010e+02s^3 - 3.2765e+01s^2 + 1.1032e+04s - 1.5582e+03)$

where $NR_{ij}/DENR$ represents the transfer function of the reduced order model from the torque τ_i to the angle θ_j. The step responses of the original model and the reduced order model are compared in Figure 2.

5 Summaries

We presented a simple, efficient method for model reduction of flexible manipulators based on the fact that the translation transformation in s-plane preserves input-output property of the system. By translation transformation in frequency domain, we can change the stability of the system without lose of its input-output property. The original balancing method requires that the model be asymptotically stable while this method is applicable to a marginally stable system. We employed the proposed method to reduce the order of flexible manipulator models such that model-based scheme can be implemented. The method has been examined by two examples, an infinite-dimensional model and a finite-dimensional model.

References

[1] B.C. Moore, Principal component analysis in linear systems: controllability, observability, and model reduction, *IEEE Trans. Automat. Contr.*, vol. AC26, p.17, 1981.

[2] J.A. De Abreu-Garcia and F.W. Fairman, On using permutation symmetric Jordan realizations to achieve SISO balancing, *Int. J. Syst. Sci.*, vol.18, p.441, 1987.

[3] J.A. De Abreu-Garcia and F.W. Fairman, Balanced realization of orthogonally symmetric transfer function matrices, *IEEE Trans. Circuits and Systems*, vol.CAS-34, NO.9, p. 997, 1987.

[4] J.R. Sveinsson and F.W. Fairman, Minimal balanced realization of transfer function matrices using Markov parameters, *IEEE Trans. Automat. Contr.*, vol. AC-30, p.1014, No.10, 1985.

[5] A.J. Laub, Computation of 'balancing' transformations, *Proc. Joint Automat. Contr. Conf.*, San Francisco, CA, Aug. 1980.

[6] E.A. Jonckheere and Silverman, Singular value analysis of deformable system, *Proc. 20th Conf. CDC*, San Diego, CA, Dec. 1981, p.660.

[7] C.Z. Gregory, Reduction of large flexible spacecraft models using internal balancing theory, *J. Guidance Contr. Dynamics*, vol.7, p.725,1984.

[8] E. A. Jonckheere, Principle component analysis of flexible systems - open-loop case, *IEEE Trans. Automat. Contr.*, vol. AC-29, p.1095,1984.

[9] T. Williams, Closed-form Grammians and model reduction for flexible space structures, *IEEE Trans. Automat. Contr.*, vol. AC-35, p.397, 1990.

[10] S. Cetinkunt and W. Yu, Closed-loop Behavior of a Feedback-controlled flexible Arm: A Comparative Study, *The Int. J of Robotics Res.*, vol. 10, No.3, p.263, 1991.

[11] A.G. Butkovskiy, Structure theory of distributed systems, *Ellis-Horwood*, 1983.

[12] Y. Xu and H. Ueno, Configuration-independent control of self-mobile space manipulator (Submitted to Journal of Intelligent and Robotic Systems).

[13] Y. Xu, B. Brown and F. Friedman, Control System of Self-Mobile Space Manipulator, *Proceedings of IEEE Inter. Conf. on Robotics and Automation*, 1992.

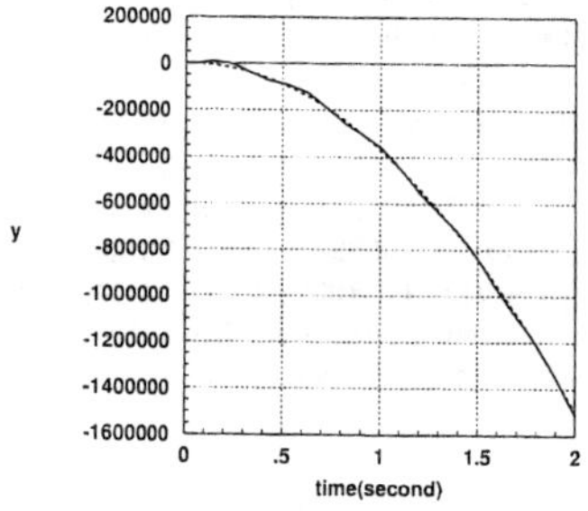

— Original model ..Reduced order model

Figure 1 Step responses of the full and reduced order model

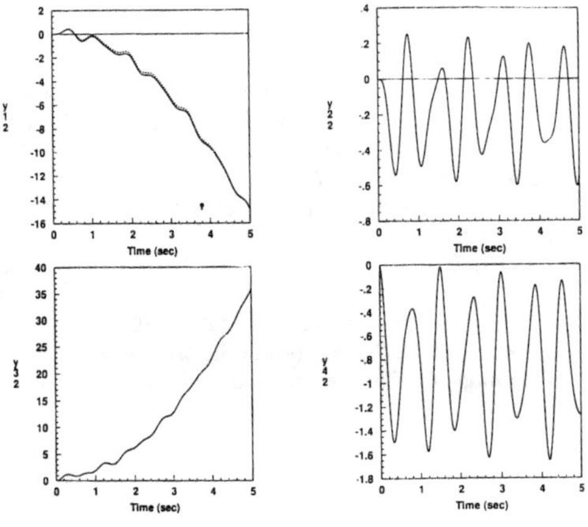

— Original model ...Reduced order model

Figure 2 Step responses of the full and reduced order model

PART 7 : FUZZY LOGIC APPLICATIONS IN ROBOTICS AND MANUFACTURING

TORQUE CONVERTER MODULATING CLUTCH CONTROLLER USING FUZZY LOGIC

ADNAN SHAOUT and PAUL KUEBER
The Electrical and Computer Engineering Dept.
The University of Michigan - Dearborn
Dearborn, Michigan 48128

ABSTRACT

This paper describes a fuzzy logic controller and its simulation for a slipping clutch on a transmission torque converter. The fuzzy control logic added a level of uncertainty, which is an invaluable component to the slipping clutch control system.

INTRODUCTION

The torque converter is used in all automatic transmissions on the market today. It employs a clutch mechanism whereby when the clutch engages, or locks up, the wheels and engine of the vehicle are connected in a mechanical system. When the clutch disengages, the wheels are driven by the engine through a fluid coupling. This disengagement results in slip between the engine rotation and the axle rotation. The axle is connected to the wheels of the vehicle. This is known as a lock/unlock converter problem.

A new type of torque converter clutch engagement/disengagement assembly is being utilized in today's vehicles whereby the slip in the converter is controlled depending on the conditions exhibited by the vehicle. Specifications, such as low slip at a light crowd in fourth gear, are outlined to improve the shift quality and "feel" of the vehicle. A target slip is established which is computed from throttle position, vehicle speed, clutch position and transmission oil temperature. This slipping clutch needs an efficient control system to match the actual slip required by the system.

The controller is based on conventional control methods which rely on fixed parameters. One main reason exists why a more efficient controller is needed. The conventional control methods generally will not adapt to the unavoidable variability of clutch wear and operating conditions, which will lead to transmission shifting instability and ultimately driving instability. For example, the conventional control system cannot maintain the target slip speed if the slip speed is between the clutch open position and the clutch closed position. Because of this lack of repeatability, it is difficult for the controller to be effective. A much more robust controller is needed that could take into account all shifting and crowding conditions and provide quick response in meeting target speeds.

This paper descrips a fuzzy logic controller and its simulation for a slipping clutch on a transmission torque converter on an IBM Model 70 386. The fuzzy control model will be compared to the conventional controller in terms of responses in meeting target slips as seen in the simulation program. Particular attention in this comparison has been paid to improve the slip speed stability.

CONTROL CONFIGURATION

A closed loop feedback control system utilizes an additional measure of the actual output in order to compare the actual output with the desired output response will be used to implement the torque converter problem in this paper. Figure 1 demonstrates this configuration.

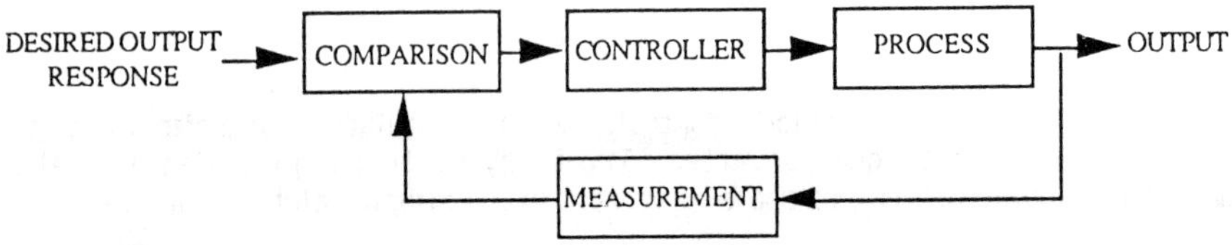

Figure 1. Feedback loop control configuration.

When in the slipping clutch mode, the convergence of actual slip speed to target slip speed is very important. The system uses a method whereby the target slip speed needed is read from a table based on vales of throttle position and vehicle speed. The system then attempts to model the slip speed by decreasing or increasing the pressure on the clutch This change in pressure is what causes slip.

The feedback in this example is the amount of throttle position change that is needed. This value is "fed back" to the original value to change it only in the amount needed.

While modern control theory is well founded mathematically, it does have its limitations. First, applications require the availability of an accurate mathematical model of the process to be controlled. The control configuration used here does not have a mathematical model of the process to be controlled. In fact, the process is not even completely understood. As a result, the clutch control configuration has difficulty in satisfying all the matching for various driving conditions that cause engaging and disengaging of the clutch which initiate slip. Second, parameter variation due to environmental factors change the model in generally unpredictable and undesirable ways over time. Because of this, the target slip speed is rarely attained because the system always endure wear. The friction material used by the clutch will wear and the oil in the converter tends to deteriorate over time.

THE FUZZY CONTROLLER

The implementation of fuzzy logic allows the use of rule based control, whereby the controller is defined by abstract rules that give them a great flexibility. In other words, one can set up the control at a high level where phrases are used to describe the needed action by the vehicle. This level of control is mode descriptive than the level of signals, and is closer to the actual level of behavior. As a result, if a control model is known in qualitative terms and a mathematical model is unclear, fuzzy logic quite possibly can solve the control problem.

In our controller, values for inputs and outputs are defined as fuzzy subsets (fuzzy linguistic variables)[3]. Typical membership function shapes are triangular or exponential. In this case, the triangular type was used because it can be handled easier than exponential and because it does not differ much in control capability when compare to exponential[4].

The fuzzy values used for the membership functions and in the rule table are: Negative Big(NB), Negative Medium(NM), Negative Small(NS), Zero(ZO), Positive Small (PS), Positive Medium(PM), and Positive Large(PL). Figure 2 shows the fuzzy values as they are used in the membership functions in the triangular form.

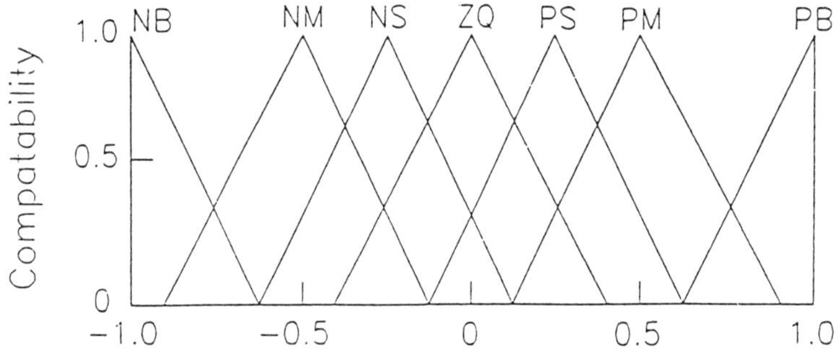

Figure 2. Membership functions used in the fuzzy controller.

These fuzzy values are applied in the control flow. Figure 3 shows this flow. First, in order to determine dU, the change of throttle position, the deviation of actual slip speed V from target slip speed V_0 is computed. This term is called E. The rate of change $dE=(V - V_0)/dT$ is also computed. The fuzzy rules are then checked for "believability". The main importance of this flow is not to understand specifically why calculations are performed or how they affect the whole car. The importance is to notice the following: How dE and E are calculated from the actual and target slip speeds; That dE and E are used to get dU, which is found from the fuzzy rules; And that dU actually changes the system, so that a recalculation of E and dE can occur.

The control strategy is expressed by a set of control rules of the form:

IF (condition on the controller inputs)
THEN (action taken on the controller output)

Figure 4 shows the rule table for the controller. The inputs are both E and dE and the output is dU.

All the rules are needed in the control system. Thirteen of the rules fix E or dE at ZO while allowing the value not fixed at zero, either E or dE, to take on the values of all other fuzzy values. Two rules are added to quickly fix slip speed when actual slip speed comes close to target slip speed. For example, one rule states that "If E is PS and dE is NS, THEN dU is ZO". By setting dU to zero the slip speed will set itself quickly. However, these rules may induce over or under shoot. Two other rules were added to minimize this effect. For example, one of the rules states that "IF E is PS and dE is NB, THEN dU is PB". This rule shows that if deviation of actual slip speed from target speed is small (E = PS), but the time rate of change of E is fast (dE = NB), the amount of change in the throttle position should be large (dU = PB). In other words, when the conditions are right for an overshoot - the value E is close to zero and dE is moving fast, the lever to reduce or increase the slip speed should change in an abrupt way. Two more rules were added to rise the response in an abrupt manner. The rules are "IF E is NB and dE is PS, THEN dU is PB" and "IF E is PB and dE is NS, THEN dU is NB". These two rules will quickly get the slip speed up or down to its requested speed[1,2].

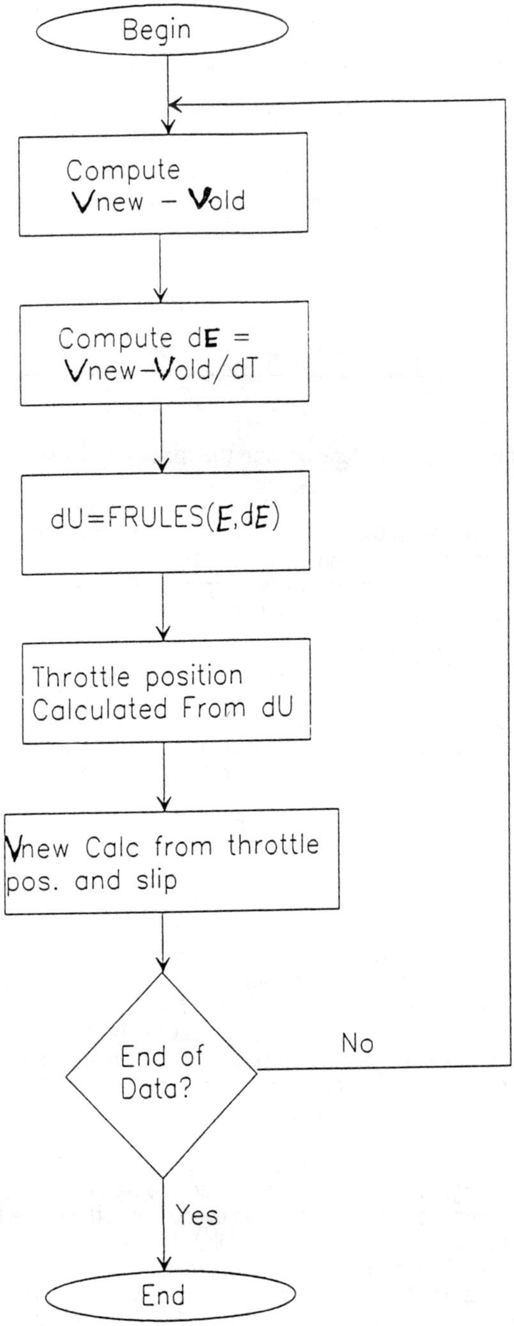

Figure 3. Flowchart showing the fuzzy control flow.

Delta E

		NB	NM	NS	ZO	PS	PM	PB
	NB				PB	PB		
	NM				PM			
E	NS				PS	ZO		NB
	ZO	PB	PM	PS	ZO	N	NM	NB
	PS	PB		ZO	NS			
	PM				NM			
	PB			NB	NB			

E: Deviation of Actual Speed from Target Speed
dE: Time Rate of Change of E

Figure 4. The fuzzy controller rule base.

The premise or IF portion of the rule is evaluated by translating the input signal to a level of belief by imposing the input upon the membership function (which represents the association between the linguistic variable in the fuzzy production rule). The minimum value of all the levels of belief of the premise portion of the rule is used as the output to the conclusion. In the simulation program, this value is known as the alpha. If the alpha is not zero, then the rule has been fired.

The conclusion or THEN portion of the rule is evaluated by using the max product method where the inputs and outputs of the system are "crisp". This portion imposes the alpha onto the conclusion membership function such that the conclusion membership graph is clipped by the alpha value. For all of the rules which were fired the graphs are combined with the logic OR resulting in a graphical polygon. This result is then converted to a crisp output by using the centroid method of evaluation[4].

SIMULATION

The program written to simulate this system is based on all the previous fuzzy rules and fuzzy values stated in the pervious section. Figure 5 shows the target slip speed verses the actual slip speed for 296 timing intervals. The crisp output values from the system are used to calculate a new actual slip speed that is closer to the target as needed. One will notice that the system rarely achieves the desired target exactly. This is the nature of the fuzzy system. The values tends to fluctuate +/- 0.2 percent in error. A standard system for slip speed control shows an average error of 8 percent. A comparison of this fluctuation to the average fluctuation seen in actual driving conditions shows that the fuzzy control system performs much better. Speed control is also an issue. The fuzzy controller tended to reach its goal quickly. The actual control system in vehicles tended the 8 percent error quickly, but never achieved its goal.

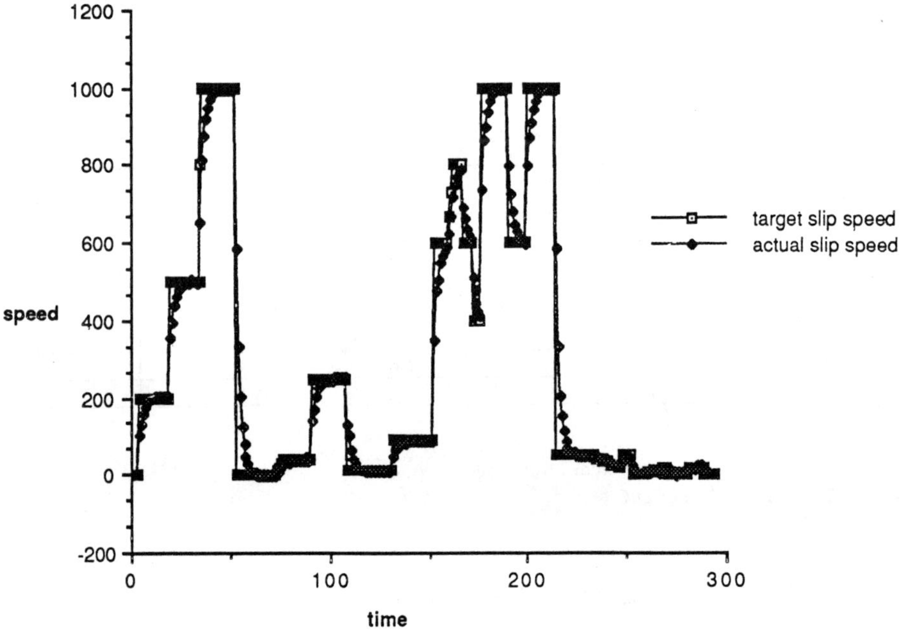

Figure 5. Target slip speed verses actual slip speed.

CONCLUSION

As a result of application of fuzzy logic to a torque converter clutch slip control, improvements are realized over the conventional control system. Fuzzy logic enables the system to respond quickly to external occurrences, which reduces the time needed to reach stability in most cases.

The fuzzy control logic adds a level of uncertainty, which is an invaluable component to the slipping clutch control system. Because conventional controllers lacked the ability to accept many changes in the system, they cannot achieve acceptable tracking. However, the fuzzy control system showed the ability to handle changes because of the high level of description of the rules for control.

REFERENCES

[1] Kayano, Masasuki, "Improvement of Automatic Cruising for Truck Using Fuzzy Control," Mitsubishi Motor Co. JSAE vol. 44 No. 2, 1990.

[2] Maurizo Abate and Norberto Dosio, "Use of Fuzzy Logic for Engine Idle Speed Control," Centro Ricerche Fiat.

[3] Zadeh, Lotfi A, "Outline of a New Approach to the Analysis of Complex Systems and Decision Processes," IEEE Transactions on Systems, Man, and Cybernetics, vol. SMC-3, No. 1, January 1973.

[4] Lee, Chuen Chien, "Fuzzy Logic in Control Systems: Fuzzy Logic Controller-Part I and Part II," IEEE Transactions on Systems, Man, and Cybernetics, vol. 20, No. 2, March/April 1990, pp. 404-435.

Building Knowledge Structure in Neural Networks Using Fuzzy Logic

Fei-Yue Wang
Department of Systems and Industrial Engineering
The University of Arizona, Tucson, Arizona 85721 USA

Abstract

This paper presents an approach of constructing structured neural networks by fuzzy logic based control systems. It builds up the neural networks by integrating the subnets of pattern recognition, fuzzy reasoning, and control synthesis, and thus creats a knowledge structure within the resulting network that will enable the transfer of human heuristics and expert skill into the neural network controllers and accelerate the learning process of the controllers. The proposed approach also suggests a way for a fuzzy logic based control system to use neural networks as a vehicle to conduct its inherent parallel computation, to refine its reasoning procedure and control rules, and to suppress the memory space required for its knowledge base and inference programs.

Introduction

The purpose of this paper is to illustrate the use of fuzzy logic based technique in the construction of a class of structured neural nets for intelligent control of complex systems. Such effort toward the combination of fuzzy logic description and neural net implementation has already been made by several researchers, e.g., network representation of membership functions [4], fuzzy associative memories [2], and hybrid neurofuzzy controllers [1]. It is observed, however, that in most of cases the implementation of rule-based fuzzy logic control systems by neural networks has led to the loss of knowledge structures embedded in the original systems. This is partly due to the fact that, under mild assumptions, a multilayer neural network with only one hidded layer of neurons or two layers of weights can generate input/output maps which will approximate any function with any desired accuracy. Thus it seems that the knowledge structure of a rule-based system is not very important as long as one is just going to capture the input/output behavior of the system by a neural network.

The knowledge structure of a system usually provides us important information about the pattern and organization of its decision-making processes. For a fuzzy logic based control system (FLCS) which selects its control action according to its sensor reading, the decision-making process can be decomposed into three steps: sensor fusion (i.e., association of sensor readings with of linguistic sensory patterns), approximate reasoning (i.e, calculation of fuzzy control actions), and control synthesis (i.e., formation of fuzzy control actions and the corresponding procedure of defuzzification). Regardless the modeling power of neural networks, there are two major reasons that the knowledge structure of a FLCS is still very important and has to be sustained when it is to be implemented by a neural network. First, it is always desired to know not only the external input/output relationship of a system, but also the structural information of its internal decision-making process, e.g., how many distinctive stages are involved during the process, and further, which is the stage(s) responsible for the system performance improvement or degeneration. The knowledge of structural information must be an indispensable part to the understanding of the system. Second, for supervised learning, one needs a large number of input/output pairs to train a network in order to achieve the desired system behavior. For control systems, however, such desired input/output pairs are generally not available at least during the early stage of the system development. But common sense, heuristics, or even expert skills can often be acquired. This kind of knowledge usually is associated with specific operations or procedures and is organized into some well-conceived structure or pattern. Thus it would be very useful to transplant this structure of knowledge into a neural network so that the heuristics can be transferred to it and be refined through the learning capability of the neural network as the system evolves. The process is very similar to that of training a new operator: at first he/she is taught with general knowledge and basic procedures in an off-line fashion and then is expected to gain sophisticated task skills later based upon his/her on-line, hand-on experience.

To build knowledge structure in a neural network, we propose to divide the whole network into three subnets, i.e., *pattern recognition*, *fuzzy reasoning*, and *control synthesis* subnets, and then train each of the subnets independently and simultaneously according to the three steps of FLCS outlined above. It should be pointed out that the proposed method also provides a mechanism to accomplish the parallel firing and continuous refinement of control rules of FLCSs.

Heuristics & Fuzzy Logic Based Control Systems

Heuristics or heuristic logic has been well recognized as a valuable source of knowledge for designing control algorithms for systems where it is infeasible, or perhaps not even practical, to formulate a mathematical model of the systems. In many industrial production processes, for example, human operators are capable of monitoring the state of a given task process, and take appropriate control actions in response to the process state. Practicing the task for a long period of time, they can understand what is happening in the task process by observing the signal displays collected through various sensors. Namely, they have acquired knowledge to interpret the pattern of sensory information and to generate appropriate commands for the task process. Generally, the knowledge acquired in such way can be represented by the following form,

IF (*signal pattern*) **THEN** (*control commands*)

The collection of such rules forms a *knowledge base* for the control of the task process.

The intrinsical uncertainties and imprecision embedded in such kind of knowledge, however, has made the classical bivalent logic no longer a proper formula for its description. On the other hand, fuzzy logic, created to characterize uncertainties and imprecision, has provided an ideal tool for this purpose. As human knowledge is usually expressed in natural languages, the key idea in fuzzy logic is to represent knowledge in some linguistic terms such as *large* or *small* and interpret these terms as fuzzy sets [6]. A fuzzy logic based control system consists of three major parts: a set of linguistic variables which are used to label signal patterns and control actions, a knowledge base which specifies IF-THEN rules in terms of linguistic variables, and a mechanism of conducting fuzzy reasoning to determine the actual control action based on the knowledge base.

More specific, for a process monitored through a signal s with m readings $s=(s_1, s_2, ..., s_m)$ and driven by a control u with n components $u=(u_1, u_2, ..., u_n)$ (s_i or u_j

itself may also be a vector) a set of linguistic variables is created for each of sensor readings and control components. Let $A_i=\{S_i^1, S_i^2, ..., S_i^{p_i}\}$ and $B_j=\{U_j^1, U_j^2, ..., U_j^{q_j}\}$ be the set of linguistic variables associated with s_i and u_j, $i = 1, ..., m$, $j = 1, ..., n$, respectively. Association of a particular value x of s_i or u_j with a linguistic variable Z in A_i or B_j is described by the concept of a membership function used in fuzzy set theory, denoted as $\mu_Z(x)$. For example, if Z is *SmallForce* and x is a reading of a force/torque sensor then $\mu_Z(x)$ describes to what degree x represents a small force.

The knowledge base of a FLCS is a set of linguistic **IF-THEN** control rules. Assume that the knowledge base has M control rules,

Rule 1: **IF** s_1 is $S_1^{k_{11}}$ and ... and s_m is $S_m^{k_{m1}}$ **THEN** u_1 is $U_1^{c_{11}}$ and ... and u_n is $U_n^{c_{n1}}$

......

Rule M: **IF** s_1 is $S_1^{k_{1M}}$ and ... s_m is $S_m^{k_{mM}}$ **THEN** u_1 is $U_1^{c_{1M}}$ and ... and u_n is $U_n^{c_{nM}}$

where $S_i^{k_{ir}} \in A_i$ and $U_j^{c_{jr}} \in B_j$ for $r = 1, ..., M$, $i = 1, ..., m$ and $j = 1, ..., n$. In general the signal patterns represented by the **IF** parts of more than one control rules will be partially matched by a particular sensory input, thus leads to the conflict of simultaneous firing or triggering of control rules. Various approximate reasoning methods have been proposed to resolve the firing conflict [3], and among them the *Max-Min compositional rule of inference* might be the most commonly used strategy. For a given sensor reading $s=(s_1, s_2, ..., s_m)$ to the FLCS, the procedure of *Max-Min* inference can be described briefly as follows,

Rule strength calculation: the strength of Rule r can be calculated by,

$$\alpha_r = \mu_{S_1^{k_{1r}}}(s_1) \wedge \mu_{S_2^{k_{2r}}}(s_2) \wedge ... \wedge \mu_{S_m^{k_{mr}}}(s_m), \qquad r = 1, ..., M$$

where $\wedge$ is the *and* operator.

Rule consequence deduction: the consequence of Rule r for *jth* control component can be deduced from,

$$\mu_{C_j^r}(u_j) = \alpha_r \wedge \mu_{U_j^{d_{jr}}}(u_j), \qquad j = 1, ..., n, \quad r = 1, ..., M$$

where C_j^r is the fuzzy set of *jth* control component according to Rule r.

Resultant control generation: now the resultant fuzzy set C_j of *jth* control component can be obtained by,

$$\mu_{C_j}(u_j) = \mu_{C_j^1}(u_j) \vee \mu_{C_j^2}(u_j) \vee ... \vee \mu_{C_j^M}(u_j), \qquad j = 1, ..., n$$

where $\vee$ is the *or* operator.

This process of conflict-resolution generates a fuzzy set for each of the control component. As a real system usually requires a nonfuzzy value of control, a *defuzzification* stage is needed to produce a single control action that best represents the membership function of an fuzzy control action. There are several ways of tackling this problem. Both *max-procedure* and *center-of-area* procedure are commonly used. According to the later one, the fuzzy control C_j is defuzzified to a single value given by,

$$u_j^* = \frac{\int_{U_i} x * \mu_{C_j}(x)dx}{\int_{U_i} \mu_{C_j}(x)dx}, \qquad j = 1, ..., n$$

where U_i is the set of admissible control for u_i.

It should be pointed out that although the *Max-Min compositional rule* has been used for approximate reasoning and the *center-of-area* procedure selected for control defuzzification, these choices are not important to our method here, as it will become clear later in the next section.

Construction of Structured Neural Nets

The task of constructing structured neural networks from a FLCS can be divided into three subtasks: pattern recognition subnets, fuzzy reasoning subnets, and control synthesis subnets. It is important to know that these three kinds of neural nets are not dependent on with each other and their construction and training can be carried out simultaneously.

Pattern Recognition Neural Nets In a FLCS, each sensor reading s_i is to be associated with the linguistic variables, or signal patterns, in A_i through the corresponding membership functions. A neural net SN_i is constructed to capture this association of sensor readings and signal patterns for s_i, In other word, the task for SN_i is to approximate the membership function $\mu_{S_i^k}(x)$ for $k = 1, ..., p_i$. Figure 1 shows a three-layer SN_i for pattern identification. It should be pointed out, however, that neural net SN_i is not required to learn the memberships very accurately. Instead, learning process of SN_i can be stopped once the shapes of the membership functions have been captured with appropriate degree. This is due to the fact that usually a membership function itself is subjective and can not be specified very accurately. In other words, the learning of SN_i is a qualitative learning, not a quantitative one. Totally, m such neural nets will be needed.

Fuzzy Reasoning Neural Nets The whole process of fuzzy reasoning should cover all three steps: rule strength calculation, rule consequence deduction, and resultant control generation. For the sake of simplicity of neural net representation, however, we will construct fuzzy reasoning neural nets only for the calculation of rule strengthes. Both consequence deduction and control generation will be included as part of control synthesis to be discussed next. The strength of a control rule is a function of truth values of the linguistic variables of its antecedent, and the task of neural net RN_k is to emulate this function for rule k, $k = 1, ..., n$. Since both truth values and strengthes are numbers ranging from 0 to 1, no input and output scaling is needed for net RN_k. Depend on the selection of evaluation method for logic operators $\wedge$, one may find certain type of neural network are especially appropriate for RN_k. For example, if the *Min* function are used for $\wedge$, the Hamming net *MAXNET* may be the best choice for RN_k. But here we will choose the standard forwardfeed network for RN_k, as shown in Figure 2. The training input/output pairs for RN_k can be generated from the reasoning procedure of the FLCS. As for SN_i, the learning process of RN_k is a qualitative, not quantitative one.

Control Synthesis Neural Nets Control synthesis is the process of determining the actual, nonfuzzy control actions according to the given values of rule strengthes and the knowledge base of a FLCS. This involves the procedures of both fuzzy reasoning and defuzzification. There are two ways to use neural nets to conduct the control synthesis. The first is to develop an individual net for each of the control components, and the second is to build a single net for all the control components. To construct a neural net CN_j to perform the synthesis for control component u_j, $j = 1, ..., n$, we will first produce sufficient number of input/output patterns through approximate reasoning and defuzzification using randomly generated values of strength for each of control rules over [0,1], and then train CN_j to record these input/output patterns. Since all n nets CN_j's share the same input, a single CN net can be used to produce the whole control vector $u=(u_1, ..., u_n)$. This may lead to longer training session but save the total number of neurons used for the process of control synthesis. Figure 3 presents typical neural nets CN_i and CN. In any case, due to the common input, all control rules are invoked simultaneously in the selection of the resultant control actions, which corresponds to the concurrent firing of control rules in a FLCS. As for pattern recognition and fuzzy reasoning, the learning of CN_i or CN is still a qualitative one.

Integration of Neural Nets Once neural nets SN_i, RN_i, and CN_i or CN have been created, the final step toward a structured neural network is to connect those nets appropriately (see Figure 4). From its construction, it is very clear that the whole computation process of the integrated neural network can be divided into three stages: pattern recognition, fuzzy reasoning, and control synthesis (Figure 4), resulting in a functional interpretation for each layer of neurons and a structure of flow of knowledge thereof. Although the structured neural network, designated as FCN, can be viewed as an ordinary multilayer network with sensor readings as its input and control actions as its output, the distinctive knowledge structure embedded within this network makes it different from others.

No additional training is required to put the structured neural network FCN to work, since its performance should be at least as good as the original FLCS from which the network is constructed. Additional training and on-line learning by the network, however, can improve its performance and thus lead to the refinement of the control rules of the original FLCS. By breaking up the structured network into subnets of pattern recognition, fuzzy reasoning, and control synthesis, the actual modifications in these aspects can be recovered individually. Therefore neural network implementation also provides a mechanism for the refinement of fuzzy logic based control systems, a problem which can not be addressed easily within its own context. This is the reason that the selection of the compositional rule of inference and procedure of defuzzification is not very important in our approach. This is also why all the learning processes are qualitative ones during the construction phase of the structured network.

Note that as a neural network, FCN is not a fully connected one. For example, its subnet SN_i is linked only to input node s_i and has no connection with other input nodes (i.e., connection weights are zero), and subnet RN_i linked only to one output node of each of subnets SN_i's. With on-line learning by the network, this partial connection can provide a way to incorporate the coupling effects among signal patterns as well as control rules.

Concluding Remarks

Due to the space limitation, several other important issues such as on-line learning and automatic generation of new signal patterns and control rules will not be discussed in this paper. A case study of controlling a two-link planar robotic arm has been conducted and presented in [5]. The simulation result has shown a very good performance of a 7-layer structured neural network, though the on-line learning is very slow.

Clearly, fuzzy logic and neural networks are complementary. From the perspective of a fuzzy logic based system, neural networks provide an ideal vehicle to carry out its inherent parallelism, refine its reasoning and rules, and suppress the space required for its knowledge base and reasoning procedure. On the other hand, fuzzy logic enables a knowledge structure to be built within a neural network. The knowledge structure in turn assigns different functions to the different parts of the network, and makes the learning of the network become distributed, i.e., different functional subnets can learn or be trained independent with their own pace.

References

[1] H. R. Berenji, Neural networks and fuzzy logic in intelligent control, *Proc. 1990 IEEE Int. Symp. Intelligent Control*, pp. 916-919, Phil., PA, 1990.

[2] Bart Kosko, *Neural Networks and Fuzzy Systems: A Dynamical Systems Approach to Machine Intelligence*, Prentice Hall, Englewood, Cliffs, NJ, 1992.

[3] M. Mizumoto, Fuzzy controls under various fuzzy reasoning methods,, *Information sciences*, vol. 45, no. 2, pp.129-152, 1988..

[4] Yoh-Han Pao, *Adaptive Pattern Recognition and Neural Networks*, Addison-Wesley Publishing Co., Inc., Reading, MA., 1989.

[5] Fei-Yue Wang, Fuzzy logic based neural nets, Technical Report # 35, Robotics and Automation Lab., Systems and Industrial Engineering Department, University of Arizona, Tucson, AZ, 1992.

[6] L. A. Zadeh, The concept of a linguistic variable and its application to approximate reasoning, *Information Sciences*, Part 1: vol. 8, pp.199-249; Part 2: vol. 8, pp.301-357; Part 3: vol. 9, pp.43-80, 1975.

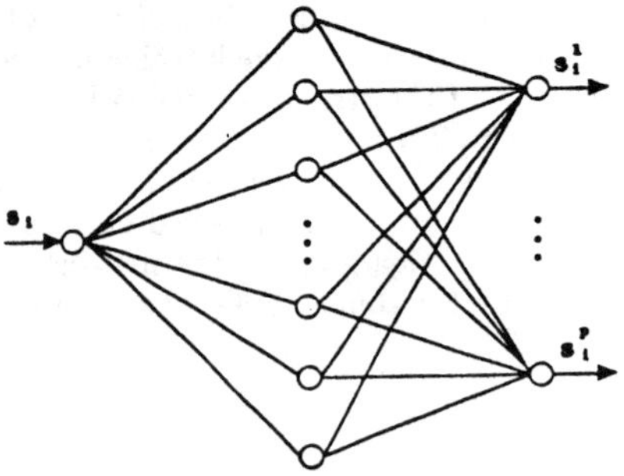

Figure 1: Pattern Recognition Neural Net

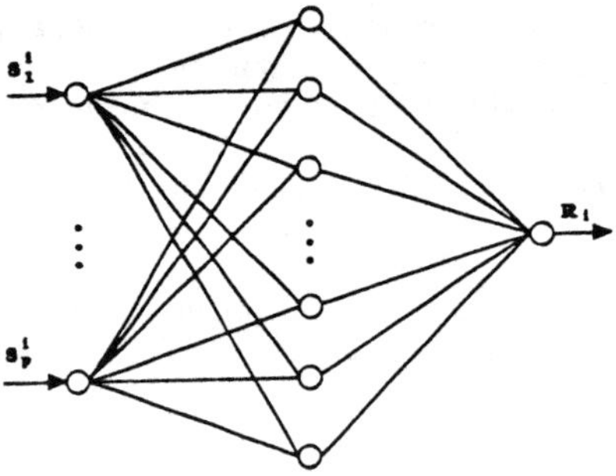

Figure 2: Fuzzy Reasoning Neural Net

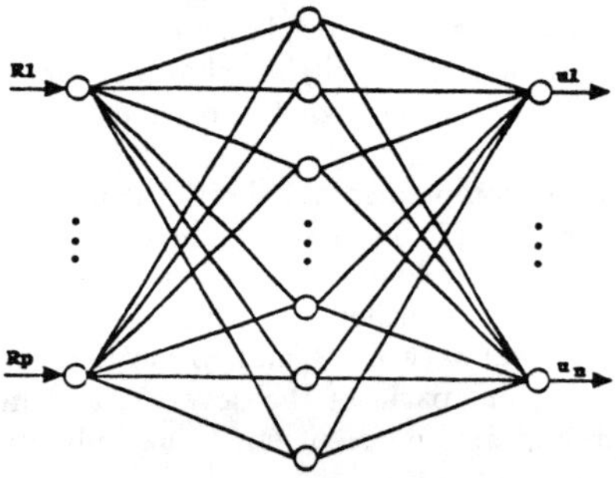

Figure 3: Control Synthesis Neural Net

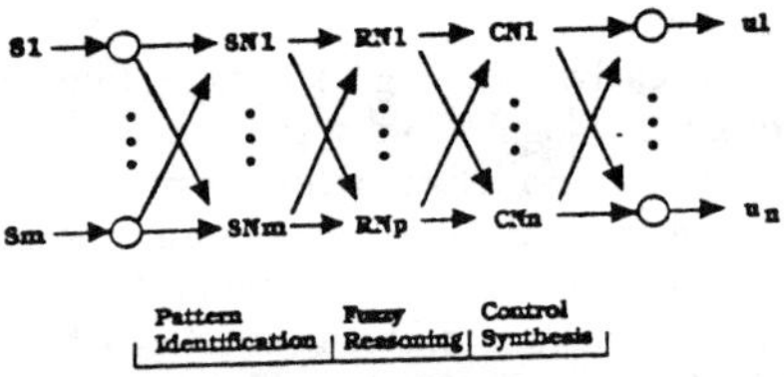

Figure 4: Structured Neural Network Based on Fuzzy Logic

SENSOR-BASED DRIVING OF A CAR WITH FUZZY INFERENCING VLSI CHIPS AND BOARDS

FRANÇOIS G. PIN and **YUTAKA WATANABE**
Autonomous Robotic Systems Group
Oak Ridge National Laboratory
P.O. Box 2008
Oak Ridge, TN 37831-6364, U.S.A.

ABSTRACT

This paper discusses the sensor-based driving of a car in a-priori unknown environments using "human-like" reasoning schemes. The schemes are implemented on custom-designed VLSI fuzzy inferencing boards and are used to investigate two control modes for driving a car on the basis of very sparse and imprecise range data. In the first mode, the car navigates fully autonomously, while in the second mode, the system acts as a driver's aid providing the driver with linguistic (fuzzy) commands to turn left or right and speed up, slow down, stop, or back up depending on the obstacles perceived by the sensors. Experiments with both modes of control are described in which the system uses only three acoustic range (sonar) sensor channels to perceive the environment. Sample results are presented which illustrate the feasibility of developing autonomous navigation systems and robust safety enhancing driver's aid using the new fuzzy inferencing VLSI hardware and "human-like" reasoning schemes.

INTRODUCTION

One of the greatest challenges in developing motion planning and control systems for vehicles operating in a-priori unknown, unpredictable, and dynamic environments is to design the methods for handling the many imprecisions, inaccuracies, and uncertainties that are present and pervasive in the perception and reasoning modules. These imprecisions typically are caused by: (1) errors in the sensor data (current sensor systems are far from perfect) which lead to inaccuracies and uncertainties in the representation of the environment, the robot's estimated position, etc., (2) imprecisions or lack of knowledge in our understanding of the system, i.e., we are unable to generate complete and exact (crisp) mathematical and/or numerical descriptions of all the phenomena contributing to the environment's and/or the system's behavior, and (3) approximations and imprecisions in the information processing schemes (e.g., discretization, numerical truncation, convergence thresholds, etc.) that are used to build environmental models and to generate decisions or control output signals. In such systems, for which it is not currently feasible to fully engineer all the uncertainties away from the perception subsystems, approximate (or "qualitative") reasoning may provide an alternative to the complexity and prohibitive computational requirements of conventional uncertainty analysis and propagation techniques.

In cooperation with MCNC, Inc. and the University of North Carolina, two types of VME-bus-compatible computer boards including custom-designed VLSI chips have been developed to add a qualitative reasoning capability to real-time control systems [1],[2],[3],[4]. The methodologies embodied on the VLSI hardware utilize the Fuzzy Set Theoretic operations [5],[6],[7],[8] to implement a production rule type of inferencing on input and output variables that can directly be specified as qualitative variables through membership functions. All rules on a chip are processed in parallel, allowing full execution of the rule base in about 30 μsec. This extremely short time of operation makes real-time reasoning feasible at speeds much faster than typical sensor data acquisition rates, therefore, making envisionable the control of very fast processes such as sensor-based "reflex-type" motions.

The basic operation of these boards and a formalism merging the fuzzy and behaviorist theories for the development of qualitative reasoning schemes emulating human-like navigation have been discussed in [4]. The approach using superposition of elemental sensor-based fuzzy behaviors has been shown to allow easy development and testing of the inferencing rule base, while providing for progressive addition of behaviors to resolve situations of increasing complexity. This fuzzy behavior formalism has been used to demonstrate the feasibility of autonomous robot navigation in a-priori unknown environments on the basis of sparse and very imprecise sensor data [9]. For these feasibility experiments, a small omnidirectional robotic platform prototype [10] equipped with a ring of acoustic range finders (sonars) was used in a laboratory environment. In this paper, we present further developments on the feasibility of autonomous navigation in a-priori unknown environments using approximate reasoning and very inaccurate sensor data. The next section describes how the "human-like reasoning" navigation rule base of the small omnidirectional platform was extended to allow for the kinematic limitations of a car (non-holonomic and steering constraints) and was applied to the autonomous navigation of a car in laboratory simulations. The following section presents the extension of the system to driver's aid mode, and discusses its operation in simulated as well as in actual outdoor environments where the entire perception and fuzzy inferencing system was positioned on the trunk of a car. The last section discusses the experimental results and presents the concluding remarks.

FUZZY BEHAVIORS FOR CAR DRIVING

In the experiments with the small omnidirectional platform, fuzzy rule bases embodying six basic navigation behaviors [9] were developed to control the turn rate (TR) and the translational speed (TS) of the platform as a function of the goal direction (GD) and obstacle proximity (OP). The single chip board [1] was used which allows inferencing on four input variables to produce two output variables. The four input variables were selected as the goal direction and obstacle proximity in sectors at the left, center, and right of the travel direction. As shown on Fig. 1, each sector encompasses five sonars. In each sector, the distance returns from each of the five sonars are weighed by a factor proportional to their firing direction, and the smallest value is utilized to indicate obstacle proximity within the sector. Effectively, this corresponds to giving the platform the equivalent of three "very wide and blurry" eyes. In the current system the navigation goal can be specified by the user as a goal point or as a heading to be maintained. When the goal is a point, the odometry system updates the position of the robot at each loop rate and calculates the relative direction to the goal point as input to the inferencing system. When the goal is a heading, a compass is used to directly provide the relative goal direction as the difference between the platform current heading and the goal heading. As explained in [4], membership functions representing the levels of uncertainty with which the values were obtained are applied to the four input values. Very robust navigation characteristics were obtained in the laboratory experiments using these very sparse and imprecise sensor data (purposefully selected as such to emphasize the feasibility demonstration), and as little as fourteen fuzzy rules representing the six basic behaviors controlling the platform's turning rate and speed (see [4] or [9]): GD → TR, GD → TS,OP → TS, "far" OP → TR, "near" OP → TR, "very near" OP → TR.

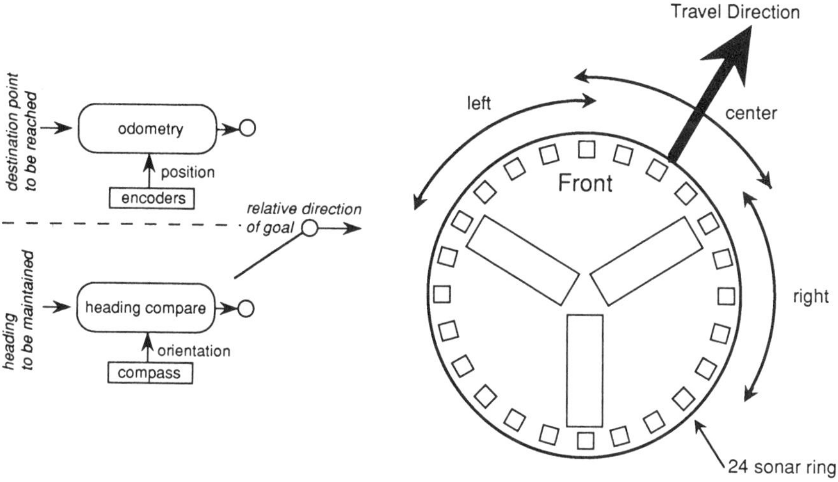

FIG. 1. Schematic of the three 5-sonar sectors providing obstacle proximity input data, and the two methods for calculating the goal direction depending on the mode of goal specification.

Application to Car Driving

One of the expected strengths of our proposed "Fuzzy-Behaviorist" approach using "human-like" behaviors is that the *linguistic logic* embodied in the behaviors should be invariant among systems of similar characteristics. In other words, for robots with similar perceptive and motion capabilities, the linguistic expression of given behaviors, and therefore their representation in the fuzzy framework, should be the same for compatible input and output. For example, a "goal tracking" behavior connecting the perceived goal direction to a rate of turn [e.g. IF (goal is to the right) THEN (apply increment of turn to the right)] should be invariant for any robot which has a means to perceive the goal direction and to perform the required turn. Using this property (and realizing that the rate of turn of a car is proportional to the steering angle of the wheels), all navigation behaviors developed for the laboratory omnidirectional platform appear directly applicable to the speed and steering control of the car, except for those requiring a rate of turn too large for the car to perform because of its limited steering angle. The "very near" OP → TC behavior, which requires the platform to perform high rates of turn (using its omnidirectional capability) when obstacles are detected at dangerously close ("very near") distances, is the only behavior which therefore could not be considered invariant from the platform to the car.

As a demonstration of the transportability of invariant behaviors from one system to another, the same behaviors (except for the "very near"

OP → TC behavior) and the very same fuzzy rules that were utilized for the omnidirectional platform were used to implement the autonomous control of a car on the basis of the same "three wide blurry eyes" and goal direction input. Figure 2 shows a simulation example of such a navigation in which the car has to reach a goal (in the upper right section) and then return to its start position (in the lower left section). Note that the out and return paths are different. Also note that a large maximum steering angle has been selected for the car in this simulation to allow very small radii of turn (e.g. see the sharp turn in the upper right section) and therefore prevent situations with "very near" obstacles.

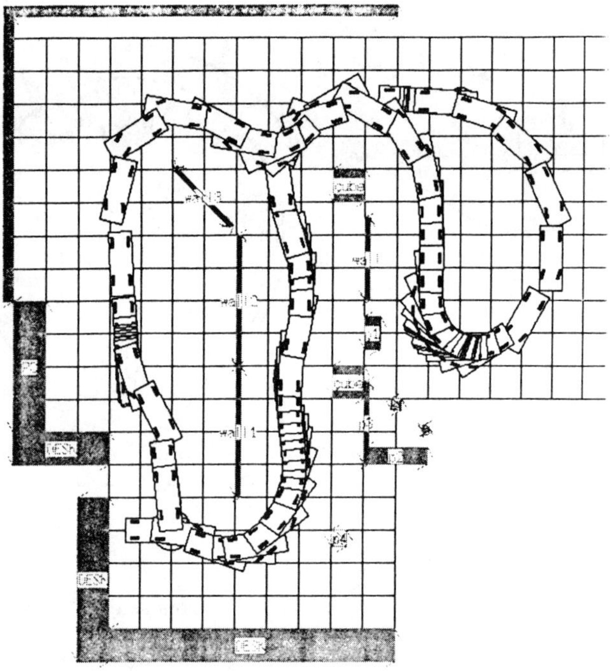

FIG. 2. Simulation example of the autonomous navigation of a car using three "wide" sonars and the same invariant navigation behaviors than for the omnidirectional platform.

Addition of a Maneuvering Behavior

To complete the navigation rule base for the driving of the car, a behavior has to be included to handle the situations where "very near" obstacles are detected. Another strength of our proposed "Fuzzy-Behaviorist" approach is its capability for superposition of elemental behaviors along a "subsumption-type" of architecture (e.g. see [11]), allowing for progressive addition of behaviors to the system to resolve situations of increasing complexity. Since the five other basic behaviors assure collision-free navigation amidst "far" and "near" frontal obstacles, the situations involving "very near" obstacles would occur when the car does not have enough space to complete a turn away from obstacles because of its limited steering angle and radius of turn, and thus would require some maneuvers using reverse gear.

By observing human reactions to such stimuli, a "human-like" response was created which can be expressed as follows: IF (obstacle is "very near" on right (left)) THEN (steer right (left)) AND (back up). This response was further divided into a steer control behavior: "very near" OP → TR, and a speed control (back up) behavior: "very near" OP → TS, to respect our approach's requirement for independence of behaviors [4]. Note that this latter behavior is intrinsically "human-like" since it implements a human reaction which implicitly utilizes the inertia present in the car in order to produce the desired effect.

Figure 3 displays sample results showing several maneuvers generated by the two "very near" OP behaviors in a simulation of the autonomous navigation of a car using the three "wide sonar" eyes as a perception system. Note that in this simulation, the "front" of the car, where the three wide-sonar perception eyes are mounted, corresponds to the axle with non-steering wheels, while the axle with the steering wheels is to the "back" of the car. This was done to closely duplicate the configuration utilized in the outdoor experiments in which the perception system was positioned on the back trunk of the vehicle, as explained in the next section.

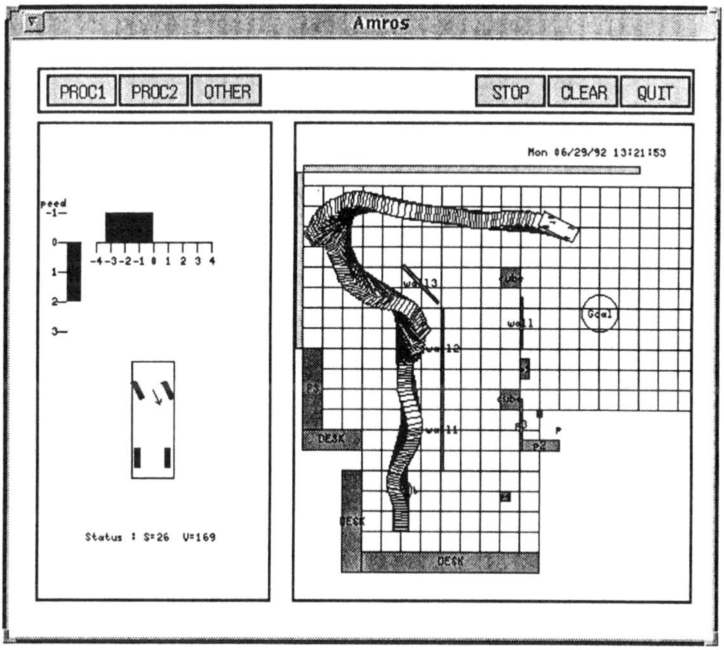

FIG. 3. Simulation example of the autonomous navigation of a car using three "wide" sonars and a maneuvering behavior to overcome the limited radius of turn.

DRIVER'S AID MODE OF OPERATION

In the simulation experiments, the output of the fuzzy inferencing can be conveniently displayed on the screen, as is shown on the left-hand side of Fig. 3. The horizontal and vertical bar scales respectively represent the steering and speed commands which are calculated by the fuzzy inferencing and sent to the controls of the vehicle emulator in the autonomous mode simulation. In the driver's aid mode, the very same commands and displays are used to guide the operator in driving the car, which he/she does using either the keyboard arrow keys in the simulations or the gas and brake pedals and steering wheel in the implementation of the system on one of the company's cars. For the testing and verification experiments, the driver was prohibited from seeing the environment while driving, either by covering the vehicle motion display part of the screen in the graphic simulations, or by positioning the sensing platform on the rear trunk of the car for the outdoor experiments and having the operator drive backwards while looking at the portable computer screen located on his/her lap. Note that the commands are not displayed to the operator as crisp control values, but as bars of variable lengths over the generic speed and steering scales, effectively providing only the direction of the command (left or right, forward or back) and the *relative strength* (i.e., more steering, faster, slower, etc.) to apply on the controls between the maximum steering and speed values. It was interesting to observe each operator develop his/her own interpretation of and response to these relative commands, which from the system's development point of view, provided a very stringent robustness test of the inferencing rule base. As described in more details in Ref. [12], the car was very successfully driven in the "blind" driver's aid mode in the diversely occupied parking lots of ORNL. Our future plans include the integration of encoders and servo controls on the wheels, steering, accelerator, and braking systems of the car to experiment with, test and demonstrate the autonomous control mode in outdoors environment.

CONCLUDING REMARKS

VLSI fuzzy inferencing chips and a "fuzzy behaviorist" approach have been used to demonstrate the feasibility of sensor-based autonomous navigation and driver's aid systems functioning with sparse data from very inaccurate sensors. The "subsumption-type" formalism proposed for the development of fuzzy behavior-based systems allows easy development of the behaviors and progressive augmentation of the fuzzy rule base to deal with situations of increasing complexity, such as in the example treated here of a need for maneuvering due to the car's limited radius of turn. Moreover, the framework has been shown to allow the same behaviors, rules, and inferencing code to be used for systems with similar characteristics, greatly enhancing code transportability among robots and systems. As shown in the driver's aid feasibility study, the straightforward linguistic interfacing capability of the fuzzy behavior-based system is also of great appeal for telerobotics and man-machine decisional systems. Our ongoing activities are focusing on the use of the recently developed multi-chip board in conjunction with additional on-board image sensors to increase the car's autonomous navigation capabilities and the safety enhancing driver's aid system for a variety of outdoor environments.

REFERENCES

1. H. Watanabe, J. R. Symon, W. D. Dettloff, and K. E. Yount, "VLSI Fuzzy Chip and Inference Accelerator Board Systems," in Proceedings of the International Symposium on Multivalued Logic, Victoria, Canada, May 1991, pp. 120–127.

2. J. R. Symon and H. Watanabe, "Single Board System for Fuzzy Inference," in Proceedings of the Workshop on Software Tools for Distributed Intelligent Control Systems (September 1990), pp. 253–261.
3. H. Watanabe, W. Dettloff, and E. Yount, "A VLSI Fuzzy Logic Controller with Reconfigurable, Cascadable Architecture," *IEEE J. of Solid State Circuits* **25**(2), 376–382 (1990).
4. F. G. Pin, H. Watanabe, J. R. Symon, and R. S. Pattay, "Using Custom-Designed VLSI Fuzzy Inferencing Chips for the Autonomous Navigation of a Mobile Robot" in Proceedings of IROS 92, the 1992 IEEE/RSJ International Conference on Intelligent Robots and Systems, Raleigh, North Carolina, July 7–10, 1992.
5. L. A. Zadeh, "Fuzzy Set," *Information and Control* **8**, 338–353 (1965).
6. L. A. Zadeh, "Outline of a New Approach to the Analysis of Complex Systems and Decision-Making Approach," *IEEE Transactions on Systems, Man, and Cybernetics* **SMC-3**(1), 28–45 (January 1973).
7. L. A. Zadeh, "Fuzzy Logic," *IEEE Computer* **21**(4), 83–93 (April 1988).
8. "Fuzzy Sets and Their Applications to Cognitive and Decision Processes," eds. L. A. Zadeh, K. S. Fu, K. Tanaka, and M. Shinmra, Academic Press, Inc., New York (1975).
9. F. G. Pin, H. Watanabe, J. R. Symon, and R. S. Pattay, "Autonomous Navigation of a Mobile Robot Using Custom-Designed Qualitative Reasoning VLSI Chips and Boards," in Proceedings of the 1992 IEEE International Conference on Robotics and Automation, Nice, France, May 10–15, 1992, pp. 123–128.
10. S. M. Killough and F. G. Pin, "Design of an Omnidirectional and Holonomic Wheeled Platform Prototype," in Proceedings of the 1992 IEEE International Conference on Robotics and Automation, May 10–15, 1992, Nice, France, pp. 84–90.
11. R. A. Brooks, "Elephants Don't Play Chess," *Robotics and Autonomous Systems* **6**(1–2), 3–15 (1990).
12. F. G. Pin and Y. Watanabe, "Driving a Car with Custom-Designed Fuzzy Inferencing VLSI Chips and Boards," in Proceedings of the 1992 International Joint Technology Workshop on Fuzzy Logic and Neural Networks, Houston, Texas, June 1–3, 1992.

Hybrid Fuzzy State Feedback Control of High Order Nonlinear Systems --- A Case Study

Lienjing Chen and Scott M. Packer

United Technologies Research Center
East Hartford, CT 06108

ABSTRACT

In this paper, a study of a high order nonlinear system is reported by employing both fuzzy logic control and state feedback control of a cart-pole system. The hybrid fuzzy state feedback control (HFSFC) demonstrated the capability to control the system in both nonlinear and linear cases when the model of the system is assumed unknown for the fuzzy logic control. The simulation results show that the pole can be balanced at a larger initial pole angle by the HFSFC than the nonlinear controller.

INTRODUCTION

Manufacturing in today's global economy demands high level of quality at minimum cost. Automated control of manufacturing processes is one of methods of achieving this goal. To control a process by modern control theory often requires an accurate model; however, due to the complexity and uncertainty of manufacturing processes, the process is often not well understood and therefore a process model is not developed. This lack of process models limits the application of modern control theory. An alternative to modern control theory is required so that these processes can be controlled.

Several advantages of fuzzy logic control (FLC) over conventional control have been demonstrated including the capability to engage the experience, heuristic and intuition of an expert operator in the control algorithm without detailed models. Significant reduction in the development time using the FLC leads to a lower cost and faster to market time. It is therefore worth investigating the suitability of the FLC for complex systems including high order nonlinear dynamic systems. It is also recognized that there is a need to take advantage of both FLC and modern control when it is appropriate [1].

In this paper, a study of high order nonlinear system control using a hybrid controller based on both fuzzy logic control and state feedback control is reported. The fuzzy control rules are determined based upon the responses of the system obtained through a nonlinear controller. The simulation results show that the pole can be balanced at a larger initial pole angle by the hybrid fuzzy state feedback control (HFSFC) than the nonlinear controller.

FUZZY LINGUISTIC CONTROL

One of the most important applications of the fuzzy set theory is in the area of control system design and implementation. The design of a control algorithm is based on a very general kind of inexact information, namely information expressed in natural language. This linguistic information may be obtained from three sources [2], namely human operator's experience and/or control engineer's knowledge, fuzzy modeling of human operator's control actions, and a fuzzy model of a process. The obtained linguistic control strategy consists of a set of linguistic decision rules in the following form:

R_1: *if x is* A_1, *and y is* B_1 *then z is* C_1,

or

R_2: *if x is* A_2, *and y is* B_2 *then z is* C_2,

or

...

or

R_n: *if x is* A_n, *and y is* B_n *then z is* C_n.

where A_i, B_i, and C_i are fuzzy subsets of U, V, W accordingly (i=1, 2, ..., n). Given (x is A') and (y is B'), the degrees of match, α_i, β_i , of (A_i and A') and (B_i and B') are given respectively by Eqs. (1) and (2) as follows:

$$\alpha_i = \max_{\mu} \min \{\mu_{A_i}(\mu), \mu_{A'}(\mu)\}, \qquad (1)$$

$$\beta_i = \max_{v} \min \{\mu_{B_i}(v), \mu_{B'}(v)\}. \qquad (2)$$

The firing strength λ_i of the ith rule can be expressed by Eq. (3):

$$\lambda_i = \min (\alpha_i, \beta_i). \qquad (3)$$

Thus, the control action inferred from the ith rule is obtained from

$$\mu_{C'_i}(w) = \min \{\lambda_i, \mu_{C_i}(w)\}. \qquad (4)$$

Finally, when the union operator is used for the sentence connective *or*, the control decision based upon individual control action from Eq. (4) can be obtained by a conflict-resolution process:

$$\mu_C(w) = \max\{\mu_{C'_1}, \mu_{C'_2}, \ldots, \mu_{C'_n}\}. \qquad (5)$$

HYBRID FUZZY STATE FEEDBACK CONTROL DESIGN

The high order nonlinear system is chosen to be the cart-pole system with friction. The mathematic model of the car-pole is given by Eqs. (6)-(11) as follows:

$$\ddot{\theta} = h_1 - h_2 \ddot{X} \quad (6)$$

$$\ddot{X} = (h_3 + F(w)/h_4) \quad (7)$$

with

$$h_1 = (3/4L) g \sin\theta \quad (8)$$

$$h_2 = (3/4L) \cos\theta \quad (9)$$

$$h_3 = m(L\sin\theta\dot{\theta}^2 - (3/8) g \sin 2\theta) - f\dot{X} \quad (10)$$

$$h_4 = M + m(1 - (3/4)\cos^2\theta) . \quad (11)$$

where $w(t)=(X(t),\dot{X},\theta(t),\dot{\theta}(t))$ is the state variable and F(w) corresponds to the control force, cart mass M=1 kg, pole mass m=0.1 kg, pole length L=1 m, friction coefficient f=5 kg/s, and acceleration of gravity $g=9.8 \text{ m/s}^2$. Previous work in the FLC of the cart-pole system [3] addresses the pole and omits the cart dynamics, which leads to a reduced second order system. The assumptions of no friction and small pole angle ($\theta \leq 10$ deg) are also usually made. In this research, the fourth order system is considered with friction force and larger pole angle ($\theta = 68.8$ deg). Immediate control action should be taken when a large pole angle occurs in order to prevent the pole from falling to the ground. Therefore, three controllers are designed for large, medium and small pole angles accordingly. The overall structure of the proposed HFSFC is depicted in Fig. 1. The first fuzzy controller is used when the pole angle θ is greater than angle θ_1 ($\theta_1=8.6$ deg). The second fuzzy controller is employed for the case when the pole angle θ is in the range of $[\theta_2,\theta_1]$ ($\theta_2=4$ deg). The universe discourses for the first and second fuzzy controllers are normalized so that the membership functions of the primary fuzzy sets and fuzzy control rules are actually identical for both fuzzy controllers, which simplifies the design of the controllers. Since the system behaves much like a linear system when the pole angle is smaller than angle θ_2, a conventional state feedback control is utilized.

Most of fuzzy controllers have been designed based on a human operator's experience and knowledge. This method is effective where human experience and knowledge play an important role. However, for many control problems such as the cart-pole system, linguistical control rules are not intuitive to the operator. In this case, fuzzy control rules can be derived based on a fuzzy model of process behavior. Specifically, the expected behavior of the process is considered in the derivation of the specific rules. The fuzzy control rules for the cart-pole system were derived from the system responses and control force generated by a feedback linearizing and decoupling transform [4]

$$F[w] = (h_4/h_2)[h_1 + K_1\theta + K_2\dot{\theta} + C_1X + C_2\dot{X}] - h_3 \quad (12)$$

where $K_1=25$, $K_2=10$, $C_1=1$, and $C_2=2.6$. A typical system response and corresponding control force with w(0)=(4m, -8m/s, 40 deg, 180 deg/s) are illustrated in Figs. 2, 3, 4. A total thirty-one fuzzy control rules were derived and are not listed here due to space limitation. Since a crisp control action is required for the control of the cart-pole system, defuzzification of the Mean of Maximum Method (MOM) is applied to the output fuzzy subsets. The state feedback controller is designed for the near-linear region, and expressed as follows:

$$F(w) = K\, w^T \qquad (13)$$

with

$$K = [171.5,\ 68.2,\ 6.8,\ 12.8].$$

SIMULATION RESULTS AND DISCUSSIONS

The simulations of the cart-pole system controlled by the HFSFC shown in Fig. 1 were conducted. The cart-pole system reaches the equilibrium point at $t=12$ seconds when the initial condition w(0)=(4 m,-8 m/s,48.1 deg,180 deg/s) is used. The cart-pole system also reaches the equilibrium point at $t=12$ seconds with initial condition w(0)=(4 m,-8 m/s,68.8 deg,180 deg/s). The system responses and control force with such initial condition are depicted in Figs. 5, 6 and 7. It should be pointed out that, although the fuzzy control rules were derived based upon the nonlinear controller responses with an initial 40 degree pole angle, the HFSFC successfully controls the system with a larger initial pole angle (68.8 deg). The nonlinear controller failed when initial condition w(0)=(4 m,-8 m/s,57.3 deg,180 deg/s) was used. However, it can be seen that there are more oscillations in the responses of the pole angle, pole angular velocity and cart velocity controlled by the FLC than those controlled by the nonlinear controller. In Figs. 5, 6, 7 less oscillations can be observed in the system responses and control force using the state feedback controller in the near-linear region.

CONCLUSIONS AND FUTURE RESEARCH

The hybrid fuzzy state feedback controller is successfully designed and used to control a fourth order nonlinear system (cart-pole system). The simulation results indicate that the HFSFC demonstrates the capability to control the system in both nonlinear and linear cases when the model of the system is assumed unknown for the fuzzy logic control. The HFSFC can also control the cart-pole system with a larger initial angle (68.8 deg) comparing with the nonlinear controller. One of advantages of the HFSFC is the increase of robustness by the FLC while improving performance including less oscillations by the state feedback control. It is realized, however, that during the design of the fuzzy logic controllers the determination of the fuzzy control rules is much more complex with increase of the system order and when the friction force is included.

It is anticipated that the results from this research will provide useful information which is essential for application of fuzzy controls in manufacturing processes.

REFERENCES

1. Y. Saito and T. Ishida, "Fuzzy PID Hybrid Control - An Application to Burner Control," *Proc. Int. Conf. on Fuzzy Logic & Neural Networks* (Iizuka, Japan), pp. 65-69 (1990).

2. M. Sugeno and G. T. Kang, "Fuzzy Modeling and Control of Multi-Layer Incinerator," *Fuzzy Sets and Systems*, 18, pp. 329-346, (1986).
3. T. Yamakawa, "Stability of An Inverted Pendulum by A High-Speed Fuzzy Logic Controller Hardware System," *Fuzzy Sets and Systems*, 32, pp. 161-180, (1989).
4. A. Guez and J. Selinsky, "A Trainable Neuromorphic Controller," *J. Robotics Systems*, Vol. 5, No. 4, pp. 368-388, (1988).

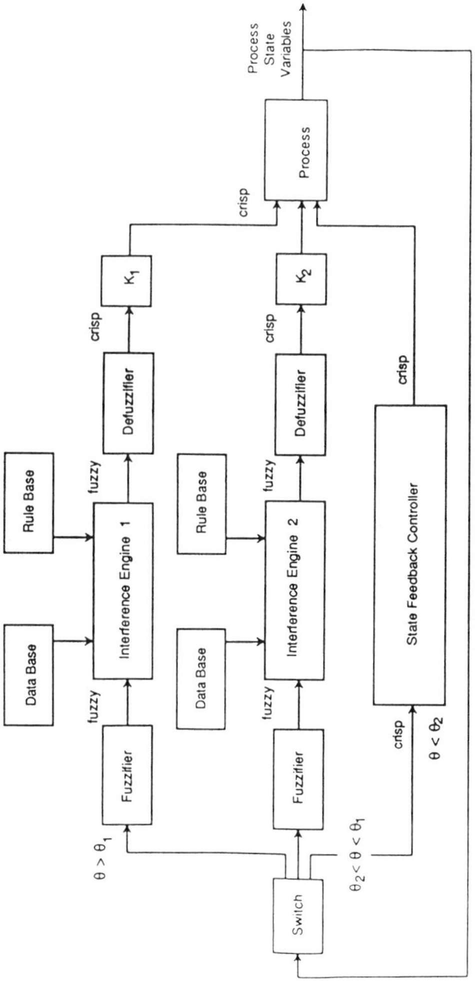

Fig. 1. Overall structure of the hybrid fuzzy state feedback controller.

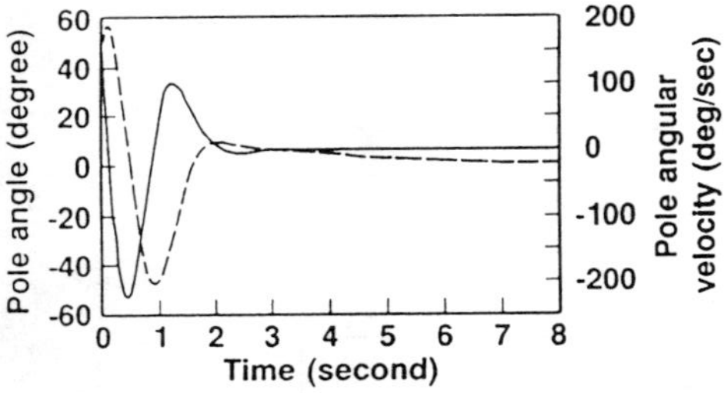

Fig. 2. Pole angle (solid) and pole angular velocity (dashed) (nonlinear controller) with w(0)=(4 m,-8 m/s,40 deg,180 deg/s).

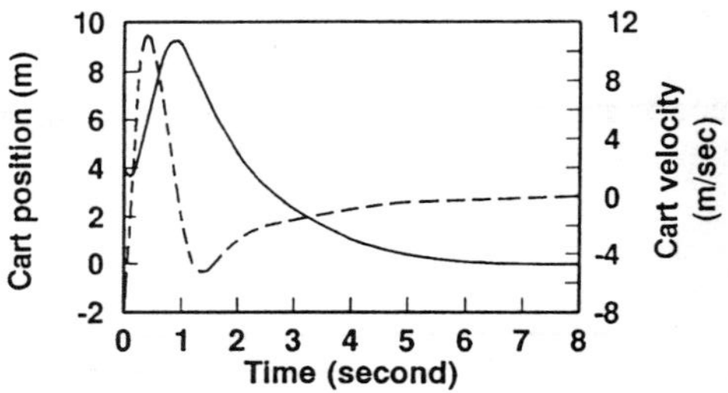

Fig. 3. Cart position (solid) and cart velocity (dashed) (nonlinear controller) with w(0)=(4 m,-8 m/s,40 deg,180 deg/s).

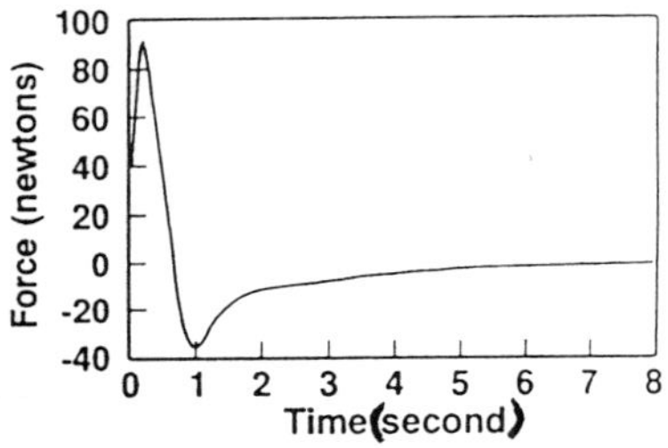

Fig. 4. Control force (nonlinear controller) with w(0)=(4 m,-8 m/s, 40 deg,180 deg/s).

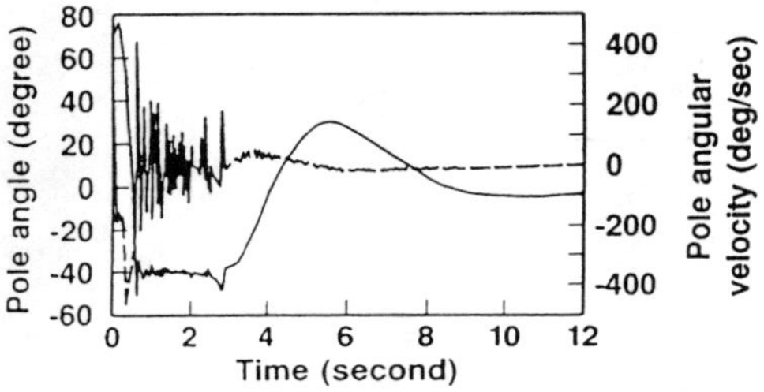

Fig. 5. Pole angle (solid) and pole angular velocity (dashed) (the HFSFC) with w(0)=(4 m,-8 m/s,68.8 deg,180 deg/s).

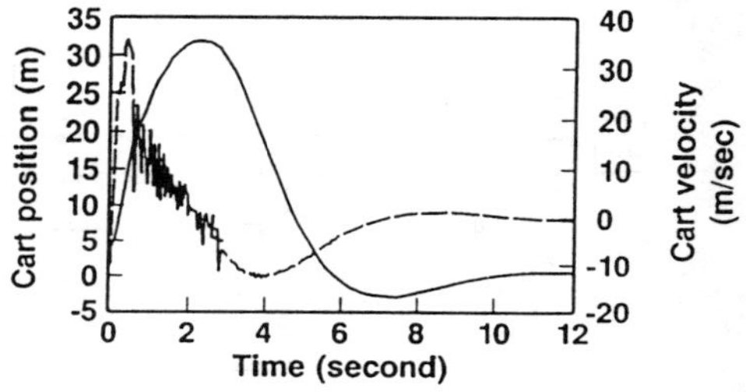

Fig. 6. Cart position (solid) and cart velocity (dashed) (the HFSFC) with w(0)=(4 m,-8 m/s,68.8 deg,180 deg/s).

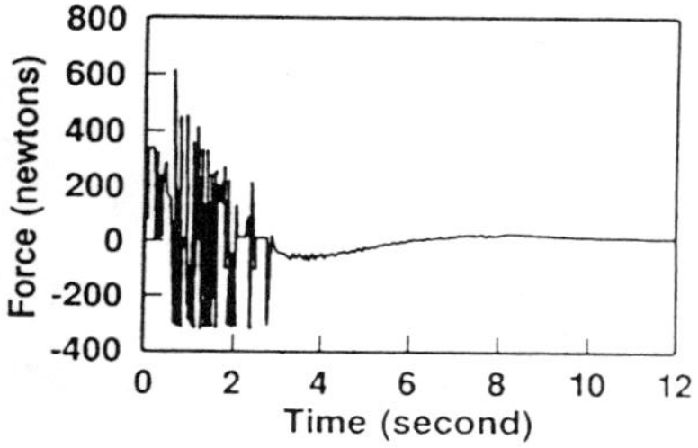

Fig. 7. Control force (the HFSFC) with w(0)=(4 m,-8 m/s, 68.8 deg,180 deg/s).

A MULTIATTRIBUTE MODEL FOR PROCESS PLANS SELECTION USING FUZZY LOGIC

F. Choobineh, P.E., Ph.D., and M. Shivani
Department of Industrial and Management Systems Engineering
University of Nebraska-Lincoln
Lincoln, NE 68588-0518

INTRODUCTION

Process plans specify the machines and auxiliary devices needed for production of parts. The most important auxiliary devices specified are cutting tools and fixtures. The set of machines, tools, and fixtures specified by a feasible process plan for a part is not necessarily unique. That is, the same part could be manufactured with an alternative plan calling for a different set of machines, tools and/or fixtures. From among the feasible alternative plans the most appropriate plan should be selected.

Selecting the most appropriate process plans for a set of parts requires an understanding and determination of the degree of appropriateness of alternative process plans for each part. The degree of appropriateness of a plan is a function of multiple attributes representing time, cost, and quality criteria. Thus, the process plans selection problem is multiattribute decision problem.

The rating of each plan with respect to most attributes can not be determined or stated with certainty. The reason is that process planners use a large amount of subjective knowledge and uncertain or incomplete data to generate process plans. Their ratings can not escape their subjectivity nor the inherent uncertainty of data. The primary cause of uncertainty in process planning is due to imprecise (vague) process knowledge. Process knowledge used in process planning is knowledge about the capabilities of manufacturing processes. It is acquired mostly by empirical methods and capabilities of a process is directly influenced by the machine tool, cutting tool and fixtures that are available in the plant. Industry guidelines based on empirical results generally specifies a feasible range of values for each decision variables. These ranges or intervals are not precise, and in addition, each value within the interval has a different degree of appropriateness. Thus, the process planner has to act according to heuristic knowledge or rule of thumb to resolve uncertainty and select appropriate values of decision variables.

The performance of a process plan with respect to a set of production attributes determines the appropriateness of the process plan. The performance with respect to an attribute may be quantitative, as measured by a crisp value, or qualitative as described by a linguistic variable. In either case each performance could be rated against a goal or a most desirable value of each attribute. This rating could be done linguistically and presented by a possibility distribution [6].

Possibility theory [6] is a viable tool for dealing with vagueness. Briefly, if X is a variable taking values in a universe of discourse U, then the possibility distribution of X, Π_x is the fuzzy set of a possible values of X. More specifically, let $\Pi_x(u)$ denote the possibility that X can take the value "u", $u \in U$. Then, the membership function of X is numerically equal to the possibility distribution function $\Pi_x(u)$: $U \rightarrow [0,1]$, which associates with each element $u \in U$, the possibility that X may take "u" as its value. A possibility distribution is called normal if it is possible to obtain an α-cut with $\alpha=1$ from the distribution. Further a possibility distribution is convex if and only if each of its α-cuts are convex. The support of a possibility distribution is defined as: SUPP ($\prod(X)$) = {X | $\prod(x) > 0$}.

Kusiak and Finke [2] developed an integer program for selecting the optimum process plans. Their model minimizes the sum of costs and weighted hamming distances among process plans. Singh et al. [4] developed a multiobjective model where they try to maximize the production rates while minimizing the costs and Hamming distances among process plans. Singh and Mohanty [5] developed a fuzzy approach to multiobjective routing problem and applied it to the process plan selection problem. They only used two objectives, time and cost, and assumed time and cost per components can be represented by triangular fuzzy numbers. They used Saaty's [3] method to obtain the weight of each objective.

This paper proposes a multiattribute fuzzy procedure for determining the optimal set of process plans. It uses an index to obtain an appropriateness value for each process plan. The appropriateness values are then used in an optimization model to obtain the set of most appropriate plan.

THE PROPOSED PROCEDURE

The steps of the proposed procedure to obtain an appropriateness value for a process plan is as follows.

1. Generate an adequate number of technologically feasible process plans for each part.
2. Identify the production performance attributes necessary for evaluating the process plans.
3. Determine the weight of each performance attribute.
4. Obtain the possibility distribution of each plan with respect to every performance attribute on the unit interval.
5. Aggregate the weighted possibility distributions to obtain a possibility distribution for each process plan.
6. Defuzzify the aggregated possibility distribution to obtain an appropriateness value for each process plan.
7. Maximize the sum of appropriateness values by an optimization model to obtain the optimal set of process plans.

Generating two or three technologically feasible process plans should pose no problem in practice. In addition, each company routinely uses a set of production performance measures (attributes) to monitor the production quality and efficiency. Therefore, implementing steps 1 and 2 of the procedure should pose no problem to practitioners.

Step 3 calls for determining the weight of each performance attribute. Analytical heirarchy process (AHP) proposed by Saaty [3] is an appropriate procedure for this purpose. Although an active debate in the management science community on the short comings of the AHP is on-going, in our opinion AHP is a viable tool for determining the attribute weights in a multiattribute decision process.

The performance of each process plan can be rated linguistically with respect to every performance attribute on the unit interval. Linguistic rating facilitates evaluation of a process plan with respect to quantitative as well as qualitative attributes. Use of five or seven linguistic terms should sufficiently covers the unit interval and provide adequate linguistic flexibility to the process planner to differentiate among process plans.

It is very important that an appropriate connective is used to obtain the aggregate possibility distribution. Generally the two most commonly used connectives are the "union" and the "intersection". From the linguistic point of view, Zadeh's operation for "or" corresponds with the union. Hence, in this case only the most important attribute is going to play a prominent role in shaping the aggregate possibility distribution of the alternative. The other most commonly used connective is the "intersection" which linguistically corresponds to "anding" operation. In this case there is a very good chance that the weaker attribute would play a dominant role in shaping the aggregate possibility distribution. For this application it is necessary to consider all the attributes, based on their relative importance. An appropriate connective to obtain the aggregate possibility distribution would be to use the generalized-means. Generalized-means are one class of averaging operation that cover the entire interval between the boundaries of the classical intersection and union operations. Weighted generalized-means could be used when it is desirable to accommodate variations in the importance of individual attributes. One such connective which has been used to obtain the aggregate possibility distribution for this application is:

$$\Pi_t(x) = \frac{1}{n} \sum_{i=1}^{n} W_i \, \Pi_{it}(x)$$

where $\Pi_t(x)$ = aggregate possibility distribution of the tth process plan.
$\Pi_{it}(x)$ = possibility distribution of process plan t with respect to the ith attribute.
W_i = weight of the ith attribute.

To obtain an appropriateness index for a process plan, its possibility distribution needs to be defuzzified. To defuzzify a possibility distribution a number of approaches are available. A most common approach is to obtain the centroid of the

possibility distribution. That is $R_i = \int_0^1 x \, \Pi_i(x) \, dx \, / \int_0^1 \Pi_i(x) \, dx$, where R_i is

the centroid of the distribution. An alternative method is to use the ranking index proposed in [1]. This ranking index utilizes a maximizing and a minimizing barriers. A maximizing barrier is a crisp number that is placed at U, and a minimizing barrier is also a crisp number that is place at L. The range U-L should contain the support of all possibility distributions under considerations. The appropriateness index as proposed in [1] is

$$R_i = 1/2 \, [h_i - (D(\Pi_i(x),U) - D(\Pi_i(x),L)) \, / \, (U-L)].$$

where h_i is the height of the possibility distribution and $D(\Pi_i(x),U) =$

$\int_0^{h_i} [U-g_i(\alpha)] \, d\alpha$ and $D(\Pi_i(x),L) = \int_0^{h_i} [f_i(\alpha)-L] \, d\alpha$. $g_i(\alpha)$ and $f_i(\alpha)$ are

respectively the right and left edges of the possibility distribution. This index in essence is based on the set difference of the area of a rectangle and the area under the membership function. Properties of this index are delineated in [1]. Once the possibility distributions are defuzzified the following integer program could be solved to obtain the optimal set of process plans.

$$\text{maximize} \sum_i \sum_j R_{ij} Y_{ij}$$

s.t.

$$\sum_i Y_{ij} = 1 \quad \forall\ j$$

$$\sum_i \sum_j t_{ijm} Y_{ij} \le T_m \quad \forall\ m$$

$$Y_{ij} = (0,1)$$

where

Y_{ij} = 1 if ith process plan of part j is selected, and 0 otherwise.
R_{ij} = Appropriateness index of ith process plan of jth part.
t_{ijm} = Amount of time needed by process plan i of part j on machine m.
T_m = Total time available on machine m.

The first constraint insures that only one process plan is selected for each part. The second constraint ensures that the process time of parts on any machine does not exceed the available capacity on that machine during the planning horizon.

CONCLUSIONS

A multiattribute procedure for selecting the optimum set of process plans over a planning horizon has been presented. The proposed procedure provides a vehicle for modeling the inherent uncertainty and subjectivity that exist in the process plan selection problem. Fuzzy sets is a natural tool for modeling uncertainty and subjectivity of this type. More detailed optimization models for this problem are under development, and we are very optimistic about the practically of our proposed procedure.

REFERENCES

1. F. Choobineh and H. Li, "An Index for Ordering Fuzzy Numbers," to appear in *Fuzzy Set and Systems* (1992).
2. A. Kusiak and G. Finke, "Selection of Process Plans in Automated Manufacturing Systems," *IEEE Journal of Robotics and Automation*, 4(4), 397-402 (1988).
3. T. L. Saaty, "Exploring the Interfaces Between Hierarchies, Multiple Objectives and Fuzzy Sets," *Fuzzy Sets and Systems*, 1(1), 57-68 (1978).
4. N. Singh, Y. P. Aneja, and S. P. Rana, "Multi-Objective Modelling and Analysis of Process Planning in Manufacturing System," *Int. J. Systems Sci.*, 21(4), 621-630 (1990).
5. N. Sing, and B. K. Mohanty, "A Fuzzy Approach to Multi-Objective Routing Problem with Applications to Process Planning in Manufacturing System," *Int. J. Prod. Res*, 29(6), 1161-1170 (1991).
6. L. A. Zadeh, "Fuzzy Sets as a Basis for a Theory of Possibility," *Fuzzy Sets and Systems*, 1(1), 3-28 (1978).

A FUZZY LOGIC BASED ROBOT CONTROLLER

Ashok Nedungadi, Ph.D
Data Systems Department
Southwest Research Institute
6220 Culebra Road
San Antonio, TX 78228-0510

Abstract

Conventional robot controllers rely on an exact inverse kinematic model to convert user specified cartesian trajectory commands to joint set points. The excessive computational burden that these equations pose provided the impetus to investigate a robot controller that did not require explicit solution of the inverse kinematic equations. In the following, a novel approach based on fuzzy logic is presented, which replaces the computationally intensive inverse kinematic equations and lays the foundation for a robot controller with a much simpler parallel architecture. A hardware implementation of a fuzzy logic based robot controller is described. The implemented fuzzy robot controller was tested on a four degree-of-freedom planar lab robot. Hardware test results of the lab robot tracking arbitrary trajectories are presented and discussed.

1.0 INTRODUCTION

Fuzzy logic provides a rule based reasoning approach that can be applied to the control of complex systems, whose dynamic model description is either too complex to develop for the application or completely unavailable. Recent applications of fuzzy logic also include the control of mathematically well described systems, such as the inverted pendulum problem. The advantages of fuzzy logic over conventional control theories are: increased robustness in spite of noise or other sensor failures, and the ability to handle nonlinearities without degradation or failure. Another important feature of fuzzy logic lies in the formulation of the fuzzy rules that use linguistic adjectives and interrelations similar to a natural language. This makes the understanding and modification of a fuzzy logic based controller much easier than is possible with conventional control techniques. The main disadvantages of fuzzy logic based systems are the unavailability of a stability analysis for the developed controller, as well as the lack of a formal procedure to describe the necessary fuzzy sets and membership functions of the fuzzy control rules.

Since the inception of fuzzy sets by Zadeh in 1965 [1], fuzzy logic has found interesting applications worldwide, such as steam engine control [2], a chemical reactor controller [3], and for far more complex systems such as trains [4], traffic light control, automatic focusing of cameras and video camcorders.

At Southwest Research Institute (SwRI), we have developed a robot controller that exploits the principles of fuzzy logic to replace the mathematically complex inverse kinematic equations that are at the heart of operations of a conventional robot controller. In this paper, we present a set of fuzzy control rules for a planar robot with an arbitrary number of serial degrees-of-freedom (DOF). The hardware implementation of the fuzzy robot controller is described. In addition, experimental results of a planar 4 degree-of-freedom (DOF) lab robot controlled by the developed fuzzy controller are presented.

2.0 A FUZZY LOGIC BASED ROBOT CONTROLLER

Figure 1 shows a block diagram of the developed fuzzy robot controller. This was implemented on a MVME-131 single board host computer with a 16 MHz 68020 microprocessor and a 68881 math coprocessor. All software for control of the robot is written in the C programming language under the Microware OS-9 real time operating system.

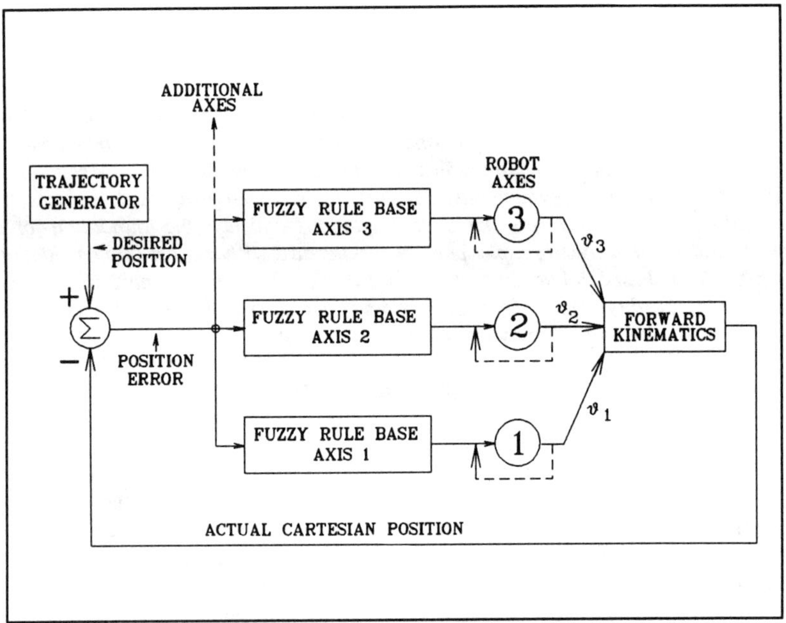

Figure 1 A Block Diagram of the Proposed Fuzzy Logic Controller

The block diagram of Figure 1 differs from traditional robot controllers in two main aspects: (i) the absence of the inverse kinematic equations, and (ii) a *fuzzy rule base* is dedicated to a single joint, thus decoupling the robot joints from one another. Therefore, the developed fuzzy logic based controller is not limited by the computationally complex inverse kinematic equations nor does it have any kinematic singularities. Additional axes or capabilities, such as obstacle avoidance can be achieved by simply adding an additional layer of fuzzy rules. The modular nature of the developed controller permits parallel execution of the individual fuzzy rule bases resulting in extremely high loop closure rates. Present experimental results indicate that the main feedback loop of Figure 1 can be closed at a rate that well exceeds 1 KHz if each fuzzy rule base was governed by a single dedicated fuzzy chip which are presently available from several vendors.

The *forward kinematic equations* are used to compute the actual end effector position. The cartesian position errors between the desired and actual position of the robot end effector drives each of the fuzzy rule bases of Figure 1. The output of each fuzzy rule base is an incremental joint command that is sent to a three axis

GALIL-DMC-530 axis motion controller card. The broken line around each robot axis indicates the loop closure of the GALIL* axis motion controller card, which is presently set at 1KHz. Theoretically, it would be feasible to completely discard the motion controller card if the output of the fuzzy rule base were an analog voltage command.

Each fuzzy rule base of Figure 1 comprises of three elements: (i) a *fuzzifier*, (ii) a *rule evaluator*, and (iii) a *defuzzifier*. The fuzzifier maps physical variables such as the cartesian position errors of the robot end effector to fuzzy set variables, which serve as inputs to the next stage: the rule evaluator. A collection of fuzzy control rules expressed in the form IF (A) THEN perform (B) and IF (A') THEN perform (B') are contained within the rule evaluator, which outputs a fuzzy command depending on the input received. The defuzzifier performs the inverse process of converting the fuzzy outputs of the fuzzy rule evaluator to a "crisp" physical variable such as robot joint increment. This defuzzified value represents the weighted average of all the fuzzy rules that were invoked within the fuzzy rule evaluator for the particular set of inputs.

3.0 FUZZY RULES FOR ROBOT CONTROL

The fuzzy control rules within the rule evaluator are critical to satisfactory and stable performance of the robot controller. The fuzzy rules may be derived from a mathematical model if one exists, or through expert knowledge, or from available experimental data that relates inputs to outputs. Naturally, there exits an infinite set of fuzzy rules that may be applicable to the robot control problem. In the following, we present a technique to derive one such acceptable set of fuzzy control rules.

Figures 2 and 3 show the fuzzy rules which determine the incremental joint motion for each individual robot joint. A_i and B_i represent the negative y and x coordinate of joint i, respectively. δx and δy correspond to the desired moves of the end effector along the X and Y axis respectively. $\delta\theta_i$ represents the incremental joint move for the i-th robot joint.

The following seven fuzzy sets were selected to describe each of the variables A_i, B_i, δx, δy, $\delta\theta_i$: Positive Big (PB), Positive Medium (PM), Positive Small (PS), Zero (Z), Negative Big (NB), Negative Medium (NM), and Negative Small (NS). Considering the first joint (i=1), each cell in the tables of Figures 2 and 3 represents one fuzzy rule of the form:

IF (A_1 is NB) AND (δx is NB) THEN $\delta\theta_1$ is PB

Note that this particular fuzzy rule results from relating the first row (A_1 = NB) with the first column (δx = NB) and reading the corresponding output for $\delta\theta_1$. Each of Figures 2 and 3 contain 49 (7x7) fuzzy rules of the above type. However, the total number of fuzzy rules can be further reduced by grouping multiple cells to form compound fuzzy rules of the form:

IF (A_1 is NEGATIVE) AND (δx is NM) THEN $\delta\theta_1$ is PB

By exploiting the pattern and forming compound fuzzy rules as the latter, the 49 rules were reduced to 14 fuzzy rules for each of Figures 2 and 3.

Rule:
If dx is NB and A_i is NB then $d\theta_i$ is PB

Rule:
If dx is PM and A_i is negative then $d\theta_i$ is NM

$d\theta_i$		dx						
		NB	NM	NS	Z	PS	PM	PB
A_i	NB	PB	PM	PS	Z	NS	NM	NB
	NM	PB	PM	PS	Z	NS	NM	NB
	NS	PB	PM	PS	Z	NS	NM	NB
	Z	Z	Z	Z	Z	Z	Z	Z
	PS	NB	NM	NS	Z	PS	PM	PB
	PM	NB	NM	NS	Z	PS	PM	PB
	PB	NB	NM	NS	Z	PS	PM	PB

Figure 2 Fuzzy Rule Base A to Determine $\delta\theta_i$

$d\theta_i$		dy						
		NB	NM	NS	Z	PS	PM	PB
A_i	NB	PB	PM	PS	Z	NS	NM	NB
	NM	PB	PM	PS	Z	NS	NM	NB
	NS	PB	PM	PS	Z	NS	NM	NB
	Z	Z	Z	Z	Z	Z	Z	Z
	PS	NB	NM	NS	Z	PS	PM	PB
	PM	NB	NM	NS	Z	PS	PM	PB
	PB	NB	NM	NS	Z	PS	PM	PB

Figure 3 Fuzzy Rule Base B to Determine $\delta\theta_i$

The fuzzy rules described in Figures 2 and 3 can be verified graphically or through the linearized forward kinematic equations. The interested reader is referred to reference [5] for details on verification of the above fuzzy rules.

During the hardware implementation of the fuzzy logic based robot controller, the Fuzzy-C Development System from Togai® was used to generate portable C code for the fuzzy rules of Figures 2 and 3, which in turn was download onto the host computer.

4.0 HARDWARE EXPERIMENTS

In the hardware experiments presented below, the fuzzy robot controller described in Section 2.0 is tested on a four DOF planar lab robot with the following link lengths: L1 = 9 inches, L2 = 9 inches, L3 = 4.5 inches, and L4 = 3.5 inches. Each robot axis is equipped with an integral rotary incremental encoder to provide joint position feedback for the axis motion controller. For the experiments that follow, the fuzzy rule bases were evaluated in software rather than using a dedicated fuzzy chip.

4.1 Experiment 1

The planar 4 DOF lab robot was commanded to move along a straight line from X_{Begin} = 10 inches, Y_{Begin} = 15 inches to X_{End} = -5 inches, Y_{End} = 8 inches. The maximum end effector velocity is 1 inch/sec taking two seconds to reach maximum velocity. The initial robot configuration and intermediate configurations as it moves from the initial location to the final location is shown in Figure 4. Figure 5 shows the actual end effector tracking errors of the robot as it moves along the desired straight line. The maximum tracking error for the specified 16.55 inch straight line trajectory was about 0.03 inches.

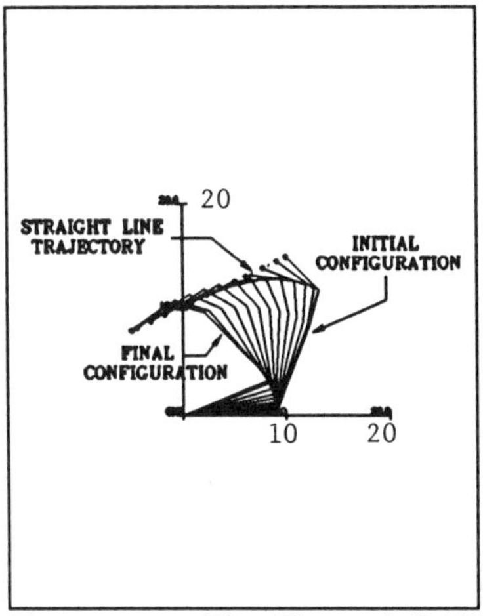

Figure 4 The Planar 4 DOF Lab Robot Tracking a Straight Line Under Fuzzy Control

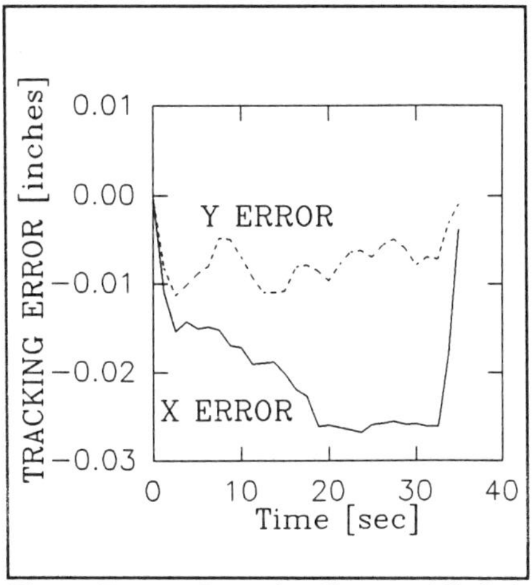

Figure 5 Cartesian Errors while Lab Robot Tracks a Straight Line

For this experiment, the trajectory position update was realized at 9 ms intervals. This update interval time is significantly less than that found in most commercially available robot controllers (28 ms for the PUMA series robots, 16 ms for the Adept Robot).

4.2 Experiment 2

In this experiment, the lab robot is required to track a circle of radius 4 inches with center point X = 5 inches and Y = 5 inches. The velocity and acceleration parameters are identical to the earlier straight line trajectory experiment. Figure 6 shows the tracking errors in both the X and Y directions, as the planar 4 DOF robot tracks the desired circular path under fuzzy control.

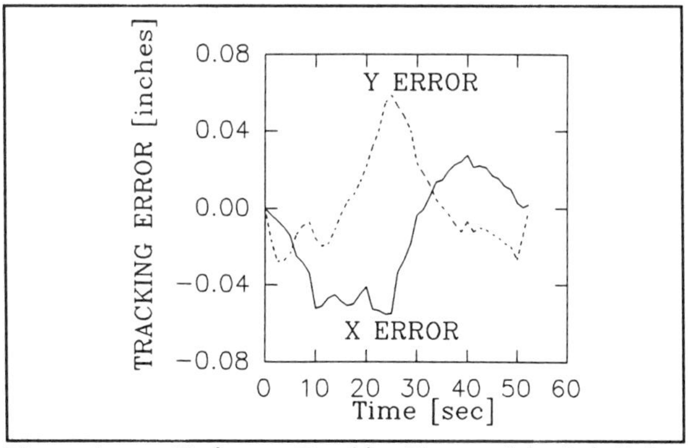

Figure 6 Trajectory Tracking Errors of the Lab Robot for a Circular Path

5.0 DISCUSSION OF RESULTS AND CONCLUSIONS

A fuzzy logic based robot controller is presented that does not rely on complex inverse kinematic equations to control a robot arm. The proposed fuzzy logic based robot controller was implemented in hardware on a VME based host computer with a real time OS-9 operating system. Experimental tests with a 4 DOF planar lab robot revealed that over a 20 inch trajectory, the maximum tracking error was 0.07 inches. The motion of each robot joint is governed by an independent fuzzy rule base, which results in a flexible, modular, parallel architecture type robot controller.

At SwRI, researchers have extended the described fuzzy robot controller to incorporate collision avoidance with obstacles that are arbitrarily placed in the robot's path. Future work will include extension of the proposed fuzzy logic based controller to control spatial robot arms.

6.0 ACKNOWLEDGMENTS

The author is grateful to the "SwRI Advisory Committee for Internal Research" for providing the funds to perform the described research. The assistance of Mr. Dennis Wenzel for development of real time software during the hardware implementation of the fuzzy robot controller is also gratefully acknowledged.

7.0 REFERENCES

1. Zadeh, L. A., Fuzzy Sets, Information and Control 8, Department of Electrical Engineering and Electronics Research Laboratory, 1965, pp. 338-353.

2. Mamdani, E. H., "Application of Fuzzy Algorithms for the Control of a Dynamic Plant," *Proc. IEEE 121, No. 12*, 1974, pp. 1585-1588.

3. Kickert, W.J.M. and Van Nauta Lemke, H.R., "Application of a Fuzzy Controller in a Warm Water Plant", *Automatica*, Vol#12, 1976 pp. 301-308.

4. Yasunabu, S. and Ihara, Miyamoto, H., "Fuzzy Control for Automatic Train Operation System," *Proc. 4th IFAC/IFIP/IFORS Int. Congress on Control in Transportation Systems*, Baden-Baden, 1983.

5. Nedungadi, A., "Application of Fuzzy Logic to Solve the Robot Inverse Kinematic Problem", to be published at the *4th World Conference on Robotics Research*.

A HEURISTIC BINARY-TREE-BASED FUZZY LOGIC CONTROLLER FOR SEMI-ACTIVE SUSPENSIONS

Dennis M. Briggs, Ka C. Cheok, Ningjian Huang, Richard E. Haskell
School of Engineering and Computer Science
Oakland University, Rochester , Michigan 48309

Abstract

The idea of controlling physical systems with artificially-intelligent control methods has great appeal, and has been investigated heavily in the past. At the very least, they offer the chance to control effectively systems that are highly non-linear and/or time-varying.

This paper describes attempts to fuse heuristic optimal control ideas and binary-decision-tree pattern recognition methods to build a fuzzy logic parametric controller for a semi-active suspension system. Simulation results show that the resulting system can provide the desired control actions to handle smoothly the irregularities in the road profile and the many non-linearities inherent in the suspension system.

Further research currently under way includes developing a system that can adapt to changing conditions encountered in the control environment..

INTRODUCTION

The problem of control of semi-active suspension systems, for automotive and non-automotive applications, has been investigated using conventional methods such as optimal and adaptive control [1]. However, it is difficult to apply these techniques to highly non-linear, time-varying systems for which there is no good closed-form model. Most control theory does not handle well the effects of natural non-linear characteristics, saturation limits, stiction, and transport delays.

Most semi-active suspension systems being considered today use shock absorbers with a finite number of damping settings, usually two. Recently, however, continuously-variable damping systems are becoming available, using electro-rheological damping fluid and other technologies. Therefore, the need has arisen to provide control systems that are able to contemplate these new capabilities.

In this paper, we present a fuzzy logic parametric control scheme that selects and implements a heuristically optimal setting for a variable-damping semi-active suspension system. The fusing of the fuzzy logic controller and the heuristic optimal solution is achieved via the use of a binary decision tree classifier.

Heuristic Optimal Parametric Control (HOPC) is a decision scheme which selects a best parameter for the system depending on the control situation [4] [6]. The decision is based on an intelligent heuristic search that evaluates predicted performance of the control system. The concept of HOPC is quite suitable for dealing with the semi-active suspension control problem.

Binary decision trees (BDTs) have long been used as a pattern recognition tool with applications in many areas. Various tree-generation alogrithms are available for use. However, very little has been done with this technique in real-time control. In one sense, BDTs can be thought of as being functionally equivalent to neural networks (NNs), in that both can be trained to recognize a finite number of patterns. However, NNs can provide a continuous output, whereas BDTs require discretizing of the output at some level. BDTs, on the other hand, provide other advantages. For instance, they lend themselves to today's von Neumann computers more readily than do the massively-parallel NNs. Also, they provide the design engineer with a set of rules describing how the state space was partitioned as the BDT was trained. These rules can be most helpful in understanding how a system is functioning.

Fuzzy Logic (FL) theory has been developing rapidly in the last two decades [5] [7]. It is currently being used in many customer products to perform control functions, such as the auto-focusing system in some video cameras. Methods for designing fuzzy control systems require the definition of rules describing how the controller should act under situations encountered during operation. These rules are usually generated by the design engineer or other system experts in a sort of knowledge engineering process. Fuzzy systems take "crisp" inputs, fuzzify them, and generate crisp outputs which have been de-fuzzified at the end of the internal process. These crisp outputs are continuous.

The main idea behind this paper is to combine HOPC, BDTs and FL and utilize the best parts of these concepts to develop a fuzzy optimal parametric controller for the semi-active suspension system. Specifically, the HOPC can select the "best" shock setting, the BDTs can automatically observe the HOPC decisions and generate the rules for the system, and the FL system can provide a continuous output to the plant being controlled.

The remainder of this paper is divided into the following sections: Problem Statement, Approach to Problem, System Description, Results, and Conclusions. It is noted that rigorous mathematical expressions are omitted in this paper due to lack of space.

PROBLEM STATEMENT

The purpose of the project is to develop a fuzzy optimal controller for a simplified but non-linear semi-active suspension system modeled for one wheel of a multi-wheeled vehicle. Interactions from other parts of the vehicle are not considered in this paper. The controller will be developed using the ideas introduced above; namely, HOPC, BDTs, and FL.

APPROACH TO PROBLEM

The basic approach to the problem is shown in Figure 1. Points in the *measure space* (i.e., those variables which can be measured and are useful) are selected by the designer for the HOPC decision program, which determines the proper shock absorber damping setting for each point. The system described here uses the following variables in its measure space: wheel absolute vertical acceleration, suspension relative (between body and wheel) vertical displacement, and suspension relative vertical velocity. The selected points and the recommended shock settings from the HOPC decision program are given to the BDT learning program. The BDT program generates a tree that divides the measure space into groups of similar shock settings, in the manner of most pattern recognition schemes. Since each path through the tree is a rule, the fuzzy logic generator can take the tree and create fuzzy rules for the fuzzy controller. It should be noted that at this point, the discretized shock settings are made continuous. The fuzzy controller then operates as do most controllers in similar schemes.

One characteristic of semi-active suspension systems is that since only the shock damping setting is controlled, little energy is put into the system by the controller. Therefore, stability problems are not of paramount concern here. However, other system maladies can occur, such as ringing or sudden jumps in acceleration, which could be felt by the passengers in the vehicle or could result in loss of vehicle control. It is the function of the HOPC decision program to provide sufficient information to prevent or overcome these maladies.

SYSTEM DESCRIPTION

This section describes each major block of the system shown in Figure 1.

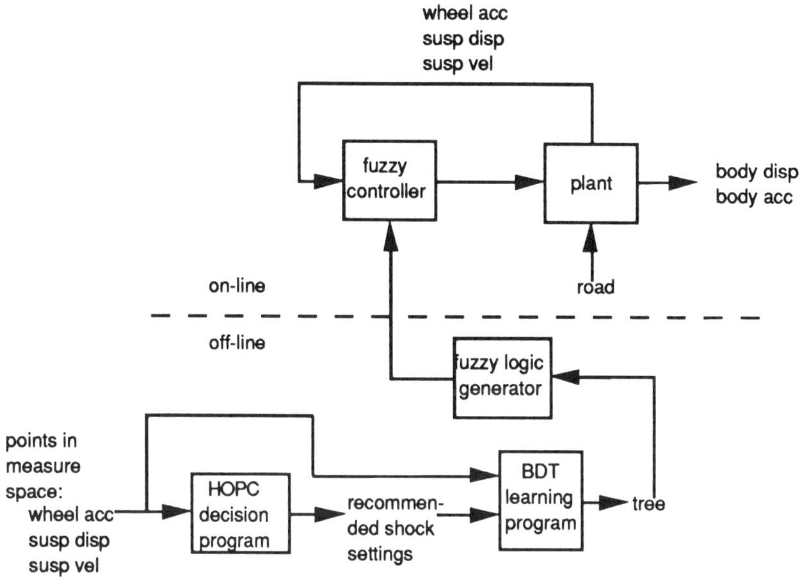

Fig. 1. HOC-BDT-Based Fuzzy Logic Controller Architecture

Heuristic Optimal Control Decision Program

The Heuristic Optimal Parametric Control Decision (HOPC) program being used currently utilizes a model of the system being controlled. In practice, this model would necessarily be an idealized version of the plant. The HOPC program takes each specified point in the state space and exercises the model through a certain amount of lookahead time The simulation runs are repeated for each possible shock setting.

Cost functions are determined for each run. For each cost function, the shock setting with the best performance is chosen and stored. Therefore, one pass through the HOPC program generates a training set for each cost function.

An assumption about the road input to the plant is made; it consists of assuming that its acceleration is the same as the given wheel acceleration, and stays the same for 10 ms, then goes to zero. This is, of course, only a guess about the road, but it is somewhat informed since the tire spring constant is rather high and the wheel does tend to follow the road. Ideally, the road acceleration itself would be used, but this must await the development of viable measurement capabilities.

The HOPC program is based on search as used in Heuristic Optimal Controllers. The search horizon for the training program is set by the designer. The model is simulated out to the horizon once for each shock setting. Even though there are multiple time steps in the simulation, it can be considered to be a one-step lookahead to the search horizon, since the shock setting does not change. In this sense, it is equivalent to the HOPC concept as applied to a one-step lookahead procedure.

BDT Learning Program

The BDT Learning Program [2] [3] creates binary decision trees by splitting the measure space into two groups, and then recursing on the groups just created, until all the groups have only one shock setting per group. This splitting of the measure space into groups is arranged into a BDT, each leaf of which consists purely of one shock setting. Normally, multiple leafs will contain the same shock setting. Some pruning of the tree usually follows, which reduces the computer load and actually increases the reliability of the pattern recognition process.

The point at which to make each split is determined by using the Kolmogorov-Smirnoff Statistic, which is the maximum distance between pairs of the probability distributions of the features in the group being considered.

The output of the program is a set rules of the form:

if wheel acc < 1000
and susp vel >= -100
and susp disp < 10
then shock setting=3

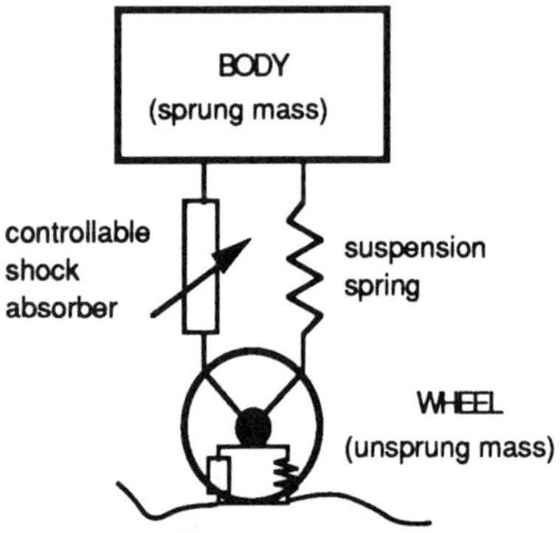

Fig. 2. PLANT TO BE CONTROLLED

Note that the BDT rules are crisp; that is, they have definite boundaries in the measure space and make definite choices. This is consistent with the discretized nature of BDTs.

Fuzzy Logic Generator

The Fuzzy Logic Generator takes the rules created by the BDT Learning Program and applies membership functions to each of the assertions in each rule. It then uses standard fuzzy logic inferencing techniques to create code that will implement the fuzzified rules.

Currently, all parameters, such as membership function shapes and inferencing techniques, are left to be specified by the designer.

Fuzzy Controller

The Fuzzy Controller implements the fuzzy rules discussed immediately above. The measured quantities are read in and the rules are applied to arrive at a crisp, but continuous, output shock absorber setting.

Plant

The plant used in this paper is shown in Figure 2. The system has four states: suspension relative velocity and displacement, and wheel relative velocity and displacement. Note that the states are not the quantities of real interest to the designer. Those quantities are the body absolute displacement and acceleration (some measures of comfort) and force between the tire and the road (a measure of handling).

Not shown in the diagram are non-linearities besides the variable shock absorber. They include the limits on suspension relative displacement (jounce or minimum, and rebound or maximum). They also include a minimum on tire relative displacement (the wheel cannot dig into the road), and the ability of the tire to leave the road. Further non-linearities include the controllable shock absorber force vs velocity curves and two other kinds of friction found in suspensions.

A Simulab model was used to represent the system in both the HOPC program and the simulation runs.

RESULTS

Figures 3 and 4 show the results obtained when using a fuzzy controller on two road disturbances: a bump and a hole. The controller was created with a training set generated by the Training Program using the following cost function:

$$J = k_1 * jerk(body\ absolute) + k_2 * velocity(body\ absolute) + k_3 * displacement(body\ absolute),\ k_1 + k_2 + k_3 = 1$$

where velocity and displacement are defined as before and jerk is the derivative of the acceleration. This function was used to try to minimize body displacement as much as possible, consistent with minimizing the harshness of impact-type forces. Other cost functions used include Average Absorbed Power (a measure of how much fatigue the human body accumulates over time as a result of experiencing vibration), and norm-max acceleration (a combination of the 2-norm and the maximum acceleration of the vehicle body).

Both figures show the results of three simulation runs: the plant with a controlled variable shock, a stiff shock and a soft shock.The stiff shock is at the high-damping end of the variable-shock range, while the soft shock is positioned similarly at the low-damping end.

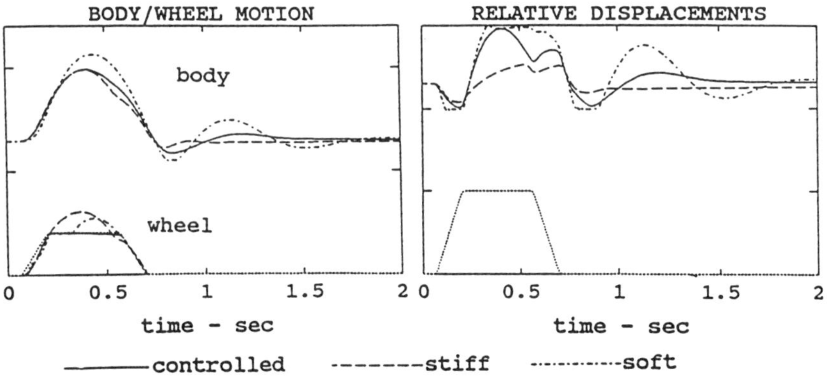

Fig. 3. Controlled shock outperforms constant shocks.

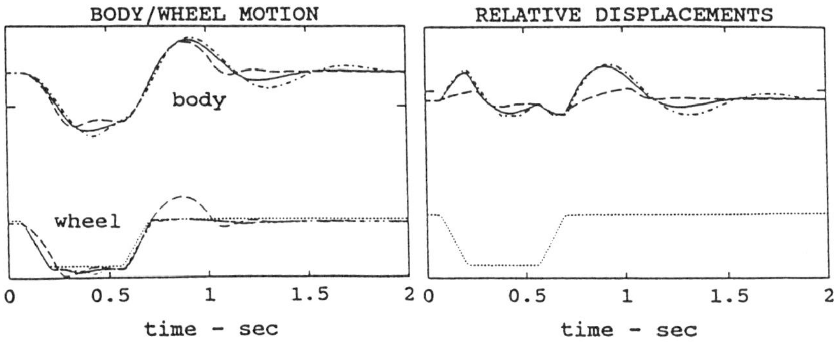

Fig. 4. Controlled shock equal to or better than constant shocks

A reasonable strategy for minimizing body movement is to allow the suspension to take up as much motion as possible without actually hitting the jounce and rebound limits. This situation is illustrated best in Figure 3. Note that the stiff shock has the least relative displacement, the soft shock has the most (it actually allows the suspension to hit both jounce and rebound limits), and the controlled shock maximizes movement without hitting limits. As a result, the body motion is comparable to that of the stiff shock, and the wheel follows the bump profile much more closely, thus improving chances for good handling.

Figure 4 shows that fixed-damping shocks may perform very well in some situations, but it is difficult to find one setting that performs well overall. Here, the soft shock appears to be comparable to the controlled shock, after performing poorly in the bump simulation.

CONCLUSIONS

This paper has shown the results of using a Fuzzy Logic parametric controller based on Heuristic Optimization and Binary Decision Tree concepts. It shows that this idea at its current very early development stage works well in some situations, but needs further work to expand the region of high performance. Further work will be done in this area, as well as in creating architectures that will allow automatic learning and improvement through self-criticism of current system performance.

REFERENCES

1. K. C. Cheok, N. K. Loh, H. D. McGee, T. Petit, "Optimal Model-Following Suspension with Microcomputerized Damping",*IEEE Trans. Ind. Elect.*, vol. 1E-32, no. 4, Nov. 1985.
2. R. Haskell, B. Mirshab, G. Castelino, "Calculating Shape Features in a Binary Image", paper presented at the Rochester Forth Conference, Rochester, NY, June 14-18,1988.
3. R. Haskell, G. Castelino, B. Mirshab,"Computer Learning using Binary Tree Classifiers", paper presented at Rochester Forth Conference, Rochester, NY,June 14-18,1988.
4. N. Huang, *Heuristic Optimal Control Systems: Theory and Applications*, Ph.D. Dissertation, Oakland University, Rochester , MI, 1991.
5. B. Kosko, *Neural Networks and Fuzzy Systems*,Prentice-Hall,Englewood Cliffs, NJ
6. J. C. Smith, K. C. Cheok, N. Huaug, "Optimal Parametric Control of a Semi-Active Suspension System Using Neural Networks", Proc. 1992 American Control Conf.,Chicago, IL, June 1992, pp963-967.
7. L. Zadeh,"Outline of a New Approach to the Analysis of Complex Systems and Decision Processes",*IEEE Trans. Syst. Man Cybern.*,vol.SMC-3,pp28-44,1973.

FUZZY CONTROL OF A DC SERVO MOTOR

Heikki Koivo Zhao Tienan
Control Engineering Laboratory
Tampere University of Technology
P.O.Box 527, 33101 Tampere, FINLAND

ABSTRACT

In this paper, application of a formula based fuzzy controller is presented. The fuzzy controller takes into account error, change of error, and change of the change of error. The consideration for the change of the change of error is an extension for the traditional fuzzy control. The representation of fuzzy logic tables in the phase plane is discussed. The fuzzy controller and PID controller are applied to a DC servo motor control. The DC motor is simulated, but includes realistic features like backlash, friction, and saturation. By comparing the simulation results of fuzzy and PID control, it can be concluded that fuzzy controller is more robust than PID controller when system working conditions are changed, therefore has industrial application potential.

INTRODUCTION

Digital control of DC servo motor has created lots of interest in the literature. The control of DC motor servos forms a formidable task, because they contain severe nonlinearities and the time requirements in digital control are quite stringent. Self-tuning and adaptive controllers are investigated in [1], [7] and [8]. State space controllers have been tried e.g. in [9] and [10]. One of the crucial issues in the implementations is, in addition to the time requirement, robustness of the controller. Especially the state space controller lacks robustness.

Fuzzy logic control has been applied to a number of applications successfully [2]. A fuzzy logic controller is a knowledge-based controller using fuzzy concepts in knowledge representation and inference. Fuzzy algorithms have been applied to servo systems by Li and Lau [3]. The problem with fuzzy logic controllers is that they are difficult to tune and calibrate.

A nice approach to overcome these deficiencies is presented by Smith and Comer [4]. In this paper the results of Li Congxin [5],[6], which resemble somewhat those in [4] are exploited and used to simulate a DC motor servo system. A simple control structure originating from fuzzy logic control is used. The tuning is reduced to determining 2-3 tuning parameters. The simulation studies of fuzzy and PID control are discussed.

PHASE PLANE REPRESENTATION OF FUZZY LOGIC TABLE

Traditionally, fuzzy control is based on error and change of error to make control decision by utilizing certain fuzzy logic relation tables. Normally the control decision is made by calculating the membership fuctions and using the fuzzy compound rules. This is a time consuming process. Li Congxin [5], [6] discussed the representation of fuzzy logic tables in the phase plane. The fuzzy logic table used by Li for a on-off hydraulic servo can be transfered to broken lines in the discrete phase plane as shown in Fig. 1. The horizontal axis is the error E. The vertical axis is the change of error CE.

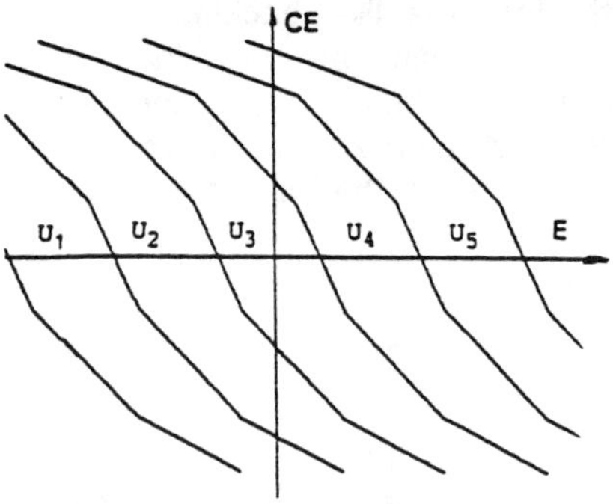

Fig. 1 Mapping of fuzzy logic tables into disctete phase plane

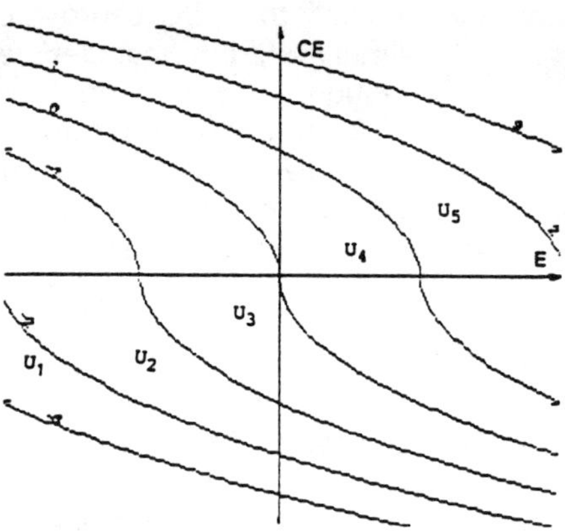

Fig. 2 Mapping of equation (1) into discrete phase plane

The broken lines in Fig. 1 can be treated as switching lines, because when the error or change of error crosses the lines, the corresponding control value U_i will be changed. In the area between the broken lines, the control input is constant. It is important to note that the control input on the switching lines is equal to the control input in the area (either left or right) between the switching lines.

It is obvious that if the broken lines are changed to curves, the control signal will be smoother. If the switching lines become switching curves, the control input can be described by an analytic expression, therefore it is easier to optimize the control parameters and another advantage of formula based controller is that the control algorithm is simple to implement, especially suitable for a fast system where the sampling time is critical. From Fig. 1, it can be observed that the broken lines are parallel and the shape is similar to that of quadratic curves. This discrete phase plane representation of fuzzy logic tables justifies the form of the following controllers.

Assuming that control input is U and

$$U = K_p*(E + K_v*CE*ABS(CE)) \quad (1)$$

then, the switching lines will be equally spaced parallel quadratic curves in discrete phase plane as shown in Fig. 2. The distance between the nearest two parallel curves in E axis is just one unit of the feedback resolution. Changing the values of K_p, K_v is similar to changing the fuzzy compound rules, but is much simpler.

By considering change in the change of error (*CCE*), formula (1) can be extended as follows:

$$U = K_p*(E + K_v*CE*ABS(CE) + K_a*CCE*ABS(CCE)) \quad (2)$$

By introducing change of the change of error, the switching lines in phase plane are transfered to phase space.

Equation (2) is the extension of fuzzy controller, which considers the *CCE* in the formula form. While using the traditional fuzzy logic table based controller, it is very difficult or even impossible to consider the *CCE*. In the simulation, equation (2) is implemented in discrete form.

SIMULATION

Fuzzy control of a DC servo motor shown in Fig. 3 was studied. The servo model is quite realistic, especially friction and backlash are modelled with more care than is common in the literature. When motor saturation is also included, tuning of a conventional PD controller becomes surprisingly hard. The principle of the system is as follows:

The set point of load angular position is the reference input of the system. Through transducer, the actual angular position of load is feedback to microcontroller. The microcontroller gives control command to the DC motor. The DC motor drives the gears with load. The constant values in Fig. 3 are respectively $K=0.488$ Nm/A; $R_a=1.6\Omega$; $L_a=0.0004$H; $i_{amax}=8.5$A; $b_m=7.6\times10^{-4}$Nms/rad^2; $J_m=200$kgm^2/rad^2; $T_{fm}=0.1$Nm; $n=250$; $K_s=10^7$Nm/rad; and $J_1=200$kgm^2/rad^2.

The simulation is based on a nonlinear system model which is comparatively near the real model of the system. The coefficients used for simulation are respectively $K_p=2000$, $K_v=1.225e3$, $K_a=7.5$ which are obtained by trial and error method.

In order to prove the robustness of the fuzzy controller, the simulations of fuzzy and PID controls for different working conditions are conducted. The simulation conditions set for the controllers are

1. Backlash is 0.01.
2. Friction is included in the simulation model.
3. A disturbance torque in the form of a unit/pulse between 0.2 and 0.3 Sec.

resisting the motion.
4. Changes in load and spring constant.

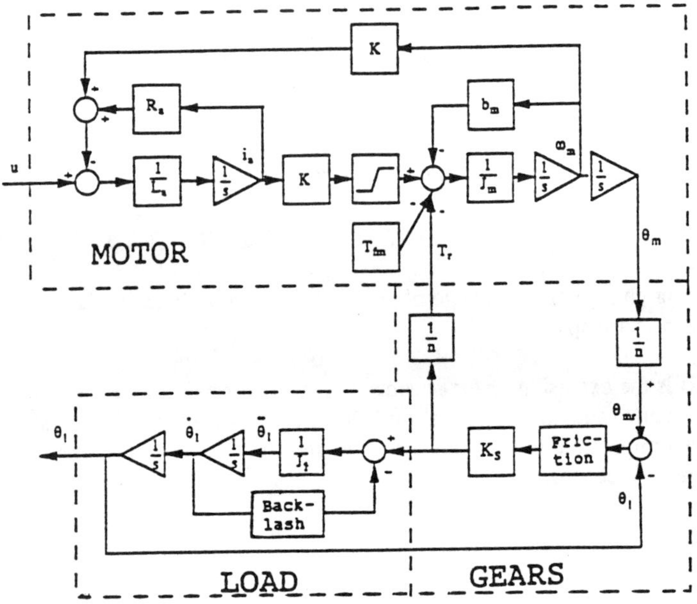

Fig. 3 Block diagram of the controlled system

Fig. 4 and Fig. 5 are respectively the dynamic responses of PID and fuzzy controllers under the ideal condition that there is no external disturbance; the load is 200kg; and the spring constant is 10^7Nm/rad. Tuning of PID controller is quite difficult. The usual rules of thumb fail, because of the system nonlinearities. The fuzzy controller, on the other hand, was relatively easy to tune. By comparing these two figures, it can be observed that the response of fuzzy control is slightly faster than PID control. The nominal case is not too indicative to make a fair comparison - the responses are quite similar. The units for the step response; torque of motor; and control voltage are respectively radian [rad]; Nm; and Volt.[V]. The unit for time is in seconds.

Fig. 6 and Fig. 7 are respectively the dynamic responses of PID and fuzzy controllers under the condition that there are external disturbances. The tuning parameters are kept constant. By comparing these two figures, it is apparent that the steady state error of fuzzy control is much smaller than that of PID control. When the spring constant is changed to 1.2×10^7Nm/rad and 0.9×10^7Nm/rad, the dynamic response of the fuzzy control has practically no change, while the steady state accuracy of the PID control is not acceptable. When the load is increased to 300kg, there is some overshoot in the response of fuzzy control but the steady state accuracy is much better than that of PID control. From above simulation results, it can be concluded that the proposed fuzzy controller is robust to the changes in system working condition, therefore it has industrial application potential.

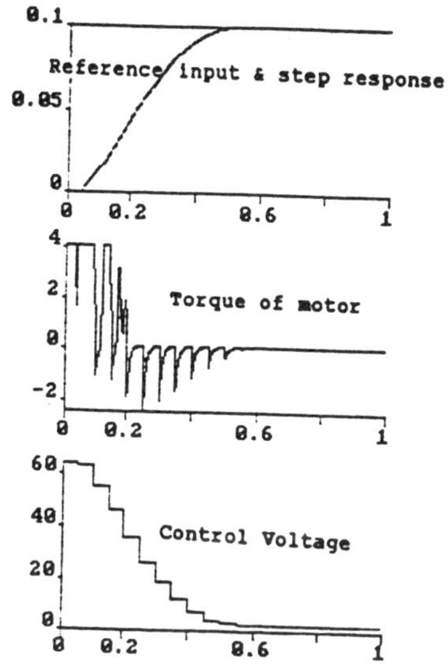

Fig. 4 PID CONTROL

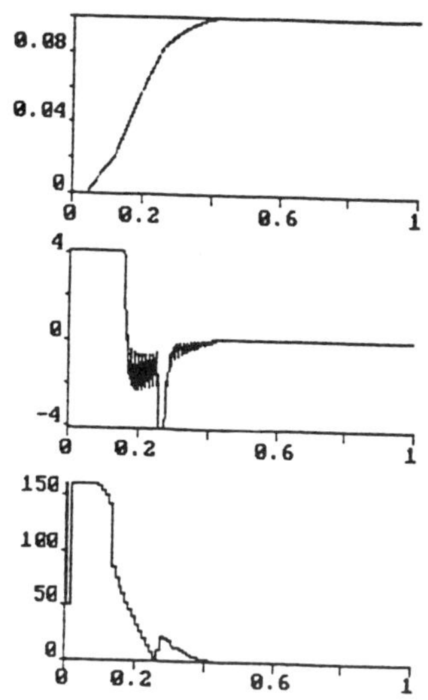

Fig. 5 FUZZY CONTROL

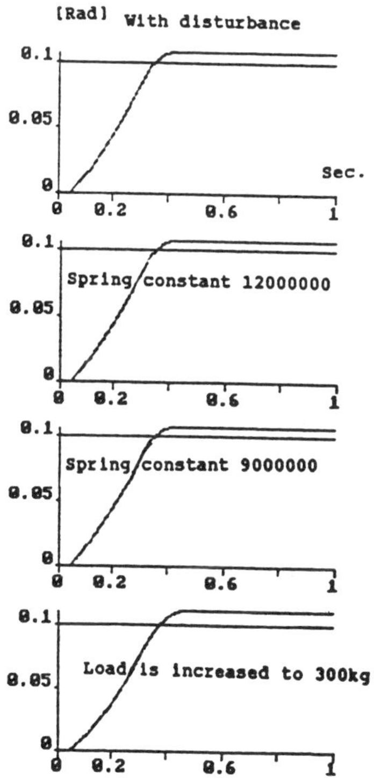

Fig. 6 PID CONTROL

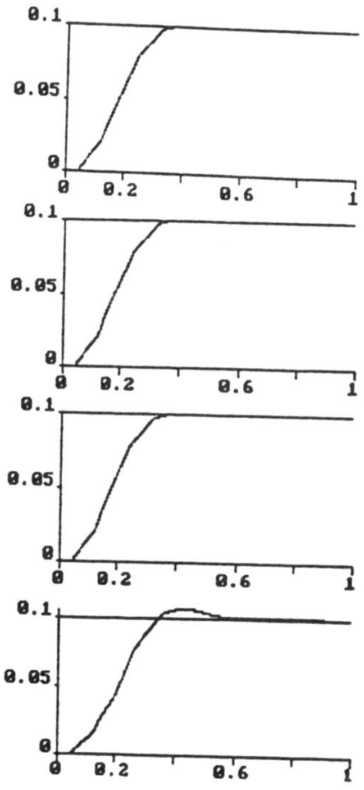

Fig. 7 FUZZY CONTROL

CONCLUSION

In this paper, a formula based fuzzy controller is presented. The origin of the fuzzy controller is discussed. The consideration for change of the change of error is a step forward compared to the traditional fuzzy logic controller. Simulation results confirmed that the proposed fuzzy controller is robust when the system working conditions are changed. The advantage of the formula based fuzzy controller is that the algorithm is simple to implement for real systems, especially suitable for fast systems where the sampling time is critical.

ACKNOWLEDGEMENT

The authors wish to thank to Dr.Li Congxin, Vickers Systems GmbH, Germany for his valuable suggestions. The financial support from Finnish Academy is acknowledged.

REFERENCES

[1] C. Canudas, K.J.Åström, K. Braun, "Adaptive Friction Compensation in DC Motor Drives", The IEEE International Conference on Robotics and Automation, San Fransisco, USA, 1986.

[2] M. Sugeno, Ed., "Industrial Applications of Fuzzy Control", New York: Elsevier, 1985.

[3] Y.F.Li and C.C.Lau, "Development of Fuzzy Algorithms for Servo Systems", IEEE International Conference on Robotics and Automation, Philadelphia, April 24-29, 1988.

[4] Samuel M. Smith and David J. Comer, "Automated Calibration of a Fuzzy Logic Controller Using a Cell State Space Algorithm", IEEE Control Systems, Aug., 1991, pp.18-28.

[5] Li Congxin, "Fuzzy Control of Hydraulic Position Servo", Machine Tool & Hydraulics, Vol 2, 1983. (In Chinese)

[6] Li Congxin, "Intelligent Hydraulic Position Control System", The 2nd Tampere International Conference on Fluid Power, March 19-21, 1991, Tampere, Finland.

[7] J. Riihilahti and H. N. Koivo, " Modelling and Control of a DC Servo Motor", Technical Report, Control Engineering Laboratory, Tampere University of Technology, Tampere, Finland, 1991.

[8] Canudas de Wit, C. Noel, " Adaptive Friction Compensation in Robot Manipulators: Low Velocities." IEEE International Conference on Robotics and Automation. Scottsdale, Arizona, USA, May 1989.

[9] J. Y. S. Luh, [1983b] "Conventional Controller Design for Industrial Robots - A Tutorial", IEEE Trans. Systems, Man, and Cybern., Vol. SMC-13, No.3, PP. 298-316.

[10] K. Ohishi, "Torque Control of DC Motor Based on Observer", IEE of Japan, Commitee of Rotating Machine, Vol. RM-82-83 (1982).

FUZZY CONTROL OF THERMAL SYSTEMS

DENIS BARAK, STEVE BAUGH, and MO JAMSHIDI
CAD Laboratory for Intelligent and Robotic Systems
Department of Electrical & Computer Engineering
The University of New Mexico
Albuquerque, NM 87131, USA

ABSTRACT

In this paper a fuzzy proportional plus integral (PI) controller is designed for a general thermal system. The general approach is being implemented for a prototype CFC-free air conditioner system produced by Refir Technologies Incorporated, which cools a volume effectively with low electrical power requirements. Experimental results demonstrate that the cooling effect can be maintained over protracted periods. The design concepts suggest applications for automobiles and over-the-road trucking cabs. This fuzzy-controlled CFC-free, non-compressor system has attractive efficiency with few mechanical or moving parts and offers the portent of increased reliability and small size at reduced costs and low maintenance requirements.

INTRODUCTION

This paper deals with an important issue of intelligent control, namely the use of fuzzy logic as an alternative method of controlling thermal processes and thereby leading to new intelligent or smart manufacturing products. The specific application or product in mind here is an air conditioning system which is not based on Chlorofluorocarbon (CFC). In this paper the potentials of fuzzy logic for controlling a thermal system are described.

The paper has the following structure: the next section will be devoted to a brief introduction to a fuzzy logic, following with the section on fuzzy control of a generic thermal system; the potentials of the approach to a non-CFC air conditioning system follows next, while the final section is devoted to some initial conclusions.

FUZZY LOGIC

The recent great interest in fuzzy logic has been, in part, due to the extent of application areas in which fuzzy control has been used. Among many possible new technologies based on artificial intelligence, fuzzy logic has emerged as a front runner judging by billions of dollars worth of sales and close to 2000 patents in Japan since 1987. In 1989 a study was made by the US Office of Technology Assessment for the US Congress. The Office studied over 12 competing technologies for cost reduction in space applications. The number one technology on their list turned out to be "expert systems," including "fuzzy expert systems."

Fuzzy logic technology is now being used in numerous consumer products such as television, camcorders, home appliances, automobiles, and such systems as elevators, trains , and even in stock market and medical diagnostics. Fuzzy logic was first proposed by Zadeh[1] in 1965 when he published his seminal paper on "Fuzzy sets." Zadeh showed that fuzzy sets are the foundation of any logic regardless of how many truth values it may have. A fuzzy set has movable boundaries, i.e. the elements of such sets represent not only black and white colors, but allow a spectrum of gray colors in between.

A fuzzy controller typically takes the form of a set of IF-THEN rules whose antecedents and consequents are membership functions. Consequents from different rules are numerically combined (typically union via MAX) and are then collapsed (typically taking the centroid of the combined distribution) to yield a single real number (binary) output. Within the framework of a fuzzy expert system, like regular expert systems, typical rules can be the result of a human operator's knowledge, e.g. :

"If the *Temperature* is *Hot* then increase the *Current* to a *Medium* level".

In this rule, *Hot* and the *Medium* are fuzzy variables. Such natural language rules can then be translated into typical computer language type statements. Using a set of rules an entire finite number of rules can be derived in the form of a natural language statements as if a human operator was performing the controlling task. In any practical system, such as an air conditioning unit, the user or an operator often fine tunes, tweaks, and adjusts the knobs until the desired cool (or hot) air with desired speed can be felt. Such operator knowledge can be utilized in the design of a fuzzy controller for an air conditioning unit system. One of the most common ways of designing fuzzy controller is through "fuzzy rule-based systems." The controller shows the processes of binary to fuzzy transformation called fuzzification, and fuzzy to binary transformation called defuzzification. An alternative way of implementing a fuzzy control regime, which is similar to a conventional adaptive control law, is through a standard 1, 2, or 3-term controller such as P, PI, PD, or PID whose gains would be adjusted through some fuzzy rules in a supervisory loop.

FUZZY CONTROLLER FOR A THERMAL SYSTEM

In this section a solution to a "real life" temperature control of a chamber, such as the inside channel of Refir Technologies Incorporated air conditioning prototype unit, is simulated by two different control systems, namely classical and fuzzy control systems. In both cases a PI compensator is used to stabilize the system and to reject the disturbance; however, in classical design K_p and K_i coefficients are kept constant, while in fuzzy control design K_p and K_i coefficients are adaptive, and they are changing depending on the error in the measured temperature, which also means that they are indirectly adapting to the applied disturbance.

Temperature control is performed through a dynamic model of a chamber whose transfer function is the following:

$$G(s) = Y(s) / U(s) = 20 / (20s + 1),$$

where the output is the chamber temperature in degree Celsius measured at specified location in the chamber, and the control input is a voltage which changes TED's current, and , consequently, heats the chamber. The sensor, in this application modeled as $H(s)=0.05$ V/°C, is a measurement device which changes its resistance with temperature, therefore providing information on chamber temperature in volts. Since we are modeling sensor as a pure gain, the sensor voltage will change very quickly compared to a chamber temperature, which allows as to use this sensor as an accurate real-time chamber temperature measurement, and, therefore, as our feedback element. We can simplify this feedback element by defining it as a unity feedback, and , therefore, changing the desired input from U(s) to (1/0.05)*U(s) which becomes 20*U(s). For this model we considered disturbance to be a random opening of the door into the chamber, which will certainly affect the temperature of the chamber. Assuming that door opening can be represented as the step input with an exponential decay, the disturbance model becomes $Gd(s)= -0.6/(s+0.1)$.

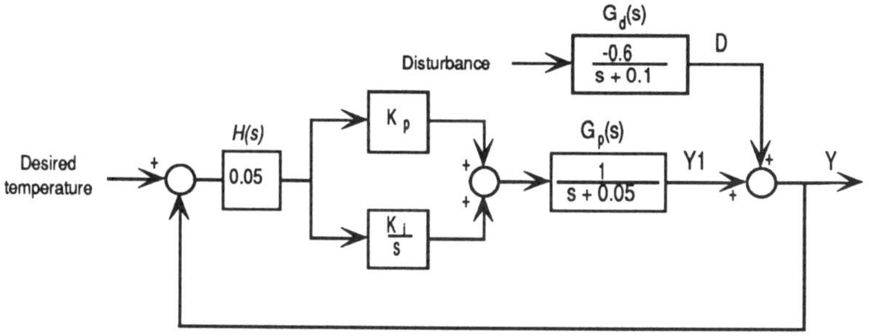

FIGURE 1 : Simulation model for the control of temperature chamber

The system's simulation results showed that the system is stable, exhibiting a classical overdamped step response. The injection of the disturbance input in each case significantly affects the systems ability to achieve a target temperature. The disturbance input alters system temperature trajectory with overdamped characteristics. Since, the system bandwidth is small, closing the loop with a P (proportional) controller will effectively increase the bandwidth, improving transient response. However, the PI (proportional-integral) controller is itself a low pass filter, which forces the steady state error to zero improving distrubance rejection, but may in fact degrade the transient response in doing so. PI controller has the following transfer function:

$$Gc(s) = K_p + K_i/s = Kp(s + Kp/Ki) / s.$$

Depending on the ratio K_i/K_p the system dynamics such as overshoot, rise time, and settling time will change. By adjusting the gain ratio, one also affects the position of the closed loop poles, which directly affects the overshoot, rise time, and stability of the system. The closed loop transfer function with no disturbance is

$$G_{ND}(s) = \frac{0.05\ K_p \left(s + \frac{K_i}{K_p} \right)}{(s^2 + 0.05(1+K_p)s + 0.05\ K_i)},$$

while the closed loop disturbance transfer function is

$$G_D(s)= \frac{0.6(s^2+0.05s)}{(s^3+0.05(3+K_p)s^2+0.005(1+K_p+10K_i)s + 0.005K_i)}$$

from which one can note the following observations :

1) If $K_p > K_i$ poles remain on the real axis, i.e no overshoot.

2) As $K_p < K_i$ poles leave real axis and overshoot will occur, and

3) As K_p is set larger than K_i the dominant pole approaches the origin leading to a slower transient response.

In order to design a fuzzy PI compensator, Control-C program was used to plot sytem's closed loop response for different values of K_i and K_p, therefore "pushing" plant to respond to different compensator proportional gains and locations of the poles and zero. By doing the above mentioned, a data knowledge base was formed, and decision was made on the limits for K_i, K_p, and error for the support of the fuzzy I/O membership functions. It was decided that error will be input variable for the fuzzy subroutine, while K_i and K_p will be output variables from the fuzzy subroutine representing adaptive PI compensator coefficients.

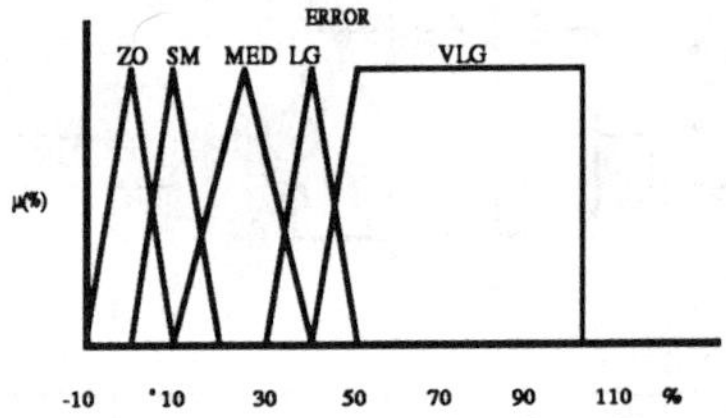

FIGURE 2 : Input membership function -- Error

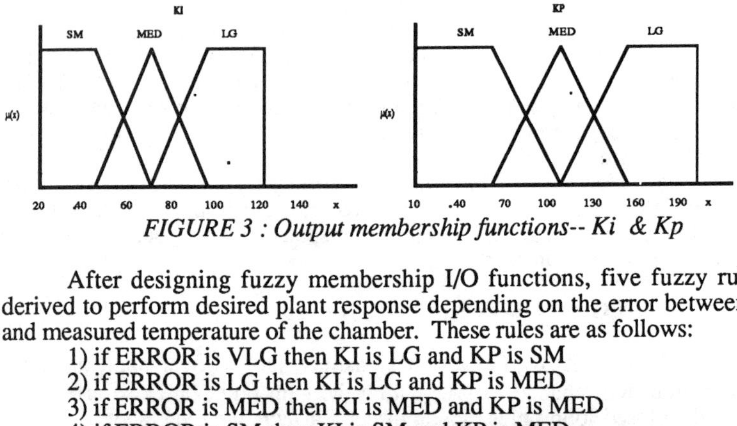

FIGURE 3 : Output membership functions-- Ki & Kp

After designing fuzzy membership I/O functions, five fuzzy rules were derived to perform desired plant response depending on the error between desired and measured temperature of the chamber. These rules are as follows:

1) if ERROR is VLG then KI is LG and KP is SM

2) if ERROR is LG then KI is LG and KP is MED

3) if ERROR is MED then KI is MED and KP is MED

4) if ERROR is SM then KI is SM and KP is MED

5) if ERROR is ZO then KI is SM and KP is LG.

From the above information, one should note that rules are designed in such a way that when the error signal is large Kp/Ki values are large to increase the control signal to the plant, and as the error signal reduces Kp/Ki values reduce to prevent overshoot.

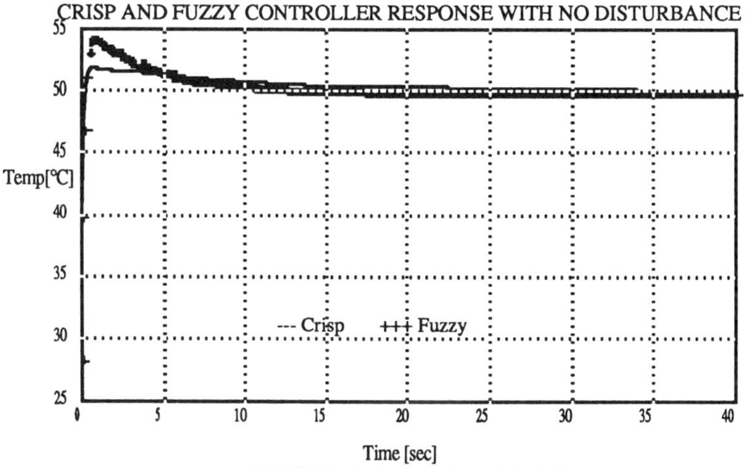

FIGURE 4 a) Results with d=0

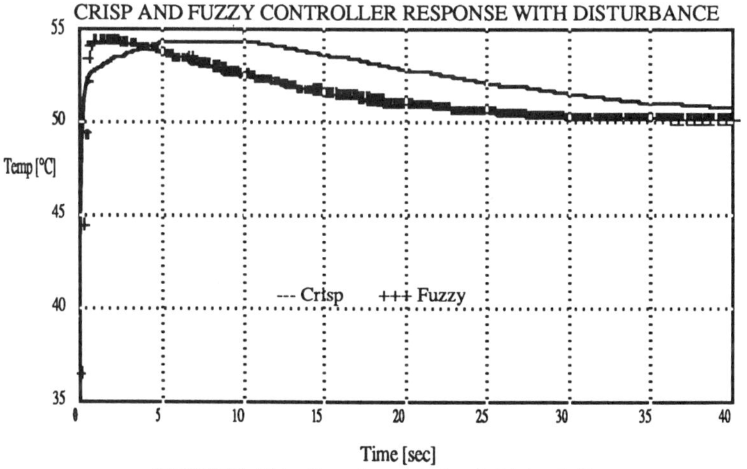

FIGURE 4 b) : Results with d=-0.6 / (s+0.1)

Despite all the effort in designing an adequate fuzzy PI compensator, without the disturbance present, replacement of classical compensator with a fuzzy one is not recommended because, for different step inputs classical compensator performs with less overshoot and faster time response than fuzzy counterpart. However, when disturbance is present in the system, fuzzy compensator has greater initial overshoot, but much faster settling time, and smaller steady state error than its classical compensator counterpart. We feel that the above are crucial characteristics in temperature control, so that control effort is reduced, and energy is saved. Moreover, the initial overshoot can be reduced by fuzzy overshoot suppression technique originally proposed by Yokogawa Incorporated[5].

CONCLUSIONS

With the impending deadline for ceasing CFC production, the need for a non-CFC-based ACs becomes indisputable. Several groups are directing their efforts toward a replacement gas, HCFC-123. However, recent findings indicate a possibility of carcinogenic effects. Start-up company Refir Technologies Incorporated has choosen to explore thermoelectric-based (TED) technology instead. Previous experiments with TEDs indicated that the technology was inefficient and laden with problems. The use of different materials and modern control technology has led to our promising prototype.

The first commercial application for this device is anticipated to be on over the road semi trucks. The device will be powered by 13.8 VDC deep cycling lead acid type battery. Total weight of both units will be 90 to 100 pounds. The unit will be used by the truck operator to provide a thermally acceptable environment for up to 10 hours without having to use the truck engine. We also anticipate that our unit will act as an ancillary A/C while the truck is in operation. Once the on-board systems have overcome any large changes in temperature the unit will turn on, relieving the primary device, which will decrease fuel consumption and increase engine efficiency. This approach finds applications in automobile, RV, small dwelling, and aircraft or other small volume situations which have an auxiliary power source.

The fuzzy control simulation of a thermal system such as the REFIR air conditioner has provided encouragement to continue ahead and implement the fuzzy PI controller for the Model II-D REFIR air conditioner which is currently housed in the Laboratory. The goal is to improve on system operational efficiency in heating, cooling, electrical power requirements, etc.

REFERENCES

[1] L.A. Zadeh, "Fuzzy Sets", Information and Control, vol. 8, 335-353 (1965).

[2] D.R. O'Geary and T.K. O'Geary, "Development of a Reliable, Efficient Non-Chlorofluorocarbon Based Air Conditioning System," in Environmental Conscious Manufacturing, M. Jamshidi editor, ECM press, pp183-191, Albuquerque, NM, 1991.

[3] M. Jamshidi, R. Marchbanks, K. Bisset, R. Kelsey, S. Baugh, and D. Barak, "Computer-Aided Design of Fuzzy Control Systems - Software and Hardware Impelmentations," in Advances in Computer-Aided Control Systems Engineering, M. Jamshidi and C.J. Herget (Eds.), Elsevier North-Holland, Amsterdam, 1993.

[4] M. Jamshidi, D. Barak, S. Baugh, and D. O'Geary, "Operations and Control of a Relaible Non-Chlorofuorocarbon Based Air Conditioning vua Fuzzy Logic, " Proc. IFAC Workshop on Intelligent Manufacturing Systems, Dearborn, MI, October, 1992.

[5] Yokogawa Corporation, "Overshoot Suppression for a New Generation of Controllers," Technical Notes, Tokyo, Japan, 1991.

AN ADAPTIVE FUZZY LOGIC CONTROLLER FOR AN AUTOMOBILE SUSPENSION SYSTEM

ROBERT C. BARASZU and ADNAN SHAOUT
Ford Motor Co. and The University of Michigan - Dearborn

ABSTRACT

This paper presents a fuzzy solution to the problem of automobile suspension control. The subjective nature of suspension system design lends itself well to the use of a heuristically developed, fuzzy controller. Fuzzy adaptive rules are used to alter a primary rule base according to the real-time performance of the controller. Performance comparisons to a mechanical and another electronic suspension system are made using a dynamical simulation.

INTRODUCTION

The goal of automotive suspension system design is to balance the tradeoff between ride quality and handling performance. Good handling requires a stiff suspension to maintain tire contact with the road. On the other hand, good ride requires a soft suspension to isolate the vehicle body from the undesirable vibrational energy imparted by the road. But a stiff, or good handling suspension, allows the road irregularities to be transmitted to the body resulting in poor ride quality. Likewise, a soft, or good riding suspension, allows the tire/wheel to oscillate uncontrollably; hence, tire-to-road contact is lost.

Many attempts have been made to quantify good ride and good handling by measuring dynamic vehicle parameters such as vertical and lateral accelerations [1], [2]. Various control systems, both mechanical and electronic, have been devised to optimize the ride/handling tradeoff. But numbers and graphs that "prove" a system is "optimal" do not necessarily correspond to real-life experience. Good ride and good handling mean different things to different people. In particular, the last word on a suspension's performance is the subjective judgement of the customer. Thus, the subjective nature of suspension design invites the use of a fuzzy, linguistically defined control policy.

This paper describes a suspension control system developed using heuristics based on expert knowledge of the fuzzy concepts: "Road Smoothness", "Road Flatness", "Poor Ride", and "Poor Handling". These terms are the linguistic variables used by the fuzzy logic controller (FLC). The proposed FLC operates on two crisp sensor inputs from each corner of the vehicle, fuzzifies them into the linguistic variables, applies them to the rule base, and defuzzifies the result into the crisp control action of shock absorber damping. Adaptive rules change the control weight of each control rule according to the real-time performance of the system as measured by "Poor Ride" and "Poor Handling".

The suspension control systems were modelled with $Matrix_x$. Simulation results are presented showing the operation of the fuzzy controller and performance comparisons of traditional control means to that of the FLC. The architecture of the FLC is shown below in Figure 1. References [3], [4], and [5] provided insight into the FLC design.

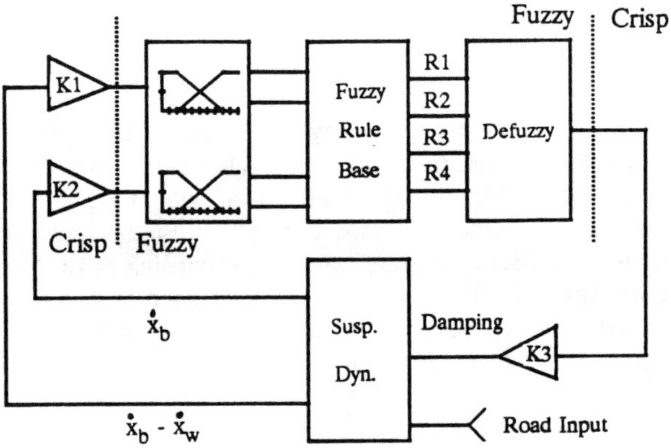

FIG. 1. Primary FLC Architecture

BASIC SUSPENSION CONTROL PRINCIPLES

The dynamical model of the suspension system used in the $Matrix_x$ simulation is a simple two degree-of-freedom model, shown below in Figure 2. It represents one-quarter of a vehicle: a wheel/tire assembly mass, and the tire spring and damper; one-quarter of the vehicle mass, the suspension spring rate, and the damping control element. The four state variables are the vertical positions and velocities of the vehicle body and the center of the wheel from the road, x_b, $\dot{x}_b$, x_w and $\dot{x}_w$, respectively. The sensor inputs to the FLC are the absolute vertical body velocity, $\dot{x}_b$, and the suspension velocity, $\dot{x}_b - \dot{x}_w$.

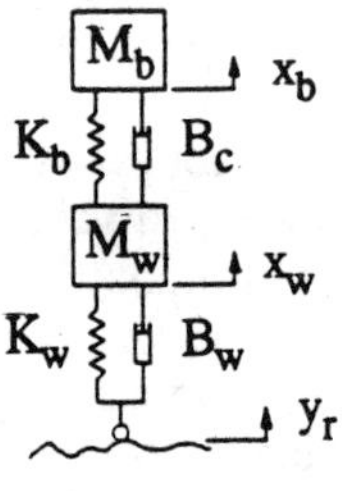

FIG. 2. Suspension Model

x_b = absolute body position
$\dot{x}_b$ = absolute body velocity
x_w = absolute wheel position
$\dot{x}_w$ = absolute wheel velocity
$\dot{x}_b - \dot{x}_w$ = suspension velocity
y_r = road position
M_b, M_w = mass of body, wheel
K_b = rate of suspension spring
K_w = rate of tire spring
B_w = damping rate of tire
B_c = controlled damping rate

The damper, conventionally a shock absorber or strut, is a device that viscously damps oscillations in the suspension system. The damping rate of conventional dampers is a fixed function of $\dot{x}_b - \dot{x}_w$. In electronically controlled dampers, the damping rate can be adjusted in real-time to manage the energy in the suspension between the body and the wheel to optimize the ride/handling tradeoff. Two heuristics illustrate this tradeoff, and are the foundation of the fuzzy knowledge base:

"*To minimize body motion, transfer energy to the wheel.*"

Requires: Low Damping
Benefit: Good ride quality.
Drawback: Wheel is allowed to oscillate. Wheel-hop occurs - contact between tire and road surface is lost. Poor handling results.

"*To minimize wheel motion, transfer energy to the body.*"

Requires: High Damping
Benefit: Good handling performance.
Drawback: Excessive body motion, both high and low frequency, results in poor ride quality.

THE SKYHOOK RIDE CONTROL LAW

The "skyhook" [6], [7] principle is considered an optimal ride control strategy since it will ideally keep the body level with the ground. This ideal can only be realized, however, with the use of active suspension components capable of both dissipating energy from and imparting energy to the suspension system. The skyhook control calls for a force to be imparted in the suspension which would be developed if a conventional damper were "hooked" between the vehicle body and the "sky", an inertial reference frame. This force is proportional to the vertical velocity of the body alone since the inertial frame is, by definition, at rest.

Using purely dissipative devices, the control law must be modified so that when ideal skyhook calls for energy to be imparted, the damping of the dissipative device is set to its minimum level. Otherwise, damping is set proportional to vertical body velocity as in the ideal case.

The skyhook law achieves optimal ride control by damping body motion at the expense of wheel motion. The wheel is given freedom to oscillate and so handling is poor. Performance comparisons made later in the paper clearly show this fact.

Traditional techniques have been employed to optimize the ride/handling tradeoff in the case of active suspensions [8]. This controller is realized with full state feedback and assumes a linear suspension model. This is reasonable for an active suspension; however, for purely dissipative dampers, the problem is nonlinear. In general, damping force is given by equations (1), (2), and (3) shown below:

$$F_{Damper} = B_{Control} \cdot (\dot{x}_B - \dot{x}_W) \tag{1}$$

where,

$$B_{Control} = f(x_B, \dot{x}_B, x_W, \dot{x}_W) \quad IF \quad \dot{x}_B \cdot (\dot{x}_B - \dot{x}_W) > 0 \tag{2}$$

$$B_{Control} = Minimim\ Damping \quad IF \quad \dot{x}_B \cdot (\dot{x}_B - \dot{x}_W) < 0. \tag{3}$$

Nonlinear optimization of this system poses a theoretical problem while the availability of all state variables poses a practical problem. Body and wheel acceleration can be measured with accelerometers; integrating the acceleration yields velocity; but, integration of velocity to obtain position is not possible since the integration constant is non-zero and varies with the weight of the vehicle, including passengers. Additional sensors could be added to measure position, but the cost is prohibitive. Thus, given these difficulties, a fuzzy approach seemed reasonable.

DESCRIPTION OF THE FUZZY SUSPENSION CONTROLLER

Two crisp variables, absolute vertical body velocity (derived from integration of an accelerometer signal) and suspension velocity (the relative velocity between the body and the wheel, derived from differentiation of a position sensor signal) are the inputs to the FLC. The absolute value of each signal is lowpass filtered to arrive at a short term average of the road condition. These signals are scaled and fuzzified into two fuzzy sets: *Road Flatness* = {*Flat Road, Undulating Road*}, and *Road Smoothness* = {*Smooth Road, Rough Road*}, respectively (Figure 3), which are linguistic measures of the road condition. Each of the two fuzzy sets contains two fuzzy variables; therefore, only four rules are required. Using the above heuristics, control rules were formulated with the damping rate of the damper as the consequence. These rules form the kernel of the FLC and will later be augmented by an adaptive strategy.

Primary Fuzzy Rule Base

Rule 1: IF *Road is Smooth* **AND** *Flat* **THEN** *Damping is Low*
Rule 2: IF *Road is Rough* **AND** *Flat* **THEN** *Damping is Medium*
Rule 3: IF *Road is Rough* **AND** *Undulating* **THEN** *Damping is Medium*
Rule 4: IF *Road is Smooth* **AND** *Undulating* **THEN** *Damping is High*

Rule 1 states that if the road disturbances are small, then little damping is required to maintain both good ride and handling. Rules 2 and 3 can be grouped together so that whenever the road is rough, damping is set to some intermediate value. Low damping here will allow excessive wheel motion while high damping will induce harshness and undesirable body motion. Rule 4 calls for maximum damping to minimize wheel motion, since body motion on an undulating road is not controllable with a damper alone.

The membership functions for the fuzzy set *Damping* = {*Low Damping*, *Medium Damping*, *High Damping*} is shown in Figure 4. The use of constant membership functions was employed to expedite computation.

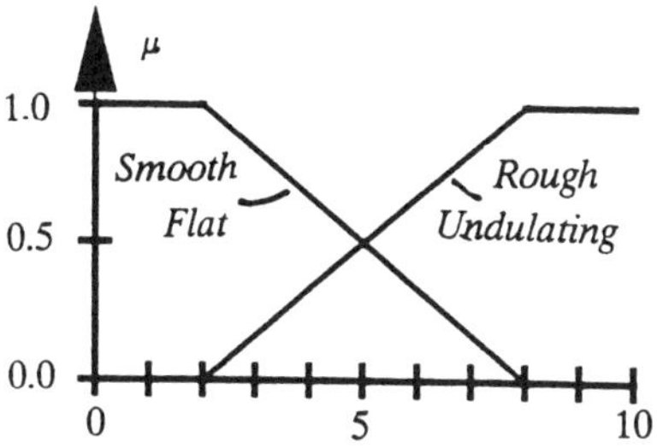

FIG. 3. Road Condition Membership

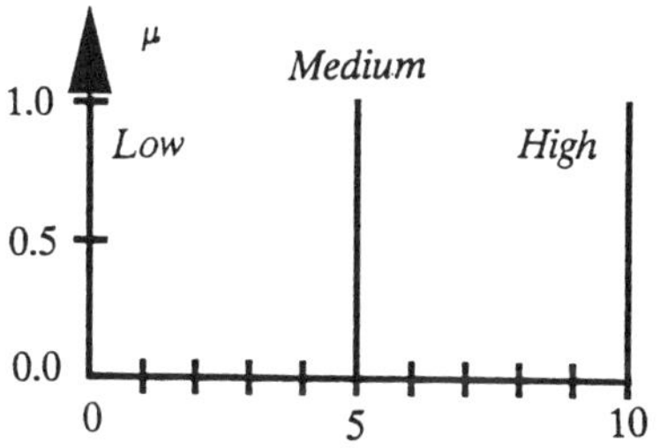

FIG. 4. Damping Level Membership

The **AND** is the operator **min**($\mu_{Road\ Smoothness}$, $\mu_{Road\ Flatness}$). The crisp value of *Damping* is obtained by defuzzification according to equation (4). R1, R2, R3, R4 are the damping values obtained from evaluation of Rules one through four; S, R, F, U refer to the fuzzy variables *Smooth*, *Rough*, *Flat*, and *Undulating Road*, respectively. The singleton membership functions for *Damping* act as weighting factors on the **min** composition of the road condition premises, simplifying defuzzification compared to fuzzy sets for the damping values.

$$Damping = \frac{R1 \cdot \min(\mu_S,\mu_F) + R2 \cdot \min(\mu_R,\mu_F) + R3 \cdot \min(\mu_R,\mu_U) + R4 \cdot \min(\mu_S,\mu_U)}{R1+R2+R3+R4} \tag{4}$$

Fuzzy Adaptive Strategy

Using only the primary rule base results in excellent controller performance, improving over the standard mechanical design. However, there is uncertainty as to the values of *Low*, *Medium*, and *High Damping* that will provide the best performance. A fuzzy strategy has been devised that monitors the effectiveness of each of the four primary rules and modifies the values of *Damping* to increase ride and handling quality.

The inputs to the fuzzy adaptive strategy are the RMS values of $\dot{x}_b$ and $\dot{x}_b - \dot{x}_w$, which are measures of ride and handling performance, respectively. The higher the value, the worse the performance. These inputs are fuzzified into the fuzzy variables *Poor Ride* and *Poor Handling*. Since no adaptation is required if performance is "good", no fuzzy variables for these conditions are required. In addition, the fuzzy variable *Long Time* measures the amount of time spent executing a primary rule. Time measurement begins when the membership grade of a rule exceeds a crisp threshold, indicating that a certain road condition, say rough and flat, is prevalent. The generalized adaptive rules are:

IF *Executing Rule N for a Long Time*
AND IF *Ride is Poor*, THEN *Decrease Rule N Damping*

and,

IF *Executing Rule N for a Long Time*
AND IF *Handling is Poor,* THEN *Increase Rule N Damping*

Augmenting the adaptive rules to each of the primary rules results in eight new rules. For example, Rule 2 becomes:

IF *Road is Rough* AND *Flat for a Long Time*
AND IF *Ride is Poor* THEN *Decrease Rule 2 Damping*

and, IF *Road is Rough* AND *Flat for a Long Time*
AND IF *Handling is Poor* THEN *Increase Rule 2 Damping*

The change in damping for each rule is summed with the previous damping value for that rule, within bounds, resulting in the new damping level. Note that if both ride and handling are poor, the net effect of the adaptive rule is to maintain the present level of damping. At this point the controller has reached an optimal state since a further change in damping would either degrade ride or handling.

OPERATION OF THE FUZZY CONTROLLER

In a practical implementation, the FLC would operate independently on each of the four corners of the vehicle; therefore, analysis of only one corner need be performed. Random road profiles provided input to the model [9]. Results presented here are from two such profiles, RD1 and RD2, with amplitude spectral densities of 0.0000375 and 0.000075 (m/sec^2/Hz), respectively, each of length 32.0 seconds. Table I summarizes the ride, handling, and combined ride/handling quality improvement or degradation of the skyhook, the primary FLC, and the adaptive FLC over the standard suspension for RD1 and RD2. A positive change indicates an improvement while a negative indicates a degradation in performance.

TABLE I. Percent Performance Change: FLC and Skyhook vs. Mechanical

	Ride Quality		Handling Quality		Ride/Handling	
	RD1	RD2	RD1	RD2	RD1	RD2
Skyhook	+12.5	+17.9	-43.2	-38.1	-39.1	-33.9
Primary Fuzzy	-6.9	+2.6	-13.8	-2.7	-15.2	-0.1
Adaptive Fuzzy	+8.0	+5.6	+13.6	+25.9	+12.0	+20.6

As expected, the skyhook strategy was superior to both FLC's in ride quality, but suffered significantly in handling quality. The performance of the primary

FLC was significantly worse than the standard suspension on RD1 (a relatively smooth road) and about the same on RD2 (a relatively rough road). Though an unexpected result, it serves to illuminate the benefits of the adaptive strategy which showed significant performance improvements in all cases, especially in handling. Further refinement of both the primary and adaptive membership functions could, perhaps, further improve performance.

The top plot of Figure 4 shows the measure of "*Poor Handling*" for the fuzzy adaptive controller over road RD2. The lower three plots show the primary rule being executed followed by the damping adaptation of Rule 1 and Rule 2. The initial damping levels of both the primary and adaptive FLC were the same. Adaptive Rule 1 changed from 1.0 to 1.2 before t = 3.0 seconds and remained there over the entire length of RD2. Rule 2 adapted to several values from its initial level of 5.0. At point A, damping is reduced and handling quality quickly decreases ("*Poor Handling*" increases). At point B, damping is raised and handling improves steadily until point C. Then, a reduction in damping begins to degrade handling performance.

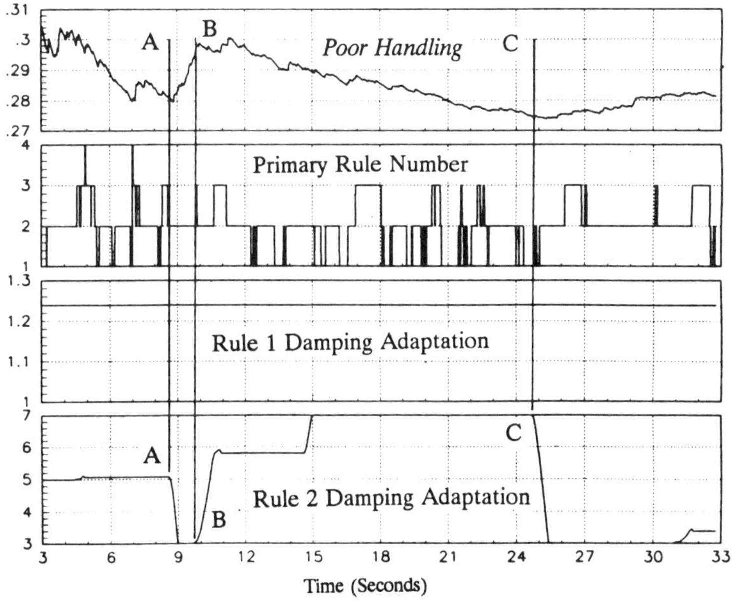

FIG. 4. Operation of Fuzzy Adaptive Strategy

CONCLUSIONS AND FUTURE WORK

Using heuristics and expert knowledge of automobile suspension systems, a simple FLC was developed and evaluated via a dynamical simulation. The FLC system exhibited some performance improvements over a standard suspension; however, the addition of simple fuzzy adaptive rules to the primary rule base increased performance significantly. The adaptive FLC provided an excellent compromise between ride and handling with only a small degradation in ride quality compared to the optimal skyhook law.

To mistake the crisp values of performance improvement data in Table I as the only end product of this study would be a grievous error. In fact, doing so would wholly disregard the spirit of fuzzy logic itself. The most important information to be gleaned from this work lies more in the approach taken than in the raw results. That is, the ability to construct a simple, yet effective, control system by the synthesis of fuzzy, linguistic concepts into a framework suitable for real-time implementation in a digital computer is the heart of this study.

The quarter-car model used in this study is only a first step in the design of a truly effective suspension controller, for a real automobile is not constrained to vertical motion nor are the suspension points independent. Ride and handling quality are functions of more than vertical motion alone. A full vehicle model, including the effects of pitch, roll, and yaw, must be considered when designing a suspension control system. Thus, future work will be toward the end of developing such a controller using the fuzzy concepts developed here as a basis.

REFERENCES

1. Butkunas, Andrius A., "Power Spectral Density and Ride Evaluation," *SAE*, 660138.
2. Leatherwood, Jack D., Dempsey, Thomas K., and Clevenson, Sherman A., "A Design Tool for Estimating Passenger Ride Discomfort Within Complex Ride Environments," *Human Factors*, vol. 22, no. 3., 1980, pp. 291-312.
3. Zadeh, Lofti A., "Outline of a New Approach to the Analysis of Complex Systems and Decision Processes," *IEEE Transactions on Systems, Man, and Cybernetics*, vol. SMC-3, No. 1, January 1973.
4. Lee, Chuen Chien, "Fuzzy Logic in Control Systems: Fuzzy Logic Controller - Parts I and II," *IEEE Transactions on Systems, Man, and Cybernetics*, vol. 20, no. 2, March/April 1990, pp. 404-435.
5. Sugeno, Michio, "An Introductory Survey of Fuzzy Control," *Information Sciences*, vol. 36, 1985, pp. 59-83.
6. Karnopp, Dean, "Active Damping in Road Vehicle Suspension Systems," *Vehicle System Dynamics*, vol 12, 1983, pp. 291-316.
7. Margolis, Donald L., "The Response of Active and Semi-Active Suspensions to Realistic Feedback Signals," *Vehicle System Dynamics*, vol. 11, 1982, pp. 267-282.
8. Armstrong, A. G., "An Active Suspension With Optimal State Feedback," *Vehicle System Dynamics*, vol. 5, 1976, pp. 187-203.
9. Kotwicki, A. J., and Bridges, G. F., "A Random Road Profile Generator," Unpublished software. Property of Ford Motor Co., 1984.

TRAINING GRAYSCALE STRUCTURING ELEMENT USING FUZZY SETS AND NEURAL NETWORKS

Sung-Jea Ko[1] and Aldo Morales[2]

[1]Department of Electronics Engineering
Korea University
Seoul, Korea

[2]College of Engineering
The Pennsylvania State University
DuBois, PA 15801

ABSTRACT

In this paper, we introduce a neural network implementation of gray scale operators. In this structure, synaptic weights are represented by a gray scale structuring element. A learning algorithm based on the averaged least mean square is used to train this fuzzy morphological neural network. It will be shown that the LMS-based algorithm has a fast rate of convergence. Two dimensional examples will be shown.

I. INTRODUCTION

Mathematical morphology involves the study of the different ways in which a structuring element interacts with a given set, modifies its shape and measures the resultant set [1,2,3]. The basic operations are erosion and dilation. Based on these operations, two compound operations, closing and opening, are defined. These operations have been successfully used in many applications including object recognition, image enhancement, texture analysis, and industrial inspection.

Artificial neural networks have become popular as a tool for computing a large mass of numerical computations in an effective, highly parallel manner [4]. These networks are composed of many nonlinear computational elements or nodes arranged in patterns reminiscent of biological neural nets. The nodes are connected via weights that are typically adapted during use to improve performance.

Fuzzy set theory is an interesting platform to cope with ambiguity present in human activity, especially in decision processes [4]. In addition, fuzzy sets have been related to Mathematical Morphology [5] to represent basic binary operators but not gray scale ones.

In this paper, we propose a neural network implementation of gray scale operators. In this structure, synaptic weights are represented by a gray scale structuring element (GSE). A learning algorithm based on the averaged least mean square is used. Numerical examples will be presented to illustrate its performance.

The organization of this paper is as follows. In section II, relationships between fuzzy sets and Mathematical Morphology are reviewed. In section III, we propose a fuzzy neural network implementation of morphological operators. In section IV, we present some numerical examples.

II. MATHEMATICAL MORPHOLOGY AND FUZZY SETS

Let the gray scale structuring element (GSE) with size N be denoted by $\mathbf{k} = \{k_1, k_2, \cdots, k_N\}$ and a set of samples covered by $\mathbf{k}$ at time i be denoted by $\mathbf{f}_i = \{f_{i1}, f_{i2}, \cdots f_{iN}\}$. The output g_i of a gray scale morphological operation is given by

$$g_i = \theta(\mathbf{f}_i, \mathbf{k}) \tag{1}$$

where θ is a Min/Max operator representing either a single operation (Min or Max) or a combined operation of Min and Max (Min of Maxima or Max of Minima). Using symmetric GSE, the output of the gray scale dilation is given by

$$g_i = \mathbf{f}_i \oplus \mathbf{k} = \max\{f_{i1} + k_1, f_{i2} + k_2, \cdots, f_{iN} + k_N\} \tag{2a}$$

and the output of the gray scale erosion is given by

$$g_i = \mathbf{f}_i \ominus \mathbf{k} = \min\{f_{i1} - k_1, f_{i2} - k_2, \cdots, f_{iN} - k_N\} \tag{2b}$$

Relationships between morphological operations with binary structuring elements and fuzzy sets were discussed by Goetcherian [5]. Next, we briefly review fuzzy set theory and membership functions induced by algebraic operations. These definitions will help us pose morphological gray scale operations as fuzzy operations.

Definition 1: If $\mathbf{F}$ is a collection of pixels located at (i, j), then a fuzzy set A in $\mathbf{F}$ is a set of ordered pairs

$$A = \{[(i,j), f_A(i,j)] | (i,j) \epsilon \ \mathbf{F}\} \tag{3}$$

where $f_A(i,j)$ is called the membership function. In images, this function can be thought of a normalized gray value, $0 \leq f_A(i,j) \leq 1$.

Performing set-theoretic or algebraic operations on fuzzy sets induces a new membership function as a result [6]. In this paper, we will utilize the following two membership functions

Definition 2: The membership function $f_{A\tilde{\oplus}B}(i,j)$ of the bounded sum $C = A\tilde{\oplus}B$ is defined by

$$f_{A\tilde{\oplus}B}(i,j) = \min[1, f_A(i,j) + f_B(i,j)] \tag{4}$$

Definition 3: The membership function $f_{A\tilde{\ominus}B}(i,j)$ of the bounded difference $C = A\tilde{\ominus}B$ is defined by

$$f_{A\tilde{\ominus}B}(i,j) = \max[0, f_A(i,j) - f_B(i,j)] \tag{5}$$

The bounded sum and bounded difference can be used to insure that the outputs of gray scale dilation and erosion are limited to the range $[0,1]$. Thus, eq. $(2a)$ may be modified to

$$g_i = \mathbf{f_i} \oplus \mathbf{k} = \max\{\min[1, f_{i1} + k_1], \cdots, \min[1, f_{iN} + k_N]\} \tag{6a}$$

and eq. $(2b)$ may be modified to

$$g_i = \mathbf{f_i} \ominus \mathbf{k} = \min\{\max[0, f_{i1} - k_1], \cdots, \max[0, f_{iN} - k_N]\} \tag{6b}$$

In sequel, these two expressions will be used to derive the training algorithm.

III. MATHEMATICAL MORPHOLOGY AND NEURAL NETWORKS

There are two factors of importance in the design of neural networks 1) an appropriate way of problem representation which leads to its formulation in terms of suitable topology and 2) an appropriate learning algorithm [4]. Referring to the first issue, morphological operators can be implemented in a structure similar to a neural network where the input $\mathbf{f_i}$ and GSE $\mathbf{k}$, respectively, form a vector of activation of all input nodes and a vector of synaptic weights which is summed to the input. The second factor in the design of neural networks is to chose an appropriate training algorithm. In this section, we use a learning algorithm based on the averaged least-mean square.

Algorithm based on Least Mean Squares

The learning rule described next is provided by the update algorithm based on least mean squares (LMS) [7]. Assume that f_{ij} and k_j take on values in the range $[0,1]$ by normalization with respect to the number of gray values of the signal $\mathbf{f}$. The updated value of $\mathbf{k}$ is given by:

$$\mathbf{k}(j+1) = \mathbf{k}(j) - \eta\Delta\mathbf{k}(j) \tag{7}$$

Here the increment $\Delta\mathbf{k}(j)$ is obtained by estimating the gradient of the mean square error between the target and actual output signal. At each iteration we have a gradient estimate of the form

$$\Delta\mathbf{k} = \begin{pmatrix} \frac{\partial E[\xi_j^2]}{\partial k_1} \\ \frac{\partial E[\xi_j^2]}{\partial k_2} \\ \vdots \\ \frac{\partial E[\xi_j^2]}{\partial k_N} \end{pmatrix} \tag{8}$$

where $E[\cdot]$ is the expected value and

$$\frac{\partial E[\xi_j^2]}{\partial k_m} = E\left[\frac{\partial \xi_j^2}{\partial k_m}\right] = 2 \cdot E\left[\xi_j \frac{\partial \xi_j}{\partial k_m}\right] \tag{9}$$

where ξ_j is the output error defined as the difference between the target and the output value, $\xi_j = t_j - \theta(\mathbf{f}_i, \mathbf{k})$. With this definition, the derivative $\frac{\partial \xi_j}{\partial k_m}$ is given by

$$\frac{\partial \xi_j}{\partial k_m} = -\frac{\partial \theta(\mathbf{f}_i, \mathbf{k})}{\partial k_m} \tag{10}$$

This derivative can be obtained using the following formula

$$\frac{\partial \theta(\mathbf{f}_i, \mathbf{k})}{\partial k_m} = \begin{cases} 1, & \text{if } g_i = f_{im} + k_m \text{ and } f_{im} + k_m \leq 1 \\ 0, & \text{otherwise.} \end{cases} \tag{11}$$

for dilation. (Note that we have used the bounded sum definition of dilation to find this derivative). For erosion the following derivative is utilized

$$\frac{\partial \theta(\mathbf{f}_i, \mathbf{k})}{\partial k_m} = \begin{cases} -1, & \text{if } g_i = f_{im} - k_m \text{ and } f_{im} - k_m \geq 0 \\ 0, & \text{otherwise.} \end{cases} \tag{12}$$

here we used the bounded difference definition to implement gray scale erosion. Next, we will make an attempt to smooth the estimate in (9) by using an averaged gradient of the mean square error. The *mth* element of the gradient vector can be estimated as

$$E\left[\xi_j \frac{\partial \xi_j}{\partial k_m}\right] = \frac{\sum_{i=1}^{L} \frac{\partial \xi_j}{\partial k_m}(t_j - \theta(\mathbf{f}_i, \mathbf{k}))}{\left|\sum_{i=1}^{L} \frac{\partial \xi_j}{\partial k_m}\right|} \tag{13}$$

here the denominator of (13) represents the total number of times that the *mth* element in the input vector becomes the output.

IV. NUMERICAL EXAMPLES

We next show the performance of the averaged LMS algorithm in 2-Dimensions. Fig. 1 and Fig. 2 illustrate the original and the target image respectively. The unknown gray scale structuring element is defined as symmetric 3x3. The momentum parameter η is set equal to 0.4. The algorithm converges after 3 iterations. The final value obtained for the gray scale structuring element is

$$k(3) = \begin{bmatrix} 2.816 & 4.800 & 2.811 \\ 2.938 & 4.800 & 2.941 \\ 2.814 & 4.800 & 2.807 \end{bmatrix}$$

Note that the values obtained in this case are not normalized with respect to the maximum pixel value (255) thus speeding up the computation process. For comparison purposes, we would like to point out that the target image was obtained with the following symmetric gray scale structuring element.

$$k = \begin{bmatrix} 3.00 & 5.00 & 3.00 \\ 3.00 & 5.00 & 3.00 \\ 3.00 & 5.00 & 3.00 \end{bmatrix}$$

We will present more examples in the conference.

V. CONCLUSION

We proposed the fuzzy neural network implementation of gray scale dilation and erosion. We used the bounded sum and difference principles to pose gray scale dilation and erosion as fuzzy operations. To train fuzzy morphological networks we considered a learning algorithm based on the averaged least mean square. Numerical examples were presented to show the performance of this training algorithm. The results obtained suggests that the averaged least mean square has a fast rate of convergence. Considerable effort remains in two aspects a) neural implementation of other morphological transformations, namely opening and closing and b) study of the convergence of the training algorithm presented. We are working on these issues.

REFERENCES

[1] G. Matheron, *Random Sets and Integral Geometry*, New York: Wiley, 1975.

[2] J. Serra, *Image Analysis and Mathematical Morphology*, New York: Academic Press, 1988.

[3] C. Giardina and E. Dougherty. *Morphological Methods in Image and Signal Processing.* New Jersey: Prentice Hall, 1988.

[4] W. Pedrycz, "Neurocomputations in Relational Systems", *IEEE Transactions on Pattern Analysis and Machine Intelligence*, Vol. PAMI 13, No. 3. March 1991.

[5] V. Goetcherian, "From Binary to Grey Tone Images Using Fuzzy Logic Concepts", *Pattern Recognition.* Vol. 12, pp 7-15, 1980.

[6] Y.-H. Pao, *Adaptive Pattern Recognition and Neural Networks.* New York: Addison-Wesley Publishing Co., 1989.

[7] B. Widrow and S.D. Stearns, *Adaptive Signal Processing.* Englewood Cliffs, New Jersey: Prentice Hall, 1985.

Fig.1 Original image.

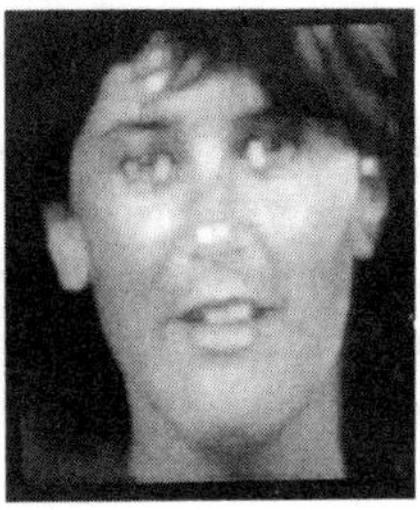

Fig. 2 Target image.

Application of Fuzzy Logic and Object Oriented Classes for Utilization in Robotic Control

Bruce S. Widmann†, Glenn R. Widmann† and Julio J. Gonzalez‡

† Hughes Aircraft Company
Missile Systems Group
Automotive Controls Engineering Directorate
Canoga Park, CA 91304-7928

‡ Colorado State University
Dept. of Electrical Engineering
Fort Collins, CO 80523

ABSTRACT: Fuzzy Logic is an emerging control methodology which has applications to robotic control. In this paper, the use of Fuzzy Logic principles and object oriented programming techniques are discussed as they can be applied to the design and implementation of control algorithms for robotic systems. Object oriented programming languages provide a powerful programming environment, through the use of abstract data types and language extensibility. This allows for the reuse of common code blocks, or classes, resulting in lower software costs and reduced algorithm design schedules. Fuzzy Logic object oriented classes have been implemented in the C++ programming language.

1. INTRODUCTION.

Fuzzy logic, a new emerging decision making methodology, has been gaining increased popularity in recent years. An outgrowth of fuzzy set theory, first conceptualized by L. Zadeh [1] in the mid 60's, fuzzy logic provides a formal means of incorporating approximate reasoning and "expert experience" into the solution of a wide range of controls and decision making problems. Current areas of fuzzy logic technology applications and research include pattern recognition, expert systems, modeling, fault isolation, neural networks, and control systems [2] [3].

In the area of control systems, the design and synthesis of conventional controllers are based on detailed, accurate models of the plant and requires extensive quantitative and numeric calculations. On the other hand, fuzzy logic provides particular advantage to applications which require the control of a plant that is poorly understood, difficult to model, and has unmodeled dynamics.

The robotics field is a perfect example where fuzzy logic control could be exploited. The detailed dynamic modeling for this class of complex nonlinear systems is very time consuming and nearly impossible to develop. Additionally, the lengthy computation time needed to specify the required control signal, based on the detailed model, is difficult to achieve within practical sampling periods for real-time control considerations. Also, the performance of robotic system that utilize finely tuned conventional controller algorithms are degraded by the presence of plant parameter variations, which can be attributed to component aging, varying loads, structural variations between robots of the same model due to imprecise manufacturing processes, etc. Fuzzy logic based controllers have been reported to more robust to plant parameter variations than classical control algorithms.

The use of a fuzzy logic control strategy seems ideally suited to resolve the control issues associated with robotics. Surprisingly, very few researchers have pursued the use of fuzzy logic as applied to robotics. These investigations have mostly been concerned with issues of position control [4] [5] and force control [6].

Examination of a fuzzy logic controller indicates that distinct partitioning is possible in the processing algorithms of the controller. The fuzzifier and defuzzifier processes can be further partitioned into distinct sub-units of membership functions and fuzzy variables. The knowledge base can be partitioned into individual fuzzy rules. This ability, to be able to partition a fuzzy logic knowledge based controller into distinct sub-units, makes it a likely candidate for a software solution utilizing object oriented programming techniques.

The combination of fuzzy logic concepts and object oriented programming has been explored by others [7] [8] [9]. However, these investigations have primarily examined the two technologies from a standpoint of the fuzziness or extent of an object meeting the criteria of, or matching, a particular class. The classes were considered "crisp", whereby a grade of membership of an instance of an object within that class being calculated.

This paper explores fuzzy logic and object oriented programming from a different perspective. The fuzzy logic classes presented in this paper are meant to provide a basis for generating a library of specialized object classes, for application to a wide variety of control problem solutions. It will, also, provide a brief overview of fuzzy logic rule-based controllers. Finally, the paper will present a description of fuzzy logic abstract data types and classes which can be extended through the use of the C++ language. It is the intent of this paper to illustrate how these classes can be used to facilitate the design, implementation, and maintainability of fuzzy logic controllers for robotic applications.

2. FUZZY LOGIC RULE-BASED CONTROLLERS.

Figure 1 presents a simplified block diagram of a typical manipulator/controller structure. In the case of position controlled robots, the system input and output would be commanded position setpoints and motion trajectory, respectively. In the case of force controlled robots, the system input and output would be commanded force setpoints and force trajectory, respectively.

The general structure of a fuzzy logic rule-based controller can be found in [10]. Defining the controller input and output variables as fuzzy variables, involves the formulation of fuzzy sets over the range of a variable. The method of choosing the fuzzy sets and their respective membership function shaping will not be described in this paper, since it is highly dependent upon the specific goals to be accomplished. But rather the intent of the paper is to discuss how object oriented classes can be applied to the components of a fuzzy logic controller in order to facilitate code re-usability and rapid software prototyping.

Exact or crisp controller inputs are applied to the fuzzy sets by the fuzzifier, in order to establish the fuzzy grades of membership. The generation of fuzzy sets and membership functions for a variable allows a particular value of the variable to satisfy a partial or full membership in one or more fuzzy sets at a time. These fuzzy inputs are then applied to the fuzzy base by the inference engine. The fuzzy rules consists of a set or collection of production rules that are lingually based and typically are of the form:

If {Antecedent Rule} **Then** {Consequent Rule}

where: Antecedent Rule: Fuzzy Rule defining decision criteria,
Consequent Rule: Fuzzy Rule defining the resulting control action.

These rules define the control surface over which the controller operates. The controller inference engine applies the controller inputs to the rule base in an attempt to establish which rules are "fired" or satisfied. The fired rules are then used to determine the controller output.

The Antecedent Rule portion of the fuzzy rule typically consists of a Boolean expression consisting of a combination of fuzzy variables. The Consequent Rule is typically one or more control actions to be implemented by the controller. For example, regarding the control of a robot and the decision for the motion of a link would be:

If {(Joint Displacement is PL) and (Joint Speed is PL)}
Then {Joint Actuator Input is NL}

where: "Joint Displacement", "Joint Speed" and "Joint Actuator Input" are lingual-based set variables, and "Positive Large (PL)" and "Negative Large (NL)" represent fuzzy quantifiers.

The potential for an input variable to have partial or full membership in several fuzzy sets simultaneously, allows the possibility for several fuzzy rules in the rule base to be fired or satisfied at a time. The control actions of each of these rules are members of fuzzy sets over the range of the controller output variables. The resultant fuzzy control actions specified by the fired fuzzy rules are combined in the defuzzifier to produce a single crisp or exact output from the controller. There are several methods by which the fuzzy outputs are combined. One common method involves the calculation of the centroid of the fired fuzzy outputs to form the crisp output.

3. FUZZY LOGIC CLASSES.

The construction of fuzzy logic object oriented classes will be discussed in this section. The classes are built from various aspects; from primitive classes such as linked lists [11] [12], through inheritance forming a hierarchy of classes, and through direct inclusion.

3.1. CONTAINER CLASSES & TEMPLATES.

One of the primary advantages of object oriented programming is the aspect of code reusability. In fact, the ability to reuse code forms the basis for the C and C++ languages, through the mechanism of function libraries in both C and C++, and the class libraries in C++. The reuse of code is appealing; it allows for rapid construction of software programs, as well as, the opportunity to utilize code which should have a high degree of maturity and reliability.

This aspect of code reusability is capitalized upon in the existence of a group of classes commonly referred to as container classes. These classes primarily consist of classes that hold, manipulate, and maintain other classes. For instance, a linked list class, holds an aggregate of data types, which can include both built-in and user defined data types, in a formalized manner. The linked list class is responsible for the maintenance of, and access to, the data types stored in the list.

A classic problem associated with container classes exists in the generalization of data types which can be held by the class. Even though the class methods are the same, the array class needs to be rewritten for each type of data which will be handled. This problem has been solved through the introduction of parameterized types or templates in Revision 3.0 of the C++ language standard. This allows a class to be type generic in definition, and type specific in usage.

The fuzzy logic classes make use of these classes through the use of the *Element* and *LinkedList* container classes, shown in Figure 2. These classes are used to maintain a collection of fuzzy data types and classes; a collection of *MemberFunctions* objects in the *FuzzyVariable* class, and the collection of *FuzzyRule* objects in the *FuzzyRuleBase* class. The classes utilize the template facility to allow the classes to be generic, becoming type specific in the application to a class.

3.2. ABSTRACT BASE CLASSES.

Another useful attribute in object oriented programming is the mechanism of inheritance. Inheritance provides the ability to build a general base class that contains the basic functioning of a set of objects, and later specialize through the construction of sub-classes that inherit the common features of the base class. This feature is utilized in the construction of the *MemberFunction* base class and subsequent *MemFunLookup* and *MemFunEquation* sub-classes shown in Figure 3. An interesting aspect associated with these classes occurs in the relationship of the base class to the sub-classes. The *MembershipFunction* class is commonly referred to as an abstract base class. The class contains the common capabilities of all the sub-classes. The class however is incomplete; no objects or instances of the class will ever be generated. The common heritage provided by the base class, allows a sub-class to be substituted into any usage instance of the base class. This feature allows functions to call as formal arguments and return the base class data type, thereby allowing all sub-classes access to the function. The use of virtual functions in prototyping the member methods in the base class, allows dynamic or late binding to resolve which version of the sub-class is being used.

The sub-classes, *MemFunLookup* and *MemFunEquation*, provide two possible instances of implementing membership functions. The first class, *MemFunLookup*, stores membership functions represented in lookup table format. The second class, *MemFunEquation*, specifically stores membership functions represented as equations stored in a function format. The use of a common base class allows for the generation of additional membership function sub-classes, which represent grade of membership information in alternate manners.

3.3. *FuzzyVariable* CLASS.

The *FuzzyVariable* class (see Figure 4) is actually a very specialized sub-class of the *LinkedList* class. The class inherits the *LinkedList's* methods for maintaining and manipulating a list of objects, specifically, *MembershipFunction* objects. In addition, the class overloads a number of operators, in order to support fuzzy rule predicate logic statements.

3.4. *FuzzyRule* AND *FuzzyRuleBase* CLASS.

The *FuzzyRule* class (see Figure 5) is fairly simple. The class structure consists primarily of a pointer to a function containing the fuzzy rule predicate logic statement. The *FuzzyRuleBase* class is again, a very specialized sub-class of the *LinkedList* class. The class is responsible for the maintenance, and manipulation of the fuzzy rules comprising the fuzzy rule base.

3.5. *InferenceEngine* CLASS.

The C++ class structure for the *InferenceEngine* class (see Figure 6) will not be presented or described in this paper, due to the complexity of the class structure. Future works will present this class structure.

4. SUMMARY AND CONCLUSION.

The use of a higher order language in the design and implementation of software applications offers numerous advantages. The programmer is elevated above computer implementation specific programming, thereby allowing the opportunity for solution oriented software programs. Utilizing software design methodologies, such as object oriented programming, allows code to be more portable across applications and processors, resulting in lower software design and maintenance costs and rapid design time. The fuzzy logic classes presented in this paper are meant to provide a basis for generating a library of specialized object classes, for application to a wide variety of control problem solutions.

REFERENCES.

[1] L. A. Zadeh, "Fuzzy sets," *Information and Control*, 8, pp. 338-353, 1965.

[2] J. Maiers & Y. S. Sherif, "Applications of Fuzzy Set Theory," *IEEE Trans. on Systems, Man, & Cybernetics*, Vol. SMC-15, No. 1, pp. 175-189, Jan./Feb. 1985.

[3] C. C. Lee, "Fuzzy Logic in Control Systems: Fuzzy Logic Controller - Part I & II," *IEEE Trans. on Systems, Man, & Cybernetics*, Vol. SMC-20, No. 2, pp. 404-435, April 1990.

[4] B.A.M. Wakileh & K.F. Gill, "Use of Fuzzy Logic in Robotics," *Computers in Industry*, Vol. 10, pp. 35-46, 1988.

[5] C.M. Lim & T. Hiyama, "Application of Fuzzy Logic Control to a Manipulator," *IEEE Trans. Robotics & Automation*, Vol. 7, No. 5, pp. 688-691, Oct. 1991.

[6] R.B. Kelly, D. Sood & M.C. Repko, "Integrated Sensing for Circuit Board Insertion," *J. Robotic Systems*, Vol. 7, No. 3, pp. 487-505, June 1990.

[7] D. C. Rine, "Design of Fuzzy Object-Oriented Software Components Data Bases," *Cybernetics & Systems: An Int. Journal*, pp. 22.531-551, 1991.

[8] Y. W. Huang & L. T. Fan, "Fuzzy Logic Rule Base System for Separation Sequence Synthesis: An Object-Oriented Approach," *Computers Chemical Engineering*, Vol. 12, No. 6, pp. 601-607, 1988.

[9] J. C. Westland, "Fuzzy Object Oriented Design Approaches to Systems," *23rd Annual Hawaii Intl Conf. on Systems Sciences*, pp. 346-352, 1990.

[10] B.S. Widmann & G.R. Widmann, "Implementation of Object Oriented Constructs for use in Fuzzy Logic Classifiers," *Proc. Fuzzy and Nerual Systems, and Vehicle Applications*, Japan, Nov. 1991.

[11] S. Lippman, C++ Primer, Addison-Wesley Publishing Co., 1990.

[12] R. Wiener & L. Pinson, An Introduction to Object Oriented Programming and C++, Addison-Wesley Publishing Co., 1988.

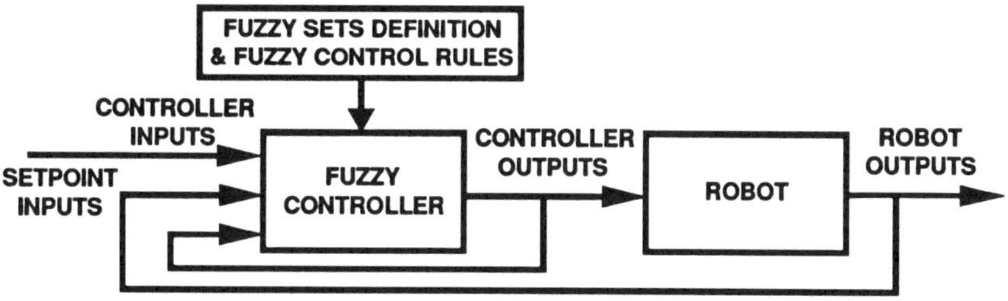

Figure 1. Simiplified Block Diagram of a Robot/Controller Configuration.

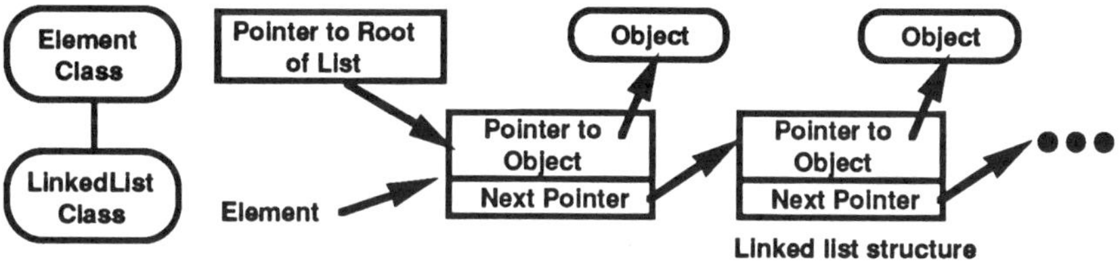

Figure 2. *LinkedList* Class Hierarchy and Associated Data Structure.

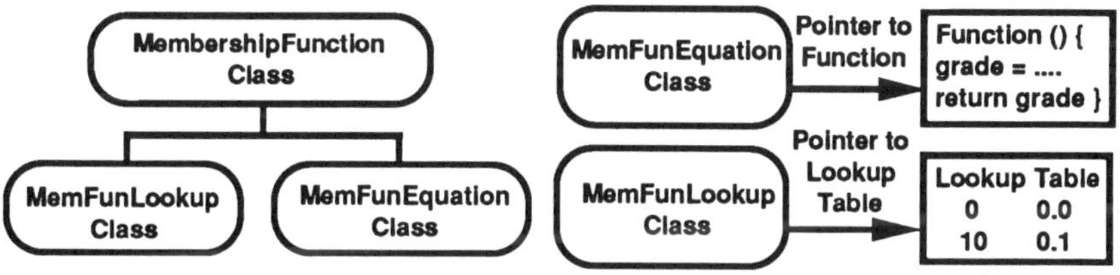

Figure 3. *MembershipFunction* Class Hierarchy and Associated Data Structure.

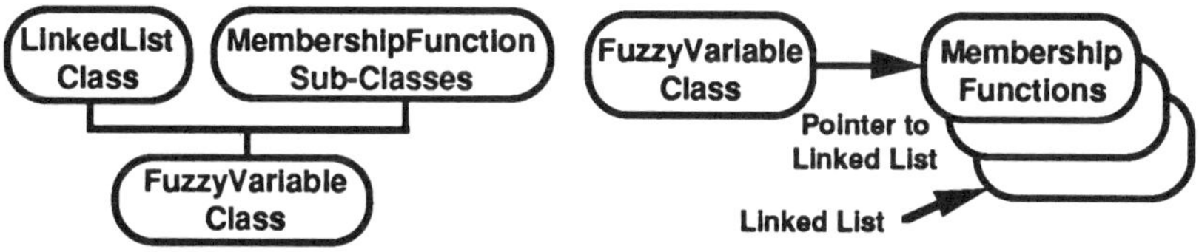

Figure 4. *FuzzyVariable* Class and Associated Component Classes.

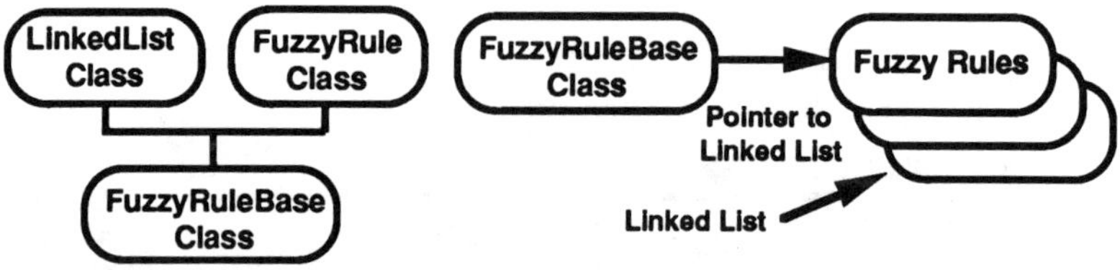

Figure 5. *FuzzyRuleBase* Class and Associated Component Classes.

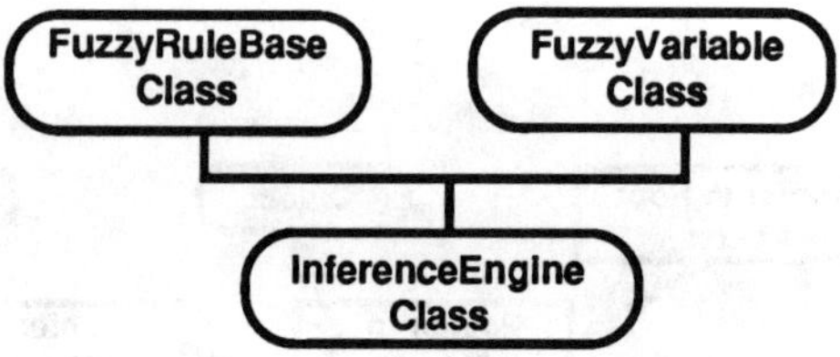

Figure 6. *InferenceEngine* Class and Associated Component Classes.

FUZZY SIMULATION IN MANUFACTURING SYSTEMS

A. DALTRINI and F. GOMIDE
UNICAMP - FEE - DCA, Brazil

ABSTRACT

This work aims at presenting an approach for manufacturing systems modeling and simulation which uses Fuzzy Set Theory and Fuzzy Logic, in special Fuzzy Petri Nets. The goal is to develop an alternative methodology for knowledge based simulation techniques. An application example of a manufacturing cell simulation is also included.

INTRODUCTION

Currently, as a result of the increasing world-wide industrial competitiveness, Manufacturing System Simulation is regarded as one of the most useful tolls for decision making in management and operating procedures [1].

Generally speaking, a simulation model is a descriptive model of a real life system, as compared to a normative model such as linear programming, which generates a candidate solution. Therefore, it is often used to predict system performance and to evaluate several strategies, and procedures. Thus, a simulation model is used as a "test-bed" to mimic the details and operational decisions of manufacturing systems under different design alternatives, and to test different tactical and operational decisions without disturbing the actual system. [2]

However, most of simulation methodologies available today do not have enough capacity to cope with uncertainty in some of the steps of simulation process (they offer resources based only on Probability Theory, which is inappropriate to solve some cases). The use of Fuzzy Petri Nets in simulation is intended to compensate this gap in the existing simulation methodologies.

In this paper, an alternative scheme to model and to simulate complex and/or ill-defined manufacturing systems with the use of concepts of Fuzzy Set Theory [3], Fuzzy Logic [4] and Petri Nets [5] is presented.

FUZZY PETRI NET - (FPN)

Fuzzy Petri Net (FPN) is a formalism used to represent fuzzy production rules of a rule based system, where a fuzzy production rule is a rule which describes the fuzzy relation among various propositions. It allows structured representation of knowledge, and has a systematic procedure for supporting fuzzy reasoning. [6]

A FPN, like a regular Petri Net, is a bipartite graph which contains two types of nodes: places and transitions, where circles represent places and bars represent transitions. Besides, the behavior and status of a FPN is determined by its marking (positions of the tokens inside the places), and each token is associated with a truth value between zero and one (this value is correspodent to the belief in the proposition that place is representing). Each transition is associated with a certainty factor value (μ) between zero and one (it means the amount you trust in that fuzzy production rule). Futhermore,corresponding to each transition, there is a transition enabling factor (λ), which is responsible for the firing of the transitions. The FPN formal description can be posed as follows:

$$FPN = (P , T , D , I , O , f , \alpha , \beta)$$

where:

- $P = \{ p1, p2, ..., pn \}$ is a finite set of places.
- $T = \{ t1, t2, ..., tn \}$ is a finite set of transitions.
- $D = \{ d1, d2, ..., dn \}$ is a finite set of propositions.
- $I : T \rightarrow P$ is the input function, a mapping from transitions to places.
- $O : T \rightarrow P$ is the output function, a mapping from transitions to places.
- $f : T \rightarrow [0,1]$ is an association function, a mapping from transitions to real values.
- $\alpha : P \rightarrow [0,1]$ is an association function, a mapping from places to real values.
- $\beta : P \rightarrow D$ is an association function, a mapping from places to propositions.

According to this definition, we can build four different types of composite fuzzy production rules using the connectives "AND" and OR". These rules will be useful in the development of a manufacturing system model. Rules examples and the processing method associated to them are summarized in the figures 1 and 2. It must be noticed that type 4 rules are unsuitable for deducing control because they do not make specific implications. Therefore, they will not be used in the manufacturing system modeling.

_ Type 1 Rules: IF d_{j1} and d_{j2} and ... and d_{jn} THEN d_k. (C.F. = μ_i)
_ Type 2 Rules: IF d_j THEN d_{k1} and d_{k2} and ... and d_{kn}. (C.F. = μ_i)

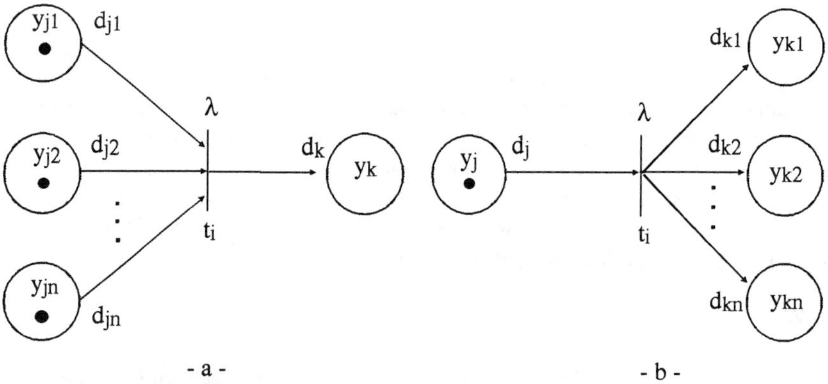

_ for i = 1 to n, if $y_{ji} \geq \lambda$:
$y_k = \text{Min}(y_{j1}, y_{j2}, ..., y_{jn}) * \mu_i$

_ if $y_j \geq \lambda$:
$y_{ki} = y_j * \mu_i$, for i = 1 to n.

FIG. 1. a: FPN representation form and reasoning method of type 1 rules.
b: FPN representation form and reasoning method of type 2 rules.

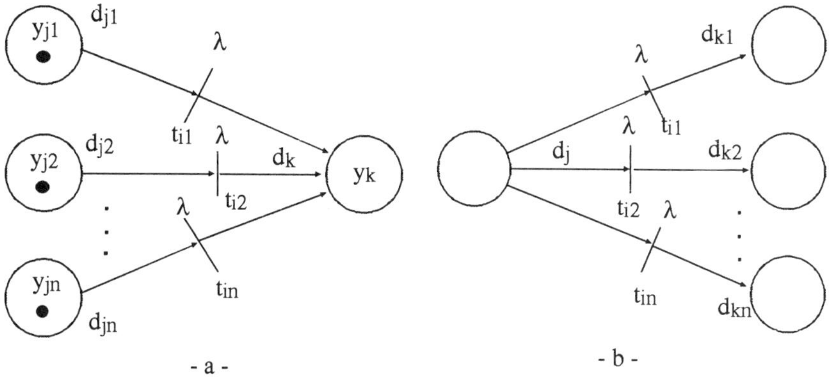

_ for i = 1 to n, if $y_{ji} \geq \lambda$:
$y_k = \text{Max}\,(y_{j1}, y_{j2}, \ldots, y_{jn}) * \mu_i$.

FIG. 2. a: FPN representation form and reasoning method of type 3 rules.
b: FPN representation form of type 4 rules.

_ Type 3 Rules: IF d_{j1} or d_{j2} or ... or d_{jn} THEN d_k. (C.F. = μ_i)
_ Type 4 Rules: IF d_j THEN d_{k1} or d_{k2} or ... or d_{kn}. (C.F. = μ_i)

SYSTEM MODELING AND SIMULATION

The manufacturing cell considered is similar to that described in [7], with some simplifications. It is composed by three machines (a, b, c). Each machine has one loading and one unloading buffer, assumed to be of infinity capacity, and a transportation system. It manufactures three family of part types (A, B, C).

Each machine has a different set-up time and each part type has a specific operation time. Transport time of the part types between two machines is disregarded because it is much lower than operation and set-up times. The routing of each part type (A, B, C) in the manufacturing cell is shown in figure 3.

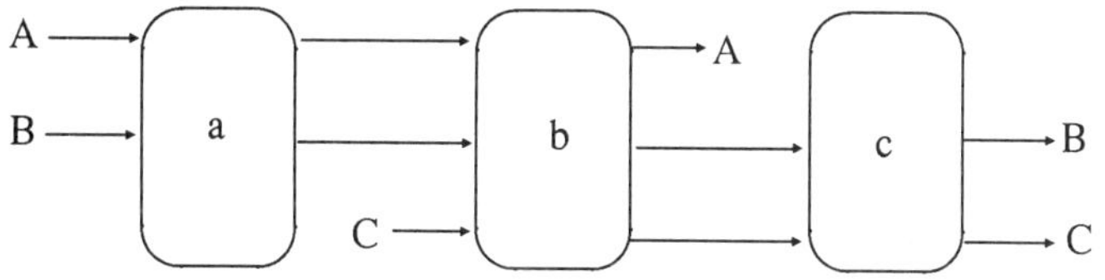

FIG. 3. Routing of each part type (A, B, C).

Similar to conventional simulation study, the model complexity level depends upon the desired outputs of the study [8]. Hence, we define as desired outputs the following: utilization rate and time spent in set-up operations of each machine. The follow input variables are needed:

- operation time of each part type (A, B, C) in each machine (a, b, c).
- set-up time of each machine.

- total number of each part type to be manufactured.
- transfer lot size of each part type.

It is convenient to distinguish between two concepts: total lot size and transfer lot size. The total lot size of each part type is splited in a convenient number, each resulting in a set of part types called transfer lot. This distinction is important because set-up operations can not be done until a machine finishes the processing of a entire transfer lot. Thus, the time spent in set-up is related to the transfer lot size. [9]

The manufacturing cell model was developed based on fuzzy production rules. These rules were estabilished using the relationship between the output and input variables. It must be noted that these rules are built upon concepts of linguistic variables [10]. These linguistic variables induce the propositions to be used by the Fuzzy Petri Net reasoning method.

Although the input variables considered are nonfuzzy, a fuzzyfication module was implemented, which assesses the membership grade of each input variable associated to the linguistic variables. More information about fuzzyfication and defuzzyfication processes can be found in [11].

Considering the knowledge acquired during many simulations experiments envolving manufacturing systems like this one, fuzzy production rules sets were proposed to evaluate the behaviour of the desired outputs.

The calculus of the utilization rate of each machine is based upon the total number of each part type to be manufactured, the part type routed into the manufacturing system, and the operation time of each part type in each machine. The calculus of the time amount spent in set-up operations is based upon the transfer lot size, the routing of each part type, the set-up time and the total number of each part type to be manufactured. Examples of some fuzzy production rules (SET A and SET B) to evaluate machine "a" utilization rate, is shown bellow:

FUZZY PRODUCTION RULES - SET A

IF ("ATN is S" OR "ATN is M") AND "BTN is S" THEN "P is S".
IF "ATN is S" AND ("BTN is S" OR "BTN is M") THEN "P is S".
IF "ATN is S" AND "BTN is H" THEN "P is M".
IF "ATN is H" AND "BTN is S" THEN "P is M"..
IF ("ATN is M" OR "ATN is H") AND "BTN is H" THEN "P is H".
IF "ATN is H" AND ("BTN is M" OR "BTN is H") THEN "P is H".

FUZZY PRODUCTION RULES - SET B

IF "P is S" AND "AOT is S" AND "BOT is S" THEN "UR is S".
IF "P is S" AND "AOT is S" AND "BOT is M" THEN "UR is S".
IF "P is S" AND "AOT is M" AND "BOT is S" THEN "UR is S".
IF "P is S" AND "AOT is H" AND "BOT is M" THEN "UR is M".
IF "P is S" AND "AOT is M" AND "BOT is H" THEN "UR is M".
IF "P is S" AND "AOT is H" AND "BOT is H" THEN "UR is H".
IF "P is M" AND "AOT is S" AND "BOT is S" THEN "UR is S".
IF "P is M" AND "AOT is M" AND "BOT is S" THEN "UR is S".
IF "P is M" AND "AOT is S" AND "BOT is M" THEN "UR is S".
IF "P is M" AND "AOT is M" AND "BOT is M" THEN "UR is M".
IF "P is M" AND "AOT is S" AND "BOT is H" THEN "UR is M".

IF "P is H" AND "AOT is S" AND "BOT is S" THEN "UR is S".
IF "P is H" AND "AOT is M" AND "BOT is S" THEN "UR is M".
IF "P is H" AND "AOT is H" AND "BOT is S" THEN "UR is M".
IF "P is H" AND "AOT is M" AND "BOT is H" THEN "UR is H".
IF "P is H" AND "AOT is H" AND "BOT is H" THEN "UR is H".

where:

S - Small. M - Medium. H - High.
ATN - total number of part type A.
BTN - total number of part type B.
P - total number of part types manufactured by machine "a".
AOT - part type A operation time in machine "a".
BOT - part type B operation time in machine "a".
UR - machine "a" utilization rate.

Every fuzzy production rule set obtained was converted into a FPN, which can be processed by the fuzzy reasoning algorithm presented in [6]. The conversion follows the method (figures 1 and 2) previously presented.

The manufactring system model (FPN) was implemented in a PC-environment. Several simulation runs were performed with the input variables taking appropriate test values. It must be noted that the simulation run results are presented in qualitative terms. Neverthless, if necessary, a deffuzification module can be implemented to generate results in quantitative terms. Table I shows a sample of results obtained from the simulation experiment using the FPN methodology.

TABLE I. Simulation run results

	Simulation run	1	2	3	4	5	6
INPUT	total lot A	10	25	48	44	14	45
	total lot B	10	25	48	44	45	15
	total lot C	10	25	48	44	60	40
	transf. lot A	4	12	20	5	3	2
	transf. lot B	4	12	20	20	15	20
	transf. lot C	4	12	20	20	22	40
	set-up time a	5	11	18	15	18	15
	set-up time b	5	11	18	15	15	8
	set-up time c	5	11	18	15	10	2
	op. time A/a	20	60	100	90	20	25
	op. time A/b	20	60	100	90	20	10
	op. time B/a	20	60	100	90	20	40
	op. time B/b	20	60	100	90	20	15
	op. time B/c	20	60	100	90	20	80
	op. time C/b	20	60	100	90	20	20
	op. time C/c	20	60	100	90	20	100
OUTPUT	util. rate a	L	M	H	H	L	H
	util. rate b	L	M	H	H	L	L
	util. rate c	L	M	H	H	L	H
	set-up util. a	M	M	M	H	H	H
	set-up util. b	M	M	M	M	H	M
	set-up util c	M	M	M	M	M	L

CONCLUSIONS

The simulation methodology using FPN and fuzzy production rules developed in this work has shown to be very useful when simulation of complex/ill-defined manufacturing systems (e.g., new products development, estrategic planning, resources allocation, etc.) have to be performed. The qualitative results obtained were satisfactory and easy to understand when compared with conventional simulation systems.

The linguistic variables concept has shown to be very useful in cell modeling because of its simplicity and intuitive meaning. The simulation process can also be performed faster. Besides, the structure utilized permits an maintenance of the manufacturing system model because it is very simple to add, delete or change the fuzzy production rules.

Tuning of the certainty factor (μ) and the transition enabling factor (λ) of each FPN rule contributes in refining the model's precision. The FPN methodology can be extended to cover a broader class of systems, and many forms of reasoning procedures can be developed to enlarge the methodology applicability.

Other important characteristic of the simulation approach here in proposed is that the knowledge representation in FPN form is simple, easy to visualize and to understand because of its graphical format and straightforward translation into fuzzy production rules form.

REFERENCES

1. R.E. Shannon, "Knowledge Based Simulation Techniques for Manufacturing", *Int. J. Prod. Res.*, Vol.26(5), 953-973, (1988).
2. I. Sabuncuoglu & D.L. Hommertzheim, "Expert Simulation Systems - Recent Developments and Applications in Flexible Manufacturing Systems", *Computers ind. Engng*, Vol.16(4), 575-585, (1989).
3. L.A. Zadeh, "Fuzzy Sets", *Information and Control*, nº8, 338-353, (1965).
4. L.A. Zadeh, "Fuzzy Logic", *Computer*, 83-93, (April, 1988)
5. T. Murata, "Petri Nets: Properties, Analysis and Applications", *Proceedings of the IEEE*, Vol.77 (4), 541-580, (April, 1989).
6. S. Chen & J. Ke & J. Chang, "Knowledge Representation using Fuzzy Petri Nets", *IEEE Trans. on Knowledge and Data Engineering*, Vol.2(3), 311-319, (September, 1990).
7. I. Hatono & K. Yamagata & H. Tamura, "Modeling and On-line Scheduling of Flexible Manufacturing Systems using Stochastic Petri Nets", *IEEE Trans. on Software Engineering*, Vol.17(2), 126-132, (February, 1991).
8. F. Neelamkavil, *Computer Simulation and Modeling*, (John Wiley & Sons, England, 1988).
9. A.M. Daltrini, "Fuzzy Petri Nets in Manufacturing Systems Simulation", *in Internal Report UNICAMP/DCA*, (June, 1991, in Portuguese).
10. L.A. Zadeh, "Outline of a New Approach to the Analysis of Complex Systems and Decision Processes", *IEEE Trans. on Systems, Man and Cybernetics*, Vol.3(1), 28-44, (January, 1973).
11. C.C. Lee, "Fuzzy Logic in Control Systems: Fuzzy Logic Controller - Part I", *IEEE Trans. on Systems, Man and Cybernetics*, Vol.20(2), 404-415, (March/April, 1990).
12. R.M. O'Keef, "The Role of Artificial Intelligence in Discrete Event Simulation", *Artificial Intelligence, Simulation & Modeling*, 359-379, (John Wiley & Sons, England, 1989).

ACKNOWLEDGEMENT

The second author thanks to CNPq, the Brazilian National Research Council for grant #300.729/86-3.

FUZZY TRACKING CONTROL OF A LASER BEAM SYSTEM

RICHARD MARCHBANKS* and MO JAMSHIDI
CAD Laboratory for Intelligent and Robotic Systems
Department of Electrical & Computer Engineering
University of New Mexico
Albuquerque, NM 87131 USA

ABSTRACT

In this paper fuzzy logic is used to control a two-axis mirror gimbal for aligning a laser beam onto the center of a quadrant detector. A PC-based controller was implemented with a simple rule set to control and stabilize the device. A comparison between a non-fuzzy proportional controller and the fuzzy controller was made. Experiments were performed to study the tracking characteristics and the effect of rule pruning on the fuzzy controller response. Initial experiments indicate a substantial improvement on the response characteristics such as overshoot in the fuzzy controller as compared to non-fuzzy control.

INTRODUCTION

Communication systems are desired that use line-of-sight transmitting and receiving. This is especially true from a military point of view with respect to submarine communication. A submarine requires the ability to communicate with other platforms but still remain "invisible." The requirement for such a system is robust tracking that will maintain the communication link between platforms given a variety of disturbances such as movement, bumps, atmospheric or hydrospheric perturbations, and obstructions. This paper presents a demonstration of fuzzy control applied to such a problem. It is an attempt to attain a proof of principle for application of fuzzy control for optical systems. Here, a two-axis mirror gimbal is set up to steer a laser beam onto a quadrant detector where the beam is desired to lie on the center of the detector. Both fuzzy and non-fuzzy control have been used on a PC-based real-time control system to compare the responses for the laser tracking problem. Initial results are rather encouraging in favor of fuzzy control.

FUZZY CONTROLLER DESIGN

Figure 1 outlines the laser beam tracking closed-loop control system. As shown, the electronic circuitry sense the error in the position of the beam relative to the center of the detector in both the x and y directions. These errors are converted to a digital format for use by the PC-based fuzzy controller . The control software processes the error information using fuzzy logic and decides on response for the motor driver. The driver supplies the required power to run the motors and reposition the beam.

---

* Currently at Los Alamos National Laboratory , Los Alamos, New Mexico, USA

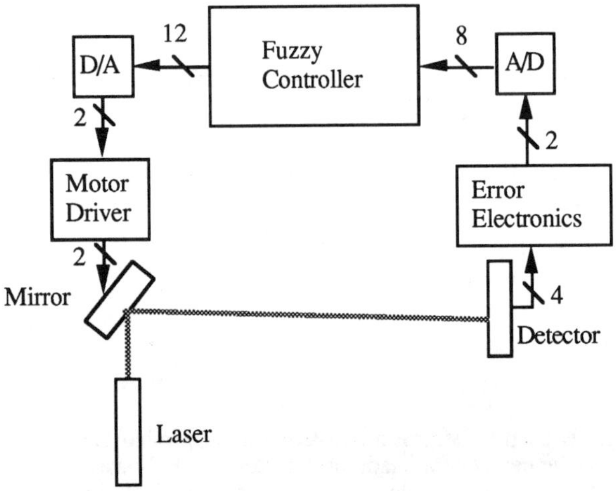

Figure 1: Block diagram of closed-loop system.

Hardware For feedback information, a quadrant detector was used to sense the error in the position of the beam from the desired position. The center of the detector is where the beam center is desired. A quadrant detector is simply four photosensitive semiconductor elements that produce a voltages proportional to the amount of photon flux incident on each one of them. A centered beam will have one quarter of its light on each quadrant and the potentials will be equal. As the beam moves off-center, the proportions change. Error in the horizontal and vertical directions can be separated by some simple electronics to reduce the problem into 2 one-dimensional ones. Based on the size and price of commercially available quadrant detectors, a decision was made to build one using solar cells. This resulted in a detector with a 16 square cm area and cost under 10 dollars.

The mirror consisted of a movable gimbal with horizontal and vertical degrees of freedom. Mounted on the mirror were 2 dc motormikes to control the elevation and azimuth of the mirror. A complication with using dc motors is that they have a nonlinear speed as a function of armature voltage. The main problem lies in the *dead band* region where the motor requires a minimum voltage to begin moving. Any voltage below this turn-on voltage will not drive the motor. A driver employing this response will always keep the armature voltage above the turn-on voltage when there is non-zero error. This response however, adds instability and can cause limit cycling near zero voltage. That is, due the the very high slope of the control response curve near the origin, the motor can switch between a forward and reverse direction- causing oscillation. Part of the objective in the design was to reduce or eliminate limit cycling with the fuzzy logic controller.

The controller was implemented using a PC with a 386 microprocessor. A Keithley-Metrabyte 8-bit analog-to-digital converter was used to put the analog error in the digital format required by the computer. The input of the A/D converter has a range from -5 volts to 5 volts with a resolution of 8-bits over this range. The error signal was conditioned with amplifiers to fill this range and acquire the full 8-bit resolution. After processing, the resulting digital response signal was converted to an analog voltage by a Keithley-Metrabyte 12-bit digital-to-analog converter.

Software Designing the required software to perform fuzzy inference can be a very involving task but fortunately there are a few packages that have been written to do this. One of these packages, Togai InfraLogic [1] Fuzzy - C System was used to write the primary code to perform the fuzzy arithmetic. This software is programmed in a language developed by the manufacturer which allows the user to easily construct membership functions and a rule base. The resulting code is compiled into C by a special compiler. This software performs all the required fuzzy operations, including defuzzification, and produces a crisp output. A driver program was written in C to communicate with the A/D and D/A converters. It also acquires data from the experiments. Listings of the Fuzzy C code, the A/D and D/A converter addresses, and driver program are listed elsewhere [2].

To represent the error input to the controller, a set of linguistic variables were chosen to represent 5 degrees of error, 3 degrees of change in error, and 5 degrees of armature voltage. Membership functions were constructed to represent the input and output values' degrees of truth for each of these linguistic variables. The input error membership functions contains a very narrow zero error (ZE) region. A faster convergence was expected by doing this since for a small error, the degree of "medium" will be large and the armature voltage will correspondingly be larger than if ZE were broad. Also, the "medium" membership functions do not intersect at zero for both the error and armature voltage. This was done to help suppress overshoot and limit cycling.

For fuzzy rule inferencing portion of the controller, two sets of rules were chosen. These "Fuzzy Associative Memories" or FAMs, are a shorthand matrix notation for listing the rule set. A linguistic armature voltage rule is fired for each pair of linguistic error and linguistic change in error variables. For example, if the error is PM and the change in error is ZE then the armature voltage is NM. Because overlap between the fuzzy variables exist, more than one rule can fire simultaneously.

The FAMs are shown in Figure 2. Figure 2a shows the full set of 15 rules. Figure 2b shows the set after pruning.

ΔError \ Error	NM	NS	ZE	PS	PM
NS	PM	PM	PS	ZE	NS
ZE	PM	PS	ZE	NS	NM
PS	PS	ZE	NS	NM	NM

(a)

ΔError \ Error	NM	NS	ZE	PS	PM
NS		PS		ZE	
ZE	PM		ZE		NM
PS		ZE		NS	

(b)

Figure 2: Fuzzy Associative Memories. a) Full set (15 rules), b) Pruned set (7 rules).

EXPERIMENTS

Two experiments were performed to test the integrity of the controller as illustrated in Figure 3. First the center finding ability was examined by placing the beam on the detector with the center of the beam offset in the positive x direction from the center of the detector. The system was started with data taken at 100 Hz for 4 seconds. The rate of convergence to the detector center was observed as well as the overshoot and oscillation. This experiment was performed for a non-fuzzy proportional feedback controller, the fuzzy controller with 15 rules, and the pruned fuzzy controller with 7 rules.

The second experiment was to test the systems tracking ability. The detector was mounted on a linear positioner with the beam centered on the detector. The positioning stage was driven at 0.2mm/s in a direction normal to the beam incidence. The distance from the mirror to the detector was 1.5 m. Again, data was taken at 100 Hz for 4 seconds while this experiment was performed. A non-fuzzy proportional feedback controller, the fuzzy controller, and the pruned fuzzy controller were tested using this scheme.

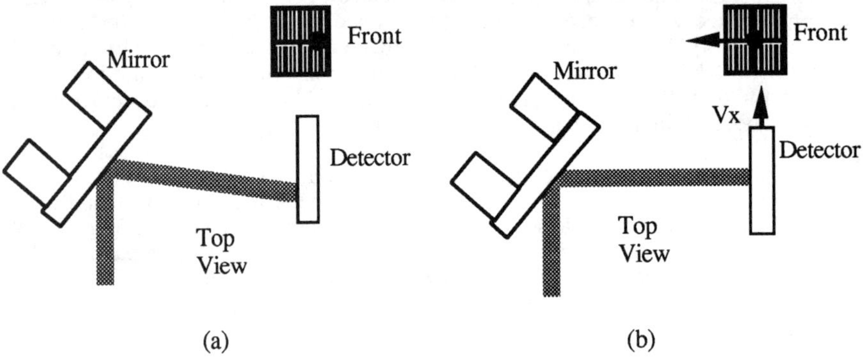

Figure 5: Experimental arrangement. a) Center finding, b) Tracking.

Results Six experiments were performed to test the systems. For Center finding ability, a non-fuzzy controller, fuzzy controller and pruned fuzzy controller were used. These 3 controllers were again used in the tracking experiment Results are shown in Figure 4.

Center Finding With the beam center initially offset in the positive x direction, the proportional feedback controller, the fuzzy controller, and the pruned fuzzy controller were all used to drive the system. The results for 4 seconds of time are shown in Figures 4a-c. When comparing the fuzzy controllers to the non-fuzzy controller, the y-error is omitted to prevent the graphs from becoming too busy.

Figure 4a shows the behavior of the non-fuzzy controller. Both x-error and y-error are shown for 4 seconds of time. The error in the x direction is maximized to 5 volts until the beam begins to overlap the center. It quickly converges, overshoots and oscillates as the controller reverses directions. The y-error remains fairly stable until the beam reaches the center. The oscillation in this error is due to

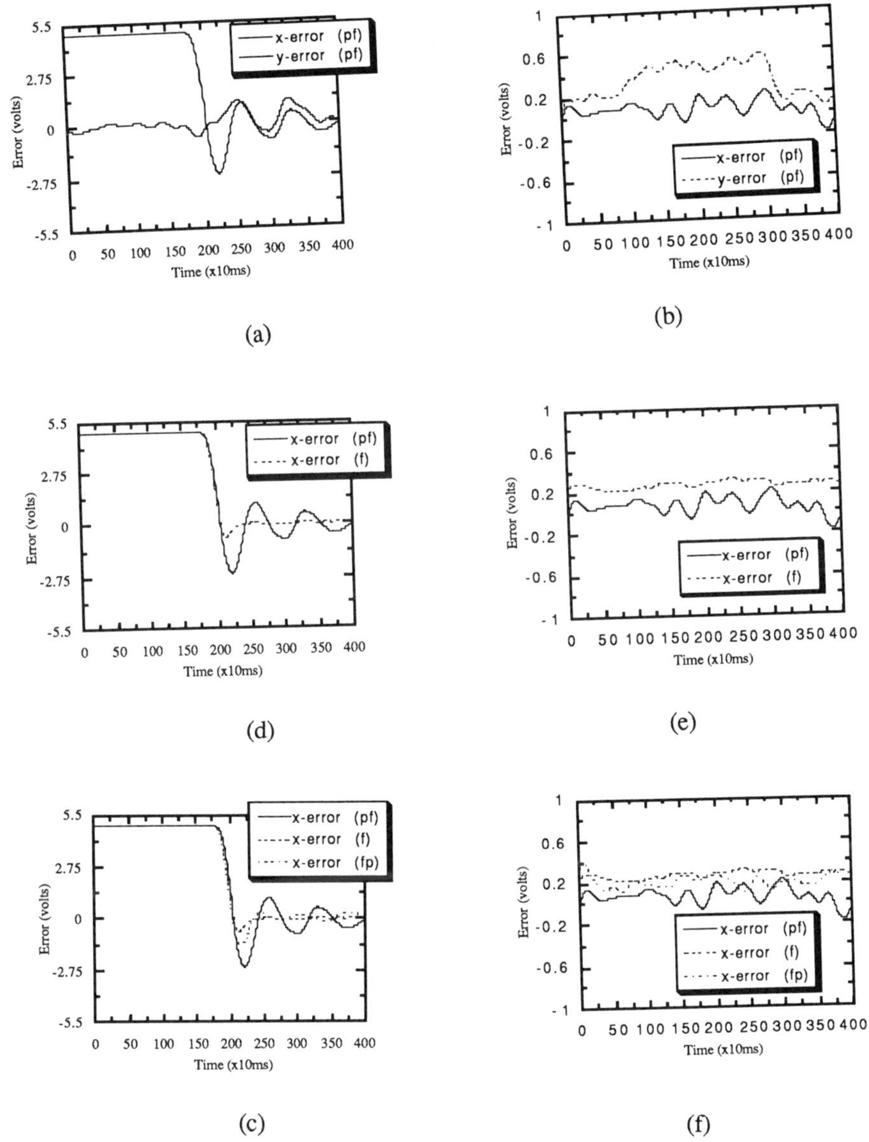

Figure 4. Experiment results. a) Non-fuzzy (proportional feedback) x and y-error, b)Non-fuzzy x-error vs fuzzy x-error, c) With pruned rules, d) Tracking non-fuzzy x and y-error, e) Tracking non-fuzzy x-error vs fuzzy x-error, f) Tracking with pruned rules.

the misalignment of the 4 quadrants. When the beam crosses over to the remaining quadrants, the abrupt change due to misalignment causes the oscillation in error.

With the fuzzy controller engaged, the overshoot is dramatically reduced and oscillation is effectively eliminated. This plot is shown in 4b. Next, the pruned fuzzy set of rules was implemented and tested. The graph in Figure 4c shows more overshoot and oscillation than the unpruned controller but the results are better than that of the non-fuzzy controller.

Tracking The second set of experiments were performed to test the tracking ability of the controllers. With the beam centered in the desired position, the detector was moved and the controllers engaged. Data was taken and is shown in Figures 4d-f. The non-fuzzy controller shows substantial oscillation in both the x-error and y-error as shown in Figure 4d. Figure 4e shows the results of the fuzzy controller x-error along with the non-fuzzy controller x-error. The fuzzy controller reduces the oscillation but a slight offset is observed. This indicates that the fuzzy controller lags slightly behind the detector. Finally, the pruned fuzzy controller is tested. As in the center-finding test, the pruned set does not reduce the oscillation as well as the full fuzzy set. However, it does perform better than the non-fuzzy controller in this respect. A slight offset was also observed here.

CONCLUSIONS

By implementing fuzzy control in this tracking system, overshoot was significantly reduced and limit cycling was virtually eliminated. However, justification of the added complexity of the micro-controller cannot be accomplished until a comparison is made with conventional proportional-plus-integral control. These results simply show the ease of implementation of fuzzy control to a problem without using a mathematical model of the system. Future study should be considered in modeling the system and implementing a PI controller to further establish or dispute the viability of fuzzy control for this particular problem.

Additional possibilities include reducing the amount of hardware. The error electronics and nonlinear motor driver circuit can be removed and replaced by the fuzzy controller . Further work can lead to a more optimum set of membership function and rules.

ACKNOWLEDGEMENTS

Special thanks to Renee Temple who helped me through the tough times by teaching me C and to the Chemical and Laser Sciences Division at Los Alamos who loaned me some equipment that made this project possible.

REFERENCES

1. Togai InfraLogic, Inc., "User's Guide - Fuzzy-C Expert Systems," Irvine, CA, 1991.

2. Marchbanks, R., "A Fuzzy Two-Axis Mirror Controller for Laser Beam Alignment," in Fuzzy Logic and Control - Software and Hardware Applications, M. Jamshidi, et al. (eds.), Prentice-Hall, 1992.

FUZZY CONTROL OF THREE LINKS OF A ROBOTIC MANIPULATOR

KISHAN KUMBLA, JOHN MOYA, RONALD BAIRD, SARALA RAJAGOPALAN, and MO JAMSHIDI

CAD Laboratory for Intelligent and Robotics Systems
Department of Electrical and Computer Engineering
The University of New Mexico, Albuquerque, NM 87131, USA

ABSTRACT

This paper presents the application of fuzzy control to three links of a Rhino robot and compares its performance to traditional PD control. The dynamics of motion of robot links are governed by nonlinear differential equations. The fuzzy controller, being an adaptive technique, gives better performance than the traditional linear PD controller over a typical operational range. The fuzzy controller reaches the desired position with no overshoot, which is unlikely with the PD controller.

1. INTRODUCTION

Robot control requires the solution of nonlinear differential equations. Unfortunately, nonlinear differential equations are plagued by substantial requirements for computation and have an incomplete theory of solution. Thus, most approaches to robot controller design have suffered due to the complications of nonlinear effects.

Because of these complications, fuzzy logic offers a very promising approach to robot controller design. Fuzzy logic offers design rules that are relatively easy to use in a wide range of applications, including nonlinear robotic equations. Fuzzy logic also allows for design in cases where models are incomplete, unlike most design techniques. In addition, microprocessor-based fuzzy controllers have performed with data streams of 8 bits or less to allow for a simple design.

The potential for improvement offered by fuzzy logic warrants further investigation into its usefulness in realistic settings. This paper presents the hardware and software design, and subsequent testing and analysis, of a project to study control of a three link Rhino robot.

2. BASELINE HARDWARE DESIGN

Figure 1 provides an overall block diagram of the robot and control hardware. The Rhino robot has five links and a gripper. Fuzzy control was applied to three of the links: the shoulder, elbow, and wrist. Each link is moved with a DC servo motor rotating about an axis at each joint. The computer system used was a Packard Bell 386PC. The remaining hardware consisted of elements for data acquisition, power control (driver), and communications.

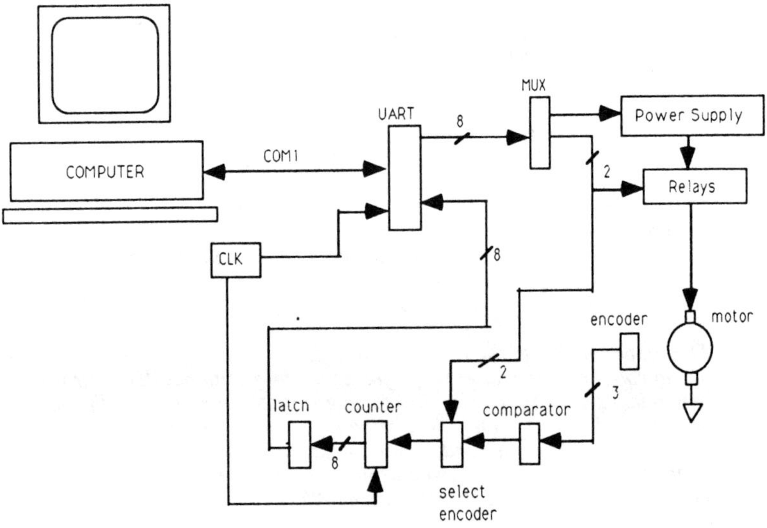

Figure 1. Schematic of the Robot Controller Hardware.

The DC motors were driven by a Fluke 4265A programmable power supply, which was interfaced to the motor through relays. These relays were computer controlled to switch power to only one motor at a time. Thus, at any point, only one of the links could be actuated. The power supply output voltage was controlled with an eight bit control word and could deliver voltages ranging from -16 to +16 volts in 0.25 volt increments.

The data acquisition hardware gathered angular position from optical encoders that pulsed a counter, giving a measurement of displacement. Each pulse corresponded to 0.12 degrees of movement. At every ten milliseconds a crystal clock triggered sampling of the counter. The count information was latched and transmitted to the computer through the RS-232C COM1 port in eight bit increments.

3. SOFTWARE IMPLEMENTATION OF THE FUZZY CONTROLLER

The controller software was divided into two primary modules written in the C language. The two modules are the fuzzy control module and the communications module.

3.1 Fuzzy Control Module

The first step in fuzzy controller design is the selection of input and output variables. The designated input variables were ERROR (robot link position error) and VELOCITY (velocity of the robot link). The output variable was chosen to be

VOLTAGE (drive voltage output by the power supply). Each variable requires a set of membership functions. The membership functions are shown in Figure 2.

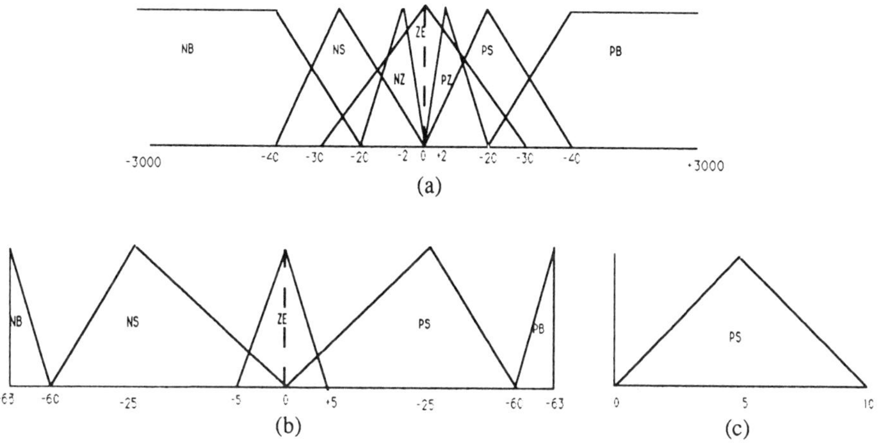

Figure 2. Fuzzy Membership Functions for (a) Position Error, (b) Output Voltage and (c) Velocity.

During actual operation, the computer reads in the robot position and then computes position error and velocity. The fuzzy controller fuzzifies the input quantities through algorithms that operate on the input data as specified by the membership functions. Next, the fuzzified input quantities pass through a series of IF-THEN decision rules that form the main body of the fuzzy controller. Thus, the fuzzy controller assesses the current state of the robot and determines which control action is most appropriate. Defuzzification is applied using the voltage output variable and a control action is selected. The control action is then passed through the interface to the external power supply driver.

The design of the fuzzy controller algorithm was accomplished with the Togai InfraLogic's fuzzy C Compiler [1]. Input to the Togai Fuzzy-C Compiler is a fuzzy source code consisting of membership functions and a set of IF-THEN decision rules. The Fuzzy-C Compiler converts the fuzzy source code into standard C source code which then passes through a C compiler to produce an executable code of the fuzzy controller. Trial experiments of the executable code were conducted to calibrate the membership functions until robot trajectory overshoot was suppressed and rise time was kept to a minimum.

3.2 Communications Module

The COM1 serial port of the PC was used to communicate between the computer and external hardware. A communications routine set the port at 9600 baud, eight data bits, two stop bits and even parity. The routine sampled the count information in the

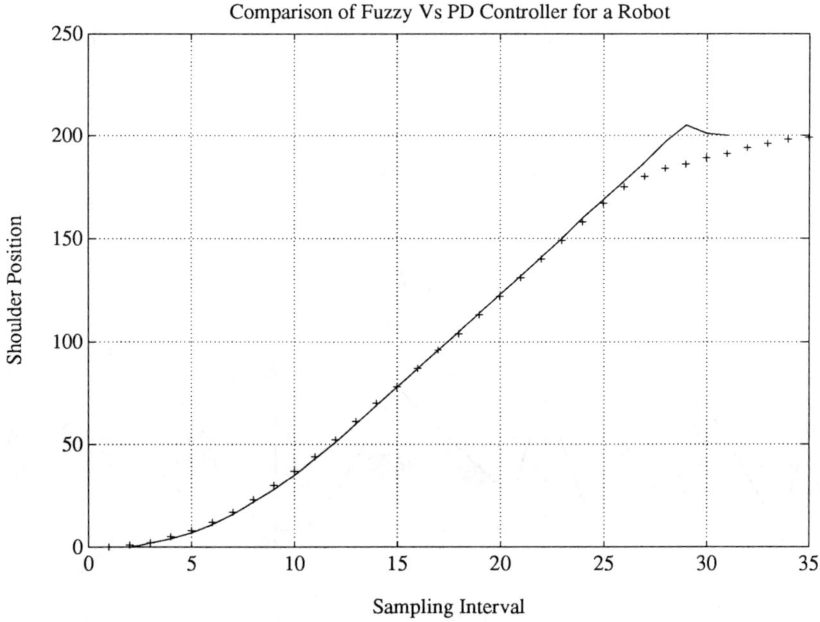

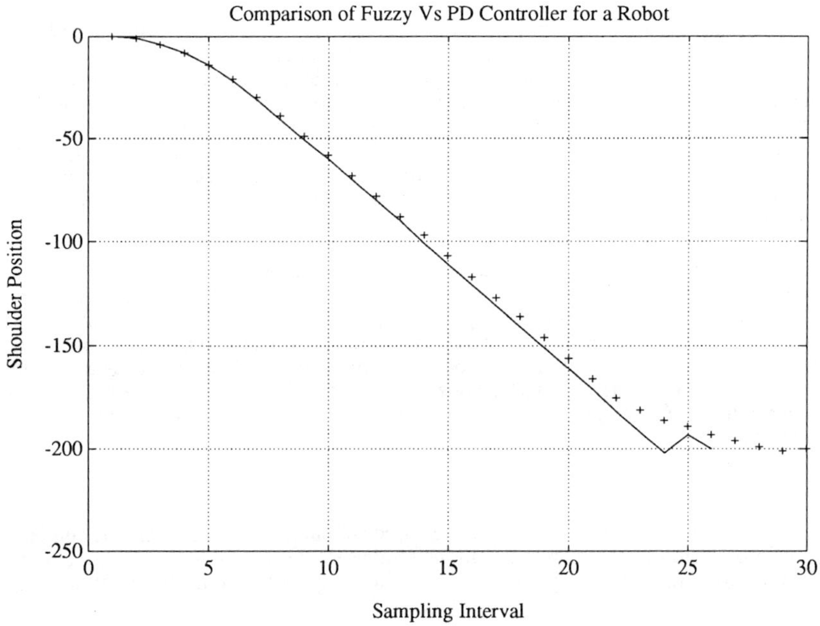

'+++' : Fuzzy Controller. '----' : PD Controller.
Figure 3a. Shoulder Motion of the Robot.

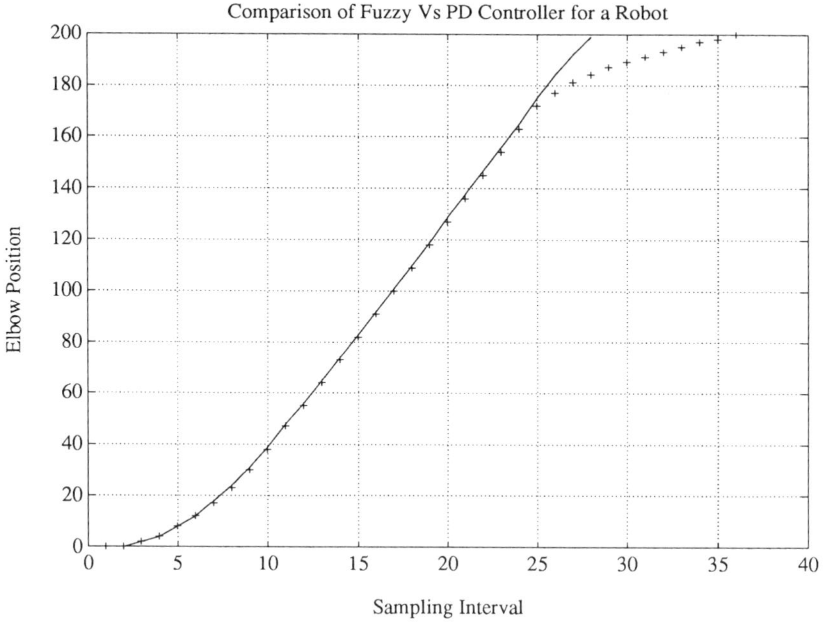

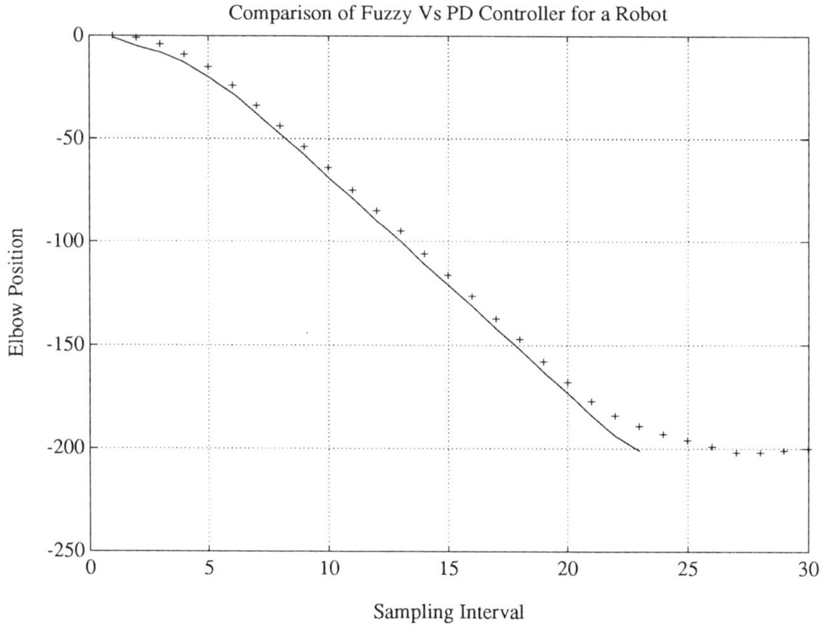

'+++' : Fuzzy Controller. '----' : PD Controller.

Figure 3b. Elbow Motion of the Robot.

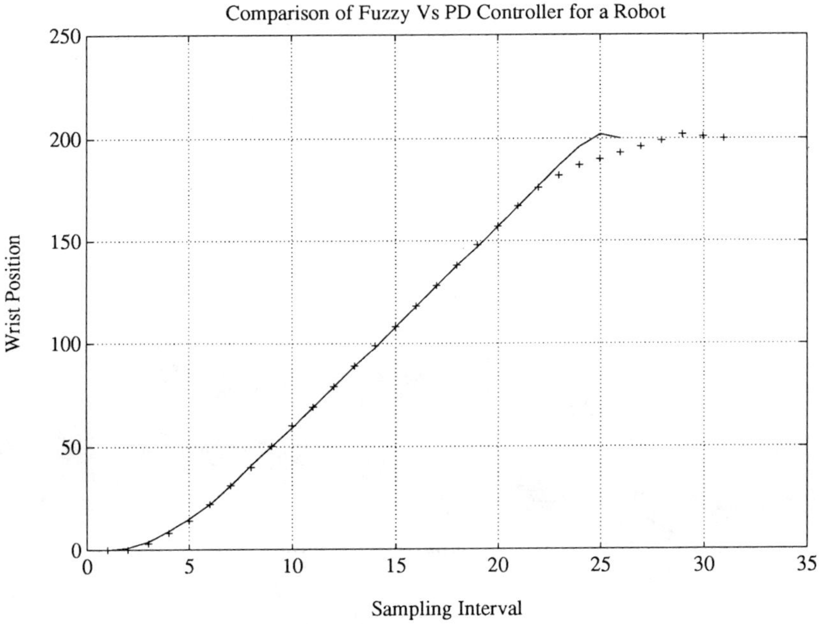

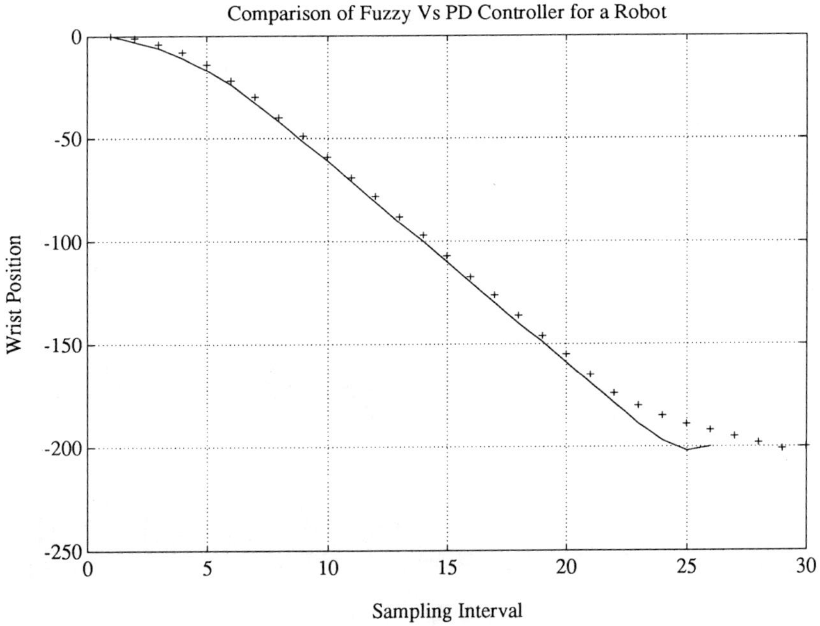

'+++' : Fuzzy Controller. '----' : PD Controller.
Figure 3c. Wrist Motion of the Robot.

receive register every ten milliseconds. The fuzzy controller operates on this data and then outputs the result into a transmit register which was used to drive the DC motors.

4. TESTING AND RESULTS

Two tests were conducted to compare controller performance on robot arm trajectory. In both tests, the gain of the PD controller was set to provide the best performance in all operating ranges. In the first test, each controller was set to traverse a trajectory of 200 counts in both positive (clockwise) and negative (counterclockwise) directions (Note: One count is equal to 0.12 degrees of rotation of the link). The test results for the trajectory of the shoulder, elbow and wrist are shown in Figure 3. The results show that the fuzzy controller is able to move the robot arm smoothly to the desired position without overshoot. The PD controller, however, generated overshoot in the trajectory. It is interesting to note that when the robot link approached its final destination, its velocity remained high (steep trajectory curve) with PD control, but began to decrease with fuzzy control.

In the second test, the controllers were compared under various ranges of operation. The PD controller with suitable gain improved trajectory response by reducing the overshoot to less than four counts. However, the PD controller was unable to move the motor when the links were initially one or two counts away from the final destination. The primary cause for this effects was the nonlinear friction in the robot joints. One could overcome this resistance with an increase in controller gain, but then the links would overshoot even further when larger distances are traversed.

The fuzzy controller was able to overcome the problems encountered by traditional PD control and perform better in all conditions tested. Results showed that the maximum overshoot with fuzzy control was held to a count of one. The fuzzy controller also was able to actuate the motor and travel a short distance of one count, unlike the PD controller.

The fuzzy controller exhibited robustness in performance against non-ideal effects like robot inertia, Coriolis effect and gravity. These effects alter the velocity, position, and acceleration of the robot links, which can degrade overall performance of conventional control. However, with fuzzy control, one set of fuzzy membership functions was sufficient to guard against these non-ideal effects and operate all links of the robot. This result is not shared by the PD controller which required different gains to accommodate for the differences between each link.

5. CONCLUSIONS

The fuzzy controller was able to suppress the robot trajectory overshoot and perform better than traditional PD control under all scenarios tested. Furthermore, one set of fuzzy membership functions was sufficient for handling the variations that occurred between the different links of the robot. This implies that fuzzy control is robust and can accommodate many unforeseen elements inevitable in any practical implementation.

6. REFERENCES

[1] Togai InfraLogic Inc. Fuzzy-C Development System User's Manual, Irvine, CA, 1991.
[2] Rhino, Inc., Rhino Robot XR Series Owner's Manual, Champaign, IL, 1988.
[3] Craig, John J. Introduction to Robotics. Addison Wesley, Reading, Mass., 1989.
[4] Larsen, D. G., Rony, P. R., Titus, J. A., and Titus, C. A. Interfacing and Scientific Data Communications Experiments. Howard W. Sams & Co., Inc., Indianapolis, Indiana, 1979.

PART 8 : INTELLIGENT SYSTEMS AND MANUFACTURING

WATCHDOG MONITOR PREVENTS MARTIAN OXYGEN PRODUCTION PLANT FROM SHUTTING ITSELF DOWN DURING STORM

F.E. CELLIER, L.C. SCHOOLEY, B.P. ZEIGLER A. DOSER, G. FARRENKOPF, J. KIM, Y. PAN and B. WILLIAMS

Dept. of Elect. & Comp. Engr., University of Arizona
Tucson, Arizona 85721, U.S.A.
E_mail: Cellier@ECE.Arizona.Edu

ABSTRACT

A distributed intelligent controller was built for high–autonomy operation of a prototype of a Martian Oxygen Production Plant. During a 100 hour test run conducted to verify the reliability of the high–autonomy control architecture, the controller broke down in the consequence of a lightning stroke. The intelligent controller was able to recognize that a fault had occurred, isolate the fault by recognizing that the fault was not to be found in the plant but had to be in the controller itself, and was able to take corrective action. The controller was able to correct this fault, although the type of fault that had occurred was not previously foreseen. This paper outlines the intelligent control architecture, describes the incident, and summarizes the lessons learned.

INTRODUCTION

The University of Arizona / NASA Space Engineering Research Center for Utilization of Local Planetary Resources (NASA/UA SERC/CULPR) is investigating means to generate oxygen from lunar and/or asteroidal rocks as well as from the Martian atmosphere. In particular, the Martian oxygen production plant has progressed beyond mere simulation to the status of rapid prototyping. Figure 1 shows diagrammatically the Martian oxygen production prototype.

The Martian atmosphere (90% CO_2 at a pressure of 6 mbar and at a temperature of 200 K) is condensed (and thereby heated) in a compressor to a pressure of 1 atm. It is then heated further to a temperature of roughly 900 K. At that temperature, carbon dioxide (CO_2) decomposes through thermal dissociation into carbon monoxide (CO) and oxygen (O_2). Unfortunately, these two gases have similar molecular weights and are therefore difficult to separate. The heart of the system is an array of zirconia tubes. These tubes separate the two gases in an electrocatalytic reaction. The oxygen is liquefied for storage, whereas the components of the CO/CO_2 gas mixture are separated in a membrane separator. The CO_2 is then rerouted, whereas the CO is further processed by a Sabatier process to generate methane (not shown on Fig.1).

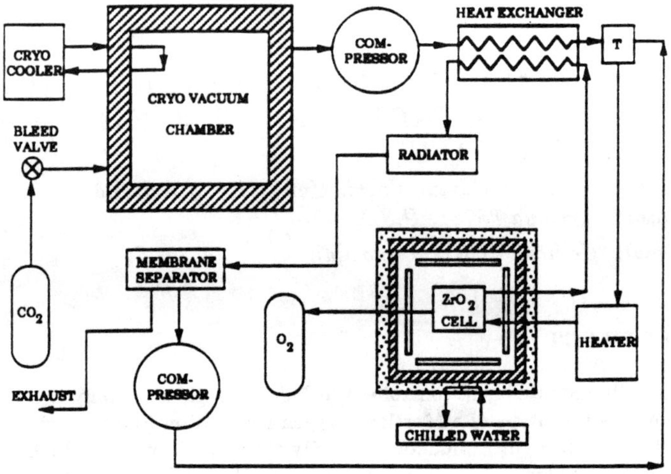

Figure 1: Martian Oxygen Production Plant

In space exploration, large amounts of oxygen are used for propulsion. In any chemically propelled spacecraft, a *fuel* reacts with an *oxidizer*. Together they constitute the *propellant*. The simplest chemical reaction that can be used for propulsion is:

$$2H_2 + O_2 \rightarrow 2H_2O \tag{1}$$

During the reaction, energy is freed that can be used to propel the spacecraft. In this propellant, the fuel (liquid hydrogen) makes up only 11% of the weight whereas the oxidizer (liquid oxygen) occupies 89%. While many different combinations of fuels and oxidizers can be employed for propulsion, oxygen is the most commonly used oxidizer. In all such reactions, the oxidizer is considerably more heavy than the fuel. For these reasons, it is economically interesting to generate oxygen on other planets.

The 16–tube breadboard generates roughly 1 kg of oxygen per day. The methane that is generated by the Sabatier process can be used as fuel. The current testbed could lead to a flight–rated plant before the turn of the century to be launched to planet Mars by NASA in an unmanned mission. A later manned mission could then use the oxygen that would meanwhile have been produced as oxidizer for the return flight to Earth. In this way, the manned mission could arrive on planet Mars with empty tanks, and the propellant for the return flight would not have to be lifted out of the gravity well of planet Earth. A manned mission to Mars could take place as early as 2014.

CONTROL ARCHITECTURE FOR HIGH–AUTONOMY OPERATION

According to NASA [4], *automation* is defined as "the ability to carry out a pre–designated function or series of actions after being initiated by an external stimulus without the necessity of further human intervention." In contrast, *autonomy* is defined as "the ability to function as an independent unit or element over an extended period of time performing a variety of actions necessary to achieve pre–designated objectives while responding to stimuli produced by integrally contained sensors."

Automation of a process is usually viewed as designing a (feedback) controller that reduces the plant sensitivity to parameter variations and/or the influence of disturbances. Parameter variations may be due to minor variations in the manufacturing process that is used to generate the plant, due to thermal effects, or to the influence of external environmental variations. Mostly, these variations are minor (a few percent). Feedback control architectures have helped to make plant operation more robust (less sensitive) to such types of variations in comparison with open–loop command architectures.

Traditional control engineers hardly ever concern themselves with abnormal situations such as failures that occur within a subsystem of the plant to be controlled or –even worse– within the controller itself. Reliability of a controlled system over long periods of time is rarely mentioned among the items on the desired performance parameter list of a control engineer. He (or she) is concerned with such properties as stability, steady–state accuracy, percent overshoot, and settling time, not failure rate, down time, or repair activities.

Such factors must be considered, however, for high–autonomy system operation. They add a new dimension of complexity to the overall system design. A high–autonomy system contains an additional hierarchy layer about the control architecture. It usually contains several alternative control architectures. It knows which of them to choose at any particular moment and when to switch between different controllers. It contains a task planning module that computes the set points for active controllers, and sequences the tasks to be executed. It reasons about both plant and controller integrity, detects faults (symptoms) as they occur, localizes faults within the system (discovers failures), and thinks about means of recovery from such faults (initiates repair activities). It furthermore collects statistics on symptoms, failures, and successful (as well as unsuccessful) repair activities, i.e., it learns from past experience.

An architecture for high–autonomy systems was recently proposed by Zeigler *et al.* [7] and Chi [1]. The oxygen production plant prototype, while still fairly primitive, contains all elements of a high–autonomy system. In the sequel, these elements will be described, and it will be shown how they cooperated during the incident.

Task Planning

Task planning refers to the activity of sequencing commands issued at task level to the control architecture. The task planner does not concern itself with the details of task execution. All it does is to determine which tasks should be executed, when, and in what sequence; it concerns itself with *resource management*, i.e., it knows which resources are required by each task, and makes sure that those resources are available during task execution; and finally, it concerns itself with *time management*, i.e., it knows how much time each task is supposed to consume and makes sure that real–time constraints are satisfied.

In this testbed, the task planner was fairly rudimentary. Since only one type of experiment was planned (the 100 hour test of continuous oxygen production without human intervention), an elaborate task planner was not required. The currently used task planner is still very simple. This part of the program will need to be drastically expanded before the full–scale plant can be launched. Task planning is not only responsible for setting the pace for normal operation, but also for recovering the plant after a fault has occurred. The Martian task planner must be able to deal with sand storms, contaminated membranes and tubes, leakage of seals, and tripped circuit breakers, to mention just a few of the types of failures that *are expected* to occur during long–term operation of this high–autonomy control system.

Chi *et al.* [2] and Chi [1] have proposed an intricate hierarchical task planning architecture that should suffice for controlling such an expedition.

Command Execution

The command executor accepts the next command to be executed from the task planner and prepares the control architecture for its execution. The command executor selects the appropriate low–level control algorithm from a set of precoded algorithms, downloads the selected controller into the smart sensor that physically houses the low–level control loop, and initiates the control activity.

The command executor is also responsible for receiving and processing sensory events that signal task completion. It is responsible for maintaining the appropriate time–window information that allows it to judge success or failure of task execution. Consequently, the command executor is responsible for fault detection during transient operational phases, such as the start–up and shut–down phases. In case of a failure (the sensory event has arrived *too–early* or *too–late*), it triggers the fault diagnoser, which starts to reason about the nature of the observed fault so as to relate the observed symptom to the failure that caused it.

Time windows are intimately linked to event–based control logic. They ensure early detection of faults during transient operational phases, and thereby provide the high–autonomy system with sufficient *reliability* to allow the system to operate adequately over an extended period of time. The time–window mechanism has been described in detail by Wang and Cellier [6].

Plant Operation

The actual low–level control is implemented in a microcontroller called a smart sensor. Once the command executor has downloaded a control program to the smart sensor, the low–level controller is activated. This can be a classical controller of any vintage. In this testbed, all control activity was strictly event–based, but there is nothing in the advocated methodology that would dictate such a solution. It should be noticed that the control programs are indeed different during different phases. For example, the voltage is carefully ramped up in the start–up phase by one control program (to avoid overshoot), whereas the voltage control program operates quite differently during steady–state operation.

Watchdog Monitors

Watchdog monitors are independent intelligent agents that monitor the high–autonomy system during steady–state operation. They have knowledge of some components of nominal system behavior during steady–state, and compare their expectations with the actually observed behavior. If a significant discrepancy is found, the "disquieted" watchdog alerts a fault diagnoser to come up with an explanation for the observed anomaly.

The watchdog monitor philosophy has been advocated by Kury [3]. In the described incident, it was one of the watchdogs that finally –and unnecessarily late– got aroused and triggered off the event chain that ultimately led to the restoration of the "distressed" temperature controller.

Fault Diagnosis

Contrary to the watchdog monitors that are active *daemons* throughout the steady–state operational phase, fault diagnosers are dormant sequential routines. They are activated only after an anomaly has been detected. The purpose of a fault diagnoser is to relate an observed symptom back to the failure most likely to have caused it.

In this testbed, only an extremely simple global rule–based fault diagnoser was employed. Those failures from which the high–autonomy architecture can recover are indicated by clear symptoms, and therefore, no complex fault diagnosers are needed. However, before an actual plant can be launched, it must be asserted that the high–autonomy system can recover from *all* foreseeable sorts of mishap. It will then become essential that the precise nature of any observed failure is well understood before an automated repair activity is initiated. For that purpose, a multi–level hierarchical model–based diagnoser is needed. Such an architecture was first proposed by Sarjoughian *et al.* [5]. It has meanwhile been elaborated upon by Chi [1].

Fault Recovery

The findings of the fault diagnoser will invariably be forwarded to a fault recovery agent. It is the task of that agent to decide whether something can be done about the failure or not.

If recovery is possible, it is the job of the recovery agent to compute a new *goal state*. It then provides the task planner with the current state and the desired goal state, and requests that a new command sequence be computed that moves the high–autonomy system from the current state to the desired goal state.

If no recovery is possible, the recovery agent will provide the task planner with the current state only and request computation of a command sequence for graceful shut–down.

THE INCIDENT

The following section uses actual data from a portion of the 100 hour test to illustrate the operation of the agents described above.

Figure 2 shows the zirconia cell temperature between 43,000 seconds and 107,000 seconds from start of the test. This is a portion of the steady–state operation.

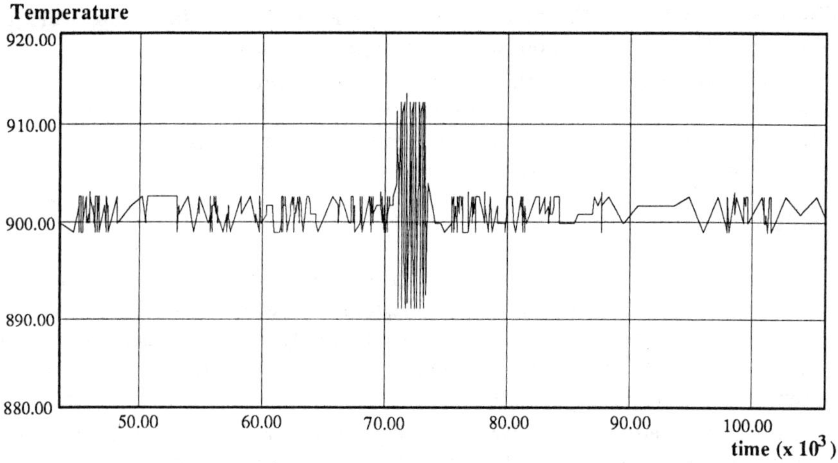

Figure 2: ZrO_2 Temperature Plotted over Time

It is immediately visible that around 71,000 seconds, something strange happened. The system recovered from the anomaly roughly 2,000 seconds later.

The anomaly started while a heavy thunderstorm took place. This was during the evening, and no human observer was at the plant site. From other curves, it can be concluded that, at that time, a short power failure (less than 2 seconds) occurred. The backup power supply took over, but there are signs of a temporary power surge. This transient upset the smart–sensor–based temperature controller; more precisely, the real–time clock of the smart sensor stopped operating.

It should have been easy to detect this anomaly immediately. However, the software being used did not require that the value of the real–time clock of the smart sensor be reported back to the PC, and thus, no watchdog monitor suspected any trouble. After a long time, a watchdog responsible for checking temperature values got suspicious, but this happened unnecessarily late. Since temperature change is normally such a slow phenomenon, that particular watchdog was executed only once every 30 minutes to save computing cycles on the PC. This is a serious drawback of the watchdog monitor philosophy *per se*: since watchdogs are daemons, they must be executed repetitively even if nothing is wrong. If they are executed rarely, they aren't very effective, but if they are executed frequently, they consume lots of computing power. Fortunately, since they are daemons, they run asynchronously and could be installed on separate CPUs, even though that was not done in this implementation.

Fault diagnosers are harmless. They are sequential routines and don't consume any computing cycles unless an anomaly has been observed. In the reported incident, the fault diagnoser, once invoked, worked beautifully. It concluded that the failure had to be in the smart–sensor program. It did not conclude that the bug was in the real–time clock, but this wasn't necessary. As an analogy: if a computer is down, the repair person will identify the fault only up to the board level and exchange the entire board. The faulty board can then be taken back to the lab where it can be further analyzed down to the chip level. There is no need to perform the second type of fault diagnosis on–line and in real time.

The recovery agent then decided to reload the control program once more into the smart sensor memory. It would have sufficed to restart the smart sensor without replacing any programs, but the additional action performed was harmless and didn't consume much time anyway.

LESSONS LEARNED

The lessons learned from this incident are enumerated in the sequel:

1. Proper functioning of the real–time clock is crucial to any event–based controller. Consequently, the smart sensor should report the value of its real–time clock back to the PC, and an additional watchdog should be installed that compares the two real–time clocks with each other and gets aroused when they start to diverge.
2. Some of the events that do not normally occur during steady–state operation can be declared as anomalous events. While no separate immediate recovery action may be needed, all anomalous events should nevertheless trigger a fault diagnoser.
3. While either of the two previous lessons could have improved results in the given situation, the watchdog that finally caught on should have been executed a little more frequently.
4. Event–based control is well suited for post–fault analysis. However, to this end it is necessary to remember which rules were fired when, why, and by whom. This was omitted in order to save computing cycles. Due to this oversight, it wasn't possible to conclude without a grain of a doubt what really had happened during the incident.

5. Similarly, while the smart sensor is equipped with means to download into it programs on the fly, nobody ever thought of the need to upload a program from the smart sensor back into the PC to save it for post–fault analysis.
6. The lessons learned from this incident should ensure that the same incident will never happen again. These sorts of experiments will help increase the reliability of the high–autonomy operation until a system results that is reliable enough for long–term operation on planet Mars. However, a good amount of research is still needed before this ultimate goal can be reached.

SUMMARY AND CONCLUSIONS

This paper outlined an intelligent high–autonomy control architecture useful for robots and manufacturing plants. The advocated methodology, once consolidated, will prove particularly valuable for space exploration in the early days of the next century, but it will also be useful for other applications where human presence at the plant location may be either undesirable, or hazardous, or impossible. Of particular interest was the fact that the architecture was able to detect and counter an anomaly in the controller itself, the potential occurrence of which had not been anticipated.

REFERENCES

1. Chi, S., (1991). *Modelling and Simulation for High Autonomy Systems*, Ph.D. Dissertation, Department of Electrical and Computer Engineering, University of Arizona, Tucson, Ariz.
2. Chi, S., B.P. Zeigler, and F.E. Cellier, (1990). "Model–Based Task Planning System for a Space Laboratory Environment," *Proceedings SPIE Conference on Cooperative Intelligent Robotics in Space*, Boston, Mass.
3. Kury, P.M., (1990). *An Intelligent Fault Diagnoser for Distributed Processing in Telescience Applications*, MS Thesis, Department of Electrical and Computer Engineering, University of Arizona, Tucson, Ariz.
4. NASA, (1985). *The Space Station Project.*
5. Sarjoughian, H.S., F.E. Cellier, and B.P. Zeigler, (1990). "Hierarchical Controllers and Diagnostic Units for Semi–Autonomous Teleoperation of a Fluid Handling Laboratory," *Proceedings IEEE Phoenix Conference on Computers and Communication*, Scottsdale, Ariz., pp. 795–802.
6. Wang, Q. and F.E. Cellier, (1991). "Time Windows: An Approach to Automated Abstraction of Continuous–Time Models into Discrete–Event Models," *International Journal of General Systems*, **19**(3), pp. 241–262.
7. Zeigler, B.P., S. Chi, and F.E. Cellier, (1991). "Model–Based Architecture for High Autonomy Systems," *Proceedings EURISCON'91 — European Robotics and Intelligent Systems Conference*, Corfu, Greece, June 23–28.

THE STATE SPACE RECURRENT NEURAL NETWORK

B. Horne, D. Hush and C. Abdallah
BSP Group, Department of Electrical and Computer Engineering
University of New Mexico
Albuquerque, NM 87131

ABSTRACT

In this paper, the state space recurrent neural network architecture (SSRNN) is reviewed and a recursive gradient descent learning algorithm is developed for the SS-RNN. An application of the SSRNN to the grammatical inference problem is presented.

1 Introduction

In the past few years, many recurrent neural network models have emerged [6, 7, 11]. Often these architectures are motivated purely by topological considerations. In this paper, we present a recurrent neural network architecture we call the *State Space Recurrent Neural Network* (SSRNN) which is motivated by the theory of nonlinear systems. It is well known that any linear system can be expressed in state-space form, i.e.,

$$\begin{aligned} \mathbf{x}(k+1) &= A\mathbf{x}(k) + B\mathbf{u}(k) \\ \mathbf{y}(k) &= C\mathbf{x}(k) + D\mathbf{u}(k) \end{aligned}$$

where $\mathbf{x}$ is a vector of *states*, and A, B, C, and D are constant coefficient matrices. The corresponding time–invariant nonlinear equivalent can be written as,

$$\begin{aligned} \mathbf{x}(k+1) &= \mathbf{f}(\mathbf{x}(k), \mathbf{u}(k)) \\ \mathbf{y}(k) &= \mathbf{g}(\mathbf{x}(k), \mathbf{u}(k)) \end{aligned}$$

It is well known that MLPs are capable of approximating arbitrary mappings under minimal constraints [1]. Thus, in theory, we could model the functions, $\mathbf{f}(\cdot)$ and $\mathbf{g}(\cdot)$, with MLPs and thus be capable of emulating a large class of nonlinear systems. The resulting architecture is the SSRNN shown in Figure 1.

Logic problems were among the first to be considered for learning in static networks because of their solid theoretical foundation. It seems natural then that Deterministic Finite Automata (DFAs) (logic gates coupled with a finite memory) be among the first problems investigated in recurrent neural networks. We use the SSRNN to infer the class of DFAs proposed by Tomita [9].

2 The SSRNN Learning Algorithm

In this section, a *real–time recurrent learning* (RTRL) learning algorithm for the SSRNN is summarized starting with some notation.

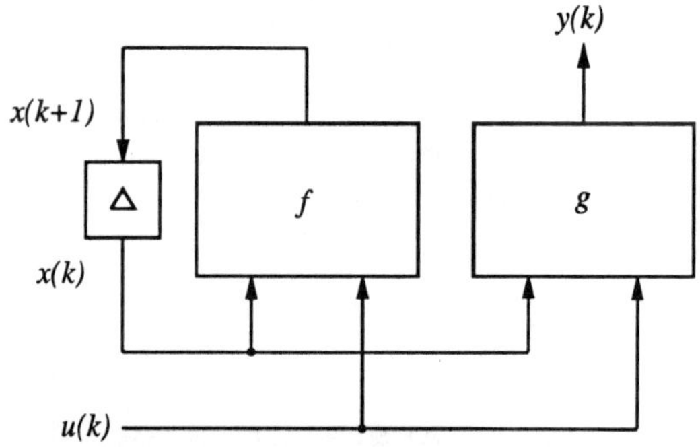

Figure 1: The State Space Recurrent Neural Network.

2.1 Notation

The number of layers (not including the input) in each of the MLPs will be denoted L. Although the number of layers in each MLP may be different, it will always be clear from the context which MLP is being referred to. The number of nodes in layer l of the MLPs will be denoted $N_l^{(f)}$ and $N_l^{(g)}$ for the **f** and **g** networks respectively.

Let $x_{l,i}(k)$ denote the output of the ith node in layer l of network **f** at time k. Let $y_{l,i}(k)$ denote the output of the ith node in in layer l of network **g** at time k.

Each of the external inputs to the network are, in general, a *sequence* of multi–dimensional vectors. Let each sequence have its own time index. Then we will denote the value of the ith component of the kth time index of the pth input sequence by $u_{p,i}(k)$. Let n denote the dimension of the input vector.

The input to the **f** network consists of the delayed state feedback and the external input. For notational convenience, we will represent this input as the "0th layer" of the network, i.e.,

$$x_{0,i}(k) = \begin{cases} x_{L,i}(k-1) & 1 \leq i \leq N_L^{(f)} \\ u_{p,i-N_L^{(f)}}(k) & N_L^{(f)} < i \leq N_L^{(f)} + n \end{cases}$$

The input to the **g** network consists only of the delayed state feedback. The ith component of these values will be denoted by $y_{0,i}(k)$ so that $y_{0,i}(k) = x_{L,i}(k-1)$.

Let $w_{l,i,j}(m)$ represent the weight which connects the jth node in layer $l-1$ to the ith node in layer l. The notation $w_{l,i,0}(m)$ will be used to refer to a node's bias weight. Note that the weights are defined with a different time scale than other variables in the network.

Each node computes a nonlinear function of a weighted sum of its inputs, i.e.,

$$\begin{aligned} x_{l,i}(k) &= \sigma\left(\sum_{j=1}^{N_l^{(f)}} w_{l,i,j}^{(f)}(m) x_{l-1,j}(k) + w_{l,i,0}^{(f)}(m)\right) \\ y_{l,i}(k) &= \sigma\left(\sum_{j=1}^{N_l^{(g)}} w_{l,i,j}^{(g)}(m) y_{l-1,j}(k) + w_{l,i,0}^{(g)}(m)\right) \end{aligned}$$

The nonlinearity $\sigma(\cdot)$ would typically be a sigmoid function, i.e., $\sigma(z) = 1/(1 + exp(-z))$. Let $\sigma_{l,i}^{\prime(f)}(k)$ and $\sigma_{l,i}^{\prime(g)}(k)$ represent the derivative of the nonlinearity with respect to its argument for the ith node in the lth layer of networks **f** and **g** respectively. For the learning algorithm a set of desired outputs will be specified for each training sequence, p. Denote the ith component of these desired outputs by $d_{p,i}(k)$.

2.2 Learning Criterion

We define an error function to evaluate the performance of the network over each of the P training patterns,

$$J = \sum_{p=1}^{P} J_p \tag{1}$$

where J_p is the Mean Squared Error (MSE) incurred on the pth training sample,

$$J_p = \frac{1}{2} \sum_{k=1}^{K_p} \sum_{h=1}^{N_L^{(g)}} [d_{p,h}(k) - y_{L,h}(k)]^2 \tag{2}$$

The weights of the network are adjusted to minimize this criterion using a gradient search, i.e.,

$$w_{l,i,j}(m+1) = w_{l,i,j}(m) - \mu \frac{\partial J}{\partial w_{l,i,j}}\bigg|_{w_{l,i,j}(m)}$$

Note that this update is performed only after one complete cycle through the training set. In practice, the gradient is estimated by approximating the criterion function by a single example sequence, i.e., $\hat{J} = J_{m \bmod P}$, or by *batching* the gradient over some fixed number of example sequences, B, i.e., $\hat{J} = \sum_{p=1}^{B} J_{(mB+p) \bmod P}$.

2.3 Learning in Network g

This weight update rule will be exactly equivalent to the standard backpropagation rule since network **g** is static. The algorithm is repeated here in a slightly different form than is usually seen to facilitate the development of the weight update rule for nodes in the **f** network. Applying the chain rule to (2) gives,

$$\frac{\partial J_p}{\partial w_{l,i,j}^{(g)}} = -\sum_{k=1}^{K_p} \sum_{h=1}^{N_L^{(g)}} (d_{p,h}(k) - y_{L,h}(k)) \frac{\partial y_{L,h}(k)}{\partial w_{l,i,j}^{(g)}} \tag{3}$$

Now, we can derive an expression for $\frac{\partial y_{L,h}(k)}{\partial w_{l,i,j}^{(g)}}$ using the chain rule applied at the output of the ith node in the lth layer of the **g** network,

$$\frac{\partial y_{L,h}(k)}{\partial w_{l,i,j}^{(g)}} = \frac{\partial y_{L,h}(k)}{\partial y_{l,i}(k)} \sigma'_{l,i}(k) y_{l-1,j}(k) \tag{4}$$

Now, $\frac{\partial y_{L,h}(k)}{\partial y_{l,i}(k)}$ can be derived recursively by applying the chain rule over all of the nodes at the next highest level as follows,

$$\frac{\partial y_{L,h}(k)}{\partial y_{l,i}(k)} = \sum_{m=1}^{N_{l+1}^{(g)}} \frac{\partial y_{L,h}(k)}{\partial y_{l+1,m}(k)} \sigma'_{l+1,m}(k) w_{l+1,m,i} \tag{5}$$

with the boundary condition, $\frac{\partial y_{L,h}(k)}{\partial y_{L,i}(k)} = \delta_{i,h}$, where $\delta_{i,h}$ is the Kronecker delta function.

2.4 Learning in Network f

In this section, we derive an expression for $\frac{\partial J_p}{\partial w_{l,i,j}}$ for a weight in the **f** network. Applying the chain rule to (2) gives,

$$\frac{\partial J_p}{\partial w^{(f)}_{l,i,j}} = -\sum_{k=1}^{K_p}\sum_{v=1}^{N_L^{(g)}} (d_{p,v}(k) - y_{L,v}(k))\frac{\partial y_{L,v}(k)}{\partial w^{(f)}_{l,i,j}}$$

By applying the chain rule at the "0th layer" of network **g** (i.e. at the input vector to **g**), we can write

$$\frac{\partial J_p}{\partial w^{(f)}_{l,i,j}} = -\sum_{k=1}^{K_p}\sum_{v=1}^{N_L^{(g)}} \left(d^{(g)}_{p,v}(k) - y_{L,v}(k)\right)\left[\sum_{h=1}^{N_L^{(f)}} \frac{\partial y_{L,v}(k)}{\partial y_{0,h}(k)}\frac{\partial x_{L,h}(k)}{\partial w^{(f)}_{l,i,j}}\right] \tag{6}$$

where $\frac{\partial y_{L,v}(k)}{\partial y_{0,h}(k)}$ is defined in the previous section. It can easily be shown that [5],

$$\frac{\partial x_{L,h}(k)}{\partial w^{(f)}_{l,i,j}} = \frac{\partial x_{L,h}(k)}{\partial x_{l,i}(k)}\sigma'_{l,i}(k)x_{l-1,j}(k) + \sum_{\beta=1}^{N_L^{(f)}} \frac{\partial x_{L,h}(k)}{\partial x_{0,\beta}(k)}\frac{\partial x_{L,\beta}(k-1)}{\partial w^{(f)}_{l,i,j}} \tag{7}$$

where $\frac{\partial x_{L,h}(k)}{\partial x_{l,i}(k)}$ is calculated recursively from the last layer backwards as follows,

$$\frac{\partial x_{L,h}(k)}{\partial x_{l,i}(k)} = \sum_{m=1}^{N_{l+1}^{(f)}} \frac{\partial x_{L,h}(k)}{\partial x_{l+1,m}(k)}\sigma'_{l+1,m}(k)w_{l+1,m,i} \tag{8}$$

with the boundary condition, $\frac{\partial x_{L,h}(k)}{\partial x_{L,i}(k)} = \delta_{i,h}$.

3 Example: Grammatical Inference

A Finite State Machine (FSM) is defined by a six–tuple, $(Q, \Sigma, \Delta, \delta, \lambda, q_0)$, where Q is a finite set of *states*, Σ and Δ are a set of symbols (typically 0 and 1) called the *input alphabet* and the *output alphabet* respectively, δ is the *transition function* which maps $Q \times \Sigma$ to Q, and q_0 is the *initial state*. An FSM always starts in state q_0 and begins processing a *string* of alphabet characters. For each character in the string it changes state according to the transition function and outputs a symbol according to the output function. A *Deterministic Finite Automaton* (DFA) is a special type of FSM in which the states are either classified as *accepting* or *rejecting*. If, after processing the entire input string, the DFA is in an accepting state then we say the string is *accepted* by the DFA, otherwise it is *rejected*. Associated with every DFA is a *regular language*, L, which consists of all the strings which are accepted by the DFA. A *grammar* is a set of rules which generate the language, L.

The problem of regular grammatical inference is as follows: Given a set of *positive strings* and a set of *negative strings*, find a DFA which accepts the positive strings and rejects the negative strings. It turns out that there are an infinite number of DFAs

	Language						
	1	2	3	4	5	6	7
% Perfect	70%	100%	70%	70%	70%	70%	80%
Avg. No. Epochs	5	257	126	135	246	447	570
Avg. Performance	100.0%	99.7%	85.0%	100.0%	96.5%	53.1%	71.1%

Table 1: Results on Tomita's languages.

which can solve this problem, some of these are trivial to construct. Many techniques have been developed for finding "good" solutions in polynomial time, for a review see [2]. Recently, there has been considerable interest in using recurrent neural networks to find solutions to this problem [3, 8, 10].

3.1 Benchmarks and Results

One set of regular languages which has become a benchmark for investigating the ability of recurrent neural networks to perform grammatical inference is the set of seven languages proposed by Tomita [9] as well as a small set of training strings.

Table 1 shows the results averaged over ten trials on each of Tomita's seven languages using a different initial condition for the weights each time.

The first row in the table shows the percentage of time the SSRNN converged to a solution for which the training set was classified perfectly. The second row in the table shows the average number of epochs it took for the SSRNN to converge to these solutions. Note that in most cases, the network only found a perfect solution seven out of ten times. In these cases, it was typical for the network to fail to learn the correct classification for only one of the strings in the training set. On some occasions, the SSRNN would only classify the empty string as a positive instance and classify all other strings as negative examples.

In order to see how well the SSRNN generalized we tested the networks which performed perfectly on the training set on a test set consisting of all strings of length 10 or less. The third row shows the average performance on this set in terms of percentage of correct classifications, where a correct classification is one in which the network outputs a value greater than 0.5 when the desired output is a 1.0 or a value less than 0.5 when the desired output is a 0.0. Note that in the cases of languages 1, 2, 4, and 5 the network achieved almost perfect performance on the test set when it was able to classify the data in the training set correctly. Although the SSRNN achieved an average performance of 71.1% on language 7, it should be noted that the total percentage of strings of length 10 or less which are negative examples is 72.6%, so simply classifying every string as negative would have given better generalization than the SSRNN. These results are encouraging in light of previously reported results on this same problem [4, 8, 10].

References

[1] G. Cybenko. Approximation by superpositions of a sigmoidal function. *Mathe-*

matics of Control, Signals, and Systems, 2(4):303–314, 1989.

[2] K.S. Fu and T.L. Booth. Grammatical inference: Introduction and survey – Part I. *IEEE Transactions on Systems, Man and Cybernetics*, 5:95–111, 1975.

[3] C.L. Giles, D. Chen, C.B. Miller, H.H. Chen, G.Z. Sun, and Y.C. Lee. Second-order recurrent neural networks for grammatical inference. In *Proceedings of the International Joint Conference on Neural Networks*, volume 2, pages 273–281, July 1991.

[4] C.L. Giles, C.B. Miller, D. Chen, H.H. Chen, G.Z. Sun, and Y.C. Lee. Learning and extracting finite state automata with second–order recurrent neural networks. *Neural Computation*, 4(3):393–405, 1992.

[5] B. Horne, D. Hush, and C. Abdallah. The state space recurrent neural network with application to regular grammatical inference. Technical Report EECE 92-002, Department of Electrical and Computer Engineering, University of New Mexico, 1992.

[6] K.S. Narendra and K. Parthasarathy. Identification and control of dynamical systems using neural networks. *IEEE Transactions on Neural Networks*, 1:4–27, March 1990.

[7] F.J. Pineda. Dynamics and architecture for neural computation. *Journal of Complexity*, 4:216–245, 1988.

[8] J.B. Pollack. The induction of dynamical recognizers. Technical Report 90–JP–AUTOMATA, Dept. of Computer and Information Science, Ohio State University, Columbus, Ohio, 1991.

[9] M. Tomita. Dynamic construction of finite automata from examples using hill-climbing. In *Proceedings of the Fourth Annual Conference of the Cognitive Science Society*, volume 4, pages 105–108, 1982.

[10] R.L. Watrous and G.M. Kuhn. Induction of finite–state automata using second–order recurrent networks. In *Advances in Neural Information Processing Systems 4*, 1992. (To appear).

[11] R.J. Williams and D. Zipser. A learning algorithm for continually running fully recurrent neural networks. *Neural Computation*, 1(2):270–280, 1989.

Making an Expert System Changes the Expert

A. Terry Bahill and K. Bharathan
Systems and Industrial Engineering

Richard F. Curlee
Speech and Hearing Sciences
University of Arizona
Tucson, AZ 85721, USA

ABSTRACT The process of extracting knowledge from a human expert can have a profound effect on the way the expert views his or her own work. This paper argues that the process of introspection ought to be encouraged, and to be used in the knowledge extraction process, thereby leading to better design and construction of expert systems. This paper is a case study covering the period from 1985 through 1992 of one domain expert and his participation in the construction of five expert systems to help with the diagnosis and prognosis of small children who might have begun to stutter.

INTRODUCTION

It is generally accepted that an expert system should mimic the input-output behavior of a human problem solver; it is also generally accepted that the main bottleneck in building expert systems lies in the process of knowledge acquisition and elicitation. Most knowledge elicitation relies on the efforts of the knowledge engineer in organizing structures that explain or predict the actions/decisions/conclusions of the human domain expert.

This paper intends to show that there is another aspect to knowledge extraction and elicitation that deserves attention: the process whereby a human domain expert becomes more cogent, in his or her own mind, about the algorithms, heuristics and strategies that he or she uses in manipulating this factual knowledge base. It is our contention that the knowledge engineer's constructs can be greatly strengthened by utilizing those constructs of the domain expert that may evolve through the knowledge extraction process. In addition, it is our contention that harnessing the domain expert's constructs by fostering such introspective processes early in the knowledge extraction stage will greatly improve the cost effectiveness of the expert system design.

This research was supported by National Institute of Child Health and Human Development, NIH, grant number 2 R44 HD26209-02A1 and by Bahill Intelligent Computer Systems.

BACKGROUND

The core arguments of this paper are based on a case study of one particular human expert who has been involved over several years -- from 1985 to the present -- in the construction of five expert systems, all of which pertain to the same domain and have the same objective: to assist in the diagnosis of stuttering among young children.

The domain expert is Richard F. Curlee, who is one of a few nationally known experts in the diagnosis of incipient stuttering among children and has been practicing in the field since the late 1960s. Beginning in 1985, he has served as the expert in five separate expert systems in this area. The authors, in conjunction with four other experts in the US, are currently building a commercial version of the fifth expert system. The following sections of this paper present background information, the expert's insights, the results of our investigation, and finally, our conclusions from an analysis of this material.

This section summarizes the history of these five expert systems. The first system was built by Richard Boyd in 1985 as part of a graduate student project for a course on expert systems being taught by Terry Bahill.

> Boyd, "using traditional question and answer interview techniques, worked with our expert and built an expert system to help give a diagnosis and prognosis for stuttering in children. In this system, the user answered yes or no to a lengthy series of questions and then received the system's recommendation. The questions were short, which seemed to make it difficult for the user to follow the system's reasoning. The knowledge base was not uniform in coverage. There were redundant questions and areas of knowledge that were left out [1]."

Moller and Bahill [1] ascribe the lacunae in this system to the failure of the knowledge extraction process. In Curlee's estimate (today and by hindsight), the problem was that the knowledge engineer in that instance did not force him to cogitate about his thought processes in undertaking a typical consultation.

Robert F. Moller built the second expert system in 1987 as a master's project in systems engineering. In tackling the problem of knowledge acquisition, Moller used the Analytic Hierarchy Process (AHP), first developed by Saaty [2]. Using this method, Moller was able to decompose the problem into object-attribute-value triplets and assign weights to combinations of these triplets (reported in detail in [1]). A major reason for using AHP was to develop an algorithm for assigning certainly factors. Traditionally, the assignment of certainty factors is intuitive and ad hoc, but AHP provided an analytic tool for deriving them. In working with AHP, Curlee had to work his way through fairly large matricies of attributes, some as large as 12 by 12. It forced "the expert to make an exhaustive comparison of attributes, every attribute [being] compared to every other attribute at each level [1]." Using the AHP process forced the expert to *systematically* evaluate the relative importance of the signs and symptoms of incipient stuttering, independently and in all possible combinations. The process forced him to clarify explicitly the significance or weight placed on specific elements of diagnostic information in arriving at a diagnosis. The resulting expert system was called Stutter.

In 1989 another graduate student in Bahill's class, Peter Vinsant, took up the task of building an expert system using the signs and symptoms that Curlee organized as a series of questions to be answered. With the set of questions already at hand, Vinsant did not spend any time extracting knowledge at a fundamental level but concentrated, instead, on the inferencing and the output of the system. They spent a lot of time determining the relative weights of the signs and symptoms represented by the answers to the questions used in arriving at a diagnosis. Using a scale from 1 through 100, Vinsant worked with Curlee in assigning weights to the answers to each question. It appears that he organized the questions in logical groups and distributed the weights within each group to approximate 100. It is not clear whether this was planned by Vinsant, or whether Curlee was applying, in his own way, overall perspective that he had acquired from working with Moller in 1987. However, the result was that assignment of weights in this manner led to several inconsistencies in this expert system, named ESP.

A fourth system, named Second Opinion, was begun in 1991 with K. Bharathan as the knowledge engineer. It built upon the set of questions that were used for ESP. However, it used object-oriented design and Fuzzy Logic. This version was written in M.1 Version 3.0, which permits object-oriented programming. In this version, there was a symptom-cluster class and a diagnostic-status class. Seven symptom-cluster objects and six diagnostic-status objects formed the core of the program. The set of possible questions was expanded to 24. The answers to each question led to increments in the certainty of one or more of the symptom-cluster objects. When all questions were answered, the program assigned a weight to each of the diagnostic-status objects on the basis of the pattern of certainties among the symptom-cluster objects. Fuzzy sets were used in deriving these weights. The system extracted up to three of the most significant diagnostic-status objects that were above a threshold and displayed these along with a set of feasible clinical management alternatives from among which the user could make the choice of treatment. Implementing the object-oriented structure forced the expert to think again about the relationships between all of the signs and symptoms. This again caused the expert to change how he thought about this knowledge. With this system we kept running out of memory. Therefore, a fifth system, named the Childhood Stuttering Consultant, was begun in May 1992. It uses C++ for data collection and screen management and M.4 for the inference engine. For this system we will use insights into the diagnostic process that have evolved from Curlee's previous expert system experiments. These insights and the role of the earlier expert systems in their generation, is the focus of the next section.

THE EXPERT'S INSIGHTS

From examining Stutter and ESP, and from Moller and Bahill's report on the use of AHP [1], several points emerged that link this expert's changed views of the diagnostic process and the expert systems.

Curlee reported that following his interaction with Moller, he changed his consultation style. Prior to 1987, he conducted most diagnostic evaluations on the basis of a single appointment with the parent(s) and the child concerned.

Starting in 1987 this diagnostic procedure was revised. The consultation is now spread over three separate sessions, each separated by several days. In the first session, he talks with the parents, discussing with them their concerns and how they perceive their child's problems and what they are doing about it. The information from this consultation is carefully reviewed for symptoms that need confirmation during the child's examination or information that is missing, inconsistent, or needs clarification. Such lacunae are filled before evaluating the child. The data that are gathered in the session with the child are based on information previously gathered in the session with the parents. Likewise the questions asked of the parents in their final interview are based on the data collected in the session with the child. Then Curlee discusses with the parents his findings and any treatment alternatives that may be needed. He is emphatic about the change in this phase; where before he would typically recommend what ought to be done, in a strictly prescriptive sense, he is concerned today with interacting with the parents to give them an informed basis on which they can make a decision about the most appropriate course for their family to follow. Because, as he puts it, "the active collaboration of the parents is an integral part of whatever treatment is undertaken. Parents (caretakers) are a major factor in shaping or changing the behavior of their children, without their complete commitment, the chances of a positive outcome are considerably diminished." It is important to point out that these changes in the expert's behavior were not caused by the evolution of new tools but rather resulted because of his participation in this expert system project.

In addition to making more precise quantitative measurements, Curlee has also modified the structure of the case-history interview. Where earlier a large number of questions were routinely posed for the parents to answer, he now has a mental model of a hierarchy of questions, with those pertaining to detail only being asked when warranted. Although this may seem trivial to an outsider, it appears that several branches of the hierarchy in his mental model evolved as he worked his way through Moller's AHP exercise.

Unfortunately, few of these insights found a place in ESP, written in 1989. What seems to have happened is that, since Vinsant did not begin his own exercise of knowledge extraction from the beginning, he did not catch the changed model that Curlee had begun using. Thus, although there is a hierarchy in ESP, with some questions being fired only when there is a pre-specified answer to an earlier question, not all of the hierarchies in Curlee's mind were incorporated.

COMMENTS OF OTHER EXPERTS

After serving as an expert in the construction of an expert system, the expert often perceives his or her knowledge differently. We have just documented this for Curlee. Moreover we have experienced similar reactions from other experts. Professor Linda Swisher of the Speech and Hearing Sciences Department at the University of Arizona, said

"Working through these production rules, and seeing the decision tree of a consultation made me realize what I actually do during a consultation. If nothing else this experience will help me teach this material better in the future. What I have been telling my students for the last 15 years, is not what I really do [1]."

We also built an expert system for a Phoenix electric company (SRP) to help homeowners reduce their electric bills. The utility energy auditors were pleased with the final expert system, but in their exit interview they said

One of the most important things you have given us is a consistent set of advice. "In the past when customers called SRP Customer Services for advice about their energy consumption, they received different recommendations, depending on the energy auditor or information specialist available at the time [4]."

CONCLUSIONS

The following results can be inferred from the foregoing discussion of this case. (1) The process of knowledge extraction, especially when done in a systematic manner, as was done by Moller with the aid of AHP, has a significant effect on the way in which an expert views his or her job. (2) Identifying and isolating the principal features of the expert's changed paradigm should help the construction of a later expert system's efficacy. That this did not happen was not due to either the expert or the knowledge engineer, but to the general failure to recognize the effect of the knowledge extraction process on the expert.

Several generalizations from the above case study are obvious and tempting: it is necessary, in any knowledge extraction process, to harness the results of the introspective process in the human expert. Had each of the five expert systems above been part of a single project, the resulting cost would have been substantial, and the final product far from ideal. Again, several of the insights, which are now being incorporated, might well have been included in ESP, making for a far more efficient process of expert system construction. Letting the processes of knowledge extraction and "paradigm evolution" grow hand-in-hand could have resulted in better results than were obtained, and in a more cost-effective manner.

In this context, a perverse possibility should also be noted, namely, that introspection could hamper some experts. The analogy is to the apocryphal story of the weak tennis player who always won his matches: all he did was to ask his opponent, just before the match, "Do you inhale or exhale at the moment when you serve the ball?" Thinking about how one does something might well have the impact of preventing one from doing that something well. Where this is the case, the introspective effort may be of little use to the knowledge engineer, and of no use to the expert. Indeed it is possible that thinking too much may reduce the quality of decisions [5]. Introspection can induce a person focus on insignificant details and subsequently place less weight on important items.

The literature indicates that it is the role of the knowledge engineer to elicit such knowledge, i.e. to build the structure that represents (by explaining or predicting) the expert's behavior. It is not the intent of this paper to belittle this role of the knowledge engineer. We only suggest that taking advantage of a process that many experts probably have to go through anyway might make things easier for the knowledge engineer and might increase the probability of having, in the end, a feasible, more cost-efficient system.

We seem to have described a reciprocal process; there is an interaction between the expert's behavior or decisions that are being modeled and the expert system that is being constructed. The expert system influences or modifies the system of the expert that is being modeled.

It is obvious that generalizing from a single case is not appropriate. However, we suggest, in conclusion, that the present case study has raised an issue that needs further exploration. More research on the manner in which cognitive processes affect the expert's mental model of his or her own work is needed before the conclusions of this paper can be confirmed or denied. One way might be to examine prototypes built during the design process to see if any improvements were due to introspection by the expert concerned.

REFERENCES

1. R.F. Moller and A.T. Bahill, "A Knowledge Extraction Technique," In *Verifying and Validating Personal Computer-Based Expert Systems,* edited by A.T. Bahill (Prentice Hall, NJ, 1991).
2. T.L. Saaty, *The Analytic Hierarchy Process* (McGraw-Hill, 1980).
3. A.T. Bahill and W.R. Ferrell, "An introductory course in expert systems," *IEEE Expert* 1(4): 59-63 (Winter 1986).
4. M.J. Jafar, A.T. Bahill and D.E. Osborn "A knowledge-based system for residential HVAC applications," *ASHRAE Journal* 33(1): 20-26 (1991).
5. T.D. Wilson and J.W. Schooler, "Thinking too much: Introspection can reduce the quality of preferences and decisions," *J. Personality and Social Psychology,* 60: 181-192 (1991).

A ROBUST LOGIC FOR RULE-BASED REASONING UNDER UNCERTAINTY

SIMON PARSONS

Department of Electronic Engineering, Queen Mary and Westfield College, Mile End Road, London, E1 4NS, United Kingdom.

MIROSLAV KUBAT

Computer Centre, Technical University of Brno, Udolni 19, 60200 Brno, Czechoslovakia.

MIRKO DOHNAL

Advanced Computation Laboratory, Imperial Cancer Research Fund, Lincoln's Inn Fields, London WC2A 3PX, United Kingdom.

ABSTRACT

Reasoning with uncertain information is a problem of key importance when reasoning about the real world. The more information required by the procedure used to handle the knowledge, the higher its chance of failure. The theory of rough sets [4] is not information intensive and is thus a good basis for handling uncertainty. We present an introduction to a logic based on rough set theory that is suitable for reasoning under uncertainty. We introduce inference rules analogous to those of classical logic, and demonstrate their effectiveness in rule-based reasoning.

INTRODUCTION

Any system designed to reason about the real world must, perforce, be capable of dealing with uncertain information, that is information whose certainty may not be completely established, and incomplete knowledge about its domain. This is a direct consequence of the complexity of the real world and the finite size of the knowledge base that such a system has at its disposal. A number of mathematical formalisms have been developed to cope with uncertainty in knowledge base systems [5], and most have been demonstrated on a number of reasoning tasks. These formalisms suffer from a number of disadvantages. The severest of these is that they are all very *information intensive*; they all require large amounts of precise information in order to deal with uncertainty. This means that the truth value of the relations between variables are required in the form of grades of membership and probability distributions. These values are often unknown, or expensive to obtain, and methods that are not information intensive are often desirable. Rough set theory seems to solve the problem of information intensity, enabling us to avoid the paradox of performing precise calculations with imprecise data.

ROUGH SET THEORY

Rough sets, originally introduced by Pawlak [4], have been further developed and applied to a number of problems by various authors, [1], [2], and [6]. Here we discuss the basic ideas behind the theory.

Basic concepts of rough sets

Consider a set of elementary objects $\mathbf{A} = \{A_1,...,A_n\}$. The members of this set are used to define a set of objects $\mathbf{E} = \{E_1,...,E_m\}$ each of the A_i corresponding to a possible attribute of the E_j. Now, since the specification of the E_j must be based on the A_i alone, it could well be the case that some of the E_j are indistinguishable since the values that distinguish them are not identified by the A_i. For instance a blue egg and a green egg will be indistinguishable if the only concepts that can be used to describe them are 'red', 'ball' and 'egg'. Thus the use of a finite set **A** implies the existence of an equivalence relation and a consequent partition on **E**:

$$\mathbf{P} = \{P_1,...,P_r\} \text{ where } \cup P_i = \mathbf{E} \text{ and } P_i \cap P_j = \emptyset \text{ for } i \neq j = 1,...,r \tag{1}$$

and each $\mathbf{P}_s = (E_s,...,E_z)$ is an equivalence class. Let $T \subset \mathbf{S}$ be an object, whose attributes are $T_A \subseteq \mathbf{A}$, that we wish to describe in terms of the set of partitioned attributes E_i. Let:

$$T^c(\mathbf{P}, \mathbf{E}) = \{e: e \in P_i, P_i \subset T \neq \emptyset\} \tag{2}$$

$$T^e(\mathbf{P}, \mathbf{E}) = \{e: e \in P_i, P_i \cap T \neq \emptyset\} \tag{3}$$

where $T^c(\mathbf{P}, \mathbf{E})$ is the core[1] of T based on **E** and **P**, the set of all partitioned attributes that T possesses, and $T^e(\mathbf{P}, \mathbf{E})$ is the envelope of T based on **E** and **P**, the set of all partitioned attributes of which at least one is possessed by T. The pair $[T^c(\mathbf{P}, \mathbf{E}), T^e(\mathbf{P}, \mathbf{E})]$ is a *rough set*. Let the set of all rough sets that may be defined using **E** partitioned as **P** be denoted by **R**. Consider $R = [R^c, R^e], R' = [R'^c, R'^e] \in \mathbf{R}$. It is simple to show that the following set theoretic relations hold where the symbol '~' stands for complement:

$$\begin{array}{llllll} (R \cup R')^c & \supseteq & R^c \cup R'^c & (R \cup R')^e & = & R^e \cup R'^e \\ (R \cap R')^c & = & R^c \cap R'^c & (R \cap R')^e & \subseteq & R^e \cap R'^e \\ (\sim R)^c & = & \sim(R^e) & (\sim R)^e & = & \sim(R^c) \end{array} \tag{4}$$

Combining rough sets

The degree to which a concept A may be defined within $(\mathbf{E}, \mathbf{A})$ depends on the cardinality of A^c and A^e. We have the following [4]:

If	$A^c = A^e$	then A is precisely defined by $(\mathbf{E}, \mathbf{A})$	
If	$A^c \neq A^e$ and $A^e \neq \emptyset$	then A is roughly defined by $(\mathbf{E}, \mathbf{A})$	
If	$A^c = \emptyset$	then A is internally undefined by $(\mathbf{E}, \mathbf{A})$	(5)
If	$A^e = E$	then A is externally undefined by $(\mathbf{E}, \mathbf{A})$	
If	$A^c = \emptyset$ and $A^e = E$	then A is totally undefined by $(\mathbf{E}, \mathbf{A})$	

We can determine the degree to which logical combinations of roughly defined objects may themselves be defined. The basic logical operations of disjunction, conjunction and negation may be defined in terms of set operations on the core and

[1] In Pawlak's original work on rough sets the envelope and core were named 'upper approximation' and 'lower approximation' respectively.

envelope of the objects concerned, and relations such as material implication defined from them:

$$\begin{array}{llllll} (A\vee B)^c & \supseteq & A^c\cup B^c & (A\wedge B)^c & = & A^c\cap B^c \\ (A\vee B)^e & = & A^e\cup B^e & (A\wedge B)^e & \subseteq & A^e\cap B^e \\ (\neg A)^c & = & \sim(A^e) & (A\rightarrow B)^c & \supseteq & \sim A^e\cup B^c \\ (\neg A)^e & = & \sim(A^c) & (A\rightarrow B)^e & = & \sim A^c\cup B^e \end{array} \quad (6)$$

A LOGIC OF ROUGH TRUTH VALUES

As stated above, it is often advantageous to deal with formalisms that are not information intensive, especially in domains in which only sparse information is typically available. In such domains we want robust reasoning mechanisms that are capable of absorbing large amounts of ill known information whilst still returning accurate answers where possible. In this section we present a simple quantified logic that meets these requirements.

Let us consider a Boolean algebra of propositions $(\mathcal{P}, \vee, \wedge, \neg)$. We define a rough measure R on $\mathcal{P}$ such that $\forall p \in \mathcal{P}, R(p) = [p^c, p^e]$. $R(p)$ is thus an estimate of the degree to which p is defined by the set of descriptors $\mathbf{A}$, and this may be related to the truth of the proposition p by careful choice of $\mathbf{A}$ [3]. We can distinguish the following limiting values of $R(p)$, for $\emptyset \subset X \subset \mathbf{A}$, and $\emptyset \subset Y \subset \mathbf{A}$, which correspond to the definitions of (5):

$$\begin{array}{lllll} \text{If} & R(p) & = & [\mathbf{A}, \mathbf{A}] & \text{then p is true} \\ \text{If} & R(p) & = & [X, \mathbf{A}] & \text{then p is roughly true} \\ \text{If} & R(p) & = & [\emptyset, \mathbf{A}] & \text{then p is of unknown value} \\ \text{If} & R(p) & = & [\emptyset, Y] & \text{then p is roughly false} \\ \text{If} & R(p) & = & [\emptyset, \emptyset] & \text{then p is false} \end{array} \quad (7)$$

These rough values form a lattice, ordered by set inclusion $\subseteq$, giving the following:

$$\begin{array}{ccccccccc} [\emptyset, \emptyset] & \subseteq & [\emptyset, X] & \subseteq & [\emptyset, \mathbf{A}] & \subseteq & [Y, \mathbf{A}] & \subseteq & [\mathbf{A}, \mathbf{A}] \\ \text{false} & & \text{roughly false} & & \text{unknown} & & \text{roughly true} & & \text{true} \end{array} \quad (8)$$

This suggests the introduction of a rough truth measure RV over $\mathcal{P}$ which identifies which of these five ordered states the rough value of each $p \in \mathcal{P}$ takes on. The advantages of such a measures are its extreme simplicity and robustness, a direct result of the simple conditions used to define the values. The axioms of the rough truth measure may be stated as:

$$\begin{array}{lll} RV(p \vee q) & = & \max(RV(p), RV(q)) \\ RV(p \wedge q) & = & \min(RV(p), RV(q)) \end{array} \quad (9)$$

These may be easily verified by considering the set operations on the rough measure for each proposition. Similar considerations will validate the following results for the negation of a rough truth value:

$RV(p)$	true	roughly true	unknown	roughly false	false
$RV\neg(p)$	false	roughly false	unknown	roughly true	true

(10)

Rough inference rules

In order to use our rough valued logic for practical reasoning purposes, we need to provide a set of rules for propagating inference. We can adapt the reasoning patterns of classical logic for rough valued logic. Firstly, modus ponens:

$$\begin{array}{ccccc} & & RV(p \rightarrow q) & = & \alpha \\ & & RV(p) & = & \beta \\ \hline \alpha & \geq & RV(q) & \geq & \min(\alpha, \beta) \end{array} \tag{11}$$

The upper limit is obtained by realising that $RV(p \rightarrow q)$ is the maximum of $RV(\neg p)$ and $RV(q)$. The lower limit stems from the fact that $\min(\alpha, \beta) = RV((p \rightarrow q) \wedge p) = RV(p \wedge q) \leq RV(q)$. A similar line of reasoning gives us the pattern for modus tollens:

$$\begin{array}{ccccc} & & RV(p \rightarrow q) & = & \alpha \\ & & RV(\neg q) & = & \beta \\ \hline \alpha & \geq & RV(\neg p) & \geq & \min(\alpha, \beta) \end{array} \tag{12}$$

AN EXAMPLE

To illustrate the kind of reasoning possible with our rough valued logic, consider the following simple example. An intelligent system has a set of rules which it uses to determine when it is appropriate to open and close a pressure regulating valve. These rules are expressed as logical sentences, each of which is quantified with a rough truth value:

C1	high(pressure)	→	open(valve)	(t)
C2	¬low(pressure)	→	low(level)	(t)
C3	low(pressure)	→	¬open(valve)	(t)
C4	high(temperature)	→	¬low(level)	(rt)

The system is also equipped with two sensors, one of which detects temperature with a high degree of accuracy, the other pressure with a lower degree of accuracy:

C5	high(pressure)	(rf)
C6	high(temperature)	(rt)

From this data the system can deduce two conflicting statements about the valve using the rules of modus ponens (11) and modus tollens (12):

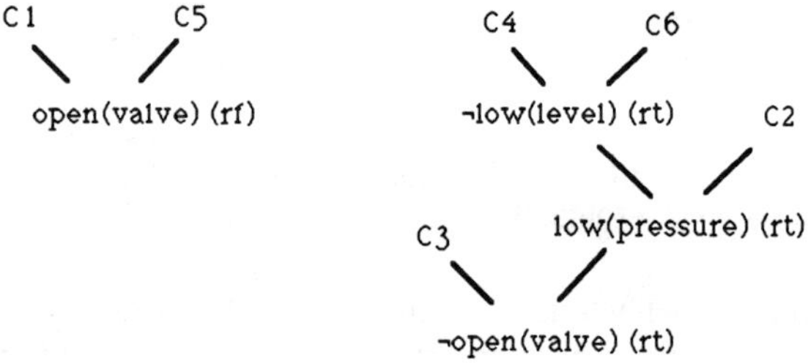

This impasse may be resolved since ¬open(value) is deduced with rough truth value rt which is a higher truth value than that with which the contradictory conclusion open(valve) is reached. The system can thus conclude that the valve should not be opened.

ROBUST REASONING IN RULE BASED SYSTEMS

For most practical purposes, knowledge-based systems are rule-based, with knowledge encoded in the form of 'if...then...' rules. In this section we consider the use of rough valued logic in such systems. In many domains detailed numerical estimates of the certainty of rules and facts may be impossible to obtain, and the reasoning mechanism adopted must be capable of dealing with vague estimates. Especially important is the robustness of the mechanism— its ability to deal with rules and facts whose certainty is unknown. In this section we analyse the robustness of rough valued logic in the context of rule based reasoning.

Forward chaining

The knowledge base of a typical rule-based system consists of a series of rules of the form 'if p then q' with a certainty value attached to each. In forward chaining inference starts with one or more facts, also with an associated certainty, which match the antecedents of particular rules. These rules are fired to obtain their consequents, with the certainty of the consequent being determined by a combination of the certainties of rule and antecedent, and the consequents used to fire more rules. This process continues until there are no facts that match the antecedents of unfired rules, or the goal fact has been deduced.

If we assume that rules of the form 'if p then q' are translated by use of material implication into logical statements of the form $p \rightarrow q$, then the mechanism of forward chaining is the rule of modus ponens (11). This allows us to establish when consequent values of unknown certainty will be generated, taking the lower bound on the value of q:

p	$p \rightarrow q$	q
t	u	u
rt	u	u
u	u	u
rf	u	rf
f	u	f

p	$p \rightarrow q$	q
u	t	u
u	rt	u
u	u	u
u	rf	rf
u	f	f

No combinations of values of antecedent and rule other than those shown, have either antecedent, consequent, or rule valued as unknown.The truth tables show that the value of the consequent can be determined if the value of the rule and the antecedent is given, or, if one has an unknown value the other has the value false (f) or roughly false (rf).

5.2 Backward chaining

A similar analysis may be performed for backward chaining. Here we are interested in determining the antecedent of a rule from the rule and its consequent. Reasoning proceeds from the goal, and continues until a known fact is identified as the antecedent of a rule that must be fired in order to generate the goal. Using the pattern for rough modus tollens (12) we obtain a similar truth tables to those above for the lowest bound on the value of p:

¬p	p → q	¬q
t	u	u
rt	u	u
u	u	u
rf	u	rf
f	u	f

¬p	p → q	¬q
u	t	u
u	rt	u
u	u	u
u	rf	rf
u	f	f

Once again, no combinations of values of antecedent and rule other than those shown, have either antecedent, consequent, or rule valued as unknown.The tables show that a fact of known value may be deduced from a fact and a rule of known value, or from a fact of unknown value and a rule that is true (t) or roughly true (rt), or from a rule of unknown value and a fact that is false or roughly false. In both these latter cases, the deuced fact has value false or roughly false.

Thus, in both forward and backward chaining, uncertainty can be absorbed by the logic at the cost of reducing the certainty value of the facts deduced. In situations in which the certainty of conclusions is secondary to the need to continue to operate in the face of degraded information such behaviour will be an advantage.

CONCLUSION

We have presented a symbolically quantified logic for reasoning under uncertainty that is based upon the concept of rough sets. This mathematical model provides a simple yet sound basis for a robust reasoning system. We have supplied rules of inference analogous to those of classical logic, and shown how they might be used by a reasoning system to determine the best action under conditions of uncertain knowledge. An analysis of the robustness of the logic in rule based reasoning has also been presented.

REFERENCES

1. Fariñas del Cerro, L. and Orlowska, E. "DAL - a logic for data analysis", *Theoretical Comp. Sci.* **36**, 251–264. 1985.
2. Orlowska, E. and Pawlak, Z. "Expressive power of knowledge representation systems", *International Journal of Man-Machine Studies*, **20**, 485–500, 1984.
3. Parsons, S., Kubat, M. and Dohnal, M. "A rough set approach to reasoning under uncertainty", *Technical Report*, Dept. Electronic Engineering, Queen Mary and Westfield College, 1991.
4. Pawlak, Z. "Rough Sets", *International Journal of Information and Computer Sciences*, **11**, 341–356, 1982.
5. Saffiotti, A. "An AI view of the treatment of uncertainty", *The Knowledge Engineering Review*, **2**, 75–97 1987.
6. Wong, S. K. M., Ziarko W. and Li Ye, R. "Comparison of rough-set and statistical methods in inductive learning", *International Journal of Man-Machine Studies*, **24**, 53–72, 1986.

A Neural Network Architecture For Three-Dimensional Manufacturing Parts Recognition

Mark B. Lynch

Department of Mechanical and Industrial Engineering

Louisiana Tech University

Ruston, Louisiana 71272-0046

ABSTRACT

Modern manufacturing is becoming more dependent on machine vision to provide the needed flexibility for automating manufacturing processes. Machine vision is being used in a variety of manufacturing tasks including the recognition, location, and orientation of manufactured parts, robotics control, and inspection. Neural networks provide new approaches for some of the machine vision processing required. This paper proposes a hybrid architecture composed of multiple neural networks and conventional programming to perform three-dimensional object recognition. The system acquires and processes images to produce a three-dimensional model of the objects found within the field of view of the cameras, and uses three-dimensional information for input into an object classifier. Stereoscopic vision techniques are used to create the three-dimensional model of the objects. Stereoscopic vision allows for a simple setup with minimal lighting requirements. A neural network approach is proposed to take advantage of the parallelism and robustness provided by artificial neural networks. Neural networks are used in stereoscopic point matching, 2-D to 3-D point conversion, and part recognition. Conventional programming is used to perform functions not implemented with neural networks. The architecture is presented along with a description of the system development.

INTRODUCTION

The machine vision field attempts to understand the surrounding environment through the acquisition and processing of images. Even with the earliest machine vision research the complexity of image understanding was seen. A fundamental problem with vision systems is the ambiguity created due to the projection process [1]. The complexity is a result of the visual task being under constrained, so there is not enough information present to uniquely solve the problem. Over the years many techniques have been developed to solve under constrained problems. The human visual system handles the under constrained nature of vision problems extremely well through use of many simple processing elements working together on the problem. This would indicate a distributed approach, such as one provided through artificial neural networks, could be successful in designing a machine vision architecture for image understanding.

Machine Vision in Manufacturing

Manufacturing is an area with great potential for machine vision. With the requirements of continuous operations performed at high speeds, machine vision can provide a means for automating monotonous tasks as well as increasing manufacturing throughput. In order to a make significant impact on the factory floor machine vision systems must increase their capabilities and operate more robustly. Neural network technology is one area that can provide increased capability and robustness in the operation of machine vision systems.

Neural Networks in Machine Vision

The use of artificial neural network approaches to machine vision problems have become increasingly popular. The massive parallelism and robustness of artificial neural networks are very desirable characteristics for machine vision processing. The ability of neural networks to learn from training examples rather than requiring the derivation of complicated non-linear equations is well suited for machine vision applications where many examples of complicated relationships can be readily generated.

One of the biggest areas of use for neural networks in machine vision is in the area of classification[2]. The networks lend themselves well to the learning of the classification criteria. Neural networks are even available which provide on-line classification without the requirement of a training process.

Stereoscopic vision has also been an area of much research within the neural network community. Optimizing neural networks have been used to minimize the error in establishing image correspondence in stereoscopic images[3]. The results have been encouraging, but only the image correspondence problem of stereoscopic vision has been addressed. Neural networks can be successfully applied to other parts of the stereoscopic vision problem, as will be shown in the proposed architecture presented in the next section.

SYSTEM ARCHITECTURE

The goal of the system is three-dimensional recognition of manufactured parts. The system is designed to recognize parts regardless of their location and orientation within the scan volume. In order to obtain three-dimensional part information, stereoscopic images are obtained and processed. Neural network technology is implemented wherever practical. The system is a combination of data transformations that transforms a set of intensity images into a set of recognized objects. Figure 1 shows the transformations that take place during the processing. A pair of stereoscopic images are acquired and the edges are enhanced through conventional processing techniques. The edges of the objects are used in finding matching points of the stereoscopic images. Edges are easy to detect and should be found in both images, unless occlusion occurs. The edges from both images are extracted and then matched by imposing several constraints on the search space. After the edges are matched the pair of two-dimensional edge points are converted into a three-dimensional object point. The three-dimensional object points are collected to form a three-dimensional model of the object's surface. It is from this three-dimensional model that a feature vector is created to be fed into the object classifier which performs the object recognition of objects within the field of view of the cameras. The next sections discuss the details of the different processing.

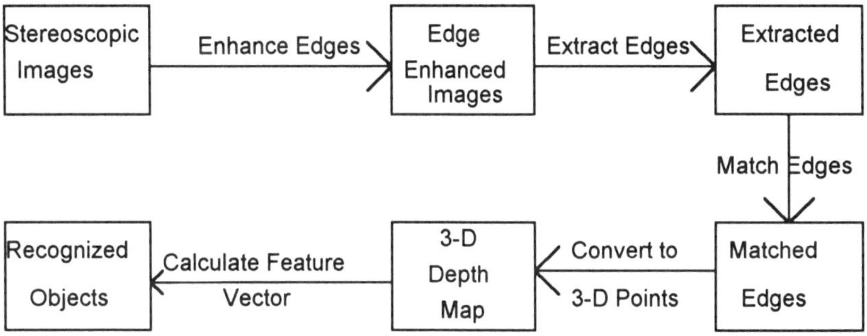

Figure 1. System Data Transformations

Data Acquisition

The stereoscopic images are acquired with two CCD cameras located over the scan volume and angled so they are pointing inward toward a common central area. This creates a shared field of view for the cameras as well as providing considerable resolution for calculating the disparity needed to obtain three-dimensional information from the stereoscopic images. The size of the images are 480 lines per image with each line being 640 pixels each.

Image Correspondence

The most critical part in obtaining three-dimensional information from stereoscopic images is the establishing of image correspondence. This entails finding the projection point in both images, corresponding to the same physical point on the object. Many factors complicate this process. One factor is that a point on the object may not appear in one or either of the images due to object occlusion. Another factor is that there are many potential matches between points in two images. Finding the correct match means eliminating many matches that seem reasonable. Fortunately, there are constraints that can be imposed to assist in solving the problem. The constraint most often imposed is that of epi-polar lines. The epi-polar line constraint corresponds to the camera geometry and states that a point found along a certain line in one image, if it can be seen in the second image, will be found along the corresponding epi-polar line. So with stereoscopic images, there is a set of corresponding epi-polar lines defined by the alignment of the cameras. Other constraints useful in the matching process include the direction of slope for the points and the disparity calculated from two potential matching points. The system design will determine a maximum height possible for detection, which can be translated into a maximum disparity possible for two matching points in the images. Any points with a disparity greater than the maximum possible cannot be a match.

The epi-polar constraint is implemented with the use of a backpropagation neural network as shown in figure 2. The network was trained by examples created from stereoscopic images. The advantage of training a neural network to learn the epi-polar line relationships is effects such as perspective distortions and system noise can be included through the training set. These effects can be very difficult to model through conventional programming approaches. The trained network thus provides an accurate estimate of the corresponding epi-polar line's vertical position given a point on an epi-polar line in the left image and an position along the horizontal axis from the right image.

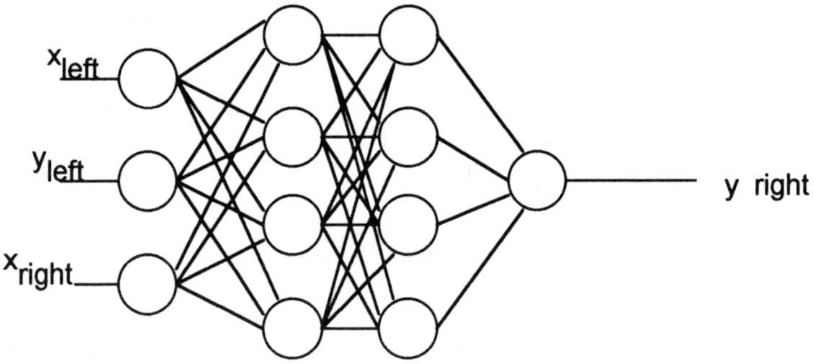

Figure 2. Epi-Polar Line Estimating Backpropagation Neural Network

Once the all the edges have been found along corresponding epi-polar lines the edge matching process utilizes the slope of the edge constraint and the maximum disparity constraint to match up the edges. Other constraints can be implemented should they be needed in the edge matching process.

Three-Dimensional Surface Estimation

With the edges matched, the next step in the process is the conversion of the matched edge points to the three-dimensional object points. The conversion of these image points to object points are dependent upon the relationship between the stereoscopic cameras and a arbitrarily established global coordinate system. The establishment of these relationships is considered the problem of system calibration. With using conventional techniques for system calibration elaborate models of the system geometry and optics must be created and verified. In the proposed architecture, neural network technology is used to learn the complicated relationships between the cameras' and global coordinate systems. The training set is created from images taken of calibration objects with known heights, placed in known locations. Figure 3 shows the matched point conversion neural network. The inputs to the network are the x and y coordinates of the matching image points, and the network is trained to output the global x, y, and z coordinates of the corresponding object point. Any non-linearity due to factors such as lens distortion or system noise are included through the training set. A backpropagation neural network can learn this complicated relationships needed for stereoscopic vision processing[4].

Once the matched image points are converted to global coordinates, they are accumulated onto a standard grid for the creation of the object's three-dimensional model. Information for the recognition of the object is obtained from the three-dimensional model.

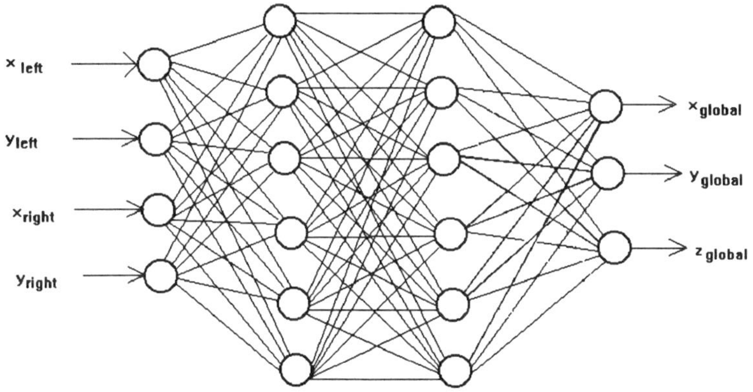

Figure 3. Matched Point Conversion Neural Network

Object Recognition

The object recognition process is completed through the creation of a feature vector that is fed into an ART2 neural network[5]. The ART2 network allows for the on-line recognition of objects and does not require a training phase. The network accepts analog input patterns and compares the input to previously stored patterns. If the input pattern matches an already stored pattern the system recognizes the object from the previous exposure. Should the input pattern not match an existing pattern the network stores the new pattern and thus learns a new object. The ART2 network provides the capability of selecting the level at which the input pattern and a stored pattern must match through a parameter called the vigilance factor. The higher the vigilance factor the closer the input pattern must be to the stored pattern. Figure 4 shows the architecture of the ART2 network used for object recognition in the system.

The feature vector, which provides the input to the recognition network, is created by using three-dimensional moment calculations[6]. The use of three-dimensional moments should provides feature vectors that are invariant to object position and orientation within the scan volume. This combination of three-dimensional moments and neural network technology provides a powerful solution to the object recognition problem[7]. Objects are recognized on-line without previous training of the network, and the system can have various recognition resolutions through the setting of the ART2 network's vigilance factor.

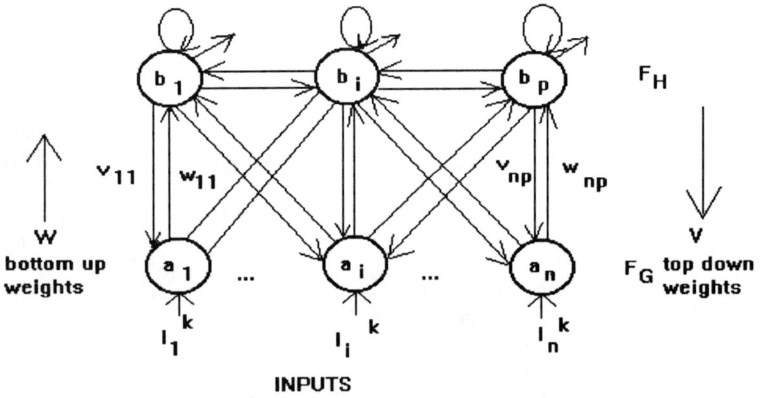

Figure 4. Object Recognition ART2 Neural Network

CONCLUSION

The hybrid neural network architecture presented provides a novel approach to solving the three-dimensional object recognition problem using stereoscopic data. The backpropagation networks are capable of learning the complicated relationships needed for estimation of the epi-polar lines and converting between matched image points to global object points. The training of the networks allows the inclusion of difficult to quantify effects such as system noise and perspective distortions.

The presented architecture also provides for on-line object classification capabilities through use of the ART2 neural network paradigm. No training cycle is required for this network and the resolution of the recognition can be controlled through a parameter known as the vigilance factor.

REFERENCES

1. H. Wechsler, Computational Vision, Academic Press, San Diego, CA, 1990
2. M. Roth, "Survey of Neural Network Technology for Automatic Target Recognition," IEEE Transactions on Neural Networks, vol. 1, no. 1, pp. 28-43, 1990
3. N. M. Nasrabadi, and C. Y. Choo (1992), "Hopfield Network for Stereo Vision Correspondence," IEEE Transactions on Neural Networks, vol. 3, no. 1, pp. 5-13
4. M. Lynch and C. Dagli , "Backpropagation neural network for stereoscopic vision calibration,", Proceedings of Intelligent Robots and Computer Vision X: Algorithms and Techniques,, pp. 289-298 , November 1991, Boston, MA
5. G. Carpenter and S. Grossberg, "ART2: self-organization of stabel category recognition codes for analog input patterns," Applied Optics, vol. 26, no. 23, pp. 4919-4930, 1987
6. C. H. Lo and H. J. Don, "Pattern Recognition Using 3-D Moments," International Conference on Neural Networks, June 17-21, 1990, San Diego Ca., vol. 1, pp. 540-544
7. M. Lynch,"Steroescopic neuro-vision for three-dimensional object recognition in a manufacturing environment," Ph.D. dissertation, University of Missouri-Rolla, 1992

Design Methodology for NASREM Software Development

Dr. Ronald Lumia
Robot Systems Division
National Institute of Standards and Technology
Bldg 220, Rm B-127
Gaithersburg, Maryland 20899

Abstract

Using a functional architecture such as the NASA/NIST Standard Reference Model for Telerobot Control System Architecture (NASREM) when implementing specific applications is helpful because much of the work resides in the infrastructure and the infrastructure is the same for all applications. Rather than recreating this infrastructure, the approach is to develop the additional 20% of the code which tailors the infrastructure for the specific application. This paper describes the process by which a system based on NASREM is developed.

1. Introduction

The Intelligent Controls Group of the Robot Systems Division at NIST has been implementing the NASREM architecture for several years. While the implementation was in support of NASA's Flight Telerobotic Servicer, the approach was to build a generic robot control system testbed. NASREM is a hierarchical system with six levels which is described in greater detail in [1-7].

Since the NASREM architecture has been developed in a technology independent manner, there are two types of modifications which must be allowed. First, the architecture must be able to support different application areas. The NASREM architecture has been use for underwater vehicles [8], mining vehicles [9], autonomous land vehicles [10], as a basis for the Next Generation Controller, as well as for its original space application. Second, the architecture must facilitate the integration of new sensors even if the application does not change so that the implementation may evolve with technology. This paper describes how a system is developed using NASREM as the functional architecture, i.e., a NASREM methodology.

The paper is organized as follows. First, the basic tenets of a NASREM implementation are given. This is followed by a description of the system development methodology. Then, a NASREM robot control system is described which is an example of using the methodology.

2. NASREM Tenets

NASREM provides fundamental real-time control system analysis, design and implementation strategies and heuristics that we will term tenets. We will use the word tenet, here, to mean guidelines and rules of thumb which characterize the philosophy of the NASREM methodology. The NASREM Methodology is based on the following empirically established method tenets:

task (or process) oriented decomposition

controller node building blocks, a controller node C is a 4-tuple C = (SP, WM, BG, HI) consisting of a sensor processing component SP, a world modeling component WM, a behavior generation control component BG and a human interface component HI. These functions are assumed to exist in each controller node.

each controller node C has one supervisor at a time and only seven + or - two subordinates

human interface is possible at all controller nodes

system controller organized as a hierarchical collection of controller nodes C.

control hierarchy organized around tasks top-down, and equipment bottom-up

order of magnitude differences partition controller nodes into adjacent hierarchical levels

design emphasis on concurrency vis a vis real-time processing

3. System Development Methodology

The process of developing a NASREM system is shown in Figure 1. A more complete description of the entire system development process is provided in [11]. The process starts with a definition of the task goals, and the performance requirements of the control system in order to develop a complete problem description task scenarios, and expectations.

In Step 2, the system developer gathers knowledge of the problem domain to define the information which is fundamental to the problem. While Step 1 was concerned with the goals for the system, Step 2 is concerned with the tools which will help achieve these goals. Consequently, the processes must be defined. This clearly depends on the specific equipment and problem but could include information such as the object and workspace geometry, machine and load kinematics, dynamics and coordinate systems, etc.

A systems analysis is performed in Step 3 to conceptualize the application in terms of Resource Management (NASREM Hierarchy), Data Management, Human Interface Management, and Communications Management requirements. The architectural design analysis of the Resource Management requirements defines and models the set of resources bottom up. Then, the controller framework, i.e., NASREM hierarchy is built to establish the relationship between resources which exhibit common spatial and temporal characteristics. This is followed by a mapping of the capabilities into NASREM hierarchial control nodes. Capabilities may be distributed across individual controller boundaries. The NASREM systems analysis continues by looking at the data management. One defines a set of objects which can be acted upon and lists their relevant attributes in terms of their hierarchical structure, e.g., points, linear features, surfaces, objects and groups.

Each application needs a vocabulary which is typically unique to the application. In Step 4, NASREM task analysis, the task tree vocabulary for each level of the hierarchy is defined. This includes the tasks, the task frames, procedures, etc. This step balances the system goals with the capabilities of the system. One iteratively refines the system controller and/or the task descriptions based on either deficiencies in the task objectives, resource capabilities, information modeling, or communication interfaces until the goals of the tasks are met by the specified system.

In Step 5, software modules are coded and tested. The code for each controller node is developed using generic NASREM coding templates in a bottom up fashion. Simulators are developed to drive each controller in a closed-loop fashion and the human interface and display simulators are developed as required. Each software module is tested and measured to determine the expected performance of each controller node. This may involve subdividing controller nodes into concurrent processes to meet real-time performance goals.

Step 6 integrates and tests the controller nodes. One maps the concurrent processes onto actual computer hardware and assesses the performance. First, controller nodes have only their communication interfaces. When these interfaces are verified, code is added module by module to test controller node functionality. As more modules are integrated into the system, it is possible that more controller nodes may be necessary to meet the design criteria.

The last step is maintenance. The above six steps are iterated in order to fix, improve, or extend the controller nodes and the application system.

4. System Development Methodology for a NASREM Robot Controller

4.1. Description of the General Approach for Real-time System Development

This section will concentrate on steps 5 and 6 of Figure 1, from code development toward integrating the code into the real-time robot control system. First, atomic units are defined. These atomic units are the smallest chunk of software, which if broken into smaller chunks, would result in a less efficient implementation. For example, robot inertia compensation should be done for all joints together. If each degree of freedom is implemented in parallel, the communication between the degrees of freedom would take more time than that gained by the parallelization of the processes. The implementation of the atomic unit, which is called a process, has five properties:

- Continuous cyclic execution
- Read-compute-write execution cycle
- Concurrency
- Interfaces through global data system
- Activation/inactivation

Once each process has been programmed and debugged as well as possible on the host system, it is cross compiled for use in the target system. Then, each process is assigned to a processor. For economic reasons, there will be fewer processors than processes. The process to processor assignment is critical to achieve good runtime performance. This assignment is revised often to hone the real-time performance of the system. Consequently, it should be easy to modify the assignments. At this point in the design process, load modules for each processor are available.

The load modules are downloaded to the computers in the target system. Depending on what is downloaded, the goal may be to obtain timing measurements for a particular algorithm, determine the performance of the algorithm during a task, or measure the performance of the entire system. At this point the testbed can be used for any research and development needs of the investigator.

4.2. Atomic Unit to Processor Assignment for the Real-time System

The result of process to processor assignment and the resulting system will be illustrated by looking at the lowest NASREM level, the Servo level. Figure 2 shows the atomic units associated with the Servo level. The square boxes depict the processes while the ovals denote the data buffers which communicate through the global data system. A diagram similar to this is developed for each part of the system so that all of the processes are identified. Then, the system designer assigns each process to a particular processor for the real-time execution phase. Figure 3 shows the current process to processor assignment for the current NIST implementation of NASREM. The manipulation backplane is shown in the upper half of the diagram. Seven processors perform the Prim and Servo functions associated with manipulation. One interface card connects the manipulation backplane to the RRC controller (power electronics for the robot). Also, two BIT-3 parallel interfaces reside in the backplane, one to connect to the SUN host, the other to connect the Manipulation and Perception backplanes. The lower half of Figure 3 shows the process to processor assignment for the perception backplane. With the current configuration, the manipulation Servo loop, which is the most time critical part of the system, supports a 2.5 ms sampling rate and a 5 ms loop time for the critical path of most control algorithms.

5. Conclusion

Using a functional architecture helps to improve efficiency in developing and integrating a system for a particular application. The contribution of this paper is to outline a methodology for system development in a way that makes it possible to implement a system.

6. References

[1] Albus, J.S., McCain, H.G., Lumia, R., "NASA/NBS Standard Reference Model Telerobot Control System Architecture (NASREM)," NIST Tech. Note 1235, NIST, Gaithersburg, MD, July, 1987.

[2] Fiala, J., "Manipulator Servo Level Task Decomposition," NIST Technical Note 1255, NIST, Gaithersburg, MD, October, 1988.

[3] Kelmar, L. "Manipulator Servo Level World Modeling," NIST Tech. Note 1258, NIST, Gaithersburg, MD, December, 1989.

[4] Fiala, J., "Note on NASREM Implementation," NISTIR 89-4215, NIST, Gaithersburg, MD, December, 1989.

[5] Chaconas, K. J., Nashman, M., "Visual Perception Processing in a Hierarchical Control System: Level 1," NIST Tech. Note 1260, June, 1988.

[6] Wavering, A. "Manipulator Primitive Level Task Decomposition," NIST Technical Note 1256, NIST, Gaithersburg, MD, October, 1988.

[7] Kelmar, L. "Manipulator Primitive Level World Modeling," NIST Tech. Note 1273, NIST, Gaithersburg, MD, December, 1989.

[8] Albus, J.S., "System Description and Design Architecture for Multiple Autonomous Undersea Vehicles," NIST Technical Note 1251, NIST, Gaithersburg, MD, September, 1988.

[9] Huang, H.M., "Hierarchical Real-Time Control System for Use with Coal Mining Automation," Fourth Conference on the Use of Computers in the Coal Industry, Morgantown, WV, June 1990.

[10] Szabo, S., Scott, H.A., Murphy, K.N., Legowik, S.A., "Control System Architecture for a Remotely Operated Unmanned Land Vehicle," Proceedings of the 5th IEEE International Symposium on Intelligent Control, Philadelphia, PA, September, 1990.

[11] Michaloski, J., "Elements of Real-time Intelligent Control System Design," NIST Technical Note, NIST, Gaithersburg, MD, March, 1992

1. Develop problem description, task scenarios and expectations.

2. Gather domain knowledge.

3. NASREM Systems Analysis. Conceptualize the application in terms of Resource Management (NASREM Hierarchy), Data Management, Human Interface Management, and Communications Management requirements

4. NASREM Task Analysis. Perform Task Decomposition vis a vis NASREM Systems Analysis.

5. Code and Test Software Modules.

6. Integrate and Test Controller Nodes

7. Maintain

Figure 1. NASREM System Development Steps

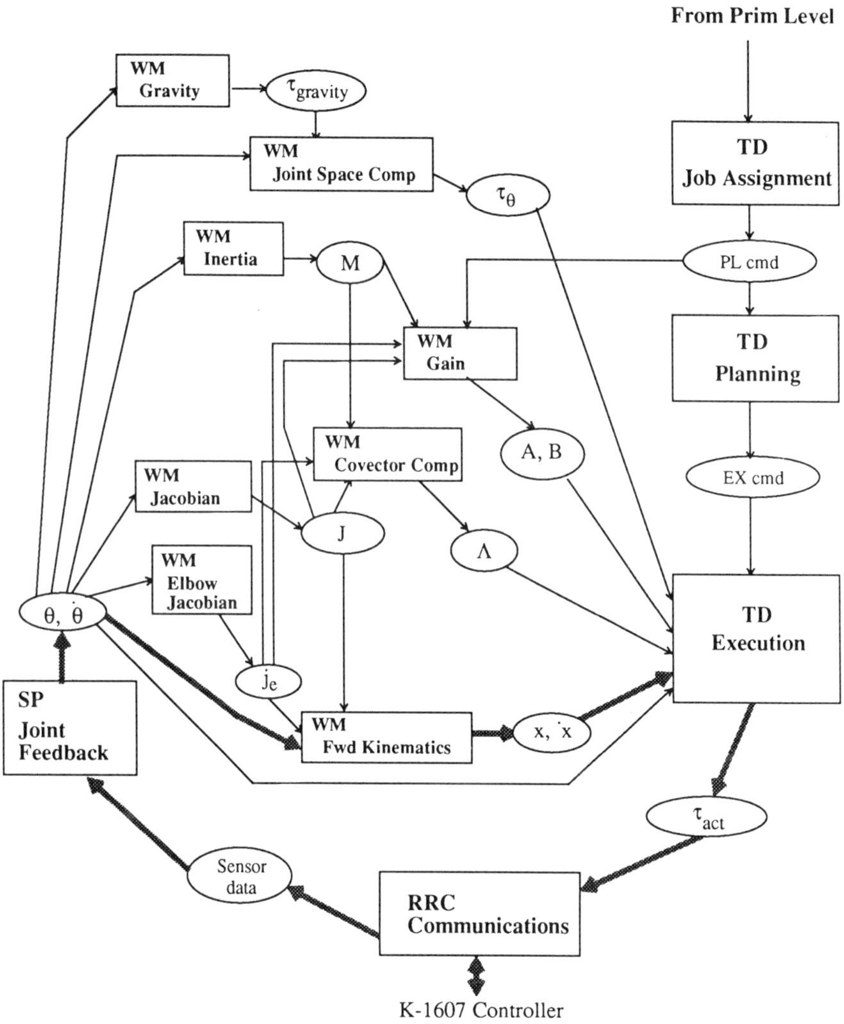

Figure 2. Servo Level Boxes Representing Atomic Units

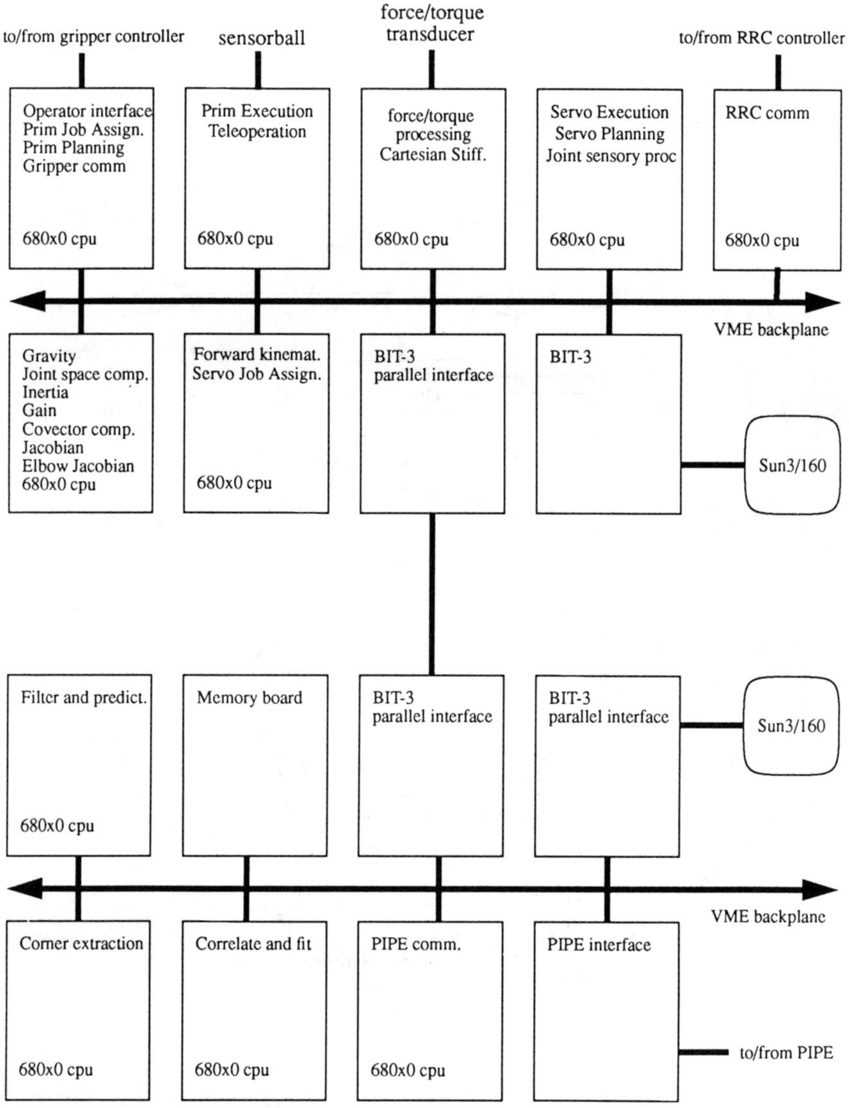

Figure 3. Process to Processor Assignment for the NIST NASREM Implementation

Effect of Thruster Dynamics and Advanced Control of Underwater Robotic Vehicles using Neural Network

J. Yuh and Kishan V. Gonugunta
Autonomous Systems Laboratory
Department of Mechanical Engineering
University of Hawaii
Honolulu, Hawaii 96822

Abstract

In this paper, we present a neural network controller for underwater robotic vehicles including the effects of thruster dynamics and saturation. It has been observed, in practice, that dominant vehicle dynamics are varying with the vehicle velocity and effect of thruster dynamics becomes significant at low velocity in the vehicle control system. It is desired to have a robust control system with respect to the vehicle velocity. When it is necessary to operate the vehicle at low velocity such as stationkeeping and hovering, the effect of thruster dynamics must be considered in the vehicle control system design. This paper discusses the effect of thruster dynamics and describes a multilayered neural network with parallel recursive adaptation algorithm. Results of computer simulation for the pitch and altitude vehicle motion indicate the effectiveness of the presented control system for the varying vehicle dynamics, especially the effects of thruster dynamics and saturation.

Introduction

Underwater Robotic Vehicles (URVs) are an important tool for undersea tasks, such as surveying, inspection, and repair, because they have greater speed, endurance and depth capability then human divers. URVs are performing several challenging functions. During the working mode, the vehicle maintains a steady position to perform a particular task (station-keeping problem); during the navigation mode, the vehicle follows a prescribed trajectory in search of an object whose location has already been determined (tracking problem). With the increased utilization of the vehicles in subsea applications, the development of autonomous URVs becomes highly desirable to enhance operator efficiency. The control system is one of the most critical subsystems that can increase autonomy of the vehicle. Control systems in current URVs are immature compared to on-land systems. The vehicles are operated by either preprogrammed controllers for repeating routine works or a human operator.

The difficult control problem with URVs stems from the following facts: the vehicle has nonlinear dynamic behavior; hydrodynamics of the vehicle are poorly known; the vehicle usually has comparable velocities along all three axes; the forces and moments generated by high density fluid motion are significant; the added mass aspect of the dense ocean medium results in time-delay control response characteristics; and a variety of unmeasurable disturbances are present due to multi-directional currents. It was also observed, in practice (Yoerger et al. 1990), that the thruster dynamics are dependent on propeller speed and the effect of thruster dynamics to the vehicle motion becomes significant at low velocity.

Most control techniques (Yoerger and Slotine 1985, Goheen et. al 1986, Kazerooni and Sheridan 1982, Yuh 1990) for underwater vehicles in the literature have addressed the control problem of the vehicle without considering the effect of thruster dynamics. The thruster dynamics usually have much smaller time constant than the vehicle's main dynamics; and therefore, the thruster dynamics can be ignored in the vehicle control system design. However, the effect of thruster dynamics becomes significant at low velocity and must be considered for the control system design. Yoerger et al. (1990) explored the effect of thruster dynamics whose time constant varies with propeller speed. They proposed adaptive sliding controller to overcome the effect of thruster dynamics at low velocity. Fjellstad et al. (1992) proposed a direct model reference adaptive controller using the reference model decomposition technique to compensate for the thruster dynamics. In their study, they assumed that time constant of thruster dynamics does not vary with propeller speed.

In this paper, we propose a neural net control system for URVs. The learning and adaptation capability of neural networks has attracted its use in underwater vehicle control (Yuh and Lakshmi 1992). We have developed a parallel recursive adaptation algorithm that can be considered as a general form of the error backpropagation algorithm, and an on-line learning approach was also developed to avoid a lengthy off-line training period. Other on-line and off-line control approaches in neural network control application are discussed in Yuh (1992). The contribution of this paper is a neural network control approach that is justified by simulation study and attracts further theoretical investigation.

This paper is organized as follows: the effect of thruster dynamics in controlling the vehicle is first discussed with a single degree-of-freedom (d.o.f) vehicle system. A neural net control system using a parallel recursive adaptation algorithm (Yuh 1992) is described, and then, the results of computer simulation are discussed. Results of the computer simulation for the altitude and pitch vehicle motion show that the presented control system can provide high performance in the presence of the effect of thruster dynamics and saturation.

Effect of Thruster Dynamics

For simplicity of analysis, a single d.o.f. vehicle model is first considered:

$$M\dot{V}(t) = -C_D V(t)|V(t)| + F_T(t) \tag{1}$$

where M is the vehicle mass plus added mass, V(t) is the vehicle velocity, F_T is the thruster force, and C_D is the drag parameter. A lumped parameter dynamic model of the thruster can be described as

$$\dot{\Omega}(t) = \beta\tau(t) - \alpha\Omega(t)|\Omega(t)| \tag{2}$$

$$F_T(t) = C_t\Omega(t)|\Omega(t)| \tag{3}$$

where Ω is the propeller angular velocity, τ is the input torque, α and β are constant parameters, and C_t is a proportionality constant. For convenience, we will call Eq. 1 the vehicle system equation and Eqs. 2 & 3 the thruster system equations. Fig. 1 shows overall diagram of the vehicle system with a simple state feedback loop where K is a constant feedback gain, and quasi-linear models of thruster system $G_T(s)$ and the vehicle system $G_V(s)$ are:

$$G_T(s) = \frac{F_T(s)}{\tau(s)} = \frac{C_t\beta|\Omega|}{S + \alpha|\Omega|}, \quad G_V(s) = \frac{V(s)}{F_T(s)} = \frac{1/M}{S + C_D|V|/M} \tag{4}$$

When the thruster dynamics can be ignored, $G_T(s)$ can be simplified to $G_{Ts}(s) =$

$C_t\beta/\alpha$.

In this analysis, we considered Jason Jr. vehicle model whoes system parameters were obtained through experiments by Yoerger et al. 1990. Figure 2 shows the pole locations of the overall system with and without thruster dynamics for high and low velocity. Since the pole of thruster is far away from the pole of vehicle at high velocity, the effect of thruster dynamics can be ignored, i.e. overall system can be represented by a first-order system $G_{Ts}(s)G_V(s)$. However, at low velocity, poles of the vehicle and thruster are very close to each other, and they are close to zero. Consequently, the effect of thruster dynamics cannot be ignored, and the overall system must be represented by a second-order system $G_T(s)G_V(s)$.

Figure 3 shows the root-locus for a simple feedback gain K. The feedback gain K value is determined for the desired performance with a first-order system model ignoring the thruster dynamics and the corresponding root locations of a second-order model to that K value are shown for low and high velocities. For low velocity, the roots are located along the vertical line in parallel and close to the imaginary axis. It will result in oscillation and the desired performance cannot be achieved for any K value. For high velocity, the roots can be located in the asymptotically stable region and desired performace may be obtained even for the K value that was determined from the first-order system model ignoring the thruster dynamics. During the vehicle operation, the vehicle velocity may vary from zero to its maximum velocity. It means that the number of the dominant poles and dynamics of the overall system are varying. Especially, during the hovering or stationkeeping of the vehicle, the effect of thruster dynamics becomes significant. Therefore, it is necessary to design a robust vehicle control system with respect to not only the system's parametric uncertainties but also the system's structural changes.

Neural Net Controller

A neural net controller considered here is shown in Figure 4. The neural network has one input layer, one output layer, and multiple hidden layers. Each layer has multiple neurons. Inputs to neurons of the input layer are position errors, velocity errors, and arbitrary constants. One neuron of each hidden layer has also arbitrary constant input signal. Neurons of the output-layer output the control input signals to the vehicle. During the feedforward process, the input signals propagate from the input layer to the output layer, being multiplied by weights between neurons of two layers in sequel and being modified at each neuron by activation functions. We consider a sigmoid activation function $f(x) = (e^{-x}-e^{x})/(e^{-x}+e^{x})$ for neurons in the hidden layers and a linear activation function $f(x) = x$ for neurons in the input and output layers. During the adaptation and learning process, the control error signals are estimated by the critic equation and then used in the adaptation algorithm to adjust the weights of the network. Each process is described in the following.

Feedforward Process

The output of the i-th neuron in the m-th layer is computed by

$$q_i^m(t) = f(W_i^m(t)^T q^{m-1}(t)) \quad (5)$$

where W_i^m is a weight vector between the i-th neuron of the m-th layer and the neurons of the (m-1)-th layer, q^{m-1} is an output vector of neurons in the (m-1)-th layer, and the superscript T implies the transpose of the matrix. The output vector of the input layer is

$$q^{input}(t) = [e^T(t)\ \ d]^T \tag{6}$$

where d is arbitrary constant and error vector e includes the position error and velocity error. The output vector of the output layer is the thruster control input U:

$$q^{output}(t) = \tau(t) \tag{7}$$

Adaptation Algorithm

A recursive algorithm for updating $W_i(t)$ can be written as follows:

$$W_i(t) = W_i(t-1) + \rho F_i(t)\ \Psi_i(t)\ \epsilon(t) \tag{8}$$

where W_i is a weight vector associated with the i-th neuron in the network, ρ is a positive constant known as a learning rate. $\Psi_i(t,W)$ is the gradient of U(t,W) with respect to W_i, and ϵ is the network output error vector. The derivation of Eq. 8 can be found in Yuh (1992). $F_i(t)$ is updated using the following equation based on constant trace method:

$$\bar{F}_i(t+1) = F_i(t) - F_i(t)\Psi_i(t+1)\ [\ \lambda\ I + \Psi_i^T(t+1)\ F_i(t)\ \Psi_i(t+1)\ ]^{-1}\ \Psi_i^T(t+1)\ F_i(t) \tag{9}$$

$$F_i(t+1) = \frac{\alpha\ \bar{F}_i\ (t+1)}{trace[\bar{F}_i(t+1)]} \tag{10}$$

where α is a positive constant trace of F_i and $0 < \lambda \leq 1$.

Critic Equation

Unlike the classic backpropagation algorithm (Rumelhart et al. 1986), the desired mapping of inputs to outputs of the network is unknown a priori. (i.e. the desired values of its outputs are not known a priori.) Therefore, the network can be trained only indirectly through a critic function (Widrow et al. 1987). The error signal ϵ is estimated at each time step by the following critic equation:

$$\epsilon(t) = C_e\ [e_V(t) + \beta e_X(t)] \tag{11}$$

where $e_V(t)=V_d(t)-V(t)$, $e_X(t)=X_d(t)-X(t)$, V and X are the actual velocity and position vectors, V_d and X_d are the desired velocity and position vectors, constant β is chosen for a reasonable compensation of position errors, and C_e is a positive constant matrix with appropriate dimension. Instead of training the network with a teacher (predetermined control law or human operator), a network is trained by the critic signal based on the system performance.

Simulation and Discussion

For the purpose of comparison, we first implemented the presented control system for the single d.o.f. model (Eqs. 1-3) used by Yoerger et al. (1990) who presented results by the lead compensator (Figs. 15 & 16 of Yoerger et al.), the pole cancellation technique (Figs. 17 & 18 of Yoerger et al.), and the adaptive sliding controller (Fig. 19 of Yoerger et al.). Their results are not repeated here; however, our results shown in Fig. 5 were obtained for the same range of the vehicle position.

Results show that the presented control system can provide good tracking performance for a wide range of the vehicle velocity after the first cycle where the adaptation occures. It is noted that the presented control system does not require any information about the vehicle and thruster system.

Results of simulation study have encouraged further research on the subject. Theoretical investigation of some crucial issues such as global stability has not been successful in the area of neural network because of the nonlinear nature of neural networks and a lack of applicable theories in neural networks. Furthermore, the real-time implementation of the neural network control system may not be realized by the current non-parallel computing hardware due to its heavy computation requirement in parallel structure. It is believed that the fast development in computer industries will eventually provide a solution at a low cost for its real-time implementation. Theoretical as well as experimental studies on the presented approach are being conducted to address issues such as global stabilty as current research.

Acknowledgment

This work was partially supported by (R/OE-13) the University of Hawaii Sea Grant College Program under Institutional Grant No. NA89AA-D-SG063 from the NOAA Office of Sea Grant, Department of Commerce and partially by National Science Foundation under contract No. BCS91-57896. This is a Sea Grant Publication UNIHI-SEAGRANT-CP-92-03.

References

Fjellstad, O.E., T.I. Fossen, and O. Egeland 1992, Adaptive Control of ROVs with Actuator Dynamics and Saturation, International Offshore and Polar Engineering, San Francisco, June, In press.

Goheen, K.R., E.R. Jefferys and D.R. Broome 1986. Robust Self-designing Control Underwater Vehicles, Proc.Conf. Offshore Mech. & Arctic Engr. Tokyo, Japan, pp. 31-38.

Kazerooni, H., and T. B. Sheridan 1982. Computer Simulation and Control of Underwater Vehicles, Report No. MITSG 82-19, Sea Grant, MIT, October.

Rumelhart, D.E., J.L. McClelland et al. 1986, Parallel Distributed Processing, MIT Press

Widrow, B., N.K. Gupta, and S. Maitra 1987, "Punish/Reward: Learning with a Critic in Adaptive Threshold Systems", IEEE Trans. Systems, Man and Cybernetics, Vol. SMC 3, No. 5, pp. 455-465

Yoerger, D.N., J.G. Cooke, and J.E. Slotine 1990. The Influence of Thruster Dynamics on Underwater Vehicle Behavior and Their Incorporation Into Control System Design, IEEE J. of Oceanic Engineering, Vol. OE-15, No. 3, pp. 167-178.

Yoerger, D.N., and J.E. Slotine 1985. Robust Trajectory Control of Underwater Vehicles, IEEE J. of Oceanic Engineering, V.OE-10, No. 4, pp.462-470.

Yuh, J. 1990. Modeling and Control of Underwater Robotic Vehicles, J. of IEEE Trans. on Systems, Man and Cybernetics, Vol. 20, No. 6, pp.1475-1482.

Yuh, J., R. Lakshmi 1992. An Intelligent Control System for Remotely Operated Vehicles, J. of IEEE Oceanic Engineering, In press.

Yuh, J. 1992. A Multilayered Neural Net Controller using Direct Learning Algorithm, International J. of Computers & Electrical Engineering, In press.

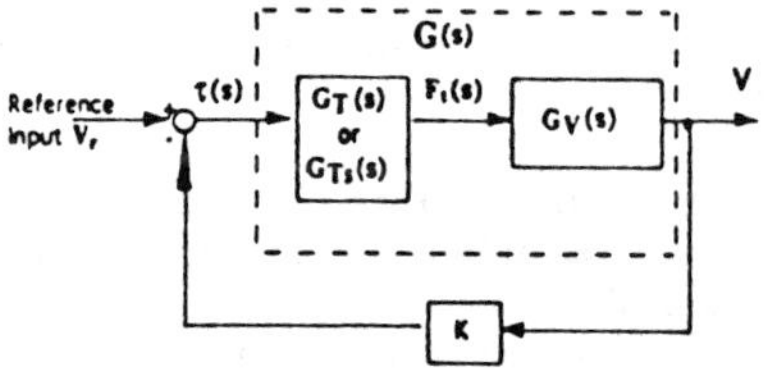

Fig.1: Vehicle dynamics

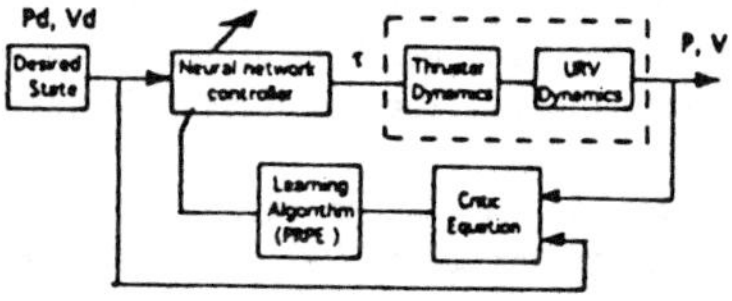

Fig.4: A neural net controller

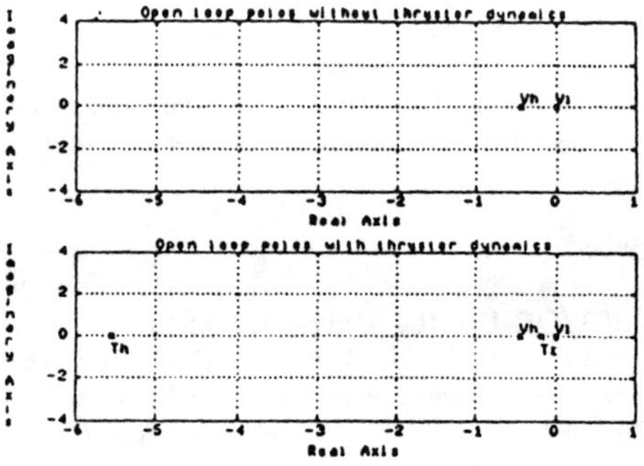

Fig.2: Open-loop poles of the vehicle system w/ & w/o thruster dynamics(h:high velocity,l: low vel)

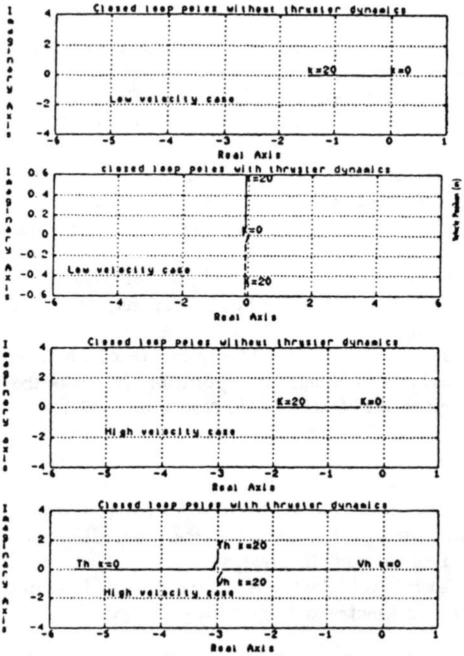

Fig.3: Closed-loop poles w/ & w/o thruster dynmaics for feedback gain K (V: vehicle pole, T:thruster pole)

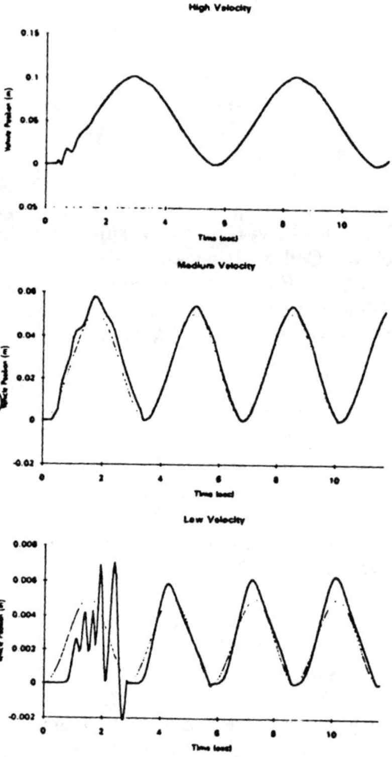

Fig. 5: Vehicle positions for high, med. & low velocities (.... desired position; ____ actual position)

INTELLIGENT SYSTEM CONTROLLER FOR REMOTE SYSTEMS

R. W. Harrigan
Sandia National Laboratories

ABSTRACT

The U.S. Department of Energy's Office of Technology Development (OTD) has sponsored the development of the Generic Intelligent System Controller (GISC) for application to the clean up of hazardous waste sites. Of primary interest to the OTD is the development of technologies which result in faster, safer, and cheaper cleanup of hazardous waste sites than possible using conventional approaches. An objective of the GISC development project is to achieve these goals by developing a modular robotics control approach which reduces the time and cost of development by allowing reuse of control system software and uses computer models to improve the safety of remote site cleanup while reducing the time and life cycle costs.

INTRODUCTION

The concept of a Generic Intelligent System Controller (GISC) resulted from a series of joint DOE/NIST workshops initiated in January 1990 [1,2] These workshops were held to determine what generalized approaches should be employed in the development of control system structures to support development of robotic systems for application to DOE waste cleanup problems. The goals of the workshops were to identify a system control environment which would

1. stimulate cooperative team based robotics system development,
2. facilitate the rapid fielding of integrated robot system demonstrations, and
3. stimulate common approaches to system control within the Robotics Technology Development Program (RTDP).

The consensus of the first Workshop was that modular open architectures to stimulate subsystem integration should be stressed. The second workshop identified the importance of multiprocessing computing environments for complex system control. The Unix operating system was identified as the operating system preferred for software development and VxWorks was identified as the preferred operating system for the real-time subsystem control environment. C and C++ were identified as the preferred programming languages. In addition, the use of graphics operator interfaces to simplify system programming was determined to be highly beneficial. Finally, the need to employ architectures and environments which could easily evolve to take advantage of improvements in computing and robotics technologies was stressed to prevent rapid obsolescence. It was felt to be very important to provide architectures which would both facilitate software development (thus the Unix based environments) and support the real-time multiprocessing computing environments (VME bus, MC68000 series single board CPUs) needed to support the development of complex robotic systems.

A GENERIC INTELLIGENT SYSTEM CONTROLLER APPROACH

The emphasis in the development of the GISC approach was functional not architectural. As a result, GISC is a strategy for integrating diverse subsystems. The emphasis, following the results of the NIST/DOE Workshops was on development of modularity and extensibility. There was an emphasis on using commercial technologies as much as possible and to extend these technologies to facilitate robotic system development. At all times during the development of the GISC approach, a goal was to produce a robotic system control environment which resulted in facilitating use of robotic technology by the site remediation technologists. As shown in Figure 1, GISC coordinates and integrates the operation of diverse subsystems to accomplish complex tasks. GISC is a modular approach for combining robots with sensors and computer models to provide robotic systems that are automatically programmed and much safer than conventional remote systems.

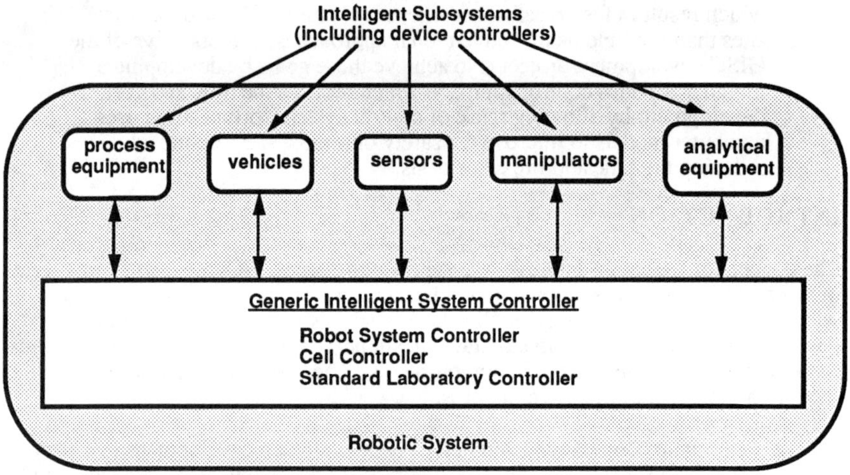

Figure 1
GISC Integrates Diverse Subsystems

GISC is termed an intelligent control system approach because computer models automate much of the system programming and condition the operator interfaces to facilitate correct decision making by a human supervisor. In addition, computer models monitor all operations (both computer planned and operator initiated) for safety. Algorithms evaluate the known environment represented by a World Model, plan collision free robot motions, and then automatically generate the robot command sequences necessary to execute the desired robot motions. Use of the computer models for error detection and recovery prevents unsafe actions which might, for example, result in collisions. The incorporation of computer models into GISC's approach to system control also enables the use of visualization technologies to provide intuitive graphic operator interfaces that allow even unskilled operators to safely program complex tasks.

Figure 2 recasts the diagram of Figure 1 in terms of the major functional blocks of GISC. As indicated, the GISC environment communicates with multiple intelligent subsystems through their respective controllers. Extensive use of computer models (labeled World Model in Figure 2) and sensors allows both operator directed task executions as well as automated operation. Direct operator control of individual subsystems is allowed only on rare occasions during emergency situations or during system development for debug purposes.

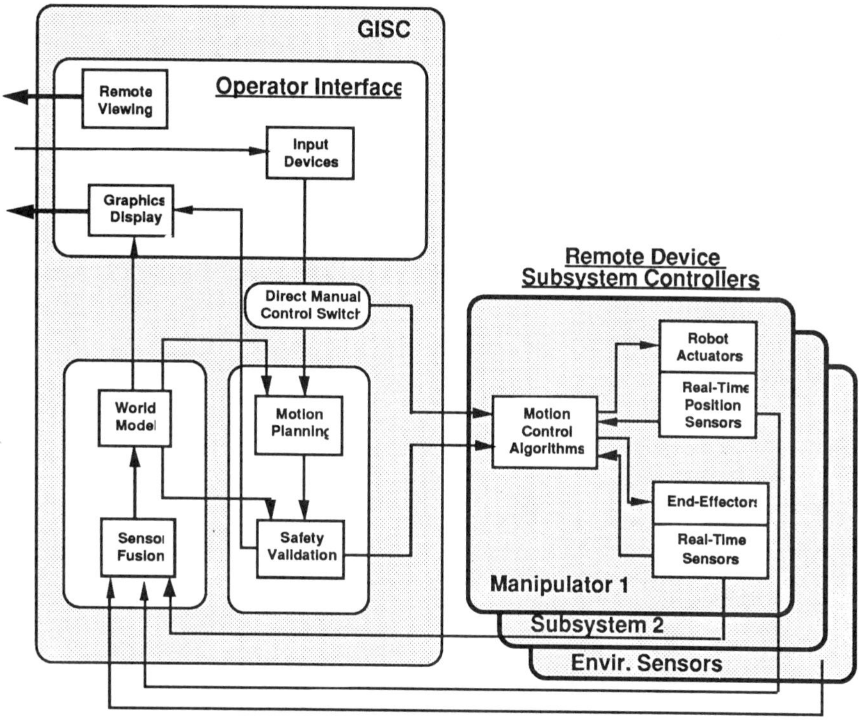

Figure 2

Generic Intelligent System Controller

The World Model is at the heart of GISC. The World Model contains the information about the robot system and its environment needed to monitor system operation and to plan the execution of tasks. The World Model is developed by combining engineering knowledge of the overall system including information from engineering drawings about known structures in the environment, robot end effectors and tools, knowledge of proper operating procedures, and knowledge of the robot system kinematics. While CAD information of the robot's environment together with the kinematic and dynamic models of the robot and other devices typically make up the major portion of the World Model, sensors also provide information to the World Model. Sensors both determine new information which is integrated into the World Model and provide updates to the World Model as operations are executed. The World Model is thus dynamically updated at all times based upon the most recent sensor information.

Visualization software technologies provide the operator of the robot system with animated graphic displays of the computer's knowledge base allowing the operator to verify that the World Model is correct and to preview intended robot movements prior to actual execution. In a way, visualization allows the operator to view what the computer is *thinking*. The close linkage of the World Model with the graphics interface allows all programming to be done in the graphics environment. Operator directed movement of devices within the graphics interface are automatically compiled into robot commands using information contained in the World Model. After graphically programming a desired sequence of robot movements, the intended programmed movements can be reviewed and modified by the operator or communicated to the robotic device for execution. At no time will the computing system allow execution of a programmed operation if the World Model determines that unsafe operation may result.

As shown in Figure 2, all operator commands to the system are checked against the knowledge contained in the World Model to ensure that the command can be executed safely and within any Standard Operating Procedures. If the operator's command can be executed safely, the command is communicated to the subsystem Motion Control module. In the case of Figure 2, this is the module which contains the real-time system servo control algorithms for movement of the remote robot subsystem. High speed sensing subsystems provide sensory inputs to the robot subsystem motion control algorithms if necessary to perform, for example, force controlled interactions with the environment. Real-time sensing also allows in-operation modification of the robot system motions to, for example, prevent collisions with previously unknown obstacles or to allow the robot end effector to track the surface of the waste. It is important to recognize that such high speed sensor based servo controlled operations are executed within the subsystem controller and not by the higher level GISC environment. This prevents delays due to bottlenecks in communication and computation. Distributed computing environments are an important feature of the GISC approach to robotic system control. This provides natural modularity and extensibility and enhances the responsiveness of large robotic systems.

Sensors used for real-time control of discrete subsystems communicate with the higher level GISC models to provide dynamic updating of the World Model. The updated models are used both to detect possible safety and operational problems as well as provide the accurate World Models needed for automated planning and programming of new operations. The importance of communicating dynamic sensory information to the GISC World Model must be stressed. Since the World Model is used to validate all commands to the active subsystems, it is important that the World Model contain measured information wherever possible. Use of measured information such as manipulator joint angles will not only provide updated World Models but allow validation that the computer models used by the World Model to automate operations accurately reflect the motions generated by the subsystem controllers. This is especially true during sensor based operations when the motion of the manipulator is governed by real-time sensor inputs not precomputed trajectories.

If an operator's command to the robot system is deemed unsafe when examined with respect to the knowledge contained in the World Model, the operator's command is either modified slightly to, for example, allow the robot to avoid a known obstacle, or interrupted completely. If the operator's command is interrupted, the operator is alerted and an indication of why the command is thought to be unsafe is sent to the operator (usually via the graphics display). One mode of operation is to display the results of all commanded robot operations graphically to the operator prior to the operator issuing the final permission for the robot system to execute the operation. This is an important operational mode since it is not always clear to an operator what the result of a commanded motion might be. Joint limitations in some commercial manipulators, for example, can result in unanticipated motions to achieve what to the operator is a very simple move command. Such unanticipated motions could result in collision if not detected prior to execution.

Notice that to this point, the description of the control system architecture is very generic. There has been no discussion of the specific robot used in the waste recovery operations nor of the characteristics of the waste environment. The general control system architecture allows variations simply by changing the knowledge contained within the World Model or the Motion Control modules. As our understanding of the problem changes, the detailed nature of the individual control system modules may change but, the basic GISC structure remains the same.

APPLICATION OF GISC

A first implementation of GISC has been completed and is currently in use at Hanford, Washington as part of the Underground Storage Tank Robotics Technology Development Program. The tasks currently being addressed by the OTD's Robotics Technology Development Program (RTDP) are mapping of internal structures and waste surfaces in underground storage tanks (USTs) as well as removal of the waste and structures such as piping from the tank both for laboratory analysis and tank cleanup. Both fully automatic and manual robot control technologies are being developed and demonstrated [3]. Computer assistance is employed whenever manual control of the robot is required.

The UST Robot Technology Testbed employs a large commercial Spar Aerospace robot as a high strength positioning system for a dextrous Schilling Titan manipulator. A Remote Tank Inspection (RTI) Robot developed by RedZone Robotics was used to deploy sensors for tank wall inspection operations. Pacific Northwest Laboratory developed a Tracking Camera which, using information from the robot's joint encoders could automatically track the motion of any part (typically the end effector) of the moving manipulators. Both ultrasonic and optical Proximity Sensors were used for localized waste surface detailed mapping and position control of the Titan manipulator. Structured Lighting was used to map the waste surface for addition to the World Model prior to the start of robotic operations. Engineering drawings of the testbed served as the Apriori Engineering Data for the initial World Model. The multilaboratory project team members included Westinghouse Hanford Company, Oak Ridge National Laboratory, Pacific Northwest Laboratory, Idaho National Engineering Laboratory, Sandia National Laboratories, as well as representatives from Spar Aerospace and RedZone Robotics [3,4,5].

The computing structure of this first implementation of GISC is shown in Figure 3 together with the ethernet communication interfaces between GISC and the intelligent subsystems. In all cases the primary interfaces are Motorola 680X0 series CPUs in a VME backplane. These CPUs interpret the generalized GISC commands and convert them to the commands understood by the individual subsystem controllers. In addition, these VME based interface units also interfaced the ethernet communication system with the subsystem communications requirements of the specific robotic devices (e.g., RS232 serial communication in the case of the Spar). The VxWorks operating system was used on the VME systems. The equipment outside the box marked GISC represent independent subsystems with diverse computing and control environments. In one case, the Tracking Camera, an IBM PC/AT was used as the subsystem controller and serial communication was used to interface this subsystem into the overall system. The modularity of the UST Robotics Testbed is apparent as is the ability of GISC to integrate diverse subsystems into a coordinated system addressing key waste cleanup issues.

Figure 3
UST Robotics Demonstration -- GISC

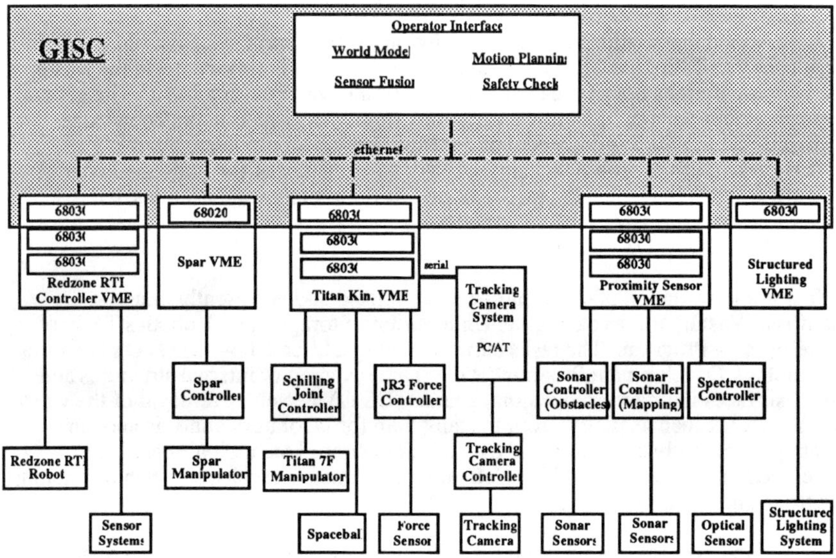

The Proximity Sensors subsystem performed two functions. In the first, sonar proximity sensors were used to provide high precision maps of the waste surface and vertical piping (risers) within the testbed environment. The Proximity Sensors module performed the statistical correlation of the multiple sonar sensors with location of the Titan manipulator (upon which the sensors were deployed) to compute map information locally to the Proximity Sensors subsystem. This computed map information was then communicated to the Sensor Fusion module of GISC and entered into the World Model. During proximity servo control of the Titan manipulator, range information was passed directly to the Titan subsystem controller without direct GISC involvement (except for the communication software) to allow high speed servo control. Proximity information for servo control could be obtained either from the sonar sensors or the Spectronics optical sensor, both of which were being tested.

The Structured Lighting used a laser based sensing system to develop maps of the waste surface to enable the robot subsystems to safely approach the waste surface for more detailed operations. The Structured Lighting subsystem analyzed all optical data locally and passed only processed map information to GISC to be combined with the initial Engineering Data to provide more complete World Models for automating robot system programming. Thus, the modular structure desired as part of GISC was implemented in a straight forward manner. The modularity enabled multilab teams to work easily together. System integration for this complex systems required less than 10 weeks.

Results

GISC and the associated subsystem controllers have been implemented and first system operation as part of the Hanford Underground Storage Tank Robotics Testbed was demonstrated in August 1991. The World Model and graphics interface modules are used in several ways. First, the animation system was used to preview robot operations to ensure that the commanded robot motions did not result in collisions. In addition, the graphics model served as the geometric world model during real-time operations. For example, during manual control of the robot (using remote video cameras), the operator's commands to the robot were evaluated in the graphics environment to ensure no inadvertent collisions would occur. Commanded motions which might have resulted in collisions were automatically interrupted by the control system to ensure safe operation.

REFERENCES

1. Richard Quintero, ed., "DOE/NIST Workshop on Common Architectures for Robotic Systems," NIST Special Publication 784, April 1990.

2. Richard Quintero, ed., "Proceedings of the Second DOE/NIST Workshop on Common Architectures for Robotic Systems," Seattle Washington, 1/23 - 1/24, 1991. (Available from NIST, Robot Systems Division, Building 220, Rm B124, Gaithersburg, MD 20899)

3. W. R. Jaquish, E. H. Shen, and, J Yount "Robotics Technology Demonstration Program for Underground Storage Tank Remediation Program," ANS Winter Meeting, San Francisco, November 10-14, 1991.

4. B. L. Griebenow, L. A. Strope, C. B. Selleck, and J. D. Burke, "Imaging, Inspection, and Characterization System for Underground Storage Tank Remediation," ANS Winter Meeting, San Francisco, November 10-14, 1991.

5. B. K. Christensen, Bl. L. Griebenow, and B. L. Burks," Graphic Model Based Control of Robotic Systems for Waste Remediation," ANS Winter Meeting, San Francisco, November 10-14, 1991.

ACKNOWLEDGEMENTS

The entire RTDP Team contributed to the development of the GISC concept in many different forms throughout the course of the technology development effort. While Sandia had the lead responsibility for the initial implementation described in this paper, the reader should be aware that GISC has been the result of the RTDP Team approach to robotics technology development. Continued GISC development is drawing heavily on the RTDP Team expertise.

INTELLIGENT SENSING AND CONTROL IN METAL CUTTING

REZA LANGARI

Department of Mechanical Engineering
Texas A&M University
College Station, TX 7783-3123

DAVID DORNFELD

Department of Mechanical Engineering
University of California,
Berkeley, CA 94720

ZACK WANG

Nyden Corporation
Santa Clara, CA 95054

ABSTRACT

In this paper we present an approach to the integration of sensing and control in metal cutting, which is meant to form the basis for intelligent control of this process. In particular, we present a strategy for modeling, and further designing a robust control algorithm, a la H_∞, that deals with some of the intricate issues associated with the control of metal cutting operation such as structural resonance mode(s) and modeling inaccuracy(ies). The results of this study are promising and indicate that active monitoring and sensor based control of metal cutting can benefit from the use of modern robust control methodologies.

INTRODUCTION

A generally agreed view of manufacturing automation is to *replace and extend the function of the human operator*[5]. In view of this idea, which in effect is an interpretation of the notion of intelligent control in the context of manufacturing automation, this article offers a framework for seamless integration of *process monitoring* and *control* in metal cutting. The key ingredients of this framework are as follows:

- *Robust process control within the H_∞ framework-* to deal with unstructured uncertainty in the process model, variations in the workpiece geometry and material characteristics, and tool wear.
- *Neural Network based sensor fusion–* to handle the limited range and variations in the sensitivity of physical sensors.

- *Fuzzy logic based diagnostic failure detection and compensation*– to detect, classify and recover from conditions of critical tool wear, and machine failure.

An outline of this framework is presented in the next section. The focus of the article, however, will be on the first of the above items and in particular on the use of the H_∞ framework in the design of a sensor based, and robust, control strategy for metal cutting, which is meant to deal with some of the intricate issues associated with this process such as structural resonance mode(s) of the machine, and the effect of external disturbances in presence of modeling uncertainty.

ARCHITECTURE FOR INTELLIGENT PROCESS CONTROL

A suitable framework for intelligent process control must accommodate heterogeneous elements employing different representational formalisms, for instance those engaged in sensor processing, process control, and diagnostics and failure compensation. For this purpose, we have adopted the hierarchical architecture proposed by Albus[1], which specially serves the real-time needs of manufacturing processes for sensing, control, and communication. A diagram depicting the use of this architecture in the present context is shown in Figure 1. The components of this architecture are as follows.

- *Process Manager:* Receives the task specification, analyzes the task and plans the execution of the task in view of hard and soft constraints.
- *Controller*: Regulates the behavior of the physical system(machine).
- *Diagnostic Modules:* Monitor the operation of the machine and its various subcomponents.

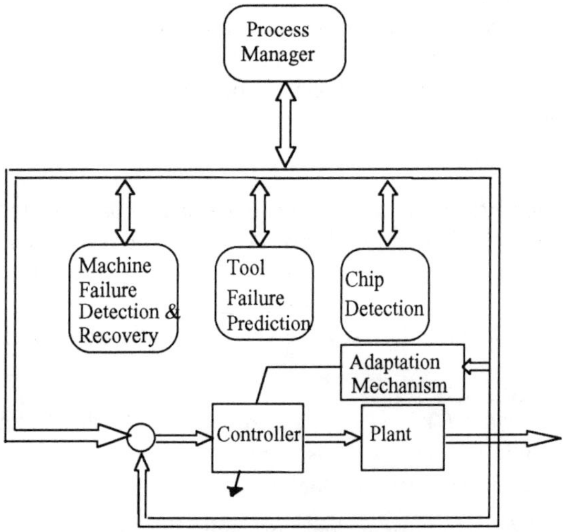

Figure 1. Architecture of an integrated control systems for machining of metals.

Arguably, the critical task within this framework is that of actively controlling, or regulating, the process. We will discuss this task from the control engineering standpoint in the next section.

ROBUST CONTROL WITHIN THE H_∞ FRAMEWORK

Manufacturing processes present an array of challenging control problems. In particular, lack of accurate process models and inadequacy of sensors and actuators, render control design a difficult task. As shown below, the difficulty stems from the basic limitation of traditional control design techniques, such as PID, to deal with the unstructured uncertainty in the plant model. This uncertainty, in presence of external disturbances, can destabilize the control system or in practice lead to poor surface finish, chatter, and even catastrophic machine failure.

The H_∞ framework on the other hand offers a practical, as well as mathematically appealing approach to handling modeling uncertainty in the control design process[4]. In particular, suppose a *static* nonlinear plant model is derived with the aid of a neural network as

$$\mathbf{y} = NL(\mathbf{u}), \tag{1}$$

where $\mathbf{u}$ and $\mathbf{y}$ are generalized plant input and output variables respectively. Further, suppose this nonlinear model is used to construct an approximate *dynamic* linear model of the plant as

$$\delta\hat{\mathbf{y}}(s) = \mathbf{J}(\mathbf{u})\big|_{\text{nom}}\mathbf{M}(s)\delta\hat{\mathbf{u}}(s), \tag{2}$$

where $\mathbf{J}(\mathbf{u})\big|_{\text{nom}}$ is the *Jacobian of* $NL(\cdot)$ evaluated at some nominal operating point; $\delta\hat{\mathbf{u}}(s)$ and $\delta\hat{\mathbf{y}}(s)$ are the Laplace Transforms of the deviations of $\mathbf{u}$ and $\mathbf{y}$, respectively, from their nominal values; and, $\mathbf{M}(s)$ denotes the modal matrix representing an estimate of the dynamic characteristics of the plant, with

$$\mathbf{M}_{i,j}(s) = \begin{cases} \dfrac{1}{\tau_{i,j}s+1}, \text{ or} \\ \dfrac{\omega_{i,j}^2}{s^2+2\xi_{i,j}\omega_{i,j}s+\omega_{i,j}^2}. \end{cases} \tag{3}$$

Note that $\tau_{i,j}$, $\omega_{i,j}$, and $\xi_{i,j}$ denote the *estimates* of the parameters associated with the various dynamic modes of the process.

Clearly, significant approximation has been introduced in the linearized plant model, a fact which needs to be considered in the control design process. Now assuming that the *true* plant, P, is described by $P = P_0 + W_\Delta\Delta$, where P_0 is the given plant model, Δ is a bounded[1] perturbation, $\|\Delta\|_\infty \leq 1$ and W_Δ is a shaping filter, the H_∞ framework can be

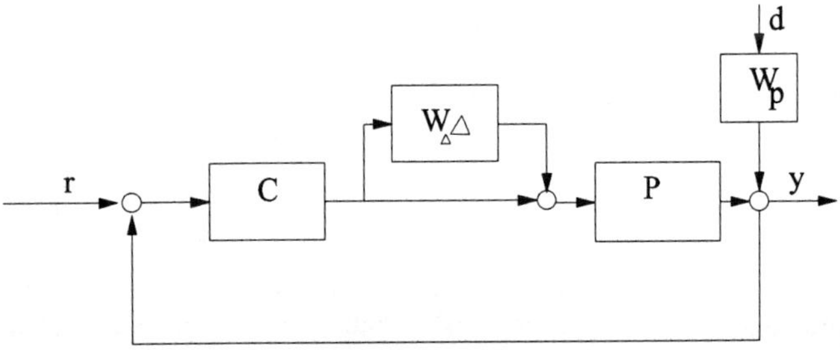

Figure 2. Architecture for Robust Control.

used to design a controller C that minimizes a cost functional, which represents a compromise between the disturbance rejection capability of the plant as represented by the system sensitivity function,

$$S(s) = \frac{1}{1+P(s)C(s)},$$

and its robustness to modeling errors as weighed by the system transmission function,

$$T(s) = \frac{P(s)C(s)}{1+P(s)C(s)}.$$

This cost functional is given by $J = \sup_{\omega}\left(|W_\Delta T| + |W_d S|\right)$, where W_d represents the disturbance weighting function. The controller C is obtained by minimizing J through a computationally intensive but conceptually straightforward process. A perceived drawback of this approach is the increased order of the controller, with controller order of the model order plus the order of the disturbance shaping filter, W_d, plus the order of the perturbation shaping filter, W_d. Model reduction is however, possible using stochastic approximation based model reduction as part of available control design packages[2].

With this in mind a controller was designed for 3 input, 3 output control of low depth of cut cutting, or deburring, and evaluated both in simulation and in practice and was shown to be capable of dealing with the issues involved in machining such as unmodeled structural resonance modes and sensor noise.

---

[1]We shall use the so called *Infinity* norm, denoted by $\|\cdot\|_\infty$, which here represents the least upper bound on the magnitude of the Fourier Transform of a signal over all frequencies.

As shown in Figure 3, the plant, that is the given machine tool is structurally flexible in the frequency range of 100 rad/sec(upper left). The plant model is assumed to be only about %50 accurate in the low frequencies and becoming less and less accurate as frequency increases(lower left). At about 20 rad/sec the plant model is no longer accurate with the Multiplicative Uncertainty reaching %100. We are, however, concerned with performance in the relatively low frequency range. Thus the Performance Weighting function(upper right) requires disturbance rejection up to approximately 10 rad/sec.

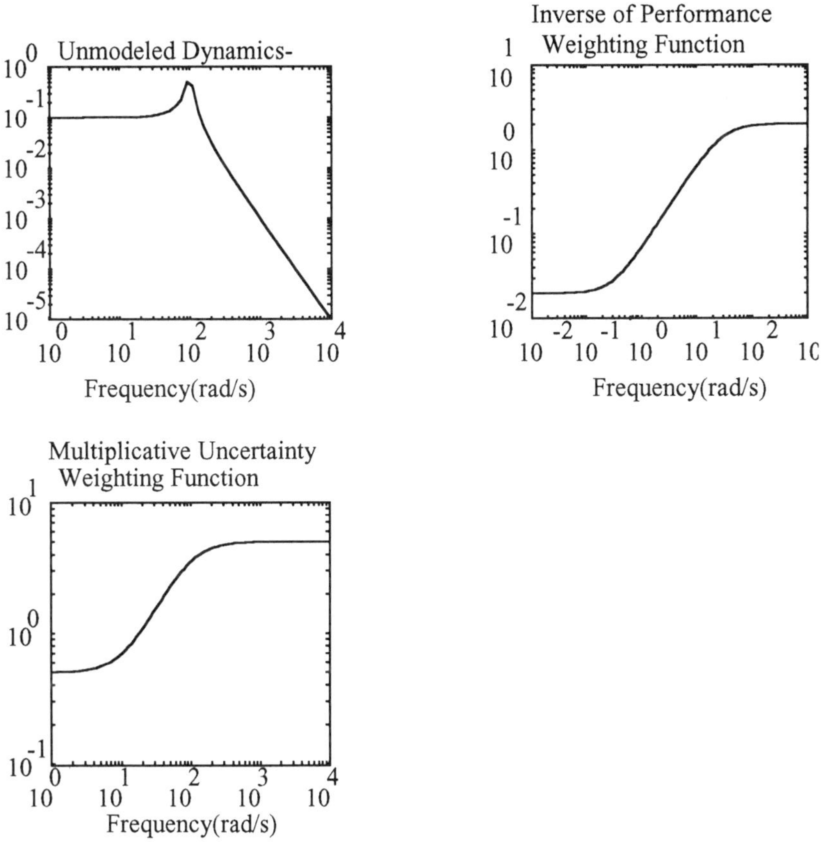

Figure 3. Disturbance and Perturbation Shaping Filters as well as Frequency Response of the Plant

Simulation results show how the H_∞ designed controller fares compared with classical(PID) controller(Figure 4). In particular, while PI control is capable of dealing with disturbances if there were no structural resonance modes, it simply becomes unstable in presence of both disturbance and structural flexibility(lower left). The H_∞ based controller, on the other hand, is capable of dealing with both disturbance and structural flexibility in the machine(upper right) .

This controller was tested in a single channel configuration on a deburring process with RMS value of the Acoustic Emission(AE) signal, measured and processed on-line, as the plant output, and relative position of the tool vs. the workpiece as the plant input. (Tests were conducted on a Matsuura Milling Machine while the first author was with the Engineering Systems Research Center, University of California. The Yaznac controller on the Matsuura machines was disabled and all positioning was performed through custom software on an i386 based AT-bus platform.) The results presented in detail in Lisciewicz[4] indicated the possibility of using this control strategy to maintain constant area of cut(constant AE-rms signal).

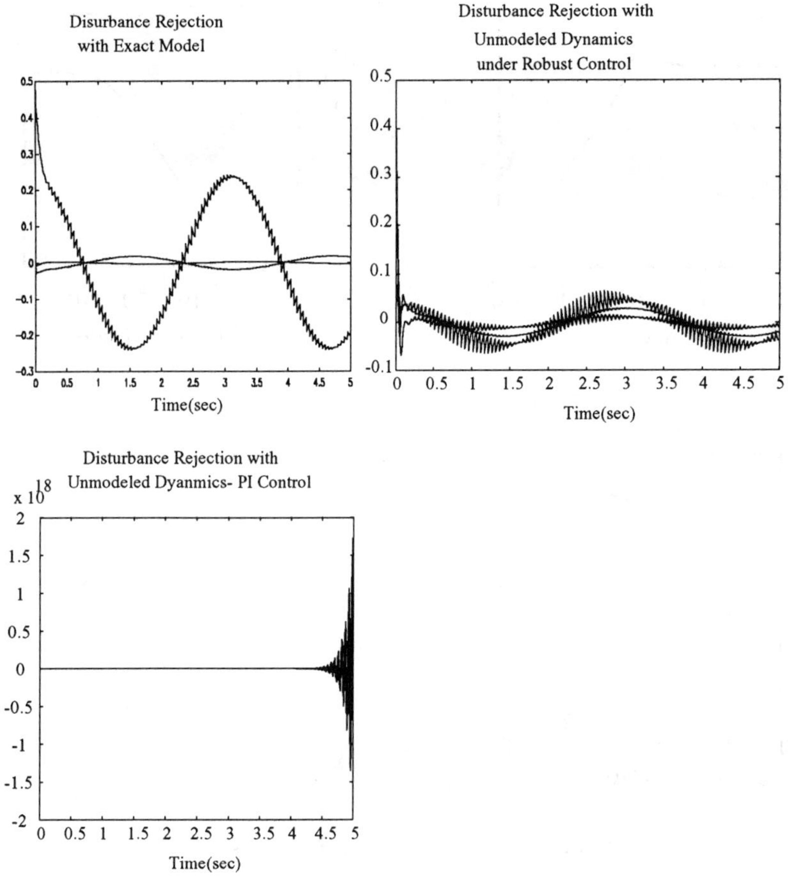

Figure 4. Response of the plant to different control algorithms.

CONCLUSION

This paper presents a framework for integrated monitoring and control of metal cutting. In particular, it presents an approach to active sensing and robust control of this process, based on the H_∞ framework, that is meant to deal with some of the intricate issues associated with metal cutting such as the structural resonance modes of the machine(due to its non-rigidity) and the effect of external disturbances in presence of unmodeled dynamics. The results of the study appear promising, but more extensive experimental work is needed before the proposed approach becomes a viable one for application in the industry.

REFERENCES

1. J. Albus, "Outline of a Theory of Intelligence," *IEEE Transactions on System, Man, and Cybernetics*, Vol. 21, No. 3,(May/June 1991).
2. G. Balas, J. Doyle, K. Glover, A. Packard and R. Smith, *μ-Analysis and Synthesis ToolBox,* MathWorks, Inc. 1991.
3. M. Morari and Zarifou, Robust Process Control, Prentice Hall, 1989
4. V. Lisciewicz, *Computer Control of Deburring*, MS Thesis, University of California, Berkeley, (December 1990).
5. A. Okada, "Automation in arc welding--- state of the art problems," *Journal of the Japan Society of Mechanical Engineers*, (March 1979, in Japanese).

On Neural Net Controllers for Robotic Manipulators

J. Yuh
Autonomous Systems Laboratory
Department of Mechanical Engineering
University of Hawaii
Honolulu, Hawaii 96822

Abstract

In this paper, we present the results of our survey on neural network application to robot control system design. Neural network approaches to robot control, discussed in the literature, can be classified into four neural design methods: supervised control method, inverse dynamic control method, critic method, and compensator method. These methods are discussed and references for each method are also cited. The objective of this paper is to provide an overview of the current state-of-art in neural network application to robot control system design and future research direction on the subject matter.

Introduction

During the last few years, application of neural networks to the robot control system has received increasing attention. The neural networks have distinct features of robust processing and adaptive capability in a changing and noisy environment. The collective processing capability of neural networks allows it to respond quickly. The network learns the correlation between the input and desired output. It can repeat these relationships and generalize it to new input. After training, the neural net can generate estimated output in response to new input. Training the network requires target signals for the output of the network. However, in its control application, the output of the network is the control signal such as each joint torque of the robot system. Therefore, one may ask "How can training signals to the network be obtained without using some other controller or supervisor who knows how to control the plant and provide the training signals?" If one controller or supervisor, who knows how to control the plant, would be used to train the network, then one might also ask "Why do we need a neural network controller when we already have an effective controller?" In spite of many attractive features of the neural network, these may not be easy questions to answer. We will see in this paper if the questions can be answered, even partially, by discussing various neural network controllers for robotic manipulators proposed by researchers. The general issues about neural networks in control systems are discussed in ref. (Antsaklis 1992, and Miller et al 1990). Approaches for the neural network robot control reported in the literature can be classified into four design methods: supervised control method, inverse dynamic control method, critic method, and compensator method. In this paper, we will focus on multilayered, feedforward neural networks trained by error backpropagation or its-like algorithm.

This paper is organized as follows: first, feedforward process and backpropagation process of the neural network are briefly described; and four different design methods are then discussed.

Multilayered Neural Network

Figure 1 shows the general multilayered neural network having one input layer, one output layer, and multiple hidden layers. Each layer has multiple neurons and one neuron of each hidden layer has one arbitrary constant input (which could be zero). Input signals of the network are coming into input neurons (i.e. neurons of the input layer). During the feedforward process, the input signals propagate from the input layer to the output layer, being multiplied by weights between neurons of two layers in sequel and being modified at each neuron by activation functions. Output neurons (i.e. neurons of the output layer) output the network output signals that are in fact control input signals to the robot. A sigmoid function such as $f(x) = (e^{-x}-e^{x})/(e^{-x}+e^{x})$ is often used as the activation function of each neuron. Discussion about the effect of activation functions is out of this paper's scpoe. However, some discussion, in general, regarding this issue can be found in reference (Stinchcombe and White 1989). During the backpropagation process, the weights of the network are adjusted based on information about error signals between the target signal and the output signal of the network. It is also called the learning process.

Feedforward Process: The output of the i-th neuron in the m-th layer is computed by

$$q_i^m(t) = f(W_i^m(t)^T q^{m-1}(t)) \tag{1}$$

where W_i^m is a weight vector between the i-th neuron of the m-th layer and the neurons of the (m-1)-th layer, q^{m-1} is an output vector of neurons in the (m-1)-th layer, and the superscript T implies the transpose of the matrix.

Backpropagation Process: The weights (W) of the network are adjusted by

$$W_i(t) = W_i(t-1) + \rho\, \Psi_i(t)\, \epsilon(t) \tag{2}$$

where W_i is a weight vector associated with the i-th neuron in the network, ρ is a positive constant known as a learning rate, $\Psi_i(t,W)$ is the gradient of $U(t,W)$ with respect to W_i, and ϵ is the network output error vector. The detailed description of the error backpropagation algorithm can be found in ref. (Rumelhart et al 1986). Chen et al (1990) and Yuh (1992) also described the parallel recursive algorithm that can be considered as a general form of the error backpropagation.

Neural Network Controller

1. Supervised Control

The neural network controller is trained to learn the supervisor. Supervisor could be conventional controller or human operator. This type of method has been used by Guez and Selinsky (1988). The implementation of this method (Fig. 2) is summarized below:

1. The same input signals are fed to the supervisor and the neural network controller;
2. The robot is controlled by the supervisor while the neural network controller is trained;
3. The output of the supervisor is used as a target value in training the neural network controller;
4. After training the network (i.e. outputs of two controllers are same.), the neural network is used to control the robot; and
5. If the neural network controller fails, then the conventional controller takes over the main control function and the network must be retrained.

This method can be implmented on-line if the overall control system has ability of carrying out steps 4 & 5 without any interruption. During the on-line implementation, steps 4 & 5 will require special attention on the issues such as when to change the main function from one to another; and how to make a smooth transition. Since the neural network is trained to imitate the supervisor, the performance of the overall control system is depending on how good the supervisor can perform. If the conventional controller is used as a supervisor and fails due to the changes in the robot system, the neural network controller cannot be trained for the desired performance.

2. Inverse Dynamic Control

The neural network controller is trained for inverse dynamics model of the robot. If the neural network is well trained for the inverse dynamics model, it is supposed to output desired joint torques for the desired joint angle input to the network. There are two methods: direct inverse modeling and indirect inverse modeling.

Direct inverse modeling

Figure 3a shows the control system diagram of this method. This type of method has been used by Zeman et al (1989). The implementation of this method can be summarized below:

1. The robot outputs such as joint angles are fed to the neural network;
2. The error signals between the control input to the robot and the output of the network are used to train the network; and
3. The trained network is then used as feedforward controller;

This method can be implemented on-line if two identical networks are used in the control system: one network is trained and its weights are copied to another network being used as feedforward controller. However, the on-line implementation of this method is not encouraged, especially when the learning takes a long period since the desired performance cannot be achieved during the learning period. Furthermore, the method can provide the desired performance only if $u \rightarrow u_d$ as $y \rightarrow y_d$.

Indirect inverse modeling

This method (Fig. 3b) uses two neural networks: one network is trained for forward dynamics model of the robot system; and another network, being used as feedforward controller, is trained for inverse dynamic model of the robot system. This type of method has been discussed by Kung and Hwang (1989). The implementation of this method can be summarized below:

1. The control inputs such as joint torques are fed to the neural network;
2. The error signals between the robot output and the network output are used to train the network for the forward dynamics model of the robot; and
3. The feedfoward neural network controller is trained by backpropagating the robot output errors through the forward dynamics model network;

This method requires two neural networks and it can be implmented on-line by training both networks at the same time. On-line implementation may not be desirable when the learning takes a long period.

3. Critic Method

The neural network controller of this method is often called reinforcement learning controller. By using a critic of evaluating robot performance, the neural network controller is trained. In this method, we need to determine how to construct a critic considering the control objective and how to use that critic in training the network. Some design methods of the reinforcement learning controller are discussed by Werbos (1991). The general implementation of this method shown in Fig. 4a is summarized below:

1. The control input and the robot output signals are measured to evaluate the robot performance that leads a critic;
2. The neural network feedforward controller is trained to maximize (or minimize) the robot performance.

This method can be implmented on-line if the network learns fast enough to provide the desired performance. This method is different from the previous two methods - the neural network is not trained to imitate the supervisor or inverse robot dynamics model itself.

Kawato et al (1990) and Yuh (1992) proposed simple designs of this method for the robot control. Both used backpropagation of the network output error (i.e. control input error) estimated by using the robot output error signals. In Kawato's approach, the control signal is a sum of neural network output and PD control signal with the robot output error. The PD control signal is then used to train the network. His approach can be summarized as shown in Fig. 4b: $u = u_1 + u_2$; as trained by $\epsilon_{u1} = u - u_1$, $u_1 \rightarrow u$; and $u \rightarrow (u + u_2)$. Only if $u \rightarrow u_d$, $y \rightarrow y_d$ and $u_2 \rightarrow 0$; and consequently $u_1 \rightarrow u_d$ and $u \rightarrow u_d$. It is called feedback learning error method. Since the feedback controller acts like nonadaptive critic, it can be seen as a critic design. In Yuh's approach, the robot output error is used as the input to the network and the network output is used as the control signal. The network is trained by a critic equation which is a function of the robot output error.

4. Compensator Method.

The conventional controller is used as a main controller to control the robot system while the neural network controller is used as a compensator for errors due to unmodeled dynamics and other system errors. Since robot dynamics are well understood, linear or nonlinear control system can be designed in the conventional way. Errors due to modeling errors and other system uncertainties such as payload are handled by a neural network compensator. The implementation of this method (Fig. 5) can be summarized below:

1. The robot system is controlled by sum of control input signal by the conventional controller and the output signal of the neural network compensator; and
2. The error between the control input torque and torque computed from the robot model used in the conventional controller design is used to train the neural network compensator.

This method can be implemented on-line since the robot is controlled by the conventional controller while the neural network compensator is trained. When the robot model used in the conventional control system design is far different from the actual system, a large burden is given to the neural network compensator and consequently, there would be a lot of errors during the learning period. This type of method has been discussed by Okumura (1990). In his approach, he used the computed torque control system with a neural network compensator.

Simulation and Discussion

We have tested some of the above methods by computer simulation with a single-link robot model. Simulation results for a high degree-of-freedom robot model are shown in reference cited for each method. Mean Square Error (MSE) of the joint angle are shown in Fig. 6. It shows that MSE gets smaller as time goes by. Other results and detailed discussion will be presented during the conference.

Acknowledgment

This work was partially supported by (R/OE-13) the University of Hawaii Sea Grant College Program under Institutional Grant No. NA89AA-D-SG063 from the NOAA Office of Sea Grant, Department of Commerce and partially by National Science Foundation under contract No. BCS91-57896. This is a Sea Grant Publication UNIHI-SEAGRANT-JC-92-04.

References

J.S. Albus (1975), "A New Approach to Manipulator Control: the Cerebeller Model Articulation Controller (CMAC)", J. Dynamic Sys. Meas. and Control, Vol 97, pp. 220-227.

P.J. Antsaklis (1992), Special Issue on Neural Networks in Control Systems, Control Systems Magazine, Vol. 12, No. 2, April.

A. Guez, and J. Selinsky (1988), "A Trainable Neuromorphic Controller", J. Robotic Systems, 5(4), pp. 363-388.

R. Hecht-Nielsen (1989), "Theory of the Back Propogation Neural Network", Proc. Int'l Joint Conference on Neural Networks, Washington D.C., pp. I593-I605.

M. Kawato (1990), "Computational Schemes and Neural Network Models for Formation and Control of Multijoint Arm Trajectory," Neural Networks for Control, eds. Miller, Sutton, and Werbos, MIT press, pp.197-228.

S. Kung and J. Hwang (1989), "Neural Network Architectures for Robotic Applications," IEEE Trans. on Robotics and Automation, 5(5), Oct., pp. 641-657.

W.T. Miller (1989), "Real-Time Application of Neural Networks for Sensor-Based Control of Robots with Vision", IEEE Transactions on Systems, Man, and Cybernetics, Vol 19, No.4, pp. 825-831, 1989.

W.T. Miller, R.S. Sutton, and P.J. Werbos, Editors, (1990) Neural Networks for Control, MIT press

D.H. Nguyen, and B. Widrow (1990), "Neural Networks for Self-Learning Control Systems", IEEE Control Systems Magazine, pp. 18-23, April.

S. Okuma, A. Ishiguro, T. Furuhashi and Y. Uchikawa (1990), "A Neural Network Compensator for Uncertainties of Robotic Manipulators", Proceedings of the 29th CDC, pp. 3303-3307.

D. Psaltis, A. Sideris, and A. A. Yamamura (1988), "A Multilayered Neural Network Controller", IEEE Control Systems Magazine, pp. 17-20, April.

D.E. Rumelhart, J.L. McClelland et al. (1986), Parallel Distributed Processing, MIT Press.

R.M. Scanner and D.L. Akin (1990), "Neuromorphic Pitch Attitude Regulation of an Underwater Telerobot", IEEE Control Sys. Mag., pp. 62-68, April

M. Stinchcombe and H. White (1989), "Universal Approximation using Feedforward Networks with Non-Sigmoid Hidden Layer Activation Functions," Proc. Int'l Joint Conference on Neural Networks, Washington D.C., pp. I613-I617.

J. Yuh (1990), "A Neural Net Controller for Underwater Robotic Vehicles", IEEE J. of Oceanic Engineering, Vol. 15, No. 3, pp. 161-166.

B. Widrow, N.K. Gupta, and S. Maitra (1987), "Punish/Reward: Learning with a Critic in Adaptive Threshold Systems", IEEE Trans. Systems, Man and Cybernetics, Vol. SMC 3, No. 5, pp. 455-465, 1987

J. Yuh, and R. Lakshmi (1992), "An Intelligent Control System for Remotely Operated Vehicles", IEEE J. of Oceanic Engineering (in print).

V. Zeman, R.V. Patel, and K. Khorasani (1989), "A Neural Network Based Control Strategy for Flexible-Joint Manipulators," IEEE Conf. on Decision and Control, Tampa, FL, December, pp. 1759-1764

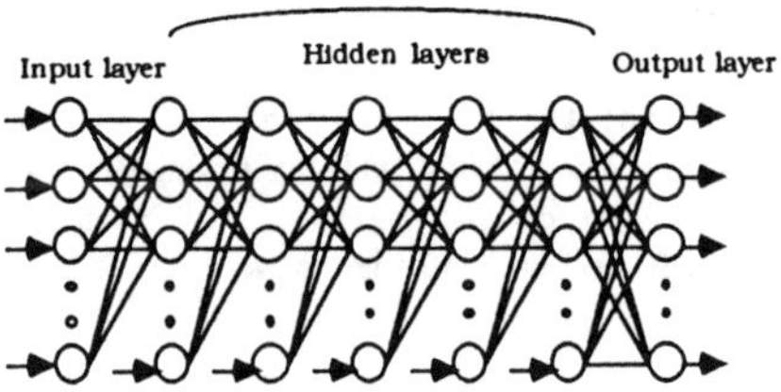

Fig.1 Multilayered Neural Network

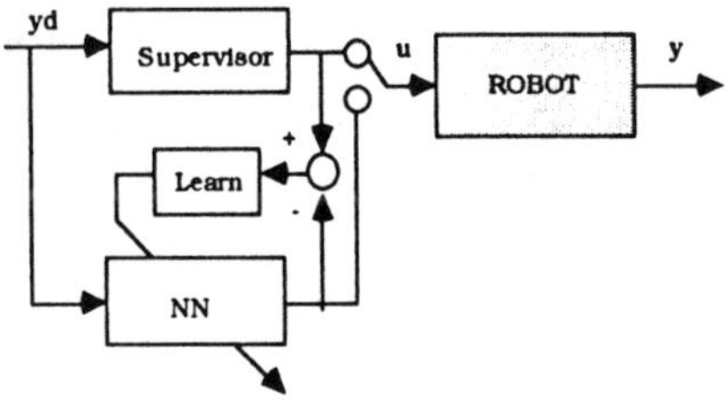

Fig. 2: Supervised Control Method

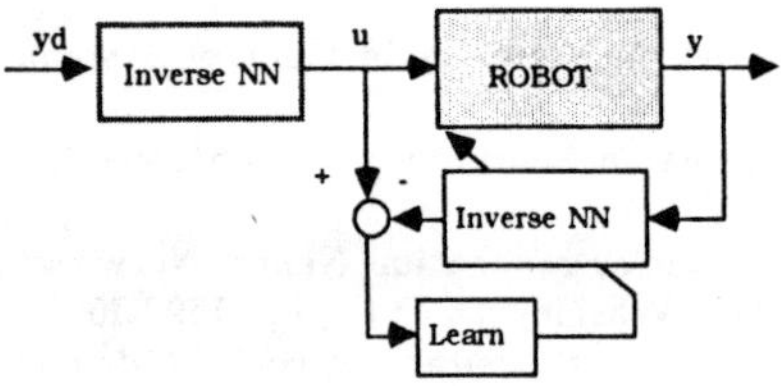

Fig. 3a: Direct Inverse Modeling Method

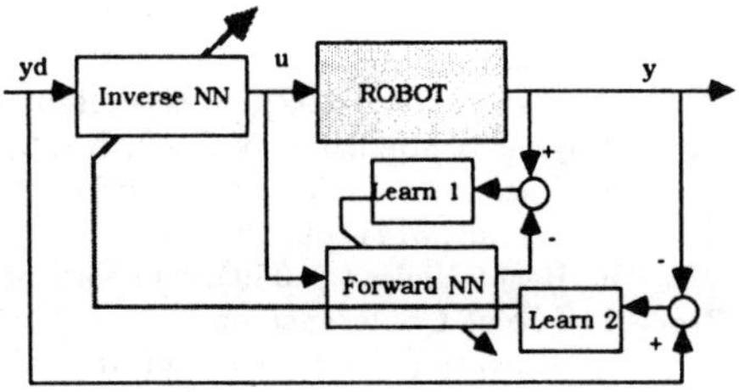

Fig. 3b: Indirect Inverse Modeling Method

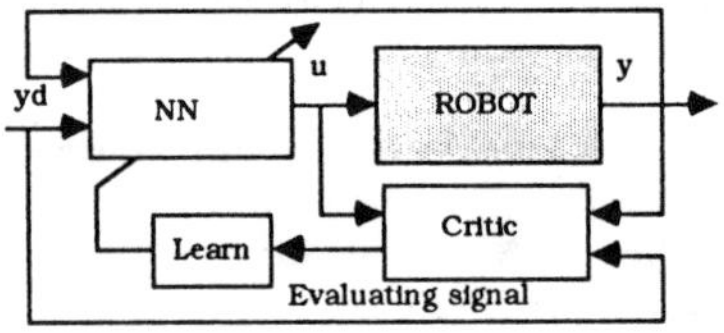

Fig. 4a: Critic Method

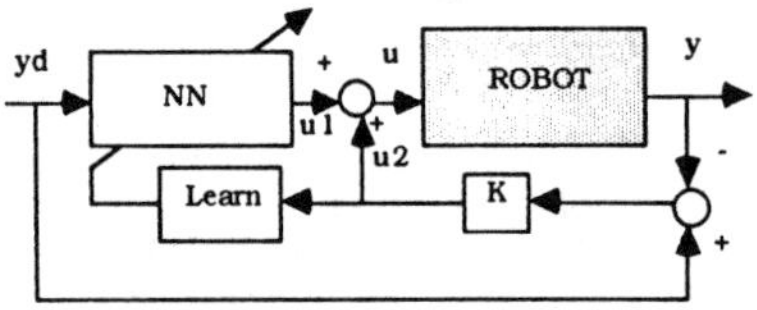

Fig. 4b: Feedback Error Method

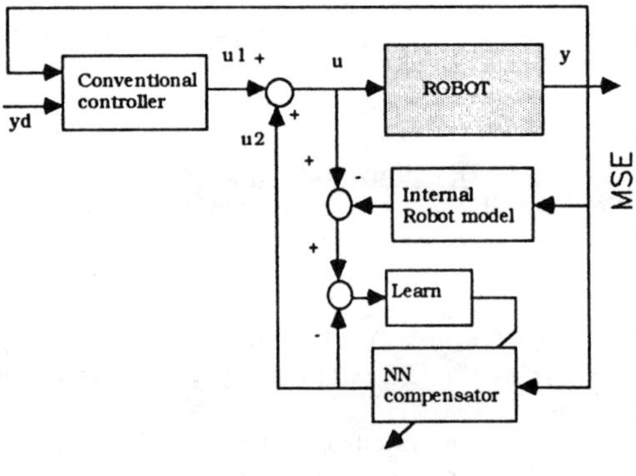

Fig. 5: Compensator Method

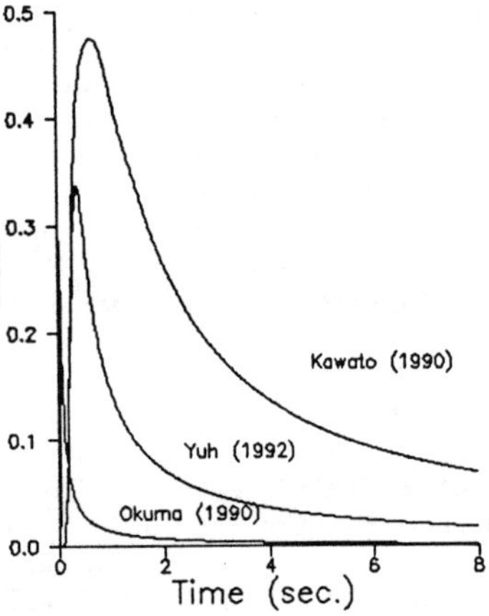

Fig. 6: Mean square errors

Collision Identification and Object-to-Obstacle Distance Calculation Using Geometric Modeling and Neural Networks

Z. Dong and H. Zhang
Department of Mechanical Engineering
University of Victoria
Victoria, B.C. Canada V8W 3P6

Abstract

A formulation for collision identification and distance calculation in motion planning, using the modified Hamming net, is presented. Derived from the homogeneous geometric transformations, the method can be used for point-to-object, line-to-object and object-to-object collision identification and distance calculation. The optimal direction for collision removal is also provided. The method can considerably decrease computation time by parallel processing, and has the potential for being applied to on-line trajectory planning.

1 Introduction

In computer automated manufacturing, collision free trajectory planning is central to programming robots, CNC machine tools, and many other manufacturing facilities. An efficient method for calculating distances between a moving object (manufacturing facility, robot) and the fixed obstacles, and interferences that may occur during the motion planning is needed for planning a collision free trajectory. The collision identification and distance calculation have been recognized occupying most computational effort in trajectory planning, and forming the bottle neck for enhancing motion planning strategies [1, 2].

Research on efficient collision identification and distance calculation methods has gained considerable attention in recent years. These approaches can be classified either by the method of distance calculation or by their underlying principles. In terms of the method of distance calculation, existing distance calculation approaches can be characterized as vertex-based approaches [3], surface-based approaches [4], and convex object-based approaches [2, 5]. In terms of the underlying principles, the research can be classified into three major categories: forbidden region identification methods [3], swept volume methods [6, 7], and constrained distance minimization methods [1, 2]. Due to the complex nature of the collision identification and distance calculation problem, most present methods require extensive computations.

The applications of intelligent robotic and manufacturing systems working in an unstructured and time-variant work environment require the on-line motion planning capability. In these applications, the work environment is either not known beforehand, or keeps changing constantly. The efficient on-line environment perception and on-line motion planning methods become critical.

*Financial support from the Institute of Robotic and Intelligent Systems (IRIS) is gratefully acknowledged.

A new approach to collision identification and distance calculation using a modified Hamming net neural networks is introduced in this work. The formulation is derived from homogeneous geometric transformations of rotation and translations among introduced local coordinates of the obstacles. The calculation also provides a quantitative evaluation of interference after collision and points the direction for eliminating the interference. The new method simplifies the object-to-obstacle distance calculation, and conducts calculations in parallel. It can considerably decreases required computational time and has the potential for being applied to on-line trajectory planning.

2 The Architecture of Hamming Net

In recent researches, neural networks have shown superior capability of solving complex pattern-recognition problems such as speech understanding and handwritten character identification, due to its capability for parallel-processing immense quantities of information. With the introduction of the Neurocomputers, this advantage of the neural networks can be fully utilized. The unique capability of neural networks make it an ideal tool for handling the large quantity of interference checks in the collision-free trajectory planning. With the parallel implementation, required computation time for interference checks is expected to drop considerably. This, in turn, makes the on-line collision free trajectory planning feasible.

An N-neuron Hamming net is a two-layer neural networks. The architecture and application of a Hamming net is well described by Lippmann [8]. A Hamming net for pattern classification in a 3-D Euclidian space is given in Figure 1, excluding the part inside the box of dashed-line. The N patterns are pre-stored in the Hamming net in terms of N sets of connection weights $\{w_{k1}, w_{k2}, w_{k3}\}_k$ $(k = 1, \ldots, N)$ and N thresholds $\{T_k\}_k$ $(k = 1, \ldots, N)$. Given an input feature vector $f = [f_x, f_y, f_y]_k^T$, the first layer (or the input layer) of the Hamming net will match this input with N patterns through N parallel inner product operations:

$$ip_k = f_x w_{k1} + f_y w_{k2} + f_z w_{k3} - T_k \quad (1 \leq k \leq N) \tag{1}$$

The resulting N inner products $\{ip_k\}_k$ $(k = 1, \ldots, N)$ are then fed into the second layer of the Hamming net (or the Hopfield N-flop). This layer makes the final classification by:

$$ip_{k^*} = \max_k \{ip_k\} \tag{2}$$

The input patterns, $f = [f_x, f_y, f_y]_k^T$ $(k = 1, \ldots, N)$, are automatically classified by the N-neuron Hamming net, according to their inner products with the N set of given weights and thresholds.

3 Collision Identification and Distance Calculation

The collision identification and distance calculation method, developed in this work, is directly applicable to convex polyhedron object models. The method can be easily extended to non-convex polyhedrons, and objects with cylindrical, spherical and sculptured surfaces.

3.1 Point-to-Object Distance Calculation

The formulation starts by attaching a local Cartesian coordinate frame to each of the N boundary planes of a polyhedron as illustrated in Figure 2. Each coordinate frame,

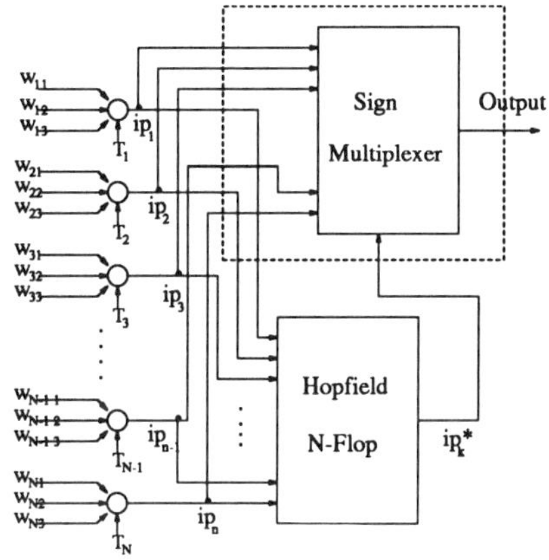

Figure 1: A Modified N-neuron Hamming Net.

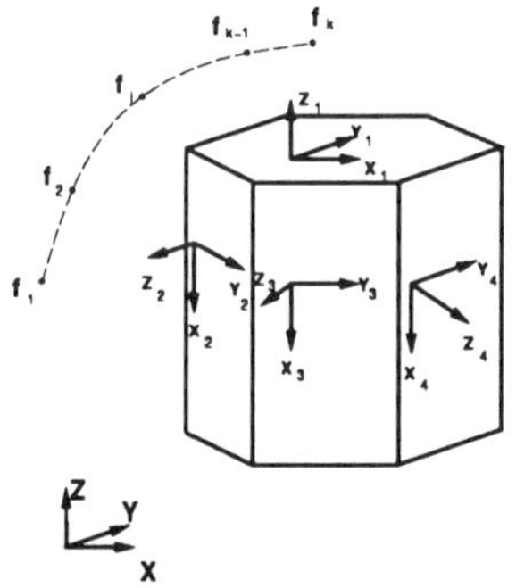

Figure 2: A Polyhedron Obstacle and Global/Local Coordinate Frames

$\{x_i, y_i, z_i\}$ $(i = 1 \ldots N)$, has the x and y axes laid within the boundary plane, the z axis along the direction of the surface normal (pointing toward the outside of the object), and the center of the coordinate frame placed anywhere within the boundary of the plane. On the other hand, the trajectory of a vertex of the moving object or all vertices of the moving object at given trajectory locations can be represented by a group of points $f_j = [f_x, f_y, f_z]_j^T$ $(j = 1, \ldots, K)$. Each of these points can be represented either in the global Cartesian coordinate $\{X, Y, Z\}$ or in the local coordinate frames of the obstacle boundary planes $\{x_i, y_i, z_i\}$.

A vertex point can be represented as ${}^{G}f = {}^{G}[f_x, f_y, f_z]^T$ with respect to the global coordinate frame G, and ${}^{L}f = {}^{L}[f_x, f_y, f_z]^T$ with respect to the local coordinate frame L. According to the theory of geometric transformation, the representation of a point in one coordinate system can be transferred into the representation in a second coordinate system by applying a series of geometric transformations. These transformations actually align the first coordinate frame with the second coordinate frame through a series of translations and rotations, and can be represented as:

$$ {}^{L}f = {}^{L}_{G}R \; {}^{G}f + {}^{L}_{G}t \tag{3} $$

where, ${}^{L}_{G}t$ is the offset vector representing the translation required for overlapping the centers of the two coordinate frames, and ${}^{L}_{G}R$ is the rotation matrix representing the rotations required for aligning the coordinate axes.

In this work, we are only concerned with which side the point is located in the halfspace, and the distance from the point to the boundary of the halfspace. In other words, only the z values in the N local coordinate frames are of interest. If the leading subscript G for the global coordinates is dropped for clarity of expression, the rotation matrix between the global coordinate system and the *ith* local coordinate frame, $(i = 1, \ldots, N)$, can thus be denoted as ${}^{i}R$. The third row of ${}^{i}R$ can be written as $\bar{w}_i = [w_{i1}, w_{i2}, w_{i3}]$ and the z component of the offset vector between the origins of the global coordinate system and the *ith* local frame attached to the *ith* plane can be denoted as T_i. The computations of all z coordinates of the vertex point f with respect to all local frames (plane boundaries) of a polyhedron obstacle can thus be represented as:

$$ {}^{i}f_z = f_x w_{i1} + f_y w_{i2} + f_z w_{i3} + T_i \qquad (1 \le i \le N) \tag{4} $$

The expression has an identical form as the equation (1), and thus can be carried out by the N neurons of the first layer of a Hamming net, through the inner product operations of a Hamming net.

If $sgn(\ {}^{i}f_z\ ) > 0$, for any $1 \le i \le N$, the point f is definitely outside of the obstacle. This criterion can be written as either $\max_i\{sgn(\ {}^{i}f_z\ )\} > 0$ or $sgn(\max_i\{{}^{i}f_z\}) > 0$.

The second layer of the Hamming net is designed to select $\max_i\{{}^{i}f_z\}$ as equation (2) indicates. A few additional connections are incorporated into the modified Hamming net to obtain the sign of ${}^{i}f_z$ for verifying $sgn(\ \max_i\{{}^{i}f_z\}\ ) > 0$. The new circuit connections are enclosed in the box of dashed-line of Figure 1, as Sign Multiplexer. The function of the Sign Multiplexer is to change the original output of the Hopfield N-flop – an address k^* that points to the winning inner product ip_{k^*}, to the sign of ip_{k^*} – $sgn(ip_{k^*})$ for collision identification.

When collision occurs, or when a point f is inside an N-plane convex polyhedron obstacle, no matter in which local coordinate frame the point is expressed, its z coordinate will always be negative. That is:

$$ sgn(\ {}^{i^*}f_z\ ) = sgn(\ \max_i\{{}^{i}f_z\}\ ) \le 0 $$

Since ${}^{i^*}f_z$ is the winning maximum inner product and has the minimum absolute value, the *ith** plane of the object is the nearest boundary of the obstacle to the point f. The shortest way to push the point f out of the obstacle is to move along the z direction in the n^*-th plane-frame. This direction is given by $[w_{n^*1}, w_{n^*2}, w_{n^*3}]$ in the global coordinate frame, and is referred as the optimal direction for collision removal.

3.2 Line-to-Object Collision Identification and Distance Calculation

The point-to-object formulation can be extended to line-to-object distance calculation. This extension can be illustrated using a simple 2-D example, shown in Figure 3.

Two lines, I-I and II-II, which represent two locations of an edge of the moving object, and a polygon obstacle with nine line boundaries, $l_1, \ldots, l_9$, are shown in Figure 3. The line I-I has no contact with the obstacle, and the line II-II cuts through the obstacle. These two lines intersect the obstacle boundary at p_1, p_2, p_3 and p_4. Using the scheme presented in the previous sections, the modified Hamming net can easily identify that point p_3 and p_4 "touch" the boundary of the obstacle (zero point-to-object distance). This, in turn, suggests that the corresponding line segment II-II either intrudes or cuts through the obstacle. On the other hand, the modified Hamming net can also find the intersecting points p_1 and p_2 located outside the obstacle

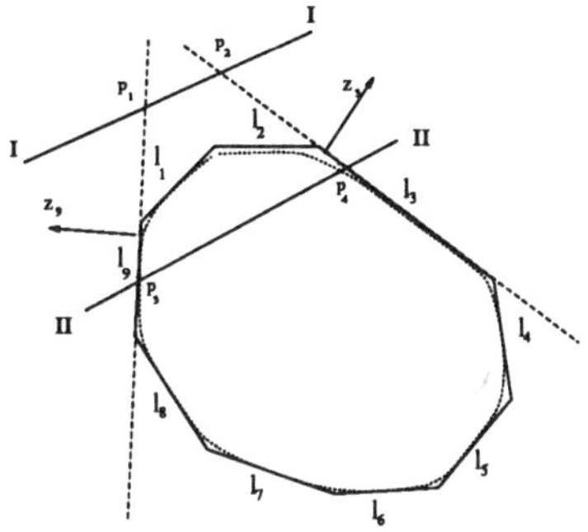

Figure 3: Line-to-Object Interference Identification

(positive point-to-object distance). This, in turn, indicates that the corresponding line I-I has no interference with the obstacle.

When the line-to-object interference is identified, the optimal direction for removing the collision can also be suggested by the modified Hamming net. This direction is the vector sum of the surface normals of the surfaces through which the line cuts. For the example shown in Figure 3, $\vec{z}_3 + \vec{z}_9$, provide an ideal direction for removing the line-to-object interference.

3.3 Object-to-Object Collision Identification and Distance Calculation

A combination of the equivalent workspace mapping method [3] with themodified Hamming net based point-to-object and line-to-object collision identification schemes leads to an efficient method for directly handling solid objects. Basically, a cube-like moving object can be modeled by a moving point and the new surrounding obstacles, of which the volumes are expanded at all boundaries. The amount of outward boundary movements of the volume expansion is equal to the radius of the circumscribed sphere to the moving object, as illustrated in Figure 4 (a). On the other hand, a beam-like moving object can be modeled by a moving line of the length of the beam, and the new surrounding obstacles, of which the volumes are expended at all boundaries. The amount of outward boundary movements of the volume expansion is equal to the radius of the circumscribed circle of the beam cross section, as illustrated in Figure 4 (b). With these two new equivalent workspace mapping methods, most of the components of robots and manufacturing facilities can be easily modeled. Detailed presentation of the method is beyond the scope of this publication.

4 Advantages

The collision identification and distance calculation method introduced in this paper can be used in tandem to the on-line workspace geometry acquisition using range scanning and geometric model generation (currently under study at the University of Victoria).

This new collision identification and distance calculation method has two major advantages over previous approaches. Firstly, the neural network operations can replace optimization-based distance calculations, thus avoiding the complex nested optimization problem. Secondly, The method can significantly improve the computational efficiency of collision identification and distance calculation by introducing

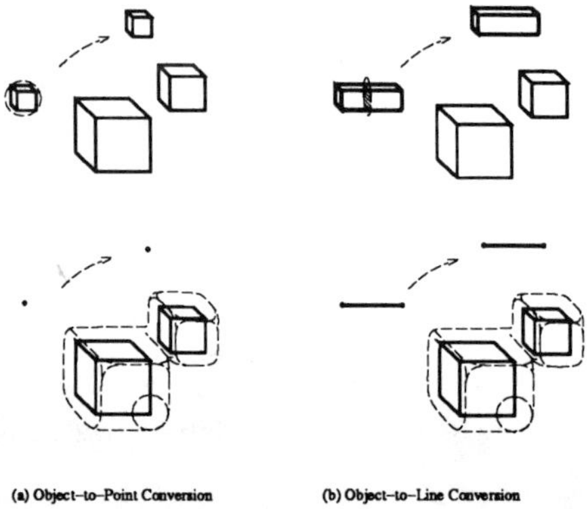

Figure 4: Object-to-Object Interference Identification

parallel processing, using a Neurocomputer. These two advantages make this method ideal for on-line motion planning of intelligent robotic and manufacturing systems.

References

[1] Bobrow, J.E., "A Direct Minimization Approach for Obtaining the Distance Between Convex Polyhedrons," *The International Journal of Robotics Research*, Vol. 8, No. 3, June, 1989, pp. 65-76.

[2] Liu, C.Y., and Mayne, R.W., "Distance Calculations in Motion Planning Problems with Interference Situations," *Proc. of the 1990 ASME Design Tech. Conf.*, DE-Vol. 23-1, 1990, pp. 145-152.

[3] Lozano-Perez, T., and Wesley, M.A., "An Algorithm for Planning Collision-Free paths Among Polyhedral Obstacles," *Communications of the ACM*, Vol. 22, No.10, October, 1979, pp. 560-570.

[4] Red, W.E., "Minimum Distance for Robot Task Simulation," *Robotica*, V.1, 1983, pp. 231-238.

[5] Gilbert, E.G., and John, D.W., "Distance Functions and Their Application to Robot Path Planning in the Presence of Obstacles," *IEEE Journal of Robotics and Automation*, Vol. RA-1, No. 1, 1985, pp. 21-30.

[6] Ganter, M.A. and Uicker, J.J., "Dynamic Collision Detection Using Swept Solid," *Journal of Mechanisms, Transmissions, and Automation in Design*, Vol. 108, 1986, pp. 549-555.

[7] Leu, M.C., Park, S.H., and Wang, K.K., "Geometric Representation of Translational Swept Volumes and Its Applications," *Journal of Engineering for Industry*, Vol. 108, 1986, pp. 113-119.

[8] Lippmann, R. P., "An Introduction to Computing with Neural Net", *IEEE ASSP Magazine*, April, 1987, pp. 4-22.

DESIGN SPECIFICATIONS FOR AN INTELLIGENT CONTROLLER OF MULTIPLE CONTACT DEXTEROUS MANIPULATION

S.P. Ananthanarayanan, A.A. Goldenberg — Robotics and Automation Laboratory
J. Mylopoulos — Department of Computer Science
University of Toronto , Toronto,
Ontario, M5S 1A4, Canada

ABSTRACT

This paper presents the fundamental issues that need to be addressed in the development of an intelligent controller for dexterous manipulation. A detailed list of functional specifications that are required to be taken into account in the design of such a controller are discussed. A comparitive analysis on the various possible methodologies of intelligent control is made in an effort to evaluate the merits and demirits of one methodology over the other, from the point of view of these functional requirements.

1. INTRODUCTION

It can be unequivocally stated that, one of the greatest legacies of robotic science to the human community has been the endeavour to replace humans by robot manipulators in hazardous environments for performing complex manipulation tasks that demand a fair deal of dexterity. From a robotic scientists point view, if such an endeavour has to blossom well beyond the realm of just being an academic pursuit into a really useful practical technology, a few fundamental problems will have to be tackled. Firstly, the robot must be capable of an *intelligent* reaction to the various types of uncertainties and anamolies that may arise (due to dispersions between the model and the actual parameters, friction and other phenomena) during the course of executing a task. *An intelligent reaction can be considered as any controlled response to an uncertainty, strategically trigerred at the time of encountering the uncertainty, by a repertoire of available knowledge of the robot, task and the environment and a presumed awareness of the manifestation of the uncertainty.* The degree of intelligence can vary from primitive levels of a controlled abortion of task execution (on encountering an uncertainty too large to be handled by the controller) to more sophisticated levels of automatic generation of a recovery procedure to accomodate the uncertainty. Secondly, the robots that replace humans in remote environments need *autonomy,* not essentially global, but at least local or supervised autonomy [1,2,3]. This issue has been strongly brought forth by researchers working in the area of telerobotics [1,3]. The time-delay involved in the communication between the human operator and the robot in the remote environment has been a major hurdle for applying teleoperated robots in unstructured or unknown environments. A local or supervised autonomy can handle uncertainties in such cases, without having to depend on control signals from the human operator at a remote control site.

The purpose of this paper is to discuss the various relevant issues regarding the development of an intelligent, autonomous control architecture for a complex dexterous manipulation process. An illustration of a complex dexterous manipulation process is given in section 2. The paper attempts to elucidate in section 3 what is expected out of an intelligent controller when it has to control such a complex dexterous manipulation task. In section 4 a comparitive analysis of various automated reasoning procedures for the intelligent controller are discussed in the pretext of choosing the best one that can match the specifications. Finally, section 5 concludes this paper.

2. AN ILLUSTRATION OF A COMPLEX MANIPULATION PROCESS - MULTIPLE AXIS ROTARY STEP

The purpose of this section is to illustrate by means of an example, a complex manipulation process that reorients the object. This manipulation process known as multiple axis rotary step, uses either push, slide or rolling contacts to rotate the object incrementally about different successive axes, with each rotation seperated by a relocation of the contacts [4]. An example of object reorientation using multiple axis rotary step is shown in fig.1. In fig. 1 (a) the object is held in a stable grasp. A relocation of finger 3 and finger 4 is shown in fig. 1(b). This is followed by a push by finger 3 and finger 4 from the new location (fig. 1(b)). As a result, the object undergoes a rotation about an axis along which finger 2 and finger 1 grasp the object. Note that this grasp is just an equilibrium grasp and not a stable grasp [5].

Finally, after a 180 deg. rotation, fingers 3 and 4 are relocated as shown in fig 1(c) to bring back the object to a stable configuration. By a similar procedure of relocation of finger 1 and finger 2, the object can be rotated about the axis collinear to the points of contact of finger 3 and finger 4. Thus by a repeated implementation of this cycle, we can perform a 3D reorientation of the object.

If an intelligent controller must be capable of doing the synthesis, control and monitoring of such a kind of manipulation autonomously, it has to satisfy certain functional requirements enumerated in the following section.

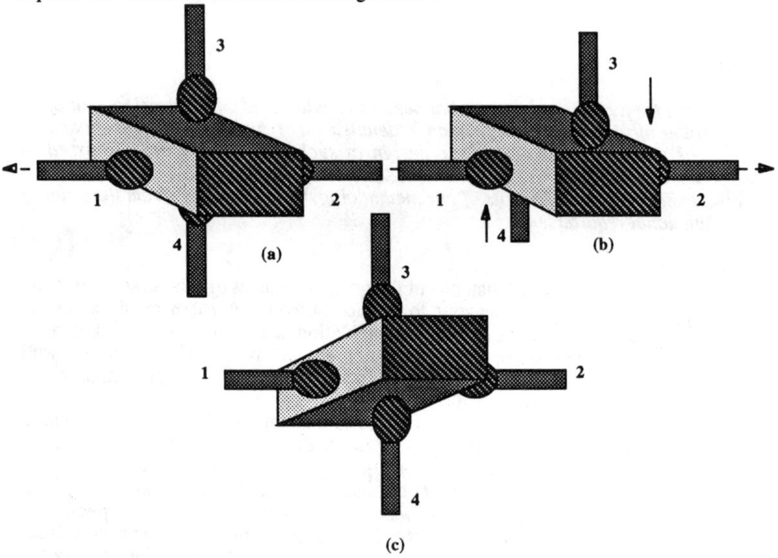

Fig. 1. A multiple axis rotary step - a complex manipulation process

3.SPECIFICATIONS FOR INTELLIGENT CONTROL OF MANIPULATION

3.1 Structural Specifications

(i) An intelligent controller for dexterous manipulation must be organized such that there exists an explicit distinction between domain and control knowledge

Domain knowledge is the knowledge of the various devices, environment, task etc. that represent a `state'. Control knowledge is that which is useful in the transformation of some particular state in the domain to another state completely distinguishable from the first. In an intelligent controller for dexterous manipulation, these two entities must be represented seperately . For example, the structure of the dynamic model of a particular robot is a part of domain knowledge specific to that robot. This structure should not be embedded in a controller that performs a manipulation synergy. By doing so, the scope of the intelligent controller becomes restricted.

(ii) Any intelligent controller for dexterous manipulation must be capable of control at three hierarchial levels viz., finger level, object level and task level, with a correspondingly decreasing order of magnitude in the control bandwidths (refer fig. 2)

The finger level control can either be a finger-joint control and/or a finger-tip control depending upon suitability, convenience etc. The control variables are either position, force or a combination of both, depending upon the task. The object level control comprises of motion and/or force control of the object/finger system as a whole. The task level plans, controls and monitors the overall performance of the task during task execution. It is in fact this level that differentiates an intelligent controller from the other types of traditional controllers. The task level in itself, can comprise of three hierarchial levels viz., reflex level, supervisory level and the planning level. The three layers from the bottom viz., the finger level/object level controllers and the reflex controllers require control bandwidths of the order of a few hundreds of hertz to 1 kilohertz (see fig. 2). The bandwidth requirement of the third layer viz., the supervisor is of the order of 5-10 hertz. The high level plan can be off-line.

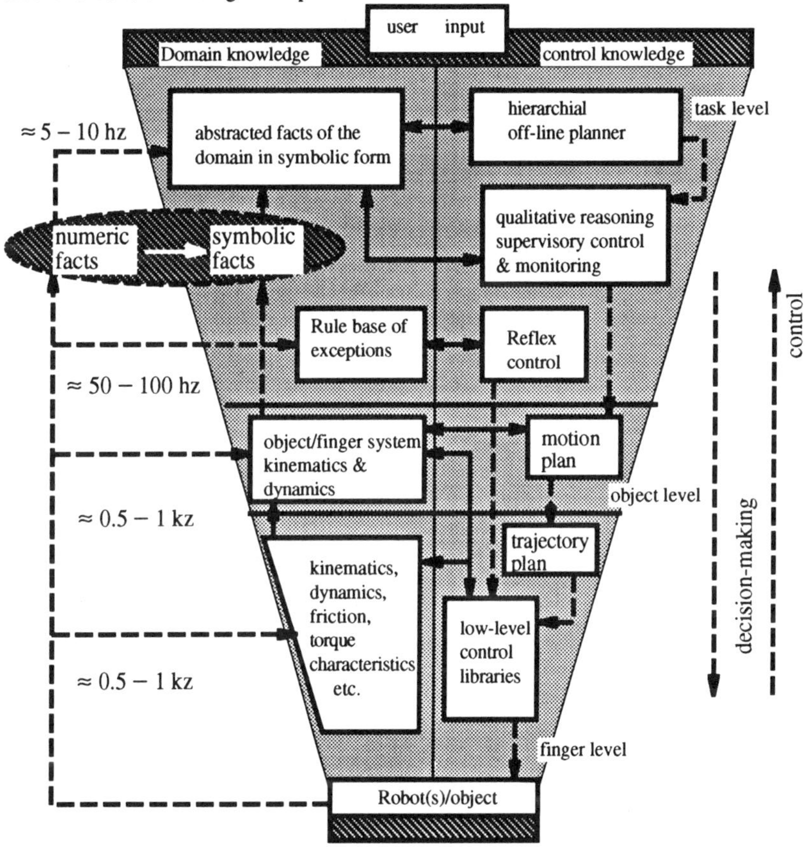

Fig. 2 Organization of an intelligent control system for dexterous manipulation

(iii) An intelligent controller for dexterous manipulation requires multiple abstracted levels of domain knowledge and reasoning, both in planning and real time control

The lowest level viz., finger level needs the knowledge of the dynamics and kinematics of each individual finger . Therefore, a knowledge base at this level must comprise of a dynamic model of the manipulator, an estimation of friction in the joints, torque input output characteristics of the individual motors at the joints, representation of frame transformations, forward and inverse kinematic models, jacobians, local contact constraints etc. At the object level, knowledge of the object/finger system kinematics and dynamics as a whole, static force constraints on the object, object attitude and location and the kinematic relationships between the object and finger-tip configurations, object inertia and other features like rigidity , shape, surface roughness etc are neccessary. The task level requires information regarding task constraints/limitations and its relationship to the finger-tip constraints and physical limitations, task performance characterestics, the task objective/goal etc.

The reasoning mechanisms (viz., planning and control algorithms) correspondingly must be capable of using the multiple abstracted levels of information. For instance, the lowest level reasoning mechanisms are the low level controllers for the individual manipulators. Each low level controller has a local objective of achieving a desired motion and/or a force in a single manipulator, a system of manipulators, or a system of manipulators inclusive of the object as a single entity, depending upon the type of manipulation task . Thus the scope of the low level controller spans the finger-tip and object levels. At the task level, three hierarchial levels of reasoning can be defined. The lowest level comprises of an expert system that functions as a reflex controller to handle exigencies like object slipping, excessive internal/external forces etc. This reasoning mechanism uses rule sets, pre-defined procedures, or look up tables. The reasoning mechanism at the next higher level is a supervisory controller. This reasoner must be able to monitor and evaluate the current state of the system in response to the low level controller, from sensory inputs using deep knowledge and generate refinement procedures or actions (on line planning). A qualitative reasoner that uses first principle laws of physics is a very suitable candidate for this type of reasoner. Finally, at the highest level viz., planning, more than one special purpose reasoners can be used.

3.2 Functional Specifications

(i) The problem of decision-making and control of an overall complex dexterous manipulation process is incremental and evolutionary and therefore requires a capability of appropriately choosing and integrating multiple control paradigms under one unified framework

A complex dexterous manipulation process can be contrived as a composite of basic synergies, wherein, each individual synergy may demand a seperate control and decision-making paradigm [4]. Therefore, an intelligent controller for a complex dexterous manipulation process must be capable of choosing appropriate control paradigms for controlling the individual synergies. The above statement presupposes the availability of a control library that is exhaustive of all the individual synergies making up the composite synergy. The intelligent controller must further be capable of integrating these individual paradigms under a single unified framework. It should be capable of addressing the issue of transitions between two different control paradigms corresponding to two successive basic synergies without a loss in the stability or performance of the controller.

(ii) An intelligent controller for dexterous manipulation must be capable of asynchronous multi-tasking

In manipulation, individual manipulators may have to be controlled simultaneously. At the same time, triggered by events during the process, decision making has to be done at different levels in the architecture on further manipulative/corrective actions. An intelligent control architecture must thus be capable of asynchronous multi-tasking.

(iii) An intelligent controller must be capable of analysing the feasibilty of achieving a given manipulation with a given set of resources

Often, it will not be possible to perform an arbitrary manipulation synergy from a given set of input configuration, using a particular hardware (e.g dexterous hand). An intelligent controller for dexterous manipulation must be able to identify such situations, using some formalised approaches.

(iv) An intelligent controller must be capable of reconciling desirability and feasibility in performing manipulation

In addition to verifying feasibility, the intelligent controller must also possess the capability of reconciling feasibility and desirability. It is often the case that more than one option will be available to the intelligent controller to perform manipulation. Under such circumstances, the intelligent controller must be able to choose the `best' amongst the available options.

(v) An intelligent controller must be capable of appropriately choosing and applying variable grain size reasoning and analysis while decision-making

In practical environments, acquisition of accurate information about the world is often very difficult. Therefore, the intelligent controller must be able to gracefully degrade under such situations, to using reasoning mechanisms that can accomodate for uncertainty and provide approximate but useful reasoning and analysis (e.g fuzzy logic, qualitative reasoners etc.) about the domain for generating further actions.

(vi) An intelligent control architecture must be capable of integrating diverse information and information represented in multiple formalisms

This is obvious from the fact that a dexterous manipulation problem is comprised of loosely coupled sub-problems. For instance let us consider any complicated dexterous manipulation process. The issues that will have to be addressed during manipulation planning, controlled execution and monitoring include the following.

(a) Assigning suitable reference frames for various entities (manipulator,object etc.)
(b) Analysing the feasibility of performing manipulation
(c) Choosing the best possible option to achieve manipulation
(d) Decomposing the complex manipulation into a sequence of procedures
(e) Designing motion plan for the individual procedures
(f) Choosing appropriate low level controllers for each procedure
(g) Checking periodically during manipulation, for exceptions like object slip, large forces etc. and appropriate activation of error recovery routines and so on.

Each problem enumerated above can be considered as independent of the others when considered seperately. Also, it is obvious that each sub-problem can be represented in a different formalism. For instance, the exception handler may be a look-up table or rule based system whereas analysing the feasibility of manipulation may be a list of procedures and so on. An intelligent controller therefore, must be equipped to handle the inherent diversity of information and integrate multiple representation formalisms.

4. A Comparitive Evaluation On The Different Possible Automated Reasoners For Intelligent Control

Any intelligent controller for dexterous manipulation must have an automated reasoning and decision-making mechanism to generate actions.The actions range from a generation of high level plans to choosing appropriate low-level control and monitoring actions during real time execution. There are various approaches to building automated reasoning in an intelligent controller in order to perform autonomous manipulation for a particular set of input and goal conditions. However each approach has its own advantages and draw-backs and is (manipulation) task dependant. In this sub-section a comparitive evaluation of different candidates for organizing an intelligent controller is made in an effort to get a better understanding of what types of organizations are suitable for the problem of intelligent control.

4.1 Reasoning based on Search Procedures

The problem of searching is to find the appropriate set of operators that transform the current state to the goal state. The set of all states (state space) can be usually represented as a graph and the various states as the nodes of the graph. The action of transforming a given initial state to a desired state then transcends to one of finding a path between the nodes that represent the two states. There exist at least five different methods of searching through the graph to find a path viz., hill climbing, depth first search, breadth first search , iterative deepening, best first search and so on . All these search procedures are suitable for theorem-proving or game-playing strategies [6]. However, they are quite inadequate to form the central decison-making mechanism in an intelligent controller. Firstly, search procedures are mechanical and would result in combinatorial explosion when confronted with a huge search space such as in intelligent control. Search procedures are not suited for generating real time monitoring actions as such actions require higher bandwidth. Also, real time control actions are often triggered by random events and therefore shift of focus between two processes needs to be done by capture. In search methods shift of control is by iteration.

4.2 Goal Oriented Decision-making

The basic idea behind this approach is based on a collection of goal-directed procedures. Each procedure is self-contained and achieves a particular goal or sub-goal and has an access to a common database. Thus the overall decision is made by decomposing the problem of decision-making into a series of sub-problems. While these procedures can be effective, there are some serious difficulties from the view point of implementing them in an intelligent controller. For instance, when a procedure gets eecuted it is unaware of the partial states of the other procedures and can only communicate with a common data base. If these procedures require intermediate results that are not placed in the data base, then they become ignorant of situations that may be important to them. Thus, this approach to decision-making in intelligent controllers can be useful only if the individual sub-problems to be considered before a decision can be made are independent of each other. It is almost impossible to contrive such decisions in an intelligent controller.

4.3 Object Oriented Decision-making

The recent development of object oriented programming offers a more rigorous approach to the problem of decision-making. In this method, procedures are called `object'. Every procedure (object) can communicate with another by message passing. The passing of messages can be thought a formalised method of structuring constraints. One of the major advantages of this method is it is possible to instantiate further procedures from a generalised class of procedures. However, the ability of using this approach in a practical environment such as in intelligent control decision-making, depends upon the possibility of decomposing the overall decision-making problem into goal oriented sub-problems. It is often the case that such a decomposition is not possible. This is true for the previous approach outlined above as well.

4.4 Rule Based Decision-making

In rule based systems, a decision can be made by forward or backward chaining through a set of rules. Automated decision-making systems based on rule bases is quite popular due to the fact that the rules interact in a more flexible manner. It is also possible to characterize uncertainty in decision-making by incorporating fuzzy logic or probability in the design of a rule base. This makes it very promising for applications like intelligent controllers.

Further, if the domain of decision-making is specific, the rule bases can encode an experts knowledge of the domain. They are fast and differ from other kinds of search procedures described previously in that, the search is not mechanical, but depends upon the current state. There are however certain inherent advantages. If the domain is highly complex, decision-making becomes complex, and some times intractible. It is not possible to exhaust the domain using rule sets and therefore, there is always a component of domain knowledge about which the decision-maker is ignorant.

4.5 Blackboard Systems

A blackboard system is an opportunistic decision-making paradigm and is usually described to consist of two main components [7] viz.,

(i) The knowledge sources : The knowledge needed to solve the problem, which are kept independent and seperate

(ii) The blackboard data structure : The data that are kept in a global data base.

Knowledge sources produce changes to the blackboard that lead incrementally towards the goal of the problem. Blackboard systems may consist of multiple panels. The purpose of these panels is to hold computational or solution state data needed by and also produced by the knowledge sources. Therefore, these blackboard panels consist of objects from the solution space. These objects can be input data, partial solutions, alternatives and final solutions. The objects in the blackboard are hierarchially organized into levels of analysis. It can be noted that, information associated with objects on one level serves as input to a set of knowledge sources which in turn, place new information in the same or other levels. Also, the solution state is partitioned into multiple hierarchies. A control panel in the blackboard monitors the changes in the blackboard and decides on the control actions that are to be pursued. The solution in a blackboard model is built one step at a time by control actions taken at the control panel. Control actions are both data driven and event driven. A knowledge source takes current information from the blackboard and updates it during every stage of the solution process. A knowledge source thus modifies the blackboard and is also responsible for knowing the conditions under which it can contribute.

From the above it is evident that blackboard systems are highly flexible in many aspects. Firstly, the knowledge sources can be of different formalisms. The knowledge can be diverse in nature. Each knowledge source can operate independently and concurrently. The domain knowledge is explicitly distinguished from the control knowledge. Therefore, they seem to be a natural choice for complicated manipulation processes as they satisfy most of the specifications stated in section 3. However, a careful examination of the suitability of the architecture to a particular task has to be investigated before implementation, in order to avoid an overkill.

5 CONCLUSIONS

This paper attempted to provide in a nut-shell, the specifications for design of an intelligent controller for a dexterous manipulation process. Various types of automated reasoning mechanisms were analysed in an effort to establish the mechanism that can best satisfy the specifications. From the analysis it is clear that a blackboard architecture is very suitable for complex manipulation processes. However, use of such an architecture is an overkill for simple synergies. Based on the principles outlined in this paper, an intelligent controller that can perform the complex manipulation process illustrated in section 2 has been successfully developed and experimentally verified [8].

REFERENCES

[1] C.R. Weisbin, M.D. Montemerio, NASA's Telerobotics Research Program, in the Proc. of IEEE Intl. Conf. on Rob. & Aut., 1992, pp. 2653 -2666

[2] R. Chatila, R. Alami, B. Degallaix, H. Laruelle, Integrated Planning and Execution Control of Autonomous Robot Actions, Proc. of IEEE Intl. Conf. on Rob. & Aut, 1992, pp. 2689-2696.

[3] P. G. Backes, Multi-sensor based Impedance Control for Task Execution, Proc. of Int. Conf. on Rob. & Aut, 1992, pp. 1245-1250

[4] S.P. Ananthanarayanan, A.A. Goldenberg, J. Mylopoulos, A Classification of Dexterous Manipulative Movements, Internal Report, RAL, University of Toronto, May 1992.

[5] Z. Li, S. Sastry, Dexterous Robot Hands : Several important issues, 1988 IEEE Workshop in Grasping & Manipulation, April, 1988.

[6] M. Lee, Intelligent Robotics, John Wiley & Sons, 1989.

[7] R. Engelmore & T. Morgan, Blackboard Systems, Addison-Wesley Publishing, 1988.

[8] S.P. Ananthanarayanan, A.A. Goldenberg, J. Mylopoulos, Intelligent Control of Dexterous Manipulation, Internal report, RAL, University of Toronto, April 1992.

A HYBRID APPROACH TO COMPOSITE STOCK CUTTING: NEURAL NETWORKS AND GENETIC ALGORITHMS

PIPATPONG POSHYANONDA AND CIHAN H. DAGLI
Department of Engineering Management
University of Missouri-Rolla
Rolla, MO 65401-0249

ABSTRACT

This paper explores the integration of artificial neural networks and genetic algorithms in automating irregular-shaped pattern generation process for cutting uncured composite materials without any pattern orientation restrictions. The proposed approach generates a packing pattern for a finite width and infinite length of uncured composite sheet.

INTRODUCTION

The problem of allocating irregular and/or rectangular patterns arises frequently in applications where it has to be determined how a set of two dimensional shapes will fit onto a large stock sheet of finite dimensions in such a way that the resulting scrap is minimized. The problem is common to many industries such as aerospace, shipbuilding and composite parts manufacturing. The savings as a result of efficient packing of irregular parts increases as the cost of material used in cutting parts becomes more expensive.

Since 1950's, the cutting stock problem has received considerable attention as the computers have appeared as fast and economic information processors and the optimization techniques have developed. Solution of this problem is of considerable practical significance. Because of the diversity of the structures of real world stock cutting problems, there exists no general standard method for solving them. Over the years, two main approaches are emerged namely; heuristics and approaches based on linear programming relaxations.

Heuristics [1,2] play an important role, since they are flexible enough to take into account various additional restrictions and objectives appearing in practice. The quality of heuristics is generally problem specific, and they can identify a pattern which is "good" for the particular problem in question. They can be integrated into intelligent decision support systems easily.

On the other hand, linear programming algorithms first solve the linear programming relaxation of the cutting pattern oriented model and search for an integer solution using branch and bound methods or simply rounding up to the next integer. In almost all linear programming models proposed [3-5], a two stage approach is used. Initially various cutting patterns are generated based on given stock sheet dimensions and small pattern shapes, then a decision variable is assigned for each cutting pattern and a linear programming model is formulated.

The Gilmore and Gomory approach [6] is classical in delaying some of the pattern generation process and generating the required patterns during the solution process using shadow prices. Generally, linear programming approaches are not widely used in practice, and most of the time, heuristics are preferred for selecting patterns. This is basically due to the fact that the large number of possible patterns combinations (in the order of hundred millions) cannot be represented in linear programming formulations. The primary reason for restricting cutting patterns used in mathematical programming formulation was basically related to technological restrictions for solving large linear programming problems.

Simulated annealing proposed by Krikpatrick and his colleagues [7] is one of the optimization method that tries to avoid pitfalls inherent in other optimization methods. Although simulated annealing could be easily mapped into parallel processing, this technique is inherently slow. Good solutions are not consistently produced by simulated annealing [8].

Due to the difficulty in formulating constraints that prevent the overlapping of patterns in the solution approaches proposed in literature cutting pattern generation is done prior to or within the solution phase. Most of the time this process is sequential and requires intense computational effort which artificial neural networks can contribute greatly to this process. In this study, a solution approach which integrates artificial neural networks and genetic algorithms for this NP-complete problem is developed. The solution model uses irregular shaped parts with different orientation requirements of fibers on uncured composite material as its inputs and generates packing pattern for the composite parts that makes the final product according to the restrictions imposed on composite cutting operations such as; manual cutting, reciprocating knife, ultrasonic vibratory cutting, die cutting, laser and water-jet cutting or broad goods cutting machine.

DESCRIPTION OF THE PROPOSED WORK

The proposed approach integrates artificial neural networks and genetic algorithms in automating irregular-shaped pattern generation process for cutting of uncured composite materials. The model generates a sequence of irregular-shaped patterns based on dynamically generated available pattern allocation shapes on the composite stock sheet of finite width and infinite length. Basic characteristics of the approach are summarized in the following sections.

Irregular-Shaped Pattern Representation

Pattern representation is an important component of the solution approach in stock cutting problems. In order to represent irregular-shaped patterns, matrix representation due to Dagli [9] is used. Each pattern is represented by a matrix of size K by L which is the minimum rectangular enclosure of the irregular shape as shown in FIG. 1 and 2. The values of K and L are determined on the desired accuracy of computations. This representation helps in reducing computational time, such as testing the degree of overlap among patterns.

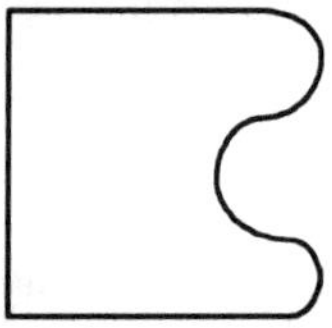

$$\begin{bmatrix} 1&1&1&1&1&1&1&1&1&1&1&0 \\ 1&1&1&1&1&1&1&1&1&1&1&1 \\ 1&1&1&1&1&1&1&1&1&1&1&1 \\ 1&1&1&1&1&1&1&1&1&1&1&0 \\ 1&1&1&1&1&1&1&0&0&0&0&0 \\ 1&1&1&1&1&1&1&0&0&0&0&0 \\ 1&1&1&1&1&1&1&0&0&0&0&0 \\ 1&1&1&1&1&1&1&1&1&0&0&0 \\ 1&1&1&1&1&1&1&1&1&1&1&1 \\ 1&1&1&1&1&1&1&1&1&1&1&1 \end{bmatrix}$$

FIG. 1. An irregular shape represented by a binary matrix of size 10x12

Nonetheless, the use of single representation for this problem is not sufficient. Once all coordinates of an irregular shape converted to the matrix representation, they will be lost. Therefore, both matrix and coordinate representations are used in this study to maintain original orientations of patterns.

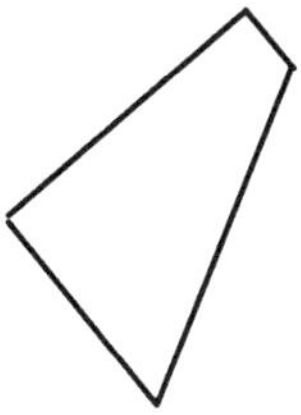

$$\begin{bmatrix} 000000000100 \\ 000000011110 \\ 000001111110 \\ 000111111110 \\ 001111111100 \\ 111111111000 \\ 011111111000 \\ 001111110000 \\ 000111110000 \\ 000011100000 \\ 000001000000 \end{bmatrix}$$

FIG. 2. An irregular shape represented by a binary matrix of size 11x12

Orientation of the input patterns

Besides the pattern representation, the orientation of the input shapes is an important issue that must be considered. Hence, all the input patterns should be reoriented in such a way that it produces the minimum scrap. As shown in FIG. 2, the input shape does not reflect the minimum size of the rectangular enclosure, it should be reoriented in order to produce the minimum rectangular enclosure. FIG. 3 shows the rotated matrix representation of the pattern given in FIG. 2 resulting a smaller rectangular enclosure. Time required in the allocation process is reduced tremendously by using the reoriented patterns since the possible allocations for a rectangle is the multiples of 90 degree, which is far less than that of irregular shapes. The pseudo code of the pattern reorientation is given below.

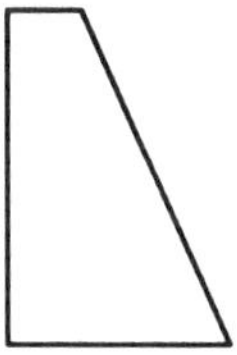

$$\begin{bmatrix} 111000000 \\ 111100000 \\ 111100000 \\ 111110000 \\ 111111000 \\ 111111100 \\ 111111100 \\ 111111110 \\ 111111111 \end{bmatrix}$$

FIG. 3. An rotated irregular shape represented by a binary matrix of size 9x9

```
FOR i = 0 TO 90
    Create the main axis which forms "i" degree with the normal X axis
    Create the perpendicular axis of the main axis
    Project all coordinates of the shape to these two axes
        This will give the distances of all points from the origin along the axes
    The length of the shape on each axis is calculated as
        The maximum projected value - The minimum projected value
    Calculate the size of the rectangular enclosure as
        Length on the main axis  X  Length on the perpendicular axis
ENDFOR
Find the smallest size of rectangular enclosure among 91 solutions found above
Rotate all coordinates of the pattern that will reflect the minimum rectangular enclosure
Create the matrix representation from these new coordinates
```

Detailed Allocation Algorithm

A composite material stock sheet of finite width and infinite length is used in this problem. The objective of the detailed allocation algorithm is to allocate all the input parts within the stock sheet with minimum scrap area. Since the algorithm uses the reoriented parts which require the minimum placement area as its inputs, "sliding pattern sequence" method is employed. Pattern allocation is done sequentially. An incoming pattern is placed to next the previously allocated one along the width of the stock sheet. If there is not enough space to allocate the pattern, an incoming pattern is placed next to the edge of the sheet, and a new row of patterns is formed (FIG. 4). All scrap areas created by the detailed allocation at each stage are kept to be utilized later. During each pattern allocation, an incoming pattern is matched with all the scrap areas kept in the previous stage. If the pattern can be allocated into one of the scrap areas, the best matched area is selected for allocation.

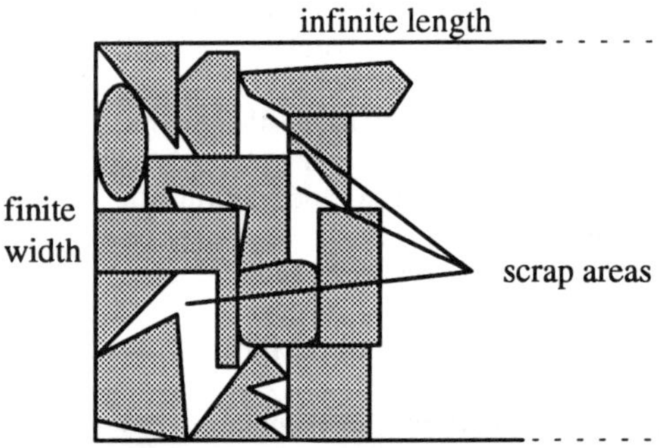

FIG. 4. Pattern allocations and scrap areas produced on the stock sheet

How do we select the best match among the scrap areas generated and the pattern being allocated? All the scrap areas, on the other hand, can be considered as irregular-shaped patterns that can be directly matched with an input pattern. If there is a possible match between these patterns, the particular scrap area is utilized as the allocation area for the input pattern. The technique of artificial neural networks can be employed during the matching process. Adaptive Resonance Theory, namely; ART1 paradigm [10-11] is selected for this purpose. ART1 learns on-line, operates in discrete time, and is capable of storing any pattern represented in binary number in its memory. All the generated scrap areas on the stock sheet at any point in time, which is the current configuration of the problem, can be represented as binary numbers and stored in the network's memory. Once an input shape needs to be allocated on the stock sheet, it is presented to the ART1 network to locate a possible match among previous generated scrap areas. All stored scrap areas compete with each other to find the best match with the input shape. The location to allocate the shape is provided if there is a possible match. The parallelism provided by artificial neural network can reduce the computation time greatly required by sequential approaches. This is achieved through the detailed allocation algorithm in which the pseudo code is given below.

```
Construct the ART1 neural network to represent all scrap areas of the configuration
FOR i = 1 TO Number of input patterns
        Pass an input to the ART1 network
        IF there is a "best-fit" scrap area
                Place the input pattern on that particular scrap area
        ELSE
                Place the input pattern next to the allocated patterns
        ENDIF
        Adjust the new pattern location in such a way that requires minimum total area
        Update the ART1 network based on the current available spaces on the stock sheet
ENDFOR
```

An Allocation Enhancement

Since the detailed allocation algorithm allocates each irregular-shaped input pattern based on pattern sequence, the performance of the algorithm depends heavily on the sequence selected. Different sequences reflect different packing allocations. At this point, this sub-problem can be considered as a travelling salesman problem (TSP) which can be solved by genetic algorithms.

Genetic algorithms, introduced by Holland [12], become an alternative solution method for optimization problems. The algorithm starts with an initial set of random configuration, called population. Each individual in the population which represents each possible solution is known as gene, in this case, a sequence of input shapes. It is usually a binary bit string representation. During each iteration, called a generation, each solution in the current population is evaluated using some measure of fitness, for example, the length of material used in the cutting process. Based on this evaluation value, individual is selected two at a time as parents to generate new individuals, call offsprings, through genetic operators. The better the individual is, the higher probability for it to be selected as a parent. Due to this stochastic selection process, the fitter parents, which are expected to contain good subsequences, should produce more offspring than the bad parents. Better the individual is, the higher probability to survive in next generations. With a fixed size of population, new population with the same size are created from the population of the previous generation as their parents. In the new generation, all of the population of the previous generation are eliminated. This inheritance also repeats for the offsprings to create new generations. As a result, the number of good subsequences in the population tends to increase through generations, on the other hand, the number of bad ones tends to decrease.

In the proposed approach, genetic algorithm is used as a solution of the travelling salesman problem. Each gene represents a possible solution of the problem which is a sequence of the input patterns to be provided to the detailed allocation algorithm. The length of an individual is the number of required shapes to be allocated. To be able to create better solution, order crossover (OX) and inversion genetic operators are applied [13]. Good characteristics within sequences of input shapes inherits through generations and creates better sequences for the detailed allocation algorithm in the next generations. The genetic algorithm is repeatedly applied until a satisfactory level of scrap produced by the allocation process is reached. The travelling salesman problem is solved through a genetic algorithm using the pseudo code is given below.

```
Generate rectangular maps for all input patterns by creating there rectangular enclosure
Generate random sequences of input patterns, called population, for a certain
        number of population size (Np)
FOR i = 1 TO Np
        Construct the solution for each individual in the population through the detailed
                allocation algorithm for the corresponding sequence for the input patterns.
        Evaluate each solution by the length of the total material used.
        Generate offsprings from the current population through genetic operators.
                The patterns are selected by the probability based on the evaluation value.
ENDFOR
Return the highest scoring configuration in the population
```

REMARKS

This paper presents an integrated approach to the uncured composite material stock cutting problem. The required parts are represented by matrix representations in order to reduce the computational time in testing for overlap parts. Both artificial neural networks and genetic algorithms used in the proposed approach provide a parallel approach to the solution in which the computational time is greatly reduced.

The results of the proposed approach is still inconclusive at this time since the complete model is currently being developed using an object oriented language. However, the preliminary results of the model without the scrap utilization in the pattern allocation is approximately 25 percent of scrap which could be reduced to 15-20 percent after the complete model is implemented.

REFERENCES

[1] Chung, J., Scott, D., and Hillman, D. J., "An Intelligent Nesting System on 2-D Highly Irregular Resources", *Proceedings of Applications of Artificial Intelligence VIII* **1293**, 472-483 (1990).

[2] Dagli, C. H., "Knowledge-based System for Cutting Stock Problem", *European Journal of Operation Research* **44**, 160-166 (1990).

[3] Christofides, N., and Whitlock, C. (1977), "An Algorithm for Two-dimensional Cutting Problems", *Operations Research* **1**, 9-16 (1977).

[4] Haessler, R. W., "A Note on Computational Modifications to the Gilmore-Gomory Cutting Stock Algorithm", *Operations Research* **28**(4), 1001-1005 (1980).

[5] Hinxman A. I., "The Trim Loss and Assortment Problems: A Survey", *European Journal of Operation Research* **5**, 8-18 (1980).

[6] Gilmore, P. C. and Gomory, R. E., "Multistage Cutting-Stock Problems of Two and More Dimensions", *Operations Research* **13**, 94-120 (1965).

[7] Krikpatrick, S., Gellatt, C. D. and Vecchi, P., "Optimization by Simulated Annealing", *Sci* **220**(4598), 671-680 (1983).

[8] Lutfiyya, H., McMillin, B., Poshaynonda, P., and Dagli, C. H., "Composite Stock Cutting Through Simulated Annealing", Mathl. Comput. Modelling **16**(1), 57-74 (1992).

[9] Dagli, C. H., "Neural Network in Manufacturing: Possible Impacts on Cutting Stock Problem", *2nd International Conference on Computer Integrated Manufacturing,* 531-537 (1990).

[10] Carpenter, G. A. and Grossberg, S., "Category Learning and Adaptive Pattern Recognition: a Neural Network Model", *Third Army Conference on Applied Mathematics and Computing, ARO Report* **86**(1), 37-56 (1985).

[11] Carpenter, G. A. and Grossberg, S., "A Massively Parallel Architecture for a Self-Organizing Neural Pattern Recognition Machine", *Comput. Vision Graphics Image process.* **37**, 54-115 (1987).

[12] Holland, J. H., "Adaptation in Natural and Artificial Systems", *University of Michigan Press,* Ann Arbor, Michigan (1985).

[13] Oliver, I. M., Smith, D. J., and Holland J. R. C., "A Study of Permutation Crossover Operators on the Travelling Salesman Problem", *Proceedings of the second international conf. on genetic algorithms*, 224-230 (1987).

LEARNING CONTROLLERS FOR A TWO-LINK ROBOT MANIPULATOR SUBJECT TO UNCERTAINTIES

G. Bojadziev and M. Bojadziev

Department of Mathematics and Statistics
Simon Fraser University, Burnaby, B.C. Canada V5A 1S6

Department of Mathematics British Columbia Institute of Technology
Burnaby, B.C. Canada V5G 3H2

ABSTRACT

A learning control of motion program is designed to guarantee that the links of a robot manipulator subject to uncertainties do not enter into an avoidance region.

1. INTRODUCTION

We study a controlled mechanical system consisting of two open-end links which rotate in a vertical plane α. A real model always contains elements which are uncertain (parameter uncertainties), or uncertainties may come from the environment (external disturbances or input uncertainties). We deal here with both, parameter uncertainties due to uncertain masses and input uncertainties; the uncertainties are assumed to be bounded.

We propose a learning procedure for the controllers which results in avoidance of a certain region by the state variables describing the manipulator motion. This procedure is used to design an avoidance strategy for the motion of the manipulator in the plane. First a Lyapunov based memoryless control program is established for the special case where the masses are assumed known constants and there are no input uncertainties (see the paper [1] by the authors). Using the experience acquired in this process, a controller for the general system with uncertainties is obtained by generating an adaptive type of control law. We utilize a Lyapunov based control methodology (Corless, Leitmann, and Skowronski [2]), and a paper by Bojadziev and Saif [3] for a planar robot manipulator with n successively connected links subject to control torque/force, with uncertain masses, but no damping and input uncertainties.

2. THE MANIPULATOR DYNAMICS

Consider the double pendulum (Fig. 1) as a model for a robot manipulator with two open-end rigid links connected at O. The links have lengths a_1 and a_2 and masses m_1 and m_2 which also represent the grippers and are centered at the points P_1 and P_2 correspondingly. The position of the links is given by the angular coordinates φ_1 and φ_2, positive in counterclock-wise direction. A damping force $R_s = -\lambda_s\dot{\varphi}_s$, $\lambda_s > 0$, s=1,2, acts on the s-th link and a damping force of the type $q = F(\dot{\varphi}_s - \dot{\varphi}_s)$ acts on the links at O (q reflects the existence of a connection between the links). Each link at the joint point O is subjected to a torque/ force $\mu_s a_s \varphi_s^2 u_s$, s=1,2 , $u_s \in R$, where $\mathbf{u}(t) = (u_1,u_2)^T \in U$ is a control vector. The unknown bounded environmental forces are $Q_s = \mu_s a_s^2 w_s$, s=1,2, $w_s \in R$, $|w_s| < c = \text{const}$, $c > 0$, where $\mathbf{w}(t) = (w_1,w_2)^T \in W$ is the vector of input uncertainties. The mass m_s is assumed to be uncertain constant in the known constant interval $[m_s^\ell, m_s^r]$, $m_s^\ell, m_s^r > 0$. The uncertain mass m_s can be expressed as

$$m_s = \bar{m}_s + \overset{*}{m}_s\,, \quad \bar{m}_s = \frac{1}{2}(m_s^\ell + m_s^r) \tag{1}$$

where $\bar{m}_s$ is an artificial mass located at the midpoint of the interval $[m_s^\ell, m_s^r]$, and $\overset{*}{m}_s$ is the difference between the actual uncertain mass m_s and the average value $\bar{m}_s$. The dynamics of the two-link robot manipulator is given by the Lagrange equations:

$$\begin{aligned}
&\dot{\varphi}_1 = y_1\,, \quad \dot{\varphi}_2 = y_2\,,\\
&(\bar{m}_1 + \overset{*}{m}_1)a_1^2\dot{y}_1 = -\lambda_1 y_1 - g(\bar{m}_1+\overset{*}{m}_1)a_1\sin\varphi_1 + F(y_2-y_1)\\
&\qquad + \bar{m}_1 a_1^2(u_1^m+u_1^a) + \bar{m}_1 a_1^2 w_1,\\
&(\bar{m}_2+\overset{*}{m}_2)a_2^2\dot{y}_2 = -\lambda_2 y_2 - g(\bar{m}_1+\overset{*}{m}_2)a_2\sin\varphi_2 - F(y_2-y_1)\\
&\qquad + \bar{m}_2 a_2^2(u_2^m+u_2^a) + \bar{m}_2 a_2^2 w_2\,,
\end{aligned} \tag{2}$$

where g is the gravitational constant. The control vector $\mathbf{u}(t)$ is presented as a sum of a memoryless control $\mathbf{u}^m(t) = (u_1^m,u_2^m)^T$ and a type of adaptive control $\mathbf{u}^a(t) = (u_1^a,u_2^a)^T$. For convenience we have set $\mu_s = \bar{m}_s$ in (2). Further we assume that:

(a) $F(y_2-y_1)$ is a C^1 function;

(b) $F(y_2-y_1) = -F(y_1-y_2) \Rightarrow F(0) = 0$ (F is an odd function);

(c) $F(\gamma_2 y_2-\gamma_1 y_1)\,F(y_2-y_1) > 0\ \forall\,|y_2-y_1|$, $\gamma_s = (\bar{m}_s a_s^2)^{-1}$, s=1,2.

By solution of (2) we mean an absolutely continuous vector function $\mathbf{z}(t) = (\boldsymbol{\varphi}(t), \mathbf{y}(t))^T \in R^4$ which satisfies (2) almost everywhere; $\boldsymbol{\varphi} = (\varphi_1,\varphi_2)^T$, $\mathbf{y} = (y_1,y_2)^T$. We refer to (2) as the general model (GM). For $\overset{*}{m}_s = 0$ and $\mathbf{u}^a = 0$ in (2) an auxiliary system with fixed artificial masses $\bar{m}_s$ (see (1)) and a memoryless control $\mathbf{u}^m$ is obtained, called here the nominal model (NM); it solution will be labeled as $\boldsymbol{\zeta}(t) \in R^4$. If in addition $\mathbf{u}^m = 0$, one gets the uncontrolled nominal model (UCNM). If $\mathbf{u}^m = \mathbf{u}^a = \mathbf{w} = 0$, then (2) reduces to the uncontrolled and undisturbed model (UCUDM).

3. PROPERTIES OF THE UCUDM

The equilibria of the UCUDM are solutions of the system $\sin\varphi_1 = 0$, $\sin\varphi_2 = 0$ which is obtained from (2) by setting $\dot{\varphi}_s = \dot{y}_s = u_s = w_s = 0$, s=1,2, and using assumption (b) (Section 2). Besides the trivial equilibrium $E_0(0,0,0,0)$, the following equilibria are of interest: $E_{1,2}(\pm\pi,0,0,0)$, $E_{3,4}(0,\pm\pi,0,0)$.

Consider the function

$$V(\varphi,y) = y_1^2 + y_2^2 + \frac{2g}{a_1}(1-\cos\varphi_1) + \frac{2g}{a_2}(1-\cos\varphi_2). \tag{3}$$

Let us denote by $\mathcal{D} \subset R^4$ the largest region which does not contain inside any other equilibrium of the UCUDM besides E_0. In $\mathcal{D}$ the function (3) has properties: $V(\mathbf{0},\mathbf{0}) = 0$, $V(\boldsymbol{\varphi},\mathbf{y})$ is positive definite in $\mathcal{D}\setminus E_0$, and

$$\frac{\partial V}{dt} = \sum_{s=1}^{2}\left(\frac{\partial V}{\partial\varphi_s}\dot{\varphi}_s + \frac{\partial V}{\partial y_s}\dot{y}_s\right) = -2\sum_{s=1}^{2}\alpha_s y_s - 2(\gamma_2 y_2 - \gamma_1 y_1)\,F(y_2-y_1) \le 0\,, \tag{4}$$

$\dot{\varphi}_s$, $\dot{y}_s$ are given in the UCUDM (2) and $\alpha_s = \lambda_s/\bar{m}_s a_s^2$, s=1,2. The inequality (4) follows from assumption (c) (Section 2). The above properties imply that (3) is a Lyapunov function for the UCUDM (2) in $\mathcal{D}$; it describes in $V \times \mathcal{D} \subset R^5$ a "cup-shaped" hypersurface with minimum at E_0. The equation $V(\boldsymbol{\varphi},\mathbf{y}) = h$, $h = \text{const} > 0$, represents a family of level surfaces in $V \times \mathcal{D}$ whose orthogonal projection onto $\mathcal{D}$ generates hypersurfaces $\mathbf{H}_h$ in R^4 which are closed, do not intersect, contain inside only the equilibrium E_0; if $h_1 < h_2$, $\mathbf{H}_{h_1}$ is inside $\mathbf{H}_{h_2}$. The largest hypersurface $\mathbf{H}_{h^*}$ with these properties passes through one or more equilibria E_i, i=1,...,4, whose coordinates minimize h: $\min h_i = h^* = \min V(E_i)$ which gives $h^* = \min(4\beta_1, 4\beta_2)$, $\beta_s = g/a_s$, s=1,2. Then $\mathbf{H}_{h^*}$ and $\mathcal{D}$ are defined as $\mathbf{H}_{h^*} = \{(\boldsymbol{\varphi},y) \in R^4\colon V(\boldsymbol{\varphi},y) = h^*\} = \partial\mathcal{D}$, and $\mathcal{D} \triangleq \{(\varphi,y) \in R\colon V(\varphi,y) \le h\}$ ($\partial\mathcal{D}$ is the boundary of $\mathcal{D}$). Further we define the values $h_A = h^* - \varepsilon$, $h_S = h^* - \delta$, $0 \le \varepsilon < \delta < h^*$, hence $h_A > h_S$; ε and δ are as small as desired for a particular study.

To clarify the control objective we introduce important regions in $\mathcal{D}$: *Admissible set* $\Omega \triangleq \{(\boldsymbol{\varphi},\mathbf{y}) \in R^4\colon V(\boldsymbol{\varphi},\mathbf{y}) < h_A \le h^*\}$; *Avoidance set* $A \triangleq \mathcal{D}\setminus\Omega$ with boundaries $\partial A \triangleq \{(\boldsymbol{\varphi},\mathbf{y}) \in \mathcal{D}\colon V(\boldsymbol{\varphi},\mathbf{y}) = h_A\}$ and $\partial\mathcal{D}$; *Safety zone* $S \triangleq \{(\varphi.\mathbf{y}) \in R\colon h_S \le V(\varphi,\mathbf{y}) < h_A\}$ with boundaries ∂A and $\partial S \triangleq \{(\boldsymbol{\varphi},\mathbf{y}) \in \mathcal{D}\colon V(\boldsymbol{\varphi},y) = h_S$; *Desirable region* $\chi \triangleq \Omega\setminus S$.

4. AVOIDANCE CONTROL PROGRAM FOR THE NOMINAL MODEL

Here we design a control program for the solution of the NM for the avoidance of the region A.

The region A is avoidable for all $\mathbf{w} \in W$ if there is a control $\mathbf{u}^m \in U^m$ such that for all states $(\boldsymbol{\varphi}(t_1), \mathbf{y}(t_1)) \in S$, $t_1 \geq t_0$, the solution of the NM (2) $\zeta(t, \boldsymbol{\varphi}(t_1), \mathbf{y}(t_1), \mathbf{u}^m, \mathbf{w})$ cannot enter A. Sufficient conditions for the avoidance are given in the following theorem.

Theorem 1. The solution $\zeta(t)$ of the NM (2) is controllable in $\Omega \subset \mathcal{D}$ for avoidance of the region A if there is a control $\mathbf{u}^m$ such that

$$\frac{dV}{dt} = \sum_{s=1}^{2}\left(\frac{\partial V}{\partial \varphi_s}\dot{\varphi}_s + \frac{\partial V}{\partial y_s}\dot{y}_s\right) \leq 0 \tag{5}$$

for all $\mathbf{w} \in W$, where $\dot{\varphi}_s$ and $\dot{y}_s$, are determined by the NM (2) and V by (3).

Proof. Following [1-3] we assume that A is not avoidable. Then for some state $(\boldsymbol{\varphi}(t_1, \mathbf{y}(t_1)) \in S$ and time $t > t_1$, the solution $\zeta(t)$ with fixed control $\mathbf{u}(t_1)$ enters A. Thus there is a $t_2 > t_1$ such that $\mathbf{z}[t_2] \in \partial A$. The point $(\boldsymbol{\varphi}(t_1), \mathbf{y}(t_1))$ lies on a hypersurface which is enclosed between ∂S and ∂A. Hence $V(\boldsymbol{\varphi}(t_1), \mathbf{y}(t_1) < V(\boldsymbol{\varphi}(t_2), \mathbf{y}(t_2))$ which implies that the function $V(\boldsymbol{\varphi},\mathbf{y})$ in increasing which contradicts (5) ($V(\boldsymbol{\varphi},\mathbf{y})$ is non-increasing along every response of the NM).

With (2) and (3) condition (5) becomes

$$\sum_{s=1}^{2} y_s u_s^m + \sum_{s=1}^{2} y_s w_s - \sum_{s=1}^{2} \alpha_s y_s^2 - (\gamma_2 y_2 - \gamma_1 y_1)\, F(y_2 - y_1) \leq 0. \tag{6}$$

The inequality (6) establishes a relationship between the control $\mathbf{u}^m$, the damping forces R_s and q, and the velocities $y_s = \dot{\varphi}_s$ of the angular coordinates φ_s, s=1,2, for all $\mathbf{w} \in W$; it is a general control law in implicit form. From (6) we see that the role of the control force is to balance the effect of the velocities y_s for all input uncertainties $\mathbf{w} \in W$. Since $w_s \in [-c,c]$, the selection of u_s^m from (6) requires consideration against the "worst case" for w_s which with property (c) (section 2) leads to $w_s = -c$ for $y_s < 0$, $w_s = c$ for $y_s > 0$.

Using Theorem 1 we introduce the following *Avoidance Control Program* for the solution $\zeta(t)$. If $\zeta(t)$ enters the safety region S, say at time t_s, $\zeta(t_s) \in S$ implying that $V(\zeta(t_s)) > h_s$, to prevent $\zeta(t)$ of entering the avoidance region A, we have to select a new control $\mathbf{u}^m(t_s)$ from (6) at $\zeta(t_s)$ (switching point). The new control will steer down $\zeta(t)$ along decreasing levels of $V = h < V(\zeta(t_s))$. Practically we monitor the level of $V(\zeta(t))$ which serves as a test.

5. AVOIDANCE CONTROL PROGRAM FOR THE GENERAL MODEL

On the basis of the experience acquired in the process of finding a memoryless control $\mathbf{u}^m(t)$ for the NM, we consider now the GM (2) with uncertain masses. We generate in addition to $\mathbf{u}^m(t)$ a control $\mathbf{u}^a(t)$ with improved performance which is to serve as a compensatory term for the model with uncertain masses and thus give the total controller $\mathbf{u}(t)$ adaptive capabilities.

Theorem 2. The solution $\mathbf{z}(t)$ of the general model (2) cannot cross the boundary ∂A and enter into the avoidance region A if there is a control $\mathbf{u}(t) = \mathbf{u}^m(t) + \mathbf{u}^a(t)$ such that the inequality (5) holds for $\dot{\varphi}_s$ and $\dot{y}_s$ determined by the GM(2).

Proof. It is the same as the proof of Theorem 1 with the difference that now $\mathbf{z}(t)$ stands for $\boldsymbol{\zeta}(t)$.

The condition (5) with (2) and (3) takes the form

$$\sum_{s=1}^{2}(u_s^m+u_s^a+w_s)y_s - \sum_{s=1}^{2}\alpha_s y_s^2 - (\gamma_2 y_2 - \gamma_1 y_1)\,F(y_2-y_1)$$
$$- \sum_{s=1}^{2}\frac{m_s^*}{\bar{m}_s}(\ddot{\varphi}_s + \beta_s \sin\varphi_s) \le 0\,. \qquad (7)$$

Here $\mathbf{u}^m$ is the memoryless control for the NM which obeys (6). The adaptive type control $\mathbf{u}^a$ is to be selected from (7). The difficulty is that (7) involves derivatives. Assuming that the state variables $\boldsymbol{\varphi}(t)$, $\mathbf{y}(t)$, hence $\mathbf{z}(t)$, are accessible, the avoidance control program in Section 4 is also valid for the GM if we manage to find $\mathbf{u}^a$.

6. AVOIDANCE STRATEGY FOR THE MANIPULATOR

It is assumed that the links can overlap ($\varphi_1=\varphi_2$). To avoid collision with stationary or moving objects, the links may not be allowed to reach or cross a specified boundary defined by the segments $0e_s$, s=1,2, which form the angles $\theta_s = \measuredangle \eta 0 e_s$ (Fig. 1). The segments Ob_s form the angles $\nu_s = \measuredangle \eta O b_s$ which are introduced to prevent the links OP_s of getting too close to the boundary segments Oe_s. For a particular study the difference $\theta_s - \nu_s$, s=1,2, can be selected as small as desired. For simplicity we assume that $|\theta_2| = \theta_1 = \theta$, $|\nu_2| = \nu_1 = \nu$, $\theta > \nu > 0$.

Now we introduce, using [1], the following angles in the plane α.

Admissible angle $\Omega_\alpha \triangleq \measuredangle(e_2Oe_1)$; $O\eta \in \Omega_\alpha$, $Oe_s \notin \Omega_\alpha$;

Avoidance angle $A_\alpha \triangleq \measuredangle(e_1Oe_2)$; $O\eta \notin A_\alpha$;

Safety angle $S_\alpha \triangleq \measuredangle(b_1\,Oe_1) \cup \measuredangle(e_2Ob_2) \in \Omega_\alpha$, $\quad Oe_s \notin S_\alpha$;

Desirable angle $\chi_\alpha \triangleq \measuredangle(b_2Ob_1)$: $O\eta \in \chi_\alpha$, $\chi_\alpha = \Omega_\alpha - S_\alpha$.

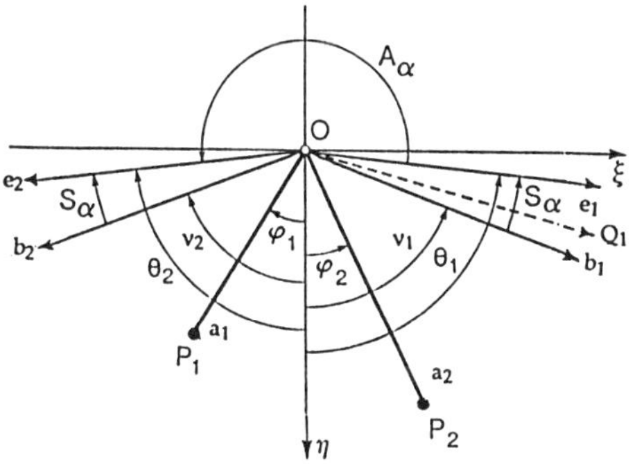

Fig. 1. Double pendulum. Angles in the plane α.

The two -link manipulator enters into the angle A_α if at least one of the connected links OP_1 and OP_2 enters into A_α.

The Control objective is to design a strategy according to which the control force **u**(t) secures that the two-link open-end manipulator does not enter into the avoidance angle A_α but preferably moves inside the angle χ_α (Fig. 1).

The following avoidance control strategy for the motion of the two-link robot manipulator based on the results derived in Section 5 concerning the solution the dynamical system (2) is proposed. Suppose that the links OP_1, OP_2 subjected to fixed control $\mathbf{u}(t_0)$ oscillate in angle $\chi_\alpha \subset R^2$. The corresponding solution $\mathbf{z}(t)$ of (2) moves in R^4. To guarantee that the links will avoid entering into angle A_α, we monitor simultaneously: (i) The value $V(\mathbf{z}(t))$ which should be smaller than h_S since we wish $\mathbf{z}(t) \in \chi$; (ii) The angular coordinates φ_1 and φ_2 so that the links OP_1, OP_2 belong to χ_α. Two situations may occur.

A. The links OP_1, $OP_2 \in$ angle χ_α at time $t_1 > t_0$ while $h_S < V(\mathbf{z}(t_1)) < h_A$, which means that the solution $\mathbf{z}(t)$ has crossed the boundary ∂S and is into the safety zone S. Our strategy is at time t_1 to change the control $\mathbf{u}(t_0)$ to a new control $\mathbf{u}(t_1) \in$ U which satisfies the control law (7). Then the solution $\mathbf{z}(t)$ would move "down" through decreasing levels of $V(\boldsymbol{\varphi}(t), \mathbf{y}(t))$.

B. One of the links enters the safety angle S_α, say $OP_2 \in$ angle(b_1Oe_1), and at time $t_1 > t_0$ coincides with the segment OQ_1 (see Fig. 1), while the solution $\mathbf{z}(t)$ reaches the state $\mathbf{z}(t_1) \equiv Q(\boldsymbol{\varphi}(t_1), \mathbf{y}(t_1)) \in \chi$, i.e. $V(\mathbf{z}(t_1)) < h_S$. To prevent OP_2 of entering into the avoidance angle A_α, a new control vector $\mathbf{u}(t_1) \in$ U which obeys (8) has to be selected at the switching point Q. The solution $\mathbf{z}(t)$ to the control $\mathbf{u}(t_1)$ will be steered "down" through decreasing levels of $V(\boldsymbol{\varphi}(t), \mathbf{y}(t))$ and the link OP_2 will move back into the desirable angle χ_α.

Acknowledgement. This work was supported by Grant No. 3967 for the Natural Sciences and Engineering Research Council of Canada.

REFERENCES

1. G. Bojadziev and M. Bojadziev, "Avoidance control program for a planar two-link robot manipulator in a uncertain environment", Proc. ANS Topical Meeting on Robotics and Remote Systems, Albuquerque (U.S.A.), pp. 71-80, 1991.
2. M. Corless, G. Leitmann and J. Skowronski, "Adaptive control for avoidance or evasion in uncertain environment", Comput. Math. Applic. Vol. 13, pp. 1-11, 1987.
3. G. Bojadziev and M. Saif, "Fault prevention control of planar manipulators with uncertain masses", Proc. AMSE Conf. "Neural Networks", San Diego (USA), May 29-31, 1991, pp. 245-251.

ART1 NEURAL NETWORK APPROACH TO LARGE SCALE GROUP TECHNOLOGY PROBLEMS

Cihan H. Dagli and Cheng-Fo Sen
Department of Engineering Management
University of Missouri-Rolla
Rolla, MO 65401

ABSTRACT

Group Technology (GT) is a manufacturing philosophy. The underlying concept is to bring together similar parts and related processes in order to take advantage of their similarities in both manufacturing and design. In general, enumerative or heuristic algorithms are used for solving the GT problem. Since the GT problem is NP complete, these algorithms tend to be slow or ineffective for large-scale problems. The ART1 paradigm has been proposed to be a binary vector classifier which can be used for machine-part family formation. In this work, large matrices (2000 x 200) are applied to the ART1 neural network and close-to-optimal solutions are achieved.

1. INTRODUCTION

Group technology is an essential tool to achieve the flexible manufacturing System (FMS) which represents one of the highest levels of achievement in automated production. Instead of using conventional process-type layout, group technology arranges the production equipment into machine cells to facilitate work flow. Similarly, parts with similar processing requirements are formed into part families. The machine-part family formation problem is, therefore, defined by identifying and grouping parts into families and machines into cells. Part families are then assigned to machine cells according to the information from route sheet and process plan. Group technology machine cells have shown benefits in reducing material handling costs, reducing the manufacturing lead times, lower set-up times, and reduce work-in-process (WIP) inventories.

The machine-part family formation problem can be represented by the machine-part incidence matrix in TABLE I(a). The element a_{ij} in the machine incidence matrix is equal to one if part j is processed on machine i, and a_{ij} is equal to zero if part j is not processed on machine i. Part families are formed by rearranging row vectors according to their similarities. Machine cells are obtained by rearranging column vectors by the same principle.

TABLE I. Machine-Part Incidence Matrix.

		PARTS				
		1	2	3	4	5
MACHINES	1			1		
	2	1	1		1	
	3			1		1
	4			1		1

(a)

		PARTS				
		1	2	4	3	5
MACHINES	1				1	
	3				1	1
	2	1	1	1		
	4		1	1		1

(b)

		PARTS				
		1	2	4	3	5
MACHINES	1				1	
	3				1	1
	4					1
	2	1	1	1		
	4		1	1		

(c)

The objective is to rearrange the order of rows and columns such that a block diagonal form of the $a_{ij} = 1$ entries is formed. This structure is given in TABLE I(b). Machine cell 1 is formed by grouping machine [1, 3]. Similarly, machine cell 2 contains machines [2, 4]. Parts [1, 2, 4] form the part family 1 and parts [3, 5] form the part family 2. Note that the matrix obtained is not a pure block diagonal form. Element (4, 5) is an exceptional element which is overlapped by two machine clusters. Machine 4, in this case, is called the bottleneck machine. Bottleneck machines can be duplicated to achieve the pure block form. TABLE I(c) is the matrix after duplicating machine 4. Many approaches have been proposed in literature to solve the machine-part family formation problem. But a large size of the matrix for realistic industrial situation (2,000 parts and 200 machines) restricts the efficiency of conventional algorithms [1]. Artificial Neural Networks employ a parallel approach which promises a solution to problems that are slow or not possible for conventional computers today. When the hardware neural network is fully realized, the problems like machine-part family formation will be easy to be solved on neural network machines. Adaptive Resonance Theory (ART) is one of the most popular paradigms in neural networks. In this work, ART1, a branch of ART family, is investigated and implemented on HP9000/750 machine. Several modifications are made to make the ART1 more suitable for the group technology problem [2].

2. ART1 NEURAL NETWORK

Adaptive Resonance Theory (ART) was first developed by Carpenter and Grossberg in 1976. ART works on-line, stable in response to arbitrarily many inputs, and effective use of full memory capacity. Most importantly, ART maintains plasticity in an unexpected world. The recent work of Carpenter and Grossberg has focused on three types of ART models. The first type, ART1, self-organizes recognition categories for arbitrary sequences of binary inputs. The second type, ART2, deals with either binary or analog inputs. The third type, ART3, includes a chemical synapse that takes care of memory search problem of ART. In general, these paradigms use either fast or slow learning. Fast learning immediately encodes the input patterns onto long term memory (LTM) and every pattern is presented to the system only once. The pattern class formation is not as statistically representative as those found by slow learning. Slow learning allows only a small portion of the input pattern to store in the long term memory. Repeated presentation is needed for the most prominent characteristics to be committed.

Two branches of ART were introduced in 1991, ART2-A [3] and Fuzzy ART [4]. ART2-A runs faster than ART2. Fuzzy ART serves for fuzzy data categorization. These two paradigms use a fast-commit slow-recode option (intermediate learning) which combines fast learning with a forgetting rule. As a result, rare events can be rapidly learned, yet previously learned events are not rapidly erased by noise. In other words, to make the category structure less dependent on input presentation order. The simulation in this work uses either fast learning or intermediate learning by the user's option. The results for these two learning algorithms are compared.

The architecture of ART1 consists of two layers of processing elements (units) and two LTM traces, a gain controller, and a reset mechanism. The ART architecture are shown in Figure 1.

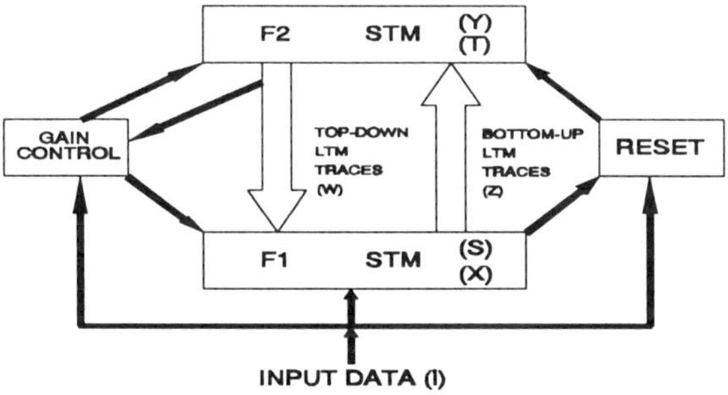

Figure 1. ART1 Architecture.

The number of units in F_1 is equal to the length of input patterns. The input pattern **I** is fed into the F_1 and vector **I** addresses itself as vector **X** to F_1. Vector **X** passes through a binary step function. The output of the step function is vector S which is than transmitted through the bottom-up LTM traces to generate the input vector **T** to F_2. The units in F_2 compete one another. There is only one winner of the competition who has the largest output value in F_2. A vector **Y** is formed by assigning the winning unit to one and others to zero among F_2 elements. The non-zero node of **Y** defines the category of the input pattern **I**. Vector **U** is formed by coping vector **Y** which is ready to be propagated back to F_1. Vector **V** at F_1 is the result of the multiplication of vector **U** by the top-down LTM traces. At this point, the input vector **I** has traveled through the bottom-up LTM traces to F_2 and through the top-down LTM traces back to F_1. An vector $\mathbf{X}^*$ is formed by the intersection of vector **V** and **I** ($\mathbf{X}^* = \mathbf{V} \cap \mathbf{I}$). The intersection operation can also be replaced by logical AND. Vector $\mathbf{X}^*$ and vector **I** are than compared to check the degree of match. If vector $\mathbf{X}^*$ and **I** are close enough, a state of resonance is achieved. A learning behavior occurs after the resonance. Both LTM traces are adapted to the new pattern. If vector **V** and **I** are not alike, a memory search is triggered. The whole procedure start all over again except that the mismatch node in F_2 is inhibited for competition (a reset occurs). The degree of match is defined by the vigilance (ρ)

parameter. Low vigilance leads to a coarse classification and high vigilance results in a fine classification. The 2/3 Rule regulates the F_1 activities which says that at least two out of three inputs are non-zero to activate a unit in F_1. Three inputs include vector **I**, $\mathbf{X}^*$, and gain control signal. The gain control signal is always turned off when both **I** and $\mathbf{X}^*$ reach F_1. Thus, the result of the 2/3 Rule controls the signal flow at F_1.

The fast learning algorithm is the most widely used in the ART network. It is proved that ART1 encoding is stable using fast learning. Fast learning enables the ART network to adapt quickly to novelty inputs which may occur rarely. Fast learning equations are provided in Equation (1).

$$Z_{Ji} = \begin{cases} \dfrac{L}{L - 1 + |S|} & \text{if } v_i \text{ is active} \\ 0 & \text{otherwise} \end{cases}$$

$$W_{iJ} = \begin{cases} 1 & \text{if } v_i \text{ is active} \\ 0 & \text{otherwise} \end{cases} \tag{1}$$

$\mathbf{Z}_{Ji}$ is the bottom-up weight vector. $\mathbf{W}_{iJ}$ is the top-down weight vector. Index J represents the winning unit. $|\mathbf{S}|$ is the number of ones in vector **S**, and L is a constant.

One drawback of fast learning is that the category structure is too dependent on input presentation order. This can be found in Huggahalli's work [5]. The input vectors are pre-sorted according to the density of the input vector. This problem can be avoided, in part, by an intermediate learning algorithm. Intermediate learning is also called the fast-commit slow-recode option which retains advantages of fast learning, such as fast responses to rarely encountered input patterns. Slow-recode prevents the already learned patterns to be deleted by noisy inputs. The top-down weight vector is updated according to Equation (2),

$$W_{iJ}^{(new)} = \begin{cases} \beta s_i + (1 - \beta) W_{iJ}^{(old)} & \text{if } v_i \text{ is active} \\ 0 & \text{otherwise} \end{cases} \tag{2}$$

where $\beta = 1$ if the first time category J becomes active. As β approaches one, this equation becomes the fast learn equation. The equation for updating bottom-up weight vector remains the same.

3. PERFORMANCE EVALUATION

An ART1 simulator is coded with modifications in learning algorithm and data pre-processing. Several test cases are used to verify the performance of the simulator. Since there is no standard method to measure the performance, especially when bottleneck machines are involved, the results are visually inspected to see how well the clusters have been formed. The basic ART1 is not suitable for the problem of machine-part family formation [2]. If sparser vectors are encountered first, those vectors that group well with the denser vectors will be classified improperly. Thus, the sequence of input presentation becomes important in this application. Slow learning can solve this problem thoroughly, but the learning time could be too long to tolerate. The first scheme, intermediate learning, can partially substitute slow learning and reduces the dependency on input presentation order. The second scheme, pre-sort the input vector, will eliminate the problem created by the sequence of input vector. The input vectors are pre-sorted according to the number of ones in each vector and apply them in the descending order to ART1. The third scheme involves the modification of ART1. Instead of storing the templet of **I** $\cap$ **V**, the one with higher number of ones is stored. By doing so, the denser patterns that have already been stored 'absorb' the sparser patterns. Sometimes, pre-sorting input vectors can not be done in real time applications making the intermediate learning the best option.

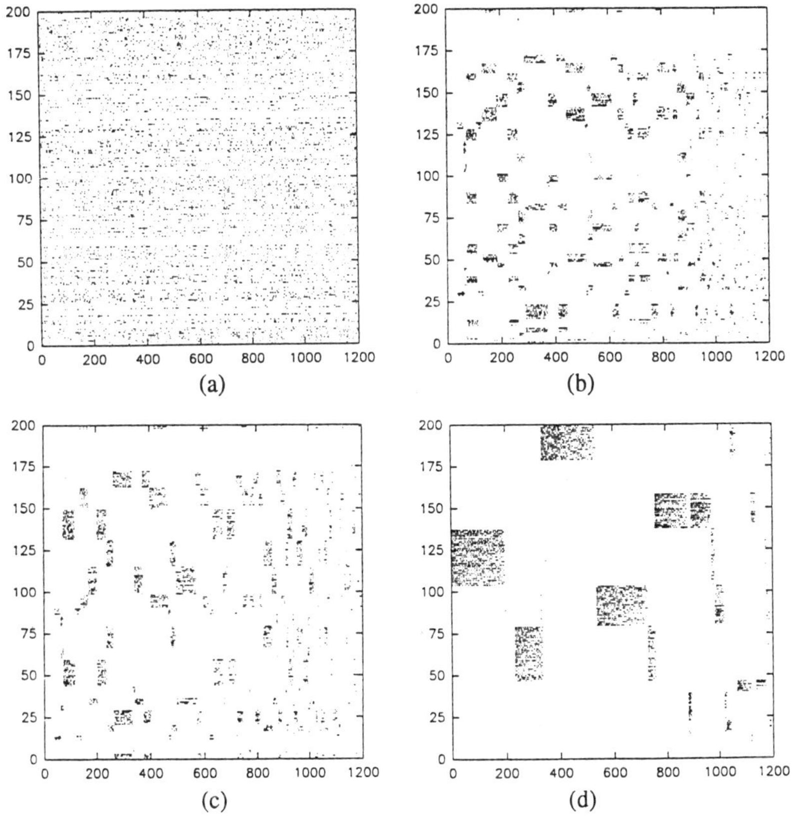

Figure 2. A 1200 × 200 Case.

Figure 2(a) is the input machine-part incidence matrix which has 1200 parts and 200 machines. In this case, column and row vigilance factors are set to 0.02. Figure 2(b) is the result obtained by the basic ART1. It is obvious that too many clusters are formed (100 part families and 46 machine cells). A lot of them could be combined. Note that the vigilance is low and there is no effect to bring down the vigilance further. Figure 2(c) uses the intermediate learning option and Figure 2(d) is obtained by applying all the schemes developed in this work. We can see the improvements. Only 30 machine cells and 15 part families are formed in Figure 2(d). The rest of the work can be done manually to achieve the pure block diagonal form if one exists. For demonstrating the ART1 can be used to solve large scale group technology problems. A realistic size with 2000 parts and 200 machines are used as our conclusion. Figure 3(a) is the input matrix which is generated by a synthetic matrix generator. Figure 3(b) is the result matrix which contains 47 part families and 14 machine cells. Both the column and row vigilance are set to 0.01.

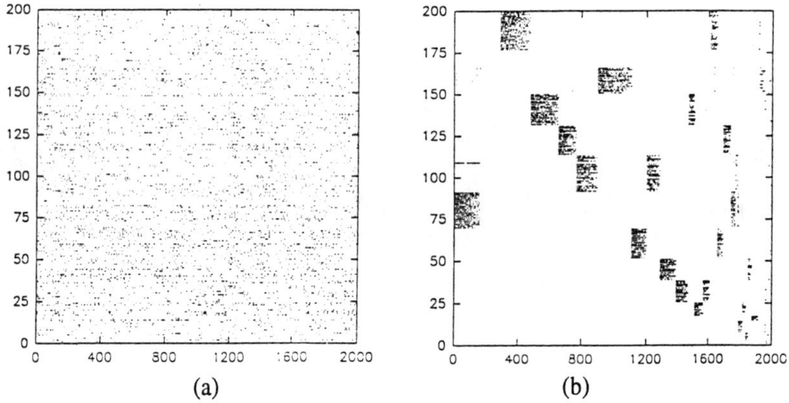

Figure 3. A Realistic Case.

4. CONCLUSION

With the implementation of ART1 on HP 9000/750 machine, ART1 has been proved to be suitable for solving large scale group technology problems. Several modifications and schemes are used to improve the performance of ART1, namely, (1) intermediate learning, (2) pre-sorting input vectors, and (3) retaining higher density vector as the templet. The first scheme is most useful when pre-processing of the input vector is not possible. The combination of the last two schemes could provide a satisfactory solution to the machine-part family formation problem.

REFERENCES

1. Roger Kerr, *Knowledge-Based Manufacturing Management* (Addison-Wesley, 1991).
2. C. Dagli, and R. Huggahalli, "Neural Network Approach to Group Technology" in *Knowledge-Based Systems and Neural Networks, Techniques and Applications*, R. Sharda, J. Y. Cheung, and N.J. Cochran, Editors, pp 213-228, Elsevier, 1991.
3. G. A. Carpenter, S. Grossberg, and D. B. Rosen "ART2-A: An Adaptive Resonance Algorithm for Rapid Category Learning and Recognition" *Neural Networks*, Vol. 4, pp 493-504.
4. G. A. Carpenter, S. Grossberg, and D. B. Rosen "Fuzzy ART: Fast Stable Learning and Categorization of Analog Pattern by an Adaptive Resonance System" *Neural Networks*, Vol. 4, pp 759-771, 1991.
5. R. Huggahalli, *A Neural Network Approach to Group Technology* (Thesis, UMR Library, 1991).

EXPERT SYSTEM FRAMEWORK FOR FAULT DETECTION AND FAULT TOLERANCE IN ROBOTICS

M. L. Visinsky, J. R. Cavallaro and I. D. Walker
Department of Electrical and Computer Engineering
Rice University, Houston, Texas 77251-1892

ABSTRACT

Fault tolerance is of increasing importance for modern autonomous or industrial robots. The ability to detect and tolerate failures will enable robots to effectively cope with internal failures and continue performing assigned tasks without the need for immediate human intervention. To monitor fault tolerance actions performed by lower level routines and to provide higher level information about a robot's recovery capabilities, we use an expert system to develop a novel fault tolerance framework combining fault detection and tolerance routines with dynamic fault tree analysis. Fault tree analysis reveals the key components for providing fault detection and tolerance within a system and details the interconnection between failures in the system. The trees can also be used quantitatively to provide a dynamic estimate of the probability of failure of the entire system or various subsystems. Using fault trees as a standard framework, the expert system package can provide fault tolerance for robots of significantly different origin and structure.

INTRODUCTION

In previous work, we developed and verified algorithms which successfully detect and tolerate sensor and motor failures in a specific robot system [9, 11]. To apply these algorithms to a wider variety of robots, we proposed an expert system based supervisor which performs robotic fault tolerance using a standardized representation of the robot's fault structure and modularized detection and recovery actions applicable to common robot components [12]. In this paper, we discuss how the expert system manipulates fault trees [8], the chosen standard representation for the robots, in order to dynamically respond to detected failures in the robot system. In fault trees, the possible failure events for the system are connected by easily understood logic symbols to produce a flow chart of failures [11]. Some failure analysis programs use digraphs [1, 6] to represent the failure interconnections. The digraphs are easily transformed into fault trees for which there exist a wide variety of research efforts [3, 8, 9, 10].

Figure 1 highlights the position of the expert system as supervisor in the fault tolerance framework discussed in [9] and [12]. The supervisor prunes failed components or branches from fault trees as failures are reported from the detection routines in the interface layer. The system immediately notifies the operator of failures in the robot during on-line operations so the operator may monitor the fault tolerance capabilities of the system and suggest alternative recovery actions. When integrated with fault detection and fault tolerance modules [9, 12], the expert system will also be able to suggest recovery actions based on the structure of the robot fault trees. The nodes in the tree will have an associated detection or recovery action module depending on the type of component or tree event. By maneuvering around the trees, the expert system will perform fault tolerant recovery actions for a specific robot structure as a sequence of these smaller, generic actions. The expert system also recomputes the probability of failure of the robot system as failures occur. The probability computations are treated as background processes which occur after more pressing recovery or tolerance actions have been completed to allow for swift recovery from failures.

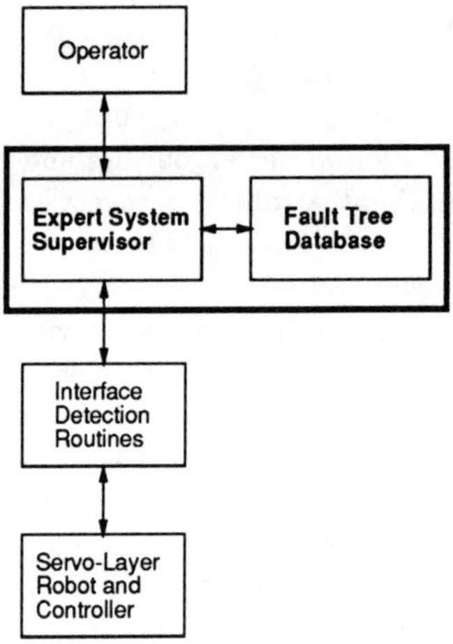

FIG. 1. Expert System Supervisor within Fault Tolerance Framework.

We are currently implementing the fault tree database supervisor in CLIPS ('C' Language Integrated Production System) [2], a NASA-developed expert system software package. With this system, we can design an intelligent and flexible framework which will provide fault detection and fault tolerance for many structurally diverse robots. Fault tree traversing schemes provide rules for the supervisor layer. The expert system is capable of generalizing the detection and tolerance algorithms into modular procedures and can be easily embedded within other programs to allow for future expansions. This framework will provide fault tolerance to the wide variety of robots which have already been developed to perform tasks in space or other hazardous environments and which may be far from human control for long periods of time.

FAULT TREE DATABASE

The operator initializes the fault tree database with facts defining the structure of a predetermined fault tree for the robot (Figure 2). In the tree, the failure events have each been given a simple letter encoding to simplify the database facts. Event **A** is the failure of the entire robot, the top event. Failure events

B, **C**, and **D** represent failures of the joints of the robot. Note that the suppressed tree under event **D** would look much the same as the subtrees under **B** and **C**. The basic component failures **u** through **z** represent failures of sensors or motors, the critical components in the joint subsystems. In defining the facts for these failure events, some of the fact fields will have default values so that their values do not have to be specifically detailed at initialization. For example, all fact "gate-status" fields have a default value of "alive". While the robot is operational, the supervisor maintains the database model of the trees, pruning branches as failures occur based on the expert system rules for failure propagation through AND- and OR-gates. As the system handles failures, new failures may be asserted by the propagation rules, thus initiating a recursive propagation of failure events up through the trees.

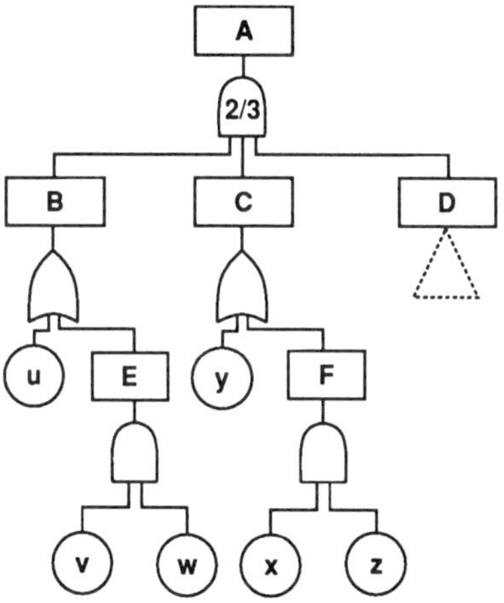

Operator Initializes Expert System with:

```
;structural facts:
(F-AND (out A) (in B C D) (fail-with 2))
(F-OR (out B) (in u E))
(F-AND (out E) (in v w))
(F-OR (out C) (in y F))
(F-AND (out F) (in x z))
....

;initial probabilities:
(prob u 2.0e-4)
(prob v 1.0e-4)
(prob w 1.0e-4)
(prob y 2.0e-4)
(prob x 1.0e-4)
(prob z 1.0e-4)
....
```

FIG. 2. Expert System Fact Initialization Example.

When a sensor or motor failure is detected by the interface layer [9, 12], a failure fact is asserted in the expert system. This fact consolidates with all structural facts containing the failed component as input to activate a specific failure propagation rule. If the component is an input to an OR-gate, the failure propagates through the gate and fails the output. The OR-gate is pruned from the tree by modifying the gate status to "failed".

Because AND-gates in a tree represent some form of functional redundancy in the robot [9], AND-gate facts have an associated "fail-with" field (see Figure 2) which informs the system of how many input failures will fail the AND-gate. The default value of a "fail-with" field is two. If an input to an AND-gate fails and the "fail-with" field of the gate is greater than one, then this level still has redundancy and only the failed input is pruned from the tree. The gate now fails with one less failure than before and has one less input. When the AND-gate fact reveals that the gate will fail with one failure, the gate is essentially an OR-gate and a failure of any remaining input will fail the output of the gate.

Facts associated with a pruned component or branch are not actually removed from the database but are simply modified to prevent certain rule activations. This allows the original tree to be rebuilt if components are repaired/replaced. Asserting the "failed" status only upwards in the tree from a failed component allows the system to continue monitoring the siblings of the faulty component (i.e.: other inputs to the gate). For example, when component **u** fails in the tree of Figure 2, the failure propagates through the OR-gate and fails **B** as well. Suppose that while **u** is down, component **v** fails. The expert system is able to handle the new failure appropriately for the AND-gate beneath **E**, which is still considered alive although it no longer contributes to system failure because node **B** has already failed. If **u** is now repaired and reinstated into the database, the failure of **v** is still known to the expert system. If this branch had been "pruned" as well when **u** originally failed, the failure of **v** would not have been correctly recorded at its onset.

Currently, fault tolerance is automatically implemented by the interface layer and the supervisor's role is limited to logging failures and reporting changes in the database to the operator. The operator has the option to signal an abort at any time if it is deemed that the assigned task has been compromised. The expert system will use suggestions made by the operator or information learned by traversing the database to develop new tolerance strategies based on the existing structure of the robot. There may be alternate sensors available or external sensors (such as cameras) that the interface layer is originally unaware of which can be used to monitor a joint for failures when the known or basic set of sensors has failed. Increasing the redundancy of the motors or sensors as in [7] and [13] would provide a wider range of survivor choices for the system.

The supervisor can be further enhanced to monitor the reachable workspace of the robot throughout its operation [5]. When failures occur and a joint is shut down, the workspace shrinks due to the immobility of the failed joint. The supervisor would compute the changes to the reachable workspace [5] and decide if the target position is still reachable by the robot. If the robot can no longer reach its assigned goal, the supervisor would signal an abort and wait for further instructions from the operator. The supervisor can also provide the operator with information on possible causes of failures by traversing the appropriate subtrees in the database to accumulate a list of components which may have caused a given failure. An operator could then use the list for future inspections and repairs. When pruning the trees of failed components, the system may also search for indications that a component is likely to fail because it is linked either structurally or functionally to a previously failed component. Advanced detection routines would then be more alert to the possibility of failures in these components while monitoring the system for failures.

PROBABILITY OF FAILURE

The expert system also maintains the probability of failure for each node within the trees. The operator initializes the probabilities of all the basic components (leaves) in the tree with appropriate probability facts (see Figure 2). The expert system then initializes the probabilities for inner nodes of the tree by combining the basic component probabilities through the gates in the tree structure. The expert system uses time-varying probabilities [8] in the above analysis to mimic component decay during the life of the robot. A small update routine monitors the system time and modifies the basic component probability facts. As detailed in [8], the probability of a component failure can be easily calculated from the standard failure rate for the component [3]. Given a constant failure rate λ and using the exponential distribution, the probability of failure at time t is:

$$p(t) = 1 - e^{-\lambda t},$$

which is typically approximated as λt. Within the fault trees, these failure probabilities are combined through the logic gates using simple multiplication and addition [8]. The probability of failure for the output event of an AND-gate is the product of all the input probabilities and a conservative estimate of the output event probability for an OR-gate is the sum of the input probabilities.

Figure 3 shows two stages of the probabilities for the subtree under component **B** in Figure 2. The probabilities are taken at an instant in time and reflect typical failure rates for robot components. In Figure 3a, a failure has occurred in motor **u** and the probabilities are modified following the rules described above. Failed components and events are shown with a grey background and failed gates and paths are shown with dashed lines. In Figure 3b, sensor **v** has failed (and **u** has not failed) and the new probabilities are as shown.

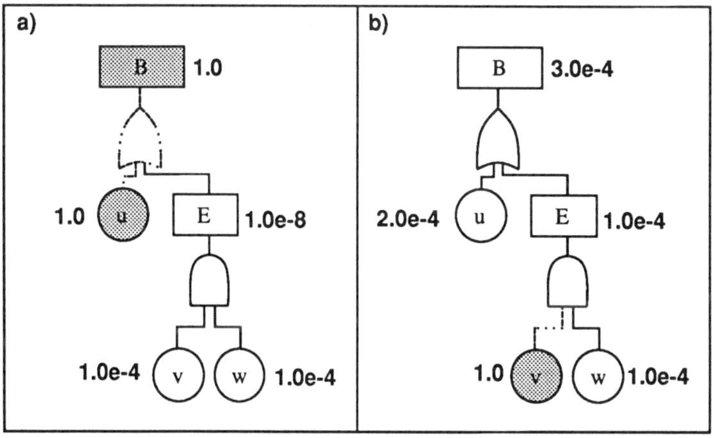

FIG. 3. Example of Probability Propagation in Fault Trees.

When a component fails, its probability of failure becomes 1 and the expert system asserts a change-probability fact to initiate the changes in the system probabilities that are affected by the failed node. The failed component is matched with all structure facts containing the component as input and new facts are asserted which detail the type of gate, the output event, and the inputs. These new facts activate probability-propagation rules which recompute the probability of the output event by recursively multiplying or adding (depending on the gate type) the probabilities of the inputs.

Currently, the branch-in factor (number of inputs) of the gates in the trees is relatively small so that each recomputation step is not too expensive time-wise. An alternative method if the branch-in factor is high for a robot's fault trees is to divide (for AND-gates) or subtract (for OR-gates) out the old value of the probability for the modified input and then multiply or add in the new value. This method limits the computation at each level to a single divide/multiply or subtract/add combination but requires additional memory space to save the old probabilities for each node in the tree. There is thus a tradeoff in time versus space which must be considered.

A quantitative analysis provides a measure of the overall chance of a complete failure for each robot system. The structure provided by the fault trees organizes the probabilities appropriately for the robot system and provides a simple map of how the events relate to each other. Using the trees, robots of significantly different origin and structure can be compared for fault tolerance and survivability. Static analysis of failure probabilities and their propagation through the fault trees allows the robot designer to focus on the components which are more likely to fail in the robot and which are thus more central to the monitoring software and fault detection routines.

By dynamically modifying the probabilities as failures occur, the expert system can alert the robot operator of any large changes in failure probabilities (several orders of magnitude) which may indicate a loss of redundancy or a loss of a component vital for the current task. Eventually, the dynamic probability analysis will also be useful in directing task-oriented decisions on which joint or component to use for a given task by promoting the use of those systems which are less likely to fail. The robot would mimic optimal configurations [4] so as not to put undo stress on systems which are injured or "sore".

CONCLUSION

Fault trees are used statically to initially develop the fault detection and fault tolerance algorithms for a specific system. The fault trees can also be used dynamically by an expert system supervisor as a database of run-time fault detection and recovery data which can reveal functional replacements for faulty components or indicate possible causes or effects of failures. The expert system can further maintain dynamic failure probabilities for the system as an indication of system weak spots. The expert system described by this paper provides a novel implementation for a flexible, intelligent, and modularly expandable fault tolerance framework for robotics.

Acknowledgements

This work was supported in part by the National Science Foundation under grants MIP-8909498 and MSS-9024391 and by DOE Sandia National Laboratory Contract #18-4379A. Further support was provided by a Mitre Corporation Graduate Fellowship and NSF Graduate Fellowship RCD-9154692.

REFERENCES

1. V. V. Dixit. EDNA: Expert Fault Digraph Analysis Using CLIPS. In *Proceedings of the First CLIPS Conference*, Houston, TX, August 1990. NASA Conference Publication 10049.

2. J. Giarratano and G. Riley. *Expert Systems: Principles and Programming.* PWS-Kent Publishing Co., Boston, MA, 1989.

3. V. H. Guthrie and D. K. Whittle. RAM Analysis Software for Optimization of Servomanipulator Designs. DOE SMALL BUSINESS INNOVATIVE RESEARCH (SBIR) PROGRAM REPORT JBFA-101-89, JBF Associates, Inc., Knoxville, TN, March 1989. Performed for Oak Ridge National Laboratory.

4. A. A. Maciejewski. Kinetic Limitations on the Use of Redundancy in Robotic Manipulators. *IEEE Transactions on Robotcs and Automation*, 7(2):205–210, April 1991.

5. A. K. Pradeep, P. J. Yoder, R. Mukundan, and R. J. Schilling. Crippled Motion in Robots. *IEEE Transactions on Aerospace and Electronic Systems*, 24(1):2–13, January 1988.

6. R. W. Stevenson and R. W. McNenny. Fault Isolation in Space Station Freedom Systems from the Space Station Control Center. In *Joint Applications in Instrumentation Process and Computer Control (JAIPCC) 1992 Symposium: Technologies for New Exploration*, Clear Lake, TX, March 1992.

7. D. Tesar, D. Sreevijayan, and C. Price. Four-Level Fault Tolerance in Manipulator Design for Space Operations. In *First International Symposium on Measurement and Control in Robotics*, volume 3, page J3.2.1, Houston, TX, June 1990.

8. W. E. Vesley, F. F. Goldberg, N. H. Roberts, and D. F. Haasi. Fault Tree Handbook. NUREG 0492, Systems and Reliability Research Office of Nuclear Regulatory Research, US Nuclear Regulatory Commission, Washington D.C. 20555, January 1981.

9. M. L. Visinsky. Fault Detection and Fault Tolerance Methods for Robotics. Master's thesis, Department of Electrical and Computer Engineering, Rice University, Houston, TX, December 1991.

10. M. L. Visinsky, I. D. Walker, and J. R. Cavallaro. Fault Detection and Fault Tolerance in Robotics. In *Proceedings of NASA Space Operations, Applications, Research Symposium*, pages 262–271, Houston, TX, July 1991.

11. M. L. Visinsky, I. D. Walker, and J. R. Cavallaro. Chapter 16, Robotic Fault Tolerance: Algorithms and Architectures. In *Robotics and Remote Systems for Hazardous Environments, Volume 1.* Prentice Hall Publishing Co., 1992. In Press.

12. M. L. Visinsky, I. D. Walker, and J. R. Cavallaro. A Dynamic Fault Tolerance Framework for Remote Robots. *IEEE Transactions on Robotics and Automation: Special Issue on Space Robotics*, 1993. Submitted.

13. E. Wu, M. Diftler, J. Hwang, and J. Chladek. A Fault Tolerant Joint Drive Systems for the Space Shuttle Remote Manipulator System. In *1991 IEEE International Conference on Robotics and Automation*, pages 2504–2509, Sacremento, CA, April 1991.

NONLINEAR SYSTEM CONTROL BASED ON NEURAL NETWORKS

SUKHAN LEE[1,2] and RHEE M. KIL[2]

Jet Propulsion Laboratory[1]
California Institute of Technology
Pasadena, CA 91109

Dept. of EE-Systems[2]
University of Southern California
Los Angeles, CA 90089-0273

Abstract

A neural network approach to the control of a nonlinear system subject to input saturation is proposed. The approach is based on 1) the training of a *Gaussian Potential Function Network* (GPFN) to model nonlinear system dynamics using the *Hierarchically Self-Organizing Learning* (HSOL) algorithm, and 2) the planning of a path from the initial state to the goal state in the *Feasible State Transition Region* (FSTR) based on a sequence of via points, considering the reachability of states under input saturation. The effectiveness of the proposed approach is demonstrated through simulation.

INTRODUCTION

Controlling a nonlinear system is difficult due to the problems involved in modeling nonlinear system dynamics as well as finding control laws based on general principles and methodologies. Neural networks can provide a means of solving the above difficulties: neural networks can be trained to model arbitrary nonlinear mappings, based on which control laws or a sequence of control inputs can be searched and defined.

The neural network approach to control can in general be classified into two: the *direct inverse approach* and the *indirect inverse approach.* In the direct inverse approach, the plant function is assumed to be invertible, and the one-to-one mapping between the control input and the plant output is stored and retrieved directly by a neural network [1, 2, 3]. In the indirect inverse approach, a neural network is trained first for modeling plant dynamics, such that control input can be derived indirectly from the learned plant model. The derivation of control input can be based on either 1) training another neural network as a controller with the teaching patterns collected from the learned plant model [4, 5], or 2) searching control input iteratively based on the gradient of a cost function defined in terms of the error between the desired output and the output of the learned plant model [6, 7, 8], or 3) a closed form of control law derived from the plant model, the parameters of which are learned during the modeling of system dynamics [9].

An important issue associated with neural-network based nonlinear system control, which has not been paid a proper attention yet, is how to generate a sequence of control inputs necessary for reaching the desired output under the constraints on possible state transitions; i.e., the desired output is not reachable by a single-step control due to plant nonlinearity and actuator saturation.

In this paper, we investigate the problem of controlling a nonlinear system from one regulatable state to another regulatable state under input saturation. The proposed approach first models the plant dynamics based on a Gaussian Potential Function Network (GPFN) with the network training by Hierarchically Self-Organizing Learning (HSOL) algorithm [10, 11]. The proposed approach also formulates the control law by planning a feasible state trajectory and generating a sequence of via points in a feasible state transition region.

PROBLEMS IN NONLINEAR SYSTEM CONTROL

A nonlinear plant can be modeled as the following discrete-time state equation:

$$\mathbf{x}_{k+1} = \mathbf{f}(\mathbf{x}_k, \mathbf{u}_k) \tag{1}$$

$$\mathbf{u}_{min} \leq \mathbf{u}_k \leq \mathbf{u}_{max} \tag{2}$$

where $\mathbf{x}_k$ and $\mathbf{u}_k$ represent respectively N dimensional state vector and M dimensional control input vector at time k, and $\mathbf{f} \equiv [f_1(\mathbf{x},\mathbf{u}), f_2(\mathbf{x},\mathbf{u}), \cdots, f_N(\mathbf{x},\mathbf{u})]^t$.

If $\mathbf{f}$ is invertible, i.e., a unique control input can be determined within its saturation limit by

$$\mathbf{u}_k = \mathbf{g}(\mathbf{x}^d_{k+1}, \mathbf{x}_k), \tag{3}$$

then, $\mathbf{g}$ can be trained by a neural network from the observations on plant states, $\mathbf{x}_k$ and $\mathbf{x}_{k+1}$, for a given control input, $\mathbf{u}_k$. Or, $\mathbf{u}_k$ can be searched iteratively in such a way as to minimize the error function, $E(\mathbf{u}_k)$,

$$E(\mathbf{u}_k) = \frac{1}{2} \parallel \mathbf{x}^d_{k+1} - \mathbf{f}(\mathbf{x}_k, \mathbf{u}_k) \parallel^2 . \tag{4}$$

It should be noted that (3) and (4) is based on the assumption that the goal state $\mathbf{x}^d_{k+1}$ can be reached by one-step state transition from $\mathbf{x}_k$. However, in many cases, it requires a multi-step state transition based on a sequence of control inputs to reach $\mathbf{x}^d_{k+1}$ from $\mathbf{x}_k$ due to actuator and plant nonlinearity. Therefore, for the control of a plant requiring multi-step state transitions, the determination of a sequence of optimal or feasible state transitions to $\mathbf{x}^d_{k+1}$ under the constraints of input saturation and plant nonlinearity becomes essential. In this paper, we cast this problem into a path planning problem in a feasible state transition region.

CONTROL UNDER PLANT NONLINEARITY AND INPUT SATURATION

Feasible State Transition Region

Not only the actuator saturation but also the plant nonlinearity limit the range of feasible state transitions at the current state. For instance, a linear plant with input saturation, e.g.,

$$x_{k+1} = 0.5x_k + u_k \qquad \text{with} -0.25 \leq u_k \leq 0.25, \tag{5}$$

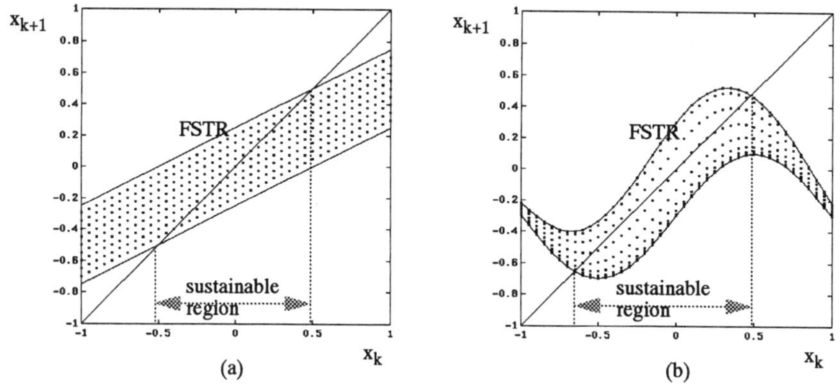

(a) (b)

Figure 1: FSTR in $(x_k,\ x_{k+1})$: The dotted areas of (a) and (b) represent respectively the FSTRs of a linear plant with input saturation, and a nonlinear plant without input saturation.

as well as a nonlinear plant without input saturation, e.g.,

$$x_{k+1} = 0.4\sin(\pi x_k) + 0.6e^{-2(x_k^2+u_k^2)} - 0.3 \tag{6}$$

equally impose a limitation on the range of feasible state transitions from x_k to x_{k+1}, as shown by the dotted region of Fig. 1(a) and (b), respectively.

Let us now define the *Feasible State Transition Region* (FSTR) as a manifold in the augmented state space, $\{\mathbf{x}_{k+1}, \mathbf{x}_k\}$ as follows:

$$FSTR = \{(\mathbf{x}_{k+1}, \mathbf{x}_k) |\ \mathbf{x}_{k+1} = \mathbf{f}(\mathbf{x}_k, \mathbf{u}_k)\ \ and\ \ \mathbf{u}_{min} \leq \mathbf{u}_k \leq \mathbf{u}_{max}\}. \tag{7}$$

A subregion inside FSTR that satisfies $\mathbf{x}_{k+1} = \mathbf{x}_k$, as illustrated by the straight lines inside FSTRs in Fig. 1, is referred to here as a sustainable region, $\mathcal{S}$:

$$\mathcal{S} = \{(\mathbf{x}, \mathbf{x}) |\ \mathbf{x} = \mathbf{f}(\mathbf{x}, \mathbf{u}_k)\ \ and\ \ \mathbf{u}_{min} \leq \mathbf{u}_k \leq \mathbf{u}_{max}\}. \tag{8}$$

Note that a sustainable state defined in a sustainable region can maintain its state indefinitely. A *regulatable region*, $\mathcal{R}$, is defined as a subset of sustainable region, such that there exists a sequence of control inputs which makes possible the transition between any pair of regulatable states inside $\mathcal{R}$.

As mentioned previously, our control goal is to transfer the initial state, $\mathbf{x}_0 \in \mathcal{S}$ to the goal state, $\mathbf{x}^d \in \mathcal{S}$, which requires planning a feasible state trajectory or a path between the two states. In what follows, we will explore an on-line method of generating such a path.

On-Line Path Planning Based on Via Points

For simplicity, let us assume that (1) can be represented by a scalar function:

$$x_{k+1} = f(x_k, x_{k-1}, \ldots, x_{k-m}, u_k), \tag{9}$$

where m represents the order of plant dynamics. Then, FSTR can be defined in $(x_{k+1}, x_k, \ldots, x_{k-m})$ space, while $\mathcal{S}$ and $\mathcal{R}$ can be defined based on $x_{k+1} = x_k = \cdots = x_{k-m}$. In this case, a feasible path *from the initial regulatable state, $\mathcal{X}_0$, to the desired regulatable state, $\mathcal{X}_d$* can be defined by a set of via points,

$$\{\mathcal{X}_1, \mathcal{X}_2, \ldots, \mathcal{X}_k\} \in \mathcal{R} \tag{10}$$

where $\mathcal{X}_1 = (x_0, x_0, \ldots, x_0) \in \mathcal{R}$ and $\mathcal{X}_k = (x_d, x_d, \ldots, x_d) \in \mathcal{R}$. The via points can be determined based on the following theorem.

Theorem 1 *The transition from a regulatable state, $\mathcal{X}_a$, to another regulatable state, $\mathcal{X}_b$, can be achieved by at least m steps of transitions, if $\mathcal{X}_b$ is located inside the hypersphere, B_x, of radius r, where B_x is centered around $\mathcal{X}_a$ and covered by FSTR.*

Lemma 1 *The next via point $\mathcal{X}_b = (x+\Delta x, x+\Delta x, \ldots, x+\Delta x)$ from the current regulatable state, $\mathcal{X}_a = (x, x, \ldots, x)$ can be determined by*

$$0 < \Delta x \leq \frac{r}{\sqrt{m+1}}. \tag{11}$$

Fig. 2 illustrates a hypersphere of radius r and the selection of Δx. The above theorem and lemma indicate that the selection of via points requires the information on the radius of hypersphere at each regulatable state. This can be predetermined based on FSTR, and stored in a neural network, e.g., Gaussian Potential Function Network (GPFN) based on Hierarchically Self-Organizing Learning (HSOL) [10,11]. The boundary of $\mathcal{R}$ can also be estimated by GPFN. Note that in the case where $\mathcal{R}$ is multi-dimensional, an optimal state trajectory needs to be determined first in $\mathcal{R}$ based on a potential field method, prior to selecting via points.

Fig. 3 summarizes the procedure of generating control inputs, as well as training networks, in the proposed approach.

Simulation

The following first order nonlinear SI/SO plant with input saturation is used for simulation:

$$x_{k+1} = f(x_k, u_k) = 0.4\sin(\pi x_k) + 0.6\cos(\pi u_k) \tag{12}$$

$$-1.0 \leq u_k \leq +1.0 \tag{13}$$

Fig. 4(a) illustrates the 3-D mesh graph of $f(x_k, u_k)$ trained by a GPFN based on HSOL. The teaching patterns are randomly generated by presenting the random control input (within the saturation limit) to the plant. The rms error of GPFN with 18 Gaussian units is estimated around 0.02 (which is 1% of the range of f).

Two cases of saturation limit are considered for simulation: $-1.0 \leq u_k \leq +1.0$ and $0.4 \leq u_k \leq 0.6$. Fig. 5(a) and (b) illustrate the FSTR of the case 1 and 2 respectively, while Fig. 5(c) and (d) illustrate respectively the regulatable regions of the case 1 and 2.

The initial and goal states are given by -0.6 and 0.7 respectively for the case 1; and -0.4 and 0.5 respectively for the case 2. In both cases, the step of next via

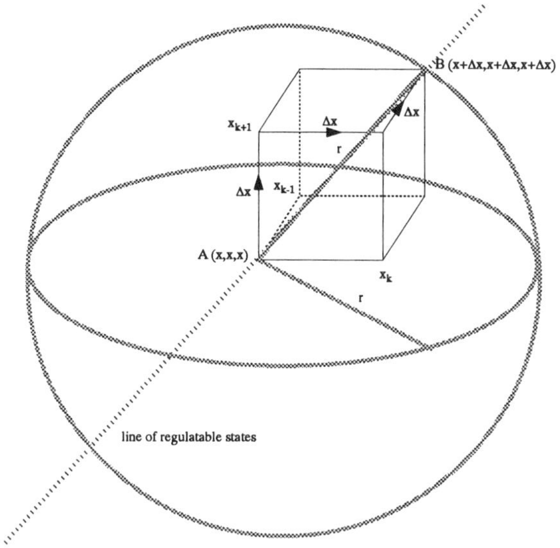

Figure 2: An example of state transition from one regulatable state to another regulatable state inside a hypersphere, $B_{\mathbf{x}}$ for a second order dynamic system.

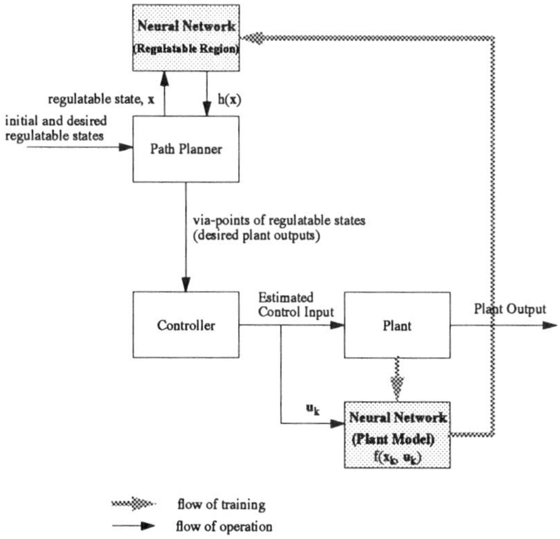

Figure 3: The block diagram describing the flow of control input generation as well as network training.

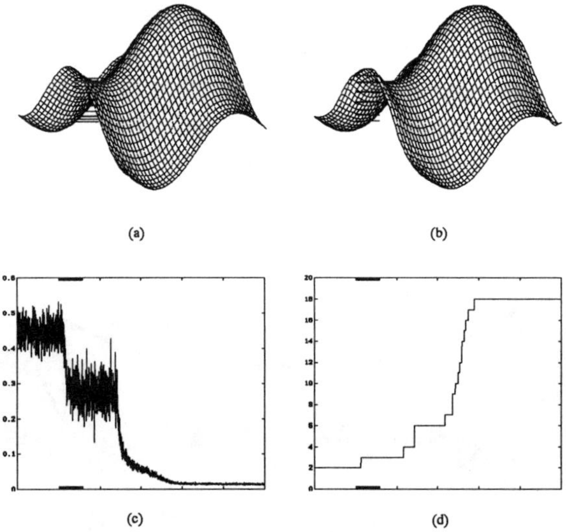

Figure 4: The learning of plant dynamics by GPFN based on HSOL: (a) and (b) represent respectively the 3-D mesh graphs of actual $f(x_k, u_k)$ and modeled $f(x_k, u_k)$ by GPFN. (c) represents the rms error vs. number of iterations, while (d) represents the number of Gaussian units vs. number of iterations.

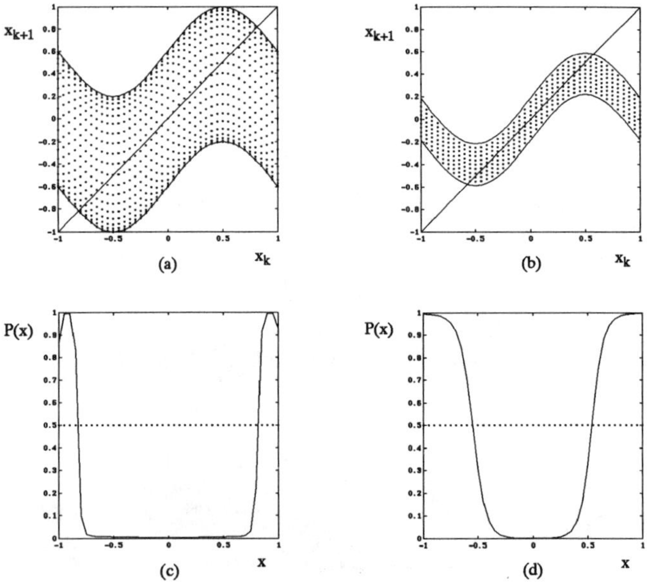

Figure 5: (a) and (b): FSTR of the nonlinear plant with $-1 \leq u_k \leq 1$ and $0.4 \leq u_k \leq 0.6$, respectively. (c) and (d): the regulatable region for $-1 \leq u_k \leq 1$ and $0.4 \leq u_k \leq 0.6$, respectively.

point, Δx_k^d, is given by 0.01. Fig. 6(a) and (b) illustrate respectively the state trajectories of the case 1 and 2. The small oscillation in the state trajectories of Fig. 6(a) and (b) are due to going through via points in regulatable region. Fig. 6(c) and (d) represent respectively the plant output vs. the number of iterations for both cases, whereas Fig. 6(e) and (f) respectively control input vs. the number of iterations for both cases. The simulation results demonstrate the successful convergence from the initial state to the goal state for both cases.

CONCLUSION

In this paper, we propose a new approach to the neural network based on-line control of a nonlinear plant subject to input saturation. Our approach is based on modeling the plant dynamics and the regulatable region by GPFN based on HSOL, and planning a path in the regulatable region by a sequence of via points along the feasible state trajectory generated by a potential field method and a sequence of via points. GPFN with HSOL has advantages over the Backpropagation network in the sense that it can overcome the structural limit of a network and the local minima problem in learning.

References

[1] B. Widrow. The original adaptive neural network broom-balancer. *IEEE International Symposium on Circuits and Systems*, pages 351–357, 1987.

[2] V. V. Tolat and B. Widrow. An adaptive broom balancer with visual inputs. *IEEE International Conference on Neural Networks*, 2:641–647, 1988.

[3] A. G. Barto, R. S. Sutton, and C. W. Anderson. Neuronlike adaptive elements that can solve difficult learning control problems. *IEEE Transactions on System, Man and Cybernetics*, 13(5):834–846, 1983.

[4] G. Wang and D. K. Miu. Unsupervised adaptation neural-network control. *IEEE International Joint Conference on Neural Networks*, 3:421–428, 1990.

[5] Y. Iiguni, H. Sakai, and H. Tokumaru. A regulator design method using multilayered neural networks. *IEEE International Joint Conference on Neural Networks*, 3:371–379, 1990.

[6] S. Lee and R. M. Kil. Robot kinematic control based on bidirectional mapping neural network. *IEEE International Joint Conference on Neural Networks*, 3:327–335, 1990.

[7] S. Lee and R. M. Kil. Redundant arm kinematic control based on liapunov function: Algorithm and connectionist model. Technical Report IRIS-271, University of Southern California, 1991.

[8] Y. Z. Tsypkin. *Adaptation and Learning in Automatic Systems*, pages 176–209. Academic Press, 1971.

[9] K. S. Narendra and K. Parthasarathy. Identification and control of dynamic systems using neural networks. *IEEE Transactions on Neural Networks*, 1(1):4–27, 1990.

[10] S. Lee and R. M. Kil. Multilayer feedforward potential function network. *IEEE International Conference on Neural Networks*, 1:161–171, 1988.

[11] S. Lee and R. M. Kil. A gaussian potential function network with hierarchically self-organizing learning. *Neural Networks*, 4(2):207–224, 1991.

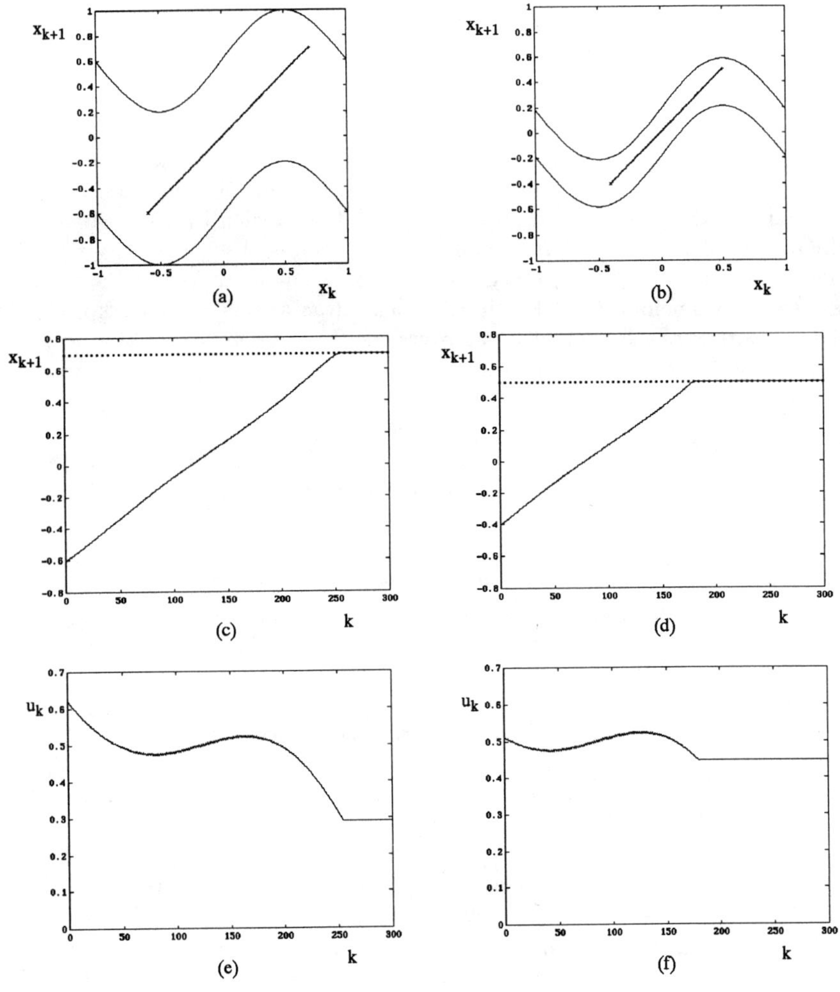

Figure 6: (a) and (b) the state trajectories for $-1 \leq u_k \leq 1$ and $-0.4 \leq u_k \leq 0.6$ respectively from the initial states to the goal states. (c) and (d) the plant output vs. number of iterations for (a) and (b) respectively. (e) and (f) the control input vs. number of iterations for (a) and (b) respectively.

PART 9 : DESIGN AND ECONOMICS OF MANUFACTURING SYSTEMS

Machine-Part Assignment for Group Technology

M. Kamel and T. Liu

Pattern Analysis and Machine Intelligence Lab
University of Waterloo
Waterloo, Ontario
Canada N2L 3G1

ABSTRACT

To date, many different criteria are used for machine-part assignment, however, none of these has used group technology (GT) concept. In this paper, two approaches to machine-part assignment are discussed. In the first approach, load balance is used to find an optimal solution, or an approximate solution then check if machines and parts can be mutually separated into cells and families. In the second approach, we use GT concept in the assignment. Instead of allocating parts to machines evenly, we try to find the commonality among the machine assignments, so that machine cells and part families can be formed. To emphasize the importance of manufacturing process sequence, a coefficient matrix based on operation sequences is used as a similarity measure between parts. After applying different clustering techniques to the matrix, we use the most reasonable partition in the GT machine assignment. Finally, the two approaches are compared and the impact of second approach is summarized.

INTRODUCTION

GT is a manufacturing philosophy or principle with far reaching implications. The basic concept is very simple: to identify and bring together related or similar parts and processes, to take advantage of the similarities which exist, during all stages of design and manufacturing. With the development of flexible manufacturing systems (FMS), there will be a greater emphasis on short run production. The evolution of FMS not only has made short runs feasible but also positively encourages the short run productions.

Many companies are now turning to GT as a means of integrating design and manufacturing [2]. Parts with design and manufacturing similarities are grouped into families. The result is very significant in reducing the number of problems confronting design and manufacturing. However, GT is not always an easy concept to accept. Most batch manufacturing will claim that their parts are always different and that they cannot be reasonably grouped.

To date, there are two basic methods used for solving the GT problem [8]. The first one is classification which is used to group parts into families, or machines into cells based upon their design characteristics. The second method is cluster analysis that decomposes a manufacturing system into homogeneous subsystems based machine features, or part characteristics. See [1] for more clustering analysis applications. However, these methods are all based on a given machine assignment. When applying GT to a given machine assignment, operation sequences, setup time, operation types, manufacturing processing time and load are no longer involved. In this case, GT is only acting as a tool to make the assignment mutually separable in machine cells and part families. Using

GT in machine assignment becomes very necessary when alternatives that may produce a good GT machine assignment exist.

This paper addresses the machine-part assignment problem for the purpose of facilitating GT. Section two discusses all possible machine-part assignments for a given manufacturing system. Section three describes an algorithm that generates an optimal or approximate machine assignment based on load balance. Section four is a discussion on using GT in machine assignment. Section five shows how to get a better GT machine assignment by partitioning parts into families based on operation sequence similarity coefficient. Finally, the impact of using GT in machine assignment is given and a conclusion is drawn from this study.

POSSIBLE MACHINE ASSIGNMENT

For any manufacturing system involving assigning parts to machines, there might be more than one way to achieve the assignment. In most cases, the number of possible machine assignments may get very large with an increase of machine capability and the number of operations required by each part. To illustrate this, a simple example, as shown in Figure 1, is used to show the relationship between parts and machines. Each part and machine is associated with a set of operations. Links indicate the operations required by parts and processed by machines.

For an operation(s) associated with a part, there may be more than one machine capable of performing the operation(s). Choosing a machine(s) to manufacture a part depends on what operations are required by the part. The standard matrix formulation enables a planner to solve the problem by direct mapping of parts to machines. Since the mapping may be arbitrary, one assignment could be better than the other in terms of grouping machines into cells and parts into families. Studying all possible machine assignments of a manufacturing system is the first step to an understanding of using GT in machine-part assignment.

By looking at how each part is processed by the machine(s), we can easily find all possible machine assignments in the matrix format for the manufacturing system in Figure 1. However, it is not easy to determine if there are good GT assignments, or none at all. In most cases, these matrices are not visually separable into machine cells and part families. This does not mean that mutually separable submatrices do not exist. To make sure of this, the cluster identification *CI* algorithm [7] can be applied to the matrix. The *CI* algorithm was originally developed by Iri [4], then implemented for GT by Kusiak *et al.* The algorithm assumes that homogeneous clusters exits. By re-arranging rows and columns of the matrix, the mutually separable submatrices may be found in the diagonal direction. Machine cells and part families eventually come from these submatrices. *CI* is an efficient algorithm to find GT solution [7]. However, it fails to recognize overlapping problem. It also does not allow a control on the size of machine cells and part families. Other recent work that deals with these problems can be found in [8].

Figure 2 shows all possible machine assignments of the system in Figure 1. Applying the *CI* algorithm, we have found two mutually separable submatrices. In this small system, there are only four possible ways to assign parts to machines. For a larger manufacturing system with six parts and eight machines with overlapping capabilities, there may be more than a million possible machine assignments. With such a great number of possible machine assignments, it is not feasible to consider all of them to select a good GT machine-part as-

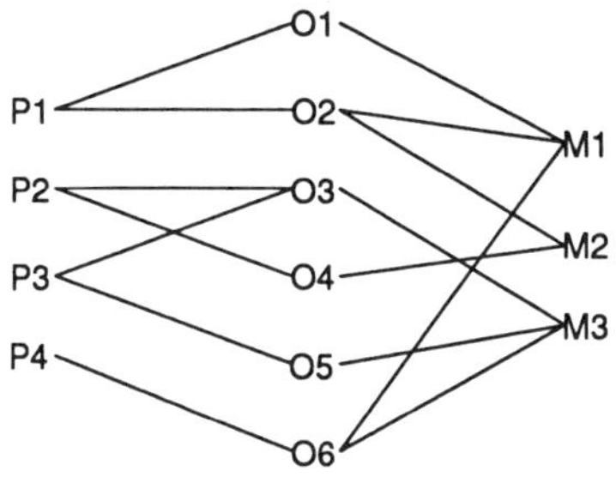

Figure 1: *A Simple Mfg System*

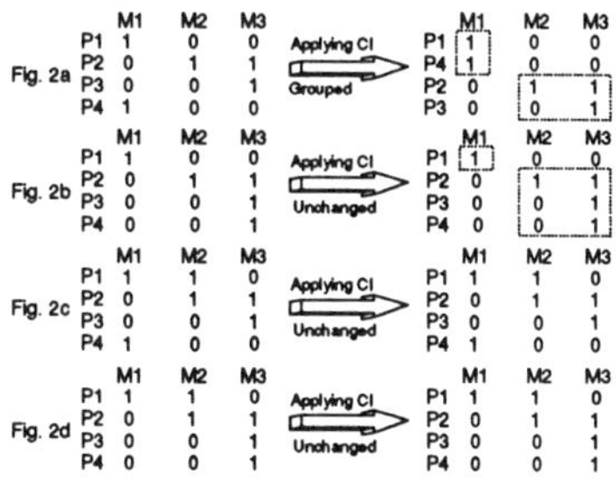

Figure 2: *Possible Assignments*

signment. An assignment criteria need to be employed. Most of the existing methods for machine-part assignment use criteria which do not incorporate GT principles. The GT problem is usually solved after an assignment has been decided. This restricts the flexibility that may exist due to having alternative assignments which could be utilized in generating a good GT solution.

MACHINE ASSIGNMENT OF MANUFACTURING SYSTEMS

Before we consider using GT in machine assignment, it is necessary to discuss current methods developed to solve machine-part assignment problems. Many different criteria are used to assign machine(s) to parts. Among them, the following are most popular: 1) Operation sequences and types, 2) Process time and setup time, and 3) Manufacturing work load.

We focus our discussion on a recent method for machine-part assignment developed by Kamel and Ghenniwa [6]. Detailed implementation and comparison to existing methods can be found in [3]. The method provides an algorithm, *POMJ* (Partially Overlapped systems with Multiple operations Jobs), that finds an optimal solution or its approximate, which satisfies the following goodness conditions:

1. Independency: minimizing the number of machines assigned to each part.

2. Load Balance: minimizing the time load difference between the machines.

POMJ uses a bipartite graph to represent manufacturing system. The algorithm approaches machine-part assignment problem by breaking the system structure into sublevels, then solves the assignment locally for each subgraph level in a way that the above two conditions are satisfied [6]. *POMJ* algorithm finds a machine-part assignment so that the machine usage is balanced.

We may find optimal solution, or a good approximation using the above two conditions. However, this machine assignment may not be mutually separable in terms of group technology. In Figure 2, for example, there are only two, out of four, good assignments from GT point of view. With a greater number of operations involved by parts and less operations processed by machines, determining good GT machine assignments becomes more difficult. Figure 3 shows a manufacturing system named *SYS* with eight machines and sixteen parts. Each part consists of more than one operation with a given process sequence. The machines have capability of handling all operations in the system.

Based on the criteria presented in *POMJ* algorithm, we have found the solution shown in Figure 4, which provides an even load to system SYS.

MACHINE	NUMBER OF OPERATIONS	MACHINE CAPABILITY
Machine 1	3	ABK
Machine 2	3	CDL
Machine 3	3	EFG
Machine 4	5	FGIHJ
Machine 5	3	KLP
Machine 6	4	JKLO
Machine 7	5	BIJSU
Machine 8	5	ABXYZ

PART	NUMBER OF OPERATIONS	OPERATION SEQUENCES
Part 1	4	ACKL
Part 2	4	ABCD
Part 3	2	JK
Part 4	3	BCD
Part 5	4	EHIJ
Part 6	3	CKL
Part 7	3	BIS
Part 8	6	JKLGFP
Part 9	3	HIJ
Part 10	4	KLGF
Part 11	5	BIOSY
Part 12	4	LUJK
Part 13	3	ABC
Part 14	7	JKLUXYZ
Part 15	3	GFO
Part 16	3	XYZ

Figure 3: *Machine-Part SYS*

	M1	M2	M3	M4	M5	M6	R1	R2
P1	1	1						
P2		1						1
P3						1		
P4		1					1	
P5			1	1				
P6		1			1			
P7							1	
P8				1	1			
P9				1				
P10			1			1		
P11						1	1	1
P12					1		1	
P13	1	1						
P14					1			1
P15			1		1		1	
P16								1

Figure 4: *POMJ Solution to SYS*

It is clear that this solution is not visually separable into machine cells and part families. Even after applying *CI* algorithm, it still can not be decomposed. In summary, *POMJ* algorithm gives a good solution to balance the work load. But, the resulting machine-part assignment may fail to generate a good GT solution.

USING GT IN MACHINE ASSIGNMENT

The conventional way of solving the GT problem is to assume that a machine assignment is given, then use GT techniques to find a partitioning of machine cells and part families. However, parts are assigned to machines based on their required operations and the operation capabilities of machines, as shown in Figure 5.

The problem is that once the machine assignment is given, no relationship associated with operations versus machines and parts can be found. Thus, solving the GT problem after the assignment is decided may not be even feasible. It is more logical to utilize the alternatives of having different assignments in order to find one that facilitates the GT partitioning. This is technically possible when more than one machine assignments exist.

In this section we develop criteria which incorporates GT in machine assignment. We also give an assignment algorithm which is based on the criteria. Let us first give some basic definitions.

Definitions:

1. Machine Combination: Machine(s) needed to process a part.

2. Machine Level: number of machine(s) required to process a part.

3. Overlapping Part: a part processed by more than one machine cells.

Comparing the goodness conditions described in the previous section, we keep the independency condition as bottleneck condition. To include GT concept in the assignment criteria, we define the commonality of a machine combination as the number of times that machine combination is selected among parts. The two conditions are used to minimize overlapping among machine(s) shared by the same parts. Therefore, to find a good GT machine assignment, the following two conditions must be satisfied:

1. Bottleneck Condition: assign parts to lowerest machine level whenever possible to avoid overlapping.

2. Commonality Condition: assign parts to most common machine combination at the same machine level with minimum overlapping machines between machine cells.

To demonstrate the first condition, we use Figure 1 as an example. In this simple manufacturing system, P_1 requires both O_1 and O_2. O_1 can be processed by M_1. However, O_2 can be processed by either M_1 or M_2. For possible machine assignment, we can assign M_1 to manufacture P_1, or both M_1 and M_2 to make P_1. To satisfy the bottleneck condition, we would assign level one machine, M_1 to P_1, rather than having level two machines, M_1 and M_2.

We also use the same example to illustrate the second condition. P_3 needs two operations, O_3 and O_5. M_3 can process O_3, O_5, and O_6. We assign M_3 to P_3. P_4 requires only O_6 which can be processed by either M_1 or M_3. There are two alternatives for machine assignment to P_4. We assign M_3 to both P_3 and P_4. This gives a machine-part matrix shown in Fig. 2b. From the first condition, P_1 is assigned to M_1. P_4 can be processed by both M_1 and M_3. According to the second condition, M_1 could be used to P_1 and P_4 shown in Fig. 2a. Comparing with assigning M_3 to P_3 and P_4, the machine combinations have appeared twice in both cases. Therefore, the planner will choose one arbitrarily. However, both machine-part matrices are nicely grouped in machine cells and part families. The following is an algorithm for using group technology in machine assignment:

GTMA Algorithm:

1. Set iteration number $k = 1$
2. While (there are still parts to be assigned) do
   (a) Find all level k machine combinations which can be used to process each part, and compute types of necessary operations required by each part.
   (b) Construct a bipartite graph at each level k to find if a part is a subset of machine combination(s) in terms of operations required.
   (c) Count the number of times that each combination appears.
   (d) Assign the machine combination(s) that appears most to the part, break ties arbitrarily.
   (e) Let $k = k + 1$.

   end while

The application of the above algorithm to the example in Figure 1 gives the result shown in Fig. 2a. Applying *GTMA* algorithm to the large problem of Figure 3, we obtain the assignment shown in Figure 6.

Applying *CI* algorithm to the resulting assignment produces the partitioning shown in Figure 7, which gives two machine cells and two part families.

USING OPERATION SEQUENNCES SIMILARITY IN GT

In the previous sections, we have discussed two algorithms used in machine assignment. *POMJ* gives an optimal solution or approximation in terms of load balance. *GTMA* uses GT concept in machine assignment to get mutually

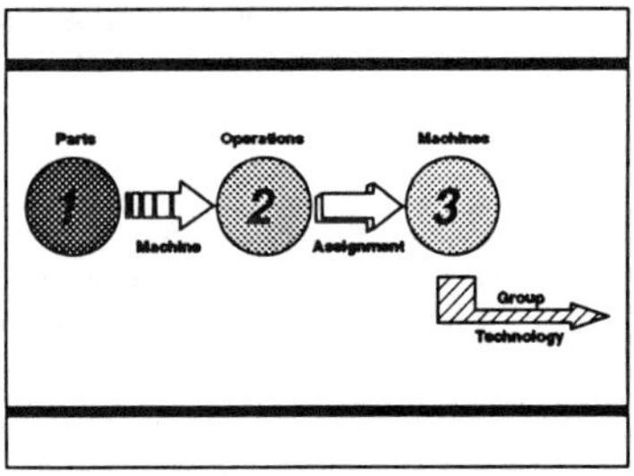

Figure 5: *Assignment vs GT*

	M1	M2	M3	M4	M5	M6	R1	R2
P1	1	1						
P2	1	1						
P3						1		
P4	1	1						
P5			1	1				
P6	1	1						
P7							1	
P8				1	1			
P9				1				
P10				1	1			
P11						1	1	1
P12					1		1	
P13	1	1						
P14						1	1	1
P15				1	1			
P16								1

Figure 6: *GTMA Solution to SYS*

	M1	M2	M3	M4	M5	M6	R1	R2
P1	1	1						
P2	1	1						
P4	1	1						
P6	1	1						
P13	1	1						
P3						1		
P5			1	1				
P7							1	
P8				1	1			
P9				1				
P10				1	1			
P11						1	1	1
P12					1		1	
P14						1	1	1
P15				1	1			
P16								1

Figure 7: *Result of Applying CI*

PART FAMILY	PARTS	OPERATIONS	SEQUENCES
PF-1	P1,P4	4	ACKL
PF-2	P2,P4,P13	4	ABCD
PF-3	P5,P9	4	EHIJ
PF-4	P8,P10,P15	6	FGJKLP
PF-5	P7,P11	5	BIOSY
PF-6	P3,P12,P14,P16	7	JKLUXYZ

Figure 8: *Part-Families of SYS*

separable machine cells and part families. In flexible manufacturing systems, setup time is important due to small-lot-size production. Frequently switching manufacturing process types may not be allowed although this situation has been improved since the evolution of numerically controlled machine tools. Grouping parts into families based operation sequence similarity is a way to reduce the setup time. We define this kind of part family as *higher level part* with all operations required by the part family.

The operation sequence part family partitioning approach is based on similarity coefficient matrix defined by Tam [10]. The basic idea is to find the minimum number of transformations required to derive one part operation sequence from the other and use a clustering technique to find part partitions. We have used the operation sequence similarity coefficient matrix, in conjunction with four different clustering methods to find the most reasonable partitioning [9]. The solution of applying the *kth* nearest neighbor clustering method [5] to the 16-part system is shown in Figure 8.

Applying *GTMA* algorithm to solve the assignment if the 6 part families and 8 machines, the mutually separable submatrices on the diagonal direction shown in Figure 9. In this case, the use of *CI* algorithm is not necessary since the matrix is already decomposed into machine cells and part families.

We have also applied *POMJ* to assign the part families to the machines. The resulting assignment is shown in Figure 10. It is obvious that such assignment can not produce further partitioning even after applying *CI* algorithm.

CONCLUSIONS

In this paper, we have presented a new approach for machine-part assignment

	M1	M2	M3	M4	M5	M6	R1	R2
PF-1	1	1						
PF-2	1	1						
PF-3			1	1				
PF-4				1	1			
PF-5						1	1	1
PF-6						1	1	1

Figure 9: *GTMA Solution to FPs*

	M1	M2	M3	M4	M5	M6	R1	R2
PF-1	1	1						
PF-2		1						1
PF-3			1	1				
PF-4				1	1			
PF-5						1	1	1
PF-6					1		1	1

Figure 10: *POMJ Solution to PFs*

to produce GT partitioning. The approach solves the GT problem at the assignment stage rather than solving it after the assignment is decided. It utilizes the possibility of having different assignments and tries to choose one which facilitates the GT partitioning. We have shown that solving the assignment problem in isolation of the GT problem by optimizing load balance, for example, may produce assignment which can not produce satisfactory GT partitioning. By using GT criteria in solving the problem machine assignment, better results can be obtained. We have provided such criteria and presented an algorithm for implementing it. We have also investigated pregrouping of parts into part families before solving the assignment problem and its effect on GT partitioning. This pregrouping is done based on operation sequence similarity which can lead to reduction in setup time of the manufacturing processes.

REFERENCES

1. M.R. Anderberg, Cluster Analysis for Application (Academic Press, New York, 1973).
2. C.C. Gallagher and W.A. Knight, Group Technology Production Methods in Manufacture (Ellis Horwood Limited, John Wiley & Sons, New York, 1986).
3. H. Ghenniwa, Solving the Assignment Problem of Multi-agent systems (Master Thesis, University of Waterloo, 1990).
4. M. Iri, "On the Synthesis of Loop and Cutset Matrices and Related Problems," *RAAG Memoirs, Research Association of Applied Geometry, Japan.* Volume 4, A-XII, pp. 376-410, (1968).
5. A.K. Jain and R.C. Dubes, Algorithms for Clustering Data (Prentice Hall, New York, 1990)
6. M. Kamel and H. Ghenniwa, "Job Allocation Scheduling for Partially Overlapped Systems," *Proceedings of International ISMM/ISTAED Conference on Parallel and Distributed Computing and Systems,* Washington D.C, (1991).
7. A. Kusiak and W.S. Chow, "Efficient Solving of the Group Technology Problem," *Journal of Manufacturing Systems,* Volume 6, No. 2, pp. 117-124, (1989).
8. A. Kusiak and W.S. Chow, "Decomposition of Manufacturing Systems," *IEEE Journal of Robotics and Automation,* Vol. 4, No. 5, pp. 457-471, (1988).
9. T.A. Liu, Algorithms for Forming Part-Families and Their Assignment to Machines in Group Technology, (Master Thesis, University of Waterloo, 1992).
10. K.Y. Tam, "An Operation Sequence Based Similarity Coefficient for Part Families Formations," *Journal of Manufacturing Systems,* Volume 9, No.1, pp. 45-68, (1990).

MICRO-AUTOMATION OF SEMICONDUCTOR FABRICATION

Ilene J. Busch-Vishniac
Department of Mechanical Engineering,
The University of Texas at Austin, Austin, TX 78712

ABSTRACT

The automation of small scale processes poses distinctly different problems from conventional automation of large scale processes and equipment. In particular, overcoming friction and operating with great delicacy assume much greater importance at small scales. This paper examines small scale automation concentrating on the automation of micron scale mechanical processes in semiconductor fabrication. Distinctions between macro-automation and micro-automation are presented and the unique challenges of micro-automation are articulated. A prototype system suitable for micro-automation of semiconductor fabrication is presented briefly.

INTRODUCTION

Most of the work on product fabrication automation has concentrated on the automated construction of large scale objects such as automobiles. However, many products, particularly in the consumer sector, are significantly enlarging their market niche by introducing substantial size reductions. Portable high fidelity tape and CD players and notebook computers are examples of such products. As a result of miniaturization, standard component size ranges have been expanded at the small size end. Manufacturers seek to either continue or introduce automated fabrication over the entire range of component sizes in order to control quality while keeping costs low. Of special significance, because of the challenges posed for automation, are microelectronics devices, which are being found in a growing number of products and in which the feature sizes are shrinking to less than a micron.

Automation of the fabrication of small components and devices poses a significantly different set of problems from automation of the production of large scale objects. To distinguish the two we refer to small scale automation as micro-automation, and large scale automation as macro-automation. The differences between micro-automation and macro-automation are most obvious in two areas: First, while friction poses a challenge for automation at any scale, friction becomes substantially more important as scales shrink. This is because friction scales with the surface area of objects while body forces such as gravity scale with volume. As a feature shrinks the ratio of surface area to volume rises, making friction a far more important force. Second, the aims of macro-automation and micro-automation tend to be quite different. In macro-automation one is concerned with the handling of large loads. In micro-automation the delicacy with which one can handle objects is much more of a design motivator. As a result of these differences between macro-automation and micro-automation, methods appropriate for one domain may not be appropriate for the other.

NEEDS OF MICRO-AUTOMATION SYSTEM FOR SEMICONDUCTOR FABRICATION

While micro-automation might be applied to a number of products, we have chosen to concentrate on semiconductor fabrication for two main reasons: First, the scales of semiconductor features are extremely small and the corresponding tolerances quite challenging. For example, typical VLSI features are currently at about 1μm and predicted to shrink to about 0.35μm in the near future. As a result, the automatic fabrication of semiconductors is probably the greatest challenge facing micro-automation. Second, the semiconductor industry has long concentrated on chemical and electrical fabrication techniques in order to achieve the necessary process control. However, some portions of the fabrication process, such as those which require motion of one part relative to another, can not be performed through purely electrical or chemical means. It is precisely these steps in the fabrication process which have prevented the full scale automation of semiconductor fabrication. Hence, the motivation to solve the micro-automation problem for mechanical processes is quite high, and there is a correspondingly high probability that breakthroughs will be moved into fabrication lines quickly.

What sets semiconductor fabrication apart from the construction of nonmicroelectronic parts are the scales of features and a number of nonconventional requirements for the production environment. The latter requirements typically include the ability to operate in a clean room, the ability to withstand high temperatures and pressures, the ability to survive exposure to a wide variety of chemicals, and the ability to operate without creation of electric fields which might damage the devices under construction.

The main components of a micro-automation system are quite similar to that of a macro-automation system[1]. Each requires an actuator or set of actuators which operate on a workpiece, sensors which monitor the actuator performance, a controller which regulates the system, and electronics which link the components together. The component sizes tend to be smaller for a micro-automation system, but the size of a component is not necessarily well correlated with the size of features of the objects on which the components operate. Hence, sensors and actuators for a micro-automation system need not themselves be micron scale in size, although it is reasonable to expect them to be smaller than their macro-automation cousins.

Semiconductor devices are constructed in an assembly line fashion although it is not uncommon for there to be long gaps in time between the completion of one part of the process and the beginning of the next. In order to automate the process the following items are required: a means of transporting the silicon wafer (workpiece) along the line, a means of loading and unloading the wafer onto the carrier used for transport, and a system suitable for precision operations in specific locations along the line. For complete automation of the line it is necessary that these items be integrated seamlessly. This makes it advantageous to use a single driving technology rather than a mixture of hydraulic, pneumatic, electrostatic, magnetic, and other mechanisms. The wafer loader/unloader design must concentrate on a means of holding a wafer along its edge or bottom since the surface being processed cannot be contacted. The precision workspaces must concentrate on positioning control, robustness and stability in small working ranges (up to the size of a wafer). The transportation between workspaces must focus on speed and smoothness of ride over long distances (up to several meters). Below we describe possible components for one prototype micro-automation system which meets these requirements.

A PROTOTYPE MICRO-AUTOMATION SYSTEM

The first issue that one must consider when designing a micro-automation system for use in semiconductor fabrication is what driving method will be used. While a number of possibilities are available, we have chosen to construct a system using magnetic drive. There are four main reasons for this choice: First, given that we wish to use a single driving mechanism, and that the needed motion will range from very small (precision workspace) to very large (inter-workspace transport), we require an approach that can be applied over extremely broad ranges. This eliminates electrostatic drive because the power needs for large scale motion are prohibitive. It also makes pneumatic and hydraulic approached unlikely choices because of the difficulty of getting very small scale motion. Second, a magnetic driving technology is compatible with clean room operation. Hydraulic or pneumatic drives, by comparison, would be unacceptable without going to great lengths to ensure clean operation. Similarly, electrostatic drive would be incompatible because of its tendency to collect fine particles. Third, magnetic approaches offer the possibility of operation in some harsh environments, including wet locations or those in vacuum. This is potentially quite important, as many of the processing steps are currently done in vacuum and most are wet. Fourth, magnetic drive is compatible with the devices being constructed. It will not introduce high electric fields which could damage the devices, nor require use of lubricants which, should they leak, could cause catastrophic failures.

Figure 1 shows a prototype wafer loader and unloader designed for semiconductor micro-automation, and based on a magnetic driving mechanism. Shown below the loader/unloader is a wafer carrier which uses clippers to hold a wafer along its edge. In the loading or unloading process, the clippers are spread by activating a pair of solenoids on a loading/unloading tool. This solenoid pair induces an attraction force between the fixed solenoids and the magnets on the moving clippers of the carrier. The loading process begins with a wafer in place on the loading/unloading tool being held by vacuum applied to the back surface. The tool positions itself just above the carrier, opens the clippers, lowers so that the wafer is gently placed onto the carrier surface, releases the clippers, then releases the vacuum. The unloading process simply reverses this operation. Further details about the loader/unloader may be found in Wang et al. [2].

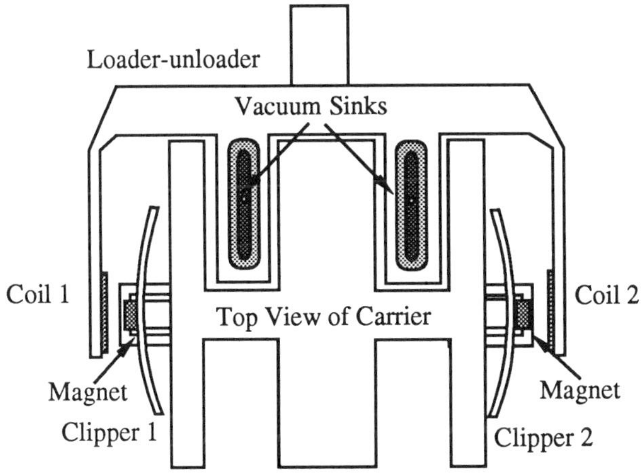

Figure 1: A prototype loader/unloader and carrier.

Figure 2 shows a view of a prototype linear transport path for the micro-automation system. This path is used to connect precision workspaces. In an ideal situation, the loading operation would begin at one end of the line. The carrier would already be located at the beginning of a transport path, and once loading was completed the wafer and carrier would progress down the line from path to precision workspace, to path, etc., until all operations which are performed on the entire wafer are completed. Then the wafer would be unloaded from the carrier which could cycle back to the line beginning to receive a new wafer.

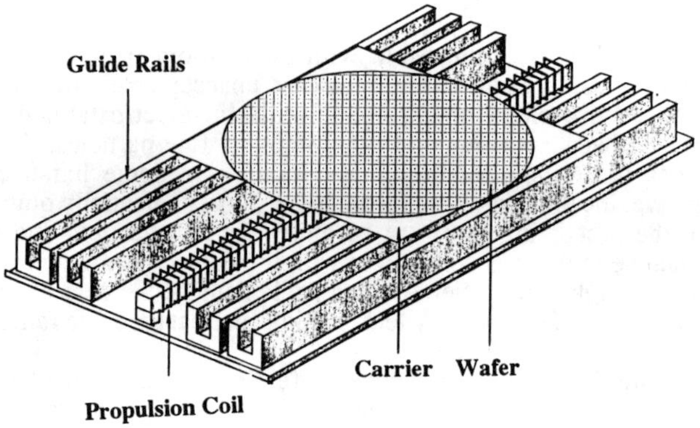

Guide Rail Detail

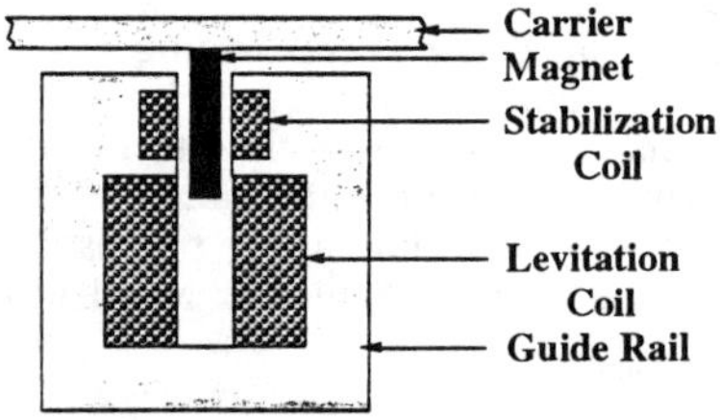

Figure 2: A prototype transport path.

The prototype path shown in Fig. 2 uses the same carrier as that shown in Fig. 1. The path consists of three independently controlled solenoids: one for levitation of the carrier in the vertical direction (large solenoid with vertical axis), one for stabilization of the levitation (small solenoid with vertical axis), and one for propulsion (central solenoid with axis in direction of propagation). With this path the carrier and wafer are moved rapidly and without any mechanical contact with other objects. We believe that a path of this sort could be manufactured in a manner similar to that currently used for toy train tracks so that great flexibility in the design of a fabrication line would be possible.

Figure 3 shows one example of a precision workspace. This part of the system, which preceeded the transport path and loader/unloader, was designed for a different carrier geometry, i.e. a triangular shape rather than rectangular. The magnets at the vertices of the triangle are driven by forces and torques induced by magnetic fields established by six electromagnets. The electromagnets are divided into two sets of three, each of which form a triangle. The two triangular shapes are located on different planes and are rotated slightly from one another. We have shown using this precision workspace that we can control four degree-of-freedom motion to the limit of our sensing system ability (0.1μm) over a range of about 400μm x 400μm [3]. Work is currently underway to expand the range of operation, to expand the number of degrees-of-freedom, and to examine rectangular carrier geometries.

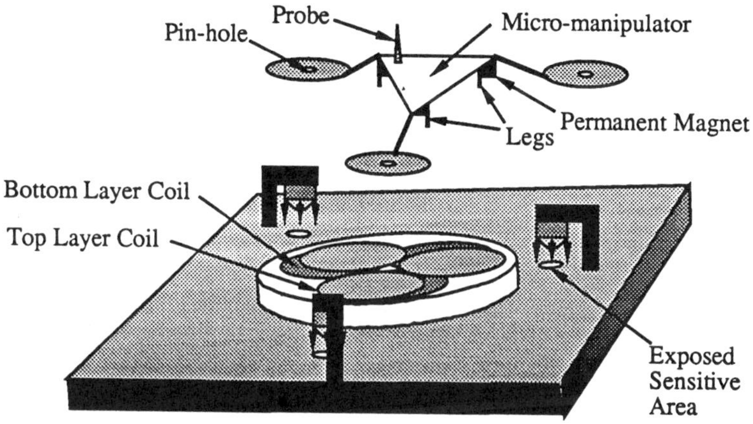

Figure 3: A prototype precision workspace system. The carrier has wings which are used to sense its location in the workspace.

By linking the above components one could develop a system suitable for the automation of mechanical processes in semiconductor fabrication. At present, the transport path and loader/unloader are fully compatible. The precision workspace defined above could be used with the loading tool and transport path, but this would require that the silicon be blocked on top and bottom. The bottom would be occupied by the rectangular carrier holding the wafer. The carrier would be held at a fixed location in the transport path. The top side of the silicon would be obscured by the workspace manipulator which would perform a task by moving to specified locations on the wafer. The advantage of this approach is that it does not require that the rectangular carrier ever leave the transport path. However, access to the silicon surface would be greatly impeded and this is not likely to be acceptable. The alternative of a precision workspace based on the rectangular carrier would permit precise positioning of the wafer in a workspace defined by solenoids below the carrier. While this would require the carrier to leave the path to enter a precision workspace, it could leave the top surface of the silicon unobscured. Work is currently underway to develop a hardware and software interface which would permit seamless handoff from the one-dimensional transport path to a multi-dimensional precision workspace.

THE FUTURE OF MICRO-AUTOMATION

A growing fraction of products rely on features that are micron scale or smaller. In order for this segment of industry to continue to develop, it is necessary that means of automating fabrication processes appropriate for such small objects be devised. To date most of the work has concentrated on the construction of micro-sensors and micro-actuators which are themselves micron scale in size. While these sensors and actuators are interesting, and will certainly find application in micro-automation and other avenues, it is not clear that it is useful to concentrate on the size of the sensors and actuators rather than their ability to position or sense quickly, accurately, and with great delicacy. Despite the successes in building micro-actuators, for instance, their applications have yet to be clearly identified and demonstrated.

In addition to the need for a shift in focus, from ever-smaller components to those that perform a function, there is a need for further work on integration of components into coordinated systems which perform a task. For example, the prototype components described here have not been integrated into a single micro-automation system and the technical obstacles to such integration are still significant.

The motivation to succeed in micro-automation is high and increasing substantially year by year. We believe that sufficient pressure to fully automate the fabrication of small scale objects will occur to interest a number of talented researchers to consider a wide variety of micro-automation approaches over the next decade. While full scale prototype micro-automation systems are likely a minimum of five to ten years away, some fully automated machines using components of interest in micro-automation might be developed sooner.

REFERENCES

1. I. J. Busch-Vishniac, "Micro-Automating Semiconductor Fabrication," *IEEE Circuits and Devices* **7**(4), 32-37 (1991).
2. W. Wang, M. Lamb, and I. J. Busch-Vishniac, "An Automated Loading and Unloading System for A Maglev Wafer Transport Path," *IEEE Trans. on Semiconductor Manufacturing* , under review.
3. M. C. Jeong and I. J. Busch-Vishniac, "A Submicron Accuracy Magnetic LevitationMicromachine with Endpoint Friction," *Sensors and Actuators,* **19**(3), 225-234 (1992).

FORMING MANUFACTURING CELLS USING GROUP TECHNOLOGY CONCEPTS

ALI K. KAMRANI, Ph.D.
Department of Engineering Management, University of Missouri-Rolla
Rolla, Missouri 65401

HAMID R. PARSAEI, Ph.D., P.E.
Department of Industrial Engineering, University of Louisville
Louisville, Kentucky 40292

ABSTRACT

This article presents a comprehensive methodology to form part families using dissimilarities between design and manufacturing attributes of parts. The methodology also forms manufacturing cells by considering several relevant operational costs. The proposed methodology employs a multi-criteria optimization model and allows decision makers to assign priorities to the objectives.

GROUP TECHNOLOGY

Group Technology (GT) is a manufacturing philosophy that identifies and exploits the similarity of parts and manufacturing processes. There are evidences that Group Technology concept has been practiced since the beginning of the century. However, it was not until 1959 that the concepts of GT were formalized in a book entitled "Scientific Principles of Group Technology" [1]. Most of the initial applications were concerned with part family manufacturing for batch-type industries. Parts requiring similar production facilities were brought together and processed in sequence in order to reduce setup times, in-process-inventories and throughput times. Over the past few years, GT concepts have broadened to encompass other manufacturing activities such as product designs and process planning using computers as an integral part of Computer Integrated Manufacturing (CIM). The philosophy of Group Technology is to group of parts according to geometrical and/or manufacturing similarities into part families. These part families once stored in a computer data base will provide the basis for productivity improvement in the design and manufacture of parts. In the design stage, geometric similarities of the parts can be exploited in order to increase standardization of parts and part features. The most common approach used to group parts into part families is classification and coding. In this method the first step is to code all manufactured components. Once this task is accomplished parts having the same codes are grouped together. These part families are then stored in a data base for subsequent use.

CELLULAR MANUFACTURING SYSTEMS

Cellular manufacturing System (CMS) also known as group technology, group production, family programming, or flexible manufacturing, can be described as a manufacturing process which produces families of parts within a single line or cell of machines operated by a machinist or by robots that only operates within the line or cell. Some of the benefits of the Cellular Manufacturing System are: 1) minimization of through-put times, 2) minimization of set-up times, 3) minimization of inventories, 4) maximization of resource (machine and labor) utilization, 5) maximization of output, 6) minimization of intra-cell and inter-cell moves of material, 7) minimization of operating costs, and 8) minimization of investment. CMS involves arranging of machines dedicated to the production of a part family. Machines are laid to allow for a continuous work flow through a series of operations. An example of the key variables in designing a manufacturing cell are the number of cells, cell size, total number of machine types, number of operations per part, and number of different part types

PROPOSED DESIGN METHODOLOGY

The proposed methodology is implemented in two phases. Phase I identifies and groups the parts into part families, and Phase II groups the families, machines, tools, and fixtures. The following sections provide detailed description of these two phases. The proposed phases are carried out sequentially.

PHASE I: Part-Families Formation - In this phase, a coding system is developed in order to categorize the parts (KAMKODE). The information for this coding system can be easily retrieved from any design and manufacturing data base which has the information concerning the parts. The coding system consists of 18 digits. The linear disagreement index between parts i and j for attribute k which is either of a binary or nominal variable type is measured by:

$$d_{ijk} = \begin{cases} 1, \text{ if } R_{ik} \neq R_{jk} \\ 0, \text{ otherwise} \end{cases} \quad (1)$$

where d_{ijk} is the disagreement index between parts i and j for attribute k and R_{ik} is the rank of part i for attribute k. The linear disagreement index for general shapes, material, and product types is calculated using equation 1. The linear disagreement index for an ordinal variable is measured using the following equation:

$$d_{ijk} = | R_{ik} - R_{jk} | / (m - 1) \quad (2)$$

where m is the number of classes for attribute k and (m - 1) is the maximum rank difference between parts i and j. The linear disagreement index for maximum diameter, overall length, and inside hole diameter is calculated using this equation. The linear disagreement index for process and end operation sequence is calculated using the following equation [2]:

$$d_{ijk} = 1 - \sum_{o} (q_{iok} * q_{jok}) / \sum_{o} (q_{iok} + q_{jok} - q_{iok} * q_{jok}) \quad (3)$$

where:

$$q_{iok} = \begin{cases} 1, \text{ Part i requires operation o for attribute k} \\ 0, \text{ otherwise} \end{cases}$$

The linear disagreement index for tools and fixture are calculated by [3]:

$$d_{ijk} = (NT_{ik} + NT_{jk} - 2NT_{ijk}) / (NT_{ik} + NT_{jk}) \quad (4)$$

where NT_{jk} is the number of attribute k required by part j and NT_{ijk} is the number of attribute k common to both parts i and j. The linear disagreement index for process and end operation machines is calculated using the Hamming Metric as follows [4]:

$$d_{ijk} = \sum_{m} \delta(X_{imk}, X_{jmk}) \quad (5)$$

where

$$X_{imk} = \begin{cases} 1, \text{ if part i uses machine m for attribute k} \\ 0, \text{ otherwise} \end{cases}$$

and

$$\delta(X_{imk}, X_{jmk}) = \begin{cases} 1, \text{ if } X_{im} \neq X_{jm} \\ 0, \text{ otherwise} \end{cases}$$

After the calculation of these values, the analyst can assign weights to represent his or her subjective evaluation of variables and finally, the weighted dissimilarity measure between parts i and j is measured by [5]:

$$DIS_{ij} = \sum_k (w_k * d_{ijk}) / \sum_k w_k \quad \textbf{(6)}$$

where, w_k is the weight assigned to attribute k, d_{ijk} is the disagreement index between parts i and j for attribute k, and d_{ij} is the weighted dissimilarity coefficient between parts i and j. To select the part-families in phase I of the methodology, the K-median model is used as follows:

<u>Coefficients</u>

p : Number of parts
K : Required number of part families
d_{ij} : Dissimilarity measure between part i and j, $d_{ij} = d_{ji}$

and, $x_{ij} = \begin{cases} 1, \text{ if part i belong to group j} \\ 0, \text{ otherwise} \end{cases}$

<u>Model</u>

$$\textbf{Minimize} \sum_i \sum_j d_{ij} * x_{ij} \quad i,j = 1,2,....p \quad \textbf{(7)}$$

subject to

$$\sum_{j=1,..p} x_{ij} = 1 \quad \text{for all } i = 1,2,...,p \quad \textbf{(8)}$$

$$\sum_{j=1,..p} x_{jj} = K \quad \textbf{(9)}$$

$$x_{ij} \leq x_{jj} \quad \text{for all } i,j = 1,2,..,p \quad \textbf{(10)}$$

PHASE II: Machine Cell Formation - In Phase II of the methodology, the main objective is to determine the number of machines, tools, fixtures, and the number of cells including the assignment of the proper families to the proper cells. The emphasis in this phase is on economic and capacity planning rather than design. This methodology takes into account several relevant operational costs and minimizes the total sum of these costs. This study uses several costs such as: 1) Machine investment cost, 2) Tool investment cost, 3) Fixture investment cost, 5) Machine production cost, 6) Inspection cost, 7) Setup cost and, 8) Intra-cell material handling cost. The formulation of the Phase II is as follows:

<u>Index Sets</u>

o : Operation	o = 1,2,...O
p : Part	p = 1,2,...P
m : Machine Type	m = 1,2,...M

g : Part Group $g = 1,2,...G$
c : Cell, $C \leq G$ $c = 1,2,...C$
t : Tool Type $t = 1,2,...T$
f : Fixture Type $f = 1,2,...F$

Decision Variables

$$X_{gc} = \begin{cases} 1, \text{ if group g is assigned to cell c} \\ 0, \text{ otherwise} \end{cases}$$

$$Y_{pc} = \begin{cases} 1, \text{ if part p is assigned to cell c} \\ 0, \text{ otherwise} \end{cases}$$

NM_{mc} = Number of type m machine assigned to cell c
NF_{fmc} = Number of type f fixture assigned to machine type m in cell c
NT_{tmc} = Number of type t tool assigned to machine type m in cell c

Cost Mathematical Model

Minimize **(11)**

$$\sum_{m}\sum_{c} CM_m * NM_{mc} + \sum_{f}\sum_{m}\sum_{c} CF_f * NF_{fmc} * \sigma_{fm} + \sum_{t}\sum_{m}\sum_{c} CT_t * NT_{tmc} * \sigma'_{tm} +$$

$$\sum_{p}\sum_{c} (NMV_p - 1) * d_p * CH_{pc} * Y_{pc} + \sum_{p}\sum_{m}\sum_{o} CI_{pm} * TI_{pm} * d_p * \mu_{omp} * NM_{mc} +$$

$$\sum_{m}\sum_{g}\sum_{c} TS_{mgc} * CS_{mgc} * \alpha'_{mg} * X_{gc} + \sum_{p}\sum_{m}\sum_{c} [CP_{pm} * R_m + CRW_m * (1 - R_m)] * \beta_{pm} * TMD_{pm} * Y_{pc}$$

Constraints

$$\sum_{m}\sum_{c} CM_m * NM_{mc} \leq BM \qquad \textbf{(12)}$$

$$\sum_{f}\sum_{m}\sum_{c} CF_f * NF_{fmc} * \sigma_{fm} \leq BF \qquad \textbf{(13)}$$

$$\sum_{t}\sum_{m}\sum_{c} CT_t * NT_{tmc} * \sigma'_{tm} \leq BT \qquad \textbf{(14)}$$

$$\sum_{p}\sum_{m}\sum_{o} CI_{pm} * TI_{pm} * d_p * \mu_{omp} * NM_{mc} \leq BI \qquad \textbf{(15)}$$

$$\sum_{p}\sum_{c} (NMV_p - 1) * d_p * CH_{pc} * Y_{pc} \leq BMH \qquad \textbf{(16)}$$

$$\sum_{p} \beta_{pm} * TMD_{pm} * Y_{pc} \leq TP_m * NM_{mc} \quad \text{all m,c} \qquad \textbf{(17)}$$

$$\sum_{p} \beta_{pm} * \beta'_{tp} * \sigma'_{tm} * TMD_{pm} * Y_{pc} \leq TL_t * NT_{tmc} \quad \text{all m,c,t} \qquad \textbf{(18)}$$

$$NF_{fmc} * \sigma_{fm} \geq NM_{mc} \quad \text{for all f,m,c} \qquad \textbf{(19)}$$

$$\sum_{d} d_p * Y_{pc} \leq IC_c \quad \text{for all c} \qquad \textbf{(20)}$$

$$\sum_{c} X_{gc} = 1 \quad \text{for all g} \qquad \textbf{(21)}$$

$$\alpha_{pg} * Y_{pc} = X_{gc} \quad \text{for all p,g,c} \qquad \textbf{(22)}$$

$$X_{gc} \; \varepsilon \; (0,1) \quad \text{for all g,c} \qquad \textbf{(23)}$$
$$Y_{pc} \; \varepsilon \; (0,1) \quad \text{for all p,c}$$

$NM_{mc} \geq 0$ and Integer for all m,c
$NF_{fmc} \geq 0$ and Integer for all f,m,c
$NT_{tmc} \geq 0$ and Integer for all t,m,c

A NUMERICAL EXAMPLE

Application of the proposed methodology is illustrated in the following example [6]. Consider a situation where the production of 15 parts required. These parts require 15 operations, 9 process operations and 6 end operations. There are 13 different machine tools, 8 process machines and 5 end operation machines available. A rating factor of 1 is assumed for all machines performing operations. This indicates that the machine selected is the most suitable for performing operations. Nine types of tools are used for operations and six types of fixture are available for process machines. The results of the Phases I and II are listed in Tables 1 through 4.

CONCLUSION

In this article, a two phase methodology is developed for forming machine cells in a computer integrated manufacturing environment using dissimilarity measure between design and manufacturing attributes' of parts. A classification and coding method is proposed which can be easily constructed using available design and manufacturing database of the firm. An 18 digits coding method is proposed. Six digits are used for coding design attributes and twelve digits are assigned for coding manufacturing attributes. The first phase of the methodology enables the decision maker to assign priority to the attributes in order to group parts into families based on the rank. A 0-1 integer linear mathematical program is developed to solve this problem. The second phase of the methodology will assign part families, and the required number of machines and fixtures, and tools to the available cells to meet the annual demands.

References

1. Snead, C.S., "**Group Technology: Foundation for Competitive Manufacturing**," Van Nostrand Reinhold, New York, 1989.
2. McAuley, J., "**Machine Grouping for Efficient Production**," Production Engineer, 1972.
3. Dutta, S.P., R.S. Lashkari, G. Nadoli, T. Ravi, "**A heuristic Procedure forDetermining Manufacturing Families From Design-Based Grouping for Flexible Manufacturing System**," Computers and Industrial Engineering, Vol. 10, No. 3, pp. 193-201, 1986.
4. Lee, R., "**Clustering Analysis and its Application**," Advances in Information System Science, Plenum Press, New York, 1981.
5. Anderberg, M.R., "**Cluster Analysis for Application**," Academic Press, New York, 1973.
6. Kamrani, A.K., "**A Methodology for Forming Machine Cells in A Computer Integrated Manufacturing Environment using Group Technology**," Doctor of Philosophy Dissertation, Department of Industrial Engineering, University of Louisville, August 1991

Table 1. Cell Configuration

Part Number	Family Number	Cell Number
1		
3		
5	1,4	3
12		
13		
-	-	-
2		
4		
6	2	1
8		
11		
14		
-	-	-
7		
9	3	2
10		
15		

Table 2. Machine Type and Assignments

Machine Type	Number of Duplicates
Cell 3	
1,3,8,9,10	1
2,12	2
11	3
Cell 1	-
5	3
8,11,13	1
9,10	2
12	4
Cell 2	-
1,4,8,9,10,12	1
3,13	2
11	3

Table 3. Tool Type and Assignments

Tool Type	Machine Type	Number of Duplicates
Cell 3		
1	1	176
2	1	194
2	2	404
2	3	91
3	2	263
5	2	137
5	3	83
9	8	49
Cell 1	-	-
2	5	196
3	5	189
5	5	136
6	5	763
9	8	51
Cell 2	-	-
1	1	71
2	1	78
2	3	318
3	3	121
4	3	117
4	4	73
9	8	227

Table 4. Fixture Type and Assignments

Fixture Type	Machine Type	Number of Duplicates
Cell 3		
1	1	1
2	3	1
3	2	2
3	3	1
5	2	2
5	3	1
6	8	1
Cell 1	-	-
2	5	3
3	5	3
6	8	1
Cell 2	-	-
1	1	1
2	3	2
3	3	2
3	4	1
5	3	2
6	8	1

ON UNSTEADY DYNAMICS OF ASSEMBLY PROCESSES

FRIEDRICH PFEIFFER and WOLFGANG SEYFFERTH
Lehrstuhl B für Mechanik, TU–München

ABSTRACT

Typical assembly tasks require the manipulator to operate either close to or in contact with the environment. This results in a combination of robot dynamics and process dynamics. The paper describes a method to match the dynamical properties of the manipulator and the assembly process by considering the corresponding time-variant constraints and the relevant constraint forces due to changing contact configurations. Some illustrating examples are given.

1 Introduction

During typical assembly tasks the manipulator always operates either close to or in contact with the environment. Small deviations in the initial conditions can therefore already lead to contact configurations completely different from those under ideal conditions. As a result, the uncertainties related to real-world assembly systems can produce excessive mating forces which affect the quality of the surfaces of the mated parts or can cause the assembly operation to fail.

From the theoretical point of view the interaction of the manipulator dynamics and the involved contact mechanics has only been studied in the past few years, mainly in the context of contour tracking problems where essentially constraints with nonchanging structures have been dealt with [2,4,5,6,7]. This paper deals with the dynamic formulation of the manipulator's motion in the presence of time-varying constraints during assembly tasks. The equations of motion for such systems are formally derived by adding the contact forces in form of Lagrange multipliers to the open loop dynamics and by considering an additional set of equations describing the acting geometrical constraints [7,8]. The current contact configuration and the respective system of differential equations are controlled by so-called kinematic and kinetic indicator magnitudes which are characteristic for such systems with time-varying topologies. Contact is controlled by the relative distance between two bodies and the constraint force perpendicular to the contact plane, while the states of sliding and stiction are indicated by the relative velocity and the difference between static friction force and constraint force. These magnitudes indicate all changes in the acting constraints and hence in the manipulator's mobility. Unsteady events like impact or stick-slip phenomena can thereby be traced during the state evolution and by means of a modified integration routine the different system equations can be evaluated piece by piece [3,11].

2 Mathematical Model

2.1 Equations of Constrained Motion

The kinematical constraints for a manipulator performing an assembly operation basically require the parts not to penetrate each other, i.e. the distance g between two bodies with respect to every possible contact configuration i (point-surface, edge – edge, ...) must satisfy:

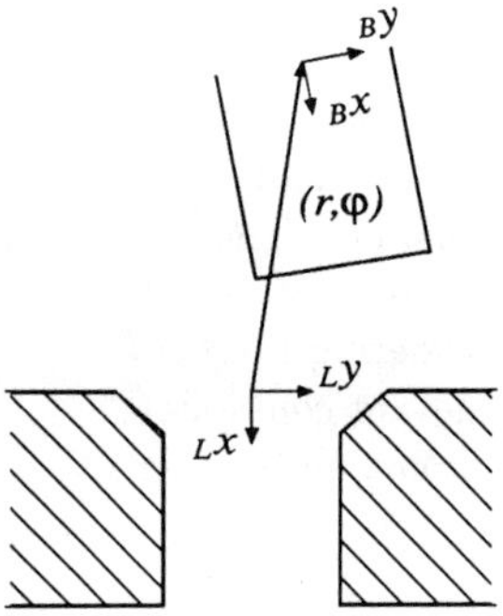

$$g(\mathbf{r}, \varphi) = \mathbf{n}^T(\mathbf{x} - \mathbf{p}) \geq 0 \tag{1}$$

$\mathbf{n}$: normal vector of the contact surface
$\mathbf{x}$: coordinates of any point within the contact surface
$\mathbf{p}$: coordinates of the corner point under consideration of the complementary body (Fig. 1)

Fig. 1: Constraint surfaces

The dynamics of an unconstrained manipulator may be represented by [7]

$$\mathbf{M}\ddot{\mathbf{q}} = \mathbf{h}(\mathbf{q}, \dot{\mathbf{q}}, t) \tag{2}$$

where $\mathbf{M}$ denotes the inertia matrix and $\mathbf{h}(\mathbf{q}, \dot{\mathbf{q}}, t)$ is a vector function which contains the Coriolis, centrifugal and gravitational load of the robot and the input generalized forces. When the robot is in contact with the environment additional forces act on the system dynamics so that its motion is consistent with all acting geometrical constraints

$$g_i(\mathbf{q}) = 0, \qquad i = 1, 2, \ldots, n_p \tag{3}$$

where n_p denotes the number of actual contact points. Differentiating eq. (3) twice yields a set of constraint conditions for the accelerations $\ddot{\mathbf{q}}$

$$\delta_{Ni} = \frac{d^2 g_i}{dt^2} = \mathbf{w}_{Ni}^T \ddot{\mathbf{q}} + \dot{\mathbf{w}}_{Ni}^T \dot{\mathbf{q}} = 0, \qquad \mathbf{w}_{Ni}^T = \frac{\partial g_i}{\partial \mathbf{q}} \tag{4}$$

where the vectors $\mathbf{w}_{Ni}$ are perpendicular to the hypersurfaces of the constraints in configuration space and physically represent the effect of a unit contact force $\mathbf{F}_n = \mathbf{n}$ in joint space

$$\mathbf{w}_{Ni} = \mathbf{J}_{Ti}^T \mathbf{n}_i \tag{5}$$

$\mathbf{J}_{Ti}$: Jacobian of translation with respect to contact point i

Friction in contacts may generate sliding or static friction forces depending on the state of the system and its influence onto the contact conditions. Sliding friction forces are active forces which in correspondance to a dry Coulomb friction model are directed opposite to the relative motion in the tangential contact plane and are proportional in magnitude to the normal reaction forces $|\mathbf{F}_N|$. The additional load imposed on the joints in configuration space can be described as

$$\mathbf{w}_{Fi}\lambda_{Ni} = -\mu \mathbf{J}_{Ti}^T \frac{\mathbf{v}_{ti}}{|\mathbf{v}_{ti}|}\lambda_{Ni} \tag{6}$$

μ : coefficient of sliding friction, $\mathbf{v}_{ti}$: relative velocity in the contact plane

where $\lambda_{Ni} > 0$ corresponds to the magnitude $|\mathbf{F}_{Ni}|$ of the constraint force perpendicular to the surface. In case of static friction in contact point i there is an additional passive force that correlates with the kinematical constraint

$$|\mathbf{v}_{ti}| = 0 \tag{7}$$

In threedimensional space (7) can be replaced by two constraints linear in $\dot{\mathbf{q}}$

$$\begin{pmatrix} v_{ti_1} \\ v_{ti_2} \end{pmatrix} = \begin{pmatrix} \boldsymbol{t}_{i1}^T \\ \boldsymbol{t}_{i2}^T \end{pmatrix} \mathbf{J}_{Ti}\,\dot{\mathbf{q}} = \mathbf{0} \tag{8}$$

$$(\mathbf{t}_{i1}, \mathbf{t}_{i2}) \perp \mathbf{n}_i \ , \ \ \mathbf{t}_{i1} \perp \mathbf{t}_{i2} \ , \ \ |\mathbf{t}_{i1}| = |\mathbf{t}_{i2}| = 1$$

which yields by analogy to (4) two more constraints in acceleration space

$$\begin{aligned} \boldsymbol{\delta}_{Ti} &= \frac{d}{dt}\begin{pmatrix} v_{ti_1} \\ v_{ti_2} \end{pmatrix} = \mathbf{W}_{Ti}^T\ddot{\mathbf{q}} + \dot{\mathbf{W}}_{Ti}^T\dot{\mathbf{q}} = \mathbf{0} \\ \mathbf{W}_{Ti} &= \frac{\partial^T}{\partial \dot{\mathbf{q}}}\begin{pmatrix} v_{ti_1} \\ v_{ti_2} \end{pmatrix} = \mathbf{J}_{Ti}^T(\mathbf{t}_{i1}\ \mathbf{t}_{i2}) \end{aligned} \tag{9}$$

The direction and magnitude of the static friction force $\mathbf{F}_{ti} = \mathbf{t}_{i_1}\lambda_{Ti} + \mathbf{t}_{i_2}\lambda_{Ti_2}$ can then be computed by introducing additional Lagrange multipliers $\boldsymbol{\lambda}_{Ti} = (\lambda_{Ti_1}\ \lambda_{Ti_2})^T$. For each contact point i we can thus summarize the effect of constraint and active forces in configuration space

a) when sliding friction is present

$$(\mathbf{w}_{Ni} + \mathbf{w}_{Fi})\lambda_{Ni} \ , \ \ \mathbf{w}_{Fi} = -\mu \mathbf{J}_{Ti}^T \frac{\mathbf{v}_{ti}}{|\mathbf{v}_{ti}|} \tag{10}$$

b) when static friction is present

$$\begin{aligned} &(\mathbf{w}_{Ni} + \mathbf{w}_{Fi})\lambda_{Ni} + \mathbf{W}_{Ti}\boldsymbol{\lambda}_{Ti} \\ &\mathbf{w}_{Fi} = \mathbf{0} \ , \ \ \mathbf{W}_{Ti} = \mathbf{J}_{Ti}^T(\mathbf{t}_{i1}\ \mathbf{t}_{i2}) \end{aligned} \tag{11}$$

Summarizing the constraints (eqs. 4, 5, 9) and considering eq. (6) in applying a Lagrange multiplier technique we come out with

$$\mathbf{M}\ddot{\mathbf{q}} = \mathbf{h}(\mathbf{q}, \dot{\mathbf{q}}, t) + (\mathbf{W}_N + \mathbf{W}_F)\boldsymbol{\lambda}_N + \mathbf{W}_T\boldsymbol{\lambda}_T \tag{12}$$

$$\boldsymbol{\delta} = \mathbf{A}\boldsymbol{\lambda} + \mathbf{b} = \mathbf{0} \tag{13}$$

and the abbreviations

$$\boldsymbol{A} = \begin{pmatrix} \boldsymbol{W}_N^T \boldsymbol{M}^{-1}(\boldsymbol{W}_N + \boldsymbol{W}_F) & \boldsymbol{W}_N^T \boldsymbol{M}^{-1} \boldsymbol{W}_T \\ \boldsymbol{W}_T^T \boldsymbol{M}^{-1}(\boldsymbol{W}_N + \boldsymbol{W}_F) & \boldsymbol{W}_T^T \boldsymbol{M}^{-1} \boldsymbol{W}_T \end{pmatrix}$$

$$\boldsymbol{b} = \begin{pmatrix} \boldsymbol{W}_N^T \boldsymbol{M}^{-1} \boldsymbol{h} + \dot{\mathbf{W}}_N^T \dot{\mathbf{q}} \\ \boldsymbol{W}_T^T \boldsymbol{M}^{-1} \boldsymbol{h} + \dot{\mathbf{W}}_T^T \dot{\mathbf{q}} \end{pmatrix}$$

$$\boldsymbol{\delta} = (\delta_{N1} \cdots \delta_{Nn_P} \boldsymbol{\delta}_{T1}^T \cdots \boldsymbol{\delta}_{Tn_{St}}^T)^T , \quad \boldsymbol{\lambda} = (\lambda_{N1} \cdots \lambda_{Nn_P} \boldsymbol{\lambda}_{T1}^T \cdots \boldsymbol{\lambda}_{Tn_{St}}^T)^T$$

$$\boldsymbol{W}_N = (\boldsymbol{w}_{N1} \cdots \boldsymbol{w}_{Nn_P}) \in \mathbb{R}^{n,n_P} , \quad \boldsymbol{W}_F = (\boldsymbol{w}_{F1} \cdots \boldsymbol{w}_{Fn_P}) \in \mathbb{R}^{n,n_P}$$

$$\boldsymbol{W}_T = (\boldsymbol{W}_{T1} \cdots \boldsymbol{W}_{Tn_{St}}) \in \mathbb{R}^{n,2n_{St}}$$

where $n_{St} \leq n_p$ is the number of contact points with static friction and (n_p+2n_{St}) is the number of overall constraints. Eqs. (12) and (13) represent $(n+n_p+2n_{St})$ scalar equations for the unknowns $(\ddot{\mathbf{q}}, \boldsymbol{\lambda})$ and automatically reduce the system by the corresponding degrees of freedom that are affected by the acting constraints.

2.2 Unilateral Constraints

The above constraints depend on the system state $(\mathbf{q}^T, \dot{\mathbf{q}}^T)^T$ and vice versa. Changing constraints are always accompanied by some kinematic and kinetic indicator magnitudes where the entry of a contact event is indicated by kinematic relations, the exit by kinetic functions. The indicator magnitudes define the unilateral constraints and are additionally complementary in kinetic respect. For example if there is geometric contact $g_i = 0$ the permissible motion of the contact point perpendicular to the constraint surface is limited to $g_i(\mathbf{q}) \geq 0$ or equivalently $\delta_{Ni} = \frac{d^2 g_i}{dt^2} \geq 0$ while the constraint force is simultaneously restricted by $\lambda_{Ni} \geq 0$ with $(\delta_{Ni} = 0,\ \lambda_{Ni} \geq 0)$ or $(\delta_{Ni} \geq 0,\ \lambda_{Ni} = 0)$. On the other hand, if the relative velocity becomes zero in some contact point i the maximum friction force according to Coulombs law is limited by the friction saturation $\Delta F_{Ti} = \mu_0 \lambda_{Ni} - |\boldsymbol{\lambda}_{Ti}| \geq 0$. If stiction is present $\boldsymbol{\delta}_{Ti} = \mathbf{0}$ and $\Delta F_{Ti} \geq 0$ must be satisfied whereas at a transition to sliding the friction force $\boldsymbol{\lambda}_{Ti}$ opposes the starting relative motion characterized by $\boldsymbol{\delta}_{Ti} \neq 0$ and hence $det(\boldsymbol{\delta}_{Ti} \boldsymbol{\lambda}_{Ti}) = 0$ and $\boldsymbol{\delta}_{Ti}^T \boldsymbol{\lambda}_{Ti} \leq 0$. Assuming a steady characteristic relating the coefficient of friction and the relative velocity (Fig. 2) the magnitude of the sliding friction force is furthermore specified by $\Delta F_{Ti} = 0$ with $\mu(v_{rel} = 0) = \mu_0$.

Thus the solution space $(\boldsymbol{\delta}^T, \boldsymbol{\lambda}^T)^T$ of the system of equations $\boldsymbol{\delta} = \mathbf{A}\boldsymbol{\lambda} + \mathbf{b}$ that relates forces and accelerations in constraint space is limited by the above requirements. The sufficient conditions for satisfying all imposed unilateral constraints can then be summarized by a system of linear and nonlinear equations,

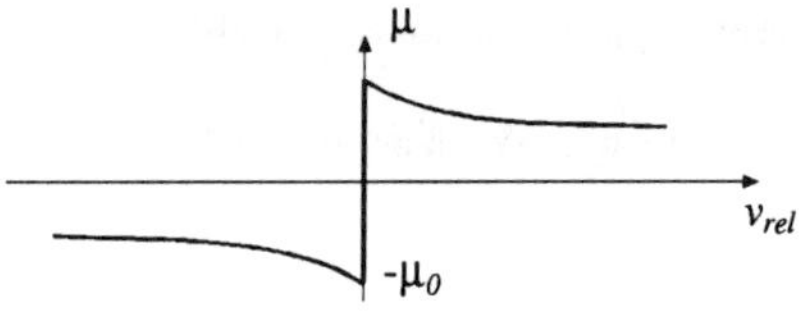

Fig. 2: Friction characteristic

a set of inequalities and additional complementarity conditions:

$$\begin{aligned}
&\boldsymbol{\delta} = \mathbf{A}\boldsymbol{\lambda} + \mathbf{b} \\
&\lambda_{Ni} \geq 0\ , \quad \delta_{Ni} \geq 0\ , \quad \lambda_{Ni}\delta_{Ni} = 0\ , \qquad i = 1, \ldots, n_p \\
&\Delta F_{Ti} = \mu_0 \lambda_{Ni} - |\boldsymbol{\lambda}_{Ti}|\ , \quad \Delta F_{Ti}\delta_{T1_i} = 0\ , \quad \Delta F_{Ti}\delta_{T2_i} = 0\ , \quad i = 1, \ldots, n_{St} \\
&\delta_{T1_i}\lambda_{T2_i} - \delta_{T2_i}\lambda_{T1_i} = 0\ , \quad \Delta F_{Ti} \geq 0\ , \quad \boldsymbol{\delta}_{Ti}^T \boldsymbol{\lambda}_{Ti} \leq 0\ , \quad i = 1, \ldots, n_{St}
\end{aligned} \tag{14}$$

The solution $(\boldsymbol{\lambda}_N^T, \boldsymbol{\lambda}_T^T, \boldsymbol{\delta}_N^T, \boldsymbol{\delta}_T^T, \Delta F_{T1}, \ldots, \Delta F_{Tn_{St}})^T$ that satisfies the nonlinear complementarity problem (14) then indicates which of the currently possible constraints are active.

Time-varying constraints always give rise to discontinuities in some derivative of the solution $\mathbf{q}(t)$. For the complete analysis of the system behaviour we therefore need to determine the points in time at which an unsteady event occurs. For that we monitor the kinematic and kinetic indicator magnitudes at macro time steps ΔT in the course of the simulation. In case a constraint has changed during the time interval $[t_1, t_2]$ we then iteratively determine the instant $t_s \in [t_1, t_2]$ when this has occured. Given the initial conditions the state history is then calculated in a piece-by-piece integration where during the respective periods with a non-changing mechanical structure the active constraints can be treated as if they acted bilaterally.

2.3 Impact Dynamics

A robot contacting its environment generates an impact, which can be modeled by the impulse/momentum equations represented by the first integral of (12)

$$\mathbf{M}(\dot{\mathbf{q}}^+ - \dot{\mathbf{q}}^-) = (\mathbf{W}_N + \mathbf{W}_F)\boldsymbol{\Lambda}_N + \mathbf{W}_T\boldsymbol{\Lambda}_T \tag{15}$$

where the superscripts $^+(t = t+0)$ and $^-(t = t-0)$ denote the state shortly after and before the impact. Typical constraints for impulsive processes are the classical impact equations which relate the velocity components perpendicular to the constraint surfaces before and after the impact [9,10]

$$\begin{aligned}
&\boldsymbol{\Delta}_N^+ = -\epsilon\boldsymbol{\Delta}_N^-\ , \qquad \boldsymbol{\Delta}_N = \mathbf{W}_N^T\dot{\mathbf{q}} \\
&\boldsymbol{\Delta}_N^+ \geq \mathbf{0}\ , \qquad \boldsymbol{\Lambda}_N \geq \mathbf{0}
\end{aligned} \tag{16}$$

with ϵ the coefficient of restitution. Since moreover a transition from sliding to sticking during the duration of impact is possible the maximum number of contact points with static friction is $n_{St} = n_p$. Let $\boldsymbol{\Delta}_{Ti} = \mathbf{W}_{Ti}^T\dot{\mathbf{q}}$ denote the tangential velocity in constraint space, then the condition for stiction in point i is

$$\boldsymbol{\Delta}_{Ti}^+ = 0\ , \qquad \Delta P_{Ti} = \mu_m \Lambda_{Ni} - |\boldsymbol{\Lambda}_{Ti}| \geq 0 \tag{17}$$

where μ_m denotes a mean coefficient of friction. Problems, however, emerge when we deal with the tangential impulse related to sliding friction as the direction of sliding can vary during the impact, including the possibility of reversed sliding [1,12]. In order to be able to establish algebraic equations for the representation of the non-smooth velocity history, we in the following make the assumption that in manipulation operations the direction of the relative velocity $\boldsymbol{\Delta}_{Ti}$ in the tangential contact plane does not change dramatically during the collision so that the sliding friction impulse can be sufficiently approximated by

$$\begin{aligned} &\boldsymbol{\Delta}_{Ti} \neq \mathbf{0}\,, \quad \Delta P_{Ti} = \mu_m \Lambda_{Ni} - |\boldsymbol{\Lambda}_{Ti}| = 0 \\ &\boldsymbol{\Lambda}_{Ti}^T \boldsymbol{\Delta}_{Ti}^+ \leq 0\,, \quad \boldsymbol{\Delta}_{Ti_1}^+ \boldsymbol{\Lambda}_{Ti_2} - \boldsymbol{\Delta}_{Ti_2}^+ \boldsymbol{\Lambda}_{Ti_1} = 0 \end{aligned} \tag{18}$$

Confining (16) to the case of perfect plastic impacts, i.e. $\epsilon = 0$, which we can assume to be dominant in assembly operations, the above conditions can be expressed by a nonlinear complementarity problem equivalent to (14)

$$\begin{aligned} &\boldsymbol{\Delta}^+ = \mathbf{A}\boldsymbol{\Lambda} + \boldsymbol{\Delta}^- \qquad (19) \\ &\Lambda_{Ni} \geq 0\,, \quad \Delta_{Ni}^+ \geq 0\,, \quad \Lambda_{Ni}\Delta_{Ni}^+ = 0\,, \quad i = 1, \ldots, n_p \\ &\Delta P_{Ti} = \mu_m \Lambda_{Ni} - |\boldsymbol{\Lambda}_{Ti}|\,, \; \Delta P_{Ti}\Delta_{T1_i}^+ = 0\,, \; \Delta P_{Ti}\Delta_{T2_i}^+ = 0\,, \; i = 1, \ldots, n_{St} \\ &\Delta_{T1_i}^+ \Lambda_{T2_i} - \Delta_{T2_i}^+ \Lambda_{T1_i} = 0\,, \; \Delta P_{Ti} \geq 0\,, \; \boldsymbol{\Lambda}_{Ti}^T \boldsymbol{\Delta}_{Ti}^+ \leq 0\,, \; i = 1, \ldots, n_{St} \end{aligned}$$

whose solution defines the subset of constraints being active during the duration of impact and with (15) yields the resulting system state $(\mathbf{q}^T, \dot{\mathbf{q}}^{+^T})^T$.

3 Results

3.1 Snap Joint

The above theory has been applied to a series of elementary processes connected with assembly activities. Two examples will be given here. The first one considers the mating of a single flexible snap joint under ideal conditions. The geometry of the parts is portrayed in Fig. 3 and the motion is given in direction of the x-axis. In this case the assembly process can be studied seperately by neglecting the manipulator dynamics since the necessary mating forces are equivalent to the extra actuator input torques that are required to fulfill the prescribed gripper motion. The respective numerical and experimental force-distance-graphs are shown in Fig. 4. The unsteady force history results from changing constraints that are characterized by the non-smooth contour of the two parts.

3.2 Peg-in-Hole Problem

The combination of robot dynamics control and contact dynamical processes has been studied considering the planar peg-in-hole insertion task. For the verification of our theoretical approach experiments with a five degree-of-freedom laboratory robot with a force-torque sensor were carried out. The fixture housing

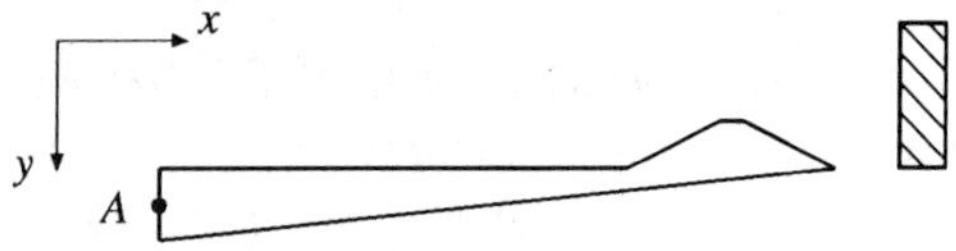

Fig. 3: Geometry of snap joint

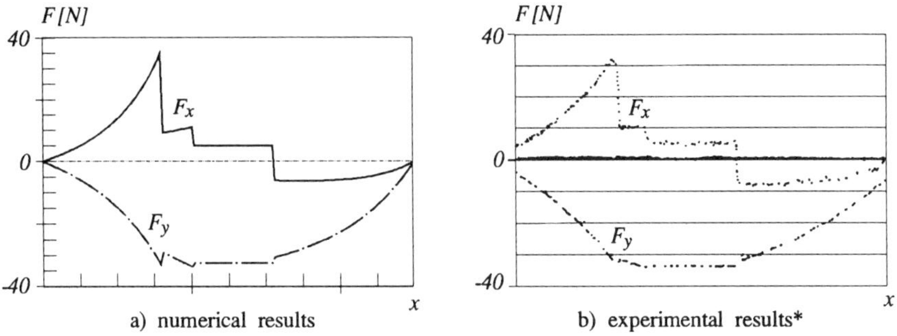

Fig. 4: Force-distance-graph of a snap joint

the complementary part is equipped with six distance sensors that measure the gripper's position and orientation and in addition disposes of two translational and three rotational adjustments which allow to produce definite positioning errors with respect to the parts that are to be mated. In experiments we investigated the insertion process when there is a lateral error between the peg and the hole. The respective numerical and experimental results are depicted in Fig. 5. The peaks in the force history at $t = 0.9s$ result from the impact when the peg hits the chamfer. When two-point-contact is reached at $t = 1.2s$ the manipulator's motion is slowed down very fast until the jamming condition is reached. Only when the controller torques become large enough the motion is continued with sliding in two and finally even one contact point. This cycle represented by the oscillations in the force history repeats until the end of the trajectory and demonstrates the phenomenon of changing constraints during an assembly operation.

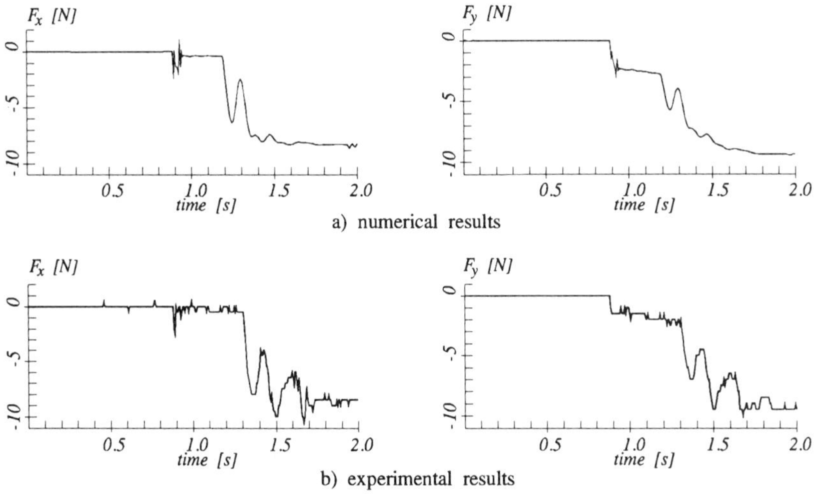

Fig. 5: Forces during insertion due to lateral offset;
F_x: insertion force, F_y: lateral force

Note, that in this experiment the sensor served only as a force measuring device for data-processing, since a simple PD joint controller was used. Controller design with force feedback in this context has been studied separately both in experiment and simulation [13].

Acknowledgement

The research work presented in this paper is partially supported by a contract with the DFG (Deutsche Forschungsgemeinschaft, SFB 336).

Notes

* Measurements of snap joints (Fig. 4b) have been conducted by Dipl.-Ing. A. Schlüter, Lehrstuhl für Konstruktion im Maschinenbau,Technische Universität München, 8000 Munich 2.

References

1. Brach, R.M.: Rigid Body Collisions, *J. of Applied Mechanics*, Vol. 56, 1989, 133 - 138
2. Cai, L., Goldenberg, A.A.: An Approach to Force and Position Control of Robot Manipulators, *Proc. IEEE Conf. on Robotics and Automation*, 1989, 86 - 91
3. Hajek, M.: Reibungsdämpfer für Turbinenlaufschaufeln. *Fortschr.-Ber. VDI*, Reihe 11, Nr. 128, Düsseldorf: VDI Verlag 1990
4. Hui, R.C., Goldenberg, A.A.: Modelling the Manipulation of Rigid Objects as a Singular System. *Proc. IEEE Conf. on Robotics and Automation*, 1988, 240 - 244
5. McClamroch, N.H.: Singular Systems of Differential Equations as Dynamic Models for Constrained Robot Systems. *Proc. IEEE Conf. on Robotics and Automation*, 1986, 21 - 27
6. McClamroch, N.H., Wang, D.: Feedback Stabilization and Tracking of Constrained Robots. *IEEE Trans. on Automatic Control*, **3**, 5, May 1988
7. Pfeiffer, F., Richter, K., Wapenhans, H.: Elastic Robot Trajectory Planning with Force Control. *Proc. of IFIP International Symp. on Theory of Robots*, Rome, Italy, 1990, 201 - 212
8. Pfeiffer, F.: Dynamical Systems with Time-Varying or Unsteady Structure. *ZAMM 71* (1991) **4**, 6 - 22
9. Pfeiffer, F.: Dynamical Systems with Impact and Stick-Slip Phenomena. *CSME Mechanical Engineering Forum*, 1990, Toronto, Canada
10. Pfeiffer, F.: On unsteady dynamics in machines with plays. *Proc. 7th world congress on the theory of machines and mechanisms*, Sevilla, 1987
11. Prestl, W.: Zahnhämmern in Rädertrieben von Dieselmotoren. *Fortschr.-Ber. VDI*, Reihe 11, Nr. 145, Düsseldorf: VDI Verlag 1991
12. Wang, Y., Mason, M.T.: Modeling Impact Dynamics for Robotic Operations, *Proc. IEEE Int. Conf. on Robotics and Automation*, 1987, 678 - 685
13. Wapenhans, H., Seyfferth, W., Pfeiffer, F.: Hybrid Position and Force Control with Unsteady Assembly Dynamics, *IMACS/SICE Int. Symp. on Robotics, Mechatronics and Manufacturing Systems*, Sept. 16 - 20, 1992, Kobe, to appear

Finding the Optimal Possible Match Between Uncertain States

M.S. Kamel, S. Leung, D. Tse, S. Wong

Pattern Analysis and Machine Intelligence Lab
University of Waterloo
Waterloo, Ontario
Canada N2L 3G1

ABSTRACT

Robot task planners that are based on state representations require a mechanism to find the difference between a current state and a goal state in order to select the appropriate operators to reduce this difference. In this paper, we develop such a mechanism for a probabilistic robot task planner. The planner utilizes a probabilistic attributed graph representation for the states. The problem is solved by first finding the optimal possible monomorphism between the graphs representing the current and goal states. A different graph representing the difference between the states is then constructed so that appropriate operators can be selected to reduce the difference. The optimal monomorphism involves a branch and bound search with cost function based on transformation costs of operations and risk of misidentification of objects.

INTRODUCTION

In order for task planners to handle real problems, the issue of uncertainty needs to be fully addressed. Most traditional planners have assumed an ideal deterministic world [6,10,12]. The real environment is not even close to that. There are many sources of uncertainty in a given domain. Uncertainty may exist due to changes in the environment (dynamic environment), lack of complete specification of the environment and/or inaccuracy introduced by sensory and other control operations. Research on planning approaches dealing with these conditions has been scarce [2,3,4,5,9,11,13].

In an effort to design and implement a planner for uncertain environment, we have proposed representation schemes for representing the uncertainties in the environment and in the robot operators [7]. In this paper, we follow this research by developing a mechanism which is used by the planner to find the difference between a current state and a goal state.

WORLD AND GOAL REPRESENTATION

Following the representation proposed in [7], each state in the world is represented using a probabilistic attributed graph. In this graph, vertices correspond to objects and edges correspond to binary relations between the objects. Each vertex has a number of attributes and each edge corresponds to a relation, but between any two vertices, there can be several edges. In order to represent the uncertainty in the planner's knowledge of the environment, these attributes and relations are treated as random variables with known ranges of values.

Attributes are classified as label attributes, which cannot be altered by any operator available in the domain (and hence identify the object), and modifiable attributes, which may be changed by the application of some operator. This classification is solely dictated by the available operators in the domain. Similarly, relations are classified into fixed relations and modifiable relations.

In the world representation, a joint probability distribution function can be defined on the Cartesian product of the ranges of all random variables so defined. This can also be viewed as a probability distribution over the state space which consists of all the possible (deterministic) world states. Since completely objective information is seldom available, this probability will not usually have a frequency interpretation. Rather, it represents the subjective knowledge of the planner concerning which state the world is currently in. This probabilistic representation (the graph structure together with the probability distribution) will be referred to as the current graph. It should be noted that in any implementation, the joint distribution is approximated by a product of distributions of lower order to reduce the computational space and time requirements. In this paper, we will only consider discrete distributions for simplicity of presentation but analogous results hold for continuous ones.

The representation of the goal is a special case of the above representation with a slightly different semantics. Each attribute and relation is associated with a subset of the range indicating the desired values of the attribute or relation, rather than a random variable. If an attribute of a relation is not present in the goal graph, it is assumed that a "don't care" condition holds for its value, i.e. any value can satisfy the goal. The goal representation is interpreted as a subset of the state space (the goal states) such that the attainment of any goal state is regarded as the attainment of the goal. It will be called the goal graph.

Operations on the Representation

We identify three basic operations to manipulate the representation: synthesis, update, and difference extraction.

The synthesis operation is used to build the world model by acquiring knowledge or information about the world. As objects are perceived by the robot, vertices are created in the current graph. The label attributes are recognized to identify the objects, then the modifiable attributes and relations are estimated. The joint probability density function can be computed using the estimations.

Updates to the current graph are made as a result of the planner operators (actions), both passive and active. Passive ones are the information gathering operators (IGOPs), through which knowledge about the environment is gained. Such operators can be realized by, for example, a computer vision system. The current graph is then updated via Bayesian learning: the posterior probability distribution is computed from the prior distribution and the IGOP measurements. Active operators are those which can change the environment: the effects of operators on the real world are modelled by procedures altering the probability distribution and structure of the current graph [7].

Another important operation on the representation is the comparison of the current graph to the goal graph to extract the difference between the current and goal states for means-ends analysis. It is this operation that will be focussed on in this paper.

FINDING THE DIFFERENCE BETWEEN UNCERTAIN STATES

The problem of finding the difference between uncertain states, represented as described above, involves matching between two probabilistic attributed graphs, and then constructing a difference graph. The matching between the two graphs that we are concerned with is actually the monomorphism between the graphs. However, here we have to develop a possible monomorphism.

Definition 1: Given the current graph X and the goal graph Y, a vertex u in X is said to be *possibly identical* to a vertex v in Y if, for every label attribute of v, there exists a non-zero probability that the value for that attribute on u lies in the range specified in the goal.

Definition 2: A *possible monomorphism* between X and Y is an injective mapping m: $V(Y) \mapsto V(X)$, where $V(G)$ is the vertex set of graph G, such that the following conditions hold: 1) for every vertex v in $V(Y), m(v)$ is possibly identical to v; 2) for every fixed relation r_1 from v_i to v_j in Y, there is a corresponding relation r_2 from $m(v_i)$ to $m(v_j)$ in X such that there is a non-zero probability that the value of r_2 lies inside the range specified for r_1.

A possible monomorphism gives a possible matching between identical objects and relations in the goal and current graphs. Existence of such a monomorphism is guaranteed if the following assumptions are made: 1) The closed world assumption; 2) The goal is actually attainable given the current world state and the available operators.

Given a possible monomorphism between the current and goal graphs, there is also a correspondence between the modifiable attributes and relations in the two graphs. The difference between the two graphs thus consists of: 1) the differences between all the modifiable attributes and relations in the current graph with the corresponding ranges in the goal, and 2) the uncertainty of the matching due to uncertainties in the label attributes of the graphs.

For each modifiable attribute or relation, its difference is classified as: 1) *match* - if $p > U$, an upper threshold; 2) *mismatch* - if $p < L$, a lower threshold; 3) *possible match* - if p is between L and U, where p is the probability that the value of a modifiable attribute (or relation) in the current graph lies in the desired range specified in the goal graph, and L, U are in [0,1] with $L < U$.

The Difference Graph

Given a possible monomorphism m, all the differences are summarized in a *difference graph*, which is defined to contain the following:

1. For every vertex v in the vertex set of the goal graph, there will be a corresponding vertex in the difference graph representing the ordered pair $(v, m(v))$ if and only if there is a mismatch or possible match in at least one of the attributes of v or the relations incident on v;
2. For every attribute or relation in the goal graph, there will be a corresponding attribute or relation on the difference graph if and only if there is a match or possible match.

Optimality of the Monomorphism

Given a monomorphism, a unique difference graph indicating the differences between the two graphs is obtained. However, there are, in general, many different monomorphisms because: 1) there are more than one object in the real world matching the requirements for a certain object in the goal, and 2) the planner is unsure about the identity of some objects in the world. This is reflected in the uncertainties in the label attributes. Thus there are more possible matches.

A cost function $C : M \mapsto R$ (M is the set of monomorphisms, and R is the set of real numbers) is needed to rank all possible monomorphisms so that only the best ones are returned to the planner. Details on the components of this function and how they are defined can be found in [8]. Given this cost function, an optimal monomorphism m_{opt} can be defined to be the one with the lowest cost. The difference between the two graphs can then be defined as the set of difference graphs whose underlying monomorphisms have cost between $C(m_{opt})$ and $C(m_{opt}) + T(C(m_{opt}))$, where T satisfies the property $T(x) \geq x$ for all real x. T dictates how many monomorphisms other than the optimal one should receive further considerations, and hence gives a measure of confidence in the accuracy of the cost function. It will be called the trust function.

SEARCHING FOR OPTIMAL MONOMORPHISMS

In this section, we describe an algorithm for finding the optimal set of monomorphisms, given the cost function (and its components), the current graph and the goal graph.

The matching problem is represented as a decision tree searching problem [1]. At each level of the tree, the nodes represent possible pairings of one particular vertex in the goal graph with different vertices in the current graph. Each path in the complete decision therefore represents an injective mapping from the vertex set of the goal graph into that of the current graph. The optimal possible monomorphisms constitute only a very small subset of these paths; the others are either sub-optimal possible monomorphisms, or are not possible monomorphisms at all because they contain at least one node pairing two completely non-identical vertices. The problem of finding the optimal set reduces to that of searching for the corresponding paths in the decision tree.

Our algorithm performs a best-first search of the decision tree,using the heuristic function to rank nodes to expand first, and prune the entire subtree under a node when the heuristic function returns a value greater than the current upper bound on the optimal set. Pruning is possible because all paths through that node must have costs higher than those of the optimal set. Using the heuristic to rank which node to expand first increases the chance to find the more optimal solutions earlier in the search, thus providing a tighter upper bound to prune a larger part of the tree in the later search. The algorithm terminates when the entire tree is searched or pruned. The optimal set of monomorphisms is guaranteed to be obtained because all paths with costs whose lower bounds are less than that of the most expensive optimal monomorphism are searched.

AN EXAMPLE

To illustrate the concepts and ideas mentioned above, consider an example of a robot moving blocks on a floor plan.

The floor plans of the current and goal states are shown in figures 1,2 and 3, 4 respectively. A cartesian plane is used to represent the robot environment, where x- and y-axes can take on discrete values. Let O_c and O_g be the sets of blocks in the current and goal states respectively. Assume O_g is a subset of O_c.

The current state is represented with a probabilistic attributed graph. Sets

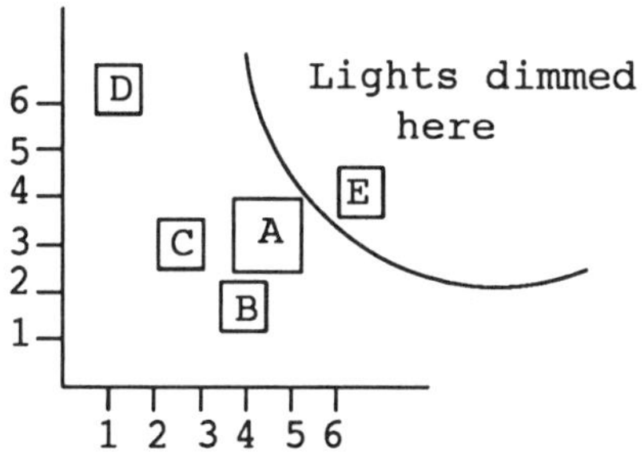

Figure 1: Current state

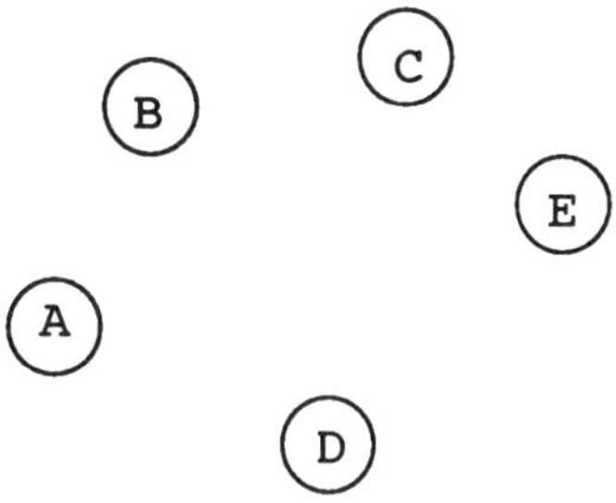

Figure 2: Current Graph

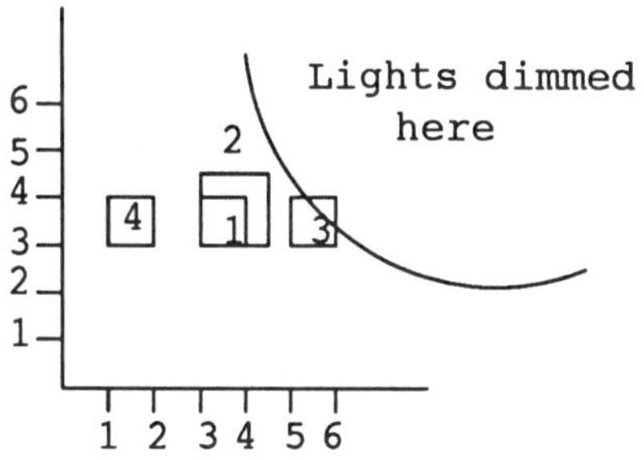

Figure 3: Goal State

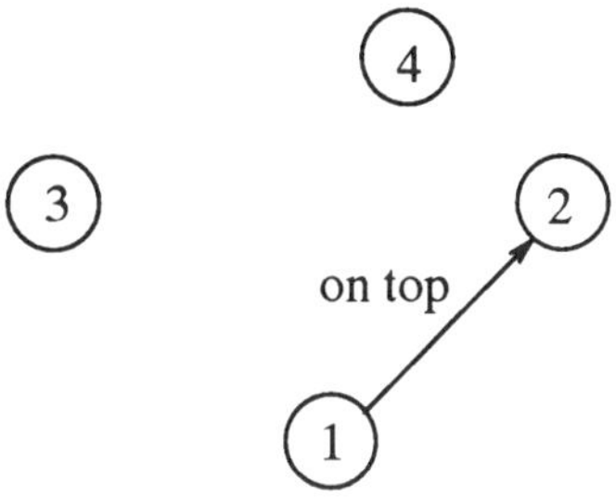

Figure 4: Goal Graph

of label, modifiable attributes, and relations of the blocks in the current state are defined below:

	Attribute	Value
Label	color	probability distribution over {green, red, orange, not applicable}
	weight	probability distribution over {1,2,3,4,5, not applicable}
Modifiable	*xpos*	discrete probability distribution over the integers 1 to 6 (co-ordinates of the x-axis of the cartesian plane)
	ypos	similar to xpos with y-axis co-ordinates
	ontop	discrete probability distribution over {yes, no, not applicable}

The goal state is also represented by an attributed graph with similar sets of attributes and relations, except that the probability distributions are replaced with range values.

Probability distributions and values for the attributes and relation in the current and goal states are shown in Tables 1, 2 and 3.

In this example, costs of transformation will be the costs associated with the robot moving the blocks around. Before transformation costs can be calculated, a set of operators must be defined. Three operators are defined below:

$movex(A, x_g)$ - to move block A to position x_g where x_g is an abscissa value.

movey(A, y_g) - to move block A to position y_g where y_g is an ordinate value.
putontop(A, B) - to put block A on top of block B

Label Attribute							
	Color			Weight			
	R	B	G	2	3	4	5
A	0.9	0.1	0	0	0.05	0.9	0.05
B	0	0	1	0.9	0.1	0	0
C	0.9	0.1	0	0.1	0.9	0	0
D	0	0.9	0.1	0	0.1	0.9	0
E	0.15	0.65	0.2	0	0.1	0.9	0

Table 1: Probability Distributions for label attributes in the current state.

Modifiable Attributes												
	x-position						y-position					
	1	2	3	4	5	6	1	2	3	4	5	6
A	0	0	0.1	0.8	0.1	0	0	0	0.9	0.05	0.05	0
B	0.05	0.05	0.1	0.75	0.05	0	0.1	0.8	0.1	0	0	0
C	0.05	0.1	0.85	0	0	0	0	0	0.9	0.1	0	0
D	0.9	0.05	0.05	0	0	0	0	0	0	0	0.1	0.9
E	0	0	0	0	0.1	0.9	0	0	0.05	0.9	0.05	0

Table 2 : Probability Distributions for modifiable attributes in the current state.

	Attributes				On Top			
	Color	Weight	x-pos	y-pos	1	2	3	4
1	G	4	4-5	3-4	n/a	yes	no	no
2	R	4	4-5	3-4	no	n/a	no	no
3	B	4	5-6	3-4	no	no	n/a	no
4	R	3	1-2	3-4	no	no	no	n/a

Table 3: Specifications for the goal state.

The distance functions are:

$$d_{xpos}(A, x_c, x_g) = \text{weight of } A * |x_c - x_g|$$

$$d_{ypos}(A, y_c, y_g) = \text{weight of } A * |y_c - y_g|$$

$$d_{ontop}(A) = \text{weight of } A$$

The cost of transformation is given by:
For *xpos* and *ypos*:

$$\sum_{i=1}^{6} Prob(X_c = i) \min_{j \text{ in } R} d_{xpos}(A, i, j)$$

$$\sum_{i=1}^{6} Prob(Y_c = i) \min_{j \text{ in } S} d_{ypos}(A, i, j)$$

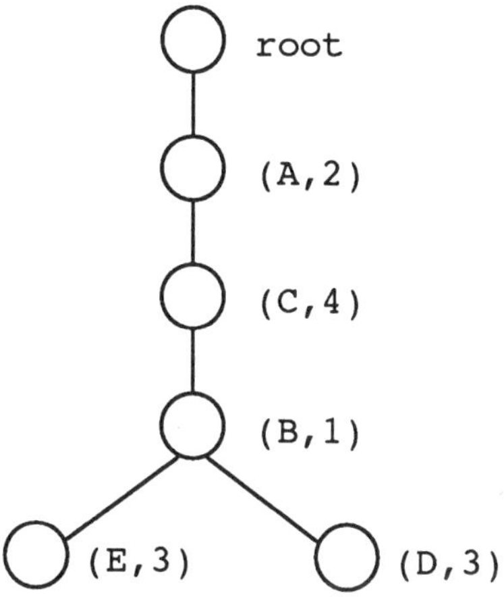

Figure 5: Matching Monomorphism

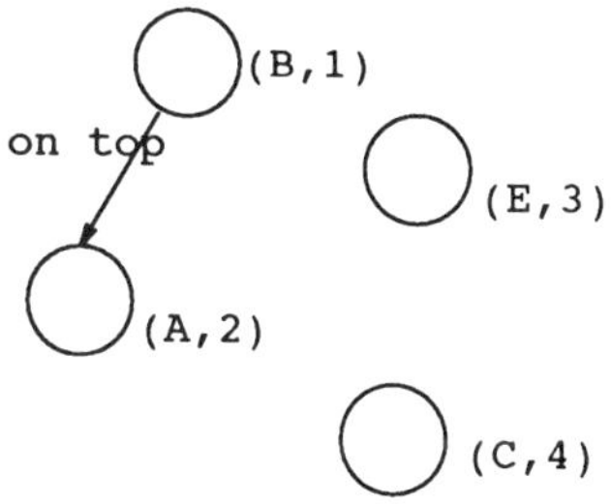

Figure 6: Difference Graph

where X_c, Y_c are the random variables for the current abscissa and ordinate values respectively

R, S are the abscissa and ordinate goal ranges of block A respectively

The risk premium of the label attribute *colour* is defined by the function $g(p) = (1/p) - 1$ where p is the probability . In this example, light of a portion of the floor plan is dimmed such that the colour of the objects in that area is identified with uncertainty. Risk premiums are calculated for matching block D with block 3 and block E with block 3.

For block E with block 3, risk premium = 0.5385, transformation cost = 0.2.
For block D with block 3 risk premium = 0.1111, transformation cost = 23

A conservative planner will match block D with 3 even that incurs a higher transformation cost. On the other hand, a risky planner will match E with 3 because this incurs a lower transformation costs even though a higher risk of matching the wrong block exists.

Upper and lower thresholds of matching are given below to decide the match, partially match, and mismatch of the blocks in the current and goal states.

Upper threshold of matching = 0.8
Lower threshold of matching = 0.1
probability of *xpos* of block A in goal range of block 2 = 0.9 (i.e. a match)
probability of *xpos* of block D in goal range of block 3 = 0 (mismatch)
probability of *xpos* of block C in goal range of block 4 = 0.15 (possibly match)

By using the branch and bound search algorithm, an optimal set of monomorphism can be found thus a difference graph can be generated. In this example, the optimal set contains only one monomorphism. The difference graph represents the differences between the current state and the goal state after the blocks are matched according to this optimal monomorphism. The differences provide the information the planner needs to select appropriate operators to reduce such differences. The monomorphism and the difference graph are shown in figures 5 and 6 respectively.

CONCLUSION

In this paper, we have presented a mechanism for finding the difference between states in planning under uncertain conditions. The mechanism is realized using an algorithm to find the optimal monomorphism between probabilistic attributed graphs representing the current and goal states. The algorithm utilizes branch and bound search. The algorithm uses cost functions related to the use of reducing the difference between the states and the risk in matching uncertain values. The algorithm uses these cost functions in the search for the optimal monomorphisms.

REFERENCES

1. A.V. Aho, J.W. Hopcroft and J.D. Ullman, "The Design and Analysis of Computer Algorithms," *Addision-Wesley*, (1974).

2. R. Brooks, "Symbolic Error Analysis and Robot Planning," *Tech. Report AI Memo 685*, (MIT, 1982).

3. D. Chapman and P.E. Agre, "Abstract Reasoning as Emergent from Concrete Activity," in *Workshop on Planning and Reasoning about Action, Portland, Oregon*, (1986).

4. A.M. Farley, "A Probabilistic Model for Uncertain Problem Solving," *IEEE Trans. on SMC-13, (4)*, 568-579, (1983).

5. R.J. Firby, "An Investigation into Reactive Planning in Complex Domains," *Proc. of Sixth National Conference on Artificial Intelligence, AAAI*, 202-206, Seattle, Washington, (1987).

6. R.E. Fikes and N.J. Nilsson, "A new approach to the application of theorem proving to problem solving", *AI 2*, 189-208, (1971).

7. M.S. Kamel and P.M. Kaufmann, "Representing Uncertainty in Robot Task Planning," *Proc. of International IEEE conf. on Robotics and Automation*, 1728-1734, (1988).

8. M.S. Kamel, S. Leung, D. Tse and S. Wong, "Probabilistic Means-end Analysis for Planning Under Uncertainty", *Technical Report, Dept. of Systems Design Engineering, University of Waterloo*, (1991).

9. J.H. Munson, "Robot Planning, Execution and Monitoring in an Uncertain Environment", *Proc. IJCAI 2*, 338, (1971).

10. E.D. Sacerdoti, "A Structure for Plans and Behaviour", *Tech. Note 109*, (AI Center, SRI International, 1975).

11. A. Tate, "Planning and Condition monitoring in a FMS", *Tech. Report AIAITR 2*, (University of Edinburgh, 1984).

12. D.E. Wilkins, "Domain-independent Planning: Representation and Plan Generation", *AI 22*, 269-301, (1984).

13. D.E. Wilkins, "Recovering from Execution Errors", in *SPIE, Computational Intelligence*, 1(1), (1985).

FLOW SHOP SCHEDULING WITH SHARED RESOURCES: A BRANCH AND BOUND APPROACH

Maria Teresa Moreira Rodrigues
Faculdade de Engenharia Química, UNICAMP.
13081 - Campinas - S.P.

Carlos Alberto Santos Passos
Instituto de Automação, CTI

Luis Gimeno Latre and Márcio Francisco Dutra e Campos
Faculdade de Engenharia Elétrica, UNICAMP.

ABSTRACT

This work is concerned with the flow shop scheduling problem when there are constraints in shared resources other than processors. A Branch and Bound approach is proposed where lower bounds are estimated considering resources constraints. Results of a simulated example show a better estimation of makespan and a reduction in the number of nodes searched.

INTRODUCTION

The flow shop scheduling problem considered here can be stated as obtaining the Gantt chart, describing operations time allocation, for a system characterized by: i- a set of N tasks each formed by M operations linked by precedence constraints, and ii- the resources shared between the tasks, operations demand profiles and available profiles of the resources. It is intended to schedule the operations minimizing the makespan using the available resources.

The application of interest is batch chemical industry where the common resources, besides processors, are utilities which are shared by equipments and have well defined available profiles. In general the solution space is limited to permutation schedules, defined as schedules where the M operations of each task are performed in the same order [1].

The flow shop problem with no restrictions in shared resources unless processors has been treated in the literature [2,3,4]. It is a NP-problem for M>2 and any regular performance measure, and for M>3 under the makespan criterion [2]. Branch and Bound techniques are often used; a node in the tree which is searched corresponds to a partial sequence of tasks and there is no need to search at the operations level because only permutation schedules are sought.

Including restrictions on available resources, other than processors, makes the problem more difficult to solve because lower bounds of partial sequences estimated considering resources restrictions are difficult to obtain. Normally restrictions are not considered and a feasibility phase follows the Branch and Bound algorithm to modify the obtained schedule to satisfy the resources's available profiles [5]. When restrictions on the resources are tight this procedure can modify strongly the schedule obtained by the Branch and Bound algorithm. In fact a resource centered scheduling perspective should be used in this situation, instead of a task/operation perspective which gives greater relevance to precedence relations than to resources availability. This approach is presented in [6] for job shops problems where heuristics techniques to allocate operations are discussed.

In this work it is proposed a Branch and Bound approach which estimates lower bounds considering resources availability.

SEQUENCING AND ALLOCATION PROCEDURES

It is well known in the literature that when there are no shared resources other than processors and there is unlimited intermediate storage the sequencing and allocation procedures can be separated. As far as the lower bound estimation considers the precedence relations between operations (e.g. Single Machine Based Bound–SMBB or Full Machine Based Bound–FMBB [2]), the Branch and Bound algorithm can be seen as a procedure to obtain the optimal sequence of tasks. To obtain the Gantt chart only recurrence relations as that given in equation 1 are necessary. For a task i in a processor j its completion time C(i,j) is given by:

$$C(i,j) = \max\{C(i,j-1), C(i-1,j)\} + t_{ij} \qquad (1)$$

where t_{ij} is the processing time of task i in processor j.

For intermediate storage policies like No-Wait(NW) and No Intermediate Storage (NIS) equivalent relations exist in the literature [1].

In the problem considered here, that is when there are limits in the available resources, a Branch and Bound algorithm cannot only be concerned with operations duration but has to define at which instant of time each operation can be started: the allocation problem has to be solved simultaneously with the sequencing problem. For a task i in the processor j the earliest start time $C_{in}(i,j)$ is given by equation 2.

$$C_{in}(i,j) = \max\left\{C(i,j-1), C(i-1,j), \max_{1 \le l \le R}\left[C^{l}_{in}(i,j)\right]\right\} \qquad (2)$$

where:

l = 1,2,...R are the shared resources, and $C^{l}_{in}(i,j)$ is the earliest time that the available profile of resource l allows to start task i in processor j.

To use a Branch and Bound procedure with start times calculated through equation 2 a procedure for obtaining $C^{l}_{in}(i,j)$ has to be defined, and a technique for estimating the lower bound, preserving its monotone property, has to be developed.

LOWER BOUND CALCULATION

To obtain a makespan lower bound estimation at each node in the search tree the problem is defined as follows:

- each node is characterized by a set of scheduled tasks $\mathbb{K}$, a set of unscheduled tasks $\mathbb{U}$ and a makespan lower bound value LB;
- resource l (l=1,2,...R) availability is given by $q^l(t)$;
- demand of resource l by task i in processor j is given by $Q^l_{ij}(t)$;
- processing time of task i in processor j is given by t_{ij}.

At each node the lower bound will be obtained, as usual, as the addition of two parts, one associated with the scheduled partial sequence $\mathbb{K}$, which is easily obtained, and the other which must be an estimation of the time needed to process the unscheduled tasks $\mathbb{U}$.

At the root node the lower bound is defined as the minimum horizon necessary to process all the tasks considering only the resources availability profiles. This time T can be obtained as:

$$T = \max_{1 \leq l \leq R} [T^l] \quad \text{and } T^l \text{ through:} \quad \int_{t=0}^{t=T^l} q^l(t)dt = \sum_{i=1}^{N} \sum_{j=1}^{M} \int_{\tau=0}^{\tau=t_{ij}} Q^l_{ij}(\tau)d\tau \qquad (3)$$

which simply states that makespan initial estimation T, will be the maximum required time T^l(l=1,..R) needed in order to process all the tasks with the hypothesis of full utilization of resource l.

At intermediate nodes the contribution of unscheduled tasks to the lower bound is estimated as follows: For each node a graph as that shown in figure 1 is considered.

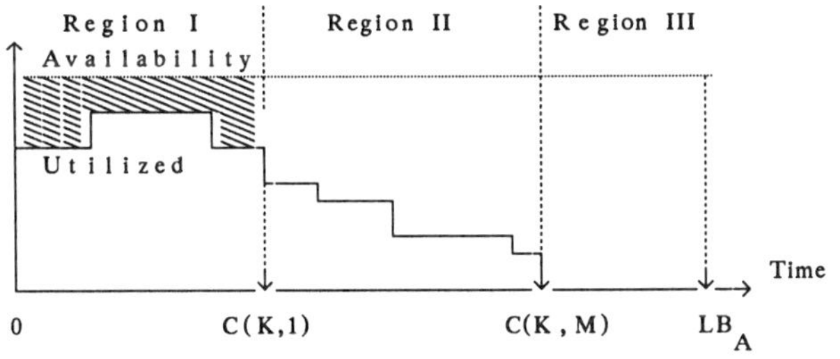

Figure 1.

In figure 1 Region I starts at initial time (t=0) and ends at C($\mathbb{K}$,1) defined as the time the first operation of the last task in partial schedule $\mathbb{K}$ is completed. For any task candidate to be sequenced after this task its first operation can only be started at a time greater or equal to C($\mathbb{K}$,1). This means that resources eventually not used in Region I are lost (dashed region in figure 1).

Region II in figure 1 ends at C($\mathbb{K}$,M), which represents the completion time of the last operation (M) of the last task in partial schedule $\mathbb{K}$. In Region II other operations will be scheduled as the Branch and Bound algorithm proceeds.

In Algorithm A, it is proposed, to estimate the lower bound of partial sequence 𝕂, by increasing the lower bound of partial sequence 𝕂-1 by the amount of time necessary to obtain the quantity of resource lost in Region I. This procedure has to be done for each resource and the lower bound will be defined by the maximum increase in time.

Algorithm A can be improved taking into consideration that the start of the first operation in the next task to be scheduled may not be C(𝕂,1). This is because some level of each resource is yet being used in Region II and the remaining profile may not allow this operation starting in C(𝕂,1). This situation is illustrated in figure 2.

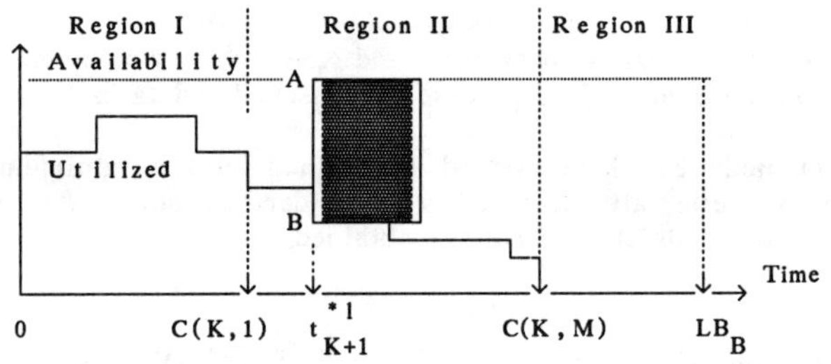

Figure 2.

The dashed region represents the resource consumption of the first operation of a task candidate to be added to partial sequence 𝕂 giving a partial sequence 𝕂+1. The demand profile of this operation (a value AB constant in the figure) does not allow starting at C(𝕂,1) but only at t^{*l}_{k+1}. When this happens resource not used between C(𝕂,1) and t^{*l}_{k+1} will be lost.

In Algorithm B this information is used to obtain a better estimate of the lower bound in the same way as in Algorithm A. The procedure is more time consuming because i) for the first operation of each task candidate t^{*l}_{k+1} has to be obtained for l=1,..R, ii)the case which gives the greater increase in lower bound has to be retained and iii) the task which leads to the minimum increase will be the one which defines the lower bound (in order to preserve the monotone property of the lower bound). A final correction can be done with little additional effort, it consists in considering eventual non complete utilization of the available resource profile by the first operation.

Algorithms A and B utilize only a resource perspective in order to define lower bounds. Precedence relations between operations also contribute to the definition of the makespan lower bound and will be important if resource constraints are not too tight. To combine the two perspectives Algorithms C_1 and C_2 are proposed, as modifications of Algorithms A and B, where in each case lower bound is calculated as the maximum between the value obtained as described above and the SMBB or FMBB values obtained through the precedence relations between operations.

A more detailed discussion of these algorithms can be found in [7].

EXAMPLE

To illustrate algorithms performance a simple example is used. Four tasks are considered, each of them consisting of three operations. Only one resource is used. Table 1 shows processing times and resource consumption; resource availability is constant equal to 10 units.

Opr	Task A	B	C	D
1	8	6	7	4
2	9	7	7	7
3	5	7	12	12

Opr	Task A	B	C	D
1	8	6	7	3
2	6	2	3	4
3	4	4	6	8

Table 1. Processing times and resource consumption

In figure 3 it is shown the consumption profiles when a Branch and Bound algorithm which does not take into account resource constraint is used, and the result obtained when the operations start times are delayed in order to maintain resource consumption under 10 units. It is seen that the makespan value given by the Branch and Bound algorithm used is strongly modified.

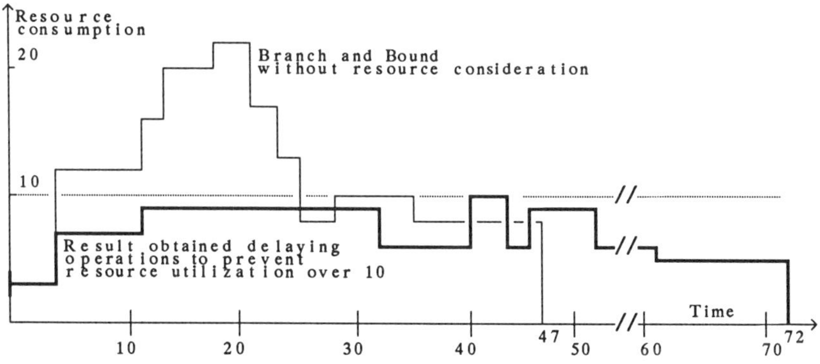

Figure 3. Resource consumption

Table 2 shows the results obtained with algorithms A,B,C_1 and C_2 in terms of nodes searched and lower bounds calculated. There is a clear reduction in searched nodes for algorithm B compared with algorithm A because the resource restriction is tight and individual resource consumption by operations is high compared with the available limit. Algorithm C_2 does not represent a big improvement with respect to B because the increase in lower bound introduced by B is, in this example, sufficient to satisfy precedence relations between operations.

CONCLUSIONS

An approach has been proposed to the calculation of lower bounds for Branch and Bound algorithms in the situation where resource constraints are tight. This resource based perspective to estimate lower bounds is a new approach as far as we know, the usual procedure being to take account of resource constraints after obtaining the final schedule by the Branch and Bound algorithm. Four alternatives are proposed and compared through a simulated example

Node	A	C_1	B	C_2	Node	A	C_1	B	C_2
A	51	53	52	53	ADB	64	68	-	-
B	52	52	53	53	ADC	61	69	-	-
C	52	52	53	53	BAC	-	-	-	-
D	53	53	53	53	BAD	-	-	-	-
AB	55	61	56	61	BCA	61	68	62	68
AC	60	67	67	67	BCD	53	53	63	63
AD	52	55	61	68	BDA	62	68	-	-
BA	64	68	64	68	BDC	63	63	-	-
BC	53	53	53	53	CAB	-	-	-	-
BD	54	59	62	66	CAD	-	-	-	-
CA	63	67	64	67	CBA	62	69	63	69
CB	53	53	55	55	CBD	55	55	61	67
CD	54	54	62	68	CDA	62	67	-	-
DA	61	64	66	66	CDB	62	65	-	-
DB	53	53	60	68	DAB	-	-	-	-
DC	61	61	62	62	DAC	-	-	-	-
ABC	56	61	56	61	DBA	60	66	67	-
ABD	57	61	66	66	DBC	61	61	70	-
ACB	60	-	-	-	DCA	-	-	-	-
ACD	62	-	-	-	DCB	-	-	-	-
ABCD	61	61	61	61	BCDA	73	73	-	-
ABDC	75	-	-	-	CBDA	67	67	-	-
ACBD	67	-	-	-	DBAC	78	-	-	-

Table 2 Lower bounds and searched nodes
(for complete sequences only searched nodes are shown)

REFERENCES

1. K.R. Baker, *Introduction to Sequencing and Scheduling* (John Wiley, 1973).
2. K.R. Baker, "A Comparative Study of Flow Shop Algorithms", *Oper. Res.*,**23** (1), 62-63, 1975.
3. B.J. Lageweg et.al., "A General Bounding Scheme for the Permutation Flow Shop Problem", *Oper. Res.*, **26**(1), 53-67, 1978.
4. F.D. Booth, J.S. Turner, "Comparing Flow Shop Scheduling Heuristics", *ORSA-TIMS Joint National Meeting,* St.Louis, 1987.
5. I. Karimi, H. Ku, "A Modified Heuristic for an Initial Sequence in Flow Shop Scheduling", *Ind. Eng. Chem. Res.*, **27**, 1654-1658, 1988.
6. K.P. Sycara et. al., "Resource Allocation in Distributed Factory Scheduling", *IEEE Expert,* 29-40, February, 1991.
7. M.T.M.Rodrigues, *Sequenciamento e Alocação de Operações em Flow Shops na Indústria Química com Restrições sobre os Recursos Compartilhados. Uma Abordagem de Busca Orientada por Restrições,* Ph D Thesis, UNICAMP, 1992 (in Portuguese).

DESIGN PROCESS MODELS FOR CONCURRENT ENGINEERING

Hyeon H. Jo, and **Hamid R. Parsaei, Ph.D., P.E.**

Dept. of Industrial Engineering
University of Louisville
Louisville, KY 40292

ABSTRACT

It has been recently acknowledged that the best opportunity to rebuild the manufacturing leadership could be found in design. However, as technological tasks in the design community have become progressively complicated, especially in a concurrent engineering environment, the entire design process has also become very complicated. A large number of highly specialized design tools has to be developed to meet the technological needs of complex design tasks, and then integrated into a *CAD framework* to provide better services to users. Development of such CAD frameworks must be based upon scientific research and practical exploration of the design process.

This paper addresses two issues in this context: significance of CAD frameworks in practicing the concurrent engineering philosophy, and models of the design process which can provide a guidance in the development of sound CAD frameworks that can support concurrent engineering practices.

Keywords : *design process, concurrent engineering, CAD frameworks*

INTRODUCTION

It has been recognized that *design must be viewed as the central pivot of all production activities* to compete in the world-wide marketplace. The philosophy of concurrent engineering addresses this issue well, which entails concurrent consideration of product design and its all related processes within an organization. It has been regarded as a manufacturing philosophy of the last decade of the 20th century that aims at the fast introduction of high-quality products to the marketplaces at lower prices [7]. The philosophy necessitates that designers take into design decision the entire issues of a product's *life-cycle* from conception to disposal. Life-cycle of a product includes every aspect involved in the development of the product, including conception of the product, functional requirements, manufacturing, assembly, cost estimation, service and maintenance, disposal, and so on. Consequently, designers will be not only confronted with the increasing complexity of product designs, but also with the constant increase in number and complexity of specialized computer-aided design (CAD) tools to analyze or evaluate their designs with respect to these multiple issues of life-cycle. Accordingly, the *design process* has inevitably become very complicated and the role of today's designers needs to be respecified. A large number of highly specialized design tools has to be developed to meet the technological needs of complex design tasks, such as design for manufacturability, engineering analysis, assemblability and testability, and then should be integrated into a *CAD framework* to provide a variety of functions and services to the design community.

The concept of CAD framework may be of special interest to those who try to practice the philosophy of concurrent engineering, because during the design phase a large number of design tools are required to examine the entire aspects of a product's life-cycle in a coherent way. The framework is a well organized software infrastructure which properly guides designers through the entire design process so that they can reach a design with maximum quality which can be produced at the lowest cost and in the shortest time. The development of such sound CAD frameworks will not be trivial and require systematic study of the design process. This paper discusses various models and aspects of the design process for the purpose of hopefully providing a guidance for the development of practical and intelligent CAD frameworks that can fully support concurrent engineering practices.

CAD FRAMEWORKS FOR CONCURRENT ENGINEERING

Concurrent engineering (CE) is stemmed from the recognition that design decisions made early in the product development cycle have significant impacts on manufacturability, quality, cost, time-to-market, and thus the ultimate marketplace success of the product.

As for the implementation of the CE philosophy, it can be inferred that computer-based approaches will be predominant. A large number of CAD tools will be required to examine the entire aspects of a product's life cycle from conception to disposal. Consequently, what we are expecting to see is an integrated design environment in which *all CAD tools interact and cooperate to find a globally optimized or compromised design* [6]. This design environment may consist of sets of various CAD tools and an underlying CAD framework.

A CAD framework is "*a software infrastructure that supports the application environment, providing a common operating environment for CAD tools*" [5]. It is a logical extension to the general computing environment that provides the functions and services demanded by the design community. A CAD framework is actually layers of code between the operating system and application environment, which provides a variety of functions and services to designers. A notable improvement in design productivity can be accomplished, because designers using the framework can concentrate on the main design rather than on other tasks such as interface.

In general, today's CAD tool users seem to be satisfied with the performance of their design tools. However, a great deal of time that could have been spent on the design is often spent in integrating these diversified tools into a single, working CAD environment [5]. The domain of the integrated CAD environment has emerged to address this problem, particularly from the area of electronic circuit design. The design process of the area has been more automated than those of any other products. The designers in the domain have been able to spend less time directly manipulating design data and more time managing CAD tools that manipulate the design data. sA for the *mechanical design community*, the concept and the value of a CAD framework have not yet been fairly understood and appreciated. However, as more and more sophisticated CAD tools required for the life cycle analyses become available, CAD frameworks should be recognized in the community as a requisite for improving the design productivity.

DESIGN PROCESS MODELS FOR CAD FRAMEWORKS

One of the major problems in developing intelligent CAD frameworks is the representation of design processes and design objects. Consequently knowledge about design processes and objects should be embedded in CAD frameworks. Especially *thorough understanding of the design process models* is very important, because the models may provide a guidance in the development of sound CAD frameworks that support concurrent engineering practices. The scientific and practical basis for such models may come from research in design theory and methodology. However, there has not existed a generally agreed theory or a set of principles to guide or explain design processes [3]. In this section, a few models and aspects of design process are discussed. It is hoped that the discussion is helpful for developing practical and intelligent CAD frameworks. The discussion is divided into two subsections: general process models and detailed process models.

General Process Models

Maher [8] describes three models of general design process : *decomposition, case-based reasoning, and transformation*. The design problem is decomposed into subproblems in a decomposition model, and case-based reasoning involves the using the design solution or solution process from previous design problems. The transformation model follows a theoretical approach to transforming the design requirements into a design solution.

In an extensive review paper, Finger and Dixon [4] classify the models of design process into three general categories : *descriptive, prescriptive, and computer-based models*. A descriptive model is usually developed from design experiments in which designers are asked to design an artifact and think 'aloud' as they solve the problems. The whole designing session is audio- and video-taped. The taped data are then analyzed to obtain a *protocol* of the session from which a descriptive model is developed. Although a prescriptive model describes how the design process should proceed, no works have verified that the better design would result in if the prescribed process were to be followed. That is, prescriptive models are premature until they can be based on validated theory, while descriptive models involve too many ill-defined variables to support a theory [2]. A computer-based model expresses a method by which a computer can perform some specified class of design tasks during the process. The general direction of design process research may be illustrated by Figure 1.

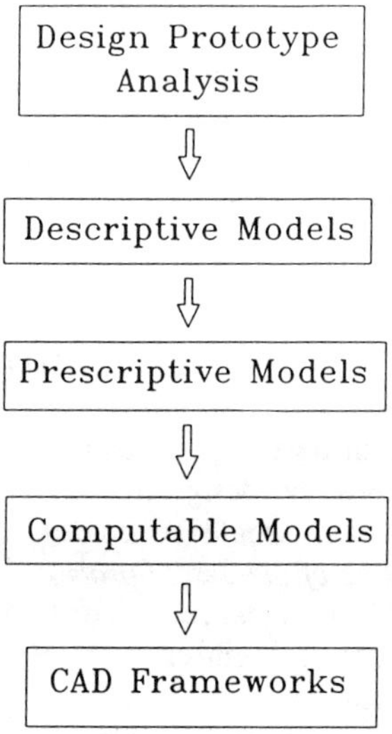

Fig. 1 General Direction of Design Process Research

Notice that descriptive models are used as the basis for prescriptive models from which computable models can be derived. These computable models of the design process are embedded in CAD frameworks which guide designers through the entire design process.

Detailed Process Models

Iterative Models. The major roles of a designer is to develop, analyze, or modify an engineering design with the use of computers. To perform these tasks, the designer may follow an iterative design process which consists of six identifiable phases : *recognition of need, definition of problem, synthesis, analysis and optimization, evaluation, and presentation* [9]. This process model shows the basic iterative nature of the design process based on single artifact.

Recursive Models. Ward [11], however, maintains that the design process should be normally recursive and allow designers to reason simultaneously about sets of design artifacts. He also argues that designers should normally not design to meet rigid specifications established at the definition of problem phase, because a wide plausible set of environments should be considered at the beginning of the design process.

Computable Models. In an attempt to build a computable design process model, Takeda et al. [10] propose *metamodel* for the central description of a design object.

A metamodel is a central description of the design object from which multiple aspect models can be derived. An example of aspect models is a geometric aspect model which contains geometric data of the design object. A metamodel is independent of a context and a aspect model is context-dependent. A metamodel is transferred to the next metamodel if the evaluation with respect to all contexts succeeds. This step-wise refinement process continues until a complete and satisfactory description of the desired artifact is obtained. Notice that any life-cycle issue of a product may be regarded as a context. Based on the frame of this descriptive model and three different types of logical reasoning, *abduction, deduction, and circumscription*, Takeda et al. discuss the development of a computable model of the design process.

Psychological Aspects. A psychological aspect of the design process should be also studied. The development of an interactive and intelligent CAD framework needs to contain knowledge about human being's mental process in order to help users facilitate their thought processes. In other words, a psychological model of the design process should be integrated into the CAD frameworks in order to *compensate any characteristic biases or cognitive limitations of human designers* [1].

Design Trade-offs. In a concurrent engineering environment, the design process model should be *interactive* and *cooperating* rather than a long iterative model. Specifically it is highly desirable for the model to incorporate a *conflict resolution scheme* since designers usually have conflicting design criteria or goals. Design is the process of recognizing, consensualizing, and resolving design conflicts [12]. Since most real-world design problems involve complex design trade-offs to satisfy multiple goals and vague constraints, it is important for the design process models to incorporate formal strategies and methodologies for identifying and resolving those design conflicts as early as possible during the design process. In this context, versatile resolution scheme for design conflicts should be integrated into the models of design process, especially in a concurrent engineering environment in which multiple life-cycle issues should be considered. One promising approach is to use fuzzy set theory which provides a flexible and formal aggregation procedure of multiple design goals. A variety of fuzzy set connectives can be used to aggregate those goals, and the membership functions of fuzzy sets may portray the designer's preferences and design strategies. Fuzzy set theory is also applicable to the evaluation of configuration level designs, because the designs can only be evaluated subjectively and implicitly.

CONCLUSIONS

It has been emphasized that *product design* must be regarded as *the central pivot of all production activities*, which corresponds to the philosophy of concurrent engineering.

A significant improvement in the design productivity may be accomplished by using a *CAD framework* that provides a variety of functions and services. The design community, especially *mechanical design community*, should be aware of the benefits of using a CAD framework that can support the concurrent engineering philosophy.

A starting point of developing a sound CAD framework should be the observation and analysis of actual design activities which are used as the basis of descriptive models of the design process. The models should, in turn, be used as the foundation of the prescriptive models upon which computable models can be based. The computable models are then embedded into the CAD framework that provides a variety of services for concurrent engineering practices.

REFERENCES

1. Baker, K. D., Ball, L. J., Culverhouse, P. F., Dennis, I., Evans, J.St.B.T., Jagodzinski, A.P., Pearce, P.D., Scothern, D.G.C., and Venner, G.M., 1989, "*A Process Oriented Approach to an Intelligent Design Aid*", Proc. 2nd Int'l Conf. on Indus. & Eng. Appl. of A.I. & Expert Sys., Tullahoma, TN, June 6-9, pp.479-485.
2. Dixon, J. R., 198, "*On Research Methodology Towards A Scientific Theory of Engineering Design*", Artificial Intelligence for Engineering, Design, Analysis, and Manufacturing, Vol.1, No.3,
3. Duffey, M.R., and Dixon, J.R., 1990, "*A Program of Research in Mechanical Design: Computer-Based Models and Representations*", Mechanism and Machine Theory, Vol.25, No.3, pp. 383-395.
4. Finger, S., and Dixon, J.R., 1989, "*A Review of Research in Mechanical Engineering Design. Part I: Descriptive, Prescriptive, and Computer-Based Models of Design Processes*", Research in Engineering Design, Vol.1, pp.51-67.
5. Graham, A., 1991, "*The CAD Framework Initiative*", IEEE Design & Test of Computers, September, pp.12-15.
6. Jo, H. H., Dong, J., and Parsaei, H. R., 1992, "*Object-oriented Paradigm for CAD Frameworks*", Proc. 2nd Int'l Conf. on Factory Auto. & Infor. Management, Falls Church, VA, June 30 - July 2.
7. Jo, H. H., and Parsaei, H. R., 1992, "*Principles of Concurrent Engineering*", Handbook of Concurrent Engineering, edited by H. R. Parsaei and W. G. Sullivan, Chapman and Hall, Chapter 1.
8. Maher, M.L., 1990, "*Process Models for Design Synthesis*", AI Magazine, Vol.11, No.4, pp.49-58.
9. Shigley, J.E., and Mischke, C.R., 1989, Mechanical Engineering Design, 5th edition, McGraw-Hill, NY, NY.
10. Takeda, H., Veerkamp, P., Tomiyama, T., and Yoshikawa, H., 1990, "*Modeling Design Processes*", AI Magazine, Vol.11, No.4, pp.37-48.
11. Ward, A.C., 1990, "*A Recursive Model for Managing the Design Process*", Design Theory and Methodology - DTM '90, ASME, pp.47-52.
12. Whitney, D. E., 1990, "*Designing the Design Process*", Research in Engineering, Vol.2, pp.3-13.

CONCURRENT ENGINEERING IN AN ACTIVITY-BASED ENVIRONMENT

Dirk Wiegmann
William G. Sullivan

Virginia Polytechnic Institute and State University

ABSTRACT

Concurrent engineering is being integrated with activity-based accounting concepts to provide designers with more accurate and life-cycle complete information. The objective of this paper is to describe how this information can be used to make better product decisions.

Key Words: concurrent engineering, activity-based costing, design-to-cost

INTRODUCTION

The responsibility of the designer is to find a product form that satisfies given functional requirements and minimizes the sum of activity costs and material costs of the product. An activity-based method appears to be an ideal tool to find this cost minimum because it calculates production costs with such high accuracy. Designers should be provided with the most accurate production cost information so they can properly calculate the costs of various design alternatives and choose the best. According to a recent survey, productivity can be improved by as much as 40% when optimizing a design with respect to material and activities, using an activity-based method [FRO91]. Classical design guidelines such as reducing the number of parts, introducing less variants, and reducing the number of required processes can then be supported by more accurate cost data [KAM89].

While all of this sounds appealing, one of the difficulties is that current activity-based systems struggle to provide information that is really required in the design process. These systems are generally unable to supply the information at the low detail level required, and they often fail to supply truly relevant costs. Accordingly, the objective of this paper is to present a model for developing detailed cost information, regarding design, for decision-making purposes.

One technique that can be applied for design-concurrent calculation using cost information from existing products is "similarity analysis" [PIC89]. The basic methodology is to find existing products that have similar characteristics to those required in the new design but for which cost data are available. These characteristics, which are specified by the designer and called product cost drivers, can be the number of parts, quantity or type of material, number of processes, or size. Similarity analysis uses several similar products as reference products and analyzes their characteristics with the help of statistical methods, such as regression, to consider several products and several characteristics. The analysis generates a function of these characteristics, and the cost of the design can be estimated by applying this function.

The design process is part of the engineering decision making process. Because similarity analysis has to be based on decision costs which frequently differ from those provided by traditional accounting systems, the design-concurrent calculation has to consider costs that are relevant from a decision perspective. Clark

[CLA23] defined relevant costs as "the costs that are incurred if that business is taken and which are not incurred if that business is not taken." While data produced by most activity-based cost systems are keyed to historical costs and activities, the design process requires future planned costs and activities for good decision-making.

Cost Model

The cost model operates according to a two stage principle. In the first stage costs and other relevant data are collected and recorded in a neutral way. "Neutral storage" means that no information is lost through any kind of proportionalization of data, which would make different data analyses impossible, or biased. In the second stage this database is made available for periodic financial accounting requirements as well as for case specific decision-making.

The cost model uses two main databases: the activity costing database and the product information database. The activity costing database ensures accurate activity cost information by storing the company's activity cost drivers, the observed and planned cost driver quantities, and the traced costs. The product information database records information about current and future products. This includes the product structure, the production plan, the main cost driving characteristics, the bill of activities, and the cost of materials. Both databases store historic data as well as planned data for several future time periods.

In the cost model the reference products are separately calculated for every time period in which the design would be produced. The reference products are calculated using the bill of activities that will be estimated for a specific future time period, i.e., the bill of activities is likely to be different in every time period due to changing product definition and activities.

As indicated in Figure 1, the present value method is applied to discount cost information of each time period to the present. Based on cost calculations for reference products, similarity analysis can be applied. A regression analysis is performed on the main cost driving characteristics, whereby product characteristics are the independent variables and present value constitutes the dependent variable. In this way a cost driver function can be determined through correlation analysis to estimate the costs of the design alternative.

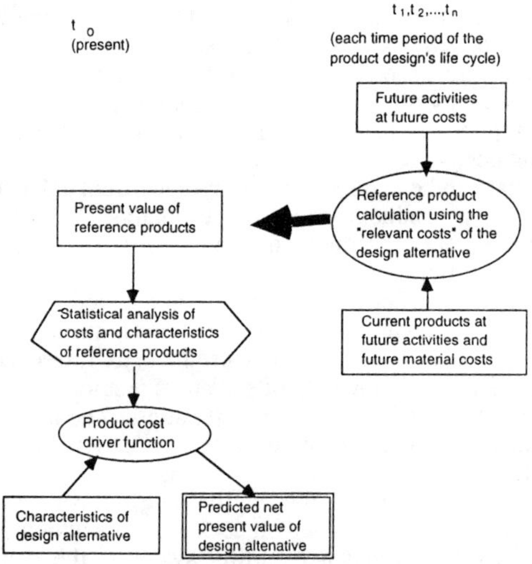

Figure 1. Concept of the design-concurrent calculation method

A hierarchical, multiple-layer functional organization is used by the cost model. For every "function" on every level of the hierarchy, a cost driver is defined to which the costs are as proportional as possible. The costs of some departments might be driven by the number of parts, e.g. in the "drilling-center," while some activities within this department might be driven by time, e.g. the activity "to perform drilling."

All costs that are traced to an activity can be categorized as either operating costs or capacity costs. Operating costs vary directly with the number of units of the cost driver performed by an activity, for example, kilowatt-hours of electricity. The slope of the operating cost function with respect to the cost driver as its independent variable is obtained by dividing the operating costs by the cost driver output (see Figure 2).

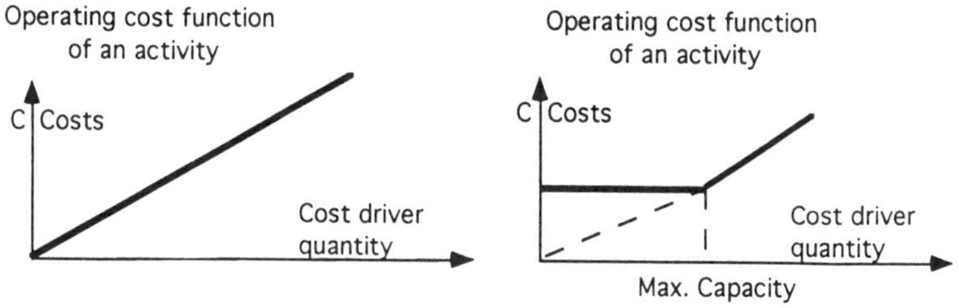

Figure 2. The operating cost and capacity functions of an activity

Special attention must be paid to the capacity consumed by every activity. This "committed" capacity includes capacity required to execute the existing production schedule in addition to unused capacity. Fluctuating capacity requirements frequently leave a considerable amount of capacity unused (see Figure 3). Information about unused capacity is important for the design-concurrent calculation because capacity-related costs have already been committed.

Capacity costs, e.g. depreciation, are incurred even if there is no production or when no cost driver is performed. Every activity's cost driver has a certain maximum capacity, and if this capacity is to be increased it will impact on all capacity costs of the related activity. For example, a new machine may require additional power supply and floor space for its operation to increase output capacity. The slope of the capacity cost function beyond the maximum capacity can be determined by dividing the existing capacity costs by the maximum capacity (see Figure 3).

For future time periods no accounting data are available yet. In order to record planned cost data, the cost model as a first step makes a cost prediction based on the existing production plan as well as on past costs. Then, in a second step the model is used to correct those predictions for expected future cost variations.

The model predicts the planned costs by multiplying the costs per unit cost driver from the most recent period by the planned activity output for each future time period. This prediction by the model initially assumes that the costs as well as the activity cost structure will remain unchanged in the future. If the user has better information about future cost developments or changes in the cost structure, he or she must adjust these predictions accordingly. Then, the cost model recalculates the costs per unit cost driver, and adjusts the operating and capacity cost functions to account for the user's insights.

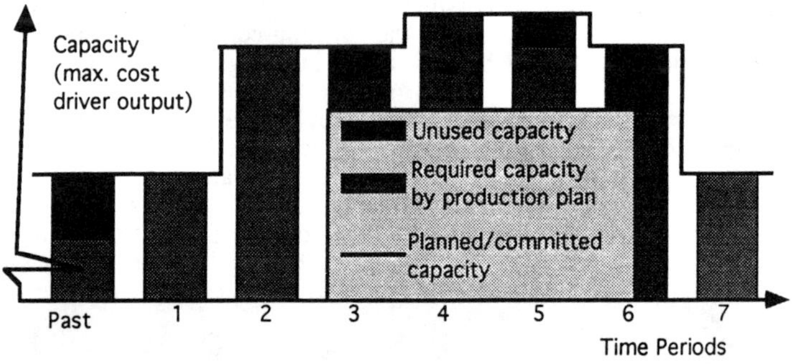

Figure 3. Capacity of an activity committed by the existing production program

The cost model always calculates the costs of all products that are similar to the design alternative. These products are called reference products because the design-concurrent calculation uses their product information as a reference point in its design-calculation. Reference products belong to the same product family as the design alternative, i.e., they have the same main characteristics and are located on the same hierarchical product level as the design alternative.

The design-concurrent calculation method utilizes the classification of the design and for every activity it determines the relevant cost driver quantity twice. First, the cost driver is determined as if the design alternative were a completely new product (P_N); then, the cost model also determines the cost driver quantity as if the design alternative were a re-used product or part (P_R). The relevant cost driver consumption is then estimated by assuming a linear combination of these two extreme cases:

$$P = (1-C)\cdot P_N + C\cdot P_R \qquad (1)$$

P = relevant cost driver quantity of an activity
C = fraction of expected re-used parts
P_R = cost driver quantity when re-using the reference product
P_N = relevant cost driver quantity when the design is new

Capacity costs committed by the design are relevant for at least one time period. The one time period case occurs when the additional capacity had been planned for the following time period. Then the costs are just incurred one time period earlier as a result of the design decision. Due to the fact that capacity costs from preceding time periods can be relevant, the design-concurrent calculation method not only has to estimate the costs for the time periods of the design life cycle, but also for several time periods thereafter.

Utilizing similarity analysis, the cost model derives a functional relationship between the characteristics of the reference products and their costs. The costs of the design alternative can be estimated from this function. Sometimes, in what can be considered a simplified version of similarity analysis, the calculations are based upon only one characteristic and the cost of one known reference product. Regression analysis can be performed on several characteristics of similar products with the product characteristics being the independent variables and the total costs being the dependent variable [EHR85]:

$$Y = \beta_0 + \beta_1 X_1 + \beta_2 X_2 + ... + \beta_Q X_Q \quad (2)$$

Y = Costs
X_i = Amount of characteristic i $i = 1,2,...,Q$
β_i = Coefficient i (values determined by ordinary least squares regression)

After their transformation, the characteristics of the design alternative are inserted into the empirical cost driver function to produce an estimate of present value of the design alternative. This information is then used to improve the decision process when selecting a best design alternative.

References

1. CLA23 Clark, Studies in the Economics of Overhead Costs, Chicago, University of Chicago Press, 1923.

2 FRO91 Frohlich, "DV-Gestutzte Prozeßkostenrechnung - Entwicklungsstand und Perspektiven," Controller Magazine, Mar. 1991, 117-127.

3. KAM89 Kammlade, "The Cost Management System: A Critical Link Between Design and Manufacturing," Dearborn, MI, Autofact'89 Proceedings, Society of Manufacturing Engineers.

4. PIC89 Pickel, "Kostenmodelle als Hilfsmittel zum Kostengunstigen Konstruieren," Dissertation Munchen 1989, Hasnser Verlag.

A SIMULATION STUDY FOR THE PERFORMANCE ANALYSIS OF COMPUTER NETWORKS

Tarun Gupta and **Amphon Phipathananunth**
Assistant Professor *Graduate Student*

Department of Industrial Engineering
Western Michigan University
Kalamazoo, MI 49008

ABSTRACT

This simulation study analyzed the performance of token-ring and carrier sense multiple access with collision detection (CSMA/CD) protocols using the NETWORK II.5 software from CACI. A simulation model was developed for a computer network in a manufacturing system. Hypothetical data, for eleven stations, was assumed. Arrival and processing rates for the packets were assumed to be exponentially distributed. Packet size, traffic intensity and transmission rate were assumed as independent variables to estimate a set of performance indicators. Several important results which can serve as useful guidelines to the network designers are also discussed.

INTRODUCTION

The development of faster and higher-capacity data transmission systems during the past decade has contributed to the phenomenal growth in the computer and computer related accessory industry. Advances in integrated circuit technology have allowed computers to become smaller, less expensive and more available. This has allowed use of computers in a variety of configurations not least the connecting of many computers to form NETWORKS. A communication network is defined as a set of data links used to interconnect a collection of autonomous computers, printers, telephones, terminals or other types of data handling devices [1]. These devices are generally known as STATIONS. The interconnection among various devices allows them to exchange information. One particular type of network is the Local Area Network or LAN as it is commonly known. This is a non-public communications system that permits various devices connected within the network to communicate with each other over distances from several feet to several miles.

Among the traditional data related applications, LAN technology has also been used in the manufacturing environment [2]. Computers in an automated factory communicate with each other through the network; the devices that can be connected to the network include computers, CNC machines, programmable controllers, robots, and vision systems amongst others. The way in which networks are geometrically arranged and connected is known as the topology of the network. The three basic topologies of interest are the bus, ring and star configurations. The purpose of this study is to analyze different combinations of LAN technologies and access techniques using simulation methods to arrive at optimum levels of performance using computer network resources in manufacturing environments.

LITERATURE REVIEW

The manufacturing industry is undergoing a dramatic change in the way computers are connected in a network environment. Recently corporate executives are coming under increasing pressure for reconsidering their options for hardware selection especially because of much stricter software licensing regulations and copyright laws. After a period in which many companies experimented with decentralizing their information systems (IS) organization, the pendulum is swinging back once again [Simson, 1990]. The major factor driving this trend toward recentralization is the continually increasing emphasis on company wide information systems that integrate business functions and support new business opportunities. There are, however, no guidelines available for determining network structures and parameters that will best utilize particular network resources in a manufacturing set-up [2]. Simulation has been used in the manufacturing environment to simulate various functions and aspects from investment decisions to new facility start-ups to evaluating system interactions in flexible assembly and flexible manufacturing systems etc. Examples of the use of simulation in these scenarios abound, [7,8]. Simulation has also been applied to the modeling and analysis of computer networks with assumptions regarding their topology, traffic characteristics and network parameters. However, it is difficult to evaluate the relative merits of differing LAN technologies for common environments or applications. Thus the purpose of this study.

METHODOLOGY

Computer Simulation has become an integral part of successful business operations. It is widely used as a quantitative method of analysis to simulate the behavior of actual systems. The main step in achieving the objectives of this study was to develop the simulation model that would represent the various network components and their inter-relationships in a computerized manufacturing system.

The software components of a simulated system are characterized in NETWORK II.5 as modules, instruction mixes, macro instructions and files. Only modules were utilized in our model. Modules are specifications of tasks to be performed by a processing element. The module description consists of four parts : scheduling conditions, host processing element options, instruction execution list and successor modules list. The stages in a module's life are : checking preconditions, requesting a host processing element, executing instructions and choosing successor modules.

The simulation model assumed ten shop controllers that were controlled individually by a main factory controller. Each of them was modeled as a processing element. The factory controller on receiving a message from a shop controller replies with an acknowledgement or instruction. The transfer device represents the cable connecting the processing elements.

Two different topologies were simulated : bus and ring. The transfer of data within the network to be simulated is done by discrete increments called packets. A packet's size can range from a few bits in length to several thousand bits in length; the actual size being determined by the network parameters; in this case the size is determined by statistical functions supported by NETWORK II.5. The parameters being controlled in the model are data transfer rate, packet size, simulation run-time and traffic load. For the purpose of performance comparison several outputs are collected. These outputs are simulation run-time, number of successful transfers, average usage time, average pre-condition time and average execution time. These are defined as follows :

Simulation run time is the length of time measured from the beginning of the simulation to the point that simulation ends and output reports are produced.

The number of successful transfers is the number of packets which have been transferred successfully.

Average usage time indicates how long a station used a transfer device from its acquisition until its release. Usage time does not include the delay time encountered waiting for this Transfer Device.

Average pre-condition time is a measure of the time from the simulation package's first attempt to activate this module until the time that the module starts execution of its first instruction.

Average execution time is a measure of the time from the start of the module's first instruction to the completion of the last inllstruction. This includes the time due to contention delays, instruction execution and interrupts encountered during execution.

The following assumptions were made for developed model:

i. There were no errors except those caused by collisions,
ii. There were 10 stations generating packets in exponential arrival times,
iii. The mean packet lengths are exponentially distributed,

varying from 500 to 5000 bits,

iv. The receiving station sends 10-bit acknowledgements to the transmit ting station,

v. The cable length is 2 kilometers,

vi. The propagation delay is 5 micro-second per kilometer,

vii. The token-holding time is a 1-bit delay,

viii. The jamming pattern for CSMA/CD bus, is 32 bits,

ix. For CSMA/CD bus, the IEEE 802.3 standard truncated binary exponential backoff algorithm has been used for the retry interval after collision, and

x. The simulation run-cycle of 100 was assumed.

The data transfer rate or baud rate, was selected as 2400 bits per second at the lower end and 10 megabits per second at the higher end. Most local area networks usually operate in the range of 1-10 Megabits per second.

The traffic load or the offered load is determined from the packet size, number of stations and baud rate. The Poisson arrival process which gives an exponential distribution of packet inter-arrival time has been used for our model. Traffic loads were controlled by varying the inter-packet arrival time of the packet. For example, if we have a system of 11 stations sending a mean packet size of 500 bits at a speed of 10 megabits per second for a 50% offered load then the inter-arrival time is given thus :

$$InterArrival\ Time = \frac{(Mean\ Packet\ size \times No.\ of\ Packets)}{(Baud\ Rate \times \%Load)} \quad \textbf{(1)}$$

The capacity of a network can also be determined by the throughput- offered load curve. A throughput-offered load curve indicates the number of packets which have been successfully transferred or throughput versus the offered load. This results are obtained from our simulation runs. As per our model we calculate an expected number of packets generated within the simulation run-time. For example, in a network operating at 10 megabits per second with an average packet length of 5000 bits, given a simulation run-time of 50,000 milliseconds under a 100% offered load, the expected number of packets generated will be :

$$No.\ of\ Packets = \frac{(\%Load \times Data\ rate \times Run\ time)}{(Average\ Packet\ length)} \quad \textbf{(2)}$$

The number of successfully transferred packets can be obtained directly from a Transfer Device output report. Assume that it is equal to 75,000 packets. Thus throughput will be :

$$Throughput = \frac{(No.\ of\ Successfully\ Transferred\ Packets)}{(NO.\ of\ Packets\ generated)} \quad \textbf{(3)}$$

Collection of the above data is followed by its conversion to more readable and understandable forms. The delay-throughput chart is of primary importance. Delays are represented by the time

measured from the generation of a packet at the source station until its successful reception at the destination station. This delay time includes the queuing and access delay at the sender station and the transmission time of the packet.
The time from the generation of a packet to its completion can be obtained from the summation of the average pre-condition time and the average execution time.

The mean packet transmission time is represented by the average usage time in transfer device output reports. Throughput measures the average number of bits per second or packets per second that can be processed through the network. Although the throughput cannot be controlled directly, it can be varied to some extent by changing the offered load. Throughput, in this model, is computed by dividing the number of successful transfers by the simulation run-time. The simulation runs were made with acknowledgement and without acknowledgement.

IMPORTANT RESULTS AND DISCUSSIONS

The performance of the two configurations simulated is analyzed. The most common performance measure is mean delay time. Delays are measured from the generation of a packet to its completion and are then compared to the mean packet transmission time. For the ideal case or the case where no delays are encountered, this ratio will be equal to 1.

The other significant measure of performance is throughput. Throughput measures the average number of bits per second or packets per second that can be processed through the network. Frequently, throughput is normalized using the channel transmission rate to obtain a value between 0 and 1. Maximum throughput for a network is also known as channel capacity [11].

It must be noted that maximizing throughput and minimizing delay for a particular network are generally conflicting goals; thus an important characteristic of concern is the delay-throughput relationship.

The results for the Delay-Throughput characteristic are as follows:

- The Token-ring is less sensitive to workload and packet size,
- CSMA/CD has a slightly shorter delay at low throughput values, but is sensitive to throughput at higher throughput values,
- Transmitting without acknowledgement can provide a lower delay in both access techniques.

The throughput of the token ring are very close to the offered loads in an ideal case. This is due to the fact that our simulation model assumes no transmission errors. In addition the token ring method is a controlled access technique which can operate more efficiently under heavy traffic load. The throughput of CSMA/CD bus is also very close to that of a token ring, however, under the high traffic loads, over 0.8 load factor, throughput of CSMA/CD decrease proportionally depending upon the data rate and the packet size.

CONCLUSION

In order to choose the right combination of protocol and topology for a network in the manufacturing environment we need to identify the requirements of the network. These are network capacity determined by traffic load and throughput. Finally, to find the best combination of network topology and access technique for a network in a manufacturing system is not easy. Every relevant parameter under every possible condition has to be taken into account. It would appear that the best way to arrive at the closest answer is to model the networks close to reality and run it through simulation.

Acknowledgments: The authors wish to acknowledge the suggestions received in the preparation of the manuscript from Mr. Sanjay Pathak who is presently a graduate student in the department of Industrial Engineering at Western Michigan University.

REFERENCES

1. Geld E Kaiser. "Local Area Networks". McGraw Hill, New York. 1989.

2. T. Gupta and B.K. Ghosh. "A multi-objective two stage approach to the design of communication networks subject to production constraints for computerized manufacturing systems". Journal of Production Research, 1989, Vol 27, No 4, pp 587-598.

3. M. Ketcham and J. Watt. "Parametric Simulation for flexible assembly systems". Journal of Manufacturing Systems, Volume 8 No.2, pp 115-125, 1989.

4. Network II.5 Manual. CACI, California. 1987.

5. L.J. Hammond and J.P.P. O'Reilly. "Performance analysis of local computer networks". Addison-Wesley, MA. 1986.

An Architecture of Computer-Aided Alternate Process Planning System (A-CAPPS) for CIM Environment

Tarun Gupta
Assistant Professor

Tata Rao Gali
Graduate Student

Department of Industrial Engineering
Western Michigan University
Kalamazoo, MI- 49008

ABSTRACT

This research presents design and development of a computer-aided process planning system (CAPP) that generates least cost alternate plans. Traditional CAPP systems generate one process plan per part. Their use in most production systems has been found to be limited due to frequent non-availability of machines and other resources recommended by the CAPP system. The proposed methodology in this research integrates order processing, aggregate planning, resource grouping, loading, process cost estimation, process planning, and finally scheduling. The developed prototype system A-CAPPS generates alternate process plans to satisfy various capacity constraints.

INTRODUCTION

Process planning is the task of transforming part design specifications from detailed engineering drawings to a set of operating instructions necessary for part manufacture. In a production system, process planning has always been the most important interface between product design and manufacturing [Lin & Bedworth, 1988]. This function is performed in virtually all industries; its significance is the greatest in small batch, discrete parts metal fabrication industries. Recent manufacturing trends indicate that a wide variety of components are produced in batches of less than fifty on facilities that are usually poor in productivity [Kimemia & Gershwin, 1985]. This has necessitated the development of efficient techniques to effectively plan and control production in these industries.

Problem Definition

Traditional CAPP systems are designed in such a way that they generate one process plan per part design. Based upon technological and/or economic considerations, these systems recommend one specific process and one machine tool for each step of production. The available machine capacities are not considered during the process and machine selection stage; as a result some machines are overloaded while others are under-utilized. The overloaded machines, usually known as bottlenecks, may cause delays in the completion of production orders. Moreover, most CIM systems are designed with a high level of flexibility and responsiveness to changing production requirements. With only one process plan per part, the advantages of system's flexibility can not be put to the fullest use. These limitations tend to make existing CAPP systems inadequate for most CIM systems.

A variant approach based CAPP system and an interface with a simulation model is developed in this research. The developed system generates alternate process plans and evaluates them for a set of performance criteria to select one for implementation.

METHODOLOGY

A recent paper by Hou and Wang (1991) reported an integration methodology for a CAPP system with an FMS. Our approach incorporates estimation of cost for alternate process plans in addition to their evaluation for a selected set of performance criteria before one is recommended for implementation. Another salient feature of our proposed architecture is the fact that there is only one standard process plan per family.

The proposed methodology for integrating CAPP in CIM planning starts from sales order processing, and integrates aggregate planning and resource grouping with process planning. The grouping of parts and machines is performed using principles of cellular manufacturing. One or more part families may be assigned to each cell. A master production schedule (MPS) is determined from the aggregate plan and work loads are assigned to each machine cell. Each cell load is split into batches of part types that are grouped into families to take advantage of group scheduling (Ham, Hitomi, Yoshida, 1987). These steps translate production planning problems from system level to problems at cell level. These steps are same as in Kusiak's FMS planning and scheduling methodology [Kusiak, 1990].

The developed system, A-CAPPS, takes part production requirements from the MPS and their detailed design as input to generate alternate process plans. Because each part may have more than one feasible process plan and there are possibly two or more part types scheduled during a planning horizon, several process plan combinations can be obtained. A-CAPPS also interacts with manufacturing cost database to determine relative costs of alternate plans. Thus, the system recommends alternate plans with their estimated unit marginal (+/-) costs when compared with the unit cost of the first plan. The cost accounting database contains unit cost of operation on each machine.

A simulation model in SLAM II with FORTRAN inserts is being developed to evaluate these process plan combinations. The developed simualtion model will evaluate a set of plan combination for predefined performance criteria. As mentioned earlier, the details of this model will be reported in the subsequent paper.

The most important step in the implementation of the proposed methodology is the development of a CAPP system, which can generate alternate process plans by interacting with production planning functions. In order to demonstrate the significance of the proposed methodology a CAPP system A-CAPPS was developed.

A-CAPPS SYSTEM DESIGN

A-CAPPS is a variant process planning system, and it uses group technology based part code to define part design information as input. Moreover, various logical divisions of process planning activity are broken into functional modules. In A-CAPPS, a part is represented by 21-digit KK-3 code. Although KK-3 is probably the most comprehensive coding scheme, it needs additional manual data input in the form of part production volume, lot size and the part feature dimensions that are required for generating process plans. Based on the part code, the family search module identifies the family to which the part belongs. The classification of parts into families is based on six vital features identified by six of the twenty-one digits of the part code. Thus, every part family has a six digit family code.

Each part family has a standard plan that is created and stored in the database. As mentioned earlier, this is contrary to the approach in [Hou and Wang, 1991] where more than one standard plan may exist for each part family. The part family standard plan is retrieved and the necessary modifications for sequence of operations are made by the process planner. At this time, the system interacts with various databases consisting of production capacity (current and avialable), machining parameters, machine capability and manufacturing cost to determine the process and the machine for each operation. This step is repeated until all operations for each part in the product mix are planned.

The plan combination can be analyzed with a simulation model the development of which will be reported in the subsequent paper.

Alternate plans for a part may differ from one another at each stage of operation in one or both of the following aspects:

a. a different machine, or

b. a different process for a specific requirement

The relative cost differential is determined assuming the first plan combination as a base. This is computed by retrieving marginal cost of each operation that is different in the revised plan.

A-CAPPS Modules

A-CAPPS is a variant approach based process planning system. The major concept used in designing A-CAPPS is to construct a generic framework for a modular system. The system interacts with capacity planning, machine loading and manufacturing cost databases and can be expanded easily. A partial list of its vital features include the following: (a) it has modular structure, (b) it allows for interactive part coding, (c) it interacts with production planning functions (capacity planning and machine loading), and (d) it interacts with order processing and costing functions.

Modular Structure

The process planning can be divided into the following sub tasks: (a) interactive part coding, (b) family search (part family identification and part family standard plan identification), (c) generation of standard plans, (d) editing standard plans, (e) process selection, (f) machine selection (candidate machine extraction and a machine selection), and (g) capacity search. In A-CAPPS, each of these listed subtasks is constructed as a software module.

Interactive Part Coding

A-CAPPS uses 21 digit KK-3 code as its input. In A-CAPPS, the coding module lets the user to code the part interactively.

Interface With Production Planning Functions

Several CAPP systems have been developed to date. These systems were designed to provide a vital link between CAD and CAM thus reducing manufacturing lead time. As mentioned earlier, their approach is static in nature and do not consider actual resource conditions in real time. As a result, the existing CAPP systems are not able to respond to the dynamics of manufacturing systems. In A-CAPPS, the process planning module interacts with other production related functions to respond to dynamic resource conditions.

A-CAPPS allows the user to interact with updated machine capacity database during the plan generation session from the capacity module. The user's request for the selection of a machine is approved if the machine has sufficient capacity to satisfy demand. In this way, A-CAPPS ensures that the process plan generated is feasible, and is based on the current shop floor conditions.

Generation of Alternative Process Plans

A-CAPPS displays the alternative processes and machines for each operation in a standard plan for a part family. It allows the user to select a specific process and a machine for each operation. The system displays alternate processes and machines alongwith their available capacities. Hence the planner can select a process and a particular machine based on the current shop floor conditions. This is the most important feature of the developed A-CAPPS, which is not available in any of the existing familiar CAPP systems. In the existing CAPP systems, the process plans generated for a part at different instants comprises the same processes and machines irrespective of the shop floor conditions and are therefore often infeasible. To circumvent this problem, A-CAPPS first presents various alternate processes in preferential order and then presents alternate machines that can perform the selected process for each step of production. Hence, the process planner can always generate a process plan which satisfies existing capacity constraints.

Operation Stages of A-CAPPS

A-CAPPS is a variant process planning system. A-CAPPS has two operation stages namely the preparatory stage and the production stage like any other variant process planning system. Hence, the developed system is described here in terms of the two operation stages.

Preparatory Stage

During the preparatory stage, the existing components are coded, classified, and subsequently grouped into families. A standard process plan for each part family is then created by the process planner and stored permanently in the standard plans database. Each processing step in a standard plan may have several feasible options for process selection. Furthermore, each of these alternate processes may be completed on a different machine. Therefore, an operation may use one of several alternate machines.

Production Stage

In the production stage, KK-3 code is first assigned to an incoming part. The code is then entered as an input to a part family searching routine to find its part family. The part family code is then used to retrieve its standard process plan. As mentioned earlier, a standard plan consists of all feasible options of processes and machines for each operation. The process planner selects a process and a machine for each operation. By using the part's detailed information provided as input and machining capability data from the database, the system computes the unit machining time for the operation. The machining capability database also contains recommended cutting parameters for the operations each machine can perform. The system then determines whether the selected machine has sufficient capacity to process the desired volume. If it does not have sufficient capacity, the system asks the user to select another machine. Once a machine with sufficient capacity is selected, then the system automatically updates the available capacity of that machine.

CONCLUSION

An architecture for the generation of alternate process plans is presented with a prototype A-CAPPS. A-CAPPS has demonstrated a definite advantage over existing CAPP systems in terms of its responsiveness to shop floor conditions. A-CAPPS can quickly determine alternate routes that are feasible and also satisfy relavant capacity constraints.

In CIM systems, frequent design and routing change requests are inevitable. These changes can be more easily accomodated in A-CAPPS. As mentioned earlier, A-CAPPS is based on variant approach. A similar project to design a generative CAPP system could be worthwhile keeping in view of the potential such an appoach has due to increased responsiveness to real time systems constraints.

REFERENCES

1. Kusiak, A. "Intelligent Manufacturing Systems", Publishers Prentice Hall, Editors W.J.Fabrycky & J.H. Mize, 1990.
2. Gupta, T. (1990). An expert system approach in process planning: Current development and its future. Computers & Industrial Engineering, 18(1), 69-80.
3. Wang, H. (1988). A layered architecture for manufacturing operation planning. Computers & Industrial Engineering, 14(2), 201-210.
4. Hou, T.H. and Wang, H.P.,"Integration of a CAPP System and an FMS", Computers in Industrial Engineering, Vol. 20, No. 2, pp. 231-242, 1991.

A METHODOLOGY FOR MANUFACTURING CELL DESIGN

LORACE L. MASSAY, Ph.D.
Department of Industrial Engineering, North Carolina A&T State University, Greensboro, NC 27411

COLIN O. BENJAMIN, Ph.D. and YILDIRIM OMURTAG, Ph.D.
Department of Engineering Management, University of Missouri-Rolla, Rolla, MO 65401

ABSTRACT

A systematic methodology for designing manufacturing cells is presented. Cell design is based on six important design axioms and proceeds in four phases viz. analysis, conceptual design, embodiment design, and detailed design. When applied to three cases, this approach yielded good results, facilitating the rapid generation and timely evaluation of design alternatives.

INTRODUCTION

Cellular Manufacturing (CM) is currently the focus of considerable interest from industry and academia [1]. This level of interest is related both to the manufacturing industry's fight for survival and to the emergence of advanced manufacturing technologies such as Computer-Integrated Manufacturing, and Flexible Manufacturing Cells and Systems. These new computer-based technologies often both support and mandate a cellular approach.

The results of a recent survey [2] show that while, labor and equipment utilization may be adversely affected, the benefits of CM far outweigh the disadvantages associated with this technique. A careful study of the survey results indicates the following. First, there were no widely accepted formal procedures for planning and designing manufacturing cells. Secondly, there was no strong evidence that adequate emphasis has been accorded to the a priori evaluation of cell performance. Finally, equipment and labor utilization clearly did not provide effective measures of performance.

Wu [3] has pointed out that the potential benefits to industrial efficiency offered by the new technologies will not be fully realized unless there is a corresponding investment in research into the design, evaluation, organization, and management aspects associated with manufacturing systems. Effective management of manufacturing operations is too complex to be measured or controlled by one performance measure. Instead a coordinated system of performance measurement must be devised so that change is facilitated and organizational goals are achieved.

In this paper a systematic methodology for the design of Manufacturing Cells is outlined. The methodology supports a concurrent engineering philosophy and is a tool for manufacturing engineers and managers, who, by the year 2000, may be required to assume the role of "Operations Integrators," and would be required to integrate Manufacturing Cells and Systems to form Computer Integrated Manufacturing Enterprises [4].

PROPOSED METHODOLOGY

The proposed cell design methodology emphasizes the axioms described in Table I. These guide the decisions required in designing manufacturing cells and involve virtually every aspect of manufacturing from part design through process planning, staffing, methods, and production planning and scheduling.

Table I DESIGN AXIOMS FOR CELL DESIGN

DESIGN AXIOMS	PROCEDURAL GUIDANCE DERIVED
(1) Top-down system definition followed by bottom-up system specification [5]	Overall system conceptualization followed by subsystems design and specification
(2) Simplification followed by integration [3][6]	Maintained by dividing the system into subsystems; by clearly identifying input, process, and output elements of each subsystem; by determining appropriate controls for monitoring achievement against predetermined standards; by integrating subsystems in a later design-integrating step.
(3) Alternative generation [7]	The system is broken into subsystems and more than one solution is sought for each part. By combining sub-solutions, a number of solutions (variants) for the system can be produced.
(4) Iterative refinement and progressive elaboration are essential [8]	Stepwise refinement and development
(5) Holistic evaluation [9]	System designs evaluated as a whole system
(6) Documentation and specification are inseparable [5]	Concurrent design documentation

The proposed methodology addresses the three fundamental aspects of systems: transformational, procedural, and structural [10], distinguishes between system logic and system structure [11], and utilizes the concepts of logical design and physical design [12]. Following the four-phased engineering design approach of Pahl and Beitz [7], the methodology is divided into four phases: (I) Analysis, (II) Conceptual design, (III) Embodiment design, and (IV) Detailed design. Figure 1 is an illustration of the proposed four phased approach for designing manufacturing cells.

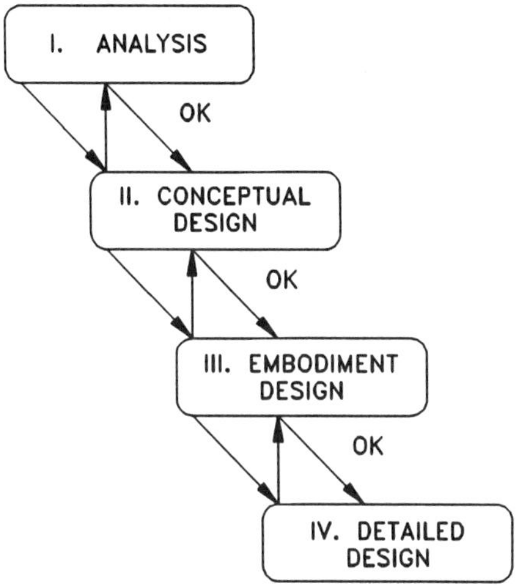

Figure 1 Proposed Four Phases of Manufacturing Cell Design

In the analysis phase, part-feature/process data are analyzed to identify part families or combinations of families that represent opportunities for exploiting CM. Many different techniques are available for identifying part families; Group Technology database analysis/process flow analysis [13], Production Flow Analysis [14]; mathematical programming methods [15],[16]; Cluster Analysis [17],[18],[19], and visual examination [20],[21].

In the conceptual design phase, "what" is to be designed is decided upon and conceptual modeling is necessary. The operations and sequence of operations required are visualized and the system description technique of Process Flow Charting is used.

Embodiment design is the phase in which an abstract concept is developed into a more concrete proposal, represented by layout drawings. In this phase, the conceptual system is divided into subsystems, solutions are sought for each subsystem, the subsystems and cells are reintegrated and adjusted, and then the whole system is evaluated. This phase comprises two stages as follows: Logical Cell Design and Physical Cell Design. Figure 2 is an outline of the proposed approach.

The purpose of the Detailed Design phase is to finalize all specifications and dimensional details of the selected "Best" design. The output from the activities in this phase will be design documentation consisting of detailed design drawings and specifications.

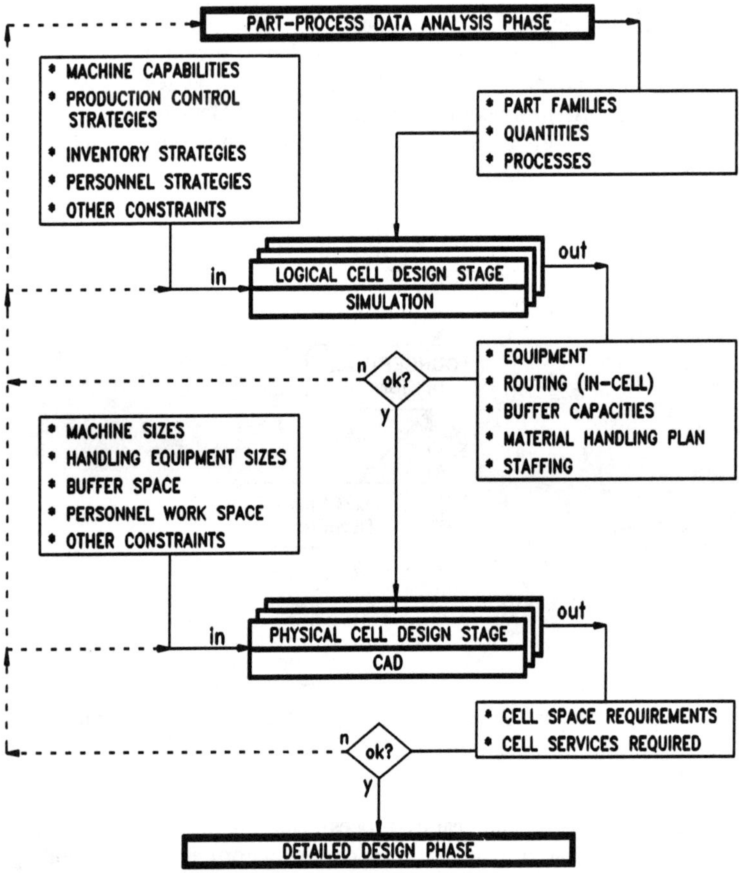

Figure 2 Two Stages of Embodiment Design Phase

APPLICATION OF METHODOLOGY

The methodology was applied to three cases which represented the range of problems and levels of complexity typically encountered by designers of manufacturing systems. Three cases were selected and classified according to (1) the area of application (academia, quasi-industry, and industry), and (2) level of system complexity (low, medium, and high). Figure 3 shows the classification of the cases included in the study.

Simulation modeling was used at the Logical Cell Design stage to develop efficient in-cell material flows. In the Physical Cell Design stage, CAD was used for drawing and arranging the machines and equipment within the cells. Workplace layout design principles were incorporated to produce effective workplace designs. In all cases, the methodology stimulated the generation of good design alternatives and facilitated their progressive refinement. However, the key benefit of the

SYSTEM COMPLEXITY	APPLICATION AREA: ACADEMIA	APPLICATION AREA: QUASI-INDUSTRY	APPLICATION AREA: INDUSTRY
High			Marathon Electric Plant
	UMR FMS		
Low		DemMaTec FCIM	

Figure 3 Classification of Cases included in Study

methodology lay in its systematic structuring of the activities that need to be considered at certain stages of the cell design process. This ensured that all design issues were attended to at the right time, and no important issues were ignored.

CONCLUSION

The methodology was applied to three cases with varying degrees of complexity and this provided preliminary confirmation of its effectiveness. Its application to more cases, with varying levels of complexity and different application areas, would further enhance its generalizability. This methodology focussed on the design of the physical production system (parts, machine tools, handling equipment). However, its capabilities should be extended to incorporate the formal information system (computer support, customer orders, work orders), and the human system (people, working environment, informal information system). Extension of the research effort in this direction would be a major contribution.

REFERENCES

1. Wemmerlov, U. and N.L. Hyer. "Research Issues In Cellular Manufacturing," *International Journal of Production Research*, 25(3),1987, 413-431.
2. Wemmerlov, U. and N.L. Hyer. "Cellular manufacturing in the U.S. industry: a survey of users," *International Journal of Production Research*, 27(9),1989, 1511-1530.
3. Wu, B. *Manufacturing systems design and analysis*. New York, NY: Van Nostrand Reinhold, 1992.
4. Koska, D.K. and J.D. Romano. "*Countdown to the Future: The Manufacturing Engineer in the 21st Century*." A.T. Kerney, Inc.,1988.
5. Nance, R.E. and J.A. Arthur. "The methodology roles in the realization of a model development environment," *Proceedings of the 1988 Winter Simulation Conference*,

(1988), 220-225.

6. Harmon, R.L. *Reinventing the factory: productivity breakthroughs in manufacturing today*. New York, NY: Free Press, 1989.
7. Pahl, G. and W. Beitz. *Engineering Design: a systematic approach*. New York, NY: Springer-Verlag, 1988.
8. Takeda, H., T. Tomiyama, and H. Yoshikawa. "Logical Formalization of Design Processes for Intelligent CAD Systems," *Intelligent CAD, II*. New York, NY: Elsevier, 1990, 325-336.
9. Black, J.T. "The Design of Manufacturing Cells (Step one to Integrated Manufacturing Systems)," *Proceedings of Manufacturing International '88*, 1988, 143-157.
10. Hitomi, K. *Manufacturing systems engineering*. Bristol, PA: Taylor and Francis, 1979.
11. Nof, S.Y. "On the structure and logic of typical material flow systems," *International Journal of Production Research*, 20(5),1982, 575-590.
12. Montreuil, B. and S.Y. Nof. "Approaches for logical vs. physical design of intelligent production facilities," *Manufacturing Research and Technology 6: Recent Developments in Production Research*. New York, NY: Elsevier, (1988), 352-360.
13. Nolen, J. *Course notes: Effective Manufacturing Cells*. SME Special Course, St. Paul, Minneapolis, MN, 1991.
14. Burbidge, J.L. *Production Flow Analysis For Planning Group Technology*. Oxford: Oxford University Press, 1989.
15. Choobineh, F. "A framework for the design of cellular manufacturing systems," *International Journal of Production Research*, 26(7),1988, 1161-1172.
16. Co, H.C. and A. Araar. "Configuring cellular manufacturing systems," *International Journal of Production Research*, 26(9),1988, 1511-1522.
17. McAuley, J. "Machine Grouping for Efficient Production," *The Production Engineer*, 21(2),1972, 53-57.
18. King, J.R. and V. Nakornchai. "Machine-component group formation in group technology: review and extension," *International Journal of Production Research*, 20(2),1982, 117-133.
19. Chan, H.M. and D.A. Milner. "Direct Clustering Algorithm for Group Formation in Cellular Manufacture," *Journal of Manufacturing Systems*, 1(1),1982, 65-74.
20. Holtz, R.D. "GT and CAPP Cut Work-in-Process Time 80%," *Assembly Engineering*, (June, 1978), 24-27.
21. Ham, I., K. Hitomi, and T. Yoshida. *Group Technology*. Kluer/Nijhoff, 1984, 9-13.

MEASURING THE RELIABILITY OF DESIGN BASED ON INFORMATION CONTENT IN A CONCURRENT ENGINEERING ENVIRONMENT

Ali Bahrami

Computer Information Systems
Department of Economics and Management
Rhode Island College
Providence, Rhode Island 02908

ABSTRACT

The concept of information content is based on the idea that a designer must design for the target values and seek to reduce the manufacturing variation around the targets rather than starting out with targets and then adding tolerance limits to facilitate manufacturing. The mindset that all products made within the tolerance limits are equally good, is clearly wrong. The use of tolerance rather than targets results in operating over the entire tolerance range, leading to an unreliable product. This paper describes a method for evaluating and validating a design solution among the candidate designs by utilizing fuzzy information content.

INTRODUCTION

The domain of reliability engineering involves considerations of all aspects of design, development, and manufacturing in order to minimize the chance of equipment breakdown. Reliability in a qualitative sense can mean a host of different things relating to the confidence in quality of the equipment, and is closely connected but often confused with the concepts of maintainability, availability, safety and even security of the system.

Information content in the product design is a measure of probability of success of achieving the required needs and wants of the end user. The total information content is achieved by summing up the information content of each of the individual components. The success probability of the individual functional requirements is regulated by how well designer specified functional requirements are satisfied by the design. The information content is simply the logarithm of the range of variable divided by its tolerance. The main difficulty in utilizing the information content is in the ambiguity and uncertainty involved in defining the design range or the system range in terms of given physical parameters. As a result, we may not be able to utilize information content as a means for design evaluation. Further, the fuzziness may result from imprecise and incomplete or subjective knowledge of an artifact to be designed. Being able to mathematically represent the phrase "*good design*," greatly simplifies the design evaluation process.

Information content and information axiom developed by Suh that defined the best design (the most reliable design) is a design that has the minimum amount of information content. The concept of information minimization has similarities to information theory, entropy and the principle of maximum entropy. The notion of information is very closely related to the probability of achieving a highly reliable product.

$$Information = I = \log_2\left(\frac{1}{p}\right) \tag{1}$$

Where p is the probability of success (in terms of producing a highly reliable product). The information content associated with the product may be thought as:

$$I = \log_2\left(\frac{range}{tolerance}\right) \qquad (2)$$

Where the ratio of (tolerance / range) defines the probability of success. For instance, if a designer specifies a very tight tolerance that cannot be produced by the manufacturing system, it would take an infinite amount of information content. Therefore in design and manufacturing, the information content is closely related to the designer's specifications and the capabilities of manufacturing system.

FUZZY LOGIC APPROACH FOR COMPUTING INFORMATION CONTENT

Nakazawa [Nakazawa87] has applied the information content measure to select the best subcompact car from the set of cars sold commercially in Japan. He has developed the satisfaction function in order to deal with the ambiguity and the uncertainty involved in defining the design range or the system range in terms of a given physical parameter.

Here, the satisfaction function has been modified by utilizing a fuzzy logic approach for computing the degree of satisfaction one may get from a design. These degrees of satisfaction are then mapped to the design range and the system range in order to compute the information content among the competing alternatives.

To compute the system range and design range, three linguistic variables, *Satisfaction* (*S*), *the Upper limit of Satisfaction* (*US*) and *the Lower limit of Satisfaction* (*LS*), must be formulated. Here, three linguistic variables (*S, LS* and *US*) are defined for technical degree of difficulty and three linguistic variables are defined for the importance rate for each class of subcomponent. See Figures 1 and 2. The designer must then enter the values for technical degree of difficulty and degree of importance for the product under design. These values are numbers between 1 and 10. The technical degree of difficulty is translated to the system range and the degree of importance is mapped into the design range. The shapes of these linguistic variables *S*, *LS* and *US* can be different (for different components) and it depends on the designer's experience and insight. In most cases triangular and trapezoidal shapes help capture the characteristics of these linguistic fuzzy sets and simplify the computations [Kosko92]. Figure 1 represents the system

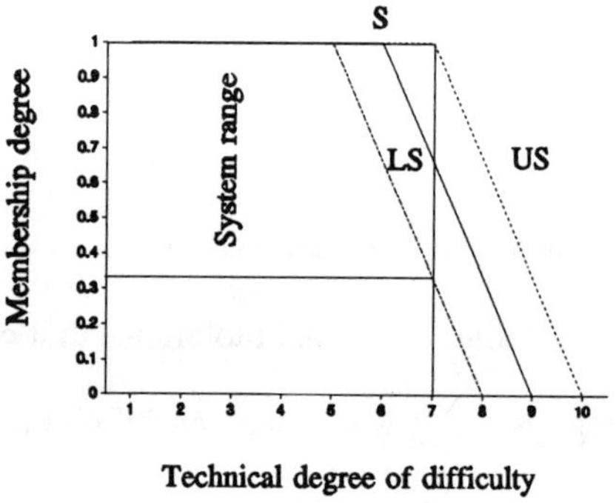

Figure 1. System range for the backrest of chair A based on the linguistic variables (LS and US) and the technical degree of difficulty.

range for a given design. Figure 2 shows the design range derivation based on the degree of importance and the fuzzy sets LS and US. According to Figure 2, the level of satisfaction can vary at a given degree of technical difficulty. The degree of satisfaction will start to decline as the parts become more difficult to manufacture. In the same fashion, the degree of satisfaction increases as the parts become more important to the customer (see Figure 2).

After mapping the degree of importance to the design range and technical degree of importance to the system range, the common range can be calculated based on the equation (2). The design with the minimum amount of information content must be selected as the best design.

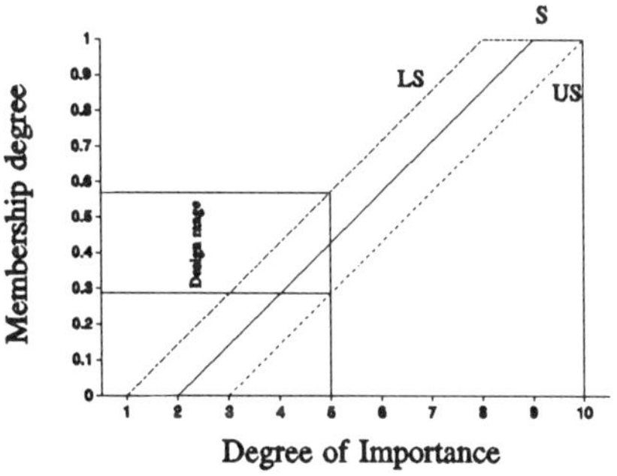

Figure 2. Design range for the backrest of chair A based on the linguistic variables (LS and US) and the degree of importance.

SUMMARY AND CONCLUSION

A proposed method for evaluating and validating design by utilizing information content and the theory of fuzzy logic has been introduced. This approach can provide us with a powerful tool for comparing various design alternatives and selecting the best design based on the information content in a concurrent engineering environment.

REFERENCES

[Kosko92] Kosko B., 1992, "Neural Networks And Fuzzy Systems, A Dynamical Systems Approach to Machine Intelligence," Prentice Hall, Englewood Cliffs, NJ.

[Nakazawa84] Nakazawa, H., and Suh, N.P., 1984, "Process Planning Based on Information Concept," *Robotics and Computer-Integrated Manufacturing*, 1(1):115-123.

[Nakazawa87] Nakazawa, H., 1987, "Information Integration Method," Corona, Tokyo.

[Suh78] Suh, N. P., Bell, A. C., and Gossard, D. C., 1978, "On an Axiomatic Approach to Manufacturing Systems," *Journal of Engineering for Industry, Transactions of ASME*, 100(2), 127-130.

[Suh84], Suh, N.P., 1984, "Development of the Science Base for the Manufacturing Field Through the Axiomatic Approach," Robotics and Computer Integrated Manufacturing 1(3/4), 399-455.

[Suh90] Suh, N, P., 1990, "The Principles of Design," Oxford University Press, New York, Oxford.

[Tyng84] Tyng, A., 1984, "Beginnings," John Wiley & Son, New York.

[Wilson80] Wilson, D.R., "An Exploratory Study of Complexity in Axiomatic Design," Ph.D. Thesis, MIT, August, 1980.

[Zadeh75a] L. A. Zadeh, "The Concept of a Linguistic Variable and Its Application to Approximate Reasoning-I," *Information Science 8*, pp. 199-249 (1975).

[Zadeh75b] L. A. Zadeh, "The Concept of a Linguistic Variable and Its Application to Approximate Reasoning-II," *Information Science 8*, pp. 301-357 (1975)

DETECTING CRITICAL FACTORS IN A FLOW LINE

F.ARCHETTI, A.SCIOMACHEN, F.STELLA

Università degli Studi di Milano,
Dip. Scienze della Informazione, Milan, Italy

G.DE MATHIA

Italtel-TLM,
Santa Maria Capua Vetere, Caserta, Italy

ABSTRACT

In this paper we are concerned with a flow line in which a conveyor is the only shared resource among serial machines. The conveyor moves capacitated pallets through the line. The aim of this work is to detect critical factors in the line by analyzing the flow of pallets and evaluating the influence of accidents of any machine on the throughput of the whole line. In particular, we first analyze the performance indices of the line by simulating its Petri Net (PN) model and subsequently we improve such indices by an optimization process. Computational results related to different weekly production plans are reported.

1.INTRODUCTION

PNs are already considered as one of the most powerful tools for modelling and analyzing complex systems, such as manufacturing and communication ones (see e.g., [6]). In this paper we propose a PN model of a flow line consisting of 10 serial machines, linked together by a conveyor that delivers pallets to the line. Starting from the PN model, we analyze the performance of each single machine in order to evaluate the influence of accidents on the throughput of the whole line. In particular, we perform a sensitivity analysis [2] for detecting the most critical machines on the basis of the rate of accidents' occurrence, given as simulation parameter. We then apply an optimization process for improving the performance indices of the line.

The purpose of the PN model optimization is to obtain the best values of some established performance indices which depend on some parameters. In the modelling of the flow line we have considered the throughput as the main performance index. However, note that the same analysis technique can be applied to the case of average waiting time of a pallet for being processed and average queue lenghts, as well. In particular, our problem consists of determining which value associated with some timed transitions of the PN is optimal for the line. It is worth remarking that we are concerned only with the possibility of accidents and not with the aspects related to macroscopic failures in the line.

The description of the telecom line is given in Section 2, while its PN model is presented in Section 3. Section 4 reports on the computational results obtained by simulating directly the PN and in the subsequent optimization process.

2. Description of the Line

Our case-study telecom line consists of 10 serial machines, linked together by a conveyor that moves pallets from a machine to the next one in order to assembly different kinds of electronic components. Each pallet has 40 locations. Any link of the conveyor can handle at most 3 pallets at a time. When a pallet begins a work cycle all its locations are empty. Following the layout of the Telecom Line depicted in *Figure1*, the pallet goes to the first machine, where a robot arm inserts some electronic components, one at a time, in the corresponding locations. When there is not any location available in the pallet, i.e., when 40 electronic components have been inserted, the pallet goes to the next machine and a new pallet arrives at the first machine. The robot arm at the second machine puts plastic components over the electronic ones in each location of the pallet. If this operation does not fail, the pallet goes to the next machine, where a robot arm inserts a spring in each location; otherwise, the robot controller pauses it. The pallet moves from the fourth assembling machine to the fifth one only if the assembly is performed without any damaging. The robot arms at the next four machines put different kinds of components in each pallet location. All such components are then assembled by the robot arm at the ninth machine. Finally, the robot arm at the last machine makes empty the 40 locations of the pallet, placing all the components into another pallet located outside the line. Once a pallet has been processed by the last machine it is moved through a conveyor to the first machine for a new work cycle. Note that after the completion of the corresponding operation the pallet is moved from machine M_i, $i=2,3,\ldots,10$, only if there is space available on the conveyor, and a new pallet, picked up from the conveyor linking M_{i-1} to M_i, arrives at the machine; otherwise, the pallet waits for being moved, thus causing a delay in the assembly process.

While a robot arm is processing a pallet some kind of accident can occur. The frequency of an accident of any machine is given by $F_i=1/MTBA_i$ $i=1,2,\ldots,10$, where $MTBA_i$ denotes the Mean Time Between Accidents of M_i. ($MTTR_i$ is the related Mean Time To Repair). The data related to each machine are reported in *Table1*, where Γ_i is the time (in seconds) required to process a pallet at machine M_i when no accident occurs. According to our weekly production data, the handling time Λ_i from M_{i-1} to M_i, $i=2,3,\ldots,10$ is fixed to 5 seconds.

The main concern here is to evaluate the influence of the accidents' rate of occurence on the performance of the line by deriving the average traversal time, i.e. the time required for a pallet to go through the line from M_1 to M_{10}. Let Ψ_S denote the expected traversal time. In [1] the authors present an analytical expression for computing such an index as a function of the values Γ_i, F_i and $MTTR_i$, in a PN model of a transfer line by exploiting the isomorphism between Markov chains and Stochastic PNs [5]; in our present case the capacity constraint on the handling system does not allow an easy characterization of Ψ_S. Moreover, the number of states required for capturing all the possible configuration of the system would result in a too large Markov chain. For this

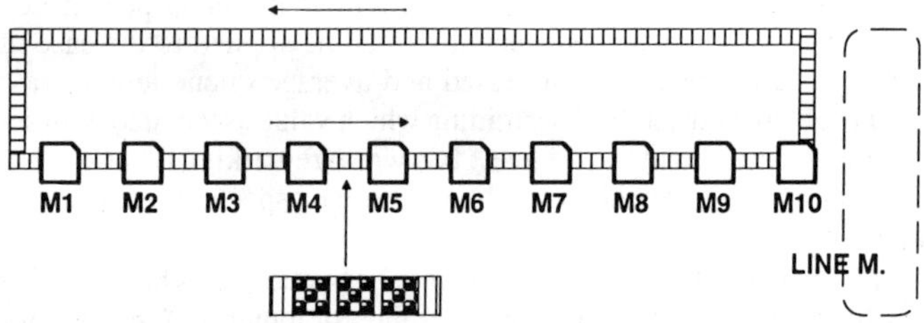

Figure1. Layout of the Telecom Line

Machine	Γ	MTBA	MTTR
M1	154.8	304.129	73.759
M2	123	473.585	106.325
M3	141.6	720.590	107.594
M4	157.8	1993.787	241.682
M5	145.2	332.010	58.885
M6	130.2	333.918	78.330
M7	121.2	9650.728	667.059
M8	168	803.580	99.647
M9	170.4	489.728	75.414
M10	171.6	1283.350	164.088

Table I. Machine Data Description

reason, a direct simulation of the system seems to be the most appropriate way for computing the value Ψ_S. Let hence Ψ_S denote our average simulated traversal time as the approximation to Ψ_S; let also Ψ_T denote the theoretical traversal time, i.e. the traversal time under the assumption that only one pallet is present and no accident occurs while flowing through the line. Ψ_T is hence given by:

$$\Psi_T = \sum_{i=1}^{10} \Gamma_i + \sum_{i=2}^{10} \Lambda_i \qquad (1)$$

An interesting performance index for evaluating the reliability of the flow line is given by the following ratio:

$$\eta = \frac{\Psi_S - \Psi_T}{\Psi_T} \qquad (2)$$

where Ψ_T is as in (1). Note that η is equal to 0 only if $\Psi_T = \Psi_S$, i.e. no accidents occur.

3. Model Generation

Let us generate the PN model of the line, focusing our attention on a single machine and its outgoing link of the conveyor which delivers pallets to the next machine.

The first step in modelling a machine by a PN is to individuate its physical components and logical conditions. Any machine is characterized by four logical conditions: "*idle*" (it is waiting for a pallet to be processed), "*busy* " (it is processing a pallet), "*down* " (an accident has occured) and "*available* " (it is working correctly).

Note that during a work cycle each machine can be in any of the following states: **"available"** AND **"idle"**, **"available"** AND **"busy"** and **"down"** AND **"busy".**

The activities, represented in the PN model of a machine by timed transitions, are related to the pallet processing, pallet handling and the occurence of accidents.

The PN model of a machine is presented in *Figure2* . Places and transitions have the following meaning: *p1*=machine "*idle*", *p2*=input buffer, *p3*=machine"*available*", *p4*=machine "*down*", *p5*=machine "*busy*", *p6*=a pallet processing is completed, *p7*=conveyor (capacity 3), *t1*=*MTBA* (exp, λ=0.003), *t2*=*MTTR* (exp,λ=0.014), *t3*=start of a pallet processing , *t4*=processing time (det, T=154.8), *t5*=a pallet enters the conveyor, *t6*=handling time (det, T=5).

In order to model correctly the beginning of a pallet processing, an immediate transition synchronizes the availability of the machine with the pallet queueing in its input buffer. Another immediate transition is required in order to characterize the so

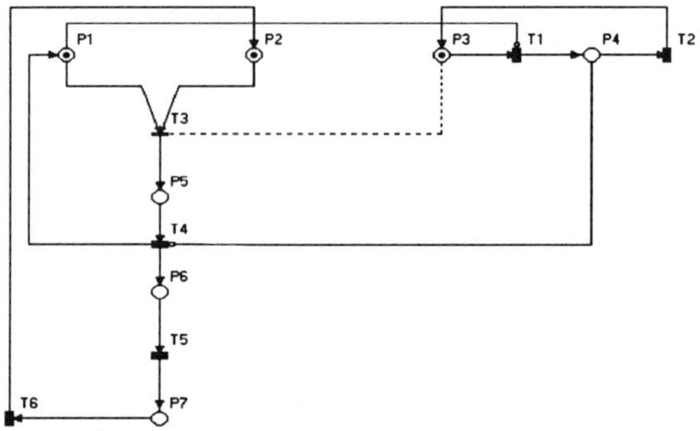

Figure2. PN model of a machine with its initial marking

called "*blocked*" state of the machine, where a pallet has been processed but it can not be moved to the next machine; this transition is enabled only if there are less than 3 tokens in the place representing the conveyor.

In the PN of *Figure2* , the inhibitor arc from *p1* to *t1* ensures that an accident can occur only while the machine is processing a pallet (i.e. no accidents occurs when M_i is idle). When an accident occurs the inhibitor arc from *p4* to *t4* disables the ongoing processing until the accident is repaired. The dashed line between *p3* and *t3* represents a loop (two directed arcs) that allows the machine to be "*available*" also when *t3* fires.

Let us now follow the evolution of this PN model, considering as initial marking **M0**=*(1,1,1,0,0,0,0)*, where *Mi* is "idle" and "available" and a pallet is in its input buffer ready to be processed. In **M0** only transition *t3*, corresponding to the starting of the pallet processing, is enabled. The firing of *t3* results in the new marking **M1**=*(0,0,1,0,1,0,0)*. In **M1** (*Mi* is processing a pallet) two events may occur: either the pallet processing terminates correctly or an accident occurs. In the first case transition *t4* fires and the new marking is **M2**=*(1,0,1,0,0,1,0)*, while if transition *t1* fires the new marking is **M3**=*(0,0,0,1,1,0,0)*. In **M2** only the immediate transition *t5* can fire; this implies that a pallet is moved from *Mi* to the conveyor between *Mi* and *Mi+1*. The firing of *t5* moves the PN in **M4**=*(1,0,1,0,0,0,1)*. From **M3** only the recovering of the accident (firing of transition *t2*) can occur, moving the net back to **M2**. Finally, from **M4** when *t6* fires the PN is newly into the initial marking **M0**.

Note that **M0** and **M4** enable only immediate transitions; in our analysis we consider only the markings **M1** (*Mi* "*busy*"), **M2** (*Mi* "*idle*") and **M3** (*Mi* "*down*"), whose probability strongly affects the performance of the line.

Starting from the above considerations the whole telecom line consisting of 10 machine can be easily modelled duplicating the net of *Figure2* and connecting the model of machine *Mi* to the model of machine *Mi+1*. Note that in this case the arc from *t6* to *p2* must be replaced by an arc from *t6* to the place corresponding to the input buffer of the next machine. The PN model of the whole flow line has 70 places and 60 transitions.

4. Simulation and Optimization Process

The first step in determining the reliability index, as defined in (2), is to compute the value Ψ_T. According to (1), and considering the processing time of each machine reported in TableI, we have Ψ_T = 1'528.8 seconds. As a second step, the average simulated traversal time has to be computed. Starting from the PN model and the specification of the initial operating conditions of the line, a direct simulation of the net can be performed. In our case, the input data are related to weekly production plans, six days per week. Three work cycles are scheduled every day, and each work cycle takes about *20'200* seconds. At the beginning of the week 28 pallets are located on the conveyor from machine *M10* to machine *M1* so that they are available for the first operation at *M1*. No other pallets are available. A pallets takes *120* seconds to go back to *M1* from *M10*.

Two software environments have been used for simulating the net, at a different stage of analysis. The first one, called *MEGAS* [8], is a software environment tailored for modelling and analyzing PN models of manufacturing systems. It provides some information related to the places of the net, such as the probability of any machine to be "*down*", "*idle*", "*busy*" and "*blocked*" and the average number of pallets waiting for being processed (average queue lenghts). The second software environment, called *NETOPT* [4] (NET OPTimization), is a stochastic optimization program for Discrete Event Dynamic Systems. In both cases, independent runs having cycle lenght of *20'200* seconds (i.e., a work cycle) have been performed, obtaining Ψ_S = 1'875.19.

Once the values of Ψ_T and Ψ_S are known, it is possible to compute the value of η as it is shown in (2). In this paper we consider the value $\underline{\eta} = (\Psi_S - \Psi_T)/\Psi_T$ as an approximation to η. For the flow line under consideration η = 0.227.

In order to verify whether an improvement on the reliability of the line can be achieved and to detect the influence of the control parameters on its main performance indices, a sensitivity analysis can be then performed. In our case, the main performance index is the throughput, while the control parameters are the $MTBA_i$ and the $MTTR_i$, i=1,...,10.

In fact, looking at the literature in this area (e.g. [3,7]), a sensitivity analysis is particularly suitable for establishing the most critical factors in improving the performance of a system. Let us hence denote by "x" the value of the control parameter under evaluation, by $F(x)$ the value of the performance index, by Δx any variation of the control parameter and by $\Delta F(x)$ the corresponding variation of the performance index. Then, the value of $\Delta F(x)$ is given by:

$$\Delta F(x) = F_X(x) \bullet \Delta x \qquad (3)$$

where $F_X(x)$ represents the sensitivity (gradient) of the performance index with respect to the controllable parameters.

Our sensitivity analysis is performed by using the software package *NETOPT* which gives an estimation of the sensitivity $F_X(x)$ together with an estimation of the performance index $F(x)$. Let us denote by S_{MTBA} and by S_{MTTR} the estimated sensitivity of the throughput on the *MTBA* and the *MTTR* of every machine, respectively. Our estimation related to the S_{MTBA} values are reported in *TableII*, while *TableIII* reports on the estimation of the S_{MTBA} values. The sensitivity estimation has

Machine	S_{MTBA}	δ_S	0.9
M1	**0.00171**	0.00017	[0.00144,0.00199]
M2	0.00046	0.00009	[0.00030,0.00061]
M3	-0.00005	0.00006	[-0.00144,0.00005]
M4	0.00005	0.00002	[0.00002,0.00008]
M5	**0.00184**	0.00014	[0.00161,0.00207]
M6	**0.00214**	0.00014	[0.00192,0.00236]
M7	-0.00001	0.00000	[-0.00002,0.00000]
M8	0.00076	0.00005	[0.00067,0.00084]
M9	**0.00280**	0.00009	[0.00265,0.00295]

TableII. Sensitivity Estimation, Std. Dev. and Confidence Interval for *MTBA*

Machine	S_{MTTRA}	δ_S	0.9
M1	-0.00086	0.00041	[-0.00154,-0.00018]
M2	-0.00085	0.00019	[-0.00116,-0.00054]
M3	-0.00123	0.00016	[-0.00149,-0.00097]
M4	-0.00082	0.00004	[-0.00089,-0.00075]
M5	**-0.00450**	0.00045	[-0.00524,-0.00376]
M6	**-0.00501**	0.00031	[-0.00552,-0.00450]
M7	-0.00022	0.00000	[-0.00022,-0.00021]
M8	**-0.00399**	0.00017	[-0.00428,-0.00371]
M9	**-0.00940**	0.00029	[-0.00988,-0.00891]

TableIII. Sensitivity Estimation, Std. Dev. and Confidence Interval for *MTTR*

Machine	*MTBA*	δp(MTBA)	*MTTR*	δp(*MTTR*)
M1	304.129	0.31	73.759	<0.1
M2	473.585	0.13	106.325	<0.1
M3	720.59	<0.1	107.594	<0.1
M4	1993.787	<0.1	241.682	0.12
M5	**332.01**	**0.37**	**58.885**	**0.16**
M6	**333.918**	**0.43**	**78.330**	**0.24**
M7	9650.728	<0.1	667.059	<0.1
M8	**803.58**	**0.36**	**99.647**	**0.24**
M9	**489.728**	**0.82**	**75.414**	**0.43**
M10	**1283.35**	**0.35**	**164.088**	**0.30**

TableIV. Average increase on the number of produced pallets

been evaluated by performing a stochastic simulation run taking *3'600* seconds. As a result, the value of the throughput is $F(x)=11.95$.

In these tables the *standard deviation*, denoted by δs, and the *0.9 conficence interval*, for evaluating the accuracy of the reported estimations, are also given. The resulting most critical machines are outlined in bold letters; in particular, we can assert that the rate of accidents' occurence at machines *M5*, *M6*, *M9* and *M10* strongly affects the throughput of the whole line. As a further analysis of the performance of the line it is possible to compute the average increase of the throughput with respect to a possible improvement of either the $MTBA_i$ or the $MTTRi$, $i=1,..., 10$. The corresponding computational results are reported in *TableIV*, where δp(*MTBA*) and δp(*MTTR*) denote the improvement of the throughput with respect to a *MTBA* decrease and to a *MTTR* decrease, respectively. These values are related to a stochastic simulation taking *21'600* seconds and to either a *10%* increase of the *MTBA* values or to a *10%* decrease of the *MTTR* ones. Considering the above table it is possible to note that in this case the most critical machine are *M5*, *M6*, *M8*, *M9*, *M10*.

References

1.F.Archetti, E.Fagiuoli, A.Sciomachen, "Computation of the Makespan in a Transfer Line with station breakdowns using Stochastic Petri Nets", Computers & Operations Research Vol. 14 N.5, pp. 409-414, 1987.

2.F.Archetti, A.Gaivoronski, A.Sciomachen, "Sensitivity Analysis and Optimization of Stochastic Petri Nets", Autofaber Research Report 7/90, 1990.

3.Ermoliev, Yu and Wets, R.J.-B, "Numerical Techniques for Stochastic Optimization", Springer-Verlag, Berlin, 1988.

4.A.Gaivoronski, "Interactive Program SQG-PC for solving Stochastic Programming problems on IBM PC/XT/AT Compatibles. User Guide", WP-88-11, IIASA, Laxenburg, 1988.

5. M.K. Molloy, "Performance Analysis using Stochastic Petri NEts", IEEE Trans. Comput., C-31, pp.913-917, 1982.

6.T.Murata, "Petri Nets: properties, analysis and applications", Proc.IEEE Vol.77, N.4, pp. 541-580, 1989.

7.G.Ch.Pflug, "Optimization of simulated discrete event processes", Preprint TR-ISI/Stamcom 87, University of Vienna, 1990.

8.A.Sciomachen, "A Software environment for modelling and simulation of manufacturing systems", Proc. Fourth Int. Conf. on CAD, CAM, Robotics and Factory of the Future, TATA Mc. Graw Hill, Vol.2, 00.690-702, 1989.

Applying Interactive Control to Waste Processing Operations

Erna L. Grasz and Roy D. Merrill
Lawrence Livermore National Laboratory

Scott A. Couture
Lawrence Livermore National Laboratory

ABSTRACT

At present waste and residue processing includes steps that require human interaction. The risk of exposure to unknown hazardous materials and the potential for radiation contamination motivates the desire to remove operators from these processes. Technologies that facilitate this include glove box robotics, modular systems for remote and automated servicing, and interactive controls that minimize human intervention. LLNL is developing an automated system which is designed to supplant the operator for glove box tasks, thus protecting the operator from the risk of radiation exposure and minimizing operator-associated waste. Although most of the processing can be automated with minimal human interaction , there are some tasks where intelligent intervention is both desirable and necessary to adapt to unexpected circumstances and events. These activities require that the operator interact with the process using a remote manipulator which provides or reflects a natural feel to the operator. The remote manipulation system which was developed incorporates sensor fusion and interactive control, and provides the operator with an effective means of controlling the robot in a potentially unknown environment. This paper describes recent accomplishments in technology development and integration, and outlines the future goals of Lawrence Livermore National Laboratory for achieving this integrated interactive control capability.

INTRODUCTION

The Advanced Processing Technology (APT) Robotics and Automation Program at Lawrence Livermore National Laboratory is an outgrowth of the automation activities of the Special Isotope Separation (SIS) program. This program is an example of DOE's commitment to the application of previously developed technologies to a broad range of programmatic needs. The automated systems which were designed and developed for servicing SIS equipment in a glove box environment are recognized as state-of-the-art for remote manufacturing of plutonium and other hazardous materials processed in glove boxes.

LLNL was chosen as the lead laboratory for the Robotics Waste Processing Operations Program by the DOE Office of Technology Development (OTD) because of LLNL's accomplishments in glove box automation. The OTD Robotics Waste Processing Operations application area is one of seven focus areas in the overall OTD Robotics Technology Development Program under the DOE Office of Environmental Restoration and Waste Management (EM). The emphasis of these

EM sponsored robotics activities is to provide practical solutions that allow remediation to be accomplished faster, safer, better and cheaper than using manual methods.

As lead lab, LLNL has responsibility for coordinating efforts across the complex aimed at using automation technologies to process waste materials following retrieval activities. The primary tools include glove box robotics technology, modular systems for remote and automated servicing, interactive controls technology (remote control of processes via telepresence and teleoperation) , and advanced material handling and transfer technologies. These technologies will be integrated and applied to waste and residue processing in support of the future DOE ER/WM waste processing facilities.

APPROACH

Present automation of glove box operations addresses a class of processing tasks that are both deterministic and within the manipulation capabilities of a robot arm. There is a much larger class of tasks that are more complex involving object manipulation, performing in a cluttered environment, requiring sensory input for control, and demanding error recovery . The performance of these tasks in a hazardous environment is normally relegated to teleoperators with human-in-the-loop controllers. The amount of sophistication required to handle most of these tasks is far beyond the capabilities of the most advanced robotics research controllers. To address these needs, teleoperation and telepresence technologies are being developed and integrated in the Interactive Controls Lab for incorporation into the glove box robot's suite of tools. By integrating autonomous and teleoperation capabilities into a robot control system, operators will be able to switch seamlessly between both modes of operation while performing maintenance activities, error recovery operations, and process intervention. The alternative is for an operator to suit up and enter the hazardous environment, and manually perform the tasks.

ACCOMPLISHMENTS

Past glove box robotics activities have focused on making improvements to the IBM Electric Cantilevered Robot (ECR). The IBM ECR has been used extensively within IBM and in industry for high speed assembly applications. Because of its size and load capability and its excellent reliability record, a joint effort between LLNL, IBM, SAIC, and other participants adapted it for use as an integral part of a plutonium glove box automation effort. Now dubbed the IBM Glove Box Robot Gantry (GR-G), the first installation took place in February 1991 in the Automated Testbed funded by the EM OTD Robotics Waste Minimization Program.(See Figure 1)

The second robotics testbed, installed in February 1992 and known as the Process Development Testbed, is the primary vehicle for incorporating advanced technologies developed in the lab into production feasibility demonstrations. (See Figure 2) A state-of-the-art controller is being developed jointly by IBM and LLNL to replace the original IBM ECR robot controller currently in use in both testbeds. The original controller, based on the IBM Series I computer, has neither the computing power nor advanced software features needed for real-time sensor-based trajectory control, improved operator interfaces including force-reflecting hand controllers, or model-based path planning. The Series II controller, coupled with a basic telerobotic capability, will be installed and operational late in summer 92. The hardware architecture for the GR-G Series II controller is seen in Figure 3.

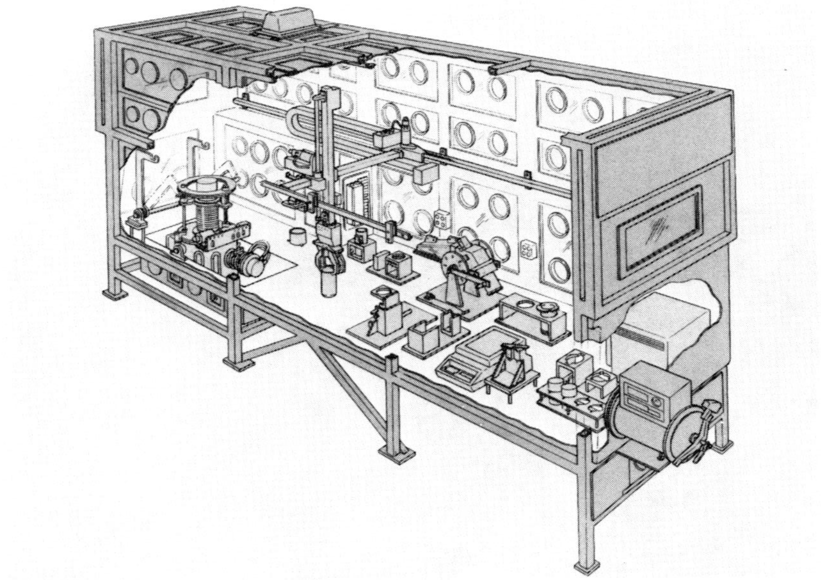

Figure 1. Automated Testbed with robot and furnace in cutaway view.

Figure 2. The Process Development Testbed that will incorporate the Series II controller.

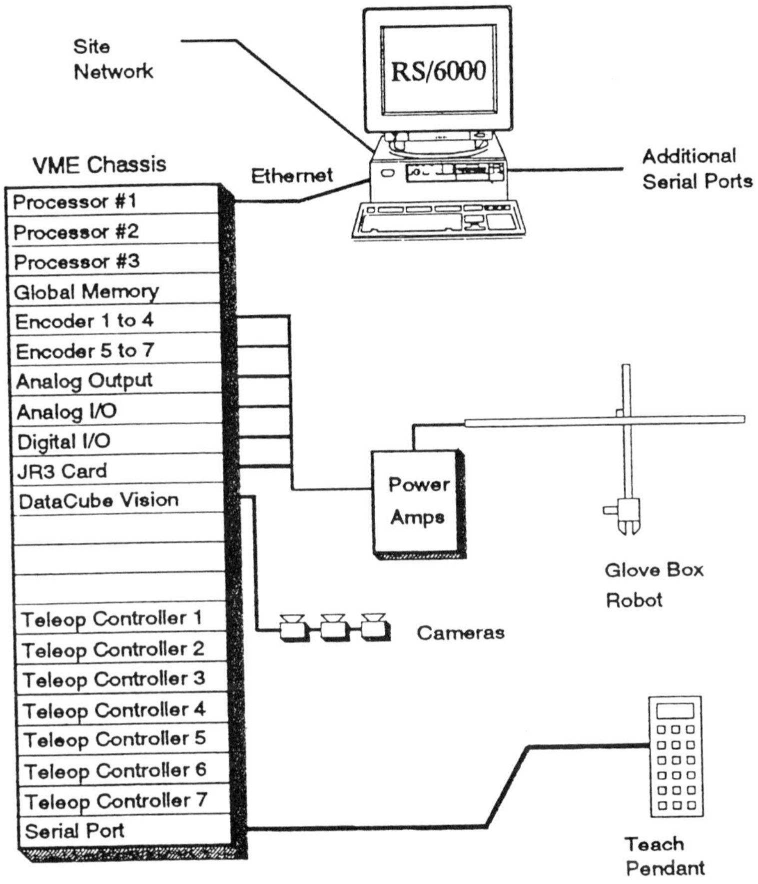

Figure 3. Hardware architecture for BGR Series II controller.

During the last year, the Interactive Controls Laboratory has successfully demonstrated an interactive telerobotic capability coupled with sensor fusion in two major demonstrations. The first is remote control of the button-breakout process in a glove box manufacturing environment. In this application, teleoperation provided an attractive alternative to manual process intervention when variations in product quality required handling in excess of the capability of the automated systems' pre-programmed response.

The second demonstration for an automated waste characterization and segregation application was performed using object recognition for unknown materials and objects. Automated path planning and seamless transfer to manual control was successfully employed here also. A sensor -based capability was developed and integrated with the autonomous and telerobotic control to evaluate and discriminate between possibly contaminated materials and waste representative of what is stored in barrels across the DOE complex. Identification and segregation was based on radiation signatures (types and levels), metallic sensing feedback, and machine vision algorithms. (Figure 4). Using this information, sample waste stream items were autonomously sorted with robot trajectories generated automatically using sensor data. This capability, when fully developed, will allow automated sorting of waste stream items based on quantifiable physical properties, thus reducing human handling and secondary/tertiary waste generation.

INTEGRATION OF INTERACTIVE CONTROL TO GLOVE BOX ROBOT

The key objectives for the interactive control effort are: 1) to provide the operator with telepresence and a means of conveying desired manipulation commands to the robot arm and it's tools , and 2) to provide the process engineer with a means of designing flexible and effective solutions by coupling autonomous and telerobotic control. Specific functional objectives are:

- Translate the operator manipulation commands to the robot arm and tool movement
- Provide the operator with feedback reflecting the forces and torques experienced by the robot arm and end-of-arm tools in the workcell
- Provide vision capabilities (2D, 3D, end-of-point tracking) to facilitate the operator in hand-eye coordination
- Collision and obstacle avoidance within the workcell
- Model the workcell for collision avoidance, and work cell calibration and off-line programming capability.

Technology and application development are underway in the Interactive Controls Laboratory in support of these objectives . These technologies will be ported to the glove box system in phases as the technology matures and is proven successful. Some of these technologies include:

- Advanced robot controller technology for sensor-based control and real-time collision avoidance
- Force-sensitive robot control to accommodate variations in expected and actual environments
- Improved man-machine interface technology including force-reflecting hand-controllers, 3-D Vision systems and context-sensitive control systems,
- Machine vision and object recognition systems to minimize hard tooling requirements and manual robot programming
- Fiducial target matching for system calibration and off-line programming
- Dexterous end-effectors for improved flexibility
- Part scanning and digitizing systems for real-time CAD model revisions.

The Series II controller currently under development by LLNL and IBM will be augmented to provide the necessary teleoperation functionality . It will use an IBM RS/6000-520 workstation host running the AIX operating system and a VME/68030 based target running the VxWorks real time operating system. The VME chassis will be sized to accommodate the teleoperational controller and machine vision hardware interface. The Series II software will be partitioned into modules as shown in Figure 5 . The dotted line separates the modules residing in the host computer above the line, and those residing in the real time system are below the line. The principal information interchange will involve controller status, error handling, and joint position measurements and setpoint commands.

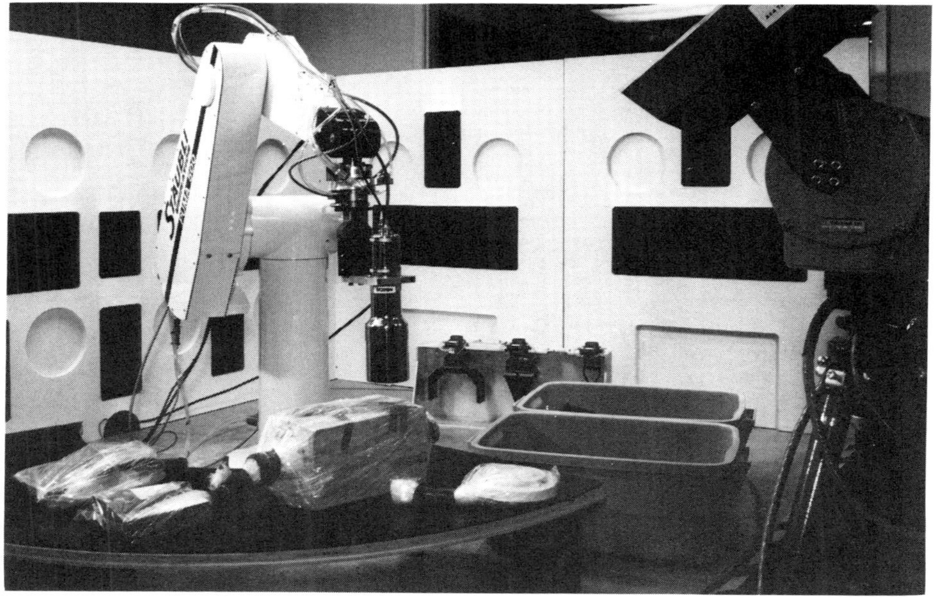

Figure 4. Identification and Segregation is based on radiation signatures.

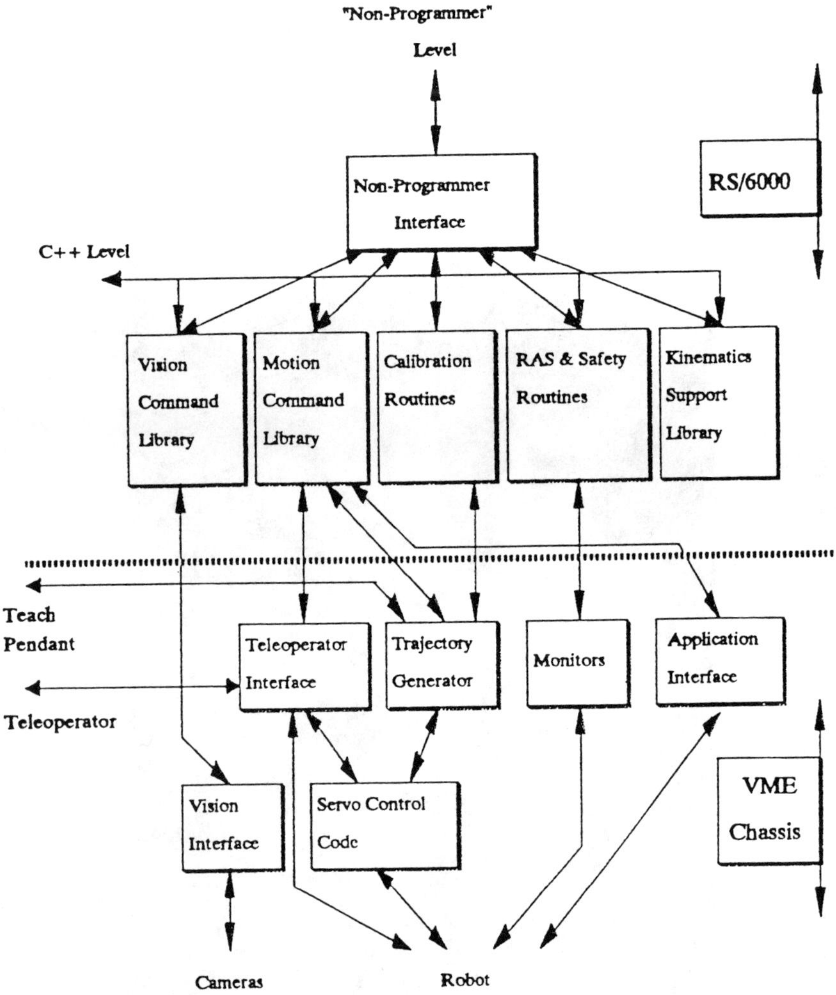

Figure 5. Software modules for BGR Series II controller.

The user interface will have three levels: non-programmer, C++, and developmental. The non-programmer level will be menu driven under X-windows and Motif so that an operator with no programming experience can do robot application development. The C++ level will enable the user to do object oriented application development in C++ in the AIX programming environment. The development level is intended for system maintenance work or for adding additional functions to the system. Work at this level will require knowledge of C++, AIX, VxWorks, and the system architecture.

A force reflecting hand controller (master controller) will be interfaced to the robot (slave) motion controller. The first objective will be to achieve master position-slave position control, and later, rate-position control with force reflection.

A seamless transfer capability will be available at the non-programmers interface (NPI) so that the operator can stop the robot while in autonomous mode and transfer to teleoperational mode (where the operator can drive the robot via the master arm), and then transfer back to autonomous mode as desired. The transfer back to the autonomous mode will be possible in either of the following ways: (1) continue at a specific step in the interrupted robot path program, or (2) select another path from a set of other autonomous programs. In each case, the reentry is to be first carried out on the graphic simulation of the robot and CAD model of the workcell to insure proper reentry before transferring to the actual robot.

The need exists to program (teach) robot paths via the graphic simulation. This need will be addressed with model-based off-line robot path programming coupled with model-based whole arm collision avoidance. Performing accuracy compensation is a requirement when correcting for differences between the graphic simulation, and the actual robot, and the workcell. Deriving this compensation information is known as workcell calibration which insures rapid set-up and teaching of a new or changed workcell configuration.

Sensor-based collision avoidance of objects in the workcell will be used to avoid objects which have changed position since the workcell model was last updated. A level of sensor integration is required to support certain robot motion paths, based on machine vision acquisition and processing. Sensor fusion will also be incorporated to support certain robot motions and for force control, (such as that required for deburring operations).

Finally, there will be supervisory level communications with other controllers to coordinate the movements of the robot with that of other mechanisms in the workcell or in adjoining workcells.

The advanced Series II controller capability will be coupled with a limited telerobotic capability and will be installed and integrated in the Glove box Robotics System in the Process Demonstration Testbed as a DOE milestone for Fiscal year 92. Throughout the next year, more advanced telerobotic and interactive control technologies will be ported to the Glove box robotics system . This flexible robotics system will be demonstrated in support of a waste and residue processing application. In order to demonstrate the advantages of such a system, an intensive performance evaluation will be conducted.

CONCLUSIONS

Development of automated technology and applications to waste processing operations are complex problems involving intelligent integration of manual, remotely controlled, and automated processes. Our approach is to facilitate concurrent development and implementation of advanced technologies required to minimize waste, and reduce hazards to operators. Technologies currently under development in the LLNL Robotics and Automation Section are addressing a wide variety of these needs and problems. Wherever possible, technology developments are based on, or tied to, existing or standard activities.

The current automated glove box development effort has demonstrated the suitability of the integrated automated glove box for automated performance of waste processing functions. Coupling of the interactive control/telerobotic capability with the advanced glove box technology has provided a high potential benefit with low risk. The completed demonstrations of interactive control technologies have illustrated the effectiveness of telerobotics for minimizing direct human intervention.

ACKNOWLEDGMENTS

This work was performed under the auspices of US DOE by LLNL under contract no. W-7405-eng-48.

We would like to acknowledge the hard work, long hours, and overall contributions of the LLNL Intelligent Controls Group, the Remote Systems and Sensors Group, and the Robot Applications Group. I would also like it noted that the IBM interactions have been dynamic, and full of synergism. All efforts and contributions are appreciated.

REFERENCES

1. Couture, Scott, *"The APT Robotics and Automation Program Plan"*, January 1992.

2. Merrill, Roy, Erna Grasz, and Ralph Sievers, *"Integrated Remote Control into a Pyrochemical Processing Glove box Robot"*, American Nuclear Society-1991 Winter Meeting, November 10-14, 1991.

3. Grasz, Erna and Ralph Sievers, *"The Application of Commercial Robotics Technology For Radioactive Material Processing"*, IEEE Conference on Robotics and Automation, April 1991.

A NEW CLASSIFICATION OF METHODOLOGIES FOR ECONOMICS JUSTIFICATION OF COMPUTER INTEGRATED MANUFACTURING SYSTEMS

Hamid R. Parsaei, Mickey R. Wilhelm, and Sai Kolli

Department of Industrial Engineering, University of Louisville, Louisville, KY 40292, U.S.A

ABSTRACT

One of the major barriers to the implementation of computer integrated manufacturing systems is the shortcoming of traditional financial justification methods. This paper presents a classification model to categorize both traditional and non-traditional economic analysis methods for capital investment justification.

INTRODUCTION

Computer integrated manufacturing systems are considered as viable tools for reducing direct and indirect manufacturing costs, improving product quality and increasing the flexibility. Although these technologies are considered to be highly beneficial, only a few companies in the U.S. have successfully implemented them.

Some of the barriers to the implementation of these technologies are the inability of traditional engineering economy and capital budgeting techniques, as well as the lack of the accounting procedures, to measure the effectiveness of new technology implementations (Meredith and Suresh, 1986; Primrose, 1985; Parsaei and Wilhelm, 1989; Wilhelm and Parsaei, 1991; Soni et al., 1992; Canada and Sullivan, 1989, 1990).

Implementation of computer integrated manufacturing systems involves high level of investment. Therefore, economic evaluation must be done using efficient techniques that incorporate qualitative benefits as well as quantitative ones in the evaluation process. However, the economic and financial justification analysis of advanced manufacturing systems is often performed using traditional economic analysis methods such as net present value, internal rate of return, payback period, etc. This paper is intended to present a classification model to categorize techniques which can be used for economic justification of computer integrated manufacturing systems.

PROPOSED CLASSIFICATION SCHEME

Various classification approaches have been proposed by researchers in the literature. Meredith and Suresh (1986) categorized economic analysis methods into three major classes, suitable for stand alone manufacturing systems, methods which are appropriate for linked systems, and methods which are more strategic in nature and can be applied to integrated systems.

Soni et al. (1992) categorizes existing economic justification methods into four groups: single objective deterministic methods, multi-objective deterministic methods, probabilistic methods, and fuzzy set methods.

Swamidass and Waller (1990) present a classification of the literature on the acquisition of new manufacturing technology. They divide the literature into planning and financial justification categories. Each of these categories is sub divided into various other themes.

The classification scheme proposed in this article categorizes existing justification methods into two major groups: single criterion and multiple criteria methods. Each major group is then divided into two subgroups termed deterministic and non-deterministic methods. Figure 1 illustrates these proposed groups and subgroups of economic justification methods. Kolli et al. (1992) present a detailed discussion of the methods and classification methodology.

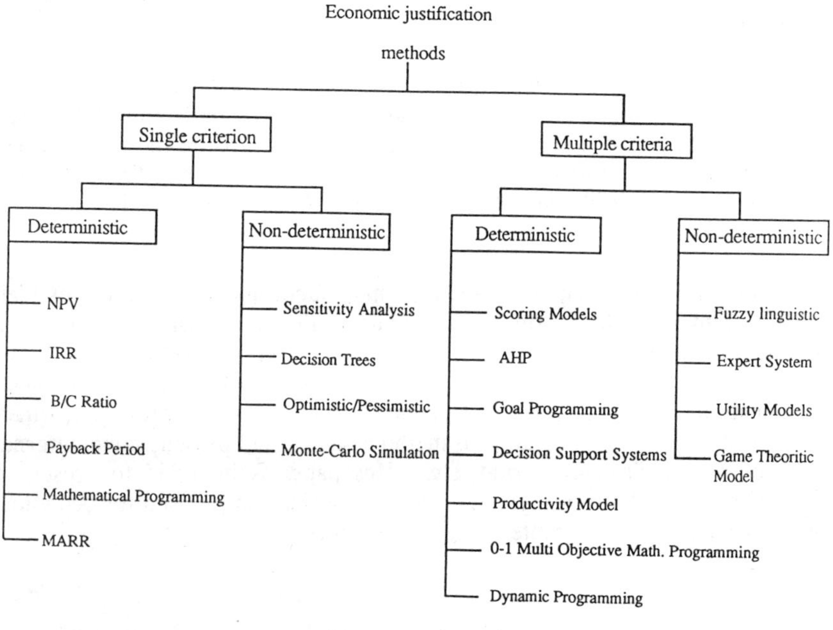

Figure 1. A proposed grouping of economic justification methods.

SINGLE CRITERION METHODS

The justification methodologies falling in this category employ a single economic objective to justify investment in advanced manufacturing technology. Various authors have used traditional financial justification methods in their justification procedure. Traditional financial methods including net present value (NPV), internal rate of return (IRR), benefit/cost ratio (B/C), payback period, mathematical programming, and minimal annual revenue requirement (MARR), belong to this category. This category also includes certain optimization methods and the productivity model as well as non-deterministic methods such as sensitivity analysis, decision trees, optimistic/pessimistic analysis, and Monte-Carlo simulation.

MULTI-CRITERIA METHODS

The decision of investing in advanced manufacturing technology frequently involves multiple and conflicting objectives, e.g., minimizing costs, maximizing flexibility, minimizing machine down times, maximizing efficiency, etc. The decision maker has to consider a various criteria and may assign certain priorities for each criterion so that the most important criterion is satisfied first. Since certain criteria are not quantitative, efficient methodologies are needed to evaluate the benefits of computer integrated manufacturing systems that are non-financial and non-quantitative in nature (Sullivan, 1991). Mathematical models have been developed to quantify performance measures such as productivity, quality, and flexibility for justifying investment in AMT (Son, 1991; Son and Park, 1987). Many multi-criteria techniques have been developed and applied to the problem of justifying advanced manufacturing technology (Falkner and Benhaijla, 1990). Sullivan (1986) discusses various models such as scoring, expert systems and AHP for automation decisions. Some of the deterministic methods for solving multi-criteria problems are multi-objective mathematical programming, scoring models, the analytic hierarchy process (AHP), outranking methods and goal programming. Non-deterministic methods include fuzzy techniques, expert systems, game theoretic methods, and multi-attribute utility and value analysis.

CONCLUSION

A new classification for categorizing traditional discounted cash flow based methods and non-traditional economic analysis methods for investment justification is proposed. This classification categorizes the existing justification methods into two major groups based on: single criterion, and multiple criteria. Each group is then branched into two subgroups of deterministic and non-deterministic methods. Purely subjective methods are not included in the proposed classification model since these methods are not quantitatively oriented and often used with some quantitative methods.

REFERENCES

Canada, J.R. and W.G. Sullivan. (1989), Economic and Multiattribute Evaluation of Advanced Manufacturing Systems, Prentice Hall, Englewood Cliffs, New Jersey.

Canada, John R. and W.G. Sullivan. (1990), Persistent Pitfalls and Applicable Approaches for Justification of Advanced Manufacturing Systems, Engineering Costs and Production Economics, Vol. 18, pp 247-253.

Falkner, Charles H. and Saida Benhaijla. (1990), Multi-attribute Decision Models in the Justification of CIM Systems, The Engineering Economist, Vol. 35, No. 2, pp 91-114.

Kolli, S., Hamid R. Parsaei, Mickey R. Wilhelm, and D.H. Liles. (1992), A Classification of Traditional and Non-Traditional Approaches for Economic Justification of Advanced Manufacturing Systems, Economic and Financial Justification of Advanced Manufacturing Systems, Elsevier Science Publishers, Holland, In press.

Meredith, Jack R. (1987), Managing Factory Automation Projects, Journal of Manufacturing Systems, Vol. 6, No. 2, pp 75-91.

Meredith, Jack R. and Nallan C. Suresh. (1986), Justification Techniques for Advanced Manufacturing Technologies, International Journal of Production Research, Vol. 24, No. 5, pp 1043-1058.

Parsaei, Hamid R. and Mickey R. Wilhelm. (1989), A Justification Methodology for Automated Manufacturing Technologies, Computers and Industrial Engineering, Vol. 19, No. 3, pp 363-373.

Park, Chan S. and Young K. Son. (1988), An Economic Evaluation Model for Advanced Manufacturing Systems, The Engineering Economist, Vol. 34, No. 1, pp 1-25.

Primrose, P.L. (1985), The Use of a Conceptual Model to Evaluate Financially Flexible Manufacturing System Projects, Proceedings of the Institute of Mechanical Engineers, Vol. 199, No. B1.

Son, Young K. (1991), A Cost Estimation Model for Advanced Manufacturing Systems, Int. J. Prod. Res., Vol. 29, No. 3, pp 441-452.

Soni, Ramesh G., Hamid Parsaei and Donald H. Liles. (1992), Economic and Financial Justification Methods for Advanced Automated Manufacturing: An Overview, Economics of Advanced Manufacturing Systems, Hamid R. Parsaei and Anil Mital (Editors), Chapman and Hall, 1992.

Sullivan, William G. (March, 1986), Models IEs Can Use for Strategic, Non-Monetary factors in Automation Decisions, Industrial Engineering, pp 42-50.

Sullivan, William G. (1991), A New Paradigm for Engineering Economy, The Engineering Economist, Vol. 36, No. 3, pp 187-200.

Swamidass, P.M. and M.A. Waller. (1990), A Classification of Approaches to Planning and Justifying New Manufacturing Technologies, Journal of Manufacturing Systems, Vol. 9, No. 3, pp 181-193.

Wilhelm, Mickey R. and Hamid R. Parsaei. (1991), A Fuzzy Linguistic Approach to Implementing a Strategy for Computer Integrated Manufacturing, Fuzzy Sets and Systems, Vol. 42, pp 191-204.

INFORMATION PROCESSING TECHNIQUES IN ECONOMIC AND FINANCIAL JUSTIFICATION OF ADVANCED MANUFACTURING TECHNOLOGIES: A DECISION MAKING APPROACH

Janardan Kulkarni
Laura Kersey Library, University of Louisville,
Louisville, KY 40292

Hamid R. Parsaei
Department of Industrial Engineering
University of Louisville, Louisville, KY 40292

ABSTRACT

In this paper we have presented a file searching problem in a decision or choice making framework. Document retrieval involves searching bibliographic databases for references to relevant documents. Usually, the record stored in a bibilio-graphic database to describe a document consists of a set of index terms or subject variables or keywords. In the field of economic and financial justification of advanced manufacturing technologies, a carefully selected set of keywords was used to obtain relevant citations from the on-line databases. Performance data was obtained from those index terms. A decision criterion that maximizes the output and increases the utility of the information to the decision maker was calculated.

INTRODUCTION

The competitive nature of manufacturing dictates that companies develop measures of performance from which to evaluate themselves and identify areas requiring improvement. These measures of performance quantify their success or failure depending on using the right information at right time. Dynamic growth depends on the appropriateness of knowing the information (i.e. knowledge) within a manufacturing environment. Only recently "Information" is considered to be viewed as a fundamental economic source that can conserve other resources, thereby enhancing productivity. Therefore, information is a critical organizational resource. To acquire the right information implies choosing or deciding what sources to select from a variety of those available. Automated document/information retrieval (IR) system or computer database searching is one of the source that is widespread in use for several decades. The majority of such commercial systems presume that requests are satisfactorily expressed in boolean logic. These IR systems require the users of the system to pose their query as a complex boolean expression of key words [1]. Such a boolean expression specifies to the system the information needs of the user and is called a user's profile.

In this paper we have constructed a profile for economic and financial justification of advanced manufacturing technologies. A set of subject terms, keywords, descriptors or pointers with cross reference relationship to other terms was selected from a thesaurus of keywords and from the index (Table 1). The goal was to choose the decision that maximized the output of useful information in the above mentioned field. Because the information improves the decision, the information is valuable. The information's value depends upon the degree to which it affects the value of the final decision. The retrieval or information processing algorithm or retrieval process is a formalization of our strategy which reflects the preferences or choices we selected in this decision model.

ACTION PLAN

To be able to plan the bibliographic search both manual and on-line, a set of indexes and databases were selected (Table 2). These sources emphasized both the depth and breadth of coverage of the topic of our paper. In addition we scanned about 50 journals in detail for the years 1975-1992. The search was not dated back before those years because there is very little information published on this topic during that time. The journals examined in detail are listed in Table 3.

The on-line search of the databases was set up with the aim of capturing the information output in the field of economic and financial justification of advanced manufacturing technologies. Searches were performed using the author names, titles, class codes, and corporate sources in addition to keywords, descriptors, or identifiers. In this case, the decision to choose a particular database or a keyword or a phrase, represents the set of alternatives available to the decision maker. Our aim was to get a precise output of information so that it can be used to its maximum.

DECISION THEORETIC MODEL

According to the statistical decision theory applied to information retrieval a document is retrieved or not retrieved depends on the expected cost (EC) of the retrieval process [3]. That is

$EC_r < EC_n$ where EC_r is the expected cost of retrieving a desirable citation and EC_n is the expected cost of not retrieving a desirable citation.

Let us call this citation profile as p which includes all the keywords and phrases that are necessary to retrieve that citation. We may formalize our rules as:

$$Pr(Rel/p)C_{Ret,Rel} + Pr(Notrel/p)C_{Ret,Notrel} <$$

$$Pr(Notrel/p)C_{Notret,Notrel} + Pr(Rel/p)C_{Notret,Rel}$$

where Pr(Rel/p) and Pr(Notrel/p) are the conditional probabilities that a document is relevant or not relevant to given profile p.

$C_{Ret,Rel}$ - Cost of retrieving a document
$C_{Notret,Rel}$ - Cost of not retrieving a document

To retrieve a document we use the following relation:

$$Pr(Rel/p) / Pr(Notrel/p) > C$$

where C is a cost constant composed of the various costs above. Let RSV(p) be the retrieval status value of a document with characteristic p. Then, a document will be retrieved if and only if RSV(p) > C were C is not known to both the searcher and system.

Computation of Retrieval Status Value (RSV)

RSV may be computed as

$$Pr(Rel/p) / Pr(Notrel/p).$$

These two distributions may be difficult to estimate, so, we may transform this expression by an application of Bayes rule [2]

$$(Pr(p/Rel) / Pr(p/Notrel)) \times (Pr(Rel)/Pr(Notrel))$$

We computed Pr(Rel)/Pr(Notrel) for those documents already retrieved in the search. The values of those computations are given in table 4. for the databases searched.

CONCLUSION

We have introduced a version of file searching problem that retrieves the best documents in the file. Our solution balances the trade-offs between the expected cost of retrieving a document and not retrieving a document. The idea of searching a file for a best document is an attempt to view a file search as decision problem.

REFERENCES

1. Salton, G. and McGill M. (1983), Introduction to Modern Information Retrieval, McGraw Hill, New York.

2. Kraft D. and Waller, W. (1981), A Bayesian Approach to User Stopping Rules for Information Retrieval Systems, Information Processing and Management, Vol. 17, pp 349-361.

3. Loose, Robert M. (1987), Predicting Document Retrieval System Performance: An Expected Precision Measure, Information Processing and Management, Vol. 23, pp 529-537.

4. Moore, James C. and Whinston, Andrew B. (1990), A Decision Theoretic Approach to Information Retrieval. ACM Transactions on Database Systems, Vol 15. pp 311-340.

Table 1. Thesaurus of the Keywords Used

Cost Justification	Economic Justification
Design to Cost	Cost Analysis
Financial Justification	FMS
Flexible Manufacturing Systems	Life Cycle Cost Analysis
Investment Management	Strategic Planning
Activity-Based Accounting	Operating Performance Management
Product Costing	Manufacturing Costs
Production Management	Cost Control
Economic and Financial Evaluation	Costing System
Computer Integrated Manufacturing	CIM
Intelligent Production Systems	Design for Manufacturablility
Design for Assembly	Design for Economic Feasibility
Computer Integrated Enterprise	Financial Management
Continuous Process Control	Life Cycle Customer Support
Long Range Planning	Risk Analysis
Managerial Economics	Investment Decisions
Quantitative Economic Analysis	Economic Comparisons
Capital Budgeting	Computer-Aided Manufacturing
CAM	Costing Methods
Cost Analysis	Cost Effectiveness
Decision Analysis	Benefit Cost Analysis
Capital Investment Analysis	Expert Systems in Economic Analysis
Engineering Economic Analysis	
Economic Models for Planning Systems	
Factory Automation Economic Justification	
Productivity Measurement Weighted Evaluation	

Table 2. List of Databases Searched

COMPENDEX
INSPEC
Harvard Business review
Japan Technology
Dissertation Abstracts International
CAD/CAM
Supertech

Table 3. Journals Examined

Harvard Business Review	Production Engineering
CIM Review	Industrial Engineering
American Metal market	Engineering Economist
Production	Tooling and Production
Annals of CIRP	Journal of Manufacturing Systems
Robotics Today	Industrial Robotics
International Labor Review	Modern Material Handling
American Machinists	Chief Executive
Computer World	Economic Bulletin for Europe
Financial World	FMS Magazine
Industrial Week	Journal of System Management
Manufacturing Review	Computers and Industrial Engineering
Production Planning and Control	Management Science
Manufacturing Systems	National Productivity Review
Quality Progress	New Scientists
Robotics World	Advanced Manufacturing Processes
Industrial Management	Material Handling Engineering
International Management	Journal of Accountancy
Management Accounting	Decision Sciences
Decision Support Systems	Journal of Business Strategy
Journal of Accounting and EDP	
International Journal of Intelligent Systems	
Journal of Manufacturing and Operations Management	
International Journal of Production Research	
Robotics and Computer Integrated Manufacturing	
IEEE Transactions on Engineering Management	
Journal of Post Keynesian Economics	
International Journal of Production Economics	
International Journal of Computer Integrated Manufacturing	

Table 4. Computation of Retreival Status Value (RSV)

Database	RSV
COMPENDEX	1.63
INSPEC	1.48
Japan Tech	0.42
NTIS	0.36
Harvard Business Review	1.96
Dissertation Abstract International	0.16

MODELING THE CONCURRENT DESIGN PROCESS WITH PETRI NETS

Andrew Kusiak and Juite Wang

Intelligent Systems Laboratory, Department of Industrial Engineering, The University of Iowa, Iowa City, Iowa 52242

ABSTRACT
The features of design process, such as distributed, concurrent, and conflicting properties, have a resemblance to Petri nets (PNs). In this paper, a PN approach is proposed to model the design process. It is observed that PNs capture typical features of the concurrent design process. In particular, PNs allow a structured definition of the model, reflecting the nature of the design problem. The design process is usually performed by several multifunctional teams. The PN approach leads to better understanding of the design effort. In particular, formal methods for determining static (structural) and dynamic (behavioral) properties of PNs, are useful.

INTRODUCTION

The design process is a source of competitive advantage in many industries, in particular if the time of product development as well as manufacturing quality and productivity issues are given adequate consideration (Nevins and Whitney 1989). To maintain a market share, methods for minimizing the cost and time consumed in the product development are considered of great importance. In recent years, the concept of concurrent engineering has emerged which is defined as a systematic approach to the integrated, simultaneous design of products and processes, including manufacturing and support, at the early design stage (Winner et al. 1988). Concurrent engineering aims at improvement of the product quality and reduction of the product development time and cost. However, with the increased level of the integration of design and manufacturing, problems have arisen that are beyond traditional solution methodologies. The increased complexity should be implemented so that the time of development and manufacturing a product is reduced. There is a perception that the only way to reduce the design cycle is to improve the efficiency of the design and analysis software. It has been identified that the design and analysis portion of engineering activities constitutes only about 1/3 of all activities (Gallagher 1991). In order to reduce the design cycle, an effort should be made to plan and manage all engineering activities. In this paper, a Petri net approach is proposed to model the concurrent design process.

CHARACTERISTICS OF THE CONCURRENT DESIGN PROCESS

To reduce the product development time, it is crucial to properly manage a design process. Poor management of design tasks may result in delays. To effectively manage, the characteristics of the design process in concurrent engineering environment have to be recognized. The characteristics of the concurrent design process are as follows:

- Complexity - A design process usually involves a large number of activities that make it difficult for a manager to acquire current information. Also, as the number and degree of interaction among activities increases, the complexity increases.
- Uncertainty - Due to the unknown state of the activities and environment, the fact that an activity is successful or fail influences the status of other activities.

• Iterations - The large number of engineering changes occurs in the design process. If a change occurs, design process has to go back to the previous step and make an adjustment.

• Concurrency - A concurrent design process usually involves multi-disciplinery participants that collaborate on problem solving.

• Conflict - It is frequent that more than one approach is used to solve a particular problem.

The traditional project management tools (Elmaghraby 1977), such as the Critical Path Method (CPM) and the Program Evaluation and Review Technique (PERT), are not suitable for managing the concurrent design process because of the poor descriptive and analytical abilities. The features of the design process, such as its distributed, concurrent, and conflicting properties, have a resemblance to Petri nets (PNs). PNs can be applied to control the design process with considerable advantages over traditional techniques. In particular, PNs allow for a structured definition of models, reflecting the nature of design problems. The engineering changes in the design process can be traced by the movement of tokens. This leads to faster implementation, maintenance, and coordination of scheduling strategies as well as consistency checking procedures by each design team. Due to the availability of a widely developed theory on PNs, both the qualitative and quantitative analysis of models can benefit from a PN-based approach. In particular, formal methods for determining static (structural) and dynamic (behavioral) properties of PNs, can be used. Also, the optimal sequence of design activities can be obtained from a PN model.

MODELING THE CONCURRENT DESIGN PROCESS

A standard PN (Peterson 1981, Murata 1989) is a directed, bipartite graph consisting of two kinds of nodes, called places and transitions. A set of directed arcs is either from a place to a transition or vice vesa. Graphically, places are represented by circles and transitions by bars. Places may contain tokens. If the concept of conditions and events is used for modeling, places represent conditions and transitions represent events. The existence of a token in a place is explained as retaining the truth of the condition corresponding to the place. If a place is interpreted as the availability of resources, then the tokens in the place indicate that the resources are available.

The behavior of the modeled system is described by the movement of tokens in the net resulting from the firing of one or more transitions. The firing rules of PNs are as follows:

- A transition is enabled if each of its input places contains at least one token.
- A transition can fire only if it is enabled.
- When a transition fires, a token is removed from each of its input places and deposited into each of its output places.

Readers interested in the theory, properties, and applications of PNs, may refer for details to Peterson (1981) and Murata (1989).

Petri Net Approach

One may define design process as a sequence of interrelated tasks that begins with an initial set of customer requirements and ends with a complete description of a product that satisfies these requirements (Brown and Chandrasekaran 1985). There are some relationships among tasks. Most network-based project management tools describe only "AND" type of precedence relationship among design tasks. However, a design process may contain more complex relationships. These complex design relationships have to be captured in the model in order to analyze the design process effectively.

In this paper, a PN approach is proposed for modeling the design process. Each design task is considered as a transition; each state can be considered as a marking. Executing a task is equivalent to firing a transition. At the completion of each task, the process leaves the original state and enters the next state.

The evolution of the design process is illustrated below in four scenarios:

• Scenario 1: Sequential design tasks (Figure 1)

Consider two design tasks t_1 and t_2. If task t_2 requires the output of task t_1, then two tasks are dependent and have to be performed sequentially. In Figure 1, transition t_2 is enabled after transition t_1 is fired and the token is removed from place p_1 to p_2.

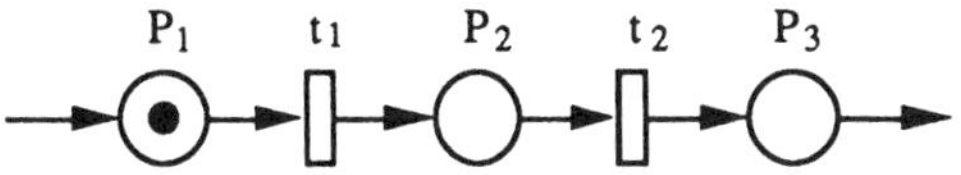

Figure 1. Sequential design tasks t_1 and t_2

- Scenario 2: Concurrent design tasks (Figure 2)

If task t_2 and t_3 are independent, then they can be performed simultaneously. In the Petri net notion, transitions t_2 and t_3 are enabled, if t_1 is fired.

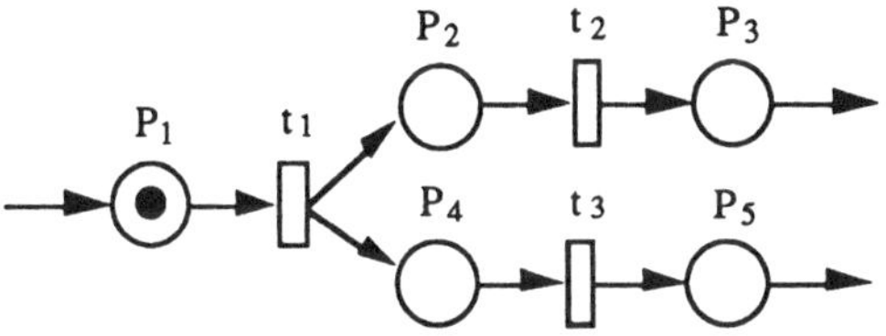

Figure 2. Concurrent design tasks t_2 and t_3

- Scenario 3: Exclusive design tasks (Figure 3)

It happens frequently that more than one approach (e.g., tasks t_2 and t_3 in Figure 3) can be used for solving a problem encountered in the design process; i.e., after transition t_1 is fired, only one of the transitions, t_2 or t_3, can be fired.

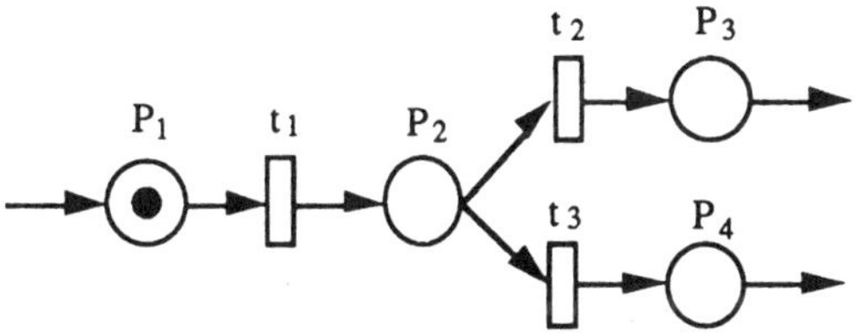

Figure 3. Exclusive design tasks t_2 and t_3

- Scenario 4: Cyclic design tasks (Figure 4)

Design process is iterative. One often starts with an initial solution and attempts to improve it. In Figure 4, transitions t_2 (product improvement) and t_3 (product test) are fired iteratively after transition t_1 (initial product development) has been fired.

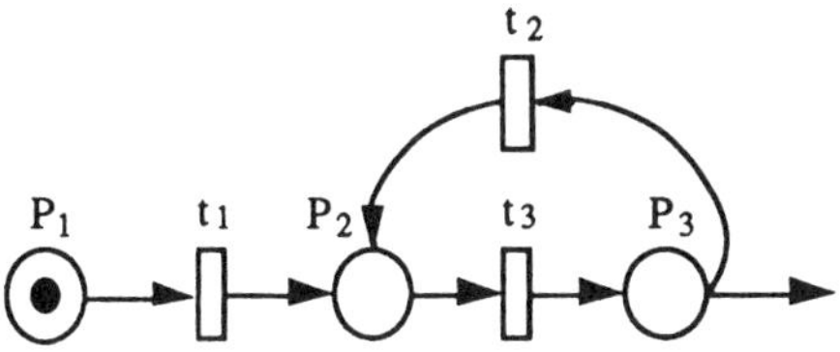

Figure 4. Cyclic design tasks t_2 and t_3

To reduce the design cycle, one may sequence the cyclic design tasks with the minimum number of the iterations (Steward 1981, Kusiak and Wang 1993). Moreover, it is possible that a task occurs at the time when not all but only one of the activities leading into this tasks has been completed. In order to model the design process, three types of relationships between design tasks are defined next.

- "AND" type

This type of relationship may occur in a design process when a design task requires the information produced by the preceding tasks, so that the task has to wait until all the preceding tasks have been completed. For example, there are three design tasks, t_1, t_2, and, t_3 (see Figure 5). Task t_3 is enabled, when the tasks t_1 and t_2 have been completed. The time of occurrence of task t_3 is the latest of the completion of the preceding tasks, t_1 and t_2.

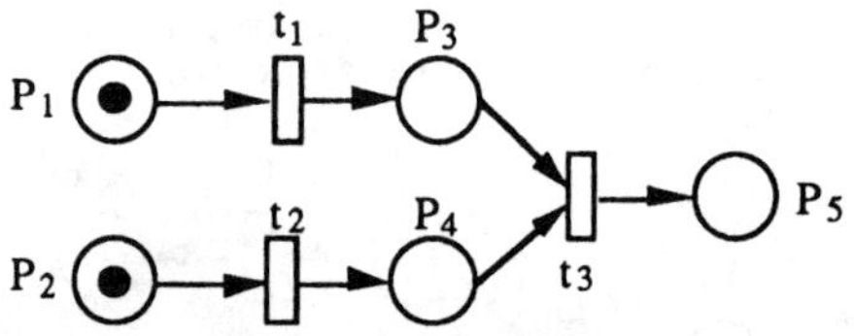

Figure 5. "AND" type of design task

- "Inclusive OR" type

This type of relationship occurs in a design process when two simultaneous design tasks are planned that are aimed at the same end result but are using different methods. When either task is successful, the process moves to the next phase. For example, in Figure 6, tasks t_1 and t_2 are been processed at the same time. However, if only one of the tasks is successful, it will disable the other task and enable task t_3. The time of occurrence of task t_3 is the earliest time of completion of the preceding tasks. An inhibitor arc is used to disable the other task, if task t_1 or t_2 is completed.

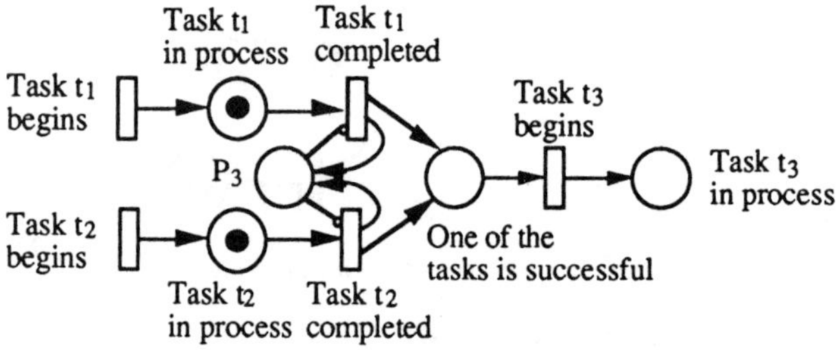

Figure 6. "Inclusive OR" type of design task

- "Exclusive OR" type

This type of relationship may occur when two parallel tasks use different approaches to accomplish the same end result. The completion of either one of the tasks starts the follow-up task. However, if both tasks are successful at about the same time, an additional analysis is required before beginning of the follow-up task, because either approach or a combination of the two is needed. In Figure 7, task t_3 occurs exactly when one of the tasks preceding task t_3 has been completed.

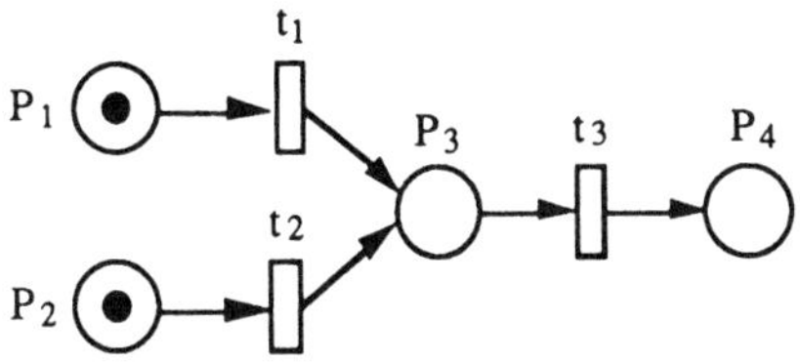

Figure 7. "Exclusive OR" type of design task

Analysis of the Concurrent Design Process

A major strength of PNs is not just the ability to represent concurrent systems graphically, but their ability to support analysis of behavioral and structural properties associated with concurrent systems (Murata 1989). The difficulty is how to interprete the properties of the concurrent design process represented as PNs. Some properties are defined and interpreted next.

Definition: Completeness

A design process is complete if and only if the final state (M) of its associated Petri net is reachable from any intermediate state M_i.

Design process involves many tasks that are performed by several teams concurrently. Each team is involved only in a part of the entire plan. One can define the design requirements (goals) as final states. Then the property of completeness ensures that tasks performed by each team contribute to the accomplishment of the goal. No irrelevant tasks are created. To remove the irrelevant design tasks, efficient graph theory algorithms can be used (Munro 1971).

Definition: Liveness

A PN net is said to be live if it is possible to ultimately fire any transition of the net by progressing through some further firing sequence no matter what marking has been reached from the initial state (M_0).

Liveness of all input transitions for a task in the design process implies that there are no deadlocks, and liveness of the output transitions means that every task can be executed in some valid task sequence.

Definition: Saveness

A PN net is safe if the number of tokens in each place does not exceed one for any marking reachable from the initial marking M_0.

Safeness of an “task-place” implies that there is no attempt to perform a task which is already in progress. Similarly, safeness of a “resource-place” means that there are no conflicts in the resource allocation, i.e., the control logic guarantees that the resources are only used when truly available.

CONCLUSION

In this paper, a PN approach was proposed to model the design process. It is observed that PNs can capture the features of the concurrent design process. The design process is usually performed by several multifunctional teams. PN models lead to efficient coordination of design teams. In particular, formal methods for determining static (structural) and dynamic (behavioral) properties of PNs, can be applied.

Design process usually involves a large number of design activities. Standard PNs are not suitable to to model the design process, due to the large size. A special class of PN, predicate-transition (PrT) net (Genrich and Lautenbach 1981, Genrich 1987), can be used to model the design process. Another extension is to incorporate time into the model. In the literature, time is usually attached either a place or transition. However, they do not seem to be general enough for modeling all of the time aspects. One may integrate PrT nets with temporal logic (Allen 1983) that provide more general operators to model the design process.

REFERENCES

1. Allen, J. F. (1983). "Maintaining Knowledge about Temporal Intervals." Communications of the ACM, Vol. 26, No. 11, pp. 832-843.
2. Brown, D. C. and B. Chandrasekaran (1985). "Expert Systems for a Class of Mechanical Design Activity." in J. S. Gero, Ed., *Knowledge Engineering in Computer-Aided Design*, Elsevier, North-Holland.
3. Elmaghraby, S. E. (1977). *Activity Networks: Project Planning and Control by Network Models*, Wiley, New York.
4. Gallagher, G. R. (1991). "ACES: Demonstrating the Infrastructure for Concurrent Engineering." *Proceedings, AUTOFACT '91 Conference*, November 10-14, Chicago, Illinois, pp. 17.27-17.34.
5. Genrich, H. J. and K. Lautenbach (1981). "System Modelling with High-Level Petri Nets." *Theoretic Computer Science*, Vol. 13, No. 1, pp. 109-136.
6. Genrich, H. J. (1987). "Predicate/Transition Nets." In W. Brauer, W. Reisig and G. Rozenberg, Eds., *Petri Nets: Central Models and Their Properties*, Lecture Notes in Computer Science, Vol. 254, Springer-Verlag, Berlin, pp. 207-247.
7. Kusiak, A. and J. Wang (1993). "Decomposition of the Design Process." in A. Kusiak, Ed., *Concurrent Engineering: Automation, Tools, and Techniques*, John Wiley, New York.
8. Munro, J. I. (1971). "Efficient Determination of the Transitive Closure of a Directed Graph." *Information Processing Letters*, Vol. 1, pp. 56-58.
9. Murata, T. (1989). "Petri Nets: Properties, Analysis and Applications." *Proceedings of the IEEE*, Vol. 77, No. 4, pp. 541-580.
10. Nevins, J. L. and D. E. Whitney (1989). *Concurrent Design of Products and Processes: A Strategies for the Next Generation in Manufacturing*, McGraw-Hill, New York.
11. Peterson, J. L. (1981). *Petri Net Theory and the Modelling of Systems*, Prentice-Hall, Englewood Cliffs, NJ.
12. Steward, D. V. (1981). *Systems Analysis and Management: Structure, Strategy, and Design*, Petrocelli Books, New York.
13. Winner, R. I., J. P. Pennell, H. E. Bertrand, and M. M. G. Slusarczuk (1988). "The Role of Concurrent Engineering in Weapons System Acquisition." *IDA Report R-338*, Institute for Defense Analyses, Alexandria, Virginia.

Using Genetic Algorithms for Equal Area Facilities Layout Design

Alice E. Smith and David M. Tate
Department of Industrial Engineering
University of Pittsburgh
Pittsburgh, PA 15261

ABSTRACT

This paper describes the use of Genetic algorithms for equal area facilities layout design. Genetic algorithms are implicitly parallel stochastic methods which avoid local minima, scale up well and generally find good solutions to nonlinear large dimensional problems. The paper describes GA encoding, fitness testing, reproduction and mutation schemes. We give results on several classical test problems, and provide evidence of robustness.

INTRODUCTION

The Facility Layout Problem is a well-known combinatorial optimization problem in which a set of n distinct **departments** is to be placed uniquely in n or more distinct **locations**. Associated with each pair of departments (i,j) is an interaction (or **traffic**) intensity, T(i,j). Associated with each pair of locations (L_k, L_l) is a **distance**, $D(L_k, L_l)$. If we let L(i) denote the location to which department i is assigned in a particular feasible layout L, the total cost of the layout L is given by:

$$\text{Cost}(L) = \sum_{\text{depts. } i} \sum_{\text{sites } j} [\, T(i,j) \bullet D(L(i), L(j)) \,]$$

This problem is arises in many manufacturing and design contexts, and is known to be NP-complete [6]. Since Koopmans and Beckmann [16] first used a quadratic assignment formulation in the context of equal area facility layout, many articles have appeared proposing heuristic procedures and enumerative schemes including Hillier and Connors [8], Picone and Wilhelm [18], Heragu and Kusiak [7], Kaku et al. [15], Scriabin and Vergin [19] and Huntley and Brown [14].

We describe below an approach for finding consistently good solutions to instances of equal area facilities layout using a Genetic algorithm (GA) framework, and show that this approach yields solutions comparable to the best previously-known heuristics when applied to test problems from the literature. We argue that this method is not overly costly in terms of computation effort, and is exceptionally robust both with respect to variations in the objective function, constraints of the problem, and the details of the GA implementation.

OVERVIEW OF GENETIC ALGORITHMS

The phrase "Genetic Algorithms" denotes a family of parallel, stochastic-search optimization heuristics which share the following features: (a) One or more "populations" of feasible solutions, (b) A mechanism for generating new feasible solutions by combining features from multiple previously-known solutions, (c) A mechanism for generating a new feasible solution by a random perturbation of a single previously-known solution, (d) A mechanism for selecting individual solutions from the population(s), giving preference to those with better objective function values, and (e) A mechanism for removing solutions from the population(s).

As first described by Holland [10], the motivation for GA is the biological paradigm of natural selection. We thus refer to the mechanism of part (b) above as a **reproduction** mechanism, that of part (c) as a **mutation** mechanism, that of part (d) as a **selection** mechanism, and that of part (e) as a **culling** mechanism. The basic flow of a typical GA implementation is thus given by the following:

1. Randomly create an initial population of solutions.
2. Repeat {
    - 2.1 **Select** parents.
    - 2.2 **Breed** offspring, and add them to the population.
    - 2.3 **Mutate** certain members of the current population.
    - 2.4 **Cull** certain members of the current population.}
3. Until (some termination criterion is met).

Since Holland's seminal work, genetic algorithms have been applied to such optimization problems as job shop scheduling by Davis [4], sizing of communications networks by Davis and Coombs [5], FMS scheduling [12], and multiprocessor task scheduling [11, 20]. A component placement problem, related to the facilities layout problem, has been addressed using genetic algorithms by Cohoon and Paris [3] and Cohoon et al. [2].

A GENETIC ALGORITHM FOR FACILITIES LAYOUT

For the equal area test problems considered, the possible location for departments are a rectangular lattice of locations in the plane, with the inter-site distance computed by Manhattan (L_1) norm. Locations were ordered in boustrophedon sequence by rows, establishing a one-to-one correspondence between permutations of {1, 2, ..., n} and feasible layouts. The number of rows and columns were pre-specified and fixed, and unused locations were assigned "dummy" departments with no traffic in or out.

Since we did not use a traditional binary encoding for solution strings, we were able to design reproduction and mutation functions which operate directly on the solution sequence. **Mutation** took the form of selecting two sites at random and reversing the order of all sites within the subsequence bounded by the two selected. For selected sites either adjacent, or one space apart in the boustrophedon sequence, this reduces to pairwise interchange.

Selected String:	B	C	F	A	G	E	H	D
Random Mutation Sites:		x			x			
Mutated String	B	G	A	F	C	E	H	D

Selection is performed with a bias towards choosing the better solutions in the current population. We selected parent strings by choosing a uniform random number between 1 and $\sqrt{m}$ (where m is the current population size), then squaring it. The result was truncated and taken to be the rank of the selected parent (where string zero is the fittest string in the population). This resulted in selection based on rank in the current population, rather than by absolute objective function value, as is done with the traditional biased roulette wheel. This approach can be generalized to give an arbitrary degree of preference or even varied dynamically during a given run by increasing the power of the root of the upper bound on the random number during subsequent generations.

Reproduction is as follows: Any department assigned to the same location in both parents will occupy that location in the offspring. For the remaining locations, one or the other department assigned to that location in one parent is chosen at random, working left to right through the sequence of locations. Any unassigned departments are then matched with the remaining unassigned locations. An example is shown below.

Parent 1:	D	C	A	H	F	E	B	G
Parent 2:	F	A	E	H	B	D	C	G
Common Locations:	-	-	-	H	-	-	-	G
Random Choice:	D	A	E	H	F	-	B	G
Leftover:	C							
Finished Child:	D	A	E	H	F	C	B	G

The number of children to be produced per generation was selected in advance, as was the probability of a given individual generating a mutant. A mutant is distinct from the individual which spawns it, and both have opportunities to survive. As children were created, the strings with worst fitness were culled to keep a constant population size. To maintain enough diversity, mutants were inserted into the population after members of the current generation had been culled. In this way, mutants, no matter what their fitness, were assured to survive at least one generation and be available for reproductive selection. The single best solution was always retained unaltered for the next generation to encourage convergence.

TEST PROBLEMS AND RESULTS

The most well known equal area facility layout problem instances for rectilinear distance are those eight defined in Nugent et al. [17], with fifteen subsequent research results summarized in Hundal [13]. These range in size from 5 departments (6 locations) to 30 departments (30 locations), where the traffic matrix is assumed symmetric. Another well known equal area test problem which is regarded as "harder" because of its increased flow dominance over the Nugent problems [14] is an adaptation of the 20 department unequal area Armour and Buffa problem [1] by Scriabin and Vergin [19].

We ran our GA on each problem 10 times since genetic algorithms are stochastic and therefore yield different searches, and potentially different results, for each random number seed used. After experimentation (see [21] for details) we determined the best mix of mutation and reproduction was 75% / 25% due to the conservative nature of our mutation operator. Each run consisted of a fixed population size of 100 solutions with a randomly generated initial population of 100 individuals. The search was allowed to continue until either the optimum had been reached (if it was known) or 2000 generations had passed. The best solution yet found and total number of solutions generated were recorded every 20 generations. The number of solutions generated was not adjusted for duplicates, and therefore represents an overestimate of the number of unique solutions investigated.

Differences, especially in search effort, existed for each different random number seed, but solution quality was within a few percentage points, as evidenced by the extremely low coefficient of variation. In stochastic search, an approach which is robust with respect to solution quality is a paramount concern. Our genetic algorithm not only performed well, it performed well consistently. Table I shows the best and average solutions, the coefficient of variation over the ten solutions, and the mean number of solutions searched to find the best solution. Asterisks denote best known solution values. Problems are denoted by first the number of departments and then the number of locations.

Table I. Solutions Found and Number Searched.

Problem	Best Solution	Mean Solution	Coefficient of Variation	Mean Searched
5 / 6 Nugent	25*	25*	0.0	277.4
6 / 6 Nugent	43*	43*	0.0	399.8
7 / 9 Nugent	74*	74*	0.0	1153.7
8 / 8 Nugent	107*	107*	0.0	1342.9
12 / 12 Nugent	289*	291.4	0.007	18892
15 / 15 Nugent	575*	585	0.011	39099
20 / 20 Nugent	1304	1321.7	0.008	75084
30 / 30 Nugent	3092	3152.7	0.011	110010
20/20 Armour/Buffa	1097.64*	1132.582	0.023	74848.5

Another important comparison is to previously published results or optimum known results. Table II shows our best found solution over the 10 runs of each problem as a percent above the best known solution [*to the best of our knowledge, our best solution to the modified Armour and Buffa problem is as yet unmatched in the literature*]. To show the reliability of our approach, Table II also shows the worst final solution found as a percent over the best known solution. Our best solutions were generally optimum, or within a percentage point, and worst solutions were within 8% of optimum.

Table II. Our Results as a Percent Over Best Known Solutions.

Problem	Best	Worst
5 / 6 Nugent	0	0
6 / 6 Nugent	0	0
7 / 9 Nugent	0	0
8 / 8 Nugent	0	0
12 / 12 Nugent	0	1.384
15 / 15 Nugent	0	3.130
20 / 20 Nugent	0.696	3.403
30 / 30 Nugent	0.422	4.190
20/20 Armour/Buffa	0	7.80

With regards to the quality of our best solutions, it must be acknowledged that for the classical rectilinear problems of Nugent et al., one approach's best solution consistently bettered ours by minute amounts. This approach, the terminal sampling procedure of Hitchings and Cottam [9], found better solutions for the 20 and 30 department problems. The improvement was less than 0.25% however, and, at least for the 30 site problem, our average solution was better than the average of the solutions they report. They did not report any results other than their best found for the 20 site problem.

DISCUSSION AND FUTURE WORK

The performance of this GA implementation is very encouraging. A simple mutation and reproduction scheme was able to match the best known heuristics from previous research, without any careful selection of parameters of the implementation. This suggests, on the one hand, that there is potential for improved performance by a genetic algorithm with a more sophisticated reproduction or mutation scheme, so that this implementation is not necessarily a "dead end" for this problem class. On the other hand, the apparent robustness of the implementation with regard to initial settings allows us to be optimistic that one can achieve near-optimal solutions quickly without having to fine-tune sensitive parameters of the implementation, or to explore a wide variety of encodings and generation algorithms. The excellent average-case performance of the GA approach across different initial populations and random number seeds reinforces this impression of robustness.

We are currently working on several extensions of the research presented in this paper. First, we are testing the sensitivity of genetic algorithms to parameters including population size, mutation probability, generation size, and number of generations. Second, we are solving the harder unequal area facilities layout problem using genetic algorithms. This has been considerably less studied in the literature due to the difficulty of obtaining good feasible solutions, but is a much more realistic formulation.

REFERENCES

1. G. C. Armour and E. S. Buffa, "A heuristic algorithm and simulation approach to relative location of facilities," *Management Science* **9**, 294-309 (1963).

2. J. P. Cohoon, S. U. Hegde, W. N. Martin and D. S. Richards, "Distributed genetic algorithms for the floorplan design problem," *IEEE Transactions on Computer-Aided Design* **10**, 483-492 (1991).

3. J. P. Cohoon and W. D. Paris, "Genetic placement," *IEEE Transactions on Computer-Aided Design* **6**, 956-964 (1987).

4. L. Davis, "Job shop scheduling with genetic algorithms," *Proceedings of an International Conference on Genetic Algorithms and Their Applications*, 136-140 (1985).

5. L. Davis and S. Coombs, "Genetic algorithms and communications link speed design: theoretical considerations," *Proceedings of the Second International Conference on Genetic Algorithms*, 252-256 (1987).

6. M. R. Garey and D. S. Johnson, *Computers and Intractability: A Guide to the Theory of NP-Completeness* (W. H. Freeman and Co., San Francisco, 1979).

7. S. S. Heragu and A. Kusiak, "Machine layout problem in flexible manufacturing systems," *Operations Research* **36**, 258-268 (1988).

8. F. S. Hillier and M. M. Connors, "Quadratic assignment problem algorithms and the location of indivisible facilities," *Management Science* **13**, 42-57 (1966).

9. G. G. Hitchings, "An efficient heuristic procedure for solving the layout design problem," *OMEGA* **4**, 205-214 (1976).

10. J. H. Holland, *Adaptation in Natural and Systems* (University of Michigan Press, Ann Arbor, 1975).

11. E. S. H. Hou, R. Hong and N. Ansari, "Efficient multiprocessor scheduling based on genetic algorithms," *Proceedings of IECON '90, Vol. 2*, 1239-1243 (1990).

12. E. S. H. Hou and H.-Y. Li, "Task scheduling for flexible manufacturing systems based on genetic algorithms," *Proceedings of the 1991 IEEE International Conference on Systems, Man, and Cybernetics*, 397-402 (1991).

13. T. S. Hundal, *The Facility Layout Problem: A Myopic Priority Heuristic*, Ph.D. Dissertation, University of Pittsburgh (1991).

14. C. L. Huntley and D. E. Brown, "A parallel heuristic for quadratic assignment problems," *Computers and Operations Research* **18**, 275-289 (1991)

15. B. K. Kaku, G. L. Thompson and T. E. Morton, "A hybrid heuristic for the facilities layout problem," *Computers in Operations Research* **18**, 241-253 (1991).

16. T. C. Koopmans and M. Beckmann, "Assignment problems and the location of economic activities," *Econometrica* **25**, 53-76 (1957).

17. C. E. Nugent, T. E. Vollmann and J. Ruml, "An experimental comparison of techniques for the assignment of facilities to locations," *Operations Research* **16**, 150-173 (1968).

18. C. J. Picone and W. E. Wilhelm, "A perturbation scheme to improve Hillier's solution to the facilities location problem," *Management Science* **30**, 1238-1249 (1984).

19. M. Scriabin and R. C. Vergin, "Comparison of computer algorithms and visual based methods for plant layout," *Management Science* **22**, 172-181 (1975).

20. H. T. Sieglemann and O. Frieder, "The allocation of documents in multiprocessor information retrieval systems: an application of genetic algorithms," *Proceedings of the 1991 IEEE International Conference on Systems, Man, and Cybernetics*, 645-650 (1991).

21. D. M. Tate and A. E. Smith, "A genetic approach to the quadratic assignment problem," under review for *Computers and Operations Research*.

COMPUTATION OF SUBASSEMBLY STABILITY FOR ASSEMBLY SEQUENCE PLANNING[1]

LUIZ S. HOMEM-DE-MELLO
Jet Propulsion Laboratory
California Institute of Technology
Pasadena CA 91109-8099

ABSTRACT

This paper presents a mathematical formulation for the computation of the set of stable orientations of a subassembly. The formulation fits well into the assembly planning framework since it also uses the computation of cut-sets of the assembly's graph of connections, as well as the computation of the polyhedral convex cones of feasible translations. When a subassembly falls apart, the motion of any part can be decomposed into a translation and a rotation. This paper only addresses stability with respect to translations.

INTRODUCTION

The automatic generation of assembly sequences for a product requires the computation of two predicates: *task-feasibility* and *subassembly-stability* [7]. The latter consists in finding whether or not there is a non-empty set of orientations such that if a part P_i of a subassembly is held fixed, the remaining parts maintain their relative positions, and do not break contact due to gravity. Furthermore, knowing what that set of orientations actually is (as opposed to only knowing whether or not the set is empty, or, perhaps, knowing just one element of that set) can be useful in selecting one assembly sequence among those that are feasible. For example, sequences that involve subassemblies that are stable for a large set of orientations are more reliable because small deviations from the nominal trajectory will not lead to the separation of the parts already assembled. Of course, the *subassembly-feasibility* predicate is a subproduct of a routine that computes the actual set of stable orientations.

Typically, in manufacturing, the joining of two subassemblies is executed by fixturing one (say S_1), and inserting the other (say S_2). The fixturing of S_1 must be in an orientation in which its parts do not move due to gravity. Furthermore, in addition to avoiding collisions, the insertion trajectory for S_2 should only use orientations in which its parts do not move due to gravity.

This paper presents a mathematical formulation for the computation of the set of stable orientations of a subassembly. The formulation fits well into the framework for assembly planning since it also uses the computation of cut-sets of the assembly's graph of connections, as well as the computation of the polyhedral convex cones of feasible translations.

When a subassembly falls apart, the motion of any part can be decomposed into a translation and a rotation. This paper addresses only stability with respect to translations. More precisely, the problem addressed can be stated as: given a subassembly $\mathcal{S}$ and one of its parts P_i, what are the orientations of $\mathcal{S}$ such that if P_i is held fixed, the remaining parts will not translate due to gravity. It is assumed that the contacts between the parts are all planar. This does not seem to be a serious limitation, and the results can be extended to other types of contacts.

The formulation presented neglects the friction in the contacts between the parts. It is assumed that friction can help the stability by preventing sliding motions, but cannot cause instability. Therefore, the results presented are actually conservative,

[1]This research was conducted at the Jet Propulsion Laboratory, California Institute of Technology, under contract with the National Aeronautics and Space Administration.

that is, the actual set of stable orientations may actually be larger than what is given by the development described in this paper. Also, there may be subassemblies that will only have stable orientations when there is friction in the contacts. For these assemblies, which seldom occur in practice, the approach presented in this paper will not find the stable orientations.

BACKGROUND

In order to compute the set of stable orientations for an assembly, one needs a representation of the set of parts, and of how they fit together. In addition, one also needs the ability to represent the directions along which parts can translate. The first subsection below reviews the main aspects of assembly representation that will be used in the developments that follow. The following subsection reviews the representation of feasible translations for a part constrained by planar contacts.

Representation of mechanical assemblies

The representation of a mechanical assemblies includes not only solid models for the individual parts, but also the geometric and mechanical relationships among the parts. (Srikanth and Turner [9] surveyed feature based models for mechanical assemblies.)

For the presentation in this paper, there is a global coordinate frame of reference, an assembly coordinate frame, and a local coordinate frame for each part. Unless otherwise specified, vectors are with respect to the global coordinate frame of reference. The *nominal* position and orientation of the assembly is such that the assembly frame coincides with the global frame. An arbitrary position and orientation of an assembly is characterized as follows: the position is given by the global coordinates a_x, a_y and a_z of the origin of its coordinate frame; and the orientation is given by the *roll-pitch-yaw* matrix [5], denoted by $\mathbf{R}_{\alpha,\beta,\gamma}$. The transformation from the global coordinate frame to the assembly coordinate frame is the concatenation of a translation by a_x, a_y and a_z, and a rotation $\mathbf{R}_{\alpha,\beta,\gamma}$. Both the set of three angles, roll (γ), pitch (β), and yaw (α), and the matrix $\mathbf{R}_{\alpha,\beta,\gamma}$ are referred to as the assembly's orientation. In the assembly's nominal position, these three angles, as well as a_x, a_y, and a_z, are zero.

A subassembly is said to be stable for translations with respect to one of its parts P_i, if and only if there is one orientation in which, if P_i is held fixed, no other part will translate due to gravity.

It is assumed that all parts are rigid, and that the product has no moving parts, that is, the relative position of one part with respect to another is constant. It is also assumed that the product can be put together by joining exactly two parts or subassemblies at a time. Wolter [10] shows an example of an assembly that can only be put together by joining 3 parts simultaneously.

The *graph of connections* [7] of an assembly is a graph for which: (1) there is a one-to-one correspondence between the nodes in the graph and the parts in the assembly; and (2) there is an edge between two nodes in the graph if and only if their corresponding parts have at least one surface contact.

Feasibility of local translations

Let P be a part that has K planar contacts with N ($N \leq K$) other parts. Let $\vec{n_1}, \vec{n_2}, \cdots, \vec{n_K}$ be the normal vectors to those K contacts, all pointing away from P. When all other parts are fixed, if P can translate along direction $\vec{t}$ then $\vec{t} \cdot \vec{n_i} \leq 0$ for $i = 1, 2, \cdots, K$. The set of vectors satisfying this system of inequalities constitutes a polyhedral convex cone. It will be referred to as the *polyhedral convex cone of feasible translations of P with respect to the set of K contacts.*

This polyhedral convex cone is the intersection of K closed half-spaces. Each of those half-spaces is actually the polyhedral convex cone of feasible translations of P with respect to one contact. Therefore, if P can translate (even only incrementally)

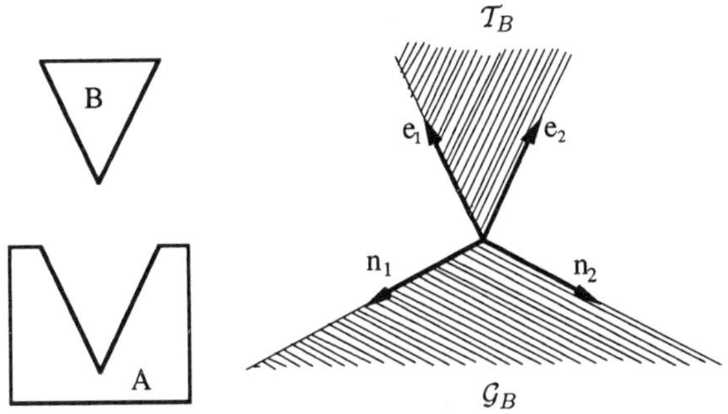

Figure 1: A two-dimensional assembly example in exploded view. $\mathcal{T}_B = \{\vec{u} | (\vec{u} \cdot \vec{n}_1 \leq 0) \wedge (\vec{u} \cdot \vec{n}_2 \leq 0)\} = \{\vec{v} | \exists a_1 \exists a_2 (a_1 \geq 0) \wedge (a_2 \geq 0) \wedge (\vec{v} = a_1 \cdot \vec{e}_1 + a_2 \cdot \vec{e}_2)\}$ is the polyhedral convex cone of feasible translations of B with respect to its contacts with A. Therefore $\mathcal{G}_B = \mathcal{T}_B^+ = \{\vec{v} | \exists a_1 \exists a_2 (a_1 \geq 0) \wedge (a_2 \geq 0) \wedge (\vec{v} = a_1 \cdot \vec{n}_1 + a_2 \cdot \vec{n}_2)\} = \{\vec{u} | (\vec{u} \cdot \vec{e}_1 \leq 0) \wedge (\vec{u} \cdot \vec{e}_2 \leq 0)\}$ is the polar of $\mathcal{T}_B$. It is straightforward to see that if part A is fixed, then no force $\vec{f} \in \mathcal{G}_B$ applied to B can cause it to move.

along $\vec{t}$, then $\vec{t}$ falls in the intersection of the polyhedral convex cones of feasible translations due to each contact.

Two distinct ways have been used to represent polyhedral cones [6]. A *tangential* representation is a set of normal vectors, such as $\{\vec{n_1}, \vec{n_2}, \cdots, \vec{n_K}\}$. A *point* representation is a set of vectors $\{\vec{e_1}, \vec{e_2}, \cdots, \vec{e_J}\}$ such that any vector in the cone can be expressed as a positive linear combination of $\vec{e_1}, \vec{e_2}, \cdots, \vec{e_J}$. Two representations are equivalent if they correspond to the same set of vectors. Furthermore, a *tangential* (*point*) representation is minimal if there is no other equivalent *tangential* (*point*) representation with fewer vectors. In the development of the next sections, it will be sometimes more convenient to use one representation, sometimes the other. It is straightforward to go back and forth between these two representations.

STABILITY OF A TWO-PART ASSEMBLY

Given an assembly made up of two parts A and B that have only planar contacts, let $\mathcal{T}_B$ be the polyhedral convex cone of feasible translations of a part B with respect to the set of planar contacts between A and B. As discussed above, a tangential representation of $\mathcal{T}_B$ is $\{\vec{n_1}, \vec{n_2}, \cdots, \vec{n_K}\}$, which is the set of normal vectors to the contacts between A and B, pointing away from B. Let $\mathcal{G}_B$ be the polar cone of $\mathcal{T}_B$, that is, $\mathcal{G}_B = \mathcal{T}_B^+ = \{\vec{u} \,|\, \forall \vec{v}\, [(\vec{v} \in \mathcal{T}_B) \Rightarrow (\vec{v} \cdot \vec{u} \leq 0)]\}$. Let $\vec{f}$ be a force applied to part B. If $\vec{f} \in \mathcal{G}_B$, and part A is kept fixed, then $\vec{f}$ will not cause part B to move. This is true because the directions along which part B can translate are those in $\mathcal{T}_B$. The force $\vec{f}$ cannot cause B to translate along any direction $\vec{t} \in \mathcal{T}_B$ because $\vec{f} \cdot \vec{t} \leq 0$.

Therefore, when A is fixed, forces that lie in $\mathcal{G}_B$ do not cause B to translate, where $\mathcal{G}_B$ is the polar of the polyhedral convex cone of feasible translations of B with respect to the set of planar contacts between A and B. Figure 1 shows a two dimensional example that clarifies this reasoning.

It should be noted that in the absence of friction and other forces, if $\vec{f} \notin \mathcal{G}_B$, then there exists at least one direction $\vec{t} \in \mathcal{T}_B$ such that $\vec{f} \cdot \vec{t} > 0$, and part B will translate.

The cone $\mathcal{G}_B$ is the polar of the intersection of the polyhedral convex cones of feasible translations of B due to each of its contacts with A. One property of polyhedral convex cones [8] is that the polar of the intersection $\mathcal{C}_1 \cap \mathcal{C}_2 \cap \cdots \cap \mathcal{C}_N$ of several cones is the sum $\mathcal{C}_1^+ + \mathcal{C}_2^+ + \cdots + \mathcal{C}_N^+$ of their polars, where the sum of two cones $\mathcal{C}_1$ and $\mathcal{C}_2$ is, by definition, the set $\mathcal{C}_1 + \mathcal{C}_2 = \{\vec{u} \mid \exists \vec{u_1}\, \exists \vec{u_2}\, (\vec{u_1} \in \mathcal{C}_1) \wedge (\vec{u_2} \in \mathcal{C}_2) \wedge (\vec{u} = \vec{u_1} + \vec{u_2})\}$. Since the polar of a closed half-space defined by $\vec{n}$ (i.e., the set $\{\, \vec{u} \mid \vec{n} \cdot \vec{u} \leq 0 \,\}$) is the ray (or half-line) defined by $\vec{n}$ (i.e., it is the set $\{\vec{v} \mid \exists \lambda\, (\lambda \geq 0) \wedge (\vec{v} = \lambda \cdot \vec{n})\}$), $\mathcal{G}_B$ is the sum of K rays. Therefore, it is straightforward to compute $\mathcal{G}_B$, given the set of contact normals. $\mathcal{G}_B$ is the polyhedral convex cone whose point representation is $\{\, \vec{n_1}\, ,\, \vec{n_2}\, ,\, \cdots\, ,\, \vec{n_K}\, \}$. It should be noted that the same set of vectors is a point representation of $\mathcal{G}_B$ and a tangential representation of $\mathcal{T}_B$.

From the discussion above, an assembly made up of two parts, A and B, is stable for translations in its nominal position and orientation, and with part A fixed, if $\vec{g} \in \mathcal{G}_B$. If the global frame is such that $\vec{g}$ has the same direction and orientation as $-\vec{z}$, the unit vector along the z axis, then the stability condition for B is $-\vec{z} \in \mathcal{G}_B$. Therefore, given the representation of the assembly $\{A\, ,\, B\}$, it is straightforward to decide whether or not, in the nominal position and orientation, part B will translate due to gravity when part A is held fixed.

As mentioned in the introduction, it is useful in assembly planning to know what is the set of orientations for the pair of parts $\{A\, ,\, B\}$ such that if A is fixed, then B will not translate due to gravity. That is, it is useful to know what are the stable orientations of the pair of parts $\{A\, ,\, B\}$, given that A is fixed.

When the orientation of the set of parts changes, $\mathcal{T}_B$, the set of feasible translations with respect to the global coordinate frame, also changes.

Given cone $\mathcal{C}$, the rotated cone $\mathcal{C}^{(\alpha,\beta,\gamma)}$ is, by definition, the set $\{\, \vec{u} \mid \exists\, \vec{v}\, [\, (\, \vec{v} \in \mathcal{C}\,) \wedge (\mathbf{R}_{\alpha,\beta,\gamma} \cdot \vec{v} = \vec{u})\,]\, \}$. Each vector $\vec{u}$ in $\mathcal{C}^{(\alpha,\beta,\gamma)}$ has a corresponding vector $\vec{v}$ in $\mathcal{C}$ such that $\vec{u} = \mathbf{R}_{\alpha,\beta,\gamma} \cdot \vec{v}$. Cone $\mathcal{C}^{(\alpha,\beta,\gamma)}$ is obtained by rotating all vectors of $\mathcal{C}$, first around x by γ, then around y by β, and then around z by α. The relation between $\mathcal{C}$ and $\mathcal{C}^{(\alpha,\beta,\gamma)}$ will be expressed by $\mathcal{C}^{(\alpha,\beta,\gamma)} = \mathbf{R}_{\alpha,\beta,\gamma} \cdot \mathcal{C}$.

Therefore, given an orientation $\mathbf{R}_{(\alpha,\beta,\gamma)}$ of the set of parts $\{A\, ,\, B\}$, the polyhedral convex cone of feasible translations of B with respect to its contacts with A is $\mathcal{T}_B^{(\alpha,\beta,\gamma)} = \mathbf{R}_{\alpha,\beta,\gamma} \cdot \mathcal{T}_B$. The polar of $\mathcal{T}_B^{(\alpha,\beta,\gamma)}$ is $\mathcal{G}_B^{(\alpha,\beta,\gamma)} = \mathbf{R}_{\alpha,\beta,\gamma} \cdot \mathcal{G}_B$. In that orientation, B will not translate due to gravity if $-\vec{z} \in \mathcal{G}_B^{(\alpha,\beta,\gamma)}$. This stability condition is equivalent to $\mathbf{R}_{\alpha,\beta,\gamma}^{-1} \cdot -\vec{z} \in \mathcal{G}_B$. Remembering that the inverse of $\mathbf{R}_{\alpha,\beta,\gamma}$ is its transpose [5], the stability condition for an arbitrary orientation can be rewritten as

$$\begin{bmatrix} \sin\beta \\ -\cos\beta \cdot \sin\gamma \\ -\cos\beta \cdot \cos\gamma \end{bmatrix} \in \mathcal{G}_B. \tag{1}$$

The point representation of $\mathcal{G}_B$ was given above. Therefore, it is straightforward to obtain its tangential representation. Let $\{\, \vec{r_1}\, ,\, \vec{r_2}\, ,\, \cdots\, ,\, \vec{r_L}\, \}$ be a tangential representation of $\mathcal{G}_B$. Let $R = [\, \vec{r_1}\, \vec{r_2}\, \cdots\, \vec{r_L}\,]$ be a $3 \times L$ matrix whose j^{th} column is $\vec{r_j}$. Equation 1 requires that

$$R^T \cdot \begin{bmatrix} \sin\beta \\ -\cos\beta \cdot \sin\gamma \\ -\cos\beta \cdot \cos\gamma \end{bmatrix} \leq \begin{bmatrix} 0 \\ 0 \\ 0 \end{bmatrix}. \tag{2}$$

It should be noted that the yaw angle (α) does not appear in equations 1 and 2. This is the mathematical representation of the fact that, as one would expect, if the gravity vector $\vec{g}$ has the same direction and orientation as $-\vec{z}$, rotations around the z axis do not affect the stability.

An example will illustrate the use of condition 2. Figure 2 shows a two-part assembly. A point representation of $\mathcal{G}_B$ is the set of three vectors $\{\, \vec{n_1}\, ,\, \vec{n_2}\, ,\, \vec{n_3}\, \}$ where $\vec{n_1} = \begin{bmatrix} -1 & 0 & 0 \end{bmatrix}^T$, $\vec{n_2} = \begin{bmatrix} 0 & 0 & -1 \end{bmatrix}^T$, and $\vec{n_3} = \begin{bmatrix} 0 & \frac{1}{2} & -\frac{\sqrt{3}}{2} \end{bmatrix}^T$. These are the normals to the contacts between A and B pointing away from B. Therefore,

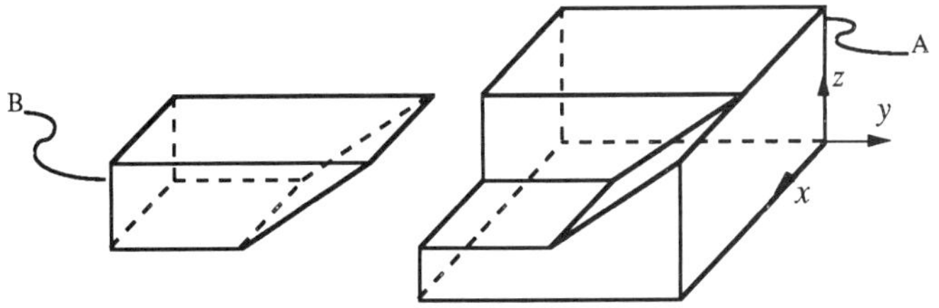

Figure 2: A two part assembly in exploded view

the tangential representation of $\mathcal{G}_B$ is the set of three vectors $\{\vec{r_1}, \vec{r_2}, \vec{r_3}\}$ where $\vec{r_1} = \begin{bmatrix} 0 & -1 & 0 \end{bmatrix}^T$, $\vec{r_2} = \begin{bmatrix} \frac{1}{2} & 0 & 0 \end{bmatrix}^T$, and $\vec{r_3} = \begin{bmatrix} 0 & \frac{\sqrt{3}}{2} & \frac{1}{2} \end{bmatrix}^T$. These are normals, the first to the plane that contains $\vec{n_1}$ and $\vec{n_2}$, the second to the plane that contains $\vec{n_2}$ and $\vec{n_3}$, and the third to the plane that contains $\vec{n_1}$ and $\vec{n_3}$. The orientations of $\vec{r_1}$, $\vec{r_2}$, and $\vec{r_3}$ are such that $\vec{r_i} \cdot \vec{n_j} \leq 0$, for all i and all j.

Using equation 2 the following conditions must be satisfied for the stability of the assembly shown in figure 2: (1) $\cos\beta \cdot \sin\gamma \leq 0$; (2) $\frac{1}{2} \cdot \sin\beta \leq 0$; and (3) $-\frac{\sqrt{3}}{2} \cdot \cos\beta \cdot \sin\gamma - \frac{1}{2} \cdot \cos\beta \cdot \cos\gamma \leq 0$. These conditions can be shown to be equivalent to $-\frac{\pi}{2} \leq \beta \leq 0$ and $-\frac{\pi}{6} \leq \gamma \leq 0$. For this simple assembly, one can easily verify, by inspection of figure 2, that these are in fact the stability conditions.

STABILITY OF AN N-PART ASSEMBLY

When the assembly has more than two parts, there are several ways in which the parts can separate from each other. Each of these ways corresponds to a cut-set of the graph of connections. Even when the assembly breaks down into several subassemblies, one can view it as a sequence of break downs into two subsets, each corresponding to a cut-set.

By considering the parts in either subset to be glued together, the analysis of the previous section can be applied to each cut-set. The subset that contains the base (i.e. the part that is fixed) is considered fixed.

Each cut-set yields a set of conditions on the angles β and γ. The intersection of all conditions corresponds to the set of orientations for which the assembly is stable.

For example, the assembly depicted in figure 3 has three parts. There are three cut-sets in its graph of connections. The first cut-set corresponds to the separation of C from the rest of the assembly. The normals to the contacts of C with the rest of the assembly are: $\vec{n_1} = \begin{bmatrix} -1 & 0 & 0 \end{bmatrix}^T$, $\vec{n_2} = \begin{bmatrix} 0 & -\frac{\sqrt{2}}{2} & -\frac{\sqrt{2}}{2} \end{bmatrix}^T$, and $\vec{n_3} = \begin{bmatrix} 0 & \frac{\sqrt{2}}{2} & -\frac{\sqrt{2}}{2} \end{bmatrix}^T$. The first two vectors correspond to the contacts of C and A, pointing away from C, and the third vector corresponds to the contact of C and B, also pointing away from C. Following the procedure used in the previous section yields the following conditions on the angles β and γ: $-\frac{\pi}{2} \leq \beta \leq 0$ and $-\frac{\pi}{4} \leq \gamma \leq \frac{\pi}{4}$. If parts A and B were glued together and held fixed, the only orientations of the assembly for which part C will not translate are those satisfying these conditions.

The second cut-set corresponds to the separation of part B from the rest of the assembly. The normals to the contacts of B with the rest of the assembly are: $\vec{n_1} = \begin{bmatrix} 0 & -\frac{\sqrt{2}}{2} & \frac{\sqrt{2}}{2} \end{bmatrix}^T$, $\vec{n_2} = \begin{bmatrix} 0 & \frac{1}{2} & -\frac{\sqrt{3}}{2} \end{bmatrix}^T$, $\vec{n_3} = \begin{bmatrix} 0 & 0 & -1 \end{bmatrix}^T$, $\vec{n_4} = \begin{bmatrix} 0 & -\frac{\sqrt{2}}{2} & -\frac{\sqrt{2}}{2} \end{bmatrix}^T$, and $\vec{n_5} = \begin{bmatrix} -1 & 0 & 0 \end{bmatrix}^T$. The first vector corresponds to the contact of B and C, pointing away from B, and the other four vectors correspond to the contacts of B and A, also pointing away from B. Following the procedure used in

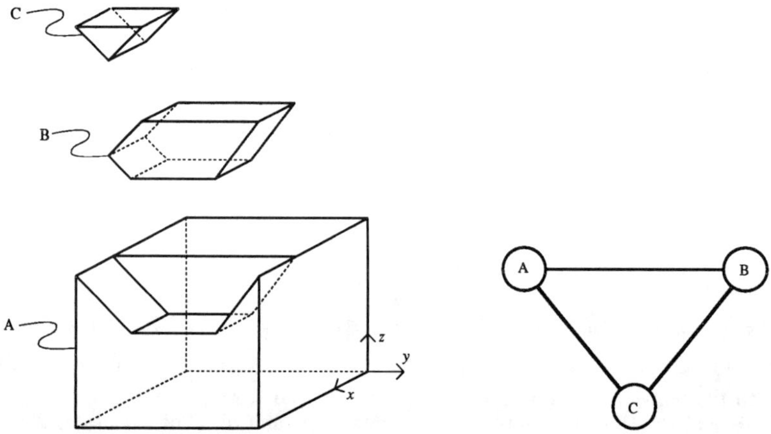

Figure 3: A three part assembly in exploded view and its graph of connections

the previous section yields the following conditions on the angles β and γ: $-\frac{\pi}{2} \leq \beta \leq 0$ and $-\frac{\pi}{3} \leq \gamma \leq \frac{3 \cdot \pi}{4}$. If parts A and C were glued together and held fixed, the only orientations of the assembly for which part B will not translate are those satisfying these conditions[2].

The third cut-set corresponds to the separation of parts B and C as a subassembly (i.e. as if they were glued together) from the rest of the assembly, which in this case has only one part. The normals to the contacts of B and C with A, pointing away from A and towards B or C are: $\vec{n_1} = \begin{bmatrix} -1 & 0 & 0 \end{bmatrix}^T$, $\vec{n_2} = \begin{bmatrix} 0 & -\frac{\sqrt{2}}{2} & -\frac{\sqrt{2}}{2} \end{bmatrix}^T$, $\vec{n_3} = \begin{bmatrix} 0 & 0 & -1 \end{bmatrix}^T$, and $\vec{n_4} = \begin{bmatrix} 0 & \frac{1}{2} & -\frac{\sqrt{3}}{2} \end{bmatrix}^T$. Following the procedure used in the previous section yields the following conditions on the angles β and γ: $-\frac{\pi}{2} \leq \beta \leq 0$ and $-\frac{\pi}{3} \leq \gamma \leq \frac{\pi}{4}$. The only orientations of the assembly for which the subset of parts $\{B, C\}$ will not translate are those satisfying these conditions.

In order to guarantee that no separation of parts will occur, these three sets of constraints must be satisfied, that is, the angles β and γ must be such that $-\frac{\pi}{2} \leq \beta \leq 0$ and $-\frac{\pi}{3} \leq \gamma \leq \frac{\pi}{4}$.

OTHER TYPES OF CONTACT

The formulation presented in the previous two sections assumed that all contacts between the parts are planar. One solution to handle other types of contacts that do not have attachments[3] is to approximate the surfaces by planar patches. But for many contacts the basic results about the stability of two-part subassemblies still can be used, that is, $\mathcal{G}_B$ is the polar of $\mathcal{T}_B$. Also, since $\mathcal{T}_B$ is the intersection of several polyhedral convex cones, $\mathcal{G}_B$ is the sum of their polars. The only thing that no longer is valid is having the same set of vectors being a tangential representation of $\mathcal{T}_B$ and a point representation of $\mathcal{G}_B$. While in the case of only planar contacts no computation was needed to obtain the point representation of $\mathcal{G}_B$, for other types of contacts some computation may be needed.

[2]This analysis addresses translations only. Assuming that A and C are glued, for $\gamma = \frac{3 \cdot \pi}{4}$ and $\beta = 0$ part B may rotate due to gravity.

[3]When contacts have attachments (e.g., pressure fit, or threaded cylindrical), the two parts may be considered to be one.

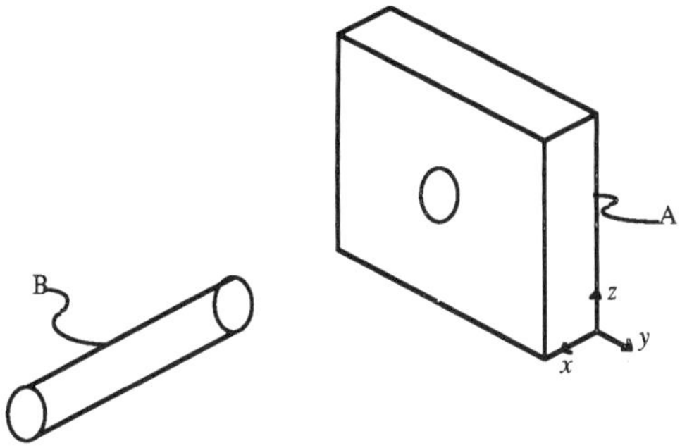

Figure 4: Cylindrical contact

As an example, consider the two-part assembly depicted in figure 4. There is only one cylindrical contact between the two parts. The polyhedral convex cone of feasible translations of B due to its contacts with A, $\mathcal{T}_B$, is the x axis. The polar of $\mathcal{T}_B$, $\mathcal{G}_B$, is the y–z plane. By the same reasoning used in the case of two-part assemblies that have only planar contacts, the stability condition for an arbitrary orientation is $\mathbf{R}^{-1}_{\alpha,\beta,\gamma} \cdot -\vec{z} \in \mathcal{G}_B$, which translates into equation 2. Since the tangential representation of $\mathcal{G}_B$ is the pair of vectors $\{ \vec{x}, -\vec{x} \}$, where $\vec{x}$ is the unit vector along the x axis, the stability conditions (from equation 2) are:

$$\begin{bmatrix} 1 & 0 & 0 \\ -1 & 0 & 0 \end{bmatrix} \cdot \begin{bmatrix} \sin\beta \\ -\cos\beta \cdot \sin\gamma \\ -\cos\beta \cdot \cos\gamma \end{bmatrix} \geq \begin{bmatrix} 0 \\ 0 \\ 0 \end{bmatrix}$$

These conditions can be shown to be equivalent to $\beta = 0$. The angle γ (and, of course, α) can take any value. Again, it is straightforward to check this result by inspecting figure 4.

Compared with the development for the case of planar contacts only, the additional complication of the analysis above stems from the fact that the point representation of $\mathcal{G}_B$ is not the same as the tangential representation of $\mathcal{T}_B$.

DISCUSSION

Unlike other approaches [3, 4], the development in this paper does not require that the subassembly be resting on a table in order to join it with another part or subassembly. It is assumed that one subassembly is fixed by using a fixturing device such as, for example, the one described by Asada and By [1]. In addition, instead of just testing whether or not an orientation is stable, the formulation described in this paper shows how to compute the set of all stable orientations. This is particularly useful in assembly planning since it allows the selection of assembly sequences in which the set of stable orientations is larger, and, therefore, the joining task is less sensitive to small inaccuracies in the fixturing and in the insertion trajectory. Furthermore, the development in this paper does not require the solution of a linear programming problem, which tends to be computationally intensive. Instead, it uses a geometric reasoning approach. Of course, a test of stability for a given orientation is a subproduct of the formulation presented in this paper.

The symbolic solution of the system of inequalities that results from condition 1 is not always straightforward as in the examples shown in this paper. But since there are only two variables, β and γ, the system of inequalities can be solved numerically without much difficulty.

The formulation presented in this paper requires the computation of the cut-sets of the graph of connections. Although this computation can be time-consuming, it is performed in assembly planning anyway [2, 7]. Therefore, it does not constitute an additional burden.

The main shorthcoming of the development described in this paper seems to be that it does not take into account that parts may rotate due to gravity. It is not clear whether or not a similarly simple formulation that includes rotational motions can be found. If one such formulation is not found, methods such as those described by Blum et al. [3] and by Boneschanscher et al. [4] can be used to verify stability with respect to rotations after a set of stable orientations is found by the method described in this paper. That is, the set of all orientations is pruned before running a computationally intensive algorithm.

CONCLUSION

This paper presented a simple method to determine, given a part P_i of a subassembly, the set of all orientations of that subassembly such that, if P_i is fixed, the other parts do not translate due to gravity. This set of orientations is needed in planning the assembly sequence for a product, and the formulation presented fits well into the framework for assembly planning.

The contribution of this paper is twofold. Unlike other methods, which require the solution of a linear programming problem, the formulation presented in this paper requires simple geometric reasoning about polyhedral convex cones. Furthermore, while other methods only test the stability for given orientations, the formulation presented in this paper computes the set of all stable orientations. The main shortcoming of this paper is that it does not take rotations into account. Future work should address the possibility that parts will separate by rotational motions.

References

[1] H. Asada and A. B. By. Kinematic analysis of workpart fixturing for flexible assembly with automatically reconfigurable fixtures. *IEEE Journal of Robotics and Automation*, RA-1(2):86–94, 1985.

[2] D. F. Baldwin et al. An Integrated Computer Aid for Generating and Evaluating Assembly Sequences for Mechanical Products. *IEEE Transactions on Robotics and Automation*, 7(1):78–94, 1991.

[3] M. Blum et al. A stability test for configurations of blocks. AI Memo 188, Massachusetts Institute of Technology – Artificial Intelligence Laboratory, 1970.

[4] N. Boneschanscher et al. Subassembly stability. In *Proceedings of AAAI-88*, pp. 780–785. Morgan Kaufmann, 1988.

[5] J. J. Craig. *Introduction to Robotics*. Addison-Wesley, 1986.

[6] L. S. Homem de Mello and A. C. Sanderson. Automatic Generation of Mechanical Assembly Sequences. Tech. Rep. CMU-RI-TR-88-19, The Robotics Institute – Carnegie Mellon University, 1988.

[7] L. S. Homem de Mello and A. C. Sanderson. A Correct and Complete Algorithm for the Generation of Mechanical Assembly Sequences. *IEEE Transactions on Robotics and Automation*, 7(2):228–240, 1991.

[8] M. Simonnard. *Linear Programming*. Prentice-Hall, 1966.

[9] S. Srikanth and J. U. Turner. Toward a Unified Representation of Mechanical Assemblies. *Engineering with Computers*, 6:103–112, 1990.

[10] J. D. Wolter. *On the Automatic Generation of Plans for Mechanical Assembly*. PhD thesis, University of Michigan, September 1988.

An Integrated Approach to Shape Design, Stress Analysis, and Cutting Pattern Generation of Tension Structures

Z. Dong, B. Tabarrok, and Z. Qin
Department of Mechanical Engineering
University of Victoria
Victoria, B.C. Canada V8W 3P6

Abstract

An integrated approach to shape design, stress analysis, and cutting pattern generation of tension structures, based on the concept of CIM, is presented. The initial shape design of a tension structure, using the "hanging cable" approximation and its implementation as a general tool in the AutoCAD system, is introduced. The fine adjustment of the initial shape to an equilibrium surface using nonlinear finite elements method is presented. Stress analysis of the designed tension structure under various loading conditions are demonstrated. Cutting pattern generation of the design shape, using 3D curved surface flattening and stress relaxation are illustrated. Methods for interfacing the developed programs to commercial CAD environments are discussed. A few examples are used to illustrate the approach and the functions of the prototype system.

1 Introduction

Tension structures are widely used due to the low construction costs, high mobile properties, and aesthetic appearances. The design and construction of a tension structure, however, is a complex task. The shape of a tension structure has to be determined based on the property of membrane material and the frame that supports the membrane. The strength and failure analysis of a tension structure has to be done using non-linear mechanics and finite element analysis. To construct a tension structure, cutting patterns of the membrane has to be obtained by curved surface flattening with stress relaxation.

Traditionally, the design and construction of a tension structure is accomplished with the help of a down-scaled physical model. This manual approach leads to a long design process, uncertain mechanical behavior, and poorly shaped structure.

Attempts are made to automate the design, analysis and cutting pattern generation tasks using computers. The design of a tension structure through computer-aided form finding have been investigated by many researchers [1-3]. Various numerical processes were used to form the shape of a tension structure under given conditions and to calculate the stress distributions on the formed structure. Numerical approach to automated cutting pattern generation, finding the shapes of fabric strips which when joined together form the curved surface of the tension structure, was reported [4]. However, the design and manufacture of tension structure is a complex task. The previous work was mostly devoted to a particular structure, and a dedicated function of automation. A systematic study on tension structure design, analysis, and cutting pattern generation, and an integrated computer system for accomplishing these tasks need to be developed.

In this paper, an integrated approach to shape design, stress analysis, and cutting pattern generation for a tension structure, based on the concept of Computer-Integrated Manufacturing (CIM), is presented. The approach breaks down the complex task of tension structure design and construction into three computer automated sub-modules: (1) form-finding through approximate shape design and/or equilibrium shape generation, (2) stress analysis under different loads, and (3) membrane material cutting pattern generation. These sub-modules of design, analysis and manufacturing share a common database and interface with the AutoCAD system, thus leading to an integrated computer system. The relations among these system modules are illustrated in Fig. 1.

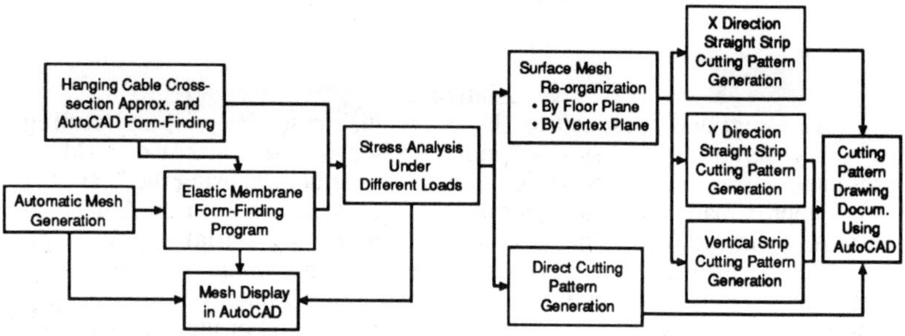

Figure 1 Integrated Design, Analysis and Cutting Pattern Generation for A Tension Structure

2 Shape Design and Form Finding

The shape of a tension structure is formed following the physical properties of the membrane material and the structure of the frame that supports the membrane. This property of tension structure separates the shape design of a tension structure from the conventional geometry design of mechanical products. A form finding process is used to transfer the intent of the designer, through the control of frame structure and membrane tension level, into the naturally formed shape. Three form finding schemes have been developed: the *hanging cable* method, the *minimum surface* method and the *elastic membrane form finding simulation* method.

2.1 From Finding Using Hanging Cable Approximation and AutoCAD

In today's mechanical design, a designer usually uses a CAD drawing to specify the design geometry. The CAD package serves as a tool for expressing the design ideas and for documenting the design. However, the ordinary CAD functions cannot be directly used to specify the shape of a tension structure.

Since the curve of a hanging cable can provides a good estimate to the cross section shape of a tension structure, a hanging cable curve generation function is added to the CAD system for the rough shape specification of a tension structure. This approximation leads to tent surfaces similar to the minimum surface with uniform stress. By controlling the tension applied to the cable (or the ratio of the distance between the two end points and the cable length), various shapes of the hanging cable can be obtained. These hanging cable curves (or soft boundaries), together with the "hard" boundaries provided by a CAD system, can be used to approximate the shape of a tension structure. The method is illustrated in Fig. 2.

The advantage of this approach is its ease of use. The user do not have to run complicated form-finding programs. Only the hard boundaries, soft boundaries and

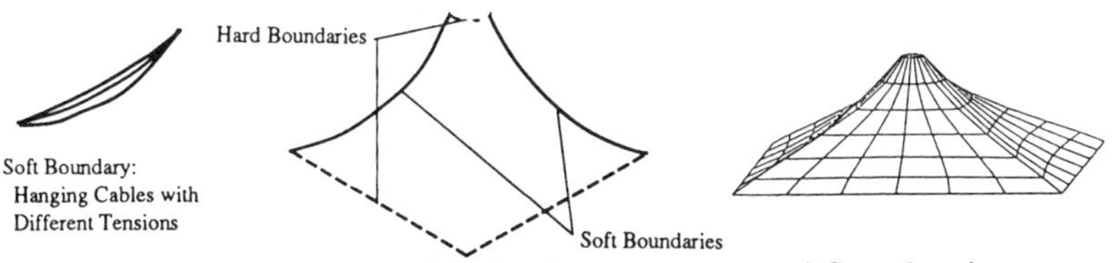

a) Cross section shapes b) Boundaries of tension structure c) Created tension structure

Figure 2 Interactive Design of A Tension Structure in AutoCAD

surface forming function of a CAD system needs to be used to form the curved surface. Drawbacks of the approach include its approximation error and its ignorance of panel interactions.

2.2 Elastic Membrane Form Finding Simulation Scheme

The procedure for form-finding, using the large displacement finite element analysis, is based on the same procedure used for stress analysis. A finite element mesh is initially generated on a flat plane. The designer then specifies a simple prestress distribution. A three-dimensional form is created by displacing the support points vertically until they attain their prescribed positions.

Minimum surfaces are often used in the design of tension structures. To create a minimum surface, a fictitious constitutive law which maintains a constant prestress, independent of any changes in strain, is used. Such a constitutive law corresponds to a zero Young's modules (in practical application, a very small Young's modules is used to avoid numerical instability). If a uniform prestress is specified, the tension structure obtained will have the specified uniform prestress in equilibrium, and form a minimum surface. The advantage of the minimum surface is its aesthetically pleasing shape and the associated uniform tensile stress in the membrane. Since the mean curvature for a minimum surface is zero, such surfaces tend to be rather flat. On the other hand, the load bearing capacity of a membrane, normal to its surface, depends on its curvatures. For this reason, minimum surfaces will not always be desirable for tension structures.

To provide the designer greater choice in form finding of tension structures, a more flexible method has been developed. This method involves the application of an appropriate external pressure load. Either a flat mesh created using the automatic mesh generation program or the surface mesh of the roughly formed shape using hanging cable scheme can be used to generate the initial mesh surface. Then the designer specifies a uniform initial stress for the membrane and a constant tension. A very small Young's modules is still used. The initial equilibrium configuration is obtained by displacing the support points and applying the specified external pressure simultaneously. The resulting shape depends on the ratio of the initial stress to the external pressure. This shape is in equilibrium under the pressure. Once a desirable shape has been found, several more iterations are needed to eliminate the external pressure and to find an equilibrating stress distribution associated with the resulting shape. For this final stage, the actual values of Young's modules for the membrane and cables are used. The resulting disequilibrium is thus eliminated through iterations and the support locations are held fixed. Because the pressure load used for form finding is usually very small, the final

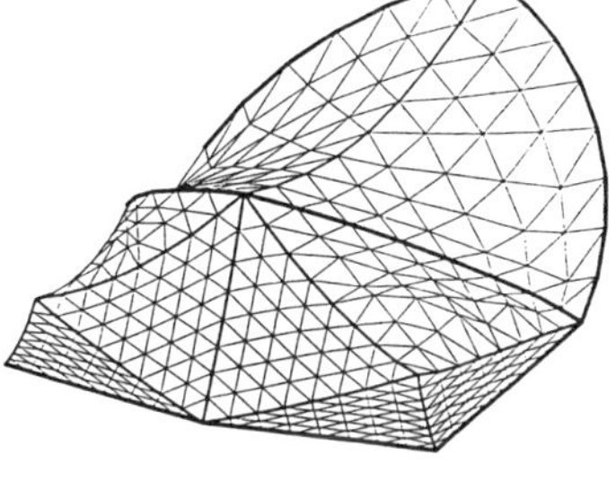

Figure 3

self-equilibrium shape, after elimination of the pressure load, will be only slightly different from the preliminary one, and the resulting non-uniform stress will be of the same order as that of the initial uniform stress. This method allows an infinite variety of feasible shapes to be generated. The procedure is equivalent to minimum surface analysis, subject to constraints of cable lengths on the surface and total volume covered by the surface. A tension structure formed using this method is illustrated in Figure 3.

3 Stress Analysis

To insure a safe and efficient design, the behavior of a formed tension structure under different loading conditions needs to be studied. In this work, five types of loading: (1) concentrated nodal loads, (2) dead loads (self-weight), (3) uniform pressure, (4) snow loads, and (5) wind loads, are defined. In addition to failure upon exceeding the ultimate stress, the lack of flexural stiffness in membrane structures leads to large deformations to changing loads. Wrinkling may occur when external loads give rise to compressive stress larger than the initial tensile stress.

The computation of wind loads is a complex problem. Wind pressure on an arbitrary surface can be expressed as:

$$P = \frac{1}{2}\rho U^2 C_p$$

where ρ is the air density, V is the wind speed and C_p is the pressure coefficient which depends on the wind direction and the geometry of the surface. The wind pressure coefficient C_p is usually determined experimentally.

A simple wind-load model has been incorporated into the present computer program. The user enters a magnitude that defines the wind pressure on a plane surface normal to the wind direction, and the direction of the wind, measured in degrees from the horizontal plane and X direction. The normal wind pressure on each membrane element is computed by projecting the wind magnitude to the direction of outward surface normal of every mesh elements. This model gives zero pressure on horizontal surfaces and suction on leeward surfaces. This approximation avoids the complex measurement for wind pressure coefficient, C_p, via experiments. An example, showing the stress distribution of the structure in Fig.3, is shown in Fig. 4.

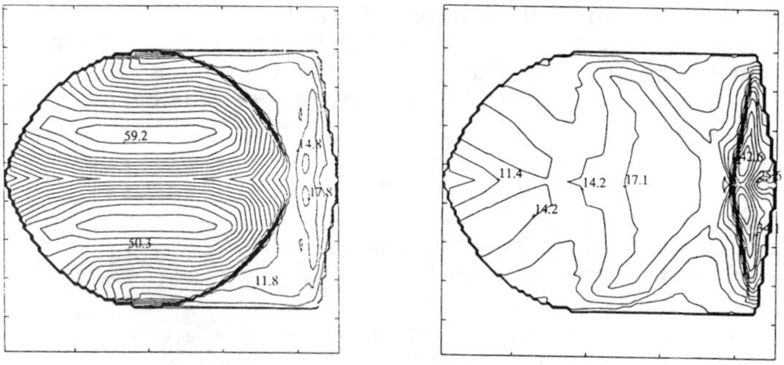

Figure 4 Stress Contours Under Snow Load and 180° Direction Wind Load

4 Cutting Pattern Generation

The cutting pattern generation module is used for finding the shapes of fabric strips which when joined together form the curved surface of the tension structure. This task is accomplished by flattening the curved surfaces of a formed tension structure.

In addition, the dimension and layout of fabric strips need to be considered to provide good aesthetic appearances for a tension structure, built using color strips.

This work allows the fiber strips of a tension structure to be lain out along two perpendicular direction, X and Y; an arbitrary direction defined by the user; and the direction perpendicular to the bottom edge of the surface panel. This flexibility allows a designer to construct a tension structure using mixed color strips to create good aesthetic appearances, which is essential for temporary exhibition halls or shelters for outdoor concert.

The cutting pattern generation starts from obtaining expected direction of fiber strip layout and the width of fiber strips from the designer. Based upon this information, new mesh surfaces of the tension structure are generated. The new mesh surfaces form a same shape as the original tension structure. But the mesh grids are re-generated to facility the cutting pattern generation along the given direction. The cutting pattern generation can thus be classified as:

- Cutting pattern generation using Floor Plane Mesh

  The tension structure, made using the cutting pattern based on the Floor Plane Mesh, presents a top view of straight fiber strips. The new mesh is created in the X-Y projection plane, with high density grids along the direction of strip layout, and low density grids (equal to fiber strip width) in its orthogonal direction. An example is illustrated in Figure 5.

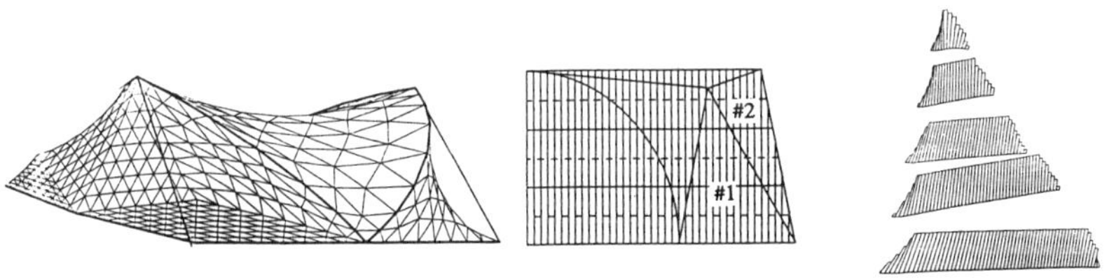

(a) Formed Structure (b) Floor Plane Mesh (c) Cutting Pattern of Panel #1

Figure 5 Cutting Pattern Generation Using Floor Plane Mesh

- Cutting pattern generation using the Vertex Plane Mesh

  The tension structure, made using the cutting pattern based on the Vertex Plane Mesh, has fiber strips along an edge of the surface panel. The plane that passes three vertexes of a surface panel is used as the mesh plane. High density grids are created along the direction of strip layout, and low density grids (equal to fiber strip width) are created along its orthogonal direction. The method is particular suitable for the almost vertical surface panels, of which the projections onto the X-Y plane will be quite narrow. An example is illustrated in Figure 6.

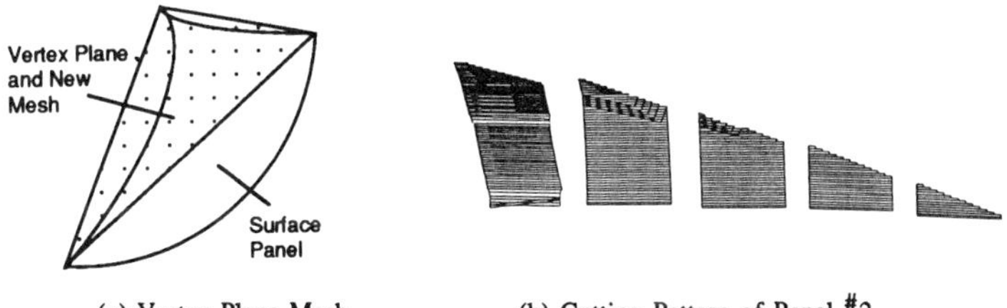

(a) Vertex Plane Mesh (b) Cutting Pattern of Panel #2

Figure 6 Cutting Pattern Generation Using Vertex Plane Mesh

5 Summary

An integrated approach to shape design, stress analysis, and cutting pattern generation of tension structures, based on the concept of Computer-Integrated Manufacturing (CIM), is presented. The approach breaks down the complex task of tension structure design and construction into three computer automated sub-modules: (1) form-finding through approximate shape design and/or equilibrium shape generation, (2) stress analysis under different loads, and (3) membrane material cutting pattern generation. These sub-modules of design, analysis and manufacturing are integrated by sharing a common database and interfacing with a general purpose CAD system.

6 Acknowledgement

Supports from the Science Council of British Columbia and the Warner Shelter Corporation are gratefully acknowledged.

References

1. Haber, R.B., and Abel, J.F., "Initial Equilibrium Solution Methods for Cable Reinforced Membranes, Part I Formulations, Part II Implementation, *Computing Methods Appl. Mech. Engrg.*, Vol. 30, 1982, 263-306.

2. Grundig, L., "Minimal Surfaces for Finding Forms of Structural Membranes," *Proceedings of Third International Conference on Civil and Structural Engineering*, Edinburgh, 1987.

3. Barnes, M.R., "Computer Aided Design of the Shade Membrane Roofs for Expr 88," *Structure Engineering Review,* Vol. 1, 1988, 3-13.

4. Moncrieft, E., and Topping, B.H.V., "Computer Methods for Generation of Membrane Cutting Patterns, *Computers and Structures*, Vol. 35, 1990, 441-450.

BASIC THEORY OF AUTOMATIC FIXTURE DESIGN FOR THIN-WALLED PARTS

Jian-Xia Liu, Doug Strong and Ostap Hawaleshka

Department of Mechanical and Industrial Engineering
University of Manitoba
Winnipeg, Manitoba, Canada R3T 2N2

ABSTRACT

In this paper the basic theory of automatic fixture design for parts with thin walls will be presented. A definition of thin-walled parts is given. A general force model of a fixture has also been developed. Supporting techniques of thin-walled parts have been exploited.

INTRODUCTION

Since Berry [1] first presented the idea of using CAD/CAM systems to automate fixture design in 1982, fixture design automation has been studied extensively and significant progress has been made. At least seventeen automated fixture design systems have been developed [2]. However, from previous literature research it has been found that automatic fixture design is still in it's youthful stage and fundamental theory regarding force models of a fixture for automatic fixture design is inadequate. Existing systems can't represent a general fixture design case because they neglect support planning within the fixture configuration. Automatic fixture design dealing with thin-walled prismatic parts has been studied by the authors. Basic theory in automatic fixture design on thin-walled parts as a workpiece model has universal representative significance. This paper presents some basic work which tries to perfect fundamental theory of fixture design automation.

PROBLEMS OF THE PARTS WITH THIN WALLS IN MANUFACTURING

In manufacturing terminology, machined parts may be classified as rotational parts or non-rotational parts. A rotational part has a symmetrical rotational axis of geometry. Shaft, bush, tray and so on are examples of rotational parts. In contrast, a non-rotational part does not have a symmetrical rotational axis of geometry. The prismatic part, plate, box, and others are examples of non-rotational parts. This classification method is very popular because it is easy to choose machining approaches for parts. For example, when it is a rotational part (shaft), turning operations would be chosen for machining the shaft. From a fixture and machining point of view, this paper introduces another parts classification method. Parts may be classified into parts with or without thin walls based on whether a part is deflected sufficiently to affect machining accuracy in manufacturing. In machining thin-walled parts the manufacturing processes are very complicated and it is difficult to obtain desired accuracy. The reason for their complexity is that thin-walled parts must consider both machining process planning and fixture design. However, each of these two aspects requires consideration by different engineering departments. Therefore this method of classification is particularly effective in the case of thin-walled parts.

The main problem with thin-walled parts in the machining state is elastic deformations. Elastic deformations are caused by many factors such as the cutting force, the cutting temperature, the frictional force, the rigidity of the workpiece, the clamping force and the internal stress. The rigidity of a workpiece with thin walls is very poor. While a part is machined and cutting forces are applied on the part, elastic deformation of the wokpiece occurs. The part was designed by consumers or engineers who are in charge of the product design. Thus, tool designers and process planners may not change directly the size of the workpiece in order to increase the rigidity of parts. It is possible to reduce cutting force, optimize the cutting process and use the appropriate supporting techniques of the fixture to reduce elastic deformations.

DEFINITION OF PARTS WITH THIN WALLS

There are several definitions for thin walled elements. Fitzgerald [3] gave a guide to differentiating between thin-walled and thick-walled cylinders based on the assumption that the stress distribution is uniform throughout the thickness of the walls. A ratio of wall thickness to diameter of about 1/10 represents a convenient dividing line between thin-walled cylinders and thick-walled cylinders. Young [4] gave a guide to differentiating among super-thin plates, thin plates and thick plates. B is the shorter length of two edges in the plate. H is the thickness of the plate.

The thin plate is defined as: $(1/80 \sim 1/100) \leq h/b \leq (1/5 \sim 1/8)$.

The super-thin plate is defined as : $h/b < (1/80 \sim 1/100)$.

The thick plate is defined as : $h/b > (1/5 \sim 1/8)$.

The theory of thin plates with small deflection (Timoshenko) is used by Young, who points out some limitations of the theory. Maximum deflection is represented by w. The thickness of the plates is h. The shorter length of two edges is b. Then, when the ratio $w / h = 1/10 \sim 1/5$, or $w / b \leq 1/50$, it may be considered that deflections are within the theory of thin plates with small deflection. The above definitions of thin walled members are suitable for their discussed topic, however, none of them can be applied directly to our study range. A definition of parts with thin walls is given by the authors based on whether or not a part is deflected sufficiently to affect machining accuracy in manufacturing.

A part with thin walls is broadly defined as one where elastic deformation of the walls is larger than or equal to allowed machining tolerance of the walls. This is represented by

$w \geq T$, or $w / T \geq 1$, where w is elastic deformation of a part, and T is the allowed machining tolerance of the part.

By knowing the deflection and tolerances of a part it is possible to identify whether or not it is a thin-walled part.

THE ANALYSIS OF A THIN WALL FROM A MACHINING AND FIXTURING POINT OF VIEW

In this section, an example of the analysis of a thin wall will be presented by using the finite element method from a machining and fixturing point of view. Fig. 1 shows a thin-walled part which is to be machined. The bold line represents machined faces of the workpiece. The thin wall of the workpiece in the FEA force model may be considered as a thin plate with one fixed edge and three free edges. Fig. 2 shows a model of finite element calculation. The cutting force in the end milling-peripheral will be assumed to be a uniformly distributed force which is equivalent in this case to a uniformly distributed concentrated load applied on eight nodes in the y direction. The cutting process will be assumed to be one cut with the cutter moving from X = 0 to X = 80 mm (length of workpiece) and cutting force along the y direction will be assumed as 30 N. Therefore, the force on each node will be 3.75 N.

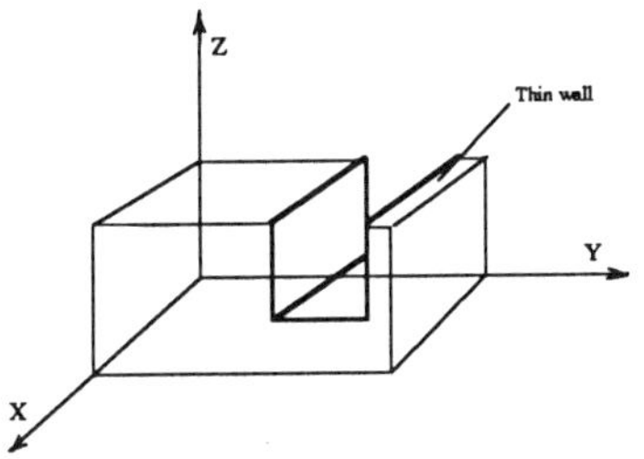

FIG. 1. A part with a thin wall.

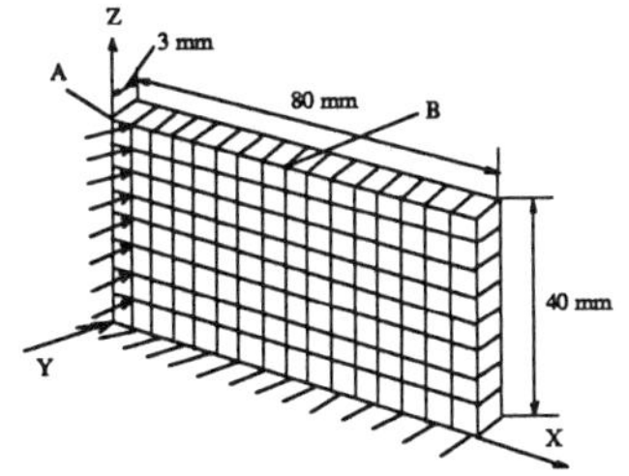

(E=72 GPa, ν=0.32, G=27 GPa)

FIG. 2. A model of finite element calculation.

Fig. 3 shows results of calculations considering cutting forces only in the y direction applied on the workpiece. Without any support, the maximum vertical displacement of the workpiece occurs at the corner of the free edges (0.05095 mm), and the minimum displacement occurs at the middle of the top free edge (0.02414 mm). According to our definition of a thin-walled part, a part with a thickness tolerance of 0.06 mm (IT11) and a flatness tolerance of 0.03 mm is not a thin-walled part. This means that we don't need to provide any support, reduce cutting force or change any cutting process, even though it is a thin member [h=3, b=40; $(1/80 \sim 1/100) \leq h/b \leq (1/5 \sim 1/8)$] according to Young's definition of thin plates. If the thickness tolerance of the thin wall is 0.01 mm (IT7) or flatness tolerance is 0.01 mm, then it is a thin-walled part. In this case, to guarantee the required accuracy of the part thin wall supporting techniques have to be used or the cutting force must be reduced by lowering cutting depth or changing the cutting process. From fig. 3 it is clearly seen that the use of two optimum point supports will reduce the displacements to within the required accuracy range. Otherwise, accuracy is still beyond tolerance range. Developing supporting techniques is important in fixture design for parts with thin walls. What is a supporting technique? Supporting techniques are techniques which are used to reduce displacement of the workpiece by adopting different kinds of support elements and different supporting schemes. They are based on the requirements of the amount of displacements to be reduced, production quantity of machined workpiece, characteristic of configuration of machined thin-walled parts, and the conditions around thin walls of the workpiece.

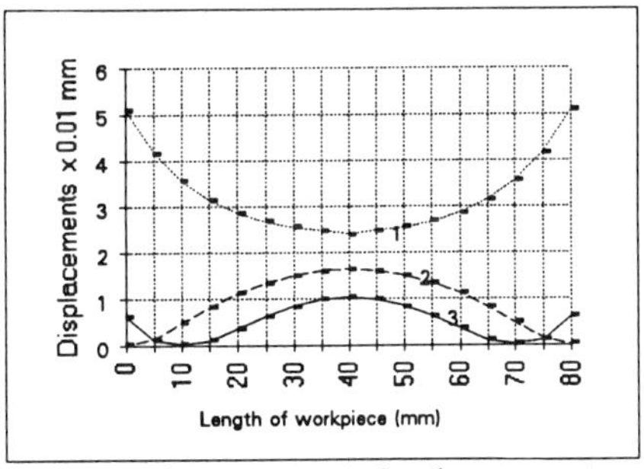

1-no support, 2-corner supports, 3-optimum supports

Fig. 3. Displacement of a thin wall.

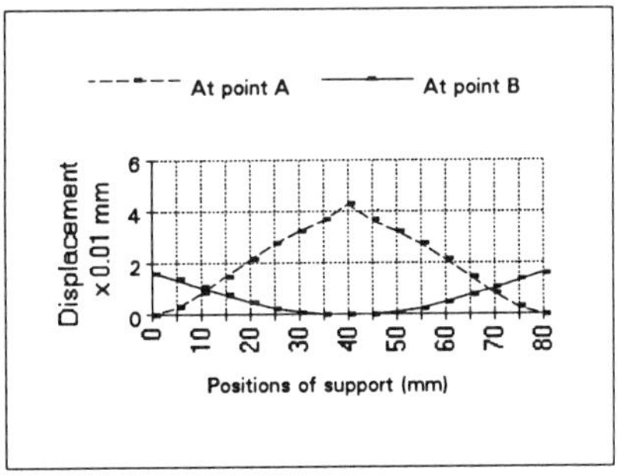

Fig. 4 Workpiece displacement vs. support position.

The approach of determining optimum supporting points was used as follows. The calculation results under our hypothesis conditions have shown that maximum deflection under load occurs at left hand and right hand side sections (X=0,X=L) and minimum at the middle. Due to the symmetry of the thin wall, the left hand side section and the middle section have been examined. The rigidity of the thin wall near the fixed edge is obviously higher than the rigidity of thin wall near free edge. Maximum deflection occurs is more likely to occur in the free edge section. Therefore, two supports are moved from two corners to the middle of the top free edge along the top free edge to look for optimum supporting points. The vertical displacements of the corner point A and the displacement of middle point B of the top edge have been obtained as shown in Fig. 4. The intersections of two curves are optimum support points, along the X axis, which is located 11.122 mm away from the side edge. Therefore, the ratio constant for optimum supporting points along the X axis is 0.139 obtained by dividing 11.122 by 80. The optimum supporting point X away from the side edge is formulated as follows:

$X = \alpha L$ where L is the length of thin wall , and α is a ratio constant which is equal to 0.139; or $X = 0.139$ L.

This time the length X is fixed on optimum supporting points determined previously. Optimum supporting points along the Z direction will be found by moving supports from the top edge of the thin wall to the bottom along direction Z. The curves represent the vertical displacements as shown in Fig. 5. The optimum support position is selected such that it corresponds to the curve with the smaller overall displacement (e.g. curve 2 in Fig. 5) and to the point of minimum local displacement. If the selected curve has two equal minima, the optimum point is chosen so that it corresponds to a smaller displacement on the other curve (curve 1). The optimum supporting points are located 5 mm away from the top edge. The optimum supporting points Z away from top edge along direction Z are found as follows:

$Z = \beta H$ where H is height of thin wall, and β is a ratio constant which is 0.125; or $Z = 0.125$ H.

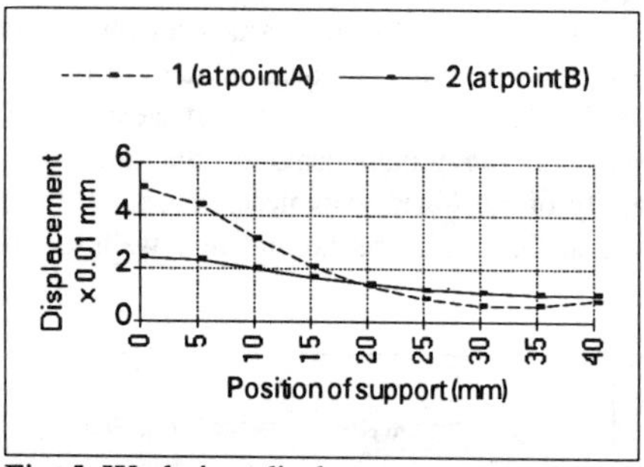

Fig. 5. Workpiece displacement vs. support position.

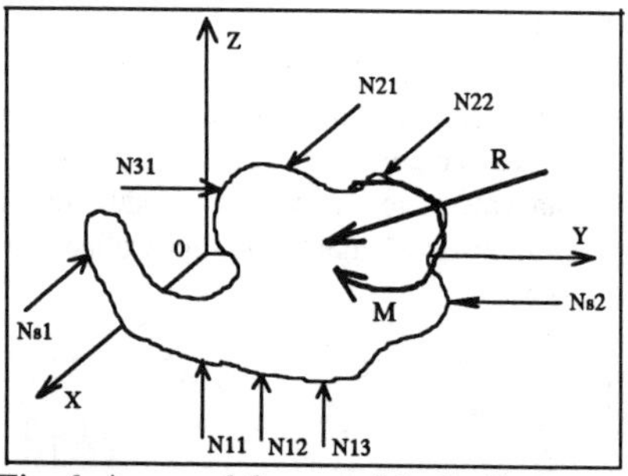

Fig. 6. A general force model of a fixture.

A THEORY OF FIXTURE CONFIGURATION

A fixturing configuration mainly contains location, clamping, and support. In order to locate and hold a workpiece accurately, securely, and control the displacement within allowed tolerances while machining, fixturing configuration must be examined for static equilibrium. Meanwhile, locating, clamping, and supporting positions and forces must be determined for a more reasonable fixturing configuration to be obtained.

In 1986, Gandhi and Thompson [5] first gave a force model for a fixtured workpiece using the 3-2-1 rule and static equilibrium. Static equilibrium equation is also expressed by the matrix equation. In 1989, Pham et al. [6] gave another force model of the fixture which expressed the resultant force and clamping force separately in one static equilibrium equation. In 1990, Cabadaj [7] gave a force model of the fixture which resolved the resultant force into six components of force applied on hexahedron. The force on the each face is equal to locating force. Two degrees of freedom of motion of the workpiece are eliminated by the face through two points reacting forces. This model serves only a prism workpiece and the forces applied in points. The above three force models of the fixture are successfully used in automated fixture configuration. However, none of them can be applied in our workpiece model. In other words, Their models have limitation, and may not serve as universal model in fixture configuration.

As a general force model of the fixture, supporting force should not be eliminated. A force model including supporting force at the present is developed by the author. The three locating planes established using the 3-2-1 locating principle of a fixture provide a force field show in Fig. 6. The external forces and torques imposed on the workpiece such as the cutting force **P**, the clamping force $\mathbf{P_c}$, the friction force **F**, and gravity **G**, can be modeled into a resultant force vector **R** and a resultant moment vector **M** about the center of mass of the workpiece. The first plane, the XY plane, provides three support points for location. The three reacting forces are denoted as N_{11}, N_{12}, and N_{13}. The second plane, the YZ plane, provides two support points for location. The corresponding reaction forces are N_{21}, N_{22}. At same time, the first supporting plane which is the opposite with the second plane provides one support point for support when it is a thin walled part. The reaction force is denoted as N_{s1}. The third plane, the ZX plane, provides one support point for location. The reaction force is denoted as N_{31}. The second supporting plane which is opposite with the third plane provides one support point for support. The reaction force is denoted as N_{s2}.

The conditions of static equilibrium are

$$\sum_{i=1}^{3}\mathbf{F}_i = 0 \qquad \text{and} \qquad \sum_{i=1}^{3}\mathbf{M}_i = 0.$$

Where $\mathbf{F}_i$ and $\mathbf{M}_i$ are the Cartesian components of the resultant force and moment vectors. Therefore, a general formula of static equilibrium for a fixtured and machined workpiece is expressed as follows:

$$\sum_{i=1}^{6}\mathbf{N}_i + \sum_{j=1}^{2}\mathbf{N}_j + \mathbf{P}_c + \mathbf{P} + \mathbf{F} + \mathbf{G} = 0.$$

In order to simplify the problems, we assume

$$\mathbf{R} = \mathbf{P}_c + \mathbf{P} + \mathbf{F} + \mathbf{G},$$

therefore,

$$\sum_{i=1}^{6}\mathbf{N}_i + \sum_{j=1}^{2}\mathbf{N}_j + \mathbf{R} = 0.$$

When $\sum X = 0$, $N_{21} + N_{22} - N_{s1} = R_x$

When the cutting force applies against support of a supported thin wall,

$$N_{21} + N_{22} = 0, \qquad -N_{s1} = R_x$$

and vice versa

$$N_{s1} = 0, \qquad N_{21} + N_{22} = R_x\,.$$

If a thin wall isn't present, the support isn't used and $N_{s1} = 0$. The method of obtaining other equations is similar. Therefore, the following matrix equation in the form of $AX = B$ is obtained.

$$\begin{bmatrix} 0 & 0 & 0 & 1 & 1 & 0 & 0 & 0 \\ 0 & 0 & 0 & 0 & 0 & 1 & 0 & 0 \\ 1 & 1 & 1 & 0 & 0 & 0 & 0 & 0 \\ r_{41} & r_{42} & r_{43} & 0 & 0 & -r_{46} & 0 & 0 \\ -r_{51} & -r_{52} & -r_{53} & r_{54} & r_{55} & 0 & 0 & 0 \\ 0 & 0 & 0 & -r_{64} & -r_{65} & r_{66} & 0 & 0 \\ 0 & 0 & 0 & 0 & 0 & 0 & -1 & 0 \\ 0 & 0 & 0 & 0 & 0 & 0 & 0 & -1 \end{bmatrix} \begin{bmatrix} N_{11} \\ N_{12} \\ N_{13} \\ N_{21} \\ N_{22} \\ N_{31} \\ N_{s1} \\ N_{s2} \end{bmatrix} = \begin{bmatrix} R_x \\ R_y \\ R_z \\ M_x \\ M_y \\ M_z \\ R_x \\ R_y \end{bmatrix}$$

where r_{ij} are the moment arm of N_{ij}.
The supporting reacting force N_s is considered as acting on the supporting plane. Normally two points supports will be put on the supporting plane. Therefore, the equations for the calculation of the two supporting forces are as follows:

$$N_{s1} = -N_{s11} - N_{s12}$$
$$N_{s1}\, r_{s1y} = -N_{s11}\, r_{s11y} - N_{s12}\, r_{s12y}$$

with determinant D, and sub-determinants D_1 and D_2. The forces that apply to supporting points N_{s11} and N_{s12} are $N_{s11} = D_1 / D$, $N_{s12} = D_2 / D$.
The arms of each optimum supporting points such as r_{s11y}, r_{s11z}, and r_{s12y}, r_{s12z} will be determined by using the method mentioned in the previous section.

CONCLUSION

The fundamental theory for automated fixture design has been expended based on considering thin walled parts. A general force model of a fixture including supporting force has been developed so that a fixture configuration in automatic fixture design is close to real production. The parts with thin walls have been defined from a fixture and machining point of view. Supporting techniques and the approach of determining optimum supporting points has been exploited.

REFERENCE

1. D. C. Berry, "Application of CAD/CAM to Fixture Design," Proc., 1st Biennial Int. Machine Tool. Technol. Conf., Chicago, pp. 43-66, Sept. 1982.
2. J. Liu & D. Strong "Survey of fixture design automation," Proc. of CSME 1992, Vol. 3, pp. 823-828, Montreal, Canada, June 1992.
3 R. W. Fitzgerald, "Mechanics of materials," 2nd edition, Addison-Wesley Publishing Company, Inc., Massachusetts, USA 1982.
4. G. Young, "Elastic and plastic mechanics," People Education Published, Inc., PRC, 1980.
5. M. V. Gandhi, B. S. Thompson, "Automated Design of Modular Fixture for Flexible Manufacturing Systems," J. of Mfg. Systems, Vol. 5/No.4,pp. 243-252, 1986.
6. D. T. Pham, M. J. Nategh & A. de Sam Lazaro, "A Knowledge-based jig-and-fixture designers' assistant," Int. J. Adv. Mfg.. Technol. 4:26-45, 1989.
7. J. Cabadaja, "Theory of Computer Aided Fixture Design," J. of Computer in Industry 15 (1990) 141-147.

DETECTION OF DEFECTS IN THE APPLICATION OF CAPS ON BABY FOOD JARS

RAFIC A. BACHNAK
Electronics Engineering Technology
Franklin University
Columbus, Ohio 43215

BARRY A. CUMMINGS
Engineering Department
Anchor Hocking Packaging
Lancaster, Ohio 43130

ABSTRACT

This paper describes a new system for inspecting caps on food jars. The system uses a laser displacement sensor operating at a wavelength of 780 nm. The voltage sensed by the laser is sampled and then digitized by a data acquisition board. The data is fed into a CPU card where, based on the inspection criteria, decisions are made to reject jars with defect caps. The system is practical and capable of scanning 1200 jars per minute. Experimental results prove that the system is robust, reliable, and suitable for real time applications.

INTRODUCTION

Closure diagnostics is a problem that has been plaguing the packaging industry. A bad application of a closure causes the flexible metal center of the closure (button) to remain in the convex position. This condition is called a "dud." The process of determining whether a closure is a dud is called "dud detection." Currently used dud detectors can properly inspect up to 800 jars per minute. They use measuring wheels to maintain a specified distance above the cap. The wheels must physically contact each package. When the processing speed increases, this may cause the conveyor carrying the packages to jam. If a faulty dud detector suddenly rejects several packages in succession, the line operator assumes that there is a problem with the packing machine. The actual problem may, however, lie in the dud detectors inability to consistently evaluate packages.

Lasers may be used for remote distance measurements [1]. Therefore, they offer the opportunity of detecting duds at higher speeds. The Laser Inspection System (LIS) described here is primarily designed to inspect "duds" on baby food jars. The system detects improperly sealed jars and generates a signal in order to eject them from the line.

DESCRIPTION OF THE SYSTEM

Fig. 1 is a block diagram showing the major components of the LIS system. It consists of a laser that operates at a wavelength of 780 nm, a laser controller that converts spatial measurements to proportional voltages, a data acquisition board that digitizes the sampled voltage, a CPU Card that employs an Intel 80386 microprocessor, a rotary pulse generator (RPG) that triggers the data acquisition board to take voltage samples from the laser controller, and a rejector system that rejects packages at predetermined locations.

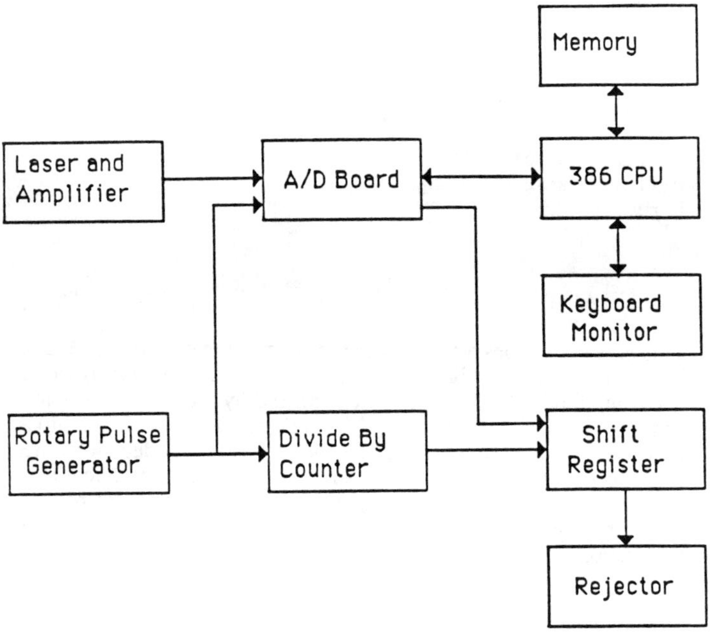

Fig. 1. System configuration

Hardware Design

Fig. 2. shows a prototype of the LIS system. It consists of the following major components:

1. Laser Displacement Sensor, model LB-72. This laser operates at a wavelength of 780 nm and is classified as a class IIIb laser. It has a standoff distance of 40 mm and a measurement range of $\pm$ 10 mm.
2. Laser controller, model LB-12. This controller has an output voltage range of -4 to +4 Vdc. The distance from the closure to the laser head is proportional to the voltage output of the controller. The resolution of the Digital-to-Analog Converter (DAC) in the LB-12 controller is 0.4 v/mm. The maximum response speed of the laser is approximately 3 KHz.
3. Data Acquisition Board, model DAS16F by Keithley-Metrabyte. This expansion board is a 12 bit analog to digital converter board with extra features and functions to enhance the manipulation of data received at the 16 channel input. The throughput is 100 KHz. It has the capability of handling 9 different voltage configurations without board modifications.
4. Rotary Pulse Generator (RPG). This is a quadrature output pulse generator using a LED as light source. Pulsing of the RPG is accomplished using a slotted disk. The output of the RPG photo sensor creates an output signal in the form of a 5

volt square wave.
5. 386-25 MHz standard CATN386 (CPU Card). This CPU Card employs an Intel 80386 microprocessor. It utilizes a PCD-890 RAM/ROM Disk Card to access program memory.
6. Rejector System, model EJH-ST. The Rejector System is the means by which packages are rejected at predetermined locations.

For more information and additional specifications concerning the major components listed above, refer to [2].

Fig. 2. Prototype of a laser inspection system

Software Design

The software includes three levels of operation: developer, installer, and user. In this paper, we will discuss only the user level operation. When the system is first turned on, it will display the Main Menu or the Run Menu depending on what the machine was doing when it was last turned off. The Main Menu has the following selections:

1. Run: this selection will invoke the Run Menu to start evaluating jars.
2. Jar Types: this selection will allow for defining which type of jars the system will be evaluating. The list of possible jar types is set by the installer.
3. Calibrate Jar: This selection is only used if the system was installed with a proximity detector. With a laser system this selection has no effect.

The Run Menu has two options:

1. Reset Totals: this resets the jar count to zero and starts the inspection process.
2. Keep Totals: this starts the inspection process. The jar count is not initialized to zero.

EXPERIMENTAL RESULTS

We present here the results of three different experiments. A high cap is defined as a cap which exceeds 1/4 of a turn from an acceptable seated position.

Experiment 1: High Caps Data

Line speed	1200 jars/minute
Samples Taken	10,000
Jars identified as okay	8,970
Jars rejected for tilt	10
Jars rejected for high caps	1,020
Number of rejects	1,030
True # of rejects	1,000
Percent of error	0.3

Percent of error represents number of rejects minus number of actual rejects divided by total packages sampled.

Experiment 2: Bad Reliability Test (High Caps. All samples are defective) (see first row in Table 1.)

Experiment 3: Bad Reliability Test (Duds. All samples are defective) (see second row in Table 1)

Table I. Data of Exp. 2 and Exp. 3.

Line speed (jars/min)	Samples taken	Rejects sensed	Rejects missed	Percent error
859	1000	1000	0	0
945	1000	999	1	0.1

CONCLUSION

This paper describes a new system for inspecting caps on baby food jars. The final production prototype proves to be successful. The system was tested with speeds up to 1200 jars per minute. It proved to be robust and reliable. Processing speed may be increased to 2000 jars/minute if the power fluctuations introduced to the unit can be eliminated. Compared with proximity based detectors, the Laser Inspection System is cutting the percent error approximately in half.

REFERENCES

1. N. L. Stauffer, "New Approach to Measuring Distance," Design News, vol 21, April 1986.
2. B. A. Cummings, Laser Detection system , Senior Project Report, Franklin University, 1991.

AN OPEN STANDARD FOR INDUSTRIAL CONTROLLERS*

BRUCE M. ANDERSON and JEFFREY R. COLE
Advanced Automation, Martin Marietta, Denver, Colorado

RICHARD G. HOLLAND
Advanced Automation, Martin Marietta, Denver, Colorado

ABSTRACT

The Next Generation Workstation/Machine Controller (NGC) program seeks to align controller manufacturers, machine tool integrators, and industrial end-users by means of a common open system standard. This standard applies to the development of controllers for a broad range of discrete part manufacturing applications that include machining, material handling, and measurement. The NGC standard must be abstract with respect to a specific application or design to cover the broad range of applications and variety of user viewpoints and methods. We address the characteristics of open systems that aid designers and developers in contributing to an integrated family of NGC products without mandating a fixed implementation topology. Adoption of an open standard will facilitate interoperability among separately developed NGC components and ease software portability during computing platform upgrades. Moreover, the standard provides for separately vended NGC components to be interchangeable and scaleable in performance and features. The strategy developed in this paper facilitates the integration of generalized system requirements into the design aspects of a working NGC system.

INTRODUCTION

The Next Generation Workstation/Machine Controller (NGC) program has as its primary objective the development and validation of a Specification for an Open System Architecture Standard (SOSAS) to govern the design and construction of a family of industrial controllers. The SOSAS facilitates commonality of components across a complete range of workstation controllers. These controllers are to be used in a wide variety of manufacturing operations which include machining, robotics, and inspection.

Available controllers are based on proprietary designs which prevent various users, machine tool builders, and system integrators from implementing creative production improvements in a truly cost effective manner. Today's controllers are single company products or product lines, and the concept of an open system is often not evident, even among and within the product lines offered by single vendors.

The SOSAS, on the other hand, allows individual companies to add their own unique capabilities, yet remain interoperable with systems supplied by other

* The work presented here is supported in part by the Manufacturing Technology Directorate, Wright Laboratory, Aeronautical Systems Center (AFMC), United States Air Force, Wright-Patterson AFB, Ohio 45433-6533, under contract F33615-89-C-5706.

vendors. The open architecture enables suppliers to customize control systems to fit users' needs more closely while the industry benefits from adopting standards from the best current commercial practices.

An NGC supports machine tools of all types, robots, electronic assembly, material handling devices, inspection devices, and virtually all types of automated equipment in both manned and unmanned environments and networked and stand-alone configurations. An NGC controls a manufacturing workstation, which is defined as a single material transformer and related material handling and inspection equipment. In the hierarchy of the overall manufacturing enterprise, several manufacturing workstations are managed by a cell.

Developing an open controller architecture standard presents significant challenges. The SOSAS must specify a family of controllers that covers an extremely broad range of manufacturing applications. Moreover, the SOSAS must readily extend to other manufacturing applications, e.g., process control. The specification must be evolvable and remain valid over the next 15 to 20 years. It must facilitate the use of advanced current technology and be flexible enough to absorb advances in manufacturing technology well into the next century. But the SOSAS also must accommodate the large investment in legacy code, written in the specialized languages for Computer Numerical Control (CNC), robotics, and automatic measurement. The SOSAS meets these challenges with the following attributes:

- Flexibility—the ability of the SOSAS to specify components needed to configure a controller for any automated equipment capable of performing a discrete part manufacturing operation or process.
- Extendibility—the ability of the SOSAS to accommodate advances in the technologies associated with manufacturing, control, and computing.
- Longevity—the ability of the SOSAS to survive as a viable standard over time periods on the order of decades.

OPEN SYSTEMS

There is an ever-expanding need for complex computing systems, composed of distributed and centralized computers and interconnecting networks. As computer usage becomes more widespread in a variety of applications by a broader class of users, open standards for computer implementation and use provide significant economic benefit to suppliers and users[1]. Organizations such as User Alliance for Open Systems, Open Software Foundation (OSF), X/Open, UniForum, and International Standards Organization (ISO) have been founded to promote open systems[2-3]. Many definitions of open system exist, but they can be grouped into two types, emphasizing the characteristics that advocates believe to be essential.

Vendor-neutral open systems† are oriented toward the end-user to standardize operations with equipment that is interchangeable and replaceable in a price competitive environment. This definition characterizes an open system as: vendor-neutral (independent of a single proprietary interest), consensus-driven (controlled by a group of suppliers and end-users), standards-based, and freely available at no cost to any interested party.

Component-integration open systems‡ are oriented toward suppliers to facilitate the building of products that integrate within a larger system. Characteristics are:

† This definition is an amalgamation of open systems definitions from X/Open, User Alliance for Open Systems, and Sun Microsystems. For example, according to X/Open, "Open systems are computer and communications environments based on formal and *de facto* interface standards. These interfaces must not be owned by any single vendor and must be freely available."

‡ IEEE P1003.0 Committee (POSIX) and Open Software Foundation contribute to this definition. According to IEEE, "An open system provides capabilities that enable properly implemented

portability, interoperability, and a common user presentation (common look and feel).

An open system architecture is a specification of the capabilities or services that provides the interconnection structure and defines the interface between interoperating components. The SOSAS is based on existing and emerging standards from computing and data communications practice, and it satisfies the vendor-neutral and component-integration definitions. This architecture standard was designed to provide for the following:

- Interoperability—the ability for independent system components to work together properly.
- Portability—the ability to operate the same system component on different platforms.
- Scaleability—the ability to change the capability of a system by upgrading or downgrading specific system components.
- Interchangeability—the ability to substitute one component with another, which was developed to provide equivalent functionality at lower cost and/or improved capability, reliability, or performance.

INDUSTRIAL CONTROL DOMAINS

Within enterprises of the future, manufacturing will be an information-intensive process, fully automating today's manual operations. Practices associated with discrete part manufacturing can be grouped into three principal domains: manufacturing practice, controller practice, and computing practice. Terminology and representations from these domains are used in the SOSAS; standards developed within the purview of these domains are incorporated in the SOSAS by reference. In this way, the considerable investment and effort devoted to the three domains, especially computing practice, can be reused effectively to support manufacturing applications.

Manufacturing practice covers the process of transforming raw materials into finished parts. The underlying concept for the discrete part manufacturing process is "art-to-part." The process begins by capturing a part design using a computer aided design (CAD) system. Then the resources for making the part are scheduled, and the machining, material handling, and measuring activities are planned using a computer aided manufacturing (CAM) system. Finally motion paths are planned and used to drive the appropriate machinery: tool paths for machining; robot end-effector paths for manipulation; and probe paths for measurement. Many of the steps in this process are done manually today. In the future, the art-to-part process is envisioned as a seamless information flow from the designer's concept to the finished part. Some related manufacturing practice standards include those for feature-based part description[4-5], part programming[6-7], and measurement[8] languages.

Controller practice covers the organization and control of manufacturing equipment. Equipment is controlled mostly by closed-loop controllers, which provide continuous (motion) control and discrete control. Computing practice covers the computing and communication technologies required to support manufacturing and controller practice. The practice includes design and use of computer hardware and software.

ARCHITECTURE

The NGC community conducted a needs analysis[9] which led to a set of requirements[10] that the SOSAS needed to encompass. The $IDEF_0$[11]

---

applications to run on a wide variety of platforms from multiple vendors, interoperate with other systems applications, and present a consistent style of interaction with the user."

methodology was applied in a top-down functional elaboration of the NGC system based on the requirements. This elaboration resulted in a set of primitive functions which were grouped into the subsystems and modules of the functional architecture (see Figure 1). An object-oriented analysis[12] of the requirements and the NASREM hierarchical control reference model[13] played key roles in grouping the primitive functions[14]. The functional architecture was created, reviewed, and approved by a consensus process within the NGC community.

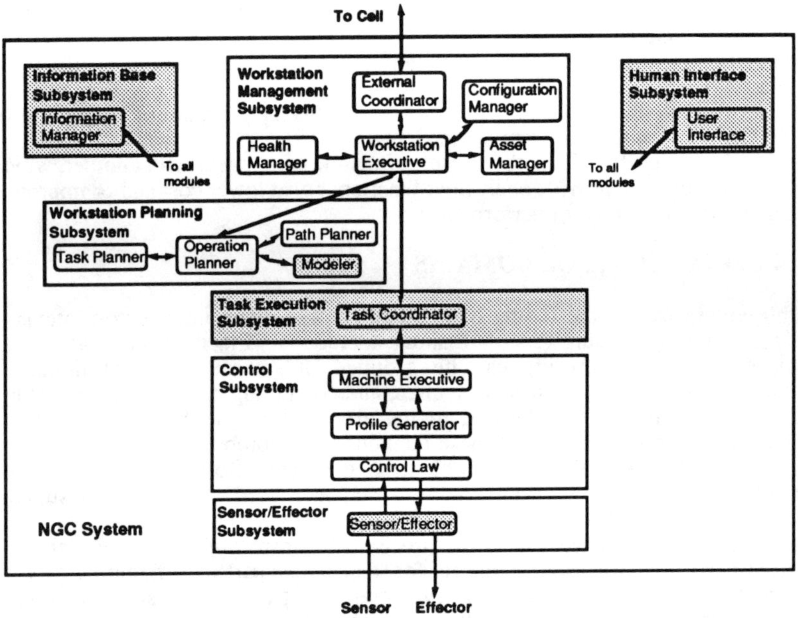

FIG. 1. NGC Functional Architecture.

To produce an enduring standard, one must describe manufactured products and manufacturing processes independent of specific equipment, methods, or topologies. A design, calling for a specific topological arrangement and interconnection of components, does not provide enough flexibility to control every application in discrete part manufacturing, nor does it accommodate technological advancements over a period of decades. Further, methods based on a single architectural paradigm are not deemed adequate for developing complex systems[15]. Functional decomposition[11], object-oriented methods[12], hierarchical control decomposition[13], structured analysis, and behaviors described in terms of states and state transitions are useful, but each of these perspectives present a topology that enhances some aspects of system development and limits others. Several topologies may be required in order to develop a system and understand its operation under a variety of situations[16]. Thus the integration architecture was created

The seven subsystems and sixteen modules shown in Figure 1 capture the functionality of the NGC. Arrows indicate the primary information paths. The Workstation Management subsystem schedules overall workstation activities, controls the modes of operation, and provides for configuration management. The Workstation Planning subsystem uses the requirements of a manufacturing process to assemble a plan to be executed by the Task Execution subsystem. The Control subsystem provides both continuous and discrete control of an individual machine, and the Sensor/Effector subsystem connects the controller to the equipment.

In addition to the subsystems, there are three elements of the functional architecture not shown explicitly in Figure 1: the Virtual Machine, the Sensor/Effector Bus, and Support Tools[17-18]. The Virtual Machine maintains portability by isolating the software from the underlying processing platform. The Sensor/Effector Bus is a standard hardware interface which improves workstation reliability and maintainability by significantly reducing the physical wiring. Support Tools aid development, installation, operation, and maintenance.

The integration architecture shown in Figure 2 is a framework that supports interoperable, portable, scaleable, and interchangeable NGC component implementations, designed and developed by competing vendors[19]. NGC functionality is partitioned between services and applications.

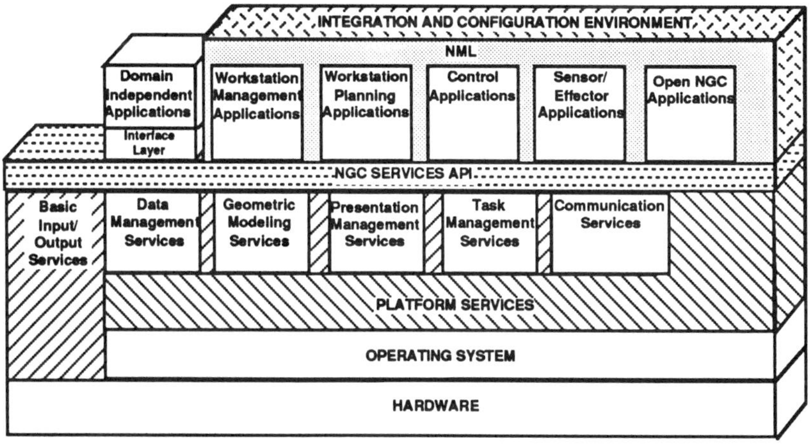

FIG. 2. Integration Architecture.

Services are based on computing practice and have a broader applicability than NGC, so they are termed "domain independent." Services are provided through a specified Application Program Interface (API), and they can be invoked by applications and other services. NGC services form seven groups: Presentation Management, Data Management, Geometric Modeling, Task Management, Communication, Platform, and Basic Input/Output. The NGC services are an abstraction layer between applications and the native operating system, computing hardware, and peripheral devices.

Applications are of three types: standard, open, and domain independent. Based on manufacturing and controller practice, both standard and open applications are interoperable within the NGC domain, they intercommunicate using Neutral Manufacturing Language (NML) messages, and they use the NGC services and data structures. Standard applications are SOSAS-specified, and open applications are not, allowing the SOSAS to be extended into other areas of manufacturing. Standard applications include Workstation Management, Workstation Planning, Control, and Sensor/Effector. Domain independent applications, e.g., graphic programs and database management systems, supply general functionality. They are available as commercial products, adapted to NGC through an interface layer.

The Integration and Configuration Environment supports machine and controller integration and facilitates the enhancement of standard functionality. The environment also supports the integration and testing of user-developed code using one or more standard languages[20-22].

The functional architecture is reconciled with the integration architecture by partitioning functionality from the former between applications and services in the latter. The shaded functional elements in Figure 1 associate most conveniently with computing practice, combining with the Virtual Machine functionality to form the NGC services in Figure 2. Thus, the Human Interface subsystem corresponds to Presentation Management services; Information Base to Data Management services; the Modeler module to Geometric Modeling services; Task Execution to Task Management services; and the Sensor/Effector module to Basic Input/Output services. The Virtual Machine provides Communication and Platform services, and Support Tools correspond to the Integration and Configuration Environment. The Sensor/Effector Bus hardware is accessed through Basic Input/Output services.

OPEN ARCHITECTURE ELEMENTS

The open system standard for industrial controllers consists of standard, open, and domain independent applications and a supporting infrastructure. The infrastructure specifies how implemented NGC components are integrated, so within constraints imposed by the infrastructure, many topologies are possible. The infrastructure includes:

- a set of services,
- a formal data standard based on reference and representation models,
- a systematic, extendible classification structure (containment tree) for all NGC entities, and
- an integration and configuration environment.

Workstation Management standard applications provide for the initiation and management of manufacturing workstation resources, establish and control standard communications[23-24] between the workstation and the cell, monitor and control safety mechanisms, detect and handle faults, diagnose the cause of system malfunctions, and manage workstation assets. Workstation Planning standard applications have the primary task of converting product descriptions and the capabilities and limitations of the manufacturing workstation into a set of manufacturing plans. Control standard applications provide direct control for all sensors and effectors associated with a particular machine or process to accomplish manufacturing goals through a user-supplied program or instruction as part of a planned task. Sensor /Effector standard applications provide the interface between NGC applications and operators that control various complex actuators or mechanisms that integrate both sensors and effectors.

Presentation management services provide standardized means for multiple applications to present data to a user in a common display representation and to receive input from shared input devices. Data management services provide an implementation-transparent mechanism for storing, retrieving, modifying, and deleting data, and they support data protection, data recovery, data consistency, access control, and security. Geometric modeling services are required for interactive and non-interactive solid modeling and geometric operations, and they are based on the Application Interface Specification[25] and the emerging STEP/PDES standards[4-5]. Task management services provide uniform mechanisms for scheduling, coordinating, and controlling manufacturing tasks created by standard and open applications. Communication services provide a means for application-to-application communication via an API that is protocol and mechanism independent. Platform services are a generic set of both real-time and non-real-time operating system and operating system utility services, and they are a subset of the portable operating system suite[26-29]. Basic input/output services are used to initialize devices and transfer data to/from them, and they are described in the specification for a basic I/O system[30].

The NGC data standard is limited in scope to the information shared among components of an NGC and between an NGC and the enterprise in which it is contained. NGC data are made up of the NGC Schema and NML. The schema specify NGC objects, the information units pertaining to those objects, and the constraints on those objects. NML is the collection of messages passed among NGC applications. The NGC data standard is based on models of two main types: reference models and representation models. A reference model is a collection of concepts and terminology for thinking and talking about some range of practice[31]. NGC reference models cover the products, processes, equipment, and resources associated with discrete part manufacturing practice. Representation models are the representations of the reference models using a formal language to define entities and types[32].

The NGC containment tree names and classifies all information structures and functions of the NGC. The containment tree is a strict inheritance hierarchy with IS-A and IS-A-KIND-OF relations between parent and child nodes[33]. Each node in the hierarchy has a symbolic and a numeric designator. A node is uniquely identified by a concatenation of designators in the inheritance chain, starting at the root. The containment tree incorporates mechanisms for extendibility, both for data and functions. Data are extended by adding new schema entities and/or adding new NML messages. Functionality is extended by augmenting the services and/or standard applications, or it is extended by adding open applications and/or domain independent applications.

CONCLUSIONS

Information-intensive domains require open system standards definitions, imitating and building on the open system standards well underway in computing and data communication technology.

Discrete part manufacturing is becoming a seamless flow of information from art to part, controlling material transformation, handling, and measuring operations. Autonomous factory floor operation is expanding, replacing today's costly manual operations. The expense of developing a new controller within this environment is reduced significantly by using and reusing off-the-shelf components.

The SOSAS facilitates the development of interoperable, portable, scaleable, and interchangeable application components by focusing on the infrastructure while maintaining the flexibility to accommodate a variety of implementation topologies. The infrastructure includes a set of services, a formal data standard, the containment tree to classify NGC entities, a configuration and integration environment, and the integration architecture to tie NGC components together. The SOSAS can be extended to other manufacturing practices, like formed component production, mechanical assembly, or electronic assembly, by adding or modifying data models, adding or modifying common functionality, and registering these changes in the containment tree.

Our approach to developing the NGC open standard can be directed at other information-intensive applications. In addition to computing and data communication, relevant domains of practice must be identified. Conventions and standards for these practices must be supported by a flexible, extendible infrastructure that accommodates designs for a wide variety of applications and unforeseeable advancements in technology.

REFERENCES

1. G. Nutt. *Open Systems*, Prentice-Hall, Inc., Englewood Cliffs, NJ, 1992.
2. M. Azzara, "Open Systems Dialectic, Part II," *UNIX Today!*, November 25, 1991.
3. J. Sheridan and J. Teresko, "Open Systems, Gateway To A New Computing Age," *Industry Week*, April 15, 1991.

4. J. Altemueller, *The STEP File Structure*, ISO TC184/SC4/WG1 Document N279, September 1988.
5. *Product Data Exchange Specification First Working Draft*, ISO TC184/SC4/WG1 Document N284, October 27, 1988.
6. *Interchangeable variable block data format for positioning, contouring, and contouring/positioning numerically controlled machines*, RS274D, February 1979.
7. *32-bit binary cutter location (BCL) exchange input format for numerically controlled machines*, RS494, 1983.
8. *Dimensional measurement interface specification (DMIS) 2.1*, Standard 101, CAM*I, Inc., July, 1989.
9. *Next Generation Workstation/Machine Controller (NGC) Needs Analysis*, Martin Marietta, Denver, CO, February 1990.§
10. *Next Generation Workstation/Machine Controller (NGC) Requirements Definition Document (RDD)*, Martin Marietta, Denver, CO, August 1990.§
11. D. Marca and C. McGowan, *SADT: Structured Analysis and Design Technique*, McGraw–Hill, New York, NY, 1988.
12. P. Coad and E. Yourdan, *Object-Oriented Analysis*, Prentice Hall, Englewood Cliffs, NJ, 1989.
13. J. Albus, H. McCain, and R. Lumia, *NASA/NBS Standard Reference Model for Telerobot Control System Architecture (NASREM)*, NIST Technical Note 1235, National Institute for Standards and Technology, Gaithersburg, MD, April 1989.
14. B. Anderson and T. Viner, "Specifying the Architecture for the Next Generation Controller," presented at Third International Conf. on Systems Engineering, Dayton, OH, August 1–3, 1991.
15. F. Brooks, "No Silver Bullets," *Computer*, Vol. 20, No. 4, April 1987, pp. 10–19.
16. D. Harel, "On Visual Formalisms," *Comm. ACM*, Vol. 31, No. 5, May 1988, pp. 514–530.
17. *Next Generation Workstation/Machine Controller Architecture Definition Document*, Martin Marietta, Denver, CO, June 1991.§
18. B. Almand, B. Anderson, and J. Graham, Jr., "Next Generation Workstation/Machine Controller (NGC)," *Proc. 22nd International Symposium on Industrial Robots*, Detroit, MI, October 21–24, 1991.
19. *Next Generation Workstation/Machine Controller (NGC) Draft Specification for an Open System Architecture Standard (SOSAS)*, Vols. I–VI, Martin Marietta, Denver, CO, March 1992.§
20. *Standard for programmable controllers*, IEC SC65A/WG6, parts 1-5.
21. *Draft standard for programmable controllers, Part 3: Programming languages*, IEC SC65B/WG7, September 1991.
22. *American National Standard for Information Systems—Programming Language C*, (ANSI X3.159–1989), 1989.
23. *Manufacturing message specification, Part 1: Service definition*, (EIA-511, Draft 5, June 1986), ISO-9506-1.
24. *Manufacturing message specification, Part 2: Protocol Specification*, (EIA-511, Draft 5, June 1986), ISO-9506-2.
25. *Application interface specification (AIS) 2.0*, Revision 2.0B, R-90-PM-03, CAM*I, Inc., October 22, 1990.
26. *Information technology-portable operating system interface (POSIX), Part 1: System application program interface (API) [C Language]*, ISO/IEC 9945-1: 1990 (IEEE 1003.1-1990), 1990.
27. *Information technology-portable operating system interface (POSIX), Part 2: Shells and utilities*, ISO/IEC 9945-2 (IEEE 1003.2-1990), 1990.
28. *Draft Real-time extension for portable operating systems*, IEEE P1003.4/DXX.
29. *Information technology-portable operating system interface (POSIX), Part 1: API network process-to-process communication*, IEEE P1003.12/DXX.
30. "OBIOS Open Basic Input Output System, Draft Specification, Version 1.15," presented at Real Time Consortium, April 10, 1991.
31. G. Maney, "NGC Controller Practice Models: Background and Requirements," private communication, Cimflex-Teknowledge, Palo Alto, CA, March 1991.
32. D. Schenck, *EXPRESS Language Reference Manual*, ISO TC184/SC4/WG5, Document N14, April 1991.
33. M. Rose, *The Simple Book.*, Prentice-Hall, Inc., Englewood Cliffs, NJ, 1991.

---

§ Document is under controlled distribution and is not generally available.

Role of Legitimization in the Group Collaborative Justification Process

Abdul Akram, Ph.D.

Dr. Akram is a Technical Specialist in the Strategic Architecture Integrated Planning department of Sprint Communications in Kansas City. Dr. Akram holds a Ph.D. in Engineering Management from the University of Missouri-Rolla, an MBA and MS in Industrial Management from the Central Missouri State University , and a BSc in Statistics from the University of Punjab, Pakistan. Dr. Akram is a member of ASEM, ACM, and DPMA. Dr. Akram's research interests include the design of collaborative work technologies, data management for workgroups, and strategic planning for Information Technologies.

Ray Kluczny, Ph.D.

Dr. Kluczny is an Associate Professor in the Engineering Management Department at the University of Missouri-Rolla. He holds a Ph.D. in Quantitative Systems from Arizona State University. His work has appeared in IEEE Computer, ASHRAE Journal, SPHE Technical Journal, Lighting Design and Application, and Journal of Systems Management. Dr. Kluczny's research interests include the design and use of systems to support decision making.

Henry Metzner, Ph.D.

Dr. Metzner is an Associate Professor in the Engineering Management Department at the University of Missouri-Rolla. He holds an MBA and Ph.D. in Business Administration from the University of Utah. In addition to short courses in Project Management and Engineering Supervision, Dr. Metzner has published in engineering management, logistics, Marketing and new product development.

Abstract

Group decision making involves the **legitimization** of a decision within a group as a necessary pre-condition for the **justification** of a decision by a group; that is, a decision must be understood and accepted within the norms of the group in order for objective criteria to be appreciated. Computer-mediated tools for group decision support and computer-based algorithms for implementing complex decision models make it seem that computer regimented decision making may be the ideal. However, research has shown that our current practice in using these tools for group decision making does not necessarily result in high quality decisions supported by participants. This paper reviews the communication process in workgroups supported by Group Decision Support Systems. It later describes the way in which the model used in the study at the University of Missouri-Rolla can be used as the basis for understanding the role of socialization in group decision process. The result of an experiment are discussed in which socialization was shown to be significant in the use of modern group decision support system in providing for high quality decisions along with high group development, and high mutual perceptual affinity.

INTRODUCTION

Cross-functional product development teams which can legitimize and successfully implement group decisions will be the key to success of the manufacturing organizations of the 1990s. The major factor for the success enjoyed by the products like Apple's Powerbook, Thomson RCA 9-inch TV, Hewlett-Packard Laser Jet Printer, Black & Decker DeWalt Power Tools, Bausch & Lomb Ray Ban, Johnson Control's METASYS Facility Management Products, Mazda Miata, was an early product definition and strategic decisions made by the product planning teams. Similar examples can be found with many domestic and international companies. The essence of success of the cross-functional teams is the speed with which the decisions are made and implemented.

As the speed of decision making becomes a high priority so will be the need to implement improved methods of working together in teams. Reaching the actual decision is only one step in the process. The subsequent critical step is the implementation of the decision. The process in collaborative teams breaks down when there is a lack of commitment to a decision made by a group. Therefore, it is necessary to adopt means that can help prevent such breakdown in the process.

Group Decision Support Systems (GDSS) have been considered a good technology to support the process of decision making groups. These systems provide computer-mediated tools and algorithms to implement complex decision models. These tools were developed to overcome the many drawbacks which were found in the face-to-face decision making process. It would seem that computer regimented decision making through these systems would be the ideal.

However, the body of experimental research in GDSS which has emerged during last few years shows that the use of GDSS provides mixed results [1]. While there is a strong evidence that the GDSS support yields higher quality results, these systems are not quite successful in generating high decision confidence in the group members. To overcome this down side of the GDSS technology has been a great challenge for the researchers in GDSS.

A recent study at the University of Missouri-Rolla found that it may be possible to overcome the above mentioned unintended consequence associated with GDSS [1]. The study analyzed the communication aspects of the decision making process. As a result of this analysis it developed a communication model which may have a means of avoiding the unintended consequences of the GDSS use in collaborative decision making. The study focused on the introduction of socialization in the decision making process supported by GDSS.

This paper first reviews the the communication model developed by one of the authors for this study. Later, the results of the experiment conducted for this study are also discussed.

LEGITIMIZATION AND JUSTIFICATION IN GDSS

Legitimization in workgroups

The term legitimization in this paper refers to the understanding and acceptance of a decision within the norms of a group.. Legitimization of a decision precedes the justification of a decision by a group. Justification occurs when the members of a group appreciate the objective criteria used in reaching the decision. The legitimization and justification of a decision in a group process are essential for building the decision confidence among the members of a workgroup.

How does a group achieve legitimization and justification for its decisions? The answer can be found in the work done by many researchers in the area of group development. The Tuckman [2] model has often been used to show the group development process.

Tuckman's model identifies four stages in group development. These stages are: Forming, Storming, Norming, and Performing. Tuckman identified that there is a "task" side and a "social" side of a group process. He characterized these two dimensions as follows:
"Forming" Task: Attempting to identify the task. Social: Testing independence.
"Storming" Task: Emotional response to task demands. Social: Emergence of intragroup conflicts.
"Norming" Task: Expression of opinions. Social: Development of group cohesion.
"Performing" Task: Emergence of solutions. Social: Understanding of roles and relationships.
Tuckman noted that groups become productive when social norms have been established within the group.

Communication and Psychological environment of workgroups

Legitimization by a participant in a workgroup depends on the idiosyncratic perception of the individual to what is happening to him/her in a group. In organizational communication this aspect of an individual is termed as "structuration." Structuration is considered to be a factor in influencing an individual's feelings and emotions related to group membership and understanding of the group process. Structuration leads to social integration initially and to system integration eventually.

According to Falcione, Sussman, and Herden [3], the individual, group and external stimuli establishes the communication climate for the group. Structuration acts as the process by which group develops a psychological climate. As such, a group through structuration develops a sense of itself as a collective entity with values and purpose.

GDSS support for workgroups

GDSS provide workgroups with computer-based tools and algorithms. These systems are successful in recording the proceedings, facilitating concurrent interaction, and passing messages. However, these systems do not facilitate the development of the psychological climate necessary for the legitimization of the decisions. The psychological climate in GDSS groups as shown in the Figure 1., is completely focused on ideas. The support necessary for Tuckman's social side of group process is lacking. Therefore, the process does not help in a group's development of social norms.

The facilitation of socioemotional communication in GDSS supported groups is needed to achieve a psychological climate which supports the communication of ideas as well as the development of group identity. The communication in a GDSS supported group will be as shown in the Figure 2.

EXPERIMENT

Experimental Design

A completely randomized 2 x 3 (task difficulty, GDSS Support) factorially designed experiment was used to study the role of structuration on affective communication in a small group GDSS environment. The decision outcome of three groups was measured using a control group consisting of non-GDSS support using

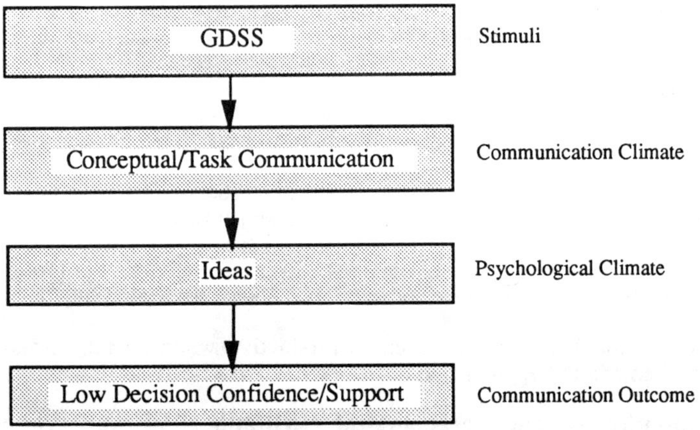

Figure 1. Communication in GDSS

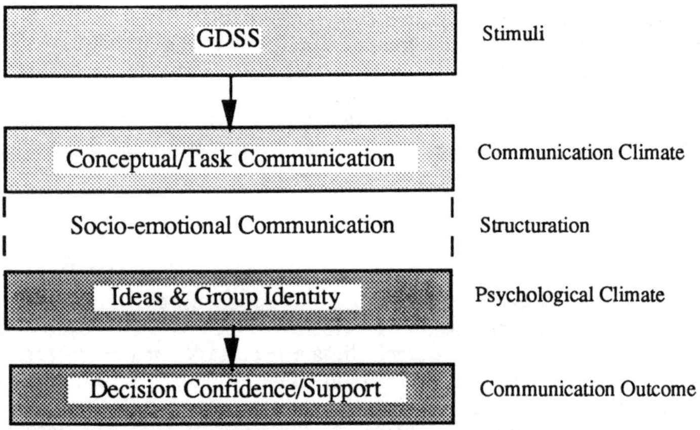

Figure 2. Role of Socioemotional communication in GDSS

paper and pencil, a GDSS support group using the "GroupSystems" GDSS software from the Ventana Corporation, and a Modified Process GDSS support (MPGDSS) group which used "GroupSystems" and took short socialization breaks between phases of the decision-making process. The groups in each of the three categories mentioned above were assigned to reach a consensus in solving a business case that had two levels of difficulty. The decision outcome was measured within the areas of decision performance, decision perception, and social development. The research design shown in Figure 4., was used for the experiment.

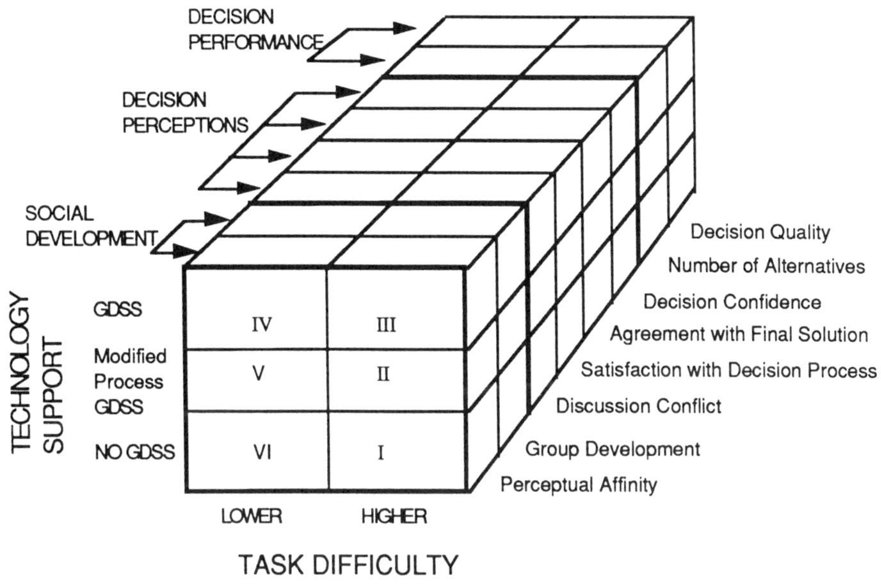

Figure 4. Research Design Showing Independent and Dependent Variables

Results

Decision Performance. Decision Quality was significantly higher for the groups who used the ordinary GDSS and were faced with the easy version of the case, and for the groups who used the GDSS along with socialization breaks and were faced with the difficult version of the case. Groups who had access to a GDSS and were faced with the difficult case were able to produce more alternatives.
Decision Perceptions. Decision confidence was found to be significantly higher for the groups who used the GDSS with socialization breaks. The socialization process seems to have provided the factor missing in the earlier studies to produce not only good decisions but decisions that are accepted by the group. Because the high decision confidence was found only for the groups faced with the difficult case version, the benefit of structuration evidently came from the way in which the groups were able to deal with the added complexity of the problem. The results also indicated significantly higher agreement with final solution among GDSS groups with socialization breaks than those without breaks when these groups worked on the difficult version of the case. Groups provided with socialization in the use of the GDSS were more satisfied with the decision process. Structuration allowed group members the equal opportunity to express their ideas and to discuss their differences during the breaks. The added opportunities for expression, evidently, made them feel good about the way the group processed the decision.
Social Development. GDSS groups aided by structuration experienced the highest group development in the study. GDSS groups aided by structuration achieved the highest levels of mutual understanding within their groups. Participants were able to gain a further understanding of each other's position on matters of mutual interest

during socialization breaks. As a result, group members were able to communicate to each other with greater sensitivity.

SUMMARY AND CONCLUSIONS

Manufacturing in 1990s and beyond will be a global phenomenon. Planners and product development teams will be dispersed globally [4]. These cross-functional teams will be a critical link for any organization. It is important that these teams be able to work together in a way that they are able to legitimize and justify their decisions. We can tell from the developments in technology that Group Decision Support Systems will be a good technology to support the collaborative work process of these teams. However, the prior experimental research in GDSS has shown that the GDSS have some intended consequences. The major concern has been related to low decision confidence found in the participants.

The recent study described in this paper analysed the process of communication in decision making groups supported by GDSS. A communication model was developed to describe the communication process in decision making groups. The study hypothesized that the lack of support for socioemotional communication in these groups is the reason for the low decision confidence observed in the participants. A laboratory experiment was conducted to test the hypotheses.

Results indicated that the addition of socialization in a group environment supported by GDSS technology can facilitate communication that influences an individual's feelings and emotions related to group membership and to an understanding of the group process. **The study showed a possible way to modify the decision making process to rectify the unintended consequences of GDSS support on the decision makers in a group.**

REFERENCES

1. A. Akram, "A Study of Structuration in Group Decision Support Systems, unpublished Ph.D. dissertation, University of Missouri-Rolla (1991).
2. B.W. Tuckman, "Development Sequence in Small Groups," *Psychological Bulletin* 63, 384-399 (1965).
3. R. Falcione, L. Sussman, and Herden, "Communication Climate in Organizations," in *Handbook of Organizational Communication*, edited by F. Jablin, L. Putnam, K. Roberts, and L. Porters (Sage, Beverly Hills, 1987).
4. A. Alter, "TEAM BOOSTERS," *CIO* , 36-40 (April 1, 1992).

ON SOLVING THE ASSIGNMENT PROBLEM FOR PARTIALLY OVERLAPPED SYSTEMS

M. Kamel and H. Ghenniwa

Pattern Analysis and Machine Intelligence Lab
University of Waterloo
Waterloo, Ontario
Canada N2L 3G1

ABSTRACT

Manufacturing systems involving more than one machine, multi-robot systems, and multi-processing systems, can all be considered as systems consisting of subsystems each of which is capable of performing specific activities. All of these subsystems are acting in some form of parallelism. These systems can be called multi-agent systems. In this paper, we give a unified classification of multi-agent systems. The multi-agent systems are classified into three main classes: general purpose systems, special purpose systems, and partially overlapped systems. Partially overlapped systems are the most general case of multi-agent systems. A graph representation is developed to model the assignment problem for partially overlapped systems. An *OR-like* graph representation is introduced to further simplify the structure of the graph representation. A method which utilizes the OR-like graph representation to produce an approximate solution is described. The generated solutions satisfy the alance and *independence* conditions. The performance of the method is demonstrated on an example of a manufacturing system.

INTRODUCTION

A great deal of research has been done to solve the scheduling problems in specialized fields such as computer system environments (parallel processors and distributed systems) [2,13], and manufacturing environments (job-shop, flow-shop, open-shop, and parallel machine systems) [8,9,12,14]. All of these systems, even though they belong to different environments, are related to a one common class. For example, parallel processors and distributed systems are systems with multi-processor structure. The same with job-shop, flow-shop, open-shop, and parallel machine systems are systems with multi-machine structure. The formulation and solution approachs for these problems use different terminology depending on the field of application [10].

MULTI-AGENT SYSTEMS: A UNIFIED APPROACH

Definition 1 *A multi-agent system is a system consisting of a set of m agents denoted by $AG = \{Ag_i \mid Ag_i \triangleq (cs_i, ps_i), 1 \leq i \leq m\}$, where each agent Ag_i, $1 \leq i \leq m$, is characterized by: a capability set, cs_i, which is a set of the operations that Ag_i can handle, $cs_i = \{op_{i,1}, .., op_{i,\alpha_i}\}$, where op_{i,k_i} is operation k_i in cs_i, $1 \leq k_i \leq \alpha_i$, and α_i is the number of operations belonging to Ag_i; and a processing speed set, ps_i, which is a set of processing speeds of the operations given in cs_i,*

$ps_i = \{sop_{i,1}, sop_{i,2}, ..., sop_{i,\alpha_i}\}$, *where* sop_{i,k_i} *is the speed of processing* op_{i,k_i} *by* Ag_i.

Definition 2 *A multi-agent system structure is the agent-job relationship which represents the possible ways of manipulating (or processing) a given set of jobs by the system's agents.*

Definition 3 *A set of n jobs is defined as:* $\mathbf{J} = \{J_j \mid J_j \triangleq (ops_j), 1 \leq j \leq n\}$, *where operations set,* ops_j, *is the set of operations required by job* J_j.

Jobs, in general, may have constraints such as ready time, time at which the job is ready to be processed; due date, time in which job should be completed; and operation precedence, order in which the operations need to be processed. In this paper we focus our attention on simple jobs which are independent and nonpreemptive. It follows from the above definitions that an agent Ag_i is capable of processing a job J_j iff $cs_i \bigcap ops_j \neq \phi$; therefore, the multi-agent system structure is determined by its agents capability.

In general, multi-agent systems can be classified into three classes according to their structure. Each class can be further classified based on the processing characteristics of the system, operations precedence and agents processing speed (ps_i). Multi-agent systems by structure, a) General Purpose ($G.P$) system: a multi-agent system with all agents have the same capability sets, thus, $cs_i = cs_j \;\; for \;\; 1 \leq i,j \leq m$; b) Special Purpose ($S.P$) system: a multi-agent system with agents having capability sets totally different from each other, thus, $cs_i \bigcap cs_j = \phi$ for $1 \leq i,j \leq m,\; j \neq i$; and c) Partially Overlapped ($P.O$) system: a multi-agent system with some or all agents' capability set partially overlapped. Thus, $cs_i = cs_i^o \bigcup cs_i^s \;\; for \;\; 1 \leq i \leq m$, where, cs_i is the capability set of agent Ag_i, cs_i^s is the special capability set which can not belong to any agent except Ag_i, and cs_i^o is the capability set of operations overlapped between agent Ag_i and some other agents in the system.

The multi-agent classes, given above, can be classified further based on the processing characteristics. First $G.P$ systems, the type of speed which agents have can be used to identify the following three classes under the G.P[1] class: a) Identical $G.P$ system ($I.G.P$), which is a general-purpose system with agents' processing speed given by $sop_{i,k} = sop_{1,k}$, $1 \leq i \leq m$, $1 \leq k \leq \alpha$; b) Uniform $G.P$ system ($U.G.P$), which is a general-purpose system with agents' processing speed given by $sop_{i,k} = IS_i \times sop_{1,k}$ where, $1 \leq i \leq m$, $1 \leq k \leq \alpha$, and the speed factor $IS_i \in N$ (the set of natural numbers); and c) Unrelated $G.P$ system *(Un.G.P)*, which is a general-purpose system with agents' processing speed given by $sop_{i,k} = RS_{i,k} \times sop_{1,k}$, where, $1 \leq i \leq m$, $1 \leq k \leq \alpha$, and the speed factor $RS_{i,k} \in \Re^+$(the set of positive real numbers). Second, $S.P$ systems, the precedence constraint between the operations of each job can be used to differentiate between the three classes under the $S.P$ class: a) $S.P$ system with *identical constraints*, which is a special-purpose system processing jobs having the same initial precedence constraint between their operations. This is commonly known as "flow-shop" system; b) $S.P$ system with *arbitrary constraints*, which is a special-purpose system processing jobs which may differ in the initial precedence constraint between their operations. This is commonly

[1]for $G.P$ systems $\alpha_i = \alpha$ (constant.)

known as "job-shop" system; and *S.P* system with *no constraints*, which is a special-purpose system processing jobs having no initial precedence constraint between their operations. This is commonly known as "open-shop" system.

GRAPH REPRESENTATION

Multi-agent systems can be modeled using a bipartite graph. Graphically, a multi-agent system's structure is the structure of the *agent-job* links in the system's graph, we call it the *logical structure* of the system. The logical structure of multi-agent systems can be classified into the three classes OR-Structure, AND-Structure, and AND-OR-Structure. The AND-OR structure of *P.O* systems makes the assignment problem hard to see from its graph model. For implementation purposes it is necessary to simplify this graph representation. One possible way of doing this is by representing the logical structure of *P.O* system using a pure OR-structure, and at the same time, keeping the characteristics of its physical structure. We call this kind of representation an *OR-like* model. The OR-like model provides *P.O* systems with a simplified graph representation at the cost of an extra level of nodes. The OR-like graph contains three different types of nodes: a) real-agent nodes, represent all actual agents given in the system; b) virtual-agent nodes, a virtual-agent node is an agent node that has a capability set equivalent to the union of capability sets of some real agents which may be needed together to process some jobs; and c) job nodes, represent all jobs given in the processed job set. To distinguish between real-agent nodes and virtual-agent nodes, real-agents are classified to be *level-1* agents, and virtual-agents are classified to be *level-k* agents, where k is equal to the number of real-agents used to construct the virtual agent, $k > 1$.

SOLVING THE ASSIGNMENT PROBLEM FOR P.O SYSTEMS

The *scheduling* problem is defined as the allocation of jobs to the appropriate agents over time in order to process a collection of jobs [1]. In other words, scheduling problems involve two sub-problems, the assignment, and sequencing problems. However, by the definition of the processed jobs given above we limit ourselves, in the rest of this paper, to the scheduling problem that doesn't contain sequencing problem, for simplicity we call it the *assignment problem*. In this paper, the performance measure of the solution of the assignment problem, with respect to the performance criterion of the completion time, is to satisfy the following conditions, we call them the *goodness conditions* [11]: load-balance condition, minimizing the time-load difference between the agents; and independence condition, minimizing the number of agents assigned to process each job. A closer look at the relationship between each of these conditions shows that they are conflicting with each other. One way to resolve this problem is to relax one of these conditions. In this paper, the load-balance condition is relaxed by assuming that any solution that satisfies a predefined boundaries for the load-balance, as will be discussed later, is optimal with respect to the load-balance condition. To achieve the load-balance condition at the operation level a third condition is introduced, called the *critical condition*: minimizing the total number of necessary operations needed from the agent within the solutions domain satisfying the independence condition.

It is impractical to try find an optimal assignment solution with respect to the completion time [3,5,6, 7]. The problem has been proven to be *NP-hard in the strong sense* for *P.O* systems [11]. Instead we develop a method to find an approximate solution that is close to the optimal. First, the independence condition cannot be applied directly for OR-like graphs because it consists of *heterogeneous* links. Thus, minimizing the number of links in a heterogeneous system of links does not lead to the objective that is sought by the independence

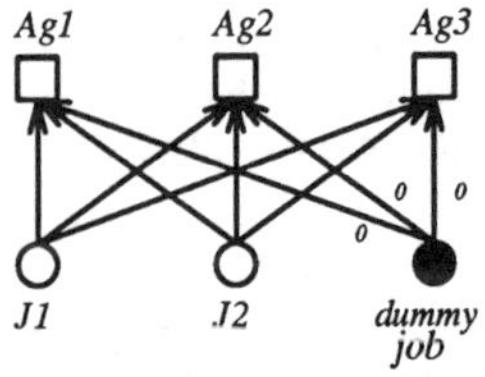

Figure 1: Converting a non-complete graph into complete graph, when $m > n$

condition. To overcome this problem the heterogeneous links are classified into groups of *homogeneous* links according to the level of the agent nodes, such that all agent nodes of level-k and their links are grouped in one group called level-k subgraph. This makes the independence condition applicable for each subgraph independently, and reduces finding the optimal assignment to finding the feasible assignment at the lowest possible level. Therefore, subgraph levels must be sorted in non-decreasing level order. Secondly, the critical condition also should be satisfied within each subgraph. This can be achieved by representing the number of necessary operations on the links between job nodes and their appropriate agent nodes as links' weights. Thus, the system critical case can be avoided by minimizing the total weight for each subgraph. Now we reach a point where we can introduce our method of obtaining an approximate solution for the assignment problem.

The Method

The main objective of the method is to solve the assignment problem incremently by breaking the system graph into subgraphs of sorted levels, and solving the assignment locally, for each subgraph, at each level in a way that the goodness conditions are satisfied. First, assume that the ideal load-balance is achievable, then the standard assignment algorithms [4] are appropriate tools to be used for satisfying the critical condition, where the total cost is the total number of the necessary operations. However, the standard assignment algorithms are designed for systems having complete bipartite graph representations. Unfortunately the graph representation of *P.O* systems may contain non-complete bipartite subgraphs because of the number of agent nodes may be not equal to the number of job nodes, or some job nodes may not be connected to all agent nodes. Hence, as a first step, we convert non-complete subgraphs into complete bipartite subgraphs. Let us consider the case when a subgraph has $m > n$. In this case some of the agents will have zero jobs assigned to them. This implies that the job nodes need to be increased to match the number of agent nodes by adding dummy job nodes, with each dummy job having $ops_j = \phi$. Thus, each dummy job node is connected by a zero weight link with each agent node, as shown in Figure 1. Now, consider the case when $m < n$. This means that there aren't enough agents to handle all the jobs simultaneously as shown in Figure 2. In this case agent nodes should be increased to match the job nodes such that the subgraph contains $(\lceil \frac{n}{m} \rceil - 1)$ copies of each existing agent node (original node), and $(m\lceil \frac{n}{m} \rceil - n)$ dummy job nodes if n is not an integer multiple of m. The set of original and its copy nodes represents one agent. Each agent copy node should have the same links as its original except for the links with the dummy job nodes. Such that zero weights are assigned to the links between dummy job nodes and original nodes, and infinity, ∞, weights to the links between copy nodes and dummy job nodes, as shown in Figure 3. To resolve the second possible case of a noncomplete bipartite graph, we simply add extra links between those unconnected nodes with each link having a weight of ∞, which reflects the infeasible assignment between these nodes.

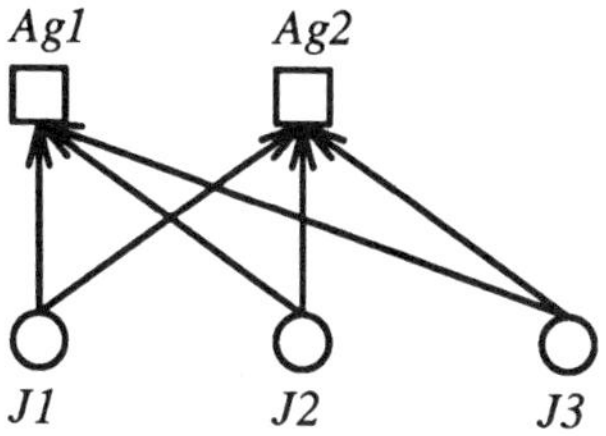

Figure 2: A non-complete graph, $m < n$

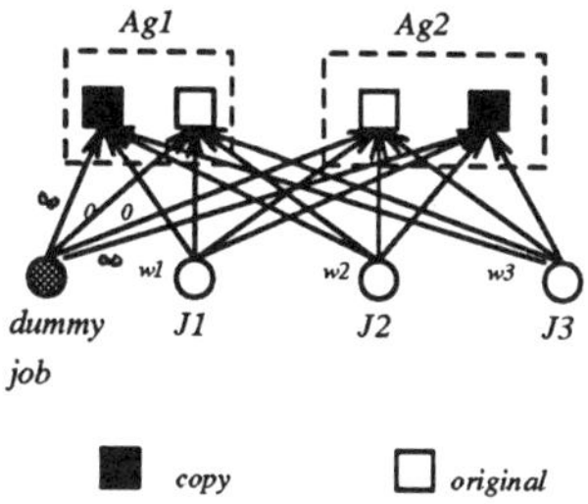

Figure 3: A complete graph version for $n > m$

Applying the standard assignment algorithms on each converted subgraph, as mentioned above, yields an ideal load balanced assignment on the jobs level, even though it may not be a feasible assignment. It is easy though, to check if the ideal balanced assignment is feasible or not; by checking if the total number of the necessary operations, we call it total *necessary-load*, of the suggested assignment is less than ∞, then a balanced necessary-load assignment does exist. If not then there are some jobs not appropriately assigned yet, these are real jobs assigned to agents with links of ∞ weights. We will call these jobs *unprocessed jobs.*

Definition 4 *An agent is called "unsaturated agent" if it has been assigned a total necessary-load of ω such that $\omega \geq \infty$. If $\omega < \infty$ and it has one dummy job in its assigned set of jobs then it is called "semisaturated agent". If it has no dummy jobs in its assigned set of jobs, then it is called " saturated agent".*

Lemma 1 *The unprocessed jobs can be reassigned to saturated or semisaturated agents only.*

Proof: is given in [11].

To control the undesirable effect of the conflict between the goodness conditions on the solution quality, we relax the load-balance condition by a predefined boundary limits for the agents' time-load. Moreover, in *P.O* systems, it is not always possible to find an assignment which minimizes the time-load difference between the agents, due to the agents' *S.P* behavior. To solve this problem, we determine the best time-load distribution within a subgraph, say at level-k'. Then if the current total time-load of each agent, for k=*1 to* k=k', is within the boundary limits, the time-load is considered to be acceptable, otherwise the suggested assignment in the current level-k' is rejected and we proceed forward by merging both level-k' and level-$(k'+1)$ subgraphs into one new subgraph. Then we solve the assignment for the new subgraph in a way that the chance of selecting the links that were belonging to the lower levels is greater than the ones belonging to the higher levels, so we do not violate the concept of homogeneous links distribution for the independence condition. One simple approach to achieve this is called the *offsetting technique* [11].

Virtual-agent nodes are used to advantage throughout the processing steps described by the method. Each virtual-agent is unfolded and its assignment solution is distributed among its real-agents. For example, assume that job J_1 is assigned to a virtual-agent Ag_{12}. In the assignment solution, Ag_{12} is unfolded to Ag_1 and Ag_2, and J_1 is added to each of Ag_1 and Ag_2 job lists. Therefore, the assignment solution includes assignment lists between real-agents and jobs only.

EXAMPLE

Consider a fully automated work cell, AWC, consisting of three multiple-operation machines named as M_1, M_2, and M_3. Each of them is characterized by its own capability set. M_1 is capable of drilling with a diameter range of (5mm-15mm), tapping with a diameter range of (5mm-20mm), and milling with an angle range of $(0^o - 45^o)$, M_2 is capable of drilling with a diameter range of (2mm-10mm), reaming with a diameter range of (2mm-10mm), and grinding with wheels diameter range of (5mm-20mm), and M_3 is capable of drilling with a diameter range of (8mm-20mm), reaming with a diameter range of (6mm-20mm), and milling with an angle range of $(30^o - 90^o)$. All the operations have processing speed of a single unit-time.
A set of five independent objects need to be processed by the AWC. Each object requires a set of operations. Object-1 requires drilling with diameters of 9mm and 12mm, tapping with a diameter of 10mm, and reaming with a diameter of 5mm, object-2 requires drilling with a diameter of 9mm, milling with an angle of 30°, grinding with an 8mm wheel, and reaming with a diameter of 7mm, object-3 requires a drilling with a diameter of 12mm, reaming with diameters of 7mm and 8mm, object-4 requires milling with angles of $30o$ and 45°, and grinding with an 8mm wheel, and object-5 requires drilling with diameters of 9mm and 12mm.

System Formulation

The AWC is considered as a multi-agent system consisting of three agents, M_1 will be named as Ag_1, M_2 as Ag_2, and M_3 as Ag_3. Due to the overlapping property between the agents capabilities AWC is classified as a $P.O$ system. For example, Ag_1 and Ag_2 both are capable of performing drilling with a diameter range of (5mm-10mm). Therefore, $AWC = \{Ag_i \;|Ag_i \triangleq (cs_i, ps_i \;|sop_{i,k_i} = 1) 1 \leq i \leq 3\}$. Each of the objects can be considered as a job where, $object_j = J_j, \;\; 1 \leq j \leq 5$. Due to the wide range of the agents capabilities, and that not all of them are being used to process the given set of jobs, we will consider the agents capabilities which are required by the jobs. As a first step we list all different operations that are required by the jobs. Drilling with diameters of 9mm and 12mm we call them op_1 and op_2 respectively. Milling with angles of 30° and 45° we call them op_3 and op_4 respectively. Tapping with a diameter of 10mm we call it op_5. Reaming with diameters of 8mm, 7mm, and 5mm we call them op_6, op_7, and op_8 respectively. Grinding with a wheel of 8mm diameter we call it op_9. Accordingly the capability sets of the agents are $cs_1 = \{op_1, op_2, op_3, op_4, op_5\}$, $cs_2 = \{op_1, op_6, op_7, op_8, op_9\}$, and $cs_3 = \{op_1, op_2, op_3, op_4, op_6, op_7\}$. $sop_k = 1$ for $1 \leq k \leq 9$. The operations sets required of the jobs are $ops_1 = \{op_1, op_2, op_5, op_8\}$, $ops_2 = \{op_1, op_3, op_7, op_9\}$, $ops_3 = \{op_2, op_6, op_7\}$, $ops_4 = \{op_3, op_4, op_9\}$, and $ops_5 = \{op_1, op_2\}$. The OR-like bipartite graph model of the system is shown in Figure 4. The suggested solution is shown in Figures 5 and 6. This solution, as obtained by the algorithm, is the optimal solution for this case.

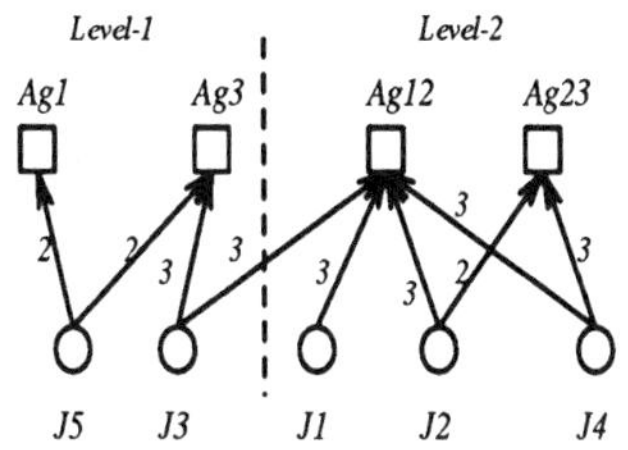

Figure 4: An *OR-like* bipartite graph model of the example.

```
Level-1
-------
 job .. J5 .. Assigned to .. Ag1 .
 job .. J3 .. Assigned to .. Ag3 .
* The agents Load condition *
---------------------------------
1 job(s) assigned to agent 1: J5
0 job(s) assigned to agent 2
1 job(s) assigned to agent 3: J3
=================================
```

Figure 5: The assignment solution for level-1 agents

```
Level-2
-------
 job .. J1 .. Assigned to .. Ag12 .
 job .. J2 .. Assigned to .. Ag23 .
 job .. J4 .. Assigned to .. Ag23 .
* The agents Load condition *
---------------------------------
2 job(s) assigned to agent 1: J1,J5
3 job(s) assigned to agent 2: J1,J2,J4
3 job(s) assigned to agent 3: J2,J3,J4
=================================
```

Figure 6: The assignment solution for level-2 agents

CONCLUSION

Most of the research done in the assignment and scheduling theory has been devoted either to scheduling single-operation jobs on *G.P* systems, or to scheduling multiple-operation jobs on *S.P* systems. Meanwhile a large number of todays real life systems are to be more efficient and applicable if they are designed as *P.O* systems. Solving the assignment and the scheduling problems in this case is challenging. In this paper, we have shown that *P.O* systems are the most general case of multi-agent systems. A graph representation has been developed to model the assignment problem of *P.O* systems. Two conditions have been developed to measure the performance of the solution of the assignment problem, namely, the load-balance and the independence conditions. An *OR-like* graph has been introduced to simplify the *AND-OR* structure of *P.O* systems. Finally, a method, which utilizes the *OR-like* representation, has been developed to generate an approximate solution to the assignment problem of *P.O* systems by satisfying the goodness conditions. The performance of the algorithm is demonstrated on an example of a manufacturing system.

REFERENCES

1. Kenneth R. Baker, *Introduction to Sequencing and Scheduling*, (John Wiley & Sons, Inc., New York, 1974).
2. Shahid H. Bokhari, *Assignment Problems in Parallel and Distributed Computing*, (Kluwer Academic, Boston, 1987).
3. R. Bulfin and R. G. Parker, "On the assignment and sequencing of operations on a crew-served project.", *NAVAL Res. Logistic Vol.(23), pp261-269, (1976).*
4. R. E. Burkard and U. Derigs, *Assignment and Matching Problems: solution methods with fortran program*, (Springer, Berlin, 1980).
5. R. Chang and R. Lee, "On a scheduling problem where a job can be executed only by a limited number of processors.", *Comput. Opns. Res. Vol.(15), No.(5), pp471-478, (1988).*
6. Y. Chen and Y. Chin, "Scheduling unit time jobs on processors with different capabilities.", *Comput. Opns. Res. Vol.(16), No.(5), pp409-417, (1989).*
7. E. G. Coffman, Jr (Ed), *Computer and Job-Shop Scheduling Theory*, (John Wiley, New York, 1976).
8. Simon French, *Sequencing and Scheduling*, (Ellis Horwood Limited, England, 1982).
9. T. Gonzalez and S. Sahni, " Opns-Shop Scheduling to Minimize Finish Time.", *J. ACM Vol.(23), No.(4), pp 665-679, (Oct.1976).*
10. M. J. Gonzalez Jr., " Deterministic Processor Scheduling., *ACM Comput. surveys Vol.(9), No.(3), pp 173-204, (Sep. 1977).*
11. M. Kamel and H. Ghenniwa, " Solving the assignment Problem for Multi-agent systems.", *to be submitted IEEE Tarns. On Robotics and Automation (1992).*
12. E. L. Lawler, J. K. Lenstra and A. H. G. Rinnooy Kan, " Recent Development in Deterministic Sequencing and Scheduling: A Survey.", M. A. H. Dempster et. al.(eds), *Deterministic and Stochastic Scheduling, pp 35-73, (1982).*
13. V. Lo and J. W. S. Liu, " Task Assignment in Distributed Multiprocessor Systems.",*Proc. of the International Conf. On Parallel Processing, pp 358-360, (1981).*
14. A.H.G. Rinnooy Kan, *Machine scheduling problems: classification, complexity and computations* (Martinus Nijhoff, The Hague, Holland, 1976.)

IDENTIFYING BOTTLENECK MACHINES AND EXCEPTIONAL PARTS FOR A CELLULAR MANUFACTURING SYSTEM

F. Choobineh, P.E., Ph.D.
Department of Industrial and Management Systems Engineering
University of Nebraska-Lincoln
Lincoln, NE 68588-0518

ABSTRACT

To economically design a cellular manufacturing system, first, the bottleneck machines and exceptional parts needs to be identified, and second, this information should be used in a cost model to finalize the design. This paper presents a heuristic procedure for identifying the minimal number of bottleneck machines and exceptional parts. The second phase of this design process is addressed in our working paper.

INTRODUCTION

A cellular manufacturing system (CMS) consists of groups of parts being produced by groups of machines. These groupings are obtained by judicious decomposition of the manufacturing system. A decomposed manufacturing system is easier to control and thus more efficient than a job shop (undecomposed) manufacturing system. Good reviews of articles about decomposition of a manufacturing system are provided by Kusiak and Chow [4] and Askin and Vakharia [1].

An ideal decomposition of a manufacturing system would result in subsystems that are totally independent (disjointed) of each other. This means that no machine is shared by any two machine cells and each part family is entirely produced in one machine cell. However, most manufacturing systems can not be ideally decomposed unless certain parts are removed and/or certain machines are duplicated or eliminated. The parts that need to be removed are referred to as <u>exceptional parts</u> and the machines that need to be duplicated or eliminated are referred to as <u>bottleneck machines</u>. The challenge is to remove the <u>minimal</u> number of exceptional parts and/or bottleneck machines so that an ideal decomposition is achieved. Identifying the minimal number of exceptions or bottlenecks is an NP-complete problem and thus heuristics that provide suboptimal solutions are used.

In this paper the cluster identification algorithm of Kusiak and Chow [5] is modified. The modified (proposed) algorithm attempts to identify the minimal number of parts and machines that are needed to be removed before an ideal decomposition is achieved.

THE PROPOSED PROCEDURE

A bipartite graph G=(P,M,A) can be used for representing the relationship among parts and machines. One set of nodes, P, represents parts and the other set of nodes, M, represents machines. The set of arcs, A, identifies (by connecting the two node sets) the set of machines that are necessary for producing each part. Figure 1 shows an example of a bipartite graph representing a part-machine relationship. In general, the objective of decomposing a manufacturing system into a CMS is to identify the disjointed subgraphs of G which may require removing minimal number of nodes.

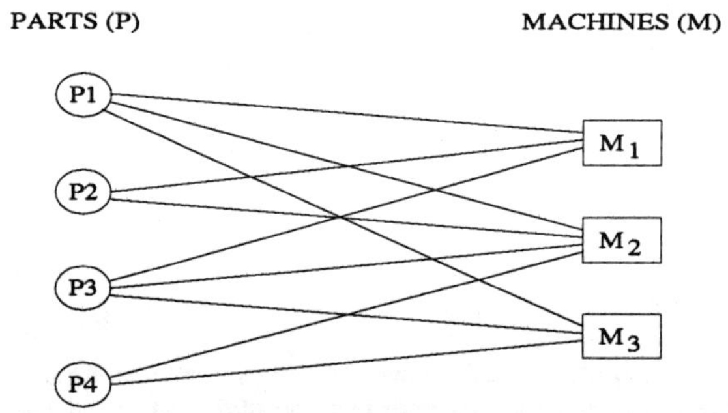

Figure 1 A Bipartite Graph Representing a Part-Machine Relationship

The cluster identification algorithm of Kusiak and Chow [5] is based on the Iri's [3] procedure and essentially checks the connectively of the graph G. The steps of the Kusiak and Chow's algorithm for identifying the disjointed subgraphs, are as follows:

1. Start from any node on the graph G.
2. Identify all the nodes that are connected to the starting node. Designate the parts, P_i's, that are connected to the starting node as a part family (group) and the machines, M_j's, that are connected to the starting node as the machine cell (group). Separate these nodes from graph G.
3. If there is any node left on the graph G, go to Step 1, otherwise stop.

In this paper this simple algorithm is modified such that the exceptional parts and bottleneck machines could be identified.

Let AP_i be the set of arcs emanating from node P_i and AM_{ij} be the set of arcs emanating from the M_jth node that is connected to node P_i. Likewise, let AM_j be the set of arcs emanating from node M_j and AP_{ji} be the set of arcs emanating from the P_ith node that is connected to node M_j. The cardinality of a node is defined as the number of arcs emanating from that node. For example in Figure 1 the cardinality of P_i is $|AP_1|=3$ and the cardinality of AM_{12} is $|AM_{12}|=|AM_2|=4$. Also, let $\Delta P_i = \sum_j |AM_{ij}| - |AP_i|$ and $\Delta M_j = \sum_i |AP_{ji}| - |AM_j|$.

The steps of the proposed procedure are as follows:

1. Start from any node on the graph G.
2. Identify the nodes that are connected to the starting node. If all nodes of G are connected, there exists an exceptional part and/or a bottleneck machine, then go to Step 4. If all nodes of G are not connected to the starting node, remove the ones that are connected from graph G and form part and machine groups, then go to Step 3.
3. If no node is left on graph G, stop, otherwise go to Step 1.
4. Determine $Q=\max\{\Delta P_i, \Delta M_j\}$. If $Q=\Delta P_i$, then P_i is an exceptional part and remove P_i and AP_i from graph G. If $Q=\Delta M_j$, then M_j is a bottleneck machine and remove M_j and AM_j from graph G. Break the ties between ΔP_i and ΔM_j at random. Go to Step 1.

Comparisons of the proposed procedure with other procedures, and a cost model for economical formations of part families and machine cells are presented in [2].

CONCLUSIONS

Identification of minimal bottleneck machines and exceptional parts is a prerequisite for an economical design of a cellular manufacturing system. Upon their identification subsequent decisions on duplication of bottleneck machines and/or subcontracting or modifying of exceptional parts can be made. In this paper an efficient heuristic procedure for identifying bottleneck machines and/or exceptional parts, as well as part families and machine cell has been proposed. The procedure has a time complexity of O(MP+P+M) where M is the number of machines and P is the number of part. This complexity makes it quite efficient for large scale manufacturing decomposition.

REFERENCES

1. R. G. Askin, A. J. Vakharia, "Group Technology - Cell Formation and Operation," *The Automated Factory Handbook: Technology and Management,* Edited by D. I. Cleland and B. Bidanda (New York: Tab Book Inc.) 317-366 (1990).

2. F. Choobineh, "Identification of Bottleneck Machines and Exceptional Parts for Cellular Manufacturing Design," Working Paper, Department of Industrial and Management Systems Engineering (1992).

3. M. Iri, "On the Synthesis of Loop and Cutset Matrices and Related Problems," K. Kondo (Editor), Raag Memoirs, Research Association of Applied Geometry, Japan, Vol. 4, A-XII, 376-410 (1968).

4. A. Kusiak, W. S. Chow, "Decomposition of Manufacturing Systems," *IEEE Journal of Robotics and Automation, 4, (5),* 457-471 (1988).

5. A. Kusiak, W. S. Chow, "Efficient Solving of the Group Technology Problem," *Journal of Manufacturing Systems, 6, (2),* 117-124 (1987).

Reflex Control for Robot System Preservation and Reliability

Thomas S. Wikman * Wyatt S. Newman †

Department of Electrical Engineering and Applied Physics
and
Center for Automation and Intelligent Systems Research ‡
Case Western Reserve University
Cleveland, OH 44106

Abstract

This paper describes the use of reflex control for robot system preservation and reliability. Reflex control is a fast, on-line and failsafe obstacle avoidance method that protects the system from erroneous higher level commands. The reflex controller inspects the immediate configuration space and does not approve commands that would result in a collision with an obstacle. Very efficient obstacle avoidance and path planning algorithms have been produced by using the reflex controller as a building block. Experimental results are described for our physical implementation of the system on the first four joints of a kinematically redundant robot.

I Introduction

As higher-level software becomes more complex, the likelihood of unsuspected, embedded programming errors becomes high. To protect against such errors, only completely trustworthy low-level software can be depended on. By analogy, this low-level software is referred to as "reflex control", since this software invokes reactions which are responsive to sensory input and are appropriate for preserving viability, yet are not complex enough to be labelled as "intelligent".

The *reflex controller* exhibits the following virtues: It does not suffer from unrealistic or overly restrictive assumptions on robot dynamics; it fits in control hierarchies and does not interfere with normal actions unless imminent danger is present; its complexity does not increase with higher dimensions; it does not fail or slow down as environment complexity increases; it does guarantee collision avoidance (at least for static obstacles) and can be computed on line; and it is failsafe i.e. if the reflex controller quits or is too slow this will still not result in a collision.

When the reflex controller easily can be incorporated with other control schemes with minimum non-essential influence on higher level controls, it serves as a natural building block in more advanced obstacle avoidance and path planning schemes. An example of this is guiding potential functions. In this paper we describe a guiding potential function which incorporates task-space based attraction towards a goal, while the reflexes permit the use of kinematic redundancy for obstacle avoidance. We also describe a geometric on-line 4D path planning algorithm that uses the reflex controller as a building block.

The reflex controller relies on a non-overshooting servo-controller for failsafe obstacle avoidance. Reflex controllers have been tested in the context of soft-PD, bang-bang, and sliding-mode control [7, 8, 10, 2, 1, 12]. We have also implemented reflex control with a so called reactive trapezoidal trajectory generator. The reactive trapezoidal trajectory generator generates in a reactive manner non-overshooting, closely time-optimal trapezoidal velocity profiles to a servo-controller.

*Graduate Student
†Associate Professor
‡A Cleveland Area Manufacturing Program/Ohio Edison Center

II The Reactive Trapezoidal Trajectory Generator

The reactive trapezoidal trajectory generator accepts position commands and generates continuous trajectory commands. These commands are issued as velocity commands when our servo-controller is a pure velocity controller, or as position, velocity, and acceleration commands in the more general case. The trajectory generator operates by planning a trapezoidal angular velocity profile for each joint. Slope and height constraints on the trapezoidal velocity profile restrict resulting trajectories to conform to the machine's joint-by-joint acceleration and velocity limits. Thus, all trajectories generated can be executed feasibly by the underlying tracking controller. At the same time, though, the trajectories generated are nearly time-optimal. A crucial new feature of our trajectory generator is that it is "reactive." That is, it is not necessary to complete any one planned trajectory before responding to new position commands. New trajectory plans are continuously recomputed with initial conditions equal to the current robot state, and goal conditions equal to the current position command. This feature is essential if the robot is working under reflex control.

The reactive trapezoidal velocity profiler first determines whether the robot should accelerate, move with a constant velocity or brake. Then it generates new trajectory commands based on the current state of the robot. In the case where the reactive trapezoidal trajectory generator generates commands to a PD-DD (position-velocity-acceleration) controller, it provides the following control law for each joint (The regions refer to Fig 1).

- Region III : If $\partial|d|/\partial t < 0$ and $|d| < d_{brake}$, where $d = \theta_{des} - \theta$ and d_{brake} is the braking distance, assert maximum braking. The maximum braking is asserted by generating the following commands to a PD-DD controller :

$$\begin{aligned} \alpha_{des} &= -\mathbf{sgn}(d)\alpha_{max} \\ \omega_{des} &= \mathbf{sgn}(d)\sqrt{2|d|\alpha_{max}} \\ \theta_{des} &= \theta + \omega_{old}\Delta t - \\ &\quad \mathbf{sgn}(d)\alpha_{max}\Delta t^2/2 \end{aligned} \tag{1}$$

  where α_{des} is the desired acceleration, α_{max} the maximum acceleration, ω_{des} the desired velocity, θ_{des} the desired position, $\omega_{old} = \omega(t - \Delta t)$ the last velocity command, and Δt is the loop time of the servo program, or in other words the time that elapses between command updates.

- Region II : else if $\omega > (1-\epsilon)\omega_{max}$ assert constant velocity, where ω_{max} is the maximum velocity and ϵ is a small constant that is adjusted for adequate noise immunity. Constant maximum velocity is asserted by generating the following commands to the PD-DD controller.

$$\begin{aligned} \alpha_{des} &= 0 \\ \omega_{des} &= \mathbf{sgn}(d)\omega_{max} \\ \theta_{des} &= \theta + \mathbf{sgn}(d)\omega_{max}\Delta t \end{aligned} \tag{2}$$

- Region I : else assert maximum acceleration. This will happen if we are moving in the opposite direction to where we want to be or the velocity is less than ω_{max} and $|d| \geq d_{brake}$. Maximum acceleration is asserted by generating the following commands to the PD-DD controller.

$$\begin{aligned} \alpha_{des} &= \mathbf{sgn}(d)\alpha_{max} \\ \omega_{des} &= \mathbf{sgn}(d)\alpha_{max}\Delta t + \omega_{old} \\ \theta_{des} &= \theta + \omega_{old}\Delta t + \\ &\quad \mathbf{sgn}(d)\alpha_{max}\Delta t^2/2 \end{aligned} \tag{3}$$

In the case where the reactive trapezoidal trajectory generator generates commands to a velocity controller, it provides the following control law for each joint.

- Region IV : If $|d| < \sigma$, where $\sigma = 0.5^o$ in our experiments, generate a velocity command that results in the same commanded torque as a fictitious PD controller. The purpose of this is to ensure that the robot is brought close to the goal. We assume that the velocity controller generates torque commands according to $\tau_{cmd} = K_v(\omega_{des} - \omega)$, and that the PD controller used in our comparison is $\tau_{cmd} = K_p(\theta_{des} - \theta) + K_v(\omega_{des} - \omega)$. The PD-equivalent velocity command is asserted by generating the following velocity command :

$$\omega_{des} = K_p d / K_v \tag{4}$$

- Region III : else if $\partial|d|/\partial t < 0$ and $|d| < d_{brake}$, assert close to maximum braking. The maximum braking is asserted by generating the following velocity command :

$$\omega_{des} = \mathbf{sgn}(d) \max(\sqrt{2|d|\alpha_{max}}, \sigma K_p / K_v) \tag{5}$$

- Region I,II : else assert maximum acceleration or maximum velocity by generating the following velocity command :

$$\omega_{des} = \mathbf{sgn}(d)\omega_{max} \tag{6}$$

The reactive trapezoidal trajectory profiler is smooth, non overshooting, nearly time optimal, stable, adjustable, it updates quickly, and it is amenable to quickly changing position commands, and is therefore ideal in the context of reflex control. For the same reasons the reactive trapezoidal trajectory generator allows construction of systems which are capable of exhibiting stimulus-responsive behaviors. The dominant class of machines outside the laboratory utilizes velocity control at the analog level, thus a reactive trapezoidal trajectory profiler generating commands to a velocity controller is applicable to a broad range of machines. In Fig 1 we illustrate the general behavior of the reactive trapezoid.

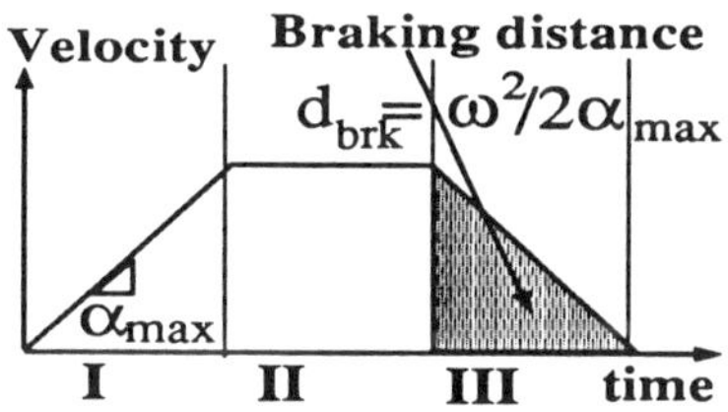

Figure 1 : **The reactive trapezoidal trajectory profiler**

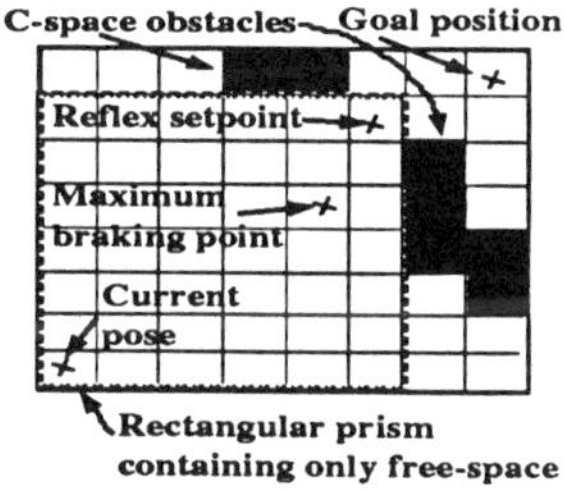

Figure 2 : **Rectangular free-space prism in C-space**

III The Reflex Controller

The primary objective of a reflex controller is to guarantee system survivability. The reflex controller accepts commands from higher, more intellient layers, monitors the state of the system, inspects the local configuration space, and anticipates potential damage to the system. The reflex layer continuously and rapidly projects whether it would be possible to prevent a collision, if necessary. On each reflex cycle, if it can be proven that a protective reaction may be safely deferred, then the higher-level motion command is accepted and executed verbatim. If, however, the reflex layer cannot prove that protective action may be postponed, then the reflexes substitute a protective action in place of the higher-level command. The protective motion approaches the commanded motion as closely as possible, subject to the constraint that no damage may occur.

A heuristic (though incomplete) description of the reflex control task may be stated as follows: the reflex controller approves a "reflex setpoint" which is the closest point to the goal pose for which the hyper-rectangular-prism defined by opposite vertices at 1) the current pose and 2) the reflex setpoint, contains only freespace. This is illustrated in Fig 2. The setpoints are continuously updated until an obstacle is found. When an obstacle is found the updating of setpoints is halted along the joint or joints for which the obstacle represents a hazard. In Fig 2 the obstacles represents hazards to both joints. A more complete description of the reflexes may be found in [6, 7, 8, 10, 9]. We note, though, that when the current position and the goal position both lie within this prism, the goal position is approved by the reflex controller and passed along unaltered to the servo controller or the trajectory generator. If the reflex controller unexpectantly quits or is too slow, it would fail to update new setpoints accordingly, which would cause the robot to unnecessarily slow down or brake. However, since no setpoints that could result in a collision with an obstacle are ever approved, the robot would still not collide with obstacles.

Reflex controllers that approve or disapprove acceleration commands or velocity commands have also been implemented. In [8] an acceleration-based reflex controller which emulates direct torque control, yet guarantees collision avoidance is described. In the case of acceleration-based or velocity-based reflex control, the reflex controller receives an acceleration or velocity command, which it converts into a corresponding virtual setpoint. The virtual setpoint is the position command that would result in the same applied torque as the received acceleration or velocity command. This conversion requires that the reflex controller is informed about the servo-control-parameters and the dynamics of the system. The virtual setpoint is approved or disapproved based on the evaluation of the local configuration space map. If approved the original acceleration or velocity command is approved, and if not the new setpoint is converted into the corresponding acceleration command. An overview of the system is illustrated in Fig 3.

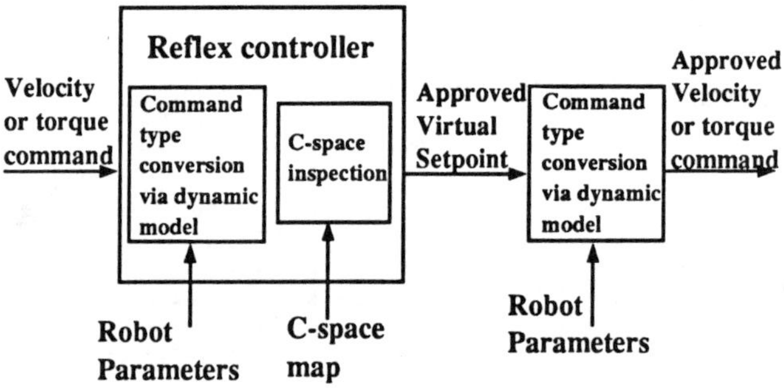

Figure 3: A velocity or torque command is received by the reflex controller, converted into a virtual setpoint, approved or disapproved by the reflex controller after comparison with the C-space map, and finally converted into an approved velocity or torque command

The reflex controller has been shown to be efficient, permitting full-speed robot operation while ensuring collision prevention. Reflex control has been tested with higher layers of both teleoperation and on-line motion planning. In the case of teleoperation, the reflex controller permitted the robot to respond to operator commands up to the full bandwidth of the actuators, while simultaneously protecting the robot from even deliberately malicious motion commands. In the case of on-line planning, the reflex layer actually improved the performance of the planner, as the planner did not need to concern itself with high resolution or with possible danger from dynamic tracking errors.

We also implemented reflex control in the context of two moving robots sharing a common work-space. In our implementation one of the robots, a Hitachi PW-10, had a fictitious protective wall located at its tool position, while the second robot, a RRC K-2107HR, avoided this wall. In this case it was not sufficient that the RRC robot reflexively avoid collision with the wall. The RRC also had to escape from the wall when the wall approached the RRC. For this purpose a second type of reflex called a "flee-reflex" was added. The flee-reflex controller inspects the C-space and generates a "flee-reflex" if the C-space obstacle corresponding to the wall advances towards the RRC robot's current location.

Finally, we note that the reflex controller require access to a configuration-space map of all reachable obstacles. Presumption of the existence of such a map, particularly in higher dimensions, should not be taken lightly, since C-space maps can be demanding of both storage space and computation time. A brief divergence into our use of C-space is given next.

IV Construction and Inspection of 4D Configuration Space

The configuration space contains static obstacles, corresponding to joint limits, static obstacles in its environment, and poses representing self-collision configurations. These C-space obstacles do not require recomputation or sensory input and can therefore be precomputed. However, sometimes obstacles are brought in or out of the robot's work-space, and at these instances the C-space map must be recomputed as fast as possible. For this purpose we constructed a 4D configuration-space cartographer that takes workspace obstacle positions and their sizes as inputs, and rapidly adds them to the C-space map. Currently, obstacles are defined either manually (by typing coordinates) or "taught" by using the robot to touch points on an obstacle. The cartographer uses a very fast method for computing C-space obstacles, using a database of precomputed transformations corresponding to simple workspace shapes. This method was developed in [7] and is further described in [2, 1, 10, 5].

The configuration-space cartographer makes it possible to add obstacles to the configuration space as soon as they appear in the robot's workspace. If the robot operator knows the location and the size of these obstacles he can add them manually, but if not he can define their location and size by letting the robot touch the obstacles and record the touchpoints. The time it takes to add an obstacle to the C-space is a few seconds up to several minutes, depending on the obstacle. The cartographer can also be interfaced to external sensors that search the robot's work-space for obstacles. We are currently working on interfacing an ultrasonic sensor to our system to accomplish this. In our current implementation, configuration space is discretized into 64^4 "voxels" in 4-D, and these voxels are bit-mapped in RAM.

The inspection of the configuration space can be made very fast when only local regions of the configuration space are inspected. To improve the speed of the inspection routine, copies of the configuration space with lower resolution are created. Using these low resolution maps, it is only necessary to continuously inspect one low-resolution pixel, and when this pixel contains an obstacle, the higher resolution pixels are inspected.

V Guiding Potential Functions

While reflex control is capable of providing some low-level potential-function type motion planning, we found that the reflex controller was often not as competent as deliberately chosen potential functions in finding simple paths around obstacles.

Our guiding potential function augments the planning behavior of our reflex controller. The current improvement can be explained best in terms of energy. If a joint-space goal point, $\boldsymbol{q}_{goal}$, is represented as an attractive potential function, then we can evaluate the energy associated with any pose of the robot, $\boldsymbol{q}$, as $E = \sum_{i=1}^{N} E_i$ where $E_i = \frac{1}{2}K_i(q_{des,i} - q_i)^2$. Under our reflex control, joints are controlled in a manner which enforces $d(E_i)/dt \leq 0$ for *each* joint i. Alternatively, a typical potential-function controller would execute a path corresponding to the steepest descent of the *total* energy, E. At an obstacle boundary, moving into an obstacle is equivalent to a steep positive energy gradient. As a result, the steepest descent of E near an obstacle might well require an *increase* in one or more individual joint energies, E_i. Thus, potential-function controllers can often continue to make progress towards a goal ($dE/dt < 0$) in cases where reflex control alone would stall ($d(E_j)/dt > 0$ for some j).

This problem can be alleviated by adding to the reflex controller a so called "guiding potential function" that generates subgoals that are achievable by the reflexes.

The purpose of the guiding potential function is solely to guide the reflex controller towards the goal, not to avoid collisions. Collisions are prevented by the reflexes. Thus it is not necessary for the guiding potential function to consider actuator effort saturation or robot dynamics when generating subgoals. Through this combination of reflexes and potential functions, we retain the speed and safety virtues of the reflex controller while inheriting the local-planning nature of potential functions.

We have investigated a few potential functions: potential functions exerted in joint space, in workspace with an additional augmented constraint for redundancy, and potential functions expressed in *augmented* task-space coordinates, as defined in [11]. These potential functions are described as follows:

- A *configuration-space guiding potential function* that computes energies as a function of distance in joint-space to a specified joint-space goal, $\boldsymbol{q}_{des,i}$, assuming a virtual joint stiffness matrix K_q:

$$E_{CS} = \sum_{i=1}^{N} K_{q,ii}(q_{des,i} - q_i)^2 \tag{7}$$

  In this case the guiding potential function results in the same path as the joint by joint reflex controller when no obstacles are present. When at the boundary of an obstacle, though, the guiding potential function leads the robot around obstructions when such motion would decrease total energy.

- A *workspace guiding potential function* that computes energies as a function of distance to a specified hand pose, $\boldsymbol{x}_{des}$ (with a large workspace stiffness K_x), plus a weak attraction to a preferred joint pose, $\boldsymbol{q}_{des}$ (with weak joint-space stiffness K_q):

$$\begin{aligned} E_{WS} &= \sum_{i=1}^{M} K_{x,ii}(x_{des,i} - x_i)^2 \\ &\quad + \sum_{i=1}^{N} K_{q,ii}(q_{des,i} - q_i)^2 \end{aligned} \tag{8}$$

- An *augmented task-space potential function* that computes energies as a function of distance to a specified task-space goal, $\boldsymbol{y}_{des}$ consisting of the hand pose and an elbow angle :

$$E_{TS} = \sum_{i=1}^{N} K_{y,ii}(y_{des,i} - y_i)^2 \tag{9}$$

  By using augmented task-space coordinates, the joint space term is eliminated.

where $\boldsymbol{x}_{des}$ is an M-dimensional desired workspace goal, $\boldsymbol{y}_{des}$ is an N-dimensional augmented task-space goal, and $\boldsymbol{q}_{des}$ is an N-dimensional desired joint-space goal. A significant problem for energy-gradient based methods acting in a discretized space is that discretization causes false local energy minima and maxima. In [12] we describe how to tunnel through such local energy minima and maxima.

VI On-line Path Planning

The 4D on-line motion planning algorithm described here is an augmententation of the 3D motion planning algorithm developed for monodextrous industrial robots described in [4, 3]. The efficiency of the path planning algorithm is due to the fact that the reflex controller is able to bring the robot between so called critical points in configuration space in a nearly time optimal manner, while guaranteeing failsafe collision avoidance.

The path planning algorithm uses 2D slices of the discretized 4D C-space. We chose to slice the 4D C-space into θ_3, θ_4 slices along the θ_1 and θ_2 axes. A very important building block in our 4D path planner is our reflex based 2D-planner. The 2D-planner is able to bring the robot to a specified goal within a given 2D C-space slice by generating a small set of subgoals achievable by the reflexes. First the 2D-planner identifies so called critical points within the 2D C-space slice. These critical points are peaks of C-space obstacles as illustrated in Fig 4. These peaks are extremal points on the border curves of the C-space obstacles with respect to one of the joint coordinates. For reasons explained in section V the reflexes are unable to bring the robot around such peaks. If such peaks exists in the current 2D C-space slice and they represent impediments to the reflex controller, the 2D-planner uses these peaks to generate a set of subgoals that are given to the reflex controller. An example of how this can be done is described in [4, 3].

The first task of the 4D path planning algorithm is to identify connected regions of free-space within each 2D-slice. These regions of free-space are called "F-regions" and are illustrated in Fig 4. The second task of the planner is, for every F-region, to identify which F-regions in the neighboring 2D-slices the F-region is connected to, and generate a tabula describing these connections. This tabula is called the inter-region graph. Two F-regions are connected if they belong to two adjacent 2D slices of C-space, and have overlapping areas of free-space. Such connections are illustrated in Fig 5. If the current position of the robot is located in an F-region F_1, and F_1 is connected to an F-region F_2, the robot could be moved to F_2 by first moving the robot to an area where F_1 and F_2 overlap, and at that point cross over from F_1 to F_2. The third task of the planner is to identify the F-region where the goal is located, and the F-region where the current position is located, called F_{goal} and F_{start} respectively. Using the inter-region graph the planner then finds a path through the connected F-regions that connects F_{goal} to F_{start}, using a graph search method. The 4D-planner uses the 2D-planner to navigate within the F-regions. On execution the robot is moved from its start position within F_{start} to a point which belongs to an area overlapping with the next desired F-region. When this point is reached the robot crosses over to the next F-region, and repeats this operation until the final goal is reached.

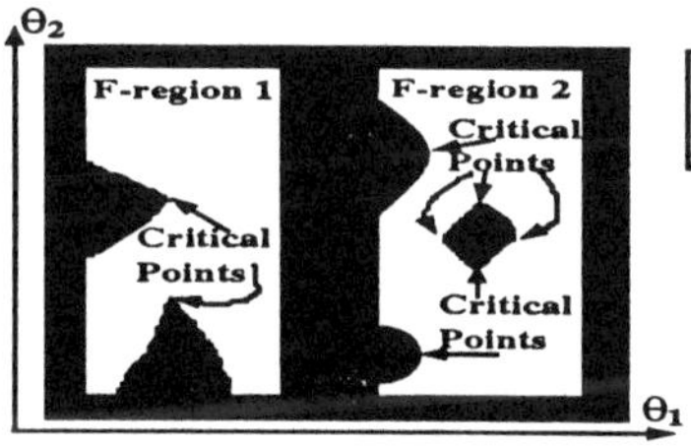

Figure 4 : **Reversal sets and F-regions**

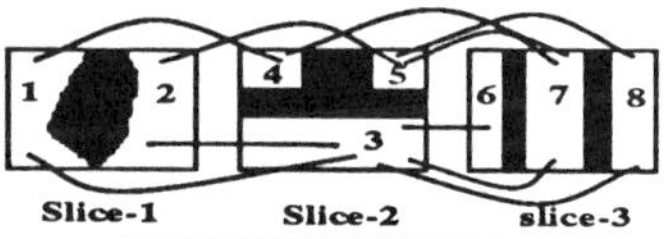

F-region	connected to
1	3,4
2	3,5
3	6,7,8,1,2
4	1,6
5	7,8
6	4,3
7	3,5
8	3,5

Figure 5 : **Inter-region graph for 8 F-regions located in 3 adjacent C-space slices**

Our 4D-motion planning algorithm finds a solution whenever a solution is possible. It is on-line in the sense that when new goals are requested, new paths are generated while the robot is moving. If new obstacles are brought in to the robot's environment, a new C-space map must be computed and the corresponding F-regions identfied. However, with our configuration space cartographer described in section IV this can be done fairly quickly. Another important virtue of our path-planner is that it generates subgoals to the reflex controller instead of paths to a servo controller, thus making it possible to utilize the full speed and acceleration capabilities of the robot.

VII Experiments on a Kinematically Redundant Industrial Robot

The algorithms described here were implemented on the first four joints of the Robotics Research Corp K-2107HR robot arm. The reflex controller was tested and shown to provide a protective behavior of the robot. When the robot accidentally or purposely was given bad commands, the reflex controller prevented the robot from exceeding its joint limits, or from colliding with itself or its environment. Our reflex controller continuously inspected a local 4D configuration space, and delivered approved set-points to the reactive trajectory generator at 300Hz or faster. Since this rate was fast with respect to the servo system position-control bandwidth, the reflex layer was virtually transparent, except in cases where necessary collision avoidance reactions were deliberately induced.

In an experiment with two moving robots having overlapping work-spaces, one of the robots was able to avoid the other robot through a combination of static reflexive obstacle avoidance and a flee-reflex. The potential functions described above were tested in coordination with reflexes, and the robot showed an ability to avoid collisions and circumvent obstacles quickly. Our task-space commands consisted of the desired Cartesian coordinates of the wrist, which left the "elbow orbit" redundant degree of freedom available for obstacle avoidance. We constructed a reflex-based 4D path planner that was able to find paths around obstacles while the robot was operating.

VIII Summary and Conclusions

To protect a system against the dangers of software errors, a reflexive self-preservation controller is proposed. The reflex controller guarantees collision avoidance in a rapid, and failsafe manner, with minimum non-essential influence on higher level controls. The reflex controller serves well as a building block in obstacle avoidance and path planning algorithms. Reflex control may be extended through future research and development as a path to grow towards increasingly autonomous systems.

IX Acknowledgments

This work was made possible through the support of the U.S. Dept. of Energy Sandia National Laboratories under contract #18-4379F. Their support is gratefully acknowledged.

References

[1] M.S. Branicky. Efficient configuration space transforms for real-time robotic reflexes. Master's thesis, Case Western Reserve University,Department of Electrical Engineering and Applied Physics, February 1990.

[2] M.S. Branicky and W.S. Newman. Rapid computation of configuration space obstacles. In *Proceedings of International Conference on Robotics and Automation*, pages 304–310, May 1990.

[3] Vinay Krishnaswamy and Wyatt S. Newman. On-line collision motion planning using critical point graphs in two-dimensional configuration space. In *Proceedings of International Conference on Robotics and Automation*, pages 2334–2339, May 1992.

[4] V.K Krishnaswamy. On-line motion planning in three dimensional configuration space for a robotic manipulator. Master's thesis, Case Western Reserve University,Department of Electrical Engineering and Applied Physics, October 1990.

[5] T. Lozano-Perez. Parallel robot motion planning. In *Proceedings of International Conference on Robotics and Automation*, pages 1000–1007, May 1991.

[6] W.S. Newman. High speed robot control and obstacle avoidance using dynamic potential functions. In *Proceedings of International Conference on Robotics and Automation*, pages 14–24, April 1987.

[7] W.S. Newman. *High Speed Robot Control in Complex Environments*. PhD thesis, Massachusetts Institute of Technology, Department of Mechanical Engineering, October 1987.

[8] W.S. Newman. Automatic obstacle avoidance at high speeds via reflex control. In *Proceedings of International Conference on Robotics and Automation*, pages 1104–1109, May 1989.

[9] W.S. Newman and M.S. Branicky. Experiments in reflex control for industrial manipulators. Report TR-89-153, Center for Automation and Intelligent Systems Research, Case Western Reserve University, Cleveland, Ohio, October 1989.

[10] W.S. Newman and M.S. Branicky. Experiments in reflex control for industrial manipulators. In *Proceedings of International Conference on Robotics and Automation*, pages 266–271, May 1990.

[11] Homayoun Seraji. Configuration control of redundant manipulators: Theory and implementaion. *IEEE Transactions on Robotics and Automation*, vol 5, no. 4:472–490, August 1989.

[12] Thomas S. Wikman and Wyatt S. Newman. A fast, on-line collision avoidance method for a kinematically redundant manipulator based on reflex control. In *Proceedings of International Conference on Robotics and Automation*, pages 261–267, May 1992.

A SMART DOCKING/BERTHING MECHANISM

J.M. Vranish
Intelligent Robotics Laboratory
NASA/Goddard Space Flight Center
Greenbelt, Maryland 20771

Abstract

A smart docking/berthing mechanism is being developed by NASA/GSFC for use by robots in space. This mechanism will permit robots to perform docking/berthing operations; both in the sense of docking/berthing themselves (as in robot walking) and in the sense of attaching another object (payload) to a fixture. The mechanism incorporates both a unique mechanical device-termed the Workpiece Attachment Mechanism/Workpiece Attachment Fixture or (WAM/WAF) and an equally unique sensing system-termed the "Capaciflector" or (capacitive reflector) to produce an intelligent system with outstanding performance and efficency and wide margins of safety in a small, light and compact package. Originated by NASA/GSFC, the mechanical WAM/WAF provides the robot a means of fastening during the docking/berthing process. This device is small and compact; yet it generates a large fastening force from minimal input torque. Also, its alignment and locating properties are outstanding. And, it passes power and signal to and from the robot to the object it is grasping. Thus, whatever it fastens to can also become a smart mechanism. The mechanical device is liberally instrumented with the NASA/GSFC-developed "Capaciflector" sensors. These simple, but effective, sensors are mounted, skin-like, at various strategic points on the WAM/WAF and on the payloads. The resulting systems will permit robots and their attached payloads to avoid collisions in space with humans and other objects and to align and dock/berth these payloads in a cluttered environment with a very soft and sure touchdown. The principles of operation and test results of the WAM/WAF and "Capaciflector" prototypes are described individually along with their combined performance as parts of an intelligent system.

1 INTRODUCTION

NASA is conducting research on berthing/docking and assembly and maintenance operations in space using robotic-type machines.The concept of a smart berthing/docking mechanism seems particularly promising for this inherently difficulty task. In this paper, the task will be explained. Following this, the smart berthing/docking mechanism will be described; both in terms of its main components and in terms of how it performs as a system. It will be shown why it s systems performance has been found to exceed all expectations to compare very favorably with more sophisticated sensors such as lasers and fiber-optics in roles once considered unique to those sensors.

Figure 1: Smart Docking/Berthing Mechanism

The task is inherently difficult. Apart from the hostile environment of gamma rays, electron clouds, oxygenization and micrometeorites and man-made debris, two main problems, micro-gravity and the presence of sensitive scientific instruments, stand out. Because of micro-gravity, items must be fastened to fixtures. This, in turn, means that in grasping an object in the berthing scenario, that fastened object cannot shift its position when it is being grasped. This, in turn, forces the stresses and compliance back into the robot arm and mandates soft, precise, berthing and grasping. In the docking scenario, the object being grasped is free to accelerate away if the robot makes a slight collision with the object when trying to grasp it. Again, a soft, precise acquisition is required. The presence of sensitive scientific instruments underlines the need for soft, sure, intelligent acquisition and grasping.The prime components of the smart berthing/docking mechanism are discussed first, followed by systems considerations. These components are a). The fastening system known as the Workpiece Attachment Mechanism/Workpiece Attachment Fixture (WAM/WAF). The WAM/WAF fasteners are used by GSFC as robot end effectors and as the fastening components for the Orbital Replacement Unit (ORU) payloads. They also provide an electrical signal and power transfer capability between the robot end effector and the ORU or the ORU and the fixture. b). The "Capaciflector" (capacitive reflector) sensors, which are added skin-like to the robot end effectors and the ORU payloads, and their supporting software, permit the end effectors and ORUs to become smart devices. Components aside, the smart berthing/docking system created by combining the WAM/WAF fastener and the "Capaciflector" sensor, with supporting softwear, is much more than the sum of the capabilities of its main components(Fig.1)[1]. Laboratory tests are showing that this smart mechanism will be able to perform berthing/docking and assembly and maintenance, in its entirety, safely and easily in space including: collision avoidance, alignment and imaging, insertion/removal, "Virtual Force" precontact precision control and soft touchdown and fastening/release.

2 THE WAM/WAF

The WAM/WAF is derived from a NASA/GSFC Spline-Locking Screw modified into a robot end effector with electrical connectors capable of transponding power and data to and from payloads. The Spline-Locking Screw, in turn, is an improvement over the standard NASA acme screw[2] from the Solar Max satellite repair mission of 1982.

Screws with low-pitch machine threads had been used successfully many times in space launch operations. But; these would cross-thread when the astronauts tried to fasten them while wearing gloves and space suits. A guide was added to the screw and this helped; but not enough. Next, the low pitch machine thread was changed to a more high pitch acme thread. This worked; but now the screw would back out during vibration tests. To overcome this, a taper interface was added at the top of the screw. However, the additions added a great deal of friction to the system and this meant that large torques, on the order of 100 ft-lbf were required to free-up the bolts used in the Solar Max mission. This, in turn, meant that very large and clumsy power tools had to be used and the astronauts (or robots) could be subjected to dangerously large reaction torques and forces. For a robot, the situation was nearly impossible. Using the large tools would severely limit robot dexterity. On the other hand, the existing robot end effectors could only produce 20 ft-lbf of force[3]

The Spline-Locking Screw was developed by returning to the common machine thread and taking a fresh approach[4].It was decided to cut the bolt in two and leave the bottom half threaded (Fig. 2a). A new interface was created in the shaft of the bolt. Thus, we have the the bolt head (or Driver) and the Object which is normally pinched by the screw section on one side of the interface and the Bolt and the threaded fixture it screws into on the other. A Spline-Locking type interface (Fig. 2b) was chosen because it was simple, direct and effective.There is an increase in size; but it is minimal. In operation, the Driver is inserted into the Bolt such that the Male Splines of the Driver fit into matching Female Splines of the Bolt. The Driver is then turned clockwise to tighten the screw system. During this process, the male Driver Spline engages in the female Bolt Spline, it cannot be pulled out and the Bolt and Driver turn together as a single complete machine screw. To unfasten the system, the Driver is turned counter-clockwise, the screw loosens and the Driver Splines relocate in the Bolt (disengage) such that the Driver can easily be pulled out of the Bolt. The Splines are very coarse so they can be designed to seat and fit together such that cross-threading is virtually impossible. And the Bolt is never unthreaded, thus, we have, in effect, a machine screw that cannot cross-thread. At the same time, it is now possible to use low pitch machine threads so large preload forces can be generated from minimum input torques (on the order of 8 ft-lbf. to generate 1,000 lbf. preload[4]). And, these Bolts would not shake loose during launch because of their low pitch. Also, robot End Effectors and Astronaut Hand Power Tools could now be made more modest in size and power, enhancing safety for the Astronaut and safety and dexterity for the robot.

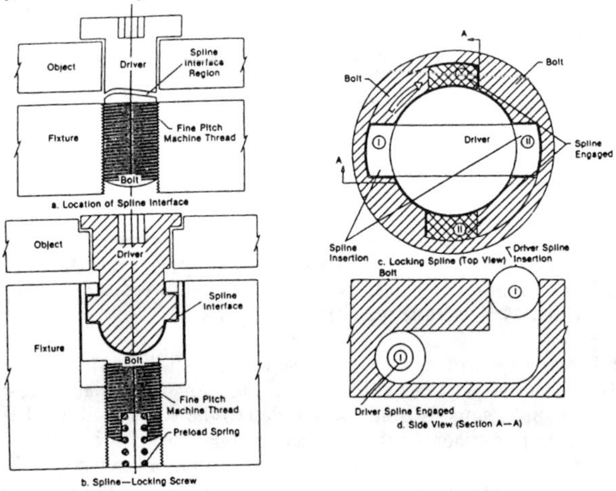

Figure 2: Evolution of Spline-Locking Screw

Of course, the Spline -Locking design requires careful and detailed treatment beyond that shown in Fig. 2. This is to make certain that the Splines engage and disengage properly and that the payload remains attached either to the Robot End Effector (or Astronaut Power Hand Tool) or to the Fixture at all times to prevent it drifting off into space. Proper so\t dock and hard dock are required during the fastening phase and "Push-Off" to prevent cold welding or jamming of the Object to the Fixture are required during the disengagement phase. Details are given in ref.[4].

For the WAM/WAF[4], An electrical connector is incorporated into the Spline- Locking Screw concept. The Spline-Locking Screw Bolt is, then, threaded on a Nut rather than into the Fixture. The Bolt is coupled to the Fixture by means of a Preload Spring (nominally 100lbf) forcing the Bolt down towards the Fixture. Contact between the Bottom Stop on the Nut and the Fixture prevents the downward translation of the Bolt. Thus, the Bolt and Nut are preloaded against the Fixture with a force equal and opposite to that of the Preload Spring. Also, the interface between the Nut and Fixture are splined so that the Nut cannot rotate; but can translate between the Fixture Top and Bottom Stops. Electrical Pins are added to the Nut, and Pin Receptors added to the Fixture. As the Driver turns clockwise, the Bolt turns with it and, since the Bolt cannot translate downwards, and the Nut cannot rotate, the Nut translates upwards. Throughout this process, the force sustaining the Spline- Locking remains constant and equal to that of the Preload Spring. And, The Object and Fixture remain strongly forced against and precisely aligned with each other. This new condition, termed ***Hard Dock***., preconditions the electrical connection. As the Driver continued to turn clockwise, the Nut translates upwards until a precision electrical connection is made. Shortly afterwards, the Nut hits its Top Stop on the Fixture. With Nut translation stopped, the Bolt once again attempts to translate downwards forcing the Object and Fixture together with preload forces. This condition is termed ***Preload***. Once again, it should be noted that Spline Locking can be done in either a ***Hard Dock*** or ***Preload*** condition. Both are adequate in ensuring that the Object is safely secured to the Fixture throughout the Fastening process. It is also, perhaps, appropriate to note that the Object is secured (or Docked) to the Fixture as soon as the Splines are Engaged, prior to Hard Dock. However, there is rattle between the Object and the Fixture during this condition so we will describe it with the term ***Soft Dock***. Turning the Driver counter-clockwise reverses the steps described above and disengages the Object and the Fixture as well as the Driver and the Bolt and forces the two Splines to be lined up such that they can easily and reliably be pulled apart.

A WAM/WAF prototype (Fig. 1) was constructed and tested as a first prototype. The prototype used a tiny motor that produced only 8 ft-lbf torque even after force multiplication through a harmonic drive. Still, it was able to repeatedly and routinely grasp a large ORU mockup and maneuver it in a 1-"g" field withstanding 100ft-lbf torque on the robot wrist during the process. Not the slightest evidence of "rattle" or gapping was seen between the WAM/WAF and the payload. Its holding force exceeded all expectations and is far superior to that we have experienced with grippers. It went through its proper fastening sequence of Soft Dock, Hard Dock, Electrical Connections (to include actuating dust covers), and to provide Preload Forces. It handed off the ORU correctly, demonstrating its berthing capabilities and it fastened the robot to a fixture in the absence of an ORU, demonstrating its ability to support a walking robot[1].

Additional minor modifications to the Spline- Locking Screw could produce any number of useful devices. One such device would be a combination Special Tool Interface and Autochanger [4]. The WAM would be attached to the robot and each special tool or end effector would be attached to the back side of a standard-sized WAF. If the tool required signal and power, it would receive these from the WAM/WAF electrical interface. It should be noted that the mating procedure by which the robot acquires a Special Tool would use common berthing practice. That is, alignment is classical "peg-in-the-hole" using alignment pins and mating receptacles. The bottom of the Nut,which constitutes the electrical connector, can be used to pinch tabs of a Tool Storage Holster and there-by store Special Tools[5]. This system would be equally effective on earth or in Zero or micro "g". Thus the system would become its own Autochanger. Although not detailed in this paper, a similar approach is used in permitting a robot to use "Spline-Locking Screw" techniques to release and fasten ORUs[1,4].

3 THE "CAPACIFLECTOR"

The "Capaciflector"[5] (capacitive reflector) is a capacitive sensor which can be mounted flush on a grounded structure and still provide extraordinary range, sensitivity and signal-to-noise ratio. It consists of a sensing element backed by a reflector element which is driven by the same voltage as the sensor to reflect all field lines away from the grounded robot arm, thus extending the range of the sensor. This approach is an extension of the technique used in instrumentation systems where a shield or guard is used to eliminate stray capacitance. Fig.3 shows the principles of operation in terms of charges and electric fields. Fig. 3a. shows a capacitive sensor not using the "Capaciflector" principle. Since we are using relatively low frequencies (approximately 20 khz) we have the quasi-static condition and static charges and fields can be used to calculate the capacitance the sensor "sees".

We can see that the smaller the stand-off from the grounded robot arm, the larger the capacitive coupling between the sensor and ground. This, of course, has the effect of reducing the relative coupling between the sensor and the object being sensed and hence reducing sensor range and sensitivity. On the other hand, increasing the stand-off increases the bulk of the robot arm and adds wires and wiring complications. And, when the insulation materials are added to support the stand-off, the ability of the robot arm to dissipate thermal energy into space is impeded. When the "capaciflector" principle is used (Fig. 3b)[5], the field lines from the sensor are prevented from returning directly to ground. The effective stand-off is approximately the width of the active shield or capacitive reflector. Thus, we can have a skin with very little thickness (on the order of 0.060 inches) and a robot arm with very little bulk and still have the performance of a large stand-off. The capacitive coupling between the sensor and the object being sensed is used as the input capacitance tuning the oscillator frequency. As an object comes closer, the capacitance increases and the oscillator frequency decreases. On the other hand, the reflector is attached to the output of the voltage follower so it is electrically isolated and prevented from affecting the tuning of the oscillator frequency. At the same time, the voltage of the reflector follows that of the oscillator. Thus, the reflector is in phase with (and reflects) the electric field of the sensor without being affected by the coupling between the sensor and an approaching object.

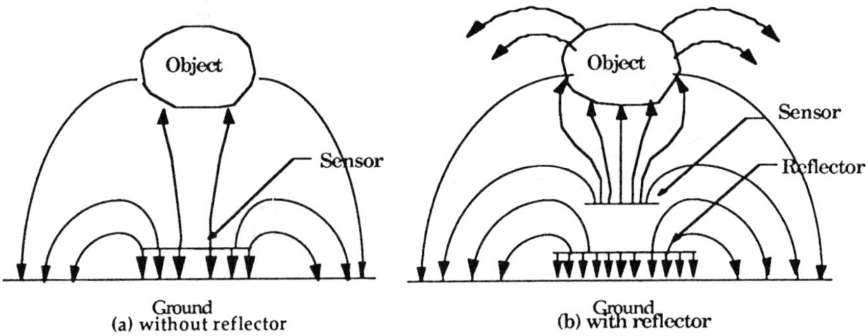

Fig. 3 : "Capaciflector" principle[5]

We will now examine the means by which the sensor detects an object[5]. The discussion will be limited to conductors for simplicity although dielectrics are also easily detected. Both the grounded and ungrounded cases will be discussed. Since we have low frequency, (approximately 20 kHz), the quasi-static case holds. Assuming a momentary positive potential V, we can see that the electric field lines emanating from the sensor towards the object induce negative charges on the object surface nearest the sensor. Thus that surface can be considered one plate of a capacitor and the sensor the other. But, an ungrounded conductive object is charge neutral so an equal amount of positive charge will form on the surface away from the sensor so as to ensure that there is no net electric field in the conductor. These charges couple back to ground which creates a second capacitor in series with the one mentioned above. These can be labeled respectively as:

$$C_{so} \text{ and } C_{og}$$

But; there also is a path where the electric fields from the sensor can go around the active shield and couple to ground directly. This is labeled as:

$$C_{sg}$$

Thus our tuning capacitance is:

$$\frac{C_{so}\,C_{og}}{C_{so}+C_{og}} + C_{sg} = C_t \qquad (1)$$

In the case where the object is grounded, equation (1) reduces to:

$$C_t = C_{sg} + C_{so} \qquad (2)$$

Eq. 2 also holds for an ungrounded conductor with any dimension more than a few inches (for example length). And, as the experimental evidence confirms,in practice, all objects of size more than a few inches are approximately grounded.

Closely-spaced "Capaciflectors" will now be be discussed. The "Capaciflector" circuit described above is limted in that sensor elements can only be as close to each other as their shield widths. But; there are many circumstances where it is desireable to have two or more sensors very close to each other. For example, in locating an alignment pin, suppose we have three sensors in our ORU pin mating hole which run up the iside of that hole and are spaced at 120 deg. increments around its circumference. Each sensor will give the minimum normal distance to a point on the suface of the bolt to great precision (on the order of 0.030 in. at a range of 1/4 in.). And, as the robot inserts the box over the alignment pin, the clearances get much closer, say on the order of 0.010 in. in dia.. And at these ranges, the resolution of each of the three "Capaciflectors" would be outstanding, say 0.002 in. If the robot control were is good enough, the box and pin could be inserted together using "Virtual Force" and without touching in the face of clearances on the order of a few Thousandths of an in. And even if contact did occur, it would be very soft and the sensor involved would know it immediately.

But how does one pack multiple sensors close together and require them to share the same driven shield, yet prevent them from cross-talking? We begin by driving the common shield with a crystal-controlled oscillator. And, we put a ground directly behind the shield, turning it into a stable parallel-plate ref. capacitor. We, then place each sensor element above the common shield (separated, of course, by a thin layer of insulation). We also drive each of these sensors by the same crystal-controlled oscillator. We then place resistors between each of the capacitors (sensors and ref. shield) such that their impedance is approximately equal to the impedance of the respective capacitor of each to form the bridge circuits. The resistor of the ref. shield is fixed and the resistors of each of the sensors are voltage-controlled resistors. Each sensor signal at the mid-points of its bridge is compared to the corresponding signal of the ref. Actually, the signals are rectified and the average dc amplitude values are compared.

In the presence of an object, the ref. signal will always be greater than the sensor seeing the object. The signal difference between each sensor and the ref. is fed into an analog closed loop simple Position/Integration controller and closed back into the circuit through its variable resistor. In this manner, the variable resistor is servoed until the sensor bridge is locked to the ref.(lab. experience shows that this locking is < 1 mv. The voltage being applied to the variable resistor is the output signal for that sensor. Modelling shows and experience confirms that resistance in the object is not usually a factor and, hence, the locking is in phase as well as amplitude. As a result, each of the sensors stays at essentially the same voltage and phase and so they do not talk to each other and they do not leak through the common shield.

4 THE SYSTEM

The WAM/WAF and the various ORU boxes can be instumented with "Capaciflector" sensors and converted into intelligent, guided, smart mechanisms with capabilities that are more than the sum of their parts. These capabilities show themselves as fundamental throughout the steps of the berthing/docking process and provide unique capabilities in each step of that process. The first step of the Berthing/Docking process involves moving the robot end effector to the vicinity of the object-to-be-grasped. In this step an array of "Capaciflector" sensors on the robot arms and the WAM/WAF(s) provides unprecedented safety via a comprehensive collision-avoidance capability. The sensors provide an all-encompassing electric field which detects objects (including humans) to a distance of 1 foot ; more than adequate for the robot to take corrective action. The second step of the Berthing/Docking process involves moving the robot end effector to the vicinity of the mating alignment pins in the ORU payload and aligning it with these pins. In this step "Capaciflector" imaging and cuing techniques are used. NASA/GSFC research has demonstrated that scanning a single sensor and correlating its output with robot joint angle information will produce clear images showing a 3-D model of the topography of the region around the sensor[6]. Since the sensor information is 3-D, the 3-D topographic model can be constructed with much less compute power than is required for a vision system. Accordingly, GSFC has been able to construct the 3-D images of the berthing alignment pins for an ORU using a single sensor in a robot automated scan routine. Returning to our example of the end effector mating with the ORU, it is clear that imaging will be able to prevent the end effector from hitting an unforseen obstacle on or near the ORU box. The use of an array of sensors will facilitate this task. And, the near range resolution is such that the robot can align the end effector with those alignment pins in the correct orientation and with sufficient resolution to enable insertion. Thus, we have what amounts to electrical guidance rather than the more traditional guidance using the mechanical action of pins hitting cones and this reaction being monitored by a 6 degree-of-freedom (6 DOF) force/torque sensor. Electrical guidance is, of course, much smoother and more accurate than mechanical guidance because friction and collisions do not exist in the process. Also, the physical sizes and weights of the pins and cones can be drastically reduced. The third step involves the precision insertion of the alignment pins and cones and final soft touchdown. In this step, "Capaciflector" sensing using cues is employed along with the concept of "Virtual Force". Since the range respose of a "Capaciflector" sensor is essentially hyperbolic and since this sensor is sufficiently rugged that the sensor can also constitute the contact surface, the range determination is extremely precise at very close ranges. For example, at about 1/16 in. range, the uncertainty is approximately 0.010 in. And, if we employ three sensors very close to each other (and sharing a common shield as described in the section on Doublet "Capaciflectors" described above, we will be able to range two separate regions of the pin's surface and, therefore, to cue on it with precision. The robot will be able to literally follow the alignment pins during the insertion process with little or no contact. The robot control system will treat the range measurement as a "Virtual Force" and will operate in the force control mode rather than position control. Just prior to bottoming, the "Capaciflectors will signal the robot that contact is imminent. The robot will then slow way down across the final 0.010 in. and make a very soft contact touchdown and, in the process, transfer from "Virtual Force" to actual force as measured by the 6-DOF force/torque sensor. The fourth step, then, will be to relax the robot joints and allow the mating members to seat using mechanical compliance and robot impedance control (with the results monitored by the 6-DOF force/torque sensor). And, the fifth step is to fasten the WAM/WAF, securing the end effector to the ORU and getting power and signal into the ORU. After the end effector is fastened to the ORU, the ORU is mechanically released from its fixture so the robot can move it to another fixture; berth it to that fixture and release the end effector from the berthed ORU to begin a new task. With "Capaciflectors" on the ORU, it becomes an instrumented intelligent payload of its own, complete with WAM/WAF mechanicals and can be (and has been) berthed to a fixture following the same steps as described for the end effector.

The berthing/docking system using WAM/WAF and "Capaciflector" technologies works; but that is it better than competing technologies? There are pursuasive arguments to suggest that it is, both fundamentally and in principle.and for a host of practical reasons. Four (4) competing technologies come to mind; vision w/targets; vision with force-reflecting teleoperations; Lasers and fiber-optic sensors. This discussion will center on the "Capaciflector" and its competitors. The arguments in favor of WAM/WAF as opposed to its competitors are described in the section on WAM/WAF above.

The first, and most common competing technology, is vision using targets (mounted on the fixture). The target is designed to give the robot-mounted camera 3-D (6 DOF) information as to where it (the camera) is with respect to the target. If we disregard the practical problems associated with operating cameras in space for prolonged periods of time and if further, we assume that the alignment between the target and the camera is error-free, there are still significant problems with this approach. First, the camera looks at the target; but contact between mating members comes at points remote from both the camera and the target . Thus, the system relies on the world model being correct which relates all points on the ORU to the robot and all points on the fixture to the target. This approach is, at heart, a computer-intensive modelling solution using the vision system as a calibration device. And, this modelling has uncertainties. Lab experience has indicated typical errors on the order of 1/4 in. to 1/8 in. even for small ORUs. This means "Virtual Force" and soft touchdown are not possible. Also, if there is a mistake in the model; such as the target being incorrectly aligned with the fixture or, for that matter, any alignment pin being misplaced, the results could be catastrophic. Furthermore, the camera cannot see behind the ORU so, in cases where ORUs are being berthed or docked, an unexpected obstacle (perhaps an astronaut) could also cause a catastrophic failure.

The second competing technology adds teleoperated force reflecting hand controllers to the vision/target system. The force reflecting hand controllers do not help until after contact is made with an object so this approach has essentially the same problems as the first.

The third competing technology would use lasers (Fig.4). This has been proposed; but has not been implemented in the lab to solve the ORU berthing/docking problem.

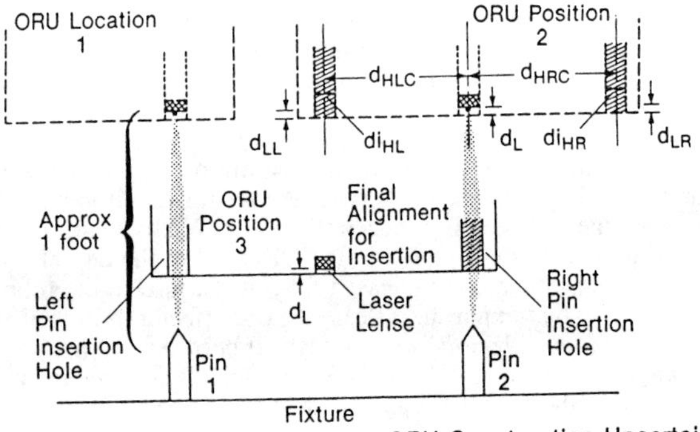

Figure 4: Virtual Force Uncertainties Using Lasers

ORU Construction Uncertainties

d_{HLC} — Distance from center of laser lense to center of left hole.

d_{HRC} — Distance from center of laser lense to center of right hole.

di_{HL} — Diameter of left hole.

di_{HR} — Diameter of right hole.

d_L — Distance laser lense is recessed below ORU surface.

Lasers can determine ranges and edges with great precision. In this respect they are superior to vision and "Capaciflectors". They normally use one of two methods for determining range; time-of-flight or interferometry techniques. Time-of-Flight is simpler in terms of mirrors, prisms and reflectors and we will, for the moment, assume it is the preferred solution. The laser would scan the berthing portion of the fixture at a range in the order of one (1) foot to permit accurate timeof-flight measurements. This would provide a precise (we shall say perfect) 3-D image of the fixture as referrenced to the laser lense. But; the lense is remote from the points on the ORU that represent the actual points of contact (note the ORU constructiøn uncertainties listed in Fig. 4). Construction uncertainty errors are typically on the order of +/- 0.005 in. for machine tolerances to +/- 0.030 in.to as much as +/- 0.125 in. for system assembly[7]. So, even if the laser provides perfect ranges and edges, the system appears inherently limited in its capability to provide near-contact electrical guidance, "Virtual Force" and soft touchdown. It must still rely on extensive prior knowledge and computer modelling to reference its location on the ORU or end effector to other remote potential points of contact on the same surface of the ORU or end effector. And, once the robot moves to the prealignment position shown in Fig. 4, and loses sight of the alignment pins, the repeatability errors of the robot , itself, comes into play (typically 0.020 in. to 0.030 in.[9]) so even if we can map everything, ORU and Fixture, there are still relatively large, inherent errors.

The fourth competing technology would use a distributed array of fiber-optic sensors. These are also purely in the proposal phase and have not been demonstrated in the lab for ORU/endeffector berthing/docking. These would be short range and local in their location on the ORU or end effector so unless there were massive numbers of them and unless they were supplemented by a longer range sensor (such as a scanning laser or camera) they would be of limited use in collision avoidance. Like vision and lasers, these fiber-optic sensors could not be co-located at the points of contact and fastening so there would a degree of computer modelling involved in estimating where these points of contact would be and when contact was likely to occur. But; because of the closeness to the points of contact the compute burden would be reduced from that of the Laser and it would be better in providing "Virtual Forces" and soft touchdown. It would not, however, be as good as the "Capaciflector".

Two subtleties make "Capaciflector" a surprisingly effective sensor for near-contact "Virtual Force" guidance and insertion/disinsertion (Fig. 5).

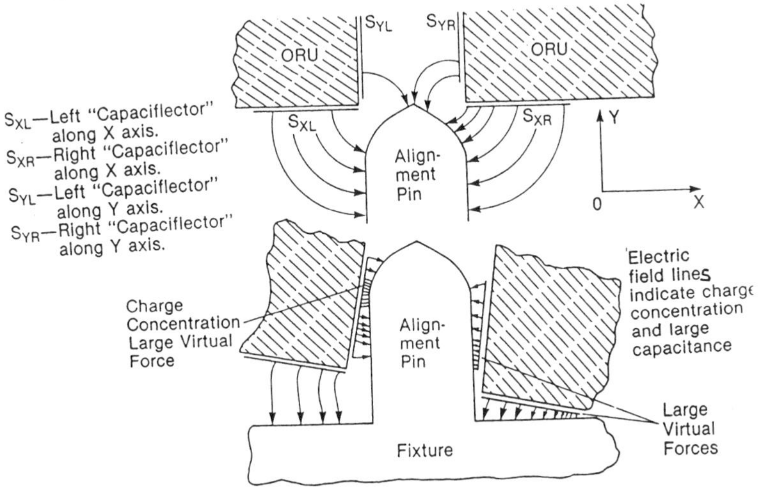

Figure 5: Virtual Force Sensing Using "Capaciflectors"

When two members make contact, they make contact across a very small area (typically <0.01X 0.01 sq. in.)[10]. It is virtually impossible to know exactly where that point is and it changes during each engagement. So, it is equally virtually impossible to locate an electro-optical sensor such that its pencil beam will land exactly on that contact point. With the "Capaciflector", we locate the sensors so that they cover the entire region in which the contact points can occur. During an engagement, the charges on the sensor plate slide along the surface of the plate and accumulate at the point where the range to the fixture is a minimum.So we know the range to contact with precision (assuring outstanding "Virtual

Force" soft contact); but we are slightly less certain as to where, on the sensor plate, this point is. These contact points can be located by implication by applying local robotic search techniques, noting sequential data points during servoing and using multiple sensors. These permit insertions where we always can be assured that we can avoid actual contact; even in close quarters and even with an inexact model. Whereas, with sharp beam electro-optical sensors, we know the range from the sensor lense to the fixture opposite it; but we estimate the contact points and ranges by prior knowledge (modelling); inherently limiting its "Virtual Force" soft contact no matter how many sensors and no matter how many sequential data points are used. It is usefull to recall that the rugged "Capaciflector" sensors can be right on the point(s) of contact ; even on the fastening pins, cones and screws. Further, "Capaciflectors" have outstanding range resolution at close ranges (hyperbolic relationship between range and output voltage[5]) and with the sensors covering the contact points, sensor/electronics drift is not a factor.

Although we have not yet begun the laboratory work, it seems clear that the GSFC smart berthing mechanism will be outstanding in docking. First, the strategy for docking would be entirely different than is practiced presently. Presently, an oject which is to be docked with a structure in orbit, is placed in the same orbit as that structure and on a collision course with it. When the two collide, the shock-absorbers of the mating bodies are activated to provide as soft a collision as possible and to prevent the bodies from bouncing apart. A capture mechanism is excited and holds the two bodies together and shortly afterwards, precision alignment and fastening occurs. A smart docking/berthing mechanism would go about this entirely differently. The object would position itself in an orbit along side that of the structure already in space and synchronize itself in speed with that structure. One of the two would likely have a small 6-DOF robotic device with a smart docking/berthing mechanism on its wrist. This device would naturally have a much higher frequency response than would the structure, so it could slowly, surely, safely and precisely use its sensors to guide the mating members together despite the undulations of the space craft. The docking process would require very little force and would be very soft. Also, it would have a large margin of safety; it would not be an all-or-nothing, one-shot event.

There are also numerous practical advantages to using "Capaciflectors" over the four competing technologies described above in the roles of collision avoidance, alignment, insertion/disinsertion and "Virtual Force" softtouchdown. "Capaciflectors" are ultra-simple; particularly the sensors (plates and shields). They are virtually indestructable in the low-earth environment. Oxygen ions cannot degrade them (these do erode and cloud the glass lenses of electro-optical devices[10]); micrometeoriite and space debris collisions do not effect them (these would be catastrophic to the lenses and mirrors of electro-optical devices[10]). And, gamma rays and electrons do not effect either plates and shields.or lenses and mirrors. The electronic signal processing capabilities of all sensors can be remoted and shielded from the corrosive elements of low-earth orbit. And, abrasions and contacts made during berthing/docking and/or fastening cannot degrade the performance of "Capaciflectors" even under repeated applications. Lenses and mirrors simply cannot tolerate this. "Capaciflectors" have been examined with respect to EMI, thermal problems and jamming from interferrence from instruments and power consumption and have been found to present no problem[11]. "Capaciflectors" also do not effect the configuration of the robot, end effector(s) or ORU payloads. Thin, flexible and conformal, they can be mounted anywhere and even covered with paint. For example, they can be made out of a thermal blanket and used to cover a payload.

5 SUMMARY/CONCLUSIONS

A smart docking/berthing mechanism, being developed by NASA/GSFC for use by robots in space, shows outstanding promise in initial laboratory tests. An innovative mechanical fastener, the WAM/WAF, is covered, skin-like, with an array of equally innovative "Capaciflector" sensors supported by extensive software to form the smart docking/berthing mechanism. This mechanism is a system which is more than the sum of its parts and which will permit robots to perform docking/berthing operations; both in the sense of docking/berthing themselves (as in robot walking) and in the sense of attaching another object (payload) to a fixture. ORU payloads also use the WAM/WAF fasteners and are covered with "Capaciflector" sensors to make them intelligent berthing/docking mechanisms. The WAM/WAF fastener clearly out-performs its competitors. Simple, compact and very strong, it fastens securely and safely with minimal input torque. It cannot cross-thread and it has excellent soft dock and hand-off features which make it safe to use without tethers. And, it mates electrical connectors to objects that it fastens with permitting power and data to be passed The "Capaciflector" array also clearly out-performs rival sensors in collision avoidance, alignment, insertion/removal, 3-D imaging and "Virtual Force" precontact/soft touchdown. It is proving to be a sensor fundamental to berthing/docking assembly and maintenance operations in space. Laboratory tests have given strong indications that the resulting systems will permit robots and their attached payloads to avoid collisions in space with humans and other objects and to align and dock/berth these payloads in a cluttered environment with a very soft and sure touchdown. Development is continuing to prove out the entire concept and ready it for space use. Certain components, the single "Capaciflector" sensor and the WAM/WAF robot end effector and ORU fastening system already are mature enough to undergo design freeze and hardening for space. The NASA/GSFC "Capaciflector", WAM/WAF and smart berthing/docking mechanism are well on the way towards becoming a central part of collision avoidance, docking and berthing and pre-contact force control in space and will be used extensively on robot arms, robot end effectors and payloads GSFC has a significant in-house capability focussed on this effort. Progress has been substantial, continuous and steadily improving.

References

[1] Private notes of John Vranish (includes laboratory tests,videos, diagrams & photos).

[2] *Portion of Shop Drawing #GL 1087209, Rockwell International Corp. Space Division*, dated 9/7/76 showing the operation of an ACME
Screw Payload Fastener for the Solar Max repair mission of 1982, courtesy of Robert Davis, NASA/GSFC.

[3] *FTS End Effector End Item Specification* dated 1990, courtesy of Paul W. Richards, NASA/GSFC.

[4] Vranish, J.M., *Fourth World Conference on Robotics Research,* [Spline-Locking Screw Fastening Strategy (SLSFS)], Robotics International/Society of Manufacturing Engineers, published as a separate paper, Pittsburgh, Pa., U.S.A., Sep. 17-19, 1991.

[5] Vranish, J. M., McConnell, R. L., Mahalingam, S., *The International Journal of Computers and Electrical Engineering*, Vol. 17, Number 3 1991 ["Capaciflector" Collision Avoidance Sensors for Robots].

[6] Campbell, Charles E. Jr., *Robotic Berthing Using Capaciflector Imaging; 4 th International Symposium on Robotics and Manufacturing (ISRAM'92), Nov. 11-13, 1992, Santa Fe, New Mexico, USA.*[Robotic Berthing Using Capaciflector Imaging].

[7] +/- 0.005 in. is common practice machining tolerance. Assembly of components across an ORU several feet across would be expected to be much less precise; +/- 0.030 to +/- 0.125 in. is the estimated range based on experience of the author.

[8] Typical range of repeatability performance values given for like-sized industrial robots and based mainly on the capabilities of joint sensors.

[9] Conservative estimate based on contact forces compressing steel against steel or against aerospace quality aluminum, say 6061 T-6.

[10] National Geographic Vol. 180, No. 5 Nov. 1991, pp. 106-130, [Satellite Rescue].

[11] Martin Marietta Study entitled, Collision Avoidance Sensor Evaluation Study by R. Onnen, dtd. Jan. 31, 1991. This study was followed up by FTS staff members evaluating the sensor for EMI, thermal and safety. It passed all evaluations.

SUBJECT INDEX

AUTHOR INDEX